Handbook of

PARAMETRIC and NONPARAMETRIC STATISTICAL PROCEDURES

Handbook of

PARAMETRIC and NONPARAMETRIC STATISTICAL PROCEDURES

David J. Sheskin

Western Connecticut State University

CRC Press

Boca Raton New York London Tokyo

Acquiring Editor:	Tim Pletscher
Assistant Managing Editor:	Paul Gottehrer
Marketing Manager:	Susie Carlisle
Direct Marketing Manager:	Becky McEldowney
Cover Design:	Dawn Boyd
Prepress:	Gary Bennett

Library of Congress Cataloging-in-Publication Data

Catalog record is available from the Library of Congress

International Standard Book Number 0-8493-3119-6
Printed in the United States of America 1 2 3 4 5 6 7 8 9 0
Printed on acid-free paper

To Vicki and Emily

Preface

My intent in writing **The Handbook of Parametric and Nonparametric Statistical Procedures** is to provide researchers, teachers, and students with a comprehensive reference book in the areas of parametric and nonparametric statistics. The book is intended for individuals who are involved in a broad spectrum of academic disciplines encompassing both the social and biological sciences, as well as the fields of mathematics/statistics, business, and education. My philosophy in writing this volume was to create a reference book on parametric and nonparametric statistical procedures that I (as well as colleagues and students I have spoken with over the years) have always wanted, yet could never find. To be more specific, my primary goal was to produce a comprehensive reference book on univariate and bivariate statistical procedures which covers a scope of material that extends far beyond that which is covered in any single available source. It was essential that the book be applications oriented, yet at the same time that it address relevant theoretical and practical issues which are of concern to the sophisticated researcher. In addition, I wanted to write a book that is accessible to people who have a limited knowledge of statistics, as well as those who are well versed in the subject. I believe I have achieved these goals, and on the basis of this I believe that **The Handbook of Parametric and Nonparametric Statistical Procedures** will serve as an invaluable resource for people in multiple academic disciplines who conduct research, are involved in teaching, or are presently in the process of learning statistics.

I am not aware of any applications oriented book that provides in-depth coverage of as many statistical procedures as the number that are covered in **The Handbook of Parametric and Nonparametric Statistical Procedures**. Inspection of the **Table of Contents** and **Index** should confirm the scope of material covered in the book. A unique feature of the Handbook, which distinguishes it from other reference books on statistics, is that it provides the reader with a practical guide that emphasizes application over theory. Although the book will be of practical value to statistically sophisticated individuals who are involved in research, it is also accessible to those who lack the theoretical and/or mathematical background required for understanding the material documented in more conventional statistics reference books. Since a major goal of the book is to serve as a practical guide, emphasis is placed on decision making with respect to which test is most appropriate to employ in evaluating a specific design, and within the framework of being "user friendly," clear computational guidelines, accompanied by easy-to-understand examples are provided for all procedures.

One should not, however, get the impression that **The Handbook of Parametric and Nonparametric Statistical Procedures** is little more than a "cookbook." In point of fact, the design of the handbook is such that within the framework of each of the statistical procedures which are covered, in addition to the basic guidelines for decision making and computation, substantial in-depth discussion is devoted to a broad spectrum of practical and theoretical issues, many of which are not discussed in conventional statistics books. Inclusion of the latter material ensures that the handbook will serve as an invaluable resource for those who are sophisticated as well as unsophisticated in statistics.

In order to facilitate its usage, most of the procedures contained in the handbook are organized within a standardized format. Specifically, for most of the procedures the following information is provided:

I. Hypothesis evaluated with test and relevant background information The first part of this section provides a general statement of the hypothesis evaluated with the test. This is followed by relevant background information on the test such as the following: a) Information regarding the experimental design for which the test is appropriate; b) Any assumptions underlying the test which, if violated, would compromise its reliability; and c) General information on other statistical procedures which are related to the test.

II. Example This section presents a description of an experiment, with an accompanying data set (or in some instances two experiments utilizing the same data set), for which the test will be employed. All examples employ **small sample sizes**, as well as **integer data** consisting of **small numbers**, in order to facilitate the reader's ability to follow the computational procedures to be described in Section IV.

III. Null versus alternative hypotheses This section contains both a **symbolic** and **verbal** description of the statistical hypotheses evaluated with the test (i.e., the **null hypothesis** versus the **alternative hypothesis**). It also states the form the data will assume when the null hypothesis is supported, as opposed to when one or more of the possible alternative hypotheses are supported.

IV. Test computations This section contains a step-by-step description of the procedure for computing the test statistic. The computational guidelines are clearly outlined in reference to the data for the example presented in Section II.

V. Interpretation of the test results This section describes the protocol for evaluating the computed test statistic. Specifically: a) It provides clear guidelines for employing the appropriate table of critical values to analyze the test statistic; b) Guidelines are provided delineating the relationship between the tabled critical values and when a researcher should retain the null hypothesis, as opposed to when the researcher can conclude that one or more of the possible alternative hypotheses are supported; c) The computed test statistic is interpreted in reference to the example presented in Section II; and d) In instances where a parametric and nonparametric test can be used to evaluate the same set of data, the results obtained using both procedures are compared with one another, and the relative power of both tests is discussed.

VI. Additional analytical procedures for the test and/or related tests Since many of the tests described in the handbook have additional analytical procedures associated with them, such procedures are described in this section. Many of these procedures are commonly employed (such as comparisons conducted following an analysis of variance), while others are used and/or discussed less frequently (such as the tie correction employed for the large sample normal approximation of many nonparametric test statistics). Many of the analytical procedures covered in Section VI are not discussed (or if so, only discussed briefly) in other books. Some representative topics which are covered in Section VI are planned versus unplanned comparison procedures, measures of association for inferential statistical tests, computation of confidence intervals, and computation of power. In addition to the aforementioned material, for many of the tests there is additional discussion of other statistical procedures that are directly related to the test under discussion. In instances where two or more tests produce equivalent results, examples are provided which clearly demonstrate the equivalency of the procedures.

VII. Additional discussion of the test Section VII discusses theoretical concepts and issues, as well as practical and procedural issues that are relevant to a specific test. In some instances where a subject is accorded brief coverage in the initial material presented on the test, the reader is alerted to the fact that the subject is discussed in greater depth in Section VII. Many of the issues discussed in this section are topics that are generally not covered in other books, or if they are, they are only discussed briefly. Among the topics covered in Section VII is additional discussion of the relationship between a specific test and other tests that are related to it. Section VII also provides bibliographic information on less commonly employed alternative procedures that can be used to evaluate the same design for which the test under discussion is used.

VIII. Additional examples illustrating the use of the test This section provides descriptions of one or more additional experiments for which a specific test is applicable. For the most part, these examples employ the **same data set** as that in the original example(s) utilized in Section II for that test. By virtue of using standardized data for most of the examples, the material for a test contained in Section IV (**Test computations**) and Section V (**Interpretation of the test results**) will be applicable to most of the additional examples. Because of this the reader is able to focus on common design elements in various experiments which indicate that a given test is appropriate for use with a specific type of design.

IX. Addendum At the conclusion of the discussion of a number of tests an **Addendum** has been included which describes one or more related tests that are not discussed in Section VI. As an example, the **Addendum** of the **between-subjects factorial analysis of variance** contains an overview and computational guidelines for the **factorial analysis of variance for a mixed design** and the **within-subjects factorial analysis of variance**.

Although it is not a prerequisite, **The Handbook of Parametric and Nonparametric Statistical Procedures** is designed to be used by those who have a basic familiarity with descriptive statistics and experimental design. Prior familiarity with the latter subject matter will facilitate one's ability to use the book efficiently. In order to insure that the reader has familiarity with these topics, an **Introduction** has been included which provides a general overview of descriptive statistics and experimental design. Following the **Introduction** the reader is provided with guidelines and decision tables for selecting the appropriate statistical test for evaluating a specific experimental design. **The Handbook of Parametric and Nonparametric Statistical Procedures** can be used as a reference book or it can be employed as a textbook in undergraduate and graduate courses that are designed to cover a broad spectrum of parametric and/or nonparametric statistical procedures.

The author would like to express his gratitude to a number of people who helped make this book a reality. First, I would like to thank Tim Pletscher of CRC Press for his confidence in and support of this project. Paul Gottehrer of CRC Press deserves thanks for overseeing the production of the final product. I am also indebted to Glena Ames who did an excellent job preparing the copy-ready manuscript. Finally, I must express my appreciation to my wife Vicki and daughter Emily who both endured and tolerated the difficulties associated with a project of this magnitude.

David Sheskin

Table of Contents with Summary of Topics

Introduction

Although it is assumed that the reader has prior familiarity with these topics, the intent of this Introduction is to provide a general overview of the basic terminology employed within the areas of descriptive statistics and experimental design. It will also review basic concepts that are required for both understanding and using the statistical procedures that are described in this book. Following the Introduction is an outline of all the procedures that are covered, as well as decision making charts to aid the reader in selecting the appropriate statistical procedure.

Descriptive versus Inferential Statistics

Statistics is a field within mathematics that involves the summary and analysis of data. The field of statistics can be divided into two general areas, **descriptive statistics** and **inferential statistics. Descriptive statistics** is a branch of statistics in which data are only used for descriptive purposes and are not employed to make predictions. Thus, descriptive statistics consists of methods and procedures for presenting and summarizing data. The procedures most commonly employed in descriptive statistics are the use of tables and graphs, and the computation of measures of central tendency and variability. **Measures of association or correlation,** which are covered in this book, are also categorized by some sources as descriptive statistical procedures, insofar as they serve to describe the relationship between two or more variables.

Inferential statistics employs data in order to draw inferences (i.e., derive conclusions) or make predictions. Typically, in inferential statistics sample data are employed to draw inferences about one or more populations from which the samples have been derived. Whereas a **population** consists of the sum total of subjects or objects that share something in common with one another, a **sample** is a set of subjects or objects which have been derived from a population. In order for a sample to be useful in drawing inferences about the larger population from which it was drawn, it must be representative of the population. Thus, typically (although there are exceptions) the ideal sample to employ in research is a **random sample.** In a random sample, each subject or object in the population has an equal likelihood of being selected as a member of that sample. In point of fact, it would be highly unusual to find an experiment that employed a truly random sample. Pragmatic and/or ethical factors make it literally impossible in most instances to obtain random samples for research. Insofar as a sample is not random, it will limit the degree to which a researcher will be able to generalize one's results. Put simply, one can only generalize to objects or subjects that are similar to the sample employed.

Statistic versus Parameter

A **statistic** refers to a characteristic of a sample, such as the average score (also known as the **mean**). A **parameter**, on the other hand, refers to a characteristic of a population (such as the average of a whole population). In inferential statistics the computed value of a statistic (e.g., a sample mean) is employed to make inferences about a parameter in the population from which the sample was derived (e.g., the population mean). The statistical procedures described in this book all employ data derived from one or more samples, in order to draw inferences or make predictions with respect to the larger population(s) from which the sample(s) were drawn.

A statistic can be employed for either descriptive or inferential purposes. An example of using a statistic for descriptive purposes is obtaining the mean of a group (which represents a sample) in order to summarize the average performance of the group. On the other hand, if we use the mean of a group to estimate the mean of a larger population the group is supposed to represent, the statistic (i.e., the group mean) is being employed for inferential purposes. The most basic statistics that are employed for both descriptive and inferential purposes are **measures of central tendency** (of which the mean is an example) and **measures of variability**.

When data from a sample are employed to estimate a population parameter, any statistic derived from the sample should be **unbiased**. An **unbiased statistic** is one that provides the most accurate estimate of a population parameter. A **biased statistic**, on the other hand, does not provide as accurate an estimate of that parameter. The subject of bias in statistics will be discussed later in reference to the **mean** (which is the most commonly employed measure of central tendency), and the **variance** (which is the most commonly employed measure of variability).

Levels of Measurement

Typically, information that is quantified in research for purposes of analysis is categorized with respect to what level of measurement the data represent. Different levels of measurement contain different amounts of information with respect to whatever the data are measuring. Statisticians generally conceptualize data as fitting within one of the following four measurement categories: **nominal data** (also known as **categorical data**), **ordinal data** (also know as **rank-order data**), **interval data**, and **ratio data**. As one moves from the lowest level of measurement, nominal data, to the highest level, ratio data, the amount of information provided by the numbers increases, as well the meaningful mathematical operations that can be performed on those numbers. Each of the levels of measurement will now be discussed in more detail.

a) **Nominal/categorical level measurement** In nominal/categorical measurement, numbers are employed merely to identify mutually exclusive categories, but cannot be manipulated in a meaningful mathematical manner. As an example, a person's social security number represents nominal measurement, since it is used purely for purposes of identification, and cannot be meaningfully manipulated in a mathematical sense (i.e., adding, subtracting, etc. the social security numbers of people does not yield anything of tangible value).

b) **Ordinal/rank-order level measurement** In an ordinal scale, the numbers represent rank-orders, and do not give any information regarding the differences between adjacent ranks. Thus, the order of finish in a horse race represents an ordinal scale. If in a race Horse A beats Horse B in a photo finish, and Horse B beats Horse C by twenty lengths, the respective order of finish of the three horses reveals nothing about the fact that the distance between the first and second place horses was minimal, while the difference between second and third place horses was substantial.

c) **Interval level measurement** An interval scale not only considers the relative order of the measures involved (as is the case with an ordinal scale) but, in addition, is characterized by the fact that throughout the length of the scale, equal differences between measurements correspond to equal differences in the amount of the attribute being measured. What this translates to is that if IQ is conceptualized as an interval scale, the one point difference between a person who has an IQ of 100 and someone who has an IQ of 101 should be equivalent to the one point difference between a person who has an IQ of 140 and someone with an IQ of 141. In actuality some psychologists might argue this point, suggesting that a greater increase in intelligence is required to jump from an IQ of 140 to 141, than to jump from an IQ of 100 to 101. If the latter, in fact is true, a one point difference does not reflect the same magnitude of difference across the full range of the IQ scale. Although in practice IQ and most other human characteristics measured by psychological tests (such as anxiety, introversion–extroversion, etc.) are treated as interval scales, many researchers would argue that they are more appropriately categorized as ordinal scales. Such an argument would be based on the fact that such measures do not really meet the requirements of an interval scale, because it cannot be demonstrated that equal numerical differences at different points of the scale are comparable.

It should also be noted that, unlike ratio scales which will discussed next, interval scales do not have a true zero point. If interval scales have a zero score that can be assigned to a person or object, it is assumed to be arbitrary. Thus, in the case of IQ we can ask the question of whether or not there is truly an IQ which is so low that it literally represents zero IQ. In reality, you probably can only say a person who is dead has a zero IQ! In point of fact, someone who has obtained an IQ of zero on an IQ test has been assigned that score because his performance on the test was extremely poor. The zero IQ designation does not necessarily mean the person could not answer any of the test questions (or, to go further, that the individual possesses none of the requisite skills or knowledge for intelligence). The developers of the test just decided to select a certain minimum score on the test and designate it as the zero IQ point.

d) **Ratio level measurement** As is the case with interval level measurement, ratio level measurement is also characterized by the fact that throughout the length of the scale, equal differences between measurements correspond to equal differences in the amount of the attribute being measured. However, ratio level measurement is also characterized by the fact that it has a true zero point. Because of the latter, with ratio measurement one is able to make meaningful ratio statements with regard to the attribute/variable being measured. To illustrate these points, most physical measures such as weight, height, blood glucose level, as well as measures of certain behaviors such as the number of times a person coughs or the number of times a child cries, represent ratio scales. For all of the aforementioned measures there is a true zero point (i.e., zero weight, zero height, zero blood glucose, zero coughs, zero episodes of crying), and for each of these measures one is able to make meaningful ratio statements (such as Ann weighs twice as much as Joan, Bill is one-half the height of Steve, Phil's blood glucose is 100 times Sam's, Mary coughs five times as often as Pete, and Billy cries three times as much as Heather).

Continuous versus Discrete Variables

When measures are obtained on people or objects, in most instances we assume that there will be variability. Since we assume variability, if we are quantifying whatever it is that is being measured, not everyone or everything will produce the same score. For this reason, when something is measured it is commonly referred to as a **variable**. As noted above, variables can be categorized with respect to the level of measurement they represent. In addition, a variable can be categorized with respect to whether it is **continuous** or **discrete**. A **continuous variable** can assume any value within the range of scores that define the limits of that variable. A **discrete variable**, on the other hand, can only assume a limited number of values. To illustrate, temperature (which can assume both integer and fractional/decimal values within a given range) is a **continuous variable**. Theoretically there are an infinite number of possible temperature values, and the number of temperature values we can measure is only limited by the precision of the instrument we are employing to obtain the measurements. The face value of a die, on the other hand, is a **discrete variable**, since it can only assume the integer values 1 through 6.

Measures of Central Tendency

Earlier in the **Introduction** it is noted that the most commonly employed statistics are measures of central tendency and measures of variability. This section will describe three measures of central tendency: the **mode,** the **median,** and the **mean.**

The mode The **mode** is the most frequently occurring score in a distribution of scores. Thus, in the following distribution of scores the mode is 5 since it occurs two times, whereas all other scores occur only once: 0, 1, 2, 5, 5, 8, 10. If more than one score occurs with the highest frequency, it is possible to have two or more modes in a distribution. Thus, in the distribution 0, 1, 2, 5, 6, 8, 10, all of the scores represent the mode, since each score occurs one time. A distribution with more than one mode is referred to as a **multimodal distribution.** If it happens that two scores both occur with the highest frequency, the distribution would be described as a **bimodal** distribution, which represents one type of multimodal distribution. The distribution 0, 5, 5, 8, 9, 9, 12 is bimodal, since the scores 5 and 9 both occur two times, and all other scores appear once. The most common situation for which the mode is employed in evaluating data is when a large body of data is presented in a tabular format listing the frequency of each score. Such a table is referred to as a **frequency distribution.** The score with the highest frequency in the distribution represent the modal score for the sample. A mode that is derived for a sample is a statistic, whereas the mode of a population is a parameter.

The median The **median** is the middle score in a distribution. If there is an odd number of scores in a distribution, in order to determine the median the following protocol should be employed: Divide the total number of scores by 2 and add .5 to the result of the division. The obtained value indicates the **ordinal position** of the score which represents the median of the distribution (note that this value does not represent the median). Thus, if we have a distribution consisting of five scores (e.g., 6, 8, 9, 13, 16), we divide the number of scores in the distribution by two, and add .5 to the result of the division. Thus, $(5/2) + .5 = 3$. The obtained value of 3 indicates that if the five scores are arranged ordinally (i.e., from lowest to highest), the median is the 3rd highest (or 3rd lowest) score in the distribution. With respect to the distribution 6, 8, 9, 13, 16, the value of the median will equal 9, since 9 is the score in the third ordinal position.

If there are an even number of scores in a distribution, there will be two middle scores. The median is the average of the two middle scores. To determine the ordinal positions of the two middle scores, divide the total number of scores in the distribution by 2. The number value obtained by that division and the number value that is one above it represent the ordinal positions of the two middle scores. To illustrate, assume we have a distribution consisting of the following six scores: 6, 8, 9, 12, 13, 16. To determine the median, we initially divide 6 by 2 which equals 3. Thus, if we arrange the scores ordinally, the 3rd and 4th (since $3 + 1 = 4$) scores are the middle scores. The average of these scores, which are respectively 9 and 12, is the median (which will be represented by the notation M). Thus, $M = (9 + 12)/2 = 10.5$. Note once again that in this example, as was the case in the previous one, the initial values computed (3 and 4) do not themselves represent the median, but instead represent the ordinal position of the scores used to compute the median. As was the case with the mode, a median value derived for a sample is a statistic, whereas the median of a whole population is a parameter.

The mean The **mean,** which is the most commonly employed measure of central tendency, is the average score in a distribution. Within the framework of the discussion to follow, the notation n will represent the number of subjects or objects in a sample, and the notation N will represent the total number of subjects or objects in the population from which the sample is derived.

Equation I.1 is employed to compute the mean of a sample. Σ, which is the upper case Greek letter **sigma**, is a summation sign. The notation ΣX indicates that the set of scores should be summed.

$$\bar{X} = \frac{\Sigma X}{n} \qquad \textbf{(Equation I.1)}$$

Sometimes Equation I.1 is written in the following more complex but equivalent form containing subscripts: $\bar{X} = \Sigma_{i=1}^{n} X_i / n$. In the latter equation, the notation $\Sigma_{i=1}^{n} X_i$ indicates that beginning with the first score, scores 1 through n (i.e., all the scores) are to be summed. X_i represents the score of the i th subject or object.

Equation I.1 will now be applied to the following distribution of five scores: 6, 8, 9, 13, 16. Since $n = 5$ and $\Sigma X = 52$, $\bar{X} = \Sigma X/n = 52/5 = 10.4$.

Whereas Equation I.1 describes how one can compute the mean of a sample, Equation I.2 describes how one can compute the mean of a population. The simplified version without subscripts is to the right of the first = sign, and the subscripted version of the equation is to the right of the second = sign. The mean of a population is represented by the notation μ, which is the lower case Greek letter **mu**. In practice, it would be highly unusual to have occasion to compute the mean of a population. Indeed, a great deal of analysis in inferential statistics is concerned with trying to estimate the mean of a population from the mean of a sample.

$$\mu = \frac{\Sigma X}{N} = \frac{\sum_{i=1}^{N} X}{N} \qquad \textbf{(Equation I.2)}$$

Note that in Equation I.2 all N scores in the population are summed and averaged, as opposed to just summing n scores when the value of $\bar{X}$ is computed. The sample mean $\bar{X}$ is an **unbiased estimate** of the population mean μ, which indicates that if one has a distribution of n scores, $\bar{X}$ provides the best possible estimate of the true value of μ. Typically, when the mean is used as a measure of central tendency, it is employed with interval or ratio level data.

Measures of Variability

In this section a number of measures of variability will be discussed. Primary emphasis, however, will be given to the **standard deviation** and the **variance**, which are the most commonly employed measures of variability.

a) **The range** The **range** is the difference between the highest and lowest score in a distribution. Thus in the distribution 2, 3, 5, 6, 7, 12, the range is the difference between 12 (the highest score) and 2 (the lowest score). Thus: Range = 12 − 2 = 10. Some sources add one to the obtained value, and would thus say that the Range = 11. Although the **range** is employed on occasion for descriptive purposes, it is of little use in inferential statistics.

b) **Percentiles, quartiles, and deciles** A percentile or percentile rank is the point in a distribution at which a given percentage of scores falls at or below. Thus if on an IQ test a score of 115 falls at the 84th percentile, it means 84% of the population has an IQ of 115 or less. A distribution can be divided into four **quartiles**, which are defined as the 25th percentile, the 50th percentile, the 75th percentile, and the 100% percentile. The **interquartile range** is the difference between the scores at the 75th percentile (which is the upper limit of the third quartile) and the 25th percentile (the upper limit of the first quartile). A distribution can be divided into ten **quantiles**, which are defined by the 10th percentile, 20th percentile, ..., 90th percentile, and 100th percentile. The **interdecile range** is the difference between the scores at the 90th percentile (the upper limit of the ninth decile) and the 10th percentile (the upper limit of the first decile).

Infrequently, the interquartile and/or interdecile ranges may be employed to represent variability. An example of a situation where a researcher might elect to employ either of these measures would be when the researcher wishes to omit a few extreme scores in a distribution. Such extreme scores are referred to as **outliers**. Specifically, an **outlier** is a score in set of data which is so extreme, that by all appearances it is not representative of the population from which the sample is ostensibly derived. Since the presence of outliers can dramatically affect variability (as well as the value of the sample mean), their presence may lead a researcher to believe that the variability of a distribution might best be expressed through use of the interquartile or interdecile range (as well as the fact that the median, instead of the mean, should be employed as the measure of central tendency for the sample).

c) **The variance and the standard deviation** The most commonly employed measures of variability in both inferential and descriptive statistics are the **variance** and the **standard deviation**. These two measures are directly related to one another, since the standard deviation is the square root of the variance (and thus the variance is the square of the standard deviation). As is the case with the mean, the standard deviation and the variance are generally only employed with interval or ratio level data.

The formal definition of the variance is that it is the mean of the squared difference scores (which are also referred to as **deviation scores**). This definition implies that in order to compute the variance of a distribution one must subtract the mean of the distribution from each score in the distribution, square each of the difference scores, sum the squared difference scores, and divide the latter value by the number of scores in the distribution. The logic of this definition is reflected in the definitional equations which will be presented later in this section for both the variance and the standard deviation.

A **definitional equation** for a statistic (or parameter) contains the specific mathematical operations that are described in the definition of that statistic (or parameter). On the other hand, a **computational equation** for the same statistic (or parameter) does not clearly reflect the definition of that statistic (or parameter). A computational equation, however, facilitates computation of the statistic (or parameter), since it is computationally less involved than the definitional equation. In this book, in instances where a definitional and computational equation are available for computing a test statistic, in order to facilitate calculations the computational equation will be employed.

The following notation will be used in the book with respect to the values of the variance and the standard deviation.

σ^2 (where σ is the lower case Greek letter **sigma**) will represent the **variance of a population.**

s^2 will represent the variance of a sample, when the variance is just being employed for descriptive purposes. s^2 will be a **biased estimate of the population variance** σ^2, and because of this s^2 will generally underestimate the true value of σ^2.

$\tilde{s}^2$ will represent the variance of a sample, when the variance is being employed for inferential purposes. $\tilde{s}^2$ will be an **unbiased estimate of the population variance** σ^2.

σ will represent the **standard deviation of a population.**

s will represent the standard deviation of a sample, when the standard deviation is just being employed for descriptive purposes. s will be a **biased estimate of the population standard deviation** σ, and because of this s will generally underestimate the true value of σ.

$\tilde{s}$ will represent the standard deviation of a sample, when the standard deviation is being employed for inferential purposes. $\tilde{s}$ will be an **unbiased estimate of the population standard deviation** σ.

Equations I.3–I.8 are employed to compute the values σ^2, s^2, $\tilde{s}^2$, σ, s, and $\tilde{s}$. Note that in each case, two equivalent methods are presented for computing the statistic or parameter in question. The formula to the left is the definitional equation, whereas the formula to the right is the computational equation.

$$\sigma^2 = \frac{\Sigma(X - \mu)^2}{N} = \frac{\Sigma X^2 - \frac{(\Sigma X)^2}{N}}{N} \qquad \textbf{(Equation I.3)}$$

$$s^2 = \frac{\Sigma(X - \bar{X})^2}{n} = \frac{\Sigma X^2 - \frac{(\Sigma X)^2}{n}}{n} \qquad \textbf{(Equation I.4)}$$

$$\tilde{s}^2 = \frac{\Sigma(X - \bar{X})^2}{n - 1} = \frac{\Sigma X^2 - \frac{(\Sigma X)^2}{n}}{n - 1} \qquad \textbf{(Equation I.5)}$$

$$\sigma = \sqrt{\frac{\Sigma(X - \mu)^2}{N}} = \sqrt{\frac{\Sigma X^2 - \frac{(\Sigma X)^2}{N}}{N}} \qquad \textbf{(Equation I.6)}$$

$$s = \sqrt{\frac{\Sigma(X - \bar{X})^2}{n}} = \sqrt{\frac{\Sigma X^2 - \frac{(\Sigma X)^2}{n}}{n}} \qquad \textbf{(Equation I.7)}$$

$$\tilde{s} = \sqrt{\frac{\Sigma(X - \bar{X})^2}{n - 1}} = \sqrt{\frac{\Sigma X^2 - \frac{(\Sigma X)^2}{n}}{n - 1}} \qquad \textbf{(Equation I.8)}$$

When the variance or standard deviation of a sample is computed within the framework of an inferential statistical test, one always wants an unbiased estimate of the population variance or the population standard deviation. Thus, the computational form of Equation I.5 will be employed throughout this book when a sample variance is employed to estimate a population variance, and the computational form of Equation I.8 will be employed when a sample standard deviation is used to estimate a population standard deviation.

The reader should take note of the fact that some sources employ subscripted versions of the above equations. Thus, the computational form of Equation I.5 is often written as.

$$\tilde{s}^2 = \frac{\sum_{i=1}^{n} X_i^2 - \frac{\left(\sum_{i=1}^{n} X_i\right)^2}{n}}{n - 1}$$

Although the subscripted version will not be employed for computing the values of $\tilde{s}^2$ and $\tilde{s}$, in the case of some equations that are presented, subscripted versions may be employed in order to clarify the mathematical operations involved in an equation.

As noted previously, for the same set of data the value of $\tilde{s}^2$ will always be larger than the value of s^2. This can be illustrated with a distribution consisting of the five scores: 6, 8, 9, 13, 16. The following values are substituted in Equations I.4 and I.5: $\Sigma X = 52$, $\Sigma X^2 = 606$, $n = 5$.

$$s^2 = \frac{606 - \frac{(52)^2}{5}}{5} = 13.04$$

$$\tilde{s}^2 = \frac{606 - \frac{(52)^2}{5}}{5 - 1} = 16.3$$

Since the standard deviation is the square root of the variance, we can quickly determine that $s = \sqrt{s^2} = \sqrt{13.04} = 3.61$ and $\tilde{s} = \sqrt{\tilde{s}^2} = \sqrt{16.3} = 4.04$. Note that $\tilde{s}^2 > s^2$ and $\tilde{s} > s$.[1]

The reader should take note of the following with respect to the standard deviation and the variance:

a) The value of a standard deviation or a variance can never be a negative number. If a negative number is ever obtained for either value, it indicates a mistake has been made in the calculations. The only time the value of a standard deviation or variance will not be a positive number is when its value equals zero. The only instance in which the value of both the standard deviation and variance of a distribution will equal zero is when all of the scores in the distribution are identical to one another.

b) As the value of the sample size (n) increases, the difference between the values of s^2 and $\tilde{s}^2$ will decrease. In the same respect, as the value of n increases, the difference between the values of s and $\tilde{s}$ will decrease. Thus, the biased estimate of the variance (or standard deviation) will be more likely to underestimate the true value of the population variance (or standard deviation) with small sample sizes than it will with large sample sizes.

c) The numerator of any of the equations employed to compute a variance or a standard deviation is often referred to as the **sum of squares**. Thus in the example in this section, the value of the sum of squares is 65.2, since $\Sigma X^2 - [(\Sigma X)^2/n] = 606 - [(52)^2/5] = 65.2$. The denominators of both Equation I.5 and Equation I.8 are often referred to as the **degrees of freedom** (a concept that is discussed later in the book within the framework of the **single-sample *t* test (Test 2)**). Based on what has been said with respect to the sum of squares and the degrees of freedom, the variance is sometimes defined as the sum of squares divided by the degrees of freedom.

The Normal Distribution

When an inferential statistical test is employed with one or more samples to draw inferences about one or more populations, such a test may make certain assumptions about the shape of an underlying population distribution. The most commonly encountered assumption in this regard is that a distribution is **normal**. When viewed from a visual perspective, the **normal distribution** (which is often referred to as the **bell-shaped curve**) is a graph of a frequency distribution which can be described mathematically and observed empirically (insofar as many variables in the real world appear to be distributed normally). The shape of the normal distribution is such that the closer a score is to the mean, the more frequently it occurs. As scores deviate more and more from the mean (i.e., become higher or lower), the more extreme the score, the lower the frequency with which that score occurs.

Obviously data can be distributed in ways that are other than normal. Two examples of other types of distributions are a **uniform** distribution and a **skewed** distribution. In a **uniform distribution** each score occurs with approximately the same frequency. If a distribution is **skewed**, it will be characterized by the presence of a disproportionately large number of high scores (in which case the distribution is **negatively skewed/skewed to the left**), or a disproportionately large number of low scores (in which case the distribution is **positively skewed/skewed to the right**).

When the normal distribution is assigned a mean value of 0 (i.e., $\mu = 0$) and a standard deviation of 1 (i.e., $\sigma = 1$), it is described as the **standard normal distribution.** The **standard normal distribution,** which is represented in Figure I.1, is employed more frequently in inferential statistics than any other theoretical probability distribution. The use of the term theoretical probability distribution in this context is based on the fact that it is known that in the standard normal distribution a certain proportion of cases will always fall within specific areas of the curve. As a result of this, if one knows how far removed a score is from the mean of the distribution, one can specify the proportion of cases that obtain that score, as well as the likelihood of randomly selecting a subject or object with that score.

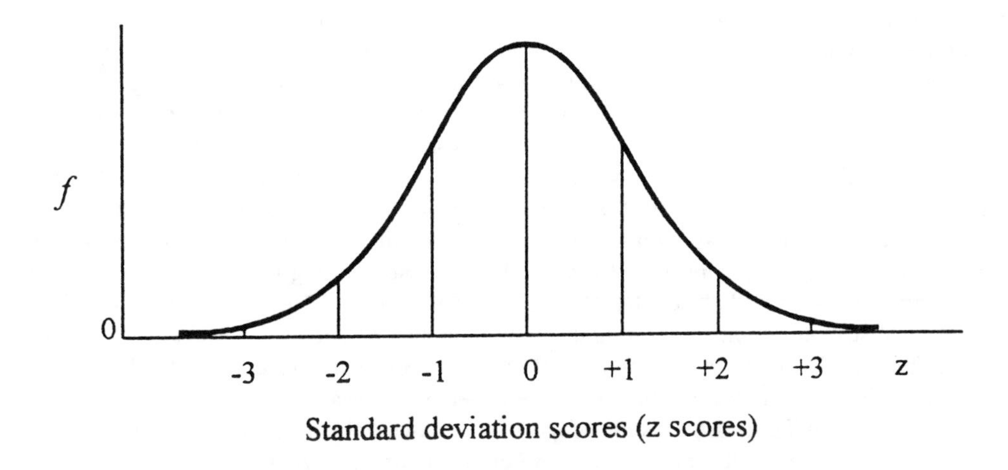

Figure I.1 The Standard Normal Distribution

The Y-axis (which is referred to as the **ordinate**) of the distribution depicted in Figure I.1 represents the frequency (f) with which scores occur in the standard normal distribution, while the X-axis (which is referred to as the **abscissa**) lists the range of possible scores. More specifically, the scores are represented on the X-axis with respect to how many standard deviation units they fall above or below the mean. For any variable that is normally distributed, regardless of the values of the population mean and population standard deviation, the distance of a score from the mean in standard deviation units can be computed with Equation I.9.

$$z = \frac{X - \mu}{\sigma} \qquad \textbf{(Equation I.9)}$$

Where: X is a specific score
μ is the value of the population mean
σ is the value of the population standard deviation

When Equation I.9 is employed, any score that is above the mean will yield a positive z score, and any score that is below the mean will yield a negative z score. Any score that is equal to the mean will yield a z score of zero.

To illustrate this, assume we have an IQ test for which it is known that the population mean is $\mu = 100$ and the population standard deviation is $\sigma = 15$. Assume three people take the test and obtain the following IQ scores: **Person A:** 135; **Person B:** 65; and **Person C:** 100. The z score (standard deviation score) for each person is computed below. The reader should take note of the fact that a z score is always computed to at least two decimal places.

$$\text{Person A: } z = \frac{135 - 100}{15} = 2.33$$

$$\text{Person B: } z = \frac{65 - 100}{15} = -2.33$$

$$\text{Person C: } z = \frac{100 - 100}{15} = 0$$

Thus, Person A obtains an IQ score that is 2.33 standard deviation units above the mean, Person B obtains an IQ score that is 2.33 standard deviation units below the mean, and Person C obtains an IQ score at the mean. If one wanted to determine the likelihood (i.e., the probability) of selecting a person (as well as the proportion of people) who obtains a specific score in a normal distribution, **Table A1 (Table of the Normal Distribution)** in the **Appendix** can provide this information. Although **Table A1** is comprised of four columns, for the analysis to be discussed in this section we will only be interested in the first three columns.

Column 1 in **Table A1** lists z scores that range from 0 to an **absolute value** of 4.[2] The use of the term **absolute value of 4** is based on the fact that since the normal distribution is symmetrical, anything we say with respect to the probability or the proportion of cases associated with a positive z score will also apply to the corresponding negative z score. Note that positive z scores will always fall to the right of the mean (often referred to as the **right tail** of the distribution), thus indicating that the score is above the mean. Negative z scores, on the other hand, will always fall to the left of the mean (often referred to as the **left tail** of the distribution), thus indicating that the score is below the mean.

Column 2 in **Table A1** lists the proportion of cases (which can also be interpreted as probability values) that falls between the mean of the distribution and the z score that appears in a specific row.

Column 3 in the table lists the proportion of cases that falls beyond the z score in that row. More specifically, the proportion listed in Column 3 is evaluated in relation to the tail of the distribution in which the score appears. Thus, if a z score is positive, the value in Column 3 will represent the proportion of cases that falls above that z score, whereas if the z score is negative, the value in Column 3 will represent the proportion of cases that falls below that z score.

Table A1 will now be employed in reference to the IQ scores of Person A and Person B. For both subjects the computed absolute value of z associated with their IQ score is $z = 2.33$. For $z = 2.33$, the tabled values in Columns 2 and 3, are respectively, .4901 and .0099. The value in Column 2 indicates that the proportion of the population that obtains a z score between the mean and $z = 2.33$ is .4901 (which expressed as a percentage is 49.01%),[3] and the proportion of the population which obtains a z score between the mean and $z = -2.33$ is .4901. We can make comparable statements with respect to the IQ values associated with these z scores. Thus, we can say that the proportion of the population that obtains an IQ score between 100 and 135 is .4901, and the proportion of the population which obtains an IQ score between 65 and 100 is .4901. Since the normal distribution is symmetrical, .5 (or 50%) represents the proportion of cases that falls both above and below the mean. Thus, we can determine that $.5 + .4901 = .9901$ (or 99.01%) is the proportion of people with an IQ of 135 or less, as well as the proportion of people who have an IQ of 65 or greater.

The value in Column 3 indicates that the proportion of the population that obtains a score of $z = 2.33$ or greater (and thus, in reference to Person A, an IQ of 135 or greater) is .0099 (which is .99%). In the same respect, the proportion of the population which obtains a score of $z = -2.33$ or less (and thus, in reference to Person B, an IQ of 65 or less) is .0099.

If one interprets the values in Columns 2 and 3 as probability values instead of proportions, we can state that if one randomly selects a person from the population, the probability of selecting someone with an IQ of 135 or greater will be approximately 1%. In the same respect, the probability of selecting someone with an IQ of 65 or less will also be approximately 1%.

In the case of Person C, whose IQ score of 100 results in the standard deviation score $z = 0$, inspection of **Table A1** reveals that the values in Columns 2 and 3 associated with $z = 0$ are, respectively, .0000 and .5000. This indicates that the proportion of the population that obtains an IQ of 100 or greater is .5 (which is equivalent to 50%), and that the proportion of the population which obtains an IQ of 100 or less is .5. Thus, if we randomly select a person from the population, the probability of selecting someone with an IQ equal to or greater than 100 will be .5, and the probability of selecting someone with an IQ equal to or less than 100 will be .5.

Figure I.2 provides a graphic summary of the results of the above analysis.

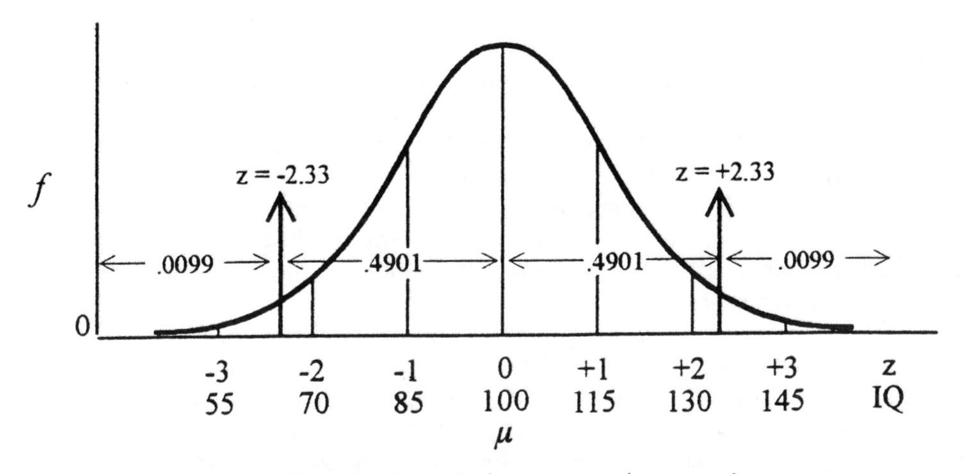

Figure I.2. Summary of Normal Curve Problem

Hypothesis Testing

Inferential statistics primarily employs sample data in two ways to draw inferences about one or more populations. The two methodologies employed in inferential statistics are **hypothesis testing** and **estimation of population parameters**. This section will discuss hypothesis testing.

Within the framework of inferential statistics, a **hypothesis** can be defined as a prediction about a single population or about the relationship between two or more populations. **Hypothesis testing** is a procedure in which sample data are employed to evaluate a hypothesis. In using the term hypothesis, some sources make a distinction between a **research hypothesis** and **statistical hypotheses**.

A **research hypothesis** is a general statement of what a researcher predicts. Two examples of a research hypothesis are: a) The average IQ of all males is some value other than 100; and b) Clinically depressed patients who take an antidepressant for six months will be less depressed than clinically depressed patients who take a placebo for six months.

In order to evaluate a research hypothesis, it is restated within the framework of two **statistical hypotheses**. Through use of a symbolic format, the statistical hypotheses summarize the research hypothesis with reference to the population parameter or parameters under study. The two statistical hypotheses are the **null hypothesis**, which is represented by the notation H_0 and, the **alternative hypothesis**, which is represented by the notation H_1.

The **null hypothesis** is a statement of **no effect** or **no difference**. Since the statement of the research hypothesis generally predicts the presence of an effect or a difference with respect to whatever it is that is being studied, the null hypothesis will generally be a hypothesis that the researcher expects to be rejected. The **alternative hypothesis**, on the other hand, represents a statistical statement indicating the **presence of an effect or a difference**. Since the research hypothesis typically predicts an effect or difference, the researcher generally expects the alternative hypothesis to be supported.

The null and alternative hypotheses will now be discussed in reference to the two research hypotheses noted earlier. Within the framework of the first research hypothesis that was presented, we will assume that a study is conducted in which an IQ score is obtained for each of n males who have been randomly selected from a population comprised of N males. The null and alternative hypotheses can be stated as follows: H_0: $\mu = 100$ and H_1: $\mu \neq 100$. The null hypothesis states that the mean (IQ score) of the population the sample represents equals 100. The alternative hypothesis states that the mean of the population the sample represents does not equal 100. The absence of an effect will be indicated by the fact that the sample mean is equal to or reasonably close to 100. If such an outcome is obtained, a researcher can be reasonably confident that the sample has come from a population with a mean value of 100. The presence of an effect, on the other hand, will be indicated by the fact that the sample mean is significantly above or below the value 100. Thus, if the sample mean is substantially larger or smaller than 100, the researcher can conclude there is a high likelihood that the population mean is some value other than 100, and thus reject the null hypothesis.

As stated above, the alternative hypothesis is **nondirectional**. A **nondirectional** (also referred to as a **two-tailed**) **alternative hypothesis** does not make a prediction in a specific direction. The alternative hypothesis H_1: $\mu \neq 100$ just states that the population mean will not equal 100, but it does not predict whether it will be less than or greater than 100. If, however, a researcher wants to make a prediction with respect to direction, the alternative hypothesis can also be stated **directionally**. Thus, with respect to the above example, either of the following two **directional** (also referred to as **one-tailed**) **alternative hypotheses** can be employed: H_1: $\mu > 100$ or H_1: $\mu < 100$.

The alternative hypothesis H_1: $\mu > 100$ states that the mean of the population the sample represents is some value greater than 100. If the directional alternative hypothesis H_1: $\mu > 100$ is employed, the null hypothesis can only be rejected if the data indicate that the population mean is some value above 100. The null hypothesis cannot, however, be rejected if the data indicate that the population mean is some value below 100.

The alternative hypothesis H_1: $\mu < 100$ states that the mean of the population the sample represents is some value less than 100. If the directional alternative hypothesis H_1: $\mu < 100$ is employed, the null hypothesis can only be rejected if the data indicate that the population mean is some value below 100. The null hypothesis cannot, however, be rejected if the data indicate that the population mean is some value above 100. The reader should take note of the fact that although there are three possible alternative hypotheses that one can employ (one that is nondirectional and two that are directional), the researcher will select only one of the alternative hypotheses.

Researchers are not in agreement with respect to the conditions under which one should employ a nondirectional or a directional alternative hypothesis. Some researchers take the position that a nondirectional alternative hypothesis should always be employed, regardless of one's prior expectations about the outcome of an experiment. Other researchers believe that a nondirectional alternative hypothesis should only be employed when one has no prior expectations about the outcome of an experiment (i.e., no expectation with respect to the direction of an effect or difference). These same researchers believe that if one does have a definite expectation about the direction of an effect or difference, a directional alternative hypothesis should be employed. One advantage of employing a directional alternative hypothesis is that in order to reject the null hypothesis, a directional alternative hypothesis does not require that there be as large an effect or difference in the sample data as will be the case if a nondirectional alternative hypothesis is employed.

The second of the research hypotheses discussed earlier in this section predicted that an antidepressant will be more effective that a placebo in treating depression. Let us assume that in order to evaluate this research hypothesis, a study is conducted which involves two groups of clinically depressed patients. One group, which will represent Sample 1, is comprised of n_1 patients, and the other group, which will represent Sample 2, is comprised of n_2 patients. The subjects in Sample 1 take an antidepressant for six months, and the subjects in Sample 2 take a placebo during the same period of time. After six months have elapsed, each subject is assigned a score with respect to his or her level of depression.

The null and alternative hypotheses can be stated as follows: H_0: $\mu_1 = \mu_2$ and H_1: $\mu_1 \neq \mu_2$. The null hypothesis states that the mean (depression score) of the population Sample 1 represents equals the mean of the population Sample 2 represents. The alternative hypothesis (which is stated **nondirectionally**) states that the mean of the population Sample 1 represents does not equal the mean of the population Sample 2 represents. In this instance the two populations are a population comprised of N_1 clinically depressed people who take an antidepressant for six months versus a population comprised of N_2 clinically depressed people who take a placebo for six months. The absence of an effect or difference will be indicated by the fact that the two sample means are exactly the same value or close to being equal. If such an outcome is obtained, a researcher can be reasonably confident that the samples do not represent two different populations.[4] The presence of an effect, on the other hand, will be indicated if a significant difference is observed between the two sample means. Thus, we can reject the null hypothesis if the mean of Sample 1 is significantly larger than the mean of Sample 2, or the mean of Sample 1 is significantly smaller than the mean of Sample 2.

As is the case with the first research hypothesis discussed earlier, the alternative hypothesis can also be stated directionally. Thus, either of the following two directional alternative hypotheses can be employed: H_1: $\mu_1 > \mu_2$ or H_1: $\mu_1 < \mu_2$.

The alternative hypothesis H_1: $\mu_1 > \mu_2$ states that the mean of the population Sample 1 represents is greater than the mean of the population Sample 2 represents. If the directional alternative hypothesis H_1: $\mu_1 > \mu_2$ is employed, the null hypothesis can only be rejected if the data indicate that the mean of Sample 1 is significantly greater than the mean of Sample 2. The null hypothesis cannot, however, be rejected if the mean of Sample 1 is significantly less than the mean of Sample 2.

The alternative hypothesis H_1: $\mu_1 < \mu_2$ states that the mean of the population Sample 1 represents is less than the mean of the population Sample 2 represents. If the directional alternative hypothesis H_1: $\mu_1 < \mu_2$ is employed, the null hypothesis can only be rejected if the data indicate that the mean of Sample 1 is significantly less than the mean of Sample 2. The null hypothesis cannot, however, be rejected if the mean of Sample 1 is significantly greater than the mean of Sample 2.

Upon collecting the data for a study, the next step in the hypothesis testing procedure is to evaluate the data through use of the appropriate inferential statistical test. An inferential statistical test yields a **test statistic**. The latter value is interpreted by employing special tables that contain information with regard to the expected distribution of the test statistic. More specifically, such tables contain extreme values of the test statistic (referred to as **critical values**) that are highly unlikely to occur if the null hypothesis is true. Such tables allow a researcher to determine whether or not the result of a study is **statistically significant.**

The term **statistical significance** implies that one is determining whether an obtained difference in an experiment is due to chance or is the result of a genuine experimental effect. To clarify this, think of a roulette wheel on which there are 38 possible numbers which may occur on any roll of the wheel. Suppose we spin a wheel 38,000 times. On the basis of chance each number should occur 1/38th of the time, and thus each value should occur 1000 times (i.e., 38000 ÷ 38). Suppose the number 32 occurs 998 times in 38,000 spins of the wheel. Since this value is close to the expected value of 1000, it is highly unlikely that the wheel is biased against the number 32, and is thus not a fair wheel (at least in reference to the number 32). This is because 998 is extremely close to 1000, and a difference of 2 outcomes isn't unlikely on the basis of the random occurrence of events (i.e., chance). On the other hand, if the number 32 only occurs 380 times in 38,000 trials (i.e., 1/100th of the time), since 380 is well below the expected value of 1000, this strongly suggests that the wheel is biased against the number 32 (and is thus probably biased in favor of one or more of the other numbers). On the basis of this, one would probably conclude that the wheel is defective and should be replaced.

When evaluating the results of an experiment, one employs a logical process similar to that involved in the above situation with the roulette wheel. The decision on whether to retain or reject the null hypothesis is based on contrasting the observed outcome of an experiment with the outcome one can expect if, in fact, the null hypothesis is true. This decision is made by using the appropriate inferential statistical test. An inferential statistical test is essentially an equation which describes a set of mathematical operations that are to be performed on the data obtained in a study. The end result of conducting such a test is a final value which is designated as the **test statistic.** A test statistic is evaluated in reference to a **sampling distribution**, which is a theoretical probability distribution of all the possible values the test statistic can assume if one were to conduct an infinite number of studies which employed a sample size equal to that used in study being evaluated. The probabilities in a sampling distribution are based on the assumption that each of the samples is randomly drawn from the population it represents.

When evaluating the study involving the use of a drug versus a placebo in treating depression, one is asking if the difference between the scores of the two groups is due to chance, or if instead, it is due to some nonchance factor (which in a well controlled study will be the differential treatment to which the groups are exposed). The larger the difference between the average scores of the two groups (just like the larger the difference between the observed and expected occurrence of a number on a roulette wheel), the less likely the difference is due to chance factors, and the more likely it is due to the experimental treatments. Thus, by declaring a difference statistically significant, the researcher is saying that based on an analysis of the sampling distribution of the test statistic, it is highly unlikely that a difference equal to or greater than that which was observed in the study could have occurred as result of chance. In view of this, the most logical decision is to conclude that the difference is due to the experimental treatments, and thus reject the null hypothesis.

Scientific convention has established that in order to declare a difference statistically significant, there can be no more than a 5% likelihood that the difference is due to chance. If a researcher believes that 5% is too high a value, one may elect to employ a 1%, or an even lower minimum likelihood, before one will be willing to conclude that a difference is significant. The notation $p > .05$ is employed to indicate that the result of an experiment is not significant. This notation indicates that there is a greater than 5% likelihood that an observed difference or effect could be due to chance. On the other hand, the notation $p < .05$ indicates that the outcome of a study is significant at the .05 level.[5] This indicates that there is less than a 5% likelihood that an obtained difference or effect can be due to chance. The notation $p < .01$ indicates a significant result at the .01 level (i.e., there is less than a 1% likelihood that the difference is due to chance).

When the normal distribution is employed for inferential statistical analysis, four tabled critical values are commonly employed. These values are summarized in Table I.1.

Table I.1 Tabled Critical Two-Tailed and One-Tailed .05 and .01 z Values

	$z_{.05}$	$z_{.01}$
Two-tailed values	1.96	2.58
One-tailed values	1.65	2.33

The value $z = 1.96$ is referred to as the tabled critical two-tailed .05 z value. This value is employed since the total proportion of cases in the normal distribution that falls above $z = +1.96$ or below $z = -1.96$ is .05. This can be confirmed by examining Column 3 of **Table A1** for the value $z = 1.96$. The value of .025 in Column 3 indicates that the proportion of cases in the right tail of the curve that falls above $z = +1.96$ is .025, and the proportion of cases in the left tail of the curve that falls below $z = -1.96$ is .025. If the two .025 values are added, the resulting proportion is .05. Note that this is a two-tailed critical value, since the proportion .05 is based on adding the extreme 2.5% of the cases from the two tails of the distribution.

The value $z = 2.58$ is referred to as the tabled critical two-tailed .01 z value. This value is employed since the total proportion of cases in the normal distribution that falls above $z = +2.58$ or below $z = -2.58$ is .01. This can be confirmed by examining Column 3 of **Table A1** for the value $z = 2.58$. The value of .0049 (which rounded off equals .005) in Column 3 indicates that the proportion of cases in the right tail of the curve that falls above $z = +2.58$ is .0049, and the proportion of cases in the left tail of the curve that falls below $z = -2.58$ is .0049. If the two .0049 values are added, the resulting proportion is .0098, which rounded off equals .01. Note that this is a two-tailed critical value, since the proportion .01 is based on adding the extreme .5% of the cases from the two tails of the distribution.

The value $z = 1.65$ is referred to as the tabled critical one-tailed .05 z value. This value is employed since the proportion of cases in the normal distribution that falls above $z = +1.65$ or below $z = -1.65$ in each tail of the distribution is .05. This can be confirmed by examining Column 3 of **Table A1** for the value $z = 1.65$. The value of .0495 (which rounded off equals .05) in Column 3 indicates that the proportion of cases in the right tail of the curve that falls above $z = +1.65$ is .0495, and the proportion of cases in the left tail of the curve that falls below $z = -1.65$ is .0495. Note that this is a one-tailed critical value, since the proportion .05 is based on the extreme 5% of the cases in one tail of the distribution.[6]

The value $z = 2.33$ is referred to as the tabled critical one-tailed .01 z value. This value is employed since the proportion of cases in the normal distribution that falls above $z = +2.33$ or below $z = -2.33$ in each tail of the distribution is .01. This can be confirmed by examining Column 3 of **Table A1** for the value $z = 2.33$. The value of .0099 (which rounded off equals .01) in Column 3 indicates that the proportion of cases in the right tail of the curve that falls above $z = +2.33$ is .0099, and the proportion of cases in the left tail of the curve that falls below $z = -2.33$ is .0099. Note that this is a one-tailed critical value, since the proportion .01 is based on the extreme 1% of the cases in one tail of the distribution.

Although in practice it is not scrupulously adhered to, the conventional hypothesis testing model employed in inferential statistics assumes that prior to conducting a study a researcher stipulates whether a directional or nondirectional alternative hypothesis will be employed, as well as at what level of significance the null hypothesis will be evaluated. The probability value which identifies the level of significance is represented by the notation α, which is the lower case Greek letter **alpha**. Throughout the book the latter value will be referred to as the **prespecified alpha value** (or **prespecified level of significance**), since it will be assumed that the value was specified prior to the data collection phase of a study.

When one employs the term **significance** in the context of scientific research, it is instructive to make a distinction between **statistical significance** and **practical significance**. Statistical significance only implies that the outcome of a study is highly unlikely to have occurred as a result of chance. It does not necessarily suggest that any difference or effect detected in a set of data is of any practical value. As an example, assume that the Scholastic Aptitude Test (SAT) scores of two school districts that employ different teaching methodologies are contrasted. Assume that the teaching methodology of each school district is based on specially designed classrooms. The results of the study indicate that the SAT average in School District A is one point higher than the SAT average in School District B, and this difference is statistically significant at the .01 level. Common sense suggests that it would be illogical for School District B to invest the requisite time and money in order to redesign its physical environment for the purpose of increasing the average SAT score in the district by one point. Thus, in this example, even though the obtained difference is statistically significant, in the final analysis it is of little or no practical significance.

Type I and Type II errors in hypothesis testing Within the framework of hypothesis testing, it is possible for a researcher to commit two types of errors. These errors are referred to as a **Type I error** and a **Type II error**.

A **Type I error** is when a true null hypothesis is rejected (i.e., one concludes that a false alternative hypothesis is true). The likelihood of committing a Type I error is specified by the alpha level a researcher employs in evaluating an experiment. The more concerned a researcher is with committing a Type I error, the lower the value of alpha the researcher should employ. Thus, the likelihood of committing a Type I error if $\alpha = .01$, is 1%, as compared with a 5% likelihood if $\alpha = .05$.

A **Type II error** is when a false null hypothesis is retained (i.e., one concludes that a true alternative hypothesis is false). The likelihood of committing a Type II error is represented by β, which is the lower case Greek letter **beta**. The likelihood of rejecting a false null hypothesis represents what is known as the **power** of a statistical test. The power of a test is determined by subtracting the value of beta from 1 (i.e., Power $= 1 - \beta$). The likelihood of committing a Type II error is inversely related to the likelihood of committing a Type I error. In other words, as the likelihood of committing one type of error decreases, the likelihood of committing the other type of error increases. Thus, with respect to the alternative hypothesis one employs, there is a higher likelihood of committing a Type II error when alpha is set equal to .01 than when it is set equal to .05. The likelihood of committing a Type II error is also inversely related to the power of a statistical test. In other words, as the likelihood of committing a Type II error decreases, the power of the test increases. Consequently, the higher the alpha value (i.e., the higher the likelihood of committing a Type I error), the more powerful the test.

Although the hypothesis testing model as described here is based on conducting a single study in order to evaluate a research hypothesis, throughout the book the author emphasizes the importance of replication in research. This recommendation is based on the fact that inferential statistical tests make certain assumptions, many of which a researcher can never be sure have been met. Since the accuracy of the probability values in tables of critical values for test statistics are contingent upon the validity of the assumptions underlying the test, if any of the assumptions have been violated, the accuracy of the tables can be compromised. In view of this, the most effective way of determining the truth with regard to a particular question, especially if practical decisions are to be made on the basis of the results of research, is to conduct multiple studies which evaluate the same hypothesis. When multiple studies yield consistent results, one is less likely to be challenged that the correct decision has been made with respect to the hypothesis under study.

Estimation in Inferential Statistics

In addition to hypothesis testing, inferential statistics can also be employed for estimating the value of one or more population parameters. Within this framework there are two types of estimation. **Point estimation** (which is the less commonly employed of the two methods) involves estimating the value of a parameter from the computed value of a statistic. The more commonly employed method of estimation is **interval estimation**, which involves computing a range of values within which a researcher can state, with a high degree of confidence, the true value of the parameter, as estimated by a statistic, falls. Such a range of values is referred to as a **confidence interval**. As an example, a 95% confidence interval for a population mean stipulates the range of values within which a researcher can be 95% confident that the true value of the population mean falls.

Basic Concepts and Terminology Employed in Experimental Design

Inferential statistical tests can be employed to evaluate data that are generated from a broad range of experimental designs. This section will review some basic terminology in the area of experimental design that will be employed throughout this book.

Typically, experiments involve two or more **experimental conditions**. These conditions are often referred to as **treatments**, and in experiments where different subjects serve in each of the conditions, the term **groups** is commonly employed to differentiate the conditions from one another. At this point it is instructive to review the distinction between a **between-subjects design** and a **within-subjects design**. A **between-subjects** design (also known as an **independent-groups design**) is one in which different subjects serve in each of the experimental conditions. In a **within-subjects design** (also referred to as a **repeated-measures design, dependent samples design**, and **correlated samples design**), each subject serves in only one of the experimental conditions. A design involving **matched subjects** is also treated as a **within-subjects design**. In a **matched-subjects design** (which is discussed in detail under the ***t* test for two dependent samples (Test 12)**), each subject is paired with one or more other subjects who are similar with respect to one or more characteristics that are highly correlated with the dependent variable (which will be discussed shortly).

A basic distinction in experimental design exists between an **independent** and a **dependent variable**. In any experiment involving two or more experimental conditions, the **independent variable** is the experimental manipulation or preexisting subject characteristic that distinguishes the different experimental conditions from one another. Thus, in the antidepressant drug study discussed earlier, the independent variable is whether or not a subject receives a drug or a placebo. Since the number of **levels** of an independent variable corresponds to the number of experimental conditions or treatments, the independent variable in the drug study is comprised of two levels.

A **dependent variable** is the specific measure which is hypothesized to be influenced by or associated with the independent variable. Thus in the drug study, the depression scores of subjects represent the dependent variable, since it is hypothesized that the depression score of a subject will be a function of which treatment the subject receives. When a null hypothesis is rejected in an experiment involving two or more treatments, the researcher is concluding that the subjects' scores on the dependent variable are dependent upon which level of the independent variable they were assigned.

It is possible to have more than one independent variable in an experiment. Experimental designs that involve more than one independent variable are referred to as **factorial designs**. In such experiments, the number of independent variables will correspond to the number of factors in the experiment, and each independent variable/factor will be comprised of two or more levels. It is also possible to have more than two dependent variables in an experiment. Typically, experiments involving two or more dependent variables are evaluated with **multivariate statistical procedures**, a topic that will not be covered in detail in this book.

In describing a **between-subjects design**, a distinction is commonly made between a **true experiment** as opposed to a **natural experiment** (which is also referred to as an **ex post facto study**). This distinction is predicated on the fact that in a **true experiment** the following applies: a) Subjects are **randomly assigned** to a group; and b) The independent variable is **manipulated** by the experimenter. The antidepressant drug study illustrates an example of an experiment in which the independent variable is manipulated, since in that study the experimenter determined who received the drug and who received the placebo. In a **natural experiment** random assignment of subjects to groups is impossible, since the independent variable is not manipulated by the experimenter, but instead is some preexisting subject characteristic (such as gender, race, etc.). Thus, if we compare the overall health of smokers and nonsmokers, the independent variable in such a study is whether or not a person smokes, which is something that is determined by "nature" prior to the experiment.

The advantage of a **true experiment** over a **natural experiment** is that the **true experiment** allows a researcher to exercise much greater control over the experimental situation. Since the experimenter randomly assigns subjects to groups in the **true experiment**, it is assumed that the groups which are formed are equivalent to one another, and as a result of this any differences between the groups with respect to the dependent variable can be directly attributed to the manipulated independent variable. The end result of all this is that the **true experiment** allows a researcher to draw conclusions with regard to cause and effect.

The **natural experiment**, on the other hand, does not allow one to draw conclusions with regard to cause and effect. Essentially the type of information that results from a **natural experiment** is **correlational** in nature. Such experiments can only tell a researcher that a statistical association exists between the independent and dependent variables. The reason why **natural experiments** do not allow a researcher to draw conclusions with regard to cause and effect is that such experiments do not control for the potential effects of **confounding variables** (also known as **extraneous variables**). A **confounding variable** is any variable that systematically varies with the different levels of the independent variable. To illustrate, assume that in a study comparing the overall health of smokers and nonsmokers, unbeknownst to the researcher all of the smokers in the study are people who have high stress jobs and all the nonsmokers are people with low stress jobs. If the outcome of such a study indicates that smokers are in poorer health than nonsmokers, the researcher will have no way of knowing whether the inferior health of the smokers is due to smoking and/or job stress, or even to some other confounding variable of which he is unaware.

Correlational Research

As noted above, **natural experiments** only provide correlational information. In point of fact, there are a large number of correlational measures that have been developed which are appropriate for use with different kinds of data. Measures of correlation are not inferential statistical tests, but are instead, descriptive measures which indicate the degree to which two or more variables are related to one another. Upon computing a measure of correlation, it is common practice to employ one or more inferential statistical tests to evaluate one or more hypotheses concerning the degree to which the variables are related to one another. Although correlations can be computed for more than two variables, the focus in this book will be on **bivariate correlational procedures**, which are procedures that measure the degree of association between two variables.

In the typical correlational study, scores on two measures/variables are available for *n* subjects. A major goal of correlational research is to determine the degree to which a subject's score on one variable can be predicted if one knows the score of the subject on the second variable. As a general rule (although there are exceptions), the value computed for a measure of correlation/association (often referred to as a **correlation coefficient**) will usually fall within a range of values between 0 and an absolute value of 1. Whereas a value of 0 indicates that no statistical relationship exists between the variables, an absolute value of 1 indicates the presence of a maximal relationship. Consequently, the closer the absolute value of a correlation is to 1, the stronger the relationship between the two variables. To state it another way, the closer the absolute value of a correlation is to 1, the more accurately a subject's score on one variable can be predicted from the subject's score on the second variable.

Many measures of correlation can assume both positive and negative values, and typically, in such cases the range of values the coefficient of correlation can assume is between -1 and $+1$. Whereas the absolute value of a correlation coefficient indicates the strength of the relationship between the two variables, the sign of the correlation coefficient indicates the nature of the relationship. A positive correlation indicates the presence of what is referred to as a **direct relationship**. In a **direct relationship** a change in one variable is associated with a change in the other variable in the **same direction**. On the other hand, a negative correlation indicates the presence of an **indirect** or **inverse relationship** between the variables. In an **indirect relationship** a change in one variable is associated with a change in the other variable in the **opposite direction**. A more comprehensive discussion of the subject of correlation can be found in Section I of the **Pearson product-moment correlation coefficient (Test 22)**.

Parametric versus Nonparametric Inferential Statistical Tests

The inferential statistical procedures discussed in this book have been categorized as being **parametric** versus **nonparametric tests**. Some sources distinguish between parametric and nonparametric tests on the basis that **parametric tests** make specific assumptions with regard to one or more of the population parameters that characterize the underlying distribution(s) for which the test is employed. These same sources describe **nonparametric tests** as making no such assumptions about population parameters. In truth, nonparametric tests are really not assumption free, and in view of this some sources (e.g., Marascuilo and McSweeney (1977)) suggest that it might be more appropriate to employ the term **"assumption freer"** rather than nonparametric in relation to such tests.

The distinction employed in this book for categorizing a procedure as a parametric versus a nonparametric test is primarily based on the level of measurement represented by the data that are being analyzed. As a general rule, inferential statistical tests which evaluate **categorical/nominal data** and **ordinal/rank-order data** are categorized as **nonparametric tests**, while those tests that evaluate **interval data** or **ratio data** are categorized as **parametric tests**. Although the appropriateness of employing level of measurement as a criterion in this context has been debated, its usage provides a reasonably simple and straightforward schema for categorization that facilitates the decision making process for selecting an appropriate statistical test.

There is general agreement among most researchers that as long as there is no reason to believe that one or more of the assumptions of a parametric test has been violated, when the level of measurement for a set of data is interval or ratio, the data should be evaluated with the appropriate parametric test. However, if one or more of the assumptions of a parametric test is violated, some (but not all) sources believe it is prudent to transform the data into a format that makes it compatible for analysis with the appropriate nonparametric test. Related to this is that even though parametric tests generally provide a more powerful test of an alternative hypothesis than their nonparametric analogs, the power advantage of a parametric test may be negated if one or more of its assumptions are violated.

The reluctance among some sources to transform interval/ratio data[7] into an ordinal/rank-order or categorical/nominal format for the purpose of analyzing it with a nonparametric test, is based on the fact that interval/ratio data contain more information than either of the latter two forms of data. Because of their reluctance to sacrifice information, these sources take the position that even when there is reason to believe that one or more of the assumptions of a parametric test has been violated, it is still more prudent to employ the appropriate parametric test. Generally, when a parametric test is employed under such conditions, certain adjustments are made in evaluating the test statistic in order to improve its reliability.

In the final analysis, the debate concerning whether one should employ a parametric or nonparametric test for a specific experimental design turns out to be of little consequence in most instances. The reason for this is that most of the time a parametric test and its nonparametric analog are employed to evaluate the same set of data, they lead to identical or similar conclusions. This latter observation is demonstrated throughout this book with numerous examples. In those instances where the two types of test yield conflicting results, the truth can best be determined by conducting multiple experiments which evaluate the hypothesis under study. Although it will not be covered in this book, a procedure called **meta-analysis** (which is described in detail in Mullen and Rosenthal (1985) and Rosenthal (1991)) can be employed to pool the results of multiple studies which evaluate the same general hypothesis.

Selection of the Appropriate Statistical Procedure

The Handbook of Parametric and Nonparametric Statistical Procedures is intended to be a comprehensive resource on inferential statistical tests and measures of correlation/association. The section to follow presents an outline of the statistical procedures covered in the book. Following the outline the reader is provided with guidelines and accompanying decision tables to facilitate the selection of the appropriate statistical procedure for a specific experimental design.

References

Marascuilo, L. A. and McSweeney, M. (1977). **Nonparametric and distribution-free methods for the social sciences.** Monterey, CA: Brooks/Cole Publishing Company.

Mullen, B. and Rosenthal, R. (1985). **Basic meta-analysis: Procedures and programs.** Hillsdale, NJ: Lawrence Erlbaum Associates, Publishers.

Rosenthal, R. (1991). **Meta-analytic procedures for social research.** Newbury Park, CA: Sage Publications.

Endnotes

1. The inequality sign $>$ means **greater than.** Some other inequality signs used throughout the book are $<$, which means **less than;** $\geq$, which means **greater than or equal to;** and $\leq$, which means **less than or equal to.**

2. The **absolute value** is the magnitude of a number value irrespective of the sign.

3. A proportion is converted into a percentage by moving the decimal point two places to the right.

4. In actuality, the values of the sample means do not have to be identical in order for the null hypothesis to be supported. Due to **sampling error**, which is a discrepancy between the value of a statistic and the parameter it estimates, even when two samples come from the same population, the value of the two sample means will usually not be identical. The larger the sample size employed in a study the less the influence of sampling error, and consequently the closer one can expect two sample means to be to one another if, in fact, they do represent the same population. With small sample sizes, however, a large difference between sample means is not unusual even when the samples come from the same population, and because of this a large difference may not be grounds for rejecting the null hypothesis.

5. Some sources employ the notation $p \leq .05$, indicating a probability of equal to or less than .05. The latter notation will not be used unless the computed value of a test statistic is the exact value of the tabled critical value.

6. Inspection of Column 3 in **Table A1** reveals that the proportion for $z = 1.64$ is .0505. This latter value is the same distance from the proportion .05 as the value .0495 derived for $z = 1.65$. If **Table A1** documented proportions to five decimal places, it would turn out that $z = 1.65$ yields a value that is slightly closer to .05 than does $z = 1.64$. Some books, however, do employ $z = 1.64$ as the tabled critical one-tailed .05 z value.

7. Since interval and ratio data are viewed the same within the decision making process with respect to test selection, the expression interval/ratio will be used throughout to indicate that either type of data is appropriate for use with a specific test.

Outline of Inferential Statistical Tests and Measures of Correlation/Association

I. Inferential statistical tests employed with a single sample

A. Inferential statistical tests employed with interval/ratio data

1. Inferential statistical tests employed with interval/ratio data for evaluating a hypothesis about the mean of a single population

Test 1: The Single-Sample *z* Test
Test 2: The Single-Sample *t* Test

2. Inferential statistical tests employed with interval/ratio data for evaluating a hypothesis about the variance of a single population

Test 3: The Single-Sample Chi-Square Test for a Population Variance

B. Inferential statistical tests employed with ordinal/rank-order data

1. Inferential statistical tests employed with ordinal/rank-order data for evaluating a hypothesis about the median of a single population

Test 4: The Wilcoxon Signed-Ranks Test
Test 6b: The Single-Sample Test for the Median

C. Inferential statistical tests employed with categorical/nominal data

1. Inferential statistical tests employed with categorical/nominal data for evaluating a hypothesis about the distribution of data in a single population

Test 5: The Chi-Square Goodness-of-Fit Test
Test 6: The Binomial Sign Test for a Single Sample
Test 6a: The *z* Test for a Population Proportion
Test 7: The Single-Sample Runs Test (and Other Tests of Randomness)
Test 11b: The Chi-Square Test of Independence

II. Inferential statistical tests employed with two independent samples

A. Inferential statistical tests employed with interval/ratio data

1. Inferential statistical tests employed with interval/ratio data for evaluating a hypothesis about the means of two independent populations

Test 8: The *t* Test for Two Independent Samples
Test 8c: The *z* Test for Two Independent Samples

2. Inferential statistical tests employed with interval/ratio data for evaluating a hypothesis about variability in two independent populations

Test 8a: Hartley's F_{max} Test for Homogeneity of Variance/*F* Test for Two Population Variances

B. Inferential statistical tests employed with ordinal/rank-order data

1. Inferential statistical tests employed with ordinal/rank-order data for evaluating a hypothesis about the medians of two independent populations

Test 9: The Mann–Whitney *U* Test
Test 11e: The Median Test for Independent Samples

2. Inferential statistical tests employed with ordinal/rank-order data for evaluating a hypothesis about variability of two independent populations

Test 10: The Siegel–Tukey Test for Equal Variability

C. Inferential statistical tests employed with categorical/nominal data

1. Inferential statistical tests employed with categorical/nominal data for evaluating a hypothesis about the distribution of data in two independent populations

Test 11a. The Chi-Square Test for Homogeneity

Test 11c: The Fisher Exact Test

Test 11d: The *z* Test for Two Independent Proportions

III. Inferential statistical tests employed with two dependent samples

A. Inferential statistical tests employed with interval/ratio data

1. Inferential statistical tests employed with interval/ratio data for evaluating a hypothesis about the means of two dependent populations

Test 12: The *t* Test for Two Dependent Samples

Test 12c: Sandler's *A* Test

Test 12d: The *z* Test for Two Dependent Samples

2. Inferential statistical tests employed with interval/ratio data for evaluating a hypothesis about variability in two dependent populations

Test 12a: The *t* Test for Homogeneity of Variance for Two Dependent Samples

B. Inferential statistical tests employed with ordinal/rank-order data

1. Inferential statistical tests employed with ordinal/rank-order data for evaluating a hypothesis about the distribution of data in two dependent populations

Test 13: The Wilcoxon Matched-Pairs Signed-Ranks Test

Test 14: The Binomial Sign Test for Two Dependent Samples

C. Inferential statistical tests employed with categorical/nominal data

1. Inferential statistical tests employed with categorical/nominal data for evaluating a hypothesis about the distribution of data in two dependent populations

Test 15: The McNemar Test

IV. Inferential statistical tests employed with two or more independent samples

A. Inferential statistical tests employed with interval/ratio data

1. Inferential statistical tests employed with interval/ratio data for evaluating a hypothesis about the means of two or more independent populations which involve one independent variable/factor

Test 16: The Single-Factor Between-Subjects Analysis of Variance

Test 16a: Multiple *t* Tests/Fisher's LSD Test

Test 16b: The Bonferroni–Dunn test

Test 16c: Tukey's HSD Test

Test 16d: The Newman–Keuls Test

Test 16e: The Scheffé Test

Test 16f: The Dunnett Test

2. Inferential statistical tests employed with interval/ratio data for evaluating a hypothesis about variability in two or more independent populations

Test 8a: Hartley's F_{max} Test for Homogeneity of Variance/*F* Test for Two Population Variances

B. Inferential statistical tests employed with ordinal/rank-order data

1. Inferential statistical tests employed with ordinal/rank data for evaluating a hypothesis about the medians of two or more independent populations

Test 17: The Kruskal-Wallis One-Way Analysis of Variance by Ranks
Test 11e: The Median Test for Independent Samples

C. Inferential statistical tests employed with categorical/nominal data

1. Inferential statistical tests employed with categorical/nominal data for evaluating a hypothesis about the distribution of data in two or more independent populations

Test 11a: The Chi-Square Test for Homogeneity

V. Inferential statistical tests employed with two or more dependent samples

A. Inferential statistical tests employed with interval/ratio data

1. Inferential statistical tests employed with interval/ratio data for evaluating a hypothesis about the means of two or more dependent populations which involve one independent variable/factor.

Test 18: The Single-Factor Within-Subjects Analysis of Variance
Test 18a: Multiple *t* Tests/Fisher's LSD Test
Test 18b: The Bonferroni-Dunn Test
Test 18c: Tukey's HSD Test
Test 18d: The Newman-Keuls Test
Test 18e: The Scheffé Test
Test 18f: The Dunnett Test

B. Inferential statistical tests employed with ordinal/rank-order data

1. Inferential statistical tests employed with ordinal/rank-order data for evaluating a hypothesis about the medians of two or more dependent populations

Test 19: The Friedman Two-Way Analysis of Variance by Ranks

C. Inferential statistical tests employed with categorical/nominal data

1. Inferential statistical tests employed with categorical/nominal data for evaluating a hypothesis about the distribution of data in two or more dependent populations

Test 20: The Cochran *Q* Test

VI. Inferential statistical tests employed with factorial designs

A. Inferential statistical tests employed with interval/ratio data

1. Inferential statistical tests employed with interval/ratio data for evaluating a hypothesis about the means of two or more populations in a design involving two independent variables/factors

Test 21: The Between-Subjects Factorial Analysis of Variance
Test 21a: Fisher's LSD Test
Test 21b: The Bonferroni-Dunn Test
Test 21c: Tukey's HSD Test
Test 21d: The Newman-Keuls Test
Test 21e: The Scheffé Test
Test 21f: The Dunnett Test
Test 21h: The Factorial Analysis of Variance for a Mixed Design
Test 22i: The Within-Subjects Factorial Analysis of Variance

VII. Measures of correlation/association

A. Measures of correlation/association employed with interval/ratio data

1. Bivariate measures

Test 22: The Pearson Product–Moment Correlation Coefficient (and tests for evaluating various hypotheses concerning the value of one or more product-moment correlation coefficients **(Tests 20a–20g)**

2. Multivariate measures

Test 22k:. The Multiple Correlation Coefficient (and the test for evaluating the significance of a multiple correlation coefficient **(Test 22k-a)**)

Test 22l: The Partial Correlation Coefficient (and the test for evaluating the significance of a partial correlation coefficient **(Test 22l-a)**)

Test 22m: The Semipartial Correlation Coefficient (and the test for evaluating the significance of a semi-partial correlation coefficient **(Test 22m-a)**)

B. Measures of correlation/association employed with ordinal/rank order data

1. Bivariate measures/Two sets of ranks

Test 23: Spearman's Rank-Order Correlation Coefficient (and the test for evaluating the significance of Spearman's rank-order correlation coefficient **(Test 23a)**

Test 24: Kendall's Tau (and the test for evaluating the of significance of Kendall's tau **(Test 24a)**)

Test 26: Goodman and Kruskal's Gamma (and the test for evaluating the significance of gamma **(Test 26a)**)

2. Ordinal measure of association for three or more samples/sets of ranks

Test 25: Kendall's Coefficient of Concordance (and the test for evaluating the significance of the coefficient of concordance **(Test 25a)**)

C. Measures of correlation/association employed with categorical/nominal data

Test 11f: The Contingency Coefficient
Test 11g: The Phi Coefficient
Test 11h: Cramér's Phi Coefficient
Test 11i: Yule's *Q*
Test 11j: The Odds Ratio

D. Other bivariate measures of correlation/association employed when interval/ratio data are used or implied for at least one variable

Test 22h: The Point-Biserial Correlation Coefficient (and the test for evaluating the significance of a point-biserial correlation coefficient **(Test 22h-a)**)

Test 22i: The Biserial Correlation Coefficient (and the test for evaluating the significance of a biserial correlation coefficient **(Test 22i-a)**)

Test 22j: The Tetrachoric Correlation Coefficient (and the test for evaluating the significance of a tetrachoric correlation coefficient **(Test 22j-a)**)

Tests 8b/12b/16g/18g/21g: Omega Squared
Test 16h: Eta Squared

Guidelines and Decision Tables for Selecting the Appropriate Statistical Procedure

Tables I.2–I.5 are designed to facilitate the selection of the appropriate statistical test. Tables I.2–I.4 list the major inferential statistical procedures described in the book, based on the level of measurement the data being evaluated represent. Specifically, Table I.2 lists inferential statistical tests employed with interval/ratio data, Table I.3 lists inferential statistical tests employed with ordinal/rank-order data, and Table I.4 lists inferential statistical tests employed with categorical/nominal data. Table I.5 lists the measures of correlation/association that are described in the book. Using the aforementioned tables, the following guidelines should be employed in selecting the appropriate statistical test.

1. Determine if the analysis involves computing a correlation coefficient/measure of association, and if it does go to Table I.5. The selection of the appropriate measure in Table I.5 is based on the level of measurement represented by each of the variables for which the measure of correlation/association is computed.
2. If the analysis does not involve computing a measure of correlation/association, it will be assumed that the data will be evaluated through use of an inferential statistical test. To select the appropriate inferential statistical test, the following protocol should be employed.
   a) State the general hypothesis that is being evaluated.
   b) Determine if the study involves a single sample or more than one sample.
   c) If the study involves a single sample, the appropriate test will be one of the tests for a single sample in Tables I.2, I.3, or I.4. In order to determine which table to employ, determine the level of measurement represented by the data that are being evaluated. If the level of measurement is interval/ratio, Table I.2 is employed. If the level of measurement is ordinal/rank-order, Table I.3 is appropriate. If the level of measurement is categorical/nominal, Table I.4 is utilized.
   d) If there is more than one sample, determine how many samples/treatments there are and whether they are independent or dependent. Determine the the level of measurement represented by the data that are being evaluated (which represents the dependent variable in the study).
      1) If the level of measurement is interval/ratio, go to Table I.2. Identify the test or tests that are appropriate for that level of measurement with respect to the number and type of samples employed in the study.
      2) If the level of measurement is ordinal/rank-order, go to Table I.3. Identify the test or tests that are appropriate for that level of measurement with respect to the number and type of samples employed in the study.
      3) If the level of measurement is categorical/nominal, go to Table I.4. Identify the test or tests that are appropriate for that level of measurement with respect to the number and type of samples employed in the study.

Table I.2 Decision Table for Inferential Statistical Tests Employed with Interval/Ratio Data

Number of samples: One independent variable		Hypothesis evaluated	Test
Single sample		Hypothesis about a population mean	The single-sample z test (Test 1) (σ known) The single-sample t test (Test 2) (σ unknown)
Single sample		Hypothesis about a population variance	The single-sample chi-square test for a population variance (Test 3)
Two samples	Two independent samples	Hypothesis about difference between two independent population means	The t test for two independent samples (Test 8) The z test for two independent samples (Test 8c) The single-factor between-subjects analysis of (Test 16)
Two samples	Two independent samples	Hypothesis about two independent population variances	Hartley's F_{max} test for homogeneity of variance/ F test for two population variances (Test 8a)
Two samples	Two dependent samples	Hypothesis about difference between two dependent population means	The t test for two dependent samples (Test 12) Sandler's A test (Test 12c) The z test for two dependent samples (Test 12d) The single-factor within-subjects analysis of variance (Test 18)
Two samples	Two dependent samples	Hypothesis about two dependent population variances	The t test for homogeneity of variance for two dependent samples (Test 12a)
Two or more samples	Two or more independent samples	Hypothesis about difference between two or more independent population means	The single-factor between-subjects analysis of variance (Test 16) See Test 16 for discussion of the analysis of covariance
Two or more samples	Two or more independent samples	Hypothesis about two or more dependent population variances	Hartley's F_{max} test for homogeneity of variance/ F test for two population variances (Test 8a)
Two or more samples	Two or more dependent samples	Hypothesis about difference between two or more dependent population means	The single-factor within-subjects analysis of variance (Test 18)
Two or more samples	Two or more dependent samples	Hypothesis about two or more dependent population variances	See discussion of sphericity assumption under the single-factor within-subjects analysis of variance (Test 18)
Two independent variables		Hypothesis about difference between two or more population means	The between-subjects factorial analysis of variance (Test 21) The factorial analysis of variance for a mixed design (Test 21h) The within-subjects factorial analysis of variance (Test 21i)

Table I.3 Decision Table for Inferential Statistical Tests Employed with Ordinal/Rank-Order Data

Number of samples		Hypothesis evaluated	Test
Single sample		Hypothesis about a population median	The Wilcoxon signed-ranks test (Test 4) The single-sample test for the median (Test 6b)
Two samples	Two independent samples	Hypothesis about two independent population medians	The Mann–Whitney U test (Test 9) The median test for independent samples (Test 11e)
Two samples	Two independent samples	Hypothesis about variability in two independent populations	The Siegel–Tukey test for equal variability (Test 10)
Two samples	Two dependent samples	Hypothesis about medians or ordering of data in two dependent populations	The Wilcoxon matched-pairs signed-ranks test (Test 13) The binomial sign test for two dependent samples (Test 14)
Two or more samples	Two or more independent samples	Hypothesis about two or more independent population medians	The Kruskal–Wallis one-way analysis of variance by ranks (Test 17) The median test for independent samples (Test 11e)
Two or more samples	Two or more dependent samples	Hypothesis about two or more dependent population medians	The Friedman two-way analysis of variance by ranks (Test 19)

Table I.4 Decision Table for Inferential Statistical Tests Employed with Categorical/Nominal Data

Number of samples		Hypothesis evaluated	Test
Single sample		Hypothesis about distribution of data in a single population	The chi-square goodness-of-fit test (Test 5) The binomial sign test for a single sample (Test 6) The z test for a population proportion (Test 6a) The chi-square test of independence (Test 11b) The single-sample runs test (and other tests of randomness) (Test 7)
Two samples	Two independent samples	Hypothesis about distribution of data in two independent populations	The chi-square test for homogeneity (Test 11a) The Fisher exact test (Test 11c) The z test for two independent proportions (Test 11d)
Two samples	Two dependent samples	Hypothesis about distribution of data in two dependent populations	The McNemar test (Test 15)
Two or more samples	Two or more independent samples	Hypothesis about distribution of data in two or more independent populations	The chi-square test for homogeneity (Test 11a)
Two or more samples	Two or more dependent samples	Hypothesis about distribution of of data in two or more dependent populations	The Cochran Q test (Test 20)

Table I.5 Decision Table for Measures of Correlation/Association

Level of measurement		Test
Interval/ratio data	Bivariate	The Pearson product-moment correlation coefficient (Test 22)
	Multivariate	The multiple correlation coefficient (Test 22k) The partial correlation coefficinet (Test 22l) The semipartial correlation coefficient (Test 22m)
Ordinal/rank order data	Bivariate/two sets of ranks	Spearman's rank-order correlation coefficient (Test 23) Kendall's tau (Test 24) Goodman and Kruskal's gamma (for ordered contingency tables) (Test 26)
	More than two samples/sets of ranks	Kendall's coefficient of concordance (Test 25)
Categorical/ nominal data	Two dichotomous variables	The contingency coefficient (Test 11f) The phi coefficient (Test 11g) Yule's Q (Test 11i) The odds ratio (Test 11j)
	Two nondichotomous variables	The contingency coefficient (Test 11f) Cramér's phi coefficient (Test 11h) The odds ratio (Test 11j)
Other bivariate correlational measures for which interval/ ratio/data are employed or implied for at least one of the variables		Omega squared (One variable, interval/ratio data; second variable, two or more nominal levels) Tests (8b/12b/16g/18g/21g) Eta squared (One variable, interval/ratio data; second variable, two or more nominal levels) (Test 16h) The point-biserial correlation coefficient (One variable, interval/ratio data; second variable represented by dichotomous categories) (Test 22h) The biserial correlation coefficient (One variable, interval/ratio data; second variable, an interval/ratio variable expressed in form of dichotomous categories) (Test 22i) The tetrachoric correlation coefficient (Two interval/ratio variables, both of which are expressed in the the form of dichotomous categories) (Test 22j)

Inferential Statistical Tests Employed with a Single Sample

Inferential Statistical Tests

Employed with a Single Sample

Test 1: The Single-Sample z Test

Test 2: The Single-Sample t Test

Test 3: The Single-Sample Chi-Square Test for a Population Variance

Test [illegible]: [illegible]

Test 8: The Chi-Square Goodness-of-Fit Test

Test 9: The Binomial Sign Test for a Single Sample

Test 10: The Single-Sample Runs Test and Other Tests of Randomness

Test 1

The Single-Sample z Test
(Parametric Test Employed with Interval/Ratio Data)

I. Hypothesis Evaluated with Test and Relevant Background Information

Hypothesis evaluated with test Does a sample of n subjects come from a population in which the mean (μ) equals a specified value?

Relevant background information on test The **single-sample z test** is employed in a hypothesis testing situation involving a single sample in order to determine whether a sample with a mean of $\bar{X}$ is derived from a population with a mean of μ. If the result of the **single-sample z test** yields a significant difference, the researcher can conclude there is a high likelihood the sample is derived from a population with a mean value other than μ. The test statistic for the **single-sample z test** is based on the normal distribution. A general discussion of the normal distribution can be found in the **Introduction.**

The **single-sample z test** is used with interval/ratio data. The test should only be employed if the value of the population standard deviation (σ) is known. In the event the value of σ is unknown, the data should be evaluated with the **single-sample t test (Test 2)**. The reader should take note of the fact that some sources argue that even when one knows the value of σ, if the sample size is very small, the **single-sample t test** provides a more accurate estimate of the underlying sampling distribution for the data. Sources that take the latter position are not in agreement with respect to the minimum sample size above which it is acceptable to employ the **single-sample z test** (although it is usually $n \geq 25$).

The **single-sample z test** is based on the following assumptions: a) The sample has been randomly selected from the population it represents; and b) The distribution of data in the underlying population the sample represents is normal. If either of the aforementioned assumptions is saliently violated, the reliability of the z test statistic may be compromised.

II. Example

Example 1.1. *Thirty subjects take a test of visual-motor coordination for which the value of the population mean is $\mu = 8$, and the value of the population standard deviation is $\sigma = 2$. If the average score of the sample of 30 subjects equals 7.4 (i.e., $\bar{X} = 7.4$), can one conclude that the sample, in fact, came from a population in which the mean is $\mu = 8$?*

III. Null versus Alternative Hypotheses

Null hypothesis H_0: $\mu = 8$

(The mean of the population the sample represents equals 8.)

Alternative hypothesis H_1: $\mu \neq 8$

(The mean of the population the sample represents does not equal 8. This is a **nondirectional alternative hypothesis**, and it is evaluated with a **two-tailed test.** In order to be supported, the absolute value of z must be equal to or greater than the tabled critical two-tailed z value at the prespecified level of significance. Thus, either a significant positive z value or a significant negative z value will provide support for this alternative hypothesis.)

or

$$H_1\text{: } \mu > 8$$

(The mean of the population the sample represents is greater than 8. This is a **directional alternative hypothesis**, and it is evaluated with a **one-tailed test**. It will only be supported if the sign of z is positive, and the absolute value of z is equal to or greater than the tabled critical one-tailed z value at the prespecified level of significance.)

or

$$H_1\text{: } \mu < 8$$

(The mean of the population the sample represents is less than 8. This is a **directional alternative hypothesis**, and it is evaluated with a **one-tailed test**. It will only be supported if the sign of z is negative, and the absolute value of z is equal to or greater than the tabled critical one-tailed z value at the prespecified level of significance.)

Note: Only one of the above noted alternative hypotheses is employed. If the alternative hypothesis the researcher selects is supported, the null hypothesis is rejected.

IV. Test Computations

Assume that the following values represent the scores of the sample of $n = 30$ subjects who take the test of visual-motor coordination in Example 1.1: 9, 10, 6, 4, 8, 11, 10, 5, 5, 6, 13, 12, 4, 4, 3, 9, 12, 5, 6, 6, 8, 9, 8, 5, 7, 9, 10, 9, 5, 4.

Since X_i can be employed to represent the score of the ith subject, by adding all thirty scores we obtain: $\Sigma X_i = \Sigma X = 222$.

Equation 1.1 is used to compute the mean of the sample.

$$\bar{X} = \frac{\Sigma X}{n} \qquad \textbf{(Equation 1.1)}$$

Employing Equation 1.1, we confirm that the mean of the sample is $\bar{X} = 7.4$, the value stated in Example 1.1.

$$\bar{X} = \frac{222}{30} = 7.4$$

Before the test statistic can be computed, it is necessary to compute a value that is referred to as the **standard error of the population mean**. This value, which is represented by the notation $\sigma_{\bar{X}}$, is computed with Equation 1.2. A full explanation of what $\sigma_{\bar{X}}$ represents can be found in Section VII.

$$\sigma_{\bar{X}} = \frac{\sigma}{\sqrt{n}} \qquad \textbf{(Equation 1.2)}$$

Substituting the values $\sigma = 2$ and $n = 30$ in Equation 1.2, the value of $\sigma_{\bar{X}} = .36$ is computed.

$$\sigma_{\bar{X}} = \frac{2}{\sqrt{30}} = .36$$

It should be noted that $\sigma_{\bar{X}}$ can never be a negative value. If a negative value is obtained for $\sigma_{\bar{X}}$ it indicates a computational error has been made.

Equation 1.3 is employed to compute the value of z, which is the test statistic for the **single-sample *z* test.** Note that in Equation 1.3, the value that represents μ is the value $\mu = 8$ which is stated in the null hypothesis.

$$z = \frac{\bar{X} - u}{\sigma_{\bar{X}}} \qquad \textbf{(Equation 1.3)}$$

Employing Equation 1.3, the value $z = -1.67$ is computed for Example 1.1.

$$z = \frac{7.4 - 8}{.36} = -1.67$$

Note that Equation 1.3 will always yield a positive z value when the sample mean is greater than the hypothesized value of μ. The value of z will always be negative when the sample mean is less than the hypothesized value of μ. When the sample mean is equal to the hypothesized value of μ, z will equal zero.

V. Interpretation of the Test Results

The obtained value $z = -1.67$ is evaluated with **Table A1 (Table of the Normal Distribution)** in the **Appendix**. Table 1.1 summarizes the tabled critical two-tailed and one-tailed .05 and .01 z values listed in **Table A1**.

Table I.1 Tabled Critical Two-Tailed and One-Tailed .05 and .01 *z* Values

	$z_{.05}$	$z_{.01}$
Two-tailed values	1.96	2.58
One-tailed values	1.65	2.33

The following guidelines are employed in evaluating the null hypothesis for the **single-sample *z* test.**

a) If the alternative hypothesis employed is nondirectional, the null hypothesis can be rejected if the obtained absolute value of z is equal to or greater than the tabled critical two-tailed value at the prespecified level of significance.

b) If the alternative hypothesis employed is directional and predicts a population mean larger than the value stated in the null hypothesis, the null hypothesis can be rejected if the sign of z is positive, and the value of z is equal to or greater than the tabled critical one-tailed value at the prespecified level of significance.

c) If the alternative hypothesis employed is directional and predicts a population mean smaller than the value stated in the null hypothesis, the null hypothesis can be rejected if the sign of z is negative, and the absolute value of z is equal to or greater than the tabled critical one-tailed value at the prespecified level of significance.

Employing the above guidelines, we can only reject the null hypothesis if the directional alternative hypothesis H_1: $\mu < 8$ is employed, and the null hypothesis can only be rejected at the .05 level. This is the case, since the obtained value of z is a negative number, and the absolute value of z is greater than the tabled critical one-tailed .05 value $z_{.05} = 1.65$.[1] The alternative hypothesis H_1: $\mu < 8$ is not supported at the .01 level, since the absolute value $z = 1.67$ is not greater than the tabled critical one-tailed .01 value $z_{.01} = 2.33$.

The nondirectional alternative hypothesis H_1: $\mu \neq 8$ is not supported, since the obtained absolute value $z = 1.67$ is less than the tabled critical two-tailed .05 value $z_{.05} = 1.96$.

The directional alternative hypothesis H_1: $\mu > 8$ is not supported, since the obtained value $z = -1.67$ is a negative number. In order for the alternative hypothesis H_1: $\mu > 8$ to be supported, the computed value of z must be a positive number (as well as the fact that it must be equal to or greater than the tabled critical one-tailed value at the prespecified level of significance).

A summary of the analysis of Example 1.1 with the **single-sample z test** follows: With respect to the test of visual-motor coordination, we can conclude that the sample of 30 subjects comes from a population with a mean value other than 8 only if we employ the directional alternative hypothesis H_1: $\mu < 8$, and prespecify as our level of significance $\alpha = .05$. This result can be summarized as follows: $z = -1.67$, $p < .05$.

A more in-depth discussion of the interpretation of the z value computed with the **single-sample z test** is contained in Section VII.

VI. Additional Analytical Procedures for the Single-Sample z Test and/or Related Tests

Procedures are available for computing **power** and **confidence intervals** for the **single-sample z test**. These computations are discussed in Section VI of the **single-sample t test** (which employs the same protocol for such computations as does the **single-sample z test**).

VII. Additional Discussion of the Single-Sample z Test

1. The interpretation of a negative z value The actual range of scores on the abscissa (i.e., the X-axis) of the normal distribution is $-\infty \leq z \leq +\infty$. The guidelines outlined in Section V for interpreting negative z values are intended to provide the reader with the simplest and least confusing protocol for interpreting such values. In terms of the actual distribution of z values, it should be noted that although the tabled critical z values listed in Table 1.1 are positive numbers, they are also applicable to interpreting negative z values. Since the critical values recorded in Table 1.1 represent absolute values, the corresponding negative z values are listed in Table 1.2.

Table 1.2 Tabled Critical Two-Tailed and One-Tailed .05 and .01 Negative z Values

	$z_{.05}$	$z_{.01}$
Two-tailed values	-1.96	-2.58
One-tailed values	-1.65	-2.33

Within the framework of the values noted in Table 1.2, if one employs the directional (one-tailed) alternative hypothesis H_1: $\mu < 8$, in order to reject the null hypothesis the obtained value of z must be a negative number that is **equal to or less than the prespecified tabled critical value**. Thus, to be significant at the .05 level, the obtained z value would have to be equal to or less than $z = -1.65$. The reader should take note of the fact that any negative number which has an absolute value greater than 1.65 is less than −1.65. In the same respect, in order for the alternative hypothesis H_1: $\mu < 8$ to be supported at the .01 level, the obtained z value would have to be equal to or less than $z = -2.33$, since any negative number which has an absolute value greater than 2.33 is less than −2.33. The important thing for the reader to understand is that when one is dealing with a negative number, the larger the absolute value of the negative number, the lower the value of that number.

2. The standard error of the population mean and graphical representation of the results of the single-sample *z* test The intent of this section is to provide further clarification with respect to what the z value computed with the **single-sample *z* test** represents. In order to do this, it is necessary to understand what is represented by the **standard error of the population mean** ($\sigma_{\bar{X}}$), which is the denominator of Equation 1.3. The standard error of the population mean represents a standard deviation of a **sampling distribution of means.** Although such a sampling distribution is theoretical and is based on an infinite number of samples, it is possible to construct an empirical sampling distribution that is based on a smaller number of sample means. In order to construct such a sampling distribution of means, a random sample consisting of n subjects is drawn from a population of N subjects. Upon doing this, the mean of the sample of n subjects is computed. Once again, employing the whole population of N subjects, a second random sample consisting of n subjects is selected, and the mean of that sample is computed. This process is repeated over and over again. At whatever point one decides to terminate the process one will have computed a large number of sample means, each of which is based on a sample size of n subjects. The frequency distribution of these sample means (which will be distributed normally) is known as a sampling distribution of means. The mean of a sampling distribution (represented by the notation $\mu_{\bar{X}}$) that is based on an infinite number of sample means will be the same value as the population mean (μ). As the number of sample means used to construct a sampling distribution increases, the greater the likelihood that the computed value of the mean of the sampling distribution equals the value of μ. The standard deviation of a sampling distribution (i.e., the standard deviation of all of the sample means), is the **standard error of the population mean** ($\sigma_{\bar{X}}$), which in many sources is referred to as the **standard error of the mean.**[2]

The z value computed with Equation 1.3 represents the number of standard deviation units (based on the value of $\sigma_{\bar{X}}$) that the sample mean deviates from the hypothesized population mean. Thus in Example 1.1, the value $\sigma_{\bar{X}} = .36$ represents the standard deviation of a sampling distribution of means in which in each sample $n = 30$. The obtained value $z = -1.67$ indicates that $\bar{X} = 7.4$ (the sample mean) is 1.67 sampling distribution standard deviation units below the hypothesized population mean $\mu = 8$ (which as noted earlier has the same value as $\mu_{\bar{X}}$). The difference is statistically significant since a sample mean of 7.4 obtained with 30 subjects is a relatively unlikely occurrence in a sampling distribution that has a mean of 8 and a standard deviation of .36. If we make the assumption that the distribution of means in the sampling distribution is normal, use of the **single-sample *z* test** will lead to the conclusion that if, in fact, the true value of the population mean is 8, the likelihood of obtaining a sample mean equal to or less than 7.4 is less than .05 (to be exact, it equals .0475).

Figure 1.1 provides a visual description of the sampling distribution of means for Example 1.1. In Figure 1.1, the numbers 6.92, 7.28, ..., 9.08 along the abscissa identify the values a sample mean will assume if it is 1, 2, and 3 standard deviation (sd) units below and above the mean of the sampling distribution. Since the value of one standard deviation unit for the sampling distribution under discussion is equal to $\sigma_{\bar{X}} = .36$, the value 8.36, which is one standard deviation unit above the mean of the distribution, is obtained simply by adding the value $\sigma_{\bar{X}} = .36$ to $\mu_{\bar{X}} = 8$. The value 8.72, which is two standard deviations above the mean, is obtained by adding two times the value of $\sigma_{\bar{X}}$ to $\mu_{\bar{X}} = 8$, and so on.

Since almost 100% of the cases in a normal distribution fall within three standard deviation units above or below the mean,[3] if a researcher has a sample of 30 subjects that is derived from a population in which $\mu = 8$ and $\sigma = 2$, he can be almost 100% sure that the mean of the sample will fall between the values 6.92 and 9.08 (which are the values that correspond to the −3 sd and +3 sd points in Figure 1.1). A mean value outside of this range is highly unlikely to occur and if a more extreme mean value does occur, it is reasonable to conclude that the sample is derived from a population which had a mean value other than 8.

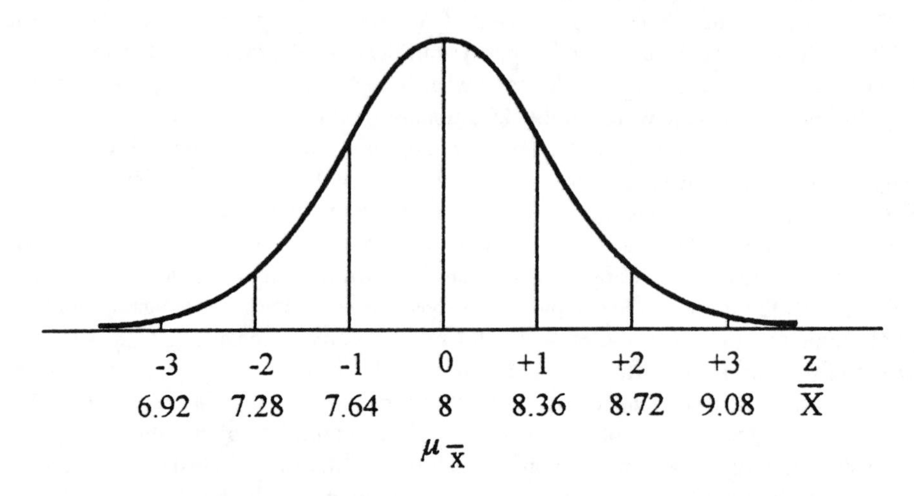

Figure 1.1 Sampling Distribution for Example 1.1

Earlier in the discussion of a sampling distribution it is noted that it is assumed that the sample means are normally distributed. The basis for this statement is a general principle in mathematics known as the **central limit theorem.** The central limit theorem states that in a population with a mean value of μ and a standard deviation of σ, the following will be true with respect to the sampling distribution of means: a) The sampling distribution will have a mean value equal to μ and a standard deviation equal to $\sigma_{\bar{X}} = \sigma/\sqrt{n}$; and b) The sampling distribution approaches being normal as the size (n) of each of the samples employed in generating the sampling distribution increases, and as the total number of means used to generate the sampling distribution increases. Although the underlying population each of the samples is derived from does not in itself have to be normal, the more it approximates normality the lower the value of n required for the sampling distribution to be normal. In addition, the more the underlying population each of the samples is derived from approaches normality, the fewer sample means will be required before the sampling distribution becomes normal.

Based on what has been said with respect to the standard error of the population mean, one can determine the value that a sample mean will have to be equal to or more extreme than in order to reject a null hypothesis at a prespecified level of significance. Figure 1.2 depicts these values for Example 1.1 in reference to the tabled critical one- and two-tailed .05 and .01 z values. Note that in each of the graphs, the value that is written directly below the tabled critical z value for the relevant level of significance is the value the sample mean will have to be equal to or more extreme in order to reject the null hypothesis H_0: $\mu = 8$.

The values that the sample mean must be equal to or more extreme than in Figure 1.2 are computed with Equation 1.4, which is the result of algebraically transposing the terms in Equation 1.3 in order to solve for the value of $\bar{X}$.

$$\bar{X} = \mu + z\,\sigma_{\bar{X}} \qquad \textbf{(Equation 1.4)}$$

The value employed to represent z in Equation 1.4 is the relevant tabled critical z value at the prespecified level of significance. By multiplying the latter value by $\sigma_{\bar{X}}$ and adding and subtracting the product from the value of the population mean, one is able to compute the upper limit that the sample mean must be equal to or greater than and the lower limit that the sample mean must be equal to or less than in order for a result to be

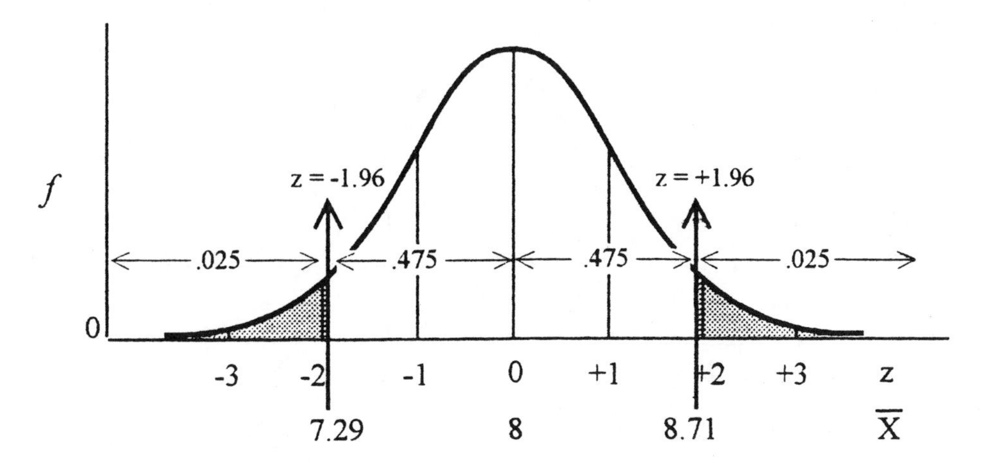

Figure 1.2a Distribution of Critical Two-Tailed .05 z Values for H_1: $\mu \neq 8$

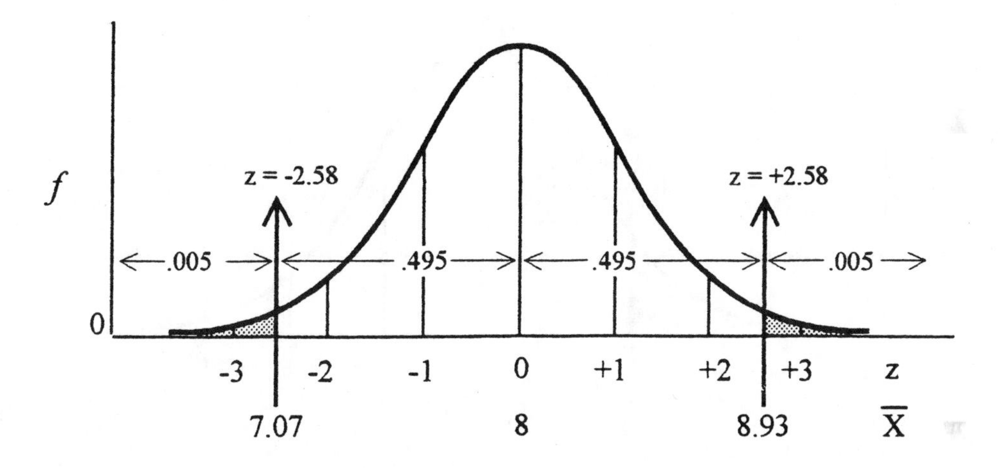

Figure 1.2b Distribution of Critical Two-Tailed .01 z Values for H_1: $\mu \neq 8$

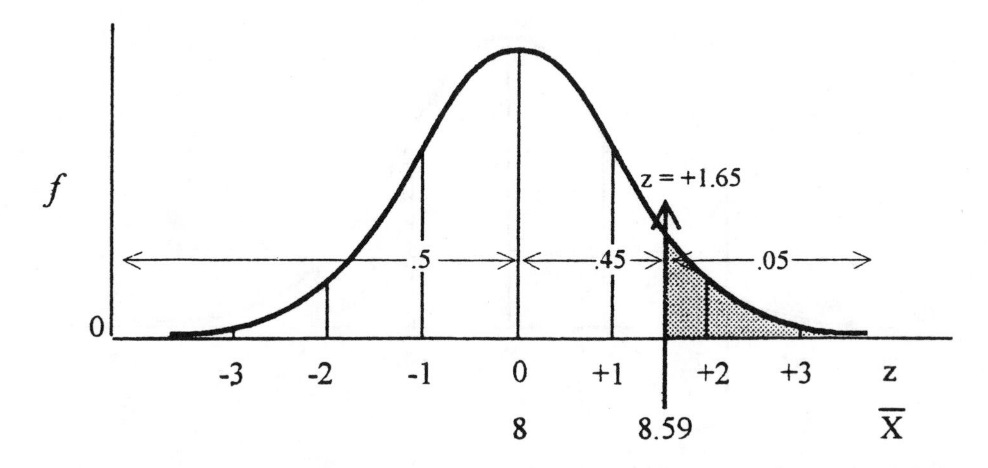

Figure 1.2c Distribution of Critical One-Tailed .05 z Values for H_1: $\mu > 8$

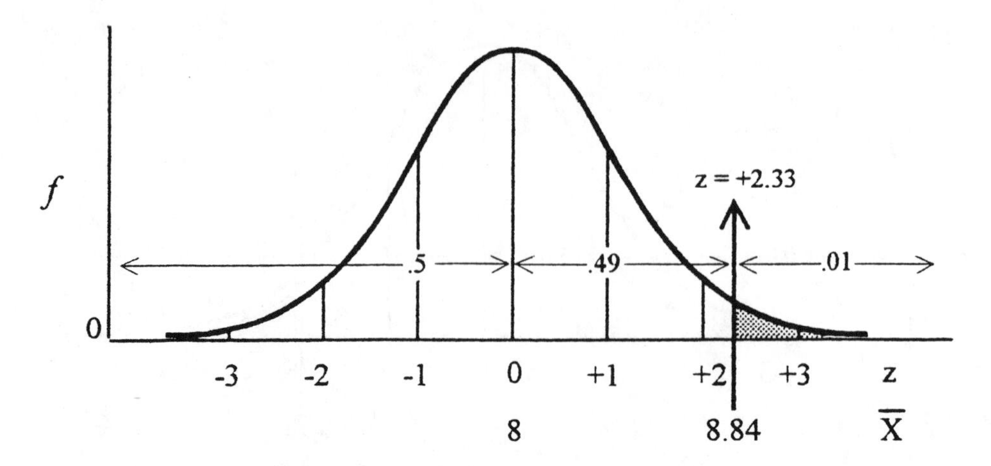

Figure 1.2d Distribution of Critical One-Tailed .01 z Values for H_1: $\mu > 8$

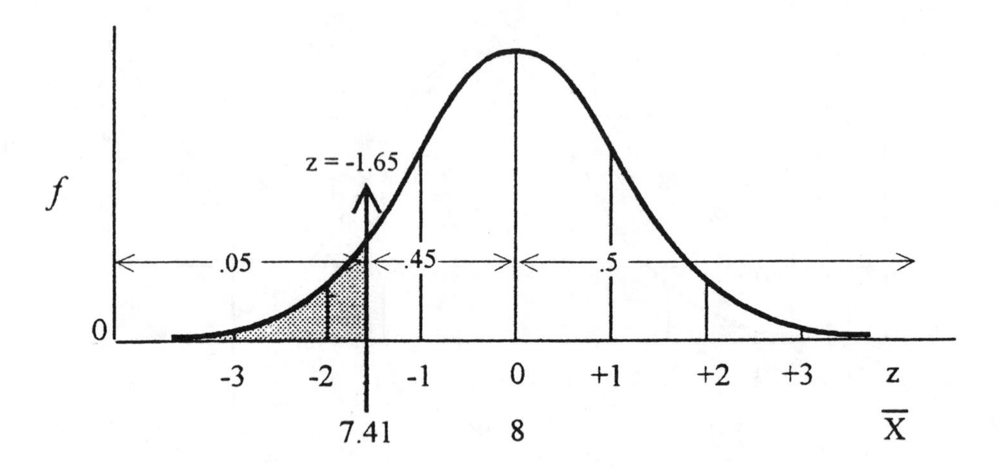

Figure I.2e Distribution of Critical One-Tailed .05 z Values for H_1: $\mu < 8$

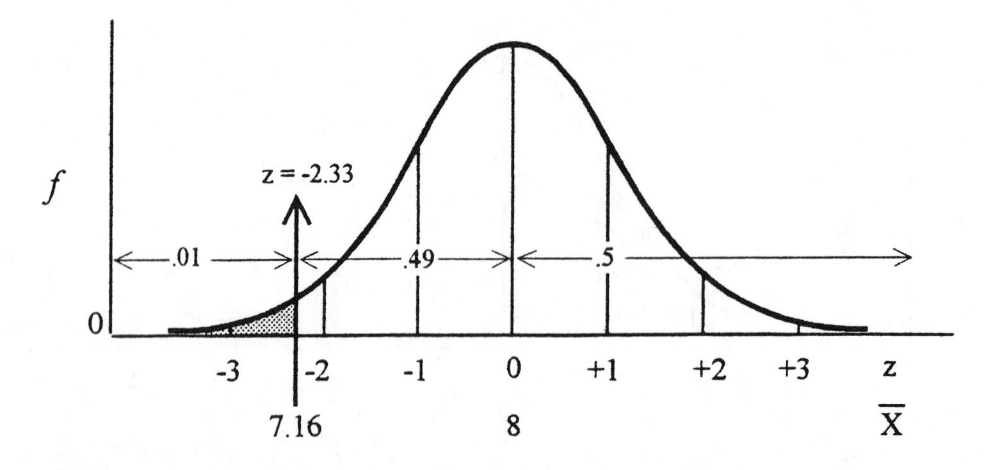

Figure 1.2f Distribution of Critical One-Tailed .01 z Values for H_1: $\mu < 8$

significant. This is illustrated below for the case depicted in Figure 1.2a, which describes the upper and lower limits for the sample mean when the nondirectional alternative hypothesis H_1: $\mu \neq 8$ is employed, with $\alpha = .05$.

$$\bar{X} = 8 \pm (1.96)(.36) = 8 \pm .71$$

Since $8 + .71 = 8.71$ and $8 - .71 = 7.29$, in order to be significant at the .05 level, a sample mean will have to be equal to or greater than 8.71, or equal to or less than 7.29. This result can be summarized as follows: $7.29 \geq \bar{X} \geq 8.71$.

A summary of the results depicted in Figure 1.2 follows:

Figure 1.2a: If the nondirectional alternative hypothesis H_1: $\mu \neq 8$ is employed, with $\alpha = .05$; in order to reject the null hypothesis, the obtained value of the sample mean will have to be equal to or greater than 8.71 or be equal to or less than 7.29.

Figure 1.2b: If the nondirectional alternative hypothesis H_1: $\mu \neq 8$ is employed, with $\alpha = .01$; in order to reject the null hypothesis, the obtained value of the sample mean will have to be equal to or greater than 8.93 or be equal to or less than 7.07.

Figure 1.2c: If the directional alternative hypothesis H_1: $\mu > 8$ is employed, with $\alpha = .05$; in order to reject the null hypothesis, the obtained value of the sample mean will have to be equal to or greater than 8.59.

Figure 1.2d: If the directional alternative hypothesis H_1: $\mu > 8$ is employed, with $\alpha = .01$; in order to reject the null hypothesis, the obtained value of the sample mean will have to be equal to or greater than 8.84.

Figure 1.2e: If the directional alternative hypothesis H_1: $\mu < 8$ is employed, with $\alpha = .05$; in order to reject the null hypothesis, the obtained value of the sample mean will have to be equal to or less than 7.41.

Figure 1.2f: If the directional alternative hypothesis H_1: $\mu < 8$ is employed, with $\alpha = .01$; in order to reject the null hypothesis, the obtained value of the sample mean will have to be equal to or less than 7.16.

Note that with respect to a specific alternative hypothesis in the above examples, the lower the value of alpha, the larger the value computed for an upper limit and the lower the value computed for a lower limit. Additionally, if the value of alpha is fixed, the computed value for an upper limit will be higher and the computed value for a lower limit will be lower when a nondirectional alternative hypothesis is employed, as opposed to when a directional alternative hypothesis is used.

3. Additional examples illustrating the interpretation of a computed *z* value To further clarify the interpretation of z values, Table 1.3 lists three additional z values that could have been obtained for Example 1.1 if a different set of data had been employed. Table 1.3 notes the decisions that would be made with reference to the three possible alternative hypotheses a researcher could employ on the basis of each of these z values. The table assumes that H_0: $\mu = 8$.

4. The *z* test for a population proportion Another test that employs the normal distribution in order to analyze data derived from a single sample is the ***z* test for a population proportion (Test 6a)**. Equation 6.6, the equation for computing the test statistic for the ***z* test for a population proportion,** is a special case of Equation 1.3. The use of Equation 6.6 is reserved for evaluating a set of scores for a binomially distributed variable (for which the values of μ and σ can be determined). The ***z* test for a population proportion** is discussed after a full discussion of the binomial distribution (which can be found under the **binomial sign test for a single sample (Test 6)**).

Table 1.3 Decision Table for z Values

Obtained z value	Alternative hypothesis	Decision
1.75	H_1: $\mu \neq 8$	The null hypothesis cannot be rejected since the obtained value $z = 1.75$ is less than the tabled critical two-tailed .05 and .01 values $z_{.05} = 1.96$ and $z_{.01} = 2.58$.
	H_1: $\mu > 8$	The null hypothesis can be rejected at the .05 level of significance since the obtained value $z = 1.75$ is a positive number which is greater than the tabled critical one-tailed .05 value $z_{.05} = 1.65$. The null hypothesis cannot be rejected at the .01 level since it is less than the tabled critical one-tailed .01 value $z_{.01} = 2.33$.
	H_1: $\mu < 8$	The null hypothesis cannot be rejected since the obtained value $z = 1.75$ is a positive number.
-2.75	H_1: $\mu \neq 8$	The null hypothesis can be rejected at both the .05 and .01 levels of significance since the obtained absolute value $z = 2.75$ is greater than the tabled critical two-tailed .05 and .01 values $z_{.05} = 1.96$ and $z_{.01} = 2.58$.
	H_1: $\mu > 8$	The null hypothesis cannot be rejected since the obtained value $z = -2.75$ is a negative number.
	H_1: $\mu < 8$	The null hypothesis can be rejected at both the .05 and .01 levels of significance since the obtained value $z = -2.75$ is a negative number and the absolute value $z = 2.75$ is greater than the tabled critical one-tailed .05 and .01 values $z_{.05} = 1.65$ and $z_{.01} = 2.33$.
.75	H_1: $\mu \neq 8$	The null hypothesis cannot be rejected since the obtained value $z = .75$ is less than the tabled critical two-tailed .05 and .01 values $z_{.05} = 1.96$ and $z_{.01} = 2.58$.
	H_1: $\mu > 8$	The null hypothesis cannot be rejected since the obtained value $z = .75$ is less than the tabled critical one-tailed .05 and .01 values $z_{.05} = 1.65$ and $z_{.01} = 2.33$.
	H_1: $\mu < 8$	The null hypothesis cannot be rejected since the obtained value $z = .75$ is a positive number.

VIII. Additional Examples Illustrating the Use of the Single-Sample z Test

Five additional examples that can be evaluated with the **single-sample z test** are presented in this section. Since Examples 1.2–1.4 employ the same population parameters and data set used in Example 1.1, they yield the identical result. Note that Examples 1.2 and 1.3 employ objects in lieu of subjects. Examples 1.5 and 1.6 illustrate the application of the **single-sample z test** when the size of the sample is $n = 1$.

Example 1.2. *The Brite battery company manufactures batteries which are programmed by a computer to have an average life span of 8 months and a standard deviation of 2 months. If the average life of a random sample of 30 Brite batteries purchased from 30 different stores is 7.4 months, are the data consistent with the mean value parameter programmed into the computer?*

Example 1.3. *The Smooth Road cement company stores large quantities of its cement in* 30 *storage tanks. A state law says that the machine which fills the tanks must be calibrated so as not to deviate substantially from a mean load of* 8 *tons. It is known that the standard deviation of the loads delivered by the machine is* 2 *tons. An inspector visits the storage facility and determines that the mean number of tons in the* 30 *storage tanks is* 7.4 *tons. Does this conform to the requirements of the state law?*

Example 1.4. *A study involving* 30 *subjects is conducted in order to determine the subjects' ability to accurately judge weight. In the study subjects are required (by adding or subtracting sand) to adjust the weight of a cylinder, referred to as the variable stimulus, until it is judged equal in weight to a standard comparison cylinder whose weight is fixed. The weight of the standard comparison stimulus is* 8 *ounces. Prior research has indicated that the standard deviation of the subjects' judgements in such a task is* 2 *ounces. Prior to testing the subjects, the experimenter decides she will conclude that a kinesthetic illusion occurs if the mean of the subjects' judgements differs significantly from the value of the standard stimulus. If the average weight assigned by the* 30 *subjects to the variable stimulus is* 7.4 *ounces, can the experimenter conclude that a kinesthetic illusion has occurred?*

Example 1.5. *A meteorologist determines that during the current year there were* 80 *major storms recorded in the Western Hemisphere. He claims that* 80 *storms represent a significantly greater number than the annual average. Based on data that have been accumulated over the past* 100 *years, it is known that on the average there have been* 70 *major storms in the Western Hemisphere, and the standard deviation is* 2. *Do* 80 *storms represent a significant deviation from the mean value of* 70*?*

Example 1.5 illustrates a problem that would be evaluated with the **single-sample *z* test** in which the value of n is equal to one. In the example, the sample size of one represents the single year during which there were 80 storms. When the value of $n = 1$, Equation 1.3 becomes Equation 1.5 (which is the same as Equation I.9 in the **Introduction**).

$$z = \frac{X - \mu}{\sigma} \qquad \textbf{(Equation 1.5)}$$

Equation 1.5, the equation for converting a raw score into a standard deviation (z) score, allows one to determine the likelihood of a specific score occurring within a normally distributed population. Thus, within the context of Example 1.5, Equation 1.5 will allow the meteorologist to determine the likelihood that a score of 80 will occur in a normally distributed population in which $\mu = 70$ and $\sigma = 2$. The analysis assumes that within the total population there are $N = 100$ scores (where each score represents the number of storms in a given year during the 100 year period). Since the frequency of storms during one year is being compared to the population mean, the value of $n = 1$. Note that when $n = 1$, the value of the sample mean ($\bar{X}$) in Equation 1.3 reduces to the value X in Equation 1.5. Additionally, since $n = 1$, the value of $\sigma_{\bar{X}}$ in Equation 1.3 becomes σ in Equation 1.5, since $\sigma_{\bar{X}} = \sigma/\sqrt{n} = \sigma/\sqrt{1} = \sigma$.

Employing Equations 1.3/1.5 with the data for Example 1.5, the value $z = 5$ is computed.

$$z = \frac{80 - 70}{2} = 5$$

The null hypothesis employed for the above analysis is H_0: $\mu = 70$. Example 1.5 implies that either the directional alternative hypothesis H_1: $\mu > 70$ or the nondirectional alternative hypothesis H_1: $\mu \neq 70$ can be employed. Regardless of which of these alternative hypotheses is employed, since the computed value $z = 5$ is greater than all of the tabled critical values in Table 1.1, the null hypothesis can be rejected at both the .05 and .01 levels. Thus, the meteorologist can conclude that a significantly greater number of storms were recorded during the current year than the mean value recorded for the past 100 years. The directional alternative hypothesis H_1: $\mu < 70$ is not supported, since in order to support the latter alternative hypothesis the computed value of z must be a negative number (which will only be the case if the number of storms observed during the year is less than the population mean of $\mu = 70$).

Example 1.6. *A physician assesses the level of a specific chemical in a* 40*-year-old male patient's blood. Assume that the average level of the chemical in adult males is* 70 *milligrams (per* 100 *milliliters), with a standard deviation of* 2 *milligrams (per* 100 *milliliters). If the patient has a blood reading of* 80, *will the patient be viewed as abnormal?*

As is the case in Example 1.5, Example 1.6 also employs a sample size of $n = 1$. Since this example uses the same data as Example 1.5, the computed value $z = 5$ is obtained, thus allowing the physician to reject the null hypothesis. The value $z = 5$ indicates that the patient has a blood reading that is five standard deviation units above the population mean. The proportion of cases in the normal distribution associated with a z score of 5 or greater is so small that there is no tabled value listed for $z = 5$ in **Table A1**.

Reference

Freund, J. E. (1984). **Modern elementary statistics** (6th ed.). Englewood Cliffs, NJ: Prentice–Hall, Inc.

Endnotes

1. The exact probability value recorded for $z = 1.67$ in Column 3 of **Table A1** is .0475 (which is equivalent to 4.75%). This indicates that the proportion of cases that falls above the value $z = 1.67$ is .0475, and the proportion of cases that falls below the value $z = -1.67$ is .0475. Since this indicates that in the left tail of the distribution there is less than a 5% chance of obtaining a z value equal to or less than $z = -1.67$, we can reject the null hypothesis at the .05 level if we employ the nondirectional alternative hypothesis H_1: $\mu < 8$, with $\alpha = .05$.

2. Equation 1.2 is employed to compute the standard error of the population mean when the size of the underlying population is infinite. In practice, it is employed when the size of the underlying population is large and the size of the sample is believed to constitute less than 5% of the population. However, among others, Freund (1984) notes that in a finite population, if the size of a sample constitutes more than 5% of the population, a correction factor is introduced into Equation 1.2. The computation of the standard error of the mean with the **finite population correction factor** is noted below:

$$\sigma_{\bar{X}} = \frac{\sigma}{\sqrt{n}} \sqrt{\frac{N - n}{N - 1}}$$

Where: N represents the total number of subjects/objects that comprise the population.

The finite population corrected equation will result in a smaller value for $\sigma_{\bar{X}}$. This is the case since as the proportion of a population represented by a sample increases, the less variability there will be among the means that comprise the sampling distribution, and thus the smaller the expected difference between the sample mean obtained for a set of data and the value of the population mean. Thus when $n = N$, employing the finite corrected equation, the value of $\sigma_{\bar{X}}$ will always equal zero. This is the case since when $n = N$, the sample mean and population mean will always be the same value, and thus no error is involved in estimating the value of $\mu_{\bar{X}}$. Since it is usually assumed that the size of a sample is less than 5% of the population it represents, Equation 1.2 is the only equation listed in most sources for computing the value of $\sigma_{\bar{X}}$.

3. Inspection of **Table A1** reveals that the exact percentage of cases in a normal distribution that falls within three standard deviations above or below the mean is 99.74%.

Test 2

The Single-Sample *t* Test
(Parametric Test Employed with Interval/Ratio Data)

I. Hypothesis Evaluated with Test and Relevant Background Information

Hypothesis evaluated with test Does a sample of n subjects come from a population in which the mean (μ) equals a specified value?

Relevant background information on test The **single-sample *t* test** is one of a number of inferential statistical tests that are based on the t distribution. Like the normal distribution, the t distribution is a bell-shaped, continuous, symmetrical distribution, which to the statistically unsophisticated eye is almost indistinguishable from the normal distribution. t, which is the computed test statistic for the **single-sample *t* test**, represents a standard deviation score, and is interpreted in the same manner as the z value computed for the **single-sample *z* test.** The only difference between a z value and a t value is that for a given standard deviation score, the proportion of cases that falls between the mean and the standard deviation score will be a function of which of the two distributions one employs. Except when $n = \infty$, for a given standard deviation score, a larger proportion of cases falls between the mean of the normal distribution and the standard deviation score than the proportion of cases that falls between the mean and that same standard deviation score in the t distribution. In point of fact, there are actually an infinite number of t distributions — each distribution being based on the number of subjects/objects in the sample. As the size of the sample increases, the proportions (and consequently the critical values) in the t distribution approach the proportions (and critical values) in the normal distribution, and in fact, when $n = \infty$, the normal and t distributions are identical. A more detailed discussion (as well as visual illustrations) of the t distribution can be found in Section VII.

The **single-sample *t* test** is employed in a hypothesis testing situation involving a single sample in order to determine whether a sample with a mean of $\bar{X}$ is derived from a population with a mean of μ. If the result of the **single-sample *t* test** yields a significant difference, the researcher can conclude there is a high likelihood the sample is derived from a population with a mean value other than μ.

The **single-sample *t* test** is used with interval/ratio data. The test is employed when a researcher does not know the value of the population standard deviation (σ), and therefore must estimate it by computing the sample standard deviation ($\tilde{s}$). As is noted in the discussion of the **single-sample *z* test (Test 1)**, some sources argue that even if one knows the value of σ, when the sample size is very small (generally less than 25), the **single sample *t* test** provides a more accurate estimate of the underlying sampling distribution for the data.

The following two assumptions which are noted for the **single-sample *z* test**, also apply to the **single-sample *t* test**: a) The sample has been randomly selected from the population it represents; and b) The distribution of data in the underlying population the sample represents is normal. If either of the aforementioned assumptions is saliently violated, the reliability of the t test statistic may be compromised.

II. Example

Example 2.1 *A physician states that the average number of times he sees each of his patients during the year is five. In order to evaluate the validity of this statement, he randomly selects ten of his patients and determines the number of office visits each of them made during the past year. He obtains the following values for the ten patients in his sample:* 9, 10, 8, 4, 8, 3, 0, 10, 15, 9. *Do the data support his contention that the average number of times he sees a patient is five?*

III. Null versus Alternative Hypotheses

Null hypothesis H_0: $\mu = 5$

(The mean of the population the sample represents equals 5.)

Alternative hypothesis H_1: $\mu \neq 5$

(The mean of the population the sample represents does not equal 5. This is a **nondirectional alternative hypothesis**, and it is evaluated with a **two-tailed test**. In order to be supported, the absolute value of t must be equal to or greater than the tabled critical two-tailed t value at the prespecified level of significance. Thus, either a significant positive t value or a significant negative t value will provide support for this alternative hypothesis.)

or

H_1: $\mu > 5$

(The mean of the population the sample represents is greater than 5. This is a **directional alternative hypothesis**, and it is evaluated with a **one-tailed test**. It will only be supported if the sign of t is positive, and the absolute value of t is equal to or greater than the tabled critical one-tailed t value at the prespecified level of significance.)

or

H_1: $\mu < 5$

(The mean of the population the sample represents is less than 5. This is a **directional alternative hypothesis**, and it is evaluated with a **one-tailed test**. It will only be supported if the sign of t is negative, and the absolute value of t is equal to or greater than the tabled critical one-tailed t value at the prespecified level of significance.)

Note: Only one of the above noted alternative hypotheses is employed. If the alternative hypothesis the researcher selects is supported, the null hypothesis is rejected.

IV. Test Computations

Table 2.1 summarizes the number of visits recorded for the $n = 10$ subjects in Example 2.1. In order to compute the test statistic for the **single-sample *t* test**, it is necessary to determine the mean of the sample and to obtain an unbiased estimate of the population standard deviation.

Employing Equation 1.1 (which is the same as Equation I.1 in the **Introduction**) the mean of the sample is computed to be $\bar{X} = 7.6$.

$$\bar{X} = \frac{\Sigma X}{n} = \frac{76}{10} = 7.6$$

Table 2.1 Summary of Data for Example 2.1

Battery	X	X^2
1	9	81
2	10	100
3	8	64
4	4	16
5	8	64
6	3	9
7	0	0
8	10	100
9	15	225
10	9	81
	$\Sigma X = 76$	$\Sigma X^2 = 740$

Equation 2.1 (which is the same as Equation I.8 in the **Introduction**) is employed to compute $\tilde{s}$, which represents an unbiased estimate of the value of the population standard deviation.

$$\tilde{s} = \sqrt{\frac{\Sigma X^2 - \frac{(\Sigma X)^2}{n}}{n - 1}} \qquad \textbf{(Equation 2.1)}$$

Using Equation 2.1, the value $\tilde{s} = 4.25$ is computed.

$$\tilde{s} = \sqrt{\frac{740 - \frac{(76)^2}{10}}{9 - 1}} = 4.25$$

As is the case with the **single-sample *z* test**, computation of the test statistic for the **single-sample *t* test** requires that the value of the **standard error of the population mean** be computed. $s_{\bar{X}}$, which represents an estimate of $\sigma_{\bar{X}}$, is computed with Equation 2.2. Note that $s_{\bar{X}}$, which is referred to as the **estimated standard error of the population mean,** is based on the value of $\tilde{s}$ computed with Equation 2.1. A discussion of the theoretical meaning of the standard error of the population mean can be found in Section VII of the **single-sample *z* test.** Further discussion of $s_{\bar{X}}$ (which represents a standard deviation of a sampling distribution of means and is interpreted in the same manner as $\sigma_{\bar{X}}$) can be found in Section V.

$$s_{\bar{X}} = \frac{\tilde{s}}{\sqrt{n}} \qquad \textbf{(Equation 2.2)}$$

Employing Equation 2.2, the value $s_{\bar{X}} = 1.34$ is computed.

$$s_{\bar{X}} = \frac{4.25}{\sqrt{10}} = 1.34$$

It should be noted that neither $\tilde{s}$ or $s_{\bar{X}}$ can ever be a negative value. If a negative value is obtained for either $\tilde{s}$ or $s_{\bar{X}}$, it indicates a computational error has been made.

Equation 2.3 is the test statistic for the **single-sample *t* test.**[1]

$$t = \frac{\bar{X} - \mu}{s_{\bar{X}}} \qquad \textbf{(Equation 2.3)}$$

Inspection of Equation 2.3 reveals that it is similar in structure to Equation 1.3, the equation for the **single-sample *z* test**. The only differences between the two equations are that: a) Equation 2.3 employs the t distribution as opposed to the z distribution; and b) The value of the standard error of the population mean is estimated in Equation 2.3 from the value of $\tilde{s}$.[2]

Employing Equation 2.3, the value $t = 1.94$ is computed. Note that in Equation 2.3, the value that represents μ is the value $\mu = 5$ which is stated in the null hypothesis.

$$t = \frac{7.6 - 5}{1.34} = 1.94$$

V. Interpretation of the Test Results

Since like a z value a t value represents a standard deviation score, except for the fact that a different distribution is employed, it is interpreted in the same manner. The obtained value $t = 1.94$ is evaluated with **Table A2 (Table of Student's *t* Distribution)** in the **Appendix.**[3] In **Table A2** the critical t values are listed in relation to the proportion of cases (which are recorded at the top of each column) that falls below a specified t score in the t distribution, and the number of **degrees of freedom** for the sampling distribution that is being evaluated (which are recorded in the left hand column of each row). Equation 2.4 is employed to compute the degrees of freedom for the **single-sample *t* test**. A full explanation of the meaning of the degrees of freedom can be found in Section VII.

$$df = n - 1 \qquad \textbf{(Equation 2.4)}$$

Employing Equation 2.4, we compute that $df = 10 - 1 = 9$. Thus, the tabled critical t values that are employed in evaluating the results of Example 2.1 are the values recorded in the cells of **Table A2** that fall in the row for $df = 9$ and the columns with probabilities/proportions that correspond to the one- and two-tailed .05 and .01 values. These critical t values are summarized in Table 2.1.

Table 2.1 Tabled Critical Two-Tailed and One-Tailed .05 and .01 *t* Values

	$t_{.05}$	$t_{.01}$
Two-tailed values	2.26	3.25
One-tailed values	1.83	2.82

Note that the tabled critical two-tailed value $t_{.05} = 2.26$ is the value in the row $df = 9$ and the column $p = .975$, since $t_{.05} = 2.26$ is the standard deviation score above which (as well as below which in the case of $t = -2.26$) a proportion equivalent to .025 of the cases in the distribution falls. The tabled critical two-tailed value $t_{.01} = 3.25$ is the value in the row $df = 9$ and the column $p = .995$, since $t_{.01} = 3.25$ is the standard deviation score above which (as well as below which in the case of $t = -3.25$) a proportion equivalent to .005 of the cases in the distribution falls. The tabled critical one-tailed value $t_{.05} = 1.83$ is the value in the row $df = 9$ and the column $p = .95$, since $t_{.05} = 1.83$ is the standard deviation score above which (as well as below which in the case of $t = -1.83$) a proportion equivalent to .05 of the cases in the distribution falls. The tabled critical one-tailed value $t_{.01} = 2.82$ is the value in the row $df = 9$ and the column $p = .99$, since $t_{.01} = 2.82$ is the standard deviation score above which (as well as below which in the case of $t = -2.82$) a proportion equivalent to .01 of the cases in the distribution falls.

The following guidelines are employed in evaluating the null hypothesis for the **single-sample *t* test.**

a) If the alternative hypothesis employed is nondirectional, the null hypothesis can be rejected if the obtained absolute value of t is equal to or greater than the tabled critical two-tailed value at the prespecified level of significance.

b) If the alternative hypothesis employed is directional and predicts a population mean larger than the value stated in the null hypothesis, the null hypothesis can be rejected if the sign of t is positive and the value of t is equal to or greater than the tabled critical one-tailed value at the prespecified level of significance.

c) If the alternative hypothesis employed is directional and predicts a population mean smaller than the value stated in the null hypothesis, the null hypothesis can be rejected if the sign of t is negative and the absolute value of t is equal to or greater than the tabled critical one-tailed value at the prespecified level of significance.

Employing the above guidelines, we can only reject the null hypothesis if the directional alternative hypothesis H_1: $\mu > 5$ is employed, and the null hypothesis can only be rejected at the .05 level. This is the case since the obtained value of t is a positive number which is greater than the tabled critical one-tailed .05 value $t_{.05} = 1.83$. Note that the alternative hypothesis H_1: $\mu > 5$ is not supported at the .01 level, since the obtained value $t = 1.94$ is less than the tabled critical one-tailed .01 value $t_{.01} = 2.82$.

The nondirectional alternative hypothesis H_1: $\mu \neq 5$ is not supported since the obtained value $t = 1.94$ is less than the tabled critical two-tailed .05 value $t_{.05} = 2.26$.

The directional alternative hypothesis H_1: $\mu < 5$ is not supported since the obtained value $t = 1.94$ is a positive number. In order for the alternative hypothesis H_1: $\mu < 5$ to be supported, the computed value of t must to be a negative number (as well as the fact that the absolute value of t must be equal to or greater than the tabled critical one-tailed value at the prespecified level of significance).

In Section IV it is noted that the **estimated standard error of the population mean** ($s_{\bar{X}}$) computed for the **single-sample *t* test** represents a standard deviation of a sampling distribution of means. The use of the t distribution for Example 2.1 is based on the fact that when the population standard deviation is unknown, the latter distribution provides a better approximation of the underlying sampling distribution than does the normal distribution. Figure 2.1 depicts the sampling distribution employed for Example 2.1. This sampling distribution is interpreted in the same manner as the sampling distribution for the **single-sample *z* test** which is depicted in Figure 1.1.

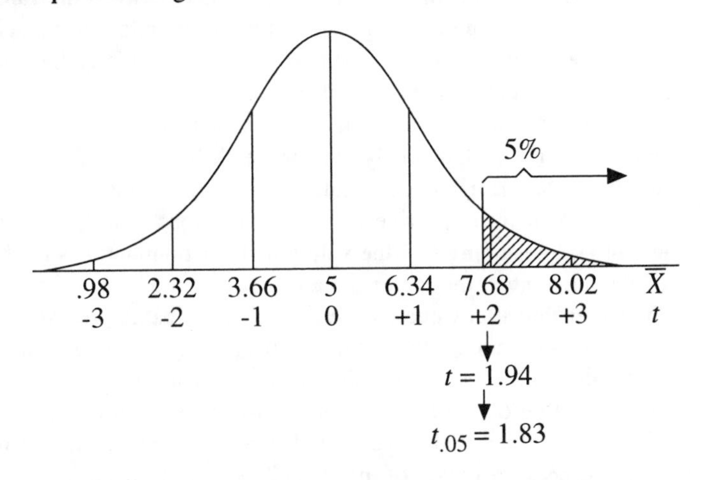

Figure 2.1 Sampling Distribution for Example 2.1

Inspection of the sampling distribution depicted in Figure 2.1 reveals that the obtained value $t = 1.94$ falls to the right of the tabled critical one-tailed value $t_{.05} = 1.83$. At this point it should be noted that $t_{.05} = 1.83$ is greater than $z_{.05} = 1.65$, the tabled critical one-tailed .05 value employed with the normal distribution. Both of these values demarcate the upper 5% of their respective distributions. If one elects to employ the **single-sample *z* test,** as opposed to the **single-sample *t* test,** for an analysis in which the value of σ is unknown, it will inflate the likelihood of committing a Type I error. This is the case since, except when the sample size is very large (in which case the corresponding values of t and z are identical), a tabled critical z value at a prespecified level of significance will always be smaller than the corresponding tabled critical t value at that same level of significance. Thus, in the case of Example 2.1, if we employ the tabled critical one-tailed value $z_{.05} = 1.65$, the likelihood of committing a Type I error will be greater than .05. This can be confirmed by inspection of **Table A2** which indicates that for $df = 9$, the proportion of cases in the t distribution that falls at or above the value of $t = 1.65$ is greater than .05 (i.e., a t score of 1.65 falls below the 95th percentile of the distribution).[4]

A summary of the analysis of Example 2.1 with the **single-sample *t* test** follows: With respect to the average number of times the doctor sees a patient, we can conclude that the sample of 10 subjects comes from a population with a mean value other than 5 only if we employ the directional alternative hypothesis H_1: $\mu > 5$, and prespecify as our level of significance $\alpha = .05$. This result can be summarized as follows: $t(9) = 1.94, p < .05$. (The degrees of freedom employed in the analysis are noted in parentheses after the t.)

VI. Additional Analytical Procedures for the Single-Sample *t* Test and/or Related Tests

1. Determination of the power of the single-sample *t* test and the single-sample *z* test The power of either the **single-sample *z* test** or the **single-sample *t* test** will represent the probability of the test identifying a difference between the value for the population mean stipulated in the null hypothesis and a specific value that represents the true value of the mean of the population represented by the experimental sample. In order to compute the power of a test, it is necessary for the researcher to stipulate the latter value which will be identified with the notation μ_1. In practice, a researcher can compute the power of a test for any value of μ_1.

The power of the test will be a function of the difference between the value of μ stated in the null hypothesis and the value of μ_1. The test's power will increase as the absolute value of the difference between the values μ and μ_1 increases. This is the case since if the sample is derived from a population with a mean value that is substantially above or below the value of μ stated in the null hypothesis, it is likely that this would be reflected in the value of the sample mean ($\bar{X}$). Obviously, the more the value of $\bar{X}$ deviates from the hypothesized value of μ, the greater the absolute value of the numerators (i.e., $\bar{X} - \mu$) in Equations 1.3 and 2.3 (which are respectively the equations for the **single-sample *z* test** and the **single-sample *t* test**). Assuming that the value of the denominator is held constant, the larger the value of the numerator, the larger the absolute value of the computed test statistic (i.e., z or t). The larger the latter value the more likely it is that the researcher will be able to reject the null hypothesis (assuming the obtained difference is in the direction predicted in the alternative hypothesis), and consequently the more powerful the test.

Since the obtained value of the test statistic is also a function of the denominator of Equations 1.3 and 2.3 (i.e., the actual or estimated standard error of the population mean), the latter value also influences the power of the test. Specifically, as the value of the denominator decreases, the computed absolute value of the test statistic increases. It happens to be the case that the value of the standard error of the mean is a function of the population standard deviation (which is estimated in the case of the t test) and the sample size. Inspection of Equations 1.3 and 2.3 reveals that the standard error of the mean will decrease if the value of the standard deviation is decreased and the sample size is increased. Thus, by employing an accurate estimate of the population standard deviation (more specifically, in the case of the t test, one that is not spuriously inflated due to sampling error) and a large sample size, one can minimize the value of the denominator in Equations 1.3 and 2.3, and consequently maximize the absolute value of the test statistic. As a result one can increase the likelihood that the null hypothesis will be rejected, which increases the power of the test.

The power of a statistical test can be represented both mathematically and graphically. Figures 2.2 and 2.3 illustrate the concept of power and its relationship to the Type I and Type II error rates. Both figures contain two distributions which represent the sampling distributions of means for two populations.[5] In each figure, the sampling distribution on the left represents a population with the mean μ (i.e., the value stated in the null hypothesis). The sampling distribution on the right represents a population with the mean μ_1, which we will assume is the true value of the mean of the population from which the experimental sample is derived. Both Figures 2.2 and 2.3 assume a fixed value for the sample size upon which the sampling distributions are based, and that each of the underlying populations represented by the sampling distributions has the same standard deviation. Figure 2.2 represents a case in which there is a large difference between the values of μ_1 and μ, whereas Figure 2.3 represents a case in which there is a small difference between the two values. When expressed in standard deviation units, the magnitude of the absolute value of the difference between μ_1 and μ is referred to as the **effect size**. Thus, in Figure 2.2 a **large effect size** is present, whereas Figure 2.3 depicts a **small effect size**.

The reader should note the following with respect to Figures 2.2 and 2.3.

a) The closer the values μ and μ_1 are to one another, the more the sampling distributions of the two populations overlap.

b) In the case of a one-tailed analysis, the value of alpha (α) is represented by area (///) in the distribution on the left. Recollect that α represents the likelihood of committing a Type I error (i.e., rejecting a true null hypothesis). Numerically, α represents the proportion of the left distribution that comprises area (///). In the case of a two-tailed analysis, the proportion of the left distribution represented by area (///) will be equal to $\alpha/2$.

c) The value of beta (β) is represented by area ($\equiv$) in the distribution on the right. β represents the likelihood of committing a Type II error (i.e., retaining a false null hypothesis). Numerically, β represents the proportion of the right distribution that comprises area ($\equiv$).

d) The power of the test is represented by area (\\\) in the right distribution. Note that this is the area in the right distribution that falls to the right of the area delineating β. Numerically, the power of the test represents the proportion of the right distribution that comprises area (\\\). The power of the test can also be represented by subtracting the value of β from 1 (i.e., Power = $1 - \beta$). Note that the area in the left distribution that represents α overlaps the area in the right distribution representing the power of the test.

e) In order to increase the value of α, one must move the boundary in the left distribution that delineates α to the left. By doing the latter one will decrease the value of β, since the area in the right distribution that corresponds to β will decrease. By increasing the value of α, one also increases the area in the right distribution which represents the power of the test. This illustrates the fact that if one increases the likelihood of committing a Type I error (α), one decreases the likelihood of committing a Type II error (β), and at the same time increases the power of the test ($1 - \beta$). In the same respect, to decrease the value of α one must move the boundary in the left distribution that delineates alpha to the right. By doing the latter one will increase the value of β, since the area in the right distribution that

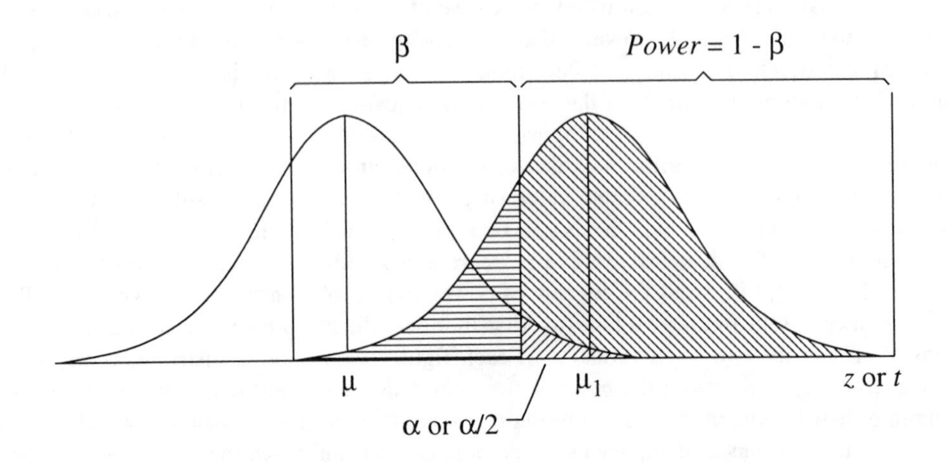

Figure 2.2 Sampling Distributions Employed in Determining the Power of a Test Involving a Large Effect Size

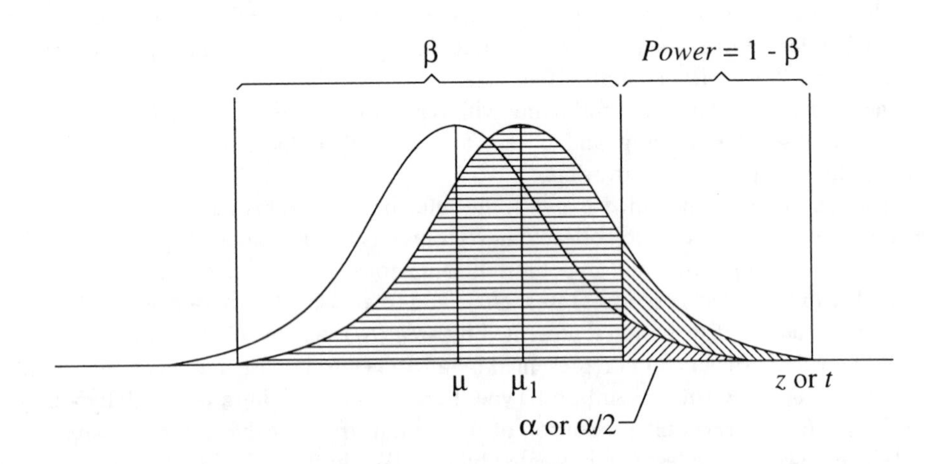

Figure 2.3 Sampling Distributions Employed in Determining the Power of a Test Involving a Small Effect Size

corresponds to β will increase. By decreasing the value of α, one also decreases the area in the right distribution which represents the power of the test. This illustrates the fact that if one decreases the likelihood of committing a Type I error, the likelihood of committing a Type II error increases, and at the same time the power of the test decreases.

f) It should also be noted that at a given level of significance, a one-tailed test will be more powerful than a two-tailed test. This is the case since with a one-tailed test the point that delineates α will be farther to the left in the left distribution than will be the case with a two-tailed test. As an example, when $\alpha = .05$, the tabled critical one-tailed value for z is $z_{.05} = 1.65$, whereas the tabled critical two-tailed value is $z_{.05} = 1.96$. Since both of these critical values are in the left distribution, the former value will be farther to the left, thus expanding the area in the right distribution which represents the power of the test.

Two methods will now be demonstrated which can be used to determine the power of either the **single-sample *t* test** or the **single-sample *z* test.** The first method (which is more time consuming) reveals all of the logical operations involved in computing the power of a test. The second method, which employs **Table A3 (Power Curves for Student's *t* Distribution)** in the **Appendix**, requires fewer computations. It should be emphasized that whenever possible a power analysis should be conducted prior to the data collection phase of a study. By computing power beforehand, one is able to design a study with a sample size that is large enough to detect the specific effect size predicted by the researcher.

Method 1 for computing the power of the single-sample *t* test and the single sample *z* test The first method will initially be demonstrated with reference to the **single-sample *z* test.** The reason for employing the latter test is that it will allow us to use **Table A1 (Table of the Normal Distribution)** in the **Appendix**, which lists probabilities for all z values between 0 and 4. Detailed tables of the t distribution that list probabilities for all t values within this range are generally not available.

Let us assume that in our example we are employing the same null hypothesis that is employed in Example 2.1 (i.e., H_0: $\mu = 5$). It will be assumed that the researcher wishes to evaluate the power of the **single-sample *z* test** in reference to the alternative hypothesis H_1: $\mu_1 = 6$. Note that in conducting a power analysis the alternative hypothesis states that the population mean is a specific value that is different from the value stated in H_0. In conducting the power analysis it will be assumed that the null hypothesis will be evaluated with a two-tailed test, with $\alpha = .05$. For purposes of illustration, it will also be assumed that the researcher evaluates the null hypothesis employing a sample size of $n = 121$. In addition, we will assume that the value of the population standard deviation is known to be $\sigma = 4.25$.

Employing Equation 1.2, the value of the standard error of the population mean is computed. Thus: $\sigma_{\bar{X}} = 4.25/\sqrt{121} = .39$.

Figure 2.4, which depicts the analysis graphically, is comprised of two overlapping normal distributions. Each distribution is a sampling distribution of population means. The distribution on the left, which is the sampling distribution of means of a population with a mean of 5, will be referred to as Distribution A. The distribution on the right, which is the sampling distribution of means of a population with a mean of 6, will be referred to as Distribution B. We have already determined above that $\sigma_{\bar{X}} = .39$, and we will assume that this value represents the standard deviation of each of the sampling distributions. The area (///) delineates the proportion of Distribution A that corresponds to the value $\alpha/2$, which equals .025. This is the case, since $\alpha = .05$ and a two-tailed test is being used. In such an instance, the proportion of the curve comprising the critical area in each of the tails of Distribution A will be .05/2 = .025. Area (≡) delineates the proportion of Distribution B that corresponds to the probability of committing a Type II error (β). Area (\\\) delineates the proportion of Distribution B that represents the power of the test.

The procedure for computing the proportions in Figure 2.4 will now be described. The first step in computing the power of the test requires one to determine how far above the value $\mu = 5$ the sample mean will have to be in order to reject the null hypothesis. Equation 1.3 is employed to determine this minimum required difference. By algebraically transposing the terms in Equation 1.3 we can demonstrate that $\bar{X} - \mu = (z_{.05})(\sigma_{\bar{X}})$. Thus, by substituting the values $z_{.05} = 1.96$ (which is the tabled critical two-tailed .05 z value) and $\sigma_{\bar{X}} = .39$ in the latter equation we can determine that the minimum required difference is $\bar{X} - \mu = (1.96)(.39) = .76$.

Thus, any sample mean .76 units above or below the value $\mu = 5$ will allow the researcher to reject the null hypothesis at the .05 level (if a two-tailed analysis is employed). With respect to evaluating the power of the test in reference to the alternative hypothesis

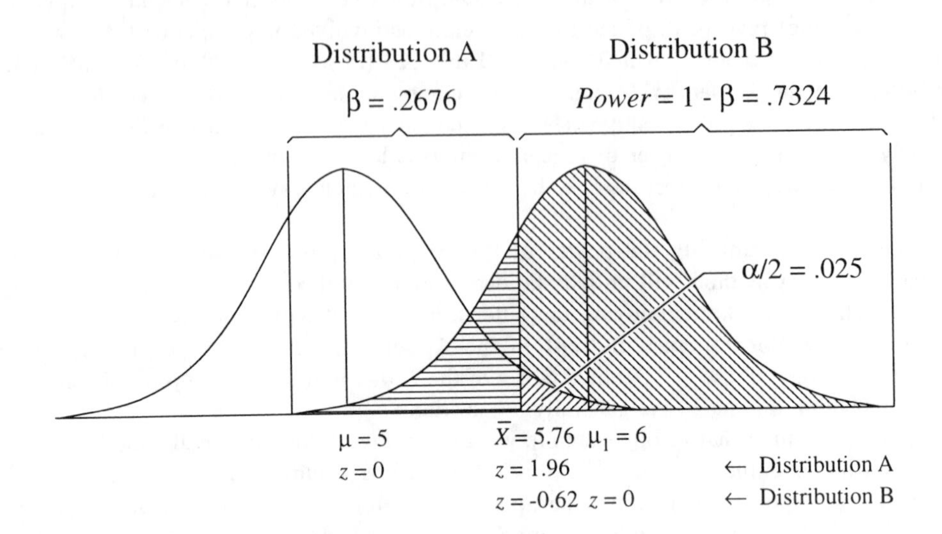

Figure 2.4 Visual Representation of Power when H_0: $\mu = 5$ and H_1: $\mu = 6$ for $n = 121$

H_1: $\mu_1 = 6$, the researcher is only concerned with a mean value above 5 (which will fall in the right tail of Distribution A).[6] Thus, a mean value of $\bar{X} = 5.76$ or greater will allow the researcher to reject the null hypothesis (since $\mu + (z_{.05})(\sigma_{\bar{X}}) = 5 + .76 = 5.76$).

The next step in the analysis requires that the area in Distribution B that falls between the mean $\mu_1 = 6$ and the value 5.76 be computed. This is accomplished by employing Equation 1.3 and substituting 5.76 to represent the value of $\bar{X}$ and $\mu_1 = 6$ to represent the value of μ.

$$z = \frac{\bar{X} - \mu}{\sigma_{\bar{X}}} = \frac{5.76 - 6}{.39} = -.62$$

Utilizing **Table A1**, we determine that the proportion of Distribution B that lies between $\mu_1 = 6$ and 5.76 (i.e., between the mean and a z score of $-.62$ or $+.62$) is .2324. Since the value 5.76 is below the mean of Distribution B, if .5 (which is the proportion of Distribution B that falls above the mean $\mu_1 = 6$) is added to .2324, the resulting value of .7324 will represent the power of the test. This latter value is represented by area (\\\) in Figure 2.4. The likelihood of committing a Type II error (i.e., β) is represented by area (≡). The proportion of Distribution B that constitutes this latter area is determined by subtracting the value .7324 from 1. Thus: $\beta = 1 - .7324 = .2676$. Based on the results of the power analysis we can state that if the alternative hypothesis H_1: $\mu_1 = 6$ is true, the likelihood that the null hypothesis will be rejected is .7324, and at the same time there is a .2676 likelihood that it will be retained. If the researcher considered the computed value for the power too low (which we are assuming is determined prior to implementing the study), she can increase the power of the test by employing a larger sample size.

If the value of σ is not known and has to be estimated from the sample data, the power analysis will be based on the t distribution instead of the normal distribution. In such a case the identical protocol described above for computing power is employed, except for the fact that a tabled critical t value is used in place of the tabled critical z value. Unless the sample size is extremely large, the tabled critical t value will be larger than the tabled critical z value used for the same data. As a result of this, the power of the test computed for the t distribution will be lower than the value computed for the normal distribution.

In the case of the example under discussion, if the t distribution is employed one would use the tabled critical two-tailed .05 t value for $df = n - 1 = 121 - 1 = 120$. From **Table A2** it can be determined that this value is $t_{.05} = 1.98$. Using the latter value and the value $s_{\bar{X}} = .39$ it can be determined that .77 is the minimum required difference in order to achieve significance. When .77 is added to the value $\mu = 5$, it indicates that a sample mean of 5.77 or greater (as well as 4.23 or lower) will allow the researcher to reject the null hypothesis. The t value required to complete the power calculations is determined by utilizing Equation 2.3 and substituting 5.77 to represent the value of $\bar{X}$ and $\mu_1 = 6$ to represent the value of μ. The calculations are noted below.

$$(t_{.05})(s_{\bar{X}}) = \bar{X} - \mu$$

$$(1.98)(.39) = .77$$

$$t = \frac{5.77 - 6}{.39} = -.59$$

Detailed tables of the t distribution indicate that for $df = 120$ the proportion of cases between the mean and a t score of $-.59$ or $+.59$ is approximately .22.[7] The power of the test is derived by adding .5 to the latter value. Thus, Power $= .22 + .5 = .72$, which is slightly lower than the value .7324 obtained for the normal distribution.

It was previously noted that the size of the sample employed in a study is directly related to the power of a statistical test. Thus, in the example under discussion, if instead of using a sample size of $n = 121$ we employ a sample size of $n = 10$, the power of both the **single-sample z test** and the **single-sample t test** will be considerably less than the computed values .7324 and .72. In point of fact, when $n = 10$ the power of the **single-sample z test** equals .1122. The dramatic decrease in power for the small sample size can be understood by determining the minimum amount by which $\bar{X}$ and μ must differ from one another in order to reject the null hypothesis. Since we are still employing the normal distribution, when $n = 10$ the same tabled critical two-tailed value $z_{.05} = 1.96$ is used. However, the value of $\sigma_{\bar{X}}$ is increased substantially, since $\sigma_{\bar{X}} = 4.25/\sqrt{10} = 1.34$. Employing the values $z_{.05} = 1.96$ and $\sigma_{\bar{X}} = 1.34$, we can compute that the minimum required difference in order to reject H_0 when $n = 10$ is 2.63. Specifically: $\bar{X} - \mu = (1.96)(1.34) = 2.63$.

A sample mean that is 2.63 units above $\mu = 5$ is equal to 7.63. The latter value will fall farther to the right in the right tail of the Distribution B than the value 5.76 which is computed when $n = 121$. Substituting the values $\bar{X} = 7.63$ and $\mu = 6$ in Equation 1.3, we determine that $\bar{X} = 7.63$ is 1.22 standard deviation units above the mean of Distribution B.

$$z = \frac{7.63 - 6}{1.34} = 1.22$$

If one examines Figure 2.5, which depicts the analysis graphically, it can be seen that in Distribution B the value $z = 1.22$ lies to the right of the mean of the distribution. Thus, when $n = 10$ the power of the **single-sample z test** will be represented by the proportion of Distribution B that comprises area (\\\). Employing the table for the normal distribution, it can be determined that the proportion of the curve to the right of $z = 1.22$ is .1122, which represents the power of the test. On the basis of this, we can determine that the likelihood of committing a Type II error (represented by area (≡)) will be $\beta = 1 - .1122 = .8878$, which is substantially greater than the value .2676 obtained when $n = 121$. Note that area (\\\) in Distribution B is much smaller than the corresponding area depicted in Figure 2.4 (when $n = 121$). By virtue of area (\\\) being smaller, the proportion of Distribution B in

Figure 2.5 representing area (≡) is substantially larger than the corresponding proportion/area in Figure 2.4.

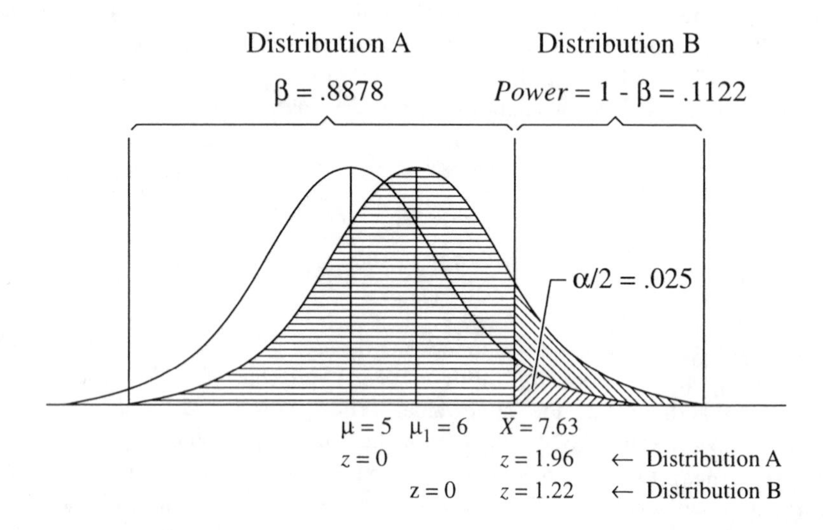

Figure 2.5 Visual Representation of Power when H_0: μ = 5 and H_1: μ = 6 for n = 10

The t distribution will now be applied to the above problem. Let us assume that $n = 10$ and the value of σ is unknown. The latter value, however, is estimated from the sample data to be $\tilde{s} = 4.25$. The aforementioned values $n = 10$ and $\tilde{s} = 4.25$ correspond to those employed in Example 2.1. If one wants to compute the power of the **single-sample *t* test** for Example 2.1 with reference to the alternative hypothesis H_1: $\mu_1 = 6$, the same protocol as described above for the normal distribution is employed, except for the fact that the value $t_{.05} = 2.26$ (which is the tabled critical two-tailed .05 t value for $df = 9$) is used in the analysis. Thus:

$$t_{.05}\, s_{\bar{X}} = \bar{X} - \mu$$

$$(2.26)(1.34) = 3.03$$

$$t = \frac{8.03 - 6}{1.34} = 1.51$$

The use of the value $\bar{X} = 8.03$ in the t test equation above is predicated on the fact that a mean of 8.03 is 3.03 units above the value $\mu = 5$ stated in the null hypothesis. Detailed tables of the t distribution indicate that for $df = 9$, the proportion of cases that falls above a t score of 1.51 is approximately .085 (which corresponds to area (\\\) in Figure 2.5 if the latter represented the t distribution).[8] The value .085 represents the power of the test. The likelihood of committing a Type II error (which corresponds to area (≡) is $\beta = 1 - .085 = .915$.

A comparison of the values obtained for the power of the **single-sample z test** and the **single-sample t test** for the two sample sizes employed in the discussion of power (i.e., for $n = 121$ and $n = 10$), reveals that when the values of both n and the standard deviation are fixed, the **single-sample z test** provides a more powerful test of an alternative hypothesis than does the **single-sample t test** (keeping in mind, however, that the use of the **single-sample z test** is only be justified if the value of σ is known).

Method 2 for computing the power of the single-sample t test and the single-sample z test It was noted previously that when the magnitude of the absolute value of the difference between μ_1 and μ is expressed in standard deviation units, the resulting value is referred to as the **effect size**. The computation of effect size, represented by the notation d, can be summarized by Equation 2.5.

$$d = \frac{|\mu_1 - \mu|}{\sigma} \qquad \textbf{(Equation 2.5)}$$

In the above equation in the case of the **single-sample z test**; the value of σ will be known; whereas, in the case of the **single-sample t test**, the latter value will have to be estimated (either from the sample data or from prior research). Cohen (1988) has proposed the following (admittedly arbitrary) d values as criteria for identifying the magnitude of an effect size: a) A **small effect size** is one that is greater than 0 but not more than .2 standard deviation units; b) A **medium effect size** is one that is greater than .2 but not more than .5 standard deviation units; and c) A **large effect size** is greater than .5 standard deviation units.

Note that in Equation 2.5 the effect size is based on population parameters and does not take into account the size of the sample. Since the power of a test is also a function of sample size, it is necessary to convert the value of d into a measure that takes into account both the population parameters and the sample size. This measure, represented by the notation δ (which is the lower case Greek letter **delta**), is referred to as the **noncentrality parameter**. The value of δ is computed with Equation 2.6.[9]

$$\delta = d\sqrt{n} \qquad \textbf{(Equation 2.6)}$$

If the value of δ is computed for a specific sample size, the power of both the **single-sample z test** and the **single-sample t test** can be determined by using **Table A3** in the **Appendix**, which consists of four sets of power curves in which the value of δ is plotted in reference to the power of a test. Each set of power curves is based on a different level of significance. Specifically: **Table A3-A** is employed for either a two-tailed analysis with $\alpha = .01$ or a one-tailed analysis with $\alpha = .005$. Table **A3-B** is employed for either a two-tailed analysis with $\alpha = .02$ or a one-tailed analysis with $\alpha = .01$. Table **A3-C** is employed for either a two-tailed analysis with $\alpha = .05$ or a one-tailed analysis with $\alpha = .025$. Table **A3-D** is employed for either a two-tailed analysis with $\alpha = .10$ or a one-tailed analysis with $\alpha = .05$. Note that each set of power curves is comprised of either eight or ten curves, each of which represents a different degrees of freedom value. When the degrees of freedom computed for an experiment do not equal one of the df values represented by the curves, the researcher must interpolate to approximate the power of the test. Regardless of the sample size, the curve for $df = \infty$ should always be used in determining the power of the **single-sample z test**. The latter curve is also used for the **single-sample t test** for large sample sizes.

The protocol for employing the curves in **Table A3** is as follows: a) Compute the value of δ; b) Upon locating δ on the horizontal axis of the appropriate set of curves, draw a line that is perpendicular to the axis which intersects the curve that represents the appropriate df value; and c) At the point the line intersects the curve, drop a perpendicular to the vertical axis on which power values are noted. The point at which the latter line intersects the vertical axis will indicate the power of the test.

The noncentrality parameter will now be employed to compute the power of the **single-sample z test** and the **single-sample t test** using the same data as that for Method 1. Thus, the power analysis will assume that the null hypothesis will be evaluated with a two-tailed test, with $\alpha = .05$. In addition:

$$H_0\text{: } \mu = 5 \quad H_1\text{: } \mu_1 = 6 \quad \sigma = \tilde{s} = 4.25 \quad n = 121$$

Employing Equation 2.5, the value $d = .235$ is computed.

$$d = \frac{6 - 5}{4.25} = .235$$

Note that using Cohen's (1988) criteria for effect size, the value $d = .235$ indicates that we are attempting to detect a medium effect size (albeit barely).

For $n = 121$, Equation 2.6 is employed to calculate the value $\delta = 2.59$.

$$\delta = (.235)\sqrt{121} = 2.59$$

Employing the power curve for $df = \infty$ in Table **A3-C**, the power of the **single-sample z test** is determined to be approximately .73. Since there is no curve for $df = 120$, the power of the **single-sample t test** is based on a curve that falls between the $df = \infty$ and $df = 24$ power curves. Through interpolation, the power of the **single-sample t test** is determined to be approximately .72. Note that these are the same values that are computed with Method 1.

For the same example, with $n = 10$, $\delta = (.235)\sqrt{10} = .743$. Employing the power curve for $df = \infty$, the power of the **single-sample z test** is determined to be approximately .11. Since $df = 9$, using the $df = 6$ and $df = 12$ power curves as reference points, we determine that the power of the **single-sample t test** is approximately .085. These values are consistent with those computed with Method 1.

It should be emphasized again that whenever possible prior to the data collection phase of a study, a researcher should stipulate the minimum effect size that she is attempting to detect. The smaller the effect size, the larger the sample size that will be required in order to have a test of sufficient power that will allow one to reject a false null hypothesis. As long as a researcher knows or is able to estimate (from the sample data) the population standard deviation, by employing trial and error one can substitute various values of n in Equation 2.6 until the computed value of δ corresponds to the desired value for the power of the test. An alternative mechanism for determining the minimum sample size necessary in order to achieve a specific level of power in reference to a specific effect size is the set of tables developed by Cohen (1988).

In closing the discussion of power, it should be pointed out that if a researcher employs a large enough sample size a significant difference can be obtained almost 100% of the time — even when the null hypothesis is true. This is the case, since due to sampling error the value of a sample mean will rarely be exactly the same as the value of μ stated in the null hypothesis. Obviously a researcher must discern whether a statistically significant difference that reflects such a minimal effect size is of any practical or theoretical significance. In instances where it is not, for all practical purposes, if one rejects the null hypothesis under such circumstances, one is committing a Type I error.

2. Computation of a confidence interval for the mean of the population represented by a sample The hypothesis testing procedure described for the **single-sample *t* test** and the **single-sample *z* test** merely allows the researcher to determine whether it is reasonable to conclude that the mean of a population is equal to a specific value. **Interval estimation,** another methodology used in inferential statistics (which is discussed briefly in the **Introduction**), allows a researcher to specify a range of values within which she can be confident the true value of a population parameter falls. Such an interval, which is always computed from sample data, is referred to as a **confidence interval.**

In this section the procedure for computing a confidence interval for the mean of a population will be described. The confidence interval for the mean is a range of values within which a researcher can be confident to a specified degree that the true value of the population mean falls. When the value of the population standard deviation is unknown, computation of a confidence interval for a single sample involving interval/ratio data utilizes the t distribution. The following confidence intervals are the ones that are most commonly computed: a) The 95% confidence interval which stipulates the range of values within which one can be 95% confident the true population mean falls; b) The 99% confidence interval which stipulates the range of values within which one can be 99% confident the true population mean falls.

Equation 2.7 is the general equation for computing a confidence interval for a population mean.

$$CI_{(1-\alpha)} = \bar{X} \pm (t_{\alpha/2})(s_{\bar{X}}) \qquad \textbf{(Equation 2.7)}$$

Where: $t_{\alpha/2}$ represents the tabled critical two-tailed value in the t distribution, for $df = n - 1$, below which a proportion equal to $[1 - (\alpha/2)]$ of the cases falls. If the percentage of the distribution that falls within the confidence interval is subtracted from 1, it will equal the value of α.

Equation 2.8 is employed to compute the 95% confidence interval, which will be represented by the notation $CI_{.95}$.

$$CI_{.95} = \bar{X} \pm (t_{.05})(s_{\bar{X}}) \qquad \textbf{(Equation 2.8)}$$

In Equation 2.8, $t_{.05}$ represents the tabled critical two-tailed .05 t value for $df = n - 1$. By employing the latter critical t value, one will be able to identify the range of values within the sampling distribution that define the middle 95% of the distribution. Only 5% of the scores in the sampling distribution will fall outside that range. Specifically, 2.5% of the scores will fall above the upper limit of the range, and 2.5% of the scores will fall below the lower limit of the range.

Equation 2.9 is employed to compute the 99% confidence interval, which will be represented by the notation $CI_{.99}$.

$$CI_{.99} = \bar{X} \pm (t_{.01})(s_{\bar{X}}) \qquad \textbf{(Equation 2.9)}$$

Note that the only difference between Equation 2.9 and Equation 2.8 is that in Equation 2.9 the critical value $t_{.01}$ is employed. The latter value represents the tabled critical two-tailed .01 value for $df = n - 1$. By using the two-tailed $t_{.01}$ value, one will be able to identify the range of values within the sampling distribution that define the middle 99% of the distribution. Only 1% of the scores in the sampling distribution will fall outside that range. Specifically, .5% of the scores will fall above the upper limit of the range, and .5% of the scores will fall below the lower limit of the range.

The values $\bar{X} = 7.6$, $s_{\bar{X}} = 1.34$, and $t_{.05} = 2.26$ will now be substituted in Equation 2.8 to compute the 95% confidence interval for the mean of the population employed in Example 2.1.

$$CI_{.95} = 7.6 \pm (2.26)(1.34) = 7.6 \pm 3.03$$

The above result can be summarized as follows: $4.57 \leq \mu \leq 10.63$. The notation $4.57 \leq \mu \leq 10.63$ means that the value of μ is greater than or equal to 4.57 and less than or equal to 10.63. This result tells us that if a mean of $\bar{X} = 7.6$ is computed for a sample size of $n = 10$, we can be 95% confident that the true value of the mean of the population the sample represents falls between the values 4.57 and 10.63. Thus, with respect to Example 2.1, the physician can be 95% confident that the average number of visits per patient is between 4.57 and 10.63.

Equation 2.9 will now be employed to compute the 99% confidence interval for the population mean in Example 2.1.

$$CI_{.99} = 7.6 \pm (3.25)(1.34) = 7.6 \pm 4.36$$

The above result can be summarized as follows: $3.24 \leq \mu \leq 11.96$. This result tells us that if a mean of $\bar{X} = 7.6$ is computed for a sample size of $n = 10$, we can be 99% confident that the true value of the mean of the population the sample represents falls between the values 3.24 and 11.96. Thus, with respect to Example 2.1, the physician can be 99% confident that the average number of visits per patient is between 3.24 and 11.96.

Note that the range of values which defines the 99% confidence interval is larger than the range of values that defines the 95% confidence interval. This will always be the case, since it is only logical that by stipulating a larger range of values one will be able to have a higher degree of confidence that the true value of the population mean has been included within that range. It is also the case that the larger the sample size employed in computing a confidence interval, the smaller the range of values that will define the confidence interval.

Figures 2.6 and 2.7 provide a graphical summary of the computation of the 95% and 99% confidence intervals.

Note that in Figure 2.6 the following is true: a) 47.5% of the scores in the sampling distribution fall between the sample mean $\bar{X} = 7.6$ and 4.57, the lower limit of $CI_{.95}$; and b) 47.5% of the scores in the sampling distribution fall between the sample mean $\bar{X} = 7.6$ and 10.63, the upper limit of $CI_{.95}$. Thus, the area of the curve between the scores 4.57 and 10.63 represents the middle 95% of the sampling distribution. Two and one-half half percent of the scores in the distribution fall below 4.57 and 2.5% of the scores fall above 10.63. The scores which are below 4.57 or greater than 10.63 comprise the extreme 5% of the sampling distribution.

In Figure 2.7 the following is true: a) 49.5% of the scores in the sampling distribution fall between the sample mean $\bar{X} = 7.6$ and 3.24, the lower limit of $CI_{.99}$; and b) 49.5% of the scores in the sampling distribution fall between the sample mean $\bar{X} = 7.6$ and 11.96, the upper limit of $CI_{.99}$. Thus, the area of the curve between the scores 3.24 and 11.96 represents the middle 99% of the sampling distribution. One-half of one percent of the scores in the distribution fall below 3.24 and .5% of the scores fall above 11.96. The scores which are below 3.24 or greater than 11.96 comprise the extreme 1% of the sampling distribution.

The reader should take note of the fact that in Figures 2.6 and 2.7 the sample mean is employed to represent the mean of the sampling distribution. This is in contrast to the sampling distribution depicted in Figure 2.1, where the hypothesized value of the population

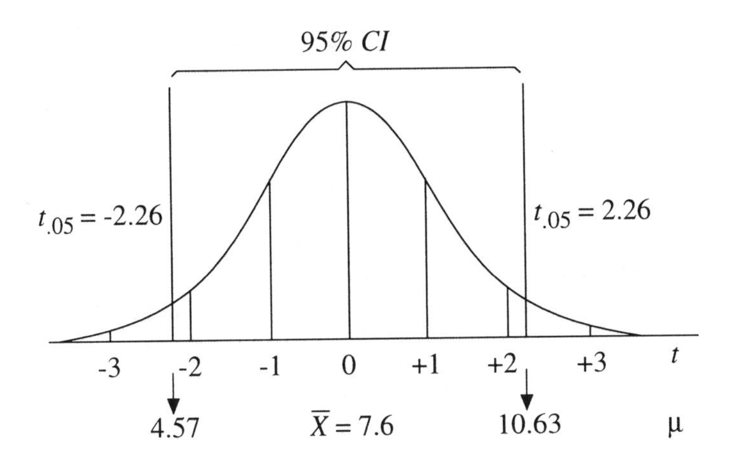

Figure 2.6 Graphical Representation of 95% Confidence Interval for Example 2.1

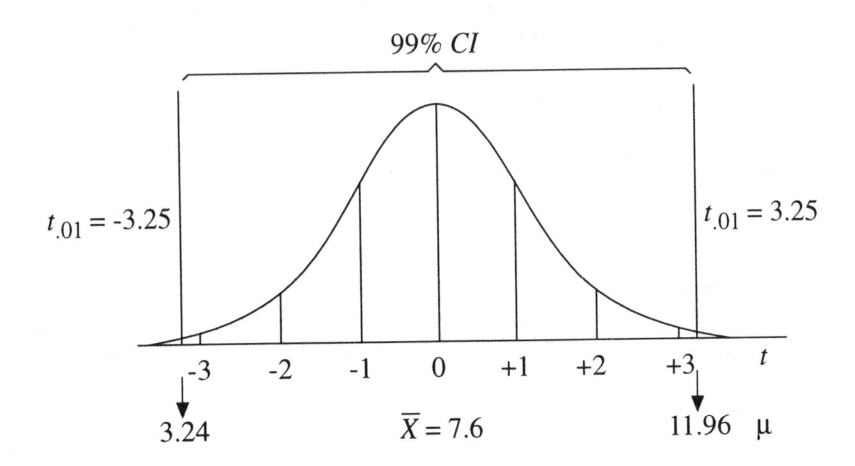

Figure 2.7 Graphical Representation of 99% Confidence Interval for Example 2.1

mean is employed to represent the mean of the sampling distribution. The reason for using different means for the two sampling distributions is that the sampling distribution depicted in Figure 2.1 is used to determine the likelihood of a sample mean deviating from the hypothesized value of the population mean, while the sampling distribution depicted in Figures 2.6 and 2.7 reflects the fact that one is engaged in interval estimation, and is thus employing the sample mean to predict the true value of the population mean.

Although, as noted previously, $CI_{.95}$ and $CI_{.99}$ are the most commonly computed confidence intervals, it is possible to calculate a confidence interval at any level of confidence. Thus, if one wanted to compute the 90% confidence interval, the equation $CI_{.90} = \bar{X} \pm (t_{.10})(s_{\bar{X}})$ is employed. In the latter equation $t_{.10}$ represents the tabled critical one-tailed .05 t value (which is also the tabled critical two-tailed .10 t value), since the latter value establishes the boundaries for the middle 90% of the sampling distribution. Only 10% of the scores in the sampling distribution will fall outside that range (5% of the scores will fall above the upper limit of the range, and 5% of the scores will fall below the lower limit of the range). Thus, for Example 2.1:

$$CI_{.90} = 7.6 \pm (1.83)(1.34) = 7.6 \pm 2.45 \quad \text{Thus:} \quad 5.15 \leq \mu \leq 10.05$$

Employing the same logic, the 98% confidence interval can computed by using the tabled critical one-tailed .01 value $t_{.01} = 2.82$ in the confidence interval equation. The latter t value is employed since it establishes the boundaries for the middle 98% of the sampling distribution. Thus:

$$CI_{.98} = 7.6 \pm (2.82)(1.34) = 7.6 \pm 3.78 \quad \text{Thus:} \quad 3.82 \leq \mu \leq 11.38$$

It should be noted that in order to accurately compute a confidence interval, one must have access to tables of the t distribution which provide the appropriate tabled value for the confidence interval in question. Since most published tables of the t distribution only provide the tabled critical one- and two-tailed .05 and .01 values, they only allow for accurate computation of the following confidence intervals: $CI_{.90}$, $CI_{.95}$, $CI_{.98}$, $CI_{.99}$. **Table A2** is more detailed than most tables of the t distribution, and thus it allows for accurate computation of a greater number of confidence intervals than those noted above. In instances where an exact t value is not tabled, interpolation can be used to estimate that value.

Although the computation of confidence intervals is not described in the discussion of the **single-sample *z* test**, when the value of the population standard deviation is known the normal distribution (as opposed to the t distribution) is employed to compute a confidence interval. If a researcher knows the value of the population standard deviation, Equation 2.10 is the general equation for computing a confidence interval.

$$CI_{(1-\alpha)} = \bar{X} \pm (z_{\alpha/2})(\sigma_{\bar{X}}) \qquad \textbf{(Equation 2.10)}$$

Where: $z_{\alpha/2}$ represents the tabled critical two-tailed value in the normal distribution below which a proportion equal to $[1 - (\alpha/2)]$ of the cases falls. If the percentage of the distribution that falls within the confidence interval is subtracted from 1, it will equal the value of α.

Note that the basic difference between Equation 2.10 and Equation 2.7, is that Equation 2.10 employs a tabled critical z value instead of the corresponding t value for the same percentile. Additionally, since use of Equation 2.10 assumes that the value of σ is known, the actual value of the standard error of the population mean can be computed. Thus, $\sigma_{\bar{X}}$ is used in place of the estimated value $s_{\bar{X}}$ employed in Equation 2.7.

Generalizing from Equation 2.10, Equation 2.11 is employed to compute the 95% confidence interval for the mean of a population when the normal distribution is used.

$$CI_{.95} = \bar{X} \pm (z_{.05})(\sigma_{\bar{X}}) \qquad \textbf{(Equation 2.11)}$$

If Equation 2.11 is employed with Example 2.1, the only value that will be different from those in Equation 2.8 is the tabled critical two-tailed value $z_{.05} = 1.96$, which is used in place of $t_{.05} = 2.26$. Since we are assuming that $\sigma = \tilde{s}$, the values of $s_{\bar{X}}$ and $\sigma_{\bar{X}}$ are equivalent. Thus, $\sigma_{\bar{X}} = \sigma/\sqrt{n} = 4.25/\sqrt{10} = 1.34$.

Equation 2.11 will now be utilized to compute the 95% confidence interval for Example 2.1.

$$CI_{.95} = 7.6 \pm (1.96)(1.34) = 7.6 \pm 2.63$$

The above result can be summarized as: $4.97 \leq \mu \leq 10.23$. Thus, by using the normal distribution to compute the 95% confidence interval, the physician can be 95% confident that the average number of visits per patient is between 4.97 and 10.23. Note that when the normal distribution is employed with the data for Example 2.1, the range of values that defines the 95% confidence interval is smaller than the range of the values that is computed with the t distribution. This will always be the case, since at the same level of confidence a tabled z value will always be smaller than the corresponding tabled t value.[10] As a result of this, the product resulting from multiplying the tabled value by the standard error of the population mean will be smaller when the normal distribution is employed.

If the value of σ is known, Equation 2.12 (as opposed to Equation 2.9) is used to calculate the 99% confidence interval for the mean of a population.

$$CI_{.99} = \bar{X} \pm (z_{.01})(\sigma_{\bar{X}}) \qquad \textbf{(Equation 2.12)}$$

Note that in contrast to Equation 2.9, Equation 2.12 employs the tabled critical two-tailed value $z_{.01} = 2.58$ instead of the corresponding value $t_{.01} = 3.25$. Equation 2.12 will now be used to compute the 99% confidence interval for Example 2.1.

$$CI_{.99} = 7.6 \pm (2.58)(1.34) = 7.6 \pm 3.46$$

The above result can be summarized as: $4.14 \leq \mu \leq 11.06$. Thus, by using the normal distribution to compute the 99% confidence interval, the physician can be 99% confident that the average number of visits per patient is between 4.14 and 11.06. Note once again that the range of values obtained with Equation 2.12 (which utilizes the normal distribution) for the 99% confidence interval is smaller than the range of the values that is obtained with Equation 2.9 (which utilizes the t distribution).

Figures 2.8 and 2.9 provide a graphical summary of the computation of the 95% and 99% confidence intervals with the normal distribution.

It should be noted that even if the value of σ is known, some researchers would challenge the use of the normal distribution in the above example. The rationale for such a challenge is that if as a result of employing a **single-sample *z* test** it is determined that a significant difference exists between the hypothesized value of the population mean and $\bar{X}$, one can question the logic of employing the normal distribution in the computation of a confidence interval. This is the case, since if we conclude that our sample is derived from a population with a different mean value than the hypothesized population mean, it is also possible that the population standard deviation is different than the value of σ we have employed in the analysis. In such an instance, the best strategy would probably be to employ the sample data to estimate the population standard deviation. Thus, Equation 2.1 is used to compute $\tilde{s}$, and consequently Equations 2.8 and 2.9 (employing the t distribution) are used to compute the 95% and 99% confidence intervals.

It is worth noting that it is unlikely that a researcher will employ the normal distribution to compute a confidence interval, for the simple reason that it is improbable that one will know the value of a population standard deviation and not know the value of the population mean. For this reason most sources only illustrate confidence interval computations in reference to the t distribution.

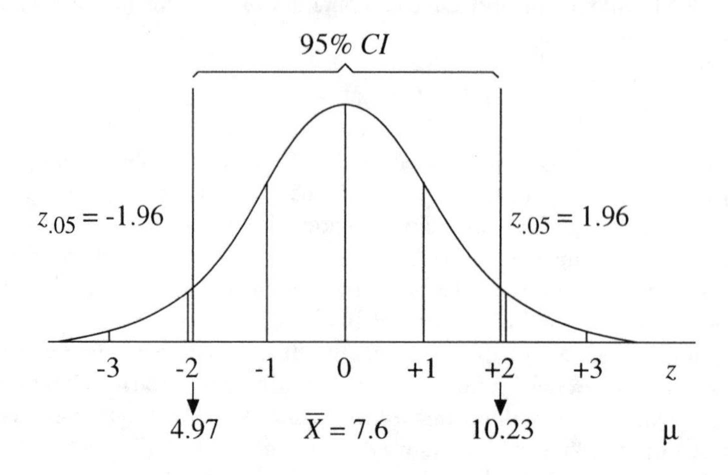

Figure 2.8 Graphical Representation of 95% Confidence Interval for Example 2.1 Through Use of the Normal Distribution

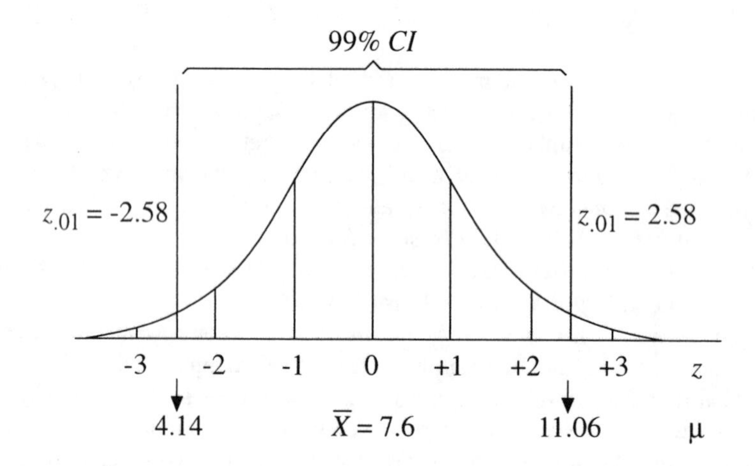

Figure 2.9 Graphical Representation of 99% Confidence Interval for Example 2.1 Through Use of the Normal Distribution

In closing the discussion of confidence intervals, the reader should take note of the fact that the range of values which defines the 95% and 99% confidence intervals for Example 2.1 are extremely large. Because of this the researcher will not be able stipulate with great precision a single value that is likely to represent the true value of the population mean. This illustrates that a confidence interval may be quite limited in its ability to accurately estimate a population parameter, especially when the sample size upon which it is based is small. The reader should also note the fact that the reliability of Equation 2.10 will be compromised if one or more of the assumptions of the **single-sample *t* test** is saliently violated.

VII. Additional Discussion of the Single-Sample *t* Test

Degrees of freedom The concept of **degrees of freedom**, which is frequently encountered in statistical analysis, represents the number of values in a set of data that are free to vary after certain restrictions have been placed upon the data. The concept of degrees of freedom will be illustrated through the use of the following example.

Assume that one is trying to construct a set consisting of three scores which are derived from a single sample, and it is known that the mean of the sample is $\bar{X} = 5$. Under these conditions, two of the three scores that comprise the set can assume a variety of values, as long as the sum of the two scores does not exceed 15. This is the case since, if $\bar{X} = 5$ and $n = 3$, it is required that $\Sigma X = 15$. Thus, some representative values that two of the three scores may assume are: 1 and 4; 0 and 6; 8 and 6; and 1 and 1. Note that in all four cases the sum of the two scores is less than 15. The value of the third score in all four instances is predetermined based on the values of the other two scores. Thus, if we know that two of the three scores that comprise a set are 1 and 4, the third score must equal 10, since it is the only value that will yield $\Sigma X = 15$.

The rationale for employing $df = n - 1$ in computing the degrees of freedom for the **single-sample *t* test** can be understood on the basis of the above discussion. Specifically, once the size of a sample is set and the mean assumes a specific value, only $n - 1$ scores will be free to vary. In the case of the **single-sample *t* test** (as well as a variety of other inferential statistical tests) degrees of freedom are a function of sample size. Specifically, as the sample size increases, the degrees of freedom increases. However, this is not always the case. When evaluating categorical data, degrees of freedom are generally a function of the number of categories involved in the analysis rather than the size of the sample.

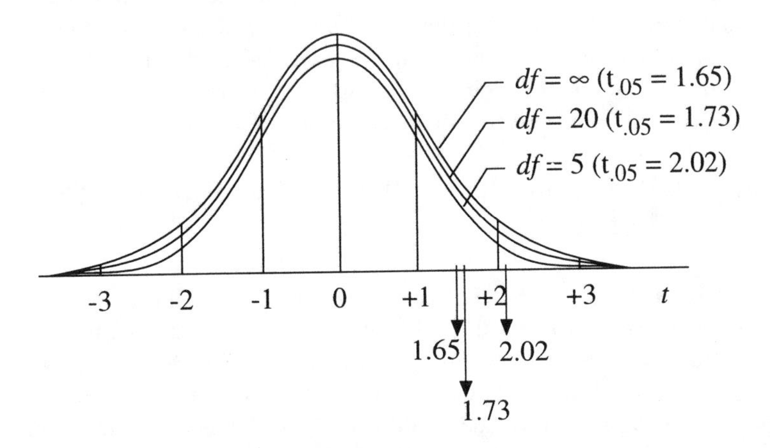

Figure 2.10. Representative *t* Distributions for Different *df* Values

Inspection of **Table A2** reveals that as degrees of freedom increase, the lower the value of the tabled critical value that the computed absolute value of t must be equal to or greater than in order to reject the null hypothesis at a given level of significance. Once again, this is not always the case. For instance, when employing the chi-square distribution (discussed later in the book) there is a direct relationship between degrees of freedom and the magnitude of the tabled critical value. In other words, as the number of degrees of freedom increases, the larger the magnitude of the tabled critical chi-square value at a given level of significance. for three different degrees of freedom values. Note that the t distribution (which is always symmetrical) closely resembles the normal distribution.

As noted in Section I, in actuality a separate t distribution exists for each sample size, and consequently for each degrees of freedom value. Figure 2.10 depicts the t distribution for three different degrees of freedom value. Note that the t distribution (which is always symmetrical) closely resembles the normal distribution. As is noted in Section I, except when $n = \infty$ for a given standard deviation score, a smaller proportion of cases will fall between the mean of the t distribution and that standard deviation score than the proportion of cases that fall between the mean and that same standard deviation score in the normal distribution. As sample size (and thus degrees of freedom) increases, the shape of the t distribution becomes increasingly similar in appearance to the normal distribution, and, in fact, when $n = \infty$ becomes identical to it. As a result of this, when the sample size employed in a study is large (usually $n \geq 200$), for all practical purposes, the tabled critical values for the normal and t distributions are identical.

Inspection of the three t distributions depicted in Figure 2.10 reveals that the lower the degrees of freedom, the larger the absolute value of t that will be required in order to reject a null hypothesis at a given level of significance. As an example, the distance between the mean and the one-tailed .05 critical t value (which corresponds to the 95th percentile of a given curve) is greatest for the $df = 5$ distribution (where $t_{.05} = 2.02$) and smallest for the $df = \infty$ distribution (where $t_{.05} = 1.65$).

Most tables of the t distribution list selected degrees of freedom values ranging from 1 through 120, and then list a final row of values for $df = \infty$. The latter set of values are identical to those in the normal distribution, since when $df = \infty$ the two distributions are identical. As a general rule, for sample sizes substantially above 121 (which correspond to $df = 120$) the critical values for $df = \infty$ can be employed. Tables of the t distribution do not include tabled critical values for all possible degrees of freedom below 120. The protocol that is generally used if the exact df value is not listed is to either interpolate the critical value or to employ the tabled df value that is closest to it. Some sources qualify the latter by stating that one should employ the tabled df value that is closest to but not above the computed df value. The intent of the this strategy is to insure that the likelihood of committing a Type I error does not exceed the prespecified alpha value.

VIII. Additional Examples Illustrating the Use of the Single-Sample *t* Test

If in the case of Examples 1.1–1.4 (all of which are employed to illustrate the **single-sample *z* test**), a researcher does not know the value of σ and has to estimate it from the sample data, the **single-sample *t* test** is the appropriate test to use. The 30 scores noted in Section IV of the **single-sample *z* test** in reference to Example 1.1 can be employed to compute the estimated population standard deviation. Utilizing the 30 scores, the following values are computed: $\Sigma X = 222$, $\bar{X} = 7.4$, and $\Sigma X^2 = 1866$. Equations 2.1–2.3 can now be employed to conduct the necessary calculations for the **single-sample *t* test**. The null hypothesis that is evaluated is H_0: $\mu = 8$.

$$\tilde{s} = \sqrt{\frac{1866 - \frac{(222)^2}{30}}{30 - 1}} = 2.77 \qquad s_{\bar{X}} = \frac{2.77}{\sqrt{30}} = .506$$

$$t = \frac{7.4 - 8}{.506} = -1.19$$

Using the t distribution for Examples 1.1–1.4, the degrees of freedom that are employed are $df = 30 - 1 = 29$. For $df = 29$, the tabled critical two-tailed .05 and .01 values are $t_{.05} = 2.05$ and $t_{.01} = 2.76$, and the tabled critical one-tailed .05 and .01 values are $t_{.05} = 1.70$ and $t_{.01} = 2.46$. Since the computed absolute value $t = 1.19$ is less than all of the aforementioned critical values, the null hypothesis cannot be rejected. Note that when the **single-sample z test** is used to evaluate the same set of data, the directional alternative hypothesis H_1: $\mu < 8$ is supported at the .05 level. The discrepancy between the results of the two tests can be attributed to the fact that the estimated population standard deviation $\tilde{s} = 2.77$ employed for the **single-sample t test** is larger than the value $\sigma = 2$ used when the data are evaluated with the **single-sample z test.**

The **single-sample t test** cannot be employed for Examples 1.5 and 1.6 (which are also used to illustrate the **single-sample z test**), since when $n = 1$ the estimated value of a population standard deviation will be indeterminate (since at least two scores are required to estimate a population standard deviation). This is confirmed by the fact that no tabled critical t values are listed for zero degrees of freedom (which is the case when $n = 1$).

Example 2.2, which is based on the same data set as Example 2.1, is an additional example that can be evaluated with the **single-sample t test.**

Example 2.2. *The Sugar Snack candy company claims that each package of candy it sells contains five bonus coupons which a consumer can use toward future purchases. Responding to a complaint by a consumer who says the company is shortchanging people on coupons, a consumer advocate purchases* 10 *bags of candy from a variety of stores. The advocate counts the number of coupons in each bag and obtains the following values:* 9, 10, 8, 4, 8, 3, 0, 10, 15, 9. *Do the data support the claim of the complainant?*

Since the data for Example 2.2 are identical to that employed for Example 2.1, analysis with the **single-sample t test** yields the value $t = 1.94$. The value $t = 1.94$, which is consistent with the directional alternative hypothesis H_1: $\mu > 5$, is totally unexpected in view of the nature of the consumer's complaint. If anything, the researcher evaluating the data is most likely to employ either the directional alternative hypothesis H_1: $\mu < 5$ or the nondirectional alternative hypothesis H_1: $\mu \neq 5$ (neither of which are supported at the .05 level). Thus, even though the directional alternative hypothesis H_1: $\mu > 5$ is supported at the .05 level, the latter alternative hypothesis would not have been stipulated prior to collecting the data.

References

Cohen, J. (1988). **Statistical power analysis for the behavioral sciences** (2nd ed.). New York: Academic Press.

Guenther, W. C. (1965). **Concepts of statistical inference.** New York: McGraw–Hill Book Company.

Endnotes

1. In order to be solvable, Equation 2.3 requires that there be variability in the sample. If all of the subjects in a sample have the same score, the computed value of $\tilde{s}$ will equal zero. When $\tilde{s} = 0$, the value of $s_{\bar{X}}$ will always equal zero. When $s_{\bar{X}} = 0$, Equation 2.3 becomes unsolvable, thus making it impossible to compute a value for t. It is also the case that when the sample size is $n = 1$, Equation 2.1 becomes unsolvable, thus making it impossible to employ Equation 2.3 to solve for t.

2. In the event that σ is known and $n < 25$, and one elects to employ the **single-sample t test**, the value of σ should be used in computing the test statistic. Given the fact that the value of σ is known, it would be foolish to employ $\tilde{s}$ as an estimate of it.

3. The t distribution was derived by William Gossett, a British statistician who published under the pseudonym of Student.

4. It is worth noting that if the value of the population standard deviation in Example 2.1 is known to be $\sigma = 4.25$, the data can be evaluated with the **single-sample z test**. When employed it yields the value $z = 1.94$, which is identical to the value obtained with Equation 2.3. Specifically, since $\sigma = \tilde{s} = 4.25$, $\sigma_{\bar{X}} = s_{\bar{X}} = 4.25/\sqrt{10} = 1.34$. Employing Equation 1.3 yields $z = (7.6 - 5)/1.34 = 1.94$. As is the case for the **single-sample t test**, the latter value only supports the directional alternative hypothesis H_1: $u > 5$ at the .05 level. This is the case since $z = 1.94$ is greater than the tabled critical one-tailed value $z_{.05} = 1.65$ in **Table A1**. The value $z = 1.94$, which is less than the tabled critical two-tailed value $z_{.05} = 1.96$, falls just short of supporting the nondirectional alternative hypothesis H_1: $\mu \neq 5$ at the .05 level.

5. A sampling distribution of means for the t distribution when employed in the context of the **single-sample t test** is interpreted in the same manner as the sampling distribution of means for the **single-sample z test** as depicted in Figure 1.1.

6. In the event the researcher is evaluating the power of the test in reference to a value of μ_1 that is less than $\mu = 5$, Distribution B will overlap the left tail of Distribution A.

7. It is really not possible to determine this value with great accuracy by interpolating the entries in **Table A2**.

8. If the table for the normal distribution is used to estimate the power of the **single-sample t test** in this example, it can be determined that the proportion of cases that falls above a z value of 1.51 is .0655. Although the value .0655 is close to .085, it slightly underestimates the power of the test.

9. The value of δ can be computed directly through use of the following equation: $\delta = (\mu_1 - \mu)/(\sigma/\sqrt{n})$. Note that the equation expresses effect size in standard deviation units of the sampling distribution.

10. As noted in Section V, the only exception to this will be when the sample size is extremely large, in which case the normal and t distributions are identical. Under such conditions the appropriate values for z and t employed in the confidence interval equation will be identical.

Test 3

The Single-Sample Chi-Square Test for a Population Variance

(Parametric Test Employed with Interval/Ratio Data)

I. Hypothesis Evaluated with Test and Relevant Background Information

Hypothesis evaluated with test Does a sample of n subjects come from a population in which the variance (σ^2) equals a specified value?

Relevant background information on test The **single-sample chi-square test for a population variance** is employed in a hypothesis testing situation involving a single sample in order to determine whether a sample with an estimated population variance of $\tilde{s}^2$ is derived from a population with a variance of σ^2. If the result of the test is significant, the researcher can conclude there is a high likelihood the sample is derived from a population in which the variance is some value other than σ^2. The **single-sample chi-square test for a population variance** is based on the chi-square distribution. The test statistic is represented by the notation χ^2 (where χ represents the lower case Greek letter **chi**).

The **single-sample chi-square test for a population variance** is used with interval/ratio level data and is based on the following assumptions: a) The distribution of data in the underlying population from which the sample is derived is normal; and b) The sample has been randomly selected from the population it represents. If either of the assumptions is saliently violated, the reliability of the test statistic may be compromised.[1]

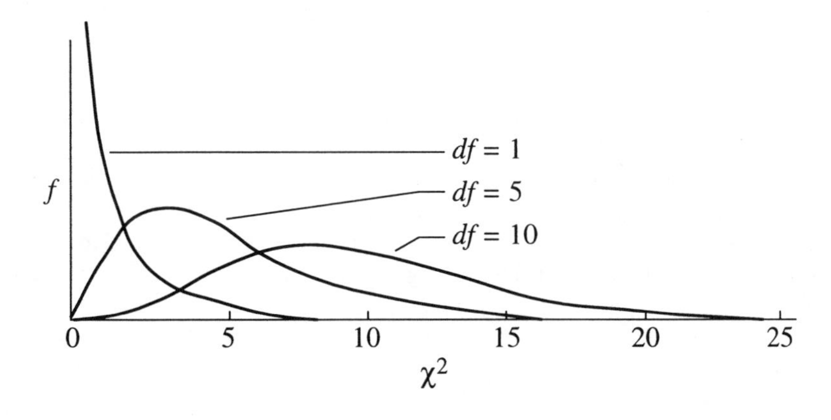

Figure 3.1 Representative Chi-Square Distributions for Different *df* Values

The chi-square distribution is a continuous asymmetrical theoretical probability distribution. A chi-square value must fall within the range $0 \leq \chi^2 \leq \infty$, and thus (unlike values for the normal and t distributions) can never be a negative number. As is the case with the t distribution, there are an infinite number of chi-square distributions — each distribution being a function of the number of degrees of freedom employed in an analysis.

Figure 3.1 depicts the chi-square distribution for three different degrees of freedom values. Inspection of the three distributions reveals: a) The lower the degrees of freedom, the more positively skewed the distribution (i.e., the larger the proportion of scores at the lower end of the distribution); and b) The greater the degrees of freedom, the more symmetrical the distribution. A thorough discussion of the chi-square distribution can be found in Section V.

II. Example

Example 3.1 *The literature published by a company which manufactures hearing aid batteries claims that a certain model battery has an average life of* 7 *hours* (μ = 7) *and a variance of* 5 *hours* (σ^2 = 5). *A customer who uses the hearing aid battery believes that the value stated in the literature for the variance is too low. In order to test his hypothesis the customer records the following times (in hours) for ten batteries he purchases during the month of September:* 5, 6, 4, 3, 11, 12, 9, 13, 6, 8. *Do the data indicate that the variance for battery time is some value other than* 5?

III. Null versus alternative hypotheses

Null hypothesis H_0: $\sigma^2 = 5$

(The variance of the population the sample represents equals 5.)

Alternative hypothesis H_0: $\sigma^2 \neq 5$

(The variance of the population the sample represents does not equal 5. This is a **nondirectional alternative hypothesis**, and it is evaluated with a **two-tailed test**. In order to be supported, the obtained chi-square value must be equal to or greater than the tabled critical two-tailed chi-square value at the prespecified level of significance in the upper tail of the chi-square distribution, or equal to or less than the tabled critical two-tailed chi-square value at the prespecified level of significance in the lower tail of the chi-square distribution. A full explanation of the protocol for interpreting chi-square values within the framework of the **single-sample chi-square test for a population variance** can be found in Section V.

or

H_0: $\sigma^2 > 5$

(The variance of the population the sample represents is greater than 5. This is a **directional alternative hypothesis**, and it is evaluated with a **one-tailed test**. In order to be supported, the obtained chi-square value must be equal to or greater than the tabled critical one-tailed chi-square value at the prespecified level of significance in the upper tail of the chi-square distribution.)

or

H_0: $\sigma^2 < 5$

(The variance of the population the sample represents is less than 5. This is a **directional alternative hypothesis**, and it is evaluated with a **one-tailed test**. In order to be supported, the obtained chi-square value must be equal to or less than the tabled critical one-tailed chi-square value at the prespecified level of significance in the lower tail of the chi-square distribution.)

Note: Only one of the above noted alternative hypotheses is employed. If the alternative hypothesis the researcher selects is supported, the null hypothesis is rejected.[2]

IV. Test Computations

Table 3.1 summarizes the data for Example 3.1.

Table 3.1 Summary of Data for Example 2.1

Battery	X	X^2
1	5	25
2	6	36
3	4	16
4	3	9
5	11	121
6	12	144
7	9	81
8	13	169
9	6	36
10	8	64
	$\Sigma X = 77$	$\Sigma X^2 = 701$

In order to compute the test statistic for the **single-sample chi-square test for a population variance,** it is necessary to use the sample data to calculate an unbiased estimate of the population variance. $\tilde{s}^2$, the unbiased estimate of σ^2, is computed with Equation 3.1 (which is the same as Equation I.5 in the **Introduction**).

$$\tilde{s}^2 = \frac{\Sigma X^2 - \frac{(\Sigma X)^2}{n}}{n - 1} \qquad \textbf{(Equation 3.1)}$$

Employing Equation 3.1, the value $\tilde{s}^2 = 12.01$ is computed.

$$\tilde{s}^2 = \frac{701 - \frac{(77)^2}{10}}{10 - 1} = 12.01$$

Equation 3.2 is the test statistic for the **single-sample chi-square test for a population variance.**

$$\chi^2 = \frac{(n - 1)\,\tilde{s}^2}{\sigma^2} \qquad \textbf{(Equation 3.2)}$$

Employing the values $n = 10$, $\tilde{s}^2 = 12.01$, and $\sigma^2 = 5$ (which is the hypothesized value of σ^2 stated in the null hypothesis), Equation 3.2 is employed to compute the value $\chi^2 = 21.62$.

$$\chi^2 = \frac{(10 - 1)(12.01)}{5} = 21.62$$

V. Interpretation of the Test Results

The computed value $\chi^2 = 21.62$ is evaluated with **Table A4 (Table of the Chi-Square Distribution)** in the **Appendix.** In **Table A4,** chi-square values are listed in relation to the proportion of cases (which are recorded at the top of each column) that falls below a tabled χ^2 value in the sampling distribution, and the number of **degrees of freedom** (which are recorded in the left hand column of each row).[3] Equation 3.3 is employed to compute the degrees of freedom for the **single-sample chi-square test for a population variance.**

$$df = n - 1 \qquad \textbf{(Equation 3.3)}$$

Employing Equation 3.3, we compute that $df = 10 - 1 = 9$. For $df = 9$, the tabled critical two-tailed .05 values are $\chi^2_{.025} = 2.70$ and $\chi^2_{.975} = 19.02$. These values are the tabled critical two-tailed .05 values, since the proportion of the distribution that falls between $\chi^2 = 0$ and $\chi^2_{.025} = 2.70$ is .025, and the proportion of the distribution that falls above $\chi^2_{.975} = 19.02$ is .025. Thus, the extreme 5% of the distribution is comprised of chi-square values that fall below $\chi^2_{.025} = 2.70$ and above $\chi^2_{.975} = 19.02$. In the same respect, the tabled critical two-tailed .01 values are $\chi^2_{.005} = 1.73$ and $\chi^2_{.995} = 23.59$. These values are the tabled critical two-tailed .01 values, since the proportion of the distribution that falls between $\chi^2 = 0$ and $\chi^2_{.005} = 1.73$ is .005, and the proportion of the distribution that falls above $\chi^2_{.995} = 23.59$ is .005. Thus, the extreme 1% of the distribution is comprised of chi-square values that fall below $\chi^2_{.005} = 1.73$ and above $\chi^2_{.995} = 23.59$.

For $df = 9$, the tabled critical one-tailed .05 values are $\chi^2_{.05} = 3.33$ and $\chi^2_{.95} = 16.92$. These values are the tabled critical one-tailed .05 values, since the proportion of the distribution that falls between $\chi^2 = 0$ and $\chi^2_{.05} = 3.33$ is .05, and the proportion of the distribution that falls above $\chi^2_{.95} = 16.92$ is .05. In the same respect, the tabled critical one-tailed .01 values are $\chi^2_{.01} = 2.09$ and $\chi^2_{.99} = 21.67$. These values are the tabled critical one-tailed .01 values, since the proportion of the distribution that falls between $\chi^2 = 0$ and $\chi^2_{.01} = 2.09$ is .01, and the proportion of the distribution that falls above $\chi^2_{.99} = 21.67$ is .01.

Figures 3.2 and 3.3 depict the tabled critical .05 and .01 values for the chi-square sampling distribution when $df = 9$. The mean of a chi-square sampling distribution will always equal the degrees of freedom for the distribution. Thus, $\mu_{\chi^2} = df = n - 1$. The standard deviation of the sampling distribution will always be $\sigma_{\chi^2} = \sqrt{2df}$. Consequently, the variance will be $\sigma^2_{\chi^2} = 2df$. In the case of Example 3.1 where $df = 9$, $\mu_{\chi^2} = 9$ and $\sigma^2_{\chi^2} = 18$.

The following guidelines are employed in evaluating the null hypothesis for the **single-sample chi-square test for a population variance.**

a) If the alternative hypothesis employed is nondirectional, the null hypothesis can be rejected if the obtained chi-square value is equal to or greater than the tabled critical two-tailed chi-square value at the prespecified level of significance in the upper tail of the chi-square distribution, or equal to or less than the tabled critical two-tailed chi-square value at the prespecified level of significance in the lower tail of the chi-square distribution.[4]

b) If the alternative hypothesis employed is directional and predicts a population variance larger than the value stated in the null hypothesis, the null hypothesis can only be rejected if the obtained chi-square value is equal to or greater than the tabled critical one-tailed chi-square value at the prespecified level of significance in the upper tail of the chi-square distribution.

c) If the alternative hypothesis employed is directional and predicts a population variance smaller than the value stated in the null hypothesis, the null hypothesis can only be rejected if the obtained chi-square value is equal to or less than the tabled critical one-tailed chi-square value at the prespecified level of significance in the lower tail of the chi-square distribution.

Employing the above guidelines, we can conclude the following.

The nondirectional alternative hypothesis H_1: $\sigma^2 \neq 5$ is supported at the .05 level. This is the case since the obtained value $\chi^2 = 21.62$ is greater than the tabled critical two-tailed .05 value $\chi^2_{.975} = 19.02$ in the upper tail of the distribution. The nondirectional alternative hypothesis H_1: $\sigma^2 \neq 5$ is not supported at the .01 level, since $\chi^2 = 21.62$ is less than the tabled critical two-tailed .01 value $\chi^2_{.995} = 23.59$ in the upper tail of the distribution.

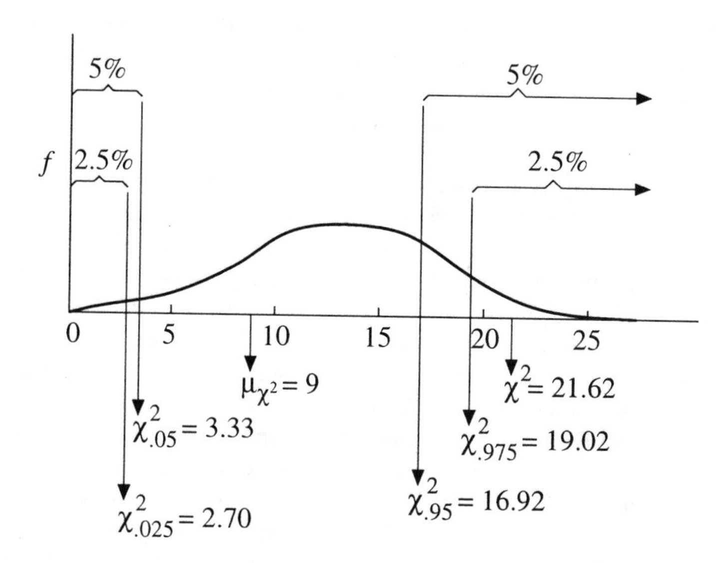

Figure 3.2 Tabled Critical Two-Tailed and One-Tailed .05 χ^2 Values for df = 9

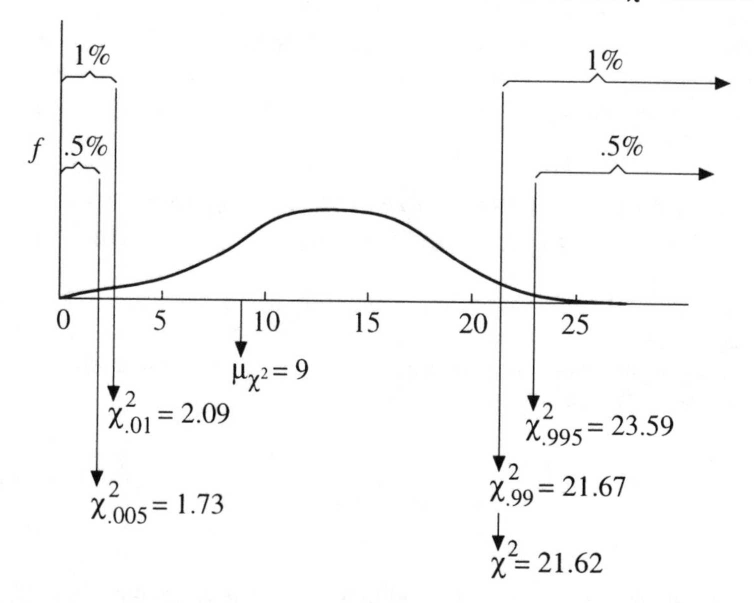

Figure 3.3 Tabled Critical Two-Tailed and One-Tailed .01 χ^2 Values for df = 9

The directional alternative hypothesis H_1: $\sigma^2 > 5$ is supported at the .05 level. This is the case since the obtained value $\chi^2 = 21.62$ is greater than the tabled critical one-tailed .05 value $\chi^2_{.95} = 16.92$ in the upper tail of the distribution. The directional alternative hypothesis H_1: $\sigma^2 > 5$ is not supported at the .01 level, since $\chi^2 = 21.62$ is less than the tabled critical one-tailed .01 value $\chi^2_{.99} = 21.67$ in the upper tail of the distribution. Note that in order for the data to be consistent with the directional alternative hypothesis H_1: $\sigma^2 > 5$, the computed value of $\tilde{s}^2$ must be greater than the value $\sigma^2 = 5$ stated in the null hypothesis.

The directional alternative hypothesis H_1: $\sigma^2 < 5$ is not supported. This is the case since in order for the latter alternative hypothesis to be supported, the obtained chi-square value must be equal to or less than the tabled critical one-tailed .05 value $\chi^2_{.05} = 3.33$ in the lower tail of the distribution. If $\alpha = .01$ is employed, in order for the directional alternative hypothesis H_1: $\sigma^2 < 5$ to be supported, the obtained chi-square value must be equal to or less than the tabled critical one-tailed value $\chi^2_{.01} = 2.09$ in the lower tail of the distribution. In order for the data to be consistent with the alternative hypothesis H_1: $\sigma^2 < 5$, the computed value of $\tilde{s}^2$ must be less than the value $\sigma^2 = 5$ stated in the null hypothesis.

A summary of the analysis of Example 3.1 with the **single-sample chi-square test for a population variance** follows: We can conclude that it is unlikely that the sample of 10 batteries comes from a population with a variance equal to 5. The data suggest that the variance of the population is greater than 5. This result can be summarized as follows: $\chi^2 (9) = 21.62, p < .05$.

Although it is not required in order to evaluate the null hypothesis H_0: $\sigma^2 = 5$, through use of Equations I.1/1.1 it can be determined that the mean number of hours the 10 batteries functioned is $\bar{X} = \Sigma X/n = 77/10 = 7.7$. The latter value is greater than $\mu = 7$, which is the mean number of hours claimed in the company's literature. If the researcher wants to determine whether the value $\bar{X} = 7.7$ is significantly larger than the value $\mu = 7$, it can be argued that one should employ the **single-sample *t* test (Test 2)** for the analysis. The rationale for employing the latter test is that if the null hypothesis H_0: $\sigma^2 = 5$ is rejected, the researcher is concluding that the true value of the population variance is unknown, and consequently, the latter value should be estimated from the sample data. If, on the other hand, one does not employ the **single-sample chi-square test for a population variance** to evaluate the null hypothesis H_0: $\sigma^2 = 5$ (and is thus unaware of the fact that the data are not consistent with the latter null hypothesis), one will assume $\sigma^2 = 5$ and employ the **single-sample *z* test (Test 1)** to evaluate the null hypothesis H_0: $\mu = 7$ (since the latter test is employed when the value of σ^2 is known).

VI. Additional Analytical Procedures for the Single-Sample Chi-Square Test for a Population Variance and/or Related Tests

1. Large sample normal approximation of the chi-square distribution When the sample size is larger than 30, the normal distribution can be employed to approximate the test statistic for the **single-sample chi-square test for a population variance.** Equation 3.4 is employed to compute the normal approximation.

$$z = \frac{\tilde{s} - \sigma}{\dfrac{\sigma}{\sqrt{2n}}} \qquad \textbf{(Equation 3.4)}$$

To illustrate the use of Equation 3.4, let us assume that in Example 3.1 the computed value $\tilde{s}^2 = 12.01$ is based on a sample size of $n = 30$. Employing Equation 3.2, the value $\chi^2 = 69.66$ is computed.

$$\chi^2 = \frac{(30 - 1)(12.01)}{5} = 69.66$$

Employing **Table A4**, for $df = 30 - 1 = 29$ the tabled critical two-tailed .05 and .01 values in the upper tail of the chi-square distribution are $\chi^2_{.975} = 45.72$ and $\chi^2_{.995} = 52.34$, and the tabled critical one-tailed .05 and .01 values in the upper tail of the distribution are $\chi^2_{.95} = 42.56$ and $\chi^2_{.99} = 49.59$. Since the obtained value $\chi^2 = 69.66$ is greater than all of the aforementioned critical values, the nondirectional alternative hypothesis H_1: $\sigma^2 \neq 5$ and the directional alternative hypothesis H_1: $\sigma^2 > 5$ are supported at both the .05 and .01 levels.

Equation 3.4 will now be employed to evaluate the data for Example 3.1, if $n = 30$. Since $\sigma^2 = 5$ and $\tilde{s}^2 = 12.01$, it follows that $\sigma = 2.24$ and $\tilde{s} = 3.47$. These values are substituted in Equation 3.4.

$$z = \frac{3.47 - 2.24}{\dfrac{2.24}{\sqrt{(2)(30)}}} = 4.24$$

Employing **Table A1 (Table of the Normal Distribution)** in the **Appendix**, we determine that the tabled critical two-tailed .05 and .01 values are $z_{.05} = 1.96$ and $z_{.01} = 2.58$, and the tabled critical one-tailed .05 and .01 values are $z_{.05} = 1.65$ and $z_{.01} = 2.33$. Since the obtained value $z = 4.24$ is greater than all of the aforementioned critical values, the nondirectional alternative hypothesis H_1: $\sigma^2 \neq 5$ and the directional alternative hypothesis H_1: $\sigma^2 > 5$ are supported at both the .05 or .01 levels. The conclusions derived through use of the normal approximation are identical to those reached with Equation 3.2.

If Equation 3.4 is employed with Example 3.1 for $n = 10$, it results in the value $z = 2.46$. Specifically: $z = [3.47 - 2.24] / [2.24/\sqrt{(2)(10)}] = 2.46$. Since the latter value is greater than the tabled critical two-tailed value $z_{.05} = 1.96$ and the tabled critical one-tailed values $z_{.05} = 1.65$ and $z_{.01} = 2.33$, the nondirectional alternative hypothesis H_1: $\sigma^2 \neq 5$ is supported at the .05 level and the directional alternative hypothesis H_1: $\sigma^2 > 5$ is supported at both the .05 and .01 levels. Recollect that when Equation 3.2 is employed with Example 3.1, the nondirectional alternative hypothesis H_1: $\sigma^2 \neq 5$ and the directional alternative hypothesis H_1: $\sigma^2 > 5$ are both supported, but only at the .05 level. Thus, when the normal approximation is employed with a small sample size, it appears to inflate the likelihood of committing a Type I error (since the normal approximation supports the nondirectional alternative hypothesis H_1: $\sigma^2 > 5$ at the .01 level).

2. Computation of a confidence interval for the variance of a population represented by a sample An equation for computing a confidence interval for the variance of a population (as well as the population standard deviation) can be derived by algebraically transposing the terms in Equation 3.2. As is noted in the discussion of the **single-sample *t* test**, a confidence interval is a range of values within which a researcher can be confident to a specified degree that the true value of a population parameter falls. Equation 3.5 is the general equation for computing a confidence interval for a population variance.

$$\frac{(n - 1)\,\tilde{s}^2}{\chi^2_{[1 - (\alpha/2)]}} \leq \sigma^2 \leq \frac{(n - 1)\,\tilde{s}^2}{\chi^2_{(\alpha/2)}} \qquad \textbf{(Equation 3.5)}$$

Where: $\chi^2_{(\alpha/2)}$ is the tabled critical two-tailed value in the chi-square distribution below which a proportion equal to $[1 - (\alpha/2)]$ of the cases falls. If the percentage of the distribution that falls within the confidence interval is subtracted from 1, it will equal the value of α.

Equations 3.6 and 3.7 are employed to compute the 95% and 99% confidence intervals for a population variance. The critical values employed in Equation 3.6 demarcate the middle 95% of the chi-square distribution, while the critical values employed in Equation 3.7 demarcate the middle 99% of the distribution.

$$\frac{(n - 1)\,\tilde{s}^2}{\chi^2_{.975}} \leq \sigma^2 \leq \frac{(n - 1)\,\tilde{s}^2}{\chi^2_{.025}} \qquad \textbf{(Equation 3.6)}$$

$$\frac{(n - 1)\,\tilde{s}^2}{\chi^2_{.995}} \le \sigma^2 \le \frac{(n - 1)\,\tilde{s}^2}{\chi^2_{.005}} \qquad \textbf{(Equation 3.7)}$$

Using the data for Example 3.1, Equation 3.6 is employed to compute the 95% confidence interval for the population variance.

$$\frac{(10 - 1)\,(12.01)}{19.02} \le \sigma^2 \le \frac{(10 - 1)\,(12.01)}{2.70}$$

$$5.68 \le \sigma^2 \le 40.03$$

Thus, we can be 95% confident that the true value of the variance of the population the sample represents falls between the values 5.68 and 40.03. By taking the square root of the latter values, we can determine the 95% confidence interval for the population standard deviation. Thus, $2.38 \le \sigma \le 6.33$. In other words, we can be 95% confident that the true value of the standard deviation of the population the sample represents falls between the values 2.38 and 6.33.

Equation 3.7 is employed below to compute the 99% confidence interval.

$$\frac{(10 - 1)\,(12.01)}{23.59} \le \sigma^2 \le \frac{(10 - 1)\,(12.01)}{1.73}$$

$$4.58 \le \sigma^2 \le 62.48$$

Thus, we can be 99% confident that the true value of the variance of the population the sample represents falls between the values 4.58 and 62.48. By taking the square root of the latter values, we can determine the 99% confidence interval for the population standard deviation. Thus, $2.14 \le \sigma \le 7.90$. In other words, we can be 99% confident that the true value of the standard deviation of the population the sample represents falls between the values 2.14 and 7.90. Note that (as is the case for confidence intervals for a population mean) a larger range of values defines the 99% confidence interval for a population variance than the 95% confidence interval.

When $n \ge 30$, the normal distribution can be employed to approximate the confidence interval for a population standard deviation. Equation 3.8 is the general equation for computing a confidence interval using the normal approximation.

$$\frac{\tilde{s}}{1 + \dfrac{z_{(\alpha/2)}}{\sqrt{2n}}} \le \sigma \le \frac{\tilde{s}}{1 - \dfrac{z_{(\alpha/2)}}{\sqrt{2n}}} \qquad \textbf{(Equation 3.8)}$$

Where: $z_{(\alpha/2)}$ is the tabled critical two-tailed value in the normal distribution below which a proportion equal to $[1 - (\alpha/2)]$ of the cases falls. If the percentage of the distribution that falls within the confidence interval is subtracted from 1, it will equal the value of α.

Equations 3.9 and 3.10 employ the normal approximation to compute the 95% and 99% confidence intervals for a population standard deviation. The values $z_{.05}$ and $z_{.01}$ used in the latter equations are the tabled critical two-tailed .05 and .01 values $z_{.05} = 1.96$ and $z_{.01} = 2.58$. By squaring a value obtained for the confidence interval for a population standard deviation, one can determine the confidence interval for the population variance.

$$\frac{\tilde{s}}{1 + \frac{z_{.05}}{\sqrt{2n}}} \leq \sigma \leq \frac{\tilde{s}}{1 - \frac{z_{.05}}{\sqrt{2n}}} \qquad \textbf{(Equation 3.9)}$$

$$\frac{\tilde{s}}{1 + \frac{z_{.01}}{\sqrt{2n}}} \leq \sigma \leq \frac{\tilde{s}}{1 - \frac{z_{.01}}{\sqrt{2n}}} \qquad \textbf{(Equation 3.10)}$$

For purposes of illustration, let us assume that $n = 30$ in Example 3.1. Using $n = 30$ and $\tilde{s}^2 = 12.01$ for Example 3.1, Equation 3.9 is employed to compute the 95% confidence interval for the population standard deviation and variance.

$$\frac{3.47}{1 + \frac{1.96}{\sqrt{(2)(30)}}} \leq \sigma \leq \frac{3.47}{1 - \frac{1.96}{\sqrt{(2)(30)}}}$$

$$2.77 \leq \sigma \leq 4.65$$

$$7.67 \leq \sigma^2 \leq 21.62$$

Using $n = 30$ and $\tilde{s}^2 = 12.01$ for Example 3.1, Equation 3.10 is employed to compute the 99% confidence interval for the population standard deviation and variance.

$$\frac{3.47}{1 + \frac{2.58}{\sqrt{(2)(30)}}} \leq \sigma \leq \frac{3.47}{1 - \frac{2.58}{\sqrt{(2)(30)}}}$$

$$2.61 \leq \sigma \leq 5.18$$

$$6.81 \leq \sigma^2 \leq 26.83$$

If Equations 3.6 and 3.7 are employed with Example 3.1 for $n = 30$ and $\tilde{s}^2 = 12.01$, they yield values close to those obtained with Equations 3.9 and 3.10. As the size of the sample increases, the values that define a confidence interval based on the use of the normal versus chi-square distributions converge, and for large sample sizes the two distributions yield the same values.

Equation 3.6 is employed to compute the 95% confidence interval. Note that for $df = 29$, the tabled critical values $\chi^2_{.025} = 16.05$ and $\chi^2_{.975} = 45.72$ are employed in Equation 3.6.

$$\frac{(30 - 1)\ (12.01)}{45.72} \leq \sigma^2 \leq \frac{(30 - 1)\ (12.01)}{16.05}$$

$$7.62 \leq \sigma^2 \leq 21.70$$

$$2.76 \leq \sigma \leq 4.66$$

Equation 3.7 is employed to compute the 99% confidence interval. Note that for $df = 29$, the tabled critical values $\chi^2_{.005} = 13.21$ and $\chi^2_{.995} = 52.34$ are employed in Equation 3.7.

$$\frac{(30 - 1)(12.01)}{52.34} \leq \sigma^2 \leq \frac{(30 - 1)(12.01)}{13.21}$$

$$6.65 \leq \sigma^2 \leq 26.37$$

$$2.58 \leq \sigma \leq 5.13$$

In closing the discussion of confidence intervals, the reader should take note of the fact that the range of values which defines the 95% and 99% confidence intervals for Example 3.1 is extremely large. Because of this the researcher will not be able stipulate with great precision a single value that is likely to represent the true value of the population variance/standard deviation. This illustrates that a confidence interval may be quite limited in its ability to accurately estimate a population parameter, especially when the sample size upon which it is based is small.[5] The reader should also take note of the fact that the reliability of Equations 3.5 and 3.8 will be compromised if one or more of the assumptions of the **single-sample chi-square test for a population variance** is saliently violated.

3. Computation of the power of the single-sample chi-square test for a population variance Although it will not be described in this book, the protocol for computing the power of the **single-sample chi-square test for a population variance** is described in Cohen (1988) and Guenther (1965).

VII. Additional Discussion of the Single-Sample Chi-Square Test for a Population Variance

No additional material will be discussed in this section.

VIII. Additional Examples Illustrating the Use of the Single-Sample Chi-Square Test for a Population Variance

With the exception of Examples 1.5 and 1.6, the **single-sample chi-square test for a population variance** can be employed to test a hypothesis about a population variance (or standard deviation) with any of the examples that are employed to illustrate the **single-sample z test** (Examples 1.1–1.4) and the **single-sample *t* test** (Examples 2.1 and 2.2).

Examples 3.2–3.5 are four additional examples that can be evaluated with the **single-sample chi-square test for a population variance.** Since these examples employ the same population parameters and sample data used in Example 3.1, they yield the same result. The reader should take note of the fact that although in Examples 3.4 and 3.5 the value $\sigma = 2.24$ is given for the population standard deviation, when the latter value is squared it results in $\sigma^2 = 5$. A different set of data are employed in Example 3.6, which is the last example presented in this section.

Example 3.2 *A manufacturer of a machine that makes ball bearings claims that the variance of the diameter of the ball bearings produced by the machine is 5 millimeters. A company that has purchased the machine measures the diameter of a random sample of ten ball bearings. The computed estimated population variance of the diameters of ten ball bearings is 12.01 millimeters. Is the obtained value $\tilde{s}^2 = 12.01$ consistent with the null hypothesis H_0: $\sigma^2 = 5$?*

Example 3.3 A meteorologist develops a theory which predicts a variance of five degrees for the temperature recorded at noon each day in a canyon situated within a mountain range 100 *miles east of the South Pole. Over a ten-day period the following temperatures are recorded:* 5, 6, 4, 3, 11, 12, 9, 13, 6, 8. *Do the data support the meteorologist's theory?*

Example 3.4 *A chemical company claims that the standard deviation for the number of tons of waste it discards annually is* $\sigma = 2.24$. *Assume that during a ten-year period the following annual values are recorded for the number of tons of waste discarded:* 5, 6, 4, 3, 11, 12, 9, 13, 6, 8. *Are the data consistent with the company's claim?*

Example 3.5 *During more than three decades of teaching, a college professor determines that the mean and standard deviation on a chemistry final examination are respectively* 7 *and* 2.24. *During the fall semester of a year in which he employs a new teaching method, ten students who take the final examination obtain the following scores:* 5, 6, 4, 3, 11, 12, 9, 13, 6, 8. *Do the data suggest that the new teaching method results in an increase in variability in performance?*

Example 3.6 *A pharmaceutical company claims that upon ingesting its cough medicine a person ceases to cough almost immediately. The company claims that the standard deviation of the number of coughs emitted by a person after ingesting the medicine is* 5. *In order to evaluate the company's claim, a physician evaluates a random sample of ten patients who come into his office coughing excessively. The physician records the following number of coughs emitted by each of the patients after he is given a therapeutic dosage of the medicine:* 9, 10, 8, 4, 8, 3, 0, 10, 15, 9. *Is the variability of the data consistent with the company's claim for the product?*

The data for Example 3.6 are identical to that employed for Example 2.1, which is used to illustrate the **single-sample *t* test**. In the computations for the latter test, through use of Equation 2.1, it is determined that the estimated population standard deviation for the data is $\tilde{s} = 4.25$. Since the population variance is the square of the standard deviation, we can determine that $\sigma^2 = (5)^2 = 25$ and $\tilde{s}^2 = (4.25)^2 = 18.06$. Since the company claims that the population standard deviation is 5, the null hypothesis for Example 3.6 can be stated as either H_0: $\sigma = 5$ or H_0: $\sigma^2 = 25$. The nondirectional and directional alternative hypotheses (stated in reference to H_0: $\sigma = 5$) that can be employed for Example 3.6 are as follows: H_1: $\sigma \neq 5$; H_1: $\sigma > 5$; and H_1: $\sigma < 5$. When Equation 3.2 is employed to evaluate the null hypothesis H_0: $\sigma = 5$, the value $\chi^2 = 6.50$ is computed.

$$\chi^2 = \frac{(10 - 1)(18.06)}{25} = 6.50$$

Since $n = 10$, the tabled critical values for $df = 9$ are employed to evaluate the computed value $\chi^2 = 6.50$. Since the latter value is less than the tabled critical two-tailed .05 value $\chi^2_{.975} = 19.02$ in the upper tail of the chi-square distribution and greater than the tabled critical two-tailed .05 value $\chi^2_{.025} = 2.70$ in the lower tail of the distribution, the nondirectional alternative hypothesis H_1: $\sigma \neq 5$ is not supported. Since $\chi^2 = 6.50$ is less than the tabled critical one-tailed .05 value $\chi^2_{.95} = 16.92$ in the upper tail of the distribution, the directional alternative hypothesis H_1: $\sigma > 5$ is not supported. Since $\chi^2 = 6.50$ is greater than the tabled critical one-tailed .05 value $\chi^2_{.05} = 3.33$ in the lower tail of the distribution, the directional alternative hypothesis H_1: $\sigma < 5$ is not supported. Thus, regardless of which alternative hypothesis one employs, the data do not contradict the company's statement that the population standard deviation is $\sigma = 5$.

References

Cohen, J. (1988). **Statistical power analysis for the behavioral sciences** (2nd ed.). New York: Academic Press.

Crisman, R. (1975). Shortest confidence interval for the standard deviation of a normal distribution. **Journal of Undergraduate Mathematics**, 7, 57.

Freund, J. E. (1984). **Modern elementary statistics** (6th ed.). Englewood Cliffs, NJ: Prentice–Hall, Inc.

Guenther, W. C. (1965). **Concepts of statistical inference.** New York: McGraw–Hill Book Company.

Hogg, R. V. and Tanis, E. A. (1988). **Probability and statistical inference** (3rd ed.). New York: Macmillan Publishing Company.

Endnotes

1. Most sources note that violation of the normality assumption is much more serious for the **single-sample chi-square test for a population variance** than it is for tests concerning the mean of a single sample (i.e., the **single-sample z test** and the **single-sample t test**). Especially in the case of small sample sizes, violation of the normality assumption can severely compromise the accuracy of the tabled values employed to evaluate the chi-square statistic.

2. One can also state the null and alternative hypotheses in reference to the population standard deviation (which is the square root of the population variance). Since in Example 3.1 $\sigma = \sqrt{\sigma^2} = \sqrt{5} = 2.24$, one can state the null hypothesis and nondirectional and directional alternative hypotheses as follows: H_0: $\sigma = 2.24$; H_1: $\sigma \neq 2.24$; H_1: $\sigma > 2.24$; and H_1: $\sigma < 2.24$.

3. The use of the chi-square distribution in evaluating the variance is based on the fact that for any value of n, the sampling distribution of $\tilde{s}^2$ has a direct linear relationship to the chi-square distribution for $df = n - 1$. As is the case for the chi-square distribution, the sampling distribution of $\tilde{s}^2$ is positively skewed. Although the average of the sampling distribution for $\tilde{s}^2$ will equal σ^2, because of the positive skew of the distribution a value of $\tilde{s}^2$ is more likely to underestimate rather than overestimate the value of σ^2.

4. When the chi-square distribution is employed within the framework of the **single-sample chi-square test for a population variance,** it is common practice to employ critical values derived from both tails of the distribution. However, when the chi-square distribution is used with other statistical tests, one generally employs critical values that are only derived from the right tail of the distribution. Examples of chi-square tests which generally only employ the right tail of the distribution are **the chi-square goodness-of-fit test (Test 5) and the chi-square test for $r \times c$ tables (Test 11).**

5. Although the procedure described in this section for computing a confidence interval for a population variance is the one that is most commonly described in statistics books, it does not result in the shortest possible confidence interval that can be computed. Hogg and Tanis (1988) describe a method (based on Crisman (1975)) requiring more advanced mathematical procedures that allows one to compute the shortest possible confidence interval for a population variance. For large sample sizes the difference between the latter method and the method described in this section will be trivial.

Test 4

The Wilcoxon Signed-Ranks Test
(Nonparametric Test Employed with Ordinal Data)

I. Hypothesis Evaluated with Test and Relevant Background Information

Hypothesis evaluated with test Does a sample of n subjects come from a population in which the median value (θ) equals a specified value?

Relevant background information on test The **Wilcoxon signed-ranks test** (Wilcoxon (1945, 1949)) is a nonparametric procedure employed in a hypothesis testing situation involving a single sample in order to determine whether a sample is derived from a population with a median of θ. (The population median will be represented by the notation θ, which is the lower case Greek letter **theta**.) If the **Wilcoxon signed-ranks test** yields a significant result, the researcher can conclude there is a high likelihood the sample is derived from a population with a median value other than θ.

The **Wilcoxon signed-ranks test** is based on the following assumptions:[1] a) The sample has been randomly selected from the population it represents; b) The original scores obtained for each of the subjects/objects are in the format of interval/ratio data; and c) The underlying population distribution is symmetrical. When there is reason to believe that this latter assumption is violated, Daniel (1990), among others, recommends that the **binomial sign test for a single sample (Test 6)** be employed in place of the **Wilcoxon signed-ranks test.**[2] Proponents of nonparametric tests recommend that the **Wilcoxon signed-ranks test** be employed in place of the **single-sample *t* test (Test 2)** when there is reason to believe that the normality assumption of the latter test has been saliently violated.[3] It should be noted that all of the other tests in this text which rank data (with the exception of the **Wilcoxon matched-pairs signed-ranks test (Test 13)**) rank the original interval/ratio scores of subjects. The **Wilcoxon signed-ranks test**, however, does not rank subjects' original interval/ratio scores, but instead ranks difference scores — specifically, the obtained difference between each subject's score and the hypothesized value of the population median. For this reason, some sources categorize the **Wilcoxon signed-ranks test** as a test of interval/ratio data. Most sources, however, (including this book) categorize the test as one involving ordinal data, by virtue of the fact that a ranking procedure is part of the test protocol.

II. Example

Example 4.1 *A physician states that the median number of times he sees each of his patients during the year is five. In order to evaluate the validity of this statement, he randomly selects ten of his patients and determines the number of office visits each of them made during the past year. He obtains the following values for the ten patients in his sample:* 9, 10, 8, 4, 8, 3, 0, 10, 15, 9. *Do the data support his contention that the median number of times he sees a patient is five?*

III. Null versus alternative hypotheses

Null hypothesis H_0: $\theta = 5$

(The median of the population the sample represents equals 5. With respect to the sample data, this translates into the sum of the sum of the ranks of the positive difference scores being equal to the sum of the ranks of the negative difference scores (i.e., $\Sigma R+ = \Sigma R-$).

Alternative hypothesis H_1: $\theta \neq 5$

(The median of the population the sample represents does not equal 5. With respect to the sample data, this translates into the sum of the ranks of the positive difference scores not being equal to the sum of the ranks of the negative difference scores (i.e., $\Sigma R+ \neq \Sigma R-$). This is a **nondirectional alternative hypothesis** and it is evaluated with a **two-tailed test.**)

or

H_1: $\theta > 5$

(The median of the population the sample represents is some value greater than 5. With respect to the sample data, this translates into the sum of the ranks of the positive difference scores being greater than the sum of the ranks of the negative difference scores (i.e., $\Sigma R+ > \Sigma R-$). This is a **directional alternative hypothesis** and it is evaluated with a **one-tailed test.**)

or

H_1: $\theta < 5$

(The median of the population the sample represents is some value less than 5. With respect to the sample data, this translates into the sum of the ranks of the positive difference scores being less than the sum of the ranks of the negative difference scores (i.e., $\Sigma R+ < \Sigma R-$). This is a **directional alternative hypothesis** and it is evaluated with a **one-tailed test.**)

Note: Only one of the above noted alternative hypotheses is employed. If the alternative hypothesis the researcher selects is supported, the null hypothesis is rejected.

IV. Test Computations

The data for Example 4.1 are summarized in Table 4.1.

Table 4.1 Data for Example 4.1

Subject	X	$D = X - \theta$	Rank of $\|D\|$	Signed rank of $\|D\|$
1	9	4	5.5	5.5
2	10	5	8	8
3	8	3	3.5	3.5
4	4	-1	1	-1
5	8	3	3.5	3.5
6	3	-2	2	-2
7	0	-5	8	-8
8	10	5	8	8
9	15	10	10	10
10	9	4	5.5	5.5
				$\Sigma R+ = 44$
				$\Sigma R- = 11$

The scores of the 10 subjects are recorded in Column 2 of Table 4.1. In Column 3, a D score is computed for each subject. This score, which is referred to as a **difference score,** is the difference between a subject's score and the hypothesized value of the population median, $\theta = 5$. Column 4 contains the ranks of the difference scores. In ranking the difference scores for the **Wilcoxon signed-ranks test,** the following guidelines are employed:

a) The **absolute values** of the difference scores ($|D|$) are ranked (i.e., the sign of a difference score is not taken into account).

b) Any difference score that equals zero is not ranked. This translates into eliminating from the analysis any subject who yields a difference score of zero.

c) In ranking the absolute values of the difference scores, the following protocol should be employed: Assign a rank of 1 to the difference score with the lowest absolute value, a rank of 2 to the difference score with the second lowest absolute value, and so on until the highest rank is assigned to the difference score with the highest absolute value. When there are tie scores present, the average of the ranks involved is assigned to all difference scores tied for a given rank. Because of this latter fact, when there are tie scores for either the lowest or highest difference scores, the rank assigned to the lowest difference score will be some value greater than 1, and the rank assigned to the highest difference score will be some value less than n. To further clarify how ties are handled, examine Table 4.2 which lists the **difference scores** of the 10 subjects. In the table, the difference scores (based on their absolute values) are arranged ordinally, after which they are ranked employing the protocol described above.

Table 4.2 Ranking Procedure for Wilcoxon Signed-Ranks Test

Subject number	4	6	3	5	1	10	2	7	8	9
Subject's difference score	–1	–2	3	3	4	4	5	–5	5	10
Absolute value of difference score	1	2	3	3	4	4	5	5	5	10
Rank of $\|D\|$	1	2	3.5	3.5	5.5	5.5	8	8	8	10

The difference score of Subject 4 has the lowest absolute value (i.e., 1), and because of this it is assigned a rank of 1. The next lowest absolute value for a difference score (2) is that of Subject 6, and thus it is assigned a rank of 2.[4] The difference score of 3 (which is obtained for both Subjects 3 and 5) is the score that corresponds to the third rank-order. Since, however, there are two instance of this difference score, it will also use up the position reserved for the fourth rank-order (i.e., 3 and 4 are the two ranks that would be employed if, in fact, these two subjects did not have the identical difference score). Instead of arbitrarily assigning one of the subjects with a difference score of 3 a rank-order of 3 and the other subject a rank-order of 4, we compute the average of the two ranks that are involved (i.e., $(3 + 4)/2 = 3.5$), and assign that value as the rank-order for the difference scores of both subjects. The next rank-order in the sequence of the 10 rank-orders is 5. Once again, however, two subjects (Subjects 1 and 10) are tied for the difference score in the fifth ordinal position (which happens to involve a difference score of 4). Since if not equal to one another these two difference scores would involve the fifth and sixth ranks, we compute the average of these two ranks (i.e., $(5 + 6)/2 = 5.5$), and assign that value as the rank for the difference scores of Subjects 1 and 10. With respect to the next difference score (5), there is a three-way tie involving Subjects 2, 7, and 8 (keeping in mind that the absolute value of the difference score for Subject 7 is 5). The average of the three ranks which would be involved if the subjects had obtained different difference scores is computed (i.e., $(7 + 8 + 9)/3 = 8$), and that average value is assigned to the difference scores of Subjects 2, 7, and 8. Since the remaining difference score of 10 (obtained by Subject 9) is the highest difference score, it is assigned the highest rank which equals 10.

It should be emphasized that in the **Wilcoxon signed-ranks test** it is essential that a rank of 1 be assigned to the difference score with the lowest absolute value, and that the highest rank be assigned to the difference score with the highest absolute value. In most other tests that involve ranking, the ranking procedure can be reversed (i.e., the same test statistic will be obtained if one assigns a rank of 1 to the highest score and the highest rank to the lowest score). However, if one reverses the ranking procedure in conducting the **Wilcoxon signed-ranks test**, it will invalidate the results of the test.

d) After ranking the absolute values of the difference scores, the sign of each difference score is placed in front of its rank. The signed ranks of the difference scores are listed in Column 5 of Table 4.1.

The sum of the ranks that have a positive sign (i.e., $\Sigma R+ = 44$) and the sum of the ranks that have a negative sign (i.e., $\Sigma R- = 11$) are recorded at the bottom of the Column 5 in Table 4.1. Equation 4.1 allows one to check the accuracy of these values. If the relationship indicated by Equation 4.1 is not obtained, it indicates an error has been made in the calculations. In Equation 4.1, n represents the number of signed ranks (i.e., the number of difference scores that are ranked).

$$\Sigma R+ + \Sigma R- = \frac{n(n+1)}{2} \qquad \textbf{(Equation 4.1)}$$

Employing the values $\Sigma R+ = 44$ and $\Sigma R- = 11$ in Equation 4.1, we confirm that the relationship described by the equation is true.

$$44 + 11 = \frac{(10)(11)}{2} = 55$$

It is important to note that in the event one or more subjects obtains a difference score of zero, such scores are not employed in the analysis. In such a case, the value of n in Equation 4.1 will only represent the number of scores that have been assigned ranks. Example 4.2 in Section VIII illustrates the use of the **Wilcoxon signed-ranks test** with data in which difference scores of zero are present.

V. Interpretation of the Test Results

As noted in Section III, if the sample is derived from a population with a median value equal to the hypothesized value of the population median (i.e., the null hypothesis is true), the values of $\Sigma R+$ and $\Sigma R-$ will be equal to one another. When $\Sigma R+$ and $\Sigma R-$ are equivalent, both of these values will equal $[n(n+1)]/4$, which in the case of Example 4.1 will be $[(10)(11)]/4 = 27.5$. This latter value is commonly referred to as the **expected value** of the **Wilcoxon *T* statistic.**

If the value of $\Sigma R+$ is significantly greater than the value of $\Sigma R-$, it indicates there is a high likelihood the sample is derived from a population with a median value which is larger than the hypothesized value of the population median. On the other hand, if $\Sigma R-$ is significantly greater than $\Sigma R+$, it indicates there is a high likelihood the sample is derived from a population with a median value which is less than the hypothesized value of the population median. The fact that $\Sigma R+ = 44$ is greater than $\Sigma R- = 11$ indicates that the data are consistent with the directional alternative hypothesis $H_1\colon \theta > 5$. The question is, however, whether the difference is significant — i.e., whether it large enough to conclude that it is unlikely to be the result of chance.

The absolute value of the **smaller** of the two values $\Sigma R+$ versus $\Sigma R-$ is designated as the **Wilcoxon *T* test statistic.** Since $\Sigma R- = 11$ is smaller than $\Sigma R+ = 44$, $T = 11$. The T value is interpreted by employing **Table A5 (Table of Critical *T* Values for Wilcoxon's**

Signed-Ranks and Matched-Pairs Signed-Ranks Tests) in the **Appendix**. **Table A5** lists the critical two-tailed and one-tailed .05 and .01 T values in relation to the number of signed ranks in a set of data. In order to be significant, the obtained value of T must be **equal to or less than** the tabled critical T value at the prespecified level of significance.[5] Table 4.3 summarizes the tabled critical two-tailed and one-tailed .05 and .01 Wilcoxon T values for $n = 10$ signed ranks.

Table 4.3 Tabled Critical Wilcoxon T Values for $n = 10$ Signed Ranks

	$T_{.05}$	$T_{.01}$
Two-tailed values	8	3
One-tailed values	10	5

Since the null hypothesis can only be rejected if the computed value $T = 11$ is equal to or less than the tabled critical value at the prespecified level of significance, we can conclude the following.

In order for the nondirectional alternative hypothesis H_1: $\theta \neq 5$ to be supported, it is irrelevant whether $\Sigma R+ > \Sigma R-$ or $\Sigma R- > \Sigma R+$. In order for the result to be significant, the computed value of T must be equal to or less than the tabled critical two-tailed value at the prespecified level of significance. Since the computed value $T = 11$ is greater than the tabled critical two-tailed .05 value $T_{.05} = 8$, the nondirectional alternative hypothesis H_1: $\theta \neq 5$ is not supported at the .05 level. It is also not supported at the .01 level, since $T = 11$ is greater than the tabled critical two-tailed .01 value $T_{.01} = 3$.

In order for the directional alternative hypothesis H_1: $\theta > 5$ to be supported, $\Sigma R+$ must be greater than $\Sigma R-$. Since $\Sigma R+ > \Sigma R-$, the data are consistent with the directional alternative hypothesis H_1: $\theta > 5$. In order for the result to be significant, the computed value of T must be equal to or less than the tabled critical one-tailed value at the prespecified level of significance. Since the computed value $T = 11$ is greater than the tabled critical one-tailed .05 value $T_{.05} = 10$, the directional alternative hypothesis is not supported at the .05 level. It is also not supported at the .01 level, since $T = 11$ is greater than the tabled critical one-tailed .01 value $T_{.01} = 5$.

In order for the directional alternative hypothesis H_1: $\theta < 5$ to be supported, the following two conditions must be met: a) $\Sigma R-$ must be greater than $\Sigma R+$; and b) the computed value of T must be equal to or less than the tabled critical one-tailed value at the prespecified level of significance. Since the first of these conditions is not met, the directional alternative hypothesis H_1: $\theta < 5$ is not supported.

A summary of the analysis of Example 4.1 with the **Wilcoxon signed-ranks test** follows: With respect to the median number of times the doctor sees a patient, we can conclude that the data do not indicate that the sample of 10 subjects comes from a population with a median value other than 5.

Except for the fact that the mean rather than the median is employed as the population parameter stated in the null and alternative hypotheses, Example 2.1 is identical to Example 4.1 (i.e., the two examples employ the same set of data). Since Example 2.1 states the null hypothesis with reference to the population mean, it is evaluated with the **single-sample *t* test**. At this point we will compare the results of the two tests. When the same data are evaluated with the **single-sample *t* test**, the null hypothesis can be rejected when the directional alternative hypothesis H_1: $\mu > 5$ is employed, but only at the .05 level. With reference to the latter alternative hypothesis, the obtained t value exceeds the tabled critical $t_{.05}$ value by a comfortable margin. When the **single-sample *t* test** is employed, the nondirectional alternative hypothesis H_1: $\mu \neq 5$ is not supported at the .05 level.

When Example 4.1 is evaluated with the **Wilcoxon signed-ranks test**, the null hypothesis cannot be rejected regardless of which alternative hypothesis is employed. However, when the directional alternative hypothesis H_1: $\theta > 5$ is employed, the **Wilcoxon signed-ranks test** falls just short of being significant at the .05 level. Directly related to this is the fact that in some sources the tabled critical values published for the Wilcoxon test statistic are not identical to the values listed in **Table A5**. These differences are the result of rounding off protocol. The critical T values in **Table A5** listed for a given level of significance are associated with the probability that is closest to but not greater than the value of alpha. In some instances a T value listed in an alternative table may be one point higher than the value listed in **Table A5**, thus making it easier to reject the null hypothesis. Although these alternative critical values are actually closer to the value of alpha than the values listed in **Table A5**, the probability associated with a tabled critical T value in the alternative table is, in fact, larger than the value of alpha. With reference to Example 4.1, the exact probability associated with $T = 10$ is .0420 (i.e., this represents the likelihood of obtaining a T value of 10 or less). The probability associated with $T = 11$, which is the critical value of T listed in the alternative table, is .0527. Although the latter probability is closer to $\alpha = .05$ than is .0420, it falls above 05. Thus, if one employs the alternative table that contains the tabled critical one-tailed .05 value $T_{.05} = 11$, the alternative hypothesis H_1: $\theta > 5$ is supported at the .05 level. Obviously in a case such as this where the likelihood of obtaining a value equal to or less than the computed value of T is just sightly above .05, it would seem prudent to conduct further studies in order to clarify the status of the alternative hypothesis H_1: $\theta > 5$.

In the case of Examples 4.1 and 2.1, the results of the **Wilcoxon signed-ranks** and **single-sample *t* test** are fairly consistent for the same set of data. Support for the analogous alternative hypotheses H_1: $\mu > 5$ and H_1: $\theta > 5$ is either clearly indicated (in the case of the t test) or falls just short of significance (in the case of the Wilcoxon test). The slight discrepancy between the two tests reflects the fact that, as a general rule, nonparametric tests are not as powerful as their parametric analogs. In the case of the two tests under consideration, the lower power of the **Wilcoxon signed-ranks test** can be attributed to the loss of information which results from expressing interval/ratio data in a rank-order format (specifically, rank-ordering the difference scores). As noted earlier, when two or more inferential statistical tests are applied to the same set of data and yield contradictory results, it is prudent to replicate the study. In the final analysis, replication is the most powerful tool a researcher has at his disposal for determining the status of a null hypothesis.

VI. Additional Analytical Procedures for the Wilcoxon Signed-Ranks Test and/or Related Tests

1. The normal approximation of the Wilcoxon *T* statistic for large sample sizes If the sample size employed in a study is relatively large, the normal distribution can be used to approximate the Wilcoxon T statistic. Although sources do not agree on the value of the sample size that justifies employing the normal approximation of the Wilcoxon distribution, they generally state it should be used for sample sizes larger than those documented in the Wilcoxon table contained within the source. Equation 4.2 provides the normal approximation for Wilcoxon T. In the equation T represents the computed value of Wilcoxon T, which for Example 4.1 is $T = 11$. n, as noted previously, represents the number of signed ranks. Thus, in our example, $n = 10$. Note that in the numerator of Equation 4.2, the term $[n(n+1)]/4$ represents the expected value of T (often summarized with the symbol T_E), which is defined in Section V. The denominator of Equation 4.2 represents the expected standard deviation of the sampling distribution of the T statistic.

$$z = \frac{T - \frac{n(n+1)}{4}}{\sqrt{\frac{n(n+1)(2n+1)}{24}}} \qquad \textbf{(Equation 4.2)}$$

Although Example 4.1 involves only ten signed ranks (a value most sources would view as too small to use with the normal approximation), it will be employed to illustrate Equation 4.2. The reader will see that in spite of employing Equation 4.2 with a small sample size, it will yield essentially the same result as that obtained when the exact table of the Wilcoxon distribution is employed. When the values $T = 11$ and $n = 10$ are substituted in Equation 4.2, the value $z = -1.68$ is computed.

$$z = \frac{11 - \frac{(10)(11)}{4}}{\sqrt{\frac{(10)(11)(21)}{24}}} = -1.68$$

The obtained value $z = -1.68$ is evaluated with **Table A1 (Table of the Normal Distribution)** in the **Appendix.** In **Table A1** the tabled critical two-tailed .05 and .01 values are $z_{.05} = 1.96$ and $z_{.01} = 2.58$, and the tabled critical one-tailed .05 and .01 values are $z_{.05} = 1.65$ and $z_{.01} = 2.33$.

Since the smaller of the two values $\Sigma R+$ versus $\Sigma R-$ is selected to represent T, the value of z computed with Equation 4.2 will always be a negative number (unless $\Sigma R+ = \Sigma R-$, in which case z will equal zero). This is the case since by selecting the smaller value T will always be less than the expected value T_E. As a result of this, the following guidelines are employed when evaluating the null hypothesis.

a) If a nondirectional alternative hypothesis is employed, the null hypothesis can be rejected if the obtained absolute value of z is equal to or greater than the tabled critical two-tailed value at the prespecified level of significance.

b) When a directional alternative hypothesis is employed, one of the two possible directional alternative hypotheses will be supported if the obtained absolute value of z is equal to or greater than the tabled critical one-tailed value at the prespecified level of significance. Which alternative hypothesis is supported depends on the prediction regarding which of the two values $\Sigma R+$ versus $\Sigma R-$ is larger. The null hypothesis can only be rejected if the directional alternative hypothesis that is consistent with the data is supported.

Employing the above guidelines, when the normal approximation is used with Example 4.1 the following conclusions can be reached.

The nondirectional alternative hypothesis H_1: $\theta \neq 5$ is not supported. This is the case since the computed absolute value $z = 1.68$ is less than the tabled critical two-tailed .05 value $z_{.05} = 1.96$. This decision is consistent with the decision that is reached when the exact table of the Wilcoxon distribution is employed to evaluate the nondirectional alternative hypothesis H_1: $\theta \neq 5$.

The directional alternative hypothesis H_1: $\theta > 5$ is supported at the .05 level. This is the case since the data are consistent with the latter alternative hypothesis (i.e., $\Sigma R+ > \Sigma R-$), and the computed absolute value $z = 1.68$ is greater than the tabled critical one-tailed .05 value $z_{.05} = 1.65$. The directional alternative hypothesis H_1: $\theta > 5$ is not supported at the .01 level, since the absolute value $z = 1.68$ is less than the tabled critical one-tailed .01 value $z_{.01} = 2.33$. When the exact table of the Wilcoxon distribution is employed, the directional alternative hypothesis H_1: $\theta > 5$ is not supported at the .05 level. However, it was noted that if an alternative table of Wilcoxon critical values is employed, the alternative hypothesis H_1: $\theta > 5$ is supported at the .05 level.

The directional alternative hypothesis H_1: $\theta < 5$ is not supported, since the data are not consistent with the latter alternative hypothesis (which requires that $\Sigma R- > \Sigma R+$).

In closing the discussion of the normal approximation, it should be noted that, in actuality, either $\Sigma R+$ or $\Sigma R-$ can be employed to represent the value of T in Equation 4.2. Either value will yield the same absolute value for z. The smaller of the two values will always yield a negative z value, and the larger of the two values will always yield a positive z value (which in this instance will be $z = 1.68$ if $\Sigma R+ = 44$ is employed in Equation 4.2 to represent T). In evaluating a nondirectional alternative hypothesis, the sign of z is irrelevant. In the case of a directional alternative hypothesis, one must determine whether the data are consistent with the alternative hypothesis that is stipulated. If the data are consistent, one then determines whether the absolute value of z is equal to or greater than the tabled critical one-tailed value at the prespecified level of significance.

2. The correction for continuity for the normal approximation of the Wilcoxon signed-ranks test Although not described in most sources, Marascuilo and McSweeney (1977) employ a correction factor known as the **correction for continuity** for the normal approximation of the Wilcoxon test statistic. The **correction for continuity** is recommended by some sources for use with a number of nonparametric tests that employ a **continuous distribution** (such as the normal distribution) to estimate a **discrete distribution** (such as in this instance the Wilcoxon distribution). As noted in the **Introduction** of the book, in a continuous distribution there is an infinite number of values a variable may assume, whereas in a discrete distribution the number of possible values a variable may assume is limited in number. The correction for continuity is based on the premise that if a continuous distribution is employed to estimate a discrete distribution, such an approximation will inflate the Type I error rate. By employing the correction for continuity, the Type I error rate is ostensibly adjusted to be more compatible with the prespecified alpha value designated by the researcher. When the correction for continuity is applied to a normal approximation of an underlying discrete distribution, it results in a slight reduction in the absolute value computed for z. In the case of the normal approximation of the Wilcoxon test statistic, the correction for continuity requires that .5 be subtracted from the absolute value of the numerator of Equation 4.2. Thus, Equation 4.3 represents the continuity corrected normal approximation of the Wilcoxon test statistic.

$$z = \frac{\left| T - \frac{n(n+1)}{4} \right| - .5}{\sqrt{\frac{n(n+1)(2n+1)}{24}}} \qquad \textbf{(Equation 4.3)}$$

If the correction for continuity is employed with Example 4.1, the value of the numerator of Equation 4.3 is 16, in contrast to the absolute value of 16.5 computed with Equation 4.2. Employing Equation 4.3, the continuity corrected value $z = 1.63$ is computed. Note that as a result of the absolute value conversion, the numerator of Equation 4.3 will always be a positive number, thus yielding a positive z value.

$$z = \frac{\left| 11 - \frac{(10)(11)}{4} \right| - .5}{\sqrt{\frac{(10)(11)(21)}{24}}} = 1.63$$

Since the absolute value $z = 1.63$ is less than the tabled critical one-tailed .05 value $z_{.05} = 1.65$, the directional hypothesis H_1: $\theta > 5$ is not supported. Note that since the obtained absolute value $z = 1.63$ is slightly below the tabled critical one-tailed value $z_{.05} = 1.65$, it is just short of being significant (in contrast to the continuity uncorrected absolute value $z = 1.68$ computed with Equation 4.2, which barely achieves significance at the .05 level). In a case such as this where the issue of whether to reject or retain the null hypothesis depends on whether or not one employs the correction for continuity, additional research should be conducted in order to clarify the status of the null hypothesis.

3. Tie correction for the normal approximation of the Wilcoxon test statistic Equation 4.4 is an adjusted version of Equation 4.2 that is recommended in some sources (e.g., Daniel (1990) and Marascuilo and McSweeney (1977)) when tied difference scores are present in the data. The tie correction results in a slight increase in the absolute value of z. Unless there are a substantial number of ties, the difference between the values of z computed with Equations 4.2 and 4.4 will be minimal.

$$z = \frac{T - \dfrac{n(n+1)}{4}}{\sqrt{\dfrac{n(n+1)(2n+1)}{24} - \dfrac{\Sigma t^3 - \Sigma t}{48}}} \qquad \textbf{(Equation 4.4)}$$

Table 4.4 illustrates the application of the tie correction with Example 4.1.

Table 4.4 Correction for Ties with Normal Approximation

Subject	Rank	t	t^3
4	1		
6	2		
3	3.5	2	8
5	3.5		
1	5.5	2	8
10	5.5		
2	8		
7	8	3	27
8	8		
9	10		
		$\Sigma t = 7$	$\Sigma t^3 = 43$

In the data for Example 4.1 there are three sets of tied ranks: Set 1 involves two subjects (Subjects 3 and 5); Set 2 involves two subjects (Subjects 1 and 10); Set 3 involves three subjects (Subjects 2, 7, and 8). The number of subjects involved in each set of tied ranks represents the values of t in the third column of Table 4.4. The three t values are cubed in the last column of the table, after which the values Σt and Σt^3 are computed. The appropriate values are now substituted in Equation 4.4.[6]

$$z = \frac{11 - \dfrac{(10)(11)}{4}}{\sqrt{\dfrac{(10)(11)(21)}{24} - \dfrac{43 - 7}{48}}} = -1.69$$

The absolute value $z = 1.69$ is slightly larger than the absolute value $z = 1.68$ obtained without the tie correction. The difference between the two methods is trivial, and in this instance, regardless of which alternative hypothesis is employed, the decision the researcher makes with respect to the null hypothesis is not affected.[7]

Conover (1980) and Daniel (1990) discuss and/or cite sources on the subject of alternative ways of handling tied difference scores. Conover (1980) also notes that in some instances retaining and ranking zero difference scores may actually provide a more powerful test of an alternative hypothesis than the more conventional method employed in this book (which eliminates zero difference scores from the data).

VII. Additional Discussion of the Wilcoxon Signed-Ranks Test

Conover (1980) and Daniel (1990) note that if, in fact, the population from which the sample is derived is symmetrical, the conclusions one draws with regard to the population median are also true with respect to the population mean (since in a symmetrical population the values of the mean and median will be identical). This amounts to saying that if in Example 4.1 we retain (or reject) H_0: $\theta = 5$, we are also reaching the same conclusion with respect to the null hypothesis H_0: $\mu = 5$. There is, however, no guarantee that the results obtained with the **Wilcoxon signed-ranks test** will be entirely consistent with the results derived when the **single-sample *t* test** is employed to evaluate the same set of data.

VIII. Additional Examples Illustrating the Wilcoxon Signed-Ranks Test

With the exception of Examples 1.5 and 1.6, the **Wilcoxon signed-ranks test** can be employed to evaluate a hypothesis about a population median with any of the examples that are employed to illustrate the **single-sample *z* test (Test 1)** and the **single-sample *t* test**. As noted in Section I, unless the normality assumption of the aforementioned tests is saliently violated, most researches would employ a parametric test in lieu of a nonparametric alternative. In all instances in which the **Wilcoxon signed-ranks test** is employed, difference scores are obtained by subtracting the hypothesized value of the population median from each score in the sample. All difference scores are then ranked and evaluated in accordance with the ranking protocol described in Section IV.

Example 4.2 (which is a restatement of Example 4.1 with a different set of data) illustrates the use of the **Wilcoxon signed-ranks test** with the presence of zero difference scores.

Example 4.2 *A physician states that the median number of times he sees each of his patients during the year is five. In order to evaluate the validity of this statement he randomly selects* 13 *of his patients and determines the number of office visits each of them made during the past year. He obtains the following values for the* 13 *patients in his sample:* 5, 9, 10, 8, 4, 8, 5, 3, 0, 10, 15, 9, 5. *Do the data support his contention that the median number of times he sees a patient is five?*

Examination of the data for Example 4.2 reveals that three of the 13 patients visited the doctor five times during the year. Since each of these three scores is equal to the hypothesized value of the population median, they will all produce difference scores of zero. In employing the ranking protocol for the **Wilcoxon signed-ranks test**, all three of these scores will be eliminated from the data analysis. Upon elimination of the three scores, the following ten scores remain: 9, 10, 8, 4, 8, 3, 0, 10, 15, 9. Since the ten remaining scores are identical to the ten scores employed in Example 4.1, the solution to Example 4.2 will be identical to that of Example 4.1.

If, on the other hand, the **single-sample *t* test** is employed to evaluate Example 4.2, all 13 scores are included in the calculations resulting in the value $t = 1.87$,[8] which is less than the value $t = 1.94$ obtained for Example 2.1 (which employs the 10 scores used in Example 4.1).[9] The point to be made here is that by not employing the zero difference scores, the same T value is computed for the **Wilcoxon signed-ranks test** for both Examples 4.1 and 4.2. Yet in the case of the **single-sample t test**, which employs all 13 scores for the analysis of Example 4.2, the computed value of t for the latter example is not the same as the computed value of t for Example 2.1 (which employs the same data as Example 4.1). Thus, the presence of zero difference scores may serve to increase the likelihood of a discrepancy between the results obtained with the **Wilcoxon signed-ranks test** and the **single-sample t test**.

Example 4.3 *A college English instructor reads in an educational journal that the median number of times a student is absent from a class that meets for fifty minutes three times a week during a* 15 *week semester is* $\theta = 5$. *During the fall semester she keeps a record of the number of times each of the* 10 *students in her writing class is absent. She obtains the following values:* 9, 10, 8, 4, 8, 3, 0, 10, 15, 9. *Do the data suggest that the class is representative of a population that has a median of* 5*?*

Since Example 4.3 employs the same data as Example 4.1, it yields the identical result.

References

Conover, W. J. (1980). **Practical nonparametric statistics** (2nd ed.). New York: John Wiley and Sons.

Daniel, W. W. (1990). **Applied nonparametric statistics** (2nd ed.). Boston: PWS–Kent Publishing Company.

Marascuilo, L. A. and McSweeney, M. (1977). **Nonparametric and distribution-free method for the social sciences.** Monterey, CA: Brooks/Cole Publishing Company.

Wilcoxon, F. (1945). Individual comparisons by ranking methods. **Biometrics**, 1, 80–83.

Wilcoxon, F. (1949). **Some rapid approximate statistical procedures.** Stamford, CT: Stamford Research Laboratories, American Cyanamid Corporation.

Endnotes

1. Some sources note that one assumption of the **Wilcoxon signed-ranks test** is that the variable being measured is based on a continuous distribution. In practice, however, this assumption is often not adhered to.

2. The **binomial sign test for a single sample** is employed with data that is in the form of a dichotomous variable (i.e., a variable represented by two categories). Each subject's score is assigned to one of the following two categories: Above the value of the hypothesized population median versus Below the value of the hypothesized population median. The test allows a researcher to compute the probability of obtaining the proportion of subjects in each of the two categories, as well as more extreme distributions with respect to the two categories.

3. The **Wilcoxon signed-ranks test** can also be employed in place of the **single-sample z test** when the value of σ is known, but the normality assumption of the latter test is saliently violated.

4. It is just coincidental in this example that the absolute value of some of the difference scores corresponds to the value of the rank assigned to that difference score.

5. The reader should take note of the fact that no critical values are recorded in **Table A5** for very small sample sizes. In the event a sample size is employed for which a critical value is not listed at a given level of significance, the null hypothesis cannot be evaluated at that level of significance. This is the case since with small sample sizes the distribution of ranks will not allow one to generate probabilities equal to or less than the specified alpha value.

6. The term $(\Sigma t^3 - \Sigma t)$ in Equation 4.4 can also be written as $\Sigma_{i=1}^{s}(t_i^3 - t_i)$. The latter notation indicates the following: a) For each set of ties, the number of ties in the set is subtracted from the cube of the number of ties in that set; and b) the sum of all the values computed in a) is obtained. Thus, in the example under discussion (in which there are $s = 3$ sets of ties):

$$\sum_{i=1}^{s} (t_i^3 - t_i) = [(2)^3 - 2] + [(2)^3 - 2] + [(3)^3 - 3] = 36$$

The above computed value of 36 is the same as the corresponding value $(\Sigma t^3 - t) = 43 - 7 = 36$ computed in Equation 4.4 through use of Table 4.4.

7. A correction for continuity can be used in conjunction with the tie correction by subtracting .5 from the absolute value computed for the numerator of Equation 4.4. Use of the correction for continuity will reduce the tie corrected absolute value of z.

8. If the **single-sample** t **test** is employed with the 13 scores listed for Example 4.2, $\Sigma X = 91$, $\bar{X} = 91/13 = 7$, and $\Sigma X^2 = 815$. Thus, $\tilde{s} = \sqrt{815 - [[(91)^2/13]/(13 - 1)]} = 3.85$, $s_{\bar{X}} = 3.85/\sqrt{13} = 1.07$, and $t = (7 - 5)/1.07 = 1.87$.

9. Even though the obtained value $t = 1.87$ is smaller than the value $t = 1.94$ obtained for Example 2.1, it is still significant at the .05 level if one employs the directional alternative hypothesis H_1: $\mu > 5$. This is the case, since for $df = 12$ the tabled critical one-tailed .05 value is $t_{.05} = 1.78$, and $t = 1.87$ exceeds the latter tabled critical value.

Test 5

The Chi-Square Goodness-of-Fit Test
(Nonparametric Test Employed with Categorical/Nominal Data)

I. Hypothesis Evaluated with Test and Relevant Background Information

Hypothesis evaluated with test In the underlying population represented by a sample are the observed cell frequencies different from the expected cell frequencies?

Relevant background information on test The **chi-square goodness-of-fit test**, also referred to as the **chi-square test for a single sample**, is employed in a hypothesis testing situation involving a single sample. Based on some preexisting characteristic or measure of performance, each of n observations (subjects/objects) that is randomly selected from a population consisting of N observations (subjects/objects) is assigned to one of k mutually exclusive categories.[1] The data are summarized in the form of a table consisting of k cells, each cell representing one of the k categories. Table 5.1 summarizes the general model for the **chi-square goodness-of-fit test**. In Table 5.1, C_i represents the ith cell/category and O_i represents the number of observations in the ith cell. The number of observations recorded in each cell of the table is referred to as the **observed frequency** of a cell.

Table 5.1 General Model for Chi-Square Goodness-of-Fit Test

							Total number of observations
Cell/Category	C_1	C_2	$\cdots$	C_i	$\cdots$	C_k	
Observed frequency	O_1	O_2	$\cdots$	O_i	$\cdots$	O_k	n

The experimental hypothesis evaluated with the **chi-square goodness-of-fit test** is whether there is a difference between the **observed frequencies** of the k cells and their **expected frequencies** (also referred to as the **theoretical frequencies**). The expected frequency of a cell is determined through the use of probability theory or is based on some preexisting empirical information about the variable under study. If the result of the **chi-square goodness-of-fit test** is significant, the researcher can conclude that in the underlying population represented by the sample there is a high likelihood that the observed frequency for at least one of the k cells is not equal to the expected frequency of the cell. It should be noted that in actuality, the test statistic for the **chi-square goodness-of-fit test** provides an approximation of a binomially distributed variable (when $k = 2$) and a multinomially distributed variable (when $k > 2$). The larger the value of n the more accurate the chi-square approximation of the **binomial** and **multinomial distributions**.[2]

The **chi-square goodness-of-fit test** is based on the following assumptions: a) Categorical/nominal data are employed in the analysis. This assumption reflects the fact that the test data should represent frequencies for k mutually exclusive categories; b) The data that are evaluated consist of a random sample of n independent observations. This assumption reflects the fact that each observation can only be represented once in the data; and c) The expected frequency of each cell is 5 or greater. When this assumption is violated, it is recommended that if $k = 2$, the **binomial sign test for a single sample (Test 6)** be employed to evaluate the data. When the expected frequency of one or more cells is less than 5 and $k > 2$, the multinomial distribution should be employed to evaluate the data. The reader should be aware of the fact that sources are not in agreement with respect to the minimum acceptable value for an expected frequency. Many sources employ criteria suggested by Cochran (1952), who stated that none of the expected frequencies should be less than 1 and that no more than 20% of the expected frequencies should be less than 5. However, many sources suggest the latter criteria may be overly conservative. In instances where a researcher believes that one or more expected cell frequencies is too small, two or more cells can be combined with one another to increase the values of the expected frequencies.

II. Examples

Two examples will be employed to illustrate the use of the **chi-square goodness-of-fit test.** Since the two examples employ identical data, they will result in the same conclusions with respect to the null hypothesis.

Example 5.1 *A die is rolled* 120 *times in order to determine whether or not it is fair (unbiased). The value* **1** *appears* 20 *times, the value* **2** *appears* 14 *times, the value* **3** *appears* 18 *times, the value* **4** *appears* 17 *times, the value* **5** *appears* 22 *times, and the value* **6** *appears* 29 *times. Do the data suggest that the die is biased?*

Example 5.2 *A librarian wishes to determine if it is equally likely that a person will take a book out of the library each of the six days of the week the library is open (assume the library is closed on Sundays). She records the number of books signed out of the library during one week and obtains the following frequencies: Monday,* 20; *Tuesday,* 14; *Wednesday,* 18; *Thursday,* 17; *Friday,* 22; *and Saturday,* 29. *Assume no person is permitted to take out more than one book during the week. Do the data indicate there is a difference with respect to the number of books taken out on different days of the week?*

III. Null versus Alternative Hypotheses

In the statement of the null and alternative hypotheses, the lower case Greek letter **omicron** (o) is employed to represent the observed frequency of a cell in the underlying population, and the lower case Greek letter **epsilon** (ε) is employed to represent the expected frequency of the cell in the population. Thus, o_i and ε_i respectively represent the observed and expected frequency of the ith cell in the underlying population. With respect to the observed and expected frequencies for the sample data, the notation O_i is employed to represent the observed frequency of a cell, and E_i the expected frequency of a cell.

Null hypothesis H_0: $o_i = \varepsilon_i$ for all cells.

(In the underlying population the sample represents, for each of the k cells, the observed frequency of a cell is equal to the expected frequency of the cell. With respect to the sample data this leads to the prediction that for all k cells $O_i = E_i$.)

Alternative hypothesis H_i: $o_i \neq \varepsilon_i$ for at least one cell.

(In the underlying population the sample represents, for at least one of the k cells the observed frequency of a cell is not equal to the expected frequency of the cell. With respect to the sample data this leads to the prediction that for at least one cell $O_i \neq E_i$. The reader should take note of the fact that the alternative hypothesis does not state that in order to reject the null hypothesis there must be a discrepancy between the observed and expected frequencies of all k cells. Rejection of the null hypothesis can be the result of a discrepancy between the observed and expected frequencies for one cell, two cells, ..., $(k - 1)$ cells, or all k cells.) As a general rule, sources always state the alternative hypothesis for the **chi-square goodness-of-fit test** nondirectionally. Although the latter protocol will be adhered to in this book, in actuality it is possible to state the alternative hypothesis directionally. The issue of the directionality of alternative hypothesis is discussed in Section VII.

IV. Test Computations

Table 5.2 summarizes the data and computations for Examples 5.1 and 5.2.

Table 5.2 Chi-square Summary Table for Examples 5.1 and 5.2

Cell	O_i	E_i	$(O_i - E_i)$	$(O_i - E_i)^2$	$\frac{(O_i - E_i)^2}{E_i}$
1/Monday	20	20	0	0	0
2/Tuesday	14	20	−6	36	1.8
3/Wednesday	18	20	−2	4	.2
4/Thursday	17	20	−3	9	.45
5/Friday	22	20	2	4	.2
6/Saturday	29	20	9	81	4.05
	$\Sigma O_i = 120$	$\Sigma E_i = 120$	$\Sigma(O_i - E_i) = 0$		$\chi^2 = 6.7$

In Table 5.2, the observed frequency of each cell (O_i) is listed in Column 2, and the expected frequency of each cell (E_i) is listed in Column 3. The computations for the **chi-square goodness-of-fit test** require that the observed and expected cell frequencies be compared with one another. In order to determine the expected frequency of a cell one must either: a) Employ the appropriate theoretical probability for the test model; or b) Employ a probability that is based on existing empirical data.

In Examples 5.1 and 5.2, computation of the expected cell frequencies is based on the theoretical probabilities for the test model.[3] Specifically, if the die employed in Example 5.1 is fair, it is equally likely that in a given trial any one of the six face values will appear. Thus, it follows that each of the six face values should occur one-sixth of the time. The probability associated with each of the possible outcomes (represented by the notation π, which is the lower case Greek letter **pi**) can be computed as follows: $\pi = r/k$ (**where**: r represents the number of outcomes that will allow an observation to be placed in a specific category, and k represents the total number of possible outcomes in any trial). Since in each trial only one face value will result in an observation being assigned to any one of the six categories, the value of the numerator for each of the six categories will equal 1. Since in each trial there are six possible outcomes, the value of the denominator for each of the six categories will equal six. Thus, for each category, $\pi_i = 1/6$.[4] Note that the sum of the k probabilities must equal 1, since if the value 1/6 is added six times it sums to 1 (i.e., $\sum_{i=1}^{k} \pi_i = 1$).

The same logic employed for Example 5.1 can be applied to Example 5.2. If it is equally likely that a person will take a book out of the library on any one of the six days of the week the library is open, it is logical to predict that on each day of the week one-sixth of the books will be taken out. Consequently, the value 1/6 will represent the expected probability for each of the six cells in Example 5.2. The expected frequency of each cell in Examples 5.1 and 5.2 is computed by multiplying the total number of observations by the probability associated with the cell. Equation 5.1 summarizes the computation of an expected frequency.

$$E_i = (n)(\pi_i) \qquad \textbf{(Equation 5.1)}$$

Where: n represents the total number of observations
π_i represents the probability that an observation will fall within the ith cell

Since in both Example 5.1 and 5.2 the total number of observations is $n = 120$, the expected frequency for each cell can be computed as follows: $E_i = (120)(1/6) = 20$.

Upon determining the expected cell frequencies, Equation 5.2 is employed to compute the test statistic for the **chi-square goodness-of-fit test.**

$$\chi^2 = \sum_{i=1}^{k} \left[\frac{(O_i - E_i)^2}{E_i} \right] \qquad \textbf{(Equation 5.2)}$$

The operations described by Equation 5.2 are as follows: a) The expected frequency of each cell is subtracted from its observed frequency. This is summarized in Column 4 of Table 5.2; b) For each cell, the difference between the observed and expected frequency is squared. This is summarized in Column 5 of Table 5.2; c) For each cell, the squared difference between the observed and expected frequency is divided by the expected frequency of the cell. This is summarized in Column 6 of Table 5.2; and d) The value of chi-square is computed by summing all of the values in Column 6. For both Examples 5.1 and 5.2, Equation 5.2 yields the value $\chi^2 = 6.7$.

Note that in Table 5.2 the sums of the observed and expected frequencies are identical. This must always be the case, and any time these sums are not equivalent it indicates that a computational error has been made.[5] It is also required that the sum of the differences between the observed and expected frequencies equals zero (i.e., $\Sigma(O_i - E_i) = 0$). Any time the latter value does not equal zero, it indicates an error has been made. Since all of the $(O_i - E_i)$ values are squared in Column 5, the sum of Column 6, which represents the value of χ^2, must always be a positive number. If a negative value is obtained for chi-square, it indicates an error has been made. The only time χ^2 will equal zero is when $O_i = E_i$ for all k cells.

V. Interpretation of the Test Results

The obtained value $\chi^2 = 6.7$ is evaluated with **Table A4 (Table of the Chi-Square Distribution)** in the **Appendix**. A general overview of the chi-square distribution and guidelines for interpreting the values in **Table A4** can be found in Sections I and V of the **single-sample chi-square test for a population variance (Test 3)**.

The degrees of freedom that are employed in evaluating the results of the **chi-square goodness-of-fit test** are computed with Equation 5.3.[6]

$$df = k - 1 \qquad \textbf{(Equation 5.3)}$$

When **Table A4** is employed to evaluate a chi-square value computed for the **chi-square goodness-of-fit test**, the following protocol is employed. The tabled critical values for the **chi-square goodness-of-fit test** are always derived from the right tail of the distribution. Thus, the tabled critical .05 chi-square value (to be designated $\chi^2_{.05}$) will be the tabled chi-square value at the 95th percentile. In the same respect, the tabled critical .01 chi-square value (to be designated $\chi^2_{.01}$) will be the tabled chi-square value at the 99th percentile. The general rule is that the tabled critical chi-square value for a given level of alpha will be the tabled chi-square value at the percentile that corresponds to the value of $(1 - \alpha)$. In order to reject the null hypothesis, the obtained value of chi-square must be equal to or greater than the tabled critical value at the prespecified level of significance. The aforementioned guidelines for determining tabled critical chi-square values are employed when the alternative hypothesis is stated nondirectionally (which as noted earlier is usually the case). The determination of tabled critical chi-square values in reference to a directional alternative hypothesis is discussed in Section VII.

Applying the guidelines for a nondirectional analysis to Examples 5.1 and 5.2, the degrees of freedom are computed to be $df = 6 - 1 = 5$. The tabled critical .05 chi-square value for $df = 5$ is $\chi^2_{.05} = 11.07$, which as noted above is the tabled chi-square value at the 95th percentile. The tabled critical .01 chi-square value for $df = 5$ is $\chi^2_{.01} = 15.09$, which as noted above is the tabled chi-square value at the 99th percentile. Since the computed value $\chi^2 = 6.7$ is less than $\chi^2_{.05} = 11.07$, the null hypothesis cannot be rejected at the .05 level. This result can be summarized as follows: $\chi^2\ (5) = 6.7, p > .05$. Although there are some deviations between the observed and expected frequencies in Table 5.2, the result of the **chi-square goodness-of-fit test** indicates there is a reasonably high likelihood that the deviations in the sample data can be attributed to chance.

A summary of the analysis of Examples 5.1 and 5.2 with the **chi-square goodness-of-fit test** follows: a) In Example 5.1 the data do not suggest that the die is biased; and b) In Example 5.2 the data do not suggest that there is any difference with respect to the number of books that are taken out of the library on different days of the week.

VI. Additional Analytical Procedures for the Chi-Square Goodness-of-Fit Test and/or Related Tests

1. Comparisons involving individual cells when $k > 2$ Within the framework of the **chi-square goodness-of-fit test** it is possible to compare individual cells with one another. To illustrate this, assume that we wish to address the following questions in reference to Examples 5.1 and 5.2.

a) In Example 5.1, is the observed frequency of 29 for the face value 6 higher than the combined observed frequency of the other five face values? Note that this is not the same thing as asking whether the face value 6 is more likely to occur when compared individually with any of the other five face values. In order to answer the latter question, the observed frequency for the face value 6 must be contrasted with the observed frequency for the specific face value in which one is interested.

b) In Example 5.2, is the observed frequency of 29 books for Saturday higher than the combined observed frequency of the other five days of the week? Note that this is not the same thing as asking whether a person is more likely to take a book out of the library on Saturday when compared individually with any one of the other five days of the week. In order to answer the latter question, the observed frequency for Saturday must be contrasted with the observed frequency of the specific day of the week in which one is interested.

In order to answer the question of whether **6/Saturday** occurs a disproportionate amount of the time, the observed frequency for **6/Saturday** must be contrasted with the combined observed frequencies of the other five face values/days of the week. In order to do this, the original six-cell chi-square table is collapsed into a two-cell table, with one cell representing **6/Saturday** (Cell 1) and the other cell representing **1, 2, 3, 4, 5/M, T, W, Th, F** (Cell 2). The expected frequency of Cell 1 remains $\pi_1 = 1/6$, since if we are dealing with a random process, there is still a one in six chance that in any trial the face value 6 will occur, or that a person will take a book out of the library on Saturday. Thus: $E_1 = (120)(1/6) = 20$. The expected frequency of Cell 2 is computed as follows: $E_2 = (120)(5/6) = 100$. Note that the probability $\pi_2 = 5/6$ for Cell 2 is the sum of the probabilities of the other five cells. In other words, if it is randomly determined what face values appears on the die or on what day of the week a person takes a book out of the library, there is a five in six chance that a face value other than 6 will appear on any role of the die, and a five in six chance that a book is taken out of the library on a day of the week other than Saturday. Table 5.3 summarizes the data for the problem under discussion.

Table 5.3 Chi-Square Summary Table When $\pi_1 = 1/6$ and $\pi_2 = 5/6$

Cell	O_i	E_i	$(O_i - E_i)$	$(O_i - E_i)^2$	$\frac{(O_i - E_i)^2}{E_i}$
6/Saturday	29	20	9	81	4.05
1,2,3,4,5/M,T,W,Th,F	91	100	–9	81	.81
	$\Sigma O_i = 120$	$\Sigma E_i = 120$	$\Sigma(O_i - E_i) = 0$		$\chi^2 = 4.86$

Since there are $k = 2$ cells, $df = 2 - 1 = 1$. Employing **Table A4** for $df = 1$, $\chi^2_{.05} = 3.84$ and $\chi^2_{.01} = 6.63$. Since the obtained value $\chi^2 = 4.86$ is larger than $\chi^2_{.05} = 3.84$, the null hypothesis can be rejected at the .05 level (i.e., $\chi^2\ (1) = 4.86, p < .05$). The null hypothesis cannot be rejected at the .01 level since $\chi^2 = 4.86 < \chi^2_{.01} = 6.63$. Note that by stating the problem in reference to one face value or one day of the week, the researcher is able to reject the null hypothesis at the .05 level. Recollect that the analysis is Section V does not allow the researcher to reject the null hypothesis.[7]

If the original null hypothesis a researcher intends to study deals with the frequency of **Cell 6/Saturday** versus the other five cells, the researcher is not obliged to defend the analysis described above. However, let us assume that the original null hypothesis under study is the one stipulated in Section III. Let us also assume that upon evaluating the data, the null hypothesis cannot be rejected. Because of this the researcher then decides to reconceptualize the problem as summarized in Table 5.3. To go even further, the researcher can extend the type of analysis depicted in Table 5.3 to all six cells (i.e., compare the observed frequency of each of the cells with the combined observed frequency of the other five cells — e.g., **Cell 1/Monday** versus **Cells 2, 3, 4, 5/T, W, Th, F, S,** as well as **Cell 2/Tuesday** versus **Cell 1, 3, 4, 5, 6/M, W, Th, F, S,** and so on for the other three cells). If $\alpha = .05$ is employed for each of the six comparisons, the overall likelihood of committing at least one Type I error within the set of six comparisons will be substantially above .05 (to be exact, it will equal $1 - (1 - .05)^6 = .26$). If within the set of six comparisons the researcher does not want more than a 5% chance of committing a Type I error, it is required that the alpha level employed for each comparison be adjusted. Specifically, by employing a probability of $.05/6 = .0083$ per comparison, the researcher will insure that the overall Type I error rate will not exceed 5%. It should be noted however, that by employing a smaller alpha level per comparison, the researcher is reducing the power associated with each comparison. A detailed discussion of the protocol for adjusting the alpha level when conducting multiple comparisons can be found in Section VI of the **single-factor between-subjects analysis of variance (Test 16)**.

It is also possible to conduct other comparisons in addition to the ones noted above. For example one can compare the observed frequencies for face values/days of the week **1, 2, 3/M, T, W** with the observed frequencies for **4, 5, 6/Th, F, S.** In such an instance there again will be two cells, with a probability of $\pi_i = 1/2$ for each cell (since $\pi_i = 3/6 = 1/2$). A researcher can also break down the original six cell table into three cells — e.g., **1, 2/M, T** versus **3, 4/W, Th** versus **5, 6/F, S.** In this instance, the probability for each cell will equal $\pi_i = 1/3$ (since $\pi_i = 2/6 = 1/3$).

Another type of comparison that can be conducted is to contrast just two of the original six cells with one another. Specifically, let us assume we want to compare **Cell 1/Monday** with **Cell 2/Tuesday**. Table 5.4 is employed to summarize the data for such a comparison.

Table 5.4 Chi-Square Summary Table for Comparison

Cell	O_i	E_i	$(O_i - E_i)$	$(O_i - E_i)^2$	$\frac{(O_i - E_i)^2}{E_i}$
1/Monday	20	17	3	9	.53
2/Tuesday	14	17	3	9	.53
	$\Sigma O_i = 34$	$\Sigma E_i = 34$	$\Sigma(O_i - E_i) = 0$		$\chi^2 = 1.06$

Note that in the above example, since we employ only two cells, the probability for each cell will be $\pi_i = 1/2$. The expected frequency of each cell is obtained by multiplying $\pi_i = 1/2$ by the total number of observations in the two cells (which equals 34). As noted previously, in conducting a comparison such as the one above, a critical issue the researcher must address is what value of alpha to employ in evaluating the null hypothesis. If $\alpha = .05$ is used, for $df = 1$, $\chi^2_{.05} = 3.84$. The null hypothesis cannot be rejected, since the obtained value $\chi^2 = 1.06 < \chi^2_{.05} = 3.84$.

A major point that has been emphasized throughout the discussion in this section is that, depending upon how one initially conceptualizes a problem, there will generally be a number of different ways in which a set of data can be analyzed. Furthermore, after analyzing the full set of data additional comparisons involving two or more categories can be conducted. The various types of comparisons that one can conduct can either be planned or unplanned. The term **planned comparison** is employed throughout the book to refer to a comparison that is planned prior to the data collection phase of a study. In contrast, an **unplanned comparison** is one that a researcher decides to conduct after the experimental data have been collected and scrutinized. Whenever possible comparisons should be planned, and most sources take the position that when a researcher plans a limited number of comparisons before the data collection phase of a study, one is not obliged to control the overall Type I error rate. However, when comparisons are not planned, most sources believe that some adjustment of the Type I error rate should be made in order to avoid inflating it excessively. As noted earlier in this section, one way of achieving the latter is to divide the maximum overall Type I error rate one is willing to tolerate by the total number of comparisons one conducts. The resulting probability value will represent the alpha level employed in evaluating each of the comparisons. A comprehensive discussion of the subject of comparisons (which is also germane to the issue of alternate ways of conceptualizing a set of data) can be found in Section VI of the **single-factor between-subjects analysis of variance.**

In closing the discussion of comparisons for the **chi-square goodness-of-fit test**, it should be noted that some sources present alternative comparison procedures that may yield results that are not in total agreement with those obtained in this section. In instances where different methodologies yield substantially different results (which will usually not be the case), a replication study evaluating the same hypothesis is in order. As noted throughout the book, replication is the most effective way to demonstrate the validity of a hypothesis. Obviously, the use of large sample sizes in both original and replication studies further increases the likelihood of obtaining reliable results. An alternative approach for conducting comparisons is presented in the next section.

2. The analysis of standardized residuals An alternative procedure for conducting comparisons (developed by Haberman (1973) and cited in sources such as Siegel and Castellan (1988)) involves the computation of **standardized residuals.** By computing the latter values, one is able to determine which cells are the major contributors to a significant chi-square value. Equation 5.4 is employed to compute a standardized residual (R_i) for each cell in a chi-square table.

$$R_i = \frac{(O_i - E_i)}{\sqrt{E_i}} \qquad \textbf{(Equation 5.4)}$$

A value computed for a residual (which is interpreted as a normally distributed variable) is evaluated with **Table A1 (Table of the Normal Distribution)** in the **Appendix.** Any residual with an absolute value that is equal to or greater than the tabled critical two-tailed .05 value $z_{.05} = 1.96$ is significant at the .05 level. Any residual with an absolute value that is equal to or greater than the tabled critical two-tailed .01 value $z_{.01} = 2.58$ is significant at the .01 level. Any cell in a chi-square table which has a significant residual makes a significant contribution to the obtained chi-square value. For any cell that has a significant residual, one can conclude that the observed frequency of the cell differs significantly from its expected frequency. The sign of the standardized residual indicates whether the observed frequency of the cell is above (+) or below (–) the expected frequency. The sum of the squared residuals for all k cells will equal the obtained value of chi-square. Although the result of the chi-square analysis for Examples 5.1 and 5.2 is not significant, the standardized residuals for the chi-square table are computed and summarized in Table 5.5.

Table 5.5 Analysis of Residuals for Examples 5.1 and 5.2

Cell	O_i	E_i	$(O_i - E_i)$	$R_i = \frac{(O_i - E_i)^2}{\sqrt{E_i}}$	$R_i^2 = \left[\frac{(O_i - E_i)}{\sqrt{E_i}}\right]^2$
1/Monday	20	20	0	0	0
2/Tuesday	14	20	–6	–1.34	1.80
3/Wednesday	18	20	–2	–.45	.20
4/Thursday	17	20	–3	–.67	.45
5/Friday	22	20	2	.45	.20
6/Saturday	29	20	9	2.01	4.05
	$\Sigma O_i = 120$	$\Sigma E_i = 120$	$\Sigma(O_i - E_i) = 0$		$\chi^2 = 1.06$

Note that the only cell with a standardized residual with an absolute value above 1.96 is **Cell 6/Saturday**. Thus, one can conclude that the observed frequency of **Cell 6/Saturday** is significantly above its expected frequency, and as such, the cell would be viewed as a major contributor in obtaining a significant chi-square value (if, in fact, the computed chi-square value had been significant). It should be noted that this result is consistent with the first comparison that was conducted in the previous section, since the latter comparison indicates that the observed frequency of 29 for **Cell 6/Saturday** deviates significantly from its expected frequency when the cell is contrasted with the combined frequencies of the other five cells.

3. Computation of a confidence interval for the chi-square goodness-of-fit test The procedure to be described in this section allows one to compute a confidence interval for the proportion of cases in the underlying population that falls within any cell in a one-dimensional chi-square table.[8] The true population proportion for a cell will be represented by the notation π_i. The procedure to be described below is a large sample approximation of the confidence interval for a binomially distributed variable (which applies to the **chi-square goodness-of-fit test** model when $k = 2$). The analysis to be described in this section will assume that if $k > 2$, the original chi-square table is converted into a table consisting of two cells.

Equation 5.5. is the general equation for computing a confidence interval for a population proportion for a specific cell, when there are $k = 2$ cells

$$\left[p_1 - z_{(\alpha/2)}\sqrt{\frac{p_1\,p_2}{n}}\right] \le \pi_1 \le \left[p_1 + z_{(\alpha/2)}\sqrt{\frac{p_1\,p_2}{n}}\right] \qquad \textbf{(Equation 5.5)}$$

Where: p_1 represents the proportion of observations in Cell 1. In the analysis under discussion, Cell 1 will represent the single cell whose observed frequency is being compared with the combined observed frequencies of the remaining five cells. The value of p_1 is computed by dividing the number of observations in Cell 1 (which will be represented by the notation x) by n (which represents the total number of observations). Thus, $p_1 = x/n$.

$p_2 = 1 - p_1$ The value p_2 represents the proportion of observations in Cell 2. In the analysis under discussion, Cell 2 will represent the combined frequencies of the other five cells. p_2 can be computed by dividing the number of observations that are not in Cell 1 by the total number of observations. Thus, $p_2 = (n - x)/n$.

$z_{\alpha/2}$ represents the tabled critical value in the normal distribution below which a proportion equal to $[1 - (\alpha/2)]$ of the cases falls. If the percentage of the distribution that falls within the confidence interval is subtracted from 1, it will equal the value of α.

If one wants to determine the 95% confidence interval, the tabled critical two-tailed .05 value $z_{.05} = 1.96$ is employed in Equation 5.5. The tabled critical two-tailed .01 value $z_{.01} = 2.58$ is employed to compute the 99% confidence interval. The value $\sqrt{(p_1 p_2)/n}$ in Equation 5.5 represents the estimated standard error of the population proportion. The latter value is an estimated standard deviation of a sampling distribution of a proportion.

If (as is done in Table 5.3) the data for Examples 5.1 and 5.2 are expressed in a format consisting of two cells, Equation 5.5 can be employed to compute a confidence interval for each of the six cells. Thus, if we wish to compute a confidence interval for the **Cell 6/Saturday** we can determine that $p_1 = x/n = 29/120 = .242$ and $p_2 = (n-x)/n = (120 - 29)/120 = .758$. Substituting the latter values and the value $z_{.05} = 1.96$ in Equation 5.5, the 95% confidence interval is computed below.

$$.242 - (1.96)\sqrt{\frac{(.242)(.758)}{120}} \le \pi_1 \le .242 + (1.96)\sqrt{\frac{(.242)(.758)}{120}}$$

$$\pi_1 = .242 \pm .077$$

$$.165 \le \pi_1 \le .319$$

Thus, the researcher can be 95% confident that the true proportion of cases in the underlying population that falls in **Cell 6/Saturday** is a value between .165 and .319.

The 99% confidence interval, which has a larger range, is computed below by employing $z_{.01} = 2.58$ in Equation 5.5.

$$.242 - (2.58)\sqrt{\frac{(.242)(.758)}{120}} \leq \pi_1 \leq .242 + (2.58)\sqrt{\frac{(.242)(.758)}{120}}$$

$$\pi_1 = .242 \pm .101$$

$$.141 \leq \pi_1 \leq .343$$

Thus, the researcher can be 99% confident that the true proportion of cases in the underlying population that falls in **Cell 6/Saturday** is a value between .141 and .343.

The above described procedure can be repeated for the other five cells in Examples 5.1 and 5.2. In each instance the observed frequency of a cell is evaluated in relation to the combined observed frequencies of the remaining five cells.[9]

4. Brief discussion of the z test for a population proportion and the single-sample test for the median In Section I it is noted that when $k = 2$ and the value of n is large, the **chi-square goodness-of-fit test** provides a good approximation of the binomial distribution. Under the discussion of the **binomial sign test for a single sample** two tests are described which yield equivalent results to those obtained with the **chi-square goodness-of-fit test** when $k = 2$. The two tests are the **z test for a population proportion (Test 6a)** and the **single-sample test for the median (Test 6b)**. In the latter test, the two cells of the chi-square table are comprised of scores that fall above the median of some distribution and scores that fall below the median of the distribution. For a full discussion of these tests the reader should consult the discussion of the **binomial sign test for a single sample.**

5. The correction for continuity for the chi-square goodness-of-fit test Although it is not generally discussed in reference to the **chi-square goodness-of-fit test**, a **correction for continuity** (which is discussed under the **Wilcoxon signed-ranks test (Test 4)**) can be applied to Equation 5.2. The basis for employing the correction for continuity with the **chi-square goodness-fit-test** is that the test employs a continuous distribution to approximate a discrete distribution (specifically, the binomial or multinomial distributions). The correction for continuity is based on the premise that if a continuous distribution is employed to estimate a discrete distribution, such an approximation will inflate the Type I error rate. By employing the correction for continuity the Type I error rate is ostensibly adjusted to be more compatible with the prespecified alpha value designated by the researcher. Equation 5.6 is the continuity corrected chi-square equation for the **chi-square goodness-of-fit test.**

$$\chi^2 = \sum_{i=1}^{k}\left[\frac{(|O_i - E_i| - .5)^2}{E_i}\right] \qquad \textbf{(Equation 5.6)}$$

Note that by subtracting .5 from the absolute value of the difference between each set of observed and expected frequencies, the chi-square value derived with Equation 5.6 will be lower than the value computed with Equation 5.2. The magnitude of the correction for continuity will be inversely related to the size of the sample. The use of the correction for continuity with the **chi-square goodness-of-fit test** is generally limited to designs in which there are $k = 2$ cells. This latter application of the correction is discussed under the **z test for a population proportion (Test 6a)**. The use of the correction for continuity with other designs that employ the chi-square statistic is discussed under the **chi-square test for $r \times c$ tables (Test 11).**

6. Sources for computing the power of the chi-square goodness-of-fit test Although the protocol for computing the power of the **chi-square goodness-of-fit test** will not be described in this book, a comprehensive discussion of the subject as well as the necessary tables can be found in Cohen (1988).

VII. Additional Discussion of the Chi-Square Goodness-of-Fit Test

1. Directionality of the chi-square goodness-of-fit test In Section III it is noted that most sources state the alternative hypothesis for the **chi-square goodness-of-fit test** nondirectionally, but that in actuality it is possible to state the alternative hypothesis directionally. This is most obvious when there are $k = 2$ cells and the expected probability associated with each cell is $\pi_i = 1/2$. Under the latter conditions a researcher can make two directional predictions, either one of which can represent the alternative hypothesis. Specifically, the following can be predicted with respect to the sample data: a) The observed frequency of Cell 1 will be significantly higher than the observed frequency of Cell 2 (which translates into the observed frequency of Cell 1 being higher than its expected frequency, and the observed frequency of Cell 2 being lower than its expected frequency); and b) The observed frequency of Cell 2 will be significantly higher than the observed frequency of Cell 1 (which translates into the observed frequency of Cell 2 being higher than its expected frequency, and the observed frequency of Cell 1 being lower than its expected frequency).

If a researcher wants to evaluate either of the aforementioned directional alternative hypotheses at the .05 level, the appropriate critical value to employ is the tabled chi-square value (for $df = 1$) at the .10 level of significance. The latter value is represented by the tabled chi-square value at the 90th percentile (which demarcates the extreme 10% in the right tail of the chi-square distribution). This latter critical value will be designated as $\chi^2_{.10}$ throughout this discussion. The rationale for employing $\chi^2_{.10}$ in evaluating the directional alternative hypothesis at the .05 level is as follows. When $k = 2$ and the alternative hypothesis is stated nondirectionally, if a computed chi-square value is equal to or greater than $\chi^2_{.05}$ (for $df = 1$) the researcher can reject the null hypothesis if the data are consistent with either of the outcomes associated with the two possible directional alternative hypotheses. If that same tabled critical chi-square value is employed to evaluate one of the two possible directional alternative hypotheses, a directional alternative hypothesis would be evaluated not at the .05 level but at one-half that value — in other words at the .025 level. Thus, if one wants to employ $\alpha = .05$ and states the alternative hypothesis directionally, the alpha level for the directional alternative hypothesis should be .05 multiplied by the number of possible directional alternative hypotheses (which in this instance equals 2). By employing the tabled critical value for $\chi^2_{.10}$, which is a lower value than $\chi^2_{.05}$, an alpha level (and Type I error rate) of .05 is established for the specific one-tailed alternative hypothesis that one employs. The area of the chi-square distribution that corresponds to the alpha level for the latter directional alternative hypothesis will be one-half of the area that comprises the extreme 10% of the right tail of the distribution. The area that corresponds to the alpha level for the other directional alternative hypothesis will be the remaining 5% of the extreme 10% in the right tail of the distribution.

If we turn our attention to Examples 5.1 and 5.2, in both examples there are, in fact, 720 possible directional predictions the researcher can make![10] The latter value is determined as follows: $k! = 6! = (6)(5)(4)(3)(2)(1) = 720$.[11] In other words, a researcher can predict any one of 720 ordinal configurations with respect to the observed frequencies of the six cells. As an example, the researcher might predict that **1/Monday** will have the highest observed frequency, followed in order by **2/Tuesday**, **3/Wednesday**, **4/Thursday**, **5/Friday**, and **6/Saturday**, and only be willing to reject the null hypothesis if the data are consistent with this specific ordering of the observed cell frequencies.

Later in this section it will be explained that when $k = 6$ it is not possible to evaluate any one of the 720 directional alternative hypotheses at either the .05 or .01 level of significance. Indeed, under these conditions the highest value of alpha that can be employed to evaluate a directional alternative hypothesis is approximately .001. In point of fact, when $k = 6$, the tabled critical $\chi^2_{.001}$ value is approximately 2.7, which happens to be the tabled chi-square value at the 28th percentile. This is the case, since if the prespecified alpha value that is employed to evaluate a nondirectional alternative hypothesis (which in this case we will assume is .001, since it is the highest value that will work with $k = 6$) is multiplied by the number of possible directional alternative hypotheses (720) we obtain: $(.001)(720) = .72$. The value .72 demarcates the extreme 72% in the right tail of the chi-square distribution when $df = 5$. Thus, it corresponds to the 28th percentile of the distribution, since $(1 - .72) = .28$. Consequently, in order to reject the null hypothesis with reference to one of the 720 possible directional alternative hypotheses, both of the following conditions will have to be met: a) The obtained value of chi-square will have to be equal to or greater than the tabled chi-square value at the 28th percentile; and b) The data will have to be consistent with the directional alternative hypothesis that is employed. In other words, the ordinal relationship between the observed frequencies of the six cells should be in the exact order stated in the directional alternative hypothesis.

As noted previously, none of the 720 possible directional alternative hypotheses can be evaluated at either the .05 or .01 levels. This is the case since if either .05 or .01 is multiplied by 720, the resulting product exceeds unity (1) — i.e., $(.05)(720) = 36$ and $(.01)(720) = 7.2$. Since both 36 and 7.2 exceed 1, they cannot be used as probability values. Thus, it is impossible to evaluate any of the directional alternative hypotheses at either the .05 or .01 level. In fact, the largest alpha level at which any of the directional alternative hypotheses can be evaluated is .001388, since $1/720 = .001388$ (i.e., the value .001388 is the maximum number which when multiplied by 720 falls short of 1. Specifically, $(.001388)(720) = .99936$).

On initial inspection it might appear that by employing $\chi^2_{.72}$ to evaluate one of the directional alternative hypotheses, a researcher is employing an inflated alpha level.[12] But as just noted, in this instance the alpha level for a directional alternative hypothesis is, in fact, .001. Of course a researcher may elect to employ a larger critical chi-square value, and if one elects to do so, the actual alpha level for a directional alternative hypothesis will be even lower than .001. For example, if the tabled critical value $\chi^2_{.05} = 11.07$ (which is the tabled value for $df = 5$ at the 95th percentile that is employed in evaluating a nondirectional alternative hypothesis) is employed to evaluate one of the 720 possible directional alternative hypotheses, the actual alpha level that one will be using in evaluating the directional alternative hypothesis will be $.05/720 = .00007$. In such a case there is obviously a minuscule likelihood of committing a Type I error in reference to the directional alternative hypothesis. Yet at the same time, the power of the analysis with respect to that alternative hypothesis will be minimal (thus resulting in a high likelihood of committing a Type II error).

It is also possible in Examples 5.1 and 5.2 to state an alternative hypothesis that predicts two or more, but less than 720, of the possible ordinal configurations with respect to the observed frequencies of the $k = 6$ cells. In other words, a directional alternative hypothesis might state that the null hypothesis can only be rejected if the magnitude of the observed cell frequencies in descending order is either **Cell 1, Cell 2, Cell 3, Cell 4, Cell 5, Cell 6** or **Cell 2, Cell 1, Cell 3, Cell 4, Cell 5, Cell 6**. In such a case, to evaluate the null hypothesis at the .001 level with respect to an alternative hypothesis involving two of the 720 possible ordinal configurations, the tabled critical chi-square value at the 64th percentile is employed (i.e., $\chi^2_{.36}$ since the area of the distribution involved is the extreme 36% $((1 - .64) = .36)$ that falls in the right tail). The general procedure to use for computing the percentile rank in the chi-square distribution to employ in determining the critical value when evaluating one or more configurations is as follows: a) Divide the total number of possible configurations by the number of acceptable configurations stated in the directional alternative hypothesis; b) Multiply the result of the division by the prespecified alpha level; and c) Subtract the value obtained in part b) from 1.

Applying this protocol to a directional alternative hypothesis in which only 2 out of 720 configurations are acceptable, we derive: a) $720/2 = 360$; b) $(360)(.001) = .36$; and c) $(1 - .36) = .64$. The resulting value of .64 (which can be converted to 64%) corresponds to the percentile rank in the chi-square distribution to employ in determining the critical value. The value .36 obtained in part b) represents the overall proportion of the right tail of the distribution which contains a proportion of the distribution equivalent to .001 that represents the rejection zone for the directional alternative hypothesis under study.

It should be noted that since $2/720 = .0028$, an alternative hypothesis involving 2 out of 720 possible configurations can be evaluated at a level above .001. In point of fact, such an alternative hypothesis can be evaluated at any level equal to or less than .0028. Thus if one elects to employ $\alpha = .002$, using the protocol described in the previous paragraph, $(720/2)(.002) = .72$, and $(1 - .72) = .28$. The latter result indicates that $\chi^2_{.72} = 2.7$ is once again employed as the critical value. This is the case, since the latter value represents the tabled value at 28th percentile for $df = 5$.

In closing this discussion, the reader should take note of the fact that all of the critical values for the **chi-square goodness-of-fit test** are derived from the right tail of the chi-square distribution. In point of fact, with the exception of the **single-sample chi-square test for a population variance** (in which case critical values are derived from both tails of the distribution), all of the tests in the book that employ the chi-square distribution only use critical values from the right tail of the distribution.

2. Modification of procedure for computing the degrees of freedom for the chi-square goodness-of-fit test There are some instances when Equation 5.3 (i.e., $df = k - 1$) is not appropriate for computing the degrees of freedom for the **chi-square goodness-of-fit test**. When one wishes to determine whether a distribution of sample data conforms to a specific theoretical distribution (such as the normal, binomial, or Poisson distributions), it may be necessary to estimate one or more population parameters prior to computing the expected frequency of each cell. In such a case, the equation for computing degrees of freedom is: $df = k - 1 - w$, where w represents the number of parameters that must be estimated.[13]

As an example, if one employs the **chi-square goodness-of-fit test** to determine whether the distribution of scores in a set of data are compatible with the normal distribution, in order to do this one must estimate the values of the population mean (μ) and standard deviation (σ) by computing the values $\bar{X}$ and $\tilde{s}$ from the sample data. Since in this instance two population parameters are estimated, the appropriate degrees of freedom to employ for the **chi-square goodness-of-fit test** are: $df = k - 1 - 2$ (which requires there be at least four cells, since df must be equal to or greater than one). The value of k will depend on how many cells are used for categorizing the n scores in sample data. Each of the cells (often referred to as **class intervals** in this context) will contain a limited range of scores on the variable being evaluated. For each cell, the proportion of cases that are expected to fall within that class interval (i.e., the expected frequency) is computed through use of **Table A1** in the **Appendix**. The observed frequency of each cell is the actual number of scores in the sample data that falls in that interval. The population parameters μ and σ have to be estimated in order to establish the cells that define the chi-square summary table, as well as to assign scores to specific cells. Among those sources that describe this special case of the **chi-square goodness-of-fit test** are Daniel (1990) and Siegel and Castellan (1988).

3. **Additional goodness-of-fit tests** In addition to the **chi-square goodness-of-fit test,** there are a number of other tests that have been developed for determining whether sample data are likely to have been derived from a population in which the distribution of scores is consistent with a specific theoretical probability distribution. Most of these tests evaluate scores that are assigned to ordered categories.[14] Among the goodness-of-fit tests that have been developed for evaluating ordered categorical data are the **Kolmogorov–Smirnov goodness-of-fit test** (Kolmogorov (1933)), the **Lilliefors test** (Lilliefors (1967)), **David's empty cell test** (David (1950)), and the **Cramér–von Mises goodness-of-fit test** (attributed to Cramér (1928), von Mises (1931), and Smirnov (1936)). Some or all of the aforementioned tests are described in books that specialize in nonparametric statistics. The general subject of goodness-of-fit tests for randomness is discussed in Section VII of the **single-sample runs tests (Test 7)**, and one specific procedure referred to as **autocorrelation** is discussed in detail in Section VII of the **Pearson product–moment correlation coefficient (Test 22)**.

VIII. Additional Examples Illustrating the Use of the Chi-Square Goodness-of-Fit Test

Three additional examples that can be evaluated with the **chi-square goodness-of-fit test** are presented in this section. Example 5.3 employs the same data set as Examples 5.1 and 5.2, and thus yields the same results. Examples 5.4 and 5.5 illustrate the application of the **chi-square goodness-of-fit test** to data in which the expected frequencies are based on existing empirical information or theoretical conjecture rather than on expected theoretical probabilities.

Example 5.3 *The owner of the Big Wheel Speedway, a stock car racetrack, asks a researcher to determine whether or not there is any bias associated with the lane to which a car is assigned at the beginning of a race. Specifically, the owner wishes to determine if there is an equal likelihood of winning a race associated with each of the six lanes of the track. The researcher examines the results of* 120 *races and determines the following number of first place finishes for the six lanes: Lane* 1 – 20; *Lane* 2 – 14; *Lane* 3 – 18; *Lane* 4 – 17; *Lane* 5 – 22; *Lane* 6 – 29.

Example 5.4 *A country in which four ethnic groups make up the population establishes affirmative action guidelines for medical school admissions. The country has one medical school, and it is mandated that each new class of medical students proportionally represents the four ethnic groups that comprise the country's population. The four ethnic groups that make up the population and the proportion of people in each ethnic group are: Balzacs* (.4), *Crosacs* (.25), *Murads* (.3), *and Isads* (.05).[15] *The number of students from each ethnic group admitted into the medical school class for the new year are: Balzacs* (300), *Crosacs* (220), *Murads* (400), *and Isads* (80). *Is there a significant discrepancy between the proportions mandated in the affirmative action guidelines and the actual proportion of the four ethnic groups in the new medical school class?*

Except for the fact that empirical data are used as a basis for determining the expected cell frequencies, this example is solved in the same manner as Examples 5.1 and 5.2. There are $k = 4$ cells — each cell representing one of the four mutually exclusive ethnic groups. The observed frequencies are the number of students from each of the four ethnic groups out of the total of 1000 students who are admitted to the medical school (i.e., $n = 300 + 220 + 400 + 80 = 1000$). The expected frequencies are computed based upon the proportion of each ethnic group in the population. Each of these values is obtained by multiplying the total number of medical school admissions (1000) by the proportion of a specific ethnic group in the population. Thus in the case of the **Balzacs**, employing Equation 5.1 the expected frequency is computed as follows: $E_1 = (1000)(.4) = 400$. In the same respect, the expected frequencies for the other three ethnic groups are: **Crosacs:** $E_2 = (1000)(.25) = 250$; **Murads:** $E_3 = (1000)(.3) = 300$; and **Isads:** $E_4 = (1000)(.05) = 50$. Table 5.6 summarizes the observed and expected frequencies and the resulting values employed in the computation of the chi-square value through use of Equation 5.2.

Table 5.6 Chi-Square Summary Table for Example 5.4

Cell	O_i	E_i	$(O_i - E_i)$	$(O_i - E_i)^2$	$\frac{(O_i - E_i)^2}{E_i}$
Balzacs	300	400	–100	10000	25
Crosacs	220	250	–30	900	3.6
Murads	400	300	100	10000	33.3
Isads	80	50	30	900	18
	$\Sigma O_i = 1000$	$\Sigma E_i = 1000$	$\Sigma(O_i - E_i) = 0$		$\chi^2 = 79.9$

Employing **Table A4**, we determine that for $df = 4 - 1 = 3$ the tabled critical .05 and .01 values are $\chi^2_{.05} = 7.81$ and $\chi^2_{.01} = 11.34$. (A nondirectional alternative hypothesis is assumed.) Since the computed value $\chi^2 = 79.9$ is greater than both of the aforementioned critical values, the null hypothesis can be rejected at both the .05 or .01 levels. Based on the chi-square analysis it can be concluded that the medical school admissions data do not adhere to the proportions mandated in the affirmative action guidelines. Inspection of Table 5.5 suggests that the significant difference is primarily due to the presence of too many **Murads** and **Isads** and too few **Balzacs**. This observation can be confirmed by employing the appropriate comparison procedure described in Section VI .

Example 5.5 *A physician who specializes in genetic diseases develops a theory which predicts that two-thirds of the people who develop a disease called cyclomeiosis will be males. She randomly selects* 300 *people who are afflicted with cyclomeiosis and observes that* 140 *of them are females. Is the physician's theory supported?*

In Example 5.5 there are two cells, one representing each gender. Since the expected frequencies are computed on the basis of the probabilities hypothesized in the physician's theory, two-thirds of the sample are expected to be males and the remaining one-third females. Thus, the respective expected frequencies for males and females are determined as follows: **Males:** $E_1 = (300)(2/3) = 200$; **Females:** $E_2 = (300)(1/3) = 100$. Since 140 females are observed with the disease, the remaining 160 people who have the disease must be males. Table 5.7 summarizes the observed and expected frequencies and the resulting values that are employed in the computation of the chi-square value with Equation 5.2.

Table 5.7 Chi-Square Summary Table for Example 5.5

Cell	O_i	E_i	$(O_i - E_i)$	$(O_i - E_i)^2$	$\frac{(O_i - E_i)^2}{E_i}$
Males	160	200	–40	1600	8
Females	140	100	40	1600	16
	$\Sigma O_i = 300$	$\Sigma E_i = 300$	$\Sigma(O_i - E_i) = 0$		$\chi^2 = 24$

Employing **Table A4**, we determine that for $df = 2 - 1 = 1$ the tabled critical .05 and .01 values are $\chi^2_{.05} = 3.84$ and $\chi^2_{.01} = 6.63$.[16] Since the computed value $\chi^2 = 24$ is greater than both of the aforementioned critical values, the null hypothesis can be rejected at both the .05 or .01 levels. Based on the chi-square analysis, it can be concluded that the observed distribution of males and females for the disease is not consistent with the doctor's theory.

References

Cochran, W. G. (1952). The chi-square goodness-of-fit test. **Annals of Mathematical Statistics**, 23, 315–345.

Cohen, J. (1988). **Statistical power analysis for the behavioral sciences** (2nd ed.). Hillsdale, NJ: Lawrence Erlbaum Associates, Publishers.

Cramér, H. (1928). On the composition of elementary errors. **Skandinavisk Aktaurietidskrift**, 11, 13–74, 141–180.

Daniel, W. W. (1990). **Applied nonparametric statistics** (2nd ed.). Boston: PWS–Kent Publishing Company.

David, F. N. (1950). Two combinatorial tests of whether a sample has come from a given population. **Biometrika**, 37, 97–110.

Haberman, S. J. (1973). The analysis of residuals in cross-classified tables. **Biometrics**, 29, 205–220.

Howell, D. C. (1992). **Statistical methods for psychology** (3rd ed.). Boston: PWS–Kent Publishing Company.

Keppel, G. and Saufley, W. H., Jr. (1992). **Introduction to design and analysis: A student's handbook** (2nd ed.). New York: W. H. Freeman and Company.

Kolmogorov, A. N. (1933). Sulla determinazione empirica di una legge di distribuzione. **Giorn dell'Inst. Ital. degli. Att.**, 4, 89–91.

Lilliefors, H. W. (1967). On the Kolmogorov–Smirnov test for normality with mean and variance unknown. **Journal of the American Statistical Association**, 62, 399–402.

Marascuilo, L. A. and McSweeney, M. (1977). **Nonparametric and distribution-free methods for the social sciences.** Monterey, CA: Brooks/Cole Publishing Company.

Siegel, S. and Castellan, N. J., Jr. (1988). **Nonparametric statistics for the behavioral sciences** (2nd ed.). New York: McGraw–Hill Book Company.

Smirnov, N. V. (1936). Sur la distribution de W^2 (criterium de M. R. v. Mises). **Comptes Rendus** (Paris), 202, 449–452.

von Mises, R. (1931). **Wahrscheinlichkeitsrechnung und ihre anwendung in derstatistik and theoretishen Physik.** Leipzig: Deuticke.

Endnotes

1. Categories are **mutually exclusive** if assignment to one of the k categories precludes a subject/object from being assigned to any one of the remaining $(k - 1)$ categories.

2. The reason why the exact probabilities associated with the binomial and multinomial distributions are generally not computed is because, except when the value of n is very small, the computations become prohibitive.

3. Example 5.4 in Section VIII illustrates an example in which the expected frequencies are based on prior empirical information.

4. It is possible for the value of the numerator of a probability ratio to be some value other than 1. For instance, if one is evaluating the number of odd versus even numbers that

appear on n rolls of a die, in each trial there are $k = 2$ categories. Three face values (1, 3, 5) will result in an observation being categorized as an odd number, and three face values (2, 4, 6) will result in an observation being categorized as an even number. Thus, the probability associated with each of the two categories will be 3/6 = 1/2. It is also possible for each of the categories to have different probabilities. Thus, if one is evaluating the relative occurrence of the face values 1 and 2 versus the face values 3, 4, 5, and 6, the probability associated with the former category will be 2/6 = 1/3 (since two outcomes fall within the category 1/2), while the probability associated with the latter will be 4/6 = 2/3 (since four outcomes fall within the category 3/4/5/6). Examples 5.4 and 5.5 in Section VIII illustrate examples where the probabilities for two or more categories are not equal to one another.

5. When decimal values are involved there may be a minimal difference between the sums of the expected and observed frequencies due to a rounding off error.

6. There are some instances when Equation 5.3 should be modified to compute the degrees of freedom for the **chi-square goodness-of-fit test**. The modified degrees of freedom equation is discussed in Section VII.

7. Sometimes when one or more cells in a set of data have an expected frequency of less than five, by combining cells (as is done in this analysis) a researcher can reconfigure the data so that the expected frequency of all the resulting cells is greater than five. Although this is one way of dealing with the violation of the assumption concerning the minimum acceptable value for an expected cell frequency, the null hypothesis evaluated with the reconfigured data will not be identical to the null hypothesis stipulated in Section III.

8. In a one-dimensional chi-square table, subjects/objects are assigned to categories which reflect their status on a single variable. In a two-dimensional table, two variables are involved in the categorization of subjects/objects. As an example, if each of n subjects is assigned to a category based on one's gender and whether one is married or single, a two-dimensional table can be constructed involving the following four cells: **Male-Married; Female-Married; Male-Not married; Female-Not married.** Note that people assigned to a given cell fall into one of the two categories on each of the two dimensions/variables (which are gender and marital status). Analysis of two-dimensional tables is discussed under the **chi-square test for $r \times c$ tables.**

9. Daniel (1990) notes that the procedure described in this section will only yield reliable results when $n\pi_1$ and $n(1 - \pi_1)$ are both greater than 5. It is assumed that the researcher estimates the value of π_1 prior to collecting the data. The researcher bases the latter value either on probability theory or preexisting empirical information. Generally speaking, if the value of n is large, the value of p_1 should provide a reasonable approximation of the value of π_1 for calculating the values $n\pi_1$ and $n(1 - \pi_1)$.

10. Since when $k = 2$ it is possible to state two directional alternative hypotheses, some sources refer to an analysis of such a nondirectional alternative hypothesis as a two-tailed test. Using the same logic, when $k > 2$ one can conceptualize a test of a nondirectional alternative hypothesis as a **multi-tailed test** (since when $k > 2$ it is possible to have more than two directional alternative hypotheses). It should be pointed out that since the **chi-square goodness-of-fit test** only utilizes the right tail of the distribution, it is questionable to use the terms two-tailed or multi-tailed in reference to the analysis of a nondirectional alternative hypothesis.

11. $k!$ is referred to as k **factorial**. The notation indicates that the integer number preceding the ! is multiplied by all integer values below it. Thus, $k! = (k)(k - 1) \ldots (1)$.

12. The subscript .72 in the notation $\chi^2_{.72}$ represents the .72 level of significance. The value .72 is based on the fact that the extreme 72% of the right tail of the chi-square distribution is employed in evaluating the directional alternative hypothesis. The value $\chi^2_{.72}$ falls at the 28th percentile of the distribution.

13. In actuality $df = k - 1 - w$ is the generic equation for computing the degrees of freedom for the **chi-square goodness-of-fit test**. The equation $df = k - 1$, which is used for the examples that are discussed to illustrate the test, represents the form the equation $df = k - 1 - w$ assumes when $w = 0$.

14. When categories are ordered there is a direct (or inverse) relationship between the magnitude of the score of a subject on the variable being measured and the ordinal position of the category to which that score has been assigned. An example of ordered categories which can be employed with the **chi-square goodness-of-fit test** are the following four categories which can be used to indicate the magnitude of a person's IQ: **Cell 1 – 1st quartile; Cell 2 – 2nd quartile; Cell 3 – 3rd quartile;** and **Cell 4 – 4th quartile.** The aforementioned categories can be employed if one wants to determine whether within a sample an equal number of subjects are observed in each of the four quartiles. Note that in Examples 5.1 and 5.2, the fact that an observation is assigned to Cell 6 is not indicative of a higher level of performance or superior quality than an observation assigned to Cell 1 (or vice versa). However, in the IQ example, there is a direct relationship between the number used to identify each cell and the magnitude of IQ scores for subjects who have been assigned to that cell.

15. Note that the sum of the proportions must equal 1.

16. Even though a nondirectional alternative hypothesis will be assumed, this example illustrates a case in which some researchers might view it more prudent to employ a directional alternative hypothesis.

Test 6

The Binomial Sign Test for a Single Sample
(Nonparametric Test Employed with Categorical/Nominal Data)

I. Hypothesis Evaluated with Test and Relevant Backward Information

Hypothesis evaluated with test In an underlying population comprised of two categories that is represented by a sample, is the proportion of observations in one of the two categories equal to a specific value?

Relevant background information on test The **binomial sign test for a single sample** is based on the **binomial distribution**, which assumes that each of n independent observations is randomly selected from a population, and that each observation can be classified in one of $k = 2$ mutually exclusive categories/cells.[1] Within the population the likelihood that an observation will fall in Category 1 will equal π_1, and the likelihood that an observation will fall in Category 2 will equal π_2. Since it is required that $\pi_1 + \pi_2 = 1$, it follows that $\pi_2 = 1 - \pi_1$. The mean (μ, which is also referred to as the **expected value**) and standard deviation (σ) of a binomially distributed variable are computed with Equations 6.1 and 6.2.[2]

$$\mu = n\pi_1 \qquad \text{(Equation 6.1)}$$

$$\sigma = \sqrt{n\pi_1\pi_2} \qquad \text{(Equation 6.2)}$$

When $\pi_1 = \pi_2 = .5$, the binomial distribution is symmetrical. When $\pi_1 < .5$, the distribution is positively skewed, with the degree of positive skew increasing as the value of π_1 approaches zero. When $\pi_1 > .5$, the distribution is negatively skewed, with the degree of negative skew increasing as the value of π_1 approaches one. The sampling distribution of a binomially distributed variable can be approximated by the normal distribution. The closer the value of π_1 is to .5 and the larger the value of n, the better the normal approximation. Because of the **central limit theorem** (which is discussed in Section VII of the **single-sample *z* test (Test 1)**), even if the value of n is small and/or the value of π_1 is close to either 0 or 1, the normal distribution still provides a reasonably good approximation of a sampling distribution for a binomially distributed variable.

The **binomial sign test for a single sample** employs the binomial distribution to determine the likelihood that x or more (or x or less) of n observations that comprise a sample will fall in one of two categories (to be designated as Category 1), if in the underlying population the true proportion of observations in Category 1 equals π_1. When there are $k = 2$ categories, the hypothesis evaluated with the **binomial sign test for a single sample** is identical to that evaluated with the **chi-square goodness-of-fit test (Test 5)**. Since the two tests evaluate the same hypothesis, the hypothesis for the **binomial sign test for a single sample** can also be stated as follows: In the underlying population represented by a sample are the observed frequencies for the two categories different from their expected frequencies? As noted in Section I of the **chi-square goodness-of-fit test**, the **binomial sign test for a single sample** is generally employed for small sample sizes since, when the value of n is large, the computation of exact binomial probabilities becomes prohibitive without access to specialized tables or the appropriate computer software.

II. Examples

Two examples will be employed to illustrate the use of the **binomial sign test for a single sample.** Since both examples employ identical data, they will result in the same conclusions with respect to the null hypothesis.

Example 6.1 *An experiment is conducted to determine whether a coin is biased. The coin is flipped ten times resulting in eight heads and two tails. Do the results indicate that the coin is biased?*

Example 6.2 *Ten women are asked to judge which of two brands of perfume has a more fragrant odor. Eight of the women select* Perfume A *and two of the women select* Perfume B. *Is there a significant difference with respect to preference for the perfumes?*

III. Null versus Alternative Hypotheses

Null hypothesis $H_0\colon \pi_1 = .5$

(In the underlying population the sample represents, the true proportion of observations in Category 1 equals .5.)

Alternative hypothesis $H_1\colon \pi_1 \neq .5$

(In the underlying population the sample represents, the true proportion of observations in Category 1 is not equal to .5. This is a **nondirectional alternative hypothesis,** and it is evaluated with a **two-tailed test.** In order to be supported, the observed proportion of observations in Category 1 in the sample data (which will be represented with the notation p_1) can be either significantly larger than the hypothesized population proportion $\pi_1 = .5$ or significantly smaller than $\pi_1 = .5$.)[3]

or

$H_1\colon \pi_1 > .5$

(In the underlying population the sample represents, the true proportion of observations in Category 1 is greater than .5. This is a **directional alternative hypothesis,** and it is evaluated with a **one-tailed test.** In order to be supported, the observed proportion of observations in Category 1 in the sample data must be significantly larger than the hypothesized population proportion $\pi_1 = .5$.)

or

$H_1\colon \pi_1 < .5$

(In the underlying population the sample represents, the true proportion of observations in Category 1 is less than .5. This is a **directional alternative hypothesis,** and it is evaluated with a **one-tailed test.** In order to be supported, the observed proportion of observations in Category 1 in the sample data must be significantly smaller than the hypothesized population proportion $\pi_1 = .5$.)

Note: Only one of the above noted alternative hypotheses is employed. If the alternative hypothesis the researcher selects is supported, the null hypothesis is rejected.

IV. Test Computations

In Example 6.1 the null and alternative hypotheses reflect the fact that it is assumed that if one is employing a fair coin, the probability of obtaining a **Heads** in any trial will equal .5 (which is equivalent to 1/2). Thus, the expected/theoretical probability for **Heads** is represented by $\pi_1 = .5$. If the coin is fair, the probability of obtaining a **Tails** in any trial will also equal .5, and thus the expected/theoretical probability for **Tails** is represented by $\pi_2 = .5$. Note that $\pi_1 + \pi_2 = 1$. In Example 6.2, it is assumed that if there is no difference with regard to preference for the two brands of perfume, the likelihood of a woman selecting **Perfume A** will equal $\pi_1 = .5$ and the likelihood of her selecting **Perfume B** will equal $\pi_2 = .5$. In both Examples 6.1 and 6.2 the question that is being asked is as follows: If $n = 10$ and $\pi_1 = \pi_2 = .5$, what is the probability of 8 or more observations in one of the two categories?[4] Table 6.1 summarizes the outcome of Examples 6.1 and 6.2. The notation x is employed to represent the number of observations in Category 1 and the notation $(n - x)$ is employed to represent the number of observations in Category 2.

Table 6.1 Model for Binomial Sign Test for a Single Sample for Examples 6.1 and 6.2

Category		
1 (Heads/Perfume A)	**2 (Tails/Perfume B)**	**Total**
$x = 8$	$n - x = 10 - 8 = 2$	$n = 10$

In Examples 6.1 and 6.2, the proportion of observations in **Category/Cell 1** is $p_1 = 8/10 = .8$, and the proportion of observations in **Category/Cell 2** is $p_2 = 2/10 = .2$. Equation 6.3 can be employed to compute the probability that exactly x out of a total of n observations will fall in one of the two categories.

$$P(x) = \binom{n}{x} (\pi_1)^x (\pi_2)^{(n - x)} \qquad \textbf{(Equation 6.3)}$$

The term $\binom{n}{x}$ in Equation 6.3 is referred to as the **binomial coefficient** and is computed with Equation 6.4. $\binom{n}{x}$ is also referred to as the **combination of n things taken x at a time.**[5]

$$\binom{n}{x} = \frac{n!}{x!\,(n - x)!} \qquad \textbf{(Equation 6.4)}$$

In the case of Examples 6.1 and 6.2 the binomial coefficient will be $\binom{10}{8}$, which is the combination of 10 things taken 8 at a time. In the combination expression, $n = 10$ represents the total number of coin tosses/women and $x = 8$ represents the observed frequency for Category 1(**Heads/Perfume A**). When the latter value (which equals $\binom{10}{8} = \frac{10!}{8!\,2!} = 45$) is multiplied by $(.5)^8\,(.5)^2$, it yields the probability of obtaining exactly 8 **Heads/Perfume A** if there are 10 observations. The probability of 8 observations in 10 trials will be represented by the notation $P(8/10)$.[6] The value $P(8/10) = .0439$ is computed below.

$$P(8/10) = \binom{10}{8} (.5)^8\,(.5)^2 = (45)(.5)^8(.5)^2 = .0439$$

Since the computation of binomial probabilities can be quite tedious, such probabilities are more commonly derived through the use of tables. By employing **Table A6 (Table of the Binomial Distribution, Individual Probabilities)** in the **Appendix**, the value .0439 can be obtained without any computations. The probability value .0439 is identified by employing the section of the **Table A6** for $n = 10$. Within this section, the value .0439 is the entry in the cell that is the intersection of the row $x = 8$ and the column $\pi = .5$ (where $\pi_1 = .5$ is employed to represent the value of π).

The probability .0439, however, does not provide enough information to allow one to evaluate the null hypothesis. The actual probability that is required is the likelihood of obtaining a value that is equal to or more extreme than the number of observations in Category 1. Thus, in the case of Examples 6.1 and 6.2, one must determine the probability of obtaining a frequency of 8 or greater for Category 1. In other words, we want to determine the likelihood of obtaining 8, 9, or 10 **Heads/Perfume A** if the total number of observations is $n = 10$. Since we have already determined that the probability of obtaining 8 **Heads/Perfume A** is .0439, we must now determine the probability associated with the values 9 and 10. Although each of these probabilities can be computed with Equation 6.3, it is quicker to use **Table A6**. Employing the table, we determine that for $\pi = .5$ and $n = 10$, the probability of obtaining exactly $x = 9$ observations in Category 1 is $P(9/10) = .0098$, and the probability of obtaining exactly $x = 10$ observations is $P(10/10) = .0010$. The sum of the three probabilities $P(8/10)$, $P(9/10)$, and $P(10/10)$ represents the likelihood of obtaining 8 or more **Heads/Perfume A** in 10 observations. Thus: $P(8, 9, \text{or } 10/10) = .0439 + .0098 + .0010 = .0547$. Equation 6.5 summarizes the computation of a cumulative probability such as that represented by $P(8, 9, \text{or } 10/10) = .0547$.

$$P(\geq x) = \sum_{r=x}^{n} \binom{n}{x} (\pi_1)^x (\pi_2)^{(n-x)} \qquad \textbf{(Equation 6.5)}$$

Where: $\sum_{r=x}^{n}$ indicates that probability values should be summed beginning with the designated value of x up through the value n

An even more efficient way of obtaining the probability $P(8, 9, \text{or } 10/10) = .0547$ is to employ **Table A7 (Table of the Binomial Distribution, Cumulative Probabilities)** in the **Appendix**. When employing **Table A7** we again find the section for $n = 10$, and locate the cell that is the intersection of the row $x = 8$ and the column $\pi = .5$. The entry .0547 in that cell represents the probability of 8 or more (i.e., 8, 9, or 10) **Heads/Perfume A**, if there is a total of $n = 10$ observations. Thus, the entry in any cell of **Table A7** represents (for the appropriate value of π_1) the probability of obtaining a number of observations that is equal to or greater than the value of x in the left margin of the row in which the cell is located.

Table A7 can be used to determine the likelihood of x being equal to or less than a specific value. In such a case, the cumulative probability associated with the value of $(x + 1)$ is subtracted from 1. To illustrate this, let us assume that in Examples 6.1 and 6.2 we want to determine the probability of obtaining 2 or less observations in one of the two categories (which applies to Category 2). In such an instance the value $x = 2$ is employed, and thus, $x + 1 = 3$. The cumulative probability associated with $x = 3$ (for $\pi = \pi_1 = .5$) is .9453. If the latter value is subtracted from 1 it yields $1 - .9453 = .0547$, which represents the likelihood of obtaining 2 or less observations in a cell when $n = 10$.[7] The value .0547 can also be obtained from **Table A6** by adding up the probabilities for the values $x = 2$ (.0439), $x = 1$ (.0098), and $x = 0$ (.0010).

It should be noted that none of the values listed for π in **Tables A6** and **A7** exceeds .5. In order to employ the tables when the value for π_1 stated in the null hypothesis is greater than .5, the following protocol is employed: a) Use the value of π_2 (i.e., $\pi_2 = 1 - \pi_1$) to represent the value of π; and b) Each of the values of x is subtracted from the value of n, and the resulting values are employed to represent x in using the table for the analysis. To illustrate, let us assume that $n = 10$, $\pi_1 = .7$, and $x = 9$, and that we wish to determine the probability that there are 9 or more observations in one of the categories. Employing the above guidelines, the tabled value to use for π is $\pi_2 = .3$ (since $1 - .7 = .3$). Since each value of x is subtracted from n, the values $x = 9$ and $x = 10$ are respectively converted into $x = 1$ and $x = 0$. In **Table A6** (for $\pi = .3$) the probabilities associated with $x = 1$ and $x = 0$ will respectively represent those probabilities associated with $x = 9$ and $x = 10$. The sum of the tabled probabilities for $x = 1$ and $x = 0$ represents the likelihood that there will be 9 or more observations in one of the categories. From **Table A6** we determine that for $\pi = .3$, $P(1/10) = .1211$ and $P(0/10) = .0282$. Thus, $P(0 \text{ or } 1/10) = .1211 + .0282 = .1493$ (which also represents $P(9 \text{ or } 10/10)$ when $\pi_1 = .7$). The value .1493 can also be obtained from **Table A7** by subtracting the tabled probability value for $(x + 1)$ from 1 (make sure that in computing $(x + 1)$, the value of x that results from subtracting the original value of x from n is employed). Thus, if the converted value of $x = 1$, then $x + 1 = 2$. The tabled value in **Table A7** for $x = 2$ and $\pi = .3$ is .8507. When the latter value is subtracted from 1, it yields .1493.[8]

V. Interpretation of the Test Results

When the **binomial sign test for a single sample** is applied to Examples 6.1 and 6.2, it provides a probabilistic answer to the question of whether $p_1 = 8/10 = .8$ (i.e., the observed proportion of cases for Category 1) deviates significantly from the value $\pi_1 = .5$ stated in the null hypothesis.[9] The following guidelines are employed in evaluating the null hypothesis.[10]

a) If a nondirectional alternative hypothesis is employed, the null hypothesis can be rejected if the probability of obtaining a value equal to or more extreme than x is equal to or less than $\alpha/2$ (where α represents the prespecified value of α). If the proportion $p_1 = x/n$ is greater than π_1, a value that is more extreme than x will be any value that is greater than the observed value of x, whereas if the proportion $p_1 = x/n$ is less than π_1, a value that is more extreme than x will be any value that is less than the observed value of x.

b) If a directional alternative hypothesis is employed which predicts that the underlying population proportion is above a specified value, in order to reject the null hypothesis both of the following conditions must be met: 1) The proportion of cases observed in Category 1 (p_1) must be greater than the value of π_1 stipulated in the null hypothesis; and 2) The probability of obtaining a value equal to or greater than x is equal to or less than the prespecified value of α.

c) If a directional alternative hypothesis is employed which predicts that the underlying population proportion is below a specified value, in order to reject the null hypothesis both of the following conditions must be met: 1) The proportion of cases observed in Category 1 (p_1) must be less than the value of π_1 stipulated in the null hypothesis; and 2) The probability of obtaining a value equal to or less than x is equal to or less than the prespecified value of α.

Applying the above guidelines to the results of the analysis of Examples 6.1 and 6.2, we can conclude the following.

If $\alpha = .05$, the nondirectional alternative hypothesis H_1: $\pi_1 \neq .5$ is not supported, since the obtained probability .0547 is greater than $\alpha/2 = .05/2 = .025$. In the same respect, if $\alpha = .01$, the nondirectional alternative hypothesis H_1: $\pi_1 \neq .5$ is not supported, since the obtained probability .0547 is greater than $\alpha/2 = .01/2 = .005$.

If $\alpha = .05$, the directional alternative hypothesis H_1: $\pi_1 > .5$ is not supported, since the obtained probability .0547 is greater than $\alpha = .05$. In the same respect, if $\alpha = .01$, the directional alternative hypothesis H_1: $\pi_1 > .5$ is not supported, since the obtained probability .0547 is greater than $\alpha = .01$.

The directional alternative hypothesis H_1: $\pi_1 < .5$ is not supported, since $p_1 = .8$ is larger than the value $\pi_1 = .5$ predicted in the null hypothesis. Of course, if the alternative hypothesis H_1: $\pi_1 < .5$ is employed and the sample data are consistent with it, in order to be supported the obtained probability must be equal to or less than the prespecified value of alpha.

To summarize, the results of the analysis of Examples 6.1 and 6.2 do not allow a researcher to conclude that the true population proportion is some value other than .5. In view of this, in Example 6.1 the data do not allow one to conclude that the coin is biased. In Example 6.2, the data do not allow one to conclude that women exhibit a preference for one of the two brands of perfume.

It should be noted that if the obtained proportion $p_1 = .8$ had been obtained with a larger sample size, the null hypothesis could be rejected. To illustrate, if for $\pi_1 = .5$, $n = 15$, $x = 12$ and thus $p_1 = .8$, the likelihood of obtaining 12 or more observations in one of the two categories is .0176. The latter value is significant at the .05 level if the directional alternative hypothesis H_1: $\pi_1 > .5$ is employed, since it is less than the value $\alpha = .05$. It is also significant at the .05 level if the nondirectional alternative hypothesis H_1: $\pi_1 \neq .5$ is employed, since .0176 is less than $\alpha/2 = .05/2 = .025$.

VI. Additional Analytical Procedures for the Binomial Sign Test for a Single Sample and/or Related Tests

1. Test 6a: The *z* test for a population proportion When the size of the sample is large the test statistic for the **binomial sign test for a single sample** can be approximated with the chi-square distribution — specifically, through use of the **chi-square goodness-of-fit test**. An alternative and equivalent approximation can be obtained by using the normal distribution. When the latter distribution is employed to approximate the test statistic for the **binomial sign test for a single sample**, the test is referred to as the ***z* test for a population proportion**. The null and alternative hypotheses employed for the ***z* test for a population proportion** are identical to those employed for the **binomial sign test for a single sample**.

Although sources are not in agreement with respect to the minimum acceptable sample size for use with the ***z* test for a population proportion**, there is general agreement that the closer the value π_1 (or π_2) is to either 0 or 1 (i.e., the further removed it is from .5), the larger the sample size required for an accurate normal approximation. Among those sources that make recommendations with respect to the minimum acceptable sample size (regardless of the value of π_1) are Freund (1984) and Marascuilo and McSweeney (1977) who state that the values of both $n\pi_1$ and $n\pi_2$ should be greater than 5. Daniel (1990) states that n should be at least equal to 12. Siegel and Castellan (1988), on the other hand, note that when π_1 is close to .5, the test can be employed when $n > 25$, but when π_1 is close to 1 or 0, the value $n\pi_1\pi_2$ should be greater than 9. In view of the different criteria stipulated in various sources, one should employ common sense in interpreting results for the normal approximation based on small samples sizes, especially when the values of π_1 or π_2 are close to 0 or 1. Since, when the sample size is small the normal approximation tends to inflate the Type I error rate, the error rate can be adjusted by conducting a more conservative test (i.e., employ a lower alpha level). A more practical alternative, however, is to use a test statistic that is corrected for continuity. As will be demonstrated in the discussion to follow, when the correction for continuity is employed for the ***z* test for a population proportion**, the test statistic will generally provide an excellent approximation of the binomial distribution even when the size of the sample is small and/or the values of π_1 and π_2 are far removed from .5.

Examples 6.3–6.5 will be employed to illustrate the use of the z **test for a population proportion.** Since the three examples use identical data they will result in the same conclusions with respect to the null hypothesis. It will also be demonstrated that when the **chi-square goodness-of-fit test** is applied to Examples 6.3–6.5 it yields equivalent results.

Example 6.3 *An experiment is conducted to determine whether a coin is biased. The coin is flipped* 200 *times resulting in* 96 *heads and* 104 *tails. Do the results indicate that the coin is biased?*

Example 6.4 *Although a senator supports a bill which favors a woman's right to have an abortion, she realizes her vote could influence whether or not the people in her state endorse her bid for reelection. In view of this she decides that she will not vote in favor of the bill unless at least* 50% *of her constituents support a woman's right to have an abortion. A random survey of* 200 *voters in her district reveals that* 96 *people are in favor of abortion. Will the senator support the bill?*

Example 6.5 *In order to determine whether a subject exhibits extrasensory ability, a researcher employs a list of* 200 *binary digits (specifically, the values* 0 *and* 1*) which have been randomly generated by a computer. The researcher conducts an experiment in which one of his assistants concentrates on each of the digits in the order it appears on the list. While the assistant does this, the subject, who is in another room, attempts to guess the value of the number for each of the* 200 *trials. The subject correctly guesses* 96 *of the* 200 *digits. Does the subject exhibit evidence of extrasensory ability?*

As is the case for Examples 6.1 and 6.2, Examples 6.3–6.5 are evaluating the hypothesis of whether the true population proportion is .5. Thus, the null hypothesis and the nondirectional alternative hypothesis are: H_0: $\pi_1 = .5$ versus H_1: $\pi_1 \neq .5$.[11]

The test statistic for the z **test for a population proportion** is computed with Equation 6.6.

$$z = \frac{p_1 - \pi_1}{\sqrt{\dfrac{\pi_1 \pi_2}{n}}} \qquad \textbf{(Equation 6.6)}$$

The denominator of Equation 6.6 ($\sqrt{\pi_1 \pi_2 / n}$), which is the standard deviation of the sampling distribution of a proportion, is commonly referred to as the **standard error of the proportion.**

For Examples 6.3–6.5, based on the null hypothesis we know that: $\pi_1 = .5$ and $\pi_2 = 1 - \pi_1 = .5$. From the information that has been provided, we can compute the following values: $p_1 = 96/200 = .48$; $p_2 = (200 - 96)/200 = 104/200 = .52$. When the relevant values are substituted in Equation 6.6, the value $z = -.57$ is computed.

$$z = \frac{.48 - .50}{\sqrt{\dfrac{(.5)(.5)}{200}}} = -.57$$

Equation 6.7 is an alternative form of Equation 6.6 that will yield the identical z value.

$$z = \frac{x - n\pi_1}{\sqrt{n \pi_1 \pi_2}} \qquad \textbf{(Equation 6.7)}$$

In Section I it is noted that the mean and standard deviation of a binomially distributed variable are respectively $\mu = n\pi_1$ and $\sigma = \sqrt{n\pi_1\pi_2}$. These values represent the mean and standard deviation of the underlying sampling distribution. In the numerator of Equation 6.7, the value $\mu = n\pi_1$ represents the expected number of observations in Category 1 if, in fact, the population proportion is equal to $\pi_1 = .5$ (i.e., the value stipulated in the null hypothesis). Thus, for the examples under discussion, $\mu = (200)(.5) = 100$. Note that the latter expected value is subtracted from the number of observations in Category 1. The denominator of Equation 6.7 is the standard deviation of a binomially distributed variable. Thus, in the case of Examples 6.3–6.5, $\sigma = \sqrt{(100)(.5)(.5)} = 7.07$. Employing Equation 6.7, the value $z = -.57$ (which is identical to the value computed with Equation 6.6) is obtained.

$$z = \frac{96 - (200)(.5)}{\sqrt{(200)(.5)(.5)}} = -.57$$

The obtained value $z = -.57$ is evaluated with **Table A1 (Table of the Normal Distribution)** in the **Appendix**. In **Table A1** the tabled critical two-tailed .05 and .01 values are $z_{.05} = 1.96$ and $z_{.01} = 2.58$, and the tabled critical one-tailed .05 and .01 values are $z_{.05} = 1.65$ and $z_{.01} = 2.33$.

The following guidelines are employed in evaluating the null hypothesis.

a) If the alternative hypothesis employed is nondirectional, the null hypothesis can be rejected if the obtained absolute value of z is equal to or greater than the tabled critical two-tailed value at the prespecified level of significance.

b) If the alternative hypothesis employed is directional and predicts a population proportion larger than the value stated in the null hypothesis, the null hypothesis can be rejected if the sign of z is positive and the value of z is equal to or greater than the tabled critical one-tailed value at the prespecified level of significance.

c) If the alternative hypothesis employed is directional and predicts a population proportion smaller than the value stated in the null hypothesis, the null hypothesis can be rejected if the sign of z is negative and the absolute value of z is equal to or greater than the tabled critical one-tailed value at the prespecified level of significance.

Using the above guidelines, the null hypothesis cannot be rejected regardless of which of the three possible alternative hypotheses is employed. The nondirectional alternative hypothesis H_1: $\pi_1 \neq .5$ is not supported, since the absolute value $z = .57$ is less than the tabled critical two-tailed value $z_{.05} = 1.96$. The directional alternative hypothesis H_1: $\pi_1 > .5$ is not supported, since in order for it to be supported the sign of z must be positive. The directional alternative hypothesis H_1: $\pi_1 < .5$ is not supported, since although the sign of z is negative as predicted, the absolute value $z = .57$ is less than the tabled critical one-tailed value $z_{.05} = 1.65$.

It was noted previously that when there are $k = 2$ categories, the **chi-square goodness-of-fit test** also provides a large sample approximation of the test statistic for the **binomial sign test for a single sample**. In point of fact, the large sample approximation based on the **chi-square goodness-of-fit test** will yield results that are equivalent to those obtained with the **z test for a population proportion**, and the relationship between the computed chi-square value and the obtained z value for the same set of data will always be $\chi^2 = z^2$. Table 6.2 summarizes the results of the analysis of Examples 6.3–6.5 with the **chi-square goodness-of-fit test**, which as noted earlier evaluates the same hypothesis as the **binomial sign test for a single sample**. The null hypothesis and nondirectional alternative hypothesis when used in reference to the **chi-square goodness-fit-test** can also be stated employing the following format: H_0: $o_i = \varepsilon_i$ for both cells versus H_1: $o_i \neq \varepsilon_i$ for both cells.

Table 6.2 Chi-Square Summary Table for Examples 6.3–6.5

Cell	O_i	E_i	$(O_i - E_i)$	$(O_i - E_i)^2$	$\frac{(O_i - E_i)^2}{E_i}$
Heads/Pro-Abortion/ Correct Guesses	96	100	−4	16	.16
Tails/Anti-Abortion/ Incorrect Guesses	104	100	4	16	.16
	$\Sigma O_i = 200$	$\Sigma E_i = 200$	$\Sigma(O_i - E_i) = 0$		$\chi^2 = .32$

In Table 6.2. the expected frequency of each cell is computed by multiplying the hypothesized population proportion for the cell by $n = 200$. Since $k = 2$, the degrees of freedom employed for the chi-square analysis are $df = k - 1 = 2$. The value $\chi^2 = .32$ (which is obtained with Equation 5.2) is evaluated with **Table A4 (Table of the Chi-Square distribution)** in the **Appendix**. For $df = 1$, the tabled critical .05 and .01 chi-square values are $\chi^2_{.05} = 3.84$ and $\chi^2_{.01} = 6.63$. Since the obtained value $\chi^2 = .32$ is less than $\chi^2_{.05} = 3.84$, the null hypothesis cannot be rejected if the nondirectional alternative hypothesis H_1: $\pi_1 \neq .5$ is employed. If the directional alternative hypothesis H_1: $\pi_1 < .5$ is employed it is not supported, since $\chi^2 = .32$ is less than the tabled critical one-tailed .05 value $\chi^2_{.05} = 2.71$ (which corresponds to the chi-square value at the 90th percentile).

As noted previously, if the z value obtained with Equations 6.6 and 6.7 is squared, it will always equal the chi-square value computed for the same data. Thus, in the current example where $z = -.57$ and $\chi^2 = .32$: $(-.57)^2 = .32$. (The minimal discrepancy is the result of rounding off error.) It is also the case that the square of a tabled critical z value at a given level of significance will equal the tabled critical chi-square value at the corresponding level of significance. This is confirmed for the tabled critical two-tailed z and χ^2 values at the .05 and .01 levels of significance: $(z_{.05} = 1.96)^2 = (\chi^2_{.05} = 3.84)$ and $(z_{.01} = 2.58)^2 = (\chi^2_{.01} = 6.63)$.

To summarize, the results of the analysis for Examples 6.3–6.5 do not allow one to conclude that the true population proportion is some value other than .5. In view of this, in Example 6.3 the data do not allow one to conclude that the coin is biased. In Example 6.4 the data do not allow the senator to conclude that the true proportion of the population that favors abortion is some value other than .5. In Example 6.5 the data do not allow one to conclude that the subject exhibited extrasensory abilities.

It was noted previously that when the **z test for a population proportion** is employed with small sample sizes, it tends to inflate the likelihood of committing a Type I error. This is illustrated below with the data for Examples 6.1 and 6.2 which are evaluated with Equation 6.6.

$$z = \frac{.8 - .5}{\sqrt{\frac{(.5)(.5)}{10}}} = 1.90$$

Since the obtained value $z = 1.90$ is greater than the tabled critical one-tailed value $z_{.05} = 1.65$, the directional alternative hypothesis H_1: $\pi_1 > .5$ is supported at the .05 level. When the **binomial sign test for a single sample** is employed to evaluate the latter alternative hypothesis for the same data, the result just falls short of being significant at the .05 level. The nondirectional alternative hypothesis H_1: $\pi_1 \neq .5$, which is not even close to being supported with the **binomial sign test for a single sample,** just falls short of being supported at the .05 level when the **z test for a population proportion** is employed (since the tabled critical two-tailed .05 value is $z_{.05} = 1.96$). When the conclusions reached with respect to Example 6.1 and 6.2 employing the **binomial sign test for a single sample** and the **z test for a population proportion** are compared with one another, it can be seen that the **z test for a population proportion** is the less conservative of the two tests (i.e., it is more likely to reject the null hypothesis).

The correction for continuity for z test for a population proportion It is noted in the discussions of the **Wilcoxon signed-ranks test (Test 4)** and the **chi-square goodness-of-fit test** that many sources recommend that a **correction for continuity** be employed when a continuous distribution is employed to estimate a discrete probability distribution. Most sources recommend the latter correction when the normal distribution is used to approximate the binomial distribution, since the correction will adjust the Type I error rate (which will generally be inflated when the normal approximation is employed with small sample sizes). Equations 6.8 and 6.9 are respectively the continuity corrected versions of Equations 6.6 and 6.7. Each of the continuity corrected equations is applied to the data for Examples 6.3–6.5.

$$z = \frac{|p_1 - \pi_1| - \dfrac{1}{2n}}{\sqrt{\dfrac{\pi_1 \pi_2}{n}}} = \frac{|.48 - .5| - \dfrac{1}{(2)(200)}}{\sqrt{\dfrac{(.5)(.5)}{200}}} = -.49 \qquad \textbf{(Equation 6.8)}$$

$$z = \frac{|x - n\pi_1| - .5}{\sqrt{n\pi_1\pi_2}} = \frac{|96 - 100| - .5}{\sqrt{(200)(.5)(.5)}} = -.49 \qquad \textbf{(Equation 6.9)}$$

As is the case when Equations 6.6 and 6.7 are employed to compute the absolute value $z = .57$, the absolute value $z = .49$ computed with Equations 6.6 and 6.7 is less than the tabled critical two-tailed .05 value $z_{.05} = 1.96$ and the tabled critical one-tailed .05 value $z_{.05} = 1.65$. Thus, regardless of which of the possible alternative hypotheses one employs, the null hypothesis cannot be rejected. Note that the continuity corrected absolute value $z = .49$ is less than the absolute value $z = .57$ obtained with Equations 6.6 and 6.7. Since a continuity corrected equation will always result in a lower absolute value for z, it will provide a more conservative test of the null hypothesis. The smaller the sample size, the greater the difference between the values computed with the continuity corrected and uncorrected equations.

Equation 5.6, which as noted previously is the continuity corrected equation for the **chi-square goodness-of-fit test**, can also be employed with the same data and will yield an equivalent result to that obtained with Equations 6.8 and 6.9. When employing Equation 5.6 there are two cells and each cell has an expected frequency of 100. The observed frequencies of the two cells are 96 and 104. Thus for each cell $(|O_i - E_i| - .5) = 3.5$. Thus:

$$\chi^2 = \sum_{i=1}^{k} \left[\frac{(|O_i - E_i| - .5)^2}{E_i} \right] = \frac{(3.5)^2}{100} + \frac{(3.5)^2}{100} = .245$$

Note that $(.49)^2 = .245$ (once again the slight discrepancy is due to rounding off error). As is the case with $\chi^2 = .32$ (the uncorrected chi-square value computed in Table 6.2), the continuity corrected value $\chi^2 = .245$ is not significant, since it is less than the tabled critical .05 value $\chi^2_{.05} = 3.84$ (in reference to the nondirectional alternative hypothesis).

Although in the case of Examples 6.3–6.5 the use of the correction for continuity does not alter the decision one can make with respect to the null hypothesis, this will not always be the case. To illustrate this, Equation 6.8 is employed below to compute the continuity corrected value of z for Examples 6.1 and 6.2.

$$z = \frac{|.8 - .5| - \frac{1}{(2)(10)}}{\sqrt{\frac{(.5)(.5)}{10}}} = 1.58$$

Since the continuity corrected value $z = 1.58$ is less than the tabled critical one-tailed value $z_{.05} = 1.65$, the directional alternative hypothesis H_1: $\pi_1 > .5$ is not supported. This is consistent with the result that is obtained when the **binomial sign test for a single sample** is employed. Recollect that when the data are evaluated with Equation 6.6 (i.e., without the continuity correction) the directional alternative hypothesis H_1: $\pi_1 > .5$ is supported. Thus, it appears that in this instance the continuity correction yields a result that is more consistent with the result based on the exact binomial probability.

Computation of a confidence interval for the *z* test for a population proportion Equation 5.5, which is described in the discussion of the **chi-square goodness-of-fit test**, can also be employed for computing a confidence interval for the *z* **test for a population proportion.** Equation 5.5 is employed below to compute the 95% confidence interval for Examples 6.3–6.5 for Category 1.

$$p_1 - z_{(\alpha/2)} \sqrt{\frac{p_1 p_2}{n}} \leq \pi_1 \leq p_1 + z_{(\alpha/2)} \sqrt{\frac{p_1 p_2}{n}}$$

$$.48 - (1.96) \sqrt{\frac{(.48)(.52)}{200}} \leq \pi_1 \leq .48 + (1.96) \sqrt{\frac{(.48)(.52)}{200}}$$

$$\pi_1 = .48 \pm .069$$

$$.411 \leq \pi_1 \leq .549$$

Thus, the researcher can be 95% confident that the true proportion of cases in the underlying population in Category 1 is a value between .411 and .549. The confidence interval for the population proportion for Category 2 (i.e., π_2) can be obtained by adding and subtracting the value .069 to and from the value $p_2 = .52$. Thus, $.451 \leq \pi_2 \geq .589$.

Extension of *z* test for a population proportion in order to evaluate the performance of *m* subjects on *n* trials on a binomially distributed variable Example 6.6 illustrates a case in which each of m subjects is evaluated for a total of n trials on a binomially distributed variable. The example represents an extension of the analysis used for Example 6.5 to a design involving m subjects. The methodology employed in analyzing the data is basically an extension of the **single-sample *z* test** to an analysis of a population proportion that is based on a binomially distributed variable.

Example 6.6 *In order to determine whether a group of* 10 *people exhibit extrasensory ability, a researcher employs as test stimuli a list of* 200 *binary digits (specifically, the values* 0 *and* 1*) which have been randomly generated by a computer. The researcher conducts an experiment in which one of his assistants concentrates on each of the digits in the order it appears on the list. While the assistant does this, each of the* 10 *subjects, all of whom are in separate rooms, attempts to guess the value of the number for each of* 200 *trials. The number of correct guesses in* 200 *trials for each of the subjects follows:* 102, 104, 100, 98, 96, 80, 110, 120, 102, 128. *Does the group as a whole exhibit evidence of extrasensory abilities?*

Equation 6.10 is employed to evaluate Example 6.6.

$$z = \frac{\bar{X} - \mu}{\frac{\sigma}{\sqrt{m}}}$$ **(Equation 6.10)**

Where: m represents the number of subjects in the sample
$\mu = n\pi_1$
$\sigma = \sqrt{n\pi_1\pi_2}$

The basic structure of Equation 6.10 is the same as that of Equation 1.3 ($z = (\bar{X} - \mu)/\sigma_{\bar{X}}$), which is the equation for the **single-sample *z* test**. In the numerator of Equation 1.3, the hypothesized population mean is subtracted from the sample mean. Equation 6.10 employs the analogous values — employing the sample mean ($\bar{X}$) and $\mu = n\pi_1$ to represent the hypothesized population mean. The denominator of Equation 1.3 represents the standard deviation of a sampling distribution that is based on a sample size of n for what is assumed to be a normally distributed variable. The denominator of Equation 6.10 represents the standard deviation of a sampling distribution of a binomially distributed variable that is based on a sample size of m. In both equations the denominator can be summarized as follows: $\sigma/\sqrt{\text{number of subjects}}$. It should also be noted that when the number of subjects is $m = 1$, Equation 6.10 reduces to Equation 6.7.[12]

Note that in Example 6.6, each of the $m = 10$ subjects is tested for $n = 200$ trials. On each trial it is assumed that a subject has a likelihood of $\pi_1 = .5$ of being correct and a likelihood of $\pi_2 = .5$ of being incorrect. Since we are dealing with a binomially distributed variable, the expected number of correct responses for each subject, as well as the expected average number of correct responses for the group of $m = 10$ subjects, is $\mu = n\pi_1$. As previously noted, the standard deviation of the sampling distribution for a single subject is defined by $\sigma = \sqrt{n\pi_1\pi_2}$. Since there are m subjects, the standard deviation of the sampling distribution for m subjects will be $\sigma/\sqrt{m}$, which is the denominator of Equation 6.10.

The null and alternative hypotheses for Example 6.6 are identical to those employed for Example 6.5. The only difference is that, whereas in Example 6.5 H_0 and H_1 are stated in reference to the population of scores for a single subject, in Example 6.6 they are stated in reference to the population of scores for a population of subjects that is represented by the m subjects in the sample. In Example 6.6, the mean number of correct guesses by the 10 subjects is computed with Equation 1.1: $\bar{X} = 1040/10 = 104$. The values $n = 200$, $\pi_1 = .5$, $\pi_2 = .5$ are identical to those employed in Example 6.5. Thus, as is the case for Example 6.5, $\mu = (200)(.5) = 100$ and $\sigma = \sqrt{(200)(.5)(.5)} = 7.07$. When the appropriate values are substituted in Equation 6.10, the value $z = 1.79$ is computed.

$$z = \frac{104 - 100}{\frac{7.07}{\sqrt{10}}} = 1.79$$

The obtained value $z = 1.79$ is evaluated with **Table A1**. The nondirectional alternative hypothesis H_1: $\pi_1 \neq .5$ is not supported, since the value $z = 1.79$ is less than the tabled critical two-tailed value $z_{.05} = 1.96$. The directional alternative hypothesis H_1: $\pi_1 > .5$ is supported at the .05 level, since the obtained value $z = 1.79$ is a positive number that is greater than the tabled critical one-tailed value $z_{.05} = 1.65$. The latter alternative hypothesis is not supported at the .01 level, since $z = 1.79$ is less than the tabled critical one-tailed value $z_{.01} = 2.33$. The directional alternative hypothesis H_1: $\pi_1 < .5$ is not supported, since in order for it to be supported the sign of z must be negative.

The above analysis allows one to conclude that the group as a whole scores significantly above chance. The latter result can be interpreted as evidence of extrasensory perception if one is able to rule out alternative sensory and cognitive explanations of information transmission. The reader should be aware of the fact that in spite of the conclusions with regard to the group, it is entirely conceivable that the performance of one or more of the subjects in the group is not statistically significant. Inspection of the data reveals that the performance of the subject who obtains a score of 100 is at chance expectancy. Additionally, the scores of some of the other subjects (e.g. 102, 104, 96, 98) are well within chance expectancy.[13]

It is instructive to note that in the case of Example 6.6, if for some reason one is unwilling to assume that the variable under study is binomially distributed with a standard deviation of $\sigma = 7.07$, the population standard deviation must be estimated from the sample data. Under the latter conditions the **single-sample *t* test (Test 2)** is the appropriate test to employ, and the following null hypothesis is evaluated: H_0: $\mu = 100$. If each of the 10 scores in Example 6.6 are squared and the squared scores are summed, they yield the value $\Sigma X^2 = 109728$. Employing Equation 2.1, the estimated population standard deviation is computed to be $\tilde{s} = 13.2$. Substituting the latter value, along with $\bar{X} = 104$, $\mu = 100$, and $n = 10$ in Equation 2.3 yields the value $t = .96$. Since for $df = 9$, $t = .96$ falls far short of the tabled critical two-tail .05 value $t_{.05} = 2.26$ and the tabled critical one-tail .05 value $t_{.05} = 1.83$, the null hypothesis is retained. This result is the opposite of that reached when Equation 6.10 is employed with the same data. The difference between the two tests can be attributed to the fact that in the case of the **single-sample *t* test** the estimated population standard deviation $\tilde{s} = 13.2$ is almost twice the value of $\sigma = 7.07$ computed for a binomially distributed variable.

2. Test 6b: The single-sample test for the median There are occasions when the **binomial sign test for a single sample** is employed to evaluate a hypothesis regarding a population median. Specifically, the test may be used to determine the likelihood of observing a specified number of scores above versus below the median of a distribution. When the **binomial sign test for a single sample** is used within this context it is often referred to as the **single-sample test for the median.**[14] This application of the **binomial sign test for a single sample** will be illustrated with Example 6.7. Since Example 6.7 assumes $\pi_1 = \pi_2 = .5$ and has the same binomial coefficient obtained for Examples 6.1 and 6.2, it yields the same result as the latter examples.

Example 6.7 *Assume that the median blood cholesterol level for a healthy* 30*-year-old male is* 200 *mg*/100 *ml. Blood cholesterol readings are obtained for a group consisting of eleven* 30*-year-old men who have had a heart attack within the last month. The blood cholesterol scores of the eleven men are:* 230, 167, 250, 345, 442, 190, 200, 248, 289, 262, 301. *Can one conclude that the median cholesterol level of the population represented by the sample (i.e., recent male heart attack victims) is some value other than* 200?

Since the median identifies the 50th percentile of a distribution, if the population median is in fact equal to 200, one would expect one-half of the sample to have a blood cholesterol reading above 200 (i.e., $p_1 = .5$), and one-half of the sample to have a reading below 200 (i.e., $p_2 = .5$). Although the null hypothesis and the nondirectional alternative hypothesis for Example 6.7 can be stated using the format H_0: $\pi_1 = .5$ versus H_1: $\pi_1 \neq .5$, they can also be stated as follows: H_0: $\theta = 200$ versus H_1: $\theta \neq 200$. Employing the latter format, the null hypothesis states that the median of the population the sample represent equals 200, and the alternative hypothesis states that the median of the population the sample represents does not equal 200.

When the **binomial sign test for a single sample** is employed to test a hypothesis about a population median, one must determine the number of cases that fall above versus below the hypothesized population median. Any score that is equal to the median is eliminated from the data. Employing this protocol, the score of the man who has a blood cholesterol of 200 is dropped from the data, leaving 10 scores, 8 of which are above the hypothesized median value, and 2 of which are below it. Thus, as is the case in Examples 6.1 and 6.2, we want to determine the likelihood of obtaining 8 or more observations in one category (i.e., **above the median**) if there are a total of 10 observations. It was previously determined that the latter probability is equal to .0537. As noted earlier, this result does not support the nondirectional alternative hypothesis H_1: $\pi_1 \neq .5$. It just falls short of supporting the directional alternative hypothesis H_1: $\pi_1 > .5$ (which in the case of Example 6.7 can also be stated as H_1: $\theta > 200$).

The data for Example 6.6 can also be evaluated within the framework of the **single-sample test for the median.** Specifically, if we assume a binomially distributed variable for which $\pi_1 = \pi_2 = .5$ and $\mu = 100$, the population median will also equal 100. In Example 6.6 6 out of the 10 subjects score above the hypothesized median value $\theta = 100$, 3 score below it, and one subject obtains a score of 100. After the latter score is dropped from the analysis, 9 scores remain. Thus, we want to determine the likelihood that there will be 6 or more observations in one category (i.e., **above the hypothesized median**) if there are a total of 9 observations. Using **Table A7**, we determine that the latter probability equals .2539. Since the value .2539 is greater than the required two-tailed .05 probability $\alpha/2 = .025$, as well as the one-tailed probability $\alpha = .05$, the null hypothesis cannot be rejected regardless of which alternative hypothesis is employed. This is in stark contrast to the decision reached when the data are evaluated with Equation 6.10. Since the latter equation employs more information (i.e., the interval/ratio scores of each subject are employed to compute the sample mean), it provides a more powerful test of an alternative hypothesis than does the **single-sample test for the median** (which conceptualizes scores as categorical data).

The **Wilcoxon signed-ranks test (Test 4)** also provides a more powerful test of an alternative hypothesis concerning a population median than does the **binomial sign test for a single sample/single-sample test for the median.**[15] This will be demonstrated by employing the **Wilcoxon signed-ranks test** to evaluate the null hypothesis H_0: $\theta = 200$ for Example 6.7. Table 6.3 summarizes the analysis.

Table 6.3 Data for Example 6.7

Subject	X	$D = X - \theta$	Rank of $\lvert D \rvert$	Signed rank of $\lvert D \rvert$
1	230	30	2	2
2	167	−33	3	−3
3	250	50	5	5
4	345	145	9	9
5	442	242	10	10
6	190	−10	1	−1
7	200	0	–	–
8	248	48	4	4
9	289	89	7	7
10	262	62	6	6
11	301	101	8	8
				$\Sigma R+ = 51$
				$\Sigma R- = 4$

The computed Wilcoxon statistic is $T = 4$, since $\Sigma R- = 4$ (the smaller of the two values $\Sigma R-$ versus $\Sigma R+$) is employed to represent the test statistic. The value $T = 4$ is evaluated with **Table A5 (Table of Critical *T* Values for Wilcoxon's Signed-Ranks and Matched-Pairs Signed-Ranks Test)** in the **Appendix**. Employing **Table A5**, we determine that for $n = 10$ signed ranks, the tabled critical two-tailed .05 and .01 values are $T_{.05} = 8$ and $T_{.01} = 3$, and the tabled critical one-tailed .05 and .01 values are $T_{.05} = 10$ and $T_{.01} = 5$. Since the null hypothesis can only be rejected if the computed value $T = 4$ is equal to or less than the tabled critical value at the prespecified level of significance, we can conclude the following:

The nondirectional alternative hypothesis $H_1: \theta \neq 200$ is supported at the .05 level, since $T = 4$ is less than the tabled critical two-tailed value $T_{.05} = 8$. It is not supported at the .01 level, since $T = 4$ is greater than the tabled critical two-tailed value $T_{.01} = 3$.

The directional alternative hypothesis $H_1: \theta > 200$ is supported at both the .05 and .01 levels since: a) The data are consistent with the directional alternative hypothesis $H_1: \theta > 200$. In other words, the fact that $\Sigma R+ > \Sigma R-$ is consistent with the directional alternative hypothesis $H_1: \theta > 200$; and b) The obtained value $T = 4$ is less than the tabled critical one-tailed values $T_{.05} = 10$ and $T_{.01} = 5$.

The directional alternative hypothesis $H_1: \theta < 200$ is not supported, since it is not consistent with the data. In order for the latter alternative hypothesis to be supported, $\Sigma R-$ must be greater than $\Sigma R+$.

Thus, if Example 6.7 is evaluated with the **Wilcoxon signed-ranks test** the nondirectional alternative hypothesis $H_1: \theta \neq 200$ is supported at the .05 level, and the directional alternative hypothesis $H_1: \theta > 200$ is supported at both the .05 and .01 levels. When the same data are evaluated with the **binomial sign test for a single sample/single-sample test for the median**, none of the alternative hypotheses are supported (although the directional alternative hypothesis $H_1: \pi_1 > .5$ just falls short of being significant at the .05 level). From the preceding it should be apparent that the **Wilcoxon signed-ranks test** (which employs a greater amount of information) provides a more powerful test of an alternative hypothesis than does the **binomial sign test for a single sample/single-sample test for the median.**

Examination of Example 4.1 (which is identical to Example 2.1) allows us to contrast the power of the **binomial sign test for a single sample/single-sample test for the median** with the power of both the **single-sample *t* test** and the **Wilcoxon signed-ranks test.** When the latter problem is evaluated with the **single-sample *t* test**, the null hypothesis $H_0: \mu = 5$ cannot be rejected if a nondirectional alternative hypothesis is employed. However, the null hypothesis can be rejected at the .05 level if the directional alternative hypothesis $H_1: \mu > 5$ is employed. With reference to the latter alternative hypothesis, the obtained t value is greater than the tabled critical $t_{.05}$ value by a comfortable margin. When Example 4.1 is evaluated with the **Wilcoxon signed-ranks test**, the null hypothesis $H_0: \theta = 5$ cannot be rejected if a nondirectional alternative hypothesis is employed. When the directional alternative hypothesis $H_1: \theta > 5$ is employed, the analysis just falls short of being significant at the .05 level. As noted in the discussion of the **Wilcoxon signed-ranks test**, the different conclusions derived from the two tests illustrate the fact that when applied to the same data, the **single-sample *t* test** provides a more powerful test of an alternative hypothesis than the **Wilcoxon signed-ranks test.**

If Example 4.1 is evaluated with the **binomial sign test for a single sample/single-sample test for the median**, it would be expected that it would be the least powerful of the three tests. In Example 4.1, 7 of the 10 scores fall above the hypothesized population median $\theta = 5$ and 3 scores fall below it. Thus, using the binomial distribution we want to determine the likelihood of obtaining 7 or more observations in one category (i.e., **above the median**) if there are a total of 10 observations. Employing either **Table A6** or **A7**, we can determine that for $\pi_1 = \pi_2 = .5$ and $n = 10$, the likelihood of 7 of more observations in one category is .1719. Since the latter value is well above the required two-tailed .05 value $\alpha/2 = .025$ and the required one-tailed .05 value $\alpha = .05$, the directional alternative hypothesis H_1: $\theta > 5$ is not supported. Thus, when compared with the **Wilcoxon signed-ranks test,** which just falls short of significance, the **binomial sign test for a single sample** does not even come close to being significant.

The above noted differences between the **single-sample *t* test**, the **Wilcoxon signed-ranks test**, and the **binomial sign test for a single sample** illustrate that when the original data are in an interval/ratio format, the most powerful test of an alternative hypothesis is provided by the **single-sample *t* test** and the least powerful by the **binomial sign test for a single sample**. As noted in the **Introduction** of the book, most researchers would not be inclined to employ a nonparametric test with interval/ratio data unless one had reason to be believe that one or more of the assumptions of the appropriate parametric test was saliently violated. In the same respect, unless there is reason to believe that the underlying population distribution is not symmetrical, it is more logical to employ the **Wilcoxon signed-ranks test** as opposed to the **binomial sign test for a single sample** to evaluate Example 4.1.

3. Sources for computing the power of the binomial sign test for a single sample Although the protocol for computing the power of the **binomial sign test for a single sample** will not be described in this book, a comprehensive discussion of the subject as well as the necessary tables can be found in Cohen (1988).

VII. Additional Discussion of the Binomial Sign Test for a Single Sample

No additional material will be discussed in this section.

VIII. Additional Example Illustrating the Use of the Binomial Sign Test for a Single Sample

Example 6.8 employs the **binomial sign test for a single sample** in a case where the value of π_1 stated in the null hypothesis is close to 1. It also illustrates that the continuity corrected version of the *z* **test for a population proportion** can provide an excellent approximation of the binomial distribution, even if the value of π_1 is far removed from .5.

Example 6.8 *A biologist has a theory that* 90% *of the people who develop a rare disease are males and only* 10% *are females. Of* 10 *people he identifies who have the disease,* 7 *are males and* 3 *are females. Do the data support the biologist's theory?*

Since the information given indicates that we are dealing with a binomially distributed variable with $\pi_1 = .9$ and $\pi_2 = .1$, the data can be evaluated with the **binomial sign test for a single sample.** Based on the biologist's theory, the null hypothesis and the nondirectional alternative hypothesis are as follows: H_0: $\pi_1 = .9$ versus H_1: $\pi_1 \neq .9$. The data consist of the number of observations in the two categories **males** versus **females**. The respective proportion of observations in the two categories are: $p_1 = 7/10 = .7$ and $p_2 = 3/10 = .3$. Thus, given that $\pi_1 = .9$, we want to determine the likelihood of 7 or fewer observations in one category if there are a total of 10 observations. Note that since $p_1 = .7$ is less than $\pi_1 = .9$, a value that is more extreme than $x = 7$ will be any value that is less than 7.

Since $\pi = \pi_1 = .9$ is not listed in either **Table A6** or **A7**, we employ for π the probability value listed for π_2, which in the case of Example 6.8 is $\pi_2 = .1$. The probability of x being equal to or less than 7 (i.e., the number of observations in Category 1) if $\pi_1 = .9$, will be equivalent to the probability of x being equal to or greater than 3 (which is the number of observations in Category 2) if $\pi_2 = .1$. From Table **A7** it can be determined that the latter probability (which is in the cell that is the intersection of the row $x = 3$ and the column $\pi = .1$) is equal to .0702. The same value can be obtained from **Table A6** by adding the probabilities for all values of x equal to or greater than 3 (for $\pi = .1$). Thus, the probability of 7 or fewer males if there is a total of 10 observations is .0702.[16] Since the latter value is greater than the two-tailed .05 value $\alpha/2 = .025$ and the one-tailed .05 value $\alpha = .05$, neither the nondirectional alternative hypothesis H_1: $\pi_1 \neq .9$ or the directional alternative hypothesis which is consistent with the data (H_1: $\pi_1 < .9$) is supported. In other words, $p_1 = .7$ (the observed the proportion of males) is not significantly below the hypothesized value $\pi_1 = .9$. In the same respect $p_2 = .3$, the observed proportion of females, is not significantly above the expected value of $\pi_2 = .1$.

When the z **test for a population proportion** is employed to evaluate Example 6.8, Equation 6.6 (which does not employ the correction for continuity) yields the following result: $z = (.7 - .9)/\sqrt{[(.9)(.1)]/10} = -2.11$. Equation 6.8 (the continuity corrected equation) has the identical denominator as Equation 6.6, but the numerator is reduced by $1/[(2)(10)] = .05$, thus yielding the value $z = -1.58$. Since the absolute value $z = 2.11$ is greater than the tabled critical two-tailed .05 value $z_{.05} = 1.96$ and the tabled critical one-tailed .05 value $z_{.05} = 1.65$, without the correction for continuity both the nondirectional alternative hypothesis H_1: $\pi_1 \neq .9$ and the directional alternative hypothesis H_1: $\pi_1 < .9$ are supported at the .05 level. When the correction for continuity is employed, the obtained absolute value $z = 1.58$ is less than both of the aforementioned tabled critical values, and because of this, regardless of which alternative hypothesis is employed, the null hypothesis cannot be rejected. The latter conclusion is consistent with the result obtained when the exact binomial probabilities are employed. Thus, even in a case where the value of π_1 is far removed from .5, the continuity corrected equation for the z **test for a population proportion** appears to provide an excellent estimate of the exact binomial probability.

References

Cohen, J. (1988). **Statistical power analysis for the behavioral sciences** (2nd ed.). Hillsdale, NJ: Lawrence Erlbaum Associates, Publishers.

Daniel, W. W. (1990). **Applied nonparametric statistics** (2nd ed.). Boston: PWS–Kent Publishing Company.

Freund, J. E. (1984). **Modern elementary statistics** (6th ed.). Englewood Cliffs, NJ: Prentice–Hall, Inc.

Marascuilo, L. A. and McSweeney, M. (1977). **Nonparametric and distribution-free methods for the social sciences.** Monterey, CA: Brooks/Cole Publishing Company.

Siegel, S. and Castellan, N. J., Jr. (1988). **Nonparametric statistics for the behavioral sciences** (2nd ed.). New York: McGraw–Hill Book Company.

Endnotes

1. The binomial distribution is actually a special case of the **multinomial distribution.** In the latter distribution each of n independent observations can be classified in one of k mutually exclusive categories, where k can be any integer value equal to or greater than two.

2. The reader should take note of the fact that most sources employ the notations p and q to represent the population proportions π_1 and π_2. Because of this, in most sources the equations for the mean and standard deviation of a binomially distributed variable are written as follows: $\mu = np$ and $\sigma = \sqrt{npq}$. The use of the notations π_1 and π_2 in

this book for the population proportions is predicated on the fact that throughout the book Greek letters are employed to represent population parameters.

3. Using the format employed for stating the null hypothesis and the nondirectional alternative hypothesis for the **chi-square goodness-of-fit test,** H_0 and H_1 can also be stated as follows for the **binomial sign test for a single sample:** H_0: $o_i = \varepsilon_i$ for both cells; H_1: $o_i \neq \varepsilon_i$ for both cells. Thus, the null hypothesis states that in the underlying population the sample represents, for both cells/categories, the observed frequency of a cell is equal to the expected frequency of the cell. The alternative hypothesis states that in the underlying population the sample represents, for both cells/categories, the observed frequency of a cell is not equal to the expected frequency of the cell.

4. The question can also be stated as follows: If $n = 10$ and $\pi_1 = \pi_2 = .5$, what is the probability of 2 or less observations in one of the two categories? When $\pi_1 = \pi_2 = .5$, the probability of two or less observations in Category 2 will equal the probability of eight or more observations in Category 1 (or vice versa). When, however, $\pi_1 \neq \pi_2$, the probability of 2 or less observations in Category 2 will not equal the probability of 8 or more observations in Category 1.

5. The **combination of *n* things taken *x* at a time** represents the number of different ways that *n* objects can be arranged *x* at a time without regard to order. For instance, if one wants to determine the number of ways that 3 objects (which we will designated *A, B,* and *C*) can be arranged 2 at a time without regard to order, the following 3 outcomes are possible: 1) An *A* and a *B* (which can result from either the sequence *AB* or *BA*); 2) An *A* and a *C* (which can result from either the sequence *AC* or *CA*); or 3) A *B* and a *C* (which can result from the sequence *BC* or *CB*). Thus, there are 3 combinations of *ABC* taken 2 at a time. This is confirmed below through use of Equation 6.4.

$$\binom{3}{2} = \frac{3!}{2!\ (3-2)!} = \frac{3!}{2!\ 1!} = 3$$

In order to extend the concept of combinations to a coin tossing situation, let us assume that one wants to determine the exact number of ways 2 **Heads** can be obtained if a coin is tossed 3 times. If a fair coin is tossed 3 times, any one of the following 8 sequences of **Heads** (*H*) and **Tails** (*T*) is equally likely to occur: *HHH, HHT, THH, HTH, THT, HTT, TTH, TTT.* Of the 8 possible sequences, only the following 3 sequences involve 2 **Heads** and 1 **Tails:** *HHT, THH, HTH.* The latter 3 arrangements represent the combination of 3 things taken 2 at a time. This can be confirmed by $\binom{3}{2} = \frac{3!}{2!\ 1!} = 3$.

When the order of the arrangements is of interest, one can compute the **permutation of *n* things taken *x* at a time.** The latter is represented by the notation P_x^n, where $P_x^n = n!/(n-x)!$. Thus, if one is interested in the order the events *A, B,* and *C* taken 2 at a time, the following number of permutations are computed: $P_2^3 = 3!/(3-2)! = 3!/1!$. The 6 possible arrangements taking order into account are *AB, BA, AC, CA, BC, and CB.*

Although based on the definition of a permutation that has been presented above, one might conclude that the three combinations *HHT, THH, HTH* take order into account, and thus represent permutations, they can be conceptualized as combinations if one views the binomial model as follows: Within each of the three combinations *HHT, THH, HTH,* the 2 **Heads** are **distinguishable** from one another, insofar as each of the **Heads** can be assigned a subscript to designate it as distinct from the other **Heads**. To illustrate, within the combination *HHT*, H_1 and H_2 can be employed to distinguish the two **Heads** from one another. If we imagine that the two **Heads** H_1 and H_2 are randomly selected from an urn, and one **Heads** is assigned to Trial 1 and the other **Heads** is assigned to Trial 2, the following two permutations are possible: H_1H_2T, H_2H_1T. Thus, the arrangement *HHT* is a combination that summarizes the two distinct permutations H_1H_2T, H_2H_1T. Based on what has been said, it follows that the following 6 permutations comprise the 3 combinations *HHT*, *THH*, and *HTH*: H_1H_2T, H_2H_1T, TH_1H_2, TH_2H_1, H_1TH_2, H_2TH_1. In point of fact, many sources (e.g., Marascuilo and McSweeney (1977, pp., 12–13)) describe the value computed for the binomial coefficient as a permutation, since it can be viewed as representing a value based on two sets of different but identical objects.

6. The application of Equation 6.3 to every possible value of x (i.e., in the case of Examples 6.1 and 6.2, the integer values 0 through 10) will yield a probability for every value of x. The sum of these probabilities will always equal 1. The algebraic expression which summarizes the summation of the probability values for all possible values of x is referred to as the **binomial expansion.**

7. If $\pi = .5$ and one wants to determine the likelihood of x being equal to or less than a specific value, one can employ the cumulative probability listed for the value $(n - x)$. Thus, if $x = 2$, the cumulative probability for $x = 8$ (which is .0547) is employed since $n - x = 10 - 2 = 8$. The value .0547 indicates the likelihood of obtaining 2 or less observations in a cell. This procedure can only be used when $\pi_1 = .5$, since when the latter is true the binomial distribution is symmetrical.

8. If in using **Tables A6** and **A7** the value of π_2 is employed to represent π in place of π_1, and the number of observations in Category 2 is employed to represent the value of x instead of the number of observations in Category 1, then the following are true: a) In **Table A6** (if all values of π within the range from 0 to 1 are listed) the probability associated with the cell that is the intersection of the values $\pi = \pi_2$ and x (where x represents the number of observations in Category 2) will be equivalent to the probability associated with the cell that is the intersection of $\pi = \pi_1$ and x (where x represents the number of observations in Category 1); and b) In **Table A7** (if all values of π within the range from 0 to 1 are listed) for $\pi = \pi_2$, the probability of obtaining x or fewer observations (where x represents the number of observations in Category 2) will be equivalent to for $\pi = \pi_1$, the probability of obtaining x or more observations (where x represents the number of observations in Category 1). Thus if $\pi_1 = .7$, $\pi_2 = .3$, and $n = 10$ and there are 9 observations in Category 1 and 1 observation in Category 2, the following are true: a) The probability in **Table A6** for $\pi = \pi_2 = .3$ and $x = 1$ will be equivalent to the probability for $\pi = \pi_1 = .7$ and $x = 9$; and b) In **Table A7** if $\pi = \pi_2 = .3$, the probability of obtaining 1 or fewer observations will be equivalent to the probability of obtaining 9 or more observations if $\pi = \pi_1 = .7$.

   A modified protocol for employing **Table A7** when a more extreme value is defined as any value that is larger than the smaller of the two observed frequencies, or smaller than the larger of the two observed frequencies is described in Section VIII in reference to Example 6.8.

9. It will also answer at the same time whether $p_2 = 2/10 = .2$, the observed proportion of cases for Category 2, deviates significantly from $\pi_2 = .5$.

10. Since like the normal distribution the binomial distribution is a two-tailed distribution, the same basic protocol is employed in interpreting nondirectional (i.e., two-tailed) and directional (one-tailed) probabilities. Thus, in interpreting binomial probabilities one can conceptualize a distribution that is similar in shape to the normal distribution, and substitute the appropriate binomial probabilities in the distribution.

11. In Example 6.4 many researchers might prefer to employ the directional alternative hypothesis H_1: $\pi_1 < .5$, since the senator will only change her vote if the observed proportion in the sample is less than .5. In the same respect, in Example 6.5 one might employ the directional alternative hypothesis H_1: $\pi_1 > .5$, since most people would only interpret above chance performance as indicative of extrasensory perception.

12. Equation 6.7 can also be expressed in the form $z = (X - \mu)/\sigma$. Note that the latter equation is identical to Equation I.9, the equation for computing a standard deviation score for a normally distributed variable. The difference between Equations 6.7 and I.9 is that Equation 6.7 computes a normal approximation for a binomially distributed variable, whereas Equation I.9 computes an exact value for a normally distributed variable.

13. The reader may be interested in knowing that in extrasensory perception (ESP) research, evidence of ESP is not necessarily limited to above chance performance. A person who consistently scores significantly below chance or only does so under certain conditions (such as being tested by an extremely skeptical and/or hostile experimenter) may also be used to support the existence of ESP. Thus, in Example 6.6, the subject who obtains a score of 80 (which is significantly below the expected value $\mu = 100$) represents someone whose poor performance (referred to as **psi missing**) might be used to suggest the presence of extrasensory processes.

14. When the **binomial sign test for a single sample** is employed to evaluate a hypothesis regarding a population median, it is categorized by some sources as a test of ordinal data (rather than as a test of categorical/nominal data), since when data are categorized with respect to the median it implies ordering of the data within two categories (i.e., **above the median** versus **below the median**).

15. In the discussion of the **Wilcoxon signed-ranks test**, it is noted that the latter test is not recommended if there is reason to believe that the underlying population distribution is asymmetrical. Thus, if there is reason to believe that blood cholesterol levels are not distributed symmetrically in the population, the **binomial sign test for a single sample** would be recommended in lieu of the **Wilcoxon signed-ranks test**.

16. The reader should take note of the fact that the protocol in using **Table A7** to interpret a π_2 value that is less than .5 in reference to the value $\pi_2 = .1$ is different than the one described in the last paragraph of Section IV. The reason for this is that in Example 6.8 we are interested in (for $\pi_2 = .1$) the probability that the number of observations in Category 2 (**females**) are equal to or greater than 3 (which equals the probability that the number of observations in Category 1 (**males**) are equal to or less than 7 for $\pi_1 = .9$). The protocol presented in the last paragraph of Section IV in reference to the value $\pi_2 = .3$ describes the use of **Table A7** to determine the probability that the number of observations in Category 2 are equal to or less than 1 (which equals the probability that the number of observations in Category 1 are equal to or greater than 9 for $\pi_1 = .7$). Note that in Example 6.8 a more extreme score is defined as one that

is larger than the lower of the two observed frequencies or smaller than the larger of the two observed frequencies. On the other hand, in the example in the last paragraph of Section IV a more extreme score is defined as one that is smaller than the lower of the two observed frequencies or larger than the higher of the two observed frequencies. The criteria for defining what constitutes an extreme score is directly related to the alternative hypothesis the researcher employs. If the alternative hypothesis is nondirectional, an extreme score can fall both above or below an observed frequency, whereas if a directional alternative hypothesis is employed a more extreme score can only be in the direction indicated by the alternative hypothesis.

Test 7

The Single-Sample Runs Test (and Other Tests of Randomness)

(Nonparametric Test Employed with Categorical/Nominal Data)

I. Hypothesis Evaluated with Test and Relevant Background Information

Hypothesis evaluated with test Is the distribution of a series of binary events in a population random?

Relevant background information on test By definition a **random series** is one for which no algorithm (i.e., set of rules) can be generated that will allow one to predict at above chance which of the n_2 possible alternatives will occur on a given trial.[1] The **single-sample runs test** is one of a number of statistical procedures that have been developed for evaluating whether or not the distribution of a series of N numbers is **random**. The test evaluates the number of **runs** in a series in which on each trial the outcome must be one of $k = 2$ alternatives. Within the series, one of the alternatives occurs on n_1 trials and the other alternative occurs on n_2 trials. Thus, $n_1 + n_2 = N$. A **run** is a sequence within a series in which one of the k alternatives occurs on consecutive trials. On the trial prior to the first trial of a run (with the exception of Trial 1 in the series) and the trial following the last trial of a run (with the exception of the Nth trial in the series), the alternative that occurs will be different than the alternative that occurs during each of the trials of the run. The minimum length of a run is one trial, and the maximum length of a run is equal to N, the total number of trials in the series. To illustrate the computation of the length of a run, consider the three series noted in Figure 7.1. Each series is comprised of $N = 10$ trials. On each trial a coin is flipped and the outcome of a Heads (H) or a Tails (T) is recorded.

Trial	**1**	**2**	**3**	**4**	**5**	**6**	**7**	**8**	**9**	**10**
Series A:	H	H	T	H	H	T	T	T	H	T
Series B:	T	H	T	H	T	H	T	H	T	H
Series C:	H	H	H	H	H	H	H	H	H	H

Figure 7.1 Illustration of Runs

In **Series A** and **Series B** there are $n_1 = 5$ Heads and $n_2 = 5$ Tails. In **Series C** there are $n_1 = 10$ Heads and $n_2 = 0$ Tails.

In **Series A** there are six runs. Run 1 consists of Trials 1 and 2 (which are Heads). Run 2 consists of Trial 3 (which is a Tails). Run 3 consists of Trials 4 and 5 (which are Heads). Run 4 consists of Trials 6–8 (which are Tails). Run 5 consists of Trial 9 (which is a Heads). Run 6 consists of Trial 10 (which is a Tails). This can be summarized visually by underlining all of the runs as noted below. Note that all the runs are comprised of sequences involving the same alternative. Thus: $\underline{\text{H H}}\ \underline{\text{T}}\ \underline{\text{H H}}\ \underline{\text{T T T}}\ \underline{\text{H}}\ \underline{\text{T}}$.

In **Series B** there are 10 runs. Each of the trials constitutes a separate run, since on each trial a different alternative occurs. Note that on each trial the alternative for that trial is preceded by and followed by a different alternative. Thus: $\underline{\text{T}}\ \underline{\text{H}}\ \underline{\text{T}}\ \underline{\text{H}}\ \underline{\text{T}}\ \underline{\text{H}}\ \underline{\text{T}}\ \underline{\text{H}}\ \underline{\text{T}}\ \underline{\text{H}}$.

As noted in the definition of a run, Trial 1 cannot be preceded by a different alternative, since it is the first trial, and Trial 10 cannot be followed by a different alternative, since it is the last trial.

In **Series C** there is one run. This is the case, since the same alternative occurs on each trial. Thus: H H H H H H H H H H.

Intuitively, one would expect that of the three series, Series A is most likely to conform to the definition of a random series. This is the case, since it is highly unlikely that a random series will exhibit a discernible pattern that will allow one to predict at above chance which of the alternatives will appear on a given trial. Series B and C, on the other hand, are characterized by patterns that will probably bias the guess of someone who is attempting to predict what the outcome will be if there is an eleventh trial. It is logical to expect that the strength of such a bias will be a direct function of the length of any series exhibiting a consistent pattern.[2]

The test statistic for the **single-sample runs test** is based on the assumption that the number of runs in a random series will be expected to fall within a certain range of values. Thus, if for a given series the number of runs is less than some minimum value or greater than some maximum value, it is likely that the series is not random. The determination of the minimum allowable number of runs and maximum allowable number of runs in a series of N trials takes into account the number of runs, as well as the frequency of occurrence of each of the two alternatives within the series.

It should be noted that although the **single-sample runs test** is most commonly employed with a binomially distributed variable for which $\pi_1 = \pi_2 = .5$ (as is the case for a coin toss), it is not required that the values π_1 and π_2 equal .5 in the underlying population. It is important to remember that the runs test does not evaluate a hypothesis regarding the values of π_1 and π_2 in the underlying population, nor does it make any assumption with regard to the latter values. The test statistic for the runs test is a function of the proportion of times each of the alternatives occurs in the sample data/series (i.e., $p_1 = n_1/N$ and $p_2 = n_2/N$). If the observed proportion for each of the alternatives is inconsistent with its actual proportion in the underlying population, the **single-sample runs test** is not designed to detect such a difference. Example 7.6 in Section VIII illustrates a situation in which the **single-sample runs test** is employed when it is known that $\pi_1 \neq \pi_2 \neq .5$.

II. Example

Example 7.1 *In a test of extrasensory ability a coin is flipped* 20 *times by an experimenter. Prior to each flip of the coin the subject is required to guess whether it will come up* Heads *or* Tails. *After each trial the subject is told whether his guess is correct or incorrect. The actual outcomes for the 20 coin flips are listed below:*

H H H T T T H H T T H T H T H T T T H H

To rule out the possibility that the subject gained extraneous information as a result of a nonrandom pattern in the above series, the experimenter decides to evaluate it with respect to randomness. Does an analysis of the series suggest that it is nonrandom?

III. Null versus Alternative Hypotheses

Null hypothesis H_0: The events in the underlying population represented by the sample series are distributed randomly.

Alternative hypothesis H_1: The events in the underlying population represented by the sample series are distributed nonrandomly. (This is a **nondirectional alternative hypothesis** and it is evaluated with a **two-tailed test**.)

or

H_1: The events in the underlying population represented by the sample series are distributed nonrandomly due to too few runs. (This is a **directional alternative hypothesis** and it is evaluated with a **one-tailed test.**)

or

H_1: The events in the underlying population represented by the sample series are distributed nonrandomly due to too many runs. (This is a **directional alternative hypothesis** and it is evaluated with a **one-tailed test.**)

Note: Only one of the above noted alternative hypotheses is employed. If the alternative hypothesis the researcher selects is supported, the null hypothesis is rejected.

IV. Test Computations

In order to compute the test statistic for the **single-sample runs test**, one must determine the number of times each of the two alternatives appears in the series and the number of runs in the series. Thus, we determine that the series described in Example 7.1 is comprised of $n_1 = 10$ **Heads** and $n_2 = 10$ **Tails**. Note that $n_1 + n_2 = N = 20$. We also determine that there are $r = 11$ runs, which represents the test statistic for the **single-sample runs test.** Specifically as one moves from Trial 1 to Trial 20: **H H H** (Run 1); **T T T** (Run 2); **H H** (Run 3); **T T** (Run 4); **H** (Run 5); **T** (Run 6); **H** (Run 7); **T** (Run 8); **H** (Run 9); **T T T** (Run 10); and **H H** (Run 11). This can also be represented visually by underlining each of the runs. Thus:

$$\underline{HHH}\ \underline{TTT}\ \underline{HH}\ \underline{TT}\ \underline{H}\ \underline{T}\ \underline{H}\ \underline{T}\ \underline{H}\ \underline{TTT}\ \underline{HH}$$

V. Interpretation of the Test Results

The computed value $r = 11$ is interpreted by employing **Table A8 (Table of Critical Values for the Single-Sample Runs Test)** in the **Appendix.** The critical values listed in **Table A8** only allow the null hypothesis to be evaluated at the .05 level if a two-tailed/nondirectional alternative hypothesis is employed, and at the .025 level if a one-tailed/directional alternative hypothesis is employed. No critical values are recorded in **Table A8** for the **single-sample runs test** for very small sample sizes, since the levels of significance employed in the table cannot be achieved for sample sizes below a specific minimum value. More extensive tables for the **single-sample runs test** which provide critical values for other levels of significance can be found in Swed and Eisenhart (1943) and Beyer (1968).[3]

Note that in **Table A8** the critical r values are listed in reference to the values of n_1 and n_2, which represent the frequencies that each of the alternatives occurs in the series. Since in Example 7.1, $n_1 = 10$ and $n_2 = 10$, we locate the cell in **Table A8** that is the intersection of these two values. In the appropriate cell, the upper value identifies the **lower limit** for the value of r, whereas the lower value identifies the **upper limit** for the value of r. The following guidelines are employed in reference to the latter values.

a) If the nondirectional alternative hypothesis is employed, in order to reject the null hypothesis the obtained value of r must be equal to or greater than the tabled critical upper limit at the prespecified level of significance, or be equal to or less than the tabled critical lower limit at the prespecified level of significance.

b) If the directional alternative hypothesis predicting too few runs is employed, in order to reject the null hypothesis the obtained value of r must be equal to or less than the tabled critical lower limit at the prespecified level of significance.

c) If the directional alternative hypothesis predicting too many runs is employed, in order to reject the null hypothesis the obtained value of r must be equal to or greater than the tabled critical upper limit at the prespecified level of significance.[4]

Employing **Table A8**, we determine that for $n_1 = n_2 = 10$, the tabled critical lower and upper critical r values are $r = 6$ and $r = 16$. Thus, if the nondirectional alternative hypothesis is employed (with $\alpha = .05$), the obtained value of r will be significant if it is equal to or less than 6 or equal to or greater than 16. In other words, it will be significant if there are either 16 or more runs or 6 or less runs in the data. Since $r = 11$ falls inside this range, the nondirectional alternative hypothesis is not supported.

If the directional alternative hypothesis predicting too few runs is employed (with $\alpha = .025$), the obtained value of r will only be significant if it is equal to or less than 6. In other words, it will only be significant if there are 6 or less runs in the data. Since $r = 11$ is greater than 6, the directional alternative hypothesis predicting too few runs is not supported.

If the directional alternative hypothesis predicting too many runs is employed (with $\alpha = .025$), the obtained value of r will only be significant if it is equal to or greater than 16. In other words, it will only be significant if there are 16 or more runs in the data. Since $r = 11$ is less than 16, the directional alternative hypothesis predicting too many runs is not supported.

Our analysis indicates that regardless of which alternative hypothesis one employs, the null hypothesis cannot be rejected. Thus, the data do not allow the researcher to conclude that the series is nonrandom.

VI. Additional Analytical Procedures for the Single-Sample Runs Test and/or Related Tests

1. The normal approximation of the single-sample runs test for large sample sizes The normal distribution can be employed with a large sample size/series to approximate the exact distribution of the **single-sample runs test**. The large sample approximation is generally employed for sample sizes larger than those documented in **Table A8**. Equation 7.1 is employed for the normal approximation of the **single-sample runs test**.

$$z = \frac{r - \left[\dfrac{2n_1n_2}{n_1 + n_2} + 1\right]}{\sqrt{\dfrac{2n_1n_2(2n_1n_2 - n_1 - n_2)}{(n_1 + n_2)^2(n_1 + n_2 - 1)}}} \qquad \textbf{(Equation 7.1)}$$

In the numerator of the above equation the term $[(2n_1n_2)/(n_1 + n_2)] + 1$ represents the mean of the sampling distribution of runs in a random series in which there are N observations. The latter value may be summarized with the notation u_r. In other words, given $n_1 = 10$ and $n_2 = 10$, if in fact the distribution is random, the best estimate of the number of runs one can expect to observe is $\mu_r = 11$. The denominator in Equation 7.1 represents the expected standard deviation of the sampling distribution for the normal approximation of the test statistic. The latter value is summarized by the notation σ_r. If the values μ_r and σ_r are employed, Equation 7.1 can also be written as follows: $z = (r - \mu_r)/\sigma_r$.

Employing Equation 7.1 with the data for Example 7.1, the value $z = 0$ is computed.

$$z = \frac{11 - \left[\frac{(2)(10)(10)}{10 + 10} + 1\right]}{\sqrt{\frac{(2)(10)(10)[(2)(10)(10) - 10 - 10]}{(10 + 10)^2(10 + 10 - 1)}}} = \frac{0}{2.18} = 0$$

Since $\mu_r = 11$ and $\sigma_r = 2.18$, the result of the above analysis can also be summarized as follows: $z = (11 - 11)/2.18 = 0$.

The obtained value $z = 0$ is evaluated with **Table A1 (Table of the Normal Distribution) in the Appendix.** In order to be significant, the obtained absolute value of z must be equal to or greater than the tabled critical value at the prespecified level of significance. The tabled critical two-tailed .05 and .01 values are $z_{.05} = 1.96$ and $z_{.01} = 2.58$, and the tabled critical one-tailed .05 and .01 values are $z_{.05} = 1.65$ and $z_{.01} = 2.33$. The following guidelines are employed in evaluating the null hypothesis.

a) If the nondirectional alternative hypothesis is employed, in order to reject the null hypothesis the obtained absolute value of z must be equal to or greater than the tabled critical two-tailed value at the prespecified level of significance. In Example 7.1 the nondirectional alternative hypothesis is not supported, since the obtained value $z = 0$ is less than both of the aforementioned tabled critical two-tailed values.

b) If the directional alternative hypothesis predicting too few runs is employed, in order to reject the null hypothesis the following must be true: 1) The obtained value of z must be a negative number; and 2) The absolute value of z must be equal to or greater than the tabled critical one-tailed value at the prespecified level of significance. In Example 7.1 the directional alternative hypothesis predicting too few runs is not supported, since the obtained value $z = 0$ is not a negative number (as well as the fact that it is less than the tabled critical one-tailed values $z_{.05} = 1.65$ and $z_{.01} = 2.33$).

c) If the directional alternative hypothesis predicting too many runs is employed, in order to reject the null hypothesis the following must be true: 1) The obtained value of z must be a positive number; and 2) The absolute value of z must be equal to or greater than the tabled critical one-tailed value at the prespecified level of significance. In Example 7.1 the directional alternative hypothesis predicting too many runs is not supported, since the obtained value $z = 0$ is not a positive number (as well as the fact that it is less than the tabled critical one-tailed values $z_{.05} = 1.65$ and $z_{.01} = 2.33$).

Thus, when the normal approximation is employed, as is the case when the critical values in **Table A8** are used, the null hypothesis cannot be rejected regardless of which alternative hypothesis is employed. Consequently, we cannot conclude that the series is not random.

2. The correction for continuity for the normal approximation of the single-sample runs test Although it is not described by most sources, Siegel and Castellan (1988) recommend that a correction for continuity be employed for the normal approximation of the **single-sample runs test.** Equation 7.2, which is the continuity corrected equation, will always yield a smaller absolute z value than the value derived with Equation 7.1.[5]

$$z = \frac{|r - \mu_r| - .5}{\sigma_r} \qquad \textbf{(Equation 7.2)}$$

Employing Equation 7.2 with the data for Example 7.1, the value $z = -.23$ is computed.

$$z = \frac{|11 - 11| - .5}{2.18} = -.23$$

Since the absolute value $z = .23$ is lower than the tabled critical two-tailed value $z_{.05} = 1.96$ and the tabled critical one-tailed value $z_{.05} = 1.65$, the null hypothesis cannot be rejected regardless of which alternative hypothesis is employed (which is also the case when the correction for continuity is not employed). Thus, we cannot conclude that the series is not random.

VII. Additional Discussion of the Single-Sample Runs Test

Alternative tests of randomness The **single-sample runs test** is one of many tests that have been developed for assessing randomness. Unlike the **single-sample runs test**, which can only evaluate a series involving **binary events**, most of the alternative procedures for assessing randomness allow one to evaluate series in which on each trial there are more than two possible outcomes.[6] A general problem with tests of randomness is that they do not employ the same criteria for assessing randomness. As a result of this some tests are more stringent than others, and thus it is not uncommon that a series of numbers may meet the requirements of one or more of the available tests of randomness, yet not meet the requirements of one or more of the other tests. Among the more commonly cited alternative tests for evaluating randomness are the following.

a. The frequency test The **frequency test**, which is probably the least demanding of the tests of randomness, assesses randomness on the basis of whether k or more equally probable alternatives occur an equal number of times within a series. The data for the **frequency test** are evaluated with the **chi-square goodness-of-fit test (Test 5)**, and when $k = 2$ the **binomial sign test for a single sample (Test 6)** (as well as the large sample normal approximation) can be employed. Since the **frequency test** only assesses a series with respect to the frequency of occurrence of each of the outcomes, it is insensitive to systematic patterns that may exist within a series. To illustrate this limitation of the **frequency test**, consider the following two binary series consisting of Heads (H) and Tails (T), where $\pi_1 = \pi_2 = .5$.

Series A: H H H T H T T T H T H H T H T T T H T H
Series B: H T H T H T H T H T H T H T H T H T H T

Inspection of the data indicates that both Series A and B are comprised of 10 Heads and 10 Tails. Since the number of times each of the alternatives occurs is at the chance level (i.e., each occurs in 50% of the trials), if one elects to analyze either series employing either the **chi-square goodness-of-fit test** or the **binomial sign test for a single sample**, both series will meet the criterion for being random. However, visual inspection of the two series clearly suggests that, as opposed to Series A, Series B is characterized by a systematic pattern involving the alternation of Heads and Tails. This latter observation clearly suggests that the latter series is not random.

At this point, the **frequency test** and the **single-sample runs test** will be applied to the same set of data. Specifically, both tests will be employed to evaluate whether or not the series below (which consists of $N = 30$ trials) is random. In the series, each of the runs has been underlined.

HHHHH T H T HHH T H T H T HHH TTTT HHHHHHH

Since $k = 2$, the **binomial sign test for a single sample** will be employed to represent the **frequency test**. When the binomial sign test is employed the null hypothesis that is evaluated is: H_0: $\pi_1 = .5$ (since it is assumed that $\pi_1 = \pi_2 = .5$). For both the **binomial sign test for a single sample** and the **single-sample runs test**, it will be assumed that a nondirectional alternative hypothesis is evaluated.

In the series that is being evaluated, the number of Heads is $n_1 = 21$ and the number of Tails is $n_2 = 9$. Employing Equation 6.9 (which is the continuity corrected normal approximation of the **binomial sign test for a single sample**), the value $z = 2.01$ is computed.

$$z = \frac{|21 - (30)(.5)| - .5}{\sqrt{(30)(.5)(.5)}} = 2.01$$

Since the obtained value $z = 2.01$ is greater than the tabled critical two-tailed value $z_{.05} = 1.96$, the nondirectional alternative hypothesis H_1: $\pi_1 \neq .5$ is supported. Thus, based on the above analysis with the **binomial sign test for a single sample**, one can conclude that the series is not random.

In evaluating the same series with the **single-sample runs test**, we determine that there are $r = 13$ runs in the data. The expected number of runs is $[(2)(21)(9)/(21 + 9)] + 1 = 13.6$, which is barely above the observed value $r = 13$. Employing Equation 7.1, the value $z = -.27$ is computed.[7]

$$z = \frac{13 - \left[\frac{(2)(21)(9)}{21 + 9} + 1\right]}{\sqrt{\frac{(2)(21)(9)[(2)(21)(9) - 21 - 9]}{(21 + 9)^2(21 + 9 - 1)}}} = -.27$$

Since the obtained absolute value $z = .27$ is less than the tabled critical two-tailed value $z_{.05} = 1.96$, the nondirectional alternative hypothesis for the runs test is not supported. Thus, based on the above analysis with the **single-sample runs test**, one can conclude that the series is random.

The fact that a significant result is obtained when the **binomial sign test for a single sample** is employed to evaluate the series, reflects the fact that the latter test only takes into account the number of observations in each of the two categories, but does not take into consideration the ordering of the data. The **single-sample runs test**, on the other hand, is sensitive to the ordering of the data, yet will not always identify a nonrandom series if nonrandomness is a function of the number of outcomes for each of the alternatives.

There will also be instances where one may conclude a series is random based on an analysis of the data with the **single-sample runs test**, yet cannot conclude the series is random if the **binomial sign test for a single sample** is employed for the analysis. Such a series, consisting of 15 Heads and 15 Tails, is depicted below.

H H H H H H H H H H H H H H H T T T T T T T T T T T T T T T

When Equation 6.9 (representing the continuity corrected normal approximation of the **binomial sign test for a single sample**) is employed, it yields the following result: $z = [|15 - (30)(.5)| - .5]/\sqrt{(30)(.5)(.5)} = -.18$. Since the latter absolute value is less than the tabled critical two-tailed value $z_{.05} = 1.96$, the result is not significant, and thus one can conclude that the series is random. When, however, the same series, which consists of only two runs, is evaluated with Equation 7.1 (the equation for the normal approximation of the **single-sample runs test**), it yields the following result: $z = (r - \mu_r)/\sigma_r = (2 - 16)/2.69 = -5.20$. Since the absolute value $z = 5.20$ is greater than the tabled critical two-tailed values $z_{.05} = 1.96$ and $z_{.01} = 2.58$, the result is significant at both the .05 and .01 levels. Thus, if one employs the **single-sample runs test** one can conclude that the series is not random.

b. The gap test Described in Gruenberger and Jaffray (1965) and Knuth (1969), the **gap test** evaluates the number of "gaps" between the appearance of a digit in a series and the reappearance of the same digit. Thus, if we have $k = 10$ digits and each of the digits is equally likely to occur, it would be expected that if the distribution of digits in a series consisting of N digits is random, the average gap/interval for the reoccurrence of each digit will equal $k = 10$. A gap for any digit can be determined by selecting that digit and counting until the next appearance of the same digit. To illustrate the concept of a gap, consider the following series of digits: 0121046720. For the digit 0 we can count two gaps of lengths 4 and 5 respectively. This is the case, since the number of digits from the first 0 to the second 0 is 4, and the number of digits from the second 0 to the third 0 is 5. In conducting the **gap test**, all of the gaps for each of the digits in the series are counted, after which the computed gap values are evaluated. The analysis of the data for a series within the framework of the **gap test** can involve one or more of the following tests: a) The **single-sample *z* test (Test 1)** can be employed to contrast the computed mean gap versus the expected mean gap for each digit; b) The **single-sample chi-square test for a population variance (Test 3)** can be employed to contrast the observed versus expected variance of the gaps for each digit; and c) The **chi-square goodness-of-fit test** can be employed to compare the observed versus expected gap lengths of a specific value for each digit.

c. The poker test Described in Gruenberger and Jaffray (1965) and Knuth (1969), this test conceptualizes the data as a set of hands in the game of poker. Starting with the first five digits in a series of N digits, the five digits are considered as the initial poker hand (flushes are not possible, but five of a kind can occur).[8] The outcome with respect to which poker hand is represented is recorded. The analysis is repeated, employing as the second hand digits 2 through 6. The third hand will be comprised of digits 3 through 7, and so on. The analysis is carried on until the end of the series (which will be the point at which a five-digit hand is no longer possible). The total of $(N - 4)$ possible hands is commonly evaluated with the **chi-square goodness-of-fit test.**[9] The latter test compares the observed frequencies of each of the possible hands with their theoretical frequencies. Of all the tests of randomness that have been developed, the **poker test** is perhaps the most demanding. Many series of digits that are able to meet the criteria of other tests of randomness will fail the **poker test**.

d. Among the other tests of randomness that have been developed are:

1) **Autocorrelation** (also known as **serial correlation**) This procedure, which is discussed in detail in Section VII of the **Pearson product-moment correlation coefficient (Test 22)**, can be employed with series in which in each trial there are two or more possible outcomes. Within the framework of autocorrelation, one can conclude that a series is random if the correlation coefficient between successive numbers in the series is equal to zero. The **Durbin–Watson test** (1950, 1951, 1971) is one of a number of procedures that are employed for autocorrelation. The latter test is described in sources such as Chou (1989), Montgomery and Peck (1992), and Netter *et al.* (1988).

2) The **serial test** The **serial test** evaluates the occurrence of each two-digit combination ranging from 00 to kk, where k is the largest digit that can occur in the series. The serial test can be generalized to groups consisting of combinations of three or more digits.

3) The **coupon collector's test** The **coupon collector's test** evaluates the number of digits required in order to make a complete set of k digits which consist of the integer values 1 to k.

4) **Von Neumann ratio test on independence/mean square successive difference test** Attributed to Bellinson *et al.* (1941) and Von Neumann (1941) and described in Bennett and Franklin (1954) and Chou (1989), this test contrasts the mean of the squares of the differences

of $(n - 1)$ successive differences in a series of n numbers with the variance of the n numbers. Tables of the sampling distribution for the test were derived by Hart (1942).

5) **Tests of trend analysis/time series analysis** Economists often refer to a set of observations that are measured over a period of time as a **time series**. The pattern of the data in a time series may be random or may instead be characterized by patterns or trends. **Trend analysis** and **time series analysis** are terms that are used to describe a variety of statistical procedures which are employed for analyzing such data. Among those tests used for trend analysis that can be employed to identify a nonrandom series is the **Cox–Stuart test for trend** (Cox and Stuart (1955) and described in Daniel (1988)), which is a modification of the **binomial sign test for a single sample**. Tests of time series and trend analysis are commonly described in books on business and economic statistics (e.g., Chou (1989), Hoel and Jessen (1982), Montgomery and Peck (1992), and Netter *et al.* (1988)).

In closing this discussion it should be noted that a distinction is generally made between a **random** and **pseudorandom** series of numbers. It is assumed that if a series is **random** there is no algorithm that will allow one to predict at above chance which of the possible outcomes will occur on a given trial. A **pseudorandom** series, on the other hand, is typically generated through use of a computer program that employs a deterministic algorithm. As a result of this, if one is privy to the rule stated by the algorithm, one will be able to correctly predict all of the numbers in the series in the order in which they occur. **Pseudorandom** number series are commonly employed in research which tries to simulate natural processes that are assumed to be random. Such research is often referred to as **Monte Carlo research.** Use of **pseudorandom** data in simulation research provides scientists with a mechanism for studying natural processes or evaluating problems that otherwise would be impossible or more problematical to evaluate.

VIII. Additional Examples Illustrating the Single-Sample Runs Test

As is the case with Example 7.1, Examples 7.2–7.5 all involve series in which $N = 20$, $n_1 = n_2 = 10$, and $r = 11$. By virtue of employing identical data, the latter examples all yield the same result as Example 7.1. Example 7.5 illustrates the application of the **single-sample runs test** to a design involving two independent samples.[10] In Examples 7.1–7.5 it is implied that if the series involved are, in fact, random, it is probably reasonable to assume in the underlying population $\pi_1 = \pi_2 = .5$ (in other words, that each alternative has an equal likelihood of occurring in the underlying population, even if the latter is not reflected in the sample data). Example 7.6 illustrates the application of the **single-sample runs test** to a design in which it is known that in the underlying population $\pi_1 \neq \pi_2 \neq .5$.

Example 7.2 *A meteorologist conducts a study to determine whether humidity levels recorded at* 12 *noon for* 20 *consecutive days in July* 1995 *are distributed randomly with respect to whether they are above or below the average humidity recorded during the month of July during the years* 1990 *through* 1994. *Recorded below is a listing of whether the humidity for* 20 *consecutive days is above* (+) *or below* (–) *the July average.*

+ + + – – – + + – – + – + – + – – – + +

Do the data indicate that the series of temperature readings is random?

Example 7.3 *The gender of* 20 *consecutive patients who register at the emergency room of a local hospital is recorded below (where:* **M** = **Male; F** = **Female).**

F F F M M M F F M M F M F M F M M M F F

Do the data suggest that the gender distribution of entering patients is random?

Example 7.4 *A quality control study is conducted on a machine that pours milk into containers. The amount of milk (in liters) dispensed by the machine into* 21 *consecutive containers follows:* 1.90, 1.99, 2.00, 1.78, 1.77, 1.76, 1.98, 1.90, 1.65, 1.76, 2.01, 1.78, 1.99, 1,76, 1.94, 1.78, 1.67, 1.87, 1.91, 1.91, 1.89. *If the median number of liters the machine is programmed to dispense is* 1.89, *is the distribution random with respect to the amount of milk poured above versus below the median value?*

In Example 7.4 it can be assumed that if the process is random, the scores should be distributed evenly throughout the series, and that there should be no obvious pattern with respect to scores above versus below the median. Thus, initially we list the 21 scores in sequential order with respect to whether they are above (+) or below (-) the median. Since one of the scores (that of the last container) is at the median, it is eliminated from the analysis. The latter protocol is employed for all scores equal to the median when the **single-sample runs test** is used within this context. The relationship of the first 20 scores to the median is recorded below.

+ + + - - - + + - - + - + - + - - - + +

Since the above sequence of runs is identical to the sequence observed for Examples 7.1–7.3, it yields the same result. Thus, there is no evidence to indicate that the distribution is not random. Presence of a nonrandom pattern due to a defect in the machine can be reflected in a small number of large cycles (i.e., each cycle consists of many trials). Thus, one might observe 10 consecutive containers that are overfilled followed by 10 consecutive containers that are underfilled. A nonrandom pattern can also be revealed by an excess of runs attributed to multiple small cycles (i.e., each cycle consists of few trials).

The reader should take note of the fact that the although the **Wilcoxon signed-ranks test (Test 4)** can also be employed to evaluate the data for Example 7.4, it is not appropriate to employ the latter test for evaluating a hypothesis regarding randomness. The **Wilcoxon signed-ranks test** can be used to evaluate whether the data indicate that the true median value for the machine is some value other than 1.89. It does not provide information concerning the ordering of the data.

Example 7.5 *In a study on the efficacy of an antidepressant drug, each of* 20 *clinically depressed patients is randomly assigned to one of two treatment groups. For* 6 *months one group is given the antidepressant drug and the other group is give a placebo. After* 6 *months have elapsed, subjects in both groups are rated for depression by a panel of psychiatrists who are blind with respect to group membership. Each subject is rated on a* 100 *point scale (the higher the rating the greater the level of depression). The depression ratings for the two groups follow.*

Drug group: 20, 25, 30, 48, 50, 60, 70, 80, 95, 98
Placebo group: 35, 40, 42, 52, 55, 62, 72, 85, 87, 90

Do the data indicate there is a difference between the groups?

Since it is less powerful than alternative procedures for evaluating the same design (which typically contrast groups with respect to a measure of central tendency), the **single-sample runs test** is not commonly employed in evaluating a design involving two independent samples. Example 7.5 will, nevertheless, be used to illustrate its application to such a situation. In order to implement the runs test, the scores of the 20 subjects are arranged ordinally with respect to group membership as shown below (Where **D** represents the Drug group and **P** represents the Placebo group).

20	25	30	35	40	42	48	50	52	55	60	62	70	72	80	85	87	90	95	98
D	**D**	**D**	**P**	**P**	**P**	**D**	**D**	**P**	**P**	**D**	**P**	**D**	**P**	**D**	**P**	**P**	**P**	**D**	**D**

Runs are evaluated as in previous examples. In this instance, the two categories employed in the series represent the two groups from which the scores are obtained. When the scores are arranged ordinally, if there is a difference between the groups it is expected that most of the scores in one group will fall to the left of the series, and that most of the scores in the other group will fall to the right of the series. More specifically, if the drug is effective one will predict that the majority of the scores in the Drug group will fall to the left of the series. Such an outcome will result in a small number of runs. Thus, if the number of runs is equal to or less than the tabled critical lower limit at the prespecified level of significance, one can conclude that the pattern of the data is nonrandom. Such an outcome will allow the researcher to conclude that there is a significant difference between the groups. Since $n_1 = n_2 = 10$ and $r = 11$, the data for Example 7.5 are identical to that obtained for Examples 7.1–7.4. Analysis of the data do not indicate that the series is nonrandom, and thus one cannot conclude that the groups differ from one another (i.e., represent two different populations).

Let us now consider two other possible patterns for Example 7.5. The first pattern depicted below contains $r = 2$ runs. It yields a significant result since in **Table A8**, for $n_1 = n_2 = 10$, any number of runs equal to or less than 6 is significant at the .05 level. Thus, the pattern depicted below will lead the researcher to conclude that the groups represent two different populations.

D D D D D D D D D D P P P P P P P P P P

Consider next the following pattern:

D D D D D P P P P P P P P P P D D D D D

Since the above pattern contains $r = 3$ runs, it is also significant at the .05 level. Yet inspection of the pattern suggests that a test which compares the mean or median values of the two groups will probably not result in a significant difference. This is based on the observation that the group receiving the drug contains the five highest and five lowest scores. Thus, if the performance of the Drug group is summarized with a measure of central tendency, such a value will probably be close to the analogous value obtained for the placebo group (whose scores cluster in the middle of the series). Nevertheless, the pattern of the data certainly suggests that a difference with respect to the variability of scores exists between the groups. In other words, half of the people receiving the drug respond to it favorably, while the other half respond to it poorly. Most of the people in the placebo group, on the other hand, obtain scores in the middle of the distribution. The above example illustrates the fact that in certain situations the **single-sample runs test** may provide more useful information regarding two independent samples than other tests which are more commonly used for such a design — specifically, the ***t* test for two independent samples (Test 8)** and the **Mann-Whitney *U* test (Test 9)**, both of which evaluate measures of central tendency. The pattern of data depicted for the series under discussion is more likely to be identified by a test that contrasts the variability of two independent samples. In addition to the **single-sample runs test**, another procedure (which is discussed later in the book) that is even better suited to identify differences with respect to group variability is the **Siegel–Tukey test of equal variability (Test 10)**.

Example 7.6 *A quality control engineer is asked by the manager of a factory to evaluate a machine that packages glassware. The manager informs the engineer that* 90% *of the glassware processed by the machine remains intact, while the remaining* 10% *of the glassware is cracked during the packaging process. It is suspected that some cyclical environmental condition may be causing the machine to produce breakages at certain points in times. In order to assess the situation, the quality control engineer records a series comprised of* 1000 *pieces* *of glassware packaged by the machine over a two week period. It is determined that within the series 890 pieces of glassware remain intact and that 110 are cracked. It is also determined that within the series there are only 4 runs. Do the data indicate the series is nonrandom?*

In Example 7.6, $N = 1000$, $n_1 = 890$, $n_2 = 110$, and $r = 4$. Employing these values in Equation 7.1, the value $z = -3.12$ is computed. In employing the latter equation, the computed value for the expected number of runs is $\mu_r = 196.8$, which is well in excess of the observed value $r = 4$.

$$z = \frac{4 - \left[\frac{(2)(890)(110)}{890 + 110} + 1\right]}{\sqrt{\frac{(2)(890)(110)[(2)(890)(110) - 890 - 110]}{(890 + 110)^2(890 + 110 - 1)}}} = -3.12$$

Employing **Table A8**, we determine that the absolute value $z = 3.12$ is greater than the tabled critical two-tailed values $z_{.05} = 1.96$ and $z_{.01} = 2.58$, and the tabled critical one-tailed values $z_{.05} = 1.65$ and $z_{.01} = 2.33$. Thus, the nondirectional alternative hypothesis is supported at both .05 and .01 levels. Since the obtained value of z is negative, the directional alternative hypothesis predicting too few runs is supported at the both the .05 and .01 levels. Obviously, the directional alternative hypothesis predicting too many runs is not supported.

Note that in Example 7.6 the plant manager informs the quality control engineer that the likelihood of a piece of glassware cracking is $\pi_1 = .9$, whereas the likelihood of it being intact is $\pi_2 = .1$. Thus, each of the two alternatives (**Intact** versus **Cracked**) are not equally likely to occur on a given trial. The observed proportion of cases in each of the two categories $p_1 = 890/1000 = .89$ and $p_2 = 110/1000 = .11$ are quite close to the values $\pi_1 = .9$ and $\pi_2 = .1$. As noted earlier in the discussion of the **single-sample runs test**, if the observed proportions are substantially different from the values assumed for the population proportions, the test will not identify such a difference, and the analysis of runs will be based on the observed values of the proportions in the series, regardless of whether or not they are consistent with the underlying population proportions. The question of whether the proportions computed for the sample data are consistent with the population proportions is certainly relevant to the issue of whether or not the series is random. However, as noted earlier, the **binomial sign test for a single sample** and the **chi-square goodness-of-fit test** are the appropriate tests to employ to evaluate the latter question.

References

Bellinson, H. R., Von Neumann, J., Kent, R. H., and Hart, B. I. (1941). The mean square successive difference. **Annals of Mathematical Statistics**, 12, 153–162.

Bennett, C. A. and Franklin, N. L. (1954). **Statistical analysis in chemistry and the chemical industry**. New York: John Wiley and Sons, Inc.

Beyer, W. H. (1968). **Handbook of tables for probability and statistics** (2nd ed.). Cleveland, OH: The Chemical Rubber Company.

Chou, Y. (1989). **Statistical analysis for business and economics**. New York: Elsevier.

Cox, D. R. and Stuart, A. (1955). Some quick tests for trend in location and dispersion. **Biometrika**, 42, 80–95.

Daniel, W. (1990). **Applied nonparametric statistics** (2nd ed.). Boston: PWS–Kent Publishing Company.

Durbin, J. and Watson, G. S. (1950). Testing for serial correlation in least squares regression I. **Biometrika**, 37, 409–438.

Durbin, J. and Watson, G. S. (1951). Testing for serial correlation in least squares regression II. **Biometrika**, 38, 159–178.

Durbin, J. and Watson, G. S. (1971). Testing for serial correlation in least squares regression III. **Biometrika**, 58, 1–19.

Gruenberger, F. and Jaffrey, G. (1965). **Problems for computer solution**. New York: John Wiley and Sons.

Hart, B. I. (1942). Significance levels for the ratio of the mean square successive difference to the variance. **Annals of Mathematical Statistics**, 13, 445–447.

Hoel, P. G. and Jessen, R. J. (1982). **Basic statistics for business and economics** (3rd ed.). New York: John Wiley and Sons.

Hogg, R. V. and Tanis, E. A. (1988). **Probability and statistical inference** (3rd ed.). New York: Macmillan Publishing Company.

Knuth, D. (1969). **Semi-numerical algorithms: The art of computer programming (Vol. 2)**. Reading, MA: Addison–Wesley.

Montgomery, D. C. and Peck, E. A. (1992). **Introduction to linear regression analysis** (2nd ed.). New York: John Wiley and Sons, Inc.

Netter, J., Wasserman, W., and Kutner, M. H. (1983). **Applied linear regression models** (3rd ed.). Homewood, IL: Richard D. Irwin, Inc.

Siegel, S. (1956). **Nonparametric statistics for the behavioral sciences** (1st ed.). New York: McGraw–Hill Book Company.

Siegel, S. and Castellan, N. J., Jr. (1988). **Nonparametric statistics for the behavioral sciences** (2nd ed.). New York: McGraw–Hill Book Company.

Swed, F. S. and Eisenhart, C. (1943). Tables for testing randomness of grouping in a sequence of alternatives. **Annals of Mathematical Statistics**, 14, 66–87.

Von Neumann, J. (1941). Distribution of the ratio of the mean square successive difference to the variance. **Annals of Mathematical Statistics**, 12, 307–395.

Wald, A. and Wolfowitz, J. (1940). On a test whether two samples are from the same population. **Annals of Mathematical Statistics**, 11, 147–162.

Endnotes

1. An alternate definition of randomness employed by some sources is that in a random series, each of k possible alternatives is equally likely to occur on any trial, and that the outcome on each trial is independent of the outcome on any other trial. The problem with the latter definition is that it cannot be applied to a series in which on each trial there are two or more alternatives which do not have a equal likelihood of occurring (the stipulation regarding independence does, however, also apply to a series involving alternatives that do not have an equal likelihood of occurring on each trial). In point of fact, it is possible to apply the concept of randomness to a series in which $\pi_1 \neq \pi_2$. To illustrate the latter, consider the following example. Assume we have a series consisting of N trials involving a binomially distributed variable for which there are two possible outcomes **A** and **B**. The theoretical probabilities in the underlying population for each of the outcomes are $\pi_A = .75$ and $\pi_B = .25$. If a series involving the two alternatives is in fact random, on each trial the respective likelihoods of alternative **A** versus alternative **B** occurring will not be $\pi_A = \pi_B = .5$, but instead will be $\pi_A = .75$ and $\pi_B = .25$. If such a series is random it is expected that alternative **A** will occur approximately 75% of the time and alternative **B** will occur approximately 25% of the time. However, it important to note that one cannot conclude that the above series is random purely on the basis of the relative frequencies of the two alternatives. To illustrate this, consider the following series consisting of 28 trials which is characterized by the presence of an invariant pattern: **AAABAAABAAABAAABAAABAAABAAAB**.

If one is attempting to predict the outcome on the 29th trial, and if in fact the periodicity of the pattern that is depicted is invariant, the likelihood that alternative **A** will occur on the next trial is not .75, but is in fact 1. This is the case, since the occurrence of events in the series can be summarized by the simple algorithm that the series is comprised of 4 trial cycles, and within each cycle alternative **A** occurs on the first 3 trials and alternative **B** on the fourth trial. The point to be made here is that it is entirely possible to have a random series, even if each of the alternatives is not equally likely to occur on every trial. However, if the occurrence of the alternatives is consistent with their theoretical frequencies, the latter in and of itself does not insure that the series is random.

2. It should be pointed out that, in actuality, each of the three series depicted in Figure 7.1 has an equal likelihood of occurring. However, in most instances where a consistent pattern is present that persists over a large number of trials, such a pattern is more likely to be attributed to a nonrandom factor than it is to chance.

3. The computation of the values in **Table A8** is based on the following logic. If a series consists of N trials and alternative 1 occurs n_1 times and alternative 2 occurs n_2 times, the number of possible combinations involving alternative 1 occurring n_1 times and alternative 2 occurring n_2 times will be $\binom{N}{n_1} = N!/(n_1!n_2!)$. Thus, if a coin is tossed $N = 4$ times, since $\binom{4}{2} = 4!/(2!\ 2!) = 6$, there will be 6 possible ways of obtaining $n_1 = 2$ **Heads** and $n_2 = 2$ **Tails**. Specifically, the 6 ways of obtaining 2 **Heads** and 2 **Tails** are: HHTT, TTHH, THHT, HTTH, THTH, HTHT. Each of the 6 aforementioned sequences constitutes a series, and the likelihood of each of the series occurring is equal. The two series HHTT and TTHH are comprised of 2 runs, the two series THHT and HTTH are comprised of 3 runs, and the two series THTH and HTHT are comprised of 4 runs. Thus, the likelihood of observing 2 runs will equal 2/6 = .33, the likelihood of observing 3 runs will equal 2/6 = .33, and the likelihood of observing 4 runs will equal 2/6 = .33. The likelihood of observing 3 or more runs will equal .67, and the likelihood of observing 2 or more runs will equal 1. A thorough discussion of the derivation of the sampling distribution for the **single-sample runs test**, which is attributed in some sources to Wald and Wolfowitz (1940), is described in Hogg and Tanis (1988).

4. Some of the cells in **Table A8** only list a lower limit. For the sample sizes in question, there is no maximum number of runs (upper limit) that will allow the null hypothesis to be rejected.

5. A general discussion of the correction for continuity can be found under the **Wilcoxon signed-ranks test (Test 4)**. The reader should take note of the fact that the correction for continuity described in this section is intended to provide a more conservative test of the null hypothesis (i.e., make it more difficult to reject). However, when the absolute value of the numerator of Equation 7.1 is equal to or very close to zero, the z value computed with Equation 7.2 will be further removed from zero than the z value computed with Equation 7.1. Since the continuity corrected z value will be extremely close to zero, this result is of no practical consequence (i.e., the null hypothesis will still be retained). This observation regarding the correction for continuity can be generalized to the continuity correction described in the book for other nonparametric tests.

6. Although it is possible to extend the **single-sample runs test** to series that involve more than two alternatives, the critical values employed in **Table A8** (as well as the use of

Equation 7.1) are not applicable to such analyses. For each value of k greater than 2, a separate table of critical values as well as a separate large sample approximation are required.

7. Although Equation 7.2 (the continuity corrected equation for the **single-sample runs test**) yields a slightly smaller absolute value for z for this example and the example to follow, it leads to identical conclusions with respect to the null hypothesis.

8. Although it is generally employed with groups of five digits, the **poker test** can be applied to groups that consist of more or less than five digits.

9. Since the hands evaluated are not actually independent of one another (since they contain overlapping data), the assumption of independence for the **chi-square goodness-of-fit test** is violated. Nevertheless, some sources describe the use of the **chi-square goodness-of-fit test** in this context.

10. The application of the **single-sample runs test** to a design involving two independent samples is described in Siegel (1956) under the **Wald–Wolfowitz (1940) runs test**.

Inferential Statistical Tests Employed with Two Independent Samples (and Related Measures of Association/Correlation)

Test 8

The *t* Test for Two Independent Samples
(Parametric Test Employed with Interval/Ratio Data)

I. Hypothesis Evaluated with Test and Relevant Background Information

Hypothesis evaluated with test Do two independent samples represent two populations with different mean values?[1]

Relevant background information on test The ***t*** **test for two independent samples**, which is employed in a hypothesis testing situation involving two independent samples, is one of a number of inferential statistical tests that are based on the t distribution (which is discussed in detail under the **single-sample *t* test (Test 2)**). Two or more samples are independent of one another if each of the samples is comprised of different subjects.[2] In addition to being referred to as an **independent samples design**, a design involving two or more independent samples is also referred to as a **between-subjects design**, a **between-groups design**, and a **randomized-groups design**. In order to eliminate the possibility of **confounding** in an **independent samples design**, each subject should be randomly assigned to one of the k (where $k \geq 2$) experimental conditions.

In conducting the ***t*** **test for two independent samples**, the two sample means (represented by the notations $\bar{X}_1$ and $\bar{X}_2$) are employed to estimate the values of the means of the populations (μ_1 and μ_2) from which the samples are derived. If the result of the ***t*** **test for two independent samples** is significant, it indicates there is a significant difference between the two sample means, and as a result of the latter the researcher can conclude there is a high likelihood that the samples represent populations with different mean values. It should be noted that the ***t*** **test for two independent samples** is the appropriate test to employ for contrasting the means of two independent samples when the values of the underlying population variances are unknown. In instances where the latter two values are known, the appropriate test to employ is the ***z*** **test for two independent samples (Test 8c)**, which is described in Section VI.

The ***t*** **test for two independent samples** is employed with interval/ratio data, and is based on the following assumptions: a) Each sample has been randomly selected from the population it represents; b) The distribution of data in the underlying population from which each of the samples is derived is normal; and c) The third assumption, which is referred to as the **homogeneity of variance** assumption, states that the variance of the underlying population represented by Sample 1 is equal to the variance of the underlying population represented by Sample 2 (i.e., $\sigma_1^2 = \sigma_2^2$). The homogeneity of variance assumption is discussed in detail in Section VI. If any of the aforementioned assumptions is saliently violated, the reliability of the t test statistic may be compromised.

II. Example

Example 8.1 *In order to assess the efficacy of a new antidepressant drug, ten clinically depressed patients are randomly assigned to one of two groups. Five patients are assigned to* Group 1, *which is administered the antidepressant drug for a period of six months. The other five patients are assigned to* Group 2, *which is administered a placebo during the same* *six month period. Assume that prior to introducing the experimental treatments, the experimenter confirmed that the level of depression in the two groups was equal. After six months elapse all ten subjects are rated by a psychiatrist (who is blind with respect to a subject's experimental condition) on their level of depression. The psychiatrist's depression ratings for the five subjects in each group follow (the higher the rating the more depressed a subject):* **Group 1:** 11, 1, 0, 2, 0; **Group 2:** 11, 11, 5, 8, 4. *Do the data indicate that the antidepressant drug is effective?*

III. Null versus Alternative Hypotheses

Null hypothesis H_0: $\mu_1 = \mu_2$

(The mean of the population Group 1 represents equals the mean of the population Group 2 represents.)

Alternative hypothesis H_1: $\mu_1 \neq \mu_2$

(The mean of the population Group 1 represents does not equal the mean of the population Group 2 represents. This is a **nondirectional alternative hypothesis** and it is evaluated with a **two-tailed test.** In order to be supported, the absolute value of t must be equal to or greater than the tabled critical two-tailed t value at the prespecified level of significance. Thus, either a significant positive t value or a significant negative t value will provide support for this alternative hypothesis.)

or

H_1: $\mu_1 > \mu_2$

(The mean of the population Group 1 represents is greater than the mean of the population Group 2 represents. This is a **directional alternative hypothesis** and it is evaluated with a **one-tailed test.** It will only be supported if the sign of t is positive, and the absolute value of t is equal to or greater than the tabled critical one-tailed t value at the prespecified level of significance.)

or

H_1: $\mu_1 < \mu_2$

(The mean of the population Group 1 represents is less than the mean of the population Group 2 represents. This is a **directional alternative hypothesis** and it is evaluated with a **one-tailed test.** It will only be supported if the sign of t is negative, and the absolute value of t is equal to or greater than the tabled critical one-tailed t value at the prespecified level of significance.)

Note: Only one of the above noted alternative hypotheses is employed. If the alternative hypothesis the researcher selects is supported, the null hypothesis is rejected.[3]

IV. Test Computations

The data for Example 8.1 are summarized in Table 8.1. In the example there are $n_1 = 5$ subjects in Group 1 and $n_2 = 5$ subjects in Group 2. In Table 8.1 each subject is identified by a two digit number. The first digit before the comma indicates the subject's number within the group, and the second digit indicates the group identification number. Thus, Subject i, j is the ith subject in Group j. The scores of the 10 subjects are listed in the columns of Table 8.1 labelled X_1 and X_2. The adjacent columns labelled X_1^2 and X_2^2 contain the square of each subject's score.

Table 8.1 Data for Example 8.1

	Group 1			Group 2	
	X_1	X_1^2		X_2	X_2^2
Subject 1,1	11	121	**Subject 1,2**	11	121
Subject 2,1	1	1	**Subject 2,2**	11	121
Subject 3,1	0	0	**Subject 3,2**	5	25
Subject 4,1	2	4	**Subject 4,2**	8	64
Subject 5,1	0	0	**Subject 5,2**	4	16
	$\Sigma X_1 = 14$	$\Sigma X_1^2 = 126$		$\Sigma X_2 = 39$	$\Sigma X_2^2 = 347$

Employing Equations I.1 and I.5, the mean and estimated population variance for each sample is computed below.

$$\bar{X}_1 = \frac{\Sigma X_1}{n_1} = \frac{14}{5} = 2.8 \qquad \tilde{s}_1^2 = \frac{\Sigma X_1^2 - \frac{(\Sigma X_1)^2}{n_1}}{n_1 - 1} = \frac{126 - \frac{(14)^2}{5}}{5 - 1} = 21.7$$

$$\bar{X}_2 = \frac{\Sigma X_2}{n_2} = \frac{39}{5} = 7.8 \qquad \tilde{s}_2^2 = \frac{\Sigma X_2^2 - \frac{(\Sigma X_2)^2}{n_2}}{n_2 - 1} = \frac{347 - \frac{(39)^2}{5}}{5 - 1} = 10.7$$

When there are an **equal number of subjects** in each sample, Equation 8.1 can be employed to compute the test statistic for the ***t* test for two independent samples.**[4]

$$t = \frac{\bar{X}_1 - \bar{X}_2}{\sqrt{\frac{\tilde{s}_1^2}{n_1} + \frac{\tilde{s}_2^2}{n_2}}} \qquad \textbf{(Equation 8.1)}$$

Employing Equation 8.1, the value $t = -1.96$ is computed.

$$t = \frac{2.8 - 7.8}{\sqrt{\frac{21.7}{5} + \frac{10.7}{5}}} = \frac{-5}{2.55} = -1.96$$

Equation 8.2 is an alternative way of expressing Equation 8.1.

$$t = \frac{\bar{X}_1 - \bar{X}_2}{\sqrt{s_{\bar{X}_1}^2 + s_{\bar{X}_2}^2}} \qquad \textbf{(Equation 8.2)}$$

Note that in Equation 8.2, the values $s_{\bar{X}_1}^2$ and $s_{\bar{X}_2}^2$ represent the squares of the **standard error of the means** of the two groups. Employing the square of the value computed with Equation 2.2 (presented in Section IV of the **single-sample *t* test**), the squared standard error of the means of the two samples are computed: $s_{\bar{X}_1}^2 = \tilde{s}_1^2/n_1 = 21.7/5 = 4.34$ and $s_{\bar{X}_2}^2 = \tilde{s}_2^2/n_2 = 10.7/5 = 2.14$. When the values $s_{\bar{X}_1}^2 = 4.34$ and $s_{\bar{X}_1}^2 = 2.14$ are substituted in Equation 8.2, they yield the value $t = -1.96$: $t = (2.8 - 7.8)/\sqrt{4.34 + 2.14} = -1.96$.

The reader should take note of the fact that the values $\tilde{s}_1^2$, $\tilde{s}_2^2$, $s_{\bar{X}_1}^2$, and $s_{\bar{X}_2}^2$ (all of which are estimates of either the variance of a population or the variance of a sampling distribution) can never be negative numbers. If a negative value is obtained for any of the aforementioned values, it indicates a computational error has been made.

Equation 8.3 is a general equation for the **t test for two independent samples** that can be employed for both **equal and unequal samples sizes** (when $n_1 = n_2$, Equation 8.3 becomes equivalent to Equations 8.1/8.2).

$$t = \frac{\bar{X}_1 - \bar{X}_2}{\sqrt{\left[\frac{(n_1 - 1)\tilde{s}_1^2 + (n_2 - 1)\tilde{s}_2^2}{n_1 + n_2 - 2}\right]\left[\frac{1}{n_1} + \frac{1}{n_2}\right]}} \qquad \textbf{(Equation 8.3)}$$

In the case of Example 8.1, Equation 8.3 yields the identical value $t = -1.96$ obtained with Equations 8.1/8.2.

$$t = \frac{2.8 - 7.8}{\sqrt{\left[\frac{(5 - 1)(21.7) + (5 - 1)(10.7)}{5 + 5 - 2}\right]\left[\frac{1}{5} + \frac{1}{5}\right]}} = -1.96$$

The left element inside the radical of the denominator of Equation 8.3 represents a weighted average (based on the values of n_1 and n_2) of the estimated population variances of the two groups. This weighted average is referred to as a **pooled variance estimate,** represented by the notation $\tilde{s}_p^2$.[5] Thus: $\tilde{s}_p^2 = [(n_1 - 1)\tilde{s}_1^2 + (n_2 - 1)\tilde{s}_2^2]/(n_1 + n_2 - 2)$. It should be noted that if Equations 8.1/8.2 are applied to data where $n_1 \neq n_2$, the absolute value of t will be slightly higher than the value computed with Equation 8.3. Thus, use of Equations 8.1/8.2 when $n_1 \neq n_2$ makes it easier to reject the null hypothesis, and consequently inflates the likelihood of committing a Type I error. The application of Equations 8.1–8.3 to a set of data when $n_1 \neq n_2$ is illustrated in Section VII.

Regardless of which equation is employed, the denominator of the **t test for two independent samples** is referred to as the **standard error of the difference.** This latter value, which can be summarized with the notation $s_{\bar{X}_1 - \bar{X}_2}$, represents an estimated standard deviation of difference scores for two populations. Thus, in Example 8.1, $s_{\bar{X}_1 - \bar{X}_2} = 2.55$. If $s_{\bar{X}_1 - \bar{X}_2}$ is employed as the denominator of the equation for the **t test for two independent samples,** the equation can be written as follows: $t = (\bar{X}_1 - \bar{X}_2)/s_{\bar{X}_1 - \bar{X}_2}$.[6]

It should be noted that in some sources the numerator of Equations 8.1–8.3 is written as follows: $[(\bar{X}_1 - \bar{X}_2) - (\mu_1 - \mu_2)]$. The latter notation is only necessary if in stating the null hypothesis, a researcher stipulates that the difference between μ_1 and μ_2 is some value other than zero. When the null hypothesis is H_0: $\mu_1 = \mu_2$, the value $(\mu_1 - \mu_2)$ reduces to zero, leaving the term $(\bar{X}_1 - \bar{X}_2)$ as the numerator of the t test equation. The application of the **t test for two independent samples** to a hypothesis testing situation in which a value other than zero is stipulated in the null hypothesis is illustrated with Example 8.2 in Section VI.

V. Interpretation of the Test Results

The obtained value $t = -1.96$ is evaluated with **Table A2 (Table of Student's t Distribution) in the Appendix.** The degrees of freedom for the **t test for two independent samples** are computed with Equation 8.4.[7]

$$df = n_1 + n_2 - 2 \qquad \textbf{(Equation 8.4)}$$

Employing Equation 8.4, the value $df = 5 + 5 - 2 = 8$ is computed. Thus, the tabled critical t values that are employed in evaluating the results of Example 8.1 are the values recorded in the cells of **Table A2** that fall in the row for $df = 8$, and the columns with probabilities that correspond to the two-tailed and one-tailed .05 and .01 values. (The protocol for employing **Table A2** is described in Section V of the **single-sample *t* test.**) The critical t values for $df = 8$ are summarized in Table 8.2.

Table 8.2 Tabled Critical .05 and .01 *t* Values $df = 8$

	$t_{.05}$	$t_{.01}$
Two-tailed values	2.31	3.36
One-tailed values	1.86	2.90

The following guidelines are employed in evaluating the null hypothesis for the ***t* test for two independent samples.**

a) If the nondirectional alternative hypothesis H_1: $\mu_1 \neq \mu_2$ is employed, the null hypothesis can be rejected if the obtained absolute value of t is equal to or greater than the tabled critical two-tailed value at the prespecified level of significance.

b) If the directional alternative hypothesis H_1: $\mu_1 > \mu_2$ is employed, the null hypothesis can be rejected if the sign of t is positive, and the value of t is equal to or greater than the tabled critical one-tailed value at the prespecified level of significance.

c) If the directional alternative hypothesis H_1: $\mu_1 < \mu_2$ is employed, the null hypothesis can be rejected if the sign of t is negative, and the absolute value of t is equal to or greater than the tabled critical one-tailed value at the prespecified level of significance.

Employing the above guidelines, the null hypothesis can only be rejected (and only at the .05 level) if the directional alternative hypothesis H_1: $\mu_1 < \mu_2$ is employed. This is the case, since the obtained value $t = -1.96$ is a negative number, and the absolute value $t = 1.96$ is greater than the tabled critical one-tailed .05 value $t_{.05} = 1.86$. This outcome is consistent with the prediction that the group which receives the antidepressant will exhibit a lower level of depression than the placebo group. Note that the alternative hypothesis H_1: $\mu_1 < \mu_2$ is not supported at the .01 level, since the obtained absolute value $t = 1.96$ is less than the tabled critical one-tailed .01 value $t_{.01} = 2.90$.

The nondirectional alternative hypothesis H_0: $\mu_1 \neq \mu_2$ is not supported, since the obtained absolute value $t = 1.96$ is less than the tabled critical two-tailed .05 value $t_{.05} = 2.31$.

The directional alternative hypothesis H_1: $\mu_1 > \mu_2$ is not supported, since the obtained value $t = -1.96$ is a negative number. In order for the alternative hypothesis H_1: $\mu_1 > \mu_2$ to be supported, the computed value of t must be a positive number (as well as the fact that the absolute value of t must be equal to or greater than the tabled critical one-tailed value at the prespecified level of significance). It should be noted, that it is not likely the researcher would employ the latter alternative hypothesis, since it predicts that the placebo group will exhibit a lower level of depression than the group that receives the antidepressant.

A summary of the analysis of Example 8.1 with the ***t* test for two independent samples** follows: It can be concluded that the average depression rating for the group that receives the antidepressant medication is significantly less than the average depression rating for the placebo group. This conclusion can only be reached if the directional alternative hypothesis H_1: $\mu_1 < \mu_2$ is employed, and the prespecified level of significance is $\alpha = .05$. This result can be summarized as follows: $t(8) = 1.96, p < .05$.[8]

VI. Additional Analytical Procedures for the *t* Test for Two Independent Samples and/or Related Tests

1. The equation for the *t* test for two independent samples when a value for a difference other than zero is stated in the null hypothesis In some sources Equation 8.5 is presented as the equation for the *t* **test for two independent samples.**

$$t = \frac{(\bar{X}_1 - \bar{X}_2) - (\mu_1 - \mu_2)}{s_{\bar{X}_1 - \bar{X}_2}} \qquad \textbf{(Equation 8.5)}$$

It is only necessary to employ Equation 8.5 if in stating the null hypothesis, a researcher stipulates that the difference between μ_1 and μ_2 is some value other than zero. When the null hypothesis is $H_0\colon \mu_1 = \mu_2$ (which as noted previously can also be written as $H_0\colon \mu_1 - \mu_2 = 0$), the value of $(\mu_1 - \mu_2)$ reduces to zero, and thus what remains of the numerator in Equation 8.5 is $(\bar{X}_1 - \bar{X}_2)$, which constitutes the numerator of Equations 8.1–8.3. Example 8.2 will be employed to illustrate the use of Equation 8.5 in a hypothesis testing situation in which some value other than zero is stipulated in the null hypothesis.

Example 8.2 *The Accusharp Battery Company claims that the hearing aid battery it manufactures has an average life span that is two hours longer than the average lifespan of a battery manufactured by the Keenair Battery Company. In order to evaluate the claim, an independent researcher measures the life span of five randomly selected batteries from the stock of each of the two companies, and obtains the following values:* **Accusharp:** 10, 8, 10, 9, 11; **Keenair:** 8, 9, 8, 7, 9. *Do the data support the claim of the Accusharp Company?*

Since the Accusharp Company (which will be designated as Group 1) specifically predicts that the lifespan of its battery is 2 hours longer, the null hypothesis can be stated as follows: $H_0\colon \mu_1 - \mu_2 = 2$. The alternative hypothesis if stated nondirectionally is $H_1\colon \mu_1 - \mu_2 \neq 2$. If the computed absolute value of t is equal to or greater than the tabled critical two-tailed t value at the prespecified level of significance, the nondirectional alternative hypothesis is supported. If stated directionally, the appropriate alternative hypothesis to employ is $H_1\colon \mu_1 - \mu_2 < 2$. The latter directional alternative hypothesis (which predicts a negative t value) is employed, since in order for the data to contradict the claim of the Accusharp Company (and thus reject the null hypothesis), the lifespan of the latter's battery can be any value that is less than 2 hours longer than that of the Keenair battery. The alternative hypothesis $H_1\colon \mu_1 - \mu_2 > 2$ (which is only supported with a positive t value) predicts that the superiority of the Accusharp battery is greater than 2 hours. If the latter alternative hypothesis is employed, the null hypothesis can only be rejected if the lifespan of the Accusharp battery is greater than 2 hours longer than that of the Keenair battery.

The analysis for Example 8.2 is summarized below.

$$\Sigma X_1 = 48 \qquad \bar{X}_1 = \frac{48}{5} = 9.6 \qquad \Sigma X_1^2 = 466$$

$$\Sigma X_2 = 41 \qquad \bar{X}_2 = \frac{41}{5} = 8.2 \qquad \Sigma X_2^2 = 339$$

$$\tilde{s}_1^2 = \frac{466 - \frac{(48)^2}{5}}{5 - 1} = 1.3 \qquad \tilde{s}_2^2 = \frac{339 - \frac{(39)^2}{5}}{5 - 1} = .7$$

$$t = \frac{(9.6 - 8.2) - 2}{\sqrt{\frac{1.3}{5} + \frac{.7}{5}}} = \frac{-.6}{.63} = -.95$$

Since $df = 5 + 5 - 2 = 8$, the tabled critical values in Table 8.2 can be employed to evaluate the results of the analysis. Since the obtained absolute value $t = .95$ is less than the tabled critical two-tailed value $t_{.05} = 2.31$, the null hypothesis is retained. Thus, the nondirectional alternative hypothesis H_1: $\mu_1 - \mu_2 \neq 2$ is not supported. It is also true that the directional alternative hypothesis H_1: $\mu_1 - \mu_2 < 2$ is not supported. This is the case since although, as predicted, the sign of the computed t value is negative, the absolute value $t = .95$ is less than the tabled critical one-tailed value $t_{.05} = 1.86$. Thus, irrespective of whether a nondirectional or directional alternative hypothesis is employed, the data are consistent with the claim of the Accusharp company that it manufactures a battery that has a lifespan which is at least two hours longer than that of the Keenair battery. In other words, the obtained difference $(\bar{X}_1 - \bar{X}_2) = 1.4$ in the numerator of Equation 8.5 is not small enough to support the directional alternative hypothesis H_1: $\mu_1 - \mu_2 < 2$.

If H_0: $\mu_1 = \mu_2$ and H_1: $\mu_1 \neq \mu_2$ are employed as the null hypothesis and nondirectional alternative hypothesis for Example 8.2, analysis of the data yields the following result: $t = (9.6 - 8.2)/.63 = 2.22$. Since the obtained value $t = 2.22$ is less than the tabled critical two-tailed value $t_{.05} = 2.31$, the nondirectional alternative hypothesis H_1: $\mu_1 \neq \mu_2$ is not supported at the .05 level. Thus, one cannot conclude that the lifespan of the Accusharp battery is significantly different than the lifespan of the Keenair battery. The directional alternative hypothesis H_1: $\mu_1 > \mu_2$ (which can also be written as H_1: $\mu_1 - \mu_2 > 0$) is supported at the .05 level, since $t = 2.22$ is a positive number that is larger than the tabled critical one-tailed value $t_{.05} = 1.86$. Thus, if the null hypothesis H_0: $\mu_1 = \mu_2$ and the directional alternative hypothesis H_1: $\mu_1 > \mu_2$ are employed, the researcher is able to conclude that the lifespan of the Accusharp battery is significantly longer than the lifespan of the Keenair battery.

The evaluation of Example 8.2 in this section reflects that fact that the conclusions one reaches can be affected by how a researcher states the null and alternative hypotheses. In the case of Example 8.2, the fact that the nondirectional alternative hypothesis H_1: $\mu_1 - \mu_2 \neq 2$ is not supported suggests that there is a two-hour difference in favor of Accustar (since the null hypothesis H_0: $\mu_1 - \mu_2 = 2$ is retained). Yet, if the nondirectional alternative hypothesis H_1: $\mu_1 \neq \mu_2$ is employed, the fact that it is not supported suggests there is no difference between the two brands of batteries.

2. Test 8a: Hartley's F_{max} test for homogeneity of variance/*F* test for two population variances: Evaluation of the homogeneity of variance assumption of the *t* test for two independent samples It is noted in Section I that one assumption of the t test for two independent samples is **homogeneity of variance.** Specifically, the homogeneity of variance assumption evaluates whether there is evidence to indicate that an inequality exists between the variances of the populations represented by the two experimental samples. When the latter condition exists it is referred to as **heterogeneity of variance.** The null and alternative hypotheses employed in evaluating the homogeneity of variance assumption are as follows.

Null hypothesis $H_0\colon \sigma_1^2 = \sigma_2^2$

(The variance of the population Group 1 represents equals the variance of the population Group 2 represents.)

Alternative hypothesis $H_1\colon \sigma_1^2 \neq \sigma_2^2$

(The variance of the population Group 1 represents does not equal the variance of the population Group 2 represents. This is a **nondirectional alternative hypothesis** and it is evaluated with a **two-tailed test**. In evaluating the homogeneity of variance assumption for the ***t* test for two independent samples**, a nondirectional alternative hypothesis is always employed.)

One of a number of procedures that can be used to evaluate the homogeneity of variance hypothesis is **Hartley's F_{max} test** (Hartley (1940,1950)), which can be employed with a design involving two or more independent samples.[9] Although the **F_{max} test** assumes an equal number of subjects per group, Kirk (1982) and Winer *et al.* (1991) among others note that if $n_1 \neq n_2$, but are approximately the same size, one can let the value of the larger sample size represent n when interpreting the F_{max} test statistic. The latter sources, however, note that using the larger n will result in a slight increase in the Type I error rate for the F_{max} test.

The test statistic for Hartley's F_{max} test is computed with Equation 8.6.

$$F_{max} = \frac{\tilde{s}_L^2}{\tilde{s}_S^2} \qquad \textbf{(Equation 8.6)}$$

Where: $\tilde{s}_L^2$ = The larger of the two estimated population variances
$\tilde{s}_S^2$ = The smaller of the two estimated population variances

Employing Equation 8.6 with the estimated population variances computed for Example 8.1, the value $F_{max} = 2.03$ is computed. The reader should take note of the fact that the computed value for F_{max} will always be a positive number that is greater than 1 (unless $\tilde{s}_L^2 = \tilde{s}_S^2$, in which case $F_{max} = 1$).

$$F_{max} = \frac{21.7}{10.7} = 2.03$$

The computed value $F_{max} = 2.03$ is evaluated with **Table A9 (Table of the F_{max} Distribution)** in the **Appendix**. The tabled critical values for the F_{max} distribution are listed in reference to the values $(n - 1)$ and k, where n represents the number of subjects per group, and k represents the number of groups. In the case of Example 8.1, the value of $n = n_1 = n_2 = 5$. Thus, $n - 1 = 5 - 1 = 4$. Since there are two groups, $k = 2$.

In order to reject the null hypothesis and conclude that the homogeneity of variance assumption has been violated, the obtained F_{max} value must be equal to or greater than the tabled critical value at the prespecified level of significance. All values listed in **Table A9** are two- tailed values. Inspection of **Table A9** indicates that for $n - 1 = 4$ and $k = 2$, $F_{max_{.05}} = 9.6$ and $F_{max_{.01}} = 23.2$. Since the obtained value $F_{max} = 2.03$ is less than $F_{max_{.05}} = 9.6$, the homogeneity of variance assumption is not violated — in other words the data do not suggest that the variances of the populations represented by the two groups are unequal. Thus, the null hypothesis is retained.

There are a number of additional points that should be made with regard to the above analysis:

a) Some sources employ Equation 8.7 or Equation 8.8 in lieu of Equation 8.6 to evaluate the homogeneity of variance assumption. When Equation 8.8 is employed to contrast two variances, it is often referred to as an ***F* test for two population variances.**

$$F = \frac{\tilde{s}_L^2}{\tilde{s}_S^2} \qquad \textbf{(Equation 8.7)}$$

$$F = \frac{\tilde{s}_1^2}{\tilde{s}_2^2} \qquad \textbf{(Equation 8.8)}$$

Both Equations 8.7 and 8.8 compute an F ratio, which is based on the F distribution. Critical values for the latter distribution are presented in **Table A10 (Table of the *F* Distribution) in the Appendix.** The F distribution (which is discussed in greater detail under the **single-factor between-subjects analysis of variance (Test 16)**) is, in fact, the sampling distribution upon which the F_{max} distribution is based. In **Table A10**, critical values are listed in reference to the number of degrees of freedom associated with the numerator and the denominator of the F ratio. In employing the F distribution in reference to Equation 8.7, the degrees of freedom for the numerator of the F ratio is $df_{num} = n_L - 1$ (where n_L represents the number of subjects in the group with the larger estimated population variance), and the degrees of freedom for the denominator is $df_{den} = n_S - 1$ (where n_S represents the number of subjects in the group with the smaller estimated population variance). The tabled $F_{.975}$ value is employed to evaluate a two-tailed alternative hypothesis at the .05 level, and the tabled $F_{.995}$ value is employed to evaluate it at the .01 level.[10] The reason for employing the tabled $F_{.975}$ and $F_{.995}$ values instead of $F_{.95}$ (which in this analysis represents the two-tailed .10 value and the one-tailed .05 value) and $F_{.99}$ (which in this analysis represents the two-tailed .02 value and the one-tailed .01 value), is that both tails of the distribution are used in employing the F distribution to evaluate a hypothesis about two population variances. Thus, if one is conducting a two-tailed analysis with $\alpha = .05$, .025 (i.e., $.05/2 = .025$) represents the proportion of cases in the extreme left of the left tail of the F distribution, as well as the proportion of cases in the extreme right of the right tail of the distribution. With respect to a two-tailed analysis with $\alpha = .01$, .005 (i.e., $.01/2 = .005$) represents the proportion of cases in the extreme left of the left tail of the distribution, as well as the proportion of cases in the extreme right of the right tail of the distribution.

In point of fact, if $df_L = df_{num} = 4$ and $df_S = df_{den} = 4$ (which are the values employed in Example 8.1), the tabled critical two-tailed F values employed for $\alpha = .05$ and $\alpha = .01$ are $F_{.975} = 9.6$ and $F_{.995} = 23.15$. These are the same critical .05 and .01 values that are employed for the $\boldsymbol{F_{max}}$ **test.**[11] Thus, Equation 8.7 employs the same critical values and yields an identical result to that obtained with Equation 8.6. It should be noted, however, that if $n_1 \neq n_2$, Equation 8.6 and Equation 8.7 will employ different critical values, since Equation 8.7 (which uses the value of n for each group in determining degrees of freedom) can accommodate unequal sample sizes.

When Group 1 has a larger estimated population variance than Group 2, everything that has been said with respect to Equation 8.7 applies to Equation 8.8. However, when $\tilde{s}_2^2 > \tilde{s}_1^2$, the value of F computed with Equation 8.8 will be less than 1. In such an instance, one can do either of the following: a) Designate the group with the larger variance as Group 1 and the group with the smaller variance as Group 2. Upon doing this, divide the larger variance by the smaller variance (as is done in Equation 8.7), and thus obtain the same F value derived with Equation 8.7; or b) Use Equation 8.8, and employ the tabled critical $F_{.025}$ value to evaluate a two-tailed alternative hypothesis at the .05 level of significance, and the tabled critical $F_{.005}$ value to evaluate a two-tailed alternative hypothesis at the .01 level of significance.[12] In such a case, in order to be significant the computed F value must be equal to or less than the tabled critical $F_{.025}$ value (if $\alpha = .05$) or the tabled critical $F_{.005}$ value (if $\alpha = .01$). Both of the aforementioned methods will yield the same conclusions with respect to retaining or rejecting the null hypothesis at a given level of significance.

Equation 8.8 can also be used to test a directional alternative hypothesis concerning the relationship between two population variances. If a researcher specifically predicts that the variance of the population represented by Group 1 is larger than the variance of the population represented by Group 2 (i.e., H_1: $\sigma_1^2 > \sigma_2^2$), or that the variance of the population represented by Group 1 is smaller than the variance of the population represented by Group 2 (i.e., H_1: $\sigma_1^2 < \sigma_2^2$), a directional alternative hypothesis is evaluated. In such a case, the tabled critical one-tailed F value for $(n_1 - 1),(n_2 - 1)$ degrees of freedom at the prespecified level of significance is employed.

To illustrate the analysis of a one-tailed alternative hypothesis, let us assume that the alternative hypothesis H_1: $\sigma_1^2 > \sigma_2^2$ is evaluated. If $\alpha = .05$, in order for the result to be significant the computed value of F must be greater than 1. In addition, the tabled critical F value that is employed in evaluating the above alternative hypothesis is $F_{.95}$. To be significant, the obtained F value must be equal to or greater than the tabled critical $F_{.95}$ value. If $\alpha = .01$, the tabled critical F value that is employed in evaluating the alternative hypothesis is $F_{.99}$. To be significant, the obtained F value must be equal to or greater than the tabled critical $F_{.99}$ value.

Now let us assume that the alternative hypothesis being evaluated is H_1: $\sigma_1^2 < \sigma_2^2$. If $\alpha = .05$, in order for the result to be significant the computed value of F must be less than 1. The tabled critical F value that is employed in evaluating the above alternative hypothesis is $F_{.05}$. To be significant, the obtained F value must be equal to or less than the tabled critical $F_{.05}$ value. If $\alpha = .01$, the tabled critical F value that is employed in evaluating the alternative hypothesis is $F_{.01}$. To be significant, the obtained F value must be equal to or less than the tabled critical $F_{.01}$ value.

If Equation 8.8 is employed with a one-tailed alternative hypothesis with reference to two groups consisting of five subjects per group (as is the case in Example 8.1), the following tabled critical values listed for $(n_1 - 1) = 4$, $(n_2 - 1) = 4$ are employed: a) If $\alpha = .05$, $F_{.95} = 6.39$, and $F_{.05} = .157$; and b) If $\alpha = .01$, $F_{.99} = 15.98$, and $F_{.01} = .063$.[13] To illustrate this in a situation where an F value less than 1 is computed, assume for the moment that we employ the alternative hypothesis H_1: $\sigma_1^2 < \sigma_2^2$ for the two groups described in Example 8.1, and that the values of the two group variances are reversed — i.e., $\tilde{s}_1^2 = 10.7$ and $\tilde{s}_2^2 = 21.7$. Employing Equation 8.8 with this data, $F = 10.7/21.7 = .49$. The obtained value $F = .49$ is not significant, since in order for it to be significant the computed value of F must be equal to or less than $F_{.05} = .157$.

It should be noted that when the general procedure discussed in this section is employed to evaluate the homogeneity of variance assumption with reference to Example 8.1, in order to reject the null hypothesis for $\alpha = .05$, the larger of the estimated population variances must be more than 9 times the magnitude of the smaller variance, and for $\alpha = .01$ the larger variance must be more than 23 times the magnitude of the smaller variance. Within the framework of the F_{max} test, such a large discrepancy between the estimated population variances is tolerated when the number of subjects per group is small. Inspection of **Table A9** reveals that as sample size increases, the magnitude of the tabled critical F_{max} values decreases. Thus, the larger the sample size, the smaller the difference between the variances that will be acceptable.

Two assumptions common to Equations 8.6–8.8 (i.e., all of the equations that can be employed to evaluate the homogeneity of variance assumption) are: a) Each sample has been randomly selected from the population it represents; and b) The distribution of data in the underlying population from which each of the samples is derived is normal. Violation of these assumptions can compromise the reliability of the F_{max} **test** statistic, which many sources note is extremely sensitive to violation of the normality assumption. Various sources (e.g., Keppel (1991)) point out that when the F_{max} **test** is employed to evaluate the homogeneity of variance hypothesis, it is not as powerful (i.e., likely to detect heterogeneity of variance when it is present) as some alternative but computationally more involved procedures. The consequence of not detecting heterogeneity of variance is that it increases the likelihood of committing a Type I error in conducting the ***t*** **test for two independent samples**. Additional discussion of the homogeneity of variance assumption and alternative procedures that can be used to evaluate it can be found in Section VI of the **single-factor between-subjects analysis of variance (Test 16)**.

In the event the homogeneity of variance assumption is violated, a number of different strategies (which yield similar but not identical results) are recommended with reference to conducting the ***t*** **test for two independent samples**. Since heterogeneity of variance increases the likelihood of committing a Type I error, all of the strategies that are recommended result in a more conservative t test (i.e., making it more difficult for the test to reject the null hypothesis). Such strategies compute either: a) An adjusted critical t value that is larger than the unadjusted t value; or b) An adjusted degrees of freedom value which is smaller than the value computed with Equation 8.4. By decreasing the degrees of freedom, a larger tabled critical t value is employed in evaluating the computed t value.

Before describing one of procedures that can be employed when the homogeneity of variance assumption is violated, it should be pointed out that the existence of heterogeneity of variance in a set of data may in itself be noteworthy. It is conceivable that although the analysis of the data for an experiment may indicate that there is no difference between the group means, one cannot rule out the possibility that there may be a significant difference between the variances of the two groups. This latter finding may be of practical importance in clarifying the relationship between the variables under study. This general issue is addressed in Section VIII of the **single-sample runs test (Test 7)** within the framework of the discussion of Example 7.5. In the latter discussion, an experiment is described in which subjects who receive an antidepressant either improve dramatically or regress while on the drug. In contrast, the scores of a placebo group exhibit little variability and fall in between the two extreme sets of scores in the group that receives the antidepressant. Analysis of such a study with the ***t*** **test for two independent samples** will in all likelihood not yield a significant result, since the two groups will probably have approximately the same mean depression score. The fact that the group receiving the drug exhibits greater variability than the group receiving the placebo indicates that the effect of the drug is not consistent for all people who are depressed. Such an effect can be identified through use of a test such as the F_{max} **test**, which contrasts the variability of two groups.

Two statisticians (Behrens and Fisher) developed a sampling distribution for the t statistic when the homogeneity of variance assumption is violated. The latter sampling distribution is referred to as the t' distribution. Since tables of critical values developed by Behrens and Fisher can only be employed for a limited number of sample sizes, Cochran and Cox (1957) developed a methodology that allows one to compute critical values of t' for all values of n_1 and n_2. Equation 8.9 summarizes the computation of t'.

$$t' = \frac{t_1\left[\frac{\tilde{s}_1^2}{n_1}\right] + t_2\left[\frac{\tilde{s}_2^2}{n_2}\right]}{\frac{\tilde{s}_1^2}{n_1} + \frac{\tilde{s}_2^2}{n_2}}$$ **(Equation 8.9)**

Where: t_1 = The tabled critical t value at the prespecified level of significance for $df = n_1 - 1$
t_2 = The tabled critical t value at the prespecified level of significance for $df = n_2 - 2$

Equation 8.9 will be employed with the data for Example 8.1. For purposes of illustration, it will be assumed that the homogeneity of variance assumption has been violated, and that the nondirectional alternative hypothesis H_1: $\mu_1 \neq \mu_2$ is evaluated, with $\alpha = .05$. Since $n_1 = n_2 = n = 5$, $df_1 = df_2 = n - 1 = 4$. Employing **Table A2**, we determine that for $\alpha = .05$ and $df = 4$, the tabled critical two-tailed .05 value is $t_{.05} = 2.78$. Thus, the values $t_1 = 2.78$ and $t_2 = 2.78$ are substituted in Equation 8.9, along with the values of the estimated population variances and the sample sizes.[14]

$$t' = \frac{2.78\left[\frac{21.7}{5}\right] + 2.78\left[\frac{10.7}{5}\right]}{\frac{21.7}{5} + \frac{10.7}{5}} = 2.78$$

Note that the computed value $t' = 2.78$ is larger than the tabled critical two-tailed .05 value $t_{.05} = 2.31$, which is employed if the homogeneity of variance adjustment is not violated. Since the value of t' will always be larger than the tabled critical t value at the prespecified level of significance for $df = n_1 + n_2 - 2$ (except for the instance noted in Endnote 14), use of the t' statistic will result in a more conservative test. In our hypothetical example, use of the t' statistic is designed to insure that the Type I error rate will conform to the prespecified value $\alpha = .05$. If there is heterogeneity of variance and the homogeneity of variance adjustment is not employed, the actual alpha level will be greater than $\alpha = .05$. Since in our example the computed value $t' = 2.78$ is larger than the computed absolute value $t = 1.96$ obtained for the t test, the null hypothesis cannot be rejected. The methodology described in this section for dealing with heterogeneity of variance provides for a slightly more conservative ***t* test for two independent samples** than do alternative strategies developed by Satterthwaite (1946) and Welch (1947) (which are described in Howell (1992) and Winer *et al.* (1991)).

3. Computation of the power of the *t* test for two independent samples In this section two methods for computing power, which are extensions of the methods presented for computing the power of the **single-sample *t* test**, will be described. Prior to reading this section the reader should review the discussion of power in Section VI of the latter test.

The first procedure to be described is a graphical method which reveals the logic underlying the power computations for the ***t* test for two independent samples**. In the discussion to follow, it will be assumed that the null hypothesis is identical to that employed for Example 8.1 (i.e., H_0: $\mu_1 - \mu_2 = 0$, which, as previously noted, is another way of writing H_0: $\mu_1 = \mu_2$). It will also be assumed that the researcher wants to evaluate the power of the ***t* test for two independent samples** in reference to the following alternative hypothesis: H_1: $|\mu_1 - \mu_2| \geq 5$ (which is the difference obtained between the sample means in Example 8.1). In other words, it is predicted that the absolute value of the difference between the two means is equal to or greater than 5. The latter alternative hypothesis is employed in lieu of H_0: $\mu_1 - \mu_2 \neq 0$ (which can also be written as H_1: $\mu_1 \neq \mu_2$), since in order to compute the power of the test a specific value must be stated for the difference between the population means. Note that, as stated, the alternative hypothesis stipulates a nondirectional analysis, since it does not specify which of the two means will be the larger value. It will be assumed that $\alpha = .05$ is employed in the analysis.

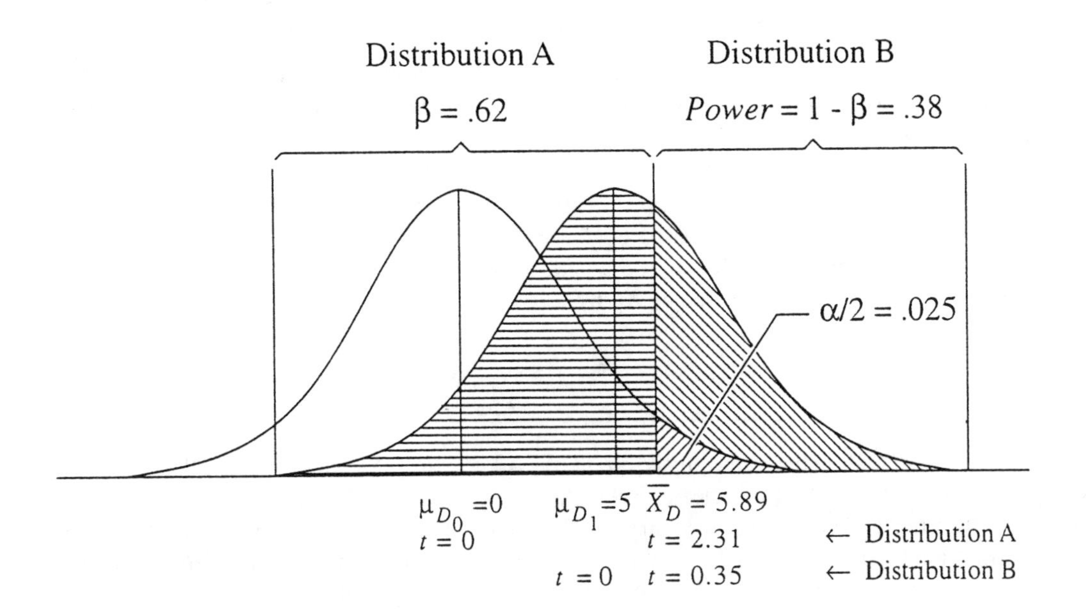

Figure 8.1 Visual Representation of Power for Example 8.1

Figure 8.1, which provides a visual summary of the power analysis, is comprised of two overlapping sampling distributions of difference scores. The distribution on the left, which will be designated as Distribution A, is a sampling distribution of difference scores that has a mean value of zero (i.e., $\mu_D = \mu_{\bar{X}_1 - \bar{X}_2} = 0$). This latter value will be represented by $\mu_{D_0} = 0$ in Figure 8.1. Distribution A represents the sampling distribution that describes the distribution of difference scores if the null hypothesis is true. The distribution on the right, which will be designated as Distribution B, is a sampling distribution of difference scores that has a mean value of 5 (i.e., $\mu_D = \mu_{\bar{X}_1 - \bar{X}_2} = 5$). This latter value will be represented by $\mu_{D_1} = 5$ in Figure 8.1. Distribution B represents the sampling distribution that describes the distribution of difference scores if the alternative hypothesis is true. It will be assumed that each of the sampling distributions has a standard deviation that is equal to the value computed for the standard error of the difference in Example 8.1 (i.e., $s_{\bar{X}_1 - \bar{X}_2} = 2.55$), since the latter value provides the best estimate of the standard deviation of the difference scores for the underlying populations.

In Figure 8.1, area (///) delineates the proportion of Distribution A that corresponds to the value $\alpha/2$, which equals .025. This is the case, since $\alpha = .05$ and a two-tailed analysis is conducted. Area (≡) delineates the proportion of Distribution B that corresponds to the probability of committing a Type II error (β). Area (\\\) delineates the proportion of Distribution B that represents the power of the test (i.e., $1 - \beta$).

The procedure for computing the proportions documented in Figure 8.1 will now be described. The first step in computing the power of the test requires one to determine how large a difference there must be between the sample means in order to reject the null hypothesis. In order to do this, we algebraically transpose the terms in Equation 8.1, using $s_{\bar{X}_1 - \bar{X}_2}$ to summarize the denominator of the equation, and $t_{.05}$ (the tabled critical two-tailed .05 t value) to represent t. Thus: $\bar{X}_1 - \bar{X}_2 = (t_{.05})(s_{\bar{X}_1 - \bar{X}_2})$. By substituting the values $t_{.05} = 2.31$ and $s_{\bar{X}_1 - \bar{X}_2} = 2.55$ in the latter equation, we determine that the minimum required difference is $\bar{X}_1 - \bar{X}_2 = (2.31)(2.55) = 5.89$. Thus, any difference between the two population means that is equal to or greater than 5.89 will allow the researcher to reject the null hypothesis at the .05 level.

The next step in the analysis requires one to compute the area in Distribution B that falls between the mean difference $\mu_{D_1} = 5$ (i.e., the mean of Distribution B) and a mean difference equal to 5.89 (represented by the notation $\bar{X}_D = 5.89$ in Figure 8.1). This is accomplished by employing Equation 8.1. In using the latter equation, the value of $\bar{X}_1$ is represented by 5.89 and the value of $\bar{X}_2$ by $\mu_{D_1} = 5$.

$$t = \frac{\bar{X}_1 - \bar{X}_2}{s_{\bar{X}_1 - \bar{X}_2}} = \frac{5.89 - 5}{2.55} = .35$$

By interpolating the values listed in **Table A2** for $df = 8$, we determine that the proportion of Distribution B that lies to the right of a t score of .35 (which corresponds to a mean difference of 5.89) is approximately .38. The latter area corresponds to area (\\\) in Distribution B. Note that the left boundary of area (\\\) is also the boundary delineating the extreme 2.5% of Distribution A (i.e., $\alpha/2 = .025$, which is the rejection zone for the null hypothesis). Since area (\\\) in Distribution B overlaps the rejection zone in Distribution A, area (\\\) represents the power of the test — i.e., it represents the likelihood of rejecting the null hypothesis if the alternative hypothesis is true. The likelihood of committing a Type II error (β) is represented by area (≡), which comprises the remainder of Distribution B. The proportion of Distribution B that constitutes this latter area is determined by subtracting the value .38 from 1. Thus: $\beta = 1 - .38 = .62$.

Based on the results of the power analysis, we can state that if the alternative hypothesis H_1: $|\mu_1 - \mu_2| \geq 5$ is true, the likelihood that the null hypothesis will be rejected is .38, and at the same time there is a .62 likelihood that it will be retained. If the researcher considers the computed value for power too low (which in actuality should be determined prior to conducting a study), she can increase the power of the test by employing a larger sample size.

Method 2, the quick computational method described for computing the power of the **single sample *t* test**, can also be extended to the ***t* test for two independent samples**. In using this latter method, the researcher must stipulate an **effect size** (d), which in the case of the ***t* test for two independent samples** is computed with Equation 8.10.

$$d = \frac{|\mu_1 - \mu_2|}{\sigma} \qquad \textbf{(Equation 8.10)}$$

The numerator of Equation 8.10 represents the hypothesized difference between the two population means. As is the case with the graphical method described previously, when a power analysis is conducted after the mean of each sample has been obtained, the difference between the two sample means (i.e., $\bar{X}_1 - \bar{X}_2$) is employed as an estimate of the value of $|\mu_1 - \mu_2|$. It is assumed that the value of the standard deviation for the variable being measured is the same in each of the populations, and the latter value is employed to represent σ in the denominator of Equation 8.10 (i.e., $\sigma = \sigma_1 = \sigma_2$). In instances where the standard deviations of the two populations are not known or cannot be estimated, the latter value can be estimated from the sample data. Because of the fact that $\tilde{s}_1$ will usually not equal $\tilde{s}_2$, a pooled estimated population standard deviation ($\tilde{s}_p$) can be computed with Equation 8.11 (which is the square root of $\tilde{s}_p^2$ discussed in Section IV with reference to Equation 8.3).

$$\tilde{s}_p = \sqrt{\frac{(n_1 - 1)\tilde{s}_1^2 + (n_2 - 1)\tilde{s}_2^2}{n_1 + n_2 - 2}} \qquad \textbf{(Equation 8.11)}$$

Since the effect size computed with Equation 8.10 is based on population parameters, it is necessary to convert the value of d into a measure that takes into account the size of the samples (which is a relevant variable in determining the power of the test). This measure, as noted in the discussion of the **single-sample *t* test,** is referred to as the **noncentrality parameter**. Equation 8.12 is employed to compute the **noncentrality parameter** (δ) for the ***t* test for two independent samples.** When the sample sizes are equal, the value of n in Equation 8.12 will be $n = n_1 = n_2$. When the sample sizes are unequal, the value of n will be represented by the **harmonic mean** of the sample sizes, which is described later in this section.

$$\delta = d\sqrt{\frac{n}{2}} \qquad \textbf{(Equation 8.12)}$$

The power of the ***t* test for two independent samples** will now be computed using the data for Example 8.1. For purposes of illustration, it will be assumed that the minimum difference between the population means the researcher is trying to detect is the observed 5 point difference between the two sample means — i.e., $|\bar{X}_1 - \bar{X}_2| = |2.8 - 5.8| = 5 = |\mu_1 - \mu_2|$. The value of σ employed in Equation 8.10 is estimated by computing a pooled value for the standard deviation using Equation 8.11. Substituting the relevant values from Example 8.1 in Equation 8.11, the value $\tilde{s}_p = 4.02$ is computed.

$$\tilde{s}_p = \sqrt{\frac{(5 - 1)(21.7) + (5 - 1)(10.7)}{5 + 5 - 2}} = 4.02$$

Substituting $|\mu_1 - \mu_2| = 5$ and $\sigma = 4.02$ in Equation 8.10, the value $d = 1.24$ is computed. Employing the guidelines noted in the discussion of power for the **single-sample *t* test,** the value $d = 1.24$, which represents 1.24 standard deviation units, is categorized as a large effect size.

$$d = \frac{5}{4.02} = 1.24$$

Along with the value $n = 5$ (since $n_1 = n_2 = n = 5$), the value $d = 1.24$ is substituted in Equation 8.12, resulting in the value $\delta = 1.96$.

$$\delta = 1.24\sqrt{\frac{5}{2}} = 1.96$$

The value $\delta = 1.96$ is evaluated with **Table A3 (Power Curves for Student's *t* Distribution)** in the **Appendix.** We will assume that for the example under discussion a two-tailed test is conducted with $\alpha = .05$, and thus **Table A3-C** is the appropriate set of power curves to employ for the analysis. Since there is no curve for $df = 8$, the power of the test will be based on a curve that falls in between the $df = 6$ and $df = 12$ power curves. Through interpolation, the power of the ***t* test for two independent samples** is determined to be approximately .38 (which is the same value that is obtained with the graphical method). Thus, by employing 5 subjects in each group the researcher has a probability of .38 of rejecting the null hypothesis if the true difference between the population means is equal to or greater than 1.24 standard deviation units (which in Example 8.1 is equivalent to a 5 point difference between the means). It should be noted that in employing Equation 8.12, the smaller the value of n the smaller the computed value of δ, and consequently the lower the power of the test.

It was noted earlier in the discussion that when the sample sizes are unequal, the value of n in Equation 8.12 is represented by the **harmonic mean** of the sample sizes (which will be represented by the notation $\bar{n}_h$). The harmonic mean is computed with Equation 8.13.

$$\bar{n}_h = \frac{k}{\sum_{j=1}^{k} \left[\frac{1}{n_j} \right]} \qquad \textbf{(Equation 8.13)}$$

Where: k = The number of groups
n_j = The number of subjects in the jth group

The use of Equation 8.13 will be illustrated for a case in which there is a total of 10 subjects, but there is an unequal number of subjects in each group. Thus, let us assume that $n_1 = 7$ and $n_2 = 3$. The harmonic mean can be computed as follows: $\bar{n}_h = 2/[(1/7) + (1/3)] = 4.20$. The reader should take note of the fact that the value $\bar{n}_h = 4.20$ computed for the harmonic mean is lower than the average number of subjects per group ($\bar{n}$), which is computed to be $\bar{n} = (7 + 3)/2 = 5$. In point of fact, $\bar{n}_h$ will always be lower than $\bar{n}$ unless $n_1 = n_2$, in which case $\bar{n}_h = \bar{n}$. Since when $n_1 \neq n_2$, $\bar{n}_h$ will always be less than $\bar{n}$, it follows that when $n_1 \neq n_2$, the value computed for the power of the test when the harmonic mean is employed in Equation 8.12 will be less than the value that is computed for the power of the test if $\bar{n}$ is employed to represent the value of n. This translates into the fact that for a specific total sample size (i.e., $n_1 + n_2$), the power of the ***t* test for two independent samples** will be maximized when $n_1 = n_2$.

As is the case with power computations for the **single-sample *t* test**, as long as a researcher knows or is able to estimate (from the sample data) the population standard deviation, by employing trial and error she can substitute various values of n in Equation 8.12, until the computed value of δ corresponds to the desired power value for the ***t* test for two independent samples** for a given effect size. This process can be facilitated by employing tables developed by Cohen (1988), which allow one to determine the minimum sample size necessary in order to achieve a specific level of power in reference to a given effect size.

4. Measure of magnitude of treatment effect for the *t* test for two independent samples: Omega squared (Test 8b) At the conclusion of an experiment a researcher may want to determine the proportion of the variability on the dependent variable that is associated with the experimental treatments (i.e., the independent variable). This latter value is commonly referred to as the **treatment effect.**[15] Unfortunately, the t value computed for the ***t* test for two independent samples** does not in itself provide information regarding the magnitude of a treatment effect. The reason for this is that the absolute value of t is not only a function of the treatment effect, but is also a function of the size of the sample employed in an experiment. Since the power of a statistical test is directly related to sample size, the larger the sample size, the more likely a significant t value will be obtained if there is any difference between the means of the underlying populations. Regardless of how small a treatment effect is present, the magnitude of the absolute value of t will increase as the size of the sample employed to detect that effect increases. Thus, a t value that is significant at any level (be it .05, .01, .001, etc.) can result from the presence of a large, medium, or small treatment effect.

Before describing measures of treatment effect for the ***t* test for two independent samples**, the distinction between **statistical significance** and **practical significance** (which is discussed briefly in the **Introduction**) will be clarified. Whereas **statistical significance** only indicates that a difference between the two group means has been detected, and that the difference is unlikely to be the result of chance, **practical significance** refers to the practical implications of the obtained difference. As just noted, by employing a large sample size a researcher will be able to detect differences between means that are extremely small. Although in some instances a small treatment effect can be of practical significance, more often than not a minimal difference will be of little or no practical value (other than perhaps allowing a researcher to get a study published, since significant results are more likely to be published than nonsignificant results). On the other hand, the larger the magnitude of treatment effect computed for an experiment, the more likely the results have practical implications. To go even further, when a researcher employs a small sample size it is possible to have a moderate or large treatment effect present, yet not obtain a statistically significant result. Obviously, in such an instance (which represents an example of a Type II error) the computed t value is misleading with respect to the truth regarding the relationship between the variables under study.

A number of indices for measuring magnitude of treatment effect have been developed. Unlike the computed t value, measures of magnitude of treatment effect provide an index of the degree of relationship between the independent and dependent variables that is independent of sample size.[16] A major problem with measures of treatment effect is that for most experimental designs two or more such measures are available which are not equivalent to one another, and researchers are often not in agreement with respect to which measure is appropriate to employ. In the case of the ***t* test for two independent samples**, the most commonly employed measure of treatment effect is **omega squared**. The statistic that is computed from the sample data to estimate the value of **omega squared** is represented by the notation $\tilde{\omega}^2$ (ω is the lower case Greek letter **omega**). This latter value provides an estimate of the underlying population parameter ω^2, which represents the proportion of variability on the dependent variable that is associated with the independent variable in the underlying population. The value of $\tilde{\omega}^2$ is computed with Equation 8.14.

$$\tilde{\omega}^2 = \frac{t^2 - 1}{t^2 + n_1 + n_2 - 1} \qquad \textbf{(Equation 8.14)}$$

Although the value of $\tilde{\omega}^2$ will generally fall in the range between 0 and 1, when $t < 1$, $\tilde{\omega}^2$ will be a negative number. The closer $\tilde{\omega}^2$ is to 1, the stronger the association between the independent and dependent variables, whereas the closer $\tilde{\omega}^2$ is to 0, the weaker the association between the two variables. A $\tilde{\omega}^2$ value equal to or less than 0 indicates that there is no association between the variables.[17]

Employing Equation 8.14 with the data for Example 8.1, the value $\tilde{\omega}^2 = .22$ is computed.

$$\tilde{\omega}^2 = \frac{(-1.96)^2 - 1}{(-1.96)^2 + 5 + 5 - 1} = .22$$

The value $\tilde{\omega}^2 = .22$ indicates that 22% (or a proportion equal to .22) of the variability on the dependent variable (the depression ratings of the subjects) is associated with variability on the levels of the independent variable (the drug versus placebo conditions). To say it another way, 22% of the variability on the depression scores can be accounted for on the basis of which group a subject is a member.

Keppel (1991) and Keppel *et al.* (1992) note that in the experimental literature in the discipline of psychology it is unusual for a $\tilde{\omega}^2$ value to exceed .25 — indeed, one review of the psychological literature yielded an average $\tilde{\omega}^2$ value of .06. The inability of researchers to control experimental error with great precision is the most commonly cited reason for the low value obtained for $\tilde{\omega}^2$ in most studies. Cohen (1988) has suggested the following values, which are employed in psychology and a number of other disciplines, as guidelines for interpreting $\tilde{\omega}^2$: **Small treatment effect:** $.01 \leq \tilde{\omega}^2 < .06$; **Medium treatment effect:** $.06 \leq \tilde{\omega}^2 < .15$; **Large treatment effect:** $\tilde{\omega}^2 \geq .15$. Many sources recommend that in reporting the results of a ***t* test for two independent samples**, in addition to reporting the t value the researcher should also report the computed value of $\tilde{\omega}^2$, since by including the latter value one is providing additional information that can further clarify the nature of the relationship between the variables under study. In the case of Example 8.1, the obtained value $\tilde{\omega}^2 = .22$ indicates the presence of a large treatment effect.

Eta squared ($\tilde{\eta}^2$) is an alternative but less commonly used measure of association that can also be employed to evaluate the magnitude of a treatment effect for a ***t* test for two independent samples.** The **eta squared (Test 16h)** statistic is described in Section VI of the **single-factor between-subjects analysis of variance.** Most sources note that **eta squared** results in a more biased estimate of the degree of association in the underlying population than does **omega squared. Eta squared** is also discussed under the **Pearson product-moment correlation coefficient (Test 22)** with reference to the **point-biserial correlation coefficient (Test 22h).** In the latter discussion, it is demonstrated that **eta squared** and the **point-biserial correlation coefficient** are equivalent measures when they are used to evaluate magnitude of treatment effect for a design involving two independent samples. In the discussion of the **single-factor between-subjects analysis of variance,** it is noted that if $\tilde{\eta}^2$ and $\tilde{\omega}^2$ are computed for the same set of data they yield different values. The issue of the lack of agreement among different magnitude of treatment effect measures is considered in more detail in Section VI of the **single-factor between-subjects analysis of variance.**

5. Computation of a confidence interval for the *t* test for two independent samples
Prior to reading this section the reader should review the discussion on the computation of confidence intervals in Section VI of the **single-sample *t* test.** When interval/ratio data are available for two independent samples, a confidence interval can be computed that identifies a range of values within which one can be confident to a specified degree that the true difference lies between the two population means. Equation 8.15 is the general equation for computing the confidence interval for the difference between the means of two independent populations.

$$CI_{(1-\alpha)} = (\bar{X}_1 - \bar{X}_2) \pm (t_{\alpha/2})(s_{\bar{X}_1 - \bar{X}_2}) \qquad \textbf{(Equation 8.15)}$$

Where: $t_{\alpha/2}$ represents the tabled critical two-tailed value in the t distribution, for $df = n_1 + n_2 - 2$, below which a proportion equal to $[1 - (\alpha/2)]$ of the cases falls. If the percentage of the distribution that falls within the confidence interval is subtracted from 1, it will equal the value of α

Employing Equation 8.15, the 95% interval for Example 8.1 is computed below. In employing Equation 8.15, $(\bar{X}_1 - \bar{X}_2)$ represents the obtained difference between the group means (which is the numerator of the equation used to compute the value of t), $t_{.05}$ represents the tabled critical two-tailed .05 value for $df = n_1 + n_2 - 2$, and $s_{\bar{X}_1 - \bar{X}_2}$ represents the standard error of the difference (which is the denominator of the equation used to compute the value of t).

$$CI_{.95} = (\bar{X}_1 - \bar{X}_2) \pm (t_{.05})(s_{\bar{X}_1 - \bar{X}_2}) = -5 \pm (2.31)(2.55) = -5 \pm 5.89$$

$$-10.89 \le (\mu_1 - \mu_2) \le .89$$

This result indicates that the researcher can be 95% confident that the true difference between the population means falls within the range −10.89 and .89. Specifically, it indicates that one can be 95% confident that the mean of the population Group 2 represents is no more than 10.89 points higher than the mean of population that Group 1 represents, and that the mean of population that Group 1 represents is no more than .89 points higher than the mean of population that Group 2 represents.[18]

Note that in using the above notation, when a confidence interval range involves both a negative and positive limit (as is the case in the above example), it indicates that it is possible for either of the two population means to be the larger value. If, on the other hand, both limits identified by the confidence interval are positive values, the mean of Population 1 will always be greater than the mean of Population 2. If both limits identified by the confidence interval are negative values, the mean of Population 2 will always be greater than the mean of Population 1.

The 99% confidence interval for Example 8.1 will also be computed to illustrate that the range of values that define a 99% confidence interval is always larger than the range which defines a 95% confidence interval.

$$CI_{.99} = (\bar{X}_1 - \bar{X}_2) \pm (t_{.01})(s_{\bar{X}_1 - \bar{X}_2}) = -5 \pm (3.36)(2.55) = -5 \pm 8.57$$

$$-13.57 \le (\mu_1 - \mu_2) \le 3.57$$

Thus, the researcher can be 99% confident that the true difference between the population means falls within the range −13.57 and 3.57. Specifically, it indicates that one can be 99% confident that the mean of the population Group 2 represents is no more than 13.57 points higher than the mean of population that Group 1 represents, and that the mean of population that Group 1 represents is no more than 3.57 points higher than the mean of population that Group 2 represents. In closing the discussion of confidence intervals, it is worth noting that the broad range of values that define the above computed confidence intervals will not allow a researcher to estimate with great precision the actual difference between the means of the underlying populations. Additionally, the reader should take note of the fact that the reliability of Equation 8.15 will be compromised if one or more of the assumptions of the ***t* test for two independent samples** is saliently violated.

6. Test 8c: The *z* test for two independent samples There are occasions (albeit infrequent) when a researcher wants to compare the means of two independent samples, and happens to know the variances of the two underlying populations. In such a case, the ***z* test for two independent samples** should be employed to evaluate the data instead of the ***t* test for two independent samples**. As is the case with the latter test, the ***z* test for two independent samples** assumes that the two samples are randomly selected from populations that have normal distributions. The effect of violation of the normality assumption on the test statistic decreases as the size of the samples employed in an experiment increase. The homogeneity of variance assumption noted for the ***t* test for two independent samples** is not an assumption of the ***z* test for two independent samples**.

The null and alternative hypotheses employed for the ***z* test for two independent samples** are identical to those employed for the ***t* test for two independent samples**. Equation 8.16 is employed to compute the test statistic for the ***z* test for two independent samples**.

$$z = \frac{\bar{X}_1 - \bar{X}_2}{\sqrt{\frac{\sigma_1^2}{n_1} + \frac{\sigma_2^2}{n_2}}}$$ **(Equation 8.16)**

The only differences between Equation 8.16 and Equation 8.1 (the equation for the ***t* test for two independent samples**) are: a) In the denominator of Equation 8.16 the population variances σ_1^2 and σ_2^2 are employed instead of the estimated population variances $\tilde{s}_1^2$ and $\tilde{s}_2^2$ (which are employed in Equation 8.1); and b) Equation 8.16 computes a z score which is evaluated with the normal distribution, while Equation 8.1 derives a t score which is evaluated with the t distribution. Unlike Equation 8.1, Equation 8.16 can be used with both equal and unequal sample sizes.[19]

If it is assumed that the two population variances are known in Example 8.1, and that $\sigma_1^2 = 21.7$ and $\sigma_2^2 = 10.7$, Equation 8.16 can be employed to evaluate the data. Note that the obtained value $z = -1.96$ is identical to the value that is computed for t when Equation 8.1 is employed.

$$z = \frac{2.8 - 7.8}{\sqrt{\frac{21.7}{5} + \frac{10.7}{5}}} = -1.96$$

The obtained value $z = -1.96$ is evaluated with **Table A1 (Table of the Normal Distribution)** in the **Appendix**. In **Table A1** the tabled critical two-tailed .05 and .01 values are $z_{.05} = 1.96$ and $z_{.01} = 2.58$, and the tabled critical one-tailed .05 and .01 values are $z_{.05} = 1.65$ and $z_{.01} = 2.33$. Since the computed absolute value $z = 1.96$ is equal to the tabled critical two-tailed value $z_{.05} = 1.96$, the nondirectional alternative hypothesis H_1: $\mu_1 \neq \mu_2$ is supported at the .05 level. Since the computed value $z = -1.96$ is a negative number and the absolute value of z is greater than the tabled critical one-tailed .05 value $z_{.05} = 1.65$, the directional alternative hypothesis H_1: $\mu_1 < \mu_2$ is also supported at the .05 level.

When the same set of data is evaluated with the ***t* test for two independent samples**, although the directional alternative hypothesis H_1: $\mu_1 < \mu_2$ is supported at the .05 level, the nondirectional alternative hypothesis H_1: $\mu_1 \neq \mu_2$ is not supported. This latter fact illustrates that if the ***z* test for two independent samples** and the ***t* test for two independent samples** are employed to evaluate the same set of data (except when the value of $n_1 + n_2 - 2$ is extremely large), the latter test will provide a more conservative test of the null hypothesis (i.e., make it more difficult to reject H_0). This is the case, since the tabled critical values listed for the ***z* test for two independent samples** will always correspond to the tabled critical values listed in **Table A2** for $df = \infty$ (which are the lowest tabled critical values listed for the t distribution).

The final part of the discussion of the ***z* test for two independent samples** will describe a special case of the test in which it is employed to evaluate the difference between the average performance of two samples for whom scores have been obtained on a binomially distributed variable. Example 8.3, which is used to illustrate this application of the test, is an extension of Example 6.6 (discussed under the ***z* test for a population proportion (Test 6a)**) to a design involving two independent samples.

Example 8.3 *An experiment is conducted in which the performance of two groups is contrasted on a test of extrasensory perception. The two groups are comprised of five subjects who believe in extrasensory perception* (Group 1) *and five subjects who do not believe in it*

(Group 2). *The researcher employs as test stimuli a list of* 200 *binary digits (specifically, the values* 0 *and* 1*) which have been randomly generated by a computer. During the experiment an associate of the researcher concentrates on each of the digits in the order it appears on the list. While the associate does this, each of the ten subjects, all of whom are in separate rooms, attempts to guess the value of the number for each of* 200 *trials. The number of correct guesses for the two groups of subjects follow:* **Group 1**: 105, 120, 130, 115, 110; **Group 2**: 104, 99, 90, 100, 107. *Is there a difference in the performance of the two groups?*

The null and alternative hypotheses evaluated in Example 8.3 are identical to those evaluated in Example 8.1. Example 8.3 is evaluated with Equation 8.17, which is the form Equation 8.16 assumes when $\sigma_1^2 = \sigma_2^2$. Note that in Equation 8.17, m_i is employed to represent the number of subjects in the ith group, since the notation n is employed with a binomially distributed variable to designate the number of trials each subject is tested. Thus, $m_1 = m_2 = 5$.

$$z = \frac{\bar{X}_1 - \bar{X}_2}{\sigma \sqrt{\dfrac{1}{m_1} + \dfrac{1}{m_2}}} \qquad \textbf{(Equation 8.17)}$$

In employing Equation 8.17, we first compute the average score of each of the groups: $\bar{X}_1 = 580/5 = 116$ and $\bar{X}_2 = 500/5 = 100$. Since scores on the binary guessing task described in Example 8.3 are assumed to be binomially distributed, as is the case in Example 6.6, the following is true: $n = 200$, $\pi_1 = .5$, and $\pi_2 = .5$. The computed value for the population standard deviation for the binomially distributed variable is $\sigma = \sqrt{n\pi_1\pi_2} = \sqrt{(200)(.5)(.5)} = 7.07$. (The computation of the latter values is discussed in Section I of the **binomial sign test for a single sample (Test 6)**.) When the appropriate values are substituted in Equation 8.17, the value $z = 3.58$ is computed.

$$z = \frac{116 - 100}{7.07 \sqrt{\dfrac{1}{5} + \dfrac{1}{5}}} = 3.58$$

Since the computed absolute value $z = 3.58$ is greater than the tabled critical two-tailed values $z_{.05} = 1.96$ and $z_{.01} = 2.58$, the nondirectional alternative hypothesis H_1: $\mu_1 \neq \mu_2$ is supported at both the .05 and .01 levels. Since the computed value $z = 3.58$ is a positive number that is greater than the tabled critical one-tailed values $z_{.05} = 1.65$ and $z_{.01} = 2.33$, the directional alternative hypothesis H_1: $\mu_1 > \mu_2$ is supported at both the .05 and .01 levels. Thus, it can be concluded that the average score of Group 1 is significantly larger than the average score of Group 2.

Note that in applying Equation 8.17 to an example such as that illustrated by Example 8.3, it is entirely possible for one of the groups to have an average score that is significantly higher than the average score of the other group, without the performance of either group deviating from the expected mean value (which in Example 8.3 is $\mu = n\pi_1 = (200)(.5) = 100$). Equation 6.10 is employed to evaluate the performance of a single group. The null hypothesis that is employed for evaluating a single group is H_0: $\pi_1 = .5$ (which for Example 8.3 is commensurate with H_0: $\mu_j = 100$), and the nondirectional alternative hypothesis is H_1: $\pi_1 \neq .5$ (which for Example 8.3 is commensurate with H_1: $\mu_j \neq 100$). Since it is obvious from inspection of the data that the performance of Group 2 is at chance expectancy, the performance of Group 1, which is above chance, will be evaluated with Equation 6.10.

$$z = \frac{\bar{X} - \mu}{\frac{\sigma}{\sqrt{m}}} = \frac{116 - 100}{\frac{7.07}{\sqrt{5}}} = 5.06$$

Since the computed value $z = 5.06$ is greater than the tabled critical two-tailed values $z_{.05} = 1.96$ and $z_{.01} = 2.58$, the nondirectional alternative hypothesis H_1: $\pi_1 \neq .5$ is supported at both the .05 and .01 levels. Since the computed value $z = 5.06$ is a positive number that is greater than the tabled critical one-tailed values $z_{.05} = 1.65$ and $z_{.01} = 2.33$, the directional alternative hypothesis H_1: $\pi_1 > .5$ is also supported at both the .05 and .01 levels. Thus, it can be concluded that the average performance of Group 1 is significantly above chance.

VII. Additional Discussion of the *t* Test for Two Independent Samples

1. Unequal sample sizes In Section IV it is noted that if Equation 8.1/8.2 (which can only be used when $n_1 = n_2$) is applied to data where $n_1 \neq n_2$, the absolute value of t will be larger than the value computed with Equation 8.3. To illustrate this point, Equations 8.1 and 8.3 will be applied to a modified form of Example 8.1. Specifically, one of the scores in Group 2 will be eliminated from the data. If the score of Subject 1 (i.e., 11) is eliminated, the four scores that remain are: 11, 5, 8, 4. Employing the latter values, $n_2 = 4$, $\Sigma X_2 = 28$, $\Sigma X_2^2 = 226$, $\bar{X}_2 = 7$, $\tilde{s}_2^2 = [226 - (28)^2/4]/(4 - 1) = 10$. For Group 1 the values $n_1 = 5$, $\bar{X}_1 = 2.8$, and $\tilde{s}_1^2 = 21.7$ remain unchanged. The relevant values are substituted in Equation 8.1.

$$t = \frac{2.8 - 7}{\sqrt{\frac{21.7}{5} + \frac{10}{4}}} = -1.61$$

The same information is now substituted in Equation 8.3.

$$t = \frac{2.8 - 7}{\sqrt{\left[\frac{(5 - 1)(21.7) + (4 - 1)(10)}{5 + 4 - 2}\right]\left[\frac{1}{5} + \frac{1}{4}\right]}} = -1.53$$

Note that the absolute value $t = 1.61$ computed with Equation 8.1 is larger than the absolute value $t = 1.53$ computed with Equation 8.3, and thus Equation 8.3 provides a more conservative test of the null hypothesis. It so happens that in this instance neither of the computed t values allows the null hypothesis to be rejected since, for $df = 5 + 4 - 2 = 7$, both absolute values are below the tabled critical two-tailed values $t_{.05} = 2.37$ and $t_{.01} = 3.50$, and the tabled critical one-tailed values $t_{.05} = 1.90$ and $t_{.01} = 3.00$.

2. Outliers An **outlier** is a score in a set of data which is so extreme that by all appearances it is not representative of the population the sample represents. The presence of one or more outliers in a set of data can dramatically influence the value of the mean of a distribution, as well as the variance. As a result of this, outliers can result in distortion of the value of a test statistic.

As an example, assume that a researcher is comparing two groups with respect to their scores on a dependent variable, and that all of the subjects in both groups except for one subject in Group 2 obtain a score between 0 and 20. The subject in Group 2 with an outlier score obtains a score of 200. It should be obvious that the presence of this one score (even if the size of the sample in Group 2 is relatively large) will inflate the mean and variance of Group 2 relative to that of Group 1, and because of this either one or both of the following consequences may result: a) A significant difference between the two group means may be obtained which would not have been obtained if the outlier score was not present in the data; and/or b) The homogeneity of variance assumption will be violated due to the higher estimated population variance computed for Group 2. By virtue of adjusting the t test statistic for violation of the homogeneity of variance assumption, a more conservative test will be conducted, thus making it more difficult to reject the null hypothesis.[20]

Tabachnick and Fiddel (1989) note that outliers present a researcher with a statistical problem if one elects to include them in the data analysis, and a moral problem if one is trying to decide whether or not to eliminate them from the analysis. In essence, how a researcher deals with outliers should depend on how one accounts for their presence. In instances when a researcher is reasonably sure that an outlier is due to any one of the following, he has a strong argument for dropping such a score from the data: a) There is reason to believe that an error was made in recording the score in question; b) There is reason to believe that the score is the result of failure on the part of the subject to follow instructions, or other behavior on the part of the subject indicating a lack of cooperation and/or attention to the experiment; and c) There is reason to believe that the score is the result of failure on the part of the experimenter to employ the correct protocol in obtaining data from the subject.

When a researcher has reservations about eliminating outliers from a data analysis, there are a number of options available to him. The remainder of this section will describe strategies (which are discussed in more detail in Howell (1992)) that can be employed when it is not possible to determine the basis for outliers being present in the data, or one believes that in spite of the fact that they may be valid scores, by retaining them in the data it will result in erroneous conclusions with regard to the populations under study.

a) **Trimming data** **Trimming** involves removing a fixed percentage of extreme scores from each of the tails of any of the distributions that are involved in the data analysis. As an example, in an experiment involving two groups, one might decide to omit the two highest and two lowest scores in each group (since by doing this, any outliers in either group would be eliminated).

b) **Winsorizing** Another option available to the researcher, known as **Winsorizing**, involves replacing a fixed number of extreme scores with the score that is closest to them in the tail of the distribution in which they occur. As an example, in the distribution 0, 1, 18, 19, 23, 26, 26, 28, 33, 35, 98, 654, one can substitute a score of 18 for both the 0 and 1 (which are the two lowest scores), and a score of 35 for the 98 and 654 (which are the two highest scores). Thus, the Winsorized distribution will be: 18, 18, 18, 19, 23, 26, 26, 28, 33, 35, 35, 35.

c) **Data transformation** Data transformation involves performing a mathematical operation on each of the scores in a set of data, and thereby converting the data into a new set of scores which are then employed in analyzing the data. In addition to being able to reduce the impact of outliers, data transformation can be used to equate heterogeneous group variances, as well as to change the shape of a nonnormal distribution into a normal one. Among the most commonly employed data transformations, all of which can reduce the impact of outliers, are converting scores into their square root, reciprocal, or logarithm. In employing data transformations, a researcher may find it necessary to compare one or more different transformation procedures until he finds the one which best accomplishes the goal he is trying to achieve.

It is not uncommon that as a result of data transformation, data that will not yield a significant effect may be modified so as to be significant (and vice versa). When employed judiciously data transformation can be productive, insofar as it allows a researcher to conduct an analysis that will provide a more accurate picture of the populations under study than will be the case if the untransformed data are used for the analysis. On the other hand, it can also be used to distort data in order to support a hypothesis that is favored by an experimenter. Among those sources that describe data transformation in more detail are Howell (1992), Tabachnick and Fidell (1989), and Winer *et al.* (1991).

3. Robustness of the *t* test for two independent samples[21] Some statisticians believe that if one or more of the assumptions of a parametric test (such as the ***t* test for two independent samples**) is saliently violated, the test results will be unreliable, and because of this under such conditions it is more prudent to employ the analogous nonparametric test, which will generally have fewer or less rigorous assumptions than its parametric analog. In the case of the ***t* test for two independent samples** the most commonly employed analogous nonparametric tests are the **Mann–Whitney *U* test (Test 9)** and the **chi-square test for $r \times c$ tables (Test 11)**. Use of the **Mann–Whitney *U* test** (which is most likely to be recommended when the normality assumption of the ***t* test for two independent samples** is saliently violated) requires that the original interval/ratio scores be transformed into a rank-order format. By virtue of rank-ordering the data, information is sacrificed (since rank-orderings do not provide information regarding the magnitude of the differences between adjacent ranks). Given the fact that the **Mann–Whitney *U* test** employs less information, many researchers if given the choice will still elect to employ the ***t* test for two independent samples**, even if there is reason to believe that the normality assumption of the latter test is violated. Under such conditions, however, most researchers would probably conduct a more conservative t test in order to avoid inflating the likelihood of committing a Type I error (i.e., one might employ the tabled critical $t_{.01}$ value to represent the $t_{.05}$ value instead of the actual value listed for $t_{.05}$). In the unlikely event that a researcher elects to employ the **chi-square test for $r \times c$ tables** in place of the ***t* test for two independent samples**, he must convert the original interval/ratio data into a categorical format. This latter type of a transformation will result in an even greater loss of information than is the case with the **Mann–Whitney *U* test**.

The justification for using a parametric test in lieu of its nonparametric analog, even when one or more of the assumptions of the former test is violated, is that the results of numerous empirical sampling studies have demonstrated that under most conditions a parametric test like the ***t* test for two independent samples** is reasonably **robust**. A **robust** test is one that still provides reliable information about the underlying sampling distribution, in spite of the fact that one or more of the test's assumptions have been violated. In addition, researchers who are reluctant to employ nonparametric tests argue that parametric tests, such as the ***t* test for two independent samples**, are more powerful than their nonparametric analogs. Proponents of nonparametric tests counter with the argument that the latter group of tests are almost equivalent in power to their parametric analogs, and because of this, they state that it is preferable to use the appropriate nonparametric test if any of the assumptions of a parametric test have been saliently violated. Throughout this book it is demonstrated that in most instances when the same set of data is evaluated with both a parametric and a nonparametric test (especially a nonparametric test employing rank-order data), the two tests yield comparable results. As a general rule, in instances where only one of the two tests is significant, the parametric test is the one that is more likely to be significant. However, in most cases where a parametric test achieves significance and the nonparametric test does not, the latter test will just fall short of being significant. In instances where both tests are significant, the alpha level at which the result is significant will be lower for the parametric test.

VIII. Additional Examples Illustrating the Use of the *t* Test for Two Independent Samples

Two additional examples that can be evaluated with the *t* **test for two independent samples** are presented in this section. Since Examples 8.4 and 8.5 employ the same data set as that employed in Example 8.1, they yield the identical result.

Example 8.4 *A researcher wants to assess the relative effect of two different kinds of punishment (loud noise versus a blast of cold air) on the emotionality of mice. Each of ten mice is randomly assigned to one of two groups. During the course of the experiment each mouse is sequestered in an experimental chamber. While in the chamber, each of the five mice in* Group 1 *is periodically presented with a loud noise, and each of the five mice in* Group 2 *is periodically presented with a blast of cold air. The presentation of the punitive stimulus for each of the animals is generated by a machine that randomly presents the stimulus throughout the duration of the time the mouse is in the chamber. The dependent variable of emotionality employed in the study is the number of times each mouse defecates while in the experimental chamber. The number of episodes of defecation for the* 10 *mice follow:* **Group 1:** 11, 1, 0, 2, 0; **Group 2:** 11, 11, 5, 8, 4. *Do subjects exhibit differences in emotionality under the different experimental conditions?*

In Example 8.4, if the one-tailed alternative hypothesis H_1: $\mu_1 < \mu_2$ is employed it can be concluded that the group presented the blast of cold air (Group 2) obtains a significantly higher emotionality score than the group presented with loud noise (Group 1). This is the case, since the computed value $t = -1.96$ indicates that the average defecation score of Group 2 is significantly higher than the average defecation score of Group 1. As is the case in Example 8.1, the nondirectional alternative hypothesis H_1: $\mu_1 \neq \mu_2$ is not supported, and thus, if the latter alternative hypothesis is employed one cannot conclude that the blast of cold air results in higher emotionality.

Example 8.5 *Each of two companies that manufacture the same size precision ball bearing claims it has better quality control than its competitor. A quality control engineer conducts a study in which he compares the precision of ball bearings manufactured by the two companies. The engineer randomly selects five ball bearings from the stock of* Company A *and five ball bearings from the stock of* Company B. *He measures how much the diameter of each of the ten ball bearings deviates from the manufacturer's specifications. The deviation scores (in micrometers) for the ten ball bearings manufactured by the two companies follow:* **Company A:** 11, 1, 0 , 2, 0; **Company B**; 11, 11, 5, 8, 4. *What can the engineer conclude about the relative quality control of the two companies?*

In Example 8.5, if the one-tailed alternative hypothesis H_1: $\mu_1 < \mu_2$ is employed it can be concluded that Company B obtains a significantly higher deviation score than Company A. This will allow the researcher to conclude that Company A has superior quality control. As is the case in Example 8.1, the nondirectional alternative hypothesis H_1: $\mu_1 \neq \mu_2$ is not supported, and thus, if the latter alternative hypothesis is employed the researcher cannot conclude that Company A has superior quality control.

References

Cochran, W. G. and Cox, G. M. (1957). **Experimental designs** (2nd ed.). New York: John Wiley.

Cohen, J. (1988). **Statistical power analysis for the behavioral sciences** (2nd ed.). New York: Academic Press.

Guenther, W. C. (1965). **Concepts of statistical inference.** New York: McGraw–Hill Book Company.

Hartley, H. O. (1940). Testing the homogeneity of a set of variances. **Biometrika**, 31, 249–255.

Hartley, H. O. (1950). The maximum *F*-ratio as a shortcut test for heterogeneity of variance. **Biometrika**, 37, 308–312.

Howell, D. C. (1992). **Statistical methods for psychology** (3rd ed.). Boston: PWS–Kent Publishing Company.

Keppel, G. (1991). **Design and analysis: A researcher's handbook** (3rd ed.). Englewood Cliffs, NJ: Prentice Hall.

Keppel, G., Saufley, W. H. Jr., and Tokunaga, H. (1992). **Introduction to design and analysis: A student's handbook.** New York: W. H. Freeman and Company.

Kirk, R. E. (1982). **Experimental design: Procedures for the behavioral sciences** (2nd ed.). Belmont, CA: Brooks/Cole Publishing Company.

Satterthwaite, F. E. (1946). An approximate distribution of estimates of variance components. **Biometrics Bulletin**, 2, 110–114.

Tabachnick, B.C. and Fidell, L. S. (1989). **Using multivariate statistics** (2nd ed.). New York: HarperCollins Publishers.

Welch, B. L. (1947). The generalization of Student's problem when several different population variances are involved. **Biometrika**, 34, 28–35.

Winer, B. J., Brown, D., and Michels, K. (1991). **Statistical principles in experimental design** (3rd ed.). New York: McGraw–Hill Publishing Company.

Endnotes

1. Alternative terms that are commonly used to describe the different samples employed in an experiment are **groups**, **experimental conditions**, and **experimental treatments**.

2. It should be noted that there is a design in which different subjects serve in each of the k experimental conditions that is categorized as a **dependent samples design**. In a **dependent samples design** each subject either serves in all of the k experimental conditions, or else is matched with a subject in each of the other $(k - 1)$ experimental conditions. When subjects are matched with one another they are equated on one or more variables that are believed to be correlated with scores on the dependent variable. The concept of matching and a general discussion of the **dependent sample design** can be found under the ***t* test for two dependent samples (Test 12)**.

3. An alternative but equivalent way of writing the null hypothesis is H_0: $\mu_1 - \mu_2 = 0$. The analogous alternative but equivalent ways of writing the alternative hypotheses in the order they are presented are: H_1: $\mu_1 - \mu_2 \neq 0$, H_1: $\mu_1 - \mu_2 > 0$, H_1: $\mu_1 - \mu_2 < 0$.

4. In order to be solvable an equation for computing the t statistic requires that there is variability in the scores for both groups of subjects. If all subjects in Group 1 have the same score and all subjects in Group 2 have the same score, the values computed for the estimated population variances will equal zero (i.e., $\tilde{s}_1^2 = \tilde{s}_2^2 = 0$). If the latter is true the denominator of any of the equations to be presented for computing the value of t will equal zero, thus rendering a solution impossible.

5. When $n_1 = n_2$, $\tilde{s}_p = \sqrt{(\tilde{s}_1^2 + \tilde{s}_2^2)/2}$.

6. The actual value that is estimated by $s_{\bar{X}_1 - \bar{X}_2}$ is $\sigma_{\bar{X}_1 - \bar{X}_2}$, which is the standard deviation of the sampling distribution of the difference scores for the two populations. The meaning of the **standard error of the difference** can be best understood by considering the following procedure for generating an empirical sampling distribution of difference scores: a) Obtain a random sample of n_1 scores from Population 1 and a random sample of n_2 scores from Population 2; b) Compute the mean of each sample; c) Obtain a difference score by subtracting the mean of Sample 2 from the mean of Sample 1 — i.e., $\bar{X}_1 - \bar{X}_2 = D$; and d) Repeat steps a) through c) m times. At the conclusion of this procedure one will have obtained m difference scores. The **standard error of the difference** represents the standard deviation of the m difference scores, and can be computed by employing Equation 2.1. Thus: $s_{\bar{X}_1 - \bar{X}_2} = \sqrt{[(\Sigma D^2 - ((\Sigma D)^2/m)]/[m - 1]}$. The standard deviation that is computed with the aforementioned equation is an estimate of $\sigma_{\bar{X}_1 - \bar{X}_2}$.

7. Equation 8.4 can also be written in the form $df = (n_1 - 1) + (n_2 - 1)$, which reflects the number of degrees of freedom for each of the groups.

8. The absolute value of t is employed to represent t in the summary statement.

9. The F_{max} test is one of a number of statistical procedures that is named after the English statistician Sir Ronald Fisher. Among Fisher's contributions to the field of statistics was the development of a sampling distribution referred to as the F distribution (which bears the first letter of his surname). The values in the F_{max} distribution are derived from the F distribution.

10. A tabled $F_{.975}$ value is the value below which 97.5% of the F distribution falls and above which 2.5% of the distribution falls. A tabled $F_{.995}$ value is the value below which 99.5% of the F distribution falls and above which .5% of the distribution falls.

11. In **Table A9** the value $F_{max_{.01}} = 23.2$ is the result of rounding off $F_{.995} = 23.15$.

12. A tabled $F_{.025}$ value is the value below which 2.5% of the F distribution falls and above which 97.5% of the distribution falls. A tabled $F_{.005}$ value is the value below which .5% of the F distribution falls and above which 99.5% of the distribution falls.

13. Most sources only list values in the upper tail of the F distribution. The values $F_{.05} = .157$ and $F_{.01} = .063$ are obtained from Guenther (1965). It so happens that when $df_{num} = df_{den}$, the value of $F_{.05}$ can be obtained by dividing 1 by the value of $F_{.95}$. Thus: $1/6.39 = .157$. In the same respect the value of $F_{.01}$ can be obtained by dividing 1 by the value of $F_{.99}$. Thus: $1/15.98 = .063$.

14. When $n = n_1 = n_2$ and $t' = t_1 = t_2$, the t' value computed with Equation 8.9 will equal the tabled critical t value for $df = n - 1$. When $n_1 \neq n_2$, the computed value of t' will fall in between the values of t_1 and t_2. It should be noted, that the effect of violation of the homogeneity of variance assumption on the t test statistic decreases as the number of subjects employed in each of the samples increases. This can be demonstrated in relation to Equation 8.9, in that if there are a large number of subjects in each group the value that is employed for both t_1 and t_2 in Equation 8.9 is $t_{.05} = 1.96$. The latter tabled critical two-tailed .05 value, which is also the tabled critical two-tailed .05 value for the normal distribution, is the value that is computed for t'. Thus, in the case of large sample sizes the tabled critical value for

$df = n_1 + n_2 - 2$ will be equivalent to the value computed for $df = n_1 - 1$ and $df = n_2 - 1$.

15. The **treatment effect** described in this section is not the same thing as d, the **effect size** computed with Equation 8.10. However, if a hypothesized effect size is present in a set of data, the computed value of d can be used as a measure of treatment effect. In such an instance the value of d will be positively correlated with the value of the treatment effect described in this section.

16. It should be noted, however, that the degree of error associated with a measure of treatment effect will decrease as the size of the sample employed to compute the measure increases.

17. The reader familiar with the concept of correlation can think of a measure of treatment effect as a correlational measure which provides information analogous to that provided by the **coefficient of determination** (designated by the notation (r^2)), which is the square of the **Pearson product-moment correlation coefficient.** A **coefficient of determination** measures the degree of variability on one variable that can be accounted for by variability on a second variable. This latter definition is consistent with the definition that is provided in this section for a treatment effect.

18. This result can also be written as: $-.89 \leq (\mu_2 - \mu_1) \leq 10.89$.

19. In instances where in stating the null hypothesis a researcher stipulates that the difference between the two population means is some value other than zero, the numerator of Equation 8.16 is the same as the numerator of Equation 8.5. The protocol for computing the value of the numerator is identical to that employed for Equation 8.5.

20. If as a result of the presence of one or more outliers the difference between the group means is also inflated, the use of a more conservative test will, in part, compensate for this latter effect.

21. The general issues discussed in this section are relevant to any case in which a parametric and nonparametric test can be employed to evaluate the same set of data.

Test 9

Mann–Whitney *U* Test
(Nonparametric Test Employed with Ordinal Data)

I. Hypothesis Evaluated with Test and Relevant Background Information

Hypothesis evaluated with test Do two independent samples represent two populations with different median values (or different distributions with respect to the rank-orderings of the scores in the two underlying population distributions)?

Relevant background information on test The **Mann–Whitney *U* test** is employed with ordinal (rank-order) data in a hypothesis testing situation involving a design with two independent samples. If the result of the **Mann–Whitney *U* test** is significant, it indicates there is a significant difference between the two sample medians, and as a result of the latter the researcher can conclude there is a high likelihood that the samples represent populations with different median values.

Two versions of the test to be described under the label of the **Mann–Whitney *U* test** were independently developed by Mann and Whitney (1947) and Wilcoxon (1949). The version to be described here is commonly identified as the **Mann–Whitney *U* test**, while the version developed by Wilcoxon (1949) is usually referred to as the **Wilcoxon–Mann–Whitney test**. Although they employ different equations and different tables, the two versions of the test yield comparable results. In employing the **Mann–Whitney *U* test**, one of the following is true with regard to the rank-order data that are evaluated: a) The data are in a rank-order format, since it is the only format in which scores are available; or b) The data have been transformed into a rank-order format from an interval/ratio format, since the researcher has reason to believe that the normality assumption (as well as, perhaps, the homogeneity of variance assumption) of the ***t* test for two independent samples** (which is the parametric analog of the **Mann–Whitney *U* test**) is saliently violated. It should be noted that when a researcher elects to transform a set of interval/ratio data into ranks, information is sacrificed. This latter fact accounts for why there is reluctance among some researchers to employ nonparametric tests such as the **Mann–Whitney *U* test**, even if there is reason to believe that one or more of the assumptions of the ***t* test for two independent samples** has been violated.

Various sources (e.g., Conover (1980), Daniel (1990), and Marascuilo and McSweeney (1977)) note that the **Mann–Whitney *U* test** is based on the following assumptions: a) Each sample has been randomly selected from the population it represents; b) The two samples are independent of one another; c) The original variable observed (which is subsequently ranked) is a continuous random variable. In truth, this assumption which is common to many nonparametric tests, is often not adhered to, in that such tests are often employed with a dependent variable that represents a discrete random variable; and d) The underlying distributions from which the samples are derived are identical in shape. The shapes of the underlying population distributions, however, do not have to be normal. Maxwell and Delaney (1990) point out the assumption of identically shaped distributions implies equal dispersion of data within each distribution. Because of this, they note that like the ***t* test for two independent samples**, the **Mann–Whitney *U* test** also assumes homogeneity of variance with respect to the underlying population distributions. Because the latter assumption is not generally acknowledged for the **Mann–Whitney *U* test**, it is not uncommon for sources to state that violation of the homogeneity of variance assumption justifies use of the **Mann–Whitney *U* test** in lieu of the ***t* test for two independent samples**. It should be pointed out, however, that there is some empirical evidence which suggests that the sampling distribution for the **Mann–Whitney *U* test** is not as affected by violation of the homogeneity of variance assumption as is the sampling distribution for ***t* test for two independent samples**. One reason cited by various sources for employing the **Mann–Whitney *U* test**, is that by virtue of ranking interval/ratio data a researcher will be able to reduce or eliminate the impact of **outliers**. As noted in Section VII of the ***t* test for two independent samples**, since **outliers** can dramatically influence variability, they can be responsible for heterogeneity of variance between two or more samples. In addition, **outliers** can have a dramatic impact on the value of a sample mean.

II. Example

Example 9.1 is identical to Example 8.1 (which is evaluated with the ***t* test for two independent samples**). In evaluating Example 9.1 it will be assumed that the interval/ratio data are rank-ordered, since one or more of the assumptions of the ***t* test for two independent samples** has been saliently violated.

Example 9.1 *In order to assess the efficacy of a new antidepressant drug, ten clinically depressed patients are randomly assigned to one of two groups. Five patients are assigned to* Group 1, *which is administered the antidepressant drug for a period of six months. The other five patients are assigned to* Group 2, *which is administered a placebo during the same six month period. Assume that prior to introducing the experimental treatments, the experimenter confirmed that the level of depression in the two groups was equal. After six months elapse all ten subjects are rated by a psychiatrist (who is blind with respect to a subject's experimental condition) on their level of depression. The psychiatrist's depression ratings for the five subjects in each group follow (the higher the rating the more depressed a subject):* **Group 1**: 11, 1, 0, 2, 0; **Group 2**: 11, 11, 5, 8, 4. *Do the data indicate that the antidepressant drug is effective?*

III. Null versus Alternative Hypotheses

Null hypothesis $H_0\text{: } \theta_1 = \theta_2$

(The median of the population Group 1 represents equals the median of the population Group 2 represents. With respect to the sample data, when both groups have an equal sample size, this translates into the sum of the ranks of Group 1 being equal to the sum of the ranks of Group 2 (i.e., $\Sigma R_1 = \Sigma R_2$). A more general way of stating this, which also encompasses designs involving unequal sample sizes, is that the means of the ranks of the two groups are equal (i.e., $\bar{R}_1 = \bar{R}_2$).

Alternative hypothesis $H_1\text{: } \theta_1 \neq \theta_2$

(The median of the population Group 1 represents does not equal the median of the population Group 2 represents. With respect to the sample data, when both groups have an equal sample size, this translates into the sum of the ranks of Group 1 not being equal to the sum of the ranks of Group 2 (i.e., $\Sigma R_1 \neq \Sigma R_2$). A more general way of stating this, which also encompasses designs involving unequal sample sizes, is that the means of the ranks of the two groups are not equal (i.e., $\bar{R}_1 \neq \bar{R}_2$). This is a **nondirectional alternative hypothesis** and it is evaluated with a **two-tailed test**.)

or

$$H_1\text{: } \theta_1 > \theta_2$$

(The median of the population Group 1 represents is greater than the median of the population Group 2 represents. With respect to the sample data, when both groups have an equal sample size (so long as a rank of 1 is given to the lowest score), this translates into the sum of the ranks of Group 1 being greater than the sum of the ranks of Group 2 (i.e., $\Sigma R_1 > \Sigma R_2$). A more general way of stating this (which also encompasses designs involving unequal sample sizes) is that the mean of the ranks of Group 1 is greater than the mean of the ranks of Group 2 (i.e., $\bar{R}_1 > \bar{R}_2$). This is a **directional alternative hypothesis** and it is evaluated with a **one-tailed test.**)

or

$$H_0\text{: } \theta_1 < \theta_2$$

(The median of the population Group 1 represents is less than the median of the population Group 2 represents. With respect to the sample data, when both groups have an equal sample size (so long as a rank of 1 is given to the lowest score), this translates into the sum of the ranks of Group 1 being less than the sum of the ranks of Group 2 (i.e., $\Sigma R_1 < \Sigma R_2$). A more general way of stating this (which also encompasses designs involving unequal sample sizes) is that the mean of the ranks of Group 1 is less than the mean of the ranks of Group 2 (i.e., $\bar{R}_1 < \bar{R}_2$). This is a **directional alternative hypothesis** and it is evaluated with a **one-tailed test.**)

Note: Only one of the above noted alternative hypotheses is employed. If the alternative hypothesis the researcher selects is supported, the null hypothesis is rejected.

IV. Test Computations

The data for Example 9.1 are summarized in Table 9.1. The total number of subjects employed in the experiment is $N = 10$. There are $n_1 = 5$ subjects in Group 1 and $n_2 = 5$ subjects in Group 2. The original interval/ratio scores of the subjects are recorded in the columns labelled X_1 and X_2. The adjacent columns R_1 and R_2 contain the rank-order assigned to each of the scores. The rankings for Example 9.1 are summarized in Table 9.2. The ranking protocol for the **Mann–Whitney *U* test** is described in this section. Note that in Table 9.1 and Table 9.2 each subject's identification number indicates the order in Table 9.1 in which a subject's score appears in a given group, followed by his/her group. Thus, Subject i, j is the ith subject in Group j.

The following protocol, which is summarized in Table 9.2, is used in assigning ranks.

a) All $N = 10$ scores are arranged in order of magnitude (irrespective of group membership), beginning on the left with the lowest score and moving to the right as scores increase. This is done in the second row of Table 9.2.

b) In the third row of Table 9.2, all $N = 10$ scores are assigned a rank. Moving from left to right, a rank of 1 is assigned to the score that is furthest to the left (which is the lowest score), a rank of 2 is assigned to the score that is second from the left (which, if there are no ties, will be the second lowest score), and so on until the score at the extreme right (which will be the highest score) is assigned a rank equal to N (if there are no ties for the highest score).

c) The ranks in the third row of Table 9.2 must be adjusted when there are tie scores present in the data. Specifically, in instances where two or more subjects have the same score, the average of the ranks involved is assigned to all scores tied for a given rank. This

Table 9.1 Data for Example 9.1

	Group 1			Group 2	
	X_1	R_1		X_2	R_2
Subject 1,1	11	9	**Subject 1,2**	11	9
Subject 2,1	1	3	**Subject 2,2**	11	9
Subject 3,1	0	1.5	**Subject 3,2**	5	6
Subject 4,1	2	4	**Subject 4,2**	8	7
Subject 5,1	0	1.5	**Subject 5,2**	4	5
		$\Sigma R_1 = 19$			$\Sigma R_2 = 36$
	$\bar{R}_1 = \frac{\Sigma R_1}{n_1} = \frac{19}{5} = 3.8$			$\bar{R}_2 = \frac{\Sigma R_2}{n_2} = \frac{36}{5} = 7.2$	

Table 9.2 Rankings for the Mann–Whitney *U* Test for Example 9.1

Subject identification number	3,1	5,1	2,1	4,1	5,2	3,2	4,2	1,1	1,2	2,2
Depression score	0	0	1	2	4	5	8	11	11	11
Rank prior to tie adjustment	1	2	3	4	5	6	7	8	9	10
Tie adjusted rank	1.5	1.5	3	4	5	6	7	9	9	9

adjustment is made in the fourth row of Table 9.2. To illustrate: Both Subjects 3,1 and 5,1 have a score of 0. Since the two scores of 0 are the lowest scores out of the total of ten scores, in assigning ranks to these scores we can arbitrarily assign one of the 0 scores a rank of 1 and the other a rank of 2. However, since both of these scores are identical it is more equitable to give each of them the same rank. To do this, we compute the average of the ranks involved for the two scores. Thus, the two ranks involved prior to adjusting for ties (i.e., the ranks 1 and 2) are added up and divided by two. The resulting value $(1 + 2)/2 = 1.5$ is the rank assigned to each of the subjects who is tied for 0. There is one other set of ties present in the data which involves three subjects. Subjects 1,1, 1,2, and 2,2 all obtain a score of 11. Since the ranks assigned to these three scores prior to adjusting for ties are 8, 9, and 10, the average of the three ranks $(8 + 9 + 10)/3 = 9$ is assigned to the scores of each of the three subjects who obtain a score of 11.

Although it is not the case in Example 9.1, it should be noted that any time each set of ties involves subjects in the same group, the tie adjustment will result in the identical sum and average for the ranks of the two groups that will be obtained if the tie adjustment is not employed. Because of this, under these conditions the computed test statistic will be identical regardless of whether or not one uses the tie adjustment. On the other hand, when one or more sets of ties involve subjects from both groups, the tie adjusted ranks will yield a value for the test statistic that will be different from that which will be obtained if the tie adjustment is not employed. In Example 9.1, although the two subjects who obtain a score of zero happen to be in the same group, in the case of the three subjects who have a score of 11, one subject is in Group 1 and the other two subjects are in Group 2.

If the ranking protocol described in this section is used with Example 9.1, and the researcher elects to employ a one-tailed alternative hypothesis, the directional alternative hypothesis H_1: $\theta_1 < \theta_2$ is employed. The latter directional alternative hypothesis is employed, since it predicts that Group 1, the group that receives the antidepressant, will have a lower median score, and thus a lower sum of ranks/average rank (both of which are indicative of a lower level of depression) than Group 2.

It should be noted that it is permissible to reverse the ranking protocol described in this section. Specifically, one can assign a rank of 1 to the highest score, a rank of 2 to the second highest score, and so on, until reaching the lowest score which is assigned a rank equal to the value of N. Although this reverse ranking protocol will yield the identical **Mann–Whitney test** statistic as the ranking protocol described in this section, it will result in ranks that are the opposite of those obtained in Table 9.2. If the protocol employed in ranking is taken into account in interpreting the results of the **Mann–Whitney *U* test**, both ranking protocols will lead to identical conclusions. Since it is less likely to cause confusion in interpreting the test statistic, it is recommended that the original ranking protocol described in this section be employed — i.e., assigning a rank of 1 to the lowest score and a rank equivalent to the value of N to the highest score. In view of this, in all future discussion of the **Mann–Whitney *U* test**, as well as other tests that involve rank-ordering data, it will be assumed (unless otherwise stipulated) that the ranking protocol employed assigns a rank of 1 to the lowest score and a rank of N to the highest score.

Once all of the subjects have been assigned a rank, the sum of the ranks for each of the groups is computed. These values, $\Sigma R_1 = 19$ and $\Sigma R_2 = 36$, are computed in Table 9.1. Upon determining the sum of the ranks for both groups, the values U_1 and U_2 are computed employing Equations 9.1 and 9.2.

$$U_1 = n_1 n_2 + \frac{n_1(n_1 + 1)}{2} - \Sigma R_1 \qquad \textbf{(Equation 9.1)}$$

$$U_2 = n_1 n_2 + \frac{n_2(n_2 + 1)}{2} - \Sigma R_2 \qquad \textbf{(Equation 9.2)}$$

Employing Equations 9.1 and 9.2, the values $U_1 = 21$ and $U_2 = 4$ are computed.

$$U_1 = (5)(5) + \frac{5(5 + 1)}{2} - 19 = 21$$

$$U_2 = (5)(5) + \frac{5(5 + 1)}{2} - 36 = 4$$

Note that U_1 and U_2 can never be negative values. If a negative value is obtained for either, it indicates an error has been made in the rankings and/or calculations.

Equation 9.3 can be employed to confirm that the correct values have been computed for U_1 and U_2.

$$n_1 n_2 = U_1 + U_2 \qquad \textbf{(Equation 9.3)}$$

If the relationship in Equation 9.3 is not confirmed, it indicates that an error has been made in ranking the scores or in the computation of the U values. The relationship described by Equation 9.3 is confirmed below for Example 9.1.

$$(5)(5) = 21 + 4 = 25$$

V. Interpretation of the Test Results

The smaller of the two values U_1 versus U_2 is designated as the obtained U statistic. Since $U_2 = 4$ is smaller than $U_1 = 21$, the value of $U = 4$. The value of U is evaluated with **Table A11 (Table of Critical Values for the Mann–Whitney *U* Statistic)** in the **Appendix**. In **Table A11**, the critical U values are listed in reference to the number of subjects in each group.[1] For $n_1 = 5$ and $n_2 = 5$, the tabled critical two-tailed .05 and .01 values are $U_{.05} = 2$ and $U_{.01} = 0$, and the tabled critical one-tailed .05 and .01 values are $U_{.05} = 4$ and $U_{.01} = 1$. In order to be significant, the obtained value of U must be **equal to or less than** the tabled critical value at the prespecified level of significance.

Since the obtained value $U = 4$ must be equal to or less than the aforementioned tabled critical values, the null hypothesis can only be rejected if the directional alternative hypothesis H_1: $\theta_1 < \theta_2$ is employed. The directional alternative hypothesis H_1: $\theta_1 < \theta_2$ is supported at the .05 level, since $U = 4$ is equal to the tabled critical one-tailed value $U_{.05} = 4$. The data are consistent with the directional alternative hypothesis H_1: $\theta_1 < \theta_2$, since the average of the ranks in Group 1 is less than the average of the ranks in Group 2 (i.e., $\bar{R}_1 < \bar{R}_2$).[2] The directional alternative hypothesis H_1: $\theta_1 < \theta_2$ is not supported at the .01 level, since the obtained value $U = 4$ is greater than the tabled critical one-tailed value $U_{.01} = 1$.

The nondirectional alternative hypothesis H_1: $\theta_1 \neq \theta_2$ is not supported, since the obtained value $U = 4$ is greater than the tabled critical two-tailed value $U_{.05} = 2$.

Since the data are not consistent with the directional alternative hypothesis H_1: $\theta_1 > \theta_2$, the latter alternative hypothesis is not supported. In order for the directional alternative hypothesis H_1: $\theta_1 > \theta_2$ to be supported, the average of the ranks in Group 1 must be greater than the average of the ranks in Group 2 (i.e., $\bar{R}_1 > \bar{R}_2$) (as well as the fact that the computed value of U must be equal to or less than the tabled critical one-tailed value at the prespecified level of significance).

The results of the **Mann–Whitney U test** are consistent with those obtained when the ***t* test for independent samples** is employed to evaluate Example 8.1 (which employs the same set of data as Example 9.1). In both instances, the null hypothesis can only be rejected if the researcher employs a directional alternative hypothesis that predicts a lower degree of depression in the group that receives the antidepressant medication (Group 1).

VI. Additional Analytical Procedures for the Mann–Whitney *U* Test and/or Related Tests

1. The normal approximation of the Mann–Whitney *U* statistic for large sample sizes
If the sample size employed in a study is relatively large, the normal distribution can be employed to approximate the **Mann–Whitney *U* statistic**. Although sources do not agree on the value of the sample size that justifies employing the normal approximation of the Mann–Whitney distribution, they generally state that it should be employed for sample sizes larger than those documented in the exact table of the U distribution contained within the source. Equation 9.4 provides the normal approximation of the **Mann–Whitney *U* test** statistic.

$$z = \frac{U - \dfrac{n_1 n_2}{2}}{\sqrt{\dfrac{n_1 n_2 (n_1 + n_2 + 1)}{12}}} \qquad \textbf{(Equation 9.4)}$$

In the numerator of Equation 9.4 the term $(n_1 n_2)/2$ is often summarized with the notation U_E, which represents the expected (mean) value of U if the null hypothesis is true. In other words, if in fact the two groups are equivalent, it is expected that $\bar{R}_1 = \bar{R}_2$. If the latter is true then $U_1 = U_2$, and both of the values U_1 and U_2 will equal the value $(n_1 n_2)/2$. The denominator in Equation 9.4 represents the expected standard deviation of the sampling distribution for the normal approximation of the U statistic.

Although Example 9.1 involves only $N = 10$ scores (a value most sources would view as too small to use with the normal approximation), it will be employed to illustrate Equation 9.4. The reader will see that in spite of employing Equation 9.4 with a small sample size, it yields a result that is consistent with the result obtained when the exact table for the **Mann–Whitney *U*** distribution is employed. It should be noted that since the smaller of the two values U_1 versus U_2 is selected to represent U, the value of z will always be negative (unless $U_1 = U_2$, in which case $z = 0$). This is the case, since by selecting the smaller value U will always be less than the expected value $U_E = (n_1 n_2)/2$.

Employing Equation 9.4, the value $z = -1.78$ is computed.[3]

$$z = \frac{4 - \frac{(5)(5)}{2}}{\sqrt{\frac{(5)(5)(5 + 5 + 1)}{12}}} = -1.78$$

The obtained value $z = -1.78$ is evaluated with **Table A1 (Table of the Normal Distribution) in the Appendix**. In order to be significant, the obtained absolute value of z must be equal to or greater than the tabled critical value at the prespecified level of significance. The tabled critical two-tailed .05 and .01 values are $z_{.05} = 1.96$ and $z_{.01} = 2.58$, and the tabled critical one-tailed .05 and .01 values are $z_{.05} = 1.65$ and $z_{.01} = 2.33$. The following guidelines are employed in evaluating the null hypothesis.

a) If the nondirectional alternative hypothesis H_1: $\theta_1 \neq \theta_2$ is employed, the null hypothesis can be rejected if the obtained absolute value of z is equal to or greater than the tabled critical two-tailed value at the prespecified level of significance.

b) If a directional alternative hypothesis is employed, one of the two possible directional alternative hypotheses is supported if the obtained absolute value of z is equal to or greater than the tabled critical one-tailed value at the prespecified level of significance. The directional alternative hypothesis that is supported is the one that is consistent with the data.

Employing the above guidelines with Example 9.1, the following conclusions are reached.

Since the obtained absolute value $z = 1.78$ must be equal to or greater than the tabled critical value at the prespecified level of significance, the null hypothesis can only be rejected if the directional alternative hypothesis H_1: $\theta_1 < \theta_2$ is employed. The directional alternative hypothesis H_1: $\theta_1 < \theta_2$ is supported at the .05 level, since the absolute value $z = 1.78$ is greater than the tabled critical one-tailed value $z_{.05} = 1.65$. As noted in Section V, the data are consistent with the directional alternative hypothesis H_1: $\theta_1 < \theta_2$. The directional alternative hypothesis H_1: $\theta_1 < \theta_2$ is not supported at the .01 level, since the obtained absolute value $z = 1.78$ is less than the tabled critical one-tailed value $z_{.01} = 2.33$.

The nondirectional alternative hypothesis H_1: $\theta_1 \neq \theta_2$ is not supported, since the obtained absolute value $z = 1.78$ is less than the tabled critical two-tailed .05 value $z_{.05} = 1.96$.

Since the data are not consistent with the directional alternative hypothesis H_1: $\theta_1 > \theta_2$, the latter alternative hypothesis is not supported. As noted in Section V, in order for the latter directional alternative hypothesis to be supported, the following condition must be met: $\bar{R}_1 > \bar{R}_2$.

It turns out that the above conclusions based on the normal approximation are identical to those reached when the exact table of the **Mann–Whitney *U* distribution** is employed.

It should be noted that, in actuality, either U_1 or U_2 can be employed in Equation 9.4 to represent the value of U. This is the case, since either value yields the same absolute value for z. Thus, if for Example 9.1 $U_1 = 21$ is employed in Equation 9.4, the value $z = 1.78$ is computed. Since the decision with respect to the status of the null hypothesis is a function of the absolute value of z, the value $z = 1.78$ leads to the same conclusions that are reached when $z = -1.78$ is employed. The decision with regard to a directional alternative hypothesis is not affected, since the data are still consistent with the directional alternative hypothesis H_1: $\theta_1 < \theta_2$.

2. The correction for continuity for the normal approximation of the Mann–Whitney *U* test[4] Although not used in most sources, Siegel and Castellan (1988) employ a correction for continuity for the normal approximation of the **Mann–Whitney test** statistic. Marascuilo and McSweeney (1977) note that the correction for continuity is generally not employed, unless the computed absolute value of z is close to the prespecified tabled critical value. The correction for continuity, which reduces the absolute value of z, requires that .5 be subtracted from the absolute value of the numerator of Equation 9.4 (as well as the absolute value of the numerator of the alternative equation described in Endnote 3). The continuity corrected version of Equation 9.4 is provided by Equation 9.5.

$$z = \frac{\left| U - \frac{n_1 n_2}{2} \right| - .5}{\sqrt{\frac{n_1 n_2 (n_1 + n_2 + 1)}{12}}} \qquad \textbf{(Equation 9.5)}$$

If the correction for continuity is employed with Example 9.1, the value computed for the numerator of Equation 9.5 is 8 (in contrast to the value 8.5 computed with Equation 9.4). Employing Equation 9.5 with Example 9.1, the value $z = -1.67$ is computed.

$$z = \frac{\left| 4 - \frac{(5)(5)}{2} \right| - .5}{\sqrt{\frac{(5)(5)(5 + 5 + 1)}{12}}} = -1.67$$

Since the absolute value $z = 1.67$ is greater than the tabled critical one-tailed value $z_{.05} = 1.65$, the directional alternative hypothesis H_1: $\theta_1 < \theta_2$ is still supported at the .05 level.

3. Tie correction for the normal approximation of the Mann–Whitney U statistic Some sources recommend that when an excessive number of ties are present in the data, a tie correction should be introduced into Equation 9.4. Equation 9.6 is the tie corrected equation for the normal approximation of the **Mann–Whitney U distribution.** The latter equation results in a slight increase in the absolute value of z.[5]

$$z = \frac{U - \frac{n_1 n_2}{2}}{\sqrt{\frac{n_1 n_2 (n_1 + n_2 + 1)}{12} - \frac{n_1 n_2 \left[\sum_{i=1}^{s} (t_i^3 - t_i) \right]}{12(n_1 + n_2)(n_1 + n_2 - 1)}}} \qquad \textbf{(Equation 9.6)}$$

The only difference between Equations 9.4 and 9.6 is the term to the right of the element $[n_1 n_2 (n_1 + n_2 + 1)]/12$ in the denominator. The result of this subtraction reduces the value of the denominator, thereby resulting in the slight increase in the absolute value of z. The term $\sum_{i=1}^{s} (t_i^3 - t_i)$ in the denominator of Equation 9.6 computes a value based on the number of ties in the data. In Example 9.1 there are $s = 2$ sets of ties. Specifically, there is a set of ties involving two subjects with the score 0, and a set of ties involving three subjects with the score 11. The notation $\sum_{i=1}^{s} (t_i^3 - t_i)$ indicates the following: a) For each set of ties, the number of ties in the set is subtracted from the cube of the number of ties in that set; and b) The sum of all the values computed in part a) is obtained. Thus, for Example 9.1:

$$\sum_{i=1}^{s} (t_i^3 - t_i) = [(2)^3 - 2] + [(3)^3 - 3] = 30$$

The tie corrected value $z = -1.80$ is now computed employing Equation 9.6.

$$z = \frac{4 - \dfrac{(5)(5)}{2}}{\sqrt{\dfrac{(5)(5)(5 + 5 + 1)}{12} - \dfrac{(5)(5)(30)}{12(5 + 5)(5 + 5 - 1)}}} = -1.80$$

The difference between $z = -1.80$ and the uncorrected value $z = -1.78$ is trivial, and consequently the decision the researcher makes with respect to the null hypothesis is not affected, regardless of which alternative hypothesis is employed.[6]

4. Sources for computing a confidence interval for the Mann–Whitney *U* test Various books that specialize in nonparametric statistics (e.g., Daniel (1990) and Marascuilo and McSweeney (1977)) describe the computational procedure for computing a confidence interval that can be used in conjunction with the **Mann–Whitney *U* test.** The confidence interval identifies a range of values that the true difference between the two population medians is likely to fall.

VII. Additional Discussion of the Mann–Whitney *U* Test

1. Power efficiency of the Mann–Whitney *U* test **Power efficiency** (also referred to as **relative efficiency**) is a statistic that is employed to indicate the power of two tests relative to one another. It is most commonly used in comparing the power of a parametric test with its nonparametric analog. As an example, assume we wish to determine the relative power of the **Mann–Whitney *U* test** (designated as Test *A*) and the ***t* test for two independent samples** (designated as Test *B*). Assume that both tests employ the same alpha level with respect to the null hypothesis being evaluated.

For a fixed power value, the statistic PE_{AB} will represent the power efficiency of Test *A* relative to Test *B*. The value of PE_{AB} is computed with Equation 9.7.

$$PE_{AB} = \frac{n_B}{n_A} \qquad \textbf{(Equation 9.7)}$$

Where: n_A is the number of subjects required for Test *A* and n_B is the number of subjects required for Test *B*, when each test is required to evaluate an alternative hypothesis at the same power

Thus, if the ***t* test for two independent samples** requires 95 subjects to evaluate an alternative hypothesis at a power of .80, and the **Mann–Whitney *U* test** requires 100 subjects to evaluate the analogous alternative hypothesis at a power of .80, the value of $PE_{AB} = 95/100 = .95$. From this result it can be determined that if 100 subjects are employed to evaluate a null hypothesis with the ***t* test for two independent samples**, in order achieve the same level of power for evaluating the analogous null hypothesis with the **Mann–Whitney *U* test,** it is necessary to employ 1/.95 = 105 subjects.

Various sources (Daniel (1990), Marascuilo and McSweeney (1977) and Siegel and Castellan (1988)) note that the power efficiency of the **Mann–Whitney *U* test** relative to the ***t* test for two independent samples** approaches .95 as the total sample size increases. Daniel (1988) and Marascuilo and McSweeney (1977) note that the latter value only holds when the distributions in the underlying populations are normal. Additional references on the power efficiency of the **Mann–Whitney *U* test** can be found in Daniel (1990). Marascuilo and McSweeney (1977, p. 87) present relative efficiency values for a large number of nonparametric tests, including the **Mann–Whitney *U* test,** relative to their parametric analogs. In describing relative efficiency values, the latter source lists them in reference to the shape of the underlying population distributions. As a general rule, proponents of nonparametric tests take the position that when a researcher has reason to believe that the normality assumption and/or the homogeneity of variance assumptions of the ***t* test for two independent samples** have been saliently violated, the **Mann–Whitney *t* test** provides a relatively powerful test of the comparable alternative hypothesis.

2. Alternative nonparametric rank-order procedures for evaluating a design involving two independent samples In addition to the **Mann–Whitney *U* test** a number of other nonparametric procedures for two independent samples have been developed that can be employed with ordinal data. Among the more commonly cited alternative procedures are the following: a) **Tukey's quick test** (Tukey (1959)); b) **Normal scores tests** developed by Terry and Hoeffding (Terry (1952)), Van der Waerden (1953/1953), and Bell and Doksum (1965); c) **The median test for independent samples (Test 11e)** which involves dichotomizing two samples with respect to their median values, and evaluating the data with the **chi-square test for $r \times c$ tables (Test 11)**; d) **The Kolmogorov–Smirnov two-sample test** (Kolmogorov (1933) and Smirnov (1939)); e) **Tests of extreme reactions** developed by Moses (1952) and Hollander (1963); f) **The Wald–Wolfowitz runs test** (Wald and Wolfowitz (1940)) (discussed under the **single-sample runs test (Test 7)** within the framework of Example 7.5); and g) **Wilks' empty-cell test for identical populations** (Wilks (1961)). In addition to various books which specialize in nonparametric statistics, Sheskin (1984) describes these tests in greater detail.

VIII. Additional Examples Illustrating the Use of the Mann–Whitney *U* Test

The **Mann–Whitney *U* test** can be employed with any of the additional examples noted for the ***t* test for two independent samples.** The interval/ratio scores in all of the examples have to be rank-ordered in order to employ the **Mann–Whitney *U* test.** Example 9.2 provides one additional example that can be evaluated with the **Mann–Whitney *U* test.** It differs from Example 9.1 in the following respects: a) The original scores are in a rank-order format. Thus, there is no need to transform the scores into ranks from an interval/ratio format (as is the case in Example 9.1). It should be noted though, that it is implied in Example 9.2 that the ranks are based on an underlying interval/ratio scale; b) The sample sizes are unequal, with $n_1 = 6$ and $n_2 = 7$.

Example 9.2 *Doctor Radical, a math instructor at Logarithm University, has two classes in advanced calculus. There are six students in* Class 1 *and seven students in* Class 2. *The instructor uses a programmed textbook in* Class 1 *and a conventional textbook in* Class 2. *At the end of the semester, in order to determine if the type of text employed influences student performance, Dr. Radical has another math instructor, Dr. Root, rank the* 13 *students in the two classes with respect to math ability. The rankings of the students in the two classes follow:* **Class 1:** 1, 3, 5, 7, 11, 13; **Class 2:** 2, 4, 6, 8, 9, 10, 12 *(assume the lower the rank the better the student).*

Employing the **Mann–Whitney *U* test** with Example 9.2 the following values are computed.

$$\Sigma R_1 = 40 \qquad \Sigma R_2 = 51$$

$$U_1 = (6)(7) + \frac{6(6+1)}{2} - 40 = 23$$

$$U_2 = (6)(7) + \frac{7(7+1)}{2} - 51 = 19$$

$$(U_1 = 23) + (U_2 = 19) = (n_1 = 6)(n_2 = 7) = 42$$

$$z = \frac{19 - \frac{(6)(7)}{2}}{\sqrt{\frac{(6)(7)(6+7+1)}{12}}} = -.29$$

Since the value $U_2 = 19$ is less than $U_1 = 23$, $U = 19$. Employing **Table A11**, for $n_1 = 6$ and $n_2 = 7$, the tabled critical two-tailed values are $U_{.05} = 6$ and $U_{.01} = 3$, and the tabled critical one-tailed values are $U_{.05} = 8$ and $U_{.01} = 4$. Since in order to be significant the obtained value $U = 19$ must be equal to or less than the tabled critical value, the null hypothesis H_0: $\theta_1 = \theta_2$ cannot be rejected regardless of which alternative hypothesis is employed. The use of Equation 9.4 for the normal approximation confirms this result, since the absolute value $z = .29$ is less than the tabled critical two-tailed values $z_{.05} = 1.96$ and $z_{.01} = 2.58$, and the tabled critical one-tailed values $z_{.05} = 1.65$ and $z_{.01} = 2.33$.

References

Bell, C. B. and Doksum, K. A. (1965). Some new distribution-free statistics. **Annals of Mathematical Statistics**, 36, 203–214.

Conover, W. J. (1980). **Practical nonparametric statistics** (2nd ed.). New York: John Wiley and Sons.

Daniel, W. W. (1990). **Applied Nonparametric statistics** (2nd ed.). Boston: PWS-Kent Publishing Company.

Hollander, M. (1963). A nonparametric test for the two-sample problem. **Psychometrika**, 28, 395–403.

Kolmogorov, A. N. (1933). Sulla determinazione empiraca di una legge di distribuzione. **Giorn dell'Inst. Ital. degli. Att.**, 4, 89–91.

Mann, H. and Whitney, D. (1947). On a test of whether one of two random variables is stochastically larger than the other. **Annals of Mathematical Statistics**, 18, 50–60.

Marascuilo, L. A. and McSweeney, M. (1977). **Nonparametric and distribution-free methods for the social sciences**. Monterey, CA: Brooks/Cole Publishing Company.

Maxwell, S. E. and Delaney, H. (1990). **Designing experiments and analyzing data.** Monterey, CA: Wadsworth Publishing Company.

Moses, L. E. (1952). A two-sample test. **Psychometrika**, 17, 234–247.

Sheskin, D. J. (1984). **Statistical tests and experimental design: A guidebook.** New York: Gardner Press.

Siegel, S. and Castellan, N. J., Jr. (1988). **Nonparametric statistics for the behavioral sciences** (2nd ed.). New York: McGraw-Hill Book Company.

Smirnov, N. V. (1939). On the estimation of the discrepancy between empirical curves of distributions for two independent samples. **Bulletin University of Moscow**, 2, 3–14.

Terry, M. E. (1952). Some rank-order tests, which are most powerful against specific parametric alternatives. **Annals of Mathematical Statistics**, 23, 346–366.

Tukey, J. W. (1959). A quick, compact, two-sample test to Duckworth's specifications. **Technometrics**, 1, 31–48.

Van der Waerden, B. L. (1952/1953). Order tests for the two-sample problem and their power. **Proceedings Koninklijke Nederlandse Akademie van Wetenshappen** (A), 55 (**Indagationes Mathematicae** 14), 453–458, and 56 (**Indagationes Mathematicae**, 15), 303–316 (corrections appear in Vol. 56, p. 80).

Wald, A. and Wolfowitz, J. (1940). On a test whether two samples are from the same population. **Annals of Mathematical Statistics**, 11, 147-162.

Wilcoxon, F. (1949). **Some rapid approximate statistical procedures**. Stamford, CT: Stamford Research Laboratories, American Cyanamid Corporation.

Wilks, S. S. (1961). A combinatorial test for the problems of two samples from continuous distributions. In J. Neyman (Ed.), **Proceedings of the Fourth Berkeley Symposium on Mathematical Statistics and Probability**. Berkeley and Los Angeles: University of California Press, Vol. I, 707–717.

Endnotes

1. The reader should take note of the following with respect to the table of critical values for the **Mann–Whitney *U* distribution**: a) No critical values are recorded in the **Mann–Whitney table** for very small sample sizes, since a level of significance of .05 or less cannot be achieved for sample sizes below a specific minimum value; b) The critical values published in **Mann–Whitney tables** by various sources may not be identical. Such differences are trivial (usually one unit), and are the result of rounding off protocol; and c) The table for the alternative version of the **Mann–Whitney *U* test** (which was developed by Wilcoxon (1949)) contains critical values that are based on the sampling distribution of the sums of ranks, which differ from the tabled critical values contained in **Table A11** (which represents the sampling distribution of U values).

2. Although for Example 9.1 we can also say that since $\Sigma R_1 < \Sigma R_2$ the data are consistent with the directional alternative hypothesis H_1: $\theta_1 < \theta_2$, the latter will not necessarily always be the case when $n_1 \neq n_2$. Since the relationship between the average of the ranks will always be applicable to both equal and unequal sample sizes, it will be employed in describing the hypothesized relationship between the ranks of the two groups.

3. Some sources employ an alternative normal approximation equation which yields the same result as Equation 9.4. The alternative equation is noted below.

$$z = \frac{\Sigma R_1 - n_1 \left[\dfrac{n_1 + n_2 + 1}{2} \right]}{\sqrt{\dfrac{n_1 n_2 (n_1 + n_2 + 1)}{12}}}$$

   Note that the only difference between the above equation and Equation 9.4 is with respect to the numerator. If the value $\Sigma R_1 = 19$ from Example 9.1 is substituted in the above equation, it yields the value -8.5 for the numerator (which is the value for the numerator computed with Equation 9.4), and consequently the value $z = -1.78$. If ΣR_2 is employed in the numerator of the above equation in lieu of ΣR_1, the numerator of the equation assumes the form $\Sigma R_2 - [n_2(n_1 + n_2 + 1)]/2$. If $\Sigma R_2 = 36$ is substituted in the revised numerator, the value 8.5 is computed for the numerator, which results in the value $z = 1.78$. (Later on in this discussion it will be noted that the same conclusions regarding the null hypothesis are reached with the values $z = 1.78$ and $z = -1.78$.) The above equation is generally employed in sources which describe the version of the **Mann–Whitney *U* test** developed by Wilcoxon (1949). The latter version of the test only requires that the sum of the ranks be computed for each group, and does not require the computation of U values. As noted in Endnote 1, the table of critical values for the alternative version of the test is based on the sampling distribution of the sums of the ranks.

4. A general discussion of the correction for continuity can be found under the **Wilcoxon signed-ranks test (Test 4)**.

5. Some sources employ the term below for the denominator of Equation 9.5. It yields the identical result.

$$\sigma_U = \sqrt{\frac{n_1 n_2}{N(N-1)}\left[\frac{N^3 - N}{12} - \sum_{i=1}^{s} \frac{(t_i^3 - t_i)}{12}\right]}$$

6. A correction for continuity can be used by subtracting .5 from the value computed in the numerator of the Equation 9.5. The continuity correction will reduce the absolute value of z.

Test 10

The Siegel–Tukey Test for Equal Variability
(Nonparametric Test Employed with Ordinal Data)

I. Hypothesis Evaluated with Test and Relevant Background Information

Hypothesis evaluated with test Do two independent samples represent two populations with different variances?

Relevant background information on test Developed by Siegel and Tukey (1960), the **Siegel–Tukey test for equal variability** is employed with ordinal (rank-order) data in a hypothesis testing situation involving two independent samples. If the result of the **Siegel–Tukey test for equal variability** is significant, it indicates there is a significant difference between the sample variances, and as a result of the latter the researcher can conclude there is a high likelihood that the samples represent populations with different variances.

The **Siegel–Tukey test for equal variability** is one of a number of tests of **dispersion** (also referred to as tests of **scale** or **spread**) that have been developed for contrasting the variances of two independent samples. A discussion of alternative nonparametric tests of dispersion can be found in Section VII. Some sources recommend the use of nonparametric tests of dispersion for evaluating the homogeneity of variance hypothesis when there is reason to believe that the normality assumption of the appropriate parametric test for evaluating the same hypothesis is violated. Sources that are not favorably disposed toward nonparametric tests recommend the use of **Hartley's F_{max} test for homogeneity of variance/F test for two population variances (Test 8a)** (or one of the alternative parametric tests that are available for evaluating homogeneity of variance), regardless of whether or not the normality assumption of the parametric test is violated. Such sources do, however, recommend that in employing a parametric test, a researcher employ a lower significance level to compensate for the fact that violation of the normality assumption can inflate the Type I error rate associated with the test. When there is no evidence to indicate that the normality assumption of the parametric test has been violated, sources are in general agreement that such a test is preferable to the **Siegel–Tukey test for equal variability** (or an alternative nonparametric test of dispersion), since a parametric test (which uses more information that a nonparametric test) provides a more powerful test of an alternative hypothesis.

Since nonparametric tests are not assumption free, the choice of which of the available tests of dispersion to employ will primarily depend on what assumptions one is willing to make with regard to the underlying distributions represented by the sample data. The **Siegel–Tukey test for equal variability** is based on the following assumptions: a) Each sample has been randomly selected from the population it represents; b) The two samples are independent of one another; c) The level of measurement the data represent is at least ordinal; and d) The two populations from which the samples are derived have equal medians. If the latter assumption is violated but the researcher does know the values of the population medians, the scores in the groups can be adjusted so as to allow the use of the **Siegel–Tukey test for equal variability**. When, however, the population medians are unknown, and one is unwilling to assume they are equal, the **Siegel–Tukey test for equal variability** is not the appropriate nonparametric test of dispersion to employ. The assumption of equality of population medians presents a practical problem, in that when evaluating two independent samples a researcher will often have no prior knowledge regarding the population medians. In point of fact, most hypothesis testing addresses the issue of whether or not the medians (or means) of two or more populations are equal. In view of this, sources such as Siegel and Castellan (1988) note that if the latter values are not known, it is not appropriate to estimate them with the sample medians.

In employing the **Siegel–Tukey test for equal variability**, one of the following is true with regard to the rank-order data that are evaluated: a) The data are in a rank-order format, since it is the only format in which scores are available; or b) The data have been transformed to a rank-order format from an interval/ratio format, since the researcher has reason to believe that the normality assumption of the analogous parametric test is saliently violated.

II. Example

Example 10.1 *In order to assess the effect of two antidepressant drugs,* 12 *clinically depressed patients are randomly assigned to one of two groups. Six patients are assigned to* Group 1, *which is administered the antidepressant drug Elatrix for a period of six months. The other six patients are assigned to* Group 2, *which is administered the antidepressant drug Euphryia during the same six-month period. Assume that prior to introducing the experimental treatments, the experimenter confirmed that the level of depression in the two groups was equal. After six months elapse all* 12 *subjects are rated by a psychiatrist (who is blind with respect to a subject's experimental condition) on their level of depression. The psychiatrist's depression ratings for the six subjects in each group follow (the higher the rating the more depressed a subject):* **Group 1**: 10, 10, 9, 1, 0, 0; **Group 2**: 6, 6, 5, 5, 4, 4.

The fact that the mean and median of each group are equivalent (specifically, both values equal 5) *is consistent with prior research which suggests that there is no difference in efficacy for the two drugs (when the latter is based on a comparison of group means and/or medians). Inspection of the data does suggest, however, that there is much greater variability in the depression scores of subjects in* Group 1. *To be more specific, the data suggest that the drug Elatrix may, in fact, decrease depression in some subjects, yet increase it in others. The researcher decides to contrast the variability in the two groups through use of the* **Siegel–Tukey test for equal variability**. *The use of the latter nonparametric test is predicated on the fact that there is reason to believe that the distributions of the posttreatment depression scores in the underlying populations are not normal. Do the data indicate there is a significant difference between the variances of the two groups?*

III. Null versus Alternative Hypotheses

Null hypothesis $H_0\text{: } \sigma_1^2 = \sigma_2^2$

(The variance of the population Group 1 represents equals the variance of the population Group 2 represents. With respect to the sample data, when both groups have an equal sample size, this translates into the sum of the ranks of Group 1 being equal to the sum of the ranks of Group 2 (i.e., $\Sigma R_1 = \Sigma R_2$). A more general way of stating this, which also encompasses designs involving unequal sample sizes, is that the means of the ranks of the two groups are equal (i.e., $\bar{R}_1 = \bar{R}_2$)).

Alternative hypothesis $H_1\text{: } \sigma_1^2 \neq \sigma_2^2$

(The variance of the population Group 1 represents does not equal the variance of the population Group 2 represents. With respect to the sample data, when both groups have an equal sample size, this translates into the sum of the ranks of Group 1 not being equal to the sum of the ranks of Group 2 (i.e., $\Sigma R_1 \neq \Sigma R_2$). A more general way of stating this, which also encompasses designs involving unequal sample sizes, is that the means of the ranks of the two groups are not equal (i.e., $\Sigma R_1 \neq \Sigma R_2$). This is a **nondirectional alternative hypothesis** and it is evaluated with a **two-tailed test.**)

or

$$H_1\colon \sigma_1^2 > \sigma_2^2$$

(The variance of the population Group 1 represents is greater than the variance of the population Group 2 represents. With respect to the sample data, when both groups have an equal sample size (so long as a rank of 1 is given to the lowest score), this translates into the sum of the ranks of Group 1 being less than the sum of the ranks of Group 2 (i.e., $\Sigma R_1 < \Sigma R_2$). A more general way of stating this, which also encompasses designs involving unequal sample sizes, is that the mean of the ranks of Group 1 is less than the mean of the ranks of Group 2 (i.e., $\bar{R}_1 < \bar{R}_2$). This is a **directional alternative hypothesis** and it is evaluated with a **one-tailed test.**)

or

$$H_1\colon \sigma_1^2 < \sigma_2^2$$

(The variance of the population Group 1 represents is less than the variance of the population Group 2 represents. With respect to the sample data, when both groups have an equal sample size (so long as a rank of 1 is given to the lowest score), this translates into the sum of the ranks of Group 1 being greater than the sum of the ranks of Group 2 (i.e., $\Sigma R_1 > \Sigma R_2$). A more general way of stating this, which also encompasses designs involving unequal sample sizes, is that the mean of the ranks of Group 1 is greater than the mean of the ranks of Group 2 (i.e., $\bar{R}_1 > \bar{R}_2$). This is a **directional alternative hypothesis** and it is evaluated with a **one-tailed test.**)

Note: Only one of the above noted alternative hypotheses is employed. If the alternative hypothesis the researcher selects is supported, the null hypothesis is rejected.

IV. Test Computations

The total number of subjects employed in the experiment is $N = 12$. There are $n_1 = 6$ subjects in Group 1 and $n_2 = 6$ subjects in Group 2. The data for the analysis are summarized in Table 10.1. The original interval/ratio scores of the subjects are recorded in the columns labelled X_1 and X_2. The adjacent columns R_1 and R_2 contain the rank-order assigned to each of the scores. The rankings for Example 10.1 are summarized in Table 10.2. The ranking protocol for the **Siegel–Tukey test for equal variability** is described in this section. Note that in Table 10.1 and Table 10.2 each subject's identification number indicates the order in Table 10.1 in which a subject's score appears in a given group, followed by his/her group. Thus, Subject i, j is the ith subject in Group j.

The computational procedure for the **Siegel–Tukey test for equal variability** is identical to that employed for the **Mann–Whitney *U* test (Test 9)**, except for the fact that the two tests employ a different ranking protocol. Recollect that in the description of the alternative hypotheses for the **Siegel–Tukey test for equal variability**, it is noted that when a directional alternative hypothesis is supported, the average of the ranks of the group with the larger variance will be less than the average of the ranks of the group with the smaller variance. On the other hand, when a directional hypothesis for the **Mann–Whitney *U* test** is supported, the average of the ranks of the group with the larger median will be greater than the average of the ranks of the group with the smaller median. The difference between the two tests with

Table 10.1 Data for Example 10.1

	Group 1			Group 2	
	X_1	R_1		X_2	R_2
Subject 1,1	10	2.5	Subject 1,2	6	8.5
Subject 2,1	10	2.5	Subject 2,2	6	8.5
Subject 3,1	9	6	Subject 3,2	5	11.5
Subject 4,1	1	5	Subject 4,2	5	11.5
Subject 5,1	0	2.5	Subject 5,2	4	8.5
Subject 6,1	0	2.5	Subject 6,2	4	8.5
		$\Sigma R_1 = 21$			$\Sigma R_2 = 57$

$$\bar{R}_1 = \frac{\Sigma R_1}{n_1} = \frac{21}{6} = 3.5 \qquad \bar{R}_2 = \frac{\Sigma R_2}{n_2} = \frac{57}{6} = 9.5$$

Table 10.2 Rankings for the Siegel–Tukey Test for Equal Variability for Example 10.1

Subject identification number	5,1	6,1	4,1	5,2	6,2	3,2	4,2	1,2	2,2	3,1	2,1	1,1
Depression score	0	0	1	4	4	5	5	6	6	9	10	10
Rank prior to tie adjustment	1	4	5	8	9	12	11	10	7	6	3	2
Tie adjusted rank	2.5	2.5	5	8.5	8.5	11.5	11.5	8.5	8.5	6	2.5	2.5

respect to the ordinal position of the average ranks reflects the fact that the tests employ different ranking protocols. Whereas the ranking protocol for the **Mann–Whitney *U* test** is designed to identify differences with respect to central tendency (specifically, the median values), the ranking protocol for the **Siegel–Tukey test for equal variability** is designed to identify differences with respect to variability. The ranking protocol for the **Siegel–Tukey test for equal variability** is based on the premise that within the overall distribution of N scores, the distribution of scores in the group with the higher variance will contain more extreme values (i.e., scores that are very high and scores that are very low) than the distribution of scores in the group with the lower variance.

The following protocol, which is summarized in Table 10.2, is used in assigning ranks.

a) All $N = 12$ scores are arranged in order of magnitude (irrespective of group membership), beginning on the left with the lowest score and moving to the right as scores increase. This is done in the second row of Table 10.2.

b) Ranks are now assigned in the following manner: A rank of 1 is assigned to the lowest score (0). A rank of 2 is assigned to the highest score (10), and a rank of 3 is assigned to the second highest score (10). A rank of 4 is assigned to the second lowest score (0), and a rank of 5 is assigned to the third lowest score (1). A rank of 6 is assigned to the third highest score (9), and a rank of 7 is assigned to the fourth highest score (6). A rank of 8 is assigned to the fourth lowest score (4), and a rank of 9 is assigned to the fifth lowest score (4). A rank of 10 is assigned to the fifth highest score (6), and a rank of 11 is assigned to the sixth highest score (5). A rank of 12 is assigned to the sixth lowest score (5). Note that the ranking protocol assigns ranks to the distribution of $N = 12$ scores by alternating from one extreme of the distribution to the other. The ranks assigned employing this protocol are listed in the third row of Table 10.2

c) The ranks in the third row of Table 10.2 must be adjusted when there are tie scores present in the data. The same procedure for handling ties that is described for the **Mann–Whitney *U* test** is also employed with for **Siegel–Tukey test for equal variability.** Specifically, in instances where two or more subjects have the same score, the average of the ranks involved is assigned to all scores tied for a given rank. This adjustment is made in the fourth row of Table 10.2. To illustrate: Both Subjects 5,1 and 6,1 have a score of 0. Since the ranks assigned to the scores of these two subjects are respectively 1 and 4, the average of the two ranks $(1 + 4)/2 = 2.5$ is assigned to the score of both subjects. Both Subjects 1,1 and 2,1 have a score of 10. Since the ranks assigned to the score of these two subjects are respectively 2 and 3, the average of the two ranks $(2 + 3)/2 = 2.5$ is assigned to the score of both subjects. For the remaining three sets of ties (which all happen to fall in Group 2) the same averaging procedure is employed.

It should be noted that in Example 10.1 each set of tied scores involves subjects who are in the same group. Any time each set of ties involves subjects in the same group, the tie adjustment will result in the identical sum and average for the ranks of the two groups that will be obtained if the tie adjustment is not employed. Because of this, under these conditions the computed test statistic will be identical regardless of whether or not one uses the tie adjustment. On the other hand, when one or more sets of ties involve subjects from both groups, the tie adjusted ranks will yield a value for the test statistic that is different from that which will be obtained if the tie adjustment is not employed.

It should be noted that it is permissible to reverse the ranking protocol described in this section. Specifically, one can assign a rank of 1 to the highest score, a rank of 2 to the lowest score, a rank of 3 to the second lowest score, a rank of 4 to the second highest score, a rank of 5 to the third highest score, and so on. This reverse ranking protocol will result in the same test statistic, and consequently the same conclusion with respect to the null hypothesis as the ranking protocol described in this section.[1]

Once all of the subjects have been assigned a rank, the sum of the ranks for each of the groups is computed. These values, $\Sigma R_1 = 21$ and $\Sigma R_2 = 57$, are computed in Table 10.1. Upon determining the sum of the ranks for both groups, the values U_1 and U_2 are computed employing Equations 9.1 and 9.2, which are employed for the **Mann–Whitney *U* test**. The basis for employing the same equations and the identical distribution as that employed for the **Mann–Whitney *U* test** is predicated on the fact that both tests employ the same sampling distribution.

$$U_1 = n_1 n_2 + \frac{n_1(n_1 + 1)}{2} - \Sigma R_1 = (6)(6) + \frac{6(6 + 1)}{2} - 21 = 36$$

$$U_2 = n_1 n_2 + \frac{n_2(n_2 + 1)}{2} - \Sigma R_2 = (6)(6) + \frac{6(6 + 1)}{2} - 57 = 0$$

Note that U_1 and U_2 can never be negative values. If a negative value is obtained for either, it indicates an error has been made in the rankings and/or calculations.

As is the case for the **Mann–Whitney *U* test**, Equation 9.3 can be employed to verify the calculations. If the relationship in Equation 9.3 is not confirmed, it indicates that an error has been made in ranking the scores or in the computation of the U values. The relationship described by Equation 9.3 is confirmed below for Example 10.1.

$$n_1 n_2 = U_1 + U_2$$

$$(6)(6) = 36 + 0 = 36$$

V. Interpretation of the Test Results

The smaller of the two values U_1 versus U_2 is designated as the obtained U statistic. Since $U_2 = 0$ is smaller than $U_1 = 36$, the value of $U = 0$. The value of U is evaluated with **Table A11 (Table of Critical Values for the Mann–Whitney *U* Statistic)** in the **Appendix**.

In order to be significant, the obtained value of U must be **equal to or less than** the tabled critical value at the prespecified level of significance. For $n_1 = 6$ and $n_2 = 6$, the tabled critical two-tailed values are $U_{.05} = 5$ and $U_{.01} = 2$, and the tabled critical one-tailed values are $U_{.05} = 7$ and $U_{.01} = 3$.[2]

Since the obtained value $U = 0$ is less than the tabled critical two-tailed values $U_{.05} = 5$ and $U_{.01} = 2$, the nondirectional alternative hypothesis H_1: $\sigma_1^2 \neq \sigma_2^2$ is supported at both the .05 and .01 levels. Since the obtained value of U is less than the tabled critical one-tailed values $U_{.05} = 7$ and $U_{.01} = 3$, the directional alternative hypothesis H_1: $\sigma_1^2 > \sigma_2^2$ is also supported at both the .05 and .01 levels. The latter directional alternative hypothesis is supported since $\bar{R}_1 < \bar{R}_2$, which indicates that the variability of scores in Group 1 is greater than the variability of scores in Group 2. The directional alternative hypothesis H_1: $\sigma_1^2 < \sigma_2^2$ is not supported, since in order for the latter alternative hypothesis to be supported $\bar{R}_1$ must be greater than $\bar{R}_2$ (which indicates that the variability of scores in Group 2 is greater than the variability of scores in Group 1).

Based on the results of the **Siegel–Tukey test for equal variability**, one can conclude that there is greater variability in the depression scores of the group that receives the drug Elatrix (Group 1) than the group that receives the drug Euphyria (Group 2).

VI. Additional Analytical Procedures for the Siegel–Tukey Test for Equal Variability and/or Related Tests

1. The normal approximation of the Siegel–Tukey test statistic for large sample sizes As is the case with the **Mann–Whitney *U* test**, the normal distribution can be employed with large sample sizes to approximate the **Siegel–Tukey test** statistic. Equation 9.4, which is employed for the large sample approximation of the **Mann–Whitney distribution**, can also be employed for the large sample approximation of **Siegel–Tukey test** statistic. As is noted in Section VI of the **Mann–Whitney *U* test**, the large sample approximation is generally employed for sample sizes larger than those documented in the exact table contained within the source one is employing.

In the discussion of the **Mann–Whitney *U* test** it is noted that the term $(n_1 n_2)/2$ in the numerator of Equation 9.4 represents the expected (mean) value of U if the null hypothesis is true. This is also the case when the normal distribution is employed to approximate the **Siegel–Tukey test statistic**. Thus, if the two population variances are in fact equal, it is expected that $\bar{R}_1 = \bar{R}_2$, and consequently $U_1 = U_2 = (n_1 n_2)/2$.

Although Example 10.1 involves only $N = 12$ scores (a value most sources would view as too small to use with the normal approximation), it will be employed to illustrate Equation 9.4. The reader will see that in spite of employing Equation 9.4 with a small sample size, it yields a result that is consistent with the result obtained when the exact table for the **Mann–Whitney *U* distribution** is employed. As is noted in Section VI of the **Mann–Whitney *U* test**, since the smaller of the two values U_1 versus U_2 is selected to represent U, the value of z will always be negative (unless $U_1 = U_2$, in which case $z = 0$).

Employing Equation 9.4, the value $z = -2.88$ is computed.

$$z = \frac{U - \dfrac{n_1 n_2}{2}}{\sqrt{\dfrac{n_1 n_2 (n_1 + n_2 + 1)}{12}}} = \frac{0 - \dfrac{(6)(6)}{2}}{\sqrt{\dfrac{(6)(6)(6 + 6 + 1)}{12}}} = -2.88$$

The obtained value $z = -2.88$ is evaluated with **Table A1 (Table of the Normal Distribution)** in the **Appendix**. In order to be significant, the obtained absolute value of z must be equal to or greater than the tabled critical value at the prespecified level of significance. The tabled critical two-tailed .05 and .01 values are $z_{.05} = 1.96$ and $z_{.01} = 2.58$, and the tabled critical one-tailed .05 and .01 values are $z_{.05} = 1.65$ and $z_{.01} = 2.33$. The following guidelines are employed in evaluating the null hypothesis.

a) If the nondirectional alternative hypothesis H_1: $\sigma_1^2 \neq \sigma_2^2$ is employed, the null hypothesis can be rejected if the obtained absolute value of z is equal to or greater than the tabled critical two-tailed value at the prespecified level of significance.

b) If a directional alternative hypothesis is employed, one of the two possible directional alternative hypotheses is supported if the obtained absolute value of z is equal to or greater than the tabled critical one-tailed value at the prespecified level of significance. The directional alternative hypothesis that is supported is the one that is consistent with the data.

Employing the above guidelines with Example 10.1, the following conclusions are reached.

Since the obtained absolute value $z = 2.88$ is greater than the tabled critical two-tailed values $z_{.05} = 1.96$ and $z_{.01} = 2.58$, the nondirectional alternative hypothesis H_1: $\sigma_1^2 \neq \sigma_2^2$ is supported at the both the .05 and .01 levels. Since the obtained absolute value $z = 2.88$ is greater than the tabled critical one-tailed values $z_{.05} = 1.65$ and $z_{.01} = 2.33$, the directional alternative hypothesis H_1: $\sigma_1^2 > \sigma_2^2$ is supported at both the .05 and .01 levels. The latter directional alternative hypothesis is supported since it is consistent with the data. The directional alternative hypothesis H_1: $\sigma_1^2 < \sigma_2^2$ is not supported, since it is not consistent with the data. Note that the conclusions reached with reference to each of the possible alternative hypotheses are consistent with those reached when the exact table of the U distribution is employed.

As is the case when normal approximation is used with the **Mann–Whitney U test**, either U_1 or U_2 can be employed in Equation 9.4 to represent the value of U, since both values will yield the same absolute value for z.

2. The correction for continuity for the normal approximation of the Siegel–Tukey test for equal variability Although not described in most sources, the correction for continuity employed for the normal approximation of the **Mann–Whitney U test** can also be applied to the **Siegel–Tukey test for equal variability**. Employing Equation 9.5 (the **Mann–Whitney** continuity corrected equation) with the data for Example 10.1, the value $z = -2.80$ is computed.

$$z = \frac{\left| U - \dfrac{n_1 n_2}{2} \right| - .5}{\sqrt{\dfrac{n_1 n_2 (n_1 + n_2 + 1)}{12}}} = \frac{\left| 0 - \dfrac{(6)(6)}{2} \right| - .5}{\sqrt{\dfrac{(6)(6)(6 + 6 + 1)}{12}}} = -2.80$$

The obtained absolute value $z = 2.80$ is greater than the tabled critical two-tailed .05 and .01 values $z_{.05} = 1.96$ and $z_{.01} = 2.58$, and the tabled critical one-tailed .05 and .01 values $z_{.05} = 1.65$ and $z_{.01} = 2.33$. Thus, as is the case when the correction for continuity is not employed, both the nondirectional alternative hypothesis H_1: $\sigma_1^2 \neq \sigma_2^2$ and the directional alternative hypothesis H_1: $\sigma_1^2 > \sigma_2^2$ are supported at both the .05 and .01 levels. Note that the absolute value of the continuity corrected z value will always be less than the absolute value computed when the correction for continuity is not used.

3. Tie correction for the normal approximation of the Siegel–Tukey test statistic It is noted in the discussion of the normal approximation of the **Mann–Whitney U test** that some sources recommend that Equation 9.4 be modified when an excessive number of ties are present in the data. Since the identical sampling distribution is involved, the same tie correction (which results in a slight increase in the absolute value of z) can be employed for the normal approximation of the **Siegel–Tukey test for equal variability**. Employing Equation 9.6 (the **Mann–Whitney** tie correction equation), the tie corrected value $z = -2.91$ is computed for Example 10.1. Note that in Example 10.1 there are $s = 5$ sets of ties, each set involving two ties. Thus, in Equation 9.6 the term $\sum_{i=1}^{s}(t_i^3 - t_i) = 5[(2)^3 - 2] = 30$.

$$z = \frac{U - \dfrac{n_1 n_2}{2}}{\sqrt{\dfrac{n_1 n_2(n_1 + n_2 + 1)}{12} - \dfrac{n_1 n_2\left[\sum_{i=1}^{s}(t_i^3 - t_i)\right]}{12(n_1 + n_2)(n_1 + n_2 - 1)}}}$$

$$= \frac{0 - \dfrac{(6)(6)}{2}}{\sqrt{\dfrac{(6)(6)(6 + 6 + 1)}{12} - \dfrac{(6)(6)(30)}{12(6 + 6)(6 + 6 - 1)}}} = -2.91$$

The difference between $z = -2.91$ and the uncorrected value $z = -2.88$ is trivial, and consequently the decision the researcher makes with respect to the null hypothesis is not affected, regardless of which alternative hypothesis is employed.

4. Adjustment of scores for the Siegel–Tukey test for equal variability when $\theta_1 \neq \theta_2$ It is noted in Section I that if the values of the population medians are known but are not equal, in order to employ the **Siegel–Tukey test for equal variability** it is necessary to adjust the scores. In such a case, prior to ranking the scores the difference between the two population medians is subtracted from each of the scores in the group that represents the population with the higher median (or added to each of the scores in the group that represents the population with the lower median). This adjustment procedure will be demonstrated with Example 10.2.

Example 10.2 *In order to evaluate whether or not two teaching methods result in different degrees of variability with respect to performance, a mathematics instructor employs two methods of instruction with different groups of students. Prior to initiating the study it is determined that the two groups are comprised of students of equal math ability.* Group 1, *which is comprised of five subjects, is taught through the use of lectures and a conventional textbook* (Method A). Group 2, *which is comprised of six subjects, is taught through the use of a computer software package* (Method B). *At the conclusion of the course the final exam scores of the two groups are compared. The final exam scores follow (the maximum possible score on the final exam is* 10 *points and the minimum* 0): **Group 1**: 7, 5, 4, 4, 3; **Group 2**: 13, 12, 7, 7, 4, 3. *The researcher elects to rank-order the scores of the subjects, since she does not believe the data are normally distributed in the underlying populations. If the* **Siegel–Tukey test for equal variability** *is employed to analyze the data, is there a significant difference in within-groups variability?*

From the sample data we can determine that the median score of Group 1 is 4, and the median score of Group 2 is 7. Although the computations will not be shown here, in spite of the three point difference between the medians of the groups, if the **Mann–Whitney U test** is employed to evaluate the data, the null hypothesis H_0: $\theta_1 = \theta_2$ (i.e., that the medians of the underlying populations are equal) cannot be rejected at the .05 level, regardless of whether a nondirectional or directional alternative hypothesis is employed. The fact that the null hypothesis cannot be rejected is largely the result of the small sample size, which limits the power of the **Mann–Whitney U test** to detect a difference between underlying populations, if in fact one exists.

Let us assume, however, that based on prior research there is reason to believe that the median of the population represented by Group 2 is, in fact, three points higher than the median of the population represented by Group 1. In order to employ the **Siegel–Tukey test for equal variability** to evaluate the null hypothesis H_0: $\sigma_1^2 = \sigma_2^2$, the groups must be equated with respect to their median values. This can be accomplished by subtracting the difference between the population medians from each score in the group with the higher median. Thus, in Table 10.3 three points have been subtracted from the score of each of the subjects in Group 2.[3] The scores in Table 10.3 are ranked in accordance with the **Siegel–Tukey test** protocol. The ranks are summarized in Table 10.4.

Table 10.3 Data for Example 10.12 Employing Adjusted X_2 Scores

	Group 1			Group 2	
	X_1	R_1		X_2	R_2
Subject 1,1	7	6	Subject 1,2	10	2
Subject 2,1	5	7	Subject 2,2	9	3
Subject 3,1	4	9.5	Subject 3,2	4	9.5
Subject 4,1	4	9.5	Subject 4,2	4	9.5
Subject 5,1	3	5	Subject 5,2	1	4
			Subject 6,2	0	1
		$\Sigma R_1 = 37$			$\Sigma R_2 = 29$
	$\bar{R}_1 = \frac{\Sigma R_1}{n_1} = \frac{37}{5} = 7.4$			$\bar{R}_2 = \frac{\Sigma R_2}{n_2} = \frac{29}{6} = 4.83$	

Table 10.4 Rankings for the Siegel–Tukey Test for Equal Variability for Example 10.2

Subject identification number	6,2	5,2	5,1	3,1	4,1	3,2	4,2	2,1	1,1	2,2	1,2
Exam score	0	1	3	4	4	4	4	5	7	9	10
Rank prior to tie adjustment	1	4	5	8	9	11	10	7	6	3	2
Tie adjusted rank	1	4	5	9.5	9.5	9.5	9.5	7	6	3	2

Equations 9.1 and 9.2 are employed to compute the values $U_1 = 8$ and $U_2 = 22$.

$$U_1 = n_1 n_2 + \frac{n_1(n_1 + 1)}{2} - \Sigma R_1 = (5)(6) + \frac{5(5 + 1)}{2} - 37 = 8$$

$$U_2 = n_1 n_2 + \frac{n_2(n_2 + 1)}{2} - \Sigma R_2 = (5)(6) + \frac{6(6 + 1)}{2} - 29 = 22$$

Employing Equation 9.3, we confirm the relationship between the sample sizes and the computed values of U_1 and U_2.

$$n_1 n_2 = U_1 + U_2$$

$$(5)(6) = 8 + 22 = 30$$

Since $U_1 = 8$ is smaller than $U_2 = 22$, the value of $U = 8$. Employing **Table A11** for $n_1 = 5$ and $n_2 = 6$, we determine that the tabled critical two-tailed .05 and .01 values are $U_{.05} = 3$ and $U_{.01} = 1$, and the tabled critical one-tailed .05 and .01 values are $U_{.05} = 5$ and

$U_{.01} = 2$. Since the obtained value $U = 8$ is greater than all of the aforementioned tabled critical values, the null hypothesis cannot be rejected at either .05 or .01 level, regardless of whether a nondirectional or directional alternative hypothesis is employed.

If the normal approximation for the **Siegel–Tukey test for equal variability** is employed with Example 10.2, it is also the case that the null hypothesis cannot be rejected, regardless of which alternative hypothesis is employed. The latter is the case, since the computed absolute value $z = 1.28$ is less than the tabled critical .05 and .01 two-tailed values $z_{.05} = 1.96$ and $z_{.01} = 2.58$, and the tabled critical .05 and .01 one-tailed values $z_{.05} = 1.65$ and $z_{.01} = 2.33$.

$$z = \frac{8 - \frac{(5)(6)}{2}}{\sqrt{\frac{(5)(6)(5 + 6 + 1)}{12}}} = -1.28$$

Thus, the data do not indicate that the two teaching methods represent populations with different variances. Of course, as is the case when the **Mann–Whitney *U* test** is employed with the same set of data, it is entirely possible that a difference does exist in the underlying populations but is not detected because of the small sample size employed in the study.

VII. Additional Discussion of the Siegel–Tukey Test for Equal Variability

1. Analysis of the homogeneity of variance hypothesis for the same set of data with both a parametric and nonparametric test As noted in Section I, the use of the **Siegel-Tukey test for equal variability** would most likely be based on the fact that a researcher has reason to believe that the data in the underlying populations are not normally distributed. If, however, in the case of Example 10.1 the normality assumption is not an issue, or if it is but in spite of it a researcher prefers to use a parametric procedure such as **Hartley's F_{max} test for homogeneity of variance/*F* test for two population variances**, she can still reject the null hypothesis H_0: $\sigma_1^2 = \sigma_2^2$ at both the .05 or .01 levels, regardless of whether a nondirectional or directional alternative hypothesis is employed. This is demonstrated by employing Equation 8.6 (the equation for **Hartley's F_{max} test for homogeneity of variance**) with the data for Example 10.1.

$$\Sigma X_1 = 30 \quad \Sigma X_1^2 = 282 \quad \Sigma X_2 = 30 \quad \Sigma X_2^2 = 154$$

$$\tilde{s}_1^2 = \frac{282 - \frac{(30)^2}{6}}{6 - 1} = 26.4 \qquad \tilde{s}_2^2 = \frac{154 - \frac{(30)^2}{6}}{6 - 1} = .8$$

$$F_{max} = \frac{\tilde{s}_L^2}{\tilde{s}_S^2} = \frac{26.4}{.8} = 33$$

Table A9 (Table of the F_{max} Distribution) in the **Appendix** is employed to evaluate the computed value $F_{max} = 33$. For $k = 2$ groups and $(n - 1) = (6 - 1) = 5$ (since $n_1 = n_2 = n = 6$), the appropriate tabled critical values for a nondirectional analysis are $F_{max_{.05}} = 7.15$ and $F_{max_{.01}} = 14.9$. Since the obtained value $F_{max} = 33$ is greater than both of the aforementioned tabled critical values, the nondirectional alternative hypothesis H_1: $\sigma_1^2 \neq \sigma_2^2$ is supported at both the .05 and .01 levels.[4]

In the case of a directional analysis, the appropriate tabled critical one-tailed .05 and .01 values must be obtained from **Table A10 (Table of the *F* Distribution)** in the **Appendix.** In **Table A10**, the values for $F_{.95}$ and $F_{.99}$ for $df_{\text{num}} = n_1 - 1 = 6 - 1 = 5$ and $df_{\text{den}} = n_2 - 1 = 6 - 1 = 5$ are employed. The appropriate values derived from **Table A10** are $F_{.95} = 5.05$ and $F_{.99} = 10.97$. Since the obtained value $F_{\max} = 33$ is greater than both of the aforementioned tabled critical values, the directional alternative hypothesis H_1: $\sigma_1^2 > \sigma_2^2$ is supported at both the .05 and .01 levels.

Note that the difference between the computed $F_{\max}$ (or F) value and the appropriate tabled critical value is more pronounced in the case of **Hartley's $F_{\max}$ test for homogeneity of variance/*F* test for two population variances** than the difference between the computed test statistic and the appropriate tabled critical value for the **Siegel–Tukey test for equal variability** (when either the exact *U* distribution or the normal approximation is employed). The actual probability associated with the outcome of the analysis is, in fact, less than .01 for both the $F_{\max}$ and **Siegel–Tukey tests**, but is even further removed from .01 in the case of the $F_{\max}$ **test**. This latter observation is consistent with the fact that when both a parametric and nonparametric test are applied to the same set of data, the former test will generally provide a more powerful test of an alternative hypothesis.

2. Alternative nonparametric tests of dispersion In Section I it is noted that the **Siegel–Tukey test for equal variability** is one of a number of nonparametric tests for ordinal data that have been developed for evaluating the hypothesis that two populations have equal variances. The determination with respect to which of these tests to employ is generally based on the specific assumptions a researcher is willing to make about the underlying population distributions. Another factor that sometimes influences which test a researcher elects to employ is the complexity of the computations required for a specific test. This section will briefly summarize a few of the alternative procedures that evaluate the same hypothesis as the **Siegel–Tukey test for equal variability.** One or more of these procedures are described in detail in various books which specialize in nonparametric statistics (e.g., Daniel (1990), Marascuilo and McSweeney (1977), Siegel and Castellan (1988)). In addition, Sheskin (1984) provides a general overview and bibliography of nonparametric tests of dispersion.

The **Ansari–Bradley test** (Ansari and Bradley (1960) and Freund and Ansari (1957)) evaluates the same hypothesis as the **Siegel–Tukey test for equal variability**, as well as sharing its assumptions. The **Moses test** (Moses, 1963), which is more computationally involved than the two aforementioned tests, can also be employed to evaluate the same hypothesis, yet unlike the **Siegel–Tukey test** and **Ansari–Bradley test**, the **Moses test** assumes that the data evaluated represent at least interval level measurement. In addition, the **Moses test** does not assume that the two populations have equal medians. Among other nonparametric tests of dispersion are procedures developed by Conover (Conover and Iman (1978), Conover (1980)), Klotz (1962), and Mood (1954). Of the aforementioned tests, the **Siegel–Tukey test for equal variability**, the **Klotz test**, and the **Mood test** can be extended to designs involving more than two independent samples. Since there is extensive literature on nonparametric tests of dispersion, the interested reader should consult sources which specialize in nonparametric statistics for a more comprehensive discussion of the subject.

VIII. Additional Examples Illustrating the Siegel–Tukey Test for Equal Variability

The **Siegel–Tukey test for equal variability** can be employed to evaluate the null hypothesis H_0: $\sigma_1^2 = \sigma_2^2$ with any of the examples noted for the ***t* test for two independent samples (Test 8)** and the **Mann–Whitney *U* test.** In order to employ the **Siegel–Tukey test for equal variability** with any of the aforementioned examples, the data must be rank-ordered employing the protocol described in Section IV. Example 10.3 is an additional example that can be evaluated with the **Siegel–Tukey test for equal variability**. It is characterized by the fact that unlike Examples 10.1 and 10.2, in Example 10.3 subjects are rank-ordered without initially obtaining scores that represent interval/ratio level measurement. Although it is implied that the ranks in Example 10.3 are based on an underlying interval/ratio scale, the data are never expressed in such a format.

Example 10.3 *A company determines that there is no difference with respect to enthusiasm for a specific product after people are exposed to a monochromatic versus a polychromatic advertisement for the product. The company, however, wants to determine whether different degrees of variability are associated with the two types of advertisement. To answer the question, a study is conducted employing twelve subjects who as a result of having no knowledge of the product are neutral towards it. Six of the subjects are exposed to a monochromatic advertisement for the product* (Group 1), *and the other six are exposed to a polychromatic version of the same advertisement* (Group 2). *One week later each subject is interviewed by a market researcher who is blind with respect to which advertisement a subject was exposed. Upon interviewing all 12 subjects, the market researcher rank-orders them with respect to their level of enthusiasm for the product. The rank-orders of the subjects in the two groups follow (assume that the lower the rank-order, the lower the level of enthusiasm for the product):*

Group 1: **Subject 1,1:** 12; **Subject 2,1:** 2; **Subject 3,1:** 4; **Subject 4,1:** 6; **Subject 5,1:** 3; **Subject 6,1:** 10
Group 2: **Subject 1,2:** 7; **Subject 2,2:** 5; **Subject 3,2:** 9; **Subject 4,2:** 8; **Subject 5,2:** 11; **Subject 6,2:** 1

Is there a significant difference in the degree of variability within each of the groups?

Employing the ranking protocol for the **Siegel–Tukey test for equal variability** with the above data, the ranks of the two groups are converted into the following new set of ranks (i.e., assigning a rank of 1 to the lowest rank, a rank of 2 to the highest rank, a rank of 3 to the second highest rank, etc.).

Group 1: **Subject 1,1:** 2; **Subject 2,1:** 4; **Subject 3,1:** 8; **Subject 4,1:** 12; **Subject 5,1:** 5; **Subject 6,1:** 6
Group 2: **Subject 1,2:** 11; **Subject 2,2:** 9; **Subject 3,2:** 7; **Subject 4,2:** 10; **Subject 5,2:** 3; **Subject 6,2:** 1

Employing the above set of ranks, $\Sigma R_1 = 37$ and $\Sigma R_2 = 41$. Employing Equations 9.1 and 9.2, the values of U_1 and U_2 are computed to be $U_1 = (6)(6) + [[6(6+1)]/2] - 37 = 20$ and $U_2 = (6)(6) + [[6(6+1)]/2 - 41] = 16$. Since $U_2 = 16$ is less than $U_1 = 20$, $U = 16$. In **Table A11**, for $n_1 = 6$ and $n_2 = 6$, the tabled critical two-tailed .05 and 01 values are $U_{.05} = 5$ and $U_{.01} = 2$, and the tabled critical one-tailed .05 and .01 values are $U_{.05} = 7$ and $U_{.01} = 3$. Since $U = 16$ is greater than all of the aforementioned critical values, the null hypothesis H_0: $\sigma_1^2 = \sigma_2^2$ cannot be rejected, regardless of whether a nondirectional or directional alternative hypothesis is employed. Thus, there is no evidence to indicate that the two types of advertisements result in different degrees of variability.

References

Ansari, A. R. and Bradley, R. A. (1960). Rank-sum tests for dispersions. **Annals of Mathematical Statistics**, 31, 1174–1189.

Conover, W. J. (1980). **Practical nonparametric statistics** (2nd. ed.). New York: John Wiley and Sons.

Conover, W. J. and Iman, R. L. (1978). Some exact tables for the squared ranks test. **Communication in Statistics: Simulation and Computation**, B7 (5), 491–513.

Daniel, W. W. (1990). **Applied nonparametric statistics** (2nd ed.). Boston: PWS-Kent Publishing Company.

Freund, J. E. and Ansari, A. R. (1957). Two-way rank-sum test for variances. Technical Report Number 34, Virginia Polytechnic Institute and State University, Blacksburg, VA.

Klotz, J. (1962). Nonparametric tests for scale. **Annals of Mathematical Statistics**, 33, 498–512.

Marascuilo, L. A. and McSweeney, M. (1977). **Nonparametric and distribution-free methods for the social sciences**. Monterey, CA: Brooks/Cole Publishing Company.

Mood, A. M. (1954). On the asymptotic efficiency of certain nonparametric two-sample tests. **Annals of Mathematical Statistics**, 25, 514–522.

Moses, L. E. (1963). Rank tests of dispersion. **Annals of Mathematical Statistics**, 34, 973–983.

Sheskin, D. J. (1984). **Statistical tests and experimental design**. New York: Gardner Press.

Siegel, S. and Castellan, N. J., Jr. (1988). **Nonparametric statistics for the behavioral sciences** (2nd ed.). New York: McGraw-Hill Book Company.

Siegel, S. and Tukey, J. W. (1960). A nonparametric sum of ranks procedure for relative spread in unpaired samples. **Journal of the American Statistical Association**, 55, 429–445.

Endnotes

1. As is the case with the **Mann–Whitney *U* test**, if the reverse ranking protocol is employed the values of U_1 and U_2 are reversed. Since the value of U, which represents the test statistic, is the lower of the two values U_1 versus U_2, the value designated U with the reverse ranking protocol will be the same U value obtained with the original ranking protocol.

2. As is the case with the **Mann–Whitney *U* test**, in describing the **Siegel–Tukey test for equal variability** some sources do not compute a U value, but rather provide tables which are based on the smaller and/or larger of the two sums of ranks. The equation for the normal approximation (to be discussed in Section VI) in these sources is also based on the sums of the ranks.

3. As previously noted, we can instead add three points to each score in Group 1.

4. If one employs Equation 8.7, and thus uses **Table A10**, the same tabled critical values are listed for $F_{.975}$ and $F_{.995}$ for $df_{\text{num}} = 6 - 1 = 5$ and $df_{\text{den}} = 6 - 1 = 5$. Thus, $F_{.975} = 7.15$ and $F_{.995} = 14.94$. (The latter value is only listed to one decimal place in Table A9.) The use of **Table A10** in evaluating homogeneity of variance is discussed in Section VI of the ***t* test for two independent samples**.

Test 11

The Chi-Square Test for $r \times c$ Tables

(Nonparametric Test Employed with Categorical/Nominal Data)

I. Hypothesis Evaluated with Test and Relevant Background Information

Hypothesis evaluated with test In the underlying population(s) represented by the sample(s) in a contingency table, are the observed cell frequencies different from the expected frequencies?

Relevant background information on test The **chi-square test for $r \times c$ tables** is one of a number of tests described in this book for which the chi-square distribution is the appropriate sampling distribution.[1] The **chi-square test for $r \times c$ tables** is an extension of the **chi-square goodness-of-fit test (Test 5)** to two-dimensional tables. Whereas the latter test can only be employed with a single sample categorized on a single dimension (the single dimension is represented by the k cells/categories that comprise the frequency distribution table), the **chi-square test for $r \times c$ tables** can be employed to evaluate designs that summarize categorical data in the form of an $r \times c$ table (which is often referred to as a **contingency table**). An $r \times c$ table consists of r rows and c columns. Both the values of r and c are integer numbers that are equal to or greater than 2. The total number of cells in an $r \times c$ table is obtained by multiplying the value of r by the value of c. The data contained in each of the cells of a contingency table represent the number of observations (i.e., subjects or objects) that are categorized in the cell.

Table 11.1 presents the general model for an $r \times c$ contingency table. There are a total of n observations in the table. Note that each cell is identified with a subscript that consists of two elements. The first element identifies the row in which the cell falls and the second element identifies the column in which the cell falls. Thus, the notation O_{ij} represents the number of observations in the cell that is in the ith row and the jth column. $O_{i.}$ represents the number of observations in the ith row and $O_{.j}$ represents the number of observations in the jth column.

Table 11.1 General Model for an $r \times c$ Contingency Table

		Column variable						Row sums
		C_1	C_2	⋯	C_j	⋯	C_c	
	R_1	O_{11}	O_{12}	⋯	O_{1j}	⋯	O_{1c}	$O_{1.}$
	R_2	O_{21}	O_{22}	⋯	O_{2j}	⋯	O_{2c}	$O_{2.}$
	⋮	⋮	⋮		⋮		⋮	⋮
Row variable	R_i	O_{i1}	O_{i2}	⋯	O_{ij}	⋯	O_{ic}	$O_{i.}$
	⋮	⋮	⋮		⋮		⋮	⋮
	R_r	O_{r1}	O_{r2}	⋯	O_{rj}	⋯	O_{rc}	$O_{r.}$
Column sums		$O_{.1}$	$O_{.2}$	⋯	$O_{.j}$	⋯	$O_{.c}$	n

In actuality, there are two chi-square tests that can be conducted with an $r \times c$ table. The two tests that will be described are the **chi-square test for homogeneity (Test 11a)** and the **chi-square test of independence (Test 11b)**. The general label **chi-square test for $r \times c$ tables** will be employed to refer to both of the aforementioned tests, since the two tests are computationally identical. Although, in actuality, the **chi-square test for homogeneity** and the **chi-square test of independence** evaluate different hypotheses, a generic hypothesis can be stated that is applicable to both tests. A brief description of the two tests follows.

The chi-square test for homogeneity (Test 11a) The **chi-square test for homogeneity** is employed when r independent samples (where $r \geq 2$) are categorized on a single dimension which consists of c categories (where $c \geq 2$). The data for the r independent samples (which are generally represented by the r rows of the contingency table) are recorded with reference to the number of observations in each of the samples that fall within each of c categories (which are generally represented by the c columns of the contingency table). It is assumed that each of the samples is randomly drawn from the underlying population it represents. The **chi-square test for homogeneity** evaluates whether or not the r samples are homogeneous with respect to the proportion of observations in each of the c categories. To be more specific, if the data are homogeneous the proportion of observations in the jth category will be equal in all of the r populations. The **chi-square test for homogeneity** assumes that the sums of the r rows (which represent the number of observations in each of the r samples) are determined by the researcher prior to the data collection phase of a study. Example 11.1 in Section II is employed to illustrate the **chi-square test for homogeneity**.

The chi-square test of independence (Test 11b) **The chi-square test of independence** is employed when a single sample is categorized on two dimensions/variables. It is assumed that the sample is randomly selected from the population it represents. One of the dimensions/variables is comprised of r categories (where $r \geq 2$) that are represented by the r rows of the contingency table, while the second dimension/variable is comprised of c categories (where $c \geq 2$) that are represented by the c columns of the contingency table. The **chi-square test of independence** evaluates the general hypothesis that the two variables are independent of one another. Another way of stating that two variables are independent of one another is to say that there is a zero correlation between them. A zero correlation indicates there is no way to predict at above chance in which category an observation will fall on one of the variables if it is known which category the observation falls on the second variable. The **chi-square test of independence** assumes that neither the sums of the r rows (which represent the number of observations in each of the r categories for Variable 1) or the sums of the c columns (which represent the number of observations in each of the c categories for Variable 2) are predetermined by the researcher prior to the data collection phase of a study. Example 11.2 in Section II is employed to illustrate the **chi-square test of independence**.

The **chi-square test for $r \times c$ tables** (i.e., both the **chi-square test for homogeneity** and the **chi-square test of independence**) is based on the following assumptions: a) Categorical/nominal data (i.e., frequencies) for $r \times c$ mutually exclusive categories are employed in the analysis; b) The data that are evaluated represent a random sample comprised of n independent observations. This assumption reflects the fact that each subject or object can only be represented once in the data; and c) The expected frequency of each cell in the contingency table is 5 or greater. When the expected frequency of one or more cells is less than 5, the probabilities in the chi-square distribution may not provide an accurate estimate of the underlying sampling distribution. As is the case for the **chi-square goodness-of-fit test**, sources are not in agreement with respect to the minimum acceptable value for an expected frequency. Many sources employ criteria suggested by Cochran (1952), who stated that none of the expected frequencies should be less than 1, and that no more than 20% of the expected frequencies should be less than 5. However, many sources suggest the latter criteria may be overly conservative. In instances where a researcher believes that one or more expected cell frequencies are too small, two or more cells can be combined with one another to increase the values of the expected frequencies.

In actuality the chi-square distribution only provides an approximation of the exact sampling distribution for a contingency table.[2] The accuracy of the chi-square approximation increases as the size of the sample increases, and except for instances involving small sample sizes, the chi-square distribution provides an excellent approximation of the exact sampling distribution. One case for which an exact probability is often computed is a 2 × 2 contingency table involving a small sample size. In the latter instance, an exact probability can be computed through use of the hypergeometric distribution. The computation of an exact probability for a 2 × 2 table using the hypergeometric distribution is described under the **Fisher exact test (Test 11c)** in Section VI.

II. Examples

Example 11.1 *A researcher conducts a study in order to evaluate the effect of noise on altruistic behavior. Each of the* 200 *subjects who participate in the experiment is randomly assigned to one of two experimental conditions. Subjects in both conditions are given a one-hour test which is ostensibly a measure of intelligence. During the test the* 100 *subjects in* Group 1 *are exposed to continual loud noise, which they are told is due to a malfunctioning generator. The* 100 *subjects in* Group 2 *are not exposed to any noise during the test. Upon completion of this stage of the experiment, each subject on leaving the room is confronted by a middle-aged man whose arm is in a sling. The man asks the subject if she would be willing to help him carry a heavy package to his car. In actuality, the man requesting help is an experimental confederate (i.e., working for the experimenter). The number of subjects in each group who help the man are recorded. Thirty of the* 100 *subjects who were exposed to noise elect to help the man, while* 60 *of the* 100 *subjects who were not exposed to noise elect to help the man. Do the data indicate that altruistic behavior is influenced by noise?*

The data for Example 11.1, which can be summarized in the form of a 2 × 2 contingency table, are presented in Table 11.2.

Table 11.2 Summary of Data for Example 11.1

	Helped the confederate	Did not help the confederate		Row sums
Noise	30	70		100
No noise	60	40		100
Column sums	90	110	Total observations	200

The appropriate test to employ for evaluating Example 11.1 is the **chi-square test for homogeneity**. This is the case, since the design of the study involves the use of categorical data (i.e., frequencies for each of the $r \times c$ cells in the contingency table) with multiple independent samples (specifically two) that are categorized on a single dimension (altruism). To be more specific, the differential treatments to which the two groups are exposed (i.e., **noise** versus **no-noise**) constitute the independent variable. The latter variable is the row variable, since it is represented by the two rows in Table 11.2. Note that the researcher assigns 100 subjects to each of the two levels of the independent variable prior to the data collection phase of the study. This is consistent with the fact that when the **chi-square test for homogeneity** is employed, the sums for the row variable are predetermined prior to collecting the data. The dependent variable is whether or not a subject exhibits altruistic behavior. The latter variable is represented by the two categories **helped the confederate** versus **did not help the confederate.** The dependent variable is the column variable, since it is represented by the two columns in Table 11.2. The hypothesis that is evaluated with the **chi-square test for homogeneity** is whether there is a difference between the two groups with respect to the proportion of subjects who help the confederate.

Example 11.2 *A researcher wants to determine if there is a relationship between the personality dimension of introversion-extroversion and political affiliation. Two hundred people are recruited to participate in the study. All of the subjects are given a personality test on the basis of which each subject is classified as an introvert or an extrovert. Each subject is then asked to indicate whether he or she is a Democrat or a Republican. The data for* Example 11.2, *which can be summarized in the form of a* 2×2 *contingency table, are presented in* Table 11.3. *Do the data indicate there is a significant relationship between one's political affiliation and whether or not one is an introvert versus an extrovert?*

Table 11.3 Summary of Data for Example 11.2

	Democrat	**Republican**		**Row sums**
Introvert	30	70		**100**
Extrovert	60	40		**100**
Column sums	**90**	**110**	**Total observations**	**200**

The appropriate test to employ for evaluating Example 11.2 is the **chi-square test of independence.** This is the case since: a) The study involves a single sample that is categorized on two dimensions; and b) The data are comprised of frequencies for each of the $r \times c$ cells in the contingency table. To be more specific, a sample of 200 subjects is categorized on the following two dimensions, with each dimension being comprised of two mutually exclusive categories: a) **introvert** versus **extrovert**; and b) **Democrat** versus **Republican.** In Example 11.2 the **introvert–extrovert** dimension is the row variable and the **Democrat–Republican** dimension is the column variable.[3] Note that in selecting the sample of 200 subjects, the researcher does not determine beforehand the number of **introverts, extroverts, Democrats,** and **Republicans** to include in the study.[4] Thus, in Example 11.2 (consistent with the use of the **chi-square test of independence**) the sums of the rows and columns (which are referred to as the **marginal sums**) are not predetermined. The hypothesis that is evaluated with the **chi-square test of independence** is whether the two dimensions are independent of one another.

III. Null versus Alternative Hypotheses

Even though the hypotheses evaluated with the **chi-square test for homogeneity** and the **chi-square test of independence** are not identical, generic null and alternative hypotheses employing common symbolic notation can be used for both tests. The generic null and alternative hypotheses employ the observed and expected cell frequencies in the underlying population(s) represented by the sample(s). The observed and expected cell frequencies for the population(s) are represented respectively by the lower case Greek letters **omicron** (o) and **epsilon** (ε). Thus, o_{ij} and ε_{ij} respectively represent the observed and expected frequency of Cell_{ij} in the underlying population.

Null hypothesis H_0: $o_{ij} = \varepsilon_{ij}$ for all cells.

(This notation indicates that in the underlying population(s) the sample(s) represents, for each of the $r \times c$ cells the observed frequency of a cell is equal to the expected frequency of the cell. With respect to the sample data, this translates into the observed frequency of each of the $r \times c$ cells being equal to the expected frequency of the cell.)

Alternative hypothesis H_1: $o_{ij} \neq \varepsilon_{ij}$ for at least one cell.

(This notation indicates that in the underlying population(s) the sample(s) represents, for at least one of the $r \times c$ cells the observed frequency of a cell is not equal to the expected frequency of the cell. With respect to the sample data, this translates into the observed frequency of at least one of the $r \times c$ cells not being equal to the expected frequency of the cell. This notation should not be interpreted as meaning that in order to reject the null hypothesis there must be a discrepancy between the observed and expected frequencies for all $r \times c$ cells. Rejection of the null hypothesis can be the result of a discrepancy between the observed and expected frequencies for one cell, two cells, ..., or all $r \times c$ cells.)

Although it is possible to employ a directional alternative hypothesis for the **chi-square test for $r \times c$ tables**, in the examples used to describe the test it will be assumed that the alternative hypothesis will always be stated nondirectionally. A discussion of the use of a directional alternative hypothesis can be found in Section VI.

The null and alternative hypotheses for each of the two tests that are described under the **chi-square test for $r \times c$ tables** can also be expressed within the framework of a different format. The alternative format for stating the null and alternative hypotheses employs the proportion of observations in the cells of the $r \times c$ contingency table. Before presenting the hypotheses in the latter format, the reader should take note of the following with respect to Tables 11.2 and 11.3. In both Tables 11.2 and 11.3 four cells can be identified: a) Cell_{11} is the upper left cell in each table (i.e., in Row 1 and Column 1 the cell with the observed frequency $O_{11} = 30$). In the case of Example 11.1, $O_{11} = 30$ represents the number of subjects exposed to **noise** who **helped the confederate**. In the case of Example 11.2, $O_{11} = 30$ represents the number of **introverts** who are **Democrats**; b) Cell_{12} is the upper right cell in each table (i.e., in Row 1 and Column 2 the cell with the observed frequency $O_{12} = 70$). In the case of Example 11.1, $O_{12} = 70$ represents the number of subjects exposed to **noise** who **did not help the confederate**. In the case of Example 11.2, $O_{12} = 70$ represents the number of **introverts** who are **Republicans**; c) Cell_{21} is the lower left cell in each table (i.e., in Row 2 and Column 1 the cell with the observed frequency $O_{21} = 60$). In the case of Example 11.1, $O_{21} = 60$ represents the number of subjects exposed to **no noise** who **helped the confederate**. In the case of Example 11.2, $O_{21} = 60$ represents the number of **extroverts** who are **Democrats**; d) Cell_{22} is the lower right cell in each table (i.e., in Row 2 and Column 2 the cell with the observed frequency $O_{22} = 40$). In the case of Example 11.1, $O_{22} = 40$ represents the number of subjects exposed to **no noise** who **did not help the confederate**. In the case of Example 11.2, $O_{22} = 40$ represents the number of **extroverts** who are **Republicans**.

Alternative way of stating the null and alternative hypotheses for the chi-square test for homogeneity If the independent variable (which represents the different groups) is employed as the row variable, the null and alternative hypotheses can be stated as follows:

H_0: **In the underlying populations the samples represent, all of the proportions in the same column of the $r \times c$ table are equal**

H_1: **In the underlying populations the samples represent, all of the proportions in the same column of the $r \times c$ table are not equal for at least one of the columns**

Viewing the above hypotheses in relation to the sample data in Table 11.2, the null hypothesis states that there are an equal proportion of observations in Cell_{11} and Cell_{21}. With respect to the sample data, the proportion of observations in Cell_{11} is the proportion of subjects who are exposed to **noise** who **helped the confederate** (which equals $O_{11}/O_{1.} = 30/100 = .3$). The proportion of observations in Cell_{21} is the proportion of subjects who are exposed to **no noise** who **helped the confederate** (which equals $O_{21}/O_{2.} = 60/100 = .6$). The null hypothesis also requires an equal proportion of observations in Cell_{12} and Cell_{22}. The proportion of observations in Cell_{12} is the proportion of subjects who are exposed to **noise** who **did not help the confederate** (which equals $O_{12}/O_{1.} = 70/100 = .7$). The proportion of observations in Cell_{22} is the proportion of subjects who are exposed to **no noise** who **did not help the confederate** (which equals $O_{22}/O_{2.} = 40/100 = .4$).

Alternative way of stating the null and alternative hypotheses for the chi-square test of independence The null and alternative hypotheses for the **chi-square test of independence** can be stated as follows:

$$H_0\text{: } \pi_{ij} = (\pi_{i.})(\pi_{.j}) \text{ for all } r \times c \text{ cells.}$$

$$H_1\text{: } \pi_{ij} \neq (\pi_{i.})(\pi_{.j}) \text{ for at least one cell.}$$

Where: π represents the value of a proportion in the population

The above notation indicates that if the null hypothesis is true, in the underlying population represented by the sample for each of the $r \times c$ cells, the proportion of observations in a cell will equal the proportion of observations in the row in which the cell appears multiplied by the proportion of observations in the column in which the cell appears. This will now be illustrated with respect to Example 11.2. In illustrating the relationship described in the null hypothesis, the notation p is employed to represent the relevant proportions obtained for the sample data. If the null hypothesis is true, in the case of Cell_{11} it is required that the proportion of observations in Cell_{11} is equivalent to the product of the proportion of observations in Row 1 (which equals $p_{1.} = O_{1.}/n = 100/200 = .5$) and the proportion of observations in Column 1 (which equals $p_{.1} = O_{.1}/n = 90/200 = .45$). The result of multiplying the row and column proportions is $(p_{1.})(p_{.1}) = (.5)(.45) = .225$. Thus, if the null hypothesis is true, the proportion of observations in Cell_{11} must equal $p_{11} = .225$.[5] Consequently, if the value .225 is multiplied by 200, which is the total number of observations in Table 11.3, the resulting value $(p_{11})(n) = (.225)(200) = 45$ is the number of observations that is expected in Cell_{11} if the null hypothesis is true. The same procedure can be used for the remaining three cells to determine the number of observations that are required in each cell in order for the null hypothesis to be supported. In Section IV these values, which in actuality correspond to the expected frequencies of the cells, are computed for each of the four cells in Table 11.3 (as well as Table 11.2).

IV. Test Computations

The computations for the **chi-square test for $r \times c$ tables** will be described for Example 11.1. The procedure to be described in this section when applied to Example 11.2 yields the identical result since: a) The computational procedure for the **chi-square test of independence** is identical to that employed for the **chi-square test for homogeneity**; and b) The identical data are employed for Examples 11.1 and 11.2. Table 11.4 summarizes the data and computations for Example 11.1.

Table 11.4 Chi-Square Summary Table for Example 11.1

Cell	O_{ij}	E_{ij}	$(O_{ij} - E_{ij})$	$(O_{ij} - E_{ij})^2$	$\frac{(O_{ij} - E_{ij})^2}{E_{ij}}$
$Cell_{11}$ — Noise/Helped the confederate	30	45	−15	225	5.00
$Cell_{12}$ — Noise/Did not help the confederate	70	55	15	225	4.09
$Cell_{21}$ — No noise/Helped the confederate	60	45	15	225	5.00
$Cell_{22}$ — No noise/Did not help the confederate	40	55	−15	225	4.09
	$\Sigma O_{ij} = 200$	$\Sigma E_{ij} = 200$	$\Sigma(O_{ij} - E_{ij}) = 0$		$\chi^2 = 18.18$

The observed frequency of each cell (O_{ij}) is listed in Column 2 of Table 11.4. Column 3 contains the expected cell frequencies (E_{ij}). In order to conduct the **chi-square test for $r \times c$ tables**, the observed frequency for each cell must be compared with its expected frequency. In order to determine the expected frequency of a cell, the data should be arranged in a contingency table that employs the format of Table 11.2. The following protocol is then employed to determine the expected frequency of a cell: a) Multiply the sum of the observations in the row in which the cell appears by the sum of the observations in the column in which the cell appears; b) Divide n, the total number of observations, into the product that results from multiplying the row and column sums for the cell.

The computation of an expected cell frequency can be summarized by Equation 11.1.

$$E_{ij} = \frac{(O_{i.})(O_{.j})}{n} \qquad \textbf{(Equation 11.1)}$$

Applying Equation 11.1 to $Cell_{11}$ in Table 11.2 (i.e., **noise/helped the confederate**), the expected cell frequency can be computed as follows. The row sum is the total number of subjects who were exposed to **noise**. Thus, $O_{1.} = 100$. The column sum is the total number of subjects who **helped the confederate**. Thus, $O_{.1} = 90$. Employing Equation 11.1, the expected frequency for $Cell_{11}$ can now be computed: $E_{11} = [(O_{1.})(O_{.1})]/n = [(100)(90)]/200 = 45$. The expected frequencies for the remaining three cells in the 2×2 contingency table that summarizes the data for Example 11.1 are computed below:

$$Cell_{12} = [(O_{1.})(O_{.2})/n = [(100)(110)]/200 = 55$$

$$Cell_{21} = [(O_{2.})(O_{.1})/n = [(100)(90)]/200 = 45$$

$$Cell_{22} = [(O_{2.})(O_{.2})/n = [(100)(110)]/200 = 55$$

Upon determining the expected cell frequencies, the test statistic for the **chi-square test for $r \times c$ tables** is computed with Equation 11.2.[6]

$$\chi^2 = \sum_{i=1}^{r}\sum_{j=1}^{c}\left[\frac{(O_{ij} - E_{ij})^2}{E_{ij}}\right] \qquad \textbf{(Equation 11.2)}$$

The operations described by Equation 11.2 (which are the same as those described for computing the chi-square statistic for the **chi-square goodness-of-fit test**) are as follows: a) The expected frequency of each cell is subtracted from its observed frequency (summarized in Column 4 of Table 11.4); b) For each cell, the difference between the observed and expected frequency is squared (summarized in Column 5 of Table 11.4); c) For each cell, the squared difference between the observed and expected frequency is divided by the expected frequency of the cell (summarized in Column 6 of Table 11.4); and d) The value of chi-square is computed by summing all of the values in Column 6. For Example 11.1, Equation 11.2 yields the value $\chi^2 = 18.18$.[7]

Note that in Table 11.4 the sums of the observed and expected frequencies are identical. This must always be the case, and any time these sums differ from one another, it indicates that a computational error has been made. It is also required that the sum of the differences between the observed and expected frequencies equals zero (i.e., $\Sigma(O_{ij} - E_{ij}) = 0$). Any time the latter value does not equal zero, it indicates an error has been made. Since all of the $(O_{ij} - E_{ij})$ values are squared in Column 5, the sum of Column 6, which represents the value of χ^2, must always be a positive number. If a negative value for chi-square is obtained, it indicates that an error has been made. The only time χ^2 will equal zero is when $O_{ij} = E_{ij}$ for all $r \times c$ cells.

V. Interpretation of the Test Results

The obtained value $\chi^2 = 18.18$ is evaluated with **Table A4 (Table of the Chi-Square Distribution) in the Appendix.** A general discussion of the values in **Table A4** can be found in Section V of the **single-sample chi-square test for a population variance (Test 3)**. When the chi-square distribution is employed to evaluate the **chi-square test for $r \times c$ tables**, the degrees of freedom employed for the analysis are computed with Equation 11.3.

$$df = (r - 1)(c - 1) \qquad \textbf{(Equation 11.3)}$$

The tabled critical values in **Table A4** for the **chi-square test for $r \times c$ tables** are always derived from the right tail of the distribution. The critical chi-square value for a specific value of alpha is the tabled value at the percentile that corresponds to the value $(1 - \alpha)$. Thus, the tabled critical .05 chi-square value (to be designated $\chi^2_{.05}$) is the tabled value at the 95th percentile. In the same respect, the tabled critical .01 chi-square value (to be designated $\chi^2_{.01}$) is the tabled value at the 99th percentile. In order to reject the null hypothesis, the obtained value of chi-square must be equal to or greater than the tabled critical value at the prespecified level of significance. The aforementioned guidelines for determining tabled critical chi-square values are employed when the alternative hypothesis is stated nondirectionally (which, as noted earlier, is generally the case for the **chi-square test for $r \times c$ tables**). The determination of tabled critical chi-square values in reference to a directional alternative hypothesis is discussed in Section VI.

The guidelines for a nondirectional analysis will now be applied to Example 11.1. Since $r = 2$ and $c = 2$, the degrees of freedom are computed to be $df = (2 - 1)(2 - 1) = 1$. The tabled critical .05 chi-square value for $df = 1$ is $\chi^2_{.05} = 3.84$, which as noted above is the tabled chi-square value at the 95th percentile. The tabled critical .01 chi-square value for

$df = 1$ is $\chi^2_{.01} = 6.63$, which as noted above is the tabled chi-square value at the 99th percentile. Since the computed value $\chi^2 = 18.18$ is greater than both of the aforementioned critical values, the null hypothesis can be rejected at both the .05 and .01 levels. Rejection of the null hypothesis at the .01 level can be summarized as follows: $\chi^2(1) = 18.18$, $p < .01$.

The significant chi-square value obtained for Example 11.1 indicates that subjects who served in the **noise** condition **helped the confederate** significantly less than subjects who served in the **no noise** condition. This can be confirmed by visual inspection of Table 11.2, which reveals that twice as many subjects who served in the **no noise** condition **helped the confederate** than subjects who served in the **noise** condition.

As noted previously, the chi-square analysis described in this section also applies to Example 11.2, since the latter example employs the same data as Example 11.1. Thus, with respect to Example 11.2, the significant $\chi^2 = 18.18$ value allows the researcher to conclude that a subject's categorization on the **introvert–extrovert** dimension is associated with (i.e., not independent of) one's political affiliation. This can be confirmed by visual inspection of Table 11.3, which reveals that **introverts** are more likely to be **Republicans** whereas **extroverts** are more likely to be **Democrats.**

It is important to note that Example 11.2 represents a correlational study, and as such does not allow a researcher to draw any conclusions with regard to cause and effect.[8] To be more specific, the study does not allow one to conclude that a subject's categorization on the personality dimension **introvert–extrovert** is the cause of one's political affiliation (**Democrat** versus **Republican**), or vice versa (i.e., that political affiliation causes one to be an **introvert** versus an **extrovert**). Although it is possible that the two variables employed in a correlational study are causally related to one another, such studies do not allow one to draw conclusions regarding cause and effect, since they fail to control for the potential influence of confounding variables. Because of this, when studies which are evaluated with the **chi-square test of independence** (such as Example 11.2) yield a significant result, one can only conclude that in the underlying population the two variables have a correlation with one another that is some value other than zero (which is not commensurate with saying that one variable causes the other).

Studies such as that represented by Example 11.2 can also be conceptualized within the framework of a **natural experiment** (also referred to as an **ex post facto study**) which is discussed in the **Introduction** of the book. In the latter type of study, one of the two variables is designated as the independent variable, and the second variable as the dependent variable. The independent variable is (in contrast to the independent variable in a **true experiment**) a nonmanipulated variable. A subject's score (or category in the case of Example 11.2) on a nonmanipulated independent variable is based on some preexisting subject characteristic, rather than being a direct result of some manipulation on the part of the experimenter. Thus, if in Example 11.2 the **introvert–extrovert** dimension is designated as the independent variable, it represents a nonmanipulated variable, since the experimenter does not determine whether or not a subject becomes an **introvert** or an **extrovert.** Which of the two aforementioned categories a subject falls in is determined beforehand by "nature" (thus the term **natural experiment**). The same logic also applies if political affiliation is employed as the independent variable, since, like **introvert–extrovert**, the **Democrat–Republican** dichotomization is a preexisting subject characteristic.

In Example 11.1, however, the independent variable, which is whether or not a subject is exposed to **noise**, is a manipulated variable. This is the case, since the experimenter randomly determines those subjects who are assigned to the **noise** condition and those who are assigned to the **no noise** condition. As noted in the **Introduction**, an experiment in which the researcher manipulates the level of the independent variable to which a subject is assigned is referred to as a **true experiment.** In the latter type of experiment, by virtue of randomly assigning subjects to the different experimental conditions, the researcher is able to control for the effects of potentially confounding variables. Because of this, if a significant result is obtained in a **true experiment**, a researcher is justified in drawing conclusions with regard to cause and effect.

VI. Additional Analytical Procedures for the Chi-Square Test for $r \times c$ Tables and/or Related Tests

1. Yates' correction for continuity In Section I it is noted that, in actuality, the **chi-square test for $r \times c$ tables** employs a continuous distribution to approximate a discrete probability distribution. Under such conditions, some sources recommend that a correction for continuity be employed. As noted previously in the book, the correction for continuity is based on the premise that if a continuous distribution is employed to estimate a discrete distribution, such an approximation will inflate the Type I error rate. By employing the correction for continuity, the Type I error rate is ostensibly adjusted to be more compatible with the prespecified alpha level designated by the researcher. Sources that recommend the correction for continuity for the **chi-square test for $r \times c$ tables** only recommend that it be employed in the case of 2×2 contingency tables. Equation 11.4 (which was developed by Yates (1934)) is the continuity corrected chi-square equation for 2×2 tables.

$$\chi^2 = \sum_{i=1}^{r} \sum_{j=1}^{c} \left[\frac{(\,|O_{ij} - E_{ij}| - .5\,)^2}{E_{ij}} \right] \qquad \textbf{(Equation 11.4)}$$

Note that by subtracting .5 from the absolute value of the difference between each set of observed and expected frequencies, the chi-square value derived with Equation 11.4 will be lower than the value computed with Equation 11.2.

Statisticians are not in agreement with respect to whether it is prudent to employ the correction for continuity described by Equation 11.4 with a 2×2 contingency table. To be more specific, various sources take the following positions with respect to what the most effective strategy is for evaluating 2×2 tables: a) Most sources agree that when the sample size for a 2×2 table is small (generally less than 20), the **Fisher exact test** (which is described later in this section) should be employed instead of the **chi-square test for $r \times c$ tables**. Cochran (1952, 1954) stated that in the case of 2×2 tables, the **chi-square test for $r \times c$ tables** should not be employed when $n < 20$, and that when $20 < n < 40$ the test should only be employed if all of the expected frequencies are at least equal to 5. Additionally, when $n > 40$ all expected frequencies should be equal to or greater than 1; b) Some sources recommend that for small sample sizes **Yates' correction for continuity** be employed. This recommendation assumes that the size of the sample is at least equal to 20 (since when $n < 20$ the **Fisher exact test** should be employed), but less than some value that defines the maximum size of a small sample size with respect to the use of Yates' correction. Sources do not agree on what value of n defines the upper limit beyond which Yates' correction is not required; c) Some sources recommend that **Yates' correction for continuity** should always be employed with 2×2 tables, regardless of the sample size; and d) To further confuse the issue, many sources take the position that **Yates' correction for continuity** should never be used, since the chi-square value computed with Equation 11.4 results in an overcorrection — i.e., it results in an overly conservative test.

Table 11.5 illustrates the application of **Yates' correction for continuity** with Example 11.1.

Thus, by employing Equation 11.4 the obtained value of chi-square is reduced to 16.98 (in contrast to the value $\chi^2 = 18.18$ obtained with Equation 11.2). Since the obtained value $\chi^2 = 16.98$ is greater than both $\chi^2_{.05} = 3.84$ and $\chi^2_{.01} = 6.83$, the null hypothesis can still

Table 11.5 Chi-Square Summary Table for Example 11.1 Employing Yates' Correction for Continuity

Cell	O_{ij}	E_{ij}	$(\lvert O_{ij} - E_{ij}\rvert - .5)$	$(\lvert O_{ij} - E_{ij}\rvert - .5)^2$	$\frac{(\lvert O_{ij} - E_{ij}\rvert - .5)^2}{E_{ij}}$
$Cell_{11}$ — Noise/ Helped the confederate	30	45	14.5	210.25	4.67
$Cell_{12}$ — Noise/ Did not help the confederate	70	55	14.5	210.25	3.82
$Cell_{21}$ — No noise/ Helped the confederate	60	45	14.5	210.25	4.67
$Cell_{22}$ — No noise/ Did not help the confederate	40	55	14.5	210.25	3.82
	$\Sigma O_{ij} = 200$	$\Sigma E_{ij} = 200$			$\chi^2 = 16.98$

be rejected at both the .05 and .01 levels. Thus, in this instance **Yates' correction for continuity** leads to the same conclusions as those reached when Equation 11.2 is employed.

2. Quick computational equation for a 2 × 2 table Equation 11.5 is a quick computational equation that can be employed for the **chi-square test for *r* × *c* tables** in the case of a 2 × 2 table. Unlike Equation 11.2, it does not require that the expected cell frequencies be computed. The notation employed in Equation 11.5 is based on the model for a 2 × 2 contingency table summarized in Table 11.6.

Table 11.6 Model for 2 × 2 Contingency Table

	Column 1	Column 2	Row sums
Row 1	a	b	$a + b = n_1$
Row 2	c	d	$c + d = n_2$
Column sums	$a + c$	$b + d$	n

$$\chi^2 = \frac{n(ad - bc)^2}{(a + b)(c + d)(a + c)(b + d)} \qquad \textbf{(Equation 11.5)}$$

Where: a, b, c, and d represent the number of observations in the relevant cell

Using the model depicted in Table 11.6, by employing the appropriate observed cell frequencies for Examples 11.1 and 11.2, we know that $a = 30$, $b = 70$, $c = 60$, and $d = 40$. Substituting these values in Equation 11.5, the value $\chi^2 = 18.18$ is computed (which is the same chi-square value that is computed with Equation 11.2).

$$\chi^2 = \frac{200[(30)(40) - (70(60)]^2}{(30 + 70)(60 + 40)(30 + 60)(70 + 40)} = 18.18$$

If **Yates' correction for continuity** is applied to a 2 × 2 table, Equation 11.6 is the continuity corrected version of Equation 11.5.

$$\chi^2 = \frac{n(|ad - bc| - .5n)^2}{(a + b)(c + d)(a + c)(b + d)}$$ **(Equation 11.6)**

Substituting the data for Examples 11.1 and 11.2 in Equation 11.6, the value $\chi^2 = 16.98$ is computed (which is the same continuity corrected chi-square value computed with Equation 11.4).

$$\chi^2 = \frac{200[|(30)(40) - (70(60)| - (.5)(200)]^2}{(30 + 70)(60 + 40)(30 + 60)(70 + 40)} = 16.98$$

3. Evaluation of a directional alternative hypothesis in the case of a 2 × 2 contingency table In the case of 2 × 2 contingency tables it is possible to employ a directional/one-tailed alternative hypothesis. Prior to reading this section, the reader may find it useful to review the relevant material on this subject in Section VII of the **chi-square goodness-of-fit test.**

In the case of a 2 × 2 contingency table, it is possible to make two directional predictions. In stating the null and alternative hypotheses, the following notation (in reference to the sample data) based on the model for a 2 × 2 contingency table described in Table 11.6 will be employed.

$$p_1 = \frac{a}{a + b} = \frac{a}{n_1} \qquad p_2 = \frac{c}{c + d} = \frac{c}{n_2}$$

The value p_1 represents the proportion of observations in Row 1 that falls in Cell a, while the value p_2 represents the proportion of observations in Row 2 that falls in Cell c. The analogous proportions in the underlying populations that correspond to p_1 and p_2 will be represented by the notation π_1 and π_2. Thus, π_1 represents the proportion of observations in Row 1 in the underlying population that falls in Cell a, while the proportion π_2 represents the proportion of observations in Row 2 in the underlying population that falls in Cell c. Employing the aforementioned notation, it is possible to make either of the two following directional predictions for a 2 × 2 contingency table.

a) In the underlying population(s) the sample(s) represent, the proportion of observations in Row 1 that falls in Cell a is greater than the proportion of observations in Row 2 that falls in Cell c. The null hypothesis and directional alternative hypothesis for this prediction are stated as follows: H_0: $\pi_1 = \pi_2$ versus H_1: $\pi_1 > \pi_2$. With respect to Example 11.1, the latter alternative hypothesis predicts that a larger proportion of subjects in the **noise** condition will **help the confederate** than subjects in the **no noise** condition. In Example 11.2, the alternative hypothesis predicts that a larger proportion of **introverts** will be **Democrats** than **extroverts.**

b) In the underlying population(s) the sample(s) represent, the proportion of observations in Row 1 that falls in Cell a is less than the proportion of observations in Row 2 that falls in Cell c. The null hypothesis and directional alternative hypothesis for this prediction are stated as follows: H_0: $\pi_1 = \pi_2$ versus H_1: $\pi_1 < \pi_2$. With respect to Example 11.1, the latter alternative hypothesis predicts that a larger proportion of subjects in the **no noise** condition will **help the confederate** than subjects in the **noise** condition. In Example 11.2, the alternative hypothesis predicts that a larger proportion of **extroverts** will be **Democrats** than **introverts.**

As is the case for the **chi-square goodness-of-fit test,** if a researcher wants to evaluate a one-tailed alternative hypothesis at the .05 level, the appropriate critical value to employ is $\chi^2_{.90}$, which is the tabled chi-square value at the .10 level of significance. The latter value is represented by the tabled chi-square value at the 90th percentile (which demarcates the extreme 10% in the right tail of the chi-square distribution). If a researcher wants to evaluate a one-tailed/directional alternative hypothesis at the .01 level, the appropriate critical value to employ is $\chi^2_{.98}$, which is the tabled chi-square value at the .02 level of significance. The latter value is represented by the tabled chi-square value at the 98th percentile (which demarcates the extreme 2% in the right tail of the chi-square distribution).

If a one-tailed alternative hypothesis is evaluated for Examples 11.1 and 11.2, from **Table A4** it can be determined that for $df = 1$ the relevant tabled critical one-tailed .05 and .01 values are $\chi^2_{.90} = 2.71$ and $\chi^2_{.98} = 5.43$.[9] Note that when one employs a one-tailed alternative hypothesis it is easier to reject the null hypothesis, since the one-tailed .05 and .01 critical values are less than the two-tailed .05 and .01 values (which for $df = 1$ are $\chi^2_{.05} = 3.84$ and $\chi^2_{.01} = 6.63$).[10] In conducting a one-tailed analysis, it is important to note, however, if the obtained value of chi-square is equal to or greater than the tabled critical value at the prespecified level of significance, only one of the two possible alternative hypotheses can be supported. The alternative hypothesis that is supported is the one that is consistent with the data.

Since for Examples 11.1 and 11.2, the computed value $\chi^2 = 18.18$ is greater than both of the one-tailed critical values $\chi^2_{.90} = 2.71$ and $\chi^2_{.98} = 5.43$, the null hypothesis can be rejected at both the .05 and .01 levels, but only if the directional alternative hypothesis $H_1\colon \pi_1 < \pi_2$ is employed. If the directional hypothesis $H_1\colon \pi_1 > \pi_2$ is employed, the null hypothesis cannot be rejected, since the data are not consistent with the latter alternative hypothesis.

When evaluating contingency tables in which the number of rows and/or columns is greater than two, it is possible to run a **multi-tailed test**, (which as noted in the discussion of the **chi-square goodness-of-fit test** is the term that is sometimes used when there are more than two possible directional alternative hypotheses). It would be quite unusual to encounter the use of a multi-tailed analysis for an $r \times c$ table, which requires that a researcher determine all possible directional patterns/ordinal configurations for a set of data, and then predict one or more of the specific patterns that will occur. In the event one elects to conduct a multi-tailed analysis, the determination of the appropriate critical values is based on the same guidelines that are discussed for multi-tailed tests under the **chi-square goodness-of-fit test.**

4. Test 11c: The Fisher exact test In Section I it is noted that the chi-square distribution provides an approximation of the exact sampling distribution for a contingency table. In the case of 2 × 2 tables, the chi-square distribution is employed to approximate the hypergeometric distribution which will be discussed in this section. As noted earlier, when $n < 20$ most sources recommend that the **Fisher exact test** (which employs exact hypergeometric probabilities) be employed to evaluate a 2 × 2 contingency table. Table 11.6, which is used earlier in this section to summarize a 2 × 2 table, will be employed to describe the model for the hypergeometric distribution upon which the **Fisher exact test** is based.

According to Daniel (1990) the **Fisher exact test,** which is also referred to as the **Fisher–Irwin test,** was simultaneously described by Fisher (1934, 1935), Irwin (1935), and Yates (1934). The test shares the same assumptions as those noted for the **chi-square test for $r \times c$ tables,** with the exception of the assumption regarding small expected frequencies (which reflects the limitations of the latter test with small sample sizes). Many sources note that an additional assumption of the **Fisher exact test** is that both the row and column sums of a 2 × 2 contingency table are predetermined by the researcher. In truth, this latter assumption is rarely met, and consequently the test is used with 2 × 2 contingency tables involving small samples sizes when one or neither of the marginal sums is predetermined by the researcher. The **Fisher exact test** is more commonly employed with the model described for the **chi-square test of homogeneity** than it is with the model described for the **chi-square test of independence.**

Equation 11.7, which is the equation for a hypergeometrically distributed variable, allows for the computation of the exact probability (P) of obtaining a specific set of observed frequencies in a 2 × 2 contingency table. Since the latter equation involves the computation of combinations, the reader may find it useful to review the discussion of combinations in Section IV of the **binomial sign test for a single sample (Test 6)**.

$$P = \frac{\binom{a+c}{a}\binom{b+d}{b}}{\binom{n}{a+b}} \qquad \textbf{(Equation 11.7)}$$

Equation 11.8 is a computationally more efficient form of Equation 11.7 which yields the same probability value.

$$P = \frac{(a+c)!\ (b+d)!\ (a+b)!\ (c+d)!}{n!\ a!\ b!\ c!\ d!} \qquad \textbf{(Equation 11.8)}$$

Example 11.3 (which is a small sample size version of Example 11.1) will be employed to illustrate the **Fisher exact test**.

Example 11.3 *A researcher conducts a study in order to evaluate the effect of noise on altruistic behavior. Each of the* 12 *subjects who participate in the experiment is randomly assigned to one of two experimental conditions. Subjects in both conditions are given a one-hour test which is ostensibly a measure of intelligence. During the test the six subjects in* Group 1 *are exposed to continual loud noise, which they are told is due to a malfunctioning generator. The six subjects in* Group 2 *are not exposed to any noise during the test. Upon completion of this stage of the experiment, each subject on leaving the room is confronted by a middle-aged man whose arm is in a sling. The man asks the subject if she would be willing to help him carry a heavy package to his car. In actuality, the man requesting help is an experimental confederate (i.e., working for the experimenter). The number of subjects in each group who help the man are recorded. One of the six subjects who were exposed to noise elects to help the man, while five of the six subjects who were not exposed to noise elect to help the man. Do the data indicate that altruistic behavior is influenced by noise?*

The data for Example 11.3, which can be summarized in the form of a 2 × 2 contingency table, are presented in Table 11.7.

Table 11.7 Summary of Data for Example 11.3

	Helped the confederate	Did not help the confederate	Row sums
Noise	$a = 1$	$b = 5$	$a + b = n_1 = 6$
No noise	$c = 5$	$d = 1$	$c + d = n_2 = 6$
Column sums	$a + c = 6$	$b + d = 6$	$n = 12$

The null and alternative hypotheses for the **Fisher exact test** are most commonly stated using the format described in the discussion of the evaluation of a directional alternative hypothesis for a 2 × 2 contingency table. Thus, the null hypothesis and nondirectional alternative hypotheses are as follows:

$$H_0\colon \pi_1 = \pi_2$$

(In the underlying populations the samples represent, the proportion of observations in Row 1 (the **noise** condition) that falls in Cell *a* is equal to the proportion of observations in Row 2 (the **no noise** condition) that falls in Cell *c*.)

$$H_1\colon \pi_1 \neq \pi_2$$

(In the underlying populations the samples represent, the proportion of observations in Row 1 (the **noise** condition) that falls in Cell *a* is not equal to the proportion of observations in Row 2 (the **no noise** condition) that falls in Cell *c*.)

The alternative hypothesis can also be stated directionally, as described in the discussion of the evaluation of a directional alternative hypothesis for a 2 × 2 contingency table — i.e., $H_1\colon \pi_1 > \pi_2$ or $H_1\colon \pi_1 < \pi_2$.[11]

Employing Equations 11.7 and 11.8, the probability of obtaining the specific set of observed frequencies in Table 11.7 is computed to be $P = .039$.

Equation 11.7:

$$P = \frac{\binom{6}{1}\binom{6}{5}}{\binom{12}{6}} = \frac{\left[\frac{6!}{1!\ 5!}\right]\left[\frac{6!}{5!\ 1!}\right]}{\frac{12!}{6!\ 6!}} = .039$$

Equation 11.8:

$$P = \frac{6!\ 6!\ 6!\ 6!}{12!\ 1!\ 5!\ 5!\ 1!} = .039$$

In order to evaluate the null hypothesis, in addition to the probability $P = .039$ (which is the probability of obtaining the set of observed frequencies in Table 11.7) it is also necessary to compute the probabilities for any sets of observed frequencies that are even more extreme than the observed frequencies in Table 11.7. The only result that is more extreme than the result summarized in Table 11.7 is if all six subjects in the **no noise** condition **helped the confederate,** while all six subjects in the **noise** condition **did not help the confederate.** Table 11.8 summarizes the observed frequencies for the latter result.

Table 11.8 Most Extreme Possible Set of Observed Frequencies for Example 11.3

	Helped the confederate	**Did not help the confederate**	**Row sums**
Noise	$a = 0$	$b = 6$	$a + b = n_1 = 6$
No noise	$c = 6$	$d = 0$	$c + d = n_2 = 6$
Column sums	$a + c = 6$	$b + d = 6$	$n = 12$

Employing Equations 11.7 and 11.8, the probability of obtaining the set of observed frequencies in Table 11.8 is computed to be $P = .001$.

Equation 11.7:

$$P = \frac{\binom{6}{0}\binom{6}{6}}{\binom{12}{6}} = \frac{\left[\frac{6!}{0!\;6!}\right]\left[\frac{6!}{6!\;0!}\right]}{\frac{12!}{6!\;6!}} = .001$$

Equation 11.8:

$$P = \frac{6!\;6!\;6!\;6!}{12!\;0!\;6!\;6!\;0!} = .001$$

When $P = .001$ (the probability of obtaining the set of observed frequencies in Table 11.8) is added to $P = .039$, the resulting probability represents the likelihood of obtaining a set of observed frequencies that is equal to or more extreme than the set of observed frequencies in Table 11.7. The notation P_T will be used to represent the latter value. Thus, in our example, $P_T = .039 + .001 = .04$.[12]

The following guidelines are employed in evaluating the null hypothesis for the **Fisher exact test.**

a) If the nondirectional alternative hypothesis H_1: $\pi_1 \neq \pi_2$ is employed, the value of P_T (i.e., the probability of obtaining a set of observed frequencies equal to or more extreme than the set obtained in the study) must be equal to or less than $\alpha/2$. Thus, if the prespecified value of alpha is $\alpha = .05$, the obtained value of P_T must be equal to or less than $.05/2 = .025$. If the prespecified value of alpha is $\alpha = .01$, the obtained value of P_T must be equal to or less than $.01/2 = .005$.

b) If a directional alternative hypothesis is employed, the observed set of frequencies for the study must be consistent with the directional alternative hypothesis, and the value of P_T must be equal to or less than the prespecified value of alpha. Thus, if the prespecified value of alpha is $\alpha = .05$, the obtained value of P_T must be equal to or less than .05. If the prespecified value of alpha is $\alpha = .01$, the obtained value of P_T must be equal to or less than .01.

Employing the above guidelines, the following conclusions can be reached.

If $\alpha = .05$, the nondirectional alternative hypothesis H_1: $\pi_1 \neq \pi_2$ is not supported, since the obtained value $P_T = .04$ is greater than $.05/2 = .025$.

The directional alternative hypothesis H_1: $\pi_1 < \pi_2$ is supported, but only at the .05 level. This is the case, since the data are consistent with the latter alternative hypothesis, and the obtained value $P_T = .04$ is less than $\alpha = .05$. The latter alternative hypothesis is not supported at the .01 level, since $P_T = .04$ is greater than $\alpha = .01$.

The directional alternative hypothesis H_1: $\pi_1 > \pi_2$ is not supported, since it is not consistent with the data. In order for the data to be consistent with the alternative hypothesis H_1: $\pi_1 > \pi_2$, it is required that a larger proportion of subjects in the **noise** condition **helped the confederate** than subjects in the **no noise** condition.

To further clarify how to interpret a directional versus a nondirectional alternative hypothesis, consider Table 11.9 which presents all seven possible outcomes of observed cell frequencies for $n = 12$ in which the marginal sums (i.e., the row and column sums) equal six (which are the values for the marginal sums in Example 11.3).

The sum of the probabilities for the seven outcomes presented in Table 11.9 equals 1. This is the case, since the seven outcomes represent all the possible outcomes for the cell frequencies if the marginal sum of each of row and column equals six. As noted earlier, if a researcher evaluates the directional alternative hypothesis H_1: $\pi_1 < \pi_2$ for Example 11.3, he will only be interested in **Outcomes 1 and 2.** The combined probability for the latter two outcomes is $P_T = .04$, which is less than the one-tailed value $\alpha = .05$ (which represents the extreme 5% of the sampling distribution in one of the two tails of the distribution). Since the data are consistent with the directional alternative hypothesis H_1: $\pi_1 < \pi_2$ and $P_T = .04$ is less than $\alpha = .05$, the latter alternative hypothesis is supported.

Table 11.9 Possible Outcomes for Observed Cell Frequencies If All Marginal Sums Equal 6, When $n = 12$

Outcome 1: $P = .001$

	Col. 1	Col. 2	Row sums
Row 1	0	6	6
Row 2	6	0	6
Column sums	6	6	12

Outcome 2: $P = .039$

	Col. 1	Col. 2	Row sums
Row 1	1	5	6
Row 2	5	1	6
Column sums	6	6	12

Outcome 3: $P = .243$

	Col. 1	Col. 2	Row sums
Row 1	2	4	6
Row 2	4	2	6
Column sums	6	6	12

Outcome 4: $P = .433$

	Col. 1	Col. 2	Row sums
Row 1	3	3	6
Row 2	3	3	6
Column sums	6	6	12

Outcome 5: $P = .243$

	Col. 1	Col. 2	Row sums
Row 1	4	2	6
Row 2	2	4	6
Column sums	6	6	12

Outcome 6: $P = .039$

	Col. 1	Col. 2	Row sums
Row 1	5	1	6
Row 2	1	5	6
Column sums	6	6	12

Outcome 7: $P = .001$

	Col. 1	Col. 2	Row sums
Row 1	6	0	6
Row 2	0	6	6
Column sums	6	6	12

If, however, the nondirectional alternative hypothesis H_1: $\pi_1 \neq \pi_2$ is employed, in addition to considering **Outcomes 1 and 2**, the researcher must also consider **Outcomes 6 and 7**, which are the analogous extreme outcomes in the opposite tail of the distribution. The latter set of outcomes also has a combined probability of .04. If the probability associated with **Outcomes 1 and 2** and **Outcomes 6 and 7** are summed, the resulting value $P_T = .04 + .04 = .08$ represents the likelihood in both tails of the distribution of obtaining an outcome equal to or more extreme than the outcome observed in Table 11.7. Since the value $P_T = .08$ is greater than the two-tailed value $\alpha = .05$, the nondirectional alternative hypothesis H_1: $\pi_1 \neq \pi_2$ is not supported. This is commensurate with saying that in order for the latter alternative hypothesis to be supported at the .05 level, in each of the tails of the distribution the maximum permissible probability value for outcomes that are equivalent to or more extreme than the observed outcome cannot be greater than the value $.05/2 = .025$. As noted earlier, since the computed probability in the relevant tail of the distribution equals .04 (which is greater than $.05/2 = .025$), the nondirectional alternative hypothesis is not supported.

If the directional alternative hypothesis H_1: $\pi_1 > \pi_2$ is employed the researcher is interested in **Outcomes 6 and 7**. As is the case for **Outcomes 1 and 2**, the combined probability for these two outcomes is $P_T = .04$, which is less than the one-tailed value $\alpha = .05$ (which represents the extreme 5% of the sampling distribution in the other tail of the distribution). However, since the data are not consistent with the directional alternative hypothesis H_1: $\pi_1 > \pi_2$, it is not supported.

In order to compare the results of the **Fisher exact test** with those that will be obtained if Example 11.3 is evaluated with the **chi-square test for $r \times c$ tables**, Equation 11.2 will be employed to evaluate the data in Table 11.7. Table 11.10 summarizes the chi-square analysis. Note that the expected frequency of each cell is 3, since when Equation 11.1 is employed, the value $E_{ij} = [(6)(6)]/12 = 3$ is computed for all $r \times c = 4$ cells.

Table 11.10 Chi-Square Summary Table for Example 11.3

Cell	O_{ij}	E_{ij}	$(O_{ij} - E_{ij})$	$(O_{ij} - E_{ij})^2$	$\frac{(O_{ij} - E_{ij})^2}{E_{ij}}$
Cell_{11} — Noise/Helped the confederate	1	3	−2	4	1.33
Cell_{12} — Noise/Did not help the confederate	5	3	2	4	1.33
Cell_{21} — No noise/Helped the confederate	5	3	2	4	1.33
Cell_{22} — No noise/Did not help the confederate	1	3	−2	4	1.33
	$\Sigma O_{ij} = 12$	$\Sigma E_{ij} = 12$	$\Sigma(O_{ij} - E_{ij}) = 0$		$\chi^2 = 5.32$

Since for $df = 1$, the obtained value $\chi^2 = 5.32$ is greater than the tabled critical two-tailed .05 value $\chi^2_{.05} = 3.84$, the nondirectional alternative hypothesis H_1: $\pi_1 \neq \pi_2$ is supported at the .05 level. It is not, however, supported at the .01 level, since $\chi^2 = 5.32$ is less than the tabled critical two-tailed .01 value $\chi^2_{.01} = 6.63$.

The directional alternative hypothesis H_1: $\pi_1 < \pi_2$ is supported at the .05 level, since the obtained value $\chi^2 = 5.32$ is greater than the tabled critical one-tailed .05 value $\chi^2_{.90} = 2.71$. The latter directional alternative hypothesis just falls short of being supported at the .01 level, since $\chi^2 = 5.32$ is less than the tabled critical one-tailed .01 value $\chi^2_{.98} = 5.43$.

Note that when the **Fisher exact test** is employed, the nondirectional alternative hypothesis H_1: $\pi_1 \neq \pi_2$ is not supported at the .05 level, yet it is supported at the .05 level when the **chi-square test for $r \times c$ tables** is used. Both the **Fisher exact test** and **chi-square test for $r \times c$ tables** allow the researcher to reject the null hypothesis at the .05 level if the directional alternative hypothesis H_1: $\pi_1 < \pi_2$ is employed. However, whereas the latter alternative hypothesis just falls short of significance at the .01 level when the **chi-quare test for $r \times c$ tables** is used, it is further removed from being significant when the **Fisher exact test** is employed. The discrepancy between the two tests when they are applied to the same set of data involving a small sample size suggests that the chi-square approximation underestimates the actual probability associated with the observed frequencies, and consequently increases the likelihood of committing a Type I error.

5. Test 11d: The *z* test for two independent proportions The ***z* test for two independent proportions** is an alternative large sample procedure for evaluating a 2×2 contingency table. In point of fact, the ***z* test for two independent proportions** yields a result that is equivalent to that obtained with the **chi-square test for $r \times c$ tables.** Later in this discussion it will be demonstrated that if both the ***z* test for two independent proportions** (which is based on the normal distribution) and the **chi-square test for $r \times c$ tables** are applied to the same set of data, the square of the z value obtained for the former test will equal the chi-square value obtained for the latter test.

The ***z* test for two independent proportions** is most commonly employed to evaluate the null and alternative hypotheses that are described for the **Fisher exact test** (which for a

2×2 contingency table are equivalent to the null and alternative hypotheses presented in Section III for the **chi-square test for homogeneity**). Thus, in reference to Example 11.1, employing the model for a 2×2 contingency table summarized by Table 11.6, the null hypothesis and nondirectional alternative hypotheses for the **z test for two independent proportions** are as follows.

$$H_0: \pi_1 = \pi_2$$

(In the underlying populations the samples represent, the proportion of observations in Row 1 (the **noise** condition) that falls in Cell a is equal to the proportion of observations in Row 2 (the **no noise** condition) that falls in Cell c.)

$$H_1: \pi_1 \neq \pi_2$$

(In the underlying populations the samples represent, the proportion of observations in Row 1 (the **noise** condition) that falls in Cell a is not equal to the proportion of observations in Row 2 (the **no noise** condition) that falls in Cell c.)

An alternate but equivalent way of stating the above noted null hypothesis and nondirectional alternative hypothesis is as follows: $H_0: \pi_1 - \pi_2 = 0$ versus $H_1: \pi_1 - \pi_2 \neq 0$. As is the case with the **Fisher exact test** (as well as the **chi-square test for $r \times c$ tables**), the alternative hypothesis can also be stated directionally. Thus, the following two directional alternative hypotheses can be employed: $H_1: \pi_1 > \pi_2$ (which can also be stated as $H_1: \pi_1 - \pi_2 > 0$) or $H_1: \pi_1 < \pi_2$ (which can also be stated as $H_1: \pi_1 - \pi_2 < 0$).

Equation 11.9 is employed to compute the test statistic for the **z test for two independent proportions.**

$$z = \frac{p_1 - p_2}{\sqrt{p(1 - p)\left[\frac{1}{n_1} + \frac{1}{n_2}\right]}} \qquad \textbf{(Equation 11.9)}$$

Where: n_1 represents the number of observations in Row 1

n_2 represents the number of observations in Row 2

$p_1 = a/(a + b) = a/n_1$ represents the proportion of observations in Row 1 that falls in Cell a. It is employed to estimate the population proportion π_1.

$p_2 = c/(c + d) = c/n_2$ represents the proportion of observations in Row 2 that falls in Cell c. It is employed to estimate the population proportion π_2.

$p = (a + c)/(n_1 + n_2) = (a + c)/n$. p is a pooled estimate of the proportion of observations in Column 1 in the underlying population.

The denominator of Equation 11.9, which represents a standard deviation of a sampling distribution of differences between proportions, is referred to as the **standard error of the difference between two proportions** (which is often summarized with the notation $s_{p_1 - p_2}$). This latter value is analogous to the **standard error of the difference** ($s_{\bar{X}_1 - \bar{X}_2}$), which is the denominator of Equations 8.1, 8.2, 8.3, and 8.5 (which are all described in reference to the **t test for two independent samples (Test 8)**). Whereas the latter value is a standard deviation of a sampling distribution of difference scores between the means of two populations, the **standard error of the difference between two proportions** is a standard deviation of a sampling distribution of difference scores between proportions for two populations.

For Example 11.1 we either know or can compute the following values:[13]

$$a = 30 \qquad c = 60 \qquad n_1 = 100 \qquad n_2 = 100$$

$$p_1 = \frac{a}{n_1} = \frac{30}{100} = .3 \qquad p_2 = \frac{c}{n_2} = \frac{60}{100} = .6$$

$$p = \frac{a + c}{n_1 + n_2} = \frac{30 + 60}{100 + 100} = .45 \qquad 1 - p = 1 - .45 = .55$$

Employing the above values in Equation 11.9, the value $z = -4.26$ is computed.

$$z = \frac{.3 - .6}{\sqrt{(.45)(.55)\left[\frac{1}{100} + \frac{1}{100}\right]}} = -4.26$$

The obtained value $z = -4.26$ is evaluated with **Table A1 (Table of the Normal Distribution)** in the **Appendix.** In **Table A1** the tabled critical two-tailed .05 and .01 values are $z_{.05} = 1.96$ and $z_{.01} = 2.58$, and the tabled critical one-tailed .05 and .01 values are $z_{.05} = 1.65$ and $z_{.01} = 2.33$. The following guidelines are employed in evaluating the null hypothesis:

a) If the nondirectional alternative hypothesis H_1: $\pi \neq \pi_2$ is employed, the null hypothesis can be rejected if the obtained absolute value of z is equal to or greater than the tabled critical two-tailed value at the prespecified level of significance.

b) If the directional alternative hypothesis H_1: $\pi_1 > \pi_2$ is employed, the null hypothesis can be rejected if $p_1 > p_2$, and the obtained value of z is a positive number that is equal to or greater than the tabled critical one-tailed value at the prespecified level of significance.

c) If the directional alternative hypothesis H_1: $\pi_1 < \pi_2$ is employed, the null hypothesis can be rejected if $p_1 < p_2$, and the obtained value of z is a negative number with an absolute value that is equal to or greater than the tabled critical one-tailed value at the prespecified level of significance.

Employing the above guidelines, the following conclusions can be reached.

Since the obtained absolute value $z = 4.26$ is greater than the tabled critical two-tailed values $z_{.05} = 1.96$ and $z_{.01} = 2.58$, the nondirectional alternative hypothesis H_1: $\pi \neq \pi_2$ is supported at both the .05 and .01 levels.

Since the obtained value of z is a negative number and absolute value $z = 4.26$ is greater than the tabled critical one-tailed values $z_{.05} = 1.65$ and $z_{.01} = 2.33$, the directional alternative hypothesis H_1: $\pi_1 < \pi_2$ is supported at both the .05 and .01 levels.

The directional alternative hypothesis H_1: $\pi_1 > \pi_2$ is not supported, since as previously noted, in order for it to be supported the following must be true: $p_1 > p_2$.

Note that the above conclusions for the analysis of Example 11.1 (as well as Example 11.2) with the **z test for two independent proportions** are consistent with the conclusions that are reached when the **chi-square test for $r \times c$ tables** is employed to evaluate the same set of data. In point of fact, the square of the z value obtained with Equation 11.9 will always be equal to the value of chi-square computed with Equation 11.2. This relationship can be confirmed by the fact that in the example under discussion, $(z = -4.26)^2 = (\chi^2 = 18.18)$. It is also the case that the square of a tabled critical z value at a given level of significance will equal the tabled critical chi-square value at the corresponding level of significance. This is confirmed for the tabled critical two-tailed z and χ^2 values at the .05 and .01 levels of significance: $(z_{.05} = 1.96)^2 = (\chi^2_{.05} = 3.84)$ and $(z_{.01} = 2.58)^2 = (\chi^2_{.01} = 6.63)$.[14]

Yates' correction for continuity can also be applied to the **z test for two independent proportions.** Equation 11.10 is the continuity corrected equation for the **z test for two independent proportions.**

$$z = \frac{[p_1 - p_2] \pm \left[\frac{1}{2}\right]\left[\frac{1}{n_1} + \frac{1}{n_2}\right]}{\sqrt{p(1 - p)\left[\frac{1}{n_1} + \frac{1}{n_2}\right]}} \qquad \textbf{(Equation 11.10)}$$

The following protocol is employed with respect to the numerator of Equation 11.10: a) If $(p_1 - p_2)$ is a positive number, the term $[1/2][(1/n_1) + (1/n_2)]$ is subtracted from $(p_1 - p_2)$; and b) If $(p_1 - p_2)$ is a negative number, the term $[1/2][(1/n_1) + (1/n_2)]$ is added to $(p_1 - p_2)$. An alternative way of computing the value of the numerator of Equation 11.10 is to subtract $[1/2][(1/n_1) + (1/n_2)]$ from the absolute value of $(p_1 - p_2)$, and then restore the original sign of the latter value.

Employing Equation 11.10, the continuity corrected value $z = -4.12$ is computed.

$$z = \frac{[.3 - .6] + \left[\frac{1}{2}\right]\left[\frac{1}{100} + \frac{1}{100}\right]}{\sqrt{(.45)(.55)\left[\frac{1}{100} + \frac{1}{100}\right]}} = -4.12$$

Note that the absolute value of the continuity corrected z value computed with Equation 11.10 will always be smaller than the absolute z value computed with Equation 11.9. As is the case when Equation 11.9 is employed, the square of the continuity corrected z value will equal the continuity corrected chi-square value computed with Equation 11.4. Thus, $(z = -4.12)^2 = (\chi^2 = 16.98)$.

By employing Equation 11.10, the obtained absolute value of z is reduced to 4.12 (when contrasted with the absolute value $z = 4.26$ computed with Equation 11.9). Since the absolute value $z = 4.12$ is greater than both $z_{.05} = 1.96$ and $z_{.01} = 2.58$, the nondirectional alternative hypothesis H_1: $\pi_1 \neq \pi_2$ is still supported at both the .05 and .01 levels. The directional alternative hypothesis H_1: $\pi_1 < \pi_2$ is also still supported at both the.05 and .01 levels, since the absolute value $z = 4.12$ is greater than both $z_{.05} = 1.65$ and $z_{.01} = 2.33$.

The protocol that has been described for the **z test for two independent proportions** assumes that the researcher employs the null hypothesis H_0: $\pi_1 = \pi_2$. If, in fact, the null hypothesis stipulates a difference other than zero between the two values π_1 and π_2, Equation 11.11 is employed to compute the test statistic for the **z test for two independent proportions.**[15]

$$z = \frac{(p_1 - p_2) - (\pi_1 - \pi_2)}{\sqrt{\frac{p_1(1 - p_1)}{n_1} + \frac{p_2(1 - p_2)}{n_2}}} \qquad \textbf{(Equation 11.11)}$$

The use of Equation 11.11 is based on the assumption that $\pi_1 \neq \pi_2$. Whenever the latter is the case, instead of computing a pooled estimate for a common population proportion, it is appropriate to estimate a separate value for the proportion in each of the underlying populations (i.e., π_1 and π_2) by using the values p_1 and p_2. This is in contrast to Equation 11.9, which computes a pooled p value (that represents the pooled estimate of the proportion of observations in Column 1 in the underlying populations). In the case of Equation 11.9, the computation of a pooled p value is based on the assumption that $\pi_1 = \pi_2$.

To illustrate the application of Equation 11.11, let us assume that the following null hypothesis and nondirectional alternative hypothesis are employed for Example 11.1: H_0: $\pi_1 - \pi_2 = -.1$ (which can also be written H_0: $\pi_2 - \pi_1 = .1$) versus

H_1: $\pi_1 - \pi_2 \neq -.1$ (which can also be written H_1: $\pi_2 - \pi_1 \neq .1$). The null hypothesis states that in the underlying populations represented by the samples, the difference between the proportion of observations in Row 1 (the **noise** condition) that falls in Cell *a* and the proportion of observations in Row 2 (the **no noise** condition) that falls in Cell *c* is $-.1$. The alternative hypothesis states that in the underlying populations represented by the samples, the difference between the proportion of observations in Row 1 (the **noise** condition) that falls in Cell *a* and the proportion of observations in Row 2 (the **no noise** condition) that falls in Cell *c* is some value other than $-.1$.

In order for the nondirectional alternative hypothesis H_1: $\pi_1 - \pi_2 \neq -.1$ to be supported, the obtained absolute value of z must be equal to or greater than the tabled critical two-tailed value at the prespecified level of significance. The directional alternative hypothesis that is consistent with the data is H_1: $\pi_1 - \pi_2 < -.1$. In order for the latter alternative hypothesis to be supported, the sign of the obtained value of z must be negative, and the obtained absolute value of z must be equal to or greater than the tabled one-tailed critical value at the prespecified level of significance. The directional alternative hypothesis H_1: $\pi_1 - \pi_2 > -.1$ is not consistent with the data. Either a significant positive z value (in which case $p_1 - p_2 > 0$) or a significant negative z value (in which case $0 > (p_1 - p_2) > -.1$) is required to support the latter alternative hypothesis.

Employing Equation 11.11 for the above analysis, the value $z = -2.99$ is computed.

$$z = \frac{(.3 - .6) - (-.1)}{\sqrt{\frac{(.3)(.7)}{100} + \frac{(.6)(.4)}{100}}} = -2.99$$

Since the obtained absolute value $z = 2.99$ is greater than the tabled critical two-tailed .05 and .01 values $z_{.05} = 1.96$ and $z_{.01} = 2.58$, the nondirectional alternative hypothesis H_1: $\pi_1 - \pi_2 \neq -.1$ is supported at both the .05 and .01 levels. Thus, one can conclude that in the underlying populations the difference $(\pi_1 - \pi_2)$ is some value other than $-.1$. The directional alternative hypothesis H_1: $\pi_1 - \pi_2 < -.1$ is supported at both the .05 and .01 levels, since the obtained value $z = -2.99$ is a negative number and the absolute value $z = -2.99$ is greater than the tabled critical one-tailed values $z_{.05} = 1.65$ and $z_{.01} = 2.33$. Thus, if the latter alternative hypothesis is employed, one can conclude that in the underlying populations the difference $(\pi_1 - \pi_2)$ is some value that is less than $-.1$ (i.e., is a negative number with an absolute value larger than .1). As noted earlier, the directional alternative hypothesis H_1: $\pi_1 - \pi_2 > -.1$ is not supported, since it is not consistent with the data.

Yates' correction for continuity can also be applied to Equation 11.11 by employing the same correction factor in the numerator that is employed in Equation 11.10. Using the correction for continuity, the numerator of Equation 11.11 becomes:

$$(p_1 - p_2) - (\pi_1 - \pi_2) \pm \left[\frac{1}{2}\right]\left[\frac{1}{n_1} + \frac{1}{n_2}\right]$$

Without the correction for continuity, the value of the numerator of Equation 11.11 is $-.2$. Since $[1/2][(1/100) + (1/100)] = .01$, using the guidelines outlined previously, the value of the numerator becomes $-2 + .01 = -.19$. When the latter value is divided by the denominator (which for the example under discussion equals .067), it yields the continuity corrected value $z = -2.84$. Thus, by employing the correction for continuity, the absolute value of z is reduced from $z = 2.99$ to $z = 2.84$. Since the absolute value $z = 2.84$ is greater than $z_{.05} = 1.96$ and $z_{.01} = 2.58$, the nondirectional alternative hypothesis H_1: $\pi_1 - \pi_2 \neq -.1$ is still supported at both the .05 and .01 levels. The directional alternative hypothesis H_1: $\pi_1 - \pi_2 < -.1$ is also still supported at both the .05 and .01 levels, since the absolute value $z = 2.84$ is greater than $z_{.05} = 1.65$ and $z_{.01} = 2.33$.

6. Computation of confidence interval for a difference between proportions With large sample sizes, a confidence interval can be computed that identifies a range of values within which one can be confident to a specified degree that the true difference lies between the two population proportions π_1 and π_2. Equation 11.12, which employs the normal distribution, is the general equation for computing the confidence interval for the difference between two population proportions. The notation employed in Equation 11.12 is identical to that used in the discussion of the **z test for two independent proportions.**

$$CI_{(1-\alpha)} = (p_1 - p_2) \pm (z_{\alpha/2})(s_{p_1 - p_2}) \qquad \textbf{(Equation 11.12)}$$

Where: $s_{p_1 - p_2} = \sqrt{[p_1(1 - p_1)/n_1] + [p_2(1 - p_2)/n_2]}$

$z_{\alpha/2}$ represents the tabled critical two-tailed value in the normal distribution, below which a proportion equal to $[1 - (\alpha/2)]$ of the cases falls. If the percentage of the distribution that falls within the confidence interval is subtracted from 1, it will equal the value of α.

Employing Equation 11.12, the 95% confidence interval for Examples 11.1/11.2 is computed below. In employing Equation 11.12, $(p_1 - p_2)$ (which is the numerator of Equation 11.9) represents the obtained difference between the sample proportions, $z_{.05}$ represents the tabled critical two-tailed .05 value $z_{.05} = 1.96$, and $s_{p_1 - p_2}$ (which is the denominator of Equation 11.11) represents the **standard error of the difference between two proportions.**[16]

$$p_1 - p_2 = .3 - .6 = -.3 \qquad s_{p_1 - p_2} = \sqrt{\frac{(.3)(.7)}{100} + \frac{(.6)(.4)}{100}} = .067$$

$$CI_{.95} = (p_1 - p_2) \pm (z_{.05})\,(s_{p_1 - p_2}) = -3 \pm (1.96)(.067) = -3 \pm .131$$

$$-.169 \geq (\pi_1 - \pi_2) \geq -.431$$

This result indicates that the researcher can be 95% confident that the true difference between the population proportions falls within the range −.431 and −.169. Specifically, it indicates that the researcher can be 95% confident that the proportion for Population 1 (π_1) is less than the proportion for Population 2 (π_2) by a value that is greater than or equal to .169, but not greater than .431. The result can also be written as $.169 \leq (\pi_2 - \pi_1) \leq .431$, which indicates that the researcher can be 95% confident that the proportion for Population 2 (π_2) is larger than the proportion for Population 1 (π_1) by a value that is greater than or equal to .169, but not greater than .431.

The 99% confidence interval, which results in a broader range of values, is computed below employing the tabled critical two-tailed .01 value $z_{.01} = 2.58$ in Equation 11.12 in lieu of $z_{.05} = 1.96$.

$$CI_{.99} = (p_1 - p_2) \pm (z_{.01})\,(s_{p_1 - p_2}) = -3 \pm (2.58)(.067) = -3 \pm .173$$

$$-.127 \geq (\pi_1 - \pi_2) \geq -.473$$

Thus, the researcher can be 99% confident that the proportion for Population 1 is less than the proportion for Population 2 by a value that is greater than or equal to .127, but not greater than .473.

7. Test 11e: The median test for independent samples The model for the **median test for independent samples** assumes that there are k independent groups, and that within each group each observation is categorized with respect to whether it is **above** or **below** a composite median value. In actuality, the **median test for independent samples** is a label that is commonly used when either the **chi-square test for $r \times c$ tables,** the **z test for two independent proportions,** or the **Fisher exact test** is employed to evaluate the hypothesis that in each of k groups there is an equal proportion of observations that are **above** versus **below** a composite median. With large sample sizes, the **median test for independent samples** is computationally identical to the **chi-square test for $r \times c$ tables** (when $k \geq 2$) and the **z test for two independent proportions** (when $k = 2$). In the case of small samples sizes, the test is computationally identical to the **Fisher exact test** (when $k = 2$). Table 11.6, which is used to summarize the model for the three aforementioned tests, can also be applied to the **median test for independent samples.** The two rows are employed to represent the two groups, and the two columns are used to represent the two categories on the dependent variable — specifically, whether a score falls **above** versus **below the median.** Example 11.4 will be employed to illustrate the **median test for two independent samples.**

Example 11.4 *A study is conducted to determine whether five-year old females are more likely than five year old males to score above the population median on a standardized test of eye-hand coordination. One hundred randomly selected females and* 100 *randomly selected males are administered the test of eye-hand coordination, and categorized with respect to whether they score above or below the overall population median (i.e., the* 50th *percentile for both males and females).* Table 11.11 *summarizes the results of the study. Do the data indicate that there are gender differences in performance?*

Table 11.11 Summary of Data for Example 11.4

	Above the median	Below the median		Row sums
Males	30	70		100
Females	60	40		100
Column sums	90	110	Total observations	200

Since Example 11.4 involves a large sample size, it can be evaluated with Equation 11.2, the equation for the **chi-square test for $r \times c$ tables** (as well as with Equation 11.9, the equation for the **z test for two independent proportions**). The reader should take note of the fact that the study described in Example 11.4 conforms to the model for the **chi-square test for homogeneity.** This is the case, since the row sums (i.e., the number of **males** and **females**) are predetermined by the researcher. Since it is consistent with the model for the latter test, the null and alternative hypotheses that are evaluated with the **median test for independent samples** are identical to those evaluated with the **chi-square test for homogeneity,** the **Fisher exact test,** and the **z test for two independent proportions.**[17]

Since the data for Example 11.4 are identical to the data for Examples 11.1/11.2, analysis of Example 11.4 with Equation 11.2 yields the value $\chi^2 = 18.18$ (which is the value obtained for Examples 11.1/11.2). Since (as noted earlier) $\chi^2 = 18.18$ is significant at both the .05 and .01 levels, one can conclude that **females** are more likely than **males** to score **above the median.**

In the case of the **median test for independent samples,** in the event that one or more subjects obtains a score that is equal to the population median, the following options are available for handling such scores: a) If the number of subjects who obtain a score that equals the median is reasonably large, a strong argument can be made for adding a third column to Table 11.11 for subjects who scored **at the median.** In such a case the contingency table is transformed into a 2 × 3 table; b) If a minimal number of subjects obtain a score at the median, such subjects can be dropped from the data; and c) Within each group, half the scores that fall at the median value are assigned to the **above the median** category and the other half to the **below the median** category. In the final analysis, the critical thing the researcher must be concerned with in employing any of the aforementioned strategies, is to make sure that the procedure he employs does not lead to misleading conclusions regarding the distribution of scores in the underlying populations.

The **median test for independent samples** can be extended to designs involving more than two groups. As an example, in Example 11.4 instead of evaluating the number of **males** and **females** who score **above** versus **below the median,** four groups of children representing different ethnic groups (e.g., **Caucasian, Asian-American, African-American, Native-American**) could be evaluated with respect to whether they score **above** versus **below the median.** In such a case, the data are summarized in the form of a 2 × 4 contingency table, with the four rows representing the four ethnic groups and the two columns representing **above** versus **below the median.**

It should be noted, that some sources categorize the **median test for independent samples** as a test of ordinal data, since categorizing scores with respect to whether they are **above** versus **below the median** involves placing scores in one of two ordered categories. The reader should also be aware of the fact that it is possible to categorize scores on more than two ordered categories. As an example, **male** and **female** children can be categorized with respect to whether they score in the **first quartile** (lower 25%), **second quartile** (25th to 50%), **third quartile** (50% to 75%), or **fourth quartile** (upper 25%) on the test of eye–hand coordination. In such a case, the data are summarized in the form of a 2 × 4 contingency table, with the two rows representing **males** and **females** and the four columns representing the four quartiles. It is also possible to have a design involving more than two groups of subjects (e.g., the four ethnic groups discussed above), and a dependent variable involving more than two ordered categories (e.g., four quartiles). The data for such a design are summarized in the form of a 4 × 4 contingency table. Finally, it is possible to have contingency tables in which both the row and the column variables are ordered. Although the **chi-square test for $r \times c$ tables** can be employed to evaluate a design in which both variables are ordered, alternative procedures may provide the researcher with more information about the relationship between the two variables. An example of such an alternative procedure is **Goodman and Kruskal's gamma (Test 26)** discussed later in the book.

8. Extension of the chi-square test for $r \times c$ tables to contingency tables involving more than two rows and/or columns, and associated comparison procedures It is noted in the previous section that the **chi-square test for $r \times c$ tables** can be employed with tables involving more than two rows and/or columns. In this section larger contingency tables will be discussed, and within the framework of the discussion additional analytical procedures that can be employed with such tables will be described. Example 11.5 will be employed to illustrate the use of the **chi-square test for $r \times c$ tables** with a larger contingency table — specifically, a 4 × 3 table.

Example 11.5 *A researcher conducts a study in order to determine if there are differences in the frequency of biting among different species of laboratory animals. He selects random samples of four laboratory species from the stock of various animal supply companies. Sixty mice,* 50 *gerbils,* 90 *hamsters, and* 80 *guinea pigs are employed in the study. Each of the animals is handled over a two-week period, and categorized in one of the following three categories with respect to biting behavior: not a biter, mild biter, flagrant biter.* Table 11.12

summarizes the data for the study. Do the data indicate that there are interspecies differences in biting behavior?

Table 11.12 Summary of Data for Example 11.5

	Not a biter	Mild biter	Flagrant biter		Row sums
Mice	20	16	24		**60**
Gerbils	30	10	10		**50**
Hamsters	50	30	10		**90**
Guinea pigs	19	11	50		**80**
Column sums	**119**	**67**	**94**	**Total observations**	**280**

The study described in Example 11.5 conforms to the model for the **chi-square test for homogeneity**. This is the case, since the row sums (i.e., the number of animals representing each of the four species) is predetermined by the researcher. The row variable represents the independent variable in the study. The independent variable, which is comprised of four levels, is nonmanipulated, since it is based on a preexisting subject characteristic (i.e., species). The reader should take note of the fact that it is not necessary to have an equal number of subjects in each of the groups/categories that constitute the row variable. The column variable, which is comprised of three categories, is the biting behavior of the animals. The latter variable represents the dependent variable in the study. Note that the marginal sums for the column variable are not predetermined by the researcher.

As is the case with a 2 × 2 contingency table, Equation 11.1 is employed to compute the expected frequency for each cell. As an example, the expected frequency of Cell_{11} **(mice/not a biter)** is computed as follows: $E_{11} = [(O_{1.})(O_{.1})]/n = [(60)(119)]/280 = 25.5$. In employing Equation 11.1, the value 60 represents the sum for Row 1 (which represents the total number of **mice**), the value 119 represents the sum for Column 1 (which represents the total number of animals categorized as **not a biter**), and 280 represents the total number of subjects/observations in the study.

After employing Equation 11.1 to compute the expected frequency for the each of the 4 × 3 = 12 cells in the contingency table, Equation 11.2 is employed to compute the value $\chi^2 = 59.16$. The analysis is summarized in Table 11.13.

Table 11.13 Chi-Square Summary Table for Example 11.5

Cell	O_{ij}	E_{ij}	$(O_{ij} - E_{ij})$	$(O_{ij} - E_{ij})^2$	$\frac{(O_{ij} - E_{ij})^2}{E_{ij}}$
Mice/Not a biter	20	25.50	-5.50	30.25	1.19
Mice/Mild biter	16	14.36	1.64	2.69	.19
Mice/Flagrant biter	24	20.14	3.86	14.90	.74
Gerbils/Not a biter	30	21.25	8.75	76.56	3.60
Gerbils/Mild biter	10	11.96	-1.96	3.84	.32
Gerbils/Flagrant biter	10	16.79	-6.79	46.10	2.75
Hamsters/Not a biter	50	38.25	11.75	138.06	3.61
Hamsters/Mild biter	30	21.54	8.46	71.57	3.32
Hamsters/Flagrant biter	10	30.21	-20.21	408.44	13.52
Guinea pigs/Not a biter	19	34.00	-15.00	225.00	6.62
Guinea pigs/Mild biter	11	19.14	-8.14	66.26	3.36
Guinea pigs/Flagrant biter	50	26.86	23.14	535.46	19.94
Column sums	280	280.00	0		$\chi^2 = 59.16$

Substituting the values $r = 4$ and $c = 3$ in Equation 11.3, the number of degrees of freedom for the analysis are $df = (4-1)(3-1) = 6$. Employing **Table A4**, the tabled critical .05 and .01 chi-square values for $df = 6$ are $\chi^2_{.05} = 12.59$ and $\chi^2_{.01} = 16.81$. Since the computed value $\chi^2 = 59.16$ is greater than both of the aforementioned critical values, the null hypothesis can be rejected at both the .05 and .01 levels. By virtue of rejecting the null hypothesis, the researcher can conclude that the four species are not homogeneous with respect to biting behavior, or to be more precise, that at least two of the species are not homogeneous.

In the case of a 2×2 contingency table, a significant result indicates that the two groups employed in the study are not homogeneous with respect to the dependent variable. However, in the case of a larger contingency table, although a significant result indicates that at least two of the r groups are not homogeneous, the chi-square analysis does not indicate which of the groups differ from one another or which of the cells are responsible for the significant effect. Visual inspection of Tables 11.12 and 11.13 suggests that the significant effect in Example 11.5 is most likely attributable to the disproportionately large number of **flagrant biters** among **guinea pigs**, and the disproportionately small number of **flagrant biters** among **hamsters**. In lieu of visual inspection (which is not a precise method for identifying the cells that are primarily responsible for a significant effect), the following two types of comparisons are among those that can be conducted.

Simple comparisons A **simple comparison** is a comparison between two of the r rows of a $r \times c$ contingency table (or two of the c columns). Table 11.14 summarizes a simple comparison that contrasts the biting behavior of **mice** with the biting behavior of **guinea pigs**. Note that in the simple comparison, the data for the other two species employed in the study (i.e., **gerbils** and **hamsters**) are not included in the analysis.

Table 11.14 Simple Comparison for Example 11.5

	Not a biter	Mild biter	Flagrant biter		Row sums
Mice	20	16	24		60
Guinea pigs	19	11	50		80
Column sums	39	27	74	Total observations	140

It should be noted that a simple comparison does not have to involve all of the columns of the contingency table. Thus, one can compare the two species **mice** and **guinea pigs**, but limit the comparison within those species to only those animals who are classified **not a biter** and **flagrant biter**. The resulting 2×2 contingency table for such a comparison will differ from Table 11.14 in that the second column (**mild biter**) is not included. As a result of omitting the latter column, the row sum for **mice** is reduced to 44 and the row sum for **guinea pigs** is reduced to 69. The total number of observations for the comparison is 113.

Complex comparisons A **complex comparison** is a comparison between two or more of the r rows of an $r \times c$ contingency table with one of the other rows or two or more of the other rows of the table.[18] Table 11.5 summarizes an example of a complex comparison which contrasts the biting behavior of **guinea pigs** with the combined biting behavior of the other three species employed in the study (i.e., **mice, gerbils**, and **hamsters**).

As is the case for a simple comparison, a complex comparison does not have to include all of the columns on which the groups are categorized. Thus, one can compare the **mice, gerbils, and hamsters** with **guinea pigs**, but limit the comparison to only those animals who are classified **not a biter** and **flagrant biter**. The resulting 2×2 contingency table for such a comparison will differ from Table 11.15, in that the second column (**mild biter**) will not be included. As a result of omitting the latter column, the row sum for **mice, gerbils, and hamsters** is reduced to 144 and the row sum for **guinea pigs** is reduced to 69. The total number of observations for the comparison is 213.

Table 11.15 Complex Comparison for Example 11.5

	Not a biter	Mild biter	Flagrant biter		Row sums
Mice, Gerbils, and Hamsters	100	56	44		**200**
Guinea pigs	19	11	50		**80**
Column sums	**119**	**67**	**94**	**Total observations**	**280**

The null and alternative hypotheses for simple and complex comparisons are identical to those that are employed for evaluating the original $r \times c$ table, except for the fact that they are stated in reference to the specific cells involved in the comparison.

Sources are not in agreement with respect to what protocol is most appropriate to employ in conducting simple and/or complex comparisons following the computation of a chi-square value for an $r \times c$ contingency table. Among those sources that describe comparison procedures for $r \times c$ contingency tables are Keppel and Saufley (1980), Keppel *et al.* (1992), Marascuilo and McSweeney (1977), and Siegel and Castellan (1988). Keppel *et al.* (1992) note that Wickens (1989) states that, as a general rule, the different protocols which have been developed for conducting comparisons yield comparable results. In discussing the general issue of partitioning contingency tables (i.e., breaking down the table for the purposes of conducting comparisons), most sources emphasize that whenever feasible, a researcher should plan a limited number of simple and/or complex comparisons prior to the data collection phase of a study, and that any comparisons one conducts should be meaningful at either a theoretical and/or practical level.[19] In the discussion of comparisons in Section VI of the **chi-square goodness-of-fit test**, it is noted that when a limited number of comparisons are planned beforehand, most sources take the position that a researcher is not obliged to control the overall Type I error rate. However, when comparisons are not planned, there is general agreement that in order to avoid inflating the Type I error rate, the latter value should be adjusted. One way of adjusting the Type I error rate is to divide the maximum overall Type I error rate one is willing to tolerate by the total number of comparisons to be conducted. The resulting probability value can then be employed as the alpha level for each comparison that is conducted. To illustrate, if one intends to conduct three comparisons and does not want the overall Type I error rate for all of the comparisons to be greater than $\alpha = .05$, the alpha level to employ for each comparison is α/Number of comparisons $= .05/3 = .0167$. There are those who would argue the latter adjustment is too severe, since it substantially reduces the power associated with each comparison. In the final analysis, the researcher must be the one who decides what per comparison alpha level strikes an equitable balance in terms of the likelihood of committing a Type I error and the power associated with a comparison. Obviously, if a researcher employs a severely reduced alpha value, it may become all but impossible to detect actual differences that exist between the underlying populations. Before continuing this section, the reader may find it useful to review the discussion of comparisons for the **chi-square goodness-of-fit test**. A thorough overview of the issues involved in conducting comparisons can be found in Section VI of the **single-factor between-subjects analysis of variance (Test 16)**. In the remainder of this section, two comparison procedures for an $r \times c$ contingency table (based on Keppel *et al.* (1992) and Keppel and Saufley (1980)) will be described. The procedures to be described can be employed for both planned and unplanned comparisons.

Method 1 The first procedure to be described (which is derived from Keppel *et al.* (1992)) employs for both simple and complex comparisons the same protocol to evaluate a comparison contingency table as the protocol that is employed when the complete $r \times c$ table is evaluated. In the case of Example 11.5, Equation 11.2 is employed to evaluate the simple comparison summarized by the 2×3 contingency table in Table 11.14. The analysis assumes there is a total of 140 observations. Equation 11.1 is employed to compute the expected frequency of each cell — i.e., the sum of the observations in the row in which the cell appears is multiplied by the sum of the observations in the column in which the cell appears, with the resulting product divided by the total number of observations in the 2×3 contingency table. As an example, the expected frequency of Cell_{11} (**mice/not a biter**) is computed as follows: $E_{11} = [(O_{1.})(O_{.1})]/n = [(60)(39)]/140 = 16.71$. In employing Equation 11.1, the value 60 represents the sum for Row 1 (which represents the total number of **mice**), the value 39 represents the sum for Column 1 (which represents the total number of **mice** and **guinea pigs** categorized as **not a biter**), and 140 represents the total number of observations in the table (i.e., **mice** and **guinea pigs**).

The sums for the two rows/species involved in the simple comparison under discussion are identical to the row sums for those species in the original 4×3 contingency table. The column sums, however, only represent the sums of the columns for the two species involved in the comparison, and are thus different from the column sums for all four species that are computed when the original 4×3 contingency table is evaluated. Table 11.16 summarizes the computation of the value $\chi^2 = 7.39$ (with Equation 11.2) for the simple comparison summarized in Table 11.14.

Table 11.16 Chi-Square Summary Table for Simple Comparison in Table 11.14 Employing Method 1

Cell	O_{ij}	E_{ij}	$(O_{ij} - E_{ij})$	$(O_{ij} - E_{ij})^2$	$\frac{(O_{ij} - E_{ij})^2}{E_{ij}}$
Mice/Not a biter	20	16.71	3.29	10.82	.65
Mice/Mild biter	16	11.57	4.43	19.62	1.70
Mice/Flagrant biter	24	31.71	−7.71	59.44	1.87
Guinea pigs/Not a biter	19	22.29	−3.29	10.82	.49
Guinea pigs/Mild biter	11	15.43	−4.43	19.62	1.27
Guinea pigs/Flagrant biter	50	42.29	7.71	59.44	1.41
Column sums	140	140.00	0		$\chi^2 = 7.39$

Substituting the values $r = 2$ and $c = 3$ in Equation 11.3, the number of degrees of freedom for the comparison are $df = (2 - 1)(3 - 1) = 2$. Employing **Table A4**, the tabled critical .05 and .01 chi-square values for $df = 2$ are $\chi^2_{.05} = 5.99$ and $\chi^2_{.01} = 9.21$. Since the computed value $\chi^2 = 7.39$ is greater than $\chi^2_{.05} = 5.99$, the null hypothesis can be rejected at the .05 level. It cannot, however, be rejected at the .01 level, since $\chi^2 = 7.39$ is less than $\chi^2_{.01} = 9.21$. By virtue of rejecting the null hypothesis the researcher can conclude that **mice** and **guinea pigs**, the two species involved in the comparison, are not homogeneous with respect to biting behavior. Inspection of Table 11.14 reveals that the significant difference can primarily be attributed to the fact that there are a disproportionately large number of **flagrant biters** among **guinea pigs**.

Method 2 An alternative and somewhat more cumbersome method for conducting both simple and complex comparisons described by Bresnahan and Shapiro (1966) and Castellan (1965) is presented in Keppel and Saufley (1980). Although most statisticians would consider the method to be described in this section as preferable to **Method 1**, in most cases the latter method will result in similar conclusions with regard to the hypothesis under study. The method to be described in this section is identical to **Method 1**, except for the fact that in computing the chi-square value, for each cell in the comparison contingency table the value $(O_{ij} - E_{ij})^2$ is divided by the expected frequency computed for the cell when the original $r \times c$ contingency table is evaluated (instead of the expected frequency for the cell based on the row and column sums in the comparison contingency table). Specifically, for each of the cells in the comparison contingency table, the following values are computed: a) The expected frequency for each cell is computed as it is for **Method 1** (i.e., using the row and column sums for the comparison contingency table and employing the number of observations in the comparison contingency table to represent the value of n); b) The values $(O_{ij} - E_{ij})$ and $(O_{ij} - E_{ij})^2$ are computed for each cell just as they are in **Method 1**; and c) In computing the values in the last column of the chi-square summary table, instead of dividing $(O_{ij} - E_{ij})^2$ by E_{ij}, it is divided by the expected frequency of the cell when all $r \times c$ cells in the original table are employed to compute the expected cell frequency. In the analysis to be described, the latter expected frequency for each cell is represented by the notation E'_{ij}. The steps noted above are illustrated in Table 11.17, which summarizes the application of **Method 2** for computing the chi-square statistic for the simple comparison presented in Table 11.14. Note that the value of E'_{ij} for each cell is the same as the value of the expected frequency for that cell in Table 11.13.

Table 11.17 Chi-Square Summary Table for Simple Comparison in Table 11.14 Employing Method 2

Cell	O_{ij}	E_{ij}	$(O_{ij} - E_{ij})$	$(O_{ij} - E_{ij})^2$	E'_{ij}	$\frac{(O_{ij} - E_{ij})^2}{E'_{ij}}$
Mice/Not a biter	20	16.71	3.29	10.82	25.50	.42
Mice/Mild biter	16	11.57	4.43	19.62	14.36	1.37
Mice/Flagrant biter	24	31.71	-7.71	59.44	20.14	2.95
Guinea pigs/Not a biter	19	22.29	-3.29	10.82	34.00	.32
Guinea pigs/Mild biter	11	15.43	-4.43	19.62	19.14	1.03
Guinea pigs/Flagrant biter	50	42.29	7.71	59.44	26.86	2.03
Column sums	140	140.00	0			$\chi^2 = 8.12$

As is the case when **Method 1** is employed, the null hypothesis can be rejected at the .05 level but not at the .01 level, since for $df = 2$, the computed value $\chi^2 = 8.12$ is greater than $\chi^2_{.05} = 5.99$, but less than $\chi^2_{.01} = 9.21$. Although the computed value $\chi^2 = 8.12$ is slightly larger than the value $\chi^2 = 7.39$ computed with **Method 1**, the difference between the two chi-square values is minimal. As noted previously, the two methods will generally yield approximately the same value.

Method 1 and **Method 2** will now be employed to evaluate the complex comparison summarized in Table 11.15. The results of these analyses are summarized in Tables 11.18 and 11.19.

Although in the case of the complex comparison, **Method 1** and **Method 2** both yield the value $\chi^2 = 42.03$, the two methods will not always yield the same value. Substituting the values $r = 2$ and $c = 3$ in Equation 11.3, the number of degrees of freedom for the comparison are $df = (2 - 1)(3 - 1) = 2$. Employing **Table A4**, the tabled critical .05 and .01 chi-square values for $df = 2$ are $\chi^2_{.05} = 5.99$ and $\chi^2_{.01} = 9.21$. Since the computed value

Table 11.18 Chi-Square Summary Table for Complex Comparison in Table 11.15 Employing Method 1

Cell	O_{ij}	E_{ij}	$(O_{ij} - E_{ij})$	$(O_{ij} - E_{ij})^2$	$\frac{(O_{ij} - E_{ij})^2}{E_{ij}}$
Mice, Gerbils, Hamsters/Not a biter	100	85.00	15.00	225.00	2.65
Mice, Gerbils, Hamsters/Mild biter	56	47.86	8.14	66.26	1.38
Mice, Gerbils, Hamsters/Flagrant biter	44	67.14	−23.14	535.46	7.98
Guinea pigs/Not a biter	19	34.00	−15.00	225.00	6.62
Guinea pigs/Mild biter	11	19.14	−8.14	66.26	3.46
Guinea pigs/Flagrant biter	50	26.86	23.14	535.46	19.34
Column sums	280	280.00	0		$\chi^2 = 42.03$

Table 11.19 Chi-Square Summary Table for Complex Comparison in Table 11.15 Employing Method 2

Cell	O_{ij}	E_{ij}	$(O_{ij} - E_{ij})$	$(O_{ij} - E_{ij})^2$	E'_{ij}	$\frac{(O_{ij} - E_{ij})^2}{E'_{ij}}$
Mice, Gerbils, Hamsters/Not a biter	100	85.00	15.00	225.00	85.00	2.65
Mice, Gerbils, Hamsters/Mild biter	56	47.86	8.14	66.26	47.86	1.38
Mice, Gerbils, Hamsters/Flagrant biter	44	67.14	−23.14	535.46	67.14	7.98
Guinea pigs/Not a biter	19	34.00	−15.00	225.00	34.00	6.62
Guinea pigs/Mild biter	11	19.14	−8.14	66.26	19.14	3.46
Guinea pigs/Flagrant biter	50	26.86	23.14	535.46	26.86	19.94
Column sums	280	280.00	0			$\chi^2 = 42.03$

$\chi^2 = 42.03$ is greater than both of the aforementioned critical values, the null hypothesis can be rejected at both the .05 and .01 levels. By virtue of rejecting the null hypothesis, the researcher can conclude that a combined population of **mice, gerbils, and hamsters** is not homogeneous with a population of **guinea pigs** with respect to biting behavior. Inspection of Table 11.15 reveals the latter can primarily be attributed to the discrepancy between the number of **guinea pigs** in the **flagrant biters** category and the number of **mice, gerbils, and hamsters** in the **not a biter** category.

As previously noted, if the two comparisons summarized in Tables 11.14 and 11.15 are not planned prior to collecting the data, most sources would argue that each comparison should be evaluated at a lower alpha level. If one does not want the likelihood of committing at least one Type I error in the set of two comparisons to be greater than .05, one can adjust the alpha level as follows: Adjusted α level = α/Number of comparisons = .05/2 = .025. Thus, in evaluating each of the comparisons, the tabled critical chi-square value at the .025 level is employed instead of the tabled critical .05 value $\chi^2_{.05} = 5.99$ (although as noted earlier some sources might consider the latter adjustment to be too severe). In **Table A4**, for the appropriate degrees of freedom, the tabled critical .025 value corresponds to value listed under $\chi^2_{.975}$. In the case of $df = 2$, $\chi^2_{.975} = 7.38$. Note that for the simple comparison discussed in this section, the obtained chi-square value $\chi^2 = 7.39$ obtained with **Method 1** barely achieves significance if the latter critical value is used, whereas without the adjustment the latter result is significant at the .05 level by a comfortable margin. This example should serve to illustrate that by employing a lower alpha level, in addition to decreasing the likelihood of committing a Type I error, one also (by virtue of reducing the power of the test) decreases the likelihood of rejecting a false null hypothesis.

The rationale for presenting two comparison methods in this section is to demonstrate that although there is not a consensus among different sources with respect to what procedure should be employed for conducting comparisons, as a general rule, if a significant effect is present it will be identified regardless of which method one employs. Although **Method 1** is the simpler of the two methods described in this section, as noted earlier, most sources would probably take the position that **Method 1** is more subject to challenge on statistical grounds. Although in some instances there will be differences with respect to the precise probability values associated with the two methods (as well as the probabilities associated with other available methods), in most cases such differences will be trivial, and will thus be of little importance in terms of their practical and/or theoretical implications. In the final analysis, the per comparison alpha level one elects to employ (rather than the use of a different comparison procedure) is the most likely reason why two or more researchers may reach dramatically different conclusions with respect to a specific comparison. In such an instance, a replication study is the best available option for clarifying the status of the null hypothesis. Alternative procedures for conducting comparisons with contingency tables are described in Marascuilo and McSweeney (1977), Marascuilo and Serlin (1988), and Siegel and Castellan (1988).

9. The analysis of standardized residuals As noted in Section VI of the **chi-square goodness-of-fit test**, an alternative procedure for conducting comparisons (developed by Haberman (1973) and cited in Siegel and Castellan (1988)) involves the computation of **standardized residuals**. By computing standardized residuals, one is able to determine which cells are the major contributors to a significant chi-square value. The computation of residuals can be useful in reinforcing or clarifying information derived from the comparison procedures described in the previous section, as well as providing additional information on the data contained in an $r \times c$ table. Through use of Equation 11.13, a standardized residual (R_{ij}) can be computed for each cell in an $r \times c$ contingency table.

$$R_{ij} = \frac{(O_{ij} - E_{ij})}{\sqrt{E_{ij}}} \qquad \textbf{(Equation 11.13)}$$

A value computed for a residual with Equation 11.13, which is interpreted as a normally distributed variable, is evaluated with **Table A1**. Any residual with an absolute value that is equal to or greater than the tabled critical two-tailed .05 value $z_{.05} = 1.96$ is significant at the .05 level. Any residual that is equal to or greater than the tabled critical two-tailed .01 value $z_{.01} = 2.58$ is significant at the .01 level. Any cell in a contingency table which has a significant residual makes a significant contribution to the obtained chi-square value. For any cell that has a significant residual, one can conclude that the observed frequency of the cell differs significantly from its expected frequency. The sign of the standardized residual indicates whether the observed frequency of the cell is above (+) or below (–) the expected frequency. The sum of the squared residuals for all $r \times c$ cells will equal the obtained value of chi-square. The analysis of the residuals for Example 11.5 is summarized in Table 11.20.

Inspection of Table 11.20 indicates that the residual computed for the following cells is significant: a) **guinea pigs/flagrant biters** — Since the absolute value $R = 4.46$ computed for the residual is greater than $z_{.05} = 1.96$ and $z_{.01} = 2.58$, the residual is significant at both the .05 and .01 levels. The positive value of the residual indicates that the observed frequency of the cell is significantly above its expected frequency. The value of the residual for this cell is consistent with the fact that in the comparisons discussed in the previous section, the cell **guinea pigs/flagrant biters** appears to play a critical role in the significant effect that is detected; b) **hamsters/flagrant biter** — Since the absolute value $R = 3.68$ computed for the residual is greater than $z_{.05} = 1.96$ and $z_{.01} = 2.58$, the residual is significant at both the .05 and .01 levels. The negative value of the residual indicates that the observed frequency of the cell is significantly below its expected frequency. The value of the residual for this cell is consistent with the fact that in the comparisons discussed in the previous section the cell **hamsters/flagrant biter** appears to play a critical role in the significant effect that is detected; and c) **guinea pigs/not a biter** — Since the absolute value $R = 2.57$ computed for the residual is greater than $z_{.05} = 1.96$, but is less (albeit barely) than $z_{.01} = 2.58$, the residual is significant at the .05 level. The negative value of the residual indicates that the observed frequency of the cell is significantly below its expected frequency.

Table 11.20 Analysis of Residuals for Example 11.5

Cell	O_{ij}	E_{ij}	$(O_{ij} - E_{ij})$	$R_{ij} = \dfrac{(O_{ij} - E_{ij})}{\sqrt{E_{ij}}}$	$R_{ij} = \left[\dfrac{(O_{ij} - E_{ij})}{\sqrt{E_{ij}}}\right]^2$
Mice/Not a biter	20	25.50	-5.50	-1.09	1.19
Mice/Mild biter	16	14.36	1.64	.43	.19
Mice/Flagrant biter	24	20.14	3.86	.86	.74
Gerbils/Not a biter	30	21.25	8.75	1.90	3.61
Gerbils/Mild biter	10	11.96	-1.96	-.57	.33
Gerbils/Flagrant biter	10	16.79	-6.79	-1.66	2.76
Hamsters/Not a biter	50	38.25	11.75	1.90	3.61
Hamsters/Mild biter	30	21.54	8.46	1.82	3.32
Hamsters/Flagrant biter	10	30.21	-20.21	-3.68**	13.52
Guinea pigs/Not a biter	19	34.00	-15.00	-2.57*	6.62
Guinea pigs/Mild biter	11	19.14	-8.14	-1.86	3.46
Guinea pigs/Flagrant biter	50	26.86	23.14	4.46**	19.93
Column sums	280	280.00	0		$\Sigma R_{ij}^2 \chi^2 = 59.28$

*Significant at the .05 level.
**Significant at the .01 level.

Four other cells in Table 11.20 approach being significant at the .05 level (**gerbils/not a biter, hamsters/not a biter, guinea pigs/mild biter, hamsters/mild biter**). The absolute value of the residual for all of the aforementioned cells is close to the tabled critical value $z_{.05} = 1.96$. Note that in Table 11.20, the sum of the squared residuals is essentially equal to the chi-square value computed in Table 11.13 for Example 11.5 (the minimal discrepancy is the result of rounding off error).

10. Sources for the computation of the power of the chi-square test for $r \times c$ tables Although the protocol for computing the power of the **chi-square test for $r \times c$ tables** will not be described in this book, a comprehensive discussion of the subject as well as the necessary tables can be found in Cohen (1988). Cohen (1988) also describes the protocol for computing the power of the **z test for two independent proportions.**

11. Measures of association for $r \times c$ contingency tables Prior to reading this section the reader may find it useful to review the discussion of the **treatment effect** (which is a measure of the degree of association between two or more variables) in Section VI of the ***t* test for two independent samples.** The computed chi-square value for an $r \times c$ contingency table does not provide a researcher with precise information regarding the size of the **treatment effect** present in the data. The reason why a chi-square value computed for a contingency table is not an accurate index of the degree of association between the two variables is because the chi-square value is a function of both the total sample size and the proportion of observations in each of the $r \times c$ cells of the table. The degree of association, on the other hand, is independent of the total sample size, and is only a function of the cell proportions. In point of fact, the magnitude of the computed chi-square statistic is directly proportional to the total sample size employed in a study.

To illustrate the latter point, consider Table 11.21 which presents a different set of data for Example 11.1. In actuality, the effect size for the data presented in Table 11.21 is the same as the effect size in Table 11.2, and the only difference between the two tables is that in Table 11.21 the total sample size and the number of observations in each of the cells are one-half of the corresponding values listed in Table 11.2. If Equation 11.2 is applied to the data presented in Table 11.21, the value $\chi^2 = 9.1$ is computed. Note that the latter value is one-half of $\chi^2 = 18.18$, which is the value computed for Table 11.2. Thus, even though the same proportion of observations appears in the corresponding cells of the two tables, the chi-square value computed for each table is directly proportional to the total sample size.

Table 11.21 Summary of Data for Example 11.1 With Reduced Sample Size

	Helped the confederate	Did not help the confederate		Row sums
Noise	15	35		50
No noise	30	20		50
Column sums	45	55	Total observations	100

A number of different measures of association/correlation that are independent of sample size can be employed as indices of the magnitude of a treatment effect for an $r \times c$ contingency table. In this section the following measures of association will be described: a) **Test 11f: The contingency coefficient**; b) **Test 11g: The phi coefficient**; c) **Test 11h: Cramér's phi coefficient**; d) **Test 11i: Yule's *Q***; and e) **Test 11j: The odds ratio.**

As a general rule (although there are some exceptions), the value computed for a measure of association/correlation will usually fall within a range of values between 0 and +1 or between −1 and +1. Whereas a value of 0 indicates no relationship between the two variables, an absolute value of 1 indicates a maximum relationship between the variables. Consequently, the closer the absolute value of a measure of association is to 1, the stronger the relationship between the variables. As noted above, some measures of association can assume values in the range between −1 and +1. In such cases, the absolute value of the measure indicates the strength of the relationship, and the sign of the measure indicates the direction of the relationship — i.e., the pattern of observations among the cells of a contingency table. Before continuing this section the reader may find it useful to read Section I of the **Pearson product-moment correlation coefficient (Test 22)**, which provides a general discussion of correlation/association.

Measures of association for $r \times c$ contingency tables can be evaluated with respect to statistical significance. In the case of a number of the measures of association to be discussed in this section, if the computed the chi-square value for the contingency table is statistically significant at a given level of significance, the measure of association computed for the contingency table will be significant at the same level of significance. In most instances, the null hypothesis and nondirectional alternative hypothesis evaluated with reference to a measure of association are as follows:[20]

Null hypothesis The correlation/degree of association between the two variables in the underlying population is zero.

Alternative hypothesis The correlation/degree of association between the two variables in the underlying population is some value other than zero.

The same guidelines discussed earlier in reference to employing a directional alternative hypothesis for the **chi-square test for $r \times c$ tables** can also be applied if one wants to state the alternative hypothesis for a measure of association directionally.

Since the different measures of association that can be computed for a contingency table do not employ the same criteria in measuring the strength of the relationship between the two variables, if two or more measures are applied to the same set of data they may not yield comparable coefficients of association. Although in the material to follow, there will be some discussion of factors that can be taken into account in considering which of the various measures of association to employ, in most cases one measure is not necessarily superior to another in terms of providing information about a contingency table. Indeed, Conover (1980) notes that the choice of which measure to employ is based more on prevailing tradition than it is on statistical considerations.

Test 11f: The contingency coefficient (C) The **contingency coefficient** (also known as **Pearson's contingency coefficient**) is a measure of association that can be computed for an $r \times c$ contingency table of any size. The value of the **contingency coefficient**, which will be represented with the notation C, is computed with Equation 11.14.

$$C = \sqrt{\frac{\chi^2}{\chi^2 + n}} \qquad \textbf{(Equation 11.14)}$$

Where: χ^2 is the computed chi-square value for the contingency table
n is the total number of observations in the contingency table

Since n can never equal zero, the value of C can never equal 1. Consequently, the range of values C may assume is $0 \le C < +1$. One limitation of the contingency coefficient is that its upper limit (i.e., the highest value it can attain) is a function of the number of rows and columns in the $r \times c$ contingency table. The upper limit of C (represented by C_{max}) can be determined with Equation 11.15.

$$C_{max} = \sqrt{\frac{k - 1}{k}} \qquad \textbf{(Equation 11.15)}$$

Where: k represents the smaller of the two values of r and c in the contingency table

Employing Equation 11.15 for a 2×2 contingency table, we can determine that the maximum value for such a table is $C_{max} = \sqrt{(2 - 1)/2} = .71$.

Employing Equation 11.14, the value $C = .29$ is computed as the value of the **contingency coefficient** for Examples 11.1/11.2.

$$C = \sqrt{\frac{18.18}{18.18 + 200}} = .29$$

Note that the value $C = .29$ is also obtained for Table 11.21 (which as noted earlier has the same effect size but half the sample size as Tables 11.2/11.3): $C = \sqrt{(9.1)/(9.1 + 100)} = .29$.

As noted in the introductory remarks on measures of association, the computed value $C = .29$ will be statistically significant at both the .05 and .01 levels, since the computed value $\chi^2 = 18.18$ for Tables 11.2/11.3 (as well as the computed value $\chi^2 = 9.1$ for Table 11.21) is significant at the aforementioned levels of significance. Thus, one can conclude that in the underlying population the **contingency coefficient** between the two variables is some value other than zero.

The reader should take note of the fact that if the value of n is reduced, but the effect size present in Tables 11.1/11.2/11.21 is maintained, at some point the chi-square value will not achieve significance. Although in such a case Equation 11.14 will yield the value $C = .29$, the latter value will not be statistically significant. This is the case, since in order for the value of C to be significant, the computed value of chi-square must be significant.

In the discussion of the **phi coefficient** (which is the next measure of association to be discussed), it is noted that Cohen (1988) interprets an effect size of .29 as just short of representing a **medium effect size.** However, since the maximum absolute value the **phi coefficient** can attain is 1, and in the case of a 2 × 2 contingency table the maximum value C can attain is .71, one can challenge applying Cohen's (1988) criteria to C.

Ott *et al.* (1992) note that among the disadvantages associated with the **contingency coefficient** is that it will always be less than 1, even when the two variables are totally dependent on one another. In addition, **contingency coefficients** that have been computed for two or more tables can only be compared with one another if all of the tables have the same number of rows and columns. One suggestion that is made to counteract the latter problems is to employ Equation 11.16 to compute an adjusted value for the **contingency coefficient.**

$$C_{\text{adj}} = \frac{C}{C_{\text{max}}} \qquad \textbf{(Equation 11.16)}$$

By employing Equation 11.16, if a perfect association between the variables exists, the value of C_{adj} will equal 1. If Equation 11.16 is employed for Examples 11.1/11.2 and Table 11.21, the value $C_{\text{adj}} = .29/.71 = .41$ is computed. It should be pointed out that although the use of C_{adj} allows for better comparison between tables of unequal size, it still does not allow one to compare such tables with complete accuracy.

Test 11g: The phi coefficient The **phi coefficient** (represented by the notation ϕ, which is the lower case Greek letter **phi**) is a measure of association that can only be employed with a 2 × 2 contingency table. The **phi coefficient** (which is discussed further in Section VII of the **Pearson product-moment correlation coefficient**) is, in actuality, a special case of the latter correlation coefficient. Specifically, it is a **Pearson product-moment correlation coefficient** that is computed if the values 0 and 1 are employed to represent the levels of two dichotomous variables. Although the value of **phi** can fall within the range −1 to +1, the latter statement must be qualified, since the lower and upper limits of **phi** are dependent on certain conditions. Carroll (1961) and Guilford (1965) note that in order for **phi** to equal −1 or +1, the following two conditions must be met with respect to the 2 × 2 contingency table described by the model in Table 11.6: $(a + b) = (c + d)$ and $(a + c) = (b + d)$.

Employing the notation in Table 11.6 for a 2 × 2 contingency table, the value of **phi** is computed with Equation 11.17.

$$\phi = \frac{ad - bc}{\sqrt{(a + b)(c + d)(a + c)(b + d)}} \qquad \textbf{(Equation 11.17)}$$

Since $\phi^2 = \chi^2/n$, many sources compute the value of **phi** through use of Equation 11.18 (which can be derived from the equation $\phi^2 = \chi^2/n$). Note that the result of Equation 11.18 will always be the absolute value of **phi** derived with Equation 11.17.[21]

$$\phi = \sqrt{\frac{\chi^2}{n}} \qquad \textbf{(Equation 11.18)}$$

Employing Equation 11.17, the value $\phi = .30$ is computed below for Examples 11.1/11.2 (as well as for the data in Table 11.21).

$$\phi = \frac{(30)(40) - (70)(60)}{\sqrt{(30 + 70)(60 + 40)(30 + 60)(70 + 40)}} = -.30$$

If Equation 11.18 is employed the absolute value of **phi** is computed to be $\phi = \sqrt{18.18/200} = .30$. As noted in the introductory remarks on measures of association, the computed absolute value $\phi = .30$ will be statistically significant at both the .05 and .01 levels, since the computed value $\chi^2 = 18.18$ is significant at the aforementioned levels of significance. Thus, one can conclude that in the underlying population the **phi** coefficient between the two variables is some value other than zero.[22] In addition to the nondirectional alternative hypothesis being supported, the directional alternative hypothesis that is consistent with the data is also supported. This is the case since the computed value of chi-square is significant at the .10 and .02 levels (i.e., the obtained value $\chi^2 = 18.18$ is greater than $\chi^2_{.90} = 2.71$ and $\chi^2_{.98} = 5.41$).

It turns out the absolute value $\phi = .30$ computed for Examples 11.1/11.2 is almost identical to $C = .29$, the value of the **contingency coefficient** computed for same set of data. As a general rule, although the two values will not be identical, the absolute value of **phi** and the **contingency coefficient** will be close to one another.

Cohen (1988) has suggested the following absolute values for **phi** be employed for identifying the magnitude of the effect size present in a set of data: **small effect**: $.10 \le \phi < .30$; **medium effect**: $.30 \le \phi < .50$; **large effect**: $\phi \ge .50$. Although admittedly, these values are somewhat arbitrary, they are often cited as guidelines in various texts. Employing Cohen's (1988) guidelines, the observed effect size for Examples 11.1/11.2 is at the lower limit of a medium effect — i.e., there is a moderate relationship between the two variables under study.

The use of the **phi coefficient** is most commonly endorsed in the case of 2×2 contingency tables involving two variables that are dichotomous in nature. Because of this, among others, Guilford (1965) and Fleiss (1981) note that one of the most useful applications of **phi** is for determining the intercorrelation between the responses of subjects on two dichotomous test items.[23] Siegel and Castellan (1988) note that when the two variables being correlated are ordered variables, computation of the **phi** coefficient sacrifices information, and because of this under such conditions it is preferable to employ alternative measures of association that are designed for ordered tables (such as **Goodman and Kruskal's gamma** which is discussed later in the book). For a more thorough overview of the **phi coefficient**, the reader should consult Guilford (1965) and Fleiss (1981), who, among other things, discuss various sources who argue in favor of employing measures other than **phi** with 2×2 tables.

Test 11h: Cramér's phi coefficient Developed by Cramér (1946), **Cramér's phi coefficient** (which will be represented by the notation ϕ_C) is an extension of the **phi coefficient** to contingency tables that are larger than 2×2 tables. **Cramér's phi coefficient**, which can assume a value between 0 and +1, is computed with Equation 11.19.

$$\phi_C = \sqrt{\frac{\chi^2}{n(k - 1)}} \qquad \textbf{(Equation 11.19)}$$

Where: k represents the smaller of the two values of r and c in the contingency table

The derivation of **Cramér's phi coefficient** is based on the fact that the maximum value chi-square can attain for a set of data is $\chi^2_{max} = n(k - 1)$. Thus, the value ϕ_C is the square root of a proportion that represents the computed value of chi-square divided by the maximum possible chi-square value for a set of data. When the computed chi-square value for a set of data equals χ^2_{max}, the value $\phi_C = 1$ will be obtained, which indicates maximum dependency between the two variables.

Since for 2 × 2 tables **Cramér's phi** and the **phi coefficient** are equivalent (i.e., when $k = 2$, $\phi_C = \phi = \sqrt{\chi^2/n}$), they will both yield the absolute value of .30 for Examples 11.1/11.2. **Cramér's phi** is computed below for the 4 × 3 contingency table presented in Example 11.5. Since $c = 3$ is less than $r = 4$, $k = c = 3$.

$$\phi_C = \sqrt{\frac{59.16}{(280)(3 - 1)}} = .325$$

The computed value $\phi_C = .325$ is significant at both the .05 and .01 levels, since the computed value $\chi^2 = 59.16$ is significant at the aforementioned levels of significance. Thus, one can conclude that in the underlying population, the value of **Cramér's phi coefficient** is some value other than zero. Employing Cohen's (1988) criteria, **Cramér's phi coefficient** indicates the presence of a medium effect size for Example 11.5.

Daniel (1990) notes that when a contingency table is square (i.e., $r = c$) and $\phi_C = 1$, there is a perfect correlation between the two variables (which will be reflected by the fact that all of the observations will be in the cells of one of the diagonals of the table). When $r \neq c$ and $\phi_C = 1$, however, the two variables will not be perfectly correlated in the same manner as is the case with a square contingency table. Conover (1980) notes that although under all conditions the possible range of values for ϕ_C will be between 0 and +1, its interpretation will depend on the values of r and c. He states that there is a tendency for the value of χ^2 (and consequently the value of ϕ_C) to increase as the values of r and c become larger. For this reason Conover (1980) suggests that ϕ_C may not be completely accurate for comparing the degree of association in different size tables. Daniel (1990) notes that the value of ϕ_C is equal to the square of the tie adjusted value of **Kendall's tau (Test 24)** (discussed later in the book) computed for the same set of data. As is the case with the **phi coefficient**, when ordered categories are employed for both variables, sources do not recommend employing **Cramér's phi** since it sacrifices information. In designs involving variables with ordered categories, it is preferable to employ an alternative measure of association such as **Goodman and Kruskal's gamma.**

Test 11i: Yule's *Q* **Yule's *Q*** (Yule (1900)) is a measure of association for a 2 × 2 contingency table. It is presented in this section to illustrate that if two or more measures of association are computed for the same set of data they may not yield comparable values. Since it employs less information than the **phi coefficient, Yule's *Q*** is less frequently recommended than **phi** as a measure of association for 2 × 2 tables. **Yule's *Q*** is actually a special case of **Goodman and Kruskal's gamma** (although unlike **gamma,** which is only used with ordered contingency tables, **Yule's *Q*** can be used for both ordered and unordered tables).

Employing the notation in Table 11.6 for a 2 × 2 contingency table, Equation 11.20 is employed to compute the value of **Yule's *Q*.**

$$Q = \frac{ad - bc}{ad + bc} \qquad \textbf{(Equation 11.20)}$$

Sources that discuss **Yule's *Q*** generally note that it tends to inflate the degree of association in the underlying population. Ott *et al.* (1992) note that an additional limitation is that if the absolute value of Q equals 1, it does not necessarily mean there is a perfect association between the two variables. In point of fact, if the observed frequency of any of the four cells in a 2 × 2 contingency table equals 0, the value of **Yule's Q** will equal either −1 or +1. For this reason, the meaning of Q can be quite misleading in cases where the frequency of one of the cells is equal to 0. Because of the latter, it is not recommended that **Yule's Q** be employed when there is a small number of frequencies in any of the four cells of a 2 × 2 contingency table.

Employing Equation 11.20, the value $Q = -.56$ is computed for Examples 11.1/11.2 (as well as for the data in Table 11.21).

$$Q = \frac{(30)(40) - (70)(60)}{(30)(40) + (70)(60)} = -.56$$

Note that although both the values of Q and **phi** suggest the presence of a negative association between the two variables, the absolute value $Q = .56$ is almost twice the value $\phi = 30$ computed previously for the same set of data. When Cohen's (1988) criteria are applied to the computed value of **phi** it indicates the presence of a medium effect size, yet if the same criteria are applied to Q it suggests the presence of a large effect (although Cohen would probably not believe that the values he lists for **phi** are appropriate to use with **Yule's Q**).

Ott *et al.* (1992) note that the significance of **Yule's Q** can be evaluated with Equation 11.21.[24]

$$z = \frac{Q}{\sqrt{\frac{1}{4}(1 - Q^2)^2 \left[\frac{1}{a} + \frac{1}{b} + \frac{1}{c} + \frac{1}{d}\right]}} \qquad \textbf{(Equation 11.21)}$$

Employing Equation 11.21 with the data for Examples 11.1/11.2, the value $z = -5.46$ is computed.

$$z = \frac{-.56}{\sqrt{\left[\frac{1}{4}\right][1 - (.56)^2]^2 \left[\frac{1}{30} + \frac{1}{70} + \frac{1}{60} + \frac{1}{40}\right]}} = -5.46$$

The obtained value $z = -5.46$ is evaluated with **Table A1**. Since the obtained absolute value $z = 5.46$ is greater than the tabled critical two-tailed .05 and .01 values $z_{.05} = 1.96$ and $z_{.01} = 2.58$, the nondirectional alternative hypothesis is supported at both the .05 and .01 levels. Since $z = 5.46$ is greater than the tabled critical one-tailed .05 and .01 values $z_{.05} = 1.65$ and $z_{.01} = 2.33$, the directional alternative hypothesis that is consistent with the data is also supported at both the .05 and .01 levels.

Test 11j: The odds ratio As is the case with **Yule's Q** (but unlike C, ϕ, and ϕ_C), the **odds ratio** (which is attributed to Cornfield (1951)) is a measure of association employed with contingency tables that is not a function of chi-square. Although the **odds ratio**, which will be represented by the notation o, can be applied to any size contingency table, it is easiest to interpret in the case of 2 × 2 tables. The **odds ratio** expresses the degree of association between the two variables in a different numerical format than all of the previously discussed measures of association. In some respects it is a more straightforward way of interpreting the results of a contingency table than are the correlational measures of association discussed previously. In the case of a 2 × 2 table, the **odds ratio** quantifies the odds of how much more likely it is that an observation in one of the categories of the row variable will be in one of the categories of the column variable than will an observation in the second category of the row variable. Thus, in the case of Example 11.1, the **odds ratio** will allow us to say how much more likely it is that a subject in the **no noise** condition will **help the confederate** than a subject in the **noise** condition.

Employing the notation in Table 11.6 for a 2 × 2 contingency table, Equation 11.22 is employed to compute the **odds ratio**.

$$o = \frac{p_b \, p_c}{p_a \, p_d}$$ **(Equation 11.22)**

From Table 11.2, we can determine that $p_a = a/n = 30/200 = .15$, $p_b = b/n = 70/200 = .35$, $p_c = c/n = 60/200 = .3$, and $p_d = d/n = 40/200 = .2$. Employing Equation 11.22 with the data for Example 11.1, the value $o = 3.5$ is computed.

$$o = \frac{(.35)(.3)}{(.15)(.2)} = 3.5$$

An alternate way to compute the value $o = 3.5$ is as follows. Divide the number of subjects in Cell a **(noise/helped the confederate)** by the number of subjects in Cell b **(noise/did not help the confederate)**. This value represents the odds that a subject in the **noise** condition will **help the confederate**. Thus: $a/b = 30/70 = .43$. We next divide the number of subjects in Cell c **(no noise/helped the confederate)** by the number of subjects in Cell d **(no noise/did not help the confederate)**. This value represents the odds that a subject in the **no noise** condition will **help the confederate**. Thus: $c/d = 60/40 = 1.5$.[25] The value of the odds ratio is computed by dividing the odds that someone in the **no noise** condition will **help the confederate** by the odds that someone in the **noise** condition will **help the confederate**. Thus: $o = 1.5/.43 = 3.49$.[26] (The same value for the **odds ratio** will be obtained for the data in Table 11.21.) The value $o = 3.5$ computed for the **odds ratio** indicates that someone in the **no noise** condition is 3.5 times more likely to **help the confederate** than someone in the **noise** condition.[27]

Although, as noted earlier, the **odds ratio** can be extended beyond 2 × 2 tables, it becomes more difficult to interpret with larger contingency tables. However, in instances where there are more than two rows but only two columns, its interpretation is still relatively straightforward. To illustrate, assume that in Example 11.1 we have three noise conditions instead of two — specifically, **loud noise, moderate noise,** and **no noise.** Table 11.22 depicts a hypothetical set of data that summarizes the results of such a study.

Table 11.22 Summary of Data for a 3 × 2 Contingency Table

	Helped the confederate	Did not help the confederate		Row sums
Loud noise	30	70		**100**
Moderate noise	50	50		**100**
No noise	80	20		**100**
Column sums	**160**	**140**	**Total observations**	**300**

Within any of the three noise conditions, we can determine the odds that someone in that condition will **help the confederate**. This is accomplished by dividing the number of subjects in a given condition who **helped the confederate** by the number of subjects in the condition who **did not help the confederate.** Thus, for the **loud noise** condition the odds that someone will **help the confederate** are 30/70 = .43. For the **moderate noise** condition the odds that someone will **help the confederate** are 50/50 = 1. For the **no noise** condition the odds that someone will **help the confederate** are 80/20 = 4. From these values we can compute the following three odds ratios: a) The relative likelihood that someone in the **no noise** condition will **help the confederate** versus someone in the **loud noise** condition: $o = 4/.43 = 9.3$; b) The relative likelihood that someone in the **no noise** condition will **help the confederate** versus someone in the **moderate noise** condition: $o = 4/1 = 4$; and c) The relative likelihood that someone in the **moderate noise** condition will **help the confederate** versus someone in the **loud noise** condition: $o = 1/.43 = 2.33$.

Thus, someone in the **no noise** condition is 9.3 times more likely to **help the confederate** than someone in the **loud noise** condition, and 4 times more likely to **help the confederate** than someone in the **moderate noise** condition. Someone in the **moderate noise** condition is 2.33 times more likely to **help the confederate** than someone in the **loud noise** condition.

VII. Additional Discussion of the Chi-Square Test for $r \times c$ Tables

1. Analysis of multidimensional contingency tables It is possible to have contingency tables which are comprised of more than two dimensions. In other words, subjects can be categorized on the basis of three or more variables. To illustrate this, the design in Example 11.2 can be modified so that subjects are categorized on three dimensions instead of two. Specifically the following three dimensions can be employed: **introvert** versus **extrovert**; **Democrat** versus **Republican**; **male** versus **female**.

Contingency tables that involve three or more dimensions are referred to as **multidimensional tables**. Analysis of such tables requires the use of procedures that are more mathematically complex than those that are employed for the **chi-square test for $r \times c$ tables**. A commonly employed methodology for evaluating multidimensional contingency tables is **log-linear analysis**. In point of fact, log-linear analysis can also be used to evaluate two-dimensional contingency tables. When the latter type of analysis is employed to evaluate $r \times c$ tables, it yields essentially the same results as those obtained with the **chi-square test for $r \times c$ tables**. Among the sources that discuss the analysis of multidimensional contingency tables are Marascuilo and McSweeney (1977) and Wickens (1989).

VIII. Additional Examples Illustrating the Chi-Square Test for $r \times c$ Tables

Examples 11.6–11.8 are additional examples that can be evaluated with the **chi-square test for $r \times c$ tables**.

Example 11.6 *A researcher conducts a study in order to evaluate the relative problem solving ability of males versus female adolescents. One hundred males and* 80 *females are randomly selected from a population of adolescents. Each subject is given a mechanical puzzle to solve. The dependent variable is whether or not a person is able to solve the puzzle. Sixty out of the* 100 *male subjects are able to solve the puzzle, while only* 30 *out of the* 80 *female subjects are able to solve the puzzle. Is there a significant difference between males and females with respect their ability to solve the puzzle?*

Table 11.23 summarizes the data for Example 11.6. Example 11.6 conforms to the requirements of the **chi-square test for homogeneity**. This is the case, since there are two independent samples/groups (**males** versus **females**) which are dichotomized with respect to the following two categories on the dimension of problem solving ability: **solved puzzle** versus **did not solve puzzle**. The grouping of subjects on the basis of gender represents a nonmanipulated independent variable, while the problem solving performance of subjects represents the dependent variable. Note that the number of people for each gender

Table 11.23 Summary of Data for Example 11.6

	Solved puzzle	Did not solve puzzle		Row sums
Males	60	40		**100**
Females	30	50		**80**
Column sums	**90**	**90**	**Total observations**	**180**

is predetermined by the experimenter. Also note that it is not necessary to have an equal number of observations in the categories of the row variable, which represents the independent variable. Since the independent variable is nonmanipulated, if the chi-square analysis is significant it will only allow the researcher to conclude that a significant association exists between gender and one's ability to solve the puzzle. The researcher cannot conclude that gender is the direct cause of any observed differences in problem solving ability between males and females. Employing Equation 11.2, the obtained chi-square value for Table 11.23 is $\chi^2 = 9$, which for $df = 1$ is greater than $\chi^2_{.05} = 3.84$ and $\chi^2_{.01} = 6.63$. Thus, the null hypothesis can be rejected at both the .05 and .01 levels. Inspection of Table 11.23 reveals that a larger proportion of **males** are able to solve the puzzle than **females**.

Example 11.7 *A pollster conducts a survey to evaluate whether Caucasians and African-Americans differ in their attitude toward gun control. Five hundred people are randomly selected from a telephone directory and called at 8 P.M. in the evening. An interview is conducted with each individual, at which time a person is categorized with respect to both race and whether one supports or opposes gun control.* Table 11.24 *summarize the results of the survey. Is there evidence of racial differences with respect to attitude toward gun control?*

This example conforms to the requirements of the **chi-square test of independence,** since a single sample is categorized on two dimensions. The two dimensions subjects are categorized with respect to are race, for which there are the two categories **Caucasian** versus **African-American,** and attitude toward gun control, for which there are the two categories **supports gun control** versus **opposes gun control.** Note that neither the number of **Caucasians** or **African-Americans** (i.e., the sums of the rows), or the people who **support gun control** or **oppose gun control** (i.e., the sums of the columns) are predetermined prior to the pollster conducting the survey. The pollster selects a single sample of 500 subjects and categorizes them on both dimensions after the data are collected. The obtained chi-square value for Table 11.24 is $\chi^2 = 59.91$, which for $df = 1$ is greater than $\chi^2_{.05} = 3.84$ and $\chi^2_{.01} = 6.63$. Thus, the null hypothesis can be rejected at both the .05 and .01 levels. Inspection of Table 11.24 reveals that a larger proportion of **Caucasians opposes gun control,** whereas a larger proportion of **African-Americans supports gun control.**

Table 11.24 Summary of Data for Example 11.7

	Supports gun control	Opposes gun control		Row sums
Caucasians	120	170		**290**
Afro-Americans	160	50		**210**
Column sums	**280**	**220**	**Total observations**	**500**

Example 11.8 *A researcher conducts a study on a college campus to examine the relationship between a student's class standing and the number of times a student visits a physician during the school year.* Table 11.25 *summarizes the responses of a random sample of* 280 *students employed in the study. Do the data indicate that the number of visits a student makes to a physician is independent of his or her class standing?*

Table 11.25 Summary of Data for Example 11.8

	0 visits	1-5 visits	More than 5 visits		Row sums
Freshman	20	16	24		**60**
Sophomore	30	10	10		**50**
Junior	50	30	10		**90**
Senior	19	11	50		**80**
Column sums	**119**	**67**	**94**	**Total observations**	**280**

This example conforms to the requirements of the **chi-square test of independence,** since a single sample is categorized on two dimensions. The two dimensions subjects are categorized with respect to are class standing, for which there are the four categories: **Freshman, Sophomore, Junior, Senior,** and the number of visits to a physician, for which there are the three categories: **0 visits, 1–5 visits, more than 5 visits.** Note that neither the sums of the rows or the columns is predetermined by the researcher. The researcher randomly selects 280 subjects and categorizes each subject on both dimensions after the data are collected. Since the data for Example 11.8 are identical to that employed in Example 11.5, it yields the same result. The null hypothesis can be rejected, since the obtained value $\chi^2 = 59.16$ is significant at both the .05 and .01 levels. Thus, the researcher can conclude that a student's class standing and the number of visits one makes to a physician are not independent of one another (i.e., the two dimensions seem to be associated/correlated with one another). As is the case with Example 11.5, a more detailed analysis of the data can be conducted through use of the comparison procedures described in Section VI.

References

Beyer, W. H. (Ed.) (1968). **CRC handbook of tables for probability and statistics** (2nd ed.). Boca Raton, FL: CRC Press.

Bresnahan, J. L. and Shapiro, M. M. (1966). A general equation and technique for the exact partitioning of chi-square contingency tables. **Psychological Bulletin**, 66, 252–262.

Carroll, J. B. (1961). The nature of data, or how to choose a correlation coefficient. **Psychometrika**, 26, 347–372.

Castellan, N. J., Jr. (1965). On the partitioning of contingency tables. **Psychological Bulletin**, 64, 330–338.

Cochran, W. G. (1952). The chi-square goodness-of-fit test. **Annals of Mathematical Statistics**, 23, 315–345.

Cochran, W. G. (1954). Some methods for strengthening the common chi-square tests. **Biometrics**, 10, 417–451.

Cohen, J. (1988). **Statistical power analysis for the behavioral sciences** (2nd ed.). Hillsdale, NJ: Erlbaum.

Conover, W. J. (1971). **Practical nonparametric statistics** (2nd ed.). New York: John Wiley and Sons.

Cornfield, J. (1951). A method of estimating comparative rates from clinical data. Applications to cancer of the lung, breast and cervix. **Journal of the National Cancer Institute**, 11, 1229–1275.

Cramér, H. (1946). **Mathematical models of statistics.** Princeton, NJ: Princeton University Press.

Daniel, W. W. (1990). **Applied nonparametric statistics** (2nd ed.). Boston: PWS–Kent Publishing Company.

Fisher, R. A. (1934). **Statistical methods for research workers** (5th ed.). Edinburgh: Oliver and Boyd.

Fisher, R. A. (1935). The logic of inductive inference. **Journal of the Royal Statistical Society, Series A**, 98, 39–54.

Fleiss, J. L. (1981). **Statistical methods for rates and proportions** (2nd ed.). New York: John Wiley and Sons.

Guilford, J. P. (1965). **Fundamental statistics in psychology and education** (4th ed.). New York: McGraw–Hill Book Company.

Haberman, S. J. (1973). The analysis of residuals in cross-classified tables. **Biometrics**, 29, 205–220.

Howell, D. C. (1992). **Statistical methods for psychology** (3rd ed.). Boston: PWS–Kent Publishing Company.

Irwin, J. O. (1935). Tests of significance for differences between percentages based on small numbers. **Metron**, 12, 83–94.

Keppel, G and Saufley, W. H., Jr. (1980). **Introduction to design and analysis: A student's handbook.** San Francisco: W. H. Freeman and Company.

Keppel, G, Saufley, W. H., Jr., and Tokunaga, H. (1992). **Introduction to design and analysis: A student's handbook** (2nd ed.). New York: W. H. Freeman and Company.

Marascuilo, L. A. and McSweeney, M. (1977). **Nonparametric and distribution-free methods for the social sciences.** Belmont, CA: Brooks/Cole Publishing Company.

Marascuilo, L. A. and Serlin, R. C. (1988). **Statistical methods for the social and behavioral sciences.** New York: W.H. Freeman and Company.

Ott, R. L, Larson, R., Rexroat, C., and Mendenhall, W. (1992). **Statistics: A tool for the social sciences** (5th ed.). Boston: PWS–Kent Publishing Company.

Owen, D. B. (1962). **Handbook of statistical tables.** Reading, MA: Addison–Wesley.

Siegel, S. and Castellan, N. J., Jr. (1988). **Nonparametric statistics for the behavioral sciences** (2nd ed.) New York: McGraw–Hill Book Company.

Wickens, T. (1989). **Multiway contingency table analysis for the social sciences.** Hillsdale, NJ: Erlbaum.

Yates, F. (1934). Contingency tables involving small numbers and the chi-square test. **Journal of the Royal Statistical Society**, 1, 217–235.

Yule, G. (1900). On the association of the attributes in statistics: With illustrations from the material of the childhood society, &c. **Philosophical Transactions of the Royal Society**, Series A, 194, 257–319.

Endnotes

1. A general discussion of the chi-square distribution can be found under the **single-sample chi-square test for a population variance.**

2. The use of the chi-square approximation (which employs a continuous probability distribution to approximate a discrete probability distribution) is based on the fact that the computation of exact probabilities requires an excessive amount of calculations.

3. In the case of both the **chi-square test of independence** and the **chi-square test of homogeneity**, the same result will be obtained regardless of which of the variables is designated as the row variable versus the column variable.

4. It is just coincidental that the number of **introverts** equals the number of **extroverts**.

5. In the context of the discussion of the **chi-square test of independence**, the proportion of observations in Cell_{11} refers to the number of observations in Cell_{11} divided by the total number of observations in the 2×2 table. In the discussion of the hypothesis for the **chi-square test for homogeneity**, the proportion of observations in Cell_{11} refers to the number of observations in Cell_{11} divided by the total number of observations in Row 1 (i.e., the row in which Cell_{11} appears).

6. Equation 11.2 is an extension of Equation 5.2 (which is employed to compute the value of chi-square for the **chi-square goodness-of-fit test**) to a two-dimensional table. In Equation 11.2, the use of the two summation expressions $\sum_{i=1}^{r}\sum_{j=1}^{c}$ indicates that the operations summarized in Table 11.4 are applied to all of the cells in the $r \times c$ table. In contrast, the single summation expression $\sum_{i=1}^{k}$ in Equation 5.2 indicates that the operations summarized in Table 5.2 are applied to all k cells in a one-dimensional table.

7. The same chi-square value will be obtained if the row and column variables are reversed — i.e., the helping variable represents the row variable and the noise variable represents the column variable.

8. Correlational studies are discussed in detail under the **Pearson product-moment correlation coefficient**.

9. The value $\chi^2_{.98} = 5.43$ is determined by interpolation. It can also be derived by squaring the tabled critical one-tailed .01 value $z_{.01} = 2.33$, since the square of the latter value is equivalent to the chi-square value at the 98th percentile. The use of z values in reference to a 2×2 contingency table is discussed later in this section under the ***z* test for two independent proportions**.

10. Within the framework of this discussion, the value $\chi^2_{.05} = 3.84$ represents the tabled chi-square value at the 95th percentile (which demarcates the extreme 5% in the right tail of the chi-square distribution). Thus, using the format employed for the one-tailed .05 and .01 values, the notation identifying the two-tailed .05 value can be written as $\chi^2_{.95} = 3.84$. In the same respect, the value $\chi^2_{.01} = 6.63$ represents the tabled chi-square value at the 99th percentile (which demarcates the extreme 1% in the right tail of the chi-square distribution). Thus, using the format employed for the one-tailed .05 and .01 values, the notation identifying the two-tailed .01 value can be written as $\chi^2_{.99} = 6.63$.

11. The null and alternative hypotheses presented for the **Fisher exact test** in this section are equivalent to the alternative form for stating the null and alternative hypotheses for the **chi-square test of homogeneity** presented in Section III (if the hypotheses in Section III are applied to a 2×2 contingency table).

12. Sourcebooks documenting statistical tables (e.g., Owen (1962) and Beyer (1968)), as well as many books that specialize in nonparametric statistics (e.g., Daniel (1990); Marascuilo and McSweeney (1977); Siegel and Castellan (1988)) contain tables of the hypergeometric distribution that can be employed with 2×2 contingency tables. Such

tables eliminate the requirement of employing Equations 11.7/11.8 to compute the value of P_T.

13. The value $(1 - p)$, which is often represented by the notation q, can also be computed as follows: $q = (1 - p) = (b + d)/(n_1 + n_2) = (b + d)/n$. The value q is a pooled estimate of the proportion of observations in Column 2 in the underlying population.

14. Due to rounding off error there may be a minimal discrepancy between the square of a z value and the corresponding chi-square value.

15. The logic for employing Equation 11.11 in lieu of Equation 11.9 is the same as that discussed in reference to the ***t* test for two independent samples**, when in the case of the latter test the null hypothesis stipulates a value other than zero for the difference between the population means (and Equation 8.5 is employed to compute the test statistic in lieu of Equations 8.1/8.2/8.3).

16. The denominator of Equation 11.11 is employed to compute $s_{p_1 - p_2}$ instead of the denominator of Equation 11.9, since in computing a confidence interval it cannot be assumed that $\pi_1 = \pi_2$ (which is assumed in Equation 11.9, and serves as the basis for computing a pooled p value in the latter equation).

17. The **median test for independent samples** can also be employed within the framework of the model for the **chi-square test of independence**. To illustrate this, assume that Example 11.4 is modified so that the researcher randomly selects a sample of 200 subjects, and does not specify beforehand that the sample is comprised of 100 females and 100 males. If it just happens by chance that the sample is comprised of 100 females and 100 males, one can state that neither the sum of the rows or the sum of the columns is predetermined by the researcher. As noted in Section I, when neither of the marginal sums is predetermined, the design conforms to the model for the **chi-square test of independence.**

18. The word **column** can be interchanged with the word **row** in the definition of a complex comparison.

19. Another consideration that should be mentioned with respect to conducting comparisons is that two or more comparisons for a set of data can be **orthogonal** (which means they are independent of one another) or comparisons can overlap with respect to the information they provide. As a general rule, when a limited number of comparisons is planned, it is most efficient to conduct orthogonal comparisons. The general subject of orthogonal comparisons is discussed in greater detail in Section VI of the **single-factor between-subjects analysis of variance.**

20. The null and alternative hypotheses stated below do not apply to the **odds ratio.**

21. Some sources note that the **phi coefficient** can only assume a range of values between 0 and +1. In these sources, the term $|ad - bc|$ is employed in the numerator of Equation 11.17. By employing the absolute value of the term in the numerator of Equation 11.17, the value of **phi** will always be a positive number. Under the latter condition the following will be true: $\phi = \sqrt{\chi^2/n}$.

22. In the case of small sample sizes, the results of the **Fisher exact test** are employed as the criterion for determining whether the computed value of **phi** is significant.

23. In such a case, the data are summarized in the form of a 2×2 contingency table documenting the proportion of subjects who answer in each of the categories of two dichotomous variables (e.g., **True** versus **False** for both variables/test items).

24. The reason why the result of the **chi-square test for $r \times c$ tables** is not employed to assess the significance of Q is because Q is not a function of chi-square. It should be noted that since Q is a special case of **Goodman and Kruskal's gamma**, it can be argued that Equation 26.2 (the significance test for **gamma**) can be employed to assess whether or not Q is significant. However, Ott *et al.* (1992) state that a different procedure is employed for evaluating the significance of Q versus **gamma**. Equation 26.2 will not yield the same result as that obtained with Equation 11.21 when it is applied to a 2×2 table. If the **gamma** statistic is computed for Examples 11.1/11.2 it yields the absolute value $\gamma = .56$ (γ is the lower case Greek letter **gamma**), which is identical to the value of Q computed for the same set of data. (The absolute value is employed since the contingency table is not ordered, and thus, depending upon how the cells are arranged, a value of either $+.56$ or $-.56$ can be derived for **gamma**.) However, when Equation 26.2 is employed to assess the significance of $\gamma = .56$, it yields the absolute value $z = 3.51$, which although significant at both the .05 and .01 levels is lower than the absolute value $z = 5.46$ obtained with Equation 11.21.

25. The odds computed in this example are more commonly stated in the form .43:1 and 1.5:1 which means .43 to 1 and 1.5 to 1. The latter odds are equivalent to 3:2.

26. The minimal discrepancy between $o = 3.49$ and the value $o = 3.50$ computed with Equation 11.22 is due to rounding off error.

27. One can also divide .43 by 1.5 and obtain the value .29, which is equivalent to employing the following equation to compute the **odds ratio**: $o = (p_a\ p_d)/(p_b\ p_c) = [(.15)(.2)]/[(.35)(.3)] = .29$. Note that in the latter equation, the numerator and denominator of Equation 11.22 are reversed. This result indicates that someone in the **noise** condition is .29 times as likely to **help the confederate** than someone in the **no noise** condition.

    It should be noted that when odds are employed, they are more commonly stated within the context of the odds of an event representing the likelihood against a specific event occurring. Using this definition, the odds that a person in the **noise** condition **did not help the confederate** are 2.33:1 (since $70/30 = 2.33$). The odds that a person in the **no noise** condition **did not help the confederate** are .667:1 or 2:3, since $40/60 = .667$. These values yield the same **odds ratio**, since $2.33/.667 = 3.49$.

Inferential Statistical Tests Employed with Two Dependent Samples (and Related Measures of Association/Correlation)

Test 12

The *t* Test for Two Dependent Samples
(Parametric Test Employed with Interval/Ratio Data)

I. Hypothesis Evaluated with Test and Relevant Background Information

Hypothesis evaluated with test Do two dependent samples represent two populations with different mean values?

Relevant background information on test The ***t* test for two dependent samples**, which is employed in a hypothesis testing situation involving two dependent samples, is one of a number of inferential statistical tests that are based on the *t* distribution (which is described in detail under the **single-sample *t* test (Test 2)**). Throughout the discussion of the ***t* test for two dependent samples**, the term **experimental conditions** will also be employed to represent the **dependent samples** employed in a study. In a **dependent samples design**, each subject either serves in all of the k (where $k \geq 2$) experimental conditions, or else is matched with a subject in each of the other $(k - 1)$ experimental conditions.[1] In designs that are evaluated with the ***t* test for two dependent samples**, the value of k will always equal 2.

In conducting the ***t* test for two dependent samples**, the means of the two experimental conditions (represented by the notations $\bar{X}_1$ and $\bar{X}_2$) are employed to estimate the values of the means of the populations (μ_1 and μ_2) the conditions represent. If the result of the ***t* test for two dependent samples** is significant, it indicates there is a significant difference between the means of the two experimental conditions, and as a result of the latter the researcher can conclude there is a high likelihood that the conditions represent populations with different mean values. It should be noted that the ***t* test for two dependent samples** is the appropriate test to employ for contrasting the means of two dependent samples when the values of the underlying population variances are unknown. In instances where the latter two values are known, the appropriate test to employ is the ***z* test for two dependent samples (Test 12c)**, which is described in Section VI.

The ***t* test for two dependent samples** is employed with interval/ratio data, and is based on the following assumptions: a) The sample of n subjects has been randomly selected from the population it represents; b) The distribution of data in the underlying populations each of the experimental conditions represents is normal; and c) The third assumption, which is referred to as the **homogeneity of variance** assumption, states that the variance of the underlying population represented by Condition 1 is equal to the variance of the underlying population represented by Condition 2 (i.e., $\sigma_1^2 = \sigma_2^2$). It should be noted that the ***t* test for two dependent samples** is more sensitive to violation of the homogeneity of variance assumption (which is discussed in Section VI) than is the ***t* test for two independent samples (Test 8)**. If any of the aforementioned assumptions of the ***t* test for two dependent samples** is saliently violated, the reliability of the test statistic may be compromised.

When a study employs a dependent samples design, the following two issues related to experimental control must be taken into account: a) In a dependent samples design in which each subject serves in both experimental conditions, it is essential that the experimenter controls for **order effects** (also known as **sequencing** or **carryover effects**). An **order effect** is where an obtained difference on the dependent variable is a direct result of the order of presentation of the experimental conditions, rather than being due to the independent variable manipulated by the experimenter. Order effects can be controlled through the use of a technique called **counterbalancing**, which is discussed in Section VII; and b) When a dependent samples design employs matched subjects, within each pair of matched subjects each of the two subjects must be randomly assigned to one of the two experimental conditions. Nonrandom assignment of subjects to the experimental conditions can compromise the **internal validity** of a study.[2] A more thorough discussion of **matching** can be found in Section VII.

II. Example

Example 12.1 *A psychologist conducts a study in order to determine whether people exhibit more emotionality when they are exposed to sexually explicit words than when they are exposed to neutral words. Each of ten subjects is shown a list of* 16 *randomly arranged words, which are projected onto a screen one at a time for a period of five seconds. Eight of the words on the list are sexually explicit and eight of the words are neutral. As each word is projected on the screen, a subject is instructed to say the word softly to him or herself. As a subject does this, sensors attached to the palms of the subject's hands record galvanic skin response (GSR), which is used by the psychologist as a measure of emotionality. The psychologist computes two scores for each subject, one score for each of the experimental conditions:* **Condition 1:** *GSR/Explicit — The average GSR score for the eight sexually explicit words;* **Condition 2:** *GSR/Neutral — The average GSR score for the eight neutral words. The GSR/Explicit and the GSR/Neutral scores of the ten subjects follow. (The higher the score the higher the level of emotionality.)* **Subject 1** (9, 8); **Subject 2** (2, 2); **Subject 3** (1, 3); **Subject 4** (4, 2); **Subject 5** (6, 3); **Subject 6** (4, 0); **Subject 7** (7, 4); **Subject 8** (8, 5); **Subject 9** (5, 4); **Subject 10** (1, 0).[3] *Do subjects exhibit differences in emotionality with respect to the two categories of words?*

III. Null versus Alternative Hypotheses

Null hypothesis $H_0\colon \mu_1 = \mu_2$

(The mean of the population Condition 1 represents equals the mean of the population Condition 2 represents.)

Alternative hypothesis $H_1\colon \mu_1 \neq \mu_2$

(The mean of the population Condition 1 represents does not equal the mean of the population Condition 2 represents. This is a **nondirectional alternative hypothesis** and it is evaluated with a **two-tailed test**. In order to be supported, the absolute value of t must be equal to or greater than the tabled critical two-tailed t value at the prespecified level of significance. Thus, either a significant positive t value or a significant negative t value will provide support for this alternative hypothesis.)

or

$$H_1\colon \mu_1 > \mu_2$$

(The mean of the population Condition 1 represents is greater than the mean of the population Condition 2 represents. This is a **directional alternative hypothesis** and it is evaluated with a **one-tailed test**. It will only be supported if the sign of t is positive, and the absolute value of t is equal to or greater than the tabled critical one-tailed t value at the prespecified level of significance.)

or

$$H_1: \mu_1 < \mu_2$$

(The mean of the population Condition 1 represents is less than the mean of the population Condition 2 represents. This is a **directional alternative hypothesis** and it is evaluated with **a one-tailed test.** It will only be supported if the sign of t is negative, and the absolute value of t is equal to or greater than the tabled critical one-tailed t value at the prespecified level of significance.)

Note: Only one of the above noted alternative hypotheses is employed. If the alternative hypothesis the researcher selects is supported, the null hypothesis (H_0) is rejected.[4]

IV. Test Computations

Two methods can be employed to compute the test statistic for the ***t*** **test for two dependent samples.** The method to be described in this section, which is referred to as the **direct-difference method**, allows for the quickest computation of the t statistic. In Section VI, a computationally equivalent but more tedious method for computing t is described.

The data for Example 12.1 and the preliminary computations for the **direct-difference method** are summarized in Table 12.1. Note that there are n = 10 subjects, and that there is a total of $2n = (2)(10) = 20$ scores, since each subject has two scores. The two scores of the 10 subjects are listed in the columns of Table 12.1 labelled X_1 and X_2. The score of a subject in the column labelled X_1 is the average GSR score of the subject for the eight sexually explicit words (Condition 1), while the score of a subject in the column labelled X_2 is the average GSR score of the subject for the eight neutral words (Condition 2). Column 4 of Table 12.1 lists a difference score for each subject (designated by the notation D), which is computed by subtracting a subject's X_2 score from his X_1 score (i.e., $D = X_1 - X_2$). Column 5 of the table lists a D^2 score for each subject, which is obtained by squaring a subject's D score.

Table 12.1 Data for Example 12.1

	Condition 1	Condition 2		
Subject	X_1	X_2	D	D^2
1	9	8	1	1
2	2	2	0	0
3	1	3	-2	4
4	4	2	2	4
5	6	3	3	9
6	4	0	4	16
7	7	4	3	9
8	8	5	3	9
9	5	4	1	1
10	1	0	1	1
	$\Sigma X_1 = 47$	$\Sigma X_2 = 31$	$\Sigma D- = -2$ $\Sigma D+ = 18$ $\Sigma D = 16$	$\Sigma D^2 = 54$
	$\bar{X}_1 = \frac{47}{10} = 4.7$	$\bar{X}_2 = \frac{31}{10} = 3.1$		

In Column 4 of Table 12.1, the summary value $\Sigma D = 16$ is obtained by adding $\Sigma D+ = 18$, the sum of the positive difference scores (i.e., all those difference scores with a + sign), and $\Sigma D- = -2$, the sum of the negative difference scores (i.e., all those difference scores with a – sign). The reader should take note of the fact that whenever $\Sigma X_1 > \Sigma X_2$ (and consequently $\bar{X}_1 > \bar{X}_2$), the value ΣD will be a positive number, whereas whenever $\Sigma X_1 < \Sigma X_2$ (and consequently $\bar{X}_1 < \bar{X}_2$), the value ΣD will be a negative number.

Equation 12.1 is the direct-difference equation for computing the test statistic for the ***t* test for two dependent samples.**

$$t = \frac{\bar{D}}{s_{\bar{D}}} \qquad \text{(Equation 12.1)}$$

Where: $\bar{D}$ represents the mean of the difference scores
$s_{\bar{D}}$ represents the **standard error of the mean difference**

The **mean of the difference scores** is computed with Equation 12.2.

$$\bar{D} = \frac{\Sigma D}{n} \qquad \text{(Equation 12.2)}$$

Employing Equation 12.2, the value $\bar{D} = 1.6$ is computed.

$$\bar{D} = \frac{16}{10} = 1.6$$

Equation 12.3 is employed to compute $\tilde{s}_D$, which represents the estimated population standard deviation of the difference scores.[5]

$$\tilde{s}_D = \sqrt{\frac{\Sigma D^2 - \frac{(\Sigma D)^2}{n}}{n - 1}} \qquad \text{(Equation 12.3)}$$

Employing Equation 12.3, the value $\tilde{s}_D = 1.78$ is computed.

$$\tilde{s}_D = \sqrt{\frac{54 - \frac{(16)^2}{10}}{10 - 1}} = 1.78$$

Equation 12.4 is employed to compute the value $s_{\bar{D}}$. The value $s_{\bar{D}}$ represents the **standard error of the mean difference**, which is an estimated population standard deviation of mean difference scores.[6]

$$s_{\bar{D}} = \frac{\tilde{s}_D}{\sqrt{n}} \qquad \text{(Equation 12.4)}$$

Employing Equation 12.4, the value $s_{\bar{D}} = .56$ is computed.

$$s_{\bar{D}} = \frac{1.78}{\sqrt{10}} = .56$$

Substituting $\bar{D} = 16$ and $s_{\bar{D}} = .56$ in Equation 12.1, the value $t = 2.86$ is computed.[7]

$$t = \frac{1.6}{.56} = 2.86$$

The reader should take note of the fact that the values $\tilde{s}_D$ and $s_{\bar{D}}$, both of which are estimates of either a population standard deviation or the standard deviation of a sampling distribution, can never be a negative number. If a negative value is obtained for either of the aforementioned values, it indicates a computational error has been made.

V. Interpretation of the Test Results

The obtained value $t = 2.86$ is evaluated with **Table A2 (Table of Student's *t* Distribution)** in the **Appendix**. The degrees of freedom for the ***t* test for two dependent samples** are computed with Equation 12.5.

$$df = n - 1 \qquad \textbf{(Equation 12.5)}$$

Employing Equation 12.5, the value $df = 10 - 1 = 9$ is computed. The tabled critical two-tailed and one-tailed .05 and .01 t values for $df = 9$ are summarized in Table 12.2. (For a review of the protocol for employing **Table A2**, the reader should review Section V of the **single-sample *t* test**.)

Table 12.2 Tabled Critical .05 and .01 *t* Values for $df = 9$

	$t_{.05}$	$t_{.01}$
Two-tailed values	2.26	3.25
One-tailed values	1.83	2.82

The following guidelines are employed in evaluating the null hypothesis for the ***t* test for two dependent samples**.

a) If the nondirectional alternative hypothesis H_1: $\mu_1 \neq \mu_2$ is employed, the null hypothesis can be rejected if the obtained absolute value of t is equal to or greater than the tabled critical two-tailed value at the prespecified level of significance.

b) If the directional alternative hypothesis H_1: $\mu_1 > \mu_2$ is employed, the null hypothesis can be rejected if the sign of t is positive, and the value of t is equal to or greater than the tabled critical one-tailed value at the prespecified level of significance.

c) If the directional alternative hypothesis H_1: $\mu_1 < \mu_2$ is employed, the null hypothesis can be rejected if the sign of t is negative, and the absolute value of t is equal to or greater than the tabled critical one-tailed value at the prespecified level of significance.

Employing the above guidelines, the following conclusions can be reached.

The nondirectional alternative hypothesis H_1: $\mu_1 \neq \mu_2$ is supported at the .05 level, since the computed value $t = 2.86$ is greater than the tabled critical two-tailed value $t_{.05} = 2.26$. The latter alternative hypothesis, however, is not supported at the .01 level, since $t = 2.86$ is less than the tabled critical two-tailed value $t_{.01} = 3.25$.

The directional alternative hypothesis H_1: $\mu_1 > \mu_2$ is supported at both the .05 and .01 levels, since the obtained value $t = 2.86$ is a positive number that is greater than the tabled critical one-tailed values $t_{.05} = 1.83$ and $t_{.01} = 2.82$. Note that when the directional alternative hypothesis H_1: $\mu_1 > \mu_2$ is supported, it is required that $\bar{X}_1 > \bar{X}_2$.

The directional alternative hypothesis H_1: $\mu_1 < \mu_2$ is not supported, since the obtained value $t = 2.86$ is a positive number. In order for the directional alternative hypothesis H_1: $\mu_1 < \mu_2$ to be supported, the computed value of t must be a negative number (as well as the fact that the absolute value of t must be equal to or greater than the tabled critical one-tailed value at the prespecified level of significance). In order for the data to be consistent with the directional alternative hypothesis H_1: $\mu_1 < \mu_2$, it is required that $\bar{X}_1 < \bar{X}_2$.

A summary of the analysis of Example 12.1 with the ***t* test for two dependent samples** follows: It can be concluded that the average GSR (emotionality) score for the sexually explicit words is significantly higher than the average GSR score for the neutral words. This result can be summarized as follows (if $\alpha = .05$ is employed): $t(9) = 2.86$, $p < .05$.

VI. Additional Analytical Procedures for the *t* Test for Two Dependent Samples and/or Related Tests

1. Alternative equation for the *t* test for two dependent samples Equation 12.6 is an alternative equation that can be employed to compute the test statistic for the ***t* test for two dependent samples.**[8]

$$t = \frac{\bar{X}_1 - \bar{X}_2}{\sqrt{s_{\bar{X}_1}^2 + s_{\bar{X}_2}^2 - 2(r_{X_1X_2})(s_{\bar{X}_1})(s_{\bar{X}_2})}} \qquad \textbf{(Equation 12.6)}$$

The computation of t with Equation 12.6 requires more computations than does Equation 12.1 (the direct difference method equation). Equation 12.6, unlike Equation 12.1, requires that the estimated population variance be computed for each of the samples (in Section VI it is noted that the latter values are required in order to evaluate the homogeneity of variance assumption of the ***t* test for two dependent samples**). Since a total understanding of Equation 12.6 requires an understanding of the concept of correlation, the reader may find it useful to read Section I of the **Pearson product-moment correlation coefficient (Test 22)** prior to continuing this section.

Except for the last term in the denominator of Equation 12.6 (i.e., $2(r_{X_1X_2})(s_{\bar{X}_1})(s_{\bar{X}_2})$), the latter equation is identical to Equation 8.2 (the equation for the ***t* test for two independent samples** when $n_1 = n_2$). The value $r_{X_1X_2}$ represents the coefficient of correlation between the two scores of subjects (or matched pairs of subjects) on the dependent variable. It is expected that as a result of using the same subjects in both conditions (or by employing matched subjects), a positive correlation will exist between pairs of scores (i.e., scores that are in the same row of Table 12.1). The closer the value of $r_{X_1X_2}$ is to $+1$, the stronger the association between the scores of subjects on the dependent variable. As the value of $r_{X_1X_2}$ approaches $+1$, the value of the denominator of Equation 12.6 decreases, which will result in an increase in the absolute value computed for t. Note that if $r_{X_1X_2} = 0$, the denominator of Equation 12.6 becomes identical to the denominator of Equation 8.2. Thus, if the scores of the n subjects under the two experimental conditions are not correlated with one another, the equation for the ***t* test for two dependent samples** (as represented by Equation 12.6) reduces to Equation 8.2.

The intent of the above discussion is to illustrate that one advantage of employing a design that can be evaluated with the ***t* test for two dependent samples,** as opposed to a design that is evaluated with the ***t* test for two independent samples,** is that if there is a positive correlation between pairs of scores, the former test will provide a more powerful test of an alternative hypothesis than will the latter test. The greater power associated with the ***t* test for two dependent samples** is a direct result of the lower value that will be computed for the denominator of Equation 12.6 when contrasted with the denominator that will be computed for Equation 8.2 for the same set of data. In the case of both equations, the denominator is an estimated measure of variability in a sampling distribution. By employing pairs of scores that are positively correlated with one another, the estimated variability in the sampling distribution will be less than will be the case if the scores are not correlated with one another.[9]

The computation of t with Equation 12.6 will now be illustrated. Note that in order to compute t with the latter equation, the following values are required: $\bar{X}_1$, $\bar{X}_2$, $s_{\bar{X}_1}$, $s_{\bar{X}_2}$, $s^2_{\bar{X}_1}$, $s^2_{\bar{X}_2}$, $r_{X_1 X_2}$. In order to compute the estimated population variances and standard deviations, the values ΣX_1^2 and ΣX_2^2 are required. The latter values are computed in Table 12.3. Employing the summary information provided in Table 12.1 and the values $\Sigma X_1^2 = 293$ and $\Sigma X_2^2 = 147$, all of the above noted values, with the exception of $r_{X_1 X_2}$, are computed.

$$\bar{X}_1 = \frac{\Sigma X_1}{n} = \frac{47}{10} = 4.7 \qquad \bar{X}_2 = \frac{\Sigma X_2}{n} = \frac{31}{10} = 3.1$$

$$\tilde{s}_1 = \sqrt{\frac{\Sigma X_1^2 - \frac{(\Sigma X_1)^2}{n}}{n-1}} = \sqrt{\frac{293 - \frac{(47)^2}{10}}{10-1}} = 2.83 \qquad \tilde{s}_2 = \sqrt{\frac{\Sigma X_2^2 - \frac{(\Sigma X_2)^2}{n}}{n-1}} = \sqrt{\frac{147 - \frac{(31)^2}{10}}{10-1}} = 2.38$$

$$s_{\bar{X}_1} = \frac{\tilde{s}_1}{\sqrt{n}} = \frac{2.83}{\sqrt{10}} = .89 \qquad s_{\bar{X}_2} = \frac{\tilde{s}_2}{\sqrt{n}} = \frac{2.38}{\sqrt{10}} = .75$$

$$s^2_{\bar{X}_1} = (.89)^2 = .79 \qquad s^2_{\bar{X}_2} = (.75)^2 = .56$$

Equation 12.7 is employed to compute the value $r_{X_1 X_2}$.

$$r_{X_1 X_2} = \frac{\Sigma X_1 X_2 - \frac{(\Sigma X_1)(\Sigma X_2)}{n}}{\sqrt{\left[\Sigma X_1^2 - \frac{(\Sigma X_1)^2}{n}\right]\left[\Sigma X_2^2 - \frac{(\Sigma X_2)^2}{n}\right]}} \qquad \textbf{(Equation 12.7)}$$

Table 12.3 Computation of ΣX_1^2, ΣX_2^2, and $\Sigma X_1 X_2$ for Example 12.1

	Condition 1		Condition 2		
Subject	X_1	X_1^2	X_2	X_2^2	$X_1 X_2$
1	9	81	8	64	72
2	2	4	2	4	4
3	1	1	3	9	3
4	4	16	2	4	8
5	6	36	3	9	18
6	4	16	0	0	0
7	7	49	4	16	28
8	8	64	5	25	40
9	5	25	4	16	20
10	1	1	0	0	0
	$\Sigma X_1 = 47$	$\Sigma X_1^2 = 293$	$X_2 = 31$	$X_2^2 = 147$	$\Sigma X_1 X_2 = 193$

The only value in Equation 12.7 that is required to compute $r_{X_1 X_2}$ which has not been computed for Example 12.1, is the term $\Sigma X_1 X_2$ in the numerator. The latter value, which is computed in Table 12.3, is obtained as follows: Each subject's X_1 score is multiplied by the subject's X_2 score. The resulting score represents an $X_1 X_2$ score for the subject. The n $X_1 X_2$ scores are summed, and the resulting value represents the term $\Sigma X_1 X_2$ in Equation 12.7.

Employing Equation 12.7, the value $r_{X_1 X_2} = .78$ is computed.

$$r_{X_1 X_2} = \frac{193 - \dfrac{(47)(31)}{10}}{\sqrt{\left[293 - \dfrac{(47)^2}{10}\right]\left[147 - \dfrac{(31)^2}{10}\right]}} = .78$$

When the relevant values are substituted in Equation 12.6, the value $t = 2.86$ is computed (which is the same value computed for t with Equation 12.1).[10] Note that the value of the numerator of Equation 12.6 is $\bar{X}_1 - \bar{X}_2 = \bar{D} = 1.6$.

$$t = \frac{4.7 - 3.1}{\sqrt{.79 + .56 - 2(.78)(.89)(.75)}} = 2.86$$

In order to illustrate that the ***t* test for two dependent samples** provides a more powerful test of an alternative hypothesis than the ***t* test for two independent samples**, the data for Example 12.1 will be evaluated with Equation 8.2 (which is the equation for the latter test). In employing Equation 8.2, the positive correlation that exists between the scores of subjects under the two experimental conditions will not be taken into account. Use of Equation 8.2 with the data for Example 12.1 assumes that in lieu of having $n = 10$ subjects, there are instead two independent groups, each group being comprised of 10 subjects. Thus, $n_1 = 10$, and the 10 X_1 scores in Table 12.1 represent the scores of the 10 subjects in Group 1, and $n_2 = 10$, and the 10 X_2 scores in Table 12.1 represent the scores of the 10 subjects in Group 2. Since the values $\bar{X}_1$, $\bar{X}_2$, $s_{\bar{X}_1}^2$, and $s_{\bar{X}_2}^2$ have already been computed, they can be substituted in Equation 8.2. When the relevant values are substituted in Equation 8.2, the value $t = 1.37$ is computed.

$$t = \frac{\bar{X}_1 - \bar{X}_2}{\sqrt{s_{\bar{X}_1}^2 + s_{\bar{X}_2}^2}} = \frac{4.7 - 3.1}{\sqrt{.79 + .56}} = 1.37$$

Employing Equation 8.4, the degrees of freedom for Equation 8.2 are $df = 10 + 10 - 2 = 18$. In **Table A2**, the tabled critical two-tailed .05 and .01 values for $df = 18$ are $t_{.05} = 2.10$ and $t_{.01} = 2.88$, and the tabled critical one-tailed .05 and .01 values are $t_{.05} = 1.73$ and $t_{.01} = 2.55$. Since the obtained value $t = 1.37$ is less than all of the aforementioned critical values, the null hypothesis cannot be rejected, regardless of whether a nondirectional or directional alternative hypothesis is employed. Recollect that when the same set of data are evaluated with Equations 12.1/12.6, the nondirectional alternative hypothesis H_1: $\mu_1 \neq \mu_2$ is supported at the .05 level and the directional alternative hypothesis H_1: $\mu_1 > \mu_2$ is supported at both the .05 and .01 levels.

Note that by employing twice the degrees of freedom, the tabled critical values employed for Equation 8.2 will always be smaller than those employed for Equations 12.1/12.6. However, if there is a reasonably high positive correlation between the pairs of scores, the lower critical values associated with the ***t* test for two independent samples** will be offset by the fact that the t value computed with Equation 8.2 will be substantially smaller than the value computed with Equations 12.1/12.6.

When the ***t* test for two dependent samples** is employed to evaluate a dependent samples design, it is assumed that a positive correlation exists between the scores of subjects on the two variables. It is, however, theoretically possible (although unlikely) that scores on the two variables will be negatively correlated. If, in fact, the correlation between the X_1 and X_2 scores of subjects is negative, the value of the numerator of Equation 12.6 will actually be larger than will be the case if Equation 8.2 is employed to evaluate the same set of data (since in the denominator of Equation 12.6, if $r_{X_1X_2}$ is a negative number, the product $(r_{X_1X_2})(s_{\bar{X}_1})(s_{\bar{X}_2})$ will be added to instead of subtracted from $s^2_{\bar{X}_1} + s^2_{\bar{X}_2}$). What's more, if the correlation between the two scores is a very low positive value that is close to 0, the slight increment in the value of t computed with Equation 12.6 may be offset by the loss of degrees of freedom (and the consequent increase in the tabled critical t value), so as to render Equation 8.2 (the ***t* test for two independent samples**) a more powerful test of an alternative hypothesis than Equation 12.6 (the ***t* test for two dependent samples**).

In the unlikely event there is a substantial negative correlation between subjects' scores on the two variables employed in a dependent samples design, it is highly improbable that evaluation of the data with Equation 12.6 will yield a significant result. However, the presence of a significant negative correlation, in and of itself, can certainly be of statistical importance. To illustrate this, consider the following example. Assume that employing a dependent samples design, each of five subjects who serve in two experimental conditions obtains the following scores: **Subject 1** (1, 5); **Subject 2** (2, 4); **Subject 3** (3, 3); **Subject 4** (4, 2); **Subject 5** (5, 1). In this hypothetical example, it turns out that the correlation between the five pairs of scores is $r_{X_1X_2} = -1$, which is the strongest possible negative correlation. Since, however, the mean and median value for both of the experimental conditions is equal to 3, evaluation of the data with the ***t* test for two dependent samples**, as well as the more commonly employed nonparametric procedures employed for a design involving two dependent samples (such as the **Wilcoxon matched-pairs signed-ranks test (Test 13)** and the **binomial sign test for two dependent samples (Test 14)**), will lead one to conclude there is no difference between the scores of subjects under the two conditions. This is the case, since such tests base the comparison of conditions on an actual or implied measure of central tendency. If it is assumed that in the above example the sample data accurately reflect what is true with respect to the underlying populations, it appears that the higher a subject's score in Condition 1, the lower the subject's score in Condition 2, and vice versa. In such an case, it can be argued that if the coefficient of correlation is statistically significant, that in itself can indicate the presence of a significant treatment effect (albeit an unusual one), in that there is a significant association between the two sets of scores. The determination of whether a correlation coefficient is significant is discussed under the **Pearson product-moment correlation coefficient,** as well as in the discussion of a number of the other correlational procedures discussed in the book.[11]

2. The equation for the *t* test for two dependent samples when a value for a difference other than zero is stated in the null hypothesis If in stating the null hypothesis for the ***t* test for two dependent samples** a researcher stipulates that the difference between μ_1 and μ_2 is some value other than zero, Equation 12.8 is employed to evaluate the null hypothesis in lieu of Equation 12.6.[12]

$$t = \frac{(\bar{X}_1 - \bar{X}_2) - (\mu_1 - \mu_2)}{\sqrt{s^2_{\bar{X}_1} + s^2_{\bar{X}_2} - 2(r_{X_1X_2})(s_{\bar{X}_1})(s_{\bar{X}_2})}} \qquad \textbf{(Equation 12.8)}$$

When the null hypothesis is H_0: $\mu_1 = \mu_2$ (which as noted previously can also be written as H_0: $\mu_1 - \mu_2 = 0$), the value of $(\mu_1 - \mu_2)$ in Equation 12.8 reduces to zero, and thus what remains of the numerator in Equation 12.8 is $(\bar{X}_1 - \bar{X}_2)$ (which represents the numerator of

Equation 12.6). In evaluating the value of t computed with Equation 12.8, the same protocol is employed that is described for evaluating a t value for the **t test for two independent samples** when the difference stated in the null hypothesis for the latter test is some value other than zero (in which case Equation 8.5 is employed to compute t).

3. Test 12a: The t test for homogeneity of variance for two dependent samples (Test 12a): Evaluation of the homogeneity of variance assumption of the t test for two dependent samples Prior to reading this section, the reader should review the discussion of homogeneity of variance in Section VI of the **t test for two independent samples.** As is the case with an independent samples design, in a dependent samples design the homogeneity of variance assumption evaluates whether there is evidence to indicate that an inequality exists between the variances of the populations represented by the two experimental conditions. The null and alternative hypotheses employed in evaluating the homogeneity of variance assumption are as follows:

Null hypothesis $H_0\text{: } \sigma_1^2 = \sigma_2^2$

(The variance of the population Condition 1 represents equals the variance of the population Condition 2 represents.)

Alternative hypothesis $H_1\text{: } \sigma_1^2 \neq \sigma_2^2$

(The variance of the population Condition 1 represents does not equal the variance of the population Condition 2 represents. This is a **nondirectional alternative hypothesis** and it is evaluated with a **two-tailed test.** In evaluating the homogeneity of variance assumption, a nondirectional alternative hypothesis is always employed.)

The test that will be described in this section for evaluating the homogeneity of variance assumption is referred to as the **t test for homogeneity of variance for two dependent samples.** The reader should take note of the fact that the **F_{max} test/F test for two population variances (Test 8a)** (employed in evaluating the homogeneity of variance assumption for the **t test for two independent samples**) is not appropriate to use with a dependent samples design, since it does not take into account the correlation between subjects' scores in the two experimental conditions (i.e., $r_{X_1X_2}$). Equation 12.9 is the equation for the **t test for homogeneity of variance for two dependent samples.**

$$t = \frac{(\tilde{s}_L^2 - \tilde{s}_S^2)\sqrt{(n-2)}}{\sqrt{4\tilde{s}_L^2\tilde{s}_S^2(1 - r_{X_1X_2}^2)}} \qquad \textbf{(Equation 12.9)}$$

Where: $\tilde{s}_L^2$ is the larger of the two estimated population variances
$\tilde{s}_S^2$ is the smaller of the two estimated population variances

Since for Example 12.1 it has already been determined that $\tilde{s}_1 = 2.83$ and $\tilde{s}_2 = 2.38$, by squaring the latter values we can determine the values of the estimated population variances. Thus: $\tilde{s}_1^2 = (2.83)^2 = 8.01 = \tilde{s}_L^2$ and $\tilde{s}_2^2 = (2.38)^2 = 5.66 = \tilde{s}_S^2$. Substituting the appropriate values in Equation 12.9, the value $t = .79$ is computed.

$$t = \frac{(8.01 - 5.66)\sqrt{(10 - 2)}}{\sqrt{4(8.01)(5.66)(1 - (.78)^2)}} = .79$$

The degrees of freedom to employ for evaluating the t value computed with Equation 12.9 are computed with Equation 12.10.

$$df = n - 2 \qquad \textbf{(Equation 12.10)}$$

Employing Equation 12.10, the degrees of freedom for the analysis are $df = 10 - 2 = 8$. For $df = 8$, the tabled critical two-tailed .05 and .01 values in **Table A2** are $t_{.05} = 2.31$ and $t_{.01} = 3.35$. In order to reject the null hypothesis the obtained value of t must be equal to or greater than the tabled critical value at the prespecified level of significance. Since the value $t = .79$ is less than both of the aforementioned critical values, the null hypothesis cannot be rejected. Thus, the homogeneity of variance assumption is not violated.

There are a number of additional points that should be noted with respect to the ***t* test for homogeneity of variance for two dependent samples.**

a) Unless $\tilde{s}_L^2 = \tilde{s}_S^2$ (in which case $t = 0$), Equation 12.8 will always yield a positive t value. This is the case, since in the numerator of the equation the smaller variance is subtracted from the larger variance.

b) In some sources Equation 12.9 is written in the form of Equation 12.11, which employs the notation $\tilde{s}_1^2$ and $\tilde{s}_2^2$ in place of $\tilde{s}_L^2$ and $\tilde{s}_S^2$.

$$t = \frac{(\tilde{s}_1^2 - \tilde{s}_2^2)\sqrt{(n - 2)}}{\sqrt{4\tilde{s}_1^2\,\tilde{s}_2^2(1 - r_{X_1X_2}^2)}} \qquad \textbf{(Equation 12.11)}$$

If Equation 12.11 is employed, the computed value of t can be a negative number. Specifically, t will be negative when $\tilde{s}_2^2 > \tilde{s}_1^2$. In point of fact, the sign of t is irrelevant, unless one is evaluating a directional alternative hypothesis. Since the homogeneity of variance assumption involves evaluation of a nondirectional alternative hypothesis, when Equation 12.11 is employed, the researcher is only concerned with the absolute value of t.

c) As is the case for a test of homogeneity of variance for two independent samples, it is possible to use the ***t* test for homogeneity of variance for two dependent samples** to evaluate a directional alternative hypothesis regarding the relationship between the variances of two populations. Thus, if a researcher specifically predicts that the variance of the population represented by Condition 1 is larger than the variance of the population represented by Condition 2 (i.e., H_1: $\sigma_1^2 > \sigma_2^2$), or that the variance of the population represented by Condition 1 is smaller than the variance of the population represented by Condition 2 (i.e., H_1: $\sigma_1^2 < \sigma_2^2$), the latter pair of directional alternative hypotheses can be evaluated with Equation 12.11. In such an case, the sign of the computed t value is relevant. If one employs Equation 12.11 to evaluate a directional alternative hypothesis, the following guidelines are employed in evaluating the null hypothesis:

1) If the directional alternative hypothesis H_1: $\sigma_1^2 > \sigma_2^2$ is employed, the null hypothesis can be rejected if the obtained absolute value of t positive, and the value of t is equal to or greater than the tabled critical one-tailed value at the prespecified level of significance.

2) If the directional alternative hypothesis H_1: $\sigma_1^2 < \sigma_2^2$ is employed, the null hypothesis can be rejected if the sign of t is negative, and the absolute value of t is equal to or greater than the tabled critical one-tailed value at the prespecified level of significance.

If the directional alternative hypothesis H_1: $\sigma_1^2 > \sigma_2^2$ is evaluated for Example 12.1, the tabled critical one-tailed .05 and .01 values employed for the analysis are $t_{.05} = 1.86$ and $t_{.01} = 2.90$ (which respectively correspond to the tabled values at the 95th and 99th percentiles). Although the data are consistent with the directional alternative hypothesis

H_1: $\sigma_1^2 > \sigma_2^2$, the null hypothesis cannot be rejected, since the obtained value $t = .79$ is less than the aforementioned one-tailed critical values. ($t = .79$ is obtained with Equation 12.11, since as previously noted, $\tilde{s}_1^2 = \tilde{s}_L^2 = 8.01$ and $\tilde{s}_2^2 = \tilde{s}_S^2 = 5.66$.)

d) Equation 12.12 is an alternative but equivalent form of Equation 12.11.

$$t = \frac{(F - 1)\sqrt{(n - 2)}}{2\sqrt{F(1 - r_{X_1 X_2}^2)}} \qquad \textbf{(Equation 12.12)}$$

In Equation 12.12, the value of F is computed with Equation 8.8 ($F = \tilde{s}_1^2/\tilde{s}_2^2$, which is described under the ***t* test for two independent samples**). Substituting the appropriate values in Equation 12.12, the value $t = .79$ is computed.

$$F = \frac{8.01}{5.66} = 1.42 \qquad t = \frac{(1.42 - 1)\sqrt{(10-2)}}{2\sqrt{(1.42(1 - .78^2)}} = .79$$

Note that Equation 12.12 can only yield a positive t value.

e) Equation 12.13 is an alternative but equivalent form of Equation 12.9 that can only be employed to evaluate a nondirectional alternative hypothesis. Since $\tilde{s}_1^2 = \tilde{s}_L^2 = 8.01$ and $\tilde{s}_2^2 = \tilde{s}_S^2 = 5.66$, Equation 12.13 yields the value $t = .79$ (since $F_{max} = 8.01/5.66 = 1.42$).

$$t = \frac{(F_{max} - 1)\sqrt{(n - 2)}}{2\sqrt{F_{max}(1 - r_{12}^2)}} \qquad \textbf{(Equation 12.13)}$$

Where: $F_{max} = \tilde{s}_L^2/\tilde{s}_S^2$ (Which is Equation 8.6)

All of the equations noted in this section for the ***t* test for homogeneity of variance for two dependent samples** are based on the following two assumptions: a) The samples have been randomly drawn from the populations they represent; and b) The distribution of data in the underlying population each of the samples represents is normal. It is noted in the discussion of homogeneity of variance in Section VI of the ***t* test for two independent samples**, that violation of the normality assumption can severely compromise the reliability of certain tests of homogeneity of variance. The ***t* test for homogeneity of variance for two dependent samples** is among those tests whose reliability can be compromised if the normality assumption is violated.

The problems associated with the use of the ***t* test for two independent samples** when the homogeneity of variance assumption is violated are also applicable to the ***t* test for two dependent samples**. Thus, if the homogeneity of variance assumption is violated, it will generally inflate the Type I error rate associated with the ***t* test for two dependent samples**. The reader should take note of the fact that when the homogeneity of variance assumption is violated with a dependent samples design, its effect on the Type I error rate will be greater than for an independent samples design. In the event the homogeneity of variance assumption is violated for a dependent samples design, either of the following strategies can be employed: a) In conducting the ***t* test for two dependent samples**, the researcher can run a more conservative test. Thus, if the researcher does not want the Type I error rate to be greater than .05, instead of employing $t_{.05}$ as the tabled critical value, she can employ $t_{.01}$ to represent the latter value; or b) In lieu of the ***t* test for two dependent samples**, a nonparametric test that does not assume homogeneity of variance can be employed to evaluate the data (such as the **Wilcoxon matched-pairs signed-ranks test**).

4. Computation of the power of the *t* test for two dependent samples In this section the two methods for computing power that are described for computing the power of the *t* **test for two independent samples** will be extended to the *t* **test for two dependent samples.** Prior to reading this section, the reader may find it useful to review the discussion of power for both the **single-sample *t* test** and the *t* **test for two independent samples.**

The first procedure to be described is the graphical method which reveals the logic underlying the power computations for the *t* **test for two dependent samples.** In the discussion to follow, it will be assumed that the null hypothesis is identical to that employed for Example 12.1 (i.e., H_0: $\mu_1 - \mu_2 = 0$, which, as previously noted, is another way of writing H_0: $\mu_1 = \mu_2$). It will also be assumed that the researcher wants to evaluate the power of the *t* **test for two dependent samples** in reference to the following alternative hypothesis: H_1: $|\mu_1 - \mu_2| \geq 1.6$ (which is the difference obtained between the means of the two experimental conditions in Example 12.1). In other words, it is predicted that the absolute value of the difference between the two means is equal to or greater than 1.6. The latter alternative hypothesis is employed in lieu of H_1: $\mu_1 - \mu_2 \neq 0$ (which can also be written as H_1: $\mu_1 \neq \mu_2$), since in order to compute the power of the test, a specific value must be stated for the difference between the population means. Note that, as stated, the alternative hypothesis stipulates a nondirectional analysis, since it does not specify which of the two means will be the larger value. It will be assumed that $\alpha = .05$ is employed in the analysis.

Figure 12.1, which provides a visual summary of the power analysis, is comprised of two overlapping sampling distributions of difference scores. The distribution on the left, which will be designated as Distribution A, is a sampling distribution of difference scores that has a mean value of zero (i.e., $\mu_{\bar{D}} = \mu_{\bar{X}_1 - \bar{X}_2} = 0$). This latter value will be represented by $\mu_{D_0} = 0$ in Figure 12.1. Distribution A represents the sampling distribution that describes the distribution of difference scores if the null hypothesis is true. The distribution on the right, which will be designated as Distribution B, is a sampling distribution of difference scores that has a mean value of 1.6 (i.e., $\mu_D = \mu_{\bar{X}_1 - \bar{X}_2} = 1.6$). This latter value will be represented by $\mu_{D_1} = 1.6$ in Figure 12.1. Distribution B represents the sampling distribution that describes the distribution of difference scores if the alternative hypothesis is true. Each of the sampling distributions has a standard deviation that is equal to the value computed for $s_{\bar{D}} = .56$, the estimated standard error of the mean difference, since the latter value provides the best estimate of the standard deviation of the mean difference in the underlying populations.

In Figure 12.1, area (///) delineates the proportion of Distribution A that corresponds to the value $\alpha/2$, which equals .025. This is the case, since $\alpha = .05$ and a two-tailed analysis is conducted. Area ($\equiv$) delineates the proportion of Distribution B that corresponds to the probability of committing a Type II error (β). Area (\\\) delineates the proportion of Distribution B that represents the power of the test (i.e., $1 - \beta$).

The procedure for computing the proportions documented in Figure 12.1 will now be described. The first step in computing the power of the test requires one to determine how large a difference there must be between the sample means in order to reject the null hypothesis. In order to do this, we algebraically transpose the terms in Equations 12.1/12.6, using $s_{\bar{D}}$ to summarize the denominator of the equation, and $t_{.05}$ (the tabled critical two-tailed .05 t value) to represent t. Thus: $\bar{X}_1 - \bar{X}_2 = (t_{.05})(s_{\bar{D}})$. By substituting the values $t_{.05} = 2.26$ and $s_{\bar{D}} = .56$ in the latter equation, we determine that the minimum required difference is $\bar{X}_1 - \bar{X}_2 = (2.26)(.56) = 1.27$ (which is represented by the notation $\bar{X}_D = 1.27$ in Figure 12.1). Thus, any difference between the two population means that is equal to or greater than 1.27 will allow the researcher to reject the null hypothesis at the .05 level.

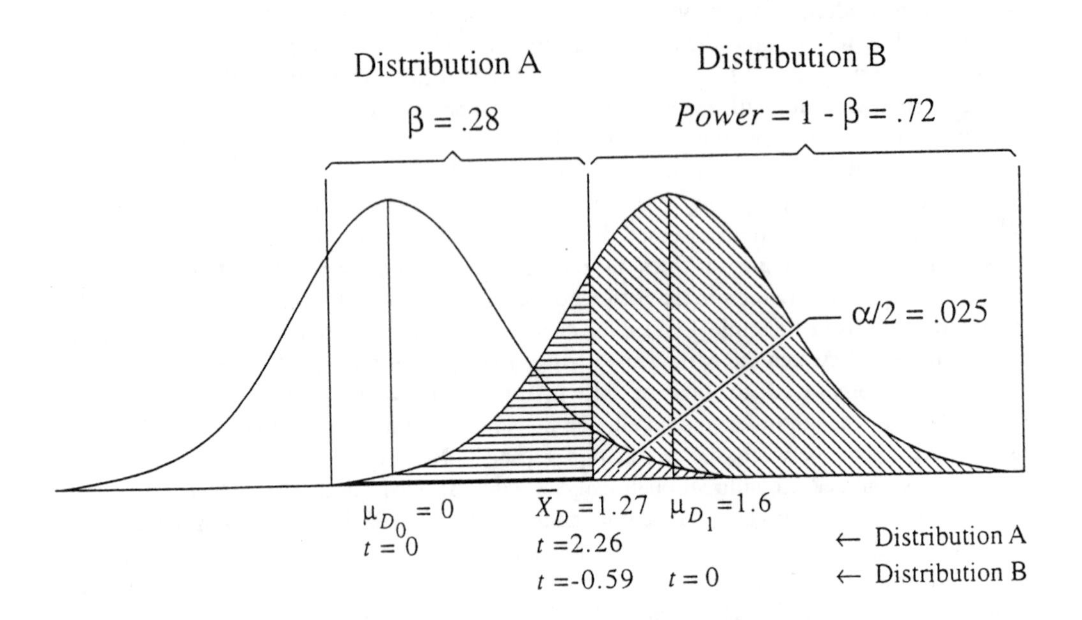

Figure 12.1 Visual Representation of Power for Example 12.1

The next step in the analysis requires one to compute the area in Distribution B that falls between the mean difference $\mu_{D_1} = 1.6$ (i.e., the mean of Distribution B) and a mean difference equal to 1.27 (represented by the notation $\bar{X}_D = 1.27$ in Figure 12.1). This is accomplished by employing Equations 12.1/12.6. In using the latter equation, the value of $\bar{X}_1$ is represented by 1.27 and the value of $\bar{X}_2$ by $\mu_{D_1} = 1.6$.

$$t = \frac{\bar{X}_1 - \bar{X}_2}{s_{\bar{D}}} = \frac{1.27 - 1.6}{.56} = -.59$$

By interpolating the values listed in **Table A2** for $df = 9$, we determine that the proportion of Distribution B that falls between the mean and a t score of $-.59$ (which corresponds to a mean difference of 1.27) is approximately .22. The latter area plus the 50% of Distribution B to the right of the mean corresponds to area (\\\) in Distribution B. Note that the left boundary of area (\\\) is also the boundary delineating the extreme 2.5% of Distribution A (i.e., $\alpha/2 = .025$, which is the rejection zone for the null hypothesis). Since area (\\\) in Distribution B overlaps the rejection zone in Distribution A, area (\\\) represents the power of the test — i.e., it represents the likelihood of rejecting the null hypothesis if the alternative hypothesis is true. The power of the test is obtained by adding .22 and .5. Thus, the power of the test equals .72. The likelihood of committing a Type II error (β) is represented by area (≡), which comprises the remainder of Distribution B. The proportion of Distribution B that constitutes this latter area is determined by subtracting the value .72 from 1. Thus: $\beta = 1 - .72 = .28$.

Based on the results of the power analysis, we can state that if the alternative hypothesis H_1: $|\mu_1 - \mu_2| \geq 1.6$ is true, the likelihood that the null hypothesis will be rejected is .72, and at the same time, there is a .28 likelihood that it will be retained. If the researcher considers the computed value for power too low (which in actuality should be determined prior to conducting a study), she can increase the power of the test by employing a larger sample size.

Method 2 described for computing the power of the ***t* test for two independent samples** can also be extended to the ***t* test for two dependent samples**. In using this latter method, the researcher must stipulate an **effect size** (d), which in the case of the ***t* test for two dependent samples** is computed with Equation 12.14.

$$d = \frac{|\mu_1 - \mu_2|}{\sigma_D} \qquad \textbf{(Equation 12.14)}$$

The numerator of Equation 12.14 represents the hypothesized difference between the two population means. As is the case with the graphical method described previously, when a power analysis is conducted after the mean of each sample has been obtained, the difference between the two sample means (i.e., $\bar{X}_1 - \bar{X}_2$) is employed as an estimate of the value of $|\mu_1 - \mu_2|$. In Equation 12.14, the value of σ_D represents the standard deviation of the difference scores in the population. In order to compute the power of the ***t* test for two dependent samples**, the latter value must either be known or be estimated by the researcher. If power is computed after the sample data have been collected, one can employ the value computed for $\tilde{s}_D$ to estimate the value of σ_D. Thus, in the case of Example 12.1 we can employ $\tilde{s}_D = 1.78$ as an estimate of σ_D.

It should be noted that if one computes the power of a test prior to collecting the data (which is what a researcher should ideally do) most researchers will have great difficulty coming up with a reasonable estimate for the value of σ_D. Since a researcher is more likely to be able to estimate the values of σ_1 and σ_2 (i.e., the population standard deviation for each of the experimental conditions), if it can be assumed that $\sigma_1 = \sigma_2$ (which is true if the population variances are homogeneous), the value of σ_D can be estimated with Equation 12.15.[13]

$$\sigma_D = \sigma\sqrt{2(1 - \rho_{X_1X_2})} \qquad \textbf{(Equation 12.15)}$$

Where: $\rho_{X_1X_2}$ is the correlation between the two variables in the underlying populations

Since the effect size computed with Equation 12.14 is only based on population parameters, it is necessary to convert the value of d into a measure that takes into account the size of the sample (which is a relevant variable in determining the power of the test). This measure, as noted in the discussions of the **single-sample *t* test** and the ***t* test for two independent samples**, is referred to as the **noncentrality parameter**. Equation 12.16 is employed to compute the **noncentrality parameter** (δ) for the ***t* test for two dependent samples**.

$$\delta = d\sqrt{n} \qquad \textbf{(Equation 12.16)}$$

The power of the ***t* test for two dependent samples** will now be computed using the data for Example 12.1. For purposes of illustration, it will be assumed that the minimum difference between the population means the researcher is trying to detect is the observed 1.6 point difference between the two sample means — i.e., $|\bar{X}_1 - \bar{X}_2| = |4.7 - 3.1| = 1.6 = |\mu_1 - \mu_2|$. The value of σ_D that will be employed in Equation 12.14 is $\tilde{s}_D = 1.78$ (which is the estimated value of the population parameter computed for the sample data). Substituting $|\mu_1 - \mu_2| = 1.6$ and $\sigma_D = 1.78$ in Equation 12.14, the value $d = .90$ is computed. Employing the guidelines noted in the discussion of power for the **single-sample *t* test,** the value $d = .90$, which represents .90 σ_D units, is categorized as a large effect size.

$$d = \frac{1.6}{1.78} = .90$$

Along with the value $n = 10$, the value $d = .90$ is substituted in Equation 12.16, resulting in the value $\delta = 2.85$.

$$\delta = .90\sqrt{10} = 2.85$$

The value $\delta = 2.85$ is evaluated with **Table A3 (Power Curves for Student's *t* Distribution)** in the **Appendix**. We will assume that for the example under discussion a two-tailed test is conducted with $\alpha = .05$, and thus **Table A3-C** is the appropriate set of power curves to employ for the analysis. Since there is no curve for $df = 9$, the power of the test will be based on a curve that falls between the $df = 6$ and $df = 12$ power curves. Through interpolation, the power of the ***t* test for two dependent samples** is determined to be approximately .72 (which is the same value that is obtained with the graphical method). Thus, by employing 10 subjects the researcher has a probability of .72 of rejecting the null hypothesis if the true difference between the population means is equal to or greater than .90 σ_D units. (which in Example 12.1 is equivalent to a 1.6 point difference between the means).

As long as a researcher knows or is able to estimate the value of σ_D, by employing trial and error she can substitute various values of n in Equation 12.16, until the computed value of δ corresponds to the desired power value for the ***t* test for two dependent samples** for a given effect size. This process can be facilitated by employing tables developed by Cohen (1988) which allow one to determine the minimum sample size necessary in order to achieve a specific level of power in reference to a given effect size.

5. Measure of magnitude of treatment effect for the *t* test for two dependent samples: Omega squared (Test 12b) Prior to reading this section, the reader should review the discussion of **magnitude of treatment effect** and the **omega squared statistic** in Section VI of the ***t* test for two independent samples.** In the latter discussion, it is noted that the computation of a t value only provides a researcher with information concerning the likelihood of the null hypothesis being false, but does not provide information on the magnitude of any treatment effect that is present. As noted in the discussion of the ***t* test for two independent samples**, a treatment effect is defined as the proportion of the variability on the dependent variable that is associated with the experimental treatments/conditions. As is the case with the ***t* test for two independent samples**, the magnitude of a treatment effect for the ***t* test for two dependent samples** can be estimated with the **omega squared** statistic ($\tilde{\omega}^2$).

Equation 12.17 (described in Kirk (1982)) will be employed to compute the value of **omega squared** for the ***t* test for two dependent samples.**[14]

$$\tilde{\omega}^2 = \frac{SS_{\text{Between conditions}} - (k-1)MS_{\text{Residual}}}{SS_{\text{Total}} + MS_{\text{Residual}}} \qquad \textbf{(Equation 12.17)}$$

Where: k equals the number of experimental conditions. In the case of the ***t* test for two dependent samples**, k will always equal 2.

In order to employ Equation 12.17, it is necessary to compute additional values that have not been determined for Example 12.1. The derivation of these values (which are summarized in Table 12.4) is described in the discussion of the **single-factor within-subjects analysis of variance (Test 18).**[15] For a description of how the values in Table 12.4 are derived, as well as what they represent, the reader should read the material in Section VI of the **single-**

factor within-subjects analysis of variance on **magnitude of treatment effect** and the **omega squared statistic.**

Table 12.4 Summary Table of Single-Factor Within-Subjects Analysis of Variance for Example 12.1

Source of variation	*SS*	*df*	*MS*	*F*
Between-subjects	108.8	9	12.09	
Between-conditions	12.8	1	12.80	8.11
Residual	14.2	9	1.58	
Total	**135.8**	**19**		

Substituting the appropriate terms from Table 12.4 in Equation 12.17, the value $\tilde{\omega}^2 = .08$ is computed.

$$\tilde{\omega}^2 = \frac{12.8 - (2 - 1)(1.58)}{135.8 + 1.58} = .08$$

Thus, 8% of the variability on the dependent variable (galvanic skin response) is associated with variability on the different levels of the independent variable (sexually explicit versus neutral words). Although 8% appears to be a relatively small value, if Cohen's (1988) guidelines are employed, it is categorized as a medium sized treatment effect.

6. Computation of a confidence interval for the *t* test for two dependent samples Prior to reading this section the reader should review the discussion of the computation of confidence intervals in Section VI of the **single-sample *t* test** and the ***t* test for two independent samples.** When interval/ratio data are available for two dependent samples, a confidence interval can be computed that identifies a range of values within which one can be confident to a specified degree that the true difference lies between the two population means. Equation 12.18 is the general equation for computing the confidence interval for the difference between two dependent population means.[16]

$$CI_{(1-\alpha)} = (\bar{X}_1 - \bar{X}_2) \pm (t_{\alpha})(s_{\bar{D}}) \qquad \textbf{(Equation 12.18)}$$

Where: $t_{\alpha/2}$ represents the tabled critical two-tailed value in the t distribution, for $df = n - 1$, below which a proportion equal to $[1 - (\alpha/2)]$ of the cases falls. If the percentage of the distribution that falls within the confidence interval is subtracted from 1, it will equal the value of α.

Employing Equation 12.18, the 95% interval for Example 12.1 is computed below. In employing Equation 12.18, $(\bar{X}_1 - \bar{X}_2)$ represents the obtained difference between the means of the two conditions (which is the numerator of the equation used to compute the value of t), $t_{.05}$ represents the tabled critical two-tailed .05 value for $df = n - 1$, and $s_{\bar{D}}$ represents the standard error of the mean difference (which is the denominator of the equation used to compute the value of t).

$$CI_{.95} = (\bar{X}_1 - \bar{X}_2) \pm (t_{.05})(s_{\bar{D}}) = 1.6 \pm (2.26)(.56) = 1.6 \pm 1.27$$

$$.33 \le (\mu_1 - \mu_2) \le 2.87$$

This result indicates that the researcher can be 95% confident that the true difference between the population means falls within the range .33 and 2.87. Specifically, it indicates that one can be 95% confident that the mean of the population Condition 1 represents is greater than the mean of population that Condition 2 represents by at least .33 words but not by more than 2.87 words.

The 99% confidence interval for Example 12.1 will also be computed to illustrate that the range of values that define a 99% confidence interval is always larger than the range which defines a 95% confidence interval.

$$CI_{.99} = (\bar{X}_1 - \bar{X}_2) \pm (t_{.01})(s_{\bar{D}}) = 1.6 \pm (3.25)(.56) = 1.6 \pm 1.82$$

$$-.22 \leq (\mu_1 - \mu_2) \leq 3.42$$

Thus, the researcher can be 99% confident that the true difference between the population means falls within the range −.22 and 3.42. Specifically, it indicates that one can be 99% confident that the mean of the population Condition 2 represents is no more than .22 words higher than the mean of population that Condition 1 represents, and that the mean of population that Condition 1 represents is no more than 3.42 words higher than the mean of population that Condition 2 represents. The reader should take note of the fact that the reliability of Equation 12.18 will be compromised if one or more of the assumptions of the ***t* test for two dependent samples** is saliently violated.

7. Test 12c: Sandler's *A* test Sandler (1955) derived a computationally simpler procedure, referred to as **Sandler's *A* test,** which is mathematically equivalent to the ***t* test for two dependent samples.** The test statistic for **Sandler's *A* test** is computed with Equation 12.19.

$$A = \frac{\Sigma D^2}{(\Sigma D)^2} \qquad \textbf{(Equation 12.19)}$$

Note that in Equation 12.19, ΣD and ΣD^2 are the same elements computed in Table 12.1 which are employed for the direct difference method for the ***t* test for two dependent samples.** When Equation 12.19 is employed for Example 12.1, the value $A = .211$ is computed.

$$A = \frac{54}{(16)^2} = .211$$

The reader should take note of the fact that except for when $\Sigma D = 0$, the value of A must be a positive number. If a negative value is obtained for A, it indicates that a computational error has been made. If $\Sigma D = 0$ (which indicates that the means of the two conditions are equal), Equation 12.19 becomes unsolvable.[17]

Table 12.5 Tabled Critical .05 and .01 *A* values for *df* = 9

	$t_{.05}$	$t_{.01}$
Two-tailed values	.276	.185
One-tailed values	.368	.213

The obtained value $A = .211$ is evaluated with **Table A12 (Table of Sandler's *A* Statistic)** in the **Appendix.** As is the case for the ***t* test for two dependent samples,** the degrees of freedom employed for **Sandler's *A* test** are computed with Equation 12.5. Thus,

$df = 10 - 1 = 9$. The tabled critical two-tailed and one-tailed .05 and .01 values for $df = 9$ are summarized in Table 12.5.

The following guidelines are employed in evaluating the null hypothesis for **Sandler's *A* test.**

a) If the nondirectional alternative hypothesis H_1: $\mu_1 \neq \mu_2$ is employed, the null hypothesis can be rejected if the obtained value of A is equal to or less than the tabled critical two-tailed value at the prespecified level of significance.

b) If the directional alternative hypothesis H_1: $\mu_1 > \mu_2$ is employed, the null hypothesis can be rejected if the sign of ΣD is positive (i.e., $\bar{X}_1 > \bar{X}_2$), and the value of A is equal to or less than the tabled critical one-tailed value at the prespecified level of significance.

c) If the directional alternative hypothesis H_1: $\mu_1 < \mu_2$ is employed, the null hypothesis can be rejected if the sign of ΣD is negative (i.e., $\bar{X}_1 < \bar{X}_2$), and the value of A is equal to or less than the tabled critical one-tailed value at the prespecified level of significance.

Employing the above guidelines, the following conclusions can be reached.

The nondirectional alternative hypothesis H_1: $\mu_1 \neq \mu_2$ is supported at the .05 level, since the computed value $A = .211$ is less than the tabled critical two-tailed value $A_{.05} = .276$. The latter alternative hypothesis, however, is not supported at the .01 level, since $A = .211$ is greater than the tabled critical two-tailed value $A_{.01} = .185$.

The directional alternative hypothesis H_1: $\mu_1 > \mu_2$ is supported at both the .05 and .01 levels, since $\Sigma D = 16$ is a positive number, and the obtained value $A = .211$ is less than the tabled critical one-tailed values $A_{.05} = .368$ and $A_{.01} = .213$.

The directional alternative hypothesis H_1: $\mu_1 < \mu_2$ is not supported, since $\Sigma D = 16$ is a positive number. In order for the directional alternative hypothesis H_1: $\mu_1 < \mu_2$ to be supported, the value of ΣD must be a negative number (as well as the fact that the computed value of A must be equal to or less than the tabled critical one-tailed value at the prespecified level of significance).

Note that the results obtained for **Sandler's *A* test** are identical to those obtained when the ***t* test for two dependent samples** is employed to evaluate Example 12.1. Equation 12.20 describes the relationship between **Sandler's *A*** statistic and the t value computed for the ***t* test for two dependent samples.**

$$A = \frac{n - 1}{nt^2} + \frac{1}{n} \qquad \textbf{(Equation 12.20)}$$

It is demonstrated below that when $t = 2.86$ (the value computed with Equations 12.1/12.6) is substituted in Equation 12.20, it yields the value $A = .211$ computed with Equation 12.19.

$$A = \frac{10 - 1}{(10)(2.86)^2} + \frac{1}{10} = .21$$

8. Test 12d: The *z* test for two dependent samples There are occasions (albeit infrequent) when a researcher wants to compare the means of two dependent samples, and happens to know the variances of the two underlying populations. In such a case, the ***z* test for two dependent samples** should be employed to evaluate the data instead of the ***t* test for two dependent samples.** As is the case with the latter test, the ***z* test for two dependent samples** assumes that the two samples are randomly selected from populations that have normal distributions. The effect of violation of the normality assumption on the test statistic decreases as the size of the sample employed in an experiment increases. The homogeneity of variance assumption noted for the ***t* test for two dependent samples** is not an assumption of the ***z* test for two dependent samples.**

The null and alternative hypotheses employed for the z **test for two dependent samples** are identical to those employed for the t **test for two dependent samples**. Equation 12.21 is employed to compute the test statistic for the z **test for two dependent samples**.[18]

$$z = \frac{\bar{X}_1 - \bar{X}_2}{\sqrt{\sigma_{\bar{X}_1}^2 + \sigma_{\bar{X}_2}^2 - 2(r_{X_1X_2})(\sigma_{\bar{X}_1})(\sigma_{\bar{X}_2})}} \qquad \textbf{(Equation 12.21)}$$

Where: $\sigma_{\bar{X}_i}^2 = \sigma_i^2/n$ and $\sigma_{\bar{X}_i} = \sigma_i/\sqrt{n}$

The only differences between Equation 12.21 and Equation 12.6 (the equation for the t **test for two dependent samples**) are: a) In the denominator of Equation 12.21, in computing the standard error of the mean for each condition, the population standard deviations σ_1 and σ_2 are employed instead of the estimated population standard deviations $\tilde{s}_1$ and $\tilde{s}_2$ (which are employed in Equation 12.6); and b) Equation 12.21 computes a z score which is evaluated with the normal distribution, while Equation 12.6 derives a t score which is evaluated with the t distribution.

If it is assumed that the two population variances are known for Example 12.1, and that $\sigma_1^2 = 8.01$ and $\sigma_2^2 = 5.66$, Equation 12.21 can be employed to evaluate the data. Note that the obtained value $z = 2.86$ is identical to the value that was computed for t when Equation 12.6 was employed.

$$z = \frac{4.7 - 3.1}{\sqrt{.79 + .56 - 2(.78)(.89)(.75)}} = 2.86$$

The obtained value $z = 2.86$ is evaluated with **Table A1 (Table of the Normal Distribution)** in the **Appendix**. In **Table A1** the tabled critical two-tailed .05 and .01 values are $z_{.05} = 1.96$ and $z_{.01} = 2.58$, and the tabled critical one-tailed .05 and .01 values are $z_{.05} = 1.65$ and $z_{.01} = 2.33$. Since the computed value $z = 2.86$ is greater than the tabled critical two-tailed values $z_{.05} = 1.96$ and $z_{.01} = 2.58$, the nondirectional alternative hypothesis H_1: $\mu_1 \neq \mu_2$ is supported at both the .05 and .01 levels. Since the computed value $z = 2.86$ is a positive number which is greater than the tabled critical one-tailed values $z_{.05} = 1.65$ and $z_{.01} = 2.33$, the directional alternative hypothesis H_1: $\mu_1 > \mu_2$ is also supported at both the .05 and .01 levels.

When the same set of data are evaluated with the t **test for two dependent samples**, although the directional alternative hypothesis H_1: $\mu_1 > \mu_2$ is supported at both the .05 and .01 levels, the nondirectional alternative hypothesis H_1: $\mu_1 \neq \mu_2$ is only supported at the .05 level. This latter fact illustrates that if the z **test for two dependent samples** and the t **test for two dependent samples** are employed to evaluate the same set of data (unless the value of n is extremely large), the latter test will provide a more conservative test of the null hypothesis (i.e., make it more difficult to reject H_0). This is the case, since the tabled critical values listed for the z **test for two dependent samples** will always correspond to the tabled critical values listed in **Table A2** for $df = \infty$ (which are the lowest tabled critical values listed for the t distribution).

The final part of the discussion of the z **test for two dependent samples** will describe a special case of the test in which it is employed to evaluate the difference between the average performance of two conditions, when the scores of subjects are based on a binomially distributed variable. Example 12.2, which is used to illustrate this application of the test, is the dependent samples analog of Example 8.3 (which illustrates the analysis of a binomially distributed variable with the z **test for two independent samples**). The null and alternative hypotheses evaluated in Example 12.2 are identical to those evaluated in Example 12.1.

Example 12.2 *An experiment is conducted in which each of five subjects is tested for extra-sensory perception under two experimental conditions. In* Condition 1 *a subject listens to a relaxation training tape, after which the subject is tested while in a relaxed state of mind. In* Condition 2 *each subject is tested while in a normal state of mind. Assume that the order of presentation of the two experimental conditions is counterbalanced, although not completely, since to do the latter would require that an even number of subjects be employed in the study. Thus, three of the five subjects initially serve in* Condition 1 *followed by* Condition 2, *while the remaining two subjects initially serve in* Condition 2 *followed by* Condition 1. *(The concept of* **counterbalancing** *is discussed in* Section VII.)

In each experimental condition a subject is tested for 200 *trials. In each condition the researcher employs as stimuli a different list of* 200 *binary digits (specifically, the values* 0 and 1*) which have been randomly generated by a computer. On each trial, an associate of the researcher concentrates on a digit in the order it appears on the list for that condition. While the associate does this, a subject is required to guess the value of the number that is employed as the stimulus for that trial. The number of correct guesses for subjects under the two experimental conditions follow. (The first score for each subject is the number of correct responses in Condition 1, and the second score is the number of correct responses in* Condition 2.): **Subject 1** (105, 90); **Subject 2** (120, 104); **Subject 3** (130, 107); **Subject 4** (115, 100); **Subject 5** (110, 99). Table 12.6 *summarizes the data for the experiment.*

Table 12.6 Data for Example 12.2

	Condition 1		Condition 2		
Subject	X_1	X_1^2	X_2	X_2^2	X_1X_2
1	105	11025	90	8100	9450
2	120	14400	104	10816	12480
3	130	16900	107	11449	13919
4	115	13225	100	10000	11500
5	110	12100	99	9801	10890
	$\Sigma X_1 = 580$	$\Sigma X_1^2 = 67650$	$\Sigma X_2 = 500$	$\Sigma X_2^2 = 50166$	$\Sigma X_1X_2 = 58230$
	$\bar{X}_1 = \frac{580}{5} = 116$		$\bar{X}_2 = \frac{500}{5} = 100$		

Note that in Example 12.2, the five scores in each of the two experimental conditions are identical to the five scores employed in the two experimental conditions in Example 8.3. The only difference between the two examples is the order in which the scores are listed. Specifically, in Example 12.2 the scores have been arranged so that the two scores in each row (i.e., the two scores of each subject) have a high positive correlation with one another. Through use of Equation 12.7, it is demonstrated that the correlation between subjects' scores in the two experimental conditions is $r_{X_1X_2} = .93$.

$$r_{X_1X_2} = \frac{58230 - \frac{(580)(500)}{5}}{\sqrt{\left[67650 - \frac{(580)^2}{10}\right]\left[50166 - \frac{(500)^2}{5}\right]}} = .93$$

Example 12.2 will be evaluated with Equation 12.22, which is the form Equation 12.21 assumes when $\sigma_1^2 = \sigma_2^2$.

$$z = \frac{\bar{X}_1 - \bar{X}_2}{\sigma \sqrt{\dfrac{1}{m} + \dfrac{1}{m} - \dfrac{2(r_{X_1 X_2})}{m}}} \qquad \textbf{(Equation 12.22)}$$

Note that in Equation 12.22, m is employed to represent the number of subjects, since the notation n is employed with the binomial variable to designate the number of trials in which each subject is tested. Since scores on the binary guessing task described in Example 12.2 are assumed to be binomially distributed, as is the case in Example 8.3, the following is true: $n = 200$, $\pi_1 = .5$, and $\pi_2 = .5$. The computed value for the population standard deviation for the binomially distributed variable is $\sigma = \sqrt{n\pi_1\pi_2} = \sqrt{(200)(.5)(.5)} = 7.07$. (The computation of the latter values is discussed in Section I of the **binomial sign test for a single sample (Test 6)**.) When the appropriate values are substituted in Equation 12.22, the value $z = 13.52$ is computed.

$$z = \frac{116 - 100}{7.07\sqrt{\dfrac{1}{5} + \dfrac{1}{5} - \dfrac{(2)(.93)}{5}}} = 13.52$$

Since the computed value $z = 13.52$ is greater than the tabled critical two-tailed values $z_{.05} = 1.96$ and $z_{.01} = 2.58$, the nondirectional alternative hypothesis H_1: $\mu_1 \neq \mu_2$ is supported at both the .05 and .01 levels. Since the computed value $z = 13.52$ is a positive number that is greater than the tabled critical one-tailed values $z_{.05} = 1.65$ and $z_{.01} = 2.33$, the directional alternative hypothesis H_1: $\mu_1 > \mu_2$ is supported at both the .05 and .01 levels. Thus, it can be concluded that the average score in Condition 1 is significantly larger than the average score in Condition 2. Note that when Equation 8.17 is employed with the same set of data, it yields the value $z = 3.58$. The fact that the value $z = 13.52$ obtained with Equation 12.22 is larger than the value $z = 3.58$ obtained with Equation 8.17, illustrates that if there is a positive correlation between the scores of subjects employed in a dependent samples design, a ***z* test for two dependent samples** will provide a more powerful test of an alternative hypothesis than will a ***z* test for two independent samples** (due to the lower value of the denominator for the former test).

VII. Additional Discussion of the *t* Test for Two Dependent Samples

1. The use of matched subjects in a dependent samples design It is noted in Section I that the ***t* test for two dependent samples** can be applied to a design involving matched subjects. Matching subjects requires that a researcher initially identify one or more variables (besides the independent variable) which she believes are positively correlated with the dependent variable employed in a study. Such a variable can be employed as a matching variable. Each subject who is assigned to one of the k experimental conditions is matched with one subject in each of the other $(k - 1)$ experimental conditions. In matching subjects it is essential that any cohort of subjects who are matched with one another are equivalent (or reasonably comparable) with respect to any matching variables employed in the study. In a design employing matched subjects there will be n cohorts (also referred to as **blocks**) of matched subjects, and within each cohort there will k subjects. Each of the k subjects should be randomly assigned to one of the k experimental conditions/levels of the independent variable. Thus, when $k = 2$, each of the two subjects within the n pairs/cohorts will be randomly assigned to one of the two experimental conditions.

By matching subjects a researcher is able to conduct a more powerful statistical analysis than will be the case if subjects in the two conditions are not matched with one another (i.e., if an independent samples design is employed). The more similar the cohorts of subjects are on the matched variable(s), the greater the power of the statistical test. In actuality, the most extreme case of matching is when each subject is matched with him or herself, and thus serves in each of the k experimental conditions. Within this framework, the design employed for Example 12.1 can be viewed as a matched-subjects design. However, as the term matching is most commonly employed, within each row of the data summary table (i.e., Table 12.1) different subjects serve in the each of the k experimental conditions. When $k = 2$, the most extreme case of matching involving different subjects in each condition is when n pairs of identical twins are employed as subjects. By virtue of their common genetic makeup, identical twins allow an experimenter to match a subject with his or her "clone." In Example 12.3 identical twins are employed as subjects in the same experiment described by Example 12.1. Analysis of Example 12.3 with the **t test for two dependent samples** yields the same result as that obtained for Example 12.1, since both examples employ the same set of data.

Example 12.3 *A psychologist conducts a study in order to determine whether people exhibit more emotionality when they are exposed to sexually explicit words than when they are exposed to neutral words. Ten sets of identical twins are employed as subjects. Within each twin pair, one of the twins is randomly assigned to* Condition 1, *in which the subject is shown a list of eight sexually explicit words, while the other twin is assigned to* Condition 2, *in which the subject is shown a list of eight neutral words. As each word is projected on the screen, a subject is instructed to say the word softly to him or herself. As a subject does this, sensors attached to the palms of the subject's hands record galvanic skin response (GSR), which is used by the psychologist as a measure of emotionality. The psychologist computes two scores for each pair of twins to represent the emotionality score for each of the experimental conditions:* **Condition 1**: *GSR/Explicit — The average GSR score for the twin presented with the eight sexually explicit words;* **Condition 2**: *GSR/Neutral — The average GSR score for the twin presented with the eight neutral words. The GSR/Explicit and the GSR/Neutral scores of the ten pairs of twins follow. (The first score for each twin pair represents the score of the twin presented with the sexually explicit words, and the second score represents the score of the twin presented with the neutral words. The higher the score the higher the level of emotionality.)* **Twin pair 1** (9, 8); **Twin pair 2** (2, 2); **Twin pair 3** (1, 3); **Twin pair 4** (4, 2); **Twin pair 5** (6, 3); **Twin pair 6** (4, 0); **Twin pair 7** (7, 4); **Twin pair 8** (8, 5); **Twin pair 9** (5, 4); **Twin pair 10** (1, 0). *Do subjects exhibit differences in emotionality with respect to the two categories of words?*

In the event $k = 3$ and a researcher wants to use identical siblings, identical triplets can be employed in a study. If $k = 4$, identical quadruplets can be used, and so on. If the critical variable(s) the researcher wants to match subjects with respect to are believed to be controlled by environmental factors, the suitability of employing identical siblings as matched subjects will be compromised to the degree that within each set of siblings the members of the set do not share common environmental experiences. Realistically, the number of available identical siblings in a human population will be quite limited. Thus, with the exception of identical twins, it would be quite unusual to encounter a study that employs identical siblings. Because of the low frequency of identical siblings in the general population, in matching subjects a researcher may elect to employ biological relatives who share less in common with one another or employ people who are not blood relatives. Example 12.4 illustrates the latter type of matching. Analysis of Example 12.4 with the **t test for two dependent samples** yields the same result as that obtained for Examples 12.1 and 12.3, since all three experiments employ the same set of data.

Example 12.4 *A psychologist conducts a study in order to determine whether people exhibit more emotionality when they are exposed to sexually explicit words than when they are exposed to neutral words. Based on previous research, the psychologist has reason to believe that the following three variables are highly correlated with the dependent variable of emotionality: a) gender; b) autonomic hyperactivity (which is measured by a series of physiological measures); and c) repression/sensitization (which is measured by a pencil and paper personality test). Ten pairs of matched subjects who are identical (or very similar) on the three aforementioned variables are employed in the study. Within each pair, one person is randomly assigned to a condition in which the subject is shown a list of eight sexually explicit words, while the other person is assigned to a condition in which the subject is shown a list of eight neutral words. As each word is projected on the screen, a subject is instructed to say the word softly to him or herself. As a subject does this, sensors attached to the palms of the subject's hands record galvanic skin response (GSR), which is used by the psychologist as a measure of emotionality. The psychologist computes two scores for each pair of matched subjects to represent the emotionality score for each of the experimental conditions:* **Condition 1**: *GSR/Explicit — The average GSR score for the subject presented with the eight sexually explicit words;* **Condition 2**: *GSR/Neutral — The average GSR score for the subject presented with the eight neutral words. The GSR/Explicit and the GSR/Neutral scores of the ten pairs of subjects follow. (The first score for each pair represents the score of the person presented with the sexually explicit words, and the second score represents the score of the person presented with the neutral words. The higher the score the higher the level of emotionality.)* **Pair 1** (9, 8); **Pair 2** (2, 2); **Pair 3** (1, 3); **Pair 4** (4, 2); **Pair 5** (6, 3); **Pair 6** (4, 0); **Pair 7** (7, 4); **Pair 8** (8, 5); **Pair 9** (5, 4); **Pair 10** (1, 0). *Do subjects exhibit differences in emotionality with respect to the two categories of words?*

One reason a researcher may elect to employ matched subjects (as opposed to employing each subject in all k experimental conditions) is because in many experiments it is not feasible to have a subject serve in more than one condition. Specifically, a subject's performance in one or more of the conditions might be influenced by his or her experience in one or more of the conditions that precede it. In some instances counterbalancing can be employed to control for such effects, but in other cases even counterbalancing does not provide the necessary control.

In spite of the fact that it can increase the power of a statistical analysis, matching is not commonly employed in experiments involving human subjects. The reason for this is that matching requires a great deal of time and effort on the part of a researcher. Not only is it necessary to identify one or more matching variables that are correlated with the dependent variable, but it is also necessary to identify and obtain the cooperation of a sufficient number of matched subjects to participate in an experiment. The latter does not present as much of a problem in animal research, where litter mates can be employed as matched cohorts. Example 12.5, which is evaluated in the next section, illustrates a design that employs animal litter mates as subjects.

Example 12.5 *A researcher wants to assess the relative effect of two different kinds of punishment (loud noise versus a blast of cold air) on the emotionality of mice. Five pairs of mice derived from five separate litters are employed as subjects. Within each pair, one of the litter mates is randomly assigned to one of two experimental conditions. During the course of the experiment each mouse is sequestered in an experimental chamber. While in the chamber, each of the five mice in* Condition 1 *is periodically presented with a loud noise, and each of the five mice in* Condition 2 *is periodically presented with a blast of cold air. The presentation of the punitive stimulus for each of the animals is generated by a machine that randomly presents the stimulus throughout the duration of the time an animal is in the* *chamber. The dependent variable of emotionality employed in the study is the number of times each mouse defecates while in the experimental chamber. The number of episodes of defecation for the five pairs of mice follows. (The first score represents the litter mate exposed to noise and the second score represents the litter mate exposed to cold air.)* **Litter 1** (11, 11); **Litter 2** (1, 11); **Litter 3** (0, 5); **Litter 4** (2, 8); **Litter 5** (0, 4). *Do subjects exhibit differences in emotionality under the different experimental conditions?*

2. Relative power of the *t* test for two dependent samples and the *t* test for two independent samples Example 12.5 will be employed to illustrate that the ***t* test for two dependent samples** provides a more powerful test of an alternative hypothesis than does the ***t* test for two independent samples**. Except for the fact that it employs a dependent samples design involving matched subjects, Example 12.5 is identical to Example 8.4 (which employs an independent samples design). Both examples employ the same set of data and evaluate the same null and alternative hypotheses. The summary values for evaluating Example 12.5 with the ***t* test for two dependent samples** (using either Equation 12.1 or 12.6) are noted below. Some of the values listed can also be found in Table 8.1 (which summarizes the same set of data for analysis with the ***t* test for two independent samples**).

$$\Sigma X_1 = 14 \qquad \Sigma X_1^2 = 126 \qquad \Sigma X_2 = 39 \qquad \Sigma X_2^2 = 347 \qquad \bar{X}_1 = 2.8 \qquad \bar{X}_2 = 5.8$$

$$\Sigma X_1 X_2 = 148 \qquad \tilde{s}_1^2 = 21.7 \qquad \tilde{s}_1^2 = 10.7 \qquad r_{X_1 X_2} = .64$$

$$\Sigma D = -25 \qquad \Sigma D^2 = 177 \qquad \bar{D} = -5 \qquad \tilde{s}_D = 3.61 \qquad s_{\bar{D}} = 1.61$$

$$t = \frac{\bar{D}}{s_{\bar{D}}} = \frac{-5}{1.61} = -3.10$$

Since $n = 5$ in Example 12.5, $df = 5 - 1 = 4$. In **Table A2**, for $df = 4$, the tabled critical two-tailed .05 and .01 values are $t_{.05} = 2.78$ and $t_{.01} = 4.60$, and the tabled critical one-tailed .05 and .01 values are $t_{.05} = 2.13$ and $t_{.01} = 3.75$.

The nondirectional alternative hypothesis H_1: $\mu_1 \neq \mu_2$ is supported at the .05 level, since the computed absolute value $t = 3.10$ is greater than the tabled critical two-tailed value $t_{.05} = 2.78$. It is not, however, supported at the .01 level, since the absolute value $t = 3.10$ is less than the tabled critical two-tailed value $t_{.01} = 4.60$.

The directional alternative hypothesis H_1: $\mu_1 < \mu_2$ is supported at the .05 level, since the computed value $t = -3.10$ is a negative number, and the absolute value $t = 3.10$ is greater than the tabled critical one-tailed value $t_{.05} = 2.13$. It is not, however, supported at the .01 level since the absolute value $t = 3.10$ is less than the tabled critical one-tailed value $t_{.01} = 3.75$.

The directional alternative hypothesis H_1: $\mu_1 > \mu_2$ is not supported, since the computed value $t = -3.10$ is a negative number.

Note that the absolute value $t = 3.10$ computed for Example 12.5 is substantially higher than the absolute value $t = 1.96$ computed for Example 8.4 with the ***t* test for two independent samples**. In the case of Example 8.4 (as well as Example 12.5), the directional alternative hypothesis H_1: $\mu_1 < \mu_2$ is supported at the .05 level. However, the nondirectional alternative hypothesis H_1: $\mu_1 \neq \mu_2$ (which is supported at the .05 level in the case of Example 12.5) is not supported in Example 8.4 when the data are evaluated with the ***t* test for two independent samples**. The difference in the conclusions reached with the two tests reflects the fact that the ***t* test for two dependent samples** provides a more powerful test of an alternative hypothesis (assuming there is a positive correlation between the scores of subjects in the two experimental conditions). In closing this discussion, it is worth noting that designs involving independent samples are more commonly employed in research than designs involving dependent samples. The reason for this is that over and above the fact that a dependent samples design allows for a more powerful test of an alternative hypothesis, it presents more practical problems in its implementation (e.g., controlling for problems that might result from subjects serving in multiple conditions; the difficulty of identifying matching variables; identifying and obtaining the cooperation of an adequate number of matched subjects).

3. Counterbalancing and order effects When each of the n subjects in an experiment serves in all k experimental conditions, it is often necessary to control for the order of presentation of the conditions.[19] Thus, if all n subjects are administered Condition 1 first followed by Condition 2 (or vice versa), factors such as practice or fatigue, which are a direct function of the order of presentation of the conditions, can differentially affect subjects' scores on the dependent variable. Specifically, subjects may perform better in Condition 2 due to practice effects or subjects may perform worse in Condition 2 as a result of fatigue. As noted earlier, **counterbalancing** is a procedure which allows a researcher to control for such order effects. In **complete counterbalancing** all possible orders for presenting the experimental conditions are represented an equal number of times with respect to the total number of subjects employed in a study. Thus, if a study with $n = 10$ subjects and $k = 2$ conditions is completely counterbalanced, five subjects will initially serve in Condition 1 followed by Condition 2, while the other five subjects will initially serve in Condition 2 followed by Condition 1. If the number of experimental conditions is $k = 3$, there will be $k! = 3! = 6$ possible presentation orders (i.e., 1,2,3; 1,3,2; 2,1,3; 2,3,1; 3,1,2; 3,2,1). Under such conditions a minimum of six subjects will be required in order to employ complete counterbalancing. If a researcher wants to assign two subjects to each of the presentation orders, $6 \times 2 = 12$ subjects must be employed. It should be obvious that in order to completely counterbalance the order of presentation of the experimental conditions, the number of subjects must equal the value of $k!$ or be some value that is evenly divisible by it.

As the number of experimental conditions increase, complete counterbalancing becomes more difficult to implement, since the number of subjects required increases substantially. Specifically, if there are $k = 5$ experimental conditions, there are $5! = 120$ presentation orders — thus requiring a minimum of 120 subjects (which can be a prohibitively large number for a researcher to use) in order that one subject serves in each of the possible presentation orders. When it is not possible to completely counterbalance the order of presentation of the conditions, alternative less complete counterbalancing procedures are available. Such procedures are described in books that specialize in experimental design.

4. Analysis of a before-after design with the *t* test for two dependent samples In a **before-after design** n subjects are administered a pretest on a dependent variable. After the pretest all n subjects are exposed to the experimental treatment. Subjects are then administered a posttest on the same dependent variable. The ***t* test for two dependent samples** can be employed to determine if there is a significant difference between the pretest versus posttest scores of subjects. Although there are published studies that employ the ***t* test for two dependent samples** to evaluate the aforementioned design, it is important to note that since it lacks a control group, a **before-after design** does not allow a researcher to conclude that the experimental treatment is responsible for a significant difference. If a significant result is obtained in such a study, it only allows the researcher to conclude that there is a significant statistical association/correlation between the experimental treatment and the dependent variable. Since correlational information does not allow one to draw conclusions with regard to cause and effect, a researcher cannot conclude that the treatment is directly responsible for the observed difference. Although it is possible that the treatment is responsible for the difference, it is also possible that the difference is due to one or more other variables that intervened between the pretest and the posttest.

In order to modify a **before-after design** so as to insure adequate experimental control, it is required that two groups of subjects be employed. In such a modification, pretest and posttest scores are obtained for both groups, but only one of the groups (the experimental group) is exposed to the experimental treatment in the time period that intervenes between the pretest and the posttest. By virtue of employing a control group which is not exposed to the experimental treatment, the researcher is able to rule out the potential influence of confounding variables. Thus, order effects, as well as other factors in the environment that subjects may have been exposed to between the pretest and the posttest, can be ruled out through use of a control group. Example 12.6 illustrates a study that employs a **before-after design** without the addition of the necessary control group.

Example 12.6 *In order to assess the efficacy of electroconvulsive therapy (ECT), a psychiatrist evaluates ten clinically depressed patients before and after a series of ECT treatments. A standardized interview is used to operationalize a patient's level of depression, and on the basis of the interview each patient is assigned a score ranging from 0 to 10 with respect to his or her level of depression prior to (pretest score) and after (posttest score) the administration of ECT. The higher a patient's score, the more depressed the patient. The pretest and posttest scores of the ten patients follow:* **Patient 1** (9, 8); **Patient 2** (2, 2); **Patient 3** (1, 3); **Patient 4** (4, 2); **Patient 5** (6, 3); **Patient 6** (4, 0); **Patient 7** (7, 4); **Patient 8** (8, 5); **Patient 9** (5, 4); **Patient 10** (1, 0). *Do the data indicate that ECT is effective?*[20]

Since the data for Example 12.6 are identical to that employed in Example 12.1, it yields the same result. Thus, analysis of the data with the ***t* test for two dependent samples** indicates that there is a significant decrease in depression following the ECT. However, as previously noted, because there is no control group, the psychiatrist cannot conclude that ECT is responsible for the decrease in depression. Inclusion of a "sham" ECT group can provide the necessary control to evaluate the impact of ECT. Such a group would be comprised of ten additional patients for whom pretest and posttest depression scores are obtained. Between the pretest and posttest, the patients in the control group undergo all of the preparations involved in ECT, but are only administered a simulated ECT treatment (i.e., they are not actually administered the shock treatment). Only by including such a control group, can one rule out the potential role of extraneous variables that might also be responsible for the lower level of depression during the posttest. An example of such an extraneous variable would be if all of the subjects who receive ECT are in psychotherapy throughout the duration of the experiment. Without a control group, a researcher cannot determine whether a lower posttest depression score is the result of the ECT, the psychotherapy, the ECT and psychotherapy interacting with one another, or some other variable of which the researcher is unaware. By including a control group, it is assumed that if any extraneous variables are present, by virtue of randomly assigning subjects to groups, the groups will be equated on such variables.

When the **before-after design** is modified by the addition of the appropriate control group, the resulting design is referred to as a **pretest-posttest control group design**. Unfortunately, researchers are not in agreement with respect to what statistical analysis is most appropriate for the latter design. Among the analytical procedures that have been recommended are the following: a) The difference scores of the two groups can be contrasted with a ***t* test for two independent samples**; b) The results can be evaluated by employing a **factorial analysis of variance for a mixed design (Test 21h)**. A factorial design has two or more independent variables (which are also referred to as factors). Thus, if the appropriate control group is employed in Example 12.6, the resulting **pretest-posttest control group design** can be conceptualized as being comprised of two independent variables. One of the independent variables is represented by the ECT versus sham ECT manipulation. The second independent variable is the pretest-posttest dichotomy. In a mixed factorial design involving two factors, one of the independent variables is a between-subjects variable (i.e., different subjects serve under different levels of that independent variable). Thus, in the example under discussion, the ECT versus sham ECT manipulation represents a between-subjects independent variable. The other independent variable in a mixed factorial design is a within-subjects variable (i.e., each subject serves under all levels of that independent variable). In the example under discussion, the pretest-posttest dichotomy represents a within-subjects independent variable; and c) An **analysis of covariance** (a procedure that is discussed in Section VI of the **single-factor between-subjects analysis of variance (Test 16)**) can also be employed to evaluate a **pretest-posttest control group design**. In conducting an analysis of covariance, the pretest scores of subjects are employed as the covariate.

VIII. Additional Example Illustrating the Use of the *t* Test for Two Dependent Samples

Example 12.7 is an additional example that can be evaluated with the ***t* test for two dependent samples**. Since Example 12.7 employs the same data as Example 12.1, it yields the same result. Note that in Example 12.7 complete counterbalancing is employed in order to control for order effects.

Example 12.7 *A study is conducted in order to evaluate the relative efficacy of two drugs (Clearoxin and Lesionoxin) on chronic psoriasis. Ten subjects afflicted with chronic psoriasis participate in the study. Each subject is exposed to both drugs for a six-month period, with a three-month hiatus between treatments. Five subjects are treated with Clearoxin initially, after which they are treated with Lesionoxin. The other five subjects are treated with Lesionoxin first and then with Clearoxin. The dependent variable employed in the study is a rating of the severity of a subject's lesions under the two drug conditions. The higher the rating the more severe a subject's psoriasis. The scores of the ten subjects under the two treatment conditions follow. (The first score represents the Clearoxin condition (which represents* Condition 1), *and the second score the Lesionoxin condition (which represents* Condition 2).) **Subject 1** (9, 8); **Subject 2** (2, 2); **Subject 3** (1, 3); **Subject 4** (4, 2); **Subject 5** (6, 3); **Subject 6** (4, 0); **Subject 7** (7, 4); **Subject 8** (8, 5); **Subject 9** (5, 4); **Subject 10** (1, 0). *Do the data indicate that subjects respond differently to the two types of medication?*

References

Cohen, J. (1988). **Statistical power analysis for the behavioral sciences** (2nd ed.). New York: Academic Press.

Conover, W. J. (1980). **Practical nonparametric statistics** (2nd ed.). New York: John Wiley and Sons.

Fisher, R. A. (1935). **The design of experiments** (7th ed.). Edinburgh–London: Oliver and Boyd.

Fisher, R. A. (1973). **Statistical methods for research workers** (14th ed.). New York: Hafner.

Kirk, R. L. (1982). **Experimental design: Procedures for the behavioral sciences** (2nd. ed.). Belmont, CA: Brooks/Cole Publishing Company.

Marascuilo, L. A. and McSweeney, M. (1977). **Nonparametric and distribution-free methods for the social sciences.** Monterey, CA: Brooks/Cole Publishing Company.

Sandler, J. (1955) A test of the significance of difference between the means of correlated measures based on a simplification of Student's *t*. **British Journal of Psychology, 46,** 225–226.

Siegel, S. and Castellan, N. J., Jr. (1988). **Nonparametric statistics for the behavioral sciences** (2nd ed.). New York: McGraw–Hill Book Company.

Endnotes

1. Alternative terms that are employed in describing a **dependent samples design** are **repeated measures design, within-subjects design, treatment-by-subjects design, correlated samples design,** and **randomized-blocks design.** The use of the terms **blocks** within the framework of a dependent samples design is discussed in Endnote 1 of the **single-factor within-subjects analysis of variance.**

2. A study has **internal validity** to the extent that observed differences between the experimental conditions on the dependent variable can be unambiguously attributed to a manipulated independent variable. Random assignment of subjects to the different experimental conditions is the most effective way to insure **internal validity** (by eliminating the possible influence of confounding/extraneous variables). In contrast to **internal validity, external validity** refers to the degree to which the results of an experiment can be generalized. The results of an experiment can only be generalized to a population of subjects, as well as environmental conditions, that are comparable to those that are employed in the experiment.

3. In actuality, when galvanic skin response (which is a measure of skin resistance) is measured, the higher a subject's GSR the less emotional the subject. In Example 12.1, it is assumed that the GSR scores have been transformed so that the higher a subject's GSR score, the greater the level of emotionality.

4. An alternative but equivalent way of writing the null hypothesis is H_0: $\mu_1 - \mu_2 = 0$. The analogous alternative but equivalent ways of writing the alternative hypotheses in the order they are presented are: H_1: $\mu_1 - \mu_2 \neq 0$; H_1: $\mu_1 - \mu_2 > 0$; and H_1: $\mu_1 - \mu_2 < 0$.

5. Note that the basic structure of Equation 12.3 is the same as Equations I.8/2.1 (the equation for the estimated population standard deviation that is employed within the framework of the **single-sample *t* test**). In Equation 12.3 a standard deviation is computed for n D scores, whereas in Equations I.8/2.1 a standard deviation is computed for n X scores.

6. The actual value that is estimated by $s_{\bar{D}}$ is $\sigma_{\bar{D}}$, which is the standard deviation of the sampling distribution of mean difference scores for the two populations. The meaning of the **standard error of the mean difference** can be best understood by considering the following procedure for generating an empirical sampling distribution of difference scores: a) Obtain n difference scores for a random sample of n subjects; b) Compute the mean difference score ($\bar{D}$) for the sample; and c) Repeat steps a) and b) m times. At the conclusion of this procedure one will have obtained m mean difference scores. The **standard error of the mean difference** represents the standard deviation of the m mean difference scores, and can be computed by substituting the term $\bar{D}$ for D in Equation 12.3. Thus: $s_{\bar{D}} = \sqrt{[\Sigma \bar{D}^2 - ((\Sigma \bar{D})^2/m)]/[m - 1]}$. The standard deviation that is computed with Equation 12.4 is an estimate of $\sigma_{\bar{D}}$.

7. In order for Equation 12.1 to be soluble, there must be variability in the n difference scores. If each subject produces the same difference score, the value of $\tilde{s}_D$ computed with Equation 12.3 will equal 0. As a result of the latter, Equation 12.4 will yield the value $s_{\bar{D}} = 0$. Since $s_{\bar{D}}$ is the denominator of Equation 12.1, when the latter value equals zero, the t test equation will be insoluble.

8. The numerator of Equation 12.6 will always equal $\bar{D}$ (i.e., the numerator of Equation 12.1). In the same respect the denominator of Equation 12.6 will always equal $s_{\bar{D}}$ (the denominator of Equation 12.1). The denominator of Equation 12.6 can also be written as follows:

$$\sqrt{\frac{\tilde{s}_1^2}{n} + \frac{\tilde{s}_2^2}{n} - 2(r_{X_1X_2})\left[\frac{\tilde{s}_1}{\sqrt{n}}\right]\left[\frac{\tilde{s}_2}{\sqrt{n}}\right]}$$

9. Note that in the case of Example 8.1 (which is employed to illustrate the ***t* test for two independent samples**), it is reasonable to assume that scores in the same row of Table 8.1 (which summarizes the data for the study) will not be correlated with one another (by virtue of the fact that two independent samples are employed in the study). When independent samples are employed, it is assumed that random factors determine the values of any pair of scores in the same row of a table summarizing the data, and consequently it is assumed that the correlation between pairs of scores in the same row will be equal to (or close to) 0.

10. Due to rounding off error, there may be a slight discrepancy between the value of t computed with Equations 12.1 and 12.6.

11. A noncorrelational procedure that allows a researcher to evaluate whether a treatment effect is present in the above described example is **Fisher's randomization procedure**, which is generally categorized as a **permutation test**. **Fisher's randomization procedure**, which is attributed to Fisher (1935, 1973) (and is described in Conover (1980), Marascuilo and McSweeney (1977), and Siegel and Castellan (1988)), requires that all possible score configurations which can be obtained for the value of the computed sum of the difference scores be determined. Upon computing the latter information, one can determine the likelihood of obtaining a configuration of scores that is equal to or more extreme than the one obtained for a set of data.

12. Equation 12.1 can be modified as follows so as to be equivalent to Equation 12.8: $t = [\bar{D} - (\mu_1 - \mu_2)]/s_{\bar{D}}$.

13. Although Equation 12.15 is intended for use prior to collecting the data, it should yield the same value for σ_D if the values computed for the sample data are substituted in it. Thus, if $\sigma = 2.60$ (which is the average of the values $\tilde{s}_1 = 2.83$ and $\tilde{s}_2 = 2.38$), and $\rho_{X_1X_2}$ (which is the population correlation coefficient estimated by the value $r_{X_1X_2} = .78$) are substituted in Equation 12.15, the value $\sigma_D = 1.72$ is computed, which is quite close to the computed value $\tilde{s}_D = 1.78$. The slight discrepancy between the two values can be attributed to the fact that the estimated population standard deviations are not identical.

14. Although other sources employ alternative equations for computing **omega squared**, the different equations result in approximately the same value. It should be noted that in

some books, Equation 8.14 is employed to compute the value of **omega squared** for the ***t* test for two dependent samples**. The latter equation, however, yields an inflated value for $\tilde{\omega}^2$.

15. In contrast to the ***t* test for two dependent samples** (which can only be employed with two dependent samples), the **single-factor within-subjects analysis of variance** can be used with a dependent samples design involving interval/ratio data in which there are k samples, where $k \geq 2$.

16. Note that the basic structure of Equation 12.18 is the same as Equation 8.15 (which is employed for computing a confidence interval for the ***t* test for two independent samples**), except that the latter equation employs $s_{\bar{X}_1 - \bar{X}_2}$ in place of $s_{\bar{D}}$.

17. It was noted earlier that if all n subjects obtain the identical difference score, Equations 12.1/12.6 become unsolvable. In the case of Equation 12.19, for a given value of n, if all n subjects obtain the same difference score the same A value will always be computed, regardless of the magnitude of the identical difference score obtained by each of the n subjects. If the value of A computed under such conditions is substituted in the equation $t = \sqrt{(n - 1)/(An - 1)}$ (which is algebraically derived from Equation 12.19), the latter equation becomes unsolvable (since the value $(An - 1)$ will always equal zero). The conclusion that results from this observation is that Equation 12.19 is insensitive to the magnitude of the difference between experimental conditions when all subjects obtain the same difference score.

18. Equation 12.21 can also be written as follows:

$$z = \frac{\bar{X}_1 - \bar{X}_2}{\sqrt{\frac{\sigma_1^2}{n} + \frac{\sigma_2^2}{n} - 2(r_{X_1 X_2})\left[\frac{\sigma_1}{\sqrt{n}}\right]\left[\frac{\sigma_2}{\sqrt{n}}\right]}}$$

In instances when a researcher stipulates in the null hypothesis that the difference between the two population means is some value other than zero, the numerator of Equation 12.21 is the same as the numerator of Equation 12.8. The protocol for computing the value of the numerator is identical to that employed for Equation 12.8.

19. In Example 12.1, the order of presentation of the conditions is controlled by randomly distributing the sexually explicit and neutral words throughout the 16 word list presented to each subject.

20. The doctor conducting the study might feel it would be unethical to employ a group of comparably depressed subjects as a control group, since patients in such a group would be deprived of a potentially beneficial treatment.

Test 13

The Wilcoxon Matched-Pairs Signed-Ranks Test

(Nonparametric Test Employed with Ordinal Data)

I. Hypothesis Evaluated with Test and Relevant Background Information

Hypothesis evaluated with test Do two dependent samples represent two different populations?

Relevant background information on test The **Wilcoxon matched-pairs signed-ranks test** (Wilcoxon (1945, 1949)) is a nonparametric procedure employed in a hypothesis testing situation involving a design with two dependent samples. Whenever one or more of the assumptions of the ***t* test for two dependent samples (Test 12)** is saliently violated, the **Wilcoxon matched-pairs signed-ranks test** (which has less stringent assumptions) may be preferred as an alternative procedure. Prior to reading the material on the **Wilcoxon matched-pairs signed-ranks test**, the reader may find it useful to review the general information regarding a dependent samples design contained in Sections I and VII of the ***t* test for two dependent samples.**

The **Wilcoxon matched-pairs signed-ranks test** is essentially an extension of the **Wilcoxon signed-ranks test (Test 4)** (which is employed for a single sample design) to a design involving two dependent samples. In order to employ the **Wilcoxon matched-pairs signed-ranks test**, it is required that each of n subjects (or n pairs of matched subjects) has two interval/ratio scores (each score having been obtained under one of the two experimental conditions). A difference score is computed for each subject (or pair of matched subjects) by subtracting a subject's score in Condition 2 from his score in Condition 1. The hypothesis evaluated with the **Wilcoxon matched-pairs signed-ranks test** is whether or not in the underlying populations represented by the samples/experimental conditions, the median of the difference scores (which will be represented by the notation θ_D) equals zero. If a significant difference is obtained, it indicates there is a high likelihood the two samples/conditions represent two different populations.

The **Wilcoxon matched-pairs signed-ranks test** is based on the following assumptions:[1] a) The sample of n subjects has been randomly selected from the population it represents; b) The original scores obtained for each of the subjects are in the format of interval/ratio data; and c) The distribution of the difference scores in the populations represented by the two samples is symmetric about the median of the population of difference scores.

As is the case for the ***t* test for two dependent samples**, in order for the **Wilcoxon matched-pairs signed-ranks test** to generate valid results, the following guidelines should be adhered to: a) In order to control for order effects, the presentation of the two experimental conditions should be random or, if appropriate, be counterbalanced; and b) If matched samples are employed, within each pair of matched subjects each of the subjects should be randomly assigned to one of the two experimental conditions.

As is the case with the ***t* test for two dependent samples**, the **Wilcoxon matched-pairs signed-ranks test** can also be employed to evaluate a **before-after design**. The limitations of the **before-after design** (which are discussed in Section VII of the ***t* test for two**

dependent samples) are also applicable when it is evaluated with the **Wilcoxon matched-pairs signed-ranks test.**

It should be noted that all of the other tests is this text which rank data (with the exception of the **Wilcoxon signed-ranks test**), rank the original interval/ratio scores of subjects. The **Wilcoxon matched-pairs signed-ranks test,** however, does not rank the original interval/ratio scores, but instead ranks the interval/ratio difference scores of subjects (or matched pairs of subjects). For this reason, some sources categorize the **Wilcoxon matched-pairs signed-ranks test** as a test of interval/ratio data. Most sources, however (including this book), categorize the **Wilcoxon matched-pairs signed-ranks test** as a test of ordinal data, by virtue of the fact that a ranking procedure is part of the test protocol.

II. Example

Example 13.1 is identical to Example 12.1 (which is evaluated with the ***t* test for two dependent samples**). In evaluating Example 13.1 it will be assumed that the ratio data are rank-ordered, since one or more of the assumptions of the ***t* test for two dependent samples** has been saliently violated.

Example 13.1 *A psychologist conducts a study in order to determine whether people exhibit more emotionality when they are exposed to sexually explicit words than when they are exposed to neutral words. Each of ten subjects is shown a list of* 16 *randomly arranged words, which are projected onto a screen one at a time for a period of five seconds. Eight of the words on the list are sexually explicit and eight of the words are neutral. As each word is projected on the screen, a subject is instructed to say the word softly to him or herself. As a subject does this, sensors attached to the palms of the subject's hands record galvanic skin response (GSR), which is used by the psychologist as a measure of emotionality. The psychologist computes two scores for each subject, one score for each of the experimental conditions:* **Condition 1:** *GSR/Explicit — The average GSR score for the eight sexually explicit words;* **Condition 2:** *GSR/Neutral — The average GSR score for the eight neutral words. The GSR/Explicit and the GSR/Neutral scores of the ten subjects follow. (The higher the score the higher the level of emotionality.)* **Subject 1** (9, 8); **Subject 2** (2, 2); **Subject 3** (1, 3); **Subject 4** (4, 2); **Subject 5** (6, 3); **Subject 6** (4, 0); **Subject 7** (7, 4); **Subject 8** (8, 5); **Subject 9** (5, 4); **Subject 10** (1, 0). *Do subjects exhibit differences in emotionality with respect to the two categories of words?*

III. Null versus Alternative Hypotheses

Null hypothesis $H_0\colon \theta_D = 0$

(In the underlying populations represented by Condition 1 and Condition 2, the median of the difference scores equals zero. With respect to the sample data, this translates into the sum of the ranks of the positive difference scores being equal to the sum of the ranks of the negative difference scores (i.e., $\Sigma R+ = \Sigma R-$).

Alternative hypothesis $H_1\colon \theta_D \neq 0$

(In the underlying populations represented by Condition 1 and Condition 2, the median of the difference scores is some value other than zero. With respect to the sample data, this translates into the sum of the ranks of the positive difference scores not being equal to the sum of the ranks of the negative difference scores (i.e., $\Sigma R+ \neq \Sigma R-$). This is a **nondirectional alternative hypothesis** and it is evaluated with a **two-tailed test.**)

or

$$H_1: \theta_D > 0$$

(In the underlying populations represented by Condition 1 and Condition 2, the median of the difference scores is some value that is greater than zero. With respect to the sample data, this translates into the sum of the ranks of the positive difference scores being greater than the sum of the ranks of the negative difference scores (i.e., $\Sigma R+ > \Sigma R-$). The latter result indicates that the scores in Condition 1 are higher than the scores in Condition 2. This is a **directional alternative hypothesis** and it is evaluated with a **one-tailed test**.)

or

$$H_1: \theta_D < 0$$

(In the underlying populations represented by Condition 1 and Condition 2, the median of the difference scores is some value that is less than zero (i.e., a negative number). With respect to the sample data, this translates into the sum of the ranks of the positive difference scores being less than the sum of the ranks of the negative difference scores (i.e., $\Sigma R+ < \Sigma R-$). The latter result indicates that the scores in Condition 2 are higher than the scores in Condition 1. This is a **directional alternative hypothesis** and it is evaluated with a **one-tailed test**.)

Note: Only one of the above noted alternative hypotheses is employed. If the alternative hypothesis the researcher selects is supported, the null hypothesis is rejected.

IV. Test Computations

The data for Example 13.1 are summarized in Table 13.1. Note that there are 10 subjects and that each subject has two scores.

Table 13.1 Data for Example 13.1

Subject	X_1	X_2	$D = X_1 - X_2$	Rank of $\|D\|$	Signed rank of $\|D\|$
1	9	8	1	2	2
2	2	2	0	–	–
3	1	3	–2	4.5	–4.5
4	4	2	2	4.5	4.5
5	6	3	3	7	7
6	4	0	4	9	9
7	7	4	3	7	7
8	8	5	3	7	7
9	5	4	1	2	2
10	1	0	1	2	2
					$\Sigma R+ = 40.5$
					$\Sigma R- = 4.5$

In Table 13.1, X_1 represents each subject's score in Condition 1 (sexually explicit words) and X_2 represents each subject's score in Condition 2 (neutral words). In Column 4 of Table 13.1 a D score is computed for each subject by subtracting a subject's score in Condition 2 from the subject's score in Condition 1 (i.e., $D = X_1 - X_2$). In Column 5 the D scores have been ranked with respect to their absolute values. Since the ranking protocol employed for the **Wilcoxon matched-pairs signed-ranks test** is identical to that employed for the **Wilcoxon signed-ranks test**, the reader may find it useful to review the ranking protocol described in Section IV of the latter test. To reiterate, the following guidelines should be adhered to when ranking the difference scores for the **Wilcoxon matched-pairs signed-ranks test.**

a) The **absolute values** of the difference scores ($|D|$) are ranked (i.e., the sign of a difference score is not taken into account).

b) Any difference score that equals zero is not ranked. This translates into eliminating from the analysis any subject who yields a difference score of zero.

c) When there are tie scores present in the data, the average of the ranks involved is assigned to all scores tied for a given rank.

d) As is the case with the **Wilcoxon signed-ranks test**, when ranking difference scores for the **Wilcoxon matched-pairs signed-ranks test** it is essential that a rank of 1 be assigned to the difference score with the lowest absolute value, and that a rank of n be assigned to the difference score with the highest absolute value (where n represents the number of signed ranks — i.e., difference scores that have been ranked).[2]

Upon ranking the absolute values of the difference scores, the sign of each difference score is placed in front of its rank. The signed ranks of the difference scores are listed in Column 6 of Table 13.1. Note that although 10 subjects participated in the experiment there are only $n = 9$ signed ranks, since Subject 2 had a difference score of zero which was not ranked. Table 13.2 summarizes the rankings of the difference scores for Example 13.1.

Table 13.2 Ranking Procedure for Wilcoxon Matched-Pairs Signed-Ranks Test

Subject number	2	1	9	10	3	4	5	7	8	6
Subject's difference score	0	1	1	1	–2	2	3	3	3	4
Absolute value of difference score	–	1	1	1	2	2	3	3	3	4
Rank of $\|D\|$	–	2	2	2	4.5	4.5	7	7	7	9

The sum of the ranks that have a positive sign (i.e., $\Sigma R+ = 40.5$) and the sum of the ranks that have a negative sign (i.e., $\Sigma R- = 4.5$) are recorded at the bottom of Column 6 in Table 13.1. Equation 13.1 (which is identical to Equation 4.1) allows one to check the accuracy of these values. If the relationship indicated by Equation 13.1 is not obtained, it indicates an error has been made in the calculations.

$$\Sigma R+ + \Sigma R- = \frac{n(n+1)}{2} \qquad \textbf{(Equation 13.1)}$$

Employing the values $\Sigma R+ = 40.5$ and $\Sigma R- = 4.5$ in Equation 13.1, we confirm that the relationship described by the equation is true.

$$40.5 + 4.5 = \frac{(9)(10)}{2} = 45$$

V. Interpretation of the Test Results

As noted in Section III, if the sample is derived from a population in which the median of the difference scores equals zero, the values of $\Sigma R+$ and $\Sigma R-$ will be equal to one another. When $\Sigma R+$ and $\Sigma R-$ are equivalent, both of these values will equal $[n(n+1)]/4$, which in the case of Example 13.1 will be $[(9)(10)]/4 = 22.5$. This latter value is commonly referred to as the **expected value of the Wilcoxon *T* statistic.**

If the value of $\Sigma R+$ is significantly greater than the value of $\Sigma R-$, it indicates there is a high likelihood that Condition 1 represents a population with higher scores than the population represented by Condition 2. On the other hand, if $\Sigma R-$ is significantly greater than $\Sigma R+$, it indicates there is a high likelihood that Condition 2 represents a population with higher scores than the population represented by Condition 1. Table 13.1 reveals that $\Sigma R+ = 40.5$ is greater than $\Sigma R- = 4.5$, and thus the data are consistent with the directional alternative hypothesis H_1: $\theta_D > 0$ (i.e., it indicates that subjects obtained higher scores in Condition 1 than Condition 2). The question is, however, whether the difference is significant — i.e., whether it is large enough to conclude that it is unlikely to be the result of chance.

The absolute value of the **smaller** of the two values $\Sigma R+$ versus $\Sigma R-$ is designated as the **Wilcoxon *T* test statistic.** Since $\Sigma R- = 4.5$ is smaller than $\Sigma R+ = 40.5$, $T = 4.5$. The T value is interpreted by employing **Table A5 (Table of Critical *T* Values for Wilcoxon's Signed-Ranks and Matched-Pairs Signed-Ranks Tests)** in the **Appendix. Table A5** lists the critical two-tailed and one-tailed .05 and .01 T values in relation to the number of signed ranks in a set of data. In order to be significant, the obtained value of T must be **equal to or less than** the tabled critical T value at the prespecified level of significance.[3] Table 13.3 summarizes the tabled critical two-tailed and one-tailed .05 and .01 Wilcoxon T values for $n = 9$ signed ranks.

Table 13.3 Tabled Critical Wilcoxon *T* Values for *n* = 9 Signed Ranks

	$T_{.05}$	$T_{.01}$
Two-tailed values	5	1
One-tailed values	8	3

Since the null hypothesis can only be rejected if the computed value $T = 4.5$ is equal to or less than the tabled critical value at the prespecified level of significance, we can conclude the following.

In order for the nondirectional alternative hypothesis H_1: $\theta_D \neq 0$ to be supported, it is irrelevant whether $\Sigma R+ > \Sigma R-$ or $\Sigma R- > \Sigma R+$. In order for the result to be significant, the computed value of T must be equal to or less than the tabled critical two-tailed value at the prespecified level of significance. Since the computed value $T = 4.5$ is less than the tabled critical two-tailed .05 value $T_{.05} = 5$, the nondirectional alternative hypothesis H_1: $\theta_D \neq 0$ is supported at the .05 level. It is not, however, supported at the .01 level, since $T = 4.5$ is greater than the tabled critical two-tailed .01 value $T_{.01} = 1$.

In order for the directional alternative hypothesis H_1: $\theta_D > 0$ to be supported, $\Sigma R+$ must be greater than $\Sigma R-$. Since $\Sigma R+ > \Sigma R-$, the data are consistent with the directional alternative hypothesis H_1: $\theta_D > 0$. In order for the result to be significant, the computed value of T must be equal to or less than the tabled critical one-tailed value at the prespecified level of significance. Since the computed value $T = 4.5$ is less than the tabled critical one-tailed .05 value $T_{.05} = 8$, the directional alternative hypothesis H_1: $\theta_D > 0$ is supported at the .05 level. It is not, however, supported at the .01 level, since $T = 4.5$ is greater than the tabled critical one-tailed .01 value $T_{.01} = 3$.

In order for the directional alternative hypothesis H_1: $\theta_D < 0$ to be supported, the following two conditions must be met: a) $\Sigma R-$ must be greater than $\Sigma R+$; and b) The computed value of T must be equal to or less than the tabled critical one-tailed value at the prespecified level of significance. Since the first of these conditions is not met, the directional alternative hypothesis H_1 : $\theta_D < 0$ is not supported.

A summary of the analysis of Example 13.1 with the **Wilcoxon matched-pairs signed-ranks test** follows: It can be concluded that subjects exhibited higher GSR (emotionality) scores with respect to the sexually explicit words than the neutral words.

The results obtained with the **Wilcoxon matched-pairs signed-ranks test** are reasonably consistent with those obtained when the ***t* test for two dependent samples** is employed to evaluate the same set of data. In the case of both tests, the analogous nondirectional alternative hypotheses H_1: $\theta_D \neq 0$ and H_1: $\mu_1 \neq \mu_2$ are supported, but only at the .05 level. In the case of the **Wilcoxon matched-pairs signed-ranks test**, the directional alternative hypothesis H_1: $\theta_D > 0$ is only supported at the .05 level, whereas the analogous directional alternative hypothesis H_1: $\mu_1 > \mu_2$ is supported at both the .05 and .01 levels when the data are evaluated with the ***t* test for two dependent samples**. The latter discrepancy between the two tests reflects the fact that when a parametric and nonparametric test are applied to the same set of data, the parametric test will generally provide a more powerful test of an alternative hypothesis. In most instances, however, similar conclusions will be reached if the same data are evaluated with the ***t* test for two dependent samples** and the **Wilcoxon matched-pairs signed-ranks test**. Siegel and Castellan (1988) note that the power efficiency of the **Wilcoxon matched-pairs signed-ranks test** relative to the ***t* test for two dependent samples** approaches .955 as the sample size increases.[4] Daniel (1990) provides additional references on the power efficiency of the **Wilcoxon matched-pairs signed-ranks test**.

VI. Additional Analytical Procedures for the Wilcoxon Matched-Signed-Ranks Test and/or Related Tests

1. The normal approximation of the Wilcoxon *T* statistic for large sample sizes As is the case with the **Wilcoxon signed-ranks test**, if the sample size employed in a study is relatively large, the normal distribution can be employed to approximate the Wilcoxon T statistic. Although sources do not agree on the value of the sample size that justifies employing the normal approximation of the Wilcoxon distribution, they generally state that it should be employed for sample sizes larger than those documented in the Wilcoxon table contained within the source. Equation 13.2 (which is identical to Equation 4.2) provides the normal approximation for Wilcoxon T. In the equation T represents the computed value of Wilcoxon T, which for Example 13.1 is $T = 4.5$. n, as noted previously, represents the number of signed ranks. Thus, in our example, $n = 9$. Note that in the numerator of Equation 13.2, the term $[n(n+1)]/4$ represents the expected value of T (often summarized with the symbol T_E), which is defined in Section V. The denominator of Equation 13.2 represents the expected standard deviation of the sampling distribution of the T statistic.

$$z = \frac{T - \dfrac{n(n+1)}{4}}{\sqrt{\dfrac{n(n+1)(2n+1)}{24}}} \qquad \textbf{(Equation 13.2)}$$

Although Example 13.1 involves only nine signed ranks (a value most sources would view as too small to use with the normal approximation), it will be employed to illustrate Equation 13.2. The reader will see that in spite of employing Equation 13.2 with a small sample size, it will yield essentially the same result as that obtained when the exact table of the Wilcoxon distribution is employed. When the values $T = 4.5$ and $n = 9$ are substituted in Equation 13.2, the value $z = -2.13$ is computed.

$$z = \frac{4.5 - \dfrac{(9)(10)}{4}}{\sqrt{\dfrac{(9)(10)(19)}{24}}} = -2.13$$

The obtained value $z = -2.13$ is evaluated with **Table A1 (Table of the Normal Distribution)** in the **Appendix**. In **Table A1** the tabled critical two-tailed .05 and .01 values are $z_{.05} = 1.96$ and $z_{.01} = 2.58$, and the tabled critical one-tailed .05 and .01 values are $z_{.05} = 1.65$ and $z_{.01} = 2.33$.

Since the smaller of the two values $\Sigma R+$ versus $\Sigma R-$ is selected to represent T, the value of z computed with Equation 13.2 will always be a negative number (unless $\Sigma R+ = \Sigma R-$, in which case z will equal zero). This is the case, since by selecting the smaller value T will always be less than the expected value T_E. As a result of this, the following guidelines are employed in evaluating the null hypothesis.

a) If a nondirectional alternative hypothesis is employed, the null hypothesis can be rejected if the obtained absolute value of z is equal to or greater than the tabled critical two-tailed value at the prespecified level of significance.

b) When a directional alternative hypothesis is employed, one of the two possible directional alternative hypotheses will be supported if the obtained absolute value of z is equal to or greater than the tabled critical one-tailed value at the prespecified level of significance. Which alternative hypothesis is supported depends on the prediction regarding which of the two values $\Sigma R+$ versus $\Sigma R-$ is larger. The null hypothesis can only be rejected if the directional alternative hypothesis that is consistent with the data is supported.

Employing the above guidelines, when the normal approximation is employed with Example 13.1 the following conclusions can be reached.

The nondirectional alternative hypothesis H_1: $\theta_D \neq 0$ is supported at the .05 level. This is the case, since the computed absolute value $z = 2.13$ is greater than the tabled critical two-tailed .05 value $z_{.05} = 1.96$. The nondirectional alternative hypothesis H_1: $\theta_D \neq 0$ is not supported at the .01 level, since the absolute value $z = 2.13$ is less than the tabled critical two-tailed .01 value $z_{.01} = 2.58$. This decision is consistent with the decision that is reached when the exact table of the Wilcoxon distribution is employed to evaluate the nondirectional alternative hypothesis H_1: $\theta_D \neq 0$.

The directional alternative hypothesis H_1: $\theta_D > 0$ is supported at the .05 level. This is the case, since the data are consistent with the latter alternative hypothesis (i.e., $\Sigma R+ > \Sigma R-$), and the computed absolute value $z = 2.13$ is greater than the tabled critical one-tailed .05 value $z_{.05} = 1.65$. The directional alternative hypothesis H_1: $\theta_D > 0$ is not supported at the .01 level, since the obtained absolute value $z = 2.13$ is less than the tabled critical one-tailed .01 value $z_{.01} = 2.33$. This decision is consistent with the decision that is reached when the exact table of the Wilcoxon distribution is employed to evaluate the directional alternative hypothesis H_1: $\theta_D > 0$.

The directional alternative hypothesis H_1: $\theta_D < 0$ is not supported, since the data are not consistent with the latter alternative hypothesis (which requires that $\Sigma R- > \Sigma R+$).

It should be noted that, in actuality, either $\Sigma R+$ or $\Sigma R-$ can be employed to represent the value of T in Equation 13.2. Either value will yield the same absolute value for z. The smaller of the two values will always yield a negative z value, and the larger of the two values will always yield a positive z value (which in this instance will be $z = 2.13$ if $\Sigma R+ = 40.5$ is employed to represent T). In evaluating a nondirectional alternative hypothesis the sign of z is irrelevant. In the case of a directional alternative hypothesis, one must determine whether the data are consistent with the alternative hypothesis that is stipulated. If the data are consistent, one then determines whether the absolute value of z is equal to or greater than the tabled critical one-tailed value at the prespecified level of significance.

2. The correction for continuity for the normal approximation of the Wilcoxon matched-pairs signed-ranks test As noted in the discussion of the **Wilcoxon signed-ranks test**, a **correction for continuity** can be employed for the normal approximation of the

Wilcoxon test statistic. The same correction for continuity can be applied to the **Wilcoxon matched-pairs signed-ranks test**. The correction for continuity (which results in a slight reduction in the absolute value computed for z) requires that .5 be subtracted from the absolute value of the numerator of Equation 13.2. Thus, Equation 13.3 (which is identical to Equation 4.3) represents the continuity corrected normal approximation of the Wilcoxon test statistic.

$$z = \frac{\left| T - \frac{n(n+1)}{4} \right| - .5}{\sqrt{\frac{n(n+1)(2n+1)}{24}}} \qquad \textbf{(Equation 13.3)}$$

Employing Equation 13.3, the continuity corrected value $z = 2.07$ is computed. Note that as a result of the absolute value conversion, the numerator of Equation 13.3 will always be a positive number, thus yielding a positive z value.

$$z = \frac{\left| 4.5 - \frac{(9)(10)}{4} \right| - .5}{\sqrt{\frac{(9)(10)(19)}{24}}} = 2.07$$

The result of the analysis with Equation 13.3 leads to the same conclusions that are reached with Equation 13.2 (i.e., when the correction for continuity is not employed). Specifically, since the absolute value $z = 2.07$ is greater than the tabled critical two-tailed .05 value $z_{.05} = 1.96$, the nondirectional alternative hypothesis H_1: $\theta_D \neq 0$ is supported at the .05 level (but not at the .01 level). Since the absolute value $z = 2.07$ is greater than the tabled critical one-tailed .05 value $z_{.05} = 1.65$, the directional alternative hypothesis H_1: $\theta_D > 0$ is supported at the .05 level (but not at the .01 level).

3. Tie correction for the normal approximation of the Wilcoxon test statistic Equation 13.4 (which is identical to Equation 4.4) is an adjusted version of Equation 13.2 that is recommended in some sources (e.g., Daniel (1990) and Marascuilo and McSweeney (1977)) when tied difference scores are present in the data. The tie correction (which is identical to the one described for the **Wilcoxon signed-ranks test**) results in a slight increase in the absolute value of z. Unless there are a substantial number of ties, the difference between the values of z computed with Equations 13.2 and 13.4 will be minimal.

$$z = \frac{T - \frac{n(n+1)}{4}}{\sqrt{\frac{n(n+1)(2n+1)}{24} - \frac{\Sigma t^3 - \Sigma t}{48}}} \qquad \textbf{(Equation 13.4)}$$

Table 13.4 illustrates the application of the tie correction with Example 13.1. In the data for Example 13.1 there are three sets of tied ranks: Set 1 involves three subjects (Subjects 1, 9, and 10); Set 2 involves two subjects (Subjects 3 and 4); Set 3 involves three subjects (Subjects 5, 7, and 8). The number of subjects involved in each set of tied ranks represents the values of t in the third column of Table 13.4. The three t values are cubed in the last column of the table, after which the values Σt and Σt^3 are computed. The appropriate values are now substituted in Equation 13.4.[5]

Table 13.4 Correction for Ties with Normal Approximation

Subject	Rank	t	t^3
1	2		
9	2	3	27
10	2		
3	4.5	2	8
4	4.5		
5	7		
7	7	3	27
8	7		
6	9		
		$\Sigma t = 8$	$\Sigma t^3 = 62$

$$z = \frac{4.5 - \frac{(9)(10)}{4}}{\sqrt{\frac{(9)(10)(19)}{24} - \frac{62 - 8}{48}}} = -2.15$$

The absolute value $z = 2.15$ is slightly larger than the absolute value $z = 2.13$ obtained without the tie correction. The difference between the two methods is trivial, and in this instance, regardless of which alternative hypothesis is employed, the decision the researcher makes with respect to the null hypothesis is not affected.[6]

Conover (1980) and Daniel (1990) discuss and/or cite sources on the subject of alternative ways of handling tied difference scores. Conover (1980) also notes that in some instances retaining and ranking zero difference scores may actually provide a more powerful test of an alternative hypothesis than the more conventional method employed in this book (which eliminates zero difference scores from the data).

4. Sources for computing a confidence interval for the Wilcoxon matched-pairs signed ranks test Daniel (1990) and Marascuilo and McSweeney (1977) describe procedures for computing a confidence interval for the **Wilcoxon matched-pairs signed-ranks test** — i.e., computing a range of values within which a researcher can be confident to a specified degree that a difference between two population medians falls.

VII. Additional Discussion of the Wilcoxon Matched-Pairs Signed-Ranks Test

1. Alternative nonparametric rank-order procedures for evaluating a design involving two dependent samples In addition to the **Wilcoxon matched-pairs signed-ranks test** there are other (albeit, less frequently employed) nonparametric procedures that can be employed in evaluating ranked data in a design involving two dependent samples. One of these tests, the **binomial sign test for two dependent samples (Test 14)**, is described later in the book. Another procedure (described in Conover (1980), Marascuilo and McSweeney (1977) and Siegel and Castellan (1988)) is **Fisher's randomization procedure** (Fisher (1935, 1973)) (which is discussed briefly in Section VI of the ***t* test for two dependent samples**). The latter test, through use of ordered permutations, determines all possible score configurations that can be obtained for the value of the sum of difference scores obtained in a set of data. Upon computing the latter value, the likelihood of obtaining a sum of difference scores that is equal to or more extreme than the one obtained for the data is determined. As a general rule, **Fisher's randomization procedure** is not practical to employ without the use of a computer, since it involves an inordinate amount of computations. Marascuilo and McSweeney (1977) also describe the **normal scores test for matched samples** (which is based on the **Van der Waerden** (1952/1953) **normal scores test** discussed in Section VII of the **Mann–Whitney *U* test (Test 9)**). Normal scores tests are procedures which involve transformation of ordinal data through use of the normal distribution. Daniel (1990) cites primary sources that describe additional nonparametric procedures for evaluating ranks in a design involving two dependent samples. In their discussion of nonparametric alternatives to the ***t* test for two dependent samples**, Marascuilo and McSweeney (1977) note that in some instances when the underlying population distribution is not normal, the power of all or some of the alternative nonparametric procedures can equal or exceed the power of the ***t* test for two dependent samples**.

VIII. Additional Examples Illustrating the Use of the Wilcoxon Matched-Pairs Signed-Ranks Test

The **Wilcoxon matched-pairs signed-ranks test** can be employed to evaluate any of the additional examples noted for the ***t* test for two dependent samples** (i.e., Examples 12.2–12.7). In all instances in which the **Wilcoxon matched-pairs signed-ranks test** is employed, difference scores are obtained for subjects (or pairs of matched subjects). All difference scores are then ranked and evaluated in accordance with the ranking protocol described in Section IV.

References

Conover, W. (1980). **Practical nonparametric statistics** (2nd ed.). New York: John Wiley and Sons.

Daniel, W. (1990). **Applied nonparametric statistics** (2nd ed.). Boston: PWS–Kent Publishing Company.

Fisher, R. A. (1935). **The design of experiments** (7th ed.). Edinburgh-London: Oliver and Boyd.

Fisher, R. A. (1973). **Statistical methods for research workers** (14th ed.). New York: Hafner.

Marascuilo, L. and McSweeney, M. (1977). **Nonparametric and distribution-free methods for the social sciences**. Monterey, CA: Brooks/Cole Publishing Company.

Siegel, S. and Castellan, N., Jr. (1988). **Nonparametric statistics for the behavioral sciences** (2nd ed.). New York: McGraw–Hill Book Company.

Van der Waerden, B. L. (1952/1953). Order tests for the two-sample problem and their power. **Proceedings Koninklijke Nederlandse Akademie van Wetenshappen** (A), 55 (**Indagationes Mathematicae** 14), 453–458, and 56 (**Indagationes Mathematicae**, 15), 303–316 (corrections appear in Vol. 56, p. 80).

Wilcoxon, F. (1945). Individual comparisons by ranking methods. **Biometrics**, 1, 80–83.

Wilcoxon, F. (1949). **Some rapid approximate statistical procedures**. Stamford, CT: Stamford Research Laboratories, American Cyanamid Corporation.

Endnotes

1. Some sources note that one assumption of the **Wilcoxon matched-pairs signed-ranks test** is that the variable being measured is based on a continuous distribution. In practice, however, this assumption is often not adhered to.

2. When there are tie scores for either the lowest or highest difference scores, as a result of averaging the ordinal positions of the tied scores, the rank assigned to the lowest difference score will be some value greater than 1, and the rank assigned to the highest difference score will be some value less than n.

3. A more thorough discussion of **Table A5** can be found in Section V of the **Wilcoxon signed-ranks test.**

4. The concept of **power efficiency** is discussed in Section VII of the **Mann-Whitney *U* test.**

5. The term $(\Sigma t^3 - \Sigma t)$ in Equation 13.4 can also be written as $\Sigma_{i=1}^{s}(t_i^3 - t_i)$. The latter notation indicates the following: a) For each set of ties, the number of ties in the set is subtracted from the cube of the number of ties in that set; and b) the sum of all the values computed in part a) is obtained. Thus, in the example under discussion (in which there are $s = 3$ sets of ties):

$$\sum_{i=1}^{s}(t_i^3 - t_i) = [(3)^3 - 3] + [(2)^3 - 2] + [(3)^3 - 3] = 54$$

   The computed value of 54 is the same as the corresponding value $(\Sigma t^3 - \Sigma t) = 62 - 8 = 54$ computed in Equation 13.4 through use of Table 13.4.

6. A correction for continuity can be used in conjunction with the tie correction by subtracting .5 from the absolute value computed for the numerator of Equation 13.4. Use of the correction for continuity will reduce the tie corrected absolute value of z.

Test 14

The Binomial Sign Test for Two Dependent Samples
(Nonparametric Test Employed with Ordinal Data)

I. Hypothesis Evaluated with Test and Relevant Background Information

Hypothesis evaluated with test Do two dependent samples represent two different populations?

Relevant background information on test The **binomial sign test for two dependent samples** is essentially an extension of the **binomial sign test for a single sample (Test 6)** to a design involving two dependent samples. Since a complete discussion of the binomial distribution (which is the distribution upon which the test is based) is contained in the discussion of the **binomial sign test for a single sample**, the reader is advised to read the material on the latter test prior to continuing this section. Whenever one or more of the assumptions of the ***t* test for two dependent samples (Test 12) or the Wilcoxon matched-pairs signed-ranks test (Test 13)** is saliently violated, the **binomial sign test for a two dependent samples** can be employed as an alternative procedure. The reader should review the assumptions of the aforementioned tests, as well as the information on a dependent samples design discussed in Sections I and VII of the ***t* test for two dependent samples**.

In order to employ the **binomial sign test for two dependent samples**, it is required that each of n subjects (or n pairs of matched subjects) has two scores (each score having been obtained under one of the two experimental conditions). The two scores are represented by the notations X_1 and X_2. For each subject (or pair of matched subjects), a determination is made with respect to whether a subject obtains a higher score in Condition 1 or Condition 2. Based on the latter, a signed difference ($D+$ or $D-$) is assigned to each pair of scores. The sign of the difference assigned to a pair of scores will be positive if a higher score is obtained in Condition 1 (i.e., $D+$ if $X_1 > X_2$), whereas the sign of the difference will be negative if a higher score is obtained in Condition 2 (i.e., $D-$ if $X_2 > X_1$). The hypothesis the **binomial sign test for two dependent samples** evaluates is whether or not in the underlying population represented by the sample, the proportion of subjects who obtain a positive signed difference (i.e., obtain a higher score in Condition 1) is some value other than .5. If the proportion of subjects who obtain a positive signed difference (which, for the underlying population, is represented by the notation $\pi+$) is some value that is either significantly above or below .5, it indicates there is a high likelihood the two dependent samples represent two different populations.

The **binomial sign test for two dependent samples** is based on the following assumptions:[1] a) The sample of n subjects has been randomly selected from the population it represents; and b) The format of the data is such that within each pair of scores the two scores can be rank-ordered.

As is the case for the ***t* test for two dependent samples** and the **Wilcoxon matched-pairs signed-ranks test**, in order for the **binomial sign test for two dependent samples** to generate valid results, the following guidelines should be adhered to: a) In order to control for order effects, the presentation of the two experimental conditions should be random or, if appropriate, be counterbalanced; and b) If matched samples are employed, within each pair of matched subjects each of the subjects should be randomly assigned to one of the two experimental conditions.

As is the case with the **t test for two dependent samples** and the **Wilcoxon matched-pairs signed-ranks test**, the **binomial sign test for two dependent samples** can also be employed to evaluate a **before-after design**. The limitations of the **before-after design** (which are discussed in Section VII of the **t test for two dependent samples**) are also applicable when it is evaluated with the **binomial sign test for two dependent samples**.

II. Example

Example 14.1 is identical to Examples 12.1 and 13.1 (which are respectively evaluated with the **t test for two dependent samples** and the **Wilcoxon matched-pairs signed-ranks test**). In evaluating Example 14.1 it will be assumed that the **binomial sign test for two dependent samples** is employed, since one or more of the assumptions of the **t test for two dependent samples** and the **Wilcoxon matched-pairs signed-ranks test** have been saliently violated.

Example 14.1 *A psychologist conducts a study in order to determine whether people exhibit more emotionality when they are exposed to sexually explicit words than when they are exposed to neutral words. Each of ten subjects is shown a list of* 16 *randomly arranged words which are projected onto a screen one at a time for a period of five seconds. Eight of the words on the list are sexually explicit in nature and eight of the words are neutral. As each word is projected on the screen, a subject is instructed to say the word softly to him or herself. As a subject does this, sensors attached to the palms of the subject's hands record galvanic skin response (GSR), which is used by the psychologist as a measure of emotionality. The psychologist computes two scores for each subject, one score for each of the experimental conditions:* **Condition 1**: *GSR/Explicit — The average GSR score for the eight sexually explicit words;* **Condition 2**: *GSR/Neutral – The average GSR score for the eight neutral words. The GSR/Explicit and the GSR/Neutral scores of the ten subjects follow. (The higher the score the higher the level of emotionality.)* **Subject 1** (9, 8); **Subject 2** (2, 2); **Subject 3** (1, 3); **Subject 4** (4, 2); **Subject 5** (6, 3); **Subject 6** (4, 0); **Subject 7** (7, 4); **Subject 8** (8, 5); **Subject 9** (5, 4); **Subject 10** (1, 0). *Do subjects exhibit differences in emotionality with respect to the two categories of words?*

III. Null versus Alternative Hypotheses

Null hypothesis H_0: $\pi + = .5$

(In the underlying population the sample represents, the proportion of subjects who obtain a positive signed difference (i.e., a higher score in Condition 1 than Condition 2) equals .5.)

Alternative hypothesis H_1: $\pi + \neq .5$

(In the underlying population the sample represents, the proportion of subjects who obtain a positive signed difference (i.e., a higher score in Condition 1 than Condition 2) does not equal .5. This is a **nondirectional alternative hypothesis**, and it is evaluated with a **two-tailed test**. In order to be supported, the observed proportion of positive signed differences in the sample data (which will be represented with the notation $p+$) can be either significantly larger than the hypothesized population proportion $\pi + = .5$ or significantly smaller than $\pi + = .5$.)

or

$$H_1\text{: } \pi+ > .5$$

(In the underlying population the sample represents, the proportion of subjects who obtain a positive signed difference (i.e., a higher score in Condition 1 than Condition 2) is greater than .5. This is a **directional alternative hypothesis**, and it is evaluated with a **one-tailed test.** In order to be supported, the observed proportion of positive signed differences in the sample data must be significantly larger than the hypothesized population proportion $\pi+ = .5$.)

or

$$H_1\text{: } \pi+ < .5$$

(In the underlying population the sample represents, the proportion of subjects who obtain a positive signed difference (i.e., a higher score in Condition 1 than Condition 2) is less than .5. This is a **directional alternative hypothesis**, and it is evaluated with a **one-tailed test.** In order to be supported, the observed proportion of positive signed differences in the sample data must be significantly smaller than the hypothesized population proportion $\pi+ = .5$.)

Note: Only one of the above noted alternative hypotheses is employed. If the alternative hypothesis the researcher selects is supported, the null hypothesis is rejected.[2]

IV. Test Computations

The data for Example 14.1 are summarized in Table 14.1. Note that there are 10 subjects and that each subject has two scores.

Table 14.1 Data for Example 14.1

Subject	X_1	X_2	$D = X_1 - X_2$	Signed Difference
1	9	8	1	+
2	2	2	0	0
3	1	3	-2	–
4	4	2	2	+
5	6	3	3	+
6	4	0	4	+
7	7	4	3	+
8	8	5	3	+
9	5	4	1	+
10	1	0	1	+
				$\Sigma D+ = 8$
				$\Sigma D- = 1$

The following information can be derived from Table 14.1: a) Eight subjects (Subjects 1, 4, 5, 6, 7, 8, 9, 10) yield a difference score with a positive sign — i.e., a positive signed difference; b) One subject (Subject 3) yields a difference score with a negative sign — i.e., a negative signed difference; and c) One subject (Subject 2) obtains the identical score in both conditions, and as a result of this yields a difference score of zero.

As is the case with the **Wilcoxon matched-pairs signed-ranks test**, in employing the **binomial sign test for two dependent samples**, any subject who obtains a zero difference score is eliminated from the data analysis. Since Subject 2 falls in this category, the size of the sample is reduced to $n = 9$, which is the same number of signed ranks employed when the **Wilcoxon matched-pairs signed-ranks test** is employed to evaluate the same set of data.

The sampling distribution of the signed differences represents a binomially distributed variable with an expected probability of .5 for each of the two mutually exclusive categories (i.e., positive signed difference versus negative signed difference). The logic underlying the **binomial sign test for two dependent samples** is that if the two experimental conditions represent equivalent populations, the signed differences should be randomly distributed. Thus, assuming that subjects who obtain a difference score of zero are eliminated from the analysis, if the remaining signed differences are, in fact, randomly distributed, one-half of the subjects should obtain a positive signed difference and one-half of the subjects should obtain a negative signed difference. In Example 14.1 the observed proportion of positive signed differences is $p+ = 8/9 = .89$ and the observed proportion of negative signed differences is $p- = 1/9 = .11$.

Equation 14.1 (which is identical to Equation 6.5, except for the fact that $\pi+$ and $\pi-$ are used in place of π_1 and π_2) is employed to determine the probability of obtaining $x = 8$ or more positive signed differences in a set of $n = 9$ scores.

$$P(\geq x) = \sum_{r=x}^{n} \binom{n}{x} (\pi+)^x \, (\pi-)^{(n-x)} \qquad \textbf{(Equation 14.1)}$$

Where: $\pi+$ and $\pi-$ respectively represent the hypothesized values for the proportion of positive and negative signed differences
n represents the number of signed differences
x represents the number of positive signed differences

In employing Equation 14.1 with Example 14.1, the following values are employed: a) $\pi+ = .5$ and $\pi- = .5$, since if the null hypothesis is true, the proportion of positive and negative signed differences should be equal. Note that the sum of $\pi+$ and $\pi-$ must always equal 1; b) $n = 9$, since there are 9 signed differences; and c) $x = 8$, since 8 subjects obtain a positive signed difference.

The notation $\Sigma_{r=x}^{n}$ in Equation 14.1 indicates that the probability of obtaining a value of x equal to the observed number of positive signed differences must be computed, as well as the probability for all values of x greater than the observed number of positive signed differences up through and including the value of n. Thus, in the case of Example 14.1, the binomial probability must be computed for the values $x = 8$ and $x = 9$. Equation 14.1 is employed below to compute the latter probability. The obtained value .0195 represents the likelihood of obtaining 8 or more positive signed differences in a set of $n = 9$ signed differences.

$$P(x \geq 8) = \binom{9}{8} (.5)^8 \, (.5)^1 + \binom{9}{9} (.5)^9 \, (.5)^0 = .0195$$

An even more efficient way of obtaining the probability $P(8 \text{ or } 9/9) = .0195$ is through use of **Table A7 (Table of the Binomial Distribution, Cumulative Probabilities)** in the **Appendix**. In employing **Table A7** we find the section for $n = 9$, and locate the cell that is the intersection of the row $x = 8$ and the column $\pi = .5$. The entry .0195 in that cell represents the probability of obtaining 8 or more (i.e., 8 and 9) positive signed differences, if there are a total of 9 signed differences.[3]

Equation 14.2 (which is identical to Equation 6.3 employed for the **binomial sign test for a single sample**, except for the fact that $\pi+$ and $\pi-$ are employed in place of π_1 and π_2) can be employed to compute each of the individual probabilities that are summed in Equation 14.1.

$$P(x) = \binom{n}{x} (\pi+)^x (\pi-)^{(n-x)} \qquad \textbf{(Equation 14.2)}$$

Since the computation of binomial probabilities can be quite tedious, in lieu of employing Equation 14.2, **Table A6 (Table of the Binomial Distribution, Individual Probabilities)** in the **Appendix** can be used to determine the appropriate probabilities. In employing **Table A6** we find the section for $n = 9$, and locate the cell that is the intersection of the row $x = 8$ and the column $\pi = .5$. The entry .0176 in that cell represents the probability of obtaining exactly 8 positive signed differences, if there are a total of 9 signed differences. Additionally, we locate the cell that is the intersection of the row $x = 9$ and the column $\pi = .5$. The entry .0020 in that cell represents the probability of obtaining exactly 9 positive signed differences, if there are a total of 9 signed differences. Summing the latter two values yields the value $P(8 \text{ or } 9/9) = .0196$, which is the likelihood of observing 8 or 9 positive signed differences in a set of $n = 9$ signed differences.[4] For a comprehensive discussion on the computation of binomial probabilities and the use of **Tables A6** and **A7**, the reader should review Section IV of the **binomial sign test for a single sample.**

V. Interpretation of the Test Results

The following guidelines are employed in evaluating the null hypothesis.

a) If a nondirectional alternative hypothesis is employed, the null hypothesis can be rejected if the probability of obtaining a value equal to or more extreme than x is equal to or less than $\alpha/2$ (where α represents the prespecified value of α). The reader should take note of the fact that if the proportion of positive signed differences in the data (i.e., $p+$) is greater than $\pi+ = .5$, a value that is more extreme than x will be any value that falls above the observed value of x, whereas if the proportion of positive signed differences in the data is less than $\pi+ = .5$, a value that is more extreme than x will be any value that falls below the observed value of x.

b) If a directional alternative hypothesis is employed which predicts that the underlying population proportion is above the hypothesized value $\pi+ = .5$, in order to reject the null hypothesis both of the following conditions must be met: 1) The proportion of positive signed differences must be greater than the value $\pi+ = .5$ stipulated in the null hypothesis; and 2) The probability of obtaining a value equal to or greater than x is equal to or less than the prespecified value of α.

c) If a directional alternative hypothesis is employed which predicts that the underlying population proportion is below the hypothesized value $\pi+ = .5$, in order to reject the null hypothesis both of the following conditions must be met: 1) The proportion of positive signed differences must be less than the value $\pi+ = .5$ stipulated in the null hypothesis; and 2) The probability of obtaining a value equal to or less than x is equal to or less than the prespecified value of α.

Applying the above guidelines to Example 14.1, we can conclude the following.

The nondirectional alternative hypothesis H_1: $\pi+ \neq .5$ is supported at the $\alpha = .05$ level, since the obtained probability .0195 is less than $\alpha/2 = .05/2 = .025$. The nondirectional alternative hypothesis H_1: $\pi+ \neq .5$ is not supported at the $\alpha = .01$ level, since the probability .0195 is greater than $\alpha/2 = .01/2 = .005$.[5]

The directional alternative hypothesis H_1: $\pi+ > .5$ is supported at the $\alpha = .05$ level. This is the case because: a) The data are consistent with the directional alternative hypothesis H_1: $\pi+ > .5$, since $p+ = .89$ is greater than the value $\pi+ = .5$ stated in the null hypothesis; and b) The obtained probability .0195 is less than $\alpha = .05$. The directional alternative hypothesis H_1: $\pi+ > .5$ is not supported at the $\alpha = .01$ level, since the probability .0195 is greater than $\alpha = .01$.

The directional alternative hypothesis H_1: $\pi+ < .5$ is not supported, since the data are not consistent with it. Specifically, $p+ = .89$ does not meet the requirement of being less than the value $\pi+ = .5$ stated in the null hypothesis.

A summary of the analysis of Example 14.1 with the **binomial sign test for two dependent samples** follows: It can be concluded that subjects exhibited higher GSR (emotionality) scores with respect to the sexually explicit words than the neutral words.

When the **binomial sign test for two dependent samples** and the **Wilcoxon matched-pairs signed-ranks test** are applied to the same set of data, the two tests yield identical conclusions. Specifically, both tests support the nondirectional alternative hypothesis and the directional alternative hypothesis that is consistent with the data at the .05 level. Although it is not immediately apparent from this example, as a general rule, when applied to the same data the **binomial sign test for two dependent samples** tends to be less powerful than the **Wilcoxon matched-pairs signed-ranks test.** This is the case, since by not considering the magnitude of the difference scores, the **binomial sign test for two dependent samples** employs less information than the **Wilcoxon matched-pairs signed-ranks test.** As is the case with the **Wilcoxon matched-pairs signed-ranks test**, the **binomial sign test for two dependent samples** utilizes less information than the ***t* test for two dependent samples**, and thus in most instances, it will provide a less powerful test of an alternative hypothesis than the latter test. In point of fact, in the case of Example 14.1, if the data are evaluated with the ***t* test for two dependent samples**, the directional alternative hypothesis that is consistent with the data is supported at both the .05 and .01 levels. It should be noted, however, that if the normality assumption of the ***t* test for two dependent samples** is saliently violated, in some instances the **binomial sign test for two dependent samples** may provide a more powerful test of an analogous alternative hypothesis.

VI. Additional Analytical Procedures for the Binomial Sign Test for Two Dependent Samples and/or Related Tests

1. The normal approximation of the binomial sign test for two dependent samples with and without a correction for continuity With large sample sizes the normal approximation for the binomial distribution (which is discussed in Section VI of the **binomial sign test for a single sample**) can provide a large sample approximation for the **binomial sign test for two dependent samples.** As a general rule, most sources recommend employing the normal approximation for sample sizes larger than those documented in the table of the binomial distribution contained in the source. Equation 14.3 (which is equivalent to Equation 6.7) is the normal approximation equation for the **binomial sign test for two dependent samples.** When a **correction for continuity** is employed, Equation 14.4 (which is equivalent to Equation 6.9) is employed.[6]

$$z = \frac{x - (n)(\pi+)}{\sqrt{(n)(\pi+)(\pi-)}} \qquad \textbf{(Equation 14.3)}$$

$$z = \frac{|x - (n)(\pi+)| - .5}{\sqrt{(n)(\pi+)(\pi-)}} \qquad \textbf{(Equation 14.4)}$$

Although Example 14.1 involves only nine signed ranks (a value most sources would view as too small to use with the normal approximation), it will be employed to illustrate Equations 14.3 and 14.4. The reader will see that in spite of employing the normal approximation with a small sample size, it yields essentially the same results as those obtained with the exact binomial probabilities.

Employing Equation 14.3, the value $z = 2.33$ is computed.

$$z = \frac{8 - (9)(.5)}{\sqrt{(9)(.5)(.5)}} = 2.33$$

Employing Equation 14.4, the value $z = 2.00$ is computed.

$$z = \frac{|\, 8 - (9)(.5)\, | - .5}{\sqrt{(9)(.5)(.5)}} = 2.00$$

The obtained z values are evaluated with **Table A1 (Table of the Normal Distribution)** in the **Appendix**. In **Table A1** the tabled critical two-tailed .05 and .01 values are $z_{.05} = 1.96$ and $z_{.01} = 2.58$, and the tabled critical one-tailed .05 and .01 values are $z_{.05} = 1.65$ and $z_{.01} = 2.33$. The following guidelines are employed in evaluating the null hypothesis.

a) If a nondirectional alternative hypothesis is employed, the null hypothesis can be rejected if the obtained absolute value of z is equal to or greater than the tabled critical two-tailed value at the prespecified level of significance.

b) If a directional alternative hypothesis is employed, only the directional alternative hypothesis that is consistent with the data can be supported. With respect to the latter alternative hypothesis, the null hypothesis can be rejected if the obtained absolute value of z is equal to or greater than the tabled critical one-tailed value at the prespecified level of significance.

Employing the above guidelines, we can conclude the following.

Since the value $z = 2.33$ computed with Equation 14.3 is greater than the tabled critical two-tailed value $z_{.05} = 1.96$ but less than the tabled critical two-tailed value $z_{.01} = 2.58$, the nondirectional alternative hypothesis H_1: $\pi + \neq .5$ is supported, but only at the .05 level. Since the value $z = 2.33$ is greater than the tabled critical one-tailed value $z_{.05} = 1.65$ and equal to the tabled critical one-tailed value $z_{.01} = 2.33$, the directional alternative hypothesis H_1: $\pi + > .5$ is supported at both the .05 and .01 levels. Note that when the exact binomial probabilities are employed, both the nondirectional alternative hypothesis H_1: $\pi + \neq .5$ and the directional alternative hypothesis H_1: $\pi + > .5$ are supported, but in both instances, only at the .05 level.

When the correction for continuity is employed, the value $z = 2.00$ computed with Equation 14.4 is greater than the tabled critical two-tailed value $z_{.05} = 1.96$ but less than the tabled critical two-tailed value $z_{.01} = 2.58$. Thus, the nondirectional alternative hypothesis H_1: $\pi + \neq .5$ is only supported at the .05 level. Since the value $z = 2.00$ is greater than the tabled critical one-tailed value $z_{.05} = 1.65$ but less than the tabled critical one-tailed value $z_{.01} = 2.33$, the directional alternative hypothesis H_1: $\pi + > .5$ is also supported at only the .05 level. The results with the correction for continuity are identical to those obtained when the exact binomial probabilities are employed. Note that the continuity corrected normal approximation provides a more conservative test of the null hypothesis than does the uncorrected normal approximation.

The **chi-square goodness-of-fit test (Test 5)** (either with or without the correction for continuity) can also be employed to provide a large sample approximation of the **binomial sign test for two dependent samples**. The **chi-square goodness-of-fit test**, which evaluates the relationship between the observed and expected frequencies in the two categories (i.e., positive signed difference versus negative signed difference), will yield a result that is equivalent to that obtained with the normal approximation. The computed chi-square value will equal the square of the z value derived with the normal approximation.

Equation 5.2 is employed for the chi-square analysis, without using a correction for continuity. Equation 5.6 is the continuity corrected equation. Table 14.2 summarizes the analysis of Example 14.1 with Equation 5.2.

Table 14.2 Chi-Square Summary Table for Example 14.1

Cell	O_i	E_i	$(O_i - E_i)$	$(O_i - E_i)^2$	$\frac{(O_i - E_i)^2}{E_i}$
Positive signed differences	8	4.5	3.5	12.25	2.72
Negative signed differences	1	4.5	-3.5	12.25	2.72
	$\Sigma O_i = 9$	$\Sigma E_i = 9$	$\Sigma(O_i - E_i) = 0$		$\chi^2 = 5.44$

In Table 14.2, the expected frequency $E_i = 4.5$ for each cell is computed by multiplying the hypothesized population proportion for the cell (.5 for both cells) by $n = 9$. Since $k = 2$, the degrees of freedom employed for the chi-square analysis are $df = k - 1 = 2$. The obtained value $\chi^2 = 5.44$ is evaluated with **Table A4 (Table of the Chi-Square Distribution)** in the **Appendix**. For $df = 1$, the tabled critical .05 and .01 chi-square values are $\chi^2_{.05} = 3.84$ (which corresponds to the chi-square value at the 95th percentile) and $\chi^2_{.01} = 6.63$ (which corresponds to the chi-square value at the 99th percentile). Since the obtained value $\chi^2 = 5.44$ is greater than $\chi^2_{.05} = 3.84$ but less than $\chi^2_{.01} = 6.63$, the nondirectional alternative hypothesis H_1: $\pi + \neq .5$ is supported, but only at the .05 level. Since $\chi^2 = 5.44$ is greater than the tabled critical one-tailed .05 value $\chi^2_{.05} = 2.71$ (which corresponds to the chi-square value at the 90th percentile) and the tabled critical one-tailed .01 value $\chi^2_{.01} = 5.43$ (which corresponds to the chi-square value at the 98th percentile), the directional alternative hypothesis H_1: $\pi + > .5$ is supported at both the .05 and .01 levels.[7] The aforementioned conclusions are identical to those reached when Equation 14.3 is employed.

As noted previously, if the z value obtained with Equation 14.3 is squared, it will always equal the chi-square value computed for the same data. Thus, in the current example where $z = 2.33$ and $\chi^2 = 5.44$, $(2.33)^2 = 5.43$.[8]

Equation 5.6 (which, as noted previously, is the continuity corrected equation for the **chi-square goodness-of-fit test**) is employed below, and yields an equivalent result to that obtained with Equation 14.4. In employing Equation 5.6, the value $(|O_i - E_i| - .5) = 3$ is employed for each cell. Thus:

$$\chi^2 = \sum_{i=1}^{k} \left[\frac{(|O_i - E_i| - .5)^2}{E_i} \right] = \frac{(3)^2}{4.5} + \frac{(3)^2}{4.5} = 4$$

Note that the obtained value $\chi^2 = 4$ is equal to the square of the value $z = 2.00$ obtained with Equation 14.4. Since $\chi^2 = 4$ is greater than $\chi^2_{.05} = 3.84$ but less than $\chi^2_{.01} = 6.63$, the nondirectional alternative hypothesis H_1: $\pi + \neq .5$ is supported, but only at the .05 level. Since $\chi^2 = 4$ is greater than the tabled critical one-tailed .05 value $\chi^2_{.05} = 2.71$ but less than the tabled critical one-tailed .01 value $\chi^2_{.01} = 5.43$, the directional alternative hypothesis H_1: $\pi + > .5$ is also supported at only the .05 level. The aforementioned conclusions are identical to those reached when Equation 14.4 is employed.

2. Computation of a confidence interval for the binomial sign test for two dependent samples Equation 14.5 (which is equivalent to Equation 5.5, except for the fact that $p+$, $p-$, and $\pi+$ are employed in place of p_1, p_2, and π_1) can be used to compute a confidence interval for the **binomial sign test for two dependent samples**. Since Equation 14.5 is based on the normal approximation, it should be employed with large sample sizes. It will, however, be used here with the data for Example 14.1. The 95% confidence interval computed with Equation 14.5 estimates the proportion of positive signed differences in the underlying population.

$$\left[p+ - z_{(\alpha/2)}\sqrt{\frac{(p+)(p-)}{n}}\right] \leq \pi+ \leq \left[p+ + z_{(\alpha/2)}\sqrt{\frac{(p+)(p-)}{n}}\right] \quad \textbf{(Equation 14.5)}$$

Employing the values computed for Example 14.1, Equation 14.5 is employed to compute the 95% confidence interval.

$$\left[.89 - (1.96)\sqrt{\frac{(.89)(.11)}{9}}\right] \leq \pi+ \leq \left[.89 + (1.96)\sqrt{\frac{(.89)(.11)}{9}}\right]$$

$$\pi+ = .89 \pm .204$$

$$.686 \leq \pi+ \leq 1.094$$

Thus, the researcher can be 95% confident that the true proportion of positive signed differences in the underlying population is a value between .686 and 1.094. Obviously, since a proportion cannot be greater than 1, the range of values identified by the confidence interval will fall between .686 and 1.

It should be noted that the above method for computing a confidence interval ignores the presence of any zero difference scores. Consequently, the range of values computed for the confidence interval assumes there are no zero difference scores in the underlying population. If, in fact, there are zero difference scores in the population, the above computed confidence interval only identifies proportions that are relevant to the total number of cases in the population that are not zero difference scores. In point of fact, when one or more zero difference scores are present in the sample data, a researcher may want to assume that zero difference scores are present in the underlying population. If the researcher makes such an assumption and employs the sample data to estimate the proportion of zero difference scores in the population, the value employed for $p+$ in Equation 14.5 will represent the number of positive signed differences in the sample divided by the total number of scores in the sample, including any zero difference scores. Thus, in the case of Example 14.1, the value $p+ = 8/10 = .8$ is computed by dividing 8 (the number of positive signed differences) by $n = 10$. The value $p-$ in Equation 14.5 will no longer represent just the negative signed differences, but will represent all signed differences that are not positive (i.e., both negative signed differences and zero difference scores). Thus, in the case of Example 14.1, $p- = 2/10 = .2$, since there is one negative signed difference and one zero difference score. If the values $n = 10$, $p+ = .8$, and $p- = .2$ are employed in Equation 14.5, the confidence interval $.552 \leq \pi+ \leq 1.048$ is computed.

$$\left[.8 - (1.96)\sqrt{\frac{(.8)(.2)}{10}}\right] \leq \pi+ \leq \left[.8 + (1.96)\sqrt{\frac{(.8)(.2)}{10}}\right]$$

$$\pi+ = .8 \pm .248$$

$$.552 \leq \pi+ \leq 1.048$$

Thus, the researcher can be 95% confident that the true proportion of positive signed differences in the underlying population is a value between .552 and 1.048. Since, as noted earlier, a proportion cannot be greater than 1, the range identified by the confidence interval will fall between .552 and 1. If the researcher wants to employ the same method for computing a confidence interval for the proportion of minus signed differences in the population (i.e., $\pi-$), the product $z_{(\alpha/2)}\sqrt{[(p+)(p-)]/n}$ is added to and subtracted from $p- = .1$. The values $p- = 1/10 = .1$ and $p+ = 9/10 = .9$ are employed in the confidence interval equation, since there is one negative signed difference and nine signed differences that are not negative (i.e., eight positive signed differences and one zero difference score).

3. Sources for computing the power of the binomial sign test for two dependent samples The protocol employed for computing the power of the **binomial sign test for a single sample** can be extended to computing the power of the **binomial sign test for two dependent samples** (since the power for both tests is computed in reference to a specific value for the proportion of observations in one of two categories). Although the protocol for computing the power of the **binomial sign test** (for both a single sample and two dependent samples) is not described in this book, a comprehensive discussion of the subject as well as the necessary tables can be found in Cohen (1988).

VII. Additional Discussion of the Binomial Sign Test for Two Dependent Samples

1. The problem of an excessive number of zero difference scores When there is an excessive number of subjects who have a zero difference score in a set of data, a substantial amount of information is sacrificed if the **binomial sign test for two dependent samples** is employed to evaluate the data. Under such conditions, it is advisable to evaluate the data with the ***t* test for two dependent samples** (assuming the interval/ratio scores of subjects are available). If one or more of the assumptions of the latter test is saliently violated, the alpha level employed for the t test should be adjusted.

2. Equivalency of the Friedman two-way analysis variance by ranks and the binomial sign test for two dependent samples when $k = 2$ In Section VII of the **Friedman two-way analysis of variance by ranks (Test 19)**, it is demonstrated that when there are two dependent samples and there are no zero difference scores, the latter test (which can be employed for two or more dependent samples) is equivalent to the chi-square approximation of the **binomial sign test for two dependent samples** (i.e., it will yield the same chi-square value computed with Equation 5.2). When employing the **Friedman two-way analysis of variance by ranks** with two dependent samples, the two scores of each subject (or pair of matched subjects) are rank-ordered. The data for Example 14.1 can be expressed in a rank-order format, if for each subject a rank of 1 is assigned to the lower of the two scores and a rank of 2 is assigned to the higher score (or vice versa). If a researcher only has such rank-order information, it is still possible to assign a signed difference to each subject, since the ordering of a subject's two ranks provides sufficient information to determine whether the difference between the two scores of a subject would yield a positive or negative value if the interval/ratio scores of the subject were available. Consequently, under such conditions one can still conduct the **binomial sign test for two dependent samples.** On the other hand, if a researcher only has the sort of rank-order information noted above, one will not be able to evaluate the data with either the ***t* test for two dependent samples** or the **Wilcoxon matched-pairs signed-ranks test**, since the latter two tests require the interval/ratio scores of subjects.

VIII. Additional Examples Illustrating the Use of the Binomial Sign Test for Two Dependent Samples

The **binomial sign test for two dependent samples** can be employed with any of the additional examples noted for the ***t* test for two dependent samples** and the **Wilcoxon matched-pairs signed-ranks test.** In each of the examples, a signed difference must be computed for each subject (or pair of matched subjects). The signed differences are then evaluated employing the protocol for the **binomial sign test for two dependent samples.**

References

Cohen, J. (1988). **Statistical power analysis for the behavioral sciences** (2nd ed.). Hillsdale, NJ: Lawrence Erlbaum Associates, Publishers.

Daniel, W. W. (1990). **Applied nonparametric statistics** (2nd ed.). Boston: PWS–Kent Publishing Company.

Marascuilo, L. A. and McSweeney, M. (1977). **Nonparametric and distribution-free methods for the social sciences.** Monterey, CA: Brooks/Cole Publishing Company.

Siegel, S. and Castellan, N. J., Jr. (1988). **Nonparametric statistics for the behavioral sciences** (2nd ed.). New York: McGraw–Hill Book Company.

Endnotes

1. Some sources note that one assumption of the **binomial sign test for two dependent samples** is that the variable being measured is based on a continuous distribution. In practice, however, this assumption is often not adhered to.

2. Another way of stating the null hypothesis is that in the underlying population the sample represents, the proportion of subjects who obtain a positive signed difference is equal to the proportion of subjects who obtain a negative signed difference.

   The null and alternative hypotheses can also be stated with respect to the proportion of people in the population who obtain a higher score in Condition 2 than Condition 1, thus yielding a negative difference score. The notation $\pi-$ represents the proportion of the population who yield a difference with a negative sign (referred to as a negative signed difference). Thus, H_0: $\pi- = .5$ can be employed as the null hypothesis, and the following nondirectional and directional alternative hypotheses can be employed: H_1: $\pi- \neq .5$; H_1: $\pi- > .5$; H_1: $\pi- < .5$.

3. It is also the likelihood of obtaining 8 or 9 negative signed differences in a set of 9 signed differences.

4. Due to rounding off protocol, the value computed with Equation 14.1 will be either .0195 or .0196, depending upon whether one employs **Table A6** or **Table A7.**

5. An equivalent way of determining whether or not the result is significant is by doubling the value of the cumulative probability obtained from **Table A7.** In order to reject the null hypothesis, the resulting value must not be greater than the value of α. Since $2 \times .0195 = .039$ is less than $\alpha = .05$, we confirm that the nondirectional alternative hypothesis is supported when $\alpha = .05$. Since .039 is greater than $\alpha = .01$, it is not supported at the .01 level.

6. Equations 6.6 and 6.8 are respectively alternate but equivalent forms of Equations 14.3 and 14.4. Note that in Equations 6.6–6.9, π_1 and π_2 are employed in place of $\pi+$ and $\pi-$ to represent the two population proportions.

7. A full discussion of the protocol for determining one-tailed chi-square values can be found in Section VII of the **chi-square goodness-of-fit test.**

8. The minimal discrepancy is the result of rounding off error.

Test 15

The McNemar Test

(Nonparametric Test Employed with Categorical/Nominal Data)

I. Hypothesis Evaluated with Test and Relevant Background Information

Hypothesis evaluated with test Do two dependent samples represent two different populations?

Relevant background information on test It is recommended that before reading the material on the **McNemar test**, the reader review the general information on a dependent samples design contained in Sections I and VII of the ***t* test for two dependent samples (Test 12)**. The **McNemar test** (McNemar, 1947) is a nonparametric procedure for categorical data employed in a hypothesis testing situation involving a design with two dependent samples. In actuality, the **McNemar test** is a special case of the **Cochran *Q* test (Test 20)**, which can be employed to evaluate a k dependent samples design involving categorical data, where $k \geq 2$. The **McNemar test** is employed to evaluate an experiment in which a sample of n subjects (or n pairs of matched subjects) is evaluated on a dichotomous dependent variable (i.e., scores on the dependent variable must fall within one of two mutually exclusive categories). The **McNemar test** assumes that each of the n subjects (or n pairs of matched subjects) contributes two scores on the dependent variable. The test is most commonly employed to analyze data derived from the two types of experimental designs described below.

a) The **McNemar test** can be employed to evaluate categorical data obtained in a **true experiment** (i.e., an experiment involving a manipulated independent variable).[1] In such an experiment, the two scores of each subject (or pair of matched subjects) represent a subject's responses under the two levels of the independent variable (i.e., the two experimental conditions). A significant result allows the researcher to conclude there is a high likelihood the two experimental conditions represent two different populations. As is the case with the ***t* test for two dependent samples**, the **Wilcoxon matched-pairs signed-ranks test (Test 13)**, and the **binomial sign test for two dependent samples (Test 14)**, when the **McNemar test** is employed to evaluate the data for a **true experiment**, in order for the test to generate valid results, the following guidelines should be adhered to: 1) In order to control for order effects, the presentation of the two experimental conditions should be random or, if appropriate, be counterbalanced; and 2) If matched samples are employed, within each pair of matched subjects each of the subjects should be randomly assigned to one of the two experimental conditions.

b) The **McNemar test** can be employed to evaluate a **before–after design** (which is described in Section VII of the ***t* test for two dependent samples**). In applying the **McNemar test** to a **before–after design**, n subjects are administered a pretest on a dichotomous dependent variable. Following the pretest, all of the subjects are exposed to an experimental treatment, after which they are administered a posttest on the same dichotomous dependent variable. The hypothesis evaluated with a **before–after design** is whether or not there is a significant difference between the pretest and posttest scores of subjects on the dependent variable. The reader is advised to review the discussion of the **before–after design** in Section

VII of the **t test for two dependent samples**, since the limitations noted for the design also apply when it is evaluated with the **McNemar test.**

The 2 × 2 table depicted in Table 15.1 summarizes the **McNemar test** model. The entries for Cells a, b, c, and d in Table 15.1 represent the number of subjects/observations in each of four possible categories that can be employed to summarize the two responses of a subject (or matched pair of subjects) on a dichotomous dependent variable. Each of the four response category combinations represents the number of subjects/observations whose response in Condition 1/Pretest falls in the response category for the row in which the cell falls, and whose response in Condition 2/Posttest falls in the response category for the column in which the cell falls. Thus, the entry in Cell a represents the number of subjects who respond in Response category 1 in both Condition 1/Pretest and Condition 2/Posttest. The entry in Cell b represents the number of subjects who respond in Response category 1 in Condition 1/Pretest and in Response category 2 in Condition 2/Posttest. The entry in Cell c represents the number of subjects who respond in Response category 2 in Condition 1/Pretest and in Response category 1 in Condition 2/Posttest. The entry in Cell d represents the number of subjects who respond in Response category 2 in both Condition 1/Pretest and Condition 2/Posttest.

Table 15.1 Model for the McNemar Test

		Condition 2/Posttest		
		Response category 1	Response category 2	Row sums
Condition 1/Pretest	Response category 1	a	b	$a + b = n_1$
	Response category 2	c	d	$c + d = n_2$
	Column sums	$a + c$	$b + d$	n

The **McNemar test** is based on the following assumptions: a) The sample of n subjects has been randomly selected from the population it represents; b) Each of the n observations in the contingency table is independent of the other observations; c) The scores of subjects are in the form of a dichotomous categorical measure involving two mutually exclusive categories; and d) Most sources state that the **McNemar test** should not be employed with extremely small sample sizes. Although the chi-square distribution is generally employed to evaluate the **McNemar test** statistic, in actuality the latter distribution is used to provide an approximation of the exact sampling distribution which is, in fact, the binomial distribution. When the sample size is small, in the interest of accuracy, the exact binomial probability for the data should be computed. Sources do not agree on the minimum acceptable sample size for computing the **McNemar test** statistic (i.e., using the chi-square distribution). Some sources endorse the use of a correction for continuity with small sample sizes (discussed in Section VI), in order to insure that the computed chi-square value provides a more accurate estimate of the exact binomial probability.

II. Examples

Since, as noted in Section I, the **McNemar test** is employed to evaluate a **true experiment** and a **before-after design**, two examples representing each of the aforementioned designs will be presented in this section. Since the two examples employ identical data, they will result in the same conclusion with respect to the null hypothesis. Example 15.1 describes a **true experiment** and Example 15.2 describes a study that employs a **before-after design.**

Example 15.1 *A psychologist wants to compare a drug for treating enuresis (bed-wetting) with a placebo. 100 enuretic children are administered both the drug (Endurin) and a placebo in a double blind study conducted over a six month period. During the duration of the study each child has six drug and six placebo treatments, with each treatment lasting one week. In order to insure that there are no carryover effects from one treatment to the another, during the week following each treatment a child is not given either the drug or the placebo. The order of presentation of the* 12 *treatment periods for each child is randomly determined. The dependent variable in the study is a parent's judgement with respect to whether or not a child improves under each of the two experimental conditions.* Table 15.2 *summarizes the results of the study. Do the data indicate the drug was effective?*

Table 15.2 Summary of Data for Example 15.1

		Favorable response to drug		
		Yes	No	Row sums
Favorable response to placebo	Yes	10	13	23
	No	41	36	77
	Column sums	51	49	100

Note that the data in Table 15.2 indicate the following: a) 10 subjects respond favorably to both the drug and the placebo; b) 13 subjects do not respond favorably to the drug but do respond favorably to the placebo; c) 41 subjects respond favorably to the drug but do not respond favorably to the placebo; and d) 36 subjects do not respond favorably to either the drug or the placebo. Of the 100 subjects, 51 respond favorably to the drug, while 49 do not. 23 of the 100 subjects respond favorably to the placebo, while 77 do not.

Example 15.2 *A researcher conducts a study to investigate whether a weekly television series that is highly critical of the use of animals as subjects in medical research influences public opinion. One hundred randomly selected subjects are administered a pretest to determine their attitude concerning the use of animals in medical research. Based on their responses, subjects are categorized as pro-animal research or anti-animal research. Following the pretest, all of the subjects are instructed to watch the television series (which last two months). At the conclusion of the series each subject's attitude toward animal research is reassessed. The results of the study are summarized in* Table 15.3. *Do the data indicate that a shift in attitude toward animal research occurred after subjects viewed the television series?*

Table 15.3 Summary of Data for Example 15.2

		Posttest		
		Anti	Pro	Row sums
Pretest	Anti	10	13	23
	Pro	41	36	77
	Column sums	51	49	100

Note that the data in Table 15.3 indicate the following: a) Ten subjects express an anti-animal research attitude on both the pretest and the posttest; b) Thirteen subjects express an anti-animal research attitude on the pretest but a pro-animal research attitude on the posttest; c) Forty-one subjects express a pro-animal research attitude on the pretest but an anti-animal research attitude on the posttest; and d) Thirty-six subjects express a pro-animal research attitude on both the pretest and the posttest. Of the 100 subjects, 23 are anti on the pretest and 77 are pro on the pretest. Fifty-one of the 100 subjects are anti on the posttest, while 49 are pro on the posttest.

Table 15.4 summarizes that data for Examples 15.1 and 15.2.

Table 15.4 Summary of Data for Examples 15.1 and 15.2

		Favorable response to drug/Posttest		
		Yes/Anti	No/Pro	Row sums
Favorable response to placebo/Pretest	Yes/Anti	$a = 10$	$b = 13$	23
	No/Pro	$c = 41$	$d = 36$	77
	Column sums	51	49	100

III. Null versus Alternative Hypotheses

In conducting the **McNemar test**, the cells of interest in Table 15.4 are Cells b and c, since the latter two cells represent those subjects who respond in different response categories under the two experimental conditions (in the case of a **true experiment**) or in the pretest versus posttest (in the case of a **before–after design**). In Example 15.1, the frequencies recorded in Cells b and c respectively represent subjects who respond **favorably to the placebo/unfavorably to the drug** and **favorably to the drug/unfavorably to the placebo.** If the drug is more effective than the placebo, one would expect the proportion of subjects in Cell c to be larger than the proportion of subjects in Cell b. In Example 15.2, the frequencies recorded in Cells b and c respectively represent subjects who are **anti-animal research in the pretest/pro-animal research in the posttest** and **pro-animal research in the pretest/anti-animal research in the posttest.** If there is a shift in attitude from the pretest to the posttest (specifically from **pro-animal research** to **anti-animal research**), one would expect the proportion of subjects in Cell c to be larger than the proportion of subjects in Cell b.

It will be assumed that in the underlying population, π_b and π_c represent the following proportions: $\pi_b = b/(b + c)$ and $\pi_c = c/(b + c)$. If there is no difference between the two experimental conditions (in the case of a **true experiment**) or between the pretest and the posttest (in the case of a **before–after design**), the following will be true: $\pi_b = \pi_c = .5$. With respect to the sample data, the values π_b and π_c are estimated with the values p_b and p_c, which in the case of Examples 15.1 and 15.2 are $p_b = b/(b + c) = 13/(13 + 41) = .24$ and $p_c = c/(b + c) = 41/(13 + 41) = .76$.

Employing the above information the null and alternative hypotheses for the **McNemar test** can now be stated.[2]

Null hypothesis $H_0\colon \pi_b = \pi_c$

(In the underlying population the sample represents, the proportion of observations in Cell b equals the proportion of observations in Cell c.)

Alternative hypothesis $H_1\colon \pi_b \neq \pi_c$

(In the underlying population the sample represents, the proportion of observations in Cell b does not equal the proportion of observations in Cell c. This is a **nondirectional alternative hypothesis** and it is evaluated with a two-tailed test. In order to be supported, the proportion of observations in Cell b (p_b) can be either significantly larger or significantly smaller than the proportion of observations in Cell c (p_c). In the case of Example 15.1, this alternative hypothesis will be supported if the proportion of subjects who respond **favorably to the placebo/unfavorably to the drug** is significantly greater than the proportion of subjects who respond **favorably to the drug/unfavorably to the placebo**, or the proportion of subjects who respond **favorably to the drug/unfavorably to the placebo** is significantly greater than the proportion of subjects who respond **favorably to the placebo/unfavorably to the drug**. In the case of Example 15.2, this alternative hypothesis will be supported if in the pretest versus posttest, a significantly larger proportion of subjects shift their response from **pro-animal research** to **anti-animal research** or a significantly larger proportion of subjects shift their response from **anti-animal research** to **pro-animal research**.)

or

$$H_1\colon \pi_b > \pi_c$$

(In the underlying population the sample represents, the proportion of observations in Cell b is greater than the proportion of observations in Cell c. This is a **directional alternative hypothesis** and it is evaluated with a one-tailed test. In order to be supported, the proportion of observations in Cell b (p_b) must be significantly larger than the proportion of observations in Cell c (p_c). In the case of Example 15.1, this alternative hypothesis will be supported if the proportion of subjects who respond **favorably to the placebo/ unfavorably to the drug** is significantly greater than the proportion of subjects who respond **favorably to the drug/ unfavorably to the placebo**. In the case of Example 15.2, this alternative hypothesis will be supported if in the pretest versus posttest, a significantly larger proportion of subjects shift their response from **anti-animal research** to **pro-animal research**.)

or

$$H_1\colon \pi_b < \pi_c$$

(In the underlying population the sample represents, the proportion of observations in Cell b is less than the proportion of observations in Cell c. This is a **directional alternative hypothesis** and it is evaluated with a one-tailed test. In order to be supported, the proportion of observations in Cell b (p_b) must be significantly smaller than the proportion of observations in Cell c (p_c). In the case of Example 15.1, this alternative hypothesis will be supported if the proportion of subjects who respond **favorably to the drug/unfavorably to the placebo** is significantly greater than the proportion of subjects who respond **favorably to the placebo/unfavorably to the drug**. In the case of Example 15.2, this alternative hypothesis will be supported if in the pretest versus posttest, a significantly larger proportion of subjects shift their responses from **pro-animal research** to **anti-animal research**.)

Note: Only one of the above noted alternative hypotheses is employed. If the alternative hypothesis the researcher selects is supported, the null hypothesis is rejected.

IV. Test Computations

The test statistic for the **McNemar test**, which is based on the chi-square distribution, is computed with Equation 15.1.[3]

$$\chi^2 = \frac{(b - c)^2}{b + c} \qquad \textbf{(Equation 15.1)}$$

Where: b and c represent the number of observations in Cells b and c of the **McNemar test** summary table

Substituting the appropriate values in Equation 15.1, the value $\chi^2 = 14.52$ is computed for Examples 15.1/15.2.

$$\chi^2 = \frac{(13 - 41)^2}{13 + 41} = 14.52$$

The computed chi-square value must always be a positive number. If a negative value is obtained, it indicates that an error has been made. The only time the value of chi-square will equal zero is when $b = c$.

V. Interpretation of the Test Results

The obtained value $\chi^2 = 14.52$ is evaluated with **Table A4 (Table of the Chi-Square Distribution)** in the **Appendix.**[4] The degrees of freedom employed in the analysis are $df = 1$.[5] Employing **Table A4**, for $df = 1$ the tabled critical two-tailed .05 and .01 chi-square values are $\chi^2_{.05} = 3.84$ (which corresponds to the chi-square value at the 95th percentile) and $\chi^2_{.01} = 6.63$ (which corresponds to the chi-square value at the 99th percentile). The tabled critical one-tailed .05 and .01 values are $\chi^2_{.05} = 2.71$ (which corresponds to the chi-square value at the 90th percentile) and $\chi^2_{.01} = 5.43$ (which corresponds to the chi-square value at the 98th percentile).[6]

The following guidelines are employed in evaluating the null hypothesis for the **McNemar test.**

a) If the nondirectional alternative hypothesis H_1: $\pi_b \neq \pi_c$ is employed, the null hypothesis can be rejected if the obtained chi-square value is equal to or greater than the tabled critical two-tailed value at the prespecified level of significance.

b) If a directional alternative hypothesis is employed, only the directional alternative hypothesis that is consistent with the data can be supported. With respect to the latter alternative hypothesis, the null hypothesis can be rejected if the obtained chi-square value is equal to or greater than the tabled critical one-tailed value at the prespecified level of significance.

Applying the above guidelines to Examples 15.1/15.2, we can conclude the following.

Since the obtained value $\chi^2 = 14.52$ is greater than the tabled critical two-tailed values $\chi^2_{.05} = 3.84$ and $\chi^2_{.01} = 6.63$, the nondirectional alternative hypothesis H_1: $\pi_b \neq \pi_c$ is supported at both the .05 and .01 levels. Since $\chi^2 = 14.52$ is greater than the tabled critical one-tailed values $\chi^2_{.05} = 2.71$ and $\chi^2_{.01} = 5.43$, the directional alternative hypothesis H_1: $\pi_b < \pi_c$ is supported at both the .05 and .01 levels (since $p_b = .24$ is less than $p_c = .76$).

A summary of the analysis of Examples 15.1 and 15.2 with the **McNemar test** follows:

Example 15.1: It can be concluded that the proportion of subjects who respond favorably to the drug is significantly greater than the proportion of subjects who respond favorably to the placebo.

Example 15.2: It can be concluded that following exposure to the television series, there is a significant change in attitude toward the use of animals as subjects in medical research. The direction of the change is from pro-animal research to anti-animal research. It is important to note, however, that since Example 15.2 is based on a **before–after design,** the researcher is not be justified in concluding that the change in attitude is a direct result of subjects watching the television series. This is the case because (as noted in Section VII of the ***t* test for two dependent samples**) a **before–after design** is an incomplete experimental design. Specifically, in order to be an adequately controlled experimental design, a **before–after design** requires the addition of a control group that is administered the identical pretest and posttest at the same time periods as the group described in Example 15.2. The control group, however, would not be exposed to the television series between the pretest and the posttest. Without inclusion of such a control group, it is not possible to determine whether an observed change in attitude from the pretest to the posttest is due to the experimental treatment (i.e., the television series), or is the result of one or more extraneous variables that may also have been present during the intervening time period between the pretest and the posttest.

VI. Additional Analytical Procedures for the McNemar Test and/or Related Tests

1. Alternative equation for the McNemar test statistic based on the normal distribution Equation 15.2 is an alternative equation that can be employed to compute the **McNemar test** statistic. It yields a result that is equivalent to that obtained with Equation 15.1.

$$z = \frac{b - c}{\sqrt{b + c}} \qquad \textbf{(Equation 15.2)}$$

The sign of the computed z value is only relevant insofar as it indicates the directional alternative hypothesis with which the data are consistent. Specifically, the z value computed with Equation 15.2 will be a positive number if the number of observations in Cell b is greater than the number of observations in Cell c, and it will be a negative number if the number of observations in Cell c is greater than the number of observations in Cell b. Since in Examples 15.1/15.2 $c > b$, the computed value of z will be a negative number. Substituting the appropriate values in Equation 15.2, the value $z = -3.81$ is computed.

$$z = \frac{13 - 41}{\sqrt{13 + 41}} = -3.81$$

The square of the z value obtained with Equation 15.2 will always equal the chi-square value computed with Equation 15.1. This relationship can be confirmed by the fact that $(z = -3.81)^2 = (\chi^2 = 14.52)$. It is also the case that the square of a tabled critical z value at a given level of significance will equal the tabled critical chi-square value at the corresponding level of significance.

The obtained z value is evaluated with **Table A1 (Table of the Normal Distribution)** in the **Appendix.** In **Table A1** the tabled critical two-tailed .05 and .01 values are $z_{.05} = 1.96$ and $z_{.01} = 2.58$, and the tabled critical one-tailed .05 and .01 values are $z_{.05} = 1.65$ and $z_{.01} = 2.33$. In interpreting the z value computed with Equation 15.2, the following guidelines are employed.

a) If the nondirectional alternative hypothesis H_1: $\pi_b \neq \pi_c$ is employed, the null hypothesis can be rejected if the obtained absolute value of z is equal to or greater than the tabled critical two-tailed value at the prespecified level of significance.

b) If a directional alternative hypothesis is employed, only the directional alternative hypothesis that is consistent with the data can be supported. With respect to the latter alternative hypothesis, the null hypothesis can be rejected if the obtained absolute value of z is equal to or greater than the tabled critical one-tailed value at the prespecified level of significance.

Employing the above guidelines with Examples 15.1/15.2, we can conclude the following.

Since the obtained absolute value $z = 3.81$ is greater than the tabled critical two-tailed values $z_{.05} = 1.96$ and $z_{.01} = 2.58$, the nondirectional alternative hypothesis H_1: $\pi_b \neq \pi_c$ is supported at both the .05 and .01 levels. Since the obtained absolute value $z = 3.81$ is greater than the tabled critical one-tailed values $z_{.05} = 1.65$ and $z_{.01} = 2.33$, the directional alternative hypothesis H_1: $\pi_b < \pi_c$ is supported at both the .05 and .01 levels. These conclusions are identical to those reached when Equation 15.1 is employed to evaluate the same set of data.

2. The correction for continuity for the McNemar test Since the **McNemar test** employs a continuous distribution to approximate a discrete probability distribution, some sources recommend that a correction for continuity be employed in computing the test statistic. Sources that recommend such a correction, either recommend it be limited to small sample sizes or that it be used in all instances.[7] Equations 15.3 and 15.4 are the continuity corrected versions of Equations 15.1 and 15.2.[8]

$$\chi^2 = \frac{(|b - c| - 1)^2}{b + c} \qquad \textbf{(Equation 15.3)}$$

$$z = \frac{|b - c| - 1}{\sqrt{b + c}} \qquad \textbf{(Equation 15.4)}$$

Substituting the appropriate values in Equations 15.3 and 15.4, the values $\chi^2 = 13.5$ and $z = 3.67$ are computed for Examples 15.1/15.2.

$$\chi^2 = \frac{(|13 - 41| - 1)^2}{13 + 41} = 13.5$$

$$z = \frac{|13 - 41| - 1}{\sqrt{13 + 41}} = 3.67$$

As is the case without the continuity correction, the square of the z value obtained with Equation 15.4 will always equal the chi-square value computed with Equation 15.3. This relationship can be confirmed by the fact that $(z = 3.67)^2 = (\chi^2 = 13.5)$. Note that the chi-square value computed with Equation 15.3 will always be less than the value computed with Equation 15.1. In the same respect, the absolute value of z computed with Equation 15.4 will always be less than the absolute value of z computed with Equation 15.2. The lower absolute values computed for the continuity corrected statistics reflect the fact that the latter analysis provides a more conservative test of the null hypothesis than does the uncorrected analysis. In this instance, the decision the researcher makes with respect to the null hypothesis is not affected by the correction for continuity, since the values $\chi^2 = 13.5$ and $z = 3.67$ are both greater than the relevant tabled critical one- and two-tailed .05 and .01 values. Thus, the nondirectional alternative hypothesis H_1: $\pi_b \neq \pi_c$ and the directional alterative hypothesis H_1: $\pi_b < \pi_c$ are supported at both the .05 and .01 levels.

3. Computation of the exact binomial probability for the McNemar test model with a small sample size In Section I it is noted that the exact probability distribution for the **McNemar test** model is the binomial distribution, and that the chi-square distribution is employed to approximate the latter distribution. Although for large sample sizes the chi-square distribution provides an excellent approximation of the binomial distribution, many sources recommend that for small sample sizes the exact binomial probabilities be computed. In order to demonstrate the computation of an exact binomial probability for the **McNemar test** model, assume that Table 15.5 is a revised summary table for Examples 15.1/15.2.

Note that although the frequencies for Cells a and d in Table 15.5 are identical to those employed in Table 15.4, different frequencies are employed for Cells b and c. Although in Table 15.5 the total sample size of $n = 56$ is reasonably large, the total number of subjects in Cells b and c is quite small, and, in the final analysis, it is when the sum of the frequencies of the latter two cells is small that computation of the exact binomial probability is recommended. The fact that the frequencies of Cells a and d are not taken into account

Table 15.5 Revised Summary Table for Examples 15.1 and 15.2 for Binomial Analysis

		Favorable response to drug/Posttest		
		Yes/Anti	No/Pro	Row sums
Favorable response to placebo/Pretest	Yes/Anti	$a = 10$	$b = 2$	12
	No/Pro	$c = 8$	$d = 36$	44
	Column sums	18	38	56

represents an obvious limitation of the **McNemar test**. In point of fact, the frequencies of Cells a and d could be 0 and 0 instead of 10 and 36, and the same result will be obtained when the **McNemar test** statistic is computed. In the same respect, frequencies of 1000 and 3600 for Cells a and d will also yield the identical result. Common sense suggests, however, that the difference $|b - c| = |2 - 8| = 6$ will be considered more important if the total sample size is small (which will be the case if the frequencies of Cells a and d are 0 and 0) than if the total sample size is very large (which will be the case if the frequencies of Cells a and d are 1000 and 3600). What the latter translates into is that a significant difference between Cells b and c may be of little or no practical significance, if the total number of observations in all four cells is very large.

Employing Equations 15.1 and 15.2 with the data in Table 15.5, the values $\chi^2 = 3.6$ and $z = 1.90$ are computed for the **McNemar test** statistic.

$$\chi^2 = \frac{(2 - 8)^2}{2 + 8} = 3.6$$

$$z = \frac{2 - 8}{\sqrt{2 + 8}} = 1.90$$

Employing Equations 15.3 and 15.4, the values $\chi^2 = 2.5$ and $z = 1.58$ are the continuity corrected values computed for the **McNemar test** statistic.

$$\chi^2 = \frac{(|2 - 8| - 1)^2}{2 + 8} = 2.5$$

$$z = \frac{|2 - 8| - 1}{\sqrt{2 + 8}} = 1.58$$

Employing **Table A1**, we determine that the exact one-tailed probability for the value $z = 1.90$ computed with Equation 15.2 (as well as for $\chi^2 = 3.6$ computed with Equation 15.1) is .0287. We also determine that the exact one-tailed probability for the value $z = 1.58$ computed with Equation 15.4 (as well as for $\chi^2 = 2.5$ computed with Equation 15.3) is .0571.[9] Note that since the continuity correction results in a more conservative test, the probability associated with the continuity corrected value will always be higher than the probability associated with the uncorrected value. Without the continuity correction, the directional alternative hypothesis H_1: $\pi_b < \pi_c$ is supported at the .05 level, since $\chi^2 = 3.6/z = 1.90$ are greater than the tabled critical one-tailed values $\chi^2_{.05} = 2.71/z_{.05} = 1.65$. The nondirectional alternative hypothesis H_1: $\pi_b \neq \pi_c$ is not supported, since $\chi^2 = 3.6/z = 1.90$ are less than the tabled critical two-tailed values $\chi^2_{.05} = 3.84/z_{.05} = 1.96$. When the continuity correction is employed, the directional alternative hypothesis H_1: $\pi_b < \pi_c$ fails to achieve significance at the .05 level, since $\chi^2 = 2.5/z = 1.58$ are less than $\chi^2_{.05} = 2.71/z_{.05} = 1.65$. The nondirectional alternative hypothesis H_1: $\pi_b \neq \pi_c$

is not supported, since $\chi^2 = 2.5/z = 1.58$ are less than the tabled critical two-tailed values $\chi^2_{.05} = 3.84/z_{.05} = 1.96$.

At this point, the exact binomial probability will be computed for the same set of data. As is the case with the equations for the **McNemar test** that are based on the chi-square and normal distributions, the binomial analysis only considers the frequencies of Cells b and c. Since only two cells are taken into account, the binomial analysis becomes identical to the analysis described for the **binomial sign test for a single sample (Test 6).**[10]

Equation 15.5 is the binomial equation that is employed to determine the likelihood of obtaining a frequency of 8 or larger in one of the two cells (or 2 or less in one of the two cells) in the McNemar model summary table, if the total frequency in the two cells is 10, where, $m = b + c = 2 + 8 = 10$. Note that Equation 15.5 is identical to Equation 6.5, except for the fact that π_b and π_c are used in place of π_1 and π_2, and the value m, which represents $b + c$, is used in place of n.

$$P(\geq x) = \sum_{r=x}^{m} \binom{m}{x} (\pi_b)^x \, (\pi_c)^{(m-x)} \qquad \textbf{(Equation 15.5)}$$

In evaluating the data in Table 15.5, the following values are employed in Equation 15.5: $\pi_b = \pi_c = .5$ (which will be the case if the null hypothesis is true), $m = 10$, $x = 8$.

$$P(x \geq 8) = \binom{10}{8} (.5)^8 \, (.5)^2 + \binom{10}{9} (.5)^9 (.5)^1 + \binom{10}{10} (.5)^{10} (.5)^0 = .0547$$

The computed probability .0547 is the likelihood of obtaining a frequency of 8 or greater in one of the two cells (as well as the likelihood of obtaining a frequency of 2 or less in one of the two cells). The value .0547 can also be obtained from **Table A7 (Table of the Binomial Distribution, Cumulative Probabilities)** in the **Appendix.** In using **Table A7** we find the section for $m = 10$ (which is represented by $n = 10$ in the table), and locate the cell that is the intersection of the row $x = 8$ and the column $\pi = .5$. The entry for the latter cell is .0547. The value .0547 computed for the exact binomial probability is quite close to the continuity corrected probability of .0571 obtained with Equations 15.3/15.4 (which suggests that even when the sample size is small, the continuity corrected chi-square/ normal approximation provides an excellent estimate of the exact probability). As is the case when the data are evaluated with Equations 15.3/15.4, the directional alternative hypothesis H_1: $\pi_b < \pi_c$ is not supported if the binomial analysis is employed. This is the case, since the probability .0547 is greater than $\alpha = .05$. In order for the directional alternative hypothesis H_1: $\pi_b < \pi_c$ to be supported, the tabled probability must be equal to or less than $\alpha = .05$. The nondirectional alternative hypothesis H_1: $\pi_b \neq \pi_c$ is also not supported, since the probability .0547 is greater than $\alpha/2 = .05/2 = .025$.[11]

4. Additional analytical procedures for the McNemar test

a) A procedure for computing a confidence interval for the difference between the marginal probabilities (i.e., $[(a + b)/n] - [(a + c)/n]$) in a **McNemar test** summary table is described in Marascuilo and McSweeney (1977) and Fleiss (1981). Daniel (1990) and Fleiss (1981) provide references that discuss the power of the **McNemar test** relative to alternative procedures.

b) Fleiss (1981), who provides a detailed discussion of the **McNemar test,** notes that an **odds ratio** (which is discussed in Section VI of the **chi-square test for $r \times c$ tables (Test 11))** can be computed for the **McNemar test** summary table. Specifically, the **odds ratio** (o) is computed with Equation 15.6. Employing the latter equation, the value $o = 3.15$ is computed for Example 15.1.

$$o = \frac{c}{b} = \frac{41}{13} = 3.15 \qquad \textbf{(Equation 15.6)}$$

In reference to Example 15.1, the computed value $o = 3.15$ indicates that a person is 3.15 times more likely to respond favorably to the drug than to the placebo.

c) Fleiss (1981) also notes that if the null hypothesis is rejected in a study such as that described by Example 15.1, Equation 15.7 can be employed to determine the relative difference (represented by the notation p_e) between the two treatments. Equation 15.7 is employed with the data for Example 15.1 to compute the value $p_e = .36$.

$$p_e = \frac{c - b}{c + d} = \frac{41 - 13}{41 + 36} = .36 \qquad \textbf{(Equation 15.7)}$$

The computed value .36 indicates that in a sample of 100 patients who do not respond favorably to the placebo, $(.36)(100) = 36$ would be expected to respond favorably to the drug. Fleiss (1981) describes the computation of the estimated standard error for the value of the relative difference computed with Equation 15.7, as well as the procedure for computing a confidence interval for the relative difference.

d) In Section IX of the **Pearson product-moment correlation coefficient (Test 22)**, the use of the **phi coefficient (Test 11g** described in Section VI of the **chi-square test for $r \times c$ tables)** as a measure of association for the **McNemar test** model is discussed within the context of the **tetrachoric correlation coefficient (Test 22j)**.

VII. Additional Discussion of the McNemar Test

1. Alternative format for the McNemar test summary table and modified test equation Although in this book Cells b and c are designated as the two cells in which subjects are inconsistent with respect to their response categories, the **McNemar test** summary table can be rearranged so that Cells a and d become the relevant cells. Table 15.6 represents such a rearrangement with respect to the response categories employed in Examples 15.1/15.2.

Table 15.6 Alternative Format for Summary of Data for Examples 15.1 and 15.2

		Favorable response to drug/Posttest		
		Yes/Anti	**No/Pro**	**Row sums**
Favorable response to placebo/Pretest	**No/Pro**	$a = 41$	$b = 36$	**77**
	Yes/Anti	$c = 10$	$d = 13$	**23**
	Column sums	**51**	**49**	**100**

Note that in Table 15.6 Cells a and d are the key cells, since subjects who are inconsistent with respect to response categories are represented in these two cells. If Table 15.6 is employed as the summary table for the **McNemar test**, Cells a and d will respectively replace Cells c and b in stating the null and alternative hypotheses. In addition, Equations 15.8 and 15.9 will respectively be employed in place of Equations 15.1 and 15.2. When the appropriate values are substituted in the aforementioned equations, Equation 15.8 yields the value $\chi^2 = 14.52$ computed with Equation 15.1, and Equation 15.9 yields the value $z = -3.81$ computed with Equation 15.2.

$$\chi^2 = \frac{(d - a)^2}{d + a} = \frac{(13 - 41)^2}{13 + 41} = 14.52 \qquad \textbf{(Equation 15.8)}$$

$$z = \frac{d - a}{\sqrt{d + a}} = \frac{13 - 41}{\sqrt{13 + 41}} = -3.81 \qquad \textbf{(Equation 15.9)}$$

2. Extension of the McNemar test model beyond 2 × 2 contingency tables The **McNemar test** model has been extended by Bowker (1948) and Stuart (1955, 1957) to a dependent samples design in which the dependent variable is a categorical measure that is comprised of more than two categories. In the **Bowker** and **Stuart test** models, a $k \times k$ contingency table (where k is the number of response categories, and $k \geq 3$) is employed to summarize the responses of n subjects (or n pairs of matched subjects) on a dependent variable under the two experimental conditions (or two time periods).[12] Whereas the **Bowker test** evaluates differences with respect to the joint probability distributions (i.e., the cells in the diagonal of a $k \times k$ table), the **Stuart test** evaluates differences between marginal probabilities. Marascuilo and McSweeney (1977) and Marascuilo and Serlin (1988) describe both the **Bowker** and **Stuart tests**. Fleiss (1981) describes the **Stuart test**, and provides references on alternative procedures that evaluate the same general hypothesis.

VIII. Additional Examples Illustrating the Use of the McNemar Test

Three additional examples that can be evaluated with the **McNemar test** are presented in this section. Since Examples 15.3–15.5 employ the same data employed in Example 15.1, they yield the identical result.

Example 15.3 *In order to determine if there is a relationship between schizophrenia and enlarged cerebral ventricles, a researcher evaluates* 100 *pairs of identical twins who are discordant with respect to schizophrenia (i.e., within each twin pair only one member of the pair has schizophrenia). Each subject is evaluated with a CAT scan in order to determine whether or not there is enlargement of the ventricles. The results of the study are summarized in* Table 15.7. *Do the data indicate there is a statistical relationship between schizophrenia and enlarged ventricles?*

Table 15.7 Summary of Data for Example 15.3

		Schizophrenic twin		Row sums
		Enlarged ventricles	Normal ventricles	
Normal twin	Enlarged ventricles	10	13	23
	Normal ventricles	41	36	77
	Column sums	51	49	100

Since Table 15.7 summarizes categorical data derived from $n = 100$ pairs of matched subjects, the **McNemar test** is employed to evaluate the data. The reader should take note of the fact that since the independent variable employed in Example 15.3 is **nonmanipulated** (specifically, it is whether or not a subject is **schizophrenic** or **normal**), analysis of the data will only provide correlational information, and thus will not allow the researcher to draw conclusions with regard to cause and effect. In other words, although the study indicates that schizophrenic subjects are significantly more likely than normal subjects to have enlarged ventricles, one cannot conclude that enlarged ventricles cause schizophrenia or that schizophrenia causes enlarged ventricles. Although either of the latter is possible, the design of the study only allows one to conclude that the presence of enlarged ventricles is associated with schizophrenia.

Example 15.4 *A company that manufactures an insecticide receives complaints from its employees about premature hair loss. An air quality analysis reveals a large concentration of a vaporous compound emitted by the insecticide within the confines of the factory. In order to determine whether or not the vaporous compound (which is known as Acherton) is related to hair loss, the following study is conducted. Each of* 100 *mice is exposed to air containing high concentrations of Acherton over a two-month period. The same mice are also exposed to air that is uncontaminated with Acherton during another two-month period. Half of the mice are initially exposed to the Acherton contaminated air followed by the uncontaminated air, while the other half are initially exposed to the uncontaminated air followed by the Acherton contaminated air. The dependent variable in the study is whether or not a mouse exhibits hair loss during an experimental condition.* Table 15.8 *summarizes the results of the study. Do the data indicate a relationship between Acherton and hair loss?*

Table 15.8 Summary of Data for Example 15.4

		Acherton contaminated air		Row sums
		Hair loss	No hair loss	
Uncontaminated air	Hair loss	10	13	23
	No hair loss	41	36	77
	Column sums	51	49	100

Analysis of the data in Table 15.8 reveals that the mice are significantly more likely to exhibit hair loss when exposed to Acherton as opposed to when they are exposed to uncontaminated air. Although the results of the study suggest that Acherton may be responsible for hair loss, one cannot assume that the results can be generalized to humans.

Example 15.5 *A market research firm is hired to determine whether a debate between the two candidates who are running for the office of Governor influences voter preference. The gubernatorial preference of* 100 *randomly selected voters is determined before and after a debate between the two candidates, Edgar Vega and Vera Myers.* Table 15.9 *summarizes the results of the voter preference survey. Do the data indicate that the debate influenced voter preference?*

Table 15.9 Summary of Data for Example 15.5

		Voter preference before debate		Row sums
		Edgar Vega	Vera Meyers	
Voter preference after debate	Edgar Vega	10	13	23
	Vera Meyers	41	36	77
	Column sums	51	49	100

When the data for Example 15.5 (which represents a **before-after design**) are evaluated with the **McNemar test**, the result indicates that following the debate there is a significant shift in voter preference in favor of Vera Myers. As noted in Section V, since a **before-after design** does not adequately control for the potential influence of extraneous variables, one cannot rule out the possibility that some factor other than the debate is responsible for the shift in voter preference.

References

Bowker, A. H. (1948). A test for symmetry in contingency tables. **Journal of the American Statistical Association**, 43, 572–574.

Daniel, W. W. (1990). **Applied nonparametric statistics** (2nd ed.). Boston: PWS–Kent Publishing Company.

Edwards, A. L. (1948). Note on the "correction for continuity" in testing the significance of the difference between correlated proportions. **Psychometrika**, 13, 185–187.

Fleiss, J. L. (1981). **Statistical methods for rates and proportions** (2nd ed.). New York: John Wiley and Sons.

Marascuilo, L. A. and McSweeney, M. (1977). **Nonparametric and distribution-free methods for the social sciences.** Monterey, CA: Brooks/Cole Publishing Company.

Marascuilo, L. A. and Serlin, R. C. (1988). **Statistical methods for the social and behavioral sciences.** New York: W. H. Freeman and Company.

Maxwell, A. E. (1970). Comparing the classification of subjects by two independent judges. **British Journal of Psychiatry**, 116, 651–655.

McNemar, Q. (1947). Note on the sampling error of the difference between correlated proportions or percentages. **Psychometrika**, 12, 153–157.

Siegel, S. and Castellan, N. J., Jr. (1988). **Nonparametric statistics for the behavioral sciences** (2nd ed.). New York: McGraw–Hill Book Company.

Stuart, A. A. (1955). A test for homogeneity of the marginal distributions in a two way classification. **Biometrika**, 42, 412–416.

Stuart, A. A. (1957). The comparison of frequencies in matched samples. **British Journal of Statistical Psychology**, 10, 29–32.

Endnotes

1. The distinction between a **true experiment** and a **natural experiment** is discussed in more detail in the **Introduction** of the book.

2. a) The reader should take note of the following with respect to the null and alternative hypotheses stated in this section:

   a) If n represents the total number of observations in Cells a, b, c, and d, the proportion of observations in Cells b and c can also be expressed as follows: b/n and c/n. The latter two values, however, are not equivalent to the values p_b and p_c that are used to estimate the values π_b and π_c employed in the null and alternative hypotheses.

   b) Many sources employ an alternative but equivalent way of stating the null and alternative hypotheses for the **McNemar test**. Assume that π_1 represents the proportion of observations in the underlying population who respond in Response category 1 in Condition 1/Pretest, and π_2 represents the proportion of observations in the underlying population who respond in Response category 1 in Condition 2/Posttest. With respect to the sample data, the values p_1 and p_2 are employed to estimate π_1 and π_2, where $p_1 = (a + b)/n$ and $p_2 = (a + c)/n$. In the case of Examples 15.1 and 15.2, $p_1 = (10 + 13)/100 = .23$ and $p_2 = (10 + 41)/100 = .51$. If there is no difference in the proportion of observations in Response category 1 in Condition 1/Pretest versus the proportion of observations in Response category 1 in Condition 2/Posttest, p_1 and p_2 would be expected to be equal, and if the latter is true one can conclude that in the underlying population $\pi_1 = \pi_2$. If, however, $p_1 \neq p_2$ (and consequently in the underlying population $\pi_1 \neq \pi_2$), it indicates a difference between the two experimental conditions in the case of a **true experiment**, and a difference between the pretest and the posttest responses of subjects in the case of a **before–after design**. Employing this information, the null hypothesis can be stated as follows: H_0: $\pi_1 = \pi_2$. The null hypothesis H_0: $\pi_1 = \pi_2$ is equivalent to the null hypothesis H_0: $\pi_b = \pi_c$. The **non-directional alternative hypothesis** can be stated as H_1: $\pi_1 \neq \pi_2$. The nondirectional alternative hypothesis H_1: $\pi_1 \neq \pi_2$ is equivalent to the nondirectional alternative hypothesis H_1 : $\pi_b \neq \pi_c$. The two **directional alternative hypotheses** that can be employed are H_1: $\pi_1 > \pi_2$ or H_1: $\pi_1 < \pi_2$. The directional alternative hypothesis H_1: $\pi_1 > \pi_2$ is equivalent to the directional alternative hypothesis H_1: $\pi_b > \pi_c$. The directional alternative hypothesis H_1: $\pi_1 < \pi_2$ is equivalent to the directional alternative hypothesis H_1: $\pi_b < \pi_c$.

3. It can be demonstrated algebraically that Equation 15.1 is equivalent to Equation 5.2 (which is the equation for the **chi-square goodness-of-fit test (Test 5)**). Specifically, if Cells a and d are eliminated from the analysis, and the **chi-square goodness-of-fit test** is employed to evaluate the observations in Cells b and c, $n = b + c$. If the expected probability for each of the cells is .5, Equation 5.2 reduces to Equation 15.1. As will be noted in Section VI, a limitation of the **McNemar test** (which is apparent from inspection of Equation 15.1) is that it only employs the data for two of the four cells in the contingency table.

4. A general overview of the chi-square distribution and interpretation of the values listed in **Table A4** can be found in Sections I and V of the **single-sample chi-square test for a population variance (Test 3)**.

5. The degrees of freedom are based on Equation 5.3, which is employed to compute the degrees of freedom for the **chi-square goodness-of-fit test**. In the case of the **McNemar test**, $df = k - 1 = 2 - 1 = 1$, since only the observations in Cells b and c (i.e., $k = 2$ cells) are evaluated.

6. A full discussion of the protocol for determining one-tailed chi-square values can be found in Section VII of the **chi-square goodness-of-fit test**.

7. A general discussion of the correction for continuity can be found under the **Wilcoxon signed-ranks test (Test 4)**. Fleiss (1981) notes that the correction for continuity for the **McNemar test** was recommended by Edwards (1948).

8. The numerator of Equation 15.4 is sometimes written as $(b - c) \pm 1$. In using the latter format, 1 is added to the numerator if the term $(b - c)$ results in a negative value, and 1 is subtracted from the numerator if the term $(b - c)$ results in a positive value. Since we are only interested in the absolute value of z, it is simpler to employ the numerator in Equation 15.4, which results in the same absolute value that is obtained when the alternative form of the numerator is employed. If the alternative form of the numerator is employed for Examples 15.1/15.2, it yields the value $z = -3.67$.

9. The values .0287 and .0571 respectively represent the proportion of the normal distribution that falls above the values $z = 1.90$ and $z = 1.58$.

10. In point of fact, it can also be viewed as identical to the analysis conducted with the **binomial sign test for two dependent samples**. In Section VII of the **Cochran *Q* test**, it is demonstrated that when the **McNemar test** (as well as the **Cochran *Q* test** when $k = 2$) and the **binomial sign test for two dependent samples** are employed to evaluate the same set of data, they yield equivalent results.

11. For a comprehensive discussion on the computation of binomial probabilities and the use of **Table A7**, the reader should review Section IV of the **binomial sign test for a single sample.**

12. In some sources the **Stuart test** is referred to as the **Stuart-Maxwell test** based on the contribution of Maxwell (1970).

Inferential Statistical Tests Employed with Two or More Independent Samples (and Related Measures of Association/Correlation)

Test 16

The Single-Factor Between-Subjects Analysis of Variance

(Parametric Test Employed with Interval/Ratio Data)

I. Hypothesis Evaluated with Test and Relevant Background Information

Hypothesis evaluated with test In a set of k independent samples (where $k \geq 2$), do at least two of the samples represent populations with different mean values?

Relevant background information on test The term **analysis of variance** (for which the acronym **ANOVA** is often employed) describes a group of inferential statistical procedures developed by the British statistician Sir Ronald Fisher. Analysis of variance procedures are employed to evaluate whether or not there is a difference between at least two means in a set of data for which two or more means can be computed. The test statistic computed for an analysis of variance is based on the F distribution (which is named after Fisher), which is a continuous theoretical probability distribution. A computed F value (commonly referred to as an ***F* ratio**) will always fall within the range $0 \leq F \leq \infty$. As is the case with the t and chi-square distributions discussed earlier in the book, there are an infinite number of F distributions — each distribution being a function of the number of degrees of freedom employed in the analysis (with degrees of freedom being a function of both the number of samples and the number of subjects per sample). A more thorough discussion of the F distribution can be found in Section V.

The **single-factor between-subjects analysis of variance** is the most basic of the analysis of variance procedures.[1] It is employed in a hypothesis testing situation involving k independent samples. In contrast to the ***t* test for two independent samples (Test 8)**, which only allows for a comparison between the means of two independent samples, the **single-factor between-subjects analysis of variance** allows for a comparison of two or more independent samples. The **single-factor between-subjects analysis of variance** is also referred to as the **completely randomized single-factor analysis of variance**, the **simple analysis of variance**, the **one-way analysis of variance**, and the **single-factor analysis of variance.**

In conducting the **single-factor between-subjects analysis of variance**, each of the k sample means is employed to estimate the value of the mean of the population the sample represents. If the computed test statistic is significant, it indicates there is a significant difference between at least two of the sample means in the set of k means. As a result of the latter, the researcher can conclude there is a high likelihood that at least two of the samples represent populations with different mean values.

In order to compute the test statistic for the **single-factor between-subjects analysis of variance**, the **total variability** in the data is divided into **between-groups variability** and **within-groups variability. Between-groups variability** (which is also referred to as **treatment variability**) is essentially a measure of the variance of the means of the k samples. **Within-groups variability** (which is essentially an average of the variance within each of the k samples) is variability that is attributable to chance factors that are beyond the control of a researcher. Since such chance factor are often referred to as experimental error, **within-**

groups variability is also referred to as **error** or **residual variability**. The ***F* ratio**, which is the test statistic for the **single-factor between-subjects analysis of variance**, is obtained by dividing **between-groups variability** by **within-groups variability**. Since **within-groups variability** is employed as a baseline measure of the variability in a set of data that is beyond a researcher's control, it is assumed that if the k samples are derived from a population with the same mean value, the amount of variability between the sample means (i.e., **between-groups variability**) will be approximately the same value as the amount of variability within any single sample (i.e., **within-groups variability**). If, on the other hand, **between-groups variability** is significantly larger than **within-groups variability** (in which case the value of the F ratio will be larger than 1), it is likely that something in addition to chance factors is contributing to the amount of variability between the sample means. In such a case, it is assumed that whatever it is that differentiates the groups from one another (i.e., the independent variable/experimental treatments) accounts for the fact that **between-groups variability** is larger than **within-groups variability**.[2] A thorough discussion of the logic underlying the **single-factor between-subjects analysis of variance** can be found in Section VII.

The **single-factor between-subjects analysis of variance** is employed with interval/ratio data and is based on the following assumptions: a) Each sample has been randomly selected from the population it represents; b) The distribution of data in the underlying population from which each of the samples is derived is normal; and c) The third assumption, which is referred to as the **homogeneity of variance** assumption, states that the variances of the k underlying populations represented by the k samples are equal to one another. The homogeneity of variance assumption is discussed in detail in Section VI.[3] If any of the aforementioned assumptions of the **single-factor between-subjects analysis of variance** is saliently violated, the reliability of the computed test statistic may be compromised.

II. Example

Example 16.1 *A psychologist conducts a study to determine whether noise can inhibit learning. Each of* 15 *subjects is randomly assigned to one of three groups. Each subject is given* 20 *minutes to memorize a list of* 10 *nonsense syllables, which they are told they will be tested on the following day. The five subjects assigned to* **Group 1**, *the* **no noise** *condition, study the list of nonsense syllables while they are in a quiet room. The five subjects assigned to* **Group 2**, *the* **moderate noise** *condition, study the list of nonsense syllables while listening to classical music. The five subjects assigned to* **Group 3**, *the* **extreme noise** *condition, study the list of nonsense syllables while listening to rock music. The number of nonsense syllables correctly recalled by the* 15 *subjects follow:* **Group 1**: 8, 10, 9, 10, 9; **Group 2**: 7, 8, 5, 8, 5; **Group 3**: 4, 8, 7, 5, 7. *Do the data indicate that noise influenced subjects' performance?*

III. Null versus Alternative Hypotheses

Null hypothesis H_0: $\mu_1 = \mu_2 = \mu_3$

(The mean of the population Group 1 represents equals the mean of the population Group 2 represents equals the mean of the population Group 3 represents.)

Alternative hypothesis H_1: Not H_0

(This indicates there is a difference between at least two of the k population means. It is important to note that the alternative hypothesis should not be written as follows: H_1: $\mu_1 \neq \mu_2 \neq \mu_3$. The reason why the latter notation for the alternative hypothesis is incorrect is because it implies that all three population means must differ from one another in order to reject the null hypothesis. In this book it will always be assumed that the alternative hypothesis for the analysis of variance is stated **nondirectionally.**[4] In order to reject the null hypothesis, the obtained F value must be equal to or greater than the tabled critical F value at the prespecified level of significance.)

IV. Test Computations

The test statistic for the **single-factor between-subjects analysis of variance** can be computed with either **computational** or **definitional equations.** Although definitional equations reveal the underlying logic behind the analysis of variance, they involve considerably more calculations than the computational equations. Because of the latter, computational equations will be employed in this section to demonstrate the computation of the test statistic. The definitional equations for the **single-factor between-subjects analysis of variance** are described in Section VII.

The data for Example 16.1 are summarized in Table 16.1. The scores of the $n_1 = 5$ subjects in Group 1 are listed in the column labelled X_1, the scores of the $n_2 = 5$ subjects in Group 2 are listed in the column labelled X_2, and the scores of the $n_3 = 5$ subjects in Group 3 are listed in the column labelled X_3. Since there are an equal number of subjects in each group, the notation n is employed to represent the number of subjects per group. In other words, $n = n_1 = n_2 = n_3$. The columns labelled X_1^2, X_2^2, and X_3^2 list the squares of the scores of the subjects in each of the three groups.

Table 16.1 Data for Example 16.1

Group 1		Group 2		Group 3	
X_1	X_1^2	X_2	X_2^2	X_3	X_3^2
8	64	7	49	4	16
10	100	8	64	8	64
9	81	5	25	7	49
10	100	8	64	5	25
9	81	5	25	7	49
$\Sigma X_1 = 46$	$\Sigma X_1^2 = 426$	$\Sigma X_2 = 33$	$X_2^2 = 227$	$X_3 = 31$	$X_3^2 = 203$
$\bar{X}_1 = \frac{\Sigma X_1}{n_1} = \frac{46}{5} = 9.2$		$\bar{X}_2 = \frac{\Sigma X_2}{n_2} = \frac{33}{5} = 6.6$		$\bar{X}_3 = \frac{\Sigma X_3}{n_3} = \frac{31}{5} = 6.2$	

The notation N represents the total number of subjects employed in the experiment. Thus:

$$N = n_1 + n_2 + \cdots + n_k$$

Since there are $k = 3$ groups:

$$N = n_1 + n_2 + n_3 = 5 + 5 + 5 = 15$$

The value ΣX_T represents the total sum of the scores of the N subjects who participate in the experiment. Thus:

$$\Sigma X_T = \Sigma X_1 + \Sigma X_2 + \cdots + \Sigma X_k$$

Since there are $k = 3$ groups, $\Sigma X_T = 110$.

$$\Sigma X_T = \Sigma X_1 + \Sigma X_2 + \Sigma X_3 = 46 + 33 + 31 = 110$$

$\bar{X}_T$ represents the grand mean, where $\bar{X}_T = \Sigma X_T / N$. Thus, $\bar{X}_T = 110/15 = 7.33$. Although $\bar{X}_T$ is not employed in the computational equations to be described in this section, it is employed in some of the definitional equations described in Section VII.

The value ΣX_T^2 represents the total sum of the squared scores of the N subjects who participate in the experiment. Thus:

$$\Sigma X_T^2 = \Sigma X_1^2 + \Sigma X_2^2 + \ldots + \Sigma X_k^2$$

Since there are $k = 3$ groups, $\Sigma X_T^2 = 856$.

$$\Sigma X_T^2 = \Sigma X_1^2 + \Sigma X_2^2 + \Sigma X_3^2 = 426 + 227 + 203 = 856$$

Although the group means are not required for computing the analysis of variance test statistic, it is recommended that they be computed since visual inspection of the group means can provide the researcher with a general idea of whether or not it is reasonable to expect a significant result. To be more specific, if two or more of the group means are far removed from one another, it is likely that the analysis of variance will be significant (especially if the number of subjects in each group is reasonably large). Another reason for computing the group means is that they are required for comparing individual groups with one another, something that is generally done following the analysis of variance on the full set of data. The latter types of comparisons are described in Section VI.

As noted in Section I, in order to compute the test statistic for the **single-factor between-subjects analysis of variance**, the **total variability** in the data is divided into **between-groups variability** and **within-groups variability**. In order to do this the following values are computed: a) The **total sum of squares** which is represented by the notation SS_T; b) The **between-groups sum of squares** which is represented by the notation SS_{BG}. The **between-groups sum of squares** is the numerator of the equation that represents **between-groups variability** (i.e., the equation that represents the amount of variability between the means of the k groups); and c) The **within-groups sum of squares** which is represented by the notation SS_{WG}. The **within-groups sum of squares** is the numerator of the equation that represents **within-groups variability** (i.e., the equation that represents the average amount of variability within each of the k groups, which, as noted earlier, represents error variability).

Equation 16.1 describes the relationship between SS_T, SS_{BG}, and SS_{WG}.

$$SS_T = SS_{BG} + SS_{WG}$$ **(Equation 16.1)**

Equation 16.2 is employed to compute SS_T.

$$SS_T = \Sigma X_T^2 - \frac{(\Sigma X_T)^2}{N}$$ **(Equation 16.2)**

Employing Equation 16.2, the value $SS_T = 49.33$ is computed.

$$SS_T = 856 - \frac{(110)^2}{15} = 49.33$$

Equation 16.3 is employed to compute SS_{BG}. In Equation 16.3, the notation n_j and ΣX_j respectively represent the values of n and ΣX for the jth group/sample.

$$SS_{BG} = \sum_{j=1}^{k} \left[\frac{(\Sigma X_j)^2}{n_j} \right] - \frac{(\Sigma X_T)^2}{N}$$ **(Equation 16.3)**

The notation $\sum_{j=1}^{k}[(\Sigma X_j)^2/n_j]$ in Equation 16.3 indicates that for each group the value $(\Sigma X_j)^2/n_j$ is computed, and the latter values are summed for all k groups. When there are an equal number of subjects in each group (as is the case in Example 16.1), the notation n can be employed in Equation 16.3 in place of n_j.[5]

With reference to Example 16.1, Equation 16.3 can be rewritten as follows:

$$SS_{BG} = \left[\frac{(\Sigma X_1)^2}{n_1} + \frac{(\Sigma X_2)^2}{n_2} + \frac{(\Sigma X_3)^2}{n_3}\right] - \frac{(\Sigma X_T)^2}{N}$$

Substituting the appropriate values from Example 16.1 in Equation 16.3, the value $SS_{BG} = 26.53$ is computed.[6]

$$SS_{BG} = \left[\frac{(46)^2}{5} + \frac{(33)^2}{5} + \frac{(31)^2}{5}\right] - \frac{(110)^2}{15} = 26.53$$

By algebraically transposing the terms in Equation 16.1, the value of SS_{WG} can be computed with Equation 16.4.

$$SS_{WG} = SS_T - SS_{BG} \qquad \textbf{(Equation 16.4)}$$

Employing Equation 16.4, the value $SS_{WG} = 22.80$ is computed.

$$SS_{WG} = 49.33 - 26.53 = 22.80$$

Since the value obtained with Equation 16.4 is a function of the values obtained with Equations 16.2 and 16.3, if the computations for either of the latter two equations are incorrect Equation 16.4 will not yield the correct value for SS_{WG}. For this reason, one may prefer to compute the value of SS_{WG} with Equation 16.5.

$$SS_{WG} = \sum_{j=1}^{k}\left[\Sigma X_j^2 - \frac{(\Sigma X_j)^2}{n_j}\right] \qquad \textbf{(Equation 16.5)}$$

The summation sign $\sum_{j=1}^{k}$ in Equation 16.5 indicates that for each group the value $\Sigma X_j^2 - [(\Sigma X_j)^2/n_j]$ is computed, and the latter values are summed for all k groups. With reference to Example 16.1, Equation 16.5 can be written as follows:

$$SS_{WG} = \left[\Sigma X_1^2 - \frac{(\Sigma X_1)^2}{n_1}\right] + \left[\Sigma X_2^2 - \frac{(\Sigma X_2)^2}{n_2}\right] + \left[\Sigma X_3^2 - \frac{(\Sigma X_3)^2}{n_3}\right]$$

Employing Equation 16.5, the value $SS_{WG} = 22.80$ is computed, which is the same value computed with Equation 16.4.[7]

$$SS_{WG} = \left[426 - \frac{(46)^2}{5}\right] + \left[227 - \frac{(33)^2}{5}\right] + \left[203 - \frac{(31)^2}{5}\right] = 22.80$$

The reader should take note of the fact that the values SS_T, SS_{BG}, and SS_{WG} must always be positive numbers. If a negative value is obtained for any of the aforementioned values, it indicates a computational error has been made.

At this point the values of the **between-groups variance** and the **within-groups variance** can be computed. In the **single-factor between-subjects analysis of variance**, the **between-groups variance** is referred to as the **mean square between-groups**, which is represented by the notation MS_{BG}. MS_{BG} is computed with Equation 16.6.

$$MS_{BG} = \frac{SS_{BG}}{df_{BG}}$$ **(Equation 16.6)**

The **within-groups variance** is referred to as the **mean square within-groups**, which is represented by the notation MS_{WG}. MS_{WG} is computed with Equation 16.7.

$$MS_{WG} = \frac{SS_{WG}}{df_{WG}}$$ **(Equation 16.7)**

Note that a total mean square is not computed.

In order to compute MS_{BG} and MS_{WG}, it is required that the values df_{BG} and df_{WG} (the denominators of Equations 16.6 and 16.7) be computed. df_{BG}, which represents the **between-groups degrees of freedom**, are computed with Equation 16.8.

$$df_{BG} = k - 1$$ **(Equation 16.8)**

df_{WG}, which represents the **within-groups degrees of freedom**, are computed with Equation 16.9.[8]

$$df_{WG} = N - k$$ **(Equation 16.9)**

Although it is not required to determine the F ratio, the **total degrees of freedom** are generally computed, since it can be used to confirm the df values computed with Equations 16.8 and 16.9, as well as the fact that it is employed in the analysis of variance summary table. The total degrees of freedom (represented by the notation df_T), are computed with Equation 16.10.[9]

$$df_T = N - 1$$ **(Equation 16.10)**

The relationship between df_{BG}, df_{WG}, and df_T is described by Equation 16.11.

$$df_T = df_{BG} + df_{WG}$$ **(Equation 16.11)**

Employing Equations 16.8–16.10, the values $df_{BG} = 2$, $df_{WG} = 12$, and $df_T = 14$ are computed. Note that $df_T = df_{BG} + df_{WG} = 2 + 12 = 14$.

$$df_{BG} = 3 - 1 = 2 \qquad df_{WG} = 15 - 3 = 12 \qquad df_T = 15 - 1 = 14$$

Employing Equations 16.6 and 16.7, the values $MS_{BG} = 13.27$ and $MS_{WG} = 1.9$ are computed.

$$MS_{BG} = \frac{26.53}{2} = 13.27 \qquad MS_{WG} = \frac{22.8}{12} = 1.9$$

The F ratio, which is the test statistic for the **single-factor between-subjects analysis of variance**, is computed with Equation 16.12.

$$F = \frac{MS_{BG}}{MS_{WG}}$$ **(Equation 16.12)**

Employing Equation 16.12, the value $F = 6.98$ is computed.

$$F = \frac{13.27}{1.9} = 6.98$$

The reader should take note of the fact that the values MS_{BG}, MS_{WG}, and F must always be positive numbers. If a negative value is obtained for any of the aforementioned values, it indicates a computational error has been made. If $MS_{WG} = 0$, Equation 16.12 will be insoluble. The only time $MS_{WG} = 0$ is when within each group all subjects obtain the same score (i.e., there is no within-groups variability). If all of the groups have the identical mean value, $MS_{BG} = 0$, and if the latter is true, $F = 0$.

V. Interpretation of the Test Results

It is common practice to summarize the results of a **single-factor between-subjects analysis of variance** with the summary table represented by Table 16.2.

Table 16.2 Summary Table of Analysis of Variance for Example 16.1

Source of variation	*SS*	*df*	*MS*	*F*
Between-groups	26.53	2	13.27	6.98
Within-groups	22.80	12	1.90	
Total	**49.33**	**14**		

The obtained value $F = 6.98$ is evaluated with **Table A10 (Table of the *F* Distribution) in the Appendix.** In **Table A10** critical values are listed in reference to the number of degrees of freedom associated with the numerator and the denominator of the F ratio (i.e., df_{num} and df_{den}).[10] In employing the F distribution in reference to Example 16.1, the degrees of freedom for the numerator are $df_{BG} = 2$ and the degrees of freedom for the denominator are $df_{WG} = 12$. In **Table A10** the tabled $F_{.95}$ and $F_{.99}$ values are respectively employed to evaluate the nondirectional alternative hypothesis H_1: Not H_0 at the .05 and .01 levels. Throughout the discussion of the analysis of variance the notation $F_{.05}$ is employed to represent the tabled critical F value at the .05 level. The latter value corresponds to the relevant tabled $F_{.95}$ value in **Table A10**. In the same respect, the notation $F_{.01}$ is employed to represent the tabled critical F value at the .01 level, and corresponds to the relevant tabled $F_{.99}$ value in **Table A10.**

For $df_{\text{num}} = 2$ and $df_{\text{den}} = 12$, the tabled $F_{.95}$ and $F_{.99}$ values are $F_{.95} = 3.89$ and $F_{.99} = 6.93$. Thus, $F_{.05} = 3.89$ and $F_{.01} = 6.93$. In order to reject the null hypothesis, the obtained F value must be equal to or greater than the tabled critical value at the prespecified level of significance. Since $F = 6.98$ is greater than $F_{.05} = 3.89$ and $F_{.01} = 6.93$, the alternative hypothesis is supported at both the .05 and .01 levels.

A summary of the analysis of Example 16.1 with the **single-factor between-subjects analysis of variance** follows: It can be concluded that there is a significant difference between at least two of the three groups exposed to different levels of noise. This result can be summarized as follows: $F\ (2,12) = 6.98, p < .01$.

VI. Additional Analytical Procedures for the Single-Factor Between-Subjects Analysis of Variance and/or Related Tests

1. Comparisons following computation of the omnibus *F* value for the single-factor between-subjects analysis of variance The F value computed with the analysis of variance is commonly referred to as the **omnibus *F* value.** The latter term implies that the obtained F value is based on an evaluation of all k group means. Recollect that in order to reject the null hypothesis, it is only required that at least two of the k group means differ significantly from one another. As a result of this, the omnibus F value does not indicate whether just two or, in fact, more than two groups have mean values that differ significantly from one another. In order to answer this question it is necessary to conduct additional tests, which are referred to as **comparisons** (since they involve comparing the means of two or more groups with one another).

Researchers are not in total agreement with respect to the appropriate protocol for conducting comparisons.[11] The basis for the disagreement revolves around the fact that each comparison one conducts increases the likelihood of committing at least one Type I error within a set of comparisons. For this reason, it can be argued that a researcher should employ a lower Type I error rate per comparison, in order to insure that the overall likelihood of committing at least one Type I error in the set of comparisons does not exceed a prespecified alpha value that is reasonably low (e.g., $\alpha = .05$). At this point in the discussion the following two terms are defined: a) The **familywise Type I error rate** (represented by the notation α_{FW}) is the likelihood that there will be at least one Type I error in a **set of *c* comparisons;**[12] and b) The **per comparison Type I error rate** (represented by the notation α_{PC}) is the likelihood that any single comparison will result in a Type I error.

Equation 16.13 defines the relationship between the **familywise Type I error rate** and the **per comparison Type I error rate**, where c = the number of comparisons.[13]

$$\alpha_{FW} = 1 - (1 - \alpha_{PC})^c \qquad \textbf{(Equation 16.13)}$$

Let us assume that upon computing the value $F = 6.98$ for Example 16.1, the researcher decides to compare each of the three group means with one another — i.e., $\bar{X}_1$ versus $\bar{X}_2$; $\bar{X}_1$ versus $\bar{X}_3$; $\bar{X}_2$ versus $\bar{X}_3$. The three aforementioned comparisons can be conceptualized as a **family/set** of comparisons, with $c = 3$. If for each of the comparisons the researcher establishes the value $\alpha_{PC} = .05$, employing Equation 16.13 it can be determined that the **familywise Type I error rate** will equal $\alpha_{FW} = .14$. This result tells the researcher that the likelihood of committing at least one Type I error in the set of three comparisons is .14.

$$\alpha_{FW} = 1 - (1 - .05)^3 = .14$$

Equation 16.14, which is computationally more efficient than Equation 16.13, provides an excellent approximation of α_{FW}.[14]

$$\alpha_{FW} = (c)(\alpha_{PC}) \qquad \textbf{(Equation 16.14)}$$

Employing Equation 16.14, the value $\alpha_{FW} = .15$ is computed for Example 16.1.

$$\alpha_{FW} = (3)(.05) = .15$$

Note that the **familywise Type I error rate** $\alpha_{FW} = .14$ is almost three times the value of the **per comparison Type I error rate** $\alpha_{PC} = .05$. Of greater importance is the fact that the value $\alpha_{FW} = .14$ is considerably higher than .05, the usual maximum value permitted for a Type I error rate in hypothesis testing. Thus, some researchers would consider the value $\alpha_{FW} = .14$ to be excessive, and if, in fact, a maximum **familywise Type I error rate** of $\alpha_{FW} = .05$ is stipulated by a researcher, it is required that the Type I error rate for each comparison (i.e., α_{PC}) be reduced. Through use of Equation 16.15 or Equation 16.16 (which are respectively the algebraic transpositions of Equation 16.13 and Equation 16.14), it can be determined that in order to have $\alpha_{FW} = .05$, the value of α_{PC} must equal .017.[15]

$$\alpha_{PC} = 1 - \sqrt[c]{1 - \alpha_{FW}} = 1 - \sqrt[3]{1 - .05} = .017 \qquad \textbf{(Equation 16.15)}$$

$$\alpha_{PC} = \frac{\alpha_{FW}}{c} = \frac{.05}{3} = .0167 \qquad \textbf{(Equation 16.16)}$$

The reader should take note of the fact that although a reduction in the value of α_{FW} reduces the likelihood of committing a Type I error, it increases the likelihood of committing a Type II error (i.e., not rejecting a false null hypothesis). Thus, as one reduces the value of α_{FW}, the power associated with each of the comparisons that is conducted is reduced. In view of this, it should be apparent that if a researcher elects to adjust the value of α_{FW}, he must consider the impact it will have on the Type I versus Type II error rates for all of the comparisons that are conducted within the set of comparisons.

A number of different strategies have been developed with regard to what a researcher should do about adjusting the **familywise Type I error rate**. These strategies are employed within the framework of the following two types of comparisons that can be conducted following an omnibus F test: **planned comparisons** versus **unplanned comparisons**. The distinction between **planned** and **unplanned comparisons** follows.

Planned comparisons (also known as **a priori comparisons**) **Planned comparisons** are comparisons a researcher plans prior to collecting the data for a study. In a well designed experiment one would expect that a researcher will probably predict differences between specific groups prior to conducting the study. As a result of this, there is general agreement that following the computation of an omnibus F value, a researcher is justified in conducting any comparisons which have been planned beforehand, regardless of whether or not the omnibus F value is significant. Although most sources state that when planned comparisons are conducted it is not necessary to adjust the **familywise Type I error rate**, under certain conditions (such as when there are a large number of planned comparisons) an argument can be made for adjusting the value of α_{FW}.

In actuality, there are two types of planned comparisons that can be conducted, which are referred to as **simple comparisons** versus **complex comparisons**. A **simple comparison** is any comparison in which two groups are compared with one another. For instance: Group 1 versus Group 2 (i.e., $\bar{X}_1$ versus $\bar{X}_2$, which allows one to evaluate the null hypothesis H_0: $\mu_1 = \mu_2$). Simple comparisons are often referred to as **pairwise comparisons**. A **complex comparison** is any comparison in which the combined performance of two or more groups is compared with the performance of one of the other groups or the combined performance of two or more of the other groups. For instance: Group 1 versus the average of Groups 2 and 3 (i.e., $\bar{X}_1$ versus $(\bar{X}_2 + \bar{X}_3)/2$, which evaluates the null hypothesis H_0: $\mu_1 = (\mu_2 + \mu_3)/2$). If there are four groups, one can conduct a complex comparison involving the average of Groups 1 and 2 versus the average of Groups 3 and 4 (i.e., $(\bar{X}_1 + \bar{X}_2)/2$ versus $(\bar{X}_3 + \bar{X}_4)/2$, which evaluates the null hypothesis H_0: $(\mu_1 + \mu_2)/2 = (\mu_3 + \mu_4)/2$).

It should be noted that if the omnibus F value is significant, it indicates there is at least one significant difference among all of the possible comparisons that can be conducted. Kirk (1982) and Maxwell and Delaney (1990), among others, note that in such a situation it is theoretically possible that none of the simple comparisons are significant, and that the one (or perhaps more than one) significant comparison is a complex comparison. It is important to note that regardless of what type of comparisons a researcher conducts, all comparisons should be meaningful within the context of the problem under study, and as a general rule comparisons should not be redundant with respect to one another.

Unplanned comparisons (also known as **post hoc comparisons, multiple comparisons, a posteriori comparisons**) An **unplanned comparison** (which can be either a simple or complex comparison) is a comparison a researcher conducts after collecting the data for a study. In conducting **unplanned comparisons**, following the data collection phase of a study a researcher examines the values of the k group means, and at that point decides which groups to compare with one another. Although for many years most researchers argued that unplanned comparisons should not be conducted unless the omnibus F value is significant, more recently many researchers (including this author) have adopted the viewpoint that it is acceptable to conduct unplanned comparisons regardless of whether a significant F value is obtained. Although there is general agreement among researchers that the **familywise Type I error rate** should be adjusted when unplanned comparisons are conducted, there is a lack of consensus with regard to the degree of adjustment that is required. This is reflected in the fact that a variety of unplanned comparison procedures have been developed, each of which employs a different method for adjusting the value of α_{FW}. More often than not, when unplanned comparisons are conducted, a researcher will compare each of the k groups with all of the other $(k - 1)$ groups (i.e., all possible comparisons between pairs of groups are made). This "shotgun" approach, which maximizes the number of comparisons conducted, represents the classic situation for which most sources argue it is imperative to control the value of α_{FW}.

The rationale behind the argument that it is more important to adjust the value of α_{FW} in the case of unplanned comparisons as opposed to planned comparisons will be illustrated with a simple example.[16] Let us assume that a set of data is evaluated with an analysis of variance, and a significant omnibus F value is obtained. Let us also assume that within the whole set of data it is possible to conduct 20 comparisons between pairs of means and/or combinations of means. If the truth were known, however, no differences exist between the means of any of the populations being compared. However, in spite of the fact that none of the population means differ, within the set of 20 possible comparisons, one comparison (specifically, the one involving $\bar{X}_1$ versus $\bar{X}_2$) results in a significant difference at the .05 level. In this example we will assume that the difference $\bar{X}_1 - \bar{X}_2$ is the largest difference between any pair of means or combination of means in the set of 20 possible comparisons. If, in fact, $\mu_1 = \mu_2$, a significant result obtained for the comparison $\bar{X}_1$ versus $\bar{X}_2$ will represent a Type I error.

Let us assume that the comparison $\bar{X}_1$ versus $\bar{X}_2$ is planned beforehand, and it is the only comparison the researcher intends to make. Since there are 20 possible comparisons, the researcher has only a 1 in 20 chance (i.e., .05) of conducting the comparison Group 1 versus Group 2, and in the process commits a Type I error. If, on the other hand, the researcher does not plan any comparisons beforehand, but after computing the omnibus F value decides to make all 20 possible comparisons, he has a 100% chance of making a Type I error, since it is certain he will compare Groups 1 and 2. Even if the researcher decides to make only one unplanned comparison — specifically, the one involving the largest difference between any pair of means or combination of means — he will also have a 100% chance of committing a Type I error, since the comparison he will make will be $\bar{X}_1$ versus $\bar{X}_2$. This example illustrates that from a probabilistic viewpoint, the **familywise Type I error rate** associated with a set of unplanned comparisons will be higher than the rate for a set of planned comparisons.

The remainder of this section will describe the most commonly recommended comparison procedures. Specifically, the following procedures will be described: a) **Linear contrasts**; b) **Multiple *t* tests/Fisher's LSD test**; c) **The Bonferroni–Dunn test**; d) **Tukey's HSD test**; e) **The Newman–Keuls test**; f) **The Scheffé test**; and g) **The Dunnett test**. Although any of the aforementioned comparison procedures can be used for both planned and unplanned comparisons, **linear contrasts** and **multiple *t* tests/Fisher's LSD test** (which do not control the value of α_{FW}) are generally described within the context of planned comparisons. Some sources, however, do employ the latter procedures for unplanned comparisons. The **Bonferroni–Dunn test**, **Tukey's HSD test**, the **Newman–Keuls test**, the **Scheffé test**, and the **Dunnett test** are generally described as unplanned comparison procedures that are employed when a researcher wants to control the value of α_{FW}. The point that will be emphasized throughout the discussion to follow is that the overriding issue in selecting a comparison procedure is whether or not the researcher wants to control the value of α_{FW}, and if so, to what degree. This latter issue is essentially what determines the difference between the various comparison procedures to be described in this section.

Linear contrasts **Linear contrasts** (which are almost always described within the framework of planned comparisons) are comparisons that involve a linear combination of population means. In the case of a simple comparison, a **linear contrast** consists of comparing two of the group means with one another (e.g., $\bar{X}_1$ versus $\bar{X}_2$). In the case of a complex comparison, the combined performance of two or more groups is compared with the performance of one of the other groups or the combined performance of two or more of the other groups (e.g., $\bar{X}_1$ versus $(\bar{X}_2 + \bar{X}_3)/2$). In conducting both simple and complex comparisons, the researcher must assign weights, which are referred to as **coefficients,** to all of the group means. These coefficients reflect the relative contribution of each of the group means to the two mean values that are being contrasted with one another in the comparison. In the case of a complex comparison, at least one of the two mean values that are contrasted in the comparison will be based on a weighted combination of the means of two or more of the groups.

The use of **linear contrasts** will be described for both simple and complex comparisons. In the examples that will be employed to illustrate **linear contrasts**, it will be assumed that all comparisons are planned beforehand, and that the researcher is making no attempt to control the value of α_{FW}. Thus, for each comparison to be conducted it will be assumed that $\alpha_{PC} = .05$. All of the comparisons (both simple and complex) to be described in this section are referred to as **single degree of freedom (*df*) comparisons.** This is the case, since one degree of freedom is always employed in the numerator of the F ratio (which represents the test statistic for a comparison).[17]

Linear contrast of a planned simple comparison Let us assume that prior to obtaining the data for Example 16.1, the experimenter hypothesizes there will be a significant difference between Group 1 (**no noise**) and Group 2 (**moderate noise**). After conducting the omnibus F test, the simple planned comparison $\bar{X}_1$ versus $\bar{X}_2$ is conducted in order to compare the performance of the two groups. The null and alternative hypotheses for the comparison follow: H_0: $\mu_1 = \mu_2$ versus H_1: $\mu_1 \neq \mu_2$.[18] Table 16.3 summarizes the information required to conduct the planned comparison $\bar{X}_1$ versus $\bar{X}_2$.

Table 16.3 Planned Simple Comparison: Group 1 Versus Group 2

Group	$\bar{X}_1$	Coefficient (c_j)	Product $(c_j)(\bar{X}_j)$	Squared Coefficient (c_j^2)
1	9.2	+1	(+1)(9.2) = +9.2	1
2	6.6	−1	(−1)(6.6) = −6.6	1
3	6.2	0	(0)(6.2) = 0	0
		$\Sigma c_j = 0$	$\Sigma(c_j)(\bar{X}_j) = 2.6$	$\Sigma c_j^2 = 2$

The following should be noted with respect to Table 16.3: a) The rows of Table 16.3 represent data for each of the three groups employed in the experiment. Even though the comparison involves two of the three groups, the data for all three groups are included to illustrate how the group which is not involved in the comparison is eliminated from the calculations; b) Column 2 contains the mean score of each of the groups; c) In Column 3 each of the groups is assigned a **coefficient**, represented by the notation c_j. The value of

c_j assigned to each group is a weight that reflects the proportional contribution of the group to the comparison. Any group not involved in the comparison (in this instance Group 3) is assigned a coefficient of zero. Thus, $c_3 = 0$. When only two groups are involved in a comparison, one of the groups (it does not matter which one) is assigned a coefficient of $+1$ (in this instance Group 1 is assigned the coefficient $c_1 = +1$) and the other group a coefficient of -1 (i.e., $c_2 = -1$). Note that Σc_j, the sum of the coefficients (which is the sum of Column 3), must always equal zero (i.e., $c_1 + c_2 + c_3 = (+1) + (-1) + 0 = 0$); d) In Column 4 a product is obtained for each group. The product for a group is obtained by multiplying the mean of the group ($\bar{X}_j$) by the coefficient that has been assigned to that group (c_j). Although it may not be immediately apparent from looking at the table, the sum of Column 4, $\Sigma(c_j)(\bar{X}_j)$, is, in fact, the difference between the two means being compared (i.e., $\Sigma(c_j)(\bar{X}_j)$ is equal to $\bar{X}_1 - \bar{X}_2 = 9.2 - 6.6 = 2.6$);[19] and e) Σc_j^2, the sum of Column 5, is the sum of the squared coefficients.

The test statistic for the comparison is an F ratio, represented by the notation F_{comp}. In order to compute the value F_{comp}, a **sum of squares** (SS_{comp}), a **degrees of freedom value** (df_{comp}), and a **mean square** (MS_{comp}) for the comparison must be computed. The **comparison sum of squares** (SS_{comp}) is computed with Equation 16.17. Note that Equation 16.17 assumes there are an equal number of subjects (n) in each group.[20]

$$SS_{\text{comp}} = \frac{n[\Sigma(c_j)(\bar{X}_j)]^2}{\Sigma c_j^2} \qquad \textbf{(Equation 16.17)}$$

Substituting the appropriate values from Example 16.1 in Equation 16.17, the value $SS_{\text{comp}} = 16.9$ is computed.

$$SS_{\text{comp}} = \frac{5(2.6)^2}{2} = 16.9$$

The **comparison mean square** (MS_{comp}) is computed with Equation 16.18. MS_{comp} represents a measure of between-groups variability which takes into account just the two group means involved in the comparison.

$$MS_{\text{comp}} = \frac{SS_{\text{comp}}}{df_{\text{comp}}} \qquad \textbf{(Equation 16.18)}$$

In a **single degree of freedom comparison**, df_{comp} will always equal 1 since the number of mean values being compared in such a comparison will always be $k_{\text{comp}} = 2$, and $df_{\text{comp}} = k_{\text{comp}} - 1 = 2 - 1 = 1$. Substituting the values $SS_{\text{comp}} = 16.9$ and $df_{\text{comp}} = 1$ in Equation 16.18, the value $MS_{\text{comp}} = 16.9$ is computed. Note that since in a **single degree of freedom comparison** the value of df_{comp} will always equal 1, the values SS_{comp} and MS_{comp} will always be equivalent.

$$MS_{\text{comp}} = \frac{16.9}{1} = 16.9$$

The test statistic F_{comp} is computed with Equation 16.19. F_{comp} is a ratio that is comprised of the variability of the two means involved in the comparison divided by the within-groups variability employed for the omnibus F test.[21]

$$F_{\text{comp}} = \frac{MS_{\text{comp}}}{MS_{WG}} \qquad \textbf{(Equation 16.19)}$$

Substituting the values $MS_{comp} = 16.9$ and $MS_{WG} = 1.9$ in Equation 16.19, the value $F_{comp} = 8.89$ is computed.

$$F_{comp} = \frac{16.9}{1.9} = 8.89$$

The value $F_{comp} = 8.89$ is evaluated with **Table A10.** Employing **Table A10,** the appropriate degrees of freedom value for the numerator is $df_{num} = 1$. This is the case, since the numerator of the F_{comp} ratio is MS_{comp}, and the degrees of freedom associated with the latter value is $df_{comp} = 1$. The denominator degrees of freedom will be the value of df_{WG} employed for the omnibus F test, which for Example 16.1 is $df_{WG} = 12$.

For $df_{num} = 1$ and $df_{den} = 12$, the tabled critical .05 and .01 F values are $F_{.05} = 4.75$ and $F_{.01} = 9.33$. Since the obtained value $F_{comp} = 8.89$ is greater than $F_{.05} = 4.75$, the nondirectional alternative hypothesis H_1: $\mu_1 \neq \mu_2$ is supported at the .05 level. Since $F_{comp} = 8.89$ is less than $F_{.01} = 9.33$, the latter alternative hypothesis is not supported at the .01 level. Thus, if the value $\alpha = .05$ is employed, the researcher can conclude that Group 1 recalled a significantly greater number of nonsense syllables than Group 2.

With respect to Example 16.1, it is possible to conduct the following additional simple comparisons: $\bar{X}_1$ versus $\bar{X}_3$; $\bar{X}_2$ versus $\bar{X}_3$. The latter two simple comparisons will be conducted later employing **multiple *t* tests/Fisher's LSD test** (which, as will be noted, are computationally equivalent to the **linear contrast** procedure described in this section).

Linear contrast of a planned complex comparison Let us assume that prior to obtaining the data for Example 16.1, the experimenter hypothesizes there will be a significant difference between the performance of Group 3 (**extreme noise**) and the combined performance of Group 1 (**no noise**) and Group 2 (**moderate noise**). Such a comparison is a complex comparison, since it involves a single group being contrasted with two other groups. As is the case with the simple comparison $\bar{X}_1$ versus $\bar{X}_2$, two means are also contrasted within the framework of the complex comparison. However, one of the two means is a composite mean that is based upon the combined performance of two groups. The complex comparison represents a single degree of freedom comparison. This is the case, since two means are being contrasted with one another — specifically, the mean of Group 3 with the composite mean of Groups 1 and 2. The fact that it is a single degree of freedom comparison is also reflected in the fact that there is one equals sign (=) in both the null and alternative hypotheses for the comparison. The null and alternative hypotheses for the complex comparison are H_0: $\mu_3 = (\mu_1 + \mu_2)/2$ versus H_1: $\mu_3 \neq (\mu_1 + \mu_2)/2$.[22] Table 16.4 summarizes the information required to conduct the planned complex comparison $\bar{X}_3$ versus $(\bar{X}_1 + \bar{X}_2)/2$.

Table 16.4 Planned Complex Comparison: Group 3 Versus Groups 1 and 2

Group	$(\bar{X}_j)$	Coefficient (c_j)	Product $(c_j)(\bar{X}_j)$	Squared Coefficient (c_j^2)
1	9.2	$-\frac{1}{2}$	$\left(-\frac{1}{2}\right)(9.2) = -4.6$	$\frac{1}{4}$
2	6.6	$-\frac{1}{2}$	$\left(-\frac{1}{2}\right)(6.6) = -3.3$	$\frac{1}{4}$
3	6.2	$+1$	$(+1)(6.2) = +6.2$	1
		$\Sigma c_j = 0$	$\Sigma(c_j)(\bar{X}_j) = -1.7$	$\Sigma c_j^2 = 1.5$

Note that the first two columns of Table 16.4 are identical to the first two columns of Table 16.3. The different values in the remaining columns of Table 16.4 result from the fact that different coefficients are employed for the complex comparison. The absolute value of $\Sigma(c_j)(\bar{X}_j) = -1.7$ represents the difference between the two sets of means contrasted in the null hypothesis — specifically, the difference between $\bar{X}_3 = 6.2$ and the composite mean of Group 1 ($\bar{X}_1 = 9.2$) and Group 2 ($\bar{X}_2 = 6.6$). The latter composite mean will be represented with the notation $\bar{X}_{1/2}$. Since $\bar{X}_{1/2} = (9.2 + 6.6)/2 = 7.9$, the difference between the two means evaluated with the comparison is $6.2 - 7.9 = -1.7$ (which is the same as the value $\Sigma(c_j)(\bar{X}_j) = -1.7$, which is the sum of Column 4 in Table 16.4).

Before conducting the computations for the complex comparison, a general protocol will be described for assigning coefficients to the groups involved in either a simple or complex comparison. Within the framework of describing the protocol, it will be employed to determine the coefficients for the complex comparison under discussion.

1) Write out the null hypothesis (i.e., H_0: $\mu_3 = [(\mu_1 + \mu_2)/2]$). Any group not involved in the comparison (i.e., not noted in the null hypothesis) is assigned a coefficient of zero. Since all three groups are included in the present comparison, none of the groups receives a coefficient of zero.

2) On each side of the equals sign of the null hypothesis write the number of group means designated in the null hypothesis. Thus:

$$H_0\text{: } \mu_3 = \frac{\mu_1 + \mu_2}{2}$$

1 mean   2 means

3) To obtain the coefficient for each of the groups included in the null hypothesis, employ Equation 16.20. The latter equation, which is applied to each side of the null hypothesis, represents the reciprocal of the number of means on a specified side of the null hypothesis.[23]

$$\text{Coefficient} = \frac{1}{\text{Number of group means}} \qquad \textbf{(Equation 16.20)}$$

Since there is only one mean to the left of the equals sign (μ_3), employing Equation 16.20 we determine that the coefficient for that group (Group 3) equals $\frac{1}{1} = 1$. Since there are two means to the right of the equals sign (μ_1 and μ_2), using Equation 16.20 we determine the coefficient for both of the groups to the right of the equals sign (Groups 1 and 2) equals $\frac{1}{2}$. Notice that all groups on the same side of the equals sign receive the same coefficient.[24]

4) The coefficient(s) on one side of the equals sign are assigned a positive sign, and the coefficient(s) on the other side of the equals sign are assigned a negative sign. Equivalent results will be obtained irrespective of which side of the equals sign is assigned positive versus negative coefficients. In the complex comparison under discussion, a positive sign is assigned to the coefficient to the left of the equals sign, and negative signs are assigned to coefficients to the right of the equals sign. Thus, the values of the coefficients are: $c_1 = -\frac{1}{2}$; $c_2 = -\frac{1}{2}$; $c_3 = +1$. Note that the sum of the coefficients must always equal zero (i.e., $c_1 + c_2 + c_3 = \left(-\frac{1}{2}\right) + \left(-\frac{1}{2}\right) + (+1)$).[25]

Equations 16.17–16.19, which are employed for the simple comparison, are also used to evaluate a complex comparison. Substituting the appropriate information from Table 16.4 in Equation 16.17, the value $SS_{\text{comp}} = 9.63$ is computed.

$$SS_{\text{comp}} = \frac{5(-1.7)^2}{1.5} = 9.63$$

Employing Equation 16.18, the value $MS_{\text{comp}} = 9.63$ is computed. Note that since the complex comparison is a single degree of freedom comparison, $df_{\text{comp}} = 1$.

$$MS_{\text{comp}} = \frac{9.63}{1} = 9.63$$

Employing Equation 16.19, the value $F_{\text{comp}} = 5.07$ is computed.

$$F_{\text{comp}} = \frac{9.63}{1.9} = 5.07$$

The protocol for evaluating the value $F_{\text{comp}} = 5.07$ computed for the complex comparison is identical to that employed for the simple comparison. In determining the tabled critical F value in **Table A10**, the same degrees of freedom values are employed. This is the case since the numerator degrees of freedom for any single degree of freedom comparison is $df_{\text{comp}} = 1$. The denominator degrees of freedom is df_{WG}, which, as in the case of the simple comparison, is the value of df_{WG} employed for the omnibus F test. Thus, the appropriate degrees of freedom for the complex comparison are $df_{\text{num}} = 1$ and $df_{\text{den}} = 12$. The tabled critical .05 and .01 F values in **Table A10** for the latter degrees of freedom are $F_{.05} = 4.75$ and $F_{.01} = 9.33$. Since the obtained value $F = 5.07$ is greater than $F_{.05} = 4.75$, the nondirectional alternative hypothesis H_1: $\mu_3 \neq (\mu_1 + \mu_2)/2$ is supported at the .05 level. Since $F = 5.07$ is less than $F_{.01} = 9.33$, the latter alternative hypothesis is not supported at the .01 level. Thus, if the value $\alpha = .05$ is employed, the researcher can conclude that Group 3 recalled a significantly fewer number of nonsense syllables than the average number recalled when the performance of Groups 1 and 2 are combined.

Orthogonal comparisons Most sources agree that intelligently planned studies involve a limited number of meaningful comparisons which the researcher plans prior to collecting the data. As a general rule, any comparisons that are conducted should address critical questions underlying the general hypothesis under study. Some researchers believe that it is not even necessary to obtain an omnibus F value if, in fact, the critical information one is concerned with is contained within the framework of the planned comparisons. It is generally recommended that the maximum number of planned comparisons one conducts should not exceed the value of df_{BG} employed for the omnibus F test. If the number of planned comparisons is equal to or less than df_{BG}, sources generally agree that a researcher is not obliged to adjust the value of α_{FW}. When, however, the number of planned comparisons exceeds df_{BG}, many sources recommend that the value of α_{FW} be adjusted. Applying this protocol to Example 16.1, since $df_{BG} = 2$, one can conduct two planned comparisons without being obliged to adjust the value of α_{FW}.

The subject of **orthogonal comparisons** is relevant to the general question of how many (and specifically which) comparisons a researcher should conduct. Orthogonal comparisons are defined as comparisons that are independent of one another. In other words, such comparisons are not redundant, in that they do not overlap with respect to the information they provide. In point of fact, the two comparisons that have been conducted (i.e., the simple comparison of Group 1 versus Group 2, and the complex comparison of Group 3 versus the combined performance of Groups 1 and 2) are orthogonal comparisons. This can be demonstrated by employing Equation 16.21, which defines the relationship that will exist between two comparisons if they are orthogonal to one another. In Equation 16.21, c_{j1} is the coefficient assigned to Group j in Comparison 1 and c_{j2} is the coefficient assigned to Group j

in Comparison 2. If, in fact, two comparisons are orthogonal, the sum of the products of the coefficients of all of k groups will equal zero.

$$\sum_{j=1}^{k} (c_{j_1})(c_{j_2}) = 0 \qquad \textbf{(Equation 16.21)}$$

Equation 16.21 is employed below with the two comparisons that have been conducted in this section. Notice that for each group, the first value in the parentheses is the coefficient for that group for the simple comparison (Group 1 versus Group 2), while the second value in parentheses is the coefficient for that group for the complex comparison (Group 3 versus Groups 1 and 2).

$$\begin{array}{ccccc} \text{Group 1} & & \text{Group 2} & & \text{Group 3} \\ (+1)\left(-\frac{1}{2}\right) & + & (-1)\left(-\frac{1}{2}\right) & + & (0)(+1) \end{array} = \left(-\frac{1}{2}\right) + \left(+\frac{1}{2}\right) + 0 = 0$$

If there are k treatments there will be $(k - 1)$ (which corresponds to df_{BG} employed for the omnibus F test) orthogonal comparisons (also known as **orthogonal contrasts**) within each complete orthogonal set. This is illustrated by the fact that two comparisons comprise the orthogonal set demonstrated above — specifically, Group 1 versus Group 2, and Group 3 versus Groups 1 and 2. Actually, when (as is the case in Example 16.1) there are $k = 3$ treatments, there are three possible sets of orthogonal comparisons — each set being comprised of one simple comparison and one complex comparison. In addition to the set noted above, the following two additional orthogonal sets can be formed: a) Group 1 versus Group 3; Group 2 versus Groups 1 and 3; b) Group 2 versus Group 3; Group 1 versus Groups 2 and 3.

When $k > 3$ there will be more than 2 contrasts in a set of orthogonal contrasts. Within that full set of orthogonal contrasts, if the coefficients from any two of the contrasts are substituted in Equation 16.21, they will yield a value of zero. It should also be noted that the number of possible sets of contrasts will increase as the value of k increases. It is important to note, however, that when all possible sets of contrasts are considered, most of them will not be orthogonal. With respect to determining those contrasts that are orthogonal, Howell (1992) describes a simple protocol that can be employed to derive most (although not all) orthogonal contrasts in a body of data. The procedure described by Howell (1992) is summarized in Figure 16.1.

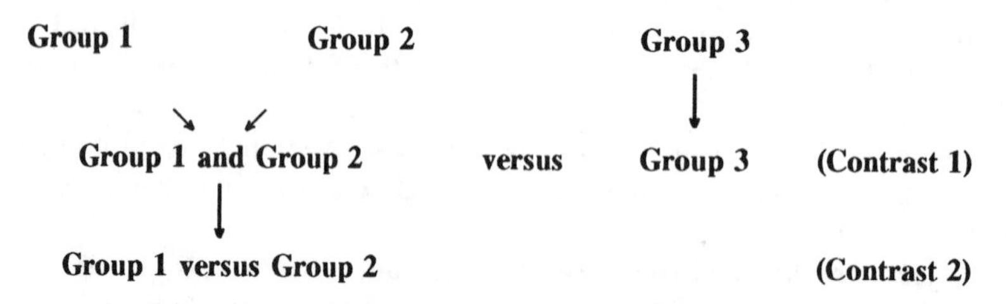

Figure 16.1 Tree Diagram for Determining Orthogonal Contrasts

In employing Figure 16.1, initially two blocks of groups are formed employing all k groups. A block can be comprised of one or more of the groups. In Figure 16.1, the first block is comprised of Groups 1 and 2, and the second block of Group 3. This will represent the first contrast, which corresponds to the complex comparison that is described in this section. Any blocks that remain which are comprised of two or more groups are broken down into smaller blocks. Thus, the block comprised of the Group 1 and Group 2 is broken down into two blocks, each consisting of one group. The contrast of these two groups (Group 1 versus Group 2) represents the second contrast in the orthogonal set.

Figure 16.1 can also be employed to derive the other two possible orthogonal sets for Example 16.1. To illustrate, the initial two blocks derived can be a block consisting of Groups 1 and 3 and a second block consisting of Group 2. This represents the complex comparison of Group 2 versus Groups 1 and 3. The remaining block consisting of Groups 1 and 3 can be broken down into two blocks consisting of Group 1 and Group 3. The comparison of these two groups, which represents a simple comparison, constitutes the second comparison in that orthogonal set.

Note that once a group has been assigned to a block, and that block is compared to an adjacent block, from that point on any other comparisons involving that group will be with other groups that fall within its own block. Thus, in our example, if the first comparison is Groups 1 and 2 versus Group 3, the researcher cannot use the comparison Group 1 versus Group 3 as the second comparison for that set, since the two groups are in different blocks. If the latter comparison is conducted, the sum of the products of the coefficients of all k groups for the two comparisons will not equal zero, and thus not constitute an orthogonal set. To illustrate this latter fact, the coefficients of the three groups for the simple comparison depicted in Table 16.3 are rearranged as follows: $c_1 = +1$, $c_2 = 0$, and $c_3 = -1$. The latter coefficients are employed if the simple comparison Group 1 versus Group 3 is conducted. Equation 16.21 is now employed to demonstrate that the Group 1 versus Group 3 comparison is not orthogonal to the complex comparison Group 3 versus Groups 1 and 2 summarized in Table 16.4.

$$\begin{array}{ccccc} \text{Group 1} & & \text{Group 2} & & \text{Group 3} \\ (+1)\left(-\frac{1}{2}\right) & + & (0)\left(-\frac{1}{2}\right) & + & (-1)(+1) \end{array} = \left(-\frac{1}{2}\right) + 0 + (-1) = -1\frac{1}{2}$$

It should be pointed out that when more than one set of orthogonal comparisons are conducted, since the different sets are not orthogonal to one another, many of the comparisons one conducts will not be independent of one another. For this reason, a researcher who does not want to conduct any nonindependent comparisons should only conduct those comparisons involving the orthogonal set which provide the most meaningful information with regard to the hypothesis under study. It should be noted, however, that there is no immutable rule that states that a researcher can only conduct orthogonal comparisons. Many sources point out there are times when the questions addressed by nonorthogonal comparisons can often contribute to a researcher's understanding of the general hypothesis under study.

Another characteristic of orthogonal comparisons is that the sum of squares for all comparisons that comprise a set of orthogonal comparisons equals the value of SS_{BG} for the omnibus F test. This reflects the fact that the variability in a set of orthogonal contrasts will account for all the between-groups variability in the full set of data. Employing the data for the simple comparison summarized in Table 16.3 and the complex comparison summarized in Table 16.4, it is confirmed below that the sum of squares for the latter two comparisons (which comprise an orthogonal set) equals the value $SS_{BG} = 26.53$ obtained with Equation 16.3.

$$SS_{BG} = 26.53 = SS_{\text{simple comparison}} + SS_{\text{complex comparison}} = 16.9 + 9.63 = 26.53$$

Test 16a: Multiple *t* tests/Fisher's LSD test One option a researcher has available after computing an omnibus F value is to run **multiple *t* tests** (specifically, the ***t* test for two independent samples**), in order to determine whether there is a significant difference between any of the pairs of means that can be contrasted within the framework of either simple or complex comparisons. In point of fact, it can be algebraically demonstrated that in the case of both simple and complex comparisons, the use of **multiple t tests** will yield a result that is equivalent to that obtained with the protocol described for conducting **linear contrasts.** When **multiple t tests** are discussed as a procedure for conducting comparisons, most sources state that: a) **Multiple t tests** should only be employed for planned comparisons; b) Since **multiple t tests** are only employed for planned comparisons, they can be conducted regardless of whether or not the omnibus F value is significant; and c) In conducting **multiple t tests** for planned comparisons, the researcher is not required to adjust the value of α_{FW}, as long as a limited number of comparisons are conducted (as noted earlier, most sources state the number of planned comparisons should not exceed df_{bG}). All of the aforementioned stipulations noted for **multiple t tests** also apply to **linear contrasts** (since as noted above, **multiple t tests** and **linear contrasts** are computationally equivalent).

When, on the other hand, comparisons are unplanned and **multiple t tests** are employed to compare pairs of means, the use of **multiple t tests** within the latter context is referred to as **Fisher's LSD test** (the term **LSD** is an abbreviation for **least significant difference**). When **Fisher's LSD test** is compared with other unplanned comparison procedures, it provides the most powerful test with respect to identifying differences between pairs of means, since it does not adjust the value of α_{FW}. Of all the unplanned comparison procedures, **Fisher's LSD test** requires the smallest difference between two means in order to conclude that a difference is significant. However, since **Fisher's LSD test** does not reduce the value of α_{FW}, it has the highest likelihood of committing one or more Type I errors in a set/family of comparisons. In the discussion to follow, since **multiple t tests** and **Fisher's LSD method** are computationally equivalent (as well as equivalent to **linear contrasts**), the term **multiple t tests/Fisher's LSD test** will refer to a computational procedure which can be employed for both planned and unplanned comparisons that do not adjust the value of α_{FW}.

Equation 16.22 can be employed to compute the test statistic (which employs the t distribution) for **multiple t tests/Fisher's LSD test.** Whereas Equation 16.22 can only be employed for simple comparisons, Equation 16.23 is a generic equation that can be employed for both simple and complex comparisons. It will be assumed that the null and alternative hypotheses for any comparisons being conducted are as follows: H_0: $\mu_a = \mu_b$ versus H_1: $\mu_a \neq \mu_b$.

$$t = \frac{\bar{X}_a - \bar{X}_b}{\sqrt{\dfrac{2MS_{WG}}{n}}} \qquad \textbf{(Equation 16.22)}$$

Where: $\bar{X}_a$ and $\bar{X}_b$ represent the two means contrasted in the comparison

$$t = \frac{\bar{X}_a - \bar{X}_b}{\sqrt{\dfrac{(\Sigma c_j^2)(MS_{WG})}{n}}} \qquad \textbf{(Equation 16.23)}$$

In the case of a **simple comparison,** the value Σc_j^2 in Equation 16.23 will always equal 2, thus resulting in Equation 16.22.[26]

The degrees of freedom employed in evaluating the t value computed with Equations 16.22 and 16.23 is the value of df_{WG} computed for the omnibus F test. Thus, in the case of Example 16.1, the value $df_{WG} = 12$ is employed. Note that in Equations 16.22/16.23, the value MS_{WG} computed for the omnibus F test is employed in computing the **standard error of the difference** in the denominator of the t test equation, as opposed to the value

$\sqrt{(\tilde{s}_a^2/n_a) + (\tilde{s}_b^2/n_b)}$, which is employed in Equation 8.1 (the equation for the ***t* test for two independent samples** when $n_a = n_b$). This is the case, since MS_{WG} is a pooled estimate of the population variance based on the full data set (i.e., the k groups for which the omnibus F value is computed).[27]

Equation 16.22 is employed below to conduct the simple comparison of Group 1 versus Group 2.

$$t = \frac{9.2 - 6.6}{\sqrt{\frac{(2)(1.9)}{5}}} = 2.99$$

The obtained value $t = 2.99$ is evaluated with **Table A2 (Table of Student's *t* Distribution)** in the **Appendix**. For $df = 12$, the tabled critical two-tailed .05 and .01 values are $t_{.05} = 2.18$ and $t_{.01} = 3.06$. Since $t = 2.99$ is greater than $t_{.05} = 2.18$, the nondirectional alternative hypothesis H_1: $\mu_a \neq \mu_b$ is supported at the.05 level. Since $t = 2.99$ is less than $t_{.01} = 3.06$, the latter alternative hypothesis is not supported at the .01 level. Thus, if $\alpha = .05$, the researcher can conclude that Group 1 recalled a significantly greater number of nonsense syllables than Group 2. This result is consistent with that obtained in the previous section using the protocol for **linear contrasts**.

Equation 16.24 employs **multiple *t* tests/Fisher's LSD test** to compute the minimum required difference in order for two means to differ significantly from one another at a prespecified level of significance. The latter value is represented by the notation CD_{LSD}, with CD being the abbreviation for **critical difference**. Whereas Equation 16.24 only applies to simple comparisons, Equation 16.25 is a generic equation that can be employed for both simple and complex comparisons.[28]

$$CD_{LSD} = \sqrt{F_{(1,WG)}}\sqrt{\frac{2MS_{WG}}{n}} \qquad \textbf{(Equation 16.24)}$$

$$CD_{LSD} = \sqrt{F_{(1,WG)}}\sqrt{\frac{(\Sigma c_j^2)(MS_{WG})}{n}} \qquad \textbf{(Equation 16.25)}$$

Where: $F_{(1,WG)}$ is the tabled critical F value for $df_{\text{num}} = 1$ and $df_{\text{den}} = df_{WG}$ at the prespecified level of significance[29]

Employing the appropriate values from Example 16.1 in Equation 16.24, the value $CD_{LSD} = 1.90$ is computed.[30]

$$CD_{LSD} = \sqrt{4.75}\sqrt{\frac{(2)(1.9)}{5}} = 1.90$$

Thus, in order to differ significantly at the .05 level, the means of any two groups must differ from one another by at least 1.90 units. Employing Table 16.5, which summarizes the differences between pairs of means involving all three experimental groups, it can be seen that the following simple comparisons are significant at the .05 level if **multiple *t* tests/Fisher's LSD test** are employed: $\bar{X}_1 - \bar{X}_2 = 2.6$; $\bar{X}_1 - \bar{X}_3 = 3$. The difference $\bar{X}_2 - \bar{X}_3 = .4$ is not significant, since it is less than $CD_{LSD} = 1.90$.

Within the framework of the discussion to follow, it will be demonstrated that the CD value computed with **multiple *t* tests/Fisher's LSD test** is the smallest CD value that can be computed with any of the comparison methods that can be employed for the analysis of variance.

Table 16.5 Differences Between Pairs of Means in Example 16.1

$$\bar{X}_1 - \bar{X}_2 = 9.2 - 6.6 = 2.6$$
$$\bar{X}_1 - \bar{X}_3 = 9.2 - 6.2 = 3.0$$
$$\bar{X}_2 - \bar{X}_3 = 6.6 - 6.2 = 0.4$$

Multiple t tests/Fisher's LSD test will now be demonstrated for a complex comparison. Equations 16.23 and 16.25 are employed to evaluate the complex comparison involving the mean of Group 3 versus the composite mean of Groups 1 and 2. Employing Equation 16.23, the absolute value $t = 2.25$ is computed.[31] The value $CD_{LSD} = 1.65$ is computed with Equation 16.25. Note that in computing the values of t and CD_{LSD}, the value $\Sigma c_j^2 = 1.5$ is employed in Equations 16.23 and 16.25, as opposed to $\Sigma c_j^2 = 2$, which is employed in Equations 16.22 and 16.24. This latter fact accounts for why the value $CD_{LSD} = 1.65$ computed for the complex comparison is smaller than the value $CD_{LSD} = 1.90$ computed for the simple comparison.

$$t = \frac{6.2 - 7.9}{\sqrt{\frac{(1.5)(1.9)}{5}}} = -2.25$$

$$CD_{LSD} = \sqrt{4.75}\sqrt{\frac{(1.5)(1.9)}{5}} = 1.65$$

Since the obtained absolute value $t = 2.25$ is greater than $t_{.05} = 2.18$ (which is the tabled critical two-tailed .05 t value for $df_{WG} = 12$), the nondirectional alternative hypothesis H_1: $\mu_3 \neq (\mu_1 + \mu_2)/2$ is supported at the.05 level. Since $t = 2.25$ is less than the tabled critical two-tailed value $t_{.01} = 3.06$, the latter alternative hypothesis is not supported at the .01 level. Thus, if $\alpha = .05$, the researcher can conclude that Group 3 recalled a significantly fewer number of nonsense syllables than the average number recalled when the performances of Groups 1 and 2 are combined. This result is consistent with that obtained in the previous section using the protocol for **linear contrasts.**

The fact that the obtained absolute value of the difference $|\bar{X}_3 - \bar{X}_{1/2}| = |6.2 - 7.9| = 1.7$ is larger than the computed value $CD_{LSD} = 1.65$, is consistent with the fact that the difference for the **complex comparison** is significant at the .05 level. The computed value $CD_{LSD} = 1.65$ indicates that for any complex comparison involving the set of coefficients employed in Table 16.4, the minimum required difference in order for the two mean values stipulated in the null hypothesis to differ significantly from one another at the .05 level is $CD_{LSD} = 1.65$.

Test 16b: The Bonferroni–Dunn test First formally described by Dunn (1961), the **Bonferroni–Dunn test** is based on the **Bonferroni inequality**, which states that the probability of the occurrence of a set of events can never be greater than the sum of the individual probabilities for each event. Although the **Bonferroni–Dunn test** is identified in most sources as a planned comparison procedure, it can also be employed for unplanned comparisons. In actuality, the **Bonferroni–Dunn test** is computationally identical to **multiple t tests/Fisher's LSD test/linear contrasts**, except for the fact that the equation for the test statistic employs an adjustment in order to reduce the value of α_{FW}. By virtue of reducing α_{FW}, the power of the **Bonferroni–Dunn test** will always be less than the power associated with **multiple t tests/Fisher's LSD test/linear contrasts** (since the latter procedure does not adjust the value of α_{FW}). As a general rule, whenever the **Bonferroni–Dunn test** is employed to conduct all possible pairwise comparisons in a set of data, it provides the least powerful test of an alternative hypothesis of all the available comparison procedures.

The **Bonferroni–Dunn test** requires a researcher to initially stipulate the highest **familywise Type I error rate** he is willing to tolerate. For purposes of illustration, let us assume that in conducting comparisons in reference to Example 16.1 the researcher does not want the value of α_{FW} to exceed .05 (i.e., he does not want more than a 5% chance of committing at least one Type I error in a set of comparisons). Let us also assume he either plans beforehand or decides after computing the omnibus F value, that he will compare each of the group means with one another. This will result in the following three simple comparisons: Group 1 versus Group 2; Group 1 versus Group 3; Group 2 versus Group 3. To insure that the **familywise Type I error rate** does not exceed .05, $\alpha_{FW} = .05$ is divided by $c = 3$, which represents the number of comparisons that comprise the set. The resulting value $\alpha_{PC} = \alpha_{FW}/c = .05/3 = .0167$ represents for each of the comparisons that are conducted, the likelihood of committing a Type I error. Thus, even if a Type I error is made for all three of the comparisons, the overall **familywise Type I error rate** will not exceed .05 (since $(3)(.0167) = .05$).

In the case of a simple comparison, Equation 16.26 is employed to compute $CD_{B/D}$, which will represent the **Bonferroni–Dunn test** statistic. $CD_{B/D}$ is the minimum required difference in order for two means to differ significantly from one another, if the familywise Type I error rate is set at a prespecified level.[32]

$$CD_{B/D} = t_{B/D} \sqrt{\frac{2MS_{WG}}{n}} \qquad \textbf{(Equation 16.26)}$$

The value $t_{B/D}$ in Equation 16.26 represents the tabled critical t value at the level of significance that corresponds to the value of α_{PC} (which in this case equals $t_{.0167}$) for df_{WG} (which for Example 16.1 is $df_{WG} = 12$). The value of $t_{B/D}$ can be obtained from detailed tables of the t distribution prepared by Dunn (1961), or can be computed with Equation 16.27 (which is described in Keppel (1991)).

$$t_{B/D} = z + \frac{z^3 + z}{4(df_{WG} - 2)} \qquad \textbf{(Equation 16.27)}$$

The value of z in Equation 16.27 is derived from **Table A1 (Table of the Normal Distribution)** in the **Appendix**. It is the z value above which $\alpha_{PC/2}$ of the cases in the normal distribution falls. Since in our example $\alpha_{PC} = .0167$, we look up the z value above which $.0167/2 = .0083$ of the distribution falls.[33] From **Table A1** we can determine that the z value which corresponds to the proportion .0083 is $z = 2.39$. Substituting $z = 2.39$ in Equation 16.27, the value $t_{B/D} = 2.79$ is computed.

$$t_{B/D} = 2.39 + \frac{(2.39)^3 + 2.39}{4(12 - 2)} = 2.79$$

Substituting the value $t_{B/D} = 2.79$ in Equation 16.26, the value $CD_{B/D} = 2.43$ is computed.

$$CD_{B/D} = 2.79\sqrt{\frac{(2)(1.9)}{5}} = 2.43$$

Thus, in order to be significant, the difference between any pair of means contrasted in a simple comparison must be at least 2.43 units. Referring to Table 16.5, we can determine that (as is the case with **multiple *t* tests/Fisher's LSD test**) the following comparisons are significant: Group 1 versus Group 2; Group 1 versus Group 3 (since the difference between the means of the aforementioned groups is larger than $CD_{B/D} = 2.43$). Note that the value $CD_{B/D} = 2.43$ is larger than the value $CD_{LSD} = 1.90$, computed with Equation 16.24. The difference between the two CD values reflects the fact that in the case of the **Bonferroni–Dunn test,** for the set of $c = 3$ simple comparisons the value of α_{FW} is .05, whereas in the case of **multiple *t* tests/Fisher's LSD test,** the value of α_{FW} (which is computed with Equation 16.13) is .14. By virtue of adjusting the value of α_{FW}, the **Bonferroni–Dunn test** will always result in a larger CD value than the value computed with **multiple *t* tests/Fisher's LSD test.**[34]

The **Bonferroni–Dunn test** can be used for both simple and complex comparisons. Earlier in this section, it was noted that both simple and complex comparisons involving two sets of means (where each set of means in a complex comparison consists of a single mean or a combination of means) represent single degree of freedom comparisons. Keppel (1991, p. 167) notes that the number of possible single degree of freedom comparisons (to be designated $c_{(1\ df)}$) in a set of data can be computed with Equation 16.28.

$$c_{(1\ df)} = 1 + \frac{(3^k - 1)}{2} - 2^k \qquad \textbf{(Equation 16.28)}$$

Employing Equation 16.28 with Example 16.1 (where $k = 3$), the value $c_{(1\ df)} = 6$ is computed.

$$c_{(1\ df)} = 1 + \frac{(3^3 - 1)}{2} - 2^3 = 6$$

The $c_{(1\ df)} = 6$ possible single degree of freedom comparisons that are possible when $k = 3$ follow: Group 1 versus Group 2; Group 1 versus Group 3; Group 2 versus Group 3; Group 1 versus Groups 2 and 3; Group 2 versus Groups 1 and 3; Group 3 versus Groups 1 and 2. Thus, if the **Bonferroni–Dunn test** is employed to conduct all six possible single degree of freedom comparisons with $\alpha_{FW} = .05$, the per comparison error rate will be $\alpha_{PC} = .05/6 = .0083$. Since $\alpha_{PC}/2 = .0083/2 = .00415$, employing **Table A1** we can determine the z value above which .00415 proportion of cases falls is approximately 2.635. Substituting $z = 2.635$ in Equation 16.27, the value $t_{B/D} = 3.16$ is computed.

$$t_{B/D} = 2.635 + \frac{(2.635)^3 + 2.635}{4(12 - 2)} = 3.16$$

Substituting $t_{B/D} = 3.16$ in Equation 16.26, the value $C_{B/D} = 2.75$ is computed.

$$C_{B/D} = 3.16\sqrt{\frac{(2)(1.9)}{5}} = 2.75$$

Since Equation 16.26 is only valid for a simple comparison, the value $C_{B/D} = 2.75$ only applies to the three simple comparisons that are possible within the full set of six single degree of freedom comparisons. Thus, in order to be significant, the difference between any two means contrasted in a simple comparison must be at least 2.75 units. Note that the latter value is larger than $CD_{B/D} = 2.43$ computed for a set of $c = 3$ comparisons.

If one or more complex comparisons are conducted, the $t_{B/D}$ value a researcher employs will depend on the total number of comparisons (both simple and complex) being conducted. As is the case with **multiple *t* tests/Fisher's LSD test,** the computed value of $CD_{B/D}$ will also be a function of the coefficients employed in a comparison. Equation 16.29 is a generic equation that can be employed for both simple and complex comparisons to compute the value of $CD_{B/D}$. In the case of a simple comparison the term $\Sigma c_j^2 = 2$, thus results in Equation 16.26.

$$CD_{B/D} = t_{B/D} \sqrt{\frac{(\Sigma c_j^2)(MS_{WG})}{n}} \qquad \textbf{(Equation 16.29)}$$

Equation 16.29 will be employed for the complex comparison of Group 3 versus the combined performance of Groups 1 and 2. On the assumption that the six possible single degree of freedom comparisons are conducted, the value $t_{B/D} = 3.16$ will be employed in the equation.

$$t_{B/D} = 3.16\sqrt{\frac{(1.5)(1.9)}{5}} = 2.39$$

Thus, in order to be significant, the absolute value of the difference $\bar{X}_3 - [(\bar{X}_1 + \bar{X}_2)/2]$ must be at least 2.39 units. Since the obtained absolute difference of 1.7 is less than the latter value, the nondirectional alternative hypothesis H_1: $\mu_3 \neq (\mu_1 + \mu_2)/2$ is not supported. Recollect that when **multiple *t* tests/Fisher's LSD test** are employed for the same comparison, the latter alternative hypothesis is supported. The difference between the results of the two comparison procedures illustrates that the **Bonferroni–Dunn test** is a more conservative/less powerful procedure.

It should be noted that in conducting comparisons (especially if they are planned) a researcher may not elect to conduct all possible comparisons between pairs of means. Obviously, the fewer comparisons that are conducted, the higher the value for α_{PC} that can be employed for the **Bonferroni–Dunn test**. Nevertheless, some researchers consider the **Bonferroni–Dunn** adjustment to be too severe, regardless of how many comparisons are conducted. In view of this, various sources (e.g., Howell (1992) and Keppel (1991)) describe a modified version of the **Bonferroni–Dunn test** that can be employed if one believes that the procedure described in this section sacrifices too much power.

It should also be pointed out that in conducting the **Bonferroni–Dunn test**, as well as any other comparison procedures which result in a reduction of the value of α_{PC}, it is not necessary that each comparison be assigned the same α_{PC} value. As long as the sum of the α_{PC} values adds up to the value stipulated for α_{FW}, the α_{PC} values can be distributed in any way the researcher deems prudent. Thus, if certain comparisons are considered more important than others, the researcher may be willing to tolerate a higher α_{PC} rate for such comparisons. As an example, assume a researcher conducts three comparisons and sets $\alpha_{FW} = .05$. If the first of the comparisons is considered to be the one of most interest and the researcher wants to maximize the power of that comparison, the α_{PC} rate for that comparison can be set equal to .04, and the α_{PC} rate for each of the other two comparisons can be set at .005. Note that since the sum of the three values is .05, the value $\alpha_{FW} = .05$ is maintained.

Test 16c: Tukey's HSD test (The term **HSD** is an abbreviation for **honestly significant difference**)[35] **Tukey's HSD test** is generally recommended for unplanned comparisons when a researcher wants to make **all possible pairwise comparisons** (i.e., simple comparisons) in a set of data. The total number of pairwise comparisons (c) that can be conducted for a set of data can be computed with the following equation: $c = [k(k - 1)]/2$. Thus, if $k = 3$, the total number of possible pairwise comparisons is $c = [3(3 - 1)]/2 = 3$ (which in the case of Example 16.1 are Group 1 versus Group 2, Group 1 versus Group 3, and Group 2 versus Group 3).

Tukey's HSD test (Tukey, 1953) controls the **familywise Type I error rate** so that it will not exceed the prespecified alpha value employed in the analysis. Many sources view it as a good compromise among the available unplanned comparison procedures, in that it maintains an acceptable level for α_{FW}, without resulting in an excessive decrease in power.[36] **Tukey's HSD test** is one of a number of comparison procedures that are based on the **Studentized range statistic** (which is represented by the notation q). Like the t distribution, the distribution for the **Studentized range statistic** is also employed to compare pairs of means. When the total number of groups/treatments involved in an experiment is greater than two, a tabled critical q value will be higher than the corresponding tabled critical t value for **multiple t tests/Fisher's LSD test** for the same comparison.[37] For a given degrees of freedom value, the magnitude of a tabled critical q value increases as the number of groups employed in an experiment increase. **Table A13** in the **Appendix** (which is discussed in more detail later in this section) is the **Table of the Studentized Range Statistic.**

Equation 16.30 is employed to compute the q statistic, which represents the test statistic for **Tukey's HSD test.**

$$q = \frac{\bar{X}_a - \bar{X}_b}{\sqrt{\dfrac{MS_{WG}}{n}}} \qquad \textbf{(Equation 16.30)}$$

Equation 16.30 will be employed with Example 16.1 to conduct the simple comparison of Group 1 versus Group 2. When the latter comparison is evaluated with Equation 16.30, the value $q = 4.22$ is computed.

$$q = \frac{9.2 - 6.6}{\sqrt{\dfrac{1.9}{5}}} = 4.22$$

The obtained value $q = 4.22$ is evaluated with **Table A13**. The latter table contains the two-tailed .05 and .01 critical q values that are employed to evaluate a nondirectional alternative hypothesis at the .05 and 01 levels or a directional alternative hypothesis at the .10 and .02 levels. As is the case with previous comparisons, the analysis will be in reference to the nondirectional alternative hypothesis H_1: $\mu_1 \neq \mu_2$, with $\alpha_{FW} = .05$. Employing the section of **Table A13** for the .05 critical values, we locate the q value that is in the cell which is the intersection of the column for $k = 3$ means (which represents the total number of groups upon which the omnibus F value is based) and the row for $df_{\text{error}} = 12$ (which represents the value $df_{\text{error}} = 12$ computed for the omnibus F test). The tabled critical $q_{.05}$ value for $k = 3$ means and $df_{\text{error}} = 12$ is $q_{.05} = 3.77$. In order to reject the null hypothesis, the obtained absolute value of q must be equal to or greater than the tabled critical value at the prespecified level of significance.[38] Since the obtained value $q = 4.22$ is greater than the tabled critical two-tailed value $q_{.05} = 3.77$, the nondirectional alternative hypothesis H_1: $\mu_1 \neq \mu_2$ is supported at the .05 level (where .05 represents the value of α_{FW}). It is not supported at the .01 level, since $q = 4.22$ is less than the tabled critical value $q_{.01} = 5.05$.

Equation 16.31, which is algebraically derived from Equation 16.30, can be employed to compute the minimum required difference (CD_{HSD}) in order for two means to differ significantly from one another at a prespecified level of significance.[39]

$$CD_{\text{HSD}} = q_{(k, df_{WG})} \sqrt{\frac{MS_{WG}}{n}} \qquad \textbf{(Equation 16.31)}$$

Where: $q_{(k,df_{WG})}$ is the tabled critical q value for k groups/means and df_{WG} is the value employed for the omnibus F test

Employing the appropriate values in Equation 16.31, the value $CD_{\text{HSD}} = 2.32$ is computed.

$$CD_{\text{HSD}} = (3.77)\sqrt{\frac{1.9}{5}} = 2.32$$

Thus, in order to be significant, the difference $\bar{X}_1 - \bar{X}_2$ must be at least 2.32 units. Note that the value $CD_{HSD} = 2.32$ is greater than $CD_{LSD} = 1.90$, but less than $CD_{B/D} = 2.43$. This reflects the fact that in conducting all pairwise comparisons, **Tukey's HSD test** will provide a more powerful test of an alternative hypothesis than the **Bonferroni-Dunn test.** It also indicates that **Tukey's HSD test** is less powerful than **multiple *t* tests/Fisher's LSD test** (which does not control the value of α_{FW}).

The use of Equations 16.30 and 16.31 for **Tukey's HSD test** is based on the following assumptions: a) The distribution of data in the underlying population from which each of the samples is derived is normal; b) The variances of the k underlying populations represented by the k samples are equal to one another (i.e. homogeneous); and c) The sample sizes of each of the groups being compared are equal. Kirk (1982), among others, discusses a number of modifications of **Tukey's HSD test** which are recommended when there is reason to believe that all or some of the aforementioned assumptions are violated.[40]

Test 16d: The Newman-Keuls test The **Newman-Keuls test** (Keuls (1952), Newman (1939)) is another procedure for pairwise unplanned comparisons that employs the **Studentized range statistic.** Although the **Newman-Keuls test** is more powerful than **Tukey's HSD test,** unlike the latter test it does not insure that in a set of pairwise comparisons the **familywise Type I error rate** will not exceed a prespecified alpha value. Equations 16.30 and 16.31 (which are employed for **Tukey's HSD test**) are also employed for the **Newman-Keuls test.** When, however, the latter equations are used for the **Newman-Keuls test,** the appropriate tabled critical q value will be a function of how far apart the two means being compared are from one another.

To be more specific, the k means are arranged ordinally (i.e., from lowest to highest). Thus, in the case of Example 16.1, the $k = 3$ means are arranged in the following order:

$$\bar{X}_3 = 6.2 \quad \bar{X}_2 = 6.6 \quad \bar{X}_1 = 9.2$$

For each pairwise comparison that can be conducted, the number of **steps** between the two means involved is determined. Because of the fact that the tabled critical q value is a function of the number of steps or **layers** that separate two mean values, the **Newman-Keuls test** is often referred to as a **stepwise** or **layered test.** The number of steps (which will be represented by the notation s) between any two means is determined as follows: Starting with the lower of the two mean values (which will be the mean to the left), count until the higher of the two mean values is reached. Each mean value employed in counting from the lower to the higher value in the pair represents one step. Thus, $s = 2$ steps are involved in the simple comparison of Group 1 versus Group 2, since we start at the left with $\bar{X}_2 = 6.6$ (the lower of the two means involved in the comparison) which is step 1, and move right to the adjacent value $\bar{X}_1 = 9.2$ (the higher of the two means involved in the comparison), which represents step 2. If Group 1 and Group 3 are compared, $s = 3$ steps separate the two means, since we start at the left with $\bar{X}_3 = 6.2$ (the lower of the two means involved in the comparison), move to the right counting $\bar{X}_2 = 6.6$ as step 2, and then move to $\bar{X}_1 = 9.2$ (the higher of the two means involved in the comparison), which is step 3.

The **Newman–Keuls test** protocol requires that the pairwise comparisons be conducted in a specific order. The first comparison conducted is between the two means which are separated by the largest number of steps (which will represent the largest absolute difference between any two means). If the latter comparison is significant, any comparisons involving the second largest number of steps are conducted. If all the comparisons in the latter subset of comparisons are significant, the subset of comparisons for the next largest number of steps is conducted, and so on. The basic rule upon which the protocol is based is that if at any point in the analysis a comparison fails to yield a significant result, no further comparisons are conducted on pairs of means that are separated by fewer steps than the number of steps involved in the nonsignificant comparison. Employing this protocol with Example 16.1, the first comparison that is conducted is $\bar{X}_1$ versus $\bar{X}_3$, which involves $s = 3$ steps. If that comparison is significant, the comparisons $\bar{X}_1$ versus $\bar{X}_2$ and $\bar{X}_2$ versus $\bar{X}_3$, both of which involve $s = 2$ steps, are conducted.

In employing **Table A13** to determine the tabled critical q value to employ with Equations 16.30 and 16.31, instead of employing the column for k (the total number of groups/means in the set of data), the column that is used is the one that corresponds to the number of steps between the two means involved in a comparison. Thus, if Group 1 and Group 3 are compared, the column for $k = 3$ groups in **Table A13** is employed in determining the value of q. Since the value of df_{error} remains $df_{WG} = 12$, the value $q_{.05} = 3.77$ is employed in Equation 16.31. Since the latter value is identical to the q value employed for **Tukey's HSD test**, the **Newman–Keuls** value computed for the minimum required difference for two means ($CD_{N/K}$) will be the same as the value CD_{HSD} computed for **Tukey's HSD test**.[41] Thus, $CD_{N/K} = (3.77)\sqrt{1.9/5} = 2.32$. The latter result indicates that if $\alpha = .05$, the minimum required difference between two means for the comparison Group 1 versus Group 3 is $CD_{N/K} = 2.32$.[42] Since the absolute difference $|\bar{X}_1 - \bar{X}_3| = 3$ is greater than $CD_{N/K} = 2.32$, the comparison is significant.

Since the result of the $s = 3$ step comparison is significant, the **Newman–Keuls test** is employed to compare Group 1 versus Group 2 and Group 2 versus Group 3, which represent $s = 2$ step comparisons. The value $q_{.05} = 3.08$ is employed in Equation 16.31, since the latter value is the tabled critical $q_{.05}$ value for $k = 2$ means and $df_{\text{error}} = 12$. Substituting $q_{.05} = 3.08$ in Equation 16.31 yields the value $CD_{N/K} = (3.08)\sqrt{1.9/5} = 1.90$. Since the absolute difference $|\bar{X}_1 - \bar{X}_2| = 2.6$ is greater than $CD_{N/K} = 1.90$, there is a significant difference between the means of Groups 1 and 2. Since the absolute difference $|\bar{X}_2 - \bar{X}_3| = .4$ is less than $CD_{N/K} = 1.90$, the researcher cannot conclude there is a significant difference between the means of Groups 2 and 3. Note that the value $CD_{N/K} = 1.90$ computed for a two-step analysis is smaller than the value $CD_{N/K} = 2.32$ computed for the three-step analysis. The general rule is that the fewer the number of steps, the smaller the computed $CD_{N/K}$ value.

The astute reader will observe that $CD_{N/K} = 1.90$ is identical to $CD_{LSD} = 1.90$ obtained with Equation 16.24. The latter result illustrates that when two steps are involved in a **Newman–Keuls** comparison, it will always yield a value that is identical to CD_{LSD} (i.e., the value computed with **multiple *t* tests/Fisher's LSD test**). In point of fact, when $s = 2$, for a given degrees of freedom value, the relationship between the values of q and t is as follows: $q = t\sqrt{2}$ and $t = q/\sqrt{2}$. If the value $q_{.05} = 3.08$ is employed in the equation $t = q/\sqrt{2}$, $t = (3.08)/\sqrt{2} = 2.18$. Note that $(t_{.05} = 2.18)^2 = (F_{.05} = 4.75)$, and that $F_{.05} = 4.75$ is the tabled critical value employed in Equation 16.24 which yields the value $CD_{LSD} = 1.90$.

The value of α_{FW} associated with the **Newman–Keuls test** will be higher than the value of α_{FW} for **Tukey's HSD test**. This is a direct result of the fact that the tabled critical **Studentized range** values employed for the **Newman–Keuls test** are smaller than those employed for **Tukey's HSD test** (with the exception of the comparison contrasting the lowest and highest means in the set of k means). Within any subset of comparisons which are an equal number of steps apart from one another, the overall Type I error rate for the **Newman–Keuls test** within that subset will not exceed the prespecified value of alpha. The latter, however, does not insure that α_{FW} for all the possible pairwise comparisons will not exceed the prespecified value of alpha. Because of its higher α_{FW} rate, the **Newman–Keuls test** is generally not held in high esteem as an unplanned comparison procedure. Excellent discussions of the **Newman–Keuls test** can be found in Maxwell and Delaney (1990) and Howell (1992).

Test 16e: The Scheffé test The **Scheffé test** (Scheffé, 1953), which is employed for unplanned comparisons, is commonly described as the most conservative of the unplanned comparison procedures. The test maintains a fixed value for α_{FW}, regardless of how many simple and complex comparisons are conducted. By virtue of controlling for a large number of potential comparisons, the error rate for any single comparison (i.e., α_{PC}) will be lower than the error rate associated with any of the other comparison procedures (assuming an alternative procedure employs the same value for α_{FW}).[43] Since in conducting unplanned pairwise comparisons **Tukey's HSD test** provides a more powerful test of an alternative hypothesis than the **Scheffé test**, most sources note that it is not prudent to employ the **Scheffé test** if only simple comparisons are being conducted. The **Scheffé test** is, however, recommended whenever a researcher wants to maintain a specific α_{FW} level, regardless of how many simple and complex comparisons are conducted. Sources note that the **Scheffé test** can accommodate unequal sample sizes and is quite robust with respect to violations of the assumptions underlying the analysis of variance (i.e., homogeneity of variance and normality of the underlying population distributions). Because of the low value the **Scheffé test** imposes on the value of α_{PC} and the consequent loss of power associated with each comparison, some sources recommend that in using the test a researcher employ a larger value for α_{FW} than would ordinarily be the case. Thus, one might employ $\alpha_{FW} = .10$ for the **Scheffé test**, instead of $\alpha_{FW} = .05$ which might be employed for a less conservative comparison procedure.

Equation 16.32 is employed to compute the minimum required difference for the **Scheffé test** (CD_S) in order for two means to differ significantly from one another at a prespecified level of significance. Whereas Equation 16.32 only applies to simple comparisons, Equation 16.33 is a generic equation that can be employed for both simple and complex comparisons.

$$CD_S = \sqrt{(k - 1)(F_{(df_{BG}, df_{WG})})} \sqrt{\frac{2MS_{WG}}{n}} \qquad \textbf{(Equation 16.32)}$$

$$CD_S = \sqrt{(k - 1)(F_{(df_{BG}, df_{WG})})} \sqrt{\frac{(\Sigma c_j^2)(MS_{WG})}{n}} \qquad \textbf{(Equation 16.33)}$$

In Equations 16.32 and 16.33, the value $F_{(df_{BG}, df_{WG})}$ is the tabled critical value that is employed for the omnibus F test for a value of alpha that corresponds to the value of α_{FW}. Thus, in Example 16.1, if $\alpha_{FW} = .05$, $F_{.05} = 3.89$ for $df = 2,12$ is used in Equations 16.32/16.33. Employing Equation 16.32, the value $CD_S = 2.43$ is computed for the simple comparison of Group 1 versus Group 2.

$$CD_S = \sqrt{(3 - 1)(3.89)}\sqrt{\frac{(2)(1.9)}{5}} = 2.43$$

Thus, in order to be significant, the difference $\bar{X}_1 - \bar{X}_2$ (as well as the difference between the means of any other two groups) must be at least 2.43 units. Since the absolute difference $|\bar{X}_1 - \bar{X}_2| = 2.6$ is greater than $CD_S = 2.43$, the nondirectional alternative hypothesis H_1: $\mu_1 \neq \mu_2$ is supported. Note that for the same comparison, the CD value computed for the **Scheffé test** is larger than the previously computed values $CD_{LSD} = CD_{N/K} = 1.90$ and $CD_{HSD} = 2.32$, but is equivalent to $CD_{B/D} = 2.43$. The fact that $CD_S = CD_{B/D}$ for the comparison under discussion illustrates that although for simple comparisons the **Scheffé test** is commonly described as the most conservative of the unplanned comparison procedures, when the **Bonferroni–Dunn test** is employed for simple comparisons it may be as or more conservative than the **Scheffé test.** Maxwell and Delaney (1990) note that in instances where a researcher conducts a small number of comparisons, the **Bonferroni–Dunn test** will provide a more powerful test of an alternative hypothesis than the **Scheffé test.** However, as the number of comparisons increase, at some point the **Scheffé test** will become more powerful than the **Bonferroni–Dunn test.** In general (although there are some exceptions), the **Bonferroni–Dunn test** will be more powerful than the **Scheffé** test when the number of comparisons conducted is less than $[k(k - 1)]/2$.

The **Scheffé test** is most commonly recommended when at least one complex comparison is conducted in a set of unplanned comparisons. Although the other comparison procedures discussed in this section can be employed for unplanned complex comparisons, the **Scheffé test** is viewed as a more desirable alternative by most sources. It will now be demonstrated how the **Scheffé test** can be employed for the complex comparison $\bar{X}_3$ versus $(\bar{X}_1 + \bar{X}_2)/2$. Substituting the value $\Sigma c_j^2 = 1.5$ (which is computed in Table 16.4) and the other relevant values in Equation 16.33, the value $CD_S = 2.11$ is computed.

$$CD_S = \sqrt{(3 - 1)(3.89)}\sqrt{\frac{(1.5)(1.9)}{5}} = 2.11$$

Thus, in order to be significant, the difference $\bar{X}_3 - [(\bar{X}_1 + \bar{X}_2)/2]$ (as well as the difference between any set of means in a complex comparison for which $\Sigma c_j^2 = 1.5$) must be equal to or greater than 2.11 units. Since the absolute difference $|\bar{X}_3 - [(\bar{X}_1 + \bar{X}_2)/2]| = 1.7$ is less than $CD_S = 2.11$, the nondirectional alternative hypothesis H_1: $\mu_3 \neq (\mu_1 + \mu_2)/2$ is not supported. Note that the CD value computed for the **Scheffé test** is larger than the previously computed value $CD_{LSD} = 1.65$, but less than $CD_{B/D} = 2.39$ computed for the same complex comparison. This reflects the fact that for the complex comparison $\bar{X}_3$ versus $(\bar{X}_1 + \bar{X}_2)/2$, the **Scheffé test** is not as powerful as **multiple *t* tests/Fisher's LSD test**, but is more powerful than the **Bonferroni–Dunn test.**[44]

Equation 16.34 can also be employed for the **Scheffé test** for both simple and complex comparisons.

$$F_S = (k - 1)F_{(BG,WG)} \qquad \textbf{(Equation 16.34)}$$

Where: $F_{(BG,WG)}$ is the tabled critical value at the prespecified level of significance employed in the omnibus F test

In order to use Equation 16.34 it is necessary to first employ Equations 16.17–16.19 to compute the value of F_{comp} for the comparison being conducted. The value computed for F_{comp} will serve as the test statistic for the **Scheffé test.** Earlier in this section (under the discussion of **linear contrasts**) the value $F_{\text{comp}} = 5.07$ is computed for the complex comparison $\bar{X}_3$ versus $(\bar{X}_1 + \bar{X}_2)/2$. When F_{comp} is used as the test statistic for the **Scheffé**

test, the critical value employed to evaluate it is different than the critical value employed in evaluating a **linear contrast**. Equation 16.34 is used to determine the **Scheffé test** critical value. In order for a comparison to be significant, the computed value of F_{comp} must be equal to or greater than the critical F value computed with Equation 16.34 (which is represented by the notation F_S). In employing Equation 16.34 to compute F_S, the tabled critical value employed for the omnibus F test (which in the case of Example 16.1 is $F_{.05} = 3.89$) is multiplied by $(k - 1)$. Obviously, the resulting value will be higher than the tabled critical value employed for the **linear contrast** for the same comparison. When the appropriate values for Example 16.1 are substituted in Equation 16.34, the value $F_S = 7.78$ is computed.

$$F_S = (3 - 1)(3.89) = 7.78$$

Since the computed value $F_{\text{comp}} = 5.07$ is less than the critical value $F_S = 7.78$ computed with Equation 16.34, it indicates that the nondirectional alternative hypothesis H_1: $\mu_3 \neq (\mu_1 + \mu_2)/2$ is not supported. Note that the value $F_S = 7.78$ is larger than the value $F_{.05} = 4.75$ (which is the tabled critical value for $df_{\text{num}} = 1$ and $df_{\text{den}} = df_{WG} = 12$) that is employed for the linear contrast for the same comparison. Recollect that the alternative hypothesis H_1: $\mu_3 \neq (\mu_1 + \mu_2)/2$ is supported when a **linear contrast** is conducted.

In closing the discussion of the **Scheffé test** some general comments will be made regarding the value of α_{FW} for the **Scheffé test**, the **Bonferroni–Dunn test**, and **Tukey's HSD test**. In the discussion to follow it will be assumed that upon computation of an omnibus F value, a researcher wishes to conduct a series of unplanned comparisons for which the **familywise error rate** does not exceed $\alpha_{FW} = .05$.[45]

a) If all possible comparisons (simple and complex) are conducted with the **Scheffé test**, the value of α_{FW} will equal exactly .05. When $k \geq 3$ there are actually an infinite number of comparisons that can be made. To illustrate this, assume $k = 3$. Beside the three pairwise/simple comparisons (Group 1 versus Group 2; Group 1 versus Group 3; Group 2 versus Group 3) and the three apparent complex comparisons (The average of Groups 1 and 2 versus Group 3; The average of Groups 1 and 3 versus Group 2; The average of Groups 2 and 3 versus Group 1), in the case of a complex comparison it is possible to combine two groups so that one group contributes more to the composite mean representing the two groups than does the other group. As an example, in comparing Groups 1 and 2 with Group 3, a coefficient of 7/8 can be assigned to Group 1 and a coefficient of 1/8 to Group 2. Employing these coefficients, a composite mean value can be computed to represent the mean of the two groups which is contrasted with Group 3. It should be obvious that if one can stipulate any combination of two coefficients/weights that add up to 1, there are potentially an infinite number of coefficient combinations that can be assigned to any two groups, and therefore an infinite number of possible comparisons can result from coefficient combinations involving the comparison of two groups with a third group. If fewer than all possible comparisons are conducted with the **Scheffé test**, the value of α_{FW} will be less than .05, thus making it an overly conservative test (since the value of α_{PC} will be lower than is necessary for α_{FW} to equal .05).

b) If all possible comparisons are conducted employing the **Bonferroni–Dunn test**, the value of α_{FW} will be less than .05. As noted earlier, as the number of comparisons conducted increases, at some point the value of α_{FW} for the **Bonferroni–Dunn test** will be less than α_{FW} for the **Scheffé test**, and thus at that point the **Bonferroni–Dunn test** will be even more conservative (and thus less powerful) than the **Scheffé test**. The decrease in the value of α_{FW} for the **Bonferroni–Dunn test** results from the fact that within the set of comparisons conducted, not all comparisons will be orthogonal with one another. Winer *et al.* (1991) note that when comparisons conducted with the **Bonferroni–Dunn test** are orthogonal, the following is true: $\alpha_{FW} = c(\alpha_{PC})$. However, when some of the comparisons conducted are not orthogonal, $\alpha_{FW} < c(\alpha_{PC})$. Thus, by virtue of some of the comparisons being nonorthogonal, the **Bonferroni–Dunn test** becomes a more conservative test (i.e., $\alpha_{FW} < .05$).

c) If **Tukey's HSD test** is employed for conducting all possible pairwise comparisons, the value of α_{FW} will be exactly .05, even though the full set of pairwise comparisons will not constitute an orthogonal set. As noted above, if the **Bonferroni–Dunn test** is employed for the full set of pairwise comparisons, due to the presence of nonorthogonal comparisons, the value of α_{FW} will be less than .05, and thus the value of $CD_{B/D}$ will be larger than the value of CD_{HSD}. If in addition to conducting all pairwise comparisons with **Tukey's HSD test**, complex comparisons are also conducted, the value of α_{FW} will exceed .05. As the number of complex comparisons conducted increases, the value of α_{FW} increases. When complex comparisons are conducted, **Tukey's HSD test** is not as powerful as the **Scheffé test**.

Test 16f: The Dunnett test The **Dunnett test** (1955, 1964) is a comparison procedure which is only employed for simple comparisons, that is designed to compare a control group with the other $(k - 1)$ groups in a set of data. Under such conditions the **Dunnett test** provides a more powerful test of an alternative hypothesis than do the **Bonferroni–Dunn test**, **Tukey's HSD test** and the **Scheffé test**. This is the case, since for the same value of α_{FW}, the α_{PC} value associated with the **Dunnett test** will be higher than the α_{PC} values associated with the aforementioned procedures (and by virtue of this provides a more powerful test of an alternative hypothesis). The larger α_{PC} value for the **Dunnett test** is predicated on the fact that by virtue of limiting the comparisons to contrasting a control group with the other groups, the **Dunnett test** statistic is based on the assumption that fewer comparisons are conducted than will be the case if all pairwise comparisons are conducted. Consequently, if a researcher specifies that $\alpha_{FW} = .05$, and the mean of the control group is contrasted with the means of each of the other $(k - 1)$ groups, the **Dunnett test** insures that the **familywise Type I error rate** will not exceed .05. It should be noted that since the control group is involved in each of the comparisons that are conducted, the comparisons will not be orthogonal to one another. In illustrating the computation of the **Dunnett test** statistic, we will assume that in Example 16.1 Group 1 (the group that is not exposed to noise) is a control group, and that Groups 2 and 3 (both of which are exposed to noise) are experimental groups. Thus, employing the **Dunnett test**, the following two comparisons will be conducted with $\alpha_{FW} = .05$: Group 1 versus Group 2; Group 1 versus Group 3.

The test statistic for the **Dunnett test** (t_D) is computed with Equation 16.35, which, except for the fact that a t_D value is computed, is identical to Equation 16.22 (which is employed to compute the test statistic for **multiple *t* tests/Fisher's LSD test**).[46]

$$t_D = \frac{\bar{X}_a - \bar{X}_b}{\sqrt{\dfrac{2MS_{WG}}{n}}} \qquad \textbf{(Equation 16.35)}$$

Equation 16.35 is employed below to compute the value $t_D = 2.99$ for the simple comparison of Group 1 versus Group 2.

$$t_D = \frac{9.2 - 6.6}{\sqrt{\dfrac{(2)(1.9)}{5}}} = 2.99$$

The computed value $t_D = 2.99$ is evaluated with **Table A14 (Table of Dunnett's Modified *t* Statistic for a Control Group Comparison)** in the **Appendix**. The latter table, which contains both two-tailed and one-tailed .05 and .01 critical values, is based on a modified t distribution derived by Dunnett (1955, 1964). Dunnett (1955) computed one-tailed critical values, since in comparing one or more treatments with a control group a researcher is often interested in the direction of the difference. The tabled critical t_D values are listed in reference to k, the total number of groups/treatments employed in the experiment, and the value of $df_{\text{error}} = df_{WG}$ computed for the omnibus F test.

For $k = 3$ and $df_{\text{error}} = df_{WG} = 12$, the tabled critical two-tailed .05 and .01 values are $t_{D_{.05}} = 2.50$ and $t_{D_{.01}} = 3.39$, and the tabled critical one-tailed .05 and .01 values are $t_{D_{.05}} = 2.11$ and $t_{D_{.01}} = 3.01$. The computed value $t_D = 2.99$ is greater than the tabled critical two-tailed and one-tailed .05 t_D values but less than the tabled critical two-tailed and one-tailed .01 t_D values. Thus, for $\alpha_{FW} = .05$ (but not $\alpha_{FW} = .01$), the nondirectional alternative hypothesis H_1: $\mu_1 \neq \mu_2$ and the directional alternative hypothesis H_1: $\mu_1 > \mu_2$ are supported. The second comparison involving the control group (Group 1) versus Group 3, yields the value $t_D = (9.2 - 6.2)/\sqrt{[(2)(1.9)]/5} = 3.45$. Since the value $t_D = 3.45$ is greater than the tabled critical two-tailed and one-tailed .05 and .01 t_D values, the nondirectional alternative hypothesis H_1: $\mu_1 \neq \mu_3$ and the directional alternative hypothesis H_1: $\mu_1 > \mu_3$ are supported for both $\alpha_{FW} = .05$ and $\alpha_{FW} = .01$.

Equation 16.36 is employed to compute the minimum required difference for the **Dunnett test** (designated CD_D) in order for two means to differ significantly from one another at a prespecified level of significance.

$$CD_D = t_{D_{(k,\, df_{WG})}} \sqrt{\frac{2MS_{WG}}{n}} \qquad \textbf{(Equation 16.36)}$$

Where: $t_{D_{(k,\, df_{WG})}}$ is the tabled critical value for **Dunnett's modified *t* statistic** for k groups and df_{WG} at the prespecified value of α_{FW}

In employing Equation 16.36, $\alpha_{FW} = .05$ will be employed, and it will be assumed that a nondirectional alternative hypothesis is evaluated for both comparisons. Substituting the two-tailed .05 value $t_{D_{.05}} = 2.50$ in Equation 16.36, the value $CD_D = 2.18$ is computed.

$$CD_D = 2.50 \sqrt{\frac{2(1.9)}{5}} = 2.18$$

Thus, in order to be significant, the differences $\bar{X}_1 - \bar{X}_2$ and $\bar{X}_1 - \bar{X}_3$ must be at least 2.18 units. Since the absolute differences $|\bar{X}_1 - \bar{X}_2| = 2.6$ and $|\bar{X}_1 - \bar{X}_3| = 3$ are greater than $CD_D = 2.18$, the nondirectional alternative hypothesis is supported for both comparisons. Note that $CD_D = 2.18$ computed for the **Dunnett test** is larger than $CD_{LSD} = 1.90$ computed for **multiple *t* tests/Fisher's LSD test** (for a simple comparson), but is less than the CD values computed for a simple comparison for the **Bonferroni–Dunn test** ($CD_{B/D} = 2.43$), **Tukey's HSD test** ($CD_{HSD} = 2.32$), and the **Scheffé test** ($CD_S = 2.43$).

Additional discussion of comparison procedures and final recommendations The accuracy of the comparison procedures described in this section may be compromised if the homogeneity of variance assumption underlying the analysis of variance (the evaluation of which is described later in Section VI) is violated. This is the case, since in such an instance MS_{WG} may not provide the best measure of error variability for a given comparison. Violation of the homogeneity of variance assumption can either increase or decrease the Type I error rate associated with a comparison, depending upon whether MS_{WG} overestimates or underestimates the pooled variability of the groups involved in a specific comparison. It is also the case that when the homogeneity of variance assumption is violated, the accuracy of a comparison may be even further compromised when there is not an equal number of subjects in each group. Sources that discuss these general issues (e.g., Howell (1992), Kirk (1982), Maxwell and Delaney (1990), Winer *et al.* (1991)) provide alternative equations which are recommended when the homogeneity of variance assumption is violated and/or sample sizes are unequal. As a general rule, the measure of within-groups variability that is employed in equations which are recommended when there is heterogeneity of variance is based on the pooled within-groups variability of just those groups which are involved in a specific comparison. Since the latter measure has a smaller degrees of freedom associated with it than MS_{WG}, the tabled critical value for the analysis will be based on fewer degrees of freedom. Although the loss of degrees of freedom can reduce the power of the test, it may be offset if the revised measure of within-groups variability is less than MS_{WG}. A full discussion of the subject of violation of the homogeneity of variance assumption with comparisons is beyond the scope of this book. The reader who refers to sources that discuss the subject in greater detail will discover that there is a lack of agreement with respect to what procedure is most appropriate to employ when the assumption is violated.

At this point the author will present some general recommendations regarding the use of comparison procedures for the analysis of variance.[47] From what has been said, it should be apparent that in conducting comparisons the minimum value required for two means to differ significantly from one another can vary dramatically, depending upon which comparison procedure a researcher employs. Although some recommendations have been made with respect to when it is viewed most prudent to employ each of the comparison procedures, in the final analysis the use of any of the procedures does not insure that a researcher will determine the truth regarding the relationship between the variables under study. Aside from the fact that researchers do not agree among themselves on which comparison procedure to employ (due largely to the fact that they do not concur with respect to the maximum acceptable value for α_{FW}), there is also the problem that one is not always able to assume with a high degree of confidence that all of the assumptions underlying a specific comparison procedure have, in fact, been met. In view of this, any probability value associated with a comparison will probably always be subject to challenge. Although most of the time a probability value may not be compromised to that great a degree, when one considers the fact that researchers may quibble over whether one should employ $\alpha_{FW} = .05$ versus $\alpha_{FW} = .10$, a minimal difference with respect to a probability value can mean a great deal, since it ultimately may determine whether a researcher elects to retain or reject a null hypothesis. If, in fact, the status of the null hypothesis is based on the result of a single study, it would seem that a researcher is obliged to arrive at a probability value in which he and others can have a high degree of confidence.[48]

In view of everything that has been discussed, this writer believes that the general strategy for conducting comparisons suggested by Keppel (1991) is the most prudent to employ. Keppel (1991) suggests that in hypothesis testing involving **unplanned comparisons**, instead of just employing the two decision categories of **retaining the null hypothesis** versus **rejecting the null hypothesis**, a third category, **suspend judgement** be added. Specifically, Keppel (1991) recommends the following:

a) If the obtained difference between two means is less than the value of CD_{LSD}, **retain the null hypothesis**. Since the value of CD_{LSD} will be the smallest CD value computed with any of the available comparison procedures, it allows for the most powerful test of an alternative hypothesis.

b) If the obtained difference between two means is equal to or greater than the CD value associated with the comparison procedure which results in the **largest α_{FW} value** one is willing to tolerate, **reject the null hypothesis**. Based on the procedures described in this book (depending upon the number of comparisons that are conducted), the largest CD value will be generated through use of either the **Bonferroni–Dunn test** or the **Scheffé test**. However, the largest α_{FW} value a researcher may be willing to tolerate may be larger than the α_{FW} value associated with either of the aforementioned procedures. In such a case, the minimum CD value required to reject the null hypothesis will be smaller than $CD_{B/D}$ or CD_S.

c) If the obtained difference between two means is greater than or equal to CD_{LSD}, but less than the CD value associated with the largest α_{FW} value one is willing to tolerate, **suspend judgement**. It is recommended that one or more replication studies be conducted employing the relevant groups/treatments for any comparisons that fall in the **suspend judgement** category.

The above guidelines will now be applied to Example 16.1. Let us assume that three simple comparisons are conducted, and that the maximum **familywise Type I error rate** the researcher is willing to tolerate is $\alpha_{FW} = .05$. Employing the above guidelines, the null hypothesis can be rejected for the comparisons of Group 1 versus Group 2 and Group 1 versus Group 3. This is the case, since $\bar{X}_1 - \bar{X}_2 = 2.6$ and $\bar{X}_1 - \bar{X}_3 = 3$, and in both instances the obtained difference between the means of the two groups is greater than $CD_{HSD} = 2.32$ (which insures that α_{FW} will not exceed .05). On the other hand, for the comparison of Group 2 versus Group 3, since the difference $\bar{X}_2 - \bar{X}_3 = .4$ is less than $CD_{LSD} = 1.90$ (which is the lowest of the computed CD values), the null hypothesis cannot be rejected. Thus, in the case of Example 16.1, none of the comparisons falls in the **suspend judgement** category. In point of fact, even if the α_{FW} value associated with the **Bonferroni-Dunn** and/or **Scheffé tests** is employed as the maximum acceptable α_{FW} rate, none of the comparisons falls in the **suspend judgement** category, since the absolute difference between the means for each comparison is either greater than or less than the relevant CD values.

As noted earlier, in the case of **planned comparisons** the issue of whether or not a researcher should control the value of α_{FW} is subject to debate. It can be argued that the strategy described for unplanned comparisons should also employed with planned comparisons — with the stipulation that the largest value for α_{FW} that one is willing to tolerate for planned comparisons be higher than the value employed for unplanned comparisons. Nevertheless, one can omit the latter stipulation and argue that the same criterion be employed for both unplanned and planned comparisons. The rationale for the latter position is as follows. Assume that two researchers independently conduct the identical study. Researcher 1 has the foresight to plan c comparisons beforehand. Researcher 2, on the other hand, conducts the same set of c comparisons, but does not plan them beforehand. The truth regarding the populations involved in the comparisons is totally independent of who conducts the study. As a result of this, one can argue that the same criterion be applied, regardless of who conducts the investigation. If Researcher 1 is allowed to conduct a less conservative analysis (i.e., tolerate a higher α_{FW} rate) than Researcher 2, it is commensurate with giving Researcher 1 a bonus for having a bit more acumen that Researcher 2 (if we consider allowing one greater latitude with respect to rejecting the null hypothesis to constitute a bonus). It would seem that if, in the final analysis, the issue at hand is the truth concerning the populations under study, each of the two researchers should be expected to adhere to the same criterion, regardless of their expectations prior to conducting the study. If one accepts this line of reasoning, it would seem that the guidelines described in this section for unplanned comparisons should also be employed for planned comparisons.

In the final analysis, regardless of whether one has conducted planned or unplanned comparisons, when there is reasonable doubt in the mind of the researcher or there is reason to believe that there will be reasonable doubt among those who scrutinize the results of a study, it is always prudent to replicate a study. In other words, anytime the result of an analysis falls within the **suspend judgement** category (or perhaps even if the result is close to falling within it), a strong argument can be made for replicating a study. Thus, regardless of which comparison procedure one employs, if the result of a comparison is not significant, yet would have been had another comparison procedure been employed, it would seem logical to conduct at least one replication study in order to clarify the status of the null hypothesis. There is also the case where the result of a comparison turns out to be significant, yet would not have been with a more conservative comparison procedure. If in such an instance the researcher (or others who are familiar with the relevant literature) has reason to believe that a Type I error may have been committed, it would seem prudent to reevaluate the null hypothesis. In the final analysis, multiple replications of a result provide the most powerful evidence regarding the status of a null hypothesis. An effective tool that can be employed to pool the results of multiple studies which evaluate the same hypothesis is a methodology called **meta-analysis**, which is described in Mullen and Rosenthal (1985) and Rosenthal (1991).

At this point in the discussion it is worth reiterating the difference between **statistical** and **practical significance** (which is discussed in both the **Introduction** and in Section VI of the ***t* test for two independent samples**). By virtue of employing a large sample size, virtually any test evaluating a difference between the means of two populations will turn out to be significant. This results from the fact that two population means are rarely identical. However, in most instances a minimal difference between two means is commensurate with there being no difference at all, since the magnitude of such a difference is of no practical or theoretical value. In conducting comparisons (or, for that matter, conducting any statistical test) one must decide what magnitude of difference is of practical and/or theoretical significance. To state it another way, one must determine the magnitude of the **effect size** one is attempting to identify.[49] If a researcher is able to stipulate a meaningful effect size prior to collecting the data for a study, he can design the study so that the test which is employed to evaluate the null hypothesis is sufficiently powerful to identify the desired effect size. As a general rule, a researcher is best able to control the power of a statistical test by employing a sample size that exceeds some minimal value. In the latter part of Section VI the computation of power for the **single-factor between-subjects analysis of variance** is discussed in reference to both the omnibus F test as well as for comparison procedures.

In the final analysis, when the magnitude of the obtained difference between the means involved in any comparison is deemed too small to be of practical or theoretical significance, it really becomes irrelevant whether a result is statistically significant. In such an instance if two or more comparison procedures yield conflicting results, replication of a study is not in order.

The computation of a confidence interval for a comparison The computation of a **confidence interval** provides a researcher with a mechanism for determining a range of values within which he can be confident the true difference between the means of two populations falls. Computation of a confidence interval for a comparison is a straightforward procedure which can be easily implemented following the computation of a CD value.[50] Specifically, to compute the range of values that define the 95% confidence interval for any comparison, one should do the following: Add to and subtract the computed value of CD from the obtained difference between the two means involved in the comparison. As an example, let us assume **Tukey's HSD test** is employed to compute the value $CD_{HSD} = 2.32$ for the comparison involving Group 1 versus Group 2. To compute the 95% confidence interval, the value 2.32 is added to and subtracted from 2.6, which is the difference between the two means. Thus, $CI_{HSD_{.95}} = 2.6 \pm 2.32$, which can also be written as $.28 \le (\mu_1 - \mu_2) \le 4.92$. In other words, the researcher can be 95% confident that the mean of the population represented by Group 1 is between .28 and 4.92 units larger than the mean of the population represented by Group 2. If the researcher wants to compute the 99% confidence interval for a comparison, the same procedure can be used, except for the fact that in computing the CD value for the comparison, the relevant .01 tabled critical q value is employed (or the tabled critical .01 t, F or t_D value if the comparison procedure happens to employ either of the aforementioned distributions).

It should be emphasized that the range of values that define a confidence interval will be a function of the tabled critical value for the relevant test statistic that a researcher elects to employ in the analysis. As noted earlier, the tabled critical value one employs will be a function of the value established for α_{FW}. In the case of the **Bonferroni–Dunn test**, the magnitude of the critical $t_{B/D}$ value one employs (and consequently the range of values that defines a confidence interval) increases as the number of comparisons one conducts increases.

If one has reason to believe that the homogeneity of variance assumption underlying the analysis of variance is violated, one can argue that the measure of error variability employed in computing a confidence interval should be a pooled measure of variability based only on the groups involved in a specific comparison. Under such circumstances, one can also argue that it is acceptable to employ Equation 8.15 (employed to compute a confidence interval for the ***t* test for two independent samples**) to compute the confidence interval for a comparison. In using the latter equation for a simple comparison, one can argue that $df = n_a + n_b - 2$ (the degrees of freedom used for the ***t* test for two independent samples**) should be employed for the analysis as opposed to df_{WG}. The use of Equation 8.15 can also be justified in circumstances where: a) The researcher views the groups involved in a comparison as distinct (with respect to both the value of μ and σ^2) from the other groups involved in the study; and b) The researcher is not attempting to control the value of α_{FW}.

In view of everything that has been said, it should be apparent that the value of a confidence interval can vary dramatically depending upon which of the comparison procedures a researcher employs, and what assumptions one is willing to make with reference to the underlying populations under study. For this reason, two researchers may compute substantially different confidence intervals as a result of employing different comparison procedures. In the final analysis, however, each of the researchers may be able to offer a persuasive argument in favor of the methodology he employs.

2. Comparing the means of three or more groups when $k \geq 4$ Within the framework of a **single-factor between-subjects analysis of variance** involving $k = 4$ or more groups, a researcher may wish to evaluate a general hypothesis with respect to the means of a subset of groups, where the number of groups in the subset is some value less than k. Although the latter type of situation is not commonly encountered in research, this section will describe the protocol for conducting such an analysis.

To illustrate, assume that a fourth group is added to Example 16.1. Assume that the scores of the five subjects who serve in Group 4 are as follows: 3, 2, 1, 4, 5. Thus, $\Sigma X_4 = 15$, $\bar{X}_4 = 3$, and $\Sigma X_4^2 = 55$. If the data for Group 4 are integrated into the data for the other three groups whose performance is summarized in Table 16.1, the following summary values are computed: $N = nk = (5)(4) = 20$, $\Sigma X_T = 125$, $\Sigma X_T^2 = 911$. Substituting the revised values for $k = 4$ groups in Equations 16.2, 16.3, and 16.4/16.5, the following sum of squares values are computed: $SS_T = 129.75$, $SS_{BG} = 96.95$, $SS_{WG} = 32.8$. Employing the values $k = 4$ and $N = 20$ in Equations 16.8 and 16.9, the values $df_{BG} = 4 - 1 = 3$ and $df_{WG} = 20 - 4 = 16$ are computed. Substituting the appropriate values for the sum of squares and degrees of freedom in Equations 16.6 and 16.7, the values $MS_{BG} = 96.95/3 = 32.32$ and $MS_{WG} = 32.8/16 = 2.05$ are computed. Equation 16.12 is employed to compute the value $F = 32.32/2.05 = 15.77$. Table 16.6 is the summary table of the analysis of variance.

Employing $df_{\text{num}} = 3$ and $df_{\text{den}} = 16$, the tabled critical .05 and .01 values are $F_{.05} = 3.24$ and $F_{.01} = 5.29$. Since the obtained value $F = 15.77$ is greater than both of the aforementioned critical values, the null hypothesis (which for $k = 4$ is H_0: $\mu_1 = \mu_2 = \mu_3 = \mu_4$) can be rejected at both the .05 and .01 levels.

Table 16.6 Summary Table of Analysis of Variance for Example 16.1 When $k = 4$

Source of variation	*SS*	*df*	*MS*	*F*
Between-groups	96.95	3	32.32	15.77
Within-groups	32.80	16	2.05	
Total	129.75	19		

Let us assume that prior to the above analysis the researcher has reason to believe that Groups 1, 2, and 3 may be distinct from Group 4. However, before he contrasts the composite mean of Groups 1, 2, and 3 with the mean of Group 4 (i.e., conduct the complex comparison which evaluates the null hypothesis H_0: $(\mu_1 + \mu_2 + \mu_3)/3 = \mu_4$), he decides to evaluate the null hypothesis H_0: $\mu_1 = \mu_2 = \mu_3$. If the latter null hypothesis is retained, he will assume that the three groups share a common mean value, and on the basis of this he will compare their composite mean with the mean of Group 4. In order to evaluate the null hypothesis H_0: $\mu_1 = \mu_2 = \mu_3$, it is necessary for the researcher to conduct a separate analysis of variance that just involves the data for the three groups identified in the null hypothesis. The latter analysis of variance has already been conducted, since it is the original analysis of variance that is employed for Example 16.1 — the results of which are summarized in Table 16.2.

Upon conducting an analysis of variance on the data for all $k = 4$ groups as well as an analysis of variance on the data for the subset comprised of $k_{\text{subset}} = 3$ groups, the researcher has the necessary information to compute the appropriate F ratio (which will be represented with the notation $F_{(1/2/3)}$) for evaluating the null hypothesis H_0: $\mu_1 = \mu_2 = \mu_3$. In computing the F ratio to evaluate the latter null hypothesis, the following values are employed: a) $MS_{BG} = 26.53$ (which is the value of MS_{BG} computed for the analysis of variance in Table 16.2 that involves only the three groups identified in null hypothesis H_0: $\mu_1 = \mu_2 = \mu_3$) is employed as the numerator of the F ratio; and b) $MS_{WG} = 2.05$ (which is the value of MS_{WG} computed in Table 16.6 for the omnibus F test when the data for all $k = 4$ groups are evaluated) is employed as the denominator of the F ratio. The reason for employing the latter value instead of $MS_{WG} = 1.9$ (which is the value of MS_{WG} computed for the analysis of variance in Table 16.2 that only employs the data for the three groups identified in the null hypothesis H_0: $\mu_1 = \mu_2 = \mu_3$) is because $MS_{WG} = 2.05$ is a pooled estimate of all $k = 4$ population variances. If, in fact, the populations represented by all four groups have equal variances, this latter value will provide the most accurate estimate of MS_{WG}. Thus:

$$F_{(1/2/3)} = \frac{MS_{BG_{(1/2/3)}}}{MS_{WG_{(1/2/3/4)}}} = \frac{13.27}{2.05} = 6.47$$

The degrees of freedom employed for the analysis are based on the mean square values employed in computing the $F_{(1/2/3)}$ ratio. Thus: $df_{\text{num}} = k_{\text{subset}} - 1 = 3 - 1 = 2$ (where $k_{\text{subset}} = 3$ groups) and $df_{\text{den}} = df_{WG_{(1/2/3/4)}} = 16$ (which is df_{WG} for the omnibus F test involving all $k = 4$ groups). For $df_{\text{num}} = 2$ and $df_{\text{den}} = 16$, $F_{.05} = 3.63$ and $F_{.01} = 6.23$. Since the obtained value $F = 6.47$ is greater than both of the aforementioned critical values, the null hypothesis can be rejected at both the.05 and .01 levels. Thus, the data do not support the researcher's hypothesis that Groups 1, 2, and 3 represent a homogenous subset. In view of this, the researcher would not conduct the contrast $(\bar{X}_1 + \bar{X}_2 + \bar{X}_3)/3$ versus $\bar{X}_4$.

It should be noted that if the researcher has reason to believe that Groups 1, 2, and 3 have homogeneous variances, and the variance of Group 4 is not homogenous with the variance of the latter three groups, it would be more appropriate to employ $MS_{WG_{(1/2/3)}} = 1.9$ as the denominator of the F ratio as opposed to $MS_{WG_{(1/2/3/4)}} = 2.05$. With respect to the problem under discussion, since the two MS_{WG} values are almost equivalent, using either value produces essentially the same result (i.e., if $MS_{WG_{(1/2/3)}} = 1.9$ is employed to compute the $F_{(1/2/3)}$ ratio, $F_{(1/2/3)} = 13.27/1.9 = 6.98$, which is greater than both the tabled critical 05 and .01 F values).

3. Evaluation of the homogeneity of variance assumption of the single-factor between-subjects analysis of variance In Section I it is noted that one assumption of the **single-factor between-subjects analysis of variance** is homogeneity of variance. As noted in the discussion of the ***t* test for two independent samples**, when there are $k = 2$ groups the homogeneity of variance assumption evaluates whether there is evidence to indicate an inequality exists between the variances of the populations represented by the two samples/groups. When there are two or more groups, the homogeneity of variance assumption evaluates whether there is evidence to indicate that an inequality exists between at least two of the population variances represented by the k samples/groups. When the latter condition exists it is referred to as **heterogeneity of variance.** In reference to Example 16.1, the null and alternative hypotheses employed in evaluating the homogeneity of variance assumption are as follows:

Null hypothesis H_0: $\sigma_1^2 = \sigma_2^2 = \sigma_3^2$

(The variance of the population Group 1 represents equals the variance of the population Group 2 represents equals the variance of the population Group 3 represents.)

Alternative hypothesis H_1: Not H_0

(This indicates that there is a difference between at least two of the three population variances.)

One of a number of procedures that can be employed to evaluate the homogeneity of variance hypothesis is **Hartley's F_{max} test (Test 8a)**, which is also employed to evaluate homogeneity of variance for the ***t* test for two independent samples.**[51] The reader is advised to review the discussion of the **F_{max} test** under the ***t* test for two independent samples** prior to continuing this section.

Equation 16.37 (which is identical to Equation 8.6) is employed to compute the **F_{max} test** statistic.

$$F_{max} = \frac{\tilde{s}_L^2}{\tilde{s}_S^2} \qquad \textbf{(Equation 16.37)}$$

Where: $\tilde{s}_L^2$ = The largest of the estimated population variances of the k groups
$\tilde{s}_S^2$ = The smallest of the estimated population variances of the k groups

Employing Equation I.5, the estimated population variances are computed for the three groups.

$$\tilde{s}_1^2 = \frac{426 - \frac{(46)^2}{5}}{5 - 1} = .7 \qquad \tilde{s}_2^2 = \frac{227 - \frac{(33)^2}{5}}{5 - 1} = 2.3 \qquad \tilde{s}_3^2 = \frac{203 - \frac{(31)^2}{5}}{5 - 1} = 2.7$$

The largest and smallest estimated population variances that are employed in Equation 16.37 are $\tilde{s}_L^2 = \tilde{s}_3^2 = 2.7$ and $\tilde{s}_S^2 = \tilde{s}_1^2 = .7$. Substituting the latter values in Equation 16.37, the value $F_{max} = 3.86$ is computed.

$$F_{max} = \frac{2.7}{.7} = 3.86$$

The value of F_{max} will always be a positive number that is greater than 1 (unless $\tilde{s}_L^2 = \tilde{s}_S^2$, in which case $F_{max} = 1$). The F_{max} value obtained with Equation 16.37 is evaluated with **Table A9 (Table of the F_{max} Distribution)** in the **Appendix**. Since in Example 16.1, there are $k = 3$ groups and $n = 5$ subjects per group, the tabled critical values in **Table A9** that are employed are the values in the cell that is the intersection of the row $n - 1 = 4$ and the column $k = 3$. In order to reject the null hypothesis, and thus conclude that the homogeneity of variance assumption is violated, the obtained F_{max} value must be equal to or greater than the tabled critical value at the prespecified level of significance. Inspection of **Table A9** indicates that for $n - 1 = 4$ and $k = 3$, $F_{max_{.05}} = 15.5$ and $F_{max_{.01}} = 37$. Since the obtained value $F_{max} = 3.86$ is less than $F_{max_{.05}} = 15.5$, the null hypothesis cannot be rejected. In other words, the alternative hypothesis indicating the presence of heterogeneity of variance is not supported.

Two assumptions of the F_{max} **test** are: a) Each of the samples has been randomly selected from the population it represents; and b) The distribution of data in the underlying populations from which each of the samples is derived is normal. Various sources (e.g., Keppel (1991), Maxwell and Delaney (1990), and Winer *et al.* (1991)) note that when the normality assumption is violated, the accuracy of the F_{max} **test** may be severely compromised. This problem becomes exacerbated when violation of the normality assumption occurs within the framework of an analysis involving small and/or unequal sample sizes. As noted in Section VI of the ***t* test for two independent samples**, the F_{max} **test** assumes that there are an equal number of subjects per group. However, if the sample sizes of the groups being compared are unequal, but are approximately the same value, the value of the larger sample size can be employed to represent n in evaluating the test statistic. Kirk (1982) and Winer *et al.* (1991) note that using the larger n will result in a slight increase in the Type I error rate for the test.

One criticism of the F_{max} **test** is that it is less powerful than some alternative but computationally more involved procedures for evaluating the homogeneity of variance assumption. Some of these procedures can be found in sources on analysis of variance (e.g., Keppel (1991), Kirk (1982), Maxwell and Delaney (1990), and Winer *et al.* (1991)). Among the more commonly cited alternatives to the F_{max} **test** are tests developed by Bartlett (1937) and Cochran (1941). Although both of these tests do use more information than the F_{max} **test**, they are also subject to distortion when the underlying populations are not normally distributed. Winer *et al.* (1991) discuss tests developed by Box (1953) and Scheffé (1959) which are not as likely to be affected by violation of the normality assumption. Keppel (1991) and Winer *et al.* (1991) note that in a review of 56 tests of homogeneity of variance, Conover *et al.* (1981) recommend a test by Brown and Forsythe (1974a, 1974b). Maxwell and Delaney (1990) and Howell (1992), on the other hand, endorse the use of a test developed by O'Brien (1981).

When all is said and done, perhaps the most reasonable approach for evaluating the homogeneity of variance hypothesis is to employ a methodology suggested by Keppel (1991) (also discussed in Keppel *et al.* (1992)). Keppel (1991) recommends the use of the F_{max} **test** in evaluating the homogeneity of variance hypothesis, but notes that regardless of the values of n or k, if $F_{max} \geq 3$ a lower level of significance (i.e., a lower α value) should be employed in evaluating the results of the analysis of variance in order to avoid inflating the Type I error rate associated with the latter test. Keppel's (1991) strategy is based on research which indicates that when $F_{max} \geq 3$, there is an increased likelihood that the accuracy of the tabled critical values in the F distribution will be compromised. The fact that the value $F_{max} = 3$ is considerably lower than most of the tabled critical F_{max} values in **Table A9**, reinforces what was noted earlier concerning the power of the F_{max} **test** — specifically, the low power of the test may increase the likelihood of committing a Type II error (i.e., not rejecting the null hypothesis when heterogeneity of variance is present).

When, in fact, the homogeneity of variance assumption is violated, there is an increased likelihood of committing a Type I error in conducting the analysis of variance evaluating the k group means. Factors that influence the degree to which the Type I error rate for the analysis of variance will be larger than the prespecified value of alpha are the size of the samples and the shapes of the underlying population distributions. Sources generally agree that the effect of violation of the homogeneity of variance assumption on the accuracy of the tabled critical F values is exacerbated when there are not an equal number of subjects in each group.

A variety of strategies have been suggested regarding how a researcher should deal with heterogeneity of variance. Among the procedures that have been suggested are the following: a) Keppel (1991) notes that one option available to the researcher is to employ an adjusted tabled critical F value in evaluating the analysis of variance. Specifically, one can employ a tabled critical value associated with a lower alpha level than the prespecified alpha level, so as to provide a more accurate estimate of the latter value (i.e., employ $F_{.025}$ or $F_{.01}$ to estimate $F_{.05}$). The problem with this strategy is that if too low an alpha value is employed, the power of the omnibus F test may be compromised to an excessive degree. Loss of power, however, can be offset by employing a large sample size; b) The data can be evaluated with a procedure other than an analysis of variance. Thus, one can employ a rank-order non-parametric procedure such as the **Kruskal–Wallis one-way analysis of variance by ranks (Test 17)**. However, by virtue of rank-ordering the data, the latter test will usually provide a less powerful test of an alternative hypothesis than the analysis of variance. A number of alternative parametric procedures developed by Brown and Forsythe (1974a, 1974c), James (1951), and Welch (1951) are discussed in various sources. Keppel (1991) notes, however, that the Brown and Forsythe and Welch procedures are not acceptable when $k > 4$, and that James' procedure is too computationally involved for conventional use. In addition, some sources believe that the aforementioned procedures have not been sufficiently researched to justify their use as an alternative to the analysis of variance, even if the homogeneity of variance assumption of the latter test is violated; and c) Another option available to the researcher is to equate the estimated population variances by employing a data transformation procedure (discussed in Section VII of the ***t* test for two independent samples**), and to conduct an analysis of variance on the transformed data. There is, however, a lack of consensus with respect to what type of data transformations are permissible, and under what conditions they should be employed.

4. Computation of the power of the single-factor between-subjects analysis of variance Prior to reading this section the reader may find it useful to review the discussion on power in Section VI of both the **single-sample *t* test (Test 2)** and the ***t* test for two independent samples**. Before conducting an analysis of variance a researcher may want to determine the minimum sample size required in order to detect a specific effect size. In order to conduct such a power analysis the researcher will have to estimate the means of all of the populations that are represented by the experimental treatments/groups. Additionally, he will have to estimate a standard deviation value which it will be assumed represents the standard deviation of all k populations. Understandably, the accuracy of one's power calculations will be a function of the researcher's ability to come up with good approximations for the means and standard deviations of the populations that are involved in the study. The basis for making such estimates will generally be prior research concerning the hypothesis under study.

Equation 16.38 is employed for computing the power of the **single-factor between-subjects analysis of variance.** The test statistic ϕ represents what is more formally known as the **noncentrality parameter,** and is based on the **noncentral F distribution.**[52]

$$\phi = \sqrt{n\left[\frac{\Sigma(\mu_j - \mu_T)^2}{k\sigma^2_{WG}}\right]} \qquad \textbf{(Equation 16.38)}$$

Where: μ_j = The estimated mean of the population represented by Group j
μ_T = The grand mean, which is the average of the k estimated population means
σ^2_{WG} = The estimated population variance for each of the k groups
n = The number of subjects per group
k = The number of groups

The computation of the minimum acceptable sample size required to achieve a specified level of power is generally determined prior to collecting the data for an experiment. Using trial and error, a researcher can determine what value of n (based on the assumption that there are an equal number of subjects per group) when substituted in Equation 16.38 will yield the desired level of power. To illustrate the use of Equation 16.38 with Example 16.1, let us assume that prior to conducting the study the researcher estimates that the means of the populations represented by the three groups are as follows: $\mu_1 = 10$, $\mu_2 = 8$, $\mu_3 = 6$. Additionally, it will be assumed that he estimates that the variance for each of the three populations the groups represent is $\sigma^2_{WG} = 2.5$. Based on this information, the value $\mu_T = 8$ can be computed: $\mu_T = (\mu_1 + \mu_2 + \mu_3)/k = (10 + 8 + 6)/3 = 8$. The appropriate values are now substituted in Equation 16.38.

$$\phi = \sqrt{n\left[\frac{(10 - 8)^2 + (8 - 8)^2 + (6 - 8)^2}{(3)(2.5)}\right]} = \sqrt{1.07n} = 1.03\sqrt{n}$$

At this point, **Table A15 (Graphs of the Power Function for the Analysis of Variance)** in the **Appendix** can be employed to determine the necessary sample size required in order to have the power stipulated by the experimenter. **Table A15** is comprised of sets of power curves that were derived by Pearson and Hartley (1951). Each set of curves is based on a different value for df_{num}, which in the case of a **single-factor between-subjects analysis of variance** is df_{BG} employed for the omnibus F test. Within each set of curves, for a given value of df_{num} there are power functions for both $\alpha = .05$ and $\alpha = .01$. For our analysis (for which it will be assumed that $\alpha = .05$) the appropriate set of curves to employ is the set for $df_{\text{num}} = df_{BG} = 2$. Let us assume we want the omnibus F test to have a power of at least .80. We now substitute what we consider to be a reasonable value for n in the equation $\phi = 1.03\sqrt{n}$ (which is the result obtained with Equation 16.38). To illustrate, the value $n = 5$ (the sample size employed for Example 16.1) is substituted in the equation. The resulting value is $\phi = 1.03\sqrt{5} = 2.30$.

The value $\phi = 2.30$ is located on the abscissa (X-axis) of the relevant set of curves in **Table A15** — specifically, the set for $df_{\text{num}} = 2$. At the point corresponding to $\phi = 2.30$, a perpendicular line is erected from the abscissa which intersects with the power curve that corresponds to the value of $df_{\text{den}} = df_{WG}$ employed for the omnibus F test. Since $df_{WG} = 12$, the curve for the latter value is employed.[53] At the point the perpendicular intersects the curve $df_{WG} = 12$, a second perpendicular line is drawn in relation to the ordinate (*Y*-axis). The point at which this perpendicular intersects the ordinate indicates the power of the test. Since $\phi = 2.30$, we determine the power equals .89. Thus, if we employ five subjects per group, there is a probability of .89 of detecting an effect size equal to or larger than the one stipulated by the researcher (which is a function of the estimated values for the population means relative to the value estimated for the variance of a population). Since the probability of committing a Type II error is $\beta = 1 - \text{power}$, $\beta = 1 - .89 = .11$. The latter value represents the likelihood of not detecting an effect size equal to or greater than the one stipulated.

Cohen (1988), who provides a detailed discussion of power computations for the analysis of variance, describes a measure of effect size for the analysis of variance based on standard deviation units that is comparable to the d value computed for the different types of t tests. In the case of the analysis of variance, d is the difference between the smallest and largest of the estimated population means divided by the standard deviation of the populations. In other words, $d = (\mu_L - \mu_S)/\sigma$. In our example the largest estimated population mean is $\mu_1 = 10$ and the smallest is $\mu_3 = 6$. The value of σ is $\sigma_{WG} = \sqrt{\sigma^2_{WG}} = \sqrt{2.5} = 1.58$. Thus, $d = (10 - 6)/1.58 = 2.53$. This result tells us that if $n = 5$, a researcher has a .89 probability of detecting a difference of about two and one-half standard deviation units.

It is also possible to conduct a power analysis for comparisons that are conducted following the computation of the omnibus F value. In fact, Keppel (1991) recommends that the sample size employed in an experiment be based on the minimum acceptable power necessary to detect the smallest effect size among all of the comparisons the researcher plans before collecting the data. The value ϕ_{comp} (described by McFatter and Gollob (1986)), which is computed with Equation 16.39, is employed to determine the power of a comparison.

$$\phi_{\text{comp}} = \sqrt{n\left[\frac{(\mu_a - \mu_b)^2}{2(\sigma^2_{WG})(\Sigma c_j^2)}\right]} \qquad \textbf{(Equation 16.39)}$$

Equation 16.39 can be used for both simple and complex single degree of freedom comparisons. As a general rule, the equation is used for planned comparisons. Although it can be extended to unplanned comparisons, published power tables for the analysis of variance generally only apply to per comparison error rates of $\alpha = .05$ and $\alpha = .01$. In the case of planned and especially unplanned comparisons which involve α_{PC} rates other than .05 or .01, more detailed tables are required.[54]

For single degree of freedom comparisons, the power curves in **Table A15** for $df_{\text{num}} = 1$ are always employed. The use of Equation 16.39 will be illustrated for the simple comparison Group 1 versus Group 2 (summarized in Table 16.3). Since $\Sigma c_j^2 = 2$, and we have estimated $\mu_a = \mu_1 = 10$, $\mu_b = \mu_2 = 8$, and $\sigma^2_{WG} = 2.5$, the following result is obtained.

$$\phi_{comp} = \sqrt{n\left[\frac{(10 - 8)^2}{(2)(2.5)(2)}\right]} = \sqrt{.4n} = .63\sqrt{n}$$

Substituting $n = 5$ in the equation $\phi_{\text{comp}} = .63\sqrt{n}$, we obtain $\phi_{\text{comp}} = .63\sqrt{5} = 1.41$. Employing the power curves for $df_{\text{num}} = 1$ with $\alpha = .05$, we use the curve for $df_{WG} = 12$ (the df_{WG} employed for the omnibus F test), and determine that when $\phi_{\text{comp}} = 1.41$, the power of the test is approximately .44.

It should be noted that if the methodology described for computing the power of the **t test for two independent samples** is employed for the above comparison or any other simple comparison, it will produce the identical result. To demonstrate that Equation 16.39 produces a result that is equivalent to that obtained with the protocol for computing the power of the **t test for two independent samples**, Equations 8.10 and 8.12 will be employed to compute the power of the comparison Group 1 versus Group 2. Employing Equation 8.10:

$$d = \frac{\mu_1 - \mu_2}{\sigma} = \frac{10 - 8}{1.58} = 1.27$$

Substituting the value of d in Equation 8.12, the value $\delta = 2.01$ is computed.

$$\delta = d\sqrt{\frac{n}{2}} = 1.27\sqrt{\frac{5}{2}} = 2.01$$

The value of $\delta = 2.01$ is evaluated with **Table A3 (Power Curves for Student's t Distribution)** in the **Appendix**. A full description of how to employ the power curves in **Table A3** can be found in Section VI of the **single-sample t test**. Employing **Table A3-C** (which is the set of curves for a two-tailed analysis with $\alpha = .05$), we use the curve for $df_{WG} = 12$. It is determined that when $\delta = 2.01$, the power of the comparison is approximately .44.

A final note regarding Equation 16.39: The value of Σc_j^2 computed for a complex comparison will always be lower than the value Σc_j^2 computed for a simple comparison (in the case of a simple comparison, Σc_j^2 will always equal 2). Because of this, for a fixed value of n the computed value of ϕ_{comp} will always be larger for a complex comparison, and consequently the power of a complex comparison will always be greater than the power of a simple comparison.

5. Measures of magnitude of treatment effect for the single-factor between-subjects analysis of variance: Omega squared (Test 16g) and eta squared (Test 16h) Prior to reading this section the reader should review the discussion of the measures of magnitude of treatment effect in Section VI of the **t test for two independent samples**. As is the case with the t value computed for the latter test, the omnibus F value computed for the **single-factor between-subjects analysis of variance** only provides a researcher with information regarding whether the null hypothesis can be rejected — i.e., whether a significant treatment effect is present. The F value (as well as the level of significance with which it is associated), however, does not provide the researcher with any information regarding the size of any treatment effect that is present. As is the case with a t value, an F value is a function of both the difference between the means of the experimental treatments and the sample size.

Omega squared (Test 16g) A number of measures of the **magnitude of treatment effect** have been developed which can be employed for the **single-factor between-subjects analysis of variance**. Such measures, which are independent of sample size, provide an estimate of the proportion of variability on the dependent variable that is associated with the independent variable/experimental treatments. Although sources are not in total agreement with respect to which measure of treatment effect is most appropriate to employ for the **single-factor between-subjects analysis of variance**, the most commonly cited measure is the **omega squared** statistic ($\tilde{\omega}^2$), which provides an estimate of the underlying population parameter ω^2. The value of $\tilde{\omega}^2$ is computed with Equation 16.40. The $\tilde{\omega}^2$ value computed with Equation 16.40 is essentially a ratio of the proportion of variability in the data that is attributed to the experimental treatments divided by the total variability in the data.

$$\tilde{\omega}^2 = \frac{SS_{BG} - (k - 1)MS_{WG}}{SS_T + MS_{WG}}$$ **(Equation 16.40)**

Although the value of $\tilde{\omega}^2$ will generally fall in the range between 0 and 1, when $F < 1$ $\tilde{\omega}^2$ will be a negative number. The closer $\tilde{\omega}^2$ is to 1 the stronger the association between the independent and dependent variables, whereas the closer $\tilde{\omega}^2$ is to 0 the weaker the association between the two variables. A $\tilde{\omega}^2$ value equal to or less than 0 indicates that there is no association between the variables. Keppel (1991) notes that in behavioral science research (which is commonly characterized by a large amount of error variability) the value of $\tilde{\omega}^2$ will rarely be close to 1.

Employing Equation 16.40 with the data for Example 16.1, the value $\tilde{\omega}^2 = .44$ is computed.

$$\tilde{\omega}^2 = \frac{26.53 - (3 - 1)(1.9)}{49.33 + 1.9} = .44$$

Equation 16.41 is an alternative equation for computing the value of $\tilde{\omega}^2$ that yields the same value as Equation 16.40.[55]

$$\tilde{\omega}^2 = \frac{(k - 1)(F - 1)}{(k - 1)(F - 1) + nk}$$ **(Equation 16.41)**

$$\tilde{\omega}^2 = \frac{(2)(6.98 - 1)}{(2)(6.98 - 1) + (5)(3)} = .44$$

The value $\tilde{\omega}^2 = .44$ indicates that 44% (or a proportion of .44) of the variability on the dependent variable (the number of nonsense syllables recalled) is associated with variability on the levels of the independent variable (noise). To say it another way, 44% of the variability on the recall scores of subjects can be accounted for on the basis of which group a subject is a member. If one employs Cohen's (1988) guidelines for magnitude of treatment effect (discussed in Section VI of the ***t* test for two independent samples**), since $\tilde{\omega}^2 = .44$ exceeds .15 a large treatment effect is detected.

Eta squared (Test 16h) Another less commonly employed measure of treatment effect for the **single-factor between-subjects analysis of variance** is the **eta squared** statistic ($\tilde{\eta}^2$) (which estimates the underlying population parameter η^2). $\tilde{\eta}^2$ is computed with Equation 16.42.[56]

$$\tilde{\eta}^2 = \frac{SS_{BG}}{SS_T}$$ **(Equation 16.42)**

Employing Equation 16.42 with the data for Example 16.1, the value $\tilde{\eta}^2 = .54$ is computed.

$$\tilde{\eta}^2 = \frac{26.53}{49.33} = .54$$

Note that the value $\tilde{\eta}^2 = .54$ is larger than $\tilde{\omega}^2 = .44$ computed with Equations 16.40/16.41. Sources note that $\tilde{\eta}^2$ is a more biased estimate of the magnitude of treatment effect in the underlying population than is $\tilde{\omega}^2$, since $\tilde{\eta}^2$ employs the values SS_{BG} and SS_T, which by themselves are biased estimates of population variability. Darlington and Carlson (1987) note that Equation 16.43 can be employed to compute a less biased estimate of the population parameter that is estimated by $\tilde{\eta}^2$.

$$\text{Adjusted } \tilde{\eta}^2 = 1 - \frac{MS_{WG}}{MS_T} \qquad \textbf{(Equation 16.43)}$$

$$\textbf{Where: } MS_T = \frac{SS_T}{N - 1}$$

Since $MS_T = 49.33/14 = 3.52$, the adjusted $\tilde{\eta}^2 = .46$ falls in between $\tilde{\omega}^2 = .44$ and $\tilde{\eta}^2 = .54$ computed with Equations 16.40/16.41 and Equation 16.42.

$$\text{Adjusted } \tilde{\eta}^2 = 1 - \frac{1.9}{3.52} = .46$$

When $\tilde{\omega}^2$ and $\tilde{\eta}^2$ are based on a small sample size, their standard error (which is a measure of error variability) will be large, and consequently the reliability of the measures of treatment effect will be relatively low. The latter will be reflected in the fact that under such conditions a confidence interval computed for a measure of treatment effect (the computation of which will not be described in this book) will have a wide range. It should be emphasized that a measure of magnitude of treatment effect is a measure of association/correlation, and in and of itself it is not a test of significance. The significance of $\tilde{\omega}^2$ and $\tilde{\eta}^2$ is based on whether or not the omnibus F value is significant.

Many sources recommend that in summarizing the results of an experiment, in addition to reporting the omnibus test statistic (e.g., an F or t value), a measure of magnitude of treatment effect also be included, since the latter can be useful in further clarifying the nature of the relationship between the independent and dependent variables. It is important to note that if the value of a measure of magnitude of treatment effect is small, it does not logically follow that the relationship between the independent and the dependent variable is trivial. There are instances when a small treatment effect may be of practical and/or theoretical value. It should be noted that when the independent variable is a nonmanipulated variable, it is possible that any treatment effect that is detected may be due to some variable other than the independent variable. Such studies (referred to as **ex post facto** studies) do not allow a researcher to adequately control for the potential effects of extraneous variables on the dependent variable. As a result of this, even if a large treatment effect is present in the data, a researcher is not justified in drawing conclusions with regard to cause and effect — specifically, the researcher cannot conclude that the independent variable is responsible for group differences on the dependent variable. Sources which provide detailed discussions of measures of magnitude of treatment effect for the **single-factor between-subjects analysis of variance** are Howell (1992), Keppel (1991), Kirk (1982), Maxwell and Delaney (1990), and Winer *et al.* (1991).[57]

6. Computation of a confidence interval for the mean of a treatment population Prior to reading this section the reader may find it useful to review the discussion of confidence intervals in Section VI of the **single-sample *t* test**. Equation 16.44 can be employed to compute a confidence interval for the mean of a population represented by one of the k treatments for which an omnibus F value has been computed.[58]

$$CI = \bar{X}_j \pm t_{df_{WG}} \sqrt{\frac{MS_{WG}}{n}} \qquad \textbf{(Equation 16.44)}$$

Where: $t_{df_{wg}}$ represents the tabled critical two-tailed value in the t distribution, for df_{WG}, below which a proportion equal to $[1 - (\alpha/2)]$ of the cases falls. If the percentage of the distribution that falls within the confidence interval is subtracted from 1, it will equal the value of α.

The use of the tabled critical t value for df_{WG} in Equation 16.44 instead of a degrees of freedom value based on the number of subjects who served in each group (i.e., $df = n - 1$) is predicated on the fact that the estimated population variance is based on the pooled variance of the k groups. If, however, there is reason to believe that the homogeneity of variance assumption is violated, it is probably more prudent to employ Equation 2.7 to compute the confidence interval. The latter equation employs the estimated population standard deviation for a specific group for which a confidence interval is computed, rather than the pooled variability (which is represented by MS_{WG}). If the standard deviation of the group is employed rather than the pooled variability, $df = n - 1$ is used for the analysis. It can also be argued that it is preferable to employ Equation 2.7 in computing the confidence interval for a population mean in circumstances where a researcher has reason to believe that the group in question represents a population that is distinct from the populations represented by the other $(k - 1)$ groups. Specifically, if the mean value of a group is significantly above or below the means of the other groups, it can be argued that one can no longer assume that the group shares a common variance with the other groups. In view of this, one can take the position that the variance of the group is the best estimate of the variance of the population represented by that group (as opposed to the pooled variance of all of the groups involved in the study). The point to be made here is that, as is the case in computing a confidence interval for a comparison, depending upon how one conceptualizes the data, more than one methodology can be employed for computing a confidence interval for a population mean.

The computation of the 95% confidence interval for the mean of the population represented by Group 1 is illustrated below employing Equation 16.44. The value $t_{.05} = 2.18$ is the tabled critical two-tailed $t_{.05}$ value for $df_{WG} = 12$.

$$CI_{.95} = 9.2 \pm 2.18\sqrt{\frac{1.9}{5}} = 9.2 \pm 1.34$$

Thus, the researcher can be 95% confident that the mean of the population represented by Group 1 falls within the range 7.86 to 10.54. Stated symbolically: $7.86 \leq u_1 \leq 10.54$.

If, on the other hand, the researcher elects to employ Equation 2.7 to compute the confidence interval for the mean of the population represented by Group 1, the standard deviation of Group 1 is employed in lieu of the pooled variability of all the groups. In addition, the tabled critical two-tailed $t_{.05}$ value for $df = n - 1 = 4$ is employed in Equation 2.7. The computations are shown below.

$$\tilde{s}_1 = \sqrt{\frac{\Sigma X_1^2 - \frac{(\Sigma X_1)^2}{n}}{n - 1}} = \sqrt{\frac{426 - \frac{(46)^2}{5}}{4}} = .837$$

$$CI_{.95} = \bar{X}_1 \pm t_{.05}\left(\frac{\tilde{s}_1}{\sqrt{n}}\right) = 9.2 \pm 2.78\left(\frac{.837}{\sqrt{5}}\right) = 9.2 \pm 1.04$$

Thus, employing Equation 2.7, the range for $CI_{.95}$ is $8.16 \leq \mu_1 \leq 10.24$.[59] Note that although the range of values computed with Equation 16.44 and Equation 2.7 are reasonably close to one another, the range of the confidence interval computed with Equation 16.44 is wider. Depending on the values of $t_{df_{WG}}$ versus $t_{df_{(n-1)}}$ and $\sqrt{MS_{WG}/n}$ versus $s_{\bar{X}_j}$, there will usually be a discrepancy between the confidence interval computed with the two equations. When the estimated population variances of all k groups are equal (or reasonably close to one another), Equation 2.7 will yield a wider confidence interval than

Equation 16.44, since $\sqrt{MS_{WG}/n}$ will equal $s_{\bar{X}_g}$ and $t_{df_{WG}} > t_{df_{(n-1)}}$. In the example illustrated in this section, the reason why use of Equation 2.7 yields a smaller confidence interval than Equation 16.44 (in spite of the fact that a larger t value is employed in Equation 2.7) is because $\tilde{s}^2 = .7$ (the estimated variance of the population represented by Group 1) is substantially less than $MS_{WG} = 1.9$ (the pooled estimate of within-groups variability).

7. The analysis of covariance **Analysis of covariance** is an analysis of variance procedure that employs a statistical adjustment (involving regression analysis which is discussed under the **Pearson product-moment correlation coefficient (Test 22)**) to control for the effect of one or more extraneous variables on a dependent variable. In this section the simplest type of analysis of covariance involving one extraneous variable will be discussed. Analysis of covariance is most commonly employed when a researcher has reason to believe that two or more experimental groups are not equivalent with respect to some extraneous variable that has a linear correlation with the dependent variable. Such an extraneous variable is referred to as a **covariate** or **concomitant variable.** By utilizing the correlation between the covariate and the dependent variable, the researcher is able to remove variability on the dependent variable that is attributable to the covariate. The effect of the latter is a reduction in the error variability employed in computing the F ratio, thereby resulting in a more powerful test of the alternative hypothesis. A second potential effect of an analysis of covariance is that by utilizing the correlation between scores on a covariate and the dependent variable, the mean scores of the different groups can be adjusted for any pre-existing differences on the dependent variable which are present prior to the administration of the experimental treatments.

When analysis of covariance is employed, the most commonly used covariates are subject variables such as intelligence, anxiety, weight, etc. Thus, in Example 16.1, if a researcher believes that the number of nonsense syllables a subject learns is a function of intelligence, and if there is a linear correlation between intelligence and one's ability to learn nonsense syllables, intelligence can be employed as a covariate. In order to employ it as a covariate it is necessary to have an intelligence score for each subject who participates in the experiment. By employing the latter scores the researcher can use an analysis of covariance to determine whether there are performance differences between the three groups that are independent of intelligence.

The optimal design for which an analysis of covariance can be employed is an experiment involving a manipulated independent variable in which subjects are randomly assigned to groups, and in which scores on the covariate are measured prior to the introduction of the experimental treatments. In such a design the analysis of covariance is able to remove variability on the dependent variable that is attributed to differences between the groups on the covariate. However, when prior to introducing the experimental treatments one knows that a strong correlation exists between the covariate and the dependent variable, and that the groups are not equal with respect to the covariate, the following two options are available to the researcher: a) Subjects can be randomly reassigned to groups, after which the researcher can check that the resulting groups are equivalent with respect to the covariate; or b) The covariate can be integrated into the study as a second independent variable. Some sources endorse the use of either or both of the aforementioned strategies as preferable to employing an analysis of covariance.

A less ideal situation for employing an analysis of covariance is an experiment involving a manipulated independent variable in which subjects are randomly assigned to groups, but in which scores on the covariate are not measured until after the experimental manipulation.

The latter situation is more problematical with respect to using the analysis of covariance, since subjects' scores on the covariate could have been influenced by the experimental treatments. This latter fact makes it more difficult to interpret the results of the analysis, since in order to draw inferences with regard to cause and effect the scores on the covariate should be independent of the treatments.

An even more problematical use of the analysis of covariance is for a design in which subjects are not randomly assigned to groups (which is often referred to as a **quasi-experimental design**). The latter can involve the use of intact groups (such as two different classes at a school) who are exposed to a manipulated independent variable or two groups which are formed on the basis of some pre-existing subject characteristic (i.e., an **ex post facto study** involving a nonmanipulated independent variable such as gender, race, etc.). It should be noted that if a substantial portion of between-groups variability can be explained on the basis of an extraneous variable (i.e., a potential covariate), it implies that there was probably some sort of systematic bias involved in forming the experimental groups. Because of the latter, it is reasonable to expect that if a researcher identifies one extraneous variable that has a substantial correlation with the dependent variable, there are probably other extraneous variables whose effects will not be controlled for even if one evaluates the data with an analysis of covariance. In view of this, some sources (Keppel and Zedeck (1989) and Lord (1967, 1969)) argue that if subjects are not randomly assigned to groups, a researcher will not be able to adequately control for the potential effects of other pre-existing extraneous variables on the dependent variable. More specifically, they argue that the analysis of covariance will not be able to produce the necessary statistical control to allow one to unambiguously interpret the effect of the independent variable on the dependent variable. Thus, in designs in which subjects are not randomly assigned to groups, sources either state that the analysis of covariance should never be employed, or that if it is employed, the results should be interpreted with extreme caution.

Kachigan (1986), among others, notes that one should never use an analysis of covariance to adjust for between-groups differences with respect to a covariate that are attributable to normal sampling error. Aside from the fact that such variability is a part of expected error variability within the framework of conducting an analysis of variance, a major reason for not employing an analysis of covariance (for which the test statistic is also an omnibus F value) is that the number of within-groups degrees of freedom required for the analysis will be one less than df_{WG} required for an analysis of variance on the same set of data. In such a case, any reduction in error variance associated with the analysis of covariance may be offset by the fact that it will require a larger critical F value to reject the null hypothesis than will an analysis of variance on the original data.

In closing the discussion of analysis of covariance, it should be noted that the procedure should not be used indiscriminately, since a considerable investment of time and effort is required to obtain subjects' scores on a covariate, as well as the fact that the analysis requires laborious computations (although admittedly, the latter will not be an issue if one has access to the appropriate computer software). In order for an analysis of covariance to generate reliable results, it is required that none of the assumptions noted for the analysis of variance as well as additional assumptions relating to correlation and regression (which will not be discussed here) have been saliently violated. The computational protocol for the analysis of covariance (which will not be described in this book) can be found in most books that specialize in analysis of variance. Keppel (1991) provides an excellent description of the computational procedure that most readers will find easy to follow. Other sources that discuss the analysis of covariance are Howell (1992), Kachigan (1986), Keppel and Zedeck (1989), and Maxwell and Delaney (1990).

VII. Additional Discussion of the Single-Factor Between-Subjects Analysis of Variance

1. Theoretical rationale underlying the single-factor between-subjects analysis of variance In the **single-factor between-subjects analysis of variance** it is assumed that any variability between the means of the k groups can be attributed to one or both of the following two elements: a) **Experimental error**; and b) **The experimental treatments.** When MS_{BG} (the value computed for between-groups variability) is significantly greater than MS_{WG} (the value computed for within-groups variability), it is interpreted as indicating that a substantial portion of between-groups variability is due to a treatment effect. The rationale for this is as follows.

Experimental error is random variability in the data that is beyond the control of the researcher. In an independent groups design the average amount of variability within each of the k groups is employed to represent experimental error. Thus, the value computed for MS_{WG} is the normal amount of variability that is expected between the scores of different subjects who serve in the same group. Within this framework, within-groups variability is employed as a baseline to represent variability which results from factors that are beyond an experimenter's control. The experimenter assumes that since such uncontrollable factors are responsible for within-groups differences, it is logical to assume that they can produce differences of a comparable magnitude between the means of the k groups. As long as the variability between the group means (MS_{BG}) is approximately the same as within-groups variability (MS_{WG}), the experimenter can attribute any between-groups variability to experimental error. When, however, between-groups variability is substantially greater than within-groups variability, it indicates that something over and above error variability is contributing to the variability between the k group means. In such a case, it is assumed that a treatment effect is responsible for the larger value of MS_{BG} relative to the value of MS_{WG}. In essence, if within-groups variability is subtracted from between-groups variability, any remaining variability can be attributed to a treatment effect. If there is no treatment effect, the result of the subtraction will be zero. Of course, one can never completely rule out the possibility that if MS_{BG} is larger than MS_{WG}, the larger value for MS_{BG} is entirely due to error variability. However, since the latter is unlikely, when MS_{BG} is significantly larger than MS_{WG} it is interpreted as indicating the presence of a treatment effect.

Tables 16.7 and 16.8 will be used to illustrate the relationship between between-groups variability and within-groups variability in the analysis of variance. Assume that both tables contain data for two hypothetical studies employing an independent groups design involving $k = 3$ groups and $n = 3$ subjects per group. In the hypothetical examples below, even though it is not employed as the measure of variability in the analysis of variance, the range (the difference between the lowest and highest score) will be used as a measure of variability. The range is employed since: a) It is simpler to employ for purposes of illustration than the variance; and b) What is derived from the range from this example can be generalized to the variance/mean squares as they are employed within the framework of the analysis of variance.

Table 16.7 presents a set of data where there is no treatment effect, since between-groups variability equals within-groups variability. In Table 16.7, in order to assess within-groups variability we do the following: a) Compute the range of each group. This is done by subtracting the lowest score in each group from the highest score in that group; and b) Compute the average range of the three groups. Since the range for all three groups equals 2, the average range equals 2 (i.e., $(2 + 2 + 2)/3 = 2$). This value will be used to represent within-groups variability (which is a function of individual differences in performance among members of the same group).

Table 16.7 Data Illustrating No Treatment Effect

Group 1	Group 2	Group 3
3	4	5
4	5	6
5	6	7
$\bar{X}_1 = 4$	$\bar{X}_2 = 5$	$\bar{X}_3 = 6$

Table 16.8 Data Illustrating Treatment Effect

Group 1	Group 2	Group 3
3	10	30
4	11	31
5	12	32
$\bar{X}_1 = 4$	$\bar{X}_2 = 11$	$\bar{X}_3 = 31$

Between-groups variability is obtained by subtracting the lowest group mean ($\bar{X}_1 = 4$) from highest group mean ($\bar{X}_3 = 6$). Thus, since $\bar{X}_3 - \bar{X}_1 = 6 - 4 = 2$, between-groups variability equals 2. As noted previously, if between-groups variability is significantly larger than within-groups variability, it is interpreted as indicating the presence of a treatment effect. Since in the example under discussion both between-groups and within-groups variability equal 2, there is no evidence of a treatment effect. To put it another way, if within-groups variability is subtracted from between-groups variability, the difference is zero. In order for there to be a treatment effect a positive difference should be present.

If, for purposes of illustration, we employ the range in the F ratio to represent variability, the value of the F will equal 1 since:

$$F = \frac{\text{Between-groups variability}}{\text{Within-groups variability}} = \frac{2}{2} = 1$$

As noted previously, when $F = 1$, there is no evidence of a treatment effect, and thus the null hypothesis is retained.[60]

Table 16.8 presents a set of data where a treatment effect is present as a result of between-groups variability being greater than within-groups variability.

In Table 16.8, if we once again use the range to assess variability, within-groups variability equals 2. This is the case since the range of each group equals 2, yielding an average range of 2. The between-groups variability, on the other hand, equals 27. The latter value is the obtained by subtracting the smallest of the three group means $\bar{X}_1 = 4$ from the largest of the group means $\bar{X}_3 = 31$. Thus, $\bar{X}_3 - \bar{X}_1 = 31 - 4 = 27$. Because the value 27 is substantially larger than 2 (which is a baseline measure of error variability that will be tolerated between the group means), there is strong evidence that a treatment effect is present. Specifically, since we will assume that only 2 of the 27 units that comprise between-groups variability can be attributed to experimental error, the remaining $27 - 2 = 25$ units can be assumed to represent the contribution of the treatment effect.

If once again, for purposes of illustration the values of the range are employed to compute the F ratio, the resulting value will be $F = 27/2 = 13.5$. As noted previously, when the value of F is substantially greater than 1, it is interpreted as indicating the presence of a treatment effect, and consequently the null hypothesis is rejected.

2. Definitional equations for the single-factor between-subjects analysis of variance In the description of the computational protocol for the **single-factor between-subjects analysis of variance** in Section IV, Equations 16.2, 16.3, and 16.5 are employed to compute the values SS_T, SS_{BG}, and SS_{WG}. The latter set of computational equations was employed, since it allows for the most efficient computation of the sum of squares values. As noted in Section IV, computational equations are derived from definitional equations which reveal the underlying logic involved in the derivation of the sums of squares. This section will describe the definitional equations for the **single-factor between-subjects analysis of variance**, and apply them to Example 16.1 in order to demonstrate that they yield the same values as the computational equations.

As noted previously, the total sum of squares (SS_T) is made up of two elements, the between-groups sum of squares (SS_{BG}) and the within-groups sum of squares (SS_{WG}). The contribution of any single subject's score to the total variability in the data can be expressed in terms of a between-groups component and a within-groups component. When the between-groups component and the within-groups component are added, the sum reflects that subject's total contribution to the overall variability in the data. The contribution of all N subjects to the total variability (SS_T) and the elements that comprise it (SS_{BG} and SS_{WG}) are summarized in Table 16.9. The definitional equations described in this section employ the following notation: X_{ij} represents the score of the ith subject in the jth group; $\bar{X}_T$ represents the grand mean (which is $\bar{X}_T = (\sum_{j=1}^{k}\sum_{i=1}^{n}X_{ij})/N = 110/15 = 7.33$); and $\bar{X}_j$ represents the mean of the jth group.

Equation 16.45 is the definitional equation for the **total sum of squares.**[61]

$$SS_T = \sum_{j=1}^{k}\sum_{i=1}^{n}(X_{ij} - \bar{X}_T)^2 \qquad \textbf{(Equation 16.45)}$$

In employing Equation 16.45 to compute SS_T, the grand mean ($\bar{X}_T$) is subtracted from each of the N scores and each of the N difference scores is squared. The total sum of squares (SS_T) is the sum of the N squared difference scores. Equation 16.45 is computationally equivalent to Equation 16.2.

Equation 16.46 is the definitional equation for the **between-groups sum of squares.**

$$SS_{BG} = n\sum_{j=1}^{k}(\bar{X}_j - \bar{X}_T)^2 \qquad \textbf{(Equation 16.46)}$$

In employing Equation 16.46 to compute SS_{BG}, the following operations are carried out for each group. The grand mean ($\bar{X}_T$) is subtracted from the group mean ($\bar{X}_j$). The difference score is squared, and the squared difference score is multiplied by the number of subjects in the group (n). After this is done for all k groups, the values that have been obtained for each group as a result of multiplying the squared difference score by the number of subjects in a group are summed. The resulting value represents the between-groups sum of squares (SS_{BG}). Equation 16.46 is computationally equivalent to Equation 16.3. An alternative but equivalent method of obtaining SS_{BG} (which is employed in deriving SS_{BG} in Table 16.9) is as follows: Within each group, for each of the n subjects the grand mean is subtracted from the group mean, each difference score is squared, and upon doing this for all k groups, the N squared difference scores are summed.

Equation 16.47 is the definitional equation for the **within-groups sum of squares.**

$$SS_{WG} = \sum_{j=1}^{k}\sum_{i=1}^{n}(X_{ij} - \bar{X}_j)^2 \qquad \textbf{(Equation 16.47)}$$

Table 16.9 Computation of Sums of Squares for Example 16.1 with Definitional Equations

	X_{ij}	$SS_{WG} = \sum_{j=1}^{k}\sum_{i=1}^{n}(X_{ij} - \bar{X}_j)^2$	$SS_{BG} = n\sum_{j=1}^{k}(\bar{X}_j - \bar{X}_T)^2$	$SS_T = \sum_{j=1}^{k}\sum_{i=1}^{n}(X_{ij} - \bar{X}_T)^2$
	8	$(8-9.2)^2 = 1.44$	$(9.2-7.33)^2 = 3.497$	$(8-7.33)^2 = .449$
	10	$(10-9.2)^2 = .64$	$(9.2-7.33)^2 = 3.497$	$(10-7.33)^2 = 7.129$
Group 1	9	$(9-9.2)^2 = .04$	$(9.2-7.33)^2 = 3.497$	$(9-7.33)^2 = 2.789$
	10	$(10-9.2)^2 = .64$	$(9.2-7.33)^2 = 3.497$	$(10-7.33)^2 = 7.129$
	9	$(9-9.2)^2 = .04$	$(9.2-7.33)^2 = 3.497$	$(9-7.33)^2 = 2.789$
	7	$(7-6.6)^2 = .16$	$(6.6-7.33)^2 = .533$	$(7-7.33)^2 = .109$
	8	$(8-6.6)^2 = 1.96$	$(6.6-7.33)^2 = .533$	$(8-7.33)^2 = .449$
Group 2	5	$(5-6.6)^2 = 2.56$	$(6.6-7.33)^2 = .533$	$(5-7.33)^2 = 5.429$
	8	$(8-6.6)^2 = 1.96$	$(6.6-7.33)^2 = .533$	$(8-7.33)^2 = .449$
	5	$(5-6.6)^2 = 2.56$	$(6.6-7.33)^2 = .533$	$(5-7.33)^2 = 5.429$
	4	$(4-6.2)^2 = 4.84$	$(6.2-7.33)^2 = 1.277$	$(4-7.33)^2 = 11.089$
	8	$(8-6.2)^2 = 3.24$	$(6.2-7.33)^2 = 1.277$	$(8-7.33)^2 = .449$
Group 3	7	$(7-6.2)^2 = .64$	$(6.2-7.33)^2 = 1.277$	$(7-7.33)^2 = .109$
	5	$(5-6.2)^2 = 1.44$	$(6.2-7.33)^2 = 1.277$	$(5-7.33)^2 = 5.429$
	7	$(7-6.2)^2 = .64$	$(6.2-7.33)^2 = 1.277$	$(7-7.33)^2 = .109$
		$SS_{WG} = 22.80$	$SS_{BG} = 26.535$	$SS_T = 49.335$

In employing Equation 16.47 to compute SS_{WG}, the following operations are carried out for each group. The group mean ($\bar{X}_j$) is subtracted from each score in the group. The difference scores are squared, after which the sum of the squared difference scores is obtained. The sum of the sum of the squared difference scores for all k groups represents the within-groups sum of squares. Equation 16.47 is computationally equivalent to Equation 16.5.

Table 16.9 illustrates the use of Equations 16.45, 16.46, and 16.47 with the data for Example 16.1.[62] The resulting values of SS_T, SS_{BG}, and SS_{WG} are identical to those obtained with the computational equations (Equations 16.2, 16.3, and 16.5). Any minimal discrepancies are the result of rounding off error.

3. Equivalency of the single-factor between-subjects analysis of variance and the *t* test for two independent samples when $k = 2$ Interval/ratio data for an experiment involving $k = 2$ independent groups can be evaluated with either a **single-factor between-subjects analysis of variance** or a ***t* test for two independent samples**. When both of the aforementioned tests are employed to evaluate the same set of data they will yield the same result. Specifically, the following will always be true with respect to the relationship between the computed F and t values for the same set of data: $F = t^2$ and $t = \sqrt{F}$. It will also be the case that the square of the tabled critical t value at a prespecified level of significance for $df = n_1 + n_2 - 2$ will be equal to the tabled critical F value at the same level of significance for $df_{BG} = 1$ and df_{WG} (which will be $df_{WG} = N - k = N - 2$, which is equivalent to the value $df = n_1 + n_2 - 2$ employed for the ***t* test for two independent samples**).

To illustrate the equivalency of the results obtained with the **single-factor between-subjects analysis of variance** and the ***t* test for two independent samples** when $k = 2$, an F value will be computed for Example 8.1. The value $t = -1.96$ ($t = 1.964$ if carried out to 3 decimal places) is obtained for the latter example when the ***t* test for two independent samples** is employed. When the same set of data is evaluated with the **single-factor between-subjects analysis of variance**, the value $F = 3.86$ is computed. Note that $(t = -1.964)^2 = (F = 3.86)$. Equations 16.2, 16.3, and 16.4 are employed below to compute the values SS_T, SS_{BG}, and SS_{WG} for Example 8.1. Since $k = 2$, $n = 5$, and

$nk = N = 10$, $df_{BG} = 2 - 1 = 1$, $df_{WG} = N - k = 10 - 2 = 8$, and $df_T = N - 1 = 10 - 1 = 9$. The full analysis of variance is summarized in Table 16.10.

$$SS_T = 473 - \frac{(53)^2}{10} = 192.1 \qquad SS_{BG} = \left[\frac{(14)^2}{5} + \frac{(39)^2}{5}\right] - \frac{(53)^2}{10} = 62.5$$

$$SS_{WG} = 192.1 - 62.5 = 129.6$$

Table 16.10 Summary Table of Analysis of Variance for Example 8.1

Source of variation	*SS*	*df*	*MS*	*F*
Between-groups	62.5	1	62.5	3.86
Within-groups	129.6	8	16.2	
Total	192.1	9		

For $df_{BG} = 1$ and $df_{WG} = 8$, the tabled critical .05 and .01 values are $F_{.05} = 5.32$ and $F_{.01} = 11.26$ (which are appropriate for a nondirectional analysis). Note that (if one takes into account rounding off error) the square roots of the aforementioned tabled critical values are (for $df = 8$) the tabled critical two-tailed values $t_{.05} = 2.31$ and $t_{.01} = 3.36$ that are employed in Example 8.1 to evaluate the value $t = -1.96$. Since the obtained value $F = 3.86$ is less than the tabled critical .05 value $F_{.05} = 5.32$, the nondirectional alternative hypothesis H_1: $\mu_1 \neq \mu_2$ is not supported. The directional alternative hypothesis H_1: $\mu_1 < \mu_2$ is supported at the .05 level, since $F = 3.86$ is greater than the tabled critical one-tailed .05 value $F_{.90} = 3.46$ (the square root of which is the tabled critical one-tailed .05 value $t_{.05} = 1.86$ employed in Example 8.1). The directional alternative hypothesis H_1: $\mu_1 < \mu_2$ is not supported at the .01 level, since $F = 3.86$ is less than the tabled critical one-tailed .01 value $F_{.98} = 8.41$ (the square root of which is the tabled critical one-tailed .01 value $t_{.01} = 2.90$ employed in Example 8.1).[63] The conclusions derived from the **single-factor between-subjects analysis of variance** are identical to those reached when the data are evaluated with the ***t* test for two independent samples**.

4. Robustness of the single-factor between-subjects analysis of variance The general comments made with respect to the robustness of the ***t* test for two independent samples** (in Section VII) are applicable to the **single-factor between-subjects analysis of variance**. Most sources state that the **single-factor between-subjects analysis of variance** is robust with respect to violation of its assumptions. Nevertheless, when either the normality and/or homogeneity of variance assumption is violated, it is recommended that a more conservative analysis be conducted (i.e., employ the tabled $F_{.01}$ value or even a tabled value for a lower alpha level to represent the $F_{.05}$ value). When the violation of one or both assumptions is extreme, some sources recommend that a researcher consider employing an alternative procedure. Alternative procedures are discussed earlier in this section under the homogeneity of variance assumption. Keppel (1991) and Maxwell and Delaney (1990) provide comprehensive discussions on the general subject of the robustness of the **single-factor between-subjects analysis of variance**.

5. Fixed-effects versus random-effects models for the single-factor between-subjects analysis of variance The terms **fixed-** versus **random-effects models** refer to the way in which a researcher selects the levels of the independent variables that are employed in an experiment. Whereas a **fixed-effects model** assumes that the levels of the independent variable are the same levels that will be employed in any attempted replication of the experiment, a **random-effects model** assumes that the levels have been randomly selected from the overall population of all possible levels that can be employed for the independent variable. The discussion of the **single-factor between-subjects analysis of variance** in this book assumes a fixed-effects model. With the exception of the computation of measures of magnitude of treatment effect, the equations employed for the fixed-effects model are identical to those that are employed for a random-effects model. However, in the case of more complex designs (within-subjects and factorial designs), the computational procedures for fixed- versus random-effects models may differ.[64] The degree to which a researcher may generalize the results of an experiment will be a direct function of which model is employed. Specifically, if a fixed-effects model is employed, one can only generalize the results of an experiment to the specific levels of the independent variable that are used in the experiment. On the other hand, if a random-effects model is employed, one can generalize the results to all possible levels of the independent variable.

VIII. Additional Examples Illustrating the Use of the Single-Factor Between-Subjects Analysis of Variance

Since the **single-factor between-subjects analysis of variance** can be employed to evaluate interval/ratio data for any independent groups design involving two or more groups, it can be used to evaluate any of the examples that are evaluated with the ***t* test for two independent samples** (with the exception of Example 8.3). Examples 16.2, 16.3, and 16.4 in this section are extensions of Examples 8.1, 8.4, and 8.5 to a design involving $k = 3$ groups. Since the data for all of the examples are identical to the data employed in Example 16.1, they yield the same result.

Example 16.2 *In order to assess the efficacy of a new antidepressant drug* 15 *clinically depressed patients are randomly assigned to one of three groups. Five patients are assigned to* Group 1, *which is administered the antidepressant drug for a period of six months. Five patients are assigned to* Group 2, *which is administered a placebo during the same six month period. Five patients are assigned to* Group 3, *which does not receive any treatment during the six month period. Assume that prior to introducing the experimental treatments, the experimenter confirmed that the level of depression in the three groups was equal. After six months elapse all* 15 *subjects are rated by a psychiatrist (who is blind with respect to a subject's experimental condition) on their level of depression. The psychiatrist's depression ratings for the five subjects in each group follow. (The higher the rating the more depressed a subject.)* **Group 1**: 8, 10, 9, 10, 9; **Group 2**: 7, 8, 5, 8, 5; **Group 3**: 4, 8, 7, 5, 7. *Do the data indicate that the antidepressant drug is effective?*

Example 16.3 *A researcher wants to assess the relative effect of three different kinds of punishment on the emotionality of mice. Each of 15 mice is randomly assigned to one of three groups. During the course of the experiment each mouse is sequestered in an experimental chamber. While in the chamber, each of the five mice in* Group 1 *is periodically presented with a loud noise, each of the five mice in* Group 2 *is periodically presented with a blast of cold air, and each of the mice in* Group 3 *is periodically presented with an electric shock. The presentation of the punitive stimulus for each of the animals is generated by a machine that randomly presents the stimulus throughout the duration of the time it is in the chamber. The dependent variable of emotionality employed in the study is the number of times each mouse defecates while in the experimental chamber. The number of episodes of defecation for the* 15 *mice follow:* **Group 1**: 8, 10, 9, 10, 9; **Group 2**: 7, 8, 5, 8, 5; **Group 3**: 4, 8, 7, 5, 7. *Do subjects exhibit differences in emotionality under the different experimental conditions?*

Example 16.4 *Each of three companies that manufacture the same size precision ball bearing claims it has better quality control than its competitor. A quality control engineer conducts a study in which he compares the precision of ball bearings manufactured by the three companies. The engineer randomly selects five ball bearings from the stock of* Company A, *five ball bearings from the stock of* Company B, *and five ball bearings from the stock of* Company C. *He measures how much the diameter of each of the* 15 *ball bearings deviates from the manufacturer's specifications. The deviation scores (in micrometers) for the 15 ball bearings manufactured by the three companies follow:* **Company A**: 8, 10, 9, 10, 9; **Company B**: 7, 8, 5, 8, 5; **Company C**: 4, 8, 7, 5, 7. *What can the engineer conclude about the relative quality control of the three companies?*

References

Bartlett, M. S. (1937). Some examples of statistical methods of research in agriculture and applied biology. **Journal of the Royal Statistical Society Supplement**, 4, 137–170.

Box, G. E. (1953). Non-normality and tests on variance. **Biometrika**, 40, 318–335.

Brown, M. B. and Forsythe, A. B. (1974a). The ANOVA and multiple comparisons for data with heterogeneous variances. **Biometrics**, 30, 719–724.

Brown, M. B. and Forsythe, A. B. (1974b). Robust tests for equality of variances. **Journal of the American Statistical Association**, 69, 364–367.

Brown, M. B. and Forsythe, A. B. (1974c). The small sample behavior of some statistics which test the equality of several means. **Technometrics**, 16, 129–132.

Cochran, W. G. (1941). The distribution of the largest of a set of estimated variances as a fraction of their total. **Annals of Eugenics**, 11, 47–52.

Cochran, W. G. and Cox, G. M. (1957). **Experimental designs** (2nd ed.). New York: John Wiley.

Cohen, J. (1988). **Statistical power analysis for the behavioral sciences** (2nd ed.). New York: Academic Press.

Conover, W. J., Johnson, M. E., and Johnson, M. M. (1981). A comparative study of tests of homogeneity of variance with applications to the outer continental shelf bidding data. **Technometrics**, 23, 351–361.

Darlington, R. B. and Carlson, P. M. (1987). **Behavioral statistics: Logic and methods.** New York: The Free Press.

Dunn, O. J. (1961). Multiple comparisons among means. **Journal of the American Statistical Association**, 56, 52–64.

Dunnett, C. W. (1955). A multiple comparison procedure for comparing several treatments with a control. **Journal of the American Statistical Association**, 50, 1096–1121.

Dunnett, C. W. (1964). New tables for multiple comparisons with a control. **Biometrics**, 20, 482–491.

Fisher, R. A. (1935). **The design of experiments**. Edinburgh and London: Oliver and Boyd.

Hartley, H. O. (1940). Testing the homogeneity of a set of variances. **Biometrika**, 31, 249–255.

Hartley, H. O. (1950). The maximum F ratio as a shortcut test for heterogeneity of variance. **Biometrika**, 37, 308–312.

Howell, D. C. (1992). **Statistical methods for psychology** (3rd ed.). Boston: PWS–Kent Publishing Company.

James, G. S. (1951). The comparison of several groups of observations when the ratios of the population variances are unknown. **Biometrika**, 38, 324–329.

Kachigan, S. K. (1986). **Statistical analysis: An interdisciplinary introduction to univariate and multivariate methods.** New York: Radius Press.

Keppel, G. (1991) **Design and analysis: A researcher's handbook** (3rd ed.). Englewood Cliffs, N.J.: Prentice Hall.

Keppel, G., Saufley, W. H. and Tokunaga, H. (1992). **Introduction to design and analysis: A student's handbook** (2nd ed.). New York: W. H. Freeman and Company.

Keppel, G. and Zedeck, S. (1989) **Data analysis for research designs.** New York: W. H. Freeman and Company.

Keuls, M. (1952). The use of studentized range in connection with an analysis of variance. **Euphytica**, 1, 112–122.

Kirk, R. E. (1982). **Experimental design: Procedures for the behavioral sciences** (2nd ed.). Belmont, CA: Brooks/Cole Publishing Company.

Lord, F. M. (1967). A paradox in the interpretation of group comparisons. **Psychological Bulletin**, 68, 304–305.

Lord, F. M. (1969). Statistical adjustments when comparing pre-existing groups. **Psychological Bulletin**, 72, 336–337.

Maxwell, S. E. and Delaney, H. D. (1990) **Designing experiments and analyzing data.** Belmont, CA: Wadsworth Publishing Company.

McFatter, R. M. and Gollob, H. F. (1986). The power of hypothesis tests for comparisons. **Educational and Psychological Measurement**, 46, 883–886.

Mullen, B. and Rosenthal, R. (1985). **Basic meta-analysis: Procedures and programs.** Hillsdale, NJ: Lawrence Erlbaum Associates, Publishers.

Newman, D. (1939). The distribution of the range in samples from a normal population, expressed in terms of an independent estimate of standard deviation. **Biometrika**, 31, 20–30.

O'Brien, R. G. (1981). A simple test for variance effects in experimental designs. **Psychological Bulletin**, 89, 570–574.

Pearson, E. S. and Hartley, H. O. (1951). Charts of the power function for analysis of variance, derived from the non-central F distribution. **Biometrika**, 38, 112–130.

Rosenthal, R. (1991). **Meta-analytic procedures for social research.** Newbury, CA: Sage Publications.

Scheffé, H. A. (1953). A method for judging all possible contrasts in the analysis of variance. **Biometrika**, 40, 87–104.

Scheffé, H. A. (1959). **The analysis of variance.** New York: John Wiley and Sons.

Tiku, M. L. (1967). Tables of the power of the F test. **Journal of the American Statistical Association**, 62, 525–539.

Tukey, J. W. (1953). The problem of multiple comparisons. Unpublished paper, Princeton University, Princeton, NJ.

Welch, B. L. (1951). On the comparison of several mean values: An alternative approach. **Biometrika**, 38, 330–336.

Winer, B. J., Brown, D. R., and Michels, K. M. (1991). **Statistical principles in experimental design** (3rd ed.). New York: McGraw-Hill Publishing Company.

Endnotes

1. The term **single-factor** refers to the fact that the design upon which the analysis of variance is based involves a single independent variable. Since factor and independent variable mean the same thing, **multifactor designs** (more commonly called **factorial designs**) that are evaluated with the analysis of variance involve more than one independent variable. Multifactor analysis of variance procedures are discussed under the **between-subjects factorial analysis of variance (Test 21)**.

2. It should be noted that if an experiment is confounded, one cannot conclude that a significant portion of **between-groups variability** is attributed to the independent variable. This is the case, since if one or more confounding variables systematically vary with the levels of the independent variable, a significant difference can be due to a confounding variable rather than the independent variable.

3. The homogeneity of variance assumption is also discussed in Section VI of the ***t* test for two independent samples** in reference to a design involving two independent samples.

4. Although it is possible to conduct a directional analysis, such an analysis will not be described with respect to the analysis of variance. A discussion of a directional analysis when $k = 2$ can be found under the ***t* test for two independent samples.** In addition, a discussion of one-tailed F values can be found in Section VI of the latter test under the discussion of the **Hartley's F_{max} test for homogeneity of variance/*F* test for two population variances (Test 8a).** A discussion of the evaluation of a directional alternative hypothesis when $k \geq 2$ can be found in Section VII of the **chi-square goodness-of-fit test (Test 5).** Although the latter discussion is in reference to analysis of a k independent samples design involving categorical data, the general principles regarding the analysis of a directional alternative hypothesis when $k \geq 2$ are applicable to the analysis of variance.

5. Some sources present an alternative method for computing SS_{BG} when the number of subjects in each group is not equal. Whereas Equation 16.3 weighs each group's contribution based on the number of subjects in the group, the alternative method (which is not generally recommended) weighs each group's contribution equally, irrespective of sample size. Keppel (1991), who describes the latter method, notes that as a general rule the value it computes for SS_{BG} is close to the value obtained with Equation 16.3, except when the sample sizes of the groups differ substantially from one another.

6. Since there are an equal number of subject in each group, the Equation for SS_{BG} can also be written as follows:

$$SS_{BG} = \frac{[(46)^2 + (33)^2 + (31)^2]}{5} - \frac{(110)^2}{15} = 26.53$$

7. SS_{WG} can also be computed with the following equation:

$$SS_{WG} = \Sigma X_T^2 - \sum_{j=1}^{k}\left[\frac{(\Sigma X_j)^2}{n_j}\right] = 856 - \left[\frac{(46)^2}{5} + \frac{(33)^2}{5} + \frac{(31)^2}{5}\right] = 22.80$$

Since there are an equal number of subjects in each group, n can be used in place of n_j in the above equation, as well as in Equation 16.5. The numerators of the term in the brackets can be combined and written over a single denominator that equals the value $n = 5$.

8. Equation 16.9 can be employed whether there are an equal or unequal number of subjects in each group. The following equation can also be employed when the number of subjects in each group is equal or unequal: $df_{WG} = (n_1 - 1) + (n_2 - 1) + \cdots + (n_k - 1)$. The equation $df_{WG} = k(n - 1) = nk - k$ can be employed to compute df_{WG}, but only when the number of subjects in each group is equal.

9. When there are an equal number of subjects in each group, since $N = nk$, $df_T = nk - 1$.

10. There is a separate F distribution for each combination of df_{num} and df_{den} values. Figure 16.2 depicts the F distribution for three different sets of degrees of freedom values. Note that in each of the distributions, 5% of the distribution falls to the right of the tabled critical $F_{.05}$ value.

    Most tables of the F distribution do not include tabled critical values for all possible values of df_{num} and df_{den}. The protocol that is generally employed for determining a critical F value for a df value that is not listed is to either employ interpolation or to employ the df value closest to the desired df value. Some sources qualify the latter by stating that in order to insure that the Type I error rate does not exceed the prespecified value of alpha, one should employ the df value that is closest to but not above the desired df value.

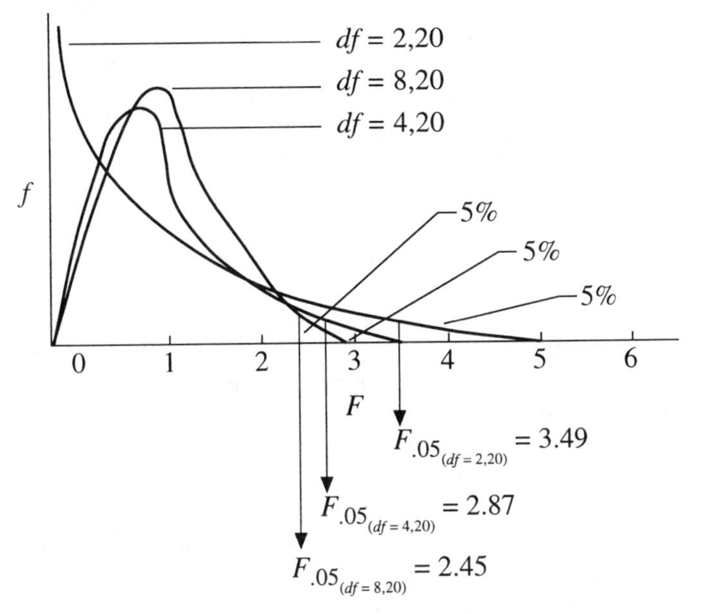

Figure 16.2 Representative *F* Distributions for Different *df* Values

11. Although the discussion of comparison procedures in this section will be limited to the analysis of variance, the underlying general philosophy can be generalized to any inferential statistical test for which comparisons are conducted.

12. The terms **family** and **set** are employed synonymously throughout this discussion.

13. The accuracy of Equation 16.13 will be compromised if all of the comparisons are not independent of one another. Independent comparisons, which are commonly referred to as **orthogonal comparisons**, are discussed later in the section.

14. Equation 16.14 tends to slightly overestimate the value of α_{FW}. The degree to which it overestimates α_{FW} increases as either the value of c or α_{PC} increase.

15. Equation 16.16 provides a computationally quick approximation of the value computed with Equation 16.15. The value computed with Equation 16.16 tends to underestimate

the value of α_{PC}. The larger the value of α_{FW} or c, the greater the degree α_{PC} will be underestimated with the latter equation. When $\alpha = .05$, however, the two equations yield values that are almost identical.

16. This example is based on a discussion of this issue in Howell (1992) and Maxwell and Delaney (1990).

17. The term **single degree of freedom comparison** reflects the fact that $k = 2$ means are contrasted with one another. Although one or both the $k = 2$ means may be a composite mean that is based on the combined scores of two or more groups, any composite mean is expressed as a single mean value. The latter is reflected in the fact that there will always one equals sign (=) in the null hypothesis for a single degree of freedom comparison.

18. Although the examples illustrating comparisons will assume a nondirectional alternative hypothesis, the alternative hypothesis can also be stated directionally. When the alternative hypothesis is stated directionally, the tabled critical one-tailed F value must be employed. Specifically, when using the F distribution in evaluating a directional alternative hypothesis, when $\alpha_{PC} = .05$, the tabled $F_{.90}$ value is employed for the one-tailed $F_{.05}$ value instead of the tabled $F_{.95}$ value (which as noted earlier is employed for the two-tailed/nondirectional $F_{.05}$ value). When $\alpha_{PC} = .01$, the tabled $F_{.98}$ value is employed for the one-tailed $F_{.01}$ value instead of the tabled $F_{.99}$ value (which is employed for the two-tailed/nondirectional $F_{.01}$ value).

19. If the coefficients of Groups 1 and 2 are reversed (i.e., $c_1 = -1$ and $c_2 = +1$), the value of $\Sigma(c_j)(\bar{X}_j)$ will equal −2.6. The fact that the sign of the latter value is negative will not affect the test statistic for the comparison. This is the case, since in computing the F value for the comparison, the value $\Sigma(c_j)(\bar{X}_j)$ is squared, and consequently it becomes irrelevant whether $\Sigma(c_j)(\bar{X}_j)$ is a positive or negative number.

20. When the sample sizes of all k groups are not equal, the value of the **harmonic mean** (which is discussed in Section VI of the ***t* test for two independent samples**) is employed to represent n in Equation 16.17. However, when the harmonic mean is employed, if there are large discrepancies between the sizes of the samples, the accuracy of the analysis may be compromised.

21. MS_{WG} is employed as the estimate of error variability for the comparison, since if the homogeneity of variance assumption is not violated, the pooled within-groups variability employed in computing the omnibus F value will provide the most accurate estimate of error variability.

22. As is the case with simple comparisons, the alternative hypothesis for a complex planned comparison can also be evaluated nondirectionally.

23. A reciprocal of a number is the value 1 divided by that number.

24. When there are equal number of subjects in each group it is possible, though rarely done, to assign different coefficients to two or more groups on the same side of the equals sign. In such an instance the composite mean reflects an unequal weighting of the groups. On the other hand, when there are an unequal number of subjects in any of the groups on the same side of the equals sign, any groups that do not have the same

sample size will be assigned a different coefficient. The coefficient a group is assigned will reflect the proportion it contributes to the total number of subjects on that side of the equals sign. Thus, if Groups 1 and 2 are compared with Group 3, and there are 4 subjects in Group 1 and 6 subjects in Group 2, there are a total of 10 subjects involved on that side of the equals sign. The absolute value of the coefficient assigned to Group 1 will be $\frac{4}{10} = \frac{2}{5}$, whereas the absolute value assigned to Group 2 will be $\frac{6}{10} = \frac{3}{5}$.

25. When any of the coefficients are fractions, in order to simplify calculations some sources prefer to convert the coefficients into integers. In order to do this, each coefficient must be multiplied by a **least common denominator**. A least common denominator is the smallest number (excluding 1) that is divisible by all of the denominators of the coefficients. With respect to the complex comparison under discussion, the least common denominator is 2, since 2 is the smallest number that can be divided by 1 and 2 (which are the denominators of the coefficients $1/1 = 1$ and $1/2$). If all of the coefficients are multiplied by 2, the coefficients are converted into the following values: $c_1 = -1$, $c_2 = -1$, $c_3 = +2$. If the latter coefficients are employed in the calculations that follow, they will produce same end result as that obtained through use of the coefficients employed in Table 16.4. It should be noted, however, that if the converted coefficients are employed, the value $\Sigma(c_j)(\bar{X}_j) = -1.7$ in Table 16.4 will become twice the value that is presently listed (i.e., it becomes 3.4). As a result of this, $\Sigma(c_j)(\bar{X}_j)$ will no longer represent the difference between the two sets of means contrasted in the null hypothesis. Instead, it will be a multiple of that value — specifically, the multiple of the value by which the coefficients are multiplied.

26. If $n_a \neq n_b$, $\sqrt{(MS_{WG}/n_a) + (MS_{WG}/n_b)}$ is employed as the denominator of Equation 16.22 and $\sqrt{([(\Sigma c_a^2)(MS_{WG})]/n_a) + ([(\Sigma c_b^2)(MS_{WG})]/n_b)}$ is employed as the denominator of Equation 16.23.

27. If the value $\sqrt{(\tilde{s}_a^2/n_a) + (\tilde{s}_b^2/n_b)}$ is employed as the denominator of Equations 16.22/16.23, or if the degrees of freedom for Equations 16.22/16.23 are computed with Equation 8.4 ($df = n_a + n_b - 2$), a different result will be obtained since: a) Unless the variance for all of the groups is equal, it is unlikely that the computed t value will be identical to the one computed with Equations 16.22/16.23; and b) If $df = n_a + n_b - 2$ is used, the tabled critical t value employed will be higher than the tabled critical value employed for Equations 16.22/16.23. This is the case, since the df value associated with MS_{WG} is larger than the df value associated with $df = n_a + n_b - 2$. The larger the value of df, the lower the corresponding tabled critical t value.

28. Equations 16.24 and 16.25 are respectively derived from Equations 16.22 and 16.23. When two groups are compared with one another, the data may be evaluated with either the t distribution or the F distribution. The relationship between the two distributions is $F = t^2$. In view of this, the term $\sqrt{F_{(1,WG)}}$ in Equations 16.24 and 16.25 can also be written as $t_{df_{WG}}$. In other words, Equation 16.24 can be written as $CD_{LSD} = t_{df_{WG}}\sqrt{(2MS_{WG})/n}$ (and in the case of Equation 16.25, the same equation is employed, except for the fact that inside the radical (Σc_j^2) is employed in place of 2). Thus, in the computations to follow, if a nondirectional alternative hypothesis is employed with $\alpha = .05$, one can employ either $F_{.05} = 4.75$ (for $df_{\text{num}} = 1$, $df_{\text{den}} = 12$) or $t_{.05} = 2.18$ (for $df = 12$), since $(t_{.05} = 2.18)^2 = (F_{.05} = 4.75)$.

29. As is noted in reference to **linear contrasts**, the $df = 1$ value for the numerator of the F ratio for a single degree of freedom comparison is based on the fact that the comparison involves $k = 2$ groups with $df = k - 1$. In the same respect, in the case of **complex comparisons** there are two sets of means, and thus $df = k_{comp} - 1$.

30. As indicated in Endnote 28, if a t value is substituted in Equation 16.24 it will yield the same result. Thus, if $t_{.05} = 2.18$ is employed in the equation $CD_{LSD} = t_{df_{WG}} \sqrt{(2MS_{WG})/n}$, $CD_{LSD} = 2.18\sqrt{[(2)(1.9)]/5} = 1.90$.

31. The absolute value of t is employed, since a nondirectional alternative hypothesis is evaluated.

32. When $n_1 \neq n_2$, $\sqrt{(MS_{WG}/n_a) + (MS_{WG}/n_b)}$ is employed in Equation 16.26 in place of $\sqrt{(2MS_{WG})/n}$.

33. Since in a two-tailed analysis we are interested in a proportion that corresponds to the most extreme .0167 cases in the distribution, one-half of the cases (i.e., 0083) falls in each tail of the distribution.

34. The only exception to this will be when one comparison is being made (i.e., $c = 1$), in which case both methods yield the identical CD value.

35. Some sources employ the abbreviation **WSD** for **wholly significant difference** instead of **HSD**.

36. The value of α_{FW} for **Tukey's HSD test** is compared with the value of α_{FW} for other comparison procedures within the framework of the discussion of the **Scheffé test** later in this section.

37. When the tabled critical q value for $k = 2$ treatments is employed, the **Studentized range statistic** will produce equivalent results to those obtained with **multiple *t* tests/Fisher's LSD test**. This will be demonstrated later in reference to the **Newman-Keuls test**, which is another comparison procedure that employs the **Studentized range statistic**.

38. As is the case with a t value, the sign of q will only be relevant if a directional alternative hypothesis is evaluated. When a directional alternative hypothesis is evaluated, in order to reject the null hypothesis the sign of the computed q value must be in the predicted direction, and the absolute value of q must be equal to or greater than the tabled critical q value at the prespecified level of significance.

39. Although Equations 16.30 and 16.31 can be employed for both simple and complex comparisons, sources generally agree that **Tukey's HSD test** should only be employed for simple comparisons. Its use with only simple comparisons is based on the fact that in the case of complex comparisons it provides an even less powerful test of an alternative hypothesis than does the **Scheffé test** (which is an extremely conservative procedure) discussed later in this section.

40. When $n_a \neq n_b$ and/or the homogeneity of variance assumption is violated, some sources recommend using the following modified form of Equation 16.31 (referred to as the **Tukey–Kramer procedure**) for computing CD_{HSD}: CD_{HSD}

$= q_{(k,df_{WG})}\sqrt{MS_{WG}[[(1/n_a) + (1/n_b)]/2]}$. Other sources, however, do not endorse the use of the latter equation and provide alternative approaches for dealing with unequal sample sizes.

In the case of unequal sample sizes, some sources (e.g., Winer *et al.*, 1991) recommend employing the **harmonic mean** of the sample sizes to compute the value of n for Equations 16.30 and 16.31 (especially if the sample sizes are approximately the same value). The harmonic mean is described in Section VI of the ***t* test for two independent samples**.

41. The only comparison for which the minimum required difference computed for the **Newman–Keuls test** will equal CD_{HSD} will be the comparison contrasting the smallest and largest means in the set of k means. One exception to this is a case in which two or more treatments have the identical mean value, and that value is either the lowest or highest of the treatment means. Although the author has not seen such a configuration discussed in the literature, one can argue that in such an instance the number of steps between the lowest and highest mean should be some value less than k. One can conceptualize all cases of a tie as constituting one step, or perhaps, for those means that are tied one might employ the average of the steps that would be involved if all those means are counted as separate steps. The larger the step value that is employed, the more conservative the test.

42. If Equation 16.30 is employed, it uses the same values that are employed for **Tukey's HSD test**, and thus yields the identical q value. Thus, $q = (9.2 - 6.6)/\sqrt{1.9/5} = 4.22$.

43. One exception to this involving the **Bonferroni–Dunn test** will be discussed later in this section.

44. Recollect that the **Bonferroni–Dunn test** assumed that a total of 6 comparisons are conducted (3 simple comparisons and 3 complex comparisons). Thus, as noted earlier, since the number of comparisons exceeds $[k(k - 1)]/2 = [3(3 - 1)]/2 = 3$, the **Scheffé test** will provide a more powerful test of an alternative hypothesis than the **Bonferroni–Dunn test**.

45. The author is indebted to Scott Maxwell for his input on the content of the discussion to follow.

46. Dunnett (1964) developed a modified test procedure (described in Winer *et al.*, (1991)) to be employed in the event that there is a lack of homogeneity of variance between the variance of the control group and the variance of the experimental groups with which it is contrasted.

47. The philosophy to be presented here can be generalized to any inferential statistical analysis.

48. Although the difference between $\alpha_{FW} = .05$ and $\alpha_{FW} = .10$ may seem trivial, a result that is declared significant is more likely to be submitted and/or accepted for publication.

49. The reader may find it useful to review the discussion of **effect size** in Section VI of the **single sample *t* test** and the ***t* test for two independent samples**.

50. The reader may want to review the discussion of confidence intervals in Section VI of both the **single sample t test** and the **t test for two independent samples.**

51. The **F test for two population variances** (discussed in conjunction with the **F_{max} test** as **Test 8a** in Section VI of the **t test for two independent samples**) can only be employed to evaluate the homogeneity of variance hypothesis when $k = 2$.

52. ϕ should not be confused with the **phi coefficient** (also represented by the notation ϕ), discussed under the **chi-square test for $r \times c$ tables (Test 11).** Although the latter measure is identified by the same symbol, it represent a different measure than the noncentrality parameter.

53. When there is no curve that corresponds exactly to df_{WG}, one can either interpolate or employ the df_{WG} value closest to it.

54. Tiku (1967) has derived more detailed power tables that include the alpha values .025 and .005.

55. When $k = 2$, Equations 16.40/16.41 and Equation 8.14 will yield the same $\tilde{\omega}^2$ value.

56. Some sources employ the notation R^2 for the **eta squared** statistic. The statistic represented by R^2 or $\tilde{\eta}^2$ is commonly referred to as the **correlation ratio,** which is the squared multiple correlation coefficient that is computed when multiple regression (discussed under the **multiple correlation coefficient (Test 22k)**) is employed to predict subjects' scores on the dependent variable based on group membership.

    When $k = 2$, the **eta squared statistic** is equivalent to r_{pb}, which represents the **point-biserial correlation coefficient (Test 22h).** Under the discussion of r_{pb} the equivalency of $\tilde{\eta}^2$ and r_{pb} is demonstrated.

57. A number of different measures have been developed to determine the magnitude of treatment effect for comparison procedures. Unfortunately, researchers do not agree with respect to which of the available measures is most appropriate to employ. Keppel (1991) provides a comprehensive discussion of this subject.

58. Some sources employ $\sqrt{F_{(1, df_{WG})}}$ in Equation 16.44 instead of $t_{df_{WG}}$. Since $t = \sqrt{F}$, the two values produce equivalent results. The tabled critical two-tailed t value at a prespecified level of significance will be equivalent to the square root of the tabled critical F value at the same level of significance for $df_{num} = 1$ and $df_{den} = df_{WG}$.

59. In using Equation 2.7, $s_1/\sqrt{n}$ is equivalent to $s_{\bar{X}_1}$.

60. In the interest of precision, Keppel and Zedeck (1989, p. 98) note that although when the null hypothesis is true the median of the sampling distribution for the value of F equals 1, the mean of the sampling distribution of F is slightly above one. Specifically, the expected value of $F = df_{WG}/df_{(WG-2)}$. It should also be noted that although it rarely occurs, it is possible that $MS_{WG} > MS_{BG}$. In such a case the value of F will be less than 1, and obviously if $F < 1$ the result cannot be significant.

61. In employing double (or even more than two) summation signs such as $\sum_{j=1}^{k}\sum_{i=1}^{n}$, the mathematical operations specified are carried out beginning with the summation

sign that is farthest to the right and continued sequentially with those operations specified by summation signs to the left. Specifically, if $k = 3$ and $n = 5$, the notation $\sum_{j=1}^{k} \sum_{i=1}^{n} X_{ij}$ indicates that the sum of the n scores in Group 1 are computed, after which the sum of the n scores in Group 2 are computed, after which the sum of the n scores in Group 3 are computed. The final result will be the sum of all the aforementioned values that have been computed.

62. For each of the $N = 15$ subjects in Table 16.9, the following is true with respect to the contribution of a subject's score to the total variability in the data:

$$(X_{ij} - \bar{X}_T) = (\bar{X}_j - \bar{X}_T) + (X_{ij} - \bar{X}_j)$$

Total deviation score = BG deviation score + WG deviation score

63. In evaluating a directional alternative hypothesis, when $k = 2$ the tabled $F_{.90}$ and $F_{.98}$ values (for the appropriate degrees of freedom) are respectively employed as the one-tailed .05 and .01 values. Since the values for $F_{.90}$ and $F_{.98}$ are not listed in **Table A10**, the values $F_{.90} = 3.46$ and $F_{.98} = 8.41$ can be obtained by squaring the tabled critical one-tailed values $t_{.05} = 1.86$ and $t_{.01} = 2.90$, by employing more extensive tables of the F distribution available in other sources, or through interpolation.

64. In the case of a factorial design it is also possible to have a **mixed-effects model**. In a **mixed-effects model** it is assumed that at least one of the independent variables is based on a fixed-effects model and that at least one is based on a random-effects model.

Test 17

The Kruskal–Wallis One-Way Analysis of Variance by Ranks
(Nonparametric Test Employed with Ordinal Data)

I. Hypothesis Evaluated with Test and Relevant Background Information

Hypothesis evaluated with test In a set of k independent samples (where $k \geq 2$), do at least two of the samples represent populations with different median values?

Relevant background information on test The **Kruskal–Wallis one-way analysis of variance by ranks** (Kruskal (1952) and Kruskal and Wallis (1952)) is employed with ordinal (rank-order) data in a hypothesis testing situation involving a design with two or more independent samples. The test is an extension of the **Mann–Whitney *U* test (Test 9)** to a design involving more than two independent samples, and when $k = 2$ the **Kruskal–Wallis one-way analysis of variance by ranks** will yield a result that is equivalent to that obtained with the **Mann–Whitney *U* test**. If the result of the **Kruskal–Wallis one-way analysis of variance by ranks** is significant, it indicates there is a significant difference between at least two of the sample medians in the set of k medians. As a result of the latter, the researcher can conclude there is a high likelihood that at least two of the samples represent populations with different median values.

In employing the **Kruskal–Wallis one-way analysis of variance by ranks** one of the following is true with regard to the rank-order data that are evaluated: a) The data are in a rank-order format, since it is the only format in which scores are available; or b) The data have been transformed into a rank-order format from an interval/ratio format, since the researcher has reason to believe that one or more of the assumptions of the **single-factor between-subjects analysis of variance (Test 16)** (which is the parametric analog of the **Kruskal–Wallis test**) is saliently violated. It should be noted that when a researcher elects to transform a set of interval/ratio data into ranks, information is sacrificed. This latter fact accounts for why there is reluctance among some researchers to employ nonparametric tests such as the **Kruskal–Wallis one-way analysis of variance by ranks**, even if there is reason to believe that one or more of the assumptions of the **single-factor between-subjects analysis of variance** has been violated.

Various sources (e.g., Conover (1980), Daniel (1990), and Marascuilo and McSweeney (1977)) note that the **Kruskal–Wallis one-way analysis of variance by ranks** is based on the following assumptions: a) Each sample has been randomly selected from the population it represents; b) The k samples are independent of one another; c) The dependent variable (which is subsequently ranked) is a continuous random variable. In truth, this assumption, which is common to many nonparametric tests, is often not adhered to, in that such tests are often employed with a dependent variable that represents a discrete random variable; and d) The underlying distributions from which the samples are derived are identical in shape. The shapes of the underlying population distributions, however, do not have to be normal. Maxwell and Delaney (1990) point out that the assumption of identically shaped distributions implies equal dispersion of data within each distribution. Because of this, they note that like the **single-factor between-subjects analysis of variance**, the **Kruskal–Wallis one-**

way analysis of variance by ranks assumes homogeneity of variance with respect to the underlying population distributions. Because the latter assumption is not generally acknowledged for the **Kruskal–Wallis one-way analysis of variance by ranks,** it is not uncommon for sources to state that violation of the homogeneity of variance assumption justifies use of the **Kruskal–Wallis one-way analysis of variance by ranks** in lieu of the **single-factor between-subjects analysis of variance.** It should be pointed out, however, that there is some empirical research which suggests that the sampling distribution for the **Kruskal–Wallis test** statistic is not as affected by violation of the homogeneity of variance assumption as is the F distribution (which is the sampling distribution for the **single-factor between-subjects analysis of variance**). One reason cited by various sources for employing the **Kruskal- Wallis one-way analysis of variance by ranks,** is that by virtue of ranking interval/ratio data a researcher can reduce or eliminate the impact of **outliers.** As noted in Section VII of the ***t* test for two independent samples,** since **outliers** can dramatically influence variability, they can be responsible for heterogeneity of variance between two or more samples. In addition, **outliers** can have a dramatic impact on the value of a sample mean.

II. Example

Example 17.1 is identical to Example 16.1 (which is evaluated with the **single-factor between-subjects analysis of variance**). In evaluating Example 17.1 it will be assumed that the ratio data are rank-ordered, since one or more of the assumptions of the **single-factor between-subjects analysis of variance** has been saliently violated.

Example 17.1 *A psychologist conducts a study to determine whether noise can inhibit learning. Each of* 15 *subjects is randomly assigned to one of three groups. Each subject is given* 20 *minutes to memorize a list of* 10 *nonsense syllables which they are told they will be tested on the following day. The five subjects assigned to* **Group 1,** *the* **no noise** *condition, study the list of nonsense syllables while they are in a quiet room. The five subjects assigned to* **Group 2,** *the* **moderate noise** *condition, study the list of nonsense syllables while listening to classical music. The five subjects assigned to* **Group 3,** *the* **extreme noise** *condition, study the list of nonsense syllables while listening to rock music. The number of nonsense syllables correctly recalled by the* 15 *subjects follow:* **Group 1:** 8, 10, 9, 10, 9; **Group 2:** 7, 8, 5, 8, 5; **Group 3:** 4, 8, 7, 5, 7. *Do the data indicate that noise influenced subjects' performance?*

III. Null versus Alternative Hypotheses

Null hypothesis H_0: $\theta_1 = \theta_2 = \theta_3$

(The median of the population Group 1 represents equals the median of the population Group 2 represents equals the median of the population Group 3 represents. With respect to the sample data, when there are an equal number of subjects in each group, the sums of the ranks will be equal for all k groups — i.e., $\Sigma R_1 = \Sigma R_2 = \Sigma R_3$. A more general way of stating this (which also encompasses designs involving unequal sample sizes) is that the means of the ranks of the k groups will be equal (i.e., $\bar{R}_1 = \bar{R}_2 = \bar{R}_3$).)

Alternative hypothesis H_1: Not H_0

(This indicates that there is a difference between at least two of the k population medians. It is important to note that the alternative hypothesis should not be written as follows: H_1: $\theta_1 \neq \theta_2 \neq \theta_3$. The reason why the latter notation for the alternative hypothesis is incorrect is because it implies that all three population medians must differ from one another in order to reject the null hypothesis. With respect to the sample data, if there are an equal number of subjects in each group, when the alternative hypothesis is true the sums of the ranks of at least two of the k groups will not be equal. A more general way of stating this (which also encompasses designs involving unequal sample sizes) is that the means of the ranks of at least two of the k groups will not be equal. In this book it will always be assumed that the alternative hypothesis for the **Kruskal–Wallis one-way analysis of variance by ranks** is stated **nondirectionally**.)[1]

IV. Test Computations

The data for Example 17.1 are summarized in Table 17.1. The total number of subjects employed in the experiment is $N = 15$. There are $n_j = n_1 = n_2 = n_3 = 5$ subjects in each group. The original interval/ratio scores of the subjects are recorded in the columns labelled X_1, X_2, and X_3. The adjacent columns R_1, R_2, and R_3 contain the rank-order assigned to each of the scores. The rankings for Example 17.1 are summarized in Table 17.2.

The ranking protocol employed for the **Kruskal–Wallis one-way analysis of variance by ranks** is the same as that employed for the **Mann–Whitney U test**. In Table 17.2 the two-digit subject identification number indicates the order in which a subject's score appears in Table 17.1 followed by his/her group. Thus, Subject i, j is the ith subject in Group j.

Table 17.1 Data for Example 17.1

Group 1		Group 2		Group 3	
X_1	R_1	X_2	R_2	X_3	R_3
8	9.5	7	6	4	1
10	14.5	8	9.5	8	9.5
9	12.5	5	3	7	6
10	14.5	8	9.5	5	3
9	12.5	5	3	7	6
	$\Sigma R_1 = 63.5$		$\Sigma R_2 = 31$		$\Sigma R_3 = 25.5$
$\bar{R}_1 = \frac{\Sigma R_1}{n_1} = \frac{63.5}{5} = 12.7$		$\bar{R}_2 = \frac{\Sigma R_2}{n_2} = \frac{31}{5} = 6.2$		$\bar{R}_3 = \frac{\Sigma R_3}{n_3} = \frac{25.5}{5} = 5.1$	

Table 17.2 Rankings for the Kruskal–Wallis Test for Example 17.1

Subject identification number	1,3	3,2	5,2	4,3	1,2	3,3	5,3	1,1	2,2	4,2	2,3	3,1	5,1	2,1	4,1
Number correct	4	5	5	5	7	7	7	8	8	8	8	9	9	10	10
Rank prior to tie adjustment	1	2	3	4	5	6	7	8	9	10	11	12	13	14	15
Tie adjusted rank	1	3	3	3	6	6	6	9.5	9.5	9.5	9.5	12.5	12.5	14.5	14.5

A brief summary of the ranking protocol employed in Table 17.2 follows:

a) All $N = 15$ scores are arranged in order of magnitude (irrespective of group membership), beginning on the left with the lowest score and moving to the right as scores increase. This is done in the second row of Table 17.2.

b) In the third row of Table 17.2, all $N = 15$ scores are assigned a rank. Moving from left to right, a rank of 1 is assigned to the score that is furthest to the left (which is the lowest score), a rank of 2 is assigned to the score that is second from the left (which, if there are no ties, will be the second lowest score), and so on until the score at the extreme right (which will be the highest score) is assigned a rank equal to N (if there are no ties for the highest score).

c) In instances where two or more subjects have the same score, the average of the ranks involved is assigned to all scores tied for a given rank. For a comprehensive discussion of how to handle tied ranks, the reader should review the description of the ranking protocol in Section IV of the **Mann–Whitney *U* test.**

It should be noted that, as is the case with the **Mann–Whitney *U* test**, it is permissible to reverse the ranking protocol described above. Specifically, one can assign a rank of 1 to the highest score, a rank of 2 to the second highest score, and so one, until reaching the lowest score which is assigned a rank equal to the value of N. This reverse ranking protocol will yield the same value for the **Kruskal–Wallis test** statistic as the protocol employed in Table 17.2.

Upon rank-ordering the scores of the $N = 15$ subjects, the sum of the ranks is computed for each group. In Table 17.1 the sum of the ranks of the jth group is represented by the notation ΣR_j. Thus, $\Sigma R_1 = 63.5$, $\Sigma R_2 = 31$, $\Sigma R_3 = 25.5$.

The chi-square distribution is generally employed to approximate the **Kruskal–Wallis test** statistic. Equation 17.1 is employed to compute the chi-square approximation of the **Kruskal–Wallis test** statistic (which is represented in most sources by the notation H).

$$H = \frac{12}{N(N+1)} \sum_{j=1}^{k} \left[\frac{(\Sigma R_j)^2}{n_j} \right] - 3(N+1) \qquad \textbf{(Equation 17.1)}$$

Note that in Equation 17.1 the term $\Sigma_{j=1}^{k}[(\Sigma R_j)^2/n_j]$ indicates that for each of the k groups, the sum of the ranks is squared and then divided by the number of subjects in the group. Upon doing this for all k groups, the resulting values are summed. Substituting the appropriate values from Example 17.1 in Equation 17.1, the value $H = 8.44$ is computed.

$$H = \frac{12}{(15)(15+1)} \left[\frac{(63.5)^2}{5} + \frac{(31)^2}{5} + \frac{(25.5)^2}{5} \right] - (3)(15+1) = 8.44$$

V. Interpretation of the Test Results

In order to reject the null hypothesis the computed value $H = \chi^2$ must be equal to or greater than the tabled critical chi-square value at the prespecified level of significance. The computed chi-square value is evaluated with **Table A4 (Table of the Chi-Square Distribution)** in the **Appendix**. For the appropriate degrees of freedom, the tabled $\chi^2_{.95}$ value (which is the chi-square value at the 95th percentile) and the tabled $\chi^2_{.99}$ value (which is the chi-square value at the 99th percentile) are employed as the .05 and .01 critical values for evaluating a nondirectional alternative hypothesis. The number of degrees of freedom employed in the analysis are computed with Equation 17.2. Thus, $df = 3 - 1 = 2$.

$$df = k - 1 \qquad \textbf{Equation 17.2)}$$

For $df = 2$, the tabled critical .05 and .01 chi-square values are $\chi^2_{.05} = 5.99$ and $\chi^2_{.01} = 9.21$. Since the computed value $H = 8.44$ is greater than $\chi^2_{.05} = 5.99$, the alternative hypothesis is supported at the .05 level. Since, however, $H = 8.44$ is less than $\chi^2_{.01} = 9.21$, the alternative hypothesis is not supported at the .01 level.[2] A summary of the analysis of

Example 17.1 with the **Kruskal–Wallis one-way analysis of variance by ranks** follows: It can be concluded that there is a significant difference between at least two of the three groups exposed to different levels of noise. This result can be summarized as follows: $H(2) = 8.44, p < .05$.

It should be noted that when the data for Example 17.1 are evaluated with a **single-factor between-subjects analysis of variance**, the null hypothesis can be rejected at both the .05 and .01 levels (although it barely achieves significance at the latter level, and in the case of the **Kruskal–Wallis test** the result just falls short of significance at the .01 level). The slight disparity between the results of the two tests reflects the fact that, as a general rule (assuming that none of the assumptions of the analysis of variance is saliently violated), the **Kruskal–Wallis one-way analysis of variance by ranks** provides a less powerful test of an alternative hypothesis than the **single-factor between-subjects analysis of variance.**

VI. Additional Analytical Procedures for the Kruskal–Wallis One-Way Analysis of Variance by Ranks and/or Related Tests

1. Tie correction for the Kruskal–Wallis one-way analysis of variance by ranks Some sources recommend that if there is an excessive number of ties in the overall distribution of N scores, the value of the **Kruskal–Wallis test** statistic be adjusted. The tie correction results in a small increase in the value of H (thus providing a slightly more powerful test of the alternative hypothesis). Equation 17.3 is employed in computing the value C, which represents the tie correction factor for the **Kruskal–Wallis one-way analysis of variance by ranks.**

$$C = 1 - \frac{\sum_{i=1}^{s} (t_i^3 - t_i)}{N^3 - N} \qquad \textbf{(Equation 17.3)}$$

Where: s = The number of sets of ties
t_i = The number of tied scores in the ith set of ties

The notation $\Sigma_{i=1}^{s}(t_i^3 - t_i)$ indicates the following: a) For each set of ties, the number of ties in the set is subtracted from the cube of the number of ties in that set; and b) The sum of all the values computed in part a) is obtained. The correction for ties will now be computed for Example 17.1. In the latter example there are $s = 5$ sets of ties (i.e., three scores of 5, three scores of 7, four scores of 8, two scores of 9, and two scores of 10). Thus:

$$\sum_{i=1}^{s} (t_i^3 - t_i) = [(3)^3 - 3] + [(3)^3 - 3] + [(4)^3 - 4] + [(2)^3 - 2] + [(2)^3 - 2] = 120$$

Employing Equation 17.3, the value $C = .964$ is computed.

$$C = 1 - \frac{120}{(15)^3 - 15} = .964$$

H_C, which represents the tie corrected value of the **Kruskal–Wallis test** statistic, is computed with Equation 17.4.

$$H_C = \frac{H}{C} \qquad \textbf{(Equation 17.4)}$$

Employing Equation 17.4, the tie corrected value $H_C = 8.76$ is computed.

$$H_C = \frac{8.44}{.964} = 8.76$$

As is the case with $H = 8.44$ computed with Equation 17.1, the value $H_C = 8.76$ computed with Equation 17.4 is significant at the .05 level (since it is greater than $\chi^2_{.05} = 5.99$), but is not significant at the .01 level (since it is less than $\chi^2_{.01} = 9.21$). Although Equation 17.4 results in a slightly less conservative test than Equation 17.1, in this instance the two equations lead to identical conclusions with respect to the null hypothesis.

2. Pairwise comparisons following computation of the test statistic for the Kruskal–Wallis one-way analysis of variance by ranks Prior to reading this section the reader should review the discussion of comparisons in Section VI of the **single-factor between-subjects analysis of variance.** As is the case with the omnibus F value computed for the **single-factor between-subjects analysis of variance,** the H value computed with Equation 17.1 is based on an evaluation of all k groups. When the value of H is significant, it does not indicate whether just two or, in fact, more than two groups differ significantly from one another. In order to answer the latter question, it is necessary to conduct comparisons contrasting specific groups with one another. This section will describe methodologies that can be employed for conducting **simple/pairwise comparisons** followingthe computation of an H value.[3]

In conducting a simple comparison, the null hypothesis and nondirectional alternative hypothesis are as follows: H_0: $\theta_a = \theta_b$ versus H_1: $\theta_a \neq \theta_b$. In the aforementioned hypotheses, θ_a and θ_b represent the medians of the populations represented by the two groups involved in the comparison. The alternative hypothesis can also be stated directionally as follows: H_1: $\theta_a > \theta_b$ or H_1: $\theta_a < \theta_b$.

Various sources (e.g., Daniel (1990) and Siegel and Castellan (1988)) describe a comparison procedure for the **Kruskal–Wallis one-way analysis of variance by ranks** (described by Dunn (1964)), which is essentially the application of the **Bonferroni–Dunn method** described in Section VI of the **single-factor between-subjects analysis of variance** to the **Kruskal–Wallis test** model. Through use of Equation 17.5, the procedure allows a researcher to identify the minimum required difference between the means of the ranks of any two groups (designated as CD_{KW}) in order for them to differ from one another at the prespecified level of significance.

$$CD_{KW} = z_{adj}\sqrt{\frac{N(N+1)}{12}\left(\frac{1}{n_a} + \frac{1}{n_b}\right)} \qquad \textbf{(Equation 17.5)}$$

Where: n_a and n_b represent the number of subjects in each of the groups involved in the simple comparison

The value of z_{adj} is obtained from **Table A1 (Table of the Normal Distribution)** in the **Appendix.** In the case of a nondirectional alternative hypothesis, z_{adj} is the z value above which a proportion of cases corresponding to the value $\alpha_{FW}/2c$ falls (where c is the total number of comparisons that are conducted). In the case of a directional alternative hypothesis, z_{adj} is the z value above which a proportion of cases corresponding to the value α_{FW}/c falls. When all possible pairwise comparisons are made $c = [k(k-1)]/2$, and thus, $2c = k(k-1)$. In Example 17.1 the number of pairwise/simple comparisons that can be conducted are $c = [3(3-1)]/2 = 3$ — specifically, Group 1 versus Group 2, Group 1 versus Group 3, and Group 2 versus Group 3.

The value of z_{adj} will be a function of both the maximum **familywise Type I error rate** (α_{FW}) the researcher is willing to tolerate and the total number of comparisons that are conducted. When a limited number of comparisons are planned prior to collecting the data, most sources take the position that a researcher is not obliged to control the value of α_{FW}. In such a case, the **per comparison Type I error rate** (α_{PC}) will be equal to the prespecified value of alpha. When α_{FW} is not adjusted, the value of z_{adj} employed in Equation 17.5 will be the tabled critical z value that corresponds to the prespecified level of significance. Thus, if a nondirectional alternative hypothesis is employed and $\alpha = \alpha_{PC} = .05$, the tabled critical two-tailed .05 value $z_{.05} = 1.96$ is used to represent z_{adj} in Equation 17.5. If $\alpha = \alpha_{PC} = .01$, the tabled critical two-tailed .01 value $z_{.01} = 2.58$ is used in Equation 17.5. In the same respect, if a directional alternative hypothesis is employed, the tabled critical .05 and .01 one-tailed values $z_{.05} = 1.65$ and $z_{.01} = 2.33$ are used for z_{adj} in Equation 17.5.

When comparisons are not planned beforehand, it is generally acknowledged that the value of α_{FW} must be controlled so as not to become excessive. The general approach for controlling the latter value is to establish a **per comparison Type I error rate** which insures that α_{FW} will not exceed some maximum value stipulated by the researcher. One method for doing this (described under the **single-factor between-subjects analysis of variance** as the **Bonferroni–Dunn method**) establishes the **per comparison Type I error rate** by dividing the maximum value one will tolerate for the **familywise Type I error rate** by the total number of comparisons conducted. Thus, in Example 17.1, if one intends to conduct all three pairwise comparisons and wants to insure that α_{FW} does not exceed .05, $\alpha_{PC} = \alpha_{FW}/c = .05/3 = .0167$. The latter proportion is used to determine the value of z_{adj}. As noted earlier, if a directional alternative hypothesis is employed for a comparison, the value of z_{adj} employed in Equation 17.5 is the z value above which a proportion equal to $\alpha_{PC} = \alpha_{FW}/c$ of the cases falls. In **Table A1**, the z value that corresponds to the proportion .0167 is $z = 2.13$. By employing $z_{adj} = 2.13$ in Equation 17.5, one can be assured that within the "family" of three pairwise comparisons, α_{FW} will not exceed .05 (assuming all of the comparisons are directional). If a nondirectional alternative hypothesis is employed for all of the comparisons, the value of z_{adj} will be the z value above which a proportion equal to $\alpha_{FW}/2c = \alpha_{PC}/2$ of the cases falls. Since $\alpha_{PC}/2 = .0167/2 = .0083$, $z = 2.39$. By employing $z_{adj} = 2.39$ in Equation 17.5, one can be assured that α_{FW} will not exceed .05.[4]

In order to employ the CD_{KW} value computed with Equation 17.5, it is necessary to determine the mean rank for each of the k groups, and then compute the absolute value of the difference between the mean ranks of each pair of groups that are compared.[5] In Table 17.1 the following values for the mean ranks of the groups are computed: $\bar{R}_1 = 12.7$, $\bar{R}_2 = 6.2$, $\bar{R}_3 = 5.1$. Employing the latter values, Table 17.3 summarizes the difference scores between pairs of mean ranks.

Table 17.3 Difference Scores Between Pairs of Mean Ranks for Example 17.1

$\lvert\bar{R}_1 - \bar{R}_2\rvert$	=	$\lvert 12.7 - 6.2\rvert$	= 6.5
$\lvert\bar{R}_1 - \bar{R}_3\rvert$	=	$\lvert 12.7 - 5.1\rvert$	= 7.6
$\lvert\bar{R}_2 - \bar{R}_3\rvert$	=	$\lvert 6.2 - 5.1\rvert$	= 1.1

If any of the differences between mean ranks is equal to or greater than the CD_{KW} value computed with Equation 17.5, a comparison is declared significant. Equation 17.5 will now be employed to evaluate the nondirectional alternative hypothesis H_1: $\theta_a \neq \theta_b$ for all three pairwise comparisons. Since it will be assumed that the comparisons are unplanned and that the researcher does not want the value of α_{FW} to exceed .05, the value $z_{adj} = 2.39$ will be used in computing CD_{KW}.

$$CD_{KW} = (2.39)\sqrt{\frac{(15)(15+1)}{12}\left(\frac{1}{5}+\frac{1}{5}\right)} = (2.39)(2.83) = 6.76$$

The obtained value $CD_{KW} = 6.76$ indicates that any difference between the mean ranks of two groups that is equal to or greater than 6.76 is significant. With respect to the three pairwise comparisons, the only difference between mean ranks which is greater than $CD_{KW} = 6.76$ is $|\bar{R}_1 - \bar{R}_3| = 7.6$. Thus, we can conclude there is a significant difference between the performance of Group 1 and Group 3. Note that although $|\bar{R}_1 - \bar{R}_2| = 6.5$ is close to $CD_{KW} = 6.76$, it is not statistically significant unless the researcher is willing to tolerate a **familywise error rate** slightly above .05.[6]

An alternative strategy that can be employed for conducting pairwise comparisons for the **Kruskal-Wallis test model** is to use the **Mann-Whitney *U* test** for each comparison. Use of the latter test requires that the data for each pair of groups to be compared be rank-ordered, and that a separate U value be computed for that comparison. The exact distribution of the **Mann-Whitney test** statistic can only be used when the value of α_{PC} is equal to one of the probabilities documented in **Table A11 (Table of Critical Values for the Mann-Whitney *U* Statistic)** in the **Appendix**. When α_{PC} is a value other than those listed in **Table A11**, the normal approximation of the **Mann-Whitney *U* test** statistic must be employed.

When the **Mann-Whitney *U* test** is employed for the three pairwise comparisons, the following U values are computed: a) Group 1 versus Group 2: $U = 1$; b) Group 1 versus Group 3: $U = .5$; and c) Group 2 versus Group 3: $U = 10$. When the aforementioned U values are substituted in Equations 9.4 and 9.5 (the uncorrected and continuity corrected normal approximations for the **Mann-Whitney *U* test**), the following absolute z values are computed: a) Group 1 versus Group 2: $z = 2.40$ and $z = 2.30$; b) Group 1 versus Group 3: $z = 2.51$ and $z = 2.40$; and c) Group 2 versus Group 3: $z = .52$ and $z = .42$. If we want to evaluate a nondirectional alternative hypothesis and insure that α_{FW} does not exceed .05, the value of α_{PC} is set equal to .0167. **Table A11** cannot be employed, since it does not list two-tailed critical U values for $\alpha_{.0167}$. In order to evaluate the result of the normal approximation, we identify the tabled critical two-tailed .0167 z value in **Table A1**. In employing Equation 17.5 earlier in this section, we determined that the latter value is $z_{.0167} = 2.39$. Since the uncorrected values $z = 2.40$ (for the comparison Group 1 versus Group 2) and $z = 2.51$ (for the comparison Group 1 versus Group 3) computed with Equation 9.4 are greater than $z_{.0167} = 2.39$, the latter two comparisons are significant if we wish to insure that α_{FW} does not exceed .05. If the correction for continuity is employed, only the value $z = 2.40$ (for the comparison Group 1 versus Group 3), computed with Equation 9.5 is significant, since it exceeds $z_{.0167} = 2.39$. The Group 1 versus Group 2 comparison falls short of significance, since $z = 2.30$ is less than $z_{.0167} = 2.39$. Recollect that when Equation 17.5 is employed to conduct the same set of comparisons, the Group 1 versus Group 3 comparison is significant, whereas the Group 1 versus Group 2 comparison just falls short of significance. Thus, the result obtained with Equation 17.5 is identical to that obtained when the continuity corrected normal approximation of the **Mann-Whitney *U* test** is employed.

In the event the researcher elects not to control the value of α_{FW} and employs $\alpha_{PC} = .05$ in evaluating the three pairwise comparisons (once again assuming a nondirectional analysis), both the Group 1 versus Group 2 and Group 1 versus Group 3 comparisons are significant at the .05 level, regardless of which comparison procedure is employed. Specifically, both the uncorrected and corrected normal approximations are significant, since $z = 2.40$ and $z = 2.30$ (computed for the comparison Group 1 versus Group 2) and $z = 2.51$ and $z = 2.40$ (computed for the comparison Group 1 versus Group 3) are greater than the tabled critical two-tailed value $z_{.05} = 1.96$. Employing **Table A11**, we also determine that both the Group 1 versus Group 2 and Group 1 versus Group 3 comparisons are significant at the .05 level, since the computed values $U = 1$ and $U = .5$ are less than the tabled critical two-tailed

.05 value $U_{.05} = 2$ (based on $n_1 = 5$ and $n_2 = 5$). If Equation 17.5 is employed for the same set of comparisons, $CD_{KW} = (1.96)(2.83) = 5.55$.[7] Thus, if the latter equation is employed, the Group 1 versus Group 2 and Group 1 versus Group 3 comparisons are significant, since in both instances the difference between the mean ranks is greater than $CD_{KW} = 5.55$.

The above discussion of comparisons illustrates that, generally speaking, the results obtained with Equation 17.5 and the **Mann–Whitney *U* test** (as well as other comparison procedures that have been developed for the **Kruskal–Wallis one-way analysis of variance**) will be reasonably consistent with one another. As noted in the discussion of comparisons in Section VI of the **single-factor between-subjects analysis of variance,** in instances where two or more comparison procedures yield inconsistent results, the most effective way to clarify the status of the null hypothesis is to replicate a study one or more times. In the final analysis, the decision regarding which of the available comparison procedures to employ is usually not the most important issue facing the researcher conducting comparisons. The main issue is what maximum value one is willing to tolerate for α_{FW}. Additional sources on comparison procedures for the **Kruskal–Wallis test** model are Wike (1978) (who provides a comparative analysis of a number of different procedures), and Marascuilo and McSweeney (1977) (who describe a methodology for conducting complex comparisons).

The same logic employed for computing a confidence interval for a comparison described in Section VI of the **single-factor between-subjects analysis of variance** can be employed to compute a confidence interval for the **Kruskal–Wallis test** model. Specifically: Add to and subtract the computed value of CD_{KW} from the obtained difference between the two mean ranks involved in a comparison. Thus, $CI_{.95}$ (based on $\alpha_{FW} = .05$) for the comparison Group 1 versus Group 3 is computed as follows: $CI_{.95} = 7.6 \pm 6.76$. In other words, the researcher can be 95% confident that the mean of the ranks in the population represented by Group 1 is between .84 and 14.36 units larger than the mean of the ranks in the population represented by Group 3. Marascuilo and McSweeney (1977) provide a more detailed discussion of the computation of a confidence interval for the **Kruskal–Wallis test** model.

VII. Additional Discussion of the Kruskal–Wallis One-Way Analysis of Variance by Ranks

1. Exact tables of the Kruskal–Wallis distribution Although an exact probability value can be computed for obtaining a configuration of ranks that is equivalent to or more extreme than the configuration observed in the data evaluated with the **Kruskal–Wallis one-way analysis of variance by ranks,** the chi-square distribution is generally employed to estimate the latter probability. As the values of k and N increase, the chi-square distribution provides a more accurate estimate of the exact Kruskal–Wallis distribution. Although most sources employ the chi-square approximation regardless of the values of k and N, some sources recommend that exact tables be employed under certain conditions. Beyer (1968), Daniel (1990), and Siegel and Castellan (1988) provide exact Kruskal–Wallis probabilities for whenever $k = 3$ and the number of subjects in any of the samples is five or less. Use of the chi-square distribution for small sample sizes will generally result in a slight decrease in the power of the test (i.e., there is a higher likelihood of retaining a false null hypothesis). Thus, for small sample sizes, the tabled critical chi-square value should, in actuality, be a little lower than the value listed in **Table A4.**

In point of fact, the exact tabled critical H values for $k = 3$ and $n_j = 5$ are $H_{.05} = 5.78$ and $H_{.01} = 7.98$. If the latter critical values are employed, the value $H = 8.44$ computed for Example 17.1 is significant at both the .05 and .01 levels, since $H = 8.44$ is greater than both $H_{.05} = 5.78$ and $H_{.01} = 7.98$. Thus, in this instance the exact tables and the chi-square approximation do not yield identical results.

2. Equivalency of the Kruskal–Wallis one-way analysis of variance by ranks and the Mann–Whitney *U* test when *k* = 2 In Section I it is noted that when $k = 2$ the **Kruskal–Wallis one-way analysis of variance by ranks** will yield a result that is equivalent to that obtained with the **Mann–Whitney *U* test.** To be more specific, the **Kruskal–Wallis test** will yield a result that is equivalent to the normal approximation for the **Mann–Whitney *U* test** when the correction for continuity is not employed (i.e., the result obtained with Equation 9.4). In order to demonstrate the equivalency of the two tests, Equation 17.1 is employed below to analyze the data for Example 9.1, which was previously evaluated with the **Mann–Whitney *U* test.**

$$H = \frac{12}{(10)(10 + 1)}\left[\frac{(19)^2}{5} + \frac{(36)^2}{5}\right] - (3)(10 + 1) = 3.15$$

Employing Equation 17.2, $df = 2 - 1 = 1$. For $df = 1$, the tabled critical .05 and .01 chi-square values are $\chi^2_{.05} = 3.84$ and $\chi^2_{.01} = 6.63$. Since the obtained value $H = 3.15$ is less than $\chi^2_{.05} = 3.84$, the null hypothesis cannot be rejected.

Equation 9.4 yields the value $z = -1.78$ for the same set of data. When the latter value is squared, it yields a value that is equal to the H (chi-square) value computed with Equation 17.1 (i.e., $(z = -1.78)^2 = (\chi^2 = 3.15)$).[8] It is also the case that the square of the tabled critical z value employed for the normal approximation of the **Mann–Whitney *U* test** will always be equal to the tabled critical chi-square value employed for the **Kruskal–Wallis test** at the same level of significance. Thus, the square of the tabled critical two-tailed value $z_{.05} = 1.96$ employed for the normal approximation of the **Mann–Whitney *U* test** equals $\chi^2_{.05} = 3.84$ employed for the **Kruskal–Wallis test** (i.e., $(z = 1.96)^2 = (\chi^2 = 3.84)$).

3. Alternative nonparametric rank-order procedures for evaluating a design involving *k* independent samples In addition to the **Kruskal–Wallis one-way analysis of variance by ranks,** a number of other nonparametric procedures for two or more independent samples have been developed that can be employed with ordinal data. Among the more commonly cited alternative procedures are the following: a) The **normal scores tests** discussed in Section VII of the **Mann–Whitney *U* test** can be extended to three or more independent samples. Marascuilo and McSweeney (1977) describe the extension of normal scores tests developed by Terry and Hoeffding (Terry (1952)) and Van der Waerden (1953/1953) to a design involving k independent samples. Conover (1980) discusses the extension of the normal scores test developed by Bell and Doksum (1965) to the latter design; b) **The Jonckheere–Terpstra test for ordered alternatives** (Jonckheere (1954); Terpstra (1952)) can be employed when the alternative hypothesis for a k independent samples design specifies the rank-order of the k population medians. The latter test is described in Daniel (1990); c) **The median test for independent samples (Test 11e)** discussed in Section VII of the **Mann–Whitney *U* test** can be extended to three or more independent samples by dichotomizing k samples with respect to their median values, and evaluating the data with the **chi-square test for *r* × *c* tables (Test 11)**; and d) **The Kolmogorov–Smirnov two-sample test** (Kolmogorov (1933) and Smirnov (1939)), alluded to in Section VII of the **Mann–Whitney *U* test,** can be extended to three or more independent samples. The **Kolmogorov–Smirnov two-sample test** is discussed in Bradley (1968) and Conover (1980). Sheskin (1984) describes some of the aforementioned tests in greater detail, as well as citing additional procedures that can be employed for k independent samples designs involving rank-order data.

VIII. Additional Examples Illustrating the Use of the Kruskal–Wallis One-Way Analysis of Variance by Ranks

The **Kruskal–Wallis one-way analysis of variance by ranks** can be employed to evaluate any of the additional examples noted for the **single-factor between-subjects analysis of variance**, if the data for the latter examples are rank-ordered. In addition, the **Kruskal–Wallis test** can be used to evaluate the data for any of the additional examples noted for the ***t* test for two independent samples (Test 8)** and the **Mann–Whitney *U* test**. Examples 17.2 and 17.3 are two additional examples that can be evaluated with the **Kruskal–Wallis one-way analysis of variance by ranks**. Example 17.2 is an extension of Example 9.2 (evaluated with the **Mann–Whitney *U* test**) to a design involving $k = 3$ groups. In Example 17.2 (as well as Example 17.3) the original data are presented as ranks, rather than in an interval/ratio format.[9] Since the rank-orderings for Example 17.2 are identical to those employed in Example 17.1, it yields the same result. Example 17.3 (which is also an extension of Example 9.2) illustrates the use of the **Kruskal–Wallis one-way analysis of variance by ranks** with a design involving $k = 4$ groups and unequal sample sizes.

Example 17.2 *Doctor Radical, a math instructor at Logarithm University, has three classes in advanced calculus. There are five students in each class. The instructor uses a programmed textbook in* Class 1, *a conventional textbook in* Class 2, *and his own printed notes in* Class 3. *At the end of the semester, in order to determine if the type of instruction employed influences student performance, Dr. Radical has another math instructor, Dr. Root, rank the* 15 *students in the three classes with respect to math ability. The rankings of the students in the three classes follow:* **Class 1**: 9.5, 14.5, 12.5, 14.5, 12.5; **Class 2**: 6, 9.5, 3, 9.5, 3; *and* **Class 3**: 1, 9.5, 6, 3, 6 *(assume the lower the rank the better the student).*

Note that whereas in Example 17.1, Group 1 (the group with the highest sum of ranks) has the best performance, in Example 17.2, Class 3 (the class with the lowest sum of ranks) is evaluated as the best class. This is the case, since in Example 17.2 the lower a student's rank the better the student, whereas in Example 17.1 the lower a subject's rank the poorer the subject performed.

Example 17.3 *Doctor Radical, a math instructor at Logarithm University, has four classes in advanced calculus. There are six students in* Class 1, *seven students in* Class 2, *eight students in* Class 3, *and six students in* Class 4. *The instructor uses a programmed textbook in* Class 1, *a conventional textbook in* Class 2, *his own printed notes in* Class 3, *and no written instructional material in* Class 4. *At the end of the semester, in order to determine if the type of instruction employed influences student performance, Dr. Radical has another math instructor, Dr. Root, rank the* 27 *students in the four classes with respect to math ability. The rankings of the students in the four classes follow:* **Class 1**: 1, 2, 4, 6, 8, 9; **Class 2**: 10, 14, 18, 20, 21, 25, 26; **Class 3**: 3, 5, 7, 11, 12 16, 17, 22; **Class 4**: 13, 15, 19, 23, 24, 27 *(assume the lower the rank the better the student).*

Example 17.3 provides us with the following information: $n_1 = 6$, $n_2 = 7$, $n_3 = 8$, $n_4 = 6$, and $N = 27$. The sums of the ranks for the four groups are: $\Sigma R_1 = 30$, $\Sigma R_2 = 134$, $\Sigma R_3 = 93$, $\Sigma R_4 = 121$. Substituting the appropriate values in Equation 17.1, the value $H = 14.99$ is computed.

$$H = \frac{12}{(27)(27+1)}\left[\frac{(30)^2}{6} + \frac{(134)^2}{7} + \frac{(93)^2}{8} + \frac{(121)^2}{6}\right] - (3)(27+1) = 14.99$$

Employing Equation 17.2, $df = 4 - 1 = 3$. For $df = 3$, the tabled critical .05 and .01 chi-square values are $\chi^2_{.05} = 7.81$ and $\chi^2_{.01} = 11.34$. Since the obtained value $H = 14.99$ is greater than both of the aforementioned critical values, the null hypothesis can be rejected at both the .05 and .01 levels. Thus, one can conclude that the rankings for at least two of the four classes differed significantly from one another. Although multiple comparisons will not be conducted for this example, visual inspection of the data suggests that the rankings for Class 1 are dramatically superior to those for the other three classes.

References

Bell, C. B. and Doksum, K. A. (1965). Some new distribution-free statistics. **Annals of Mathematical Statistics**, 36, 203–214.

Beyer, W. H. (1968). **Handbook of tables for probability and statistics** (2nd ed.). Cleveland, OH: The Chemical Rubber Company.

Bradley, J. V. (1968). **Distribution-free statistical tests**. Englewood Cliffs, NJ: Prentice–Hall.

Conover, W. J. (1980). **Practical nonparametric statistics** (2nd ed.). New York: John Wiley and Sons.

Daniel, W. W. (1990). **Applied nonparametric statistics** (2nd ed.). Boston: PWS–Kent Publishing Company.

Dunn, O. J. (1964). Multiple comparisons using rank sums. **Technometrics**, 6, 241–252.

Jonckheere, A. R. (1954). A distribution-free k sample test against ordered alternatives. **Biometrika**, 41, 133–145.

Kolmogorov, A. N. (1933). Sulla determinazione empiraca di una legge di distribuzione. **Giorn dell'Inst. Ital. degli. Att.**, 4, 89–91.

Kruskal, W. H. (1952). A nonparametric test for the several sample problem. **Annals of Mathematical Statistics**, 23, 525–540.

Kruskal, W. H. and Wallis, W. A. (1952). Use of ranks in one-criterion variance analysis. **Journal of the American Statistical Association**, 47, 583–621.

Marascuilo, L. A. and McSweeney, M. (1977). **Nonparametric and distribution-free methods for the social sciences**. Monterey, CA: Brooks/Cole Publishing Company.

Maxwell, S. and Delaney, H. (1990). **Designing experiments and analyzing data**. Belmont, CA: Wadsworth Publishing Company.

Sheskin, D. J. (1984). **Statistical tests and experimental design: A guidebook.** New York: Gardner Press.

Siegel, S. and Castellan, N. J., Jr. (1988). **Nonparametric statistics for the behavioral sciences** (2nd ed.). New York: McGraw–Hill Book Company.

Smirnov, N. V. (1939). On the estimation of the discrepancy between empirical curves of distributions for two independent samples. **Bulletin University of Moscow**, 2, 3–14.

Terpstra, T. J. (1952). The asymptotic normality and consistency of Kendall's test against trend, when ties are present in one ranking. **Indagationes Mathematicae**, 14, 327–333.

Terry, M. E. (1952). Some rank-order tests, which are most powerful against specific parametric alternatives. **Annals of Mathematical Statistics**, 23, 346–366.

Van der Waerden, B. L. (1952/1953). Order tests for the two-sample problem and their power. **Proceedings Koninklijke Nederlandse Akademie van Wetenshappen** (A), 55 (**Indagationes Mathematicae** 14), 453–458, and 56 (**Indagationes Mathematicae**, 15), 303–316 (corrections appear in Vol. 56, p. 80).

Wike, E. L. (1978). A Monte Carlo investigation of four nonparametric multiple-comparison tests for k independent samples. **Bulletin of the Psychonomic Society**, 11, 25–28.

Wike, E. L. (1985). **Numbers: A primer of data analysis**. Columbus, OH: Charles E. Merrill Publishing Company.

Endnotes

1. Although it is possible to conduct a directional analysis, such an analysis will not be described with respect to the **Kruskal–Wallis one-way analysis of variance by ranks.** A discussion of a directional analysis when $k = 2$ can be found under the **Mann–Whitney *U* test.** A discussion of the evaluation of a directional alternative hypothesis when $k \geq 2$ can be found in Section VII of the **chi-square goodness-of-fit test (Test 5).** Although the latter discussion is in reference to analysis of a k independent samples design involving categorical data, the general principles regarding analysis of a directional alternative hypothesis when $k \geq 2$ are applicable to the **Kruskal–Wallis one-way analysis of variance by ranks.**

2. As noted in Section IV, the chi-square distribution provides an approximation of the **Kruskal–Wallis test** statistic. Although the chi-square distribution provides an excellent approximation of the Kruskal–Wallis sampling distribution, some sources recommend the use of exact probabilities for small sample sizes. Exact tables of the Kruskal–Wallis distribution are discussed in Section VII.

3. In the discussion of comparisons under the **single-factor between-subjects analysis of variance,** it is noted that a **simple** (also known as a **pairwise**) **comparison** is a comparison between any two groups in a set of k groups.

4. The rationale for the use of the proportions .0167 and .0083 in determining the appropriate value for z_{adj} is as follows. In the case of a one-tailed/directional analysis, the relevant probability/proportion employed is based on only one of the two tails of the normal distribution. Consequently, the proportion of the normal curve that is used to determine the value of z_{adj} will be a proportion that is equal to the value of α_{PC} in the appropriate tail of the distribution (which is designated in the alternative hypothesis). The value $z = 2.13$ is employed, since the proportion of cases that falls above $z = 2.13$ in the right tail of the distribution is .0167, and the proportion of cases that falls below $z = -2.13$ in the left tail of the distribution is .0167. In the case of a two-tailed/nondirectional analysis, the relevant probability/proportion employed is based on both tails of the distribution. Consequently, the proportion of the normal curve that is used to determine the value of z_{adj} will be a proportion that is equal to the value of $\alpha_{PC}/2$ in each tail of the distribution. The proportion $\alpha_{PC}/2 = .0167/2 = .0083$ is employed for a two-tailed/nondirectional analysis, since one-half of the proportion that comprises $\alpha_{PC} = .0167$ comes from the left tail of the distribution and the other half from the right tail. Consequently, the value $z = 2.39$ is employed, since the proportion of cases that falls above $z = 2.39$ in the right tail of the distribution is .0083, and the proportion of cases that falls below $z = -2.39$ in the left tail of the distribution is .0083.

5. It should be noted that when a directional alternative hypothesis is employed, the sign of the difference between the two mean ranks must be consistent with the prediction stated in the directional alternative hypothesis. When a nondirectional alternative hypothesis is employed, the direction of the difference between two mean ranks is irrelevant.

6. a) Many researchers would probably be willing to tolerate a somewhat higher familywise Type I error rate than .05. In such a case the difference $|\bar{R}_1 - \bar{R}_2| = 6.5$ will be significant, since the value of z_{adj} employed in Equation 17.5 will be less than $z = 2.39$, thus resulting in a lower value for CD_{KW}; b) When there are a large number of ties in the

data, a modified version of Equation 17.5 is recommended by some sources (e.g., Daniel (1990)), which reduces the value of CD_{KW} by a minimal amount.

7. In Equation 17.5 the value $z_{.05} = 1.96$ is employed for z_{adj}, and the latter value is multiplied by 2.83, which is the value computed for the term in the radical of the equation for Example 17.1.

8. The slight discrepancy is due to rounding off error, since the actual absolute value of z computed with Equation 9.4 is 1.7756.

9. In accordance with one of the assumptions noted in Section I for the **Kruskal–Wallis one-way analysis of variance by ranks**, in both Examples 17.2 and 17.3 it is assumed that Dr. Radical implicitly or explicitly evaluates the N students on a continuous interval/ratio scale prior to converting the data into a rank-order format.

Inferential Statistical Tests Employed with Two or More Dependent Samples (and Related Measures of Association/Correlation)

[illegible]

with Two or More Dependent Samples

[illegible]

[illegible]

[illegible]

[illegible]

[illegible]

Test 18

The Single-Factor Within-Subjects Analysis of Variance
(Parametric Test Employed with Interval/Ratio Data)

I. Hypothesis Evaluated with Test and Relevant Background Information

Hypothesis evaluated with test In a set of k dependent samples (where $k \geq 2$), do at least two of the samples represent populations with different mean values?

Relevant background information on test Prior to reading this section the reader should review the general comments on the analysis of variance in Section I of the **single-factor between-subjects analysis of variance (Test 16)**. In addition, the general information regarding a dependent samples design contained in Sections I and VII of the ***t* test for two dependent samples (Test 12)** should be reviewed. The **single-factor within-subjects analysis of variance** (which is also referred to as the **single-factor repeated-measures analysis of variance** and the **randomized-blocks one-way analysis of variance**)[1] is employed in a hypothesis testing situation involving k dependent samples. In contrast to the ***t* test for two dependent samples**, which only allows for a comparison of the means of two dependent samples, the **single-factor within-subjects analysis of variance** allows for comparison of two or more dependent samples. When the number of dependent samples is $k = 2$, the **single-factor within-subjects analysis of variance** and the ***t* test for two dependent samples** yield equivalent results.

In conducting the **single-factor within-subjects analysis of variance**, each of the k sample means is employed to estimate the value of the mean of the population the sample represents. If the computed test statistic is significant, it indicates there is a significant difference between at least two of the sample means in the set of k means. As a result of the latter, the researcher can conclude there is a high likelihood that at least two of the samples represent populations with different mean values.

In order to compute the test statistic for the **single-factor within-subjects analysis of variance**, the following two variability components (which are part of the **total variability**) are contrasted with one another: **between-conditions variability** and **residual variability**. **Between-conditions variability** (which is also referred to as **treatment variability**) is essentially a measure of the variance of the means of the k experimental conditions. **Residual variability** is the amount of variability within the k scores of each of the n subjects which cannot be accounted for on the basis of a treatment effect. **Residual variability** is viewed as variability that results from chance factors that are beyond the control of a researcher, and since chance factors are often referred to as experimental error, **residual variability** is also referred to as **error variability**. The ***F* ratio**, which is the test statistic for the **single-factor within-subjects analysis of variance**, is obtained by dividing **between-conditions variability** by **residual variability**. Since **residual variability** is employed as a baseline measure of the variability in a set of data that is beyond a researcher's control, it is assumed that if the k experimental conditions represent populations with the same mean value, the amount of variability between the means of the k experimental conditions (i.e., **between-conditions variability**) will be approximately the same value as the **residual variability**. If, on the other hand, **between-conditions variability** is significantly larger than **residual variability** (in which case the value of the F ratio will be larger than 1), it is likely that something in addition to chance factors is contributing to the amount of variability between the means of the experimental conditions. In such a case, it is assumed that whatever it is that differentiates the experimental conditions from one another (i.e., the independent variable/ experimental treatments) accounts for the fact that **between-conditions variability** is larger than **residual variability**.[2] A thorough discussion of the logic underlying the **single-factor within-subjects analysis of variance** can be found in Section VII.

The **single-factor within-subjects analysis of variance** is employed with interval/ratio data and is based on the following assumptions: a) The sample of n subjects has been randomly selected from the population it represents; b) The distribution of data in the underlying populations each of the experimental conditions represents is normal; and c) The third assumption, which is referred to as the **sphericity assumption**, is the analog of the homogeneity of variance assumption of the **single-factor between-subjects analysis of variance.** The assumption of sphericity, which is mathematically more complex than the homogeneity of variance assumption, essentially revolves around the issue of whether or not the underlying population variances and covariances are equal. A full discussion of the sphericity assumption (as well as the concept of covariance) can be found in Section VI. It should also be noted that the **single-factor within-subjects analysis of variance** is more sensitive to violations of its assumptions than is the **single-factor between-subjects analysis of variance.**

As is the case for the ***t* test for two dependent samples**, in order for the **single-factor within-subjects analysis of variance** to generate valid results the following guidelines should be adhered to: a) In order to control for order effects, the presentation of the k experimental conditions should be random or, if appropriate, be counterbalanced;[3] and b) If matched samples are employed, within each set of matched subjects each of the subjects should be randomly assigned to one of the k experimental conditions.

As is the case with the ***t* test for dependent samples**, when $k = 2$ the **single-factor within-subjects analysis of variance** can be employed to evaluate a **before-after design**, as well as extensions of the latter design that involve more than two measurement periods. The limitations of the **before-after design** (which are discussed in Section VII of the ***t* test for dependent samples**) are also applicable when it is evaluated with the **single-factor within-subjects analysis of variance.**

The reader should take note of the fact that there are certain advantages associated with employing a within-subjects design as opposed to a between-subjects design. If within-subjects and between-subjects designs that evaluate the same hypothesis and involve the same number of scores in each of the experimental conditions are compared with one another, the number of subjects required for the within-subjects analysis is a fraction (specifically $1/k^{th}$) of the number required for the between-subjects analysis. Another advantage of a within-subjects analysis is that it provides for a more powerful test of an alternative hypothesis. The latter can be attributed to the fact that the error variability associated with a within-subjects analysis is less than that associated with a between-subjects analysis. In spite of the aforementioned advantages of employing a within-subjects design, the between-subjects design is more commonly employed in research, since in many experiments it is impractical for a subject to serve in more than one experimental condition.

II. Example

Example 18.1 *A psychologist conducts a study to determine whether noise can inhibit learning. Each of six subjects is tested under three experimental conditions. In each of the experimental conditions a subject is given* 20 *minutes to memorize a list of* 10 *nonsense syllables, which the subject is told he or she will be tested on the following day. The three experimental conditions each subject serves under are as follows:* **Condition 1**, *the* **no noise**

condition, requires subjects to study the list of nonsense syllables in a quiet room. **Condition 2,** *the* **moderate noise** *condition, requires subjects to study the list of nonsense syllables while listening to classical music.* **Condition 3,** *the* **extreme noise** *condition, requires subjects to study the list of nonsense syllables while listening to rock music. Although in each of the experimental conditions subjects are presented with a different list of nonsense syllables, the three lists are comparable with respect to those variables that are known to influence a person's ability to learn nonsense syllables. In order to control for order effects, the order of presentation of the three experimental conditions is completely counterbalanced.*[4] *The number of nonsense syllables correctly recalled by the six subjects under the three experimental conditions follow. (Subjects' scores are listed in the order* **Condition 1, Condition 2, Condition 3.**) **Subject 1:** 9, 7, 4; **Subject 2:** 10, 8, 7; **Subject 3:** 7, 5, 3; **Subject 4:** 10, 8, 7; **Subject 5:** 7, 5, 2; **Subject 6:** 8, 6, 6. *Do the data indicate that noise influenced subjects' performance?*

III. Null versus Alternative Hypotheses

Null hypothesis H_0: $\mu_1 = \mu_2 = \mu_3$

(The mean of the population Condition 1 represents equals the mean of the population Condition 2 represents equals the mean of the population Condition 3 represents.)

Alternative hypothesis H_1: Not H_0

(This indicates that there is a difference between at least two of the k population means. It is important to note that the alternative hypothesis should not be written as follows: H_1: $\mu_1 \neq \mu_2 \neq \mu_3$. The reason why the latter notation for the alternative hypothesis is incorrect is because it implies that all three population means must differ from one another in order to reject the null hypothesis. In this book it will always be assumed that the alternative hypothesis for the analysis of variance is stated **nondirectionally**.[5] In order to reject the null hypothesis, the obtained F value must be equal to or greater than the tabled critical F value at the prespecified level of significance.)

IV. Test Computations

The test statistic for the **single-factor within-subjects analysis of variance** can be computed with either **computational** or **definitional equations.** Although definitional equations reveal the underlying logic behind the analysis of variance, they involve considerably more calculations that do the computational equations. Because of the latter, computational equations will be employed in this section to demonstrate the computation of the test statistic. The definitional equations for the **single-factor within-subjects analysis of variance** are described in Section VII.

The data for Example 18.1 are summarized in Table 18.1. In Table 18.1 the $k = 3$ scores of the $n = 6$ subjects in Conditions 1, 2, and 3 are respectively listed in the columns labelled X_1, X_2, and X_3. The notation n is employed to represent the number of scores in each of the experimental conditions. Since there are $n = 6$ scores in each condition, $n = n_1 = n_2 = n_3 = 6$. The columns labelled X_1^2, X_2^2, and X_3^2 list the squares of the scores of the six subjects in each of the three experimental conditions. The last column labelled ΣS_i lists for each of the six subjects the sum of a subject's $k = 3$ scores. Thus, the value ΣS_i is the sum of the scores for Subject i under Conditions 1, 2, and 3.

Table 18.1 Data for Example 18.1

	Condition 1		Condition 2		Condition 3		
	X_1	X_1^2	X_2	X_2^2	X_3	X_3^2	ΣS_i
Subject 1	9	81	7	49	4	16	20
Subject 2	10	100	8	64	7	49	25
Subject 3	7	49	5	25	3	9	15
Subject 4	10	100	8	64	7	49	25
Subject 5	7	49	5	25	2	4	14
Subject 6	8	64	6	36	6	36	20
	$\Sigma X_1 = 51$	$\Sigma X_1^2 = 443$	$\Sigma X_2 = 39$	$\Sigma X_2^2 = 263$	$\Sigma X_3 = 29$	$\Sigma X_3^2 = 163$	$\Sigma X_T = 119$
	$\bar{X}_1 = \frac{\Sigma X_1}{n_1} = \frac{51}{6} = 8.5$		$\bar{X}_2 = \frac{\Sigma X_2}{n_2} = \frac{39}{6} = 6.5$		$\bar{X}_3 = \frac{\Sigma X_3}{n_3} = \frac{29}{6} = 4.83$		

The notation N represents the total number of scores in the experiment. Since there are $n = 6$ subjects and each subject has $k = 3$ scores, there are a total of $nk = N = (6)(3) = 18$ scores. The value ΣX_T represents the sum of the $N = 30$ scores (i.e., the total sum of scores). Thus:

$$\Sigma X_T = \Sigma X_1 + \Sigma X_2 + \cdots + \Sigma X_k$$

Since there are $k = 3$ experimental conditions, $\Sigma X_T = 119$.

$$\Sigma X_T = \Sigma X_1 + \Sigma X_2 + \Sigma X_3 = 51 + 39 + 29 = 119$$

The value ΣX_T can also be computed by adding up the ΣS_i scores computed for the n subjects. Thus:

$$\Sigma X_T = \Sigma S_1 + \Sigma S_2 + \cdots + \Sigma S_n$$

$$\Sigma X_T = \Sigma S_1 + \Sigma S_2 + \Sigma S_3 + \Sigma S_4 + \Sigma S_5 + \Sigma S_6$$

$$= 20 + 25 + 15 + 25 + 14 + 20 = 119$$

$\bar{X}_T$ represents the grand mean, where $\bar{X}_T = \Sigma X_T/N$. Thus, $\bar{X}_T = 119/18 = 6.61$. Although $\bar{X}_T$ is not employed in the computational equations to be described in this section, it is employed in some of the definitional equations described in Section VII.

The value ΣX_T^2 represents the total sum of the N squared scores. Thus:

$$\Sigma X_T^2 = \Sigma X_1^2 + \Sigma X_2^2 + \cdots + \Sigma X_k^2$$

Since there are $k = 3$ experimental conditions, $\Sigma X_T^2 = 869$.

$$\Sigma X_T^2 = \Sigma X_1^2 + \Sigma X_2^2 + \Sigma X_3^2 = 443 + 263 + 163 = 869$$

Although the means for each of the experimental conditions are not required for computing the analysis of variance test statistic, it is recommended that they be computed since visual inspection of the condition means can provide the researcher with a general idea of whether or not it is reasonable to expect a significant result. To be more specific, if two or more of the condition means are far removed from one another, it is likely that the analysis of variance will be significant (especially if there are a relatively large number of subjects employed in the experiment). Another reason for computing the condition means is that they are required for comparing individual conditions with one another, which is something that is generally done following the analysis of variance on the full set of data. The latter types of comparisons are described in Section VI.

In order to compute the test statistic for the **single-factor within-subjects analysis of variance,** the **total variability** in the data is divided into a number of different components. Specifically, the following variability components are computed: a) The **total sum of squares** which is represented by the notation SS_T; b) The **between-conditions sum of squares** which is represented by the notation SS_{BC}. The **between-conditions sum of squares** is the numerator of the equation that represents **between-conditions variability** (i.e., the equation that represents the amount of variability between the means of the k conditions); c) The **between-subjects sum of squares** is represented by the notation SS_{BS}. The **between-subjects sum of squares** is the numerator of the equation that represents **between-subjects variability,** which is the amount of variability between the mean scores of the n subjects (the mean of each subject being the average of a subject's k scores); and d) The **residual sum of squares** is represented by the notation SS_{res}. The **residual sum of squares** is the numerator of the equation that represents **residual variability** (i.e., error variability that is beyond the researcher's control).

Equation 18.1 describes the relationship between SS_T, SS_{BC}, SS_{BS}, and SS_{res}.[6]

$$SS_T = SS_{BC} + SS_{BS} + SS_{\text{res}} \qquad \textbf{(Equation 18.1)}$$

Equation 18.2 is employed to compute SS_T.

$$SS_T = \Sigma X_T^2 - \frac{(\Sigma X_T)^2}{N} \qquad \textbf{(Equation 18.2)}$$

Employing Equation 18.2, the value $SS_T = 82.28$ is computed.

$$SS_T = 869 - \frac{(119)^2}{18} = 82.28$$

Equation 18.3 is employed to compute SS_{BC}. In Equation 18.3 the notation ΣX_j represents the sum of the n scores in the jth condition. Note that in Equation 18.3 the notation n_j can be employed in place of n, since there are $n = n_j$ scores in the jth condition.

$$SS_{BC} = \sum_{j=1}^{k} \left[\frac{(\Sigma X_j)^2}{n}\right] - \frac{(\Sigma X_T)^2}{N} \qquad \textbf{(Equation 18.3)}$$

The notation $\Sigma_{j=1}^{k}[(\Sigma X_j)^2/n]$ in Equation 18.3 indicates that for each condition the value $(\Sigma X_j)^2/n$ is computed, and the latter values are summed for all k conditions.

With reference to Example 18.1, Equation 18.3 can be rewritten as follows:

$$SS_{BC} = \left[\frac{(\Sigma X_1)^2}{n} + \frac{(\Sigma X_2)^2}{n} + \frac{(\Sigma X_3)^2}{n}\right] - \frac{(\Sigma X_T)^2}{N}$$

Substituting the appropriate values from Example 18.1 in Equation 18.3, the value $SS_{BC} = 40.45$ is computed.[7]

$$SS_{BC} = \left[\frac{(51)^2}{6} + \frac{(39)^2}{6} + \frac{(29)^2}{6}\right] - \frac{(119)^2}{18} = 40.45$$

Equation 18.4 is employed to compute SS_{BS}.

$$SS_{BS} = \sum_{i=1}^{n} \left[\frac{(\Sigma S_i)^2}{k} \right] - \frac{(\Sigma X_T)^2}{N} \qquad \textbf{(Equation 18.4)}$$

The notation $\Sigma_{i=1}^{n}[(\Sigma S_i)^2/k]$ in Equation 18.4 indicates that for each subject the value $(\Sigma S_i)^2/k$ is computed, and the latter values are summed for all n subjects. With reference to Example 18.1, Equation 18.4 can be rewritten as follows:

$$SS_{BS} = \left[\frac{(\Sigma S_1)^2}{k} + \frac{(\Sigma S_2)^2}{k} + \frac{(\Sigma S_3)^2}{k} + \frac{(\Sigma S_4)^2}{k} + \frac{(\Sigma S_5)^2}{k} + \frac{(\Sigma S_6)^2}{k} \right] - \frac{(\Sigma X_T)^2}{N}$$

Substituting the appropriate values from Example 18.1 in Equation 18.4, the value $SS_{BS} = 36.93$ is computed.[8]

$$SS_{BS} = \left[\frac{(20)^2}{3} + \frac{(25)^2}{3} + \frac{(15)^2}{3} + \frac{(25)^2}{3} + \frac{(14)^2}{3} + \frac{(20)^2}{3} \right] - \frac{(119)^2}{18} = 36.93$$

By algebraically transposing the terms in Equation 18.1, the value of SS_{res} can be computed with Equation 18.5.

$$SS_{\text{res}} = SS_T - SS_{BC} - SS_{BS} \qquad \textbf{(Equation 18.5)}$$

Employing Equation 18.5, the value $SS_{\text{res}} = 4.9$ is computed.

$$SS_{\text{res}} = 82.28 - 40.45 - 36.93 = 4.9$$

Equation 18.6 is a computationally more complex equation for computing the value of SS_{res}.

$$SS_{\text{res}} = \Sigma X_T^2 - \sum_{j=1}^{k} \left[\frac{(\Sigma X_j)^2}{n} \right] - \sum_{i=1}^{n} \left[\frac{(\Sigma S_i)^2}{k} \right] + \frac{(\Sigma X_T)^2}{N} \qquad \textbf{(Equation 18.6)}$$

Since $\Sigma X_T^2 = 869$, $\Sigma_{j=1}^{k}[(\Sigma X_j)^2/n] = 827.17$, $\Sigma_{i=1}^{n}[(\Sigma S_i)^2/k] = 823.65$, and $(\Sigma X_T)^2/N = 786.72$, employing Equation 18.6, the value $SS_{\text{res}} = 4.9$ is computed.

$$SS_{\text{res}} = 869 - 827.17 - 823.65 + 786.72 = 4.9$$

The reader should take note of the fact that the values SS_T, SS_{BC}, SS_{BS}, and SS_{res} must always be positive numbers. If a negative value is obtained for any of the aforementioned values, it indicates a computational error has been made.

At this point the values of the **between-conditions variance**, **between-subjects variance**, and the **residual variance** can be computed. In the **single-factor within-subjects analysis of variance**, the **between-conditions variance** is referred to as the **mean square between-conditions**, which is represented by the notation MS_{BC}. MS_{BC} is computed with Equation 18.7.

$$MS_{BC} = \frac{SS_{BC}}{df_{BC}} \qquad \textbf{(Equation 18.7)}$$

The **between-subjects variance** is referred to as the **mean square between-subjects,** which is represented by the notation MS_{BS}. MS_{BS} is computed with Equation 18.8.

$$MS_{BS} = \frac{SS_{BS}}{df_{BS}} \qquad \textbf{(Equation 18.8)}$$

The **residual variance** is referred to as the **mean square residual**, which is represented by the notation MS_{res}. MS_{res} is computed with Equation 18.9.

$$MS_{\text{res}} = \frac{SS_{\text{res}}}{df_{\text{res}}} \qquad \textbf{(Equation 18.9)}$$

Note that a total mean square is not computed.

In order to compute MS_{BC}, MS_{BS}, and MS_{res}, it is required that the values df_{BC}, df_{BS}, and df_{res} (the denominators of Equations 18.7–18.9) be computed. df_{BC}, which represents the **between-conditions degrees of freedom**, are computed with Equation 18.10.

$$df_{BC} = k - 1 \qquad \textbf{(Equation 18.10)}$$

df_{BS}, which represents the **between-subjects degrees of freedom**, are computed with Equation 18.11.

$$df_{BS} = n - 1 \qquad \textbf{(Equation 18.11)}$$

df_{res}, which represents the **residual degrees of freedom**, are computed with Equation 18.12.

$$df_{\text{res}} = (n - 1)(k - 1) \qquad \textbf{(Equation 18.12)}$$

Although it is not required to determine the F ratio, the **total degrees of freedom** are generally computed, since it can be used to confirm the df values computed with Equations 18.10–18.12, as well as the fact it is employed in the analysis of variance summary table. The total degrees of freedom (represented by the notation df_T), are computed with Equation 18.13.

$$df_T = nk - 1 = N - 1 \qquad \textbf{(Equation 18.13)}$$

The relationship between df_{BC}, df_{BS}, df_{res}, and df_T is described by Equation 18.14.

$$df_T = df_{BC} + df_{BS} + df_{\text{res}} \qquad \textbf{(Equation 18.14)}$$

Employing Equations 18.10–18.13, the values $df_{BC} = 2$, $df_{BS} = 5$, $df_{\text{res}} = 10$, and $df_T = 17$ are computed. Note that $df_T = df_{BC} + df_{BS} + df_{\text{res}} = 2 + 5 + 10 = 17$.

$$df_{BC} = 3 - 1 = 2 \qquad df_{BS} = 6 - 1 = 5$$

$$df_{\text{res}} = (6 - 1)(3 - 1) = 10 \qquad df_T = 18 - 1 = 17$$

Employing Equations 18.7–18.9, the values $MS_{BC} = 20.23$, $MS_{BS} = 7.39$, and $MS_{\text{res}} = .49$ are computed.

$$MS_{BC} = \frac{40.45}{2} = 20.23 \qquad MS_{BS} = \frac{36.93}{5} = 7.39 \qquad MS_{\text{res}} = \frac{4.9}{10} = .49$$

The F ratio, which is the test statistic for the **single-factor within-subjects analysis of variance,** is computed with Equation 18.15.

$$F = \frac{MS_{BC}}{MS_{\text{res}}} \qquad \textbf{(Equation 18.15)}$$

Employing Equation 18.15, the value $F = 41.29$ is computed.

$$F = \frac{20.23}{.49} = 41.29$$

The reader should take note of the fact that the values MS_{BC}, MS_{BS}, and MS_{res} must always be positive numbers. If a negative value is obtained for any of the aforementioned values, it indicates a computational error has been made. If $MS_{\text{res}} = 0$, Equation 18.15 will be insoluble. If all of the conditions have the identical mean value, $MS_{BC} = 0$, and if the latter is true, $F = 0$.

V. Interpretation of the Test Results

It is common practice to summarize the results of a **single-factor within-subjects analysis of variance** with the summary table represented by Table 18.2.

Table 18.2 Summary Table of Analysis of Variance for Example 18.1

Source of variation	*SS*	*df*	*MS*	*F*
Between-subjects	36.93	5	7.39	
Between-conditions	40.45	2	20.23	41.29
Residual	4.90	10	.49	
Total	**82.28**	**17**		

The obtained value $F = 41.29$ is evaluated with **Table A10 (Table of the F Distribution)** in the **Appendix.** In **Table A10** critical values are listed in reference to the number of degrees of freedom associated with the numerator and the denominator of the F ratio (i.e., df_{num} and df_{den}). In employing the F distribution in reference to Example 18.1, the degrees of freedom for the numerator are $df_{BC} = 2$ and the degrees of freedom for the denominator are $df_{\text{res}} = 10$. In **Table A10** the tabled $F_{.95}$ and $F_{.99}$ values are respectively employed to evaluate the nondirectional alternative hypothesis H_1: Not H_0 at the .05 and .01 levels. As is the case for the **single-factor between-subjects analysis of variance,** the notation $F_{.05}$ is employed to represent the tabled critical F value at the .05 level. The latter value corresponds to the relevant tabled $F_{.95}$ value in **Table A10.** In the same respect, the notation $F_{.01}$ is employed to represent the tabled critical F value at the .01 level, and corresponds to the relevant tabled $F_{.99}$ value in **Table A10.**

For $df_{\text{num}} = 2$ and $df_{\text{den}} = 10$, the tabled $F_{.95}$ and $F_{.99}$ values are $F_{.95} = 4.10$ and $F_{.99} = 7.56$. Thus, $F_{.05} = 4.10$ and $F_{.01} = 7.56$. In order to reject the null hypothesis, the obtained F value must be equal to or greater than the tabled critical value at the prespecified level of significance. Since $F = 41.29$ is greater than both $F_{.05} = 4.10$ and $F_{.01} = 7.56$, the alternative hypothesis is supported at both the .05 and .01 levels.

A summary of the analysis of Example 18.1 with the **single-factor within-subjects analysis of variance** follows: It can be concluded that there is a significant difference between at least two of the three experimental conditions (i.e., different levels of noise). This result can be summarized as follows: $F(2,10) = 41.29, p < .01$.

VI. Additional Analytical Procedures for the Single-Factor Within-Subjects Analysis of Variance and/or Related Tests

1. Comparisons following computation of the omnibus F value for the single-factor within-subjects analysis of variance Prior to reading this section the reader should review the discussion of comparison procedures in Section VI of the **single-factor between-subjects analysis of variance.** As is the case with the latter test, the omnibus F value computed for a **single-factor within-subjects analysis of variance** is based on a comparison of the means of all k experimental conditions. Thus, in order to reject the null hypothesis, it is only required that the means of at least two of the k conditions differ significantly from one another.[9]

The same procedures that are employed for conducting comparisons for the **single-factor between-subjects analysis of variance** can be used for the **single-factor within-subjects analysis of variance.** Thus, the following comparison procedures discussed under the latter test can be employed for conducting comparisons within the framework of the **single-factor within-subjects analysis of variance: Test 18a: Multiple *t* tests/Fisher's LSD test** (which is equivalent to **linear contrasts**); **Test 18b: The Bonferroni–Dunn test; Test 18c: Tukey's HSD test; Test 18d: The Newman–Keuls test; Test 18e: The Scheffé test; Test 18f: The Dunnett test.** The only difference in applying the aforementioned comparison procedures to the analysis of variance under discussion is that a different measure of error variability is employed. Recollect that in the case of the **single-factor between-subjects analysis variance**, a pooled measure of within-groups variability (MS_{WG}) is employed as the measure of experimental error. In the case of the **single-factor within-subjects analysis of variance**, MS_{res} is employed as the measure of error variability. Consequently, in conducting comparisons for a **single-factor within-subjects analysis of variance**, MS_{res} is employed in place of MS_{WG} as the error term in the comparison equations described in Section VI of the **single-factor between-subjects analysis variance.** It should be noted, however, that if the sphericity assumption (which, as noted in Section I, is based on homogeneity of the underlying population variances and covariances) of the **single-factor within-subjects analysis of variance** is violated, MS_{res} may not provide the most accurate measure of error variability to employ in conducting comparisons. Because of this, an alternative measure of error variability (that is not influenced by violation of the sphericity assumption) will be presented later in this section.

At this point, employing MS_{res} as the error term, the following two **single degree of freedom comparisons** will be conducted: a) The **simple comparison** Condition 1 versus Condition 2, which is summarized in Table 18.3; and b) The **complex comparison** Condition 3 versus the combined performance of Conditions 1 and 2, which is summarized in Table 18.4.

Table 18.3 Planned Simple Comparison: Condition 1 Versus Condition 2

Condition	$\bar{X}_j$	Coefficient (c_j)	Product $(c_j)(\bar{X}_j)$	Squared Coefficient (c_j^2)
1	8.5	+1	$(+1)(8.5) = +8.5$	1
2	6.5	−1	$(-1)(6.5) = -6.5$	1
3	4.83	0	$(0)(4.83) = 0$	0
		$\Sigma c_j = 0$	$\Sigma(c_j)(\bar{X}_j) = 2$	$\Sigma c_j^2 = 2$

Since it will be assumed that the above comparisons are planned prior to collecting the data, **linear contrasts** will be conducted in which no attempt is made to control the value of the **familywise Type I error rate** (α_{FW}). The null hypothesis and nondirectional alternative hypothesis for the above comparisons are identical to those employed when the analogous comparisons are conducted in Section VI for the **single-factor between-subjects analysis variance** (i.e., H_0: $\mu_1 = \mu_2$ versus H_1: $\mu_1 \neq \mu_2$ for the simple comparison, and H_0: $\mu_3 = (\mu_1 + \mu_2)/2$ versus H_1: $\mu_3 \neq (\mu_1 + \mu_2)/2$ for the complex comparison).

Table 18.4 Planned Complex Comparison: Condition 3 Versus Conditions 1 and 2

Condition	$\bar{X}_j$	Coefficient (c_j)	Product $(c_j)(\bar{X}_j)$	Squared Coefficient (c_j^2)
1	8.5	$-\frac{1}{2}$	$\left(-\frac{1}{2}\right)(8.5) = -4.25$	$\frac{1}{4}$
2	6.5	$-\frac{1}{2}$	$\left(-\frac{1}{2}\right)(6.5) = -3.25$	$\frac{1}{4}$
3	4.83	$+1$	$(+1)(4.83) = +4.83$	1
		$\Sigma c_j = 0$	$\Sigma(c_j)(\bar{X}_j) = -2.67$	$\Sigma c_j^2 = 1.5$

Equations 16.17 and 16.18, which are employed to compute SS_{comp} and MS_{comp} for linear contrasts for the **single-factor between-subjects analysis of variance**, are also employed for linear contrasts for the **single-factor within-subjects analysis of variance**. The latter equations are employed below for the simple comparison Condition 1 versus Condition 2.

$$SS_{\text{comp}} = \frac{n\left[\Sigma(c_j)(\bar{X}_j)\right]^2}{\Sigma c_j^2} = \frac{6(2)^2}{2} = 12$$

$$MS_{\text{comp}} = \frac{SS_{\text{comp}}}{df_{\text{comp}}} = \frac{12}{1} = 12$$

Equation 18.16 (which is identical to Equation 16.19, except for the fact that it employs MS_{res} as the error term) is used to compute the value of F_{comp}.

$$F_{\text{comp}} = \frac{MS_{\text{comp}}}{MS_{\text{res}}} = \frac{12}{.49} = 24.49 \qquad \textbf{(Equation 18.16)}$$

The degrees of freedom employed in evaluating the obtained value $F_{\text{comp}} = 24.49$ are $df_{\text{num}} = df_{\text{comp}} = 1$ (since in a **single degree of freedom comparison** df_{comp} will always equal 1) and $df_{\text{den}} = df_{\text{res}} = 10$. For $df_{\text{num}} = 1$ and $df_{\text{den}} = 10$, the tabled critical .05 and .01 F values in **Table A10** are $F_{.05} = 4.96$ and $F_{.01} = 10.04$. Since the obtained value $F_{\text{comp}} = 24.49$ is greater than both of the aforementioned critical values, the nondirectional alternative hypothesis H_1: $\mu_1 \neq \mu_2$ is supported at both the .05 and .01 levels.

Applying Equations 16.17, 16.18, and 18.16 to the complex comparison Condition 3 versus Conditions 1 and 2, the following result is obtained.

$$SS_{\text{comp}} = \frac{n\left[\Sigma(c_j)(\bar{X}_j)\right]^2}{\Sigma c_j^2} = \frac{6(-2.67)^2}{1.5} = 28.52$$

$$MS_{\text{comp}} = \frac{SS_{\text{comp}}}{df_{\text{comp}}} = \frac{28.52}{1} = 28.52$$

$$F_{\text{comp}} = \frac{MS_{\text{comp}}}{MS_{\text{res}}} = \frac{28.52}{.49} = 58.20$$

As is the case for the simple comparison, the degrees of freedom employed for the complex comparison in evaluating the value $F_{\text{comp}} = 58.20$ are $df_{\text{num}} = df_{\text{comp}} = 1$ and $df_{\text{den}} = df_{\text{res}} = 10$. Since the obtained value $F_{\text{comp}} = 58.20$ is greater than the tabled critical values $F_{.05} = 4.96$ and $F_{.01} = 10.04$, the nondirectional alternative hypothesis H_1: $\mu_3 \neq (\mu_1 + \mu_2)/2$ is supported at both the .05 and .01 levels.

For both of the comparisons that have been conducted, a *CD* value can be computed that identifies a minimum required difference in order for two means (or sets of means) to differ from one another at a prespecified level of significance. For both the simple comparison Condition 1 versus Condition 2 and the complex comparison Condition 3 versus Conditions 1 and 2, a *CD* value will be computed employing **multiple *t* tests/Fisher's LSD test** (which is equivalent to a **linear contrast** in which the value of α_{FW} is not controlled) and the **Scheffé test**. Whereas **multiple *t* tests/Fisher's LSD test** allow a researcher to determine the minimum *CD* value (designated CD_{LSD}) that can be computed through use of any of the comparison procedures that are described for the **single-factor between-subjects analysis of variance**, the **Scheffé test** generally results in the maximum *CD* value (designated CD_S) that can be computed by the available procedures. Thus, if at a prespecified level of significance the obtained difference for a comparison is equal to or greater than CD_S, it will be significant regardless of which comparison procedure is employed. On the other hand, if the obtained difference for a comparison is less than CD_{LSD}, it will not achieve significance, regardless of which comparison procedure is employed. In illustrating the computation of *CD* values for both simple and complex comparisons, it will be assumed that the total number of comparisons conducted is $c = 3$.[10] It will also be assumed that when the researcher wants to control the value of the **familywise Type I error rate**, the value $\alpha_{FW} = .05$ is employed irrespective of which comparison procedure is used.

The equations employed for computing the values CD_{LSD} and CD_S are essentially the same as those used for the **single-factor between-subjects analysis of variance**. Equations 18.17 and 18.18 are employed below to compute the values of CD_{LSD} and CD_S for the simple comparison Condition 1 versus Condition 2. Note that Equations 18.17 and 18.18 are identical to Equations 16.24 and 16.32, except for the fact that MS_{res} is employed as the error term in both equations, and in Equation 18.18 df_{BC} is used in place of df_{BG} in determining the numerator degrees of freedom. In employing Equation 18.18, the tabled critical value $F_{.05} = 4.10$ is employed since $df_{\text{num}} = df_{BC} = 2$ and $df_{\text{den}} = df_{\text{res}} = 10$.

$$CD_{LSD} = \sqrt{F_{(1,\text{res})}}\sqrt{\frac{2MS_{\text{res}}}{n}} = \sqrt{4.96}\sqrt{\frac{(2)(.49)}{6}} = .90 \quad \textbf{(Equation 18.17)}$$

$$CD_S = \sqrt{(k-1)(F_{(df_{BC},\,df_{\text{res}})})}\sqrt{\frac{2MS_{\text{res}}}{n}} \quad \textbf{(Equation 18.18)}$$

$$= \sqrt{(3-1)(4.10)}\sqrt{\frac{(2)(.49)}{6}} = 1.16$$

Thus, if one employs **multiple *t* tests/Fisher's LSD test** (i.e., conducts linear contrasts with α_{FW} not adjusted), in order to differ significantly at the .05 level the means of any two conditions must differ from one another by at least .90 units. If, on the other hand, the **Scheffé test** is employed, in order to differ significantly the means of any two conditions must differ from one another by at least 1.16 units. Since the difference score for the comparison Condition 1 versus Condition 2 equals $\bar{X}_1 - \bar{X}_2 = 2$, the comparison is significant at the .05 level, regardless of which comparison procedure is employed.

Table 18.5 summarizes the differences between pairs of means involving all of the experimental conditions. Since the difference score for all three simple/pairwise comparisons is greater than $CD_S = 1.16$, all of the comparisons are significant at the .05 level, regardless of which comparison procedure is employed.

Table 18.5 Differences Between Pairs of Means in Example 18.1

$\bar{X}_1 - \bar{X}_2 = 8.5 - 6.5 = 2$
$\bar{X}_1 - \bar{X}_3 = 8.5 - 4.83 = 3.67$
$\bar{X}_2 - \bar{X}_3 = 6.5 - 4.83 = 1.67$

The complex comparison Condition 3 versus Conditions 1 and 2 is illustrated below employing Equations 18.19 and 18.20. The latter two equations (which are the generic forms of Equations 18.17 and 18.18 that can be used for both simple and complex comparisons) are identical to Equations 16.25 and 16.33 (except for the use of MS_{res} as the error term, and the use of df_{BC} in place of df_{BG} in Equation 18.20).

$$CD_{LSD} = \sqrt{F_{(1,\text{res})}} \sqrt{\frac{(\Sigma c_j^2)(MS_{\text{res}})}{n}} \qquad \textbf{(Equation 18.19)}$$

$$= \sqrt{4.96}\sqrt{\frac{(1.5)(.49)}{6}} = .78$$

$$CD_S = \sqrt{(k-1)(F_{(df_{BC},\,df_{\text{res}})})} \sqrt{\frac{(\Sigma c_j^2)(MS_{\text{res}})}{n}} \qquad \textbf{(Equation 18.20)}$$

$$= \sqrt{(3-1)(4.10)}\sqrt{\frac{(1.5)(.49)}{6}} = 1.00$$

Thus, if one employs **multiple *t* tests/Fisher's LSD test** (i.e., conducts a linear contrast with α_{FW} not adjusted), in order to differ significantly the difference between $\bar{X}_3$ and $(\bar{X}_1 + \bar{X}_2)/2$ must be at least .78 units. If, on the other hand, the **Scheffé test** is employed, in order to differ significantly the two sets of means must differ from one another by at least 1.00 units. Since the obtained difference of 2.67 is greater than $CD_S = 1.00$, the nondirectional alternative hypothesis H_1: $\mu_3 \neq (\mu_1 + \mu_2)/2$ is supported, regardless of which comparison procedure is employed.

The computation of a confidence interval for a comparison The procedure that is described for computing a confidence interval for a comparison for the **single-factor between-subjects analysis of variance** can also be used with the **single-factor within-subjects analysis of variance.** Thus, in the case of the latter analysis of variance, a confidence interval for a comparison is computed by adding to and subtracting the relevant *CD* value for the comparison from the obtained difference between the means involved in the comparison. As an example, let us assume the **Scheffé test** is employed to compute the value $CD_S = 1.16$ for the comparison Condition 1 versus Condition 2. To compute the 95% confidence interval, the value 1.16 is added to and subtracted from 2, which is the absolute value of the difference between the two means. Thus, $CI_{.95} = 2 \pm 1.16$, which can also be written as $.84 \leq (\mu_1 - \mu_2) \leq 3.16$. In other words, the researcher can be 95% confident that the mean of the population represented by Condition 1 is between .84 and 3.16 units larger than the mean of the population represented by Condition 2.

Alternative methodology for computing MS_{res} for a comparison Earlier in this section it is noted that if the sphericity assumption underlying the **single-factor within-subjects analysis of variance** is violated, MS_{res} may not provide an accurate measure of error variability for a specific comparison. Because of this, many sources recommend that when the sphericity assumption is violated a separate measure of error variability be computed for each comparison that is conducted. The procedure that will be discussed in this section (which is described in Keppel (1991)) can be employed any time a researcher has reason to believe that MS_{res} employed in computing the omnibus F value is not representative of the actual error variability for the experimental conditions involved in a specific comparison. The procedure (which will be demonstrated for both a simple and complex comparison) requires that a **single-factor within-subjects analysis of variance** be conducted employing only the data for those experimental conditions involved in a comparison. In the case of a simple comparison, the scores of subjects in the two comparison conditions are evaluated with the analysis of variance. In the case of a complex comparison, a weighted score must be computed for each subject for any composite mean that is a combination of two or more experimental conditions.

With respect to a simple planned comparison, the procedure will be employed to evaluate the difference between Condition 1 and Condition 3. The reason it will not be used for the Condition 1 versus Condition 2 comparison is because the error variability associated with the latter comparison is $MS_{res} = 0$. The reason why $MS_{res} = 0$ for the latter comparison is revealed by inspection of the scores of the six subjects in Table 18.1. Observe that the score of each of the six subjects in Condition 1 is two units higher than it is in Condition 2. Anytime all of the subjects in a within-subjects design involving two treatments obtain identical difference scores, the measure of error variability will equal zero. Consequently, whenever the value of $MS_{res} = 0$, the value of F_{comp} will be indeterminate, since the denominator of the F ratio will equal zero. In such an instance, one can either use the MS_{res} value employed in computing the omnibus F value, or elect to use the smallest MS_{res} value that can be computed for any of the other simple comparisons in the set of data.[11]

Table 18.6 summarizes the data for the comparison Condition 1 versus Condition 3. Note that the values $\Sigma X_T = 80$ and $\Sigma X_T^2 = 606$ differ from those computed in Table 18.1, since in this instance they only include the data for two of the three experimental conditions.

Table 18.6 Data for Comparison of Condition 1 Versus Condition 3

	Condition 1		Condition 3		
	X_1	X_1^2	X_3	X_3^2	ΣS_i
Subject 1	9	81	4	16	13
Subject 2	10	100	7	49	17
Subject 3	7	49	3	9	10
Subject 4	10	100	7	49	17
Subject 5	7	49	2	4	9
Subject 6	8	64	6	36	14
	$\Sigma X_1 = 51$	$\Sigma X_1^2 = 443$	$\Sigma X_3 = 29$	$\Sigma X_3^2 = 163$	$\Sigma X_T = 80$
		$\Sigma X_T = 51 + 29 = 80$		$\Sigma X_T^2 = 443 + 163 = 606$	

The sum of squares values that are required for the analysis of variance are computed with Equations 18.2–18.5. Note that since we are only dealing with two conditions, $k = 2$, and thus, $N = nk = (6)(2) = 12$.

$$SS_T = \Sigma X_T^2 - \frac{(\Sigma X_T)^2}{N} = 606 - \frac{(80)^2}{12} = 72.67$$

$$SS_{BC} = \left[\frac{(\Sigma X_1)^2}{n} + \frac{(\Sigma X_3)^2}{n}\right] - \frac{(\Sigma X_T)^2}{N} = \left[\frac{(51)^2}{6} + \frac{(29)^2}{6}\right] - \frac{(80)^2}{12} = 40.34$$

$$SS_{BS} = \left[\frac{(\Sigma S_1)^2}{k} + \cdots + \frac{(\Sigma S_6)^2}{k}\right] - \frac{(\Sigma X_T)^2}{N} = \left[\frac{(13)^2}{2} + \frac{(17)^2}{2} + \cdots + \frac{(14)^2}{2}\right] - \frac{(80)^2}{12} = 28.67$$

$$SS_{\text{res}} = SS_T - SS_{BC} - SS_{BS} = 72.67 - 40.34 - 28.67 = 3.66$$

Upon computing the sum of squares values, the appropriate degrees of freedom (employing Equations 18.10–18.13) and mean square values (employing Equations 18.7–18.9) are computed. Employing Equation 18.15, the value $F = F_{\text{comp}} = 55.26$ is computed for the comparison. The analysis of variance is summarized in Table 18.7. Note that in computing the degrees of freedom for the analysis, the values $n = 6$ and $k = 2$ are employed.

Table 18.7 Summary Table for Analysis of Variance for Comparison of Condition 1 Versus Condition 3

Source of variation	*SS*	*df*	*MS*	*F*
Between-subjects	28.67	5	5.73	
Between-conditions	40.34	1	40.34	55.26
Residual	3.66	5	.73	
Total	**72.67**	**11**		

Employing $df_{\text{num}} = df_{BC} = df_{\text{comp}} = 1$ and $df_{\text{den}} = df_{\text{res}} = 5$, the tabled critical values employed in **Table A10** are $F_{.05} = 6.61$ and $F_{.01} = 16.26$. Since the obtained value $F = 55.26$ is greater than both of the aforementioned critical values, the nondirectional alternative hypothesis H_1: $\mu_1 \neq \mu_3$ is supported at both the .05 and .01 levels. The reader should note that the value $df_{\text{res}} = 5$ employed for the comparison of Condition 1 versus Condition 3 is less than the value $df_{\text{res}} = 10$ employed when the same comparison was conducted earlier in this section employing the value $MS_{\text{res}} = .49$ obtained for the omnibus F test. The lower df_{res} value associated with the method under discussion results in a less powerful test of the alternative hypothesis. In some instances the loss of power associated with this method may be offset if the value of MS_{res} for the comparison conditions is smaller than the value of MS_{res} obtained for the omnibus F test.

Employing Equations 18.17 and 18.18, a CD_{LSD} and CD_S value can be computed for the Condition 1 versus Condition 3 comparison.

$$CD_{LSD} = \sqrt{F_{(1,\text{res})}}\sqrt{\frac{2MS_{\text{res}}}{n}} = \sqrt{6.61}\sqrt{\frac{(2)(.73)}{6}} = 1.27$$

$$CD_S = \sqrt{(k - 1)(F_{(df_{BC}, df_{\text{res}})}}\sqrt{\frac{2MS_{\text{res}}}{n}} = \sqrt{(3 - 1)(4.10)}\sqrt{\frac{(2)(.73)}{6}} = 1.41$$

In computing $CD_{LSD} = 1.27$, the tabled critical F value employed in Equation 18.17 is based on $df_{\text{num}} = df_{BC} = df_{\text{comp}} = 1$, and $df_{\text{den}} = df_{\text{res}} = 5$. Note that $df_{\text{den}} = df_{\text{res}} = 5$ is only based on the data for the two experimental conditions employed in the comparison. In computing $CD_S = 1.41$, however, the tabled critical F value employed in Equation 18.18 is based on all three conditions employed in the experiment, and thus $df_{BC} = 2$ and $df_{\text{res}} = 10$. Note that the values $CD_{LSD} = 1.27$ and $CD_S = 1.41$ are larger than the corresponding values $CD_{LSD} = .90$ and $CD_S = 1.16$, which are computed for simple comparisons when $MS_{\text{res}} = .49$ is employed. It should be obvious through inspection of Equations 18.17 and 18.18, that the larger the value of MS_{res} the greater the magnitude of the computed CD value.

At the beginning of this section it is noted that when a complex comparison is conducted employing a residual variability which is based only on the specific conditions involved in the comparison, a weighted score must be computed for each subject for any composite mean that is a combination of two or more experimental conditions. This will now be illustrated for the comparison of Condition 3 versus the combined performance of Conditions 1 and 2.

Table 18.8 contains the scores of the six subjects in Condition 3 and a weighted score for each subject for Conditions 1 and 2. The weighted score of a subject for a combination of two or more conditions is obtained as follows: a) A subject's score in each condition is multiplied by the absolute value of the coefficient for that condition (based on the coefficients for the comparison in Table 18.4); and b) The subject's weighted score is the sum of the products obtained in part a).

To clarify how the aforementioned procedure is employed to compute the weighted scores in Table 18.8, the computation of weighted scores for Subjects 1 and 2 will be described. Since the scores of Subject 1 in Conditions 1 and 2 are 9 and 7, each score is multiplied by the absolute value of the coefficient for the corresponding condition. Employing the absolute values of the coefficients noted in Table 18.4 for Conditions 1 and 2, each score is multiplied by $1/2$, yielding $(9)(1/2) = 4.5$ and $(7)(1/2) = 3.5$. The weighted score for Subject 1 is obtained by summing the latter two values. Thus, the weighted score for Subject 1 is $4.5 + 3.5 = 8$. In the case of Subject 2, $(10)(1/2) = 5$ and $(8)(1/2) = 4$. Thus, the weighted score for Subject 2 is $5 + 4 = 9$. The same procedure is used for the remaining four subjects.[12]

Table 18.8 Data for Comparison of Condition 3 Versus Conditions 1 and 2

	Condition 3		Conditions 1 & 2		
	X_3	X_3^2	$X_{1/2}$	$X_{1/2}^2$	ΣS_i
Subject 1	4	16	8	64	12
Subject 2	7	49	9	81	16
Subject 3	3	9	6	36	9
Subject 4	7	49	9	81	16
Subject 5	2	4	6	36	8
Subject 6	6	36	7	49	13
	$\Sigma X_3 = 29$	$\Sigma X_3^2 = 163$	$\Sigma X_{1/2} = 45$	$\Sigma X_{1/2}^2 = 347$	$\Sigma X_T = 74$
	$\Sigma X_T = 29 + 45 = 74$		$\Sigma X_T^2 = 163 + 347 = 510$		

The sum of squares values that are required for the analysis of variance are computed with Equations 18.2–18.5. Note that since we are only dealing with two sets of means, $k = 2$, and thus, $N = nk = (6)(2) = 12$.

$$SS_T = \Sigma X_T^2 - \frac{(\Sigma X_T)^2}{N} = 510 - \frac{(74)^2}{12} = 53.67$$

$$SS_{BC} = \left[\frac{(\Sigma X_3)^2}{n} + \frac{(\Sigma X_{1/2})^2}{n}\right] - \frac{(\Sigma X_T)^2}{N} = \left[\frac{(29)^2}{6} + \frac{(45)^2}{6}\right] - \frac{(74)^2}{12} = 21.34$$

$$SS_{BS} = \left[\frac{(\Sigma S_1)^2}{k} + \cdots + \frac{(\Sigma S_6)^2}{k}\right] - \frac{(\Sigma X_T)^2}{N} = \left[\frac{(12)^2}{2} + \frac{(16)^2}{2} + \cdots + \frac{(13)^2}{2}\right] - \frac{(74)^2}{12} = 28.67$$

$$SS_{\text{res}} = SS_T - SS_{BC} - SS_{BS} = 53.67 - 21.34 - 28.67 = 3.66$$

Upon computing the sum of squares values the appropriate degrees of freedom (employing Equations 18.10–18.13) and mean square values (employing Equations 18.7–18.9) are computed. Employing Equation 18.15, the value $F = F_{\text{comp}} = 29.23$ is computed for the comparison. The analysis of variance is summarized in Table 18.9. Note that in computing the degrees of freedom for the analysis, the values $n = 6$ and $k = 2$ are employed.

Table 18.9 Summary Table for Analysis of Variance for Comparison of Condition 3 Versus Conditions 1 and 2

Source of variation	*SS*	*df*	*MS*	*F*
Between-subjects	28.67	5	5.73	
Between-conditions	21.34	1	21.34	29.23
Residual	3.66	5	.73	
Total	**53.67**	**11**		

Employing $df_{\text{num}} = df_{BC} = df_{\text{comp}} = 1$ and $df_{\text{den}} = df_{\text{res}} = 5$, the tabled critical values employed in **Table A10** are $F_{.05} = 6.61$ and $F_{.01} = 16.26$. Since the obtained value $F = 29.23$ is greater than both of the aforementioned tabled critical values, the nondirectional alternative hypothesis H_1: $\mu_3 \neq (\mu_1 + \mu_2)/2$ is supported at both the .05 and .01 levels.

Employing Equations 18.19 and 18.20, a CD_{LSD} and CD_S value can be computed for the complex comparison.

$$CD_{LSD} = \sqrt{F_{(1,\text{res})}}\sqrt{\frac{(\Sigma c_j^2)(MS_{\text{res}})}{n}} = \sqrt{6.61}\sqrt{\frac{(1.5)(.73)}{6}} = 1.10$$

$$CD_S = \sqrt{(k-1)(F_{(df_{BC},\, df_{\text{res}})})}\sqrt{\frac{(\Sigma c_j^2)(MS_{\text{res}})}{n}}$$

$$= \sqrt{(3-1)(4.10)}\sqrt{\frac{(1.5)(.73)}{6}} = 1.22$$

Note that the values $CD_{LSD} = 1.10$ and $CD_S = 1.22$ are larger than the corresponding values $CD_{LSD} = .78$ and $CD_S = 1.00$, which are computed for the same complex comparison when $MS_{\text{res}} = .49$ is used.

2. Comparing the means of three or more conditions when $k \geq 4$ Within the framework of a **single-factor within-subjects analysis of variance** involving $k = 4$ or more conditions, a researcher may wish to evaluate a general hypothesis with respect to the means of a subset of conditions, where the number of conditions in the subset is some value less than k. Although the latter type of situation is not commonly encountered in research, this section will describe the protocol for conducting such an analysis. Specifically, the protocol described for the analogous analysis for a **single-factor between-subjects analysis of variance** will be extended to the **single-factor within-subjects analysis of variance.**

To illustrate, assume that a fourth experimental condition is added to Example 18.1. Assume that the scores of Subjects 1–6 in Condition 4 are respectively: 3, 8, 2, 6, 4, 6. Thus, $\Sigma X_4 = 29$, $\bar{X}_4 = 4.83$, and $\Sigma X_4^2 = 165$. If the data for Condition 4 are integrated into the data for the other three conditions (which are summarized in Table 18.1), the following summary values are computed: $N = nk = (6)(4) = 24$, $\Sigma X_T = 148$, $\Sigma X_T^2 = 1034$. Substituting the revised values for $k = 4$ conditions in Equations 18.2–18.5, the following sum of squares values are computed: $SS_T = 121.33$, $SS_{BC} = 54.66$, $SS_{BS} = 54.33$, $SS_{\text{res}} = 12.34$. Employing the values $k = 4$ and $n = 6$ in Equations 18.10–18.12, the values $df_{BC} = 4 - 1 = 3$, $df_{BS} = 6 - 1 = 5$, and $df_{\text{res}} = (6 - 1)(4 - 1) = 15$ are computed. Substituting the appropriate values for the sums of squares and degrees of freedom in Equations 18.7–18.9, the values $MS_{BC} = 54.66/3 = 18.22$, $MS_{BS} = 54.33/5 = 10.87$, and $MS_{\text{res}} = 12.34/15 = .82$ are computed. Equation 18.15 is employed to compute the value $F = 18.22/.82 = 22.22$. Table 18.10 is the summary table of the analysis of variance.

Table 18.10 Summary Table of Analysis of Variance for Example 18.1 When $k = 4$

Source of variation	*SS*	*df*	*MS*	*F*
Between-subjects	54.33	5	10.87	
Between-conditions	54.66	3	18.22	22.22
Residual	12.34	15	.82	
	121.33	**23**		

Employing $df_{\text{num}} = 3$ and $df_{\text{den}} = 15$, the tabled critical .05 and .01 F values are $F_{.05} = 3.29$ and $F_{.01} = 5.42$. Since the obtained value $F = 22.22$ is greater than both of the aforementioned critical values, the null hypothesis (which for $k = 4$ is H_0: $\mu_1 = \mu_2 = \mu_3 = \mu_4$) can be rejected at both the .05 and .01 levels.

Let us assume that prior to the above analysis the researcher has reason to believe that Conditions 1, 2, and 3 may be distinct from Condition 4. However, before he contrasts the composite mean of Conditions 1, 2, and 3 with the mean of Condition 4 (i.e., conduct the complex comparison which evaluates the null hypothesis H_0: $(\mu_1 + \mu_2 + \mu_3)/3 = \mu_4$), he decides to evaluate the null hypothesis H_0: $\mu_1 = \mu_2 = \mu_3$. If the latter null hypothesis is retained, he will assume that the three conditions share a common mean value, and on the basis of this he will compare their composite mean with the mean of Condition 4. In order to evaluate the null hypothesis H_0: $\mu_1 = \mu_2 = \mu_3$, it is necessary for the researcher to conduct a separate analysis of variance that just involves the data for the three conditions identified in the null hypothesis. The latter analysis of variance has already been conducted, since it is the original analysis of variance that is employed for Example 18.1 — the results of which are summarized in Table 18.2.

Upon conducting an analysis of variance on the data for all $k = 4$ conditions as well as an analysis of variance on the data for the subset comprised of $k_{\text{subset}} = 3$ conditions, the researcher has the necessary information to compute the appropriate F ratio (which will be represented with the notation $F_{(1/2/3)}$) for evaluating the null hypothesis H_0: $\mu_1 = \mu_2 = \mu_3$. If we apply the same logic employed when the analogous analysis is conducted in reference to the **single-factor between-subjects analysis of variance**, the following values are employed to compute the F ratio to evaluate the latter null hypothesis: a) $MS_{BC} = 20.23$ (which is the value of MS_{BC} computed for the analysis of variance in Table 18.2 that involves only the three conditions identified in the null hypothesis H_0: $\mu_1 = \mu_2 = \mu_3$) is employed as the numerator of the F ratio; and b) $MS_{res} = .82$ (which is the value of MS_{res} computed in Table 18.10 for the omnibus F test when the data for all $k = 4$ conditions are evaluated) is employed as the denominator of the F ratio. The use of $MS_{res} = .82$ instead of $MS_{res} = .49$ (which is the value of MS_{res} computed for the analysis of variance in Table 18.2) as the denominator of the F ratio is predicated on the assumption that $MS_{res} = .82$ provides a more accurate estimate of error variability than $MS_{res} = .49$. If the latter assumption is made, the value $F_{(1/2/3)} = 24.67$ is computed.

$$F_{(1/2/3)} = \frac{MS_{BC_{(1/2/3)}}}{MS_{res_{(1/2/3/4)}}} = \frac{20.23}{.82} = 24.67$$

The degrees of freedom employed for the analysis are based on the mean square values employed in computing the $F_{(1/2/3)}$ ratio. Thus: $df_{num} = k_{subset} - 1 = 3 - 1 = 2$ (where $k_{subset} = 3$ conditions) and $df_{den} = df_{res_{(1/2/3/4)}} = 15$ (which is df_{res} for the omnibus F test involving all $k = 4$ conditions). For $df_{num} = 2$ and $df_{den} = 15$, $F_{.05} = 3.68$ and $F_{.01} = 6.36$. Since the obtained value $F = 24.67$ is greater than both of the aforementioned critical values, the null hypothesis can be rejected at both the .05 and .01 levels. Thus, the data do not support the researcher's hypothesis that Conditions 1, 2, and 3 represent a homogeneous subset. In view of this, the researcher would not conduct the contrast $(\bar{X}_1 + \bar{X}_2 + \bar{X}_3)/3$ versus $\bar{X}_4$.

If a researcher is not willing to assume that $MS_{res} = .82$ provides a more accurate estimate of error variability than $MS_{res} = .49$, the latter value can be employed as the denominator term in computing the $F_{(1/2/3)}$ ratio. Thus, if for some reason a researcher believes that by virtue of adding a fourth experimental condition experimental error is either increased or decreased, one can justify employing the value $MS_{res} = .49$ (computed for the $k = 3$ conditions) as the denominator term in computing the value $F_{(1/2/3)}$. If $MS_{res} = .49$ is employed to compute the F ratio, the value $F_{(1/2/3)} = 20.23/.49 = 41.29$ is computed. Since the latter value is greater than $F_{.05} = 3.68$ and $F_{.01} = 6.36$, the researcher can still reject the null hypothesis at both the .05 and .01 levels. However, the fact that $F_{(1/2/3)} = 41.29$ is substantially larger than $F_{(1/2/3)} = 24.67$ illustrates that depending upon which of the two error terms is employed, it is possible that they may lead to different conclusions regarding the status of the null hypothesis. The determination of which error term to use will be based on the assumptions a researcher is willing to make concerning the data. In the final analysis, in instances where the two error terms yield inconsistent results, it may be necessary for a researcher to conduct one or more replication studies in order to clarify the status of a null hypothesis.

3. Evaluation of the sphericity assumption underlying the single-factor within-subjects analysis of variance In Section I it is noted that one of the assumptions underlying the **single-factor within-subjects analysis of variance** is the existence of a condition referred to as **sphericity**. Sphericity exists when there is homogeneity of variance among the populations of difference scores. The latter can be explained as follows: Assume that for each of the n subjects who serve under all k experimental conditions, a difference score is calculated for all pairs of conditions. The number of difference scores that can be computed for each subject will equal $[k(k - 1)]/2$. When $k = 3$, three sets of difference scores can be computed. Specifically: a) A set of difference scores that is the result of subtracting each subject's score in Condition 2 from the subject's score in Condition 1; b) A set of difference scores that is the result of subtracting each subject's score in Condition 3 from the subject's score in Condition 1; and c) A set of difference scores that is the result of subtracting each subject's score in Condition 3 from the subject's score in Condition 2.[13]

The sphericity assumption states that if the estimated population variances for the three sets of difference scores are computed, the values of the variances should be equal. The derivation of the three sets of difference scores for Example 18.1, and the computation of their estimated population variances are summarized in Table 18.11. Note that for each set of difference scores, a D value is computed for each subject. The estimated population variance of the D values (which is computed with Equation I.5) represents the estimated population variance of a set of difference scores.

Table 18.11 Computation of Estimated Population Variances of Difference Scores

Condition 1 versus Condition 2

	X_1	X_2	D	D^2
Subject 1	9	7	2	4
Subject 2	10	8	2	4
Subject 3	7	5	2	4
Subject 4	10	8	2	4
Subject 5	7	5	2	4
Subject 6	8	6	2	4
			$\Sigma D = 12$	$\Sigma D^2 = 24$

$$\tilde{s}^2_{(X_1 - X_2)} = \frac{\Sigma D^2 - \frac{(\Sigma D)^2}{n}}{n - 1} = \frac{24 - \frac{(12)^2}{6}}{6 - 1} = 0$$

Condition 1 versus Condition 3

	X_1	X_3	D	D^2
Subject 1	9	4	5	25
Subject 2	10	7	3	9
Subject 3	7	3	4	16
Subject 4	10	7	3	9
Subject 5	7	2	5	25
Subject 6	8	6	2	4
			$\Sigma D = 22$	$\Sigma D^2 = 88$

$$\tilde{s}^2_{(X_1 - X_3)} = \frac{\Sigma D^2 - \frac{(\Sigma D)^2}{n}}{n - 1} = \frac{88 - \frac{(22)^2}{6}}{6 - 1} = 1.47$$

Condition 2 versus Condition 3

	X_2	X_3	D	D^2
Subject 1	7	4	3	9
Subject 2	8	7	1	1
Subject 3	5	3	2	4
Subject 4	8	7	1	1
Subject 5	5	2	3	9
Subject 6	6	6	0	0
			$\Sigma D = 10$	$\Sigma D^2 = 24$

$$\tilde{s}^2_{(X_2 - X_3)} = \frac{\Sigma D^2 - \frac{(\Sigma D)^2}{n}}{n - 1} = \frac{24 - \frac{(10)^2}{6}}{6 - 1} = 1.47$$

Visual inspection of the estimated population variances of the difference scores reveals that the three variances are quite close to one another. This latter fact suggests that the sphericity assumption is unlikely to have been violated. Unfortunately, the tests that are discussed in this book for evaluating homogeneity of variance are not appropriate for comparing the variances of the difference scores within the framework of evaluating the sphericity assumption. The procedures that have been developed for evaluating sphericity require the use of matrix algebra and are generally conducted with the aid of a computer. Further reference to such procedures will be made later in this discussion.

Sources on analysis of variance (e.g., Myers and Well (1991)) note that there is another condition known as **compound symmetry** which is sufficient, although not necessary, in order for sphericity to exist. Compound symmetry, which represents a special case of sphericity, exists when both of the following conditions have been met: a) **Homogeneity of variance** — All of the populations that are represented by the k experimental conditions have equal variances; and b) **Homogeneity of covariance** — All of the population covariances are equal to one another.

At this point we will examine the variances of the three experimental conditions, as well as the **covariances** of each pair of conditions. Employing Equation I.5, the estimated population variance for each of the three experimental conditions is computed.

$$\tilde{s}_1^2 = \frac{\Sigma X_1^2 - \dfrac{(\Sigma X_1)^2}{n}}{n - 1} = \frac{443 - \dfrac{(51)^2}{6}}{5} = 1.9$$

$$\tilde{s}_2^2 = \frac{\Sigma X_2^2 - \dfrac{(\Sigma X_2)^2}{n}}{n - 1} = \frac{263 - \dfrac{(39)^2}{6}}{5} = 1.9$$

$$\tilde{s}_3^2 = \frac{\Sigma X_3^2 - \dfrac{(\Sigma X_3)^2}{n}}{n - 1} = \frac{163 - \dfrac{(29)^2}{6}}{5} = 4.57$$

Although, within the framework of the sphericity assumption, Equation 12.9 it is not actually employed to evaluate homogeneity of variance, for illustrative purposes it will be used. The latter equation is employed to evaluate the homogeneity of variance assumption for the ***t* test for two dependent samples** by contrasting the highest estimated population variance (which in Example 18.1 is $\tilde{s}_3^2 = 4.57$) and the lowest estimated population variance (which in Example 18.1 is $\tilde{s}_1^2 = \tilde{s}_2^2 = 1.9$). In order to employ Equation 12.9 it is necessary to compute the correlation between subjects' scores in Condition 1 (which will be used to represent the lowest variance) and Condition 3. The value of the correlation coefficient is computed with Equation 12.7. Employing Equation 12.7, the value $r_{X_1 X_3} = .85$ is computed.[14]

$$r_{X_1 X_3} = \frac{\Sigma X_1 X_3 - \dfrac{(\Sigma X_1)(\Sigma X_3)}{n}}{\sqrt{\left[\Sigma X_1^2 - \dfrac{(\Sigma X_1)^2}{n}\right]\left[\Sigma X_3^2 - \dfrac{(\Sigma X_3)^2}{n}\right]}} = \frac{259 - \dfrac{(51)(29)}{6}}{\sqrt{\left[443 - \dfrac{(51)^2}{6}\right]\left[163 - \dfrac{(29)^2}{6}\right]}} = .85$$

Substituting the appropriate values in Equation 12.9, the value $t = 1.72$ is computed.

$$t = \frac{(\tilde{s}_L^2 - \tilde{s}_S^2)\sqrt{n - 2}}{\sqrt{4\tilde{s}_L^2\tilde{s}_S^2(1 - r_{X_1X_3}^2)}} = \frac{(4.57 - 1.9)\sqrt{(6 - 2)}}{\sqrt{(4)(4.57)(1.9)(1 - (.85)^2)}} = 1.72$$

The degrees of freedom associated with the t value computed with Equation 12.9 are $df = n - 2 = 4$. Since the computed value $t = 1.72$ is less than the tabled critical two-tailed value $t_{.05} = 2.78$ (for $df = 4$), the null hypothesis H_0: $\sigma_L^2 = \sigma_S^2$ (which states there is homogeneity of variance) is retained. Thus, there is no evidence to suggest that the homogeneity of variance assumption is violated.

Earlier in the discussion it was noted that the sphericity assumption assumes equal population covariances. Whereas variance is a measure of variability of the scores of n subjects on a single variable, covariance (which is discussed in more detail in Section VII of the **Pearson product-moment correlation coefficient (Test 22)**) is a measure which represents the degree to which two variables vary together. A positive covariance is associated with variables that are positively correlated with one another, and a negative covariance is associated with variables that are negatively correlated with one another.

Equation 18.21 is the general equation for computing **covariance**. The value computed with Equation 18.21 represents the estimated covariance between **Population a** and **Population b.**

$$\text{cov}_{X_aX_b} = \frac{\Sigma X_aX_b - \frac{(\Sigma X_a)(\Sigma X_b)}{n}}{n - 1} \qquad \textbf{(Equation 18.21)}$$

Since a covariance can be computed for any pair of experimental conditions, in Example 18.1 three covariances can be computed — specifically, the covariance between Conditions 1 and 2, the covariance between Conditions 1 and 3, and the covariance between Conditions 2 and 3. To illustrate the computation of the covariance, the covariance between Conditions 1 and 2 ($\text{cov}_{X_1X_2}$) will be computed employing Equation 18.21. Table 18.12, which reproduces the data for Conditions 1 and 2, summarizes the values employed in the calculation of the covariance.

Table 18.12 Data Required for Computing Covariance of Condition 1 and Condition 2

	X_1	X_2	X_1X_2
Subject 1	9	7	63
Subject 2	10	8	80
Subject 3	7	5	35
Subject 4	10	8	80
Subject 5	7	5	35
Subject 6	8	6	48
	$\Sigma X_1 = 51$	$\Sigma X_2 = 39$	$\Sigma X_1X_2 = 341$

Employing Equation 18.21 the value $\text{cov}_{X_1X_2} = 1.9$ is computed.

$$\text{cov}_{X_1X_2} = \frac{341 - \frac{(51)(39)}{6}}{6 - 1} = 1.9$$

If the relevant data for the other two sets of scores are employed, the values $\text{cov}_{X_1X_3} = 2.5$ and $\text{cov}_{X_2X_3} = 2.5$ are computed.

$$\text{cov}_{X_1X_3} = \frac{259 - \frac{(51)(29)}{6}}{6 - 1} = 2.5$$

$$\text{cov}_{X_2X_3} = \frac{201 - \frac{(39)(29)}{6}}{6 - 1} = 2.5$$

Since the three values for covariance are extremely close to one another, on the basis of visual inspection it would appear that the data are characterized by homogeneity of covariance. Coupled with the fact that homogeneity of variance also appears to exist, it would seem reasonable to conclude that the assumptions underlying compound symmetry (and thus of sphericity) are unlikely to have been violated.

The conditions necessary for compound symmetry (which, as previously noted, is not required in order for sphericity to exist), are, in fact, more stringent than the general requirement of sphericity (i.e., that there be homogeneity of variance among the populations of difference scores). Whenever data are characterized by compound symmetry, homogeneity of variance will exist among the populations of difference scores. However, it is possible to have homogeneity of variance among the populations of difference scores, yet not have compound symmetry.

A full discussion of the tests that are employed to evaluate the sphericity assumption underlying the **single-factor within-subjects analysis of variance** is beyond the scope of this book. The interested reader can find a description of such tests in selected texts that specialize in analysis of variance (e.g., Kirk (1982)). Keppel (1991) among others, notes, however, that tests which have been developed to evaluate the sphericity assumption have their own assumptions, and when the assumptions of the latter tests are violated (which may be more often than not) their reliability will be compromised. In view of this, Keppel (1991) questions the wisdom of employing such tests for evaluating the sphericity assumption.

At this point some general comments are in order regarding the consequences of violating the sphericity assumption. In the discussion of the ***t* test for two dependent samples** it is noted that the latter test is much more sensitive to violation of the homogeneity of variance assumption than is the ***t* test for two independent samples (Test 8)**. Since this observation can be generalized to designs involving more than two treatments, the **single-factor within-subjects analysis of variance** is more sensitive to violation of its assumptions concerning homogeneity of variance and covariance (i.e., the sphericity assumption) than is the **single-factor between-subjects analysis of variance** to violation of its assumption of homogeneity of variance. In point of fact, most sources suggest that the **single-factor within-subjects analysis of variance** is extremely sensitive to violations of the sphericity assumption, and that when the latter assumption is violated the tabled critical values in **Table A10** will not be accurate. Specifically, when the sphericity assumption is violated, the tabled critical F value associated with the appropriate degrees of freedom for the analysis of variance will be too low (i.e., the Type I error rate for the analysis will actually be higher than the prespecified value). One option proposed by Geisser and Greenhouse (1958) is to employ the tabled critical F value associated with $df_{\text{num}} = 1$ and $df_{\text{den}} = n - 1$ instead of the tabled critical F value associated with the usual degrees of freedom values (i.e., $df_{\text{num}} = df_{BC} = k - 1$ and $df_{\text{den}} = df_{\text{res}} = (n - 1)(k - 1)$). However, since the Geisser–Greenhouse method tends to overcorrect the value of F (i.e., it results in too high a critical value), some sources recommend an alternative but computationally more involved method developed by Box

(1954) which does not result in as severe an adjustment of the critical F value as the Geisser-Greenhouse method.

Keppel (1991) notes that it is quite common for the sphericity assumption to be violated in experiments which utilize a within-subjects design. In view of this he recommends that in employing the **single-factor within-subjects analysis of variance** to evaluate the latter design, it is probably always prudent to run a more conservative test in order to insure that the Type I error rate is adequately controlled. An even more extreme viewpoint is articulated by other sources who recommend that when there is reason to believe that the sphericity assumption is violated, one should evaluate the data with a procedure other than the **single-factor within-subjects analysis of variance**. Specifically, these sources (e.g., Maxwell and Delaney (1990) and Howell (1992)) recommend evaluating the data for a within-subjects design with a **multivariate analysis of variance (MANOVA)**. Since a description of multivariate statistical procedures (of which the **MANOVA** is an example) is beyond the scope of this book, the interested reader should consult sources that describe such analyses (e.g., Tabachnick and Fidell (1989)).

It should be apparent from the discussion in this section that there is lack of agreement with respect to the most appropriate methodology for dealing with violation of the sphericity assumption. A cynic might conclude that regardless of which method one employs, there will always be reason to doubt the accuracy of the probability value associated with the outcome of a study. As noted throughout this book, in situations where there are doubts concerning the reliability of an analysis, the most powerful tool the researcher has at her disposal is replication. In the final analysis, the truth regarding a hypothesis will ultimately emerge if one or more researchers conducts multiple studies that evaluate the same hypothesis. In instances where replication studies have been conducted, **meta-analysis** can be employed to derive a pooled probability value for all of the published studies.[15]

4. Computation of the power of the single-factor within-subjects analysis of variance Prior to reading this section the reader should review the discussion of power in Section VI of the **single-factor between-subjects analysis of variance**, since basically the same procedure is employed to determine the power of the **single-factor within-subjects analysis of variance**. The power of the **single-factor within-subjects analysis of variance** is computed with Equation 18.22, which is identical to Equation 16.38 (which is the equation used for computing the power of the **single-factor between-subjects analysis of variance**), except for the fact that the estimated value of σ^2_{res} is employed as the measure of error variability in place of σ^2_{WG} (McFatter and Gollob (1986)).

$$\phi = \sqrt{n\left[\frac{\Sigma(\mu_j - \mu_T)^2}{k\sigma^2_{\text{res}}}\right]} \qquad \textbf{(Equation 18.22)}$$

Where: μ_j = The estimated mean of the population represented by Condition j
μ_T = The grand mean, which is the average of the k estimated population means
σ^2_{res} = The estimated measure of error variability
n = The number of subjects
k = The number of experimental conditions

To illustrate the use of Equation 18.22 with Example 18.1, let us assume that prior to conducting the study the researcher estimates that the means of the populations represented by the three conditions are as follows: $\mu_1 = 8$, $\mu_2 = 6$, $\mu_3 = 4$. Additionally, it will be assumed that he estimates the population error variance associated with the analysis will equal $\sigma_{\text{res}^2} = 1$. Based on this information, the value μ_T can be computed:

$\mu_T = (\mu_1 + \mu_2 + \mu_3)/k = (8 + 6 + 4)/3 = 6$. The appropriate values are now substituted in Equation 18.22.

$$\phi = \sqrt{n\left[\frac{(8-6)^2 + (6-6)^2 + (4-6)^2}{(3)(1)}\right]} = \sqrt{2.67n} = 1.63\sqrt{n}$$

At this point **Table A15 (Graphs of the Power Function for the Analysis of Variance)** in the **Appendix** can be employed to determine the necessary sample size required in order to have the power stipulated by the experimenter. For our analysis (for which it will be assumed $\alpha = .05$) the appropriate set of curves to employ is the set for $df_{\text{num}} = df_{BC} = 2$. Let us assume we want the omnibus F test to have a power of at least .80. We now substitute what we consider to be a reasonable value for n in the equation $\phi = 1.63\sqrt{n}$ (which is the result obtained with Equation 18.22). To illustrate, the value $n = 6$ (the sample size employed in Example 18.1) is substituted in the equation. The resulting value is $\phi = 1.63\sqrt{6} = 3.99$.

The value $\phi = 3.99$ is located on the abscissa (X-axis) of the relevant set of curves in **Table A15** — specifically, the set for $df_{\text{num}} = 2$. At the point corresponding to $\phi = 3.99$, a perpendicular line is erected from the abscissa which intersects with the power curve that corresponds to $df_{\text{den}} = df_{\text{res}}$ employed for the omnibus F test. Since $df_{\text{res}} = 10$, the curve for the latter value is employed (or closest to it if a curve for the exact value is not available). At the point the perpendicular intersects the curve $df_{\text{res}} = 10$, a second perpendicular line is drawn in relation to the ordinate (Y-axis). The point at which this perpendicular intersects the ordinate indicates the power of the test. Since $\phi = 3.99$, we determine the power equals 1.[16] Thus, if we employ six subjects in a within-subjects design, there is a 100% likelihood (which corresponds to a probability of 1) of detecting an effect size equal to or larger than the one stipulated by the researcher (which is a function of the estimated values for the population means relative to the value estimated for error variability). Since the probability of committing a Type II error is $\beta = 1 - \text{power}$, $\beta = 1 - 1 = 0$. This value represents the likelihood of not detecting an effect size equal to or greater than the one stipulated.

Equation 18.23 (described in McFatter and Gollob (1986)) can be employed to conduct a power analysis for a comparison associated with a **single-factor within-subjects analysis of variance.** Equation 18.23 is identical to Equation 16.39 (which is the equation for evaluating the power of a comparison for the **single-factor between-subjects analysis of variance**), except for the fact that σ^2_{res} is employed as the measure of error variability in place of σ^2_{WG}.

$$\phi_{\text{comp}} = \sqrt{n\left[\frac{(\mu_a - \mu_b)^2}{2(\sigma^2_{\text{res}})(\Sigma c_j^2)}\right]} \qquad \textbf{(Equation 18.23)}$$

As is the case for a **single-factor between-subjects analysis of variance,** Equation 18.23 can be used for both simple and complex single degree of freedom comparisons. As a general rule, the equation is used for planned comparisons. As noted in the discussion of the **single-factor between-subjects analysis of variance,** although the equation can be extended to unplanned comparisons, published power tables for the analysis of variance generally only apply to per comparison error rates of $\alpha = .05$ and $\alpha = .01$. In the case of planned and especially unplanned comparisons which involve α_{PC} rates other than .05 or .01, more detailed tables are required.

For single degree of freedom comparisons, the power curves in **Table A15** for $df_{\text{num}} = 1$ are always employed. The use of Equation 18.23 will be illustrated for the simple comparison Condition 1 versus Condition 2 (summarized in Table 18.3). Since $\Sigma c_j = 2$, and we have estimated $\mu_a = \mu_1 = 8$, $\mu_b = \mu_2 = 6$ and $\sigma^2_{\text{res}} = 1$, the following result is obtained:

$$\phi_{\text{comp}} = \sqrt{n\left[\frac{(8 - 6)^2}{(2)(1)(2)}\right]} = \sqrt{n}$$

Substituting $n = 6$ in the equation $\phi_{\text{comp}} = \sqrt{n}$, we obtain $\phi_{\text{comp}} = \sqrt{6} = 2.45$. Employing the power curves for $df_{\text{num}} = 1$ with $\alpha = .05$, we use the curve for $df_{\text{res}} = 10$ (df_{res} employed for the omnibus F test), and determine that when $\phi_{\text{comp}} = 2.45$, the power of the test is approximately .88.

5. Measure of magnitude of treatment effect for the single-factor within-subjects analysis of variance: Omega squared (Test 18g) Prior to reading this section the reader should review the discussion of measures of magnitude of treatment effect in Section VI of both the ***t* test for two independent samples** and the **single-factor between-subjects analysis of variance.** The latter discussions note that the computation of an omnibus F value only provides a researcher with information regarding whether the null hypothesis can be rejected — i.e., whether a significant treatment effect exists. The F value (as well as the level of significance with which it is associated), however, does not provide the researcher with any information regarding the size of any treatment effect that is present. As is noted in earlier discussions of treatment effect, the latter is defined as the proportion of the variability on the dependent variable that is associated with the experimental treatments/ conditions. Unfortunately, as Keppel (1991) notes, there is disagreement with respect to which variance components should be employed in computing the **omega squared** statistic ($\tilde{\omega}^2$) — the latter being the most commonly computed measure of treatment effect for the **single-factor within-subjects analysis of variance.**

Equation 18.24, which is presented in Kirk (1982), is one of a number of equations that have been proposed for computing $\tilde{\omega}^2$. Employing Equation 18.24 with the data for Example 18.1, the value $\tilde{\omega}^2 = .48$ is computed.

$$\tilde{\omega}^2 = \frac{SS_{BC} - (k - 1)MS_{\text{res}}}{SS_T + MS_{\text{res}}} = \frac{40.45 - (3 - 1).49}{82.28 + .49} = .48 \quad \textbf{(Equation 18.24)}$$

The value $\tilde{\omega}^2 = .48$ indicates that 48% (or a proportion of .48) of the variability on the dependent variable (the number of nonsense syllables recalled) is associated with variability on the levels of the independent variable (noise). Since alternative equations that have been proposed for computing $\tilde{\omega}^2$ (noted in such sources as Myers and Well (1991) and Winer *et al.* (1991)) may yield a different value than the one obtained with Equation 18.24, one should use caution in interpreting the value computed for $\tilde{\omega}^2$.[17] A more thorough discussion of the general issues involved in computing a measure of magnitude of treatment effect for a **single-factor within-subjects analysis of variance** can be found in the references cited in this section.

6. Computation of a confidence interval for the mean of a treatment population Prior to reading this section the reader should review the discussion of confidence intervals in Section VI of both the **single sample *t* test (Test 2)** and the **single-factor between-subjects analysis of variance.** The same procedure employed to compute a confidence interval for a treatment population for the **single-factor between-subjects analysis of variance** is employed for computing the confidence interval for the mean of a treatment population for a **single-factor within-subjects analysis of variance.** In other words, in order to compute a confidence interval for any experimental treatment/condition, one must conceptualize a within-subjects design as if it was a between-subjects design. The reason for this is that a confidence interval for any single condition will be a function of the variability of the scores of subjects who serve within that condition. Since MS_{res}, the measure of error variability for the repeated-measures analysis of variance, is a measure of within-subjects variability that is independent of any treatment effect, it cannot be employed to estimate the error variability for a specific treatment if one wants to compute a confidence interval for the mean of a treatment population. As is the case with the **single-factor between-subjects analysis of variance**, one can employ either of the following two strategies in computing the confidence interval for a treatment.

a) If one assumes that all of the treatments represent a population with the same variance, Equation 16.44 can be employed to compute the confidence interval (in our example we will assume the 95% confidence interval is being computed). In order to employ Equation 16.44 it is necessary to compute the value of MS_{WG}, which in the case of the **single-factor within-subjects analysis of variance** can be conceptualized as a **within-conditions mean square** (MS_{WC}). In order to compute MS_{WC}, it is necessary to first compute the **within-conditions sum of squares** (SS_{WC}). The latter value is computed employing Equation 18.25 (which is identical to Equation 16.5, except it employs the subscript WC in place of WG).

$$SS_{WC} = \sum_{j=1}^{k} \left[\Sigma X_j^2 - \frac{(\Sigma X_j)^2}{n} \right] \qquad \textbf{(Equation 18.25)}$$

$$= \left[443 - \frac{(51)^2}{6} \right] + \left[263 - \frac{(39)^2}{6} \right] + \left[163 - \frac{(29)^2}{6} \right] = 41.83$$

The within-conditions degrees of freedom is computed in an identical manner as is the within-groups degrees of freedom for the **single-factor between-subjects analysis of variance**. Thus, using Equation 16.9 (using the subscript WC in place of WG), $df_{WC} = N - k = 18 - 3 = 15$. The **within-conditions mean square** can now be computed: $MS_{WC} = SS_{WC}/df_{WC} = 41.83/15 = 2.79$. Employing Equation 16.44, the 95% confidence interval for the mean of the population represented by Condition 1 is computed. The value $t_{.05} = 2.13$ is the tabled critical two-tailed $t_{.05}$ value for $df_{WC} = 15$.

$$CI_{.95} = \bar{X}_j \pm t_{df_{WC}} \sqrt{\frac{MS_{WC}}{n}} = 8.5 \pm 2.13 \sqrt{\frac{2.79}{6}} = 8.5 \pm 1.45$$

Thus, the researcher can be 95% confident that the mean of the population represented by Condition 1 falls within the range 7.05 to 9.95. Stated symbolically: $7.05 \le \mu_1 \le 9.95$.

b) If one has reason to believe that the treatment in question is distinct from the other treatments, Equation 2.7 can be employed to compute the confidence interval. Specifically, if the mean value of a treatment is substantially above or below the means of the other treatments, it can be argued that one can no longer assume that the treatment shares a common variance with the other treatments. In view of this, one can take the position that the variance of the treatment is the best estimate of the variance of the population represented by that treatment (as opposed to the pooled variability of all of the treatments involved in the study). This position can also be taken even if the means of the k treatments are equal, but the treatments have substantially different estimated population variances.

If Equation 2.7 is employed to compute the 95% confidence interval for the population mean of Condition 1, the estimated variance of the population Condition 1 represents is employed in lieu of the pooled within-conditions variability. In addition, the tabled critical two-tailed $t_{.05}$ value for $df = n - 1 = 5$ is employed in the equation. The computation of the confidence interval is illustrated below. Initially the estimated population standard deviation is computed, which is then substituted in Equation 2.7.

$$\tilde{s}_1 = \sqrt{\frac{\Sigma X_1^2 - \frac{(\Sigma X_1)^2}{n}}{n - 1}} = \sqrt{\frac{443 - \frac{(51)^2}{6}}{5}} = 1.38$$

$$CI_{.95}\ \bar{X}_1 \pm t_{.05}\left(\frac{\tilde{s}_1}{\sqrt{n}}\right) = 8.5 \pm 2.57\left(\frac{1.38}{\sqrt{6}}\right) = 8.5 \pm 1.45$$

Thus, the result obtained with Equation 2.7 indicates that the range for $CI_{.95}$ is $7.05 \leq \mu_1 \leq 9.95$.[18] Note that in the case of Condition 1, the confidence intervals computed with Equations 16.44 and 2.7 are identical. This will not always be true, especially when the within-condition variability of the treatment for which the confidence interval is computed is substantially different than the within-condition variability of the other treatments.

VII. Additional Discussion of the Single-Factor Within-Subjects Analysis of Variance

1. Theoretical rationale underlying the single-factor within-subjects analysis of variance In the **single-factor within-subjects analysis of variance** the total variability can be partitioned into the following two elements: a) **Between-subjects variability** (which is represented by MS_{BS}) represents the variability between the mean scores of the n subjects. In other words, a mean value for each subject who has served in each of the k experimental conditions is computed, and MS_{BS} represents the variance of the n subject means; and b) **Within-subjects variability** (which will be represented by the notation MS_{WS}) represents variability within the k scores of each of the n subjects. In other words, for each subject the variance for that subject's k scores is computed, and the average of the n variances represents within-subjects variability. Within-subjects variability can itself be partitioned into two elements: **between-conditions variability** (which is represented by MS_{BC}) and **residual variability** (which is represented by MS_{res}). Between-conditions variability is essentially a measure of variance of the means of the k experimental conditions. In the **single-factor within-subjects analysis of variance**, it is assumed that any variability between the means of the conditions can be attributed to one or both of the following two elements: 1) **The experimental treatments**; and 2) **Experimental error**. When MS_{BC} (the value computed for between-conditions variability) is significantly greater than MS_{res} (the value computed for error variability), it is interpreted as indicating that a substantial portion of between-conditions variability is due to a treatment effect. The rationale for this is as follows.

Experimental error is random variability in the data that is beyond the control of the researcher. In a within-subjects design the average amount of variability within the k scores of each of the n subjects that cannot be accounted for on the basis of a treatment effect is employed to represent experimental error. Thus, the value computed for MS_{res} is the normal amount of variability that is expected for any subject who serves in each of k experimental conditions, if the conditions are equivalent to one another. Within this framework, residual variability is employed as a baseline to represent variability which results from factors that are beyond an experimenter's control. The experimenter assumes that even if no treatment effect is present, since such uncontrollable factors are responsible for within-subjects variability, it is logical to assume that they can produce differences of a comparable magnitude between the means of the k experimental conditions. As long as the variability between the condition means (MS_{BC}) is approximately the same as residual variability (MS_{res}), the experimenter can attribute any between-conditions variability present to experimental error. When, however, between-conditions variability is substantially greater than residual variability, it indicates that something over and above error variability is contributing to the variability between the k condition means. In such a case, it is assumed that a treatment effect is responsible for the larger value of MS_{BC} relative to the value of MS_{res}. In essence, if residual variability is subtracted from within-subjects variability, any remaining variability within the scores of subjects can be attributed to a treatment effect. If there is no treatment effect, the result of the subtraction will be zero. Of course, one can never completely rule out the possibility that if MS_{BC} is larger than MS_{res}, the larger value for MS_{BC} is entirely due to error variability. However, since the latter is unlikely, when MS_{BC} is significantly larger than MS_{res} it is interpreted as indicating the presence of a treatment effect.

Table 18.13 Alternative Summary Table for Analysis of Variance for Example 18.1

Source of variation	*SS*	*df*	*MS*	*F*
Between-subjects	36.93	5	7.39	
Within-subjects	45.35	12	3.78	
Between-conditions	40.45	2	20.23	41.29
Residual	4.9	10	.49	
Total	**82.28**	**17**		

In some sources a table employing the format depicted in Table 18.13 is used to summarize the results of a **single-factor within-subjects analysis of variance**. In contrast to Table 18.2, which does not include a row documenting within-subjects variability, Table 18.13 includes the latter variability, which is partitioned into between-conditions variability and residual variability. In point of fact, it is not necessary to compute the information documented in the row for within-subjects variability in order to compute the F ratio.

Note that in Table 18.13 the following relationships will always be true: a) $SS_T = SS_{BS} + SS_{WS}$; b) $df_T = df_{BS} + df_{WS}$; c) $SS_{WS} = SS_{BC} + SS_{\text{res}}$; and d) $df_{WS} = df_{BC} + df_{\text{res}}$. The values SS_{WS}, df_{WS}, and MS_{WS} in Table 18.13 are respectively computed with Equations 18.26, 18.27, and 18.28.

$$SS_{WS} = \Sigma X_T^2 - \sum_{i=1}^{n}\left[\frac{(\Sigma S_i)^2}{k}\right] \qquad \textbf{(Equation 18.26)}$$

$$df_{WS} = n(k - 1) \qquad \textbf{(Equation 18.27)}$$

$$MS_{WS} = \frac{SS_{WS}}{df_{WS}} \qquad \textbf{(Equation 18.28)}$$

The values $SS_{WS} = 45.35$, $df_{WS} = 12$, and $MS_{WS} = 3.78$ are computed below for Example 18.1.

$$SS_{WS} = 869 - \left[\frac{(20)^2}{3} + \frac{(25)^2}{3} + \frac{(15)^2}{3} + \frac{(25)^2}{3} + \frac{(14)^2}{3} + \frac{(20)^2}{3}\right] = 45.35$$

$$df_{WS} = 6(3 - 1) = 12 \qquad MS_{WS} = \frac{45.35}{12} = 3.78$$

Note that $SS_{WS} = SS_{BC} + SS_{res} = 40.45 + 4.9 = 45.35$ and $df_{WS} = df_{BC} + df_{res} = 2 + 10 = 12$.

2. Definitional equations for the single-factor within-subjects analysis of variance In the description of the computational protocol for the **single-factor within-subjects analysis of variance,** Equations 18.2–18.5 are employed to compute the values SS_T, SS_{BC}, SS_{BS}, and SS_{res}. The latter set of computational equations were employed, since they allow for the most efficient computation of the sum of squares values. As noted in Section IV, computational equations are derived from definitional equations which reveal the underlying logic involved in the derivation of the sums of squares. This section will describe the definitional equations for the **single-factor within-subjects analysis of variance,** and apply them to Example 18.1 in order to demonstrate that they yield the same values as the computational equations.

As noted previously, the total sum of squares (SS_T) is made up of two elements, the between-subjects sum of squares (SS_{BS}) and the within-subjects sum of squares (SS_{WS}), and that the latter sum of squares can be partitioned into the between-conditions sum of squares (SS_{BC}) and the residual sum of squares (SS_{res}). The contribution of any single subject's score to the total variability in the data can be expressed in terms of a between-subjects component and a within-subjects component. When the between-subjects component and the within-subjects component are added, the sum reflects that subject's total contribution to the overall variability in the data. The contribution of all N scores to the total variability (SS_T) and the elements that comprise it (SS_{BS} and SS_{WS}, and SS_{BC} and SS_{res} which comprise the latter) are summarized in Table 18.14. The definitional equations described in this section employ the following notation: X_{ij} represents the score of the ith subject in the jth condition, $\bar{X}_T$ represents the grand mean (which is $\bar{X}_T = (\Sigma_{j=1}^{k}\Sigma_{i=1}^{n}X_{ij})/N = 119/18 = 6.61$), $\bar{X}_j$ represents the mean of the jth condition, and $\bar{S}_i$ represents the mean of the k scores of the ith subject.

Equation 18.29 is the definitional equation for the **total sum of squares.**[19]

$$SS_T = \sum_{j=1}^{k}\sum_{i=1}^{n}(X_{ij} - \bar{X}_T)^2 \qquad \textbf{(Equation 18.29)}$$

In employing Equation 18.29 to compute SS_T, the grand mean ($\bar{X}_T$) is subtracted from each of the N scores and each of the N difference scores is squared. The total sum of squares (SS_T) is the sum of the N squared difference scores. Equation 18.29 is computationally equivalent to Equation 18.2.

Equation 18.30 is the definitional equation for the **between-subjects sum of squares.**

$$SS_{BS} = k\sum_{i=1}^{n}(\bar{S}_i - \bar{X}_T)^2 \qquad \textbf{(Equation 18.30)}$$

In employing Equation 18.30 to compute SS_{BS}, the following operations are carried out for each of the n subjects. The grand mean ($\bar{X}_T$) is subtracted from the mean of the subject's k scores. The difference score is squared and the squared difference score is multiplied by the number of experimental conditions (k). After this is done for all n subjects, the values that have been obtained for each subject as a result of multiplying the squared difference score by k are summed. The resulting value represents the between-subjects sum of squares (SS_{BS}). Equation 18.30 is computationally equivalent to Equation 18.4.

Equation 18.31 is the definitional equation for the **within-subjects sum of squares.**

$$SS_{WS} = \sum_{i=1}^{n} \sum_{j=1}^{k} (X_{ij} - \bar{S}_i)^2 \qquad \textbf{(Equation 18.31)}$$

In employing Equation 18.31 to compute SS_{WS}, the following operations are carried out for the k scores of each of the n subjects. The mean of a subject's k scores ($\bar{S}_i$) is subtracted from each of the subject's scores, and the k difference scores for that subject are squared. The sum of the k squared difference scores for all n subjects (i.e., the sum total of N squared difference scores) represents the within-subjects sum of squares (SS_{WS}). Equation 18.31 is computationally equivalent to Equation 18.26.

Equation 18.32 is the definitional equation for the **between-conditions sum of squares.**

$$SS_{BC} = n \sum_{j=1}^{k} (\bar{X}_j - \bar{X}_T)^2 \qquad \textbf{(Equation 18.32)}$$

In employing Equation 18.32 to compute SS_{BC}, the following operations are carried out for each experimental condition. The grand mean ($\bar{X}_T$) is subtracted from the condition mean ($\bar{X}_j$). The difference score is squared, and the squared difference score is multiplied by the number of scores in that condition (n). After this is done for all k conditions, the values that have been obtained for each condition as a result of multiplying the squared difference score by the number of subjects in the condition are summed. The resulting value represents the between-conditions sum of squares (SS_{BC}). Equation 18.32 is computationally equivalent to Equation 18.3. An alternative but equivalent method of obtaining SS_{BC} (which is employed in deriving SS_{BC} in Table 18.14) is as follows: Within each condition, for each of the n subjects the grand mean is subtracted from the condition mean, each difference score is squared, and upon doing this for all k conditions, the N squared difference scores are summed.

Equation 18.33 is the definitional equation for the **residual sum of squares.**

$$SS_{res} = \sum_{j=1}^{k} \sum_{i=1}^{n} \left[(X_{ij} - \bar{X}_T) - (\bar{S}_i - \bar{X}_T) - (\bar{X}_j - \bar{X}_T)\right]^2 \qquad \textbf{(Equation 18.33)}$$

In employing Equation 18.33 to compute SS_{res}, the following operations are carried out for each of the N scores: a) The grand mean ($\bar{X}_T$) is subtracted from the score (X_{ij}); b) The grand mean ($\bar{X}_T$) is subtracted from the mean of the k scores for that subject ($\bar{S}_i$); and c) The grand mean ($\bar{X}_T$) is subtracted from the mean of the condition from which the score is derived ($\bar{X}_j$). The value of the difference score obtained in b) is subtracted from the value of the difference score obtained in a), and the difference score obtained in c) is subtracted from the resulting difference. The resulting value is squared, and the sum of the squared values for all N scores represents the residual sum of squares (SS_{res}). Note that in Equation 18.33, for each subject a between-subjects and between-conditions component of variability is subtracted from the subject's contribution to the total variability, resulting in the subject's contribution to the residual variability. Equation 18.33 is computationally equivalent to Equations 18.5/18.6.

Table 18.14 illustrates the use of Equations 18.29–18.33 with the data for Example 18.1.[20] In the computations summarized in Table 18.14, the following S_i values are employed: $\bar{S}_1 = 20/3 = 6.67$, $\bar{S}_2 = 25/3 = 8.33$, $\bar{S}_3 = 15/3 = 5$, $\bar{S}_4 = 25/3 = 8.33$, $\bar{S}_5 = 14/3 = 4.67$, $\bar{S}_6 = 20/3 = 6.67$. The resulting values of SS_T, SS_{BS}, SS_{WS}, SS_{BC}, and SS_{res} are identical to those obtained with the computational equations (Equations 18.2, 18.4, 18.26, 18.3, and 18.5/18.6). Any minimal discrepancies are the result of rounding off error.

Table 18.14 Computation of Sums of Squares for Example 18.1 with Definitional Equations

	(Subject, Condition)	X_{ij}	$SS_T = \sum_{j=1}^{k}\sum_{i=1}^{n}(X_{ij} - \bar{X}_T)^2$	$SS_{BS} = k\sum_{i=1}^{n}(\bar{S}_i - \bar{X}_T)^2$	$SS_{WS} = \sum_{i=1}^{n}\sum_{j=1}^{k}(X_{ij} - \bar{S}_i)^2$
Condition 1	(1,1)	9	$(9.00-6.61)^2 = 5.71$	$(6.67-6.61)^2 = .00$	$(9.00-6.67)^2 = 5.43$
	(2,1)	10	$(10.00-6.61)^2 = 11.49$	$(8.33-6.61)^2 = 2.96$	$(10.00-8.33)^2 = 2.79$
	(3,1)	7	$(7.00-6.61)^2 = .15$	$(5.00-6.61)^2 = 2.59$	$(7.00-5.00)^2 = 4.00$
	(4,1)	10	$(10.00-6.61)^2 = 11.49$	$(8.33-6.61)^2 = 2.96$	$(10.00-8.33)^2 = 2.79$
	(5,1)	7	$(7.00-6.61)^2 = .15$	$(4.67-6.61)^2 = 3.76$	$(7.00-4.67)^2 = 5.43$
	(6,1)	8	$(8.00-6.61)^2 = 1.93$	$(6.67-6.61)^2 = .00$	$(8.00-6.67)^2 = 1.77$
Condition 2	(1,2)	7	$(7.00-6.61)^2 = .15$	$(6.67-6.61)^2 = .00$	$(7.00-6.67)^2 = .11$
	(2,2)	8	$(8.00-6.61)^2 = 1.93$	$(8.33-6.61)^2 = 2.96$	$(8.00-8.33)^2 = .11$
	(3,2)	5	$(5.00-6.61)^2 = 2.59$	$(5.00-6.61)^2 = 2.59$	$(5.00-5.00)^2 = .00$
	(4,2)	8	$(8.00-6.61)^2 = 1.93$	$(8.33-6.61)^2 = 2.96$	$(8.00-8.33)^2 = .11$
	(5,2)	5	$(5.00-6.61)^2 = 2.59$	$(4.67-6.61)^2 = 3.76$	$(5.00-4.67)^2 = .11$
	(6,2)	6	$(6.00-6.61)^2 = .37$	$(6.67-6.61)^2 = .00$	$(6.00-6.67)^2 = .45$
Condition 3	(1,3)	4	$(4.00-6.61)^2 = 6.81$	$(6.67-6.61)^2 = .00$	$(4.00-6.67)^2 = 7.13$
	(2,3)	7	$(7.00-6.61)^2 = .15$	$(8.33-6.61)^2 = 2.96$	$(7.00-8.33)^2 = 1.77$
	(3,3)	3	$(3.00-6.61)^2 = 13.03$	$(5.00-6.61)^2 = 2.59$	$(3.00-5.00)^2 = 4.00$
	(4,3)	7	$(7.00-6.61)^2 = .15$	$(8.33-6.61)^2 = 2.96$	$(7.00-8.33)^2 = 1.77$
	(5,3)	2	$(2.00-6.61)^2 = 21.25$	$(4.67-6.61)^2 = 3.76$	$(2.00-4.67)^2 = 7.13$
	(6,3)	6	$(6.00-6.61)^2 = .37$	$(6.67-6.61)^2 = .00$	$(6.00-6.67)^2 = .45$
			$SS_T = 82.24$	$SS_{BS} = 36.81$	$SS_{WS} = 45.35$

	(Subject, Condition)	X_{ij}	$SS_{BC} = n\sum_{j=1}^{k}(\bar{X}_j - \bar{X}_T)^2$	$SS_{\text{res}} = \sum_{j=1}^{k}\sum_{i=1}^{n}[(X_{ij} - \bar{X}_T) - (\bar{S}_i - \bar{X}_T) - (\bar{X}_j - \bar{X}_T)]^2$
Condition 1	(1,1)	9	$(8.50-6.61)^2 = 3.57$	$[(9.00-6.61)-(6.67-6.61)-(8.50-6.61)]^2 = .19$
	(2,1)	10	$(8.50-6.61)^2 = 3.57$	$[(10.00-6.61)-(8.33-6.61)-(8.50-6.61)]^2 = .05$
	(3,1)	7	$(8.50-6.61)^2 = 3.57$	$[(7.00-6.61)-(5.00-6.61)-(8.50-6.61)]^2 = .01$
	(4,1)	10	$(8.50-6.61)^2 = 3.57$	$[(10.00-6.61)-(8.33-6.61)-(8.50-6.61)]^2 = .05$
	(5,1)	7	$(8.50-6.61)^2 = 3.57$	$[(7.00-6.61)-(4.67-6.61)-(8.50-6.61)]^2 = .19$
	(6,1)	8	$(8.50-6.61)^2 = 3.57$	$[(8.00-6.61)-(6.67-6.61)-(8.50-6.61)]^2 = .31$
Condition 2	(1,2)	7	$(6.50-6.61)^2 = .01$	$[(7.00-6.61)-(6.67-6.61)-(6.50-6.61)]^2 = .19$
	(2,2)	8	$(6.50-6.61)^2 = .01$	$[(8.00-6.61)-(8.33-6.61)-(6.50-6.61)]^2 = .05$
	(3,2)	5	$(6.50-6.61)^2 = .01$	$[(5.00-6.61)-(5.00-6.61)-(6.50-6.61)]^2 = .01$
	(4,2)	8	$(6.50-6.61)^2 = .01$	$[(8.00-6.61)-(8.33-6.61)-(6.50-6.61)]^2 = .05$
	(5,2)	5	$(6.50-6.61)^2 = .01$	$[(5.00-6.61)-(4.67-6.61)-(6.50-6.61)]^2 = .19$
	(6,2)	6	$(6.50-6.61)^2 = .01$	$[(6.00-6.61)-(6.67-6.61)-(6.50-6.61)]^2 = .31$
Condition 3	(1,3)	4	$(4.83-6.61)^2 = 3.17$	$[(4.00-6.61)-(6.67-6.61)-(4.83-6.61)]^2 = .79$
	(2,3)	7	$(4.83-6.61)^2 = 3.17$	$[(7.00-6.61)-(8.33-6.61)-(4.83-6.61)]^2 = .20$
	(3,3)	3	$(4.83-6.61)^2 = 3.17$	$[(3.00-6.61)-(5.00-6.61)-(4.83-6.61)]^2 = .05$
	(4,3)	7	$(4.83-6.61)^2 = 3.17$	$[(7.00-6.61)-(8.33-6.61)-(4.83-6.61)]^2 = .20$
	(5,3)	2	$(4.83-6.61)^2 = 3.17$	$[(2.00-6.61)-(4.67-6.61)-(4.83-6.61)]^2 = .79$
	(6,3)	6	$(4.83-6.61)^2 = 3.17$	$[(6.00-6.61)-(6.67-6.61)-(4.83-6.61)]^2 = 1.23$
			$SS_{BC} = 40.50$	$SS_{\text{res}} = 4.86$

Table 18.15 Summary Table of Single-Factor Between-Subjects Analysis of Variance for Example 18.1

Source of variation	*SS*	*df*	*MS*	*F*
Between-groups	40.45	2	20.23	7.25
Within-groups	41.83	15	2.79	
Total	**82.28**	**17**		

3. Relative power of the single-factor within-subjects analysis of variance and the single-factor between-subjects analysis of variance The use of MS_{res} as the measure of error variability (as opposed to MS_{WC}) for the **single-factor within-subjects analysis of variance** provides for an optimally powerful test of an alternative hypothesis.[21] The reason why MS_{res} allows for a more powerful test of an alternative hypothesis than MS_{WC} is because when no treatment effect is present in the data, it is expected that the average variability of the k scores of n subjects will be less than the average variability of the scores of n different subjects who serve in any single experimental condition (in an experiment involving k experimental conditions).

To illustrate this point, let us assume that the data for Example 18.1 are obtained in an experiment employing an independent groups/between-subjects design, and as a result of the latter $MS_{WC} = MS_{WG}$ is employed as the measure of error variability. Thus, we will assume that each of $N = 18$ subjects is randomly assigned to one of $k = 3$ experimental conditions, resulting in $n = 6$ scores per condition. The data for such an experiment will be evaluated with a **single-factor between-subjects analysis of variance.** In conducting the computations for the latter analysis, the value of SS_T is computed with Equation 16.2 (which, in fact, is identical to Equation 18.2, which is employed to compute SS_T when Example 18.1 is evaluated with a **single-factor within-subjects analysis of variance**). Thus, $SS_T = 82.28$. Equation 16.3, which is employed to compute the between-groups sum of squares (SS_{BG}) is, in fact, identical to Equation 18.3 (which is employed in Section IV to compute the between-conditions sum of squares (SS_{BC})). Thus, $SS_{BG} = SS_{BC} = 40.45$. The within-groups sum of squares (SS_{WG}) can be computed with Equation 16.4. Thus, $SS_{WG} = SS_T - SS_{BG} = 82.28 - 40.45 = 41.83$. Note that the latter value is identical to the value computed with Equation 18.25 (which as noted earlier is computationally equivalent to Equation 16.5, which yields the same value as Equation 16.4). Employing the values $k = 3$ and $N = 18$ in Equations 16.8–16.10, the values $df_{BG} = 2$, $df_{WG} = 15$, and $df_T = 17$ are computed. Substituting the appropriate degrees of freedom in Equations 16.6 and 16.7, the values $MS_{BG} = 40.45/2 = 20.23$ and $MS_{WG} = 41.83/15 = 2.79$ are computed. Using Equation 16.12, $F = 20.23/2.79 = 7.25$. Table 18.15 is the summary table of the analysis of variance.

Since $df_{num} = df_{BG} = 2$ and $df_{den} = df_{WG} = 15$, $F_{.05} = 3.68$ and $F_{.01} = 6.36$ are the critical values in **Table A10** that are employed to evaluate the nondirectional alternative hypothesis. Since the obtained value $F = 7.25$ is greater than both of the aforementioned critical values, the alternative hypothesis is supported at both the.05 and .01 levels. Note, however, that the value $F = 7.25$ is substantially less than the value $F = 41.29$, which is obtained when the same set of data is evaluated with the **single-factor within-subjects analysis of variance.** Although the value $F = 7.25$ obtained for a between-subjects analysis is significant at both the .05 and .01 levels, $F = 7.25$ is not very far removed from the tabled critical value $F_{.01} = 6.36$. The value $F = 41.29$, on the other hand, is well above the tabled critical value $F_{.01} = 7.56$ (which is the tabled critical .01 value employed for the **single-factor within-subjects analysis of variance** for $df_{BC} = 2$ and $df_{res} = 10$). The fact that the difference between the computed F value and the tabled critical $F_{.01}$ value is much larger when the **single-factor within-subjects analysis of variance** is employed, illustrates that a within-subjects analysis provides a more powerful test of an alternative hypothesis than a between-subjects analysis.[22]

It should be noted that for the same set of data, the tabled critical F value at a given level of significance for a **single-factor between-subjects analysis of variance** will always be lower than the tabled critical F value for a **single-factor within-subjects analysis of variance** (unless there is an extremely large number of scores in each condition, in which case the tabled critical F values for both analyses will be equivalent). This is the case since (as long as n is not extremely large) the number of degrees of freedom associated with the denominator of the F ratio will always be larger for a **single-factor between-subjects**

analysis of variance (assuming the values of $n = n_j$ and k for both analyses are equal) than for a **single-factor within-subjects analysis of variance** — i.e., $df_{WG} > df_{res}$. It is important to note, however, that any loss of degrees of freedom associated with a within-subjects analysis will more than likely be offset as a result of employing MS_{res} as the error term in the computation of the F ratio. A final point that should be made is that if in a within-subjects design subjects' scores in the k experimental conditions are not correlated with one another (which is highly unlikely), a **single-factor within-subjects analysis of variance** and a **single-factor between-subjects analysis of variance** (as well as a ***t*** **test for two dependent samples** and a ***t*** **test for two independent samples** when $k = 2$) will yield comparable results.

4. Equivalency of the single-factor within-subjects analysis of variance and the *t* test for two dependent samples when *k* = 2 Interval/ratio data for an experiment involving $k = 2$ dependent samples can be evaluated with either a **single-factor within-subjects analysis of variance** or a t **test for two dependent samples.** When both of the aforementioned tests are employed to evaluate the same set of data they will yield the same result. Specifically, the following will always be true with respect to the relationship between the computed F and t values for the same set of data: $F = t^2$ and $t = \sqrt{F}$. It will also be the case that the square of the tabled critical t value at a prespecified level of significance for $df = n - 1$ will be equal to the tabled critical F value at the same level of significance for $df_{BC} = 1$ and df_{res} (which will be $df_{res} = (n - 1)(k - 1) = (n - 1)(2 - 1) = n - 1$, which is equivalent to the value $df = n - 1$ employed for the ***t*** **test for two dependent samples**).

To illustrate the equivalency of the results obtained with the **single-factor within-subjects analysis of variance** and the ***t*** **test for two dependent samples** when $k = 2$, an F value will be computed for Example 12.1. The value $t = 2.86$ is obtained (a more precise value $t = 2.848$ is obtained if all computations are carried out to 3 decimal places) for the latter example when the ***t*** **test for two dependent samples** is employed. When the same set of data is evaluated with the **single-factor within-subjects analysis of variance,** the value $F = 8.11$ is computed. Note that $(t = 2.848)^2 = (F = 8.11)$. Equations 18.2–18.5 are employed below to compute the values SS_T, SS_{BC}, SS_{BS}, and SS_{res} for Example 12.1. Since $k = 2$, $n = 10$, and $nk = N = 20$, $df_{BC} = 2 - 1 = 1$, $df_{BS} = 10 - 1 = 9$, $df_{res} = (10 - 1)(2 - 1) = 9$, and $df_T = 20 - 1 = 19$. The full analysis of variance is summarized in Table 18.16.

$$SS_T = 440 - \frac{(78)^2}{20} = 135.8 \qquad SS_{BC} = \left[\frac{(47)^2 + (31)^2}{10}\right] - \frac{(78)^2}{20} = 12.8$$

$$SS_{BS} = \left[\frac{(17)^2+(4)^2+(4)^2+(6)^2+(9)^2+(4)^2+(11)^2+(13)^2+(9)^2+(1)^2}{2}\right] - \frac{(78)^2}{20} = 108.8$$

$$SS_{res} = 135.8 - 12.8 - 108.8 = 14.2$$

Table 18.16 Summary Table of Analysis of Variance for Example 12.1

Source of variation	*SS*	*df*	*MS*	*F*
Between-subjects	108.8	9		
Between-conditions	12.8	1	12.80	8.11
Residual	14.2	9	1.58	
Total	**135.8**	**19**		

For $df_{BC} = 1$ and $df_{\text{res}} = 9$, the tabled critical .05 and .01 values are $F_{.05} = 5.12$ and $F_{.01} = 10.56$ (which are appropriate for a nondirectional analysis). Note that (if one takes into account rounding off error) the square roots of the aforementioned tabled critical values are (for $df = 9$) the tabled critical two-tailed values $t_{.05} = 2.26$ and $t_{.01} = 3.25$ that are employed in Example 12.1 to evaluate the value $t = 2.86$. Since the obtained value $F = 8.11$ is greater than $F_{.05} = 5.12$ but less than $F_{.01} = 10.56$, the nondirectional alternative hypothesis H_1: $\mu_1 \neq \mu_2$ is supported, but only at the .05 level. The directional alternative hypothesis H_1: $\mu_1 > \mu_2$ is supported at both the .05 and .01 levels, since $F = 8.11$ is greater than the tabled critical one-tailed .05 and .01 values $F_{.05} = 3.36$ and $F_{.01} = 7.95$ (the square roots of which are the tabled critical one-tailed .05 and .01 values $t_{.05} = 1.83$ and $t_{.01} = 2.82$ employed for Example 12.1).[23] The conclusions derived from the **single-factor within-subjects analysis of variance** are identical to those reached when the data are evaluated with the ***t* test for two dependent samples**.

VIII. Additional Examples Illustrating the Use of the Single-Factor Within-Subjects Analysis of Variance

Since the **single-factor within-subjects analysis of variance** can be employed to evaluate interval/ratio data for any dependent samples design involving two or more experimental conditions, it can be used to evaluate any of the examples that are evaluated with the ***t* test for two dependent samples** (with the exception of Example 12.2). Examples 18.2–18.6 are respectively extensions of Examples 12.1, 12.3, 12.5, 12.7, and 12.6. As is the case with Examples 12.3 and 12.5, Examples 18.3 and 18.4 employ matched subjects, and are thus evaluated as a within-subjects design. Examples 18.6 and 18.7 represent extensions of the **before-after design** to a design involving $k = 3$ experimental conditions. Since the data for all of the examples are identical to the data employed in Example 18.1, they yield the same result.

Example 18.2 *A psychologist conducts a study in order to determine whether people exhibit more emotionality when they are exposed to sexually explicit words, aggressively toned words, or neutral words. Each of six subjects is shown a list of* 15 *randomly arranged words, which are projected on a screen one at a time for a period of five seconds. Five of the words on the list are sexually explicit, five of the words are aggressively toned, and five of the words are neutral. As each word is projected on the screen, a subject is instructed to say the word softly to him or herself. As a subject does this, sensors attached to the palms of the subject's hands record galvanic skin response (GSR), which is used by the psychologist as a measure of emotionality. The psychologist computes the following three scores for each subject, one score for each of the three experimental conditions:* **Condition 1**: *GSR/Sexually explicit — The average GSR score for the five sexually explicit words;* **Condition 2**: *GSR/Aggressively toned — The average GSR score for the five aggressively toned words;* **Condition 3**: *GSR/Neutral — The average GSR score for the five neutral words. The GSR/Sexually explicit, GSR/Aggressively toned, and GSR/Neutral scores of the six subjects follow. (The higher the score the higher the level of emotionality.)* **Subject 1** (9, 7, 4); **Subject 2** (10, 8, 7); **Subject 3** (7, 5, 3); **Subject 4** (10, 8, 7); **Subject 5** (7, 5, 2); **Subject 6** (8, 6, 6). *Do subjects exhibit differences in emotionality with respect to the three categories of words?*

Example 18.3 *A psychologist conducts a study in order to determine whether people exhibit more emotionality when they are exposed to sexually explicit words, aggressively toned words, or neutral words. Six sets of identical triplets are employed as subjects and within each set of triplets one member of the set is treated as follows: a) One of the triplets is randomly assigned to* Condition 1, *in which the subject is shown a list of five sexually explicit words;*

b) One of the triplets is randomly assigned to Condition 2, *in which the subject is shown a list of five aggressively toned words; c) and One of the triplets is randomly assigned to* Condition 3, *in which the subject is shown a list of five neutral words. As each word is projected on the screen, a subject is instructed to say the word softly to him or herself. As a subject does this, sensors attached to the palms of the subject's hands record galvanic skin response (GSR), which is used by the psychologist as a measure of emotionality. The psychologist computes the following three scores for each set of triplets to represent the emotionality score for each of the experimental conditions:* **Condition 1**: *GSR/Sexually explicit — The average GSR score for the subject presented with the five sexually explicit words;* **Condition 2**: *GSR/Aggressively toned — The average GSR score for the subject presented with the five aggressively toned words;* **Condition 3**: *GSR/Neutral — The average GSR score for the subject presented with the five neutral words. The GSR/Sexually explicit, GSR/Aggressively toned, and GSR/Neutral scores of the six sets of triplets follow. (The first score for each triplet set represents the score of the subject presented with the sexually explicit words, the second score represents the score of the subject presented with the aggressively toned words, and the third score represents the score of the subject presented with the neutral words. The higher the score the higher the level of emotionality.)* **Triplet set 1** (9, 7, 4); **Triplet set 2** (10, 8, 7); **Triplet set 3** (7, 5, 3); **Triplet set 4** (10, 8, 7); **Triplet set 5** (7, 5, 2); **Triplet set 6** (8, 6, 6). *Do subjects exhibit differences in emotionality with respect to the three categories of words?*

Example 18.4 *A researcher wants to assess the impact of different types of punishment on the emotionality of mice. Six sets of mice derived from six separate litters are employed as subjects. Within each set, one of the litter mates is randomly assigned to one of the three experimental conditions. During the course of the experiment each mouse is sequestered in an experimental chamber. While in the chamber, each of the six mice in* Condition 1 *is periodically presented with a loud noise, and each of the six mice in* Condition 2 *is periodically presented with a blast of cold air. The six mice in* Condition 3 *(which is a no treatment control condition) are not exposed to any punishment. The presentation of the punitive stimulus for the animals in* Conditions 1 *and* 2 *is generated by a machine that randomly presents the stimulus throughout the duration of the time an animal is in the chamber. The dependent variable of emotionality employed in the study is the number of times each mouse defecates while in the experimental chamber. The number of episodes of defecation for the six sets of mice follows. (The higher the score the higher the level of emotionality.)* **Litter 1** (9, 7, 4); **Litter 2** (10, 8, 7); **Litter 3** (7, 5, 3); **Litter 4** (10, 8, 7); **Litter 5** (7, 5, 2); **Litter 6** (8, 6, 6). *Do subjects exhibit differences in emotionality under the different experimental conditions?*

Example 18.5 *A study is conducted in order to evaluate the relative efficacy of two drugs (Clearoxin and Lesionoxin) and a placebo on chronic psoriasis. Six subjects afflicted with chronic psoriasis participate in the study. Each subject is exposed to both drugs and the placebo for a six-month period, with a three-month hiatus between treatments. Within the six subjects, the order of presentation of the experimental treatments is completely counterbalanced. The dependent variable employed in the study is a rating of the severity of a subject's lesions under the three experimental conditions. The higher the rating the more severe a subject's psoriasis. The scores of the six subjects under the three treatment conditions follow. (The first score represents the Clearoxin condition (which represents* Condition 1), *the second score the Lesionoxin condition (which represents* Condition 2), *and the third score the placebo condition (which represents* Condition 3).) **Subject 1** (9, 7, 4); **Subject 2** (10, 8, 7); **Subject 3** (7, 5, 3); **Subject 4** (10, 8, 7); **Subject 5** (7, 5, 2); **Subject 6** (8, 6, 6). *Do the data indicate differences in subjects' responses under the three experimental conditions?*

Example 18.6 *In order to assess the efficacy of electroconvulsive therapy (ECT), a psychiatrist evaluates six clinically depressed patients who receive a series of ECT treatments. Each patient is evaluated at the following three points in time: a) One day prior to the first treatment in the ECT series; b) The day following the final treatment in the ECT series; and c) Six months after the final treatment in the ECT series. During each evaluation period a standardized interview is used to operationalize a patient's level of depression, and on the basis of the interview a patient is assigned a score ranging from* 0 *to* 10. *The higher a patient's score the more depressed the patient. The depression scores of the six patients during each of the three time periods follow:* **Patient 1** (9, 7, 4); **Patient 2** (10, 8, 7); **Patient 3** (7, 5, 3); **Patient 4** (10, 8, 7); **Patient 5** (7, 5, 2); **Patient 6** (8, 6, 6). *Do the data indicate that the ECT is effective, and, if so, is the effect maintained six months after the treatment?*

Although, as described, Example 18.6 can be evaluated with a **single-factor within-subjects analysis of variance**, the design of the study does not allow one to rule out the potential impact of confounding variables. To be more specific, Example 18.6 (which represents an extension of a **before-after design** to more than two measurement periods) does not allow a researcher to draw definitive conclusions with respect to whether any observed changes in mood are, in fact, due to the ECT treatments.[24] Thus, even if there is a significant decrease in subjects' depression scores following the final ECT treatment, and the effect is still present six months later, factors other than ECT can account for such a result. As an example, all of the patients may have been depressed about a problem related to the economy, and if, in fact, during the course of the study the economy improves dramatically, the observed changes in mood can be attributed to the improved economy rather than the ECT. In order for the design of the above study to be suitable, it is necessary to include a control group — specifically, a comparable group of depressed patients who are not given ECT (or are given "sham" ECT treatments). By contrasting the depression scores of the control group with those of the treatment group, one can determine whether any observed differences across the three time periods are in fact attributable to the ECT. Inclusion of such a control group would require that the design of the above study be modified into a **mixed factorial design**. The latter design and the analysis of variance employed to evaluate it are discussed in Section IX of the **between-subjects factorial analysis of variance (Test 21)**.

Example 18.7 *In order to assess the efficacy of a drug which a pharmaceutical company claims is effective in treating hyperactivity, six hyperactive children are evaluated during the following three time periods: a) One week prior to taking the drug; b) After a child has taken the drug for six consecutive months; and c) Six months after the drug is discontinued. The children are observed by judges who employ a standardized procedure for evaluating hyperactivity. During each time period a child is assigned a score between* 0 *and* 10, *in which the higher the score the higher the level of hyperactivity. During the evaluation process, the judges are blind with respect to whether a child is taking medication at the time he or she is evaluated. The hyperactivity scores of the six children during the three time periods follow:* **Child 1**: (9, 7, 4); **Child 2**: (10, 8, 7); **Child 3**: (7, 5, 3); **Child 4**: (10, 8, 7); **Child 5**: (7, 5, 2); **Child 6**: (8, 6, 6). *Do the data indicate that the drug is effective?*

Since it lacks a control group, Example 18.7 is subject to the same criticism that is noted for Example 18.6. Because of the lack of a control group (i.e., a group of hyperactive children who do not receive medication), any observed differences in hyperactivity between two or more of the measurement periods can be the result of extraneous factors in the external environment or physiological/maturational changes in the children that are independent of whether a child is taking the drug. In spite of its limitations, it is not unusual to encounter the use of the design employed in Example 18.7 (which is commonly referred to as an **ABA design**) in behavior modification research. Such designs are most commonly employed with individual subjects in order to assess the efficacy of a treatment protocol. The letters **A** and **B** in an **ABA design** refer to whether or not a treatment is in effect during a specific time period. In Example 18.7, Time period 1 is designated **A** since no treatment is in effect. This initial measure of the subject's behavior provides the researcher with a baseline measure of hyperactivity. During Time period 2, which is designated by the letter **B**, the treatment is in effect. If the treatment is effective, a decrease in hyperactivity in Time period 2 relative to Time period 1 is expected. Time period 3 is once again designated **A**, since the treatment is no longer employed. If, in fact, the treatment is effective it is expected that a subject's level of hyperactivity during Time period 3 will be higher than in Time period 2, and, in fact, return to the baseline level obtained during Time period 1 (unless, of course, the drug has a permanent residual effect). When an **ABA design** is employed with an individual subject, the format of the data resulting from such a study is not suitable for evaluation with an analysis of variance.

References

Box, G. E. P. (1954). Some theorems on quadratic forms applied in the study of analysis of variance problems, II. Effect of inequality of variances and correlation between error in two-way classification. **Annals of Mathematical Statistics**, 25, 484–498.

Geisser, S. and Greenhouse, S. W. (1958). An extension of Box's results to the use of the *F* distribution in multivariate analysis. **Annals of Mathematical Statistics**, 29, 885–891.

Howell, D. C (1992). **Statistical methods for psychology** (3rd ed). Boston: PWS–Kent Publishing Company.

Keppel, G. (1991) **Design and analysis: A researcher's handbook** (3rd ed.). Englewood Cliffs, NJ: Prentice Hall.

Keppel, G. and Zedeck, S. (1989) **Data analysis for research designs**. New York: W. H. Freeman and Company.

Kirk, R. E. (1982) **Experimental design: Procedures for the behavioral sciences** (2nd ed.). Belmont, CA: Brooks/Cole Publishing Company.

Maxwell, S. E. and Delaney, H. D. (1990) **Designing experiments and analyzing data.** Belmont, CA: Wadsworth Publishing Company.

McFatter, R. M. and Gollob, H. F. (1986). The power of hypothesis tests for comparisons. **Educational and Psychological Measurement**, 46, 883–886.

Myers, J. L. and Well, A. D. (1991). **Research design and statistical analysis.** New York: Harper Collins Publishers.

Tabachnick, B. C. and Fidell, L. S. (1989). **Using multivariate statistics** (2nd ed.). New York: HarperCollins Publishers.

Winer, B. J., Brown, D. R. and Michels, K. M. (1991). **Statistical principles in experimental design** (3rd ed.). New York: McGraw–Hill Publishing Company.

Endnotes

1. A **within-subjects/repeated-measures design** in which each subject serves under each of the *k* levels of the independent variable is often described as a special case of a **randomized-blocks design**. The term **randomized-blocks design** is most commonly employed to describe a dependent samples design involving matched subjects. As an example, assume that 10 sets of identical triplets are employed in a study to determine the efficacy of two drugs when compared with a placebo. Within each set of triplets one

of the members is randomly assigned to each of the three experimental conditions. Such a design is described in various sources as a **matched-subjects/samples design, a dependent samples design**, or a **randomized-blocks design.** Within the usage of the term **randomized-blocks design**, each set of triplets constitutes a **block**, and consequently 10 blocks are employed in the study with three subjects in each block.

2. It should be noted that if an experiment is confounded, one cannot conclude that a significant portion of **between-conditions** variability is attributed to the independent variable. This is the case, since if one or more confounding variables systematically vary with the levels of the independent variable, a significant difference can be due to a confounding variable rather than the independent variable.

3. A discussion of counterbalancing can be found in Section VII of the ***t* test for two dependent samples**).

4. In other words, each subject is tested under one of the six possible presentation orders for the three experimental conditions, and within the sample of six subjects each of the presentation orders is presented once. Specifically, the following six presentation orders are employed: 1,2,3; 1,3,2; 2,1,3; 2,3,1; 3,1,2; 3,2,1.

5. Although it is possible to conduct a directional analysis, such an analysis will not be described with respect to the analysis of variance. A discussion of a directional analysis when $k = 2$ can be found under the ***t* test for two dependent samples**. In addition, a discussion of one-tailed F values can be found in Section VI of the ***t* test for two independent samples** under the discussion of the **Hartley's F_{max} test for homogeneity of variance/*F* test for two population variances (Test 8a).** A discussion of the evaluation of a directional alternative hypothesis when $k \geq 2$ can be found in Section VII of the **chi-square goodness-of-fit test (Test 5).** Although the latter discussion is in reference to analysis of a k independent samples design involving categorical data, the general principles regarding analysis of a directional alternative hypothesis when $k \geq 2$ are applicable to the analysis of variance.

6. In Section VII it is noted that the sum of **between-conditions variability** and **residual variability** represents what is referred to as **within-subjects variability.** The sum of squares of **within-subjects variability** (SS_{WS}) is the sum of **between-conditions variability** and **residual variability** — i.e., $SS_{WS} = SS_{BC} + SS_{res}$.

7. Since there is an equal number of scores in each condition, the Equation for SS_{BC} can also be written as follows:

$$SS_{BC} = \left[\frac{(\Sigma X_1)^2 + (\Sigma X_2)^2 + \cdots + (\Sigma X_k)^2}{n}\right] - \frac{(\Sigma X_T)^2}{N}$$

Thus: $$SS_{BC} = \frac{[(51)^2 + (39)^2 + (29)^2]}{6} - \frac{(119)^2}{18} = 40.45$$

8. The equation for SS_{BS} can also be written as follows:

$$SS_{BS} = \frac{[(\Sigma S_1)^2 + (\Sigma S_2)^2 + \cdots + (\Sigma S_n)^2]}{k} - \frac{(\Sigma X_T)^2}{n} =$$

$$\frac{[(20)^2 + (25)^2 + (15)^2 + (14)^2 + (20)^2]}{3} - \frac{(119)^2}{18} = 36.93$$

9. In the interest of accuracy, as is the case with the **single-factor between-subjects analysis of variance**, a significant omnibus F value indicates that there is at least one significant difference among all possible comparisons that can be conducted. Thus, it is theoretically possible that none of the simple/pairwise comparisons are significant, and that the significant difference (or differences) involves one or more complex comparisons.

10. As noted in Section VI of the **single-factor between-subjects analysis of variance**, in some instances the $CD_{B/D}$ value associated with the **Bonferroni–Dunn test** will be larger than the CD_S value associated with the **Scheffé test**. However, when there are $c = 3$ comparisons, CD_S will be greater than $CD_{B/D}$.

11. One can, of course, conduct a replication study and base the estimate of MS_{res} on the value of MS_{res} obtained for the comparison in the latter study. In point of fact, one or more replication studies can serve as a basis for obtaining the best possible estimate of error variability to employ for any comparison conducted following an analysis of variance.

12. If the means of each of the conditions for which a composite mean is computed are weighted equally, an even simpler method for computing the composite score of a subject is to add the subject's scores and divide the sum by the number of conditions that are involved. Thus, the composite score of Subject 1 can be obtained by adding 9 and 7 and dividing by 2. The averaging procedure will only work if all of the means are weighted equally. The protocol described in Section VI must be employed in instances where a comparison involves unequal weighting of means.

13. The same result is obtained if (for the three difference scores) the score in the first condition noted is subtracted from the score in the second condition noted (i.e., Condition 2–Condition 1; Condition 3–Condition 1; Condition 3–Condition 2).

14. If the variance of Condition 2 is employed to represent the lowest variance, $r_{X_2 X_3}$ also equals .85.

15. It should be noted that if the accuracy of the probabilities associated with the outcome of one or more studies is subject to challenge, the accuracy of a pooled probability will be compromised. One can argue, however, that if enough replication studies are conducted, probability inaccuracies in one direction will most likely be balanced by probability inaccuracies in the opposite direction.

16. Inspection of the $df_{\text{res}} = 10$ curve reveals that for $df_{\text{res}} = 10$, a value of approximately $\phi = 3.1$ or greater will be associated with a power of 1.

17. The different equations that have been suggested for computing $\tilde{\omega}^2$ generally yield values that are close to one another.

18. In using Equation 2.7, $\tilde{s}_1/\sqrt{n}$ is equivalent to $s_{\bar{X}_1}$.

19. In employing double (or even more than two) summation signs such as $\sum_{j=1}^{k} \sum_{i=1}^{n}$, the mathematical operations specified are carried out beginning with the summation sign that is farthest to the right and continued sequentially with those operations specified by summation signs to the left. Specifically, if $k = 3$ and $n = 6$, the notation $\sum_{j=1}^{k} \sum_{i=1}^{n} X_{ij}$ indicates that the sum of the n scores in Condition 1 is computed, after which the sum

of the n scores in Condition 2 is computed, after which the sum of the n scores in Condition 3 is computed. The final result will be the sum of all the aforementioned values that have been computed. On the other hand, the notation $\sum_{i=1}^{n}\sum_{j=1}^{k}X_{ij}$ indicates that the sum of the $k = 3$ scores of Subject 1 is computed, after which the sum of the $k = 3$ scores of Subject 2 is computed, and so on until the sum of the $k = 3$ scores of Subject 6 is computed. The final result will be the sum of all the aforementioned values that have been computed. In this example the final value computed for $\sum_{j=1}^{k}\sum_{i=1}^{n}X_{ij}$ will be equal to the final value computed for $\sum_{i=1}^{n}\sum_{j=1}^{k}X_{ij}$. In obtaining the final value, however, the order in which the operations are conducted is reversed. Specifically, in computing $\sum_{j=1}^{k}\sum_{i=1}^{n}X_{ij}$, the sums of the k columns are computed and summed in order to arrive at the grand sum, while in computing $\sum_{i=1}^{n}\sum_{j=1}^{k}X_{ij}$, the sums of the n rows are computed and summed in order to arrive at the grand sum.

20. For each of the $N = 18$ scores in Table 18.14, the following is true with respect to the contribution of any score to the total variability in the data:

$$(X_{ij} - \bar{X}_T) = (\bar{S}_i - \bar{X}_T) + (X_{ij} - \bar{S}_i)$$

Total deviation score = *BS* deviation score + *WS* deviation score

and

$$(X_{ij} - \bar{S}_i) = (\bar{X}_j - \bar{X}_T) + [(X_{ij} - \bar{X}_T) - (\bar{S}_i - \bar{X}_T) - (\bar{X}_j - \bar{X}_T)]$$

WS deviation score = *BC* deviation score + Residual deviation score

21. As noted in Section VI under the discussion of computation of a confidence interval, MS_{WC} is equivalent to MS_{WG} (which is the analogous measure of variability for the **single-factor between-subjects analysis of variance**).

22. An issue discussed by Keppel (1991) that is relevant to the power of the **single-factor within-subjects analysis of variance** is that even though counterbalancing is an effective procedure for distributing practice effects evenly over the k experimental conditions in a dependent samples/within-subjects design, if practice effects are in fact present in the data, the value of MS_{res} will be inflated, and because of the latter the power of the **single-factor within-subjects analysis of variance** will be reduced. Keppel (1991) describes a methodology for computing an adjusted measure of MS_{res} that is independent of practice effects, which allows for a more powerful test of an alternative hypothesis.

23. In evaluating a directional alternative hypothesis, when $k = 2$ the tabled $F_{.90}$ and $F_{.98}$ values (for the appropriate degrees of freedom) are respectively employed as the one-tailed .05 and .01 values. Since the values for $F_{.90}$ and $F_{.98}$ are not listed in **Table A10**, the values $F_{.90} = 3.36$ and $F_{.98} = 7.95$ can be obtained by squaring the tabled critical one-tailed values $t_{.05} = 1.83$ and $t_{.01} = 2.82$, by employing more extensive tables of the F distribution available in other sources, or through interpolation.

24. Example 18.6 can also be viewed as example of what is commonly referred to as a **time series design** (although time series designs typically involve more measurement periods than are employed in the latter example). The latter design is essentially a **before-after design** involving one or more measurement periods prior to an experimental treatment, and one or more measurement periods following the experimental treatment.

Test 19

The Friedman Two-Way Analysis of Variance by Ranks
(Nonparametric Test Employed with Ordinal Data)

I. Hypothesis Evaluated with Test and Relevant Background Information

Hypothesis evaluated with test In a set of k dependent samples (where $k \geq 2$), do at least two of the samples represent populations with different median values?

Relevant background information on test Prior to reading the material on the **Friedman two-way analysis of variance by ranks**, the reader may find it useful to review the general information regarding a dependent samples design contained in Sections I and VII of the ***t* test for two dependent samples (Test 12)**. The **Friedman two-way analysis of variance by ranks** (Friedman (1937)) is employed with ordinal (rank-order) data in a hypothesis testing situation involving a design with two or more dependent samples. The test is an extension of the **binomial sign test for two dependent samples (Test 14)** to a design involving more than two dependent samples, and when $k = 2$ the **Friedman two-way analysis of variance by ranks** will yield a result that is equivalent to that obtained with the **binomial sign test for two dependent samples.**[1] If the result of the **Friedman two-way analysis of variance by ranks** is significant, it indicates there is a significant difference between at least two of the sample medians in the set of k medians. As a result of the latter, the researcher can conclude there is a high likelihood that at least two of the samples represent populations with different median values.

In employing the **Friedman two-way analysis of variance by ranks**, one of the following is true with regard to the rank-order data that are evaluated: a) The data are in a rank-order format, since it is the only format in which scores are available; or b) The data have been transformed into a rank-order format from an interval/ratio format, since the researcher has reason to believe that one or more of the assumptions of the **single-factor within-subjects analysis of variance (Test 18)** (which is the parametric analog of the **Friedman test**) is saliently violated. It should be pointed out that when a researcher elects to transform a set of interval/ratio data into ranks, information is sacrificed. This latter fact accounts for why there is reluctance among some researchers to employ nonparametric tests such as the **Friedman two-way analysis of variance by ranks**, even if there is reason to believe that one or more of the assumptions of the **single-factor within-subjects analysis of variance** has been violated.

Various sources (e.g., Conover (1980), Daniel (1990)) note that the **Friedman two-way analysis of variance by ranks** is based on the following assumptions: a) The sample of n subjects has been randomly selected from the population it represents; and b) The dependent variable (which is subsequently ranked) is a continuous random variable. In truth, this assumption, which is common to many nonparametric tests, is often not adhered to, in that such tests are often employed with a dependent variable that represents a discrete random variable.

As is the case for other tests that are employed to evaluate data involving two or more dependent samples, in order for the **Friedman two-way analysis of variance by ranks** to generate valid results the following guidelines should be adhered to:[2] a) In order to control for order effects, the presentation of the k experimental conditions should be random or, if appropriate, be counterbalanced; and b) If matched samples are employed, within each set of matched subjects each of the subjects should be randomly assigned to one of the k experimental conditions.

As is noted with respect to other tests that are employed to evaluate a design involving two or more dependent samples, the **Friedman two-way analysis of variance by ranks** can also be used to evaluate a **before-after design,** as well as extensions of the latter design that involve more than two measurement periods. The limitations of the **before-after design** (which are discussed in Section VII of the ***t* test for two dependent samples**) are also applicable when it is evaluated with the **Friedman two-way analysis of variance by ranks.**

II. Example

Example 19.1 is identical to Example 18.1 (which is evaluated with the **single-factor within-subjects analysis of variance**). In evaluating Example 19.1 it will be assumed that the ratio data are rank-ordered, since one or more of the assumptions of the **single-factor within-subjects analysis of variance** has been saliently violated.

Example 19.1 *A psychologist conducts a study to determine whether noise can inhibit learning. Each of six subjects is tested under three experimental conditions. In each of the experimental conditions a subject is given* 20 *minutes to memorize a list of* 10 *nonsense syllables, which the subject is told he or she will be tested on the following day. The three experimental conditions each subject serves under are as follows:* **Condition 1,** *the* **no noise** *condition, requires subjects to study the list of nonsense syllables in a quiet room.* **Condition 2,** *the* **moderate noise** *condition, requires subjects to study the list of nonsense syllables while listening to classical music.* **Condition 3,** *the* **extreme noise** *condition, requires subjects to study the list of nonsense syllables while listening to rock music. Although in each of the experimental conditions subjects are presented with a different list of nonsense syllables, the three lists are comparable with respect to those variables that are known to influence a person's ability to learn nonsense syllables. In order to control for order effects, the order of presentation of the three experimental conditions is completely counterbalanced. The number of nonsense syllables correctly recalled by the six subjects under the three experimental conditions follow. (Subjects' scores are listed in the order* **Condition 1, Condition 2, Condition 3.**) **Subject 1**: 9, 7, 4; **Subject 2**: 10, 8, 7; **Subject 3**: 7, 5, 3; **Subject 4**: 10, 8, 7; **Subject 5**: 7, 5, 2; **Subject 6**: 8, 6, 6. *Do the data indicate that noise influenced subjects' performance?*

III. Null versus Alternative Hypotheses

Null hypothesis H_0: $\theta_1 = \theta_2 = \theta_3$

(The median of the population Condition 1 represents equals the median of the population Condition 2 represents equals the median of the population Condition 3 represents. With respect to the sample data, when the null hypothesis is true the sums of the ranks (as well as the mean ranks) of all k conditions will be equal.

Alternative hypothesis H_1: Not H_0

(This indicates that there is a difference between at least two of the k population medians. It is important to note that the alternative hypothesis should not be written as follows: H_1: $\theta_1 \neq \theta_2 \neq \theta_3$. The reason why the latter notation for the alternative hypothesis is incorrect is because it implies that all three population medians must differ from one another in order to reject the null hypothesis. With respect to the sample data, if the alternative hypothesis is true the sum of the ranks (as well as the mean ranks) of at least two of the k conditions will not be equal. In this book it will always be assumed that the alternative hypothesis for the **Friedman two-way analysis of variance by ranks** is stated **nondirectionally**.)[3]

IV. Test Computations

The data for Example 19.1 are summarized in Table 19.1. The number of subjects employed in the experiment is $n = 6$, and thus within each condition there are $n = n_1 = n_2 = n_3 = 6$ scores. The original interval/ratio scores of the six subjects are recorded in the columns labelled X_1, X_2, and X_3. The adjacent columns R_1, R_2, and R_3 note the rank-order assigned to each of the scores.

Table 19.1 Data for Example 19.1

	Condition 1		Condition 2		Condition 3	
	X_1	R_1	X_2	R_2	X_3	R_3
Subject 1	9	3	7	2	4	1
Subject 2	10	3	8	2	7	1
Subject 3	7	3	5	2	3	1
Subject 4	10	3	8	2	7	1
Subject 5	7	3	5	2	2	1
Subject 6	8	3	6	1.5	6	1.5
	$\Sigma R_1 = 18$		$\Sigma R_2 = 11.5$		$\Sigma R_3 = 6.5$	
	$\bar{R}_1 = \frac{\Sigma R_1}{n_1} = \frac{18}{6} = 3$		$\bar{R}_2 = \frac{\Sigma R_2}{n_2} = \frac{11.5}{6} = 1.92$		$\bar{R}_3 = \frac{\Sigma R_3}{n_3} = \frac{6.5}{6} = 1.08$	

The ranking procedure employed for the **Friedman two-way analysis of variance by ranks** requires that each of the k scores of a subject be ranked within that subject.[4] Thus, in Table 19.1, for each subject a rank of 1 is assigned to the subject's lowest score, a rank of 2 to the subject's middle score, and a rank of 3 to the subject's highest score. In the event of tied scores, the same protocol described for handling ties for other rank-order tests (discussed in detail in Section IV of the **Mann–Whitney *U* test (Test 9)**) is employed. Specifically, the average of the ranks involved is assigned to all scores tied for a given rank. The only example of tied scores in Example 19.1 is in the case of Subject 6 who has a score of 6 in both Conditions 2 and 3. In Table 19.1 both of these scores are assigned a rank of 1.5, since if the scores of Subject 6 in Conditions 2 and 3 were not identical, but were still less than the subject's third score (which is 8 in Condition 1), one of the two scores that are, in fact, tied would receive a rank of 1 and the other a rank of 2. The average of these two ranks (i.e., $(1 + 2)/2 = 1.5$) is thus assigned to each of the two tied scores.

It should be noted that it is permissible to reverse the ranking protocol described above. Specifically, one can assign a rank of 1 to a subject's highest score, a rank of 2 to the subject's middle score, and a rank of 3 to subject's the lowest score. This reverse ranking protocol will yield the same value for the **Friedman test** statistic as the protocol employed in Table 19.1.

Upon rank-ordering the scores of the $n = 6$ subjects, the sum of the ranks is computed for each of the experimental conditions. In Table 19.1 the sum of the ranks of the jth condition is represented by the notation ΣR_j. Thus, $\Sigma R_1 = 18$, $\Sigma R_2 = 11.5$, $\Sigma R_3 = 6.5$.

Although they are not required for the **Friedman test** computations, the mean rank ($\bar{R}_j$) for each of the conditions is also noted in Table 19.1.

The chi-square distribution is generally employed to approximate the **Friedman test** statistic. Equation 19.1 is employed to compute the chi-square approximation of the **Friedman test** statistic (which is represented in most sources by the notation χ_r^2).

$$\chi_r^2 = \frac{12}{nk(k+1)}\left[\sum_{j=1}^{k}(\Sigma R_j)^2\right] - 3n(k+1) \qquad \textbf{(Equation 19.1)}$$

Note that in Equation 19.1 the term $[\Sigma_{j=1}^{k}(\Sigma R_j)^2]$ indicates that the sum of the ranks for each of the k experimental conditions is squared, and that the squared sums of ranks are summed. Substituting the appropriate values from Example 19.1 in Equation 19.1, the value $\chi_r^2 = 11.08$ is computed.

$$\chi_r^2 = \frac{12}{(6)(3)(3+1)}[(18)^2 + (11.5)^2 + (6.5)^2] - (3)(6)(3+1) = 11.08$$

V. Interpretation of the Test Results

In order to reject the null hypothesis the computed value χ_r^2 must be equal to or greater than the tabled critical chi-square value at the prespecified level of significance. The computed chi-square value is evaluated with **Table A4 (Table of the Chi-Square Distribution)** in the **Appendix**. For the appropriate degrees of freedom, the tabled $\chi_{.95}^2$ value (which is the chi-square value at the 95th percentile) and the tabled $\chi_{.99}^2$ value (which is the chi-square value at the 99th percentile) are employed as the .05 and .01 critical values for evaluating a nondirectional alternative hypothesis. The number of degrees of freedom employed in the analysis are computed with Equation 19.2. Thus, $df = 3 - 1 = 2$.

$$df = k - 1 \qquad \textbf{(Equation 19.2)}$$

For $df = 2$, the tabled critical .05 and .01 chi-square values are $\chi_{.05}^2 = 5.99$ and $\chi_{.01}^2 = 9.21$. Since the computed value $\chi_r^2 = 11.08$ is greater than $\chi_{.05}^2 = 5.99$ and $\chi_{.01}^2 = 9.21$, the alternative hypothesis is supported at both the .05 and .01 levels.[5] A summary of the analysis of Example 19.1 with the **Friedman two-way analysis of variance by ranks** follows: It can be concluded that there is a significant difference between at least two of the three experimental conditions exposed to different levels of noise. This result can be summarized as follows: $\chi_r^2(2) = 11.08, p < .01$.

It should be noted that when the data for Example 19.1 are evaluated with a **single-factor within-subjects analysis of variance**, the null hypothesis can also be rejected at both the .05 and .01 levels. The reader should note, however, that the difference between the value $\chi_r^2 = 11.08$ (obtained for the **Friedman test**) and $\chi_{.01}^2 = 9.21$ (the .01 tabled critical value for the **Friedman test**) is much smaller than the difference between $F = 41.29$ (obtained for the analysis of variance) and $F_{.01} = 7.56$ (the .01 tabled critical value for the analysis of variance). The smaller difference between the computed test statistic and the tabled critical value in the case of the **Friedman test** reflects the fact that, as a general rule (assuming that none of the assumptions of the analysis of variance is saliently violated), it provides a less powerful test of an alternative hypothesis than the analysis of variance.

VI. Additional Analytical Procedures for the Friedman Two-Way Analysis of Variance by Ranks and/or Related Tests

1. Tie correction for the Friedman two-way analysis of variance by ranks Some sources recommend that if there is an excessive number of ties in the overall distribution of scores, the value of the **Friedman test** statistic be adjusted. The tie correction results in a small increase in the value of χ_r^2 (thus providing a slightly more powerful test of the alternative hypothesis). Equation 19.3 (based on a methodology described in Daniel (1990) and Marascuilo and McSweeney (1977)) is employed in computing the value C, which represents the tie correction factor for the **Friedman two-way analysis of variance by ranks.**

$$C = 1 - \frac{\sum_{i=1}^{s} (t_i^3 - t_i)}{n(k^3 - k)} \qquad \textbf{(Equation 19.3)}$$

Where: s = the number of sets of ties
t_i = the number of tied scores in the ith set of ties

The notation $\Sigma_{i=1}^{s}(t_i^3 - t_i)$ indicates the following: a) For each set of ties, the number of ties in the set is subtracted from the cube of the number of ties in that set; and b) The sum of all the values computed in part a) is obtained. The tie correction will now be computed for Example 19.1. In the latter example there is $s = 1$ set of ties in which there are $t_i = 2$ ties (i.e., the two scores of 6 for Subject 6 under Conditions 2 and 3). Thus:

$$\sum_{i=1}^{s} (t_i^3 - t_i) = [(2)^3 - 2] = 6$$

Employing Equation 19.3, the value $C = .958$ is computed.

$$C = 1 - \frac{6}{6[(3)^3 - 3]} = .958$$

$\chi_{r_C}^2$, which represents the tie corrected value of the **Friedman test** statistic, is computed with Equation 19.4.

$$\chi_{r_C}^2 = \frac{\chi_r^2}{C} \qquad \textbf{(Equation 19.4)}$$

Employing Equation 19.4, the tie corrected value $\chi_{r_C}^2 = 11.57$ is computed.

$$\chi_{r_C}^2 = \frac{11.08}{.958} = 11.57$$

As is the case with $\chi_r^2 = 11.08$ computed with Equation 19.1, the value $\chi_{r_C}^2 = 11.57$ computed with Equation 19.4 is significant at both the .05 and .01 levels (since it is greater than $\chi_{.05}^2 = 5.99$ and $\chi_{.01}^2 = 9.21$). Although Equation 19.4 results in a slightly less conservative test than Equation 19.1, in this instance the two equations lead to identical conclusions with respect to the null hypothesis.

2. Pairwise comparisons following computation of the test statistic for the Friedman two-way analysis of variance by ranks Prior to reading this section the reader should review the discussion of comparisons in Section VI of the **single-factor between-subjects analysis of variance (Test 16)**. As is the case with the omnibus F value computed for an analysis of variance, the χ_r^2 value computed with Equation 19.1 is based on an evaluation of all k experimental conditions. When the value of χ_r^2 is significant, it does not indicate whether just two or, in fact, more than two conditions differ significantly from one another. In order to answer the latter question, it is necessary to conduct comparisons contrasting specific conditions with one another. This section will describe methodologies that can be employed for conducting **simple/pairwise comparisons** following the computation of a χ_r^2 value.[6]

In conducting a simple comparison, the null hypothesis and nondirectional alternative hypothesis are as follows: H_0: $\theta_a = \theta_b$ versus H_1: $\theta_a \neq \theta_b$. In the aforementioned hypotheses, θ_a and θ_b represent the medians of the populations represented by the two conditions involved in the comparison. The alternative hypothesis can also be stated directionally as follows: H_1: $\theta_a > \theta_b$ or H_1: $\theta_a < \theta_b$.

Various sources (e.g., Daniel (1990) and Siegel and Castellan (1988)) describe a comparison procedure for the **Friedman two-way analysis of variance by ranks** (which is essentially the application of the **Bonferroni-Dunn method** described in Section VI of the **single-factor between-subjects analysis of variance** to the **Friedman test** model). Through use of Equation 19.5, the procedure allows a researcher to identify the minimum required difference between the sums of the ranks of any two conditions (designated as CD_F) in order for them to differ from one another at the prespecified level of significance.[7]

$$CD_F = z_{\text{adj}} \sqrt{\frac{nk(k + 1)}{6}} \qquad \textbf{(Equation 19.5)}$$

The value of z_{adj} is obtained from **Table A1 (Table of the Normal Distribution)** in the **Appendix**. In the case of a nondirectional alternative hypothesis, z_{adj} is the z value above which a proportion of cases corresponding to the value $\alpha_{FW}/2c$ falls (where c is the total number of comparisons that are conducted). In the case of a directional alternative hypothesis, z_{adj} is the z value above which a proportion of cases corresponding to the value α_{FW}/c falls. When all possible pairwise comparisons are made $c = [k(k - 1)]/2$, and thus, $2c = k(k - 1)$. In Example 19.1 the number of pairwise/simple comparisons that can be conducted are $c = [3(3 - 1)]/2 = 3$ — specifically, Condition 1 versus Condition 2, Condition 1 versus Condition 3, and Condition 2 versus Condition 3.

The value of z_{adj} will be a function of both the maximum **familywise Type I error rate** (α_{FW}) the researcher is willing to tolerate and the total number of comparisons that are conducted. When a limited number of comparisons are planned prior to collecting the data, most sources take the position that a researcher is not obliged to control the value of α_{FW}. In such a case, the **per comparison Type I error rate** (α_{PC}) will be equal to the prespecified value of alpha. When α_{FW} is not adjusted, the value of z_{adj} employed in Equation 19.5 will be the tabled critical z value that corresponds to the prespecified level of significance. Thus, if a nondirectional alternative hypothesis is employed and $\alpha = \alpha_{PC} = .05$, the tabled critical two-tailed .05 value $z_{.05} = 1.96$ is used to represent z_{adj} in Equation 19.5. If $\alpha = \alpha_{PC} = .01$, the tabled critical two-tailed .01 value $z_{.01} = 2.58$ is used in Equation 19.5. In the same respect, if a directional alternative hypothesis is employed, the tabled critical .05 and .01 one-tailed values $z_{.05} = 1.65$ and $z_{.01} = 2.33$ are used for z_{adj} in Equation 19.5.

When comparisons are not planned beforehand, it is generally acknowledged that the value of α_{FW} must be controlled so as not to become excessive. The general approach for controlling the latter value is to establish a **per comparison Type I error rate** which insures that α_{FW} will not exceed some maximum value stipulated by the researcher. One method for doing this (described under the **single-factor between-subjects analysis of variance** as the **Bonferroni–Dunn method**) establishes the **per comparison Type I error rate** by dividing the maximum value one will tolerate for the **familywise Type I error rate** by the total number of comparisons conducted. Thus, in Example 19.1, if one intends to conduct all three pairwise comparisons and wants to insure that α_{FW} does not exceed .05, $\alpha_{PC} = \alpha_{FW}/c = .05/3 = .0167$. The latter proportion is used to determine the value of z_{adj}. As noted earlier, if a directional alternative hypothesis is employed for a comparison, the value of z_{adj} employed in Equation 19.5 is the z value above which a proportion equal to $\alpha_{PC} = \alpha_{FW}/c$ of the cases falls. In **Table A1**, the z value that corresponds to the proportion .0167 is $z = 2.13$. By employing $z_{\text{adj}} = 2.13$ in Equation 19.5, one can be assured that within the "family" of three pairwise comparisons, α_{FW} will not exceed .05 (assuming all of the comparisons are directional). If a nondirectional alternative hypothesis is employed for all of the comparisons, the value of z_{adj} will be the z value above which a proportion equal to $\alpha_{FW}/2c = \alpha_{PC}/2$ of the cases falls. Since $\alpha_{PC}/2 = .0167/2 = .0083$, $z = 2.39$. By employing $z_{adj} = 2.39$ in Equation 19.5, one can be assured that α_{FW} will not exceed .05.[8]

In order to employ the CD_F value computed with Equation 19.5, it is necessary to determine the absolute value of the difference between the sums of the ranks of each pair of experimental conditions that are compared.[9] Table 19.2 summarizes the difference scores between pairs of sums of ranks.

Table 19.2 Difference Scores Between Pairs of Sums of Ranks for Example 19.1

$$|\Sigma R_1 - \Sigma R_2| = |18 - 11.5| = 6.5$$
$$|\Sigma R_1 - \Sigma R_3| = |18 - 6.5| = 11.5$$
$$|\Sigma R_2 - \Sigma R_3| = |11.5 - 6.5| = 5$$

If any of the differences between the sums of ranks is equal to or greater than the CD_F value computed with Equation 19.5, a comparison is declared significant. Equation 19.5 will now be employed to evaluate the nondirectional alternative hypothesis H_1: $\theta_a \neq \theta_b$ for all three pairwise comparisons. Since it will be assumed that the comparisons are unplanned and that the researcher does not want the value of α_{FW} to exceed .05, the value $z_{adj} = 2.39$ will be used in computing CD_F.

$$CD_F = (2.39)\sqrt{\frac{(6)(3)(3 + 1)}{6}} = (2.39)(3.46) = 8.28$$

The obtained value $CD_F = 8.28$ indicates that any difference between the sums of ranks of two conditions that is equal to or greater than 8.28 is significant. With respect to the three pairwise comparisons, the only difference between the sum of ranks of two conditions which is greater than $CD_F = 8.28$ is $|\Sigma R_1 - \Sigma R_3| = 11.5$. Thus, we can conclude there is a significant difference between Condition 1 and Condition 3. We cannot conclude that the difference between any other pair of conditions is significant.

An alternative strategy that can be employed for conducting pairwise comparisons for the **Friedman test** model is to use one of the tests that are described for evaluating a dependent samples design involving $k = 2$ samples. Specifically, one can employ either the **Wilcoxon matched-pairs signed-ranks test (Test 13)** or the **binomial sign test for two dependent samples**. Whereas the **binomial sign test** only takes into consideration the direction of the difference of subjects' scores in the two experimental conditions, the **Wilcoxon test** rank-orders the interval/ratio difference scores of subjects. Because of the latter, the **Wilcoxon test** employs more information than the **binomial sign test**, and consequently will provide a more powerful test of an alternative hypothesis. Both the **Wilcoxon test** and **binomial sign test** will be used to conduct the three pairwise comparisons for Example 19.1.[10]

Use of the **Wilcoxon matched-pairs signed-ranks test** requires that for each comparison that is conducted the difference scores of subjects in the two experimental conditions be rank-ordered, and that the Wilcoxon T statistic be computed for that comparison. The exact distribution of the **Wilcoxon test** statistic can only be used when the value of α_{PC} is equal to one of the probabilities documented in **Table A5 (Table of Critical T Values for Wilcoxon's Signed-Ranks and Matched-Pairs Signed-Ranks Tests)** in the **Appendix.** When α_{PC} is a value other than those listed in **Table A5**, the normal approximation of the **Wilcoxon test** statistic must be employed.

When the **Wilcoxon matched-pairs signed-ranks test** is employed for the three pairwise comparisons, the following T values are computed: a) Condition 1 versus Condition 2: $T = 0$; b) Condition 1 versus Condition 3: $T = 0$; and c) Condition 2 versus Condition 3: $T = 0$.[11] When the aforementioned T values are substituted in Equations 13.2 and 13.3 (the uncorrected and continuity corrected normal approximations for the **Wilcoxon test**), the following absolute z values are computed: a) Condition 1 versus Condition 2: $z = 2.20$ and $z = 2.10$; b) Condition 1 versus Condition 3: $z = 2.20$ and $z = 2.10$; and c) Condition 2 versus Condition 3: $z = 2.02$ and $z = 1.89$. If we want to evaluate a nondirectional alternative hypothesis and insure that α_{FW} does not exceed .05, the value of α_{PC} is set equal to .0167. **Table A5** cannot be employed, since it does not list two-tailed critical T values for $\alpha_{.0167}$. In order to evaluate the result of the normal approximation, we identify the tabled critical two-tailed .0167 z value in **Table A1**. In employing Equation 19.5 earlier in this section, we determined that the latter value is $z_{.0167} = 2.39$. Since none of the z values computed for the normal approximation is equal to or greater than $z_{.0167} = 2.39$, none of the pairwise comparisons is significant. This result is not identical to that obtained with Equation 19.5, in which case a significant difference is computed for the comparison Condition 1 versus Condition 3. Although the latter comparison (as well as the Condition 1 versus Condition 2 comparison) comes close when it is evaluated with the **Wilcoxon test**, it just falls short of achieving significance.

In the event the researcher elects not to control the value of α_{FW} and employs $\alpha_{PC} = .05$ in evaluating the three pairwise comparisons (once again assuming a nondirectional analysis), both the Condition 1 versus Condition 2 and Condition 1 versus Condition 3 comparisons are significant at the .05 level if the **Wilcoxon test** is employed. Specifically, both the uncorrected and corrected normal approximations are significant, since $z = 2.20$ and $z = 2.10$ (computed for both the Condition 1 versus Condition 2 and Condition 1 versus Condition 3 comparisons) are greater than the tabled critical two-tailed value $z_{.05} = 1.96$. The Condition 2 versus Condition 3 comparison is also significant, but only if the uncorrected value $z = 2.02$ is employed. Employing **Table A5**, we also determine that both the Condition 1 versus Condition 2 and Condition 1 versus Condition 3 comparisons are significant at the .05 level, since the computed value $T = 0$ for both comparisons is equal to the tabled critical two-tailed .05 value $T_{.05} = 0$ (for $n = 6$). The Condition 2 versus Condition 3 comparison is not significant, since no two-tailed .05 critical value is listed in **Table A5** for $n = 5$. If Equation 19.5 is employed for the same set of comparisons, however, only the Condition 1 versus Condition 3 comparison is significant. This is the case, since $CD_F = (1.96)(3.46) = 6.78$, and only the difference $|\Sigma R_1 - \Sigma R_3| = 11.5$ is greater than $CD_F = 6.78$.[12] The difference $|\Sigma R_1 - \Sigma R_2| = 6.5$ (which is significant with the **Wilcoxon test**) just falls short of achieving significance. Although the result obtained with Equation 19.5 is not identical to that obtained with the **Wilcoxon test**, the two analyses are reasonably consistent with one another.

In the event the **binomial signed test for two dependent samples** is employed to conduct comparisons, a researcher must determine for each comparison the number of subjects who yield positive versus negative difference scores. With the exception of Subject 6 in the Condition 2 versus Condition 3 comparison, all of the difference scores for the three pairwise comparisons are positive (since all subjects obtain a higher score in Condition 1 than Condition 2, in Condition 1 than Condition 3, and in Condition 2 than Condition 3). For the Condition 1 versus Condition 2 and Condition 1 versus Condition 3 comparisons, we must compute $P(x = 6)$ for $n = 6$. For the Condition 2 versus Condition 3 comparison (which does not include Subject 6 in the analysis, since the latter subject has a zero difference score), we must compute $P(x = 5)$ for $n = 5$. For all three pairwise comparisons $\pi + = \pi - = .5$.

Employing **Table A6 (Table of the Binomial Distribution, Individual Probabilities)** in the **Appendix** we can determine that when $n = 6$, $P(x = 6) = .0156$. Thus, the computed two-tailed probability for the Condition 1 versus Condition 2 and Condition 1 versus Condition 3 comparisons is $(2)(.0156) = .0312$. When $n = 5$, $P(x = 5) = .0312$. The computed two-tailed probability for the Condition 2 versus Condition 3 comparison is $(2)(.0312) = .0624$.

As before, if we want to evaluate a nondirectional alternative hypothesis and insure that α_{FW} does not exceed .05, the value of α_{PC} is set equal to .0167. Thus, in order to reject the null hypothesis the computed two-tailed binomial probability for a comparison must be equal to or less than .0167. Since the computed two-tailed probabilities .0312 (for the Condition 1 versus Condition 2 and the Condition 1 versus Condition 3 comparisons) and .0624 (for the Condition 2 versus Condition 3 comparison) are greater than .0167, none of the pairwise comparisons is significant.

In the event the researcher elects not to control the value of α_{FW} and employs $\alpha_{PC} = .05$ for evaluating the three pairwise comparisons (once again assuming a nondirectional analysis), both the Condition 1 versus Condition 2 and Condition 1 versus Condition 3 comparisons are significant at the .05 level, since the computed two-tailed probability .0312 (for the Condition 1 versus Condition 2 and the Condition 1 versus Condition 3 comparisons) is less than .05. The Condition 2 versus Condition 3 comparison is not significant, since the computed two-tailed probability .0624 for the latter comparison is greater than .05.

When the results obtained with the **binomial sign test** are compared with those obtained with Equation 19.5 and the **Wilcoxon test**, it would appear that of the three procedures the **binomial sign test** results in the most conservative test (and thus, as noted previously, the least powerful test). However, if one takes into account the obtained binomial probabilities, they are, in actuality, not far removed from the probabilities obtained when Equation 19.5 and the **Wilcoxon test** are used.

In the case of Example 19.1, regardless of which comparison procedure one employs, it would appear that unless one uses a very low value for α_{PC}, the Condition 1 versus Condition 3 comparison is significant. There is some suggestion that the Condition 1 versus Condition 2 comparison may also be significant, but some researchers would recommend conducting additional studies in order to clarify whether or not the two conditions represent different populations. Although based on the analyses that have been conducted the Condition 2 versus Condition 3 comparison does not appear to be significant, it is worth noting that if the researcher uses the **Wilcoxon test** (specifically, the normal approximation not corrected for continuity) to evaluate the directional alternative hypothesis H_1: $\theta_2 > \theta_3$ with $\alpha_{PC} = .05$, the latter comparison also yields a significant result. Thus, further studies might be in order to clarify the relationship between the populations represented by Conditions 2 and 3.

The intent of presenting three comparison procedures in this section is to illustrate that, generally speaking, the results obtained with the different comparison procedures will be reasonably consistent with one another. As is noted in the discussion of comparisons in Section VI of the **single-factor between-subjects analysis of variance**, in instances where two or more comparison procedures yield inconsistent results, the most effective way to clarify the status of the null hypothesis is to replicate a study one or more times. It is also noted throughout the book that, in the final analysis, the decision regarding which of the available comparison procedures to employ is usually not the most important issue facing the researcher conducting comparisons. The main issue is what maximum value one is willing to tolerate for α_{FW}. Additional sources on comparison procedures for the **Friedman test** model are Church and Wike (1979) (who provide a comparative analysis of a number of different comparison procedures); Daniel (1990) (who describes a methodology for comparing $(k - 1)$ conditions with a control group, as well as a methodology for estimating the size of a difference between the medians of any pair of experimental conditions); Marascuilo and McSweeney (1977) (who within the framework of a comprehensive discussion of the **Friedman test** model describe a methodology for conducting complex comparisons); and Siegel and Castellan (1988) (who also describe the methodology for comparing $(k - 1)$ conditions with a control group).

Marascuilo and McSweeney (1977) also discuss the computation of a confidence interval for a comparison for the **Friedman test** model. One approach for computing a confidence interval is to add to and subtract the computed value of CD_F from the obtained difference between the sums of ranks (or mean ranks, if the equation in Endnote 7 is employed) involved in the comparison. The latter approach is based on the same logic employed for computing a confidence interval for a comparison in Section VI of the **single-factor between-subjects analysis of variance.**

VII. Additional Discussion of the Friedman Two-Way Analysis of Variance by Ranks

1. Exact tables of the Friedman distribution Although an exact probability value can be computed for obtaining a configuration of ranks that is equivalent to or more extreme than the configuration observed in the data evaluated with the **Friedman two-way analysis of variance by ranks**, the chi-square distribution is generally employed to estimate the latter probability. Although most sources employ the chi-square approximation regardless of the values of k and n, some sources recommend that exact tables be employed when the values of n and/or k are small. The exact sampling distribution for the **Friedman two-way analysis of variance by ranks** is based on the use of **Fisher's method of randomization** (which is discussed briefly in Section VI of the ***t* test for two dependent samples** and Section VII of the **Wilcoxon matched-pairs signed-ranks test**).

Tables of exact critical values, which can be viewed as adjusted chi-square values, can be found in Marascuilo and McSweeney (1977) and Siegel and Castellan (1988) (who list critical values for various values of n between 5 and 13 when the value of k is between 3 and 5). Depending upon the values of k and n, exact critical values may be either slightly larger or smaller than the critical chi-square values in **Table A4**. In point of fact, for $k = 3$ and $n = 6$ the adjusted tabled critical .05 and .01 chi-square values for the **Friedman test** statistic are respectively $\chi^2_{r_{.05}} = 7.00$ and $\chi^2_{r_{.01}} = 9.00$. Since the value $\chi^2_r = 11.08$ computed for Example 19.1 is greater than both of the aforementioned critical values, the null hypothesis can still be rejected at both the .05 and .01 levels. Although the conclusions with respect to Example 19.1 are the same regardless of whether one employs the adjusted chi-square values or the chi-square values in **Table A4**, inspection of the two sets of values indicates that the adjusted .05 chi-square value is larger than the corresponding critical value $\chi^2_{.05} = 5.99$ derived from **Table A4**, while the reverse is true with respect to the adjusted .01

chi-square value, which is less than the corresponding value $\chi^2_{.01} = 9.21$ in **Table A4.** It should be noted that for a given value of k, as the value of n increases the adjusted chi-square value approaches the tabled chi-square value in **Table A4.** An additional point of interest relevant to evaluating the **Friedman test** statistic, is that Daniel (1990) and Conover (1980) cite a study by Iman and Davenport (1980) which suggests that the F distribution can be used to approximate the sampling distribution for the **Friedman test,** and that the latter approximation may be more accurate than the more commonly employed chi-square approximation.

2. Equivalency of the Friedman two-way analysis of variance by ranks and the binomial sign test for two dependent samples when $k = 2$ In Section I it is noted that when $k = 2$ the **Friedman two-way analysis of variance by ranks** will yield a result that is equivalent to that obtained with the **binomial sign test for two dependent samples.** To be more specific, the **Friedman test** will yield a result that is equivalent to the normal approximation of the **binomial sign test for two dependent samples** when the correction for continuity is not employed (i.e., the result obtained with Equation 14.3).[13] It should be noted, however, that the two tests will only yield an equivalent result when none of the subjects has the same score in the two experimental conditions. In the case of the **binomial sign test,** any subject who has the same score in both conditions is eliminated from the data analysis. In the case of the **Friedman test,** however, such subjects are included in the analysis. In order to demonstrate the equivalency of the two tests, Equation 19.1 will be employed to analyze the data for Example 14.1 (which was previously evaluated with the **binomial sign test for two dependent samples).** In using Equation 19.1 the data for Subject 2 are not included, since the latter subject has identical scores in both conditions. Thus, in our analysis $n = 9$ and $k = 2$. Table 19.3 summarizes the rank-ordering of data for Example 14.1 within the framework of the **Friedman test** model.

Table 19.3 Summary of Data for Example 14.1 for Friedman Test Model

	Condition 1		Condition 2	
	X_1	R_1	X_2	R_2
Subject 1	9	2	8	1
Subject 2	2	1.5	2	1.5
Subject 3	1	1	3	2
Subject 4	4	2	2	1
Subject 5	6	2	3	1
Subject 6	4	2	0	1
Subject 7	7	2	4	1
Subject 8	8	2	5	1
Subject 9	5	2	4	1
Subject 10	1	2	0	1
		$\Sigma R_1 = 18.5$		$\Sigma R_2 = 11.5$

Since the data for Subject 2 are not included in the analysis, the rank-orders for Subject 2 under the two conditions are subtracted from the values $\Sigma R_1 = 18.5$ and $\Sigma R_2 = 11.5$, yielding the revised values $\Sigma R_1 = 17$ and $\Sigma R_2 = 10$ which are employed in Equation 19.1. Employing the latter equation, the value $\chi^2_r = 5.44$ is computed.

$$\chi^2_r = \frac{12}{(9)(2)(2 + 1)}\left[(17)^2 + (10)^2\right] - (3)(9)(3) = 5.44$$

Employing Equation 19.2, $df = 2 - 1 = 1$. For $df = 1$, the tabled critical .05 and .01 chi-square values are $\chi^2_{.05} = 3.84$ and $\chi^2_{.01} = 6.63$. Since the obtained value $\chi^2_r = 5.44$ is greater than $\chi^2_{.05} = 3.84$, the alternative hypothesis is supported at the .05 level. It is not, however, supported at the .01 level, since $X^2_r = 5.44$ is less than $\chi^2_{.01} = 6.63$.[14]

Equation 14.3 yields the value $z = 2.33$ for the same set of data. Since the square of a z value will equal the corresponding chi-square value computed for the same set of data, z^2 should equal χ^2_r. In point of fact, $(z = 2.33)^2 = (\chi^2_r = 5.44)$ (the minimal discrepancy is due to rounding off error). It is also the case that the square of the tabled critical z value employed for the normal approximation of the **binomial sign test for two dependent samples** will always equal the tabled critical chi-square value employed for the **Friedman test** at the same level of significance. Thus, the square of the tabled critical two-tailed value $z_{.05} = 1.96$ employed for the normal approximation of the **binomial sign test** equals $\chi^2_{.05} = 3.84$ employed for the **Friedman test** (i.e., $(z = 1.96)^2 = (\chi^2 = 3.84)$).

3. Alternative nonparametric rank-order procedures for evaluating a design involving *k* dependent samples In addition to the **Friedman two-way analysis of variance by ranks**, a number of other nonparametric procedures for two or more dependent samples have been developed that can be employed with ordinal data. Among the more commonly cited alternative procedures are the following: a) Marascuilo and McSweeney (1977) describe the extension of the **normal scores test** developed by Van der Waerden (1953/1953) (discussed in Section VII of the **Mann–Whitney *U* test** and the **Kruskal–Wallis one-way analysis of variance by ranks (Test 17)**) to a design involving k dependent samples. Conover (1980) notes that the normal scores test developed by Bell and Doksum (1965) can also be extended to the latter design; b) **Page's test for ordered alternatives** (Page (1963)) can be employed with k dependent samples to evaluate an ordered alternative hypothesis. Specifically, in stating the alternative hypothesis the ordinal position of the treatment effects is stipulated (as opposed to just stating that a difference exists between at least two of the k experimental conditions). **Page's test for ordered alternatives** is described in Daniel (1990), Marascuilo and McSweeney (1977), and Siegel and Castellan (1988); and c) Additional tests that can be employed with a k dependent samples design are either discussed or referenced in Conover (1980), Daniel (1990), Marascuilo and McSweeney (1977), and Sheskin (1984).

4. Relationship between the Friedman two-way analysis of variance by ranks and Kendall's coefficient of concordance The **Friedman two-way analysis of variance by ranks** and **Kendall's coefficient of concordance (Test 25)** (which is one of a number of measures of association that are described in this book) are based on the same statistical model. The latter measure of association is employed with three or more sets of ranks when rankings are based on the **Friedman test** protocol. A full discussion of the relationship between the **Friedman two-way analysis of variance by ranks** and **Kendall's coefficient of concordance** (which can be used as a measure of effect size for the **Friedman test**) can be found in Section VII of **Kendall's coefficient of concordance**. When there are $n = 2$ subjects/sets of matched subjects, **Spearman's rank-order correlation coefficient (Test 23)**, which is linearly related to **Kendall's coefficient of concordance**, can be conceptualized within the framework of the **Friedman test** model. The latter relationship is discussed in Section VII of **Spearman's rank-order correlation coefficient**.

VIII. Additional Examples Illustrating the Use of the Friedman Two-Way Analysis of Variance by Ranks

The **Friedman two-way analysis of variance by ranks** can be employed to evaluate any of the additional examples noted for the **single-factor within-subjects analysis of variance**, if the data for the latter examples are rank-ordered. In addition, the **Friedman test** can be used to evaluate the data for any of the additional examples noted for the ***t* test for two dependent samples/binomial sign test for two dependent samples/Wilcoxon matched-pairs signed-ranks test**. Example 19.2 is an additional example that can be evaluated with the **Friedman two-way analysis of variance by ranks**. In Example 19.2 there is no need to rank-order interval/ratio data, since the results of the study are summarized in a rank-order format.[15]

Example 19.2 *Six horses are rank-ordered by a trainer with respect to their racing form on three different surfaces. Specifically,* Track A *has a cement surface,* Track B *a clay surface, and* Track C *a grass surface. Except for the surface the three tracks are comparable to one another in all other respects.* Table 19.4 *summarizes the rankings of the horses on the three tracks. (In the case of* Horse 6, *the rank of* 1.5 *for both the clay and grass tracks reflects the fact that the horse was perceived to have equal form on both surfaces.) Do the data indicate that the form of a horse is related to the surface on which it is racing?*

Table 19.4 Data for Example 19.2

	Track A (Cement)	Track B (Clay)	Track C (Grass)
Horse 1	3	2	1
Horse 2	3	2	1
Horse 3	3	2	1
Horse 4	3	2	1
Horse 5	3	2	1
Horse 6	3	1.5	1.5

Since the ranks employed in Example 19.2 are identical to those employed for Example 19.1, the **Friedman test** will yield the identical result. Since most people would probably be inclined to employ a rank of 1 to represent a horse's best surface and a rank of 3 to represent a horse's worst surface, using such a ranking protocol the track with the lowest sum of ranks (Track C) is associated with the best racing form and the track with the highest sum of ranks (Track A) is associated with the worst racing form.

References

Bell, C. B. and Doksum, K. A. (1965). Some new distribution-free statistics. **Annals of Mathematical Statistics**, 36, 203–214.

Church, J. D. and Wike, E. L. (1979). A Monte Carlo study of nonparametric multiple-comparison tests for a two-way layout. **Bulletin of the Psychonomic Society**, 14, 95–98.

Conover, W. J. (1980). **Practical nonparametric statistics** (2nd ed.). New York: John Wiley and Sons.

Daniel, W. W. (1990). **Applied nonparametric statistics** (2nd ed.). Boston: PWS–Kent Publishing Company.

Friedman, M. (1937). The use of ranks to avoid the assumption of normality implicit in the analysis of variance. **Journal of the American Statistical Association**, 32, 675–701.

Iman, R. L. and Davenport, J. M. (1980). Approximations of the critical region of the Friedman statistic. **Communication in Statistics — Theory and Methods**, 9, 571–595.

Marascuilo, L. A. and McSweeney, M. (1977). **Nonparametric and distribution-free methods for the social sciences.** Monterey, CA: Brooks/Cole Publishing Company.

Page, E. B. (1963). Ordered hypotheses for multiple treatments: A significance test for linear ranks. **Journal of the American Statistical Association**, 58, 216–230.

Sheskin, D. J. (1984). **Statistical tests and experimental design: A guidebook.** New York: Gardner Press.

Siegel, S. and Castellan, N. J., Jr. (1988). **Nonparametric statistics for the behavioral sciences** (2nd ed.). New York: McGraw–Hill Book Company.

Van der Waerden, B. L. (1952/1953). Order tests for the two-sample problem and their power. **Proceedings Koninklijke Nederlandse Akademie van Wetenshappen** (A), 55 (**Indagationes Mathematicae** 14), 453–458, and 56 (**Indagationes Mathematicae**, 15), 303–316 (corrections appear in Vol. 56, p. 80).

Wike, E. L. (1985). **Numbers: A primer of data.** Columbus, OH: Charles E. Merrill Publishing Company.

Endnotes

1. The reader should take note of the fact that when there are $k = 2$ dependent samples, the **Wilcoxon matched-pairs signed-ranks test** (which is also described in this book as a nonparametric test for evaluating ordinal data) will not yield a result equivalent to that obtained with the **Friedman two-way analysis of variance by ranks.** Since the **Wilcoxon test** (which rank-orders interval/ratio difference scores) employs more information than the **Friedman test/binomial sign test**, it provides a more powerful test of an alternative hypothesis than the latter tests.

2. A more detailed discussion of the guidelines noted below can be found in Sections I and VII of the ***t* test for two dependent samples (Test 12)** and in Section I of the **single-factor within-subjects analysis of variance.**

3. Although it is possible to conduct a directional analysis, such an analysis will not be described with respect to the **Friedman two-way analysis of variance by ranks.** A discussion of a directional analysis when $k = 2$ can be found under the **binomial sign test for two dependent samples.** A discussion of the evaluation of a directional alternative hypothesis when $k \geq 2$ can be found in Section VII of the **chi-square goodness-of-fit test (Test 5).** Although the latter discussion is in reference to analysis of a k independent samples design involving categorical data, the general principles regarding analysis of a directional alternative hypothesis when $k \geq 2$ are applicable to the **Friedman two-way analysis of variance by ranks.**

4. Note that this ranking protocol differs from that employed for other rank-order procedures discussed in the book. In other rank-order tests, the rank assigned to each score is based on the rank-order of the score within the overall distribution of $nk = N$ scores.

5. As noted in Section IV, the chi-square distribution provides an approximation of the **Friedman test** statistic. Although the chi-square distribution provides an excellent approximation of the Friedman sampling distribution, some sources recommend the use of exact probabilities for small sample sizes. Exact tables of the Friedman distribution are discussed in Section VII.

6. In the discussion of comparisons in reference to the analysis of variance, it is noted that a **simple** (also known as a **pairwise**) **comparison** is a comparison between any two groups/conditions in a set of k groups/conditions.

7. An alternative form of the comparison equation, which identifies the minimum required difference between the **means of the ranks** of any two conditions in order for them to differ from one another at the prespecified level of significance, is noted below.

$$CD_{F_{(\bar{R}_a - \bar{R}_b)}} = z_{\text{adj}} \sqrt{\frac{k(k+1)}{6n}}$$

If the CD_F value computed with Equation 19.5 is divided by n, it yields the value $CD_{F_{(\bar{R}_a - \bar{R}_b)}}$ computed with the above equation.

8. The method for deriving the value of z_{adj} for the **Friedman two-way analysis of variance by ranks** is based on the same logic that is employed in Equation 17.5 (which is used for conducting comparisons for the **Kruskal-Wallis one-way analysis of variance by ranks**). A rationale for the use of the proportions .0167 and .0083 in determining the appropriate value for z_{adj} in Example 19.1 can be found in Endnote 4 of the **Kruskal–Wallis one-way analysis of variance by ranks.**

9. It should be noted that when a directional alternative hypothesis is employed, the sign of the difference between the two sums of ranks must be consistent with the prediction stated in the directional alternative hypothesis. When a nondirectional alternative hypothesis is employed, the direction of the difference between two sums of ranks is irrelevant.

10. In the case of both the **Wilcoxon matched-pairs signed-ranks test** and the **binomial sign test for two dependent samples**, it is assumed that for each pairwise comparison a subject's score in the second condition that is listed for a comparison is subtracted from the subject's score in the first condition that is listed for the comparison. In the case of both tests, reversing the order of subtraction will yield the same result.

11. The value $n = 6$ is employed for the Condition 1 versus Condition 2 and Condition 1 versus Condition 3 comparisons, since no subject has the same score in both experimental conditions. On the other hand, the value $n = 5$ is employed in the Condition 2 versus Condition 3 comparison, since Subject 6 has the same score in Conditions 2 and 3. The use of $n = 5$ is predicated on the fact that in conducting the **Wilcoxon matched-pairs signed-ranks test**, subjects who have a difference score of zero are not included in the computation of the test statistic.

12. In Equation 19.5 the value $z_{.05} = 1.96$ is employed for z_{adj}, and the latter value is multiplied by 3.46, which is the value computed for the term in the radical of the equation for Example 19.1.

13. It is also the case that the exact binomial probability for the **binomial sign test for two dependent samples** will correspond to the exact probability for the **Friedman test statistic.**

14. If Subject 2 is included in the analysis, Equation 19.1 yields the value $\chi_r^2 = 4.9$ which is also significant at the .05 level.

15. In Section I it is noted that in employing the **Friedman test** it is assumed that the variable which is ranked is a continuous random variable. Thus, it would be assumed that the racing form of a horse was at some point either explicitly or implicitly expressed as a continuous interval/ratio variable.

Test 20

The Cochran Q Test

(Nonparametric Test Employed with Categorical/Nominal Data)

I. Hypothesis Evaluated with Test and Relevant Background Information

Hypothesis evaluated with test In a set of k dependent samples (where $k \geq 2$), do at least two of the samples represent different populations?

Relevant background information on test It is recommended that before reading the material on the **Cochran Q test,** the reader review the general information on a dependent samples design contained in Sections I and VII of the ***t* test for two dependent samples (Test 12).** The **Cochran Q test** (Cochran (1950)) is a nonparametric procedure for categorical data employed in a hypothesis testing situation involving a design with $k = 2$ or more dependent samples. The test is employed to evaluate an experiment in which a sample of n subjects (or n sets of matched subjects) is evaluated on a dichotomous dependent variable (i.e., scores on the dependent variable must fall within one of two mutually exclusive categories). The test assumes that each of the n subjects (or n sets of matched subjects) contributes k scores on the dependent variable. The **Cochran Q test** is an extension of the **McNemar test (Test 15)** to a design involving more than two dependent samples, and when $k = 2$ the **Cochran Q test** will yield a result that is equivalent to that obtained with the **McNemar test.** If the result of the **Cochran Q test** is significant, it indicates there is a high likelihood at least two of the k experimental conditions represent different populations.

The **Cochran Q test** is based on the following assumptions: a) The sample of n subjects has been randomly selected from the population it represents; and b) The scores of subjects are in the form of a dichotomous categorical measure involving two mutually exclusive categories.

Although the chi-square distribution is generally employed to evaluate the **Cochran test** statistic, in actuality the latter distribution is used to provide an approximation of the exact sampling distribution. Sources on nonparametric analysis (e.g., Daniel (1990), Marascuilo and McSweeney (1977), and Siegel and Castellan (1988)) recommend that for small sample sizes exact tables of the Q distribution derived by Patil (1975) be employed. Use of exact tables is generally recommended when $n < 4$ and/or $nk < 24$.

As is the case for other tests that are employed to evaluate data involving two or more dependent samples, in order for the **Cochran Q test** to generate valid results the following guidelines should be adhered to:[1] a) In order to control for order effects, the presentation of the k experimental conditions should be random or, if appropriate, be counterbalanced; and b) If matched samples are employed, within each set of matched subjects each of the subjects should be randomly assigned to one of the k experimental conditions.

As is noted with respect to other tests that are employed to evaluate a design involving two or more dependent samples, the **Cochran Q test** can also be used to evaluate a **before–after design,** as well as extensions of the latter design that involve more than two measurement periods. The limitations of the **before–after design** (which are discussed in Section VII of the ***t* test for two dependent samples**) are also applicable when it is evaluated with the **Cochran Q test.**

II. Example

Example 20.1 *A market researcher asks* 12 *female subjects whether or not they would purchase an automobile manufactured by three different companies. Specifically, subjects are asked whether they would purchase a car manufactured by the following automobile manufacturers: Chenesco, Howasaki, and Gemini. The responses of the* 12 *subjects follow:* **Subject 1** *said she would purchase a Chenesco and a Howasaki but not a Gemini;* **Subject 2** *said she would only purchase a Howasaki;* **Subject 3** *said she would purchase all three makes of cars;* **Subject 4** *said she would only purchase a Howasaki;* **Subject 5** *said she would only purchase a Howasaki;* **Subject 6** *said she would purchase a Howasaki and a Gemini but not a Chenesco;* **Subject 7** *said she would not purchase any of the automobiles;* **Subject 8** *said she would only purchase a Howasaki;* **Subject 9** *said she would purchase a Chenesco and a Howasaki but not a Gemini;* **Subject 10** *said she would only purchase a Howasaki;* **Subject 11** *said she would not purchase any of the automobiles; and* **Subject 12** *said she would only purchase a Gemini. Can the market researcher conclude that there are differences with respect to car preference based on the responses of subjects?*

III. Null versus Alternative Hypotheses

In stating the null and alternative hypotheses the notation π_j will be employed to represent the proportion of **Yes** responses in the population represented by the jth experimental condition. Stated more generally, π_j represents the proportion of responses in one of the two response categories in the population represented by the jth experimental condition.

Null hypothesis H_0: $\pi_1 = \pi_2 = \pi_3$

(The proportion of **Yes** responses in the population represented by Condition 1 equals the proportion of **Yes** responses in the population represented by Condition 2 equals the proportion of **Yes** responses in the population represented by Condition 3.

Alternative hypothesis H_1: Not H_0

(This indicates that in at least two of the underlying populations represented by the $k = 3$ conditions, the proportion of **Yes** responses are not equal. It is important to note that the alternative hypothesis should not be written as follows: H_1: $\pi_1 \neq \pi_2 \neq \pi_3$. The reason why the latter notation for the alternative hypothesis is incorrect is because it implies that all three population proportions must differ from one another in order to reject the null hypothesis. In this book it will always be assumed that the alternative hypothesis for the **Cochran** Q **test** is stated **nondirectionally**.)[2]

IV. Test Computations

The data for Example 20.1 are summarized in Table 20.1. The number of subjects employed in the experiment is $n = 12$, and thus within each condition there are $n = n_1 = n_2 = n_3 = 12$ scores. The values **1** and **0** are employed to represent the two response categories in which a subject's response/categorization may fall. Specifically, a score of **1** indicates a **Yes** response and a score of **0** indicates a **No** response.

The following summary values are computed in Table 20.1 which will be employed in the analysis of the data:

a) The value ΣC_j represents the number of **Yes** responses in the jth condition. Thus, the number of **Yes** responses in Conditions 1, 2, and 3 are respectively $\Sigma C_1 = 3$, $\Sigma C_2 = 9$, and $\Sigma C_3 = 3$.

Table 20.1 Data for Example 20.1

	Chenesco C_1	Howasaki C_2	Gemini C_3	R_i	R_i^2
Subject 1	1	1	0	2	4
Subject 2	0	1	0	1	1
Subject 3	1	1	1	3	9
Subject 4	0	1	0	1	1
Subject 5	0	1	0	1	1
Subject 6	0	1	1	2	4
Subject 7	0	0	0	0	0
Subject 8	0	1	0	1	1
Subject 9	1	1	0	2	4
Subject 10	0	1	0	1	1
Subject 11	0	0	0	0	0
Subject 12	0	0	1	1	1
	$\Sigma C_1 = 3$	$\Sigma C_2 = 9$	$\Sigma C_3 = 3$	$\Sigma R_i = 15$	$\Sigma R_i^2 = 27$
	$(\Sigma C_1)^2 = (3)^2 = 9$	$(\Sigma C_2)^2 = (9)^2 = 81$	$(\Sigma C_3)^2 = (3)^2 = 9$		
	$p_1 = \frac{\Sigma C_1}{n_1} = \frac{3}{12} = .25$	$p_2 = \frac{\Sigma C_2}{n_2} = \frac{9}{12} = .75$	$p_3 = \frac{\Sigma C_3}{n_3} = \frac{3}{12} = .25$		

b) The value $(\Sigma C_j)^2$ represents the square of the ΣC_j value computed for the jth condition. The sum of the $k = 3$ $(\Sigma C_j)^2$ scores can be represented by the notation $\Sigma(\Sigma C_j)^2$. Thus, for Example 20.1, $\Sigma(\Sigma C_j)^2 = (\Sigma C_1)^2 + (\Sigma C_2)^2 + (\Sigma C_3)^2 = 9 + 81 + 9 = 99$.

c) The value R_i represents the sum of the $k = 3$ scores of the ith subject (i.e., the number of **Yes** responses for the ith subject). Note that an R_i value is computed for each of the $n = 12$ subjects. The sum of the n R_i scores is ΣR_i. Thus, for Example 20.1, $\Sigma R_i = 2 + 1 + \cdots + 0 + 1 = 15$.

d) The value R_i^2 represents the square of the R_i score of the ith subject. The sum of the n R_i^2 scores is ΣR_i^2. Thus, for Example 20.1, $\Sigma R_i^2 = 4 + 1 + \cdots + 0 + 1 = 27$.

e) The value p_j represents the proportion of **Yes** responses in the jth condition. The value of p_j is computed as follows: $p_j = \Sigma C_j / n_j$. Thus, in Table 20.1 the values of p_j for Conditions 1, 2, and 3 are respectively $p_1 = .25$, $p_2 = .75$, and $p_3 = .25$.

Equation 20.1 is employed to calculate the test statistic for the **Cochran *Q* test**. The Q value computed with Equation 20.1 is interpreted as a chi-square value. In Equation 20.1 the following notation is employed with respect to the summary values noted in this section: a) C is employed to represent the value computed for $\Sigma(\Sigma C_j)^2$; b) T is employed to represent the value computed for ΣR_i; and c) R is employed to represent the value computed for ΣR_i^2. Thus, for Example 20.1, $C = 99$, $T = 15$, and $R = 27$.[3]

$$Q = \frac{(k - 1)[(k)(C) - (T)^2]}{(k)(T) - R} \qquad \textbf{(Equation 20.1)}$$

Substituting the appropriate values from Example 20.1 in Equation 20.1, the value $Q = 8$ is computed.[4]

$$Q = \frac{(3 - 1)[(3)(99) - (15)^2]}{(3)(15) - 27} = 8$$

V. Interpretation of the Test Results

In order to reject the null hypothesis the computed value $Q = \chi^2$ must be equal to or greater than the tabled critical chi-square value at the prespecified level of significance. The computed chi-square value is evaluated with **Table A4 (Table of the Chi-Square Distribution)** in the **Appendix**. For the appropriate degrees of freedom, the tabled $\chi^2_{.95}$ value (which is the chi-square value at the 95th percentile) and the tabled $\chi^2_{.99}$ value (which is the chi-square value at the 99th percentile) are employed as the .05 and .01 critical values for evaluating a nondirectional alternative hypothesis. The number of degrees of freedom employed in the analysis are computed with Equation 20.2. Thus, $df = 3 - 1 = 2$.

$$df = k - 1 \qquad \textbf{(Equation 20.2)}$$

For $df = 2$, the tabled critical .05 and .01 chi-square values are $\chi^2_{.05} = 5.99$ and $\chi^2_{.01} = 9.21$. Since the computed value $Q = 8$ is greater than $\chi^2_{.05} = 5.99$, the alternative hypothesis is supported at the .05 level. Since, however, $Q = 8$ is less than $\chi^2_{.01} = 9.21$, the alternative hypothesis is not supported at the .01 level. A summary of the analysis of Example 20.1 with the **Cochran *C* test** follows: It can be concluded that there is a significant difference in subjects' preferences for at least two of the three automobiles. This result can be summarized as follows: $Q(2) = 8, p < .05$.

VI. Additional Analytical Procedures for the Cochran *Q* Test and/or Related Tests

1. Pairwise comparisons following computation of the test statistic for the Cochran *Q* test Prior to reading this section the reader should review the discussion of comparisons in Section VI of the **single-factor between-subjects analysis of variance (Test 16)**. As is the case with the omnibus F value computed for an analysis of variance, the Q value computed with Equation 20.1 is based on an evaluation of all k experimental conditions. When the value of Q is significant, it does not indicate whether just two or, in fact, more than two conditions differ significantly from one another. In order to answer the latter question, it is necessary to conduct comparisons contrasting specific conditions with one another. This section will describe methodologies that can be employed for conducting **simple/pairwise comparisons** following the computation of a Q value.[5]

In conducting a simple comparison, the null hypothesis and nondirectional alternative hypothesis are as follows: H_0: $\pi_a = \pi_b$ versus H_1: $\pi_a \neq \pi_b$. In the aforementioned hypotheses, π_a and π_b represent the proportion of **Yes** responses in the populations represented by the two conditions involved in the comparison. The alternative hypothesis can also be stated directionally as follows: H_1: $\pi_a > \pi_b$ or H_1: $\pi_a < \pi_b$.

A number of sources (e.g., Fleiss (1981) and Marascuilo and McSweeney (1977)) describe comparison procedures for the **Cochran *Q* test**. The procedure to be described in this section, which is one of two procedures described in Marascuilo and McSweeney (1977), is essentially the application of the **Bonferroni–Dunn method** described in Section VI of the **single-factor between-subjects analysis of variance** to the **Cochran *Q* test** model. Through use of Equation 20.3, the procedure allows a researcher to identify the minimum required difference between the observed proportion of **Yes** responses for any two experimental conditions (designated as CD_C) in order for them to differ from one another at the prespecified level of significance.

$$CD_C = z_{\text{adj}} \sqrt{2\left[\frac{(k)(T) - R}{(n^2)(k)(k-1)}\right]} \qquad \textbf{(Equation 20.3)}$$

The value of z_{adj} is obtained from **Table A1 (Table of the Normal Distribution)** in the **Appendix**. In the case of a nondirectional alternative hypothesis, z_{adj} is the z value above which a proportion of cases corresponding to the value $\alpha_{FW}/2c$ falls (where c is the total number of comparisons that are conducted). In the case of a directional alternative hypothesis, z_{adj} is the z value above which a proportion of cases corresponding to the value α_{FW}/c falls. When all possible pairwise comparisons are made $c = [k(k - 1)]/2$, and thus, $2c = k(k - 1)$. In Example 20.1 the number of pairwise/simple comparisons that can be conducted is $c = [3(3 - 1)]/2 = 3$ — specifically, Condition 1 versus Condition 2, Condition 1 versus Condition 3, and Condition 2 versus Condition 3.

The value of z_{adj} will be a function of both the maximum **familywise Type I error rate** (α_{FW}) the researcher is willing to tolerate and the total number of comparisons that are conducted. When a limited number of comparisons are planned prior to collecting the data, most sources take the position that a researcher is not obliged to control the value of α_{FW}. In such a case, the **per comparison Type I error rate** (α_{PC}) will be equal to the prespecified value of alpha. When α_{FW} is not adjusted, the value of z_{adj} employed in Equation 20.3 will be the tabled critical z value that corresponds to the prespecified level of significance. Thus, if a nondirectional alternative hypothesis is employed and $\alpha = \alpha_{PC} = .05$, the tabled critical two-tailed .05 value $z_{.05} = 1.96$ is used to represent z_{adj} in Equation 20.3. If $\alpha = \alpha_{PC} = .01$, the tabled critical two-tailed .01 value $z_{.01} = 2.58$ is used in Equation 20.3. In the same respect, if a directional alternative hypothesis is employed, the tabled critical .05 and .01 one-tailed values $z_{.05} = 1.65$ and $z_{.01} = 2.33$ are used for z_{adj} in Equation 20.3.

When comparisons are not planned beforehand, it is generally acknowledged that the value of α_{FW} must be controlled so as not to become excessive. The general approach for controlling the latter value is to establish a **per comparison Type I error rate** which insures that α_{FW} will not exceed some maximum value stipulated by the researcher. One method for doing this (described under the **single-factor between-subjects analysis of variance** as the **Bonferroni–Dunn method**) establishes the **per comparison Type I error rate** by dividing the maximum value one will tolerate for the **familywise Type I error rate** by the total number of comparisons conducted. Thus, in Example 20.1, if one intends to conduct all three pairwise comparisons and wants to insure that α_{FW} does not exceed .05, $\alpha_{PC} = \alpha_{FW}/c = .05/3 = .0167$. The latter proportion is used in determining the value of z_{adj}. As noted earlier, if a directional alternative hypothesis is employed for a comparison, the value of z_{adj} employed in Equation 20.3 is the z value above which a proportion equal to $\alpha_{PC} = \alpha_{FW}/c$ of the cases falls. In **Table A1**, the z value that corresponds to the proportion .0167 is $z = 2.13$. By employing z_{adj} in Equation 20.3, one can be assured that within the "family" of three pairwise comparisons, α_{FW} will not exceed .05 (assuming all of the comparisons are directional). If a nondirectional alternative hypothesis is employed for all of the comparisons, the value of z_{adj} will be the z value above which a proportion equal to $\alpha_{FW}/2c = \alpha_{PC}/2$ of the cases falls. Since $\alpha_{PC}/2 = .0167/2 = .0083$, $z = 2.39$. By employing z_{adj} in Equation 20.3, one can be assured that α_{FW} will not exceed .05.[6]

Table 20.2 Difference Scores Between Pairs of Proportions for Example 20.1

$\lvert p_1 - p_2 \rvert = \lvert .25 - .75 \rvert = .50$	$\lvert p_1 - p_3 \rvert = \lvert .25 - .25 \rvert = 0$	$\lvert p_2 - p_3 \rvert = \lvert .75 - .25 \rvert = .50$

In order to employ the CD_C value computed with Equation 20.3, it is necessary to determine the absolute value of the difference between the proportion of **Yes** responses for each pair of experimental conditions that are compared.[7] Table 20.2 summarizes the difference scores between pairs of proportions.

If any of the differences between two proportions is equal to or greater than the CD_C value computed with Equation 20.3, a comparison is declared significant. Equation 20.3 will now be employed to evaluate the nondirectional alternative hypothesis H_1: $\pi_a \neq \pi_b$ for all three pairwise comparisons. Since it will be assumed that the comparisons are unplanned and that the researcher does not want the value of α_{FW} to exceed .05, the value $z_{\text{adj}} = 2.39$ will be used in computing CD_C.

$$CD_C = (2.39)\sqrt{2\left[\frac{(3)(15) - 27}{(12)^2(3)(3 - 1)}\right]} = (2.39)(.204) = .49$$

The obtained value $CD_C = .49$ indicates that any difference between a pair of proportions that is equal to or greater than .49 is significant. With respect to the three pairwise comparisons, the difference between Condition 1 and Condition 2 (which equals .50) and the difference between Condition 2 and Condition 3 (which also equals .50) are significant, since they are both greater than $CD_C = .49$. We cannot conclude that the difference between Condition 1 and Condition 3 is significant, since $|p_1 - p_3| = 0$ is less than $CD_C = .49$.

An alternative strategy that can be employed for conducting pairwise comparisons for the **Cochran Q test** is to use the **McNemar test** for each comparison. In employing the **McNemar test** one can employ either the chi-square or normal approximation of the test statistic for each comparison (the continuity corrected value generally providing a more accurate estimate for a small sample size), or compute the exact binomial probability for the comparison. It will be demonstrated in Section VII, that the computed chi-square value computed with Equation 15.1 for the **McNemar test** yields the same $Q = \chi^2$ value that is obtained if a **Cochran Q test** (Equation 20.1) is employed to compare the same set of experimental conditions. In this section the exact binomial probabilities for the three pairwise comparisons for the **McNemar test** model will be computed. In order to compute the exact binomial probabilities (or, for that matter, the chi-square or normal approximations of the test statistic), the data for each comparison must be placed within a 2 × 2 table like Table 15.1 (which is the table for the **McNemar test** model). To illustrate this, the data for the Condition 1 versus Condition 2 comparison are recorded in Table 20.3. Note that of the $n = 12$ subjects involved in the comparison, only 6 of the subjects' scores are actually taken into account in computing the test statistic, since the other 6 subjects have the same score in both conditions.

Table 20.3 McNemar Test Model for Binomial Analysis of Condition 1 Versus Condition 2 Comparison

		Condition 1		Row sums
		Yes (1)	No (0)	
Condition 2	Yes (1)	$a = 3$	$b = 6$	9
	No (0)	$c = 0$	$d = 3$	3
	Column sums	3	9	12

Employing **Table A6 (Table of the Binomial Distribution, Individual Probabilities)** in the **Appendix** to compute the binomial probability, we determine that when $n = 6$, $P(x = 6) = .0156$. Thus, the two-tailed binomial probability for the Condition 1 versus Condition 2 comparison is $(2)(.0156) = .0312$. In the case of the Condition 1 versus Condition 3 comparison, the frequencies for Cells a, b, c, and d are respectively 1, 2, 2, and 7. Since the frequency of both Cells b and c is 2 (and thus, $n = 4$), the Condition 1 versus Condition 3 comparison results in no difference. In the case of the Condition 2 versus Condition 3 comparison, the frequencies for Cells a, b, c, and d are respectively 2, 1,

7, and 2. For the latter comparison, since $n = 8$, the frequencies for Cells b and c are 1 and 7. Using **Table A6** (or **Table A7** which is the **Table of the Binomial Distribution, Cumulative Probabilities**), we determine that when $n = 8$, $P(x \geq 7) = .0352$. Thus, the two-tailed binomial probability for the Condition 2 versus Condition 3 comparison is $(2)(.0352) = .0704$. Note that the binomial probabilities computed for the Condition 1 versus Condition 2 and Condition 2 versus Condition 3 comparisons are not identical, since by virtue of eliminating subjects who respond in the same category in both conditions from the analysis, the two comparisons employ different values for n (which is the sum of the frequencies for Cells b and c).

As before, if we wish to evaluate a nondirectional alternative hypothesis and insure that α_{FW} does not exceed .05, the value of α_{PC} is set equal to .0167. Thus, in order to reject the null hypothesis the computed two-tailed binomial probability for a comparison must be equal to or less than .0167. Since the computed two-tailed probabilities .0312 (for the Condition 1 versus Condition 2 comparison) and .0704 (for the Condition 2 versus Condition 3 comparison) are greater than .0167, none of the pairwise comparisons is significant.

In the event the researcher elects not to control the value of α_{FW} and employs $\alpha_{PC} = .05$ in evaluating the three pairwise comparisons (once again assuming a nondirectional analysis), only the Condition 1 versus Condition 2 comparison is significant, since the computed two-tailed binomial probability .0312 is less than .05. The Condition 2 versus Condition 3 comparison falls short of significance, since the computed two-tailed binomial probability .0704 is greater than .05. It should be noted that both of the aforementioned comparisons are significant if the directional alternative hypothesis that is consistent with the data is employed (since the one-tailed probabilities .0156 and .0352 are less than .05). If Equation 20.3 is employed for the same set of comparisons, $CD_C = (1.96)(.204) = .40$.[8] Thus, employing the latter equation, the Condition 1 versus Condition 2 and Condition 2 versus Condition 3 comparisons are significant, since in both instances the difference between the two proportions is greater than $CD_C = .40$.

Although the binomial probabilities for the **McNemar test** for the Condition 1 versus Condition 2 and Condition 2 versus Condition 3 comparisons are larger than the probabilities associated with the use of Equation 20.3, both comparison procedures yield relatively low probability values for the two aforementioned comparisons. Thus, in the case of Example 20.1, depending upon which comparison procedure one employs (as well as the value of α_{PC} and whether one evaluates a nondirectional or directional alternative hypothesis) it would appear that there is a high likelihood that the Condition 1 versus Condition 2 and Condition 2 versus Condition 3 comparisons are significant. The intent of presenting two different comparison procedures in this section is to illustrate that, generally speaking, the results obtained with different procedures will be reasonably consistent with one another.[9] As is noted in the discussion of comparisons in Section VI of the **single-factor between-subjects analysis of variance**, in instances where two or more comparison procedures yield inconsistent results, the most effective way to clarify the status of the null hypothesis is to replicate a study one or more times. It is also noted throughout the book that, in the final analysis, the decision regarding which of the available comparison procedures to employ is usually not the most important issue facing the researcher conducting comparisons. The main issue is what maximum value one is willing to tolerate for α_{FW}.

Marascuilo and McSweeney (1977) discuss the computation of a confidence interval for a comparison for the **Cochran *Q* test** model. One approach for computing a confidence interval is to add to and subtract the computed value of CD_C from the obtained difference between the proportions involved in the comparison. The latter approach is based on the same logic employed for computing a confidence interval for a comparison in Section VI of the **single-factor between-subjects analysis of variance**.

VII. Additional Discussion of the Cochran Q Test

1. Issues relating to subjects who obtain the same score under all of the experimental conditions

a) Cochran (1950) noted that since the value computed for Q is not affected by the scores of any subject (or any row, if matched subjects are employed) who obtains either all 0s or all 1s in each of the experimental conditions, the scores of such subjects can be deleted from the data analysis. If the latter is done with respect to Example 20.1, the data for Subjects 3, 7, and 11 can be eliminated from the analysis (since Subjects 7 and 11 obtain all 0s, and Subject 3 obtains all 1s). It is demonstrated below that if the scores of Subjects 3, 7, and 11 are eliminated from the analysis, the value $Q = 8$ is still obtained when the revised summary values are substituted in Equation 20.1.

$$\Sigma C_1 = 2 \quad \Sigma C_2 = 8 \quad \Sigma C_3 = 2 \quad (\Sigma C_1)^2 = 4 \quad (\Sigma C_2)^2 = 64 \quad (\Sigma C_3)^2 = 4$$

$$\Sigma(\Sigma C_j)^2 = 4 + 64 + 4 = 72 \quad \Sigma R_i = 12 \quad \Sigma R_i^2 = 18$$

$$\text{Thus:} \quad C = 72 \quad T = 12 \quad R = 18$$

$$Q = \frac{(3 - 1)[(3)(72) - (12)^2]}{(3)(12) - 18} = 8$$

It is noted in Section VI of the **McNemar test** that the latter test essentially eliminates from the analysis any subject who obtain the same score under both experimental conditions, and that this represents a limitation of the test. What was said with regard to the **McNemar test** in this respect also applies to the **Cochran Q test**. Thus, it is entirely possible to obtain a significant Q value even if the overwhelming majority of the subjects in a sample obtain the same score in each of the experimental conditions. To illustrate this, the value $Q = 8$ (obtained for Example 20.1) can be obtained for a sample of 1009 subjects, if 1000 of the subjects obtained a score of 1 in all three experimental conditions, and the remaining nine subjects had the same scores as Subjects 1, 2, 4, 5, 6, 8, 9, 10, and 12 in Example 20.1. Since the computation of the Q value in such an instance will be based on a sample size of 9 rather than on the actual sample size of 1009, it is reasonable to assume that such a result, although statistically significant, will not be of any practical significance from the perspective of the three automobile manufacturers. The latter statement is based on the fact that since all but 9 of the 1009 subjects said they would buy all three automobiles, there really does not appear to be any differences in preference that will be of any economic consequence to the manufacturers.

b) In Section I it is noted that when $n < 4$ and/or $nk < 24$, it is recommended that tables for the exact **Cochran test** statistic (derived by Patil (1975)) be employed instead of the chi-square approximation. In making such a determination, the value of n that should be used should not include any subjects who obtain all 0s or all 1s in each of the experimental conditions. Thus, in Example 20.1, the value $n = 9$ is employed, and not the value $n = 12$. Consequently $nk = (9)(3) = 27$. Since $n > 4$ and $nk > 24$, it is acceptable to employ the chi-square approximation.

c) Note, however, that Equation 20.3 (the equation employed for conducting comparisons) does employ for the value of n the total number of subjects, irrespective of whether a subject obtains the same score in all k experimental conditions. The use of the latter n value maximizes the power of the comparison procedure.

2. Equivalency of the Cochran Q test and the McNemar test when $k = 2$ In Section I it is noted that when $k = 2$ the **Cochran Q test** yields a result that is equivalent to that obtained with the **McNemar test**. To be more specific, the **Cochran Q test** will yield a result that is equivalent to the **McNemar test** statistic when the correction for continuity is not employed for the latter test (i.e., the result obtained with Equation 15.1). In order to demonstrate the equivalency of the two tests Example 20.2 will be evaluated with both tests.

Example 20.2 *A market researcher asks* 10 *female subjects whether or not they would purchase an automobile manufactured by two different companies. Specifically, subjects are asked whether they would purchase an automobile manufactured by Chenesco and Howasaki. Except for* Subjects 2 *and* 3, *all of the subjects said they would purchase a Chenesco but would not purchase a Howasaki.* Subject 2 *said she would not purchase either car, while* Subject 3 *said she would purchase a Howasaki but not a Chenesco. Based on the responses of subjects, can the market researcher conclude that there are differences with respect to car preference?*

Tables 20.4 and 20.5 respectively summarize the data for the study within the framework of the **Cochran Q test** model and the **McNemar test** model.

Table 20.4 Summary of Data for Analysis of Example 20.2 with Cochran Q Test

	Chenesco C_1	Howasaki C_2	R_i	R_i^2
Subject 1	1	0	1	1
Subject 2	0	0	0	0
Subject 3	0	1	1	1
Subject 4	1	0	1	1
Subject 5	1	0	1	1
Subject 6	1	0	1	1
Subject 7	1	0	1	1
Subject 8	1	0	1	1
Subject 9	1	0	1	1
Subject 10	1	0	1	1
	$\Sigma C_1 = 8$	$\Sigma C_2 = 1$	$\Sigma C_3 = 9$	$\Sigma R_i = 9$

Table 20.5 Summary of Data for Analysis of Example 20.2 with McNemar Test

		Condition 1 (Chenesco)		Row sums
		No (0)	Yes (1)	
Condition 2 (Howasaki)	**No (0)**	$a = 1$	$b = 8$	9
	Yes (1)	$c = 1$	$d = 0$	1
	Column sums	2	8	10

Example 20.2 is evaluated below employing both the **Cochran Q test** and the **McNemar test**. Note the computed value $Q = \chi^2 = 5.44$ is equivalent to $\chi^2 = 5.44$ computed for the **McNemar test**.

Cochran Q test:

$$\Sigma C_1 = 8 \quad \Sigma C_2 = 1 \quad (\Sigma C_1)^2 = 64 \quad (\Sigma C_2)^2 = 1$$

$$\Sigma(\Sigma C_j)^2 = 64 + 1 = 65 \quad \Sigma R_i = 9 \quad \Sigma R_i^2 = 9$$

$$\text{Thus:} \quad C = 65 \quad T = 9 \quad R = 9$$

$$Q = \frac{(2 - 1)[(2)(65) - (9)^2]}{(2)(9) - 9} = 5.44$$

McNemar test:

$$\chi^2 = \frac{(b - c)^2}{(b + c)} = \frac{(8 - 1)^2}{(8 + 1)} = 5.44$$

In the case of both tests, $df = 1$ (since in the case of the **Cochran Q test** $df = k - 1 = 2 - 1 = 1$, and in the case of the **McNemar test** the number of degrees of freedom is always $df = 1$). The tabled critical .05 and .01 chi-square values for $df = 1$ are $\chi^2_{.05} = 3.84$ and $\chi^2_{.01} = 6.63$. Since the obtained value $\chi^2 = 5.44$ is greater than $\chi^2_{.05} = 3.84$, the nondirectional alternative hypothesis is supported at the .05 level. Since $\chi^2 = 5.44$ is less than $\chi^2_{.01} = 6.63$, it is not supported at the .01 level.

In point of fact, the data in Table 20.4 are based on Example 14.1, which is employed to illustrate the **binomial sign test for two dependent samples (Test 14).** If we assume that in the case of Example 14.1 a subject is assigned a score of **1** in the condition in which she has a higher score and a score of **0** in the condition in which she has a lower score, plus the fact that a subject is assigned a score of **0** (or **1**) in both conditions if she has the same score, the data in Table 14.1 will be identical to that presented in Table 20.4.[10] When Equation 14.3 (the uncorrected normal approximation for the **binomial sign test for two dependent samples**) is employed to evaluate Example 14.1, it yields the value $z = 2.33$ which if squared equals the obtained chi-square value for Example 20.2 — i.e., $(z = 2.33)^2 = (\chi^2 = 5.44)$.[11] Thus, when $k = 2$ the **McNemar test/Cochran Q test** are equivalent to the **binomial sign test for two dependent samples.** It should also be noted that the exact binomial probability computed for the **binomial sign test for two dependent samples** will be equivalent to the exact binomial probability computed when the **McNemar test/Cochran Q test** is employed to evaluate the same data. For Examples 14.1/20.2, the two-tailed binomial probability is $P(x \geq 8) = (2)(.0196) = .0392$ (for $n = 9$).

3. Alternative nonparametric procedures for categorical data for evaluating a design involving k dependent samples Daniel (1990) and/or Fliess (1981) note that alternative procedures for comparing k or more matched samples have been developed by Bennett (1967, 1968) and Shah and Claypool (1985). Chou (1989) describes a median test that can be employed to evaluate more than two dependent samples. The latter test, which employs the chi-square distribution to approximate the exact sampling distribution, employs subject/block medians as reference points in determining whether two or more of the treatment conditions represent different populations. The test described by Chou (1989) assumes that subjects' original scores are in an interval/ratio format and are converted into categorical data.

VIII. Additional Examples Illustrating the Use of the Cochran Q Test

Since the **Cochran Q test** can be employed to evaluate any dependent samples design involving two or more experimental conditions, it can also be used to evaluate any of the examples discussed under the **McNemar test.** Examples 20.3–20.7 are additional examples that can be evaluated with the **Cochran *Q* test**. Examples 20.6 and 20.7 represent extensions of a **before–after design** to a design involving $k = 3$ experimental conditions. Since the data for all of the examples in this section (with the exception of Example 20.7) are identical to the data employed in Example 20.1, they yield the same result.

Example 20.3 *A researcher wants to assess the relative likelihood of three brands of house paint fading within two years of application. In order to make this assessment he applies the following three brands of house paint that are identical in hue to a sample of houses that have cedar shingles: Brightglow, Colorfast, and Prismalong. In selecting the houses the researcher identifies* 12 *neighborhoods which vary with respect to geographical conditions, and within each neighborhood he randomly selects* 3 *houses. Within each block of three houses, one of the houses is painted with Brightglow, a second house with Colorfast, and a third house with Prismalong. Thus, a total of* 36 *houses are painted in the study. Two years after the houses are painted, an independent judge categorizes each house with respect to whether the paint on its shingles has faded. A house is assigned the number* **1** *if there is evidence of fading and the number* **0** *if there is no evidence of fading.* Table 20.6 *summarizes the results of the study. Do the data indicate differences between the three brands of house paint with respect to fading?*

Table 20.6 Data for Example 20.3

	Brand of paint		
	Brightglow	**Colorfast**	**Prismalong**
Block 1	1	1	0
Block 2	0	1	0
Block 3	1	1	1
Block 4	0	1	0
Block 5	0	1	0
Block 6	0	1	1
Block 7	0	0	0
Block 8	0	1	0
Block 9	1	1	0
Block 10	0	1	0
Block 11	0	0	0
Block 12	0	0	1

Note that in Example 20.3 the 12 blocks, comprised of 3 houses per block, are analogous to the use of 12 sets of matched subjects with 3 subjects per set/block. The brands of house paint represent the three levels of the independent variable, and the judge's categorization for each house with respect to fading (i.e., **1** versus **0**) represents the dependent variable. Based on the analysis conducted for Example 20.1, there is a strong suggestion that Colorfast paint is perceived as more likely to fade than the other two brands.

Example 20.4 *Twelve male marines are administered a test of physical fitness which requires that an individual achieve the minimum criterion noted for the following three tasks: a) Climb a* 100 *ft. rope; b) Do* 25 *chin-ups; and c) Run a mile in under six minutes. Within the sample of* 12 *subjects the order of presentation of the three tasks is completely counterbalanced (i.e., two subjects are presented the tasks in each of the six possible presentation orders). For each of the tasks a subject is assigned a score of* **1** *if he achieves the minimum criterion and a score of* **0** *if he does not.* Table 20.7 *summarizes the results of the testing. Do the data indicate there is a difference between the three tasks with respect to subjects achieving the criterion?*

Table 20.7 Data for Example 20.4

	Task		
	Rope climb	**Chin-ups**	**Mile run**
Subject 1	1	1	0
Subject 2	0	1	0
Subject 3	1	1	1
Subject 4	0	1	0
Subject 5	0	1	0
Subject 6	0	1	1
Subject 7	0	0	0
Subject 8	0	1	0
Subject 9	1	1	0
Subject 10	0	1	0
Subject 11	0	0	0
Subject 12	0	0	1

Based on the analysis conducted for Example 20.1, the data suggest that subjects are more likely to achieve the criterion for chin-ups than they are for the other two tasks.

Example 20.5 *A horticulturist working at a university is hired to evaluate the effectiveness of three different kinds of weed killer (Zapon, Snuffout, and Shalom). Twelve athletic fields of equal size are selected as test sites. The researcher divides each athletic field into three equally sized areas, and within each field (based on random determination) he applies one kind of weed killer to one third of the field, a second kind of weed killer to another third of the field, and the third kind of weed killer to the remaining third of the field. This procedure is employed for all* 12 *athletic fields, resulting in* 36 *separate areas to which weed killer is applied. Six months after application of the weed killer an independent judge evaluates the 36 areas with respect to weed growth. The judge employs the number* **1** *to indicate that an area has evidence of weed growth and the number* **0** *to indicate that an area does not have evidence of weed growth.* Table 20.8 *summarizes the judge's categorizations. Do the data indicate there is a difference in the effectiveness between the three kinds of week killer?*

Table 20.8 Data for Example 20.5

	Weed killer		
	Zapon	**Snuffout**	**Shalom**
Field 1	1	1	0
Field 2	0	1	0
Field 3	1	1	1
Field 4	0	1	0
Field 5	0	1	0
Field 6	0	1	1
Field 7	0	0	0
Field 8	0	1	0
Field 9	1	1	0
Field 10	0	1	0
Field 11	0	0	0
Field 12	0	0	1

Note that in Example 20.5 the 12 athletic fields are analogous to 12 subjects who are evaluated under three experimental conditions. The three brands of weed killer represent the levels of the independent variable, and the judge's categorization of each area with respect to weed growth (i.e., **1** versus **0**) represents the dependent variable. Based on the analysis conducted for Example 20.1, the data suggest that Snuffout is less effective than the other two brands of weed killer.

Example 20.6 *A social scientist conducts a study assessing the impact of a federal gun control law on rioting in large cities. Assume that as a result of legislative changes the law in question, which severely limits the publics' access to firearms, was not in effect between the years* 1985–1989, *but was in effect during the five years directly preceding and following that time period (i.e., the gun control law was in effect during the periods* 1980–1984 *and* 1990–1994). *In conducting the study, the social scientist categorizes* 12 *large cities with respect to whether or not there was a major riot within each of the three designated time periods. Thus, each city is categorized with respect to whether or not a riot occurred during: a)* 1980–1984, *during which time the gun control law was in effect* (Time 1); *b)* 1985–1989, *during which time the gun control law was not in effect* (Time 2); *and c)* 1990–1994, *during which time the gun control law was in effect* (Time 3). *A code of* **1** *is employed to indicate the occurrence of at least one major riot during a specified five-year time period, and a code of* **0** *is employed to indicate the absence of a major riot during a specified time period.* Table 20.9 *summarizes the results of the study. Do the data indicate the gun control law had an effect on rioting?*

Table 20.9 Data for Example 20.6

	Time period		
	Time 1 (1980–1984)	**Time 2 (1985–1989)**	**Time 3 (1990–1994)**
New York	1	1	0
Chicago	0	1	0
Detroit	1	1	1
Philadelphia	0	1	0
Los Angeles	0	1	0
Dallas	0	1	1
Houston	0	0	0
Miami	0	1	0
Washington	1	1	0
Boston	0	1	0
Baltimore	0	0	0
Atlanta	0	0	1

Note that in Example 20.6 the 12 cities are analogous to 12 subjects who are evaluated during three time periods. Example 20.6 can be conceptualized as representing what is referred to as a **time series design.** A **time series design** is essentially a **before–after design** in which one or more blocks are evaluated one or more times both prior to and following an experimental treatment. In Example 20.6 each of the cities represents a block. **Time series** designs are most commonly employed in the social sciences when a researcher wants to evaluate social change through analysis of archival data (i.e., public records). The internal validity of a **time series design** is limited insofar as the treatment is not manipulated by the researcher, and thus any observed differences across time periods with respect to the dependent variable may be due to extraneous variables over which the researcher has no control. Thus, although when Example 20.6 is evaluated with the **Cochran Q test** the obtained Q value is significant (suggesting that more riots occurred when the gun control law was not in effect), other circumstances (such as the economy, race relations, etc.) may have varied across time periods, and such factors (as opposed to the gun control law) may have been responsible for the observed effect.[12] Another limitation of the time series design described by Example 20.6 is that since the membership of the blocks is not the result of random assignment, the various blocks will not be directly comparable to one another.

In closing the discussion of Example 20.6, it should be noted that in practice the **Cochran Q test** is not commonly employed to evaluate a **time series design**. Additionally, it is worth noting that, in the final analysis, it is probably more prudent to conceptualize Example 20.6 as a **mixed factorial design**, viewing each of the cities as a separate level of a second independent variable. In a **mixed factorial design** involving two independent variables, one independent variable is a between-subjects variable (i.e., each subject/block is evaluated under only one level of that independent variable), while the other independent variable is a within-subjects variable (i.e., each subject/block is evaluated under all levels of that independent variable). If Example 20.6 is conceptualized as a **mixed factorial design,** the different cities represent the between-subjects variable and the three time periods represent the within-subjects variable. Such a design is typically evaluated with the **factorial analysis of variance for a mixed design** which is discussed in Section IX of the **between-subjects factorial analysis of variance (Test 21)**. It should be noted, however, that in employing the latter analysis of variance, the dependent variable is generally represented by interval/ratio level data.[13]

Example 20.7 *In order to assess the efficacy of a drug which a pharmaceutical company claims is effective in treating hyperactivity,* 12 *hyperactive children are evaluated during the following three time periods: a) One week prior to taking the drug; b) After a child has taken the drug for six consecutive months; and c) Six months after the drug is discontinued. The children are observed by judges who employ a standardized procedure for evaluating hyperactivity. The procedure requires that during each time period a child be assigned a score of* **1** *if he is hyperactive and a score of* **0** *if he is not hyperactive. During the evaluation process, the judges are blind with respect to whether a child is taking medication at the time he or she is evaluated.* Table 20.10 *summarizes the results of the study. Do the data indicate the drug is effective?*

Table 20.10 Data for Example 20.7

	Time Period		
	Time 1	**Time 2**	**Time 3**
Child 1	1	0	1
Child 2	1	0	1
Child 3	1	0	1
Child 4	1	0	0
Child 5	1	0	0
Child 6	1	0	1
Child 7	1	1	1
Child 8	1	0	1
Child 9	1	0	1
Child 10	1	0	1
Child 11	1	1	1
Child 12	1	1	1

Example 20.7 employs the same experimental design to evaluate the hypothesis that is evaluated in Example 18.7 (in Section VIII of the **single-factor within-subjects analysis of variance (Test 18)**). In Example 20.7, however, categorical data are employed to represent

the dependent variable. Evaluation of the data with the **Cochran Q test** yields the value $Q = 14.89$.

$$\Sigma C_1 = 12 \quad \Sigma C_2 = 3 \quad \Sigma C_3 = 10 \quad (\Sigma C_1)^2 = 144 \quad (\Sigma C_2)^2 = 9 \quad (\Sigma C_3)^2 = 100$$

$$\Sigma(\Sigma C_j)^2 = 144 + 9 + 100 = 253 \quad \Sigma R_i = 25 \quad \Sigma R_i^2 = 57$$

$$\text{Thus:} \quad C = 253 \quad T = 25 \quad R = 57$$

$$Q = \frac{(3 - 1)[(3)(253) - (25)^2]}{(3)(25) - 57} = 14.89$$

Since $k = 3$, $df = 2$. The tabled critical .05 and .01 values for $df = 2$ are $\chi^2_{.05} = 5.99$ and $\chi^2_{.01} = 9.21$. Since the obtained value $Q = \chi^2 = 14.89$ is greater than both of the aforementioned critical values, the alternative hypothesis is supported at both the .05 and .01 levels. Inspection of Table 20.10 strongly suggests that the significant effect is due to the lower frequency of hyperactivity during the time subjects are taking the drug (Time period 2). The latter, of course, can be confirmed by conducting comparisons between pairs of time periods.

As noted in the discussion of Example 18.7, the design of the latter study does not adequately control for the effects of extraneous/confounding variables. The same comments noted in the aforementioned discussion also apply to Example 20.7.

References

Bennett, B. M. (1967). Tests of hypotheses concerning matched samples. **Journal of the Royal Statistical Society**, Ser. B., 29, 468–474.

Bennett, B. M. (1968). Notes on χ^2 tests for matched samples. **Journal of the Royal Statistical Society**, Ser. B., 30, 368–370.

Chou, Y. (1989). **Statistical analysis for business and economics**. New York: Elsevier.

Cochran, W. G. (1950). The comparison of percentages in matched samples. **Biometrika**, 37, 256–266.

Conover, W. J. (1980). **Practical nonparametric statistics** (2nd ed.). New York: John Wiley and Sons.

Daniel, W. J. (1990). **Applied nonparametric statistics** (2nd ed.). Boston: PWS-Kent Publishing Company.

Fleiss, J. L. (1981). **Statistical methods for rates and proportions** (2nd ed.). New York: John Wiley and Sons.

Marascuilo, L. A. and McSweeney, M. (1977). **Nonparametric and distribution-free methods for the social sciences**. Monterey, CA: Brooks/Cole Publishing Company.

Patil, K. D. (1975). Cochran's Q test: Exact distribution. **Journal of the American Statistical Association**, 70, 186–189.

Shah, A. K. and Claypool, P. L. (1985). Analysis of binary data in the randomized complete block design. **Communications in Statistics — Theory and Methods**, 14, 1175–1179.

Sheskin, D. J. (1984). **Statistical tests and experimental design: A guidebook**. New York: Gardner Press.

Siegel, S. and Castellan, N. J., Jr. (1988). **Nonparametric statistics for the behavioral sciences** (2nd ed.). New York: McGraw-Hill Book Company.

Winer, B. J., Brown, D. R., and Michels, K. M. (1991). **Statistical principles in experimental design** (3rd ed.). New York: McGraw-Hill, Inc.

Endnotes

1. A more detailed discussion of the guidelines noted below can be found in Sections I and VII of the **t test for two dependent samples (Test 12)** and the **single-factor within-subjects analysis of variance.**

2. Although it is possible to conduct a directional analysis, such an analysis will not be described with respect to the **Cochran Q test.** A discussion of a directional analysis when $k = 2$ can be found under the **McNemar test.** A discussion of the evaluation of a directional alternative hypothesis when $k \geq 2$ can be found in Section VII of the **chi-square goodness-of-fit test (Test 5).** Although the latter discussion is in reference to analysis of a k independent samples design involving categorical data, the general principles regarding analysis of a directional alternative hypothesis when $k \geq 2$ are applicable to the **Cochran Q test.**

3. The use of Equation 20.1 to compute the **Cochran Q test** statistic assumes that the columns in the summary table (i.e., Table 20.1) are employed to represent the k levels of the independent variable, and that the rows are employed to represent the n subjects/matched sets of subjects. If the columns and rows are reversed (i.e., the columns are employed to represent the subjects/matched sets of subjects, and the rows the levels of the independent variable), Equation 20.1 cannot be employed to compute the value of Q.

4. The same Q value is obtained if the frequencies of **No** responses (**0**) are employed in computing the summary values used in Equation 20.1 instead of the frequencies of **Yes** (**1**) responses. To illustrate this, the data for Example 20.1 are evaluated employing the frequencies of **No** (**0**) responses.

$$\Sigma C_1 = 9 \quad \Sigma C_2 = 3 \quad \Sigma C_3 = 9 \quad (\Sigma C_1)^2 = 81 \quad (\Sigma C_2)^2 = 9 \quad (\Sigma C_3)^2 = 81$$

$$\Sigma(\Sigma C_j)^2 = 81 + 9 + 81 = 171 \quad \Sigma R_i = 21 \quad \Sigma R_i^2 = 45$$

$$\text{Thus:} \quad C = 171 \quad T = 21 \quad R = 45$$

$$Q = \frac{(3 - 1)[(3)(171) - (21)^2]}{(3)(21) - 45} = 8$$

5. In the discussion of comparisons in reference to the analysis of variance, it is noted that a **simple** (also known as a **pairwise**) **comparison** is a comparison between any two groups/conditions in a set of k groups/conditions.

6. The method for deriving the value of z_{adj} for the **Cochran Q test** is based on the same logic that is employed in Equation 17.5 (which is used for conducting comparisons for the **Kruskal–Wallis one-way analysis of variance by ranks (Test 17)**). A rationale for the use of the proportions .0167 and .0083 in determining the appropriate value for z_{adj} in Example 20.1 can be found in Endnote 4 of the **Kruskal–Wallis one-way analysis of variance by ranks.**

7. It should be noted that when a directional alternative hypothesis is employed, the sign of the difference between the two proportions must be consistent with the prediction stated in the directional alternative hypothesis. When a nondirectional alternative hypothesis is employed, the direction of the difference between the two proportions is irrelevant.

8. In Equation 20.3 the value $z_{.05}$= 1.96 is employed for z_{adj}, and the latter value is multiplied by .204, which is the value computed for the term in the radical of the equation for Example 20.1.

9. In point of fact, Equation 20.3 employs more information than the **McNemar test**, and thus provides a more powerful test of an alternative hypothesis than the latter test (assuming both tests employ the same value for α_{PC}). The lower power of the **McNemar test** is directly attributed to the fact that for a given comparison, it only employs the scores of those subjects who obtain different scores under the two experimental conditions.

10. In conducting the **binomial sign test for two dependent samples**, what is relevant is in which of the two conditions a subject has a higher score, which is commensurate with assigning a subject to one of two response categories. As is the case with the **McNemar test** and the **Cochran *Q* test**, the analysis for the **binomial sign test for two dependent samples** does not include subjects who obtain the same score in both conditions.

11. The value χ^2 = 5.44 is also obtained for Example 14.1 through use of Equation 5.2, which is the equation for the **chi-square goodness-of-fit test (Test 5)**. In the case of Example 14.1 the latter equation produces an equivalent result to that obtained with Equation 14.3 (the normal approximation). The result of the binomial analysis of Example 14.1 with the **chi-square goodness-of-fit test** is summarized in Table 14.2.

12. Within the framework of a time series design, one or more blocks can be included which can serve as controls. Specifically, in Example 20.6 additional cities might have been selected in which the gun control law was always in effect (i.e., in effect during Time 2 as well as during Times 1 and 3). Differences on the dependent variable during Time 2 between the control cities and the cites in which the law was nullified between 1985–1989 could be contrasted to further evaluate the impact of the gun control law. Unfortunately, if the law in question is national, such control cities would not be available in the nation in which the study is conducted. The reader should note, however, that even if such control cities were available, the internal validity of such a study would still be subject to challenge, since it would still not ensure adequate control over extraneous variables.

13. Related to the issue of employing an analysis of variance with a design such as that described by Example 20.6, Cochran (1950) and Winer *et al.* (1991) note that if a **single-factor within-subjects analysis of variance** is employed to evaluate the data in the **Cochran *Q* test** summary table (i.e., Table 20.1), it generally leads to similar conclusions as those reached when the data are evaluated with Equation 20.1. The question of whether it is appropriate to employ an analysis of variance to evaluate the categorical data in the **Cochran *Q* test** summary table is an issue on which researchers do not agree.

Inferential Statistical Tests Employed with Factoral Designs (and Related Measures of Association/Correlation)

Test 21: The Between-Subjects Factorial Analysis of Variance

Test 21

The Between-Subjects Factorial Analysis of Variance
(Parametric Test Employed with Interval/Ratio Data)

I. Hypothesis Evaluated with Test and Relevant Background Information

The **between-subjects factorial analysis of variance** is one of a number of analysis of variance procedures that are employed to evaluate a **factorial design**. A factorial design is employed to simultaneously evaluate the effect of two or more independent variables on a dependent variable. Each of the independent variables is referred to as a **factor**. Each of the factors has two or more levels, which refer to the number of groups/experimental conditions that comprise that independent variable. If a factorial design is not employed to assess the effect of multiple independent variables on a dependent variable, separate experiments must be conducted to evaluate the effect of each of the independent variables. One major advantage of a factorial design is that it allows the same set of hypotheses to be evaluated at a comparable level of power by using only a fraction of the subjects that would be required if separate experiments were conducted to evaluate the relevant hypotheses for each of the independent variables. Another advantage of a factorial design is that it permits a researcher to evaluate whether or not there is an **interaction** between two or more independent variables — the latter being something that cannot be determined if only one independent variable is employed in a study. An **interaction** is present in a set of data when the performance of subjects on one independent variable is not consistent across all the levels of another independent variable. The concept of interaction is discussed in detail in Section V.

The **between-subjects factorial analysis of variance** (also known as a **completely randomized factorial analysis of variance**) is an extension of the **single-factor between-subjects analysis of variance (Test 16)** to experiments involving two or more independent variables. Although the **between-subjects factorial analysis of variance** can be used for more than two factors, the computational procedures described in this book will be limited to designs involving two factors. One of the factors will be designated by the letter **A**, and will have p levels, and the second factor will be designated by the letter **B**, and will have q levels. As a result of this, there will be a total of $p \times q$ groups. A $p \times q$ **between-subjects/completely randomized factorial design** requires that each of the $p \times q$ groups is comprised of different subjects who have been randomly assigned to that group. Each group serves under one of the p levels of Factor A and one of the q levels of Factor B, with no two groups serving under the same combination of levels of the two factors. All possible combinations of the levels of Factor A and Factor B are represented by the total $p \times q$ groups.

The **between-subjects factorial analysis of variance** evaluates the following hypotheses:

a) With respect to Factor A: In the set of p independent samples (where $p \geq 2$), do at least two of the samples represent populations with different mean values? The latter hypothesis can also be stated as follows: Do at least two of the levels of Factor A represent populations with different mean values?

b) With respect to Factor B: In the set of q independent samples (where $q \geq 2$), do at least two of the samples represent populations with different mean values? The latter hypothesis can also be stated as follows: Do at least two of the levels of Factor B represent populations with different mean values?

c) In addition to evaluating the above hypotheses (which assess the presence or absence of what are referred to as **main effects**),[1] the **between-subjects factorial analysis of variance** evaluates the hypothesis of whether there is a significant **interaction** between the two factors/independent variables.

A discussion of the rationale for evaluating the three sets of hypotheses for the **between-subjects factorial analysis of variance** can be found in Section VII.

The **between-subjects factorial analysis of variance** is employed with interval/ratio data and is based on the following assumptions: a) Each sample has been randomly selected from the population it represents; b) The distribution of data in the underlying population from which each of the samples is derived is normal; and c) The third assumption, which is referred to as the **homogeneity of variance** assumption, states that the variances of the $p \times q$ underlying populations represented by the $p \times q$ groups are equal to one another. The homogeneity of variance assumption (which is discussed earlier in the book in reference to the ***t* tests for two independent and two dependent samples (Tests 8 and 12)** and the **single-factor between-subjects and within-subjects analyses of variance (Tests 16 and 18)**) is discussed in greater detail in Section VI. If any of the aforementioned assumptions of the **between-subjects factorial analysis of variance** is saliently violated, the reliability of the computed test statistic may be compromised.

II. Example

Example 21.1 *A study is conducted in order to evaluate the effect of humidity (to be designated as Factor* A*) and temperature (to be designated as Factor* B*) on mechanical problem solving ability. The experimenter employs a* 2 × 3 *between-subjects factorial design. The two levels that comprise Factor* A *are* A_1*: Low humidity;* A_2*: High humidity. The three levels that comprise Factor* B *are* B_1*: Low temperature;* B_2*: Moderate temperature;* B_3*: High temperature. The study employs* 18 *subjects, each of whom is randomly assigned to one of the six experimental groups (i.e.,* $p \times q = 2 \times 3 = 6$*) resulting in three subjects per group. Each of the six experimental groups represents a different combination of the levels that comprise the two factors. The number of mechanical problems solved by the three subjects in each of the six experimental conditions/groups follow. (The notation* Group AB_{jk} *indicates the group that served under Level* j *of Factor* A *and Level* k *of Factor* B.) **Group AB_{11}**: *Low humidity/Low temperature* (11, 9, 10); **Group AB_{12}**: *Low humidity/ Moderate temperature* (7, 8, 6); **Group AB_{13}**: *Low humidity/High temperature* (5, 4, 3); **Group AB_{21}**: *High humidity/Low temperature* (2, 4, 3); **Group AB_{22}**: *High humidity/ Moderate temperature* (4, 5, 3); **Group AB_{23}**: *High humidity/High temperature* (0, 1, 2). *Do the data indicate that either humidity or temperature influences mechanical problem solving ability?*

III. Null versus Alternative Hypotheses

A **between-subjects factorial analysis of variance** involving two factors evaluates three sets of hypotheses. The first set of hypotheses evaluates the effect of Factor A on the dependent variable, the second set evaluates the effect of Factor B on the dependent variable, and the third set evaluates whether or not there is an interaction between the two factors.

Set 1: Hypotheses for Factor A

Null hypothesis H_0: $\mu_{A_1} = \mu_{A_2}$

(The mean of the population Level 1 of Factor A represents equals the mean of the population Level 2 of Factor A represents.)

Alternative hypothesis H_1: $\mu_{A_1} \neq \mu_{A_2}$

(The mean of the population Level 1 of Factor A represents does not equal the mean of the population Level 2 of Factor A represents. This is a nondirectional alternative hypothesis. In the discussion of the **between-subjects factorial analysis of variance** it will always be assumed that the alternative hypothesis is stated **nondirectionally**.[2] In order for the alternative hypothesis for Factor A to be supported, the obtained F value for Factor A (designated by the notation F_A) must be equal to or greater than the tabled critical F value at the prespecified level of significance.)

Set 2: Hypotheses for Factor B

Null hypothesis H_0: $\mu_{B_1} = \mu_{B_2} = \mu_{B_3}$

(The mean of the population Level 1 of Factor B represents equals the mean of the population Level 2 of Factor B represents equals the mean of the population Level 3 of Factor B represents.)

Alternative hypothesis H_1: Not H_0

(This indicates that there is a difference between at least two of the q population means. It is important to note that the alternative hypothesis should not be written as follows: H_1: $\mu_{B_1} \neq \mu_{B_2} \neq \mu_{B_3}$. The reason why the latter notation for the alternative hypothesis is incorrect is because it implies that all three population means must differ from one another in order to reject the null hypothesis. In order for the alternative hypothesis for Factor B to be supported, the obtained F value for Factor B (designated by the notation F_B) must be equal to or greater than the tabled critical F value at the prespecified level of significance.)

Set 3: Hypotheses for interaction

H_0: There is no interaction between Factor A and Factor B.

H_1: There is an interaction between Factor A and Factor B.

Although it is possible to state the null and alternative hypotheses for the interaction symbolically, such a format will not be employed since it requires a considerable amount of notation. It should be noted that in predicting an interaction, a researcher may be very specific with respect to the pattern of the interaction that is predicted. As a general rule, however, such predictions are not reflected in the statement of the null and alternative hypotheses. In order for the alternative hypothesis for the interaction to be supported, the obtained F value for the interaction (designated by the notation F_{AB}) must be equal to or greater than the tabled critical F value at the prespecified level of significance.

IV. Test Computations

The test statistics for the **between-subjects factorial analysis of variance** can be computed with either **computational** or **definitional equations**. Although definitional equations reveal the underlying logic behind the analysis of variance, they involve considerably more calculations than do the computational equations. Because of the latter, computational equations will be employed in this section to demonstrate the computation of the test statistic. The definitional equations for the **between-subjects factorial analysis of variance** are described in Section VII.

The data for Example 21.1 are summarized in Table 21.1. In the latter table the following notation is employed.

N represents the total number of subjects who serve in the experiment. In Example 21.1, $N = 18$.

ΣX_T represents the total sum of the scores of the $N = 18$ subjects who serve in the experiment.

$\bar{X}_T$ represents the mean of the scores of the $N = 18$ subjects who serve in the experiment. $\bar{X}_T$ will be referred to as the **grand mean.**

ΣX_T^2 represents the total sum of the squared scores of the $N = 18$ subjects who serve in the experiment.

$(\Sigma X_T)^2$ represents the square of the total sum of scores of the $N = 18$ subjects who serve in the experiment.

$n_{AB_{jk}}$ represents the number of subjects who serve in Group AB_{jk}. In Example 21.1 $n_{AB_{jk}} = 3$. In some of the equations that follow the notation n is employed to represent the value $n_{AB_{jk}}$.

$\Sigma X_{AB_{jk}}$ represents the sum of the scores of the $n_{AB_{jk}} = 3$ subjects who serve in Group AB_{jk}.

$\bar{X}_{AB_{jk}}$ represents the mean of the scores of the $n_{AB_{jk}} = 3$ subjects who serve in Group AB_{jk}.

$\Sigma X_{AB_{jk}}^2$ represents the sum of the squared scores of the $n_{AB_{jk}} = 3$ subjects who serve in Group AB_{jk}.

$(\Sigma X_{AB_{jk}})^2$ represents the square of the sum of scores of the $n_{AB_{jk}} = 3$ subjects who serve in Group AB_{jk}.

n_{A_j} represents the number of subjects who serve in level j of Factor A. In Example 21.1, $n_{A_j} = (n_{AB_{jk}})(q) = (3)(3) = 9$.

ΣX_{A_j} represents the sum of the scores of the $n_{A_j} = 9$ subjects who serve in Level j of Factor A.

$\bar{X}_{A_j}$ represents the mean of the scores of the $n_{A_j} = 9$ subjects who serve in level j of Factor A.

$\Sigma X_{A_j}^2$ represents the sum of the squared scores of the $n_{A_j} = 9$ subjects who serve in Level j of Factor A.

$(\Sigma X_{A_j})^2$ represents the square of the sum of scores of the $n_{A_j} = 9$ subjects who serve in Level j of Factor A.

n_{B_k} represents the number of subjects who serve in level k of Factor B. In Example 21.1, $n_{B_k} = (n_{AB_{jk}})(p) = (3)(2) = 6$.

ΣX_{B_k} represents the sum of the scores of the $n_{B_k} = 6$ subjects who serve in Level k of Factor B.

$\bar{X}_{B_k}$ represents the mean of the scores of the $n_{B_k} = 6$ subjects who serve in Level k of Factor B.

$\Sigma X^2_{B_k}$ represents the sum of the squared scores of the $n_{B_k} = 6$ subjects who serve in Level k of Factor B.

$(\Sigma X_{B_k})^2$ represents the square of the sum of scores of the $n_{B_k} = 6$ subjects who serve in Level k of Factor B.

Table 21.1 Data for Example 21.1

Factor A (Humidity)	Factor B (Temperature)			
	B_1 (Low)	B_2 (Moderate)	B_3 (High)	**Row sums**
A_1 (Low)	**Group AB_{11}**	**Group AB_{12}**	**Group AB_{13}**	**Level A_1**
	$X_{AB_{11}}$ $X^2_{AB_{11}}$	$X_{AB_{12}}$ $X^2_{AB_{12}}$	$X_{AB_{13}}$ $X^2_{AB_{13}}$	
	11 121	7 49	5 25	
	9 81	8 64	4 16	
	10 100	6 36	3 9	
	$n_{AB_{11}}=3$	$n_{AB_{12}}=3$	$n_{AB_{13}}=3$	$n_{A_1}=9$
	$\Sigma X_{AB_{11}}=30$	$\Sigma X_{AB_{12}}=21$	$\Sigma X_{AB_{13}}=12$	$\Sigma X_{A_1}=63$
	$\bar{X}_{AB_{11}}=\frac{\Sigma X_{AB_{11}}}{n_{AB_{11}}}=\frac{30}{3}=10$	$\bar{X}_{AB_{12}}=\frac{\Sigma X_{AB_{12}}}{n_{AB_{12}}}=\frac{21}{3}=7$	$\bar{X}_{AB_{13}}=\frac{\Sigma X_{AB_{13}}}{n_{AB_{13}}}=\frac{12}{3}=4$	$\bar{X}_{A_1}=\frac{\Sigma X_{A_1}}{n_{A_1}}=\frac{63}{9}=7$
	$\Sigma X^2_{AB_{11}}=302$	$\Sigma X^2_{AB_{12}}=149$	$\Sigma X^2_{AB_{13}}=50$	$\Sigma X^2_{A_1}=501$
	$(\Sigma X_{AB_{11}})^2=(30)^2=900$	$(\Sigma X_{AB_{12}})^2=(21)^2=441$	$(\Sigma X_{AB_{13}})^2=(12)^2=144$	$(\Sigma X_{A_1})^2=(63)^2=3969$
A_2 (High)	**Group AB_{21}**	**Group AB_{22}**	**Group AB_{23}**	**Level A_2**
	$X_{AB_{21}}$ $X^2_{AB_{21}}$	$X_{AB_{22}}$ $X^2_{AB_{22}}$	$X_{AB_{23}}$ $X^2_{AB_{23}}$	
	2 4	4 16	0 0	
	4 16	5 25	1 1	
	3 9	3 9	2 4	
	$n_{AB_{21}}=3$	$n_{AB_{22}}=3$	$n_{AB_{23}}=3$	$n_{A_2}=9$
	$\Sigma X_{AB_{21}}=9$	$\Sigma X_{AB_{22}}=12$	$\Sigma X_{AB_{23}}=3$	$\Sigma X_{A_2}=24$
	$\bar{X}_{AB_{21}}=\frac{\Sigma X_{AB_{21}}}{n_{AB_{21}}}=\frac{9}{3}=3$	$\bar{X}_{AB_{22}}=\frac{\Sigma X_{AB_{22}}}{n_{AB_{22}}}=\frac{12}{3}=4$	$\bar{X}_{AB_{23}}=\frac{\Sigma X_{AB_{23}}}{n_{AB_{23}}}=\frac{3}{3}=1$	$\bar{X}_{A_2}=\frac{\Sigma X_{A_2}}{n_{A_2}}=\frac{24}{9}=2.67$
	$\Sigma X^2_{AB_{21}}=29$	$\Sigma X^2_{AB_{22}}=50$	$\Sigma X^2_{AB_{23}}=5$	$\Sigma X^2_{A_2}=84$
	$(\Sigma X_{AB_{21}})^2=(9)^2=81$	$(\Sigma X_{AB_{22}})^2=(12)^2=144$	$(\Sigma X_{AB_{23}})^2=(3)^2=9$	$(\Sigma X_{A_2})^2=(24)^2=576$
Column sums	**Level B_1**	**Level B_2**	**Level B_3**	**Grand Total**
	$n_{B_1}=6$	$n_{B_2}=6$	$n_{B_3}=6$	$N=18$
	$\Sigma X_{B_1}=39$	$\Sigma X_{B_2}=33$	$\Sigma X_{B_3}=15$	$\Sigma X_T=87$
	$\bar{X}_{B_1}=\frac{\Sigma X_{B_1}}{n_{B_1}}=\frac{39}{6}=6.5$	$\bar{X}_{B_2}=\frac{\Sigma X_{B_2}}{n_{B_2}}=\frac{33}{6}=5.5$	$\bar{X}_{B_3}=\frac{\Sigma X_{B_3}}{n_{B_3}}=\frac{15}{6}=2.5$	$\bar{X}_T=\frac{\Sigma X_T}{N}=\frac{87}{18}=4.83$
	$\Sigma X^2_{B_1}=331$	$\Sigma X^2_{B_2}=199$	$\Sigma X^2_{B_3}=55$	$\Sigma X^2_T=585$
	$(\Sigma X_{B_1})^2=(39)^2=1521$	$(\Sigma X_{B_2})^2=(33)^2=1089$	$(\Sigma X_{B_3})^2=(15)^2=225$	$(\Sigma X_T)^2=(87)^2=7569$

As is the case for the **single-factor between-subjects analysis of variance**, the total variability for the **between-subjects factorial analysis of variance** can be divided into **between-groups variability** and **within-groups variability**. The **between-groups variability** can be divided into the following: a) Variability attributable to Factor A; b) Variability attributable to Factor B; and c) Variability attributable to any interaction that is present between Factors A and B (which will be designated as AB variability). For each of the variability components involved in the **between-subjects factorial analysis of variance**, a sum of squares is computed. Thus, the following sum of squares values are computed: a) SS_T, the **total sum of squares**; b) SS_{BG}, the **between-groups sum of squares**; c) SS_A, the **sum of squares for Factor A** (which can also be referred to as the **row sum of squares**, since in Table 21.1 Factor A is the row variable); d) SS_B, the **sum of squares for Factor B** (which can also be referred to as the **column sum of squares**, since in Table 21.1 Factor B is the column variable); e) SS_{AB}, the **interaction sum of squares**; and f) SS_{WG}, the **within-groups sum of squares**, which is also referred to as the **error sum of squares** or **residual sum of squares**, since it represents variability that is due to chance factors which are beyond the control of the researcher. Each of the aforementioned sum of squares values represents the numerator in the equation that is employed to compute the variance for that variability component (which is referred to as the **mean square** for that component).

Equations 21.1–21.3 summarize the relationship between the sum of squares components for the **between-subjects factorial analysis of variance**. Equation 21.1 summarizes the relationship between the between-groups, the within-groups, and the total sums of squares.

$$SS_T = SS_{BG} + SS_{WG} \qquad \textbf{(Equation 21.1)}$$

Because of the relationship noted in Equation 21.2, Equation 21.1 can also be written in the form of Equation 21.3.

$$SS_{BG} = SS_A + SS_B + SS_{AB} \qquad \textbf{(Equation 21.2)}$$

$$SS_T = SS_A + SS_B + SS_{AB} + SS_{WG} \qquad \textbf{(Equation 21.3)}$$

In order to compute the sums of squares for the **between-subjects factorial analysis of variance**, the following summary values are computed with Equations 21.4–21.8 which will be employed as elements in the computational equations: $[XS]$, $[T]$, $[A]$, $[B]$, $[AB]$.[3] The reader should take note of the fact that in the equations that follow the following is true: a) $n_{AB_{jk}} = n = 3$; b) $N = npq$, and thus, $N = (3)(2)(3) = 18$; c) $n_{A_j} = nq$, and thus, $n_{A_j} = (3)(3) = 9$; d) $n_{B_k} = np$, and thus, $n_{B_k} = (3)(2) = 6$.

The summary value $[XS] = 585$ is computed with Equation 21.4.

(Equation 21.4)

$$[XS] = \Sigma X_T^2 = (11)^2 + (9)^2 + (10)^2 + \cdots + (0)^2 + (1)^2 + (2)^2 = 585$$

The summary value $[T] = 420.5$ is computed with Equation 21.5.

$$[T] = \frac{(\Sigma X_T)^2}{N} = \frac{(87)^2}{18} = 420.5 \qquad \textbf{(Equation 21.5)}$$

The summary value $[A] = 505$ is computed with Equation 21.6.

$$[A] = \sum_{j=1}^{p} \left[\frac{(\Sigma X_{A_j})^2}{n_{A_j}} \right] = \frac{(63)^2}{9} + \frac{(24)^2}{9} = 505 \qquad \textbf{(Equation 21.6)}$$

The notation $\Sigma_{j=1}^{p}[(\Sigma X_{A_j})^2/n_{A_j}]$ in Equation 21.6 indicates that for each level of Factor A the scores of the $n_{A_j} = 9$ subjects who serve under that level of the factor are summed, the resulting value is squared, and the obtained value is divided by $n_{A_j} = 9$. The values obtained for each of the $p = 2$ levels of Factor A are then summed.

The summary value $[B] = 472.5$ is computed with Equation 21.7.

$$[B] = \sum_{k=1}^{q}\left[\frac{(\Sigma X_{B_k})^2}{n_{B_k}}\right] = \frac{(39)^2}{6} + \frac{(33)^2}{6} + \frac{(15)^2}{6} = 472.5 \quad \textbf{(Equation 21.7)}$$

The notation $\Sigma_{k=1}^{q}[(\Sigma X_{B_k})^2/n_{B_k}]$ in Equation 21.7 indicates that for each level of Factor B the scores of the $n_{B_k} = 6$ subjects who serve under that level of the factor are summed, the resulting value is squared, and the obtained value is divided by $n_{B_k} = 6$. The values obtained for each of the $q = 3$ levels of Factor B are then summed.

The summary value $[AB] = 573$ is computed with Equation 21.8.

$$[AB] = \sum_{k=1}^{q}\sum_{j=1}^{p}\left[\frac{(\Sigma X_{AB_{jk}})^2}{n_{AB_{jk}}}\right] = \frac{(30)^2}{3} + \frac{(9)^2}{3} + \frac{(21)^2}{3} + \frac{(12)^2}{2} + \frac{(12)^2}{3} + \frac{(3)^2}{3} = 573 \quad \textbf{(Equation 21.8)}$$

The notation $\Sigma_{k=1}^{q}\Sigma_{j=1}^{p}[(\Sigma X_{AB_{jk}})^2/n_{AB_{jk}}]$ in Equation 21.8 indicates that for each of the $pq = 6$ groups the scores of the $n_{AB_{jk}} = 3$ subjects who serve in that group are summed, the resulting value is squared, and the obtained value is divided by $n_{AB_{jk}} = 3$. The values obtained for each of the $pq = 6$ groups are then summed.

Employing the summary values computed with Equations 21.4–21.8, Equations 21.9–21.14 can be employed to compute the values SS_T, SS_{BG}, SS_A, SS_B, SS_{AB}, and SS_{WG}.

Equation 21.9 is employed to compute the value $SS_T = 164.5$.

$$SS_T = [XS] - [T] = 585 - 420.5 = 164.5 \quad \textbf{(Equation 21.9)}$$

Equation 21.10 is employed to compute the value $SS_{BG} = 152.5$.

$$SS_{BG} = [AB] - [T] = 573 - 420.5 = 152.5 \quad \textbf{(Equation 21.10)}$$

Equation 21.11 is employed to compute the value $SS_A = 84.5$.

$$SS_A = [A] - [T] = 505 - 420.5 = 84.5 \quad \textbf{(Equation 21.11)}$$

Equation 21.12 is employed to compute the value $SS_B = 52$.

$$SS_B = [B] - [T] = 472.5 - 420.5 = 52 \quad \textbf{(Equation 21.12)}$$

Equation 21.13 is employed to compute the value $SS_{AB} = 16$.

$$SS_{AB} = [AB] - [A] - [B] + [T] = 573 - 505 - 472.5 + 420.5 = 16 \quad \textbf{(Equation 21.13)}$$

Equation 21.14 is employed to compute the value $SS_{WG} = 12$.[4]

$$SS_{WG} = [XS] - [AB] = 585 - 573 = 12 \quad \textbf{(Equation 21.14)}$$

Note that $SS_{BG} = SS_A + SS_B + SS_{AB} = 84.5 + 52 + 16 = 152.5$ and $SS_T = SS_{BG} + SS_{WG} = 152.5 + 12 = 164.5$.

The reader should take note of the fact that the values SS_T, SS_{BG}, SS_A, SS_B, SS_{AB}, and SS_{WG} must always be positive numbers. If a negative value is obtained for any of the aforementioned values, it indicates a computational error has been made.

At this point the **mean square** values (which as previously noted represent variances) for the above components can be computed. In order to compute the test statistics for the **between-subjects factorial analysis of variance,** it is only required that the following mean square values be computed: MS_A, MS_B, MS_{AB}, MS_{WG}.

MS_A is computed with Equation 21.15.

$$MS_A = \frac{SS_A}{df_A} \qquad \textbf{(Equation 21.15)}$$

MS_B is computed with Equation 21.16.

$$MS_B = \frac{SS_B}{df_B} \qquad \textbf{(Equation 21.16)}$$

MS_{AB} is computed with Equation 21.17.

$$MS_{AB} = \frac{SS_{AB}}{df_{AB}} \qquad \textbf{(Equation 21.17)}$$

MS_{WG} is computed with Equation 21.18.

$$MS_{WG} = \frac{SS_{WG}}{df_{WG}} \qquad \textbf{(Equation 21.18)}$$

In order to compute MS_A, MS_B, MS_{AB}, and MS_{WG}, it is required that the values df_A, df_B, df_{AB}, and df_{WG} (the denominators of Equations 21.15–21.18) be computed.

df_A are computed with Equation 21.19.

$$df_A = p - 1 \qquad \textbf{(Equation 21.19)}$$

df_B are computed with Equation 21.20.

$$df_B = q - 1 \qquad \textbf{(Equation 21.20)}$$

df_{AB} are computed with Equation 21.21.

$$df_{AB} = (p - 1)(q - 1) \qquad \textbf{(Equation 21.21)}$$

df_{WG} are computed with Equation 21.22. As noted earlier, the value n is equivalent to the value $n_{AB_{jk}}$. The use of n in any of the equations for the **between-subjects factorial analysis of variance** assumes that there are an equal number of subjects in each of the pq groups.

$$df_{WG} = pq(n - 1) \qquad \textbf{(Equation 21.22)}$$

Although they are not required to compute the F ratios for the **between-subjects factorial analysis of variance,** the **between-groups degrees of freedom** (df_{BG}) and the **total**

degrees of freedom (df_T) are generally computed, since they can be used to confirm the df values computed with Equations 21.19–21.22, as well as the fact that they are employed in the analysis of variance summary table.

df_{BG} are computed with Equation 21.23.

$$df_{BG} = pq - 1 \qquad \textbf{(Equation 21.23)}$$

df_T are computed with Equation 21.24.

$$df_T = N - 1 \qquad \textbf{(Equation 21.24)}$$

The relationships between the various degrees of freedom values are described below.

$$df_{BG} = df_A + df_B + df_{AB} \qquad df_T = df_{BG} + df_{WG}$$

Employing Equations 21.19–21.24, the values $df_A = 1$, $df_B = 2$, $df_{AB} = 2$, $df_{WG} = 12$, $df_{BG} = 5$, and $df_T = 17$ are computed.

$$df_A = 2 - 1 = 1 \qquad df_B = 3 - 1 = 2 \qquad df_{AB} = (2 - 1)(3 - 1) = 2$$

$$df_{WG} = (2)(3)(3 - 1) = 12 \qquad df_{BG} = [(2)(3)] - 1 = 5 \qquad df_T = 18 - 1 = 17$$

Note that $df_{BG} = df_A + df_B + df_{AB} = 1 + 2 + 2 = 5$ and $df_T = df_{BG} + df_{WG} = 5 + 12 = 17$.

Employing Equations 21.15–21.18, the following the values are computed: $MS_A = 84.5$, $MS_B = 26$, $MS_{AB} = 8$, $MS_{WG} = 1$.

$$MS_A = \frac{84.5}{1} = 84.5 \qquad MS_B = \frac{52}{2} = 26 \qquad MS_{AB} = \frac{16}{2} = 8 \qquad MS_{WG} = \frac{12}{12} = 1$$

The F ratio is the test statistic for the **between-subjects factorial analysis of variance**. Since, however, there are three sets of hypotheses to be evaluated, it is required that three F ratios be computed — one for each of the components that comprise the between-groups variability. Specifically, an F ratio is computed for Factor A, for Factor B, and for the AB interaction. Equations 21.25–21.27 are respectively employed to compute the three F ratios.

$$F_A = \frac{MS_A}{MS_{WG}} \qquad \textbf{(Equation 21.25)}$$

$$F_B = \frac{MS_B}{MS_{WG}} \qquad \textbf{(Equation 21.26)}$$

$$F_{AB} = \frac{MS_{AB}}{MS_{WG}} \qquad \textbf{(Equation 21.27)}$$

Employing Equations 21.25–21.27 the values $F_A = 84.5$, $F_B = 26$, and $F_{AB} = 8$ are computed.

$$F_A = \frac{84.5}{1} = 84.5 \qquad F_B = \frac{26}{1} = 26 \qquad F_{AB} = \frac{8}{1} = 8$$

The reader should take note of the fact that any value computed for a mean square or an F ratio must always be a positive number. If a negative value is obtained for any mean square or F ratio, it indicates a computational error has been made. If $MS_{WG} = 0$,

Equations 21.25–21.27 will be insoluble. The only time $MS_{WG} = 0$ is when within each of the pq groups all subjects obtain the same score (i.e., there is no within-groups variability). If the mean values for all of the levels of any factor are identical, the mean square value for that factor will equal zero, and if the latter is true, the F value for that factor will also equal zero.

V. Interpretation of the Test Results

It is common practice to summarize the results of a **between-subjects factorial analysis of variance** with the summary table represented by Table 21.2.

Table 21.2 Summary Table of Analysis of Variance for Example 21.1

Source of variation	*SS*	*df*	*MS*	*F*
Between-groups	152.5	5		
A	84.5	1	84.5	84.5
B	52	2	26	26
AB	16	2	8	8
Within-groups	12	12	1	
Total	**164.5**	**17**		

The obtained F values are evaluated with **Table A10 (Table of the *F* Distribution)** in the **Appendix**. In **Table A10** critical values are listed in reference to the number of degrees of freedom associated with the numerator and the denominator of an F ratio. Thus, in the case of Example 21.1 the values for df_A, df_B, and df_{AB} are employed for the numerator degrees of freedom for each of the three F ratios, while df_{WG} is employed as the denominator degrees of freedom for all three F ratios. As is the case in the discussion of other analysis of variance procedures discussed in the book, the notation $F_{.05}$ is employed to represent the tabled critical F value at the .05 level. The latter value corresponds to the tabled $F_{.95}$ value in **Table A10**. In the same respect, the notation $F_{.01}$ will be employed to represent the tabled critical F value at the .01 level, and the latter value will correspond to the relevant tabled $F_{.99}$ value in **Table A10**.

The following tabled critical values are employed in evaluating the three F ratios computed for Example 21.1: a) **Factor A**: For $df_{\text{num}} = df_A = 1$ and $df_{\text{den}} = df_{WG} = 12$, $F_{.05} = 4.75$ and $F_{.01} = 9.33$; b) **Factor B**: For $df_{\text{num}} = df_B = 2$ and $df_{\text{den}} = df_{WG} = 12$, $F_{.05} = 3.89$ and $F_{.01} = 6.93$; c) **AB interaction**: For $df_{\text{num}} = df_{AB} = 2$ and $df_{\text{den}} = df_{WG} = 12$, $F_{.05} = 3.89$ and $F_{.01} = 6.93$.

In order to reject the null hypothesis in reference to a computed F ratio, the obtained F value must be equal to or greater than the tabled critical value at the prespecified level of significance. Since the computed value $F_A = 84.5$ is greater than $F_{.05} = 4.75$ and $F_{.01} = 9.33$, the alternative hypothesis for Factor A is supported at both the .05 and .01 levels. Since the computed value $F_B = 26$ is greater than $F_{.05} = 3.89$ and $F_{.01} = 6.93$, the alternative hypothesis for Factor B is supported at both the .05 and .01 levels. Since the computed value $F_{AB} = 8$ is greater than $F_{.05} = 3.89$ and $F_{.01} = 6.93$, the alternative hypothesis for an interaction between Factors A and B is supported at both the .05 and .01 levels. The aforementioned results can be summarized as follows: $F_A(1,12) = 84.5$, $p < .01$; $F_B(2,12) = 26$, $p < .01$; $F_{AB}(2,12) = 8$, $p < .01$.

The analysis of the data for Example 21.1 allows the researcher to conclude that both humidity (Factor A) and temperature (Factor B) have a significant impact on problem solving scores. Thus, both main effects are significant. As previously noted, a main effect describes the effect of one factor/independent variable on the dependent variable, ignoring any effect any of the other factors/independent variables might have on the dependent variable. There is also, however, a significant interaction present in the data. As noted in Section I, the latter indicates that the effect of one factor is not consistent across all the levels of the other factor. It is important to note that the presence of an interaction renders any significant main effects meaningless, since it requires that the relationship described by a main effect be qualified. This is the case, since when an interaction is present the nature of the relationship between the levels of a factor on which a significant main effect is detected will depend upon which level of the second factor is considered. Table 21.3, which summarizes the data for Example 21.1, will be used to illustrate this point. The six cells in the Table 21.3 contain the means of the $pq = 6$ groups. The values in the margins of the rows and columns of the table respectively represent the means of the levels of Factor A and Factor B. In Table 21.3 the average of any row or column can be obtained by adding up all of the values in that row or column, and dividing the sum by the number of cells in that row or column.[5]

Table 21.3 Group and Marginal Means for Example 21.1

		Factor B (Temperature)			
		B_1 (Low)	B_2 (Moderate)	B_3 (High)	Row averages
Factor A (Humidity)	A_1 (Low)	10	7	4	7
	A_2 (High)	3	4	1	2.67
Column averages		6.5	5.5	2.5	Grand mean = 4.83

In Table 21.3 the main effect for Factor A (Humidity) indicates that as humidity increases the number of problems solved decreases (since $(\bar{X}_{A_1} = 7) > (\bar{X}_{A_2} = 2.67)$). Similarly, the main effect for Factor B (Temperature) indicates that as temperature increases the number of problems solved decreases (since $(\bar{X}_{B_1} = 6.5) > (\bar{X}_{B_2} = 5.5) > (\bar{X}_{B_3} = 2.5)$). However, closer inspection of the data reveals that the effects of the factors on the dependent variable are not as straightforward as the main effects suggest. Specifically, the ordinal relationship depicted for the main effect on Factor B is only applicable to Level 1 of Factor A. Although under the low humidity condition (A_1) the number of problems solved decreases as temperature increases, the latter is not true for the high humidity condition (A_2). Under the latter condition the number of problems solved increases from 3 to 4 as temperature increases from low to moderate but then decreases to 1 under the high temperature condition. Thus, the main effect for Factor B is misleading, since it is based on the result of averaging the data from two rows which do not contain consistent patterns of information. In the same respect, if one examines the main effect on Factor A, it suggests that as humidity increases performance decreases. Table 21.3, however, reveals that although this ordinal relationship is observed for all three levels of Factor B, the effect is much more pronounced for Level 1 (low temperature) than it is for either Level 2 (moderate temperature) or Level 3 (high temperature). Thus, even though the ordinal relationship described by the main effect is consistent across the three levels of Factor B, the magnitude of the relationship varies depending upon which level of Factor B is considered.

Figure 21.1 summarizes the information presented in Table 21.3 in a graphical format. Each of the points depicted in the graphs described by Figures 21.1a and 21.1b represents the average score of the group that corresponds to the level of the factor represented by the line on which that point falls and the level of the factor on the abscissa (X-axis) above which the point falls. An interaction is revealed on either graph when two or more of the lines are not equidistant from one another throughout the full length of the graph, as one moves from left to right. When two or more lines on a graph intersect with one another, as is the case

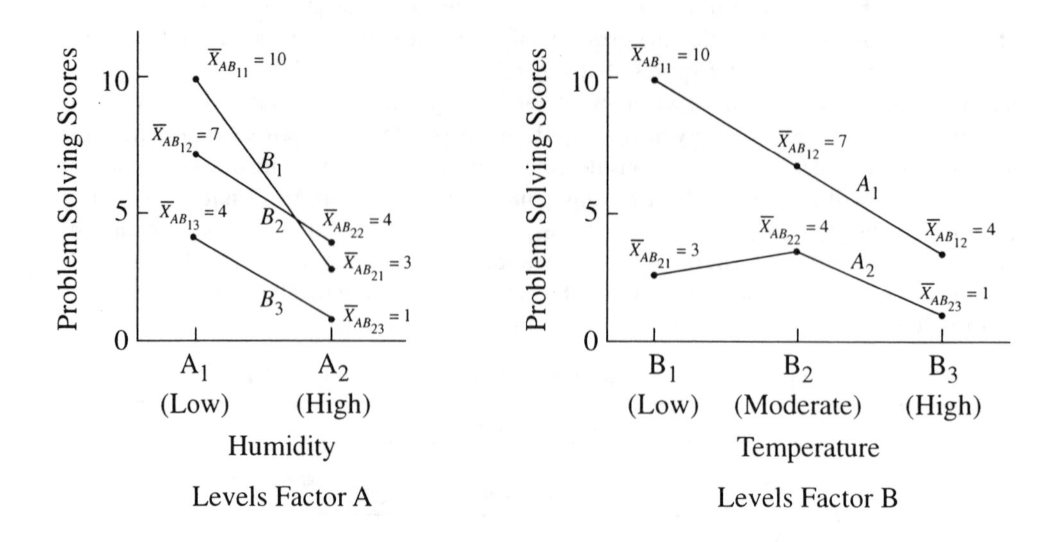

Figure 21.1 Graphical Summary of Results of Example 21.1

in Figure 21.1a, or two or more lines diverge from one another, as is the case in Figure 21.1b, it more than likely indicates the presence of an interaction. The ultimate determination, however, with respect to whether a significant interaction is present should always based on the computed value of the F_{AB} ratio.

In an experiment in which there are two factors, either of two graphs can be employed to summarize the results of the study. In Figure 21.1a the levels of Factor A are represented on the abscissa, and three lines are employed to represent subjects' performance on each of the levels of Factor B (with reference to the specific levels of Factor A). In Figure 21.1b the levels of Factor B are represented on the abscissa, and two lines are employed to represent subjects' performance on each of the levels of Factor A (with reference to the specific levels of Factor B). As noted earlier, the fact that an interaction is present is reflected in Figures 21.1a and 21.1b, since the lines are not equidistant from one another throughout the length of both graphs.[6]

Table 21.4 and Figure 21.2 summarize a hypothetical set of data (for the same experiment described by Example 21.1) in which no interaction is present. For purposes of illustration it will be assumed that in this example the computed values F_A and F_B are significant, while F_{AB} is not.

Table 21.4 Hypothetical Values for Group and Marginal Means When There Is No Interaction

		Factor B (Temperature)			
		B_1 (Low)	B_2 (Moderate)	B_3 (High)	Row averages
Factor A (Humidity)	A_1 (Low)	10	8	6	8
	A_2 (High)	6	4	2	4
Column averages		8	6	4	Grand mean = 6

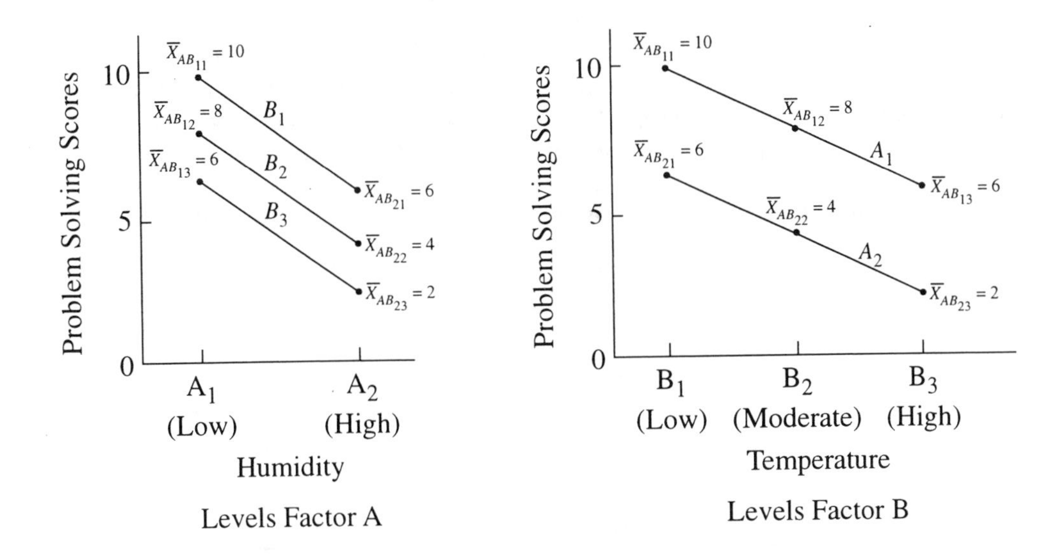

Figure 21.2 Graphical Summary of Results Described in Table 21.4

Inspection of Table 21.4 and Figure 21.2 indicates the presence of a main effect on both Factors A and B and the absence of an interaction. The presence of a main effect on Factor A is reflected in the fact that there is a reasonably large difference between $\bar{X}_{A_1} = 8$ and $\bar{X}_{A_2} = 4$. In the same respect, the significant main effect on Factor B is reflected in the discrepancy between the mean values $\bar{X}_{B_1} = 8$, $\bar{X}_{B_2} = 6$, and $\bar{X}_{B_3} = 4$. The conclusion that there is no interaction is based on the fact that the relationship described by each of the main effects is consistent across all of the levels of the second factor. To illustrate this, consider the main effect described for Factor A. In Table 21.4, the main effect for Factor A indicates that subjects solve 4 more problems under the low humidity condition than under the high humidity condition, and since this is the case regardless of which level of Factor B one considers, it indicates that there is no interaction between the two factors. The absence of an interaction is reflected in Figure 21.2a, since the three lines are equidistant from one another.[7] In Table 21.4 the main effect for Factor B indicates that the number of problems solved decreases in steps of 2 as one progresses from low to moderate to high temperature. This pattern is consistent across both of the levels of Factor A. The absence of an interaction is also reflected in Figure 21.2b, since the two lines representing each of the levels of Factor A are equidistant from one another (as well as being parallel) throughout the length of the graph.

VI. Additional Analytical Procedures for the Between-Subjects Factorial Analysis of Variance and/or Related Tests

1. Comparisons following the computation of the *F* values for the between-subjects factorial analysis of variance Upon computing the omnibus *F* values, further analysis of the data comparing one or more groups and/or factor levels with one another can provide a researcher with more detailed information regarding the relationship between the independent variables and the dependent variable. Since the procedures to be described in this section are essentially extensions of those employed for the **single-factor between-subjects analysis of variance,** the reader should review the discussion of comparison procedures in Section VI of the latter test prior to reading this section. The discussion in this section will examine additional analytical procedures that can be conducted following the computation of the F values under the following three conditions: a) **No significant main effects or interaction are present;** b) **One or both main effects are significant, but the interaction is not significant;** and c) **A significant interaction is present, with or without one or more of the main effects being significant.** Table 21.5 is a summary table for Example 21.1 depicting all of the group means for which comparison procedures will be described in this section.

Table 21.5 Summary Table of Means for Example 21.1

		Factor B (Temperature)			
		B_1	B_2	B_3	Row averages
Factor A (Humidity)	A_1	$\bar{X}_{AB_{11}}$	$\bar{X}_{AB_{12}}$	$\bar{X}_{AB_{13}}$	$\bar{X}_{A_1}$
	A_2	$\bar{X}_{AB_{21}}$	$\bar{X}_{AB_{22}}$	$\bar{X}_{AB_{23}}$	$\bar{X}_{A_2}$
Column averages		$\bar{X}_{B_1}$	$\bar{X}_{B_2}$	$\bar{X}_{B_3}$	

a) **No significant main effects or interaction are present** If in a **between-subjects factorial analysis of variance** neither of the main effects or interaction is significant, in most instances it will not be productive for a researcher to conduct additional analysis of the data. If, however, prior to the data collection phase of a study a researcher happens to have planned any of the specific types of analyses to be discussed later in this section, he can still conduct them regardless of whether or not any of the F values are significant (and not be obliged to control the value of α_{FW}). Although one can also justify conducting additional analytical procedures that are unplanned, in such a case most statisticians believe that a researcher should control the familywise Type I error rate (α_{FW}), in order that it not exceed what would be considered to be a reasonable level.

b) **One or both main effects are significant, but the interaction is not significant** When at least one of the F values is significant, the first question the researcher must ask prior to conducting any additional analytical procedures is whether or not the interaction is significant. When the interaction is not significant, a factorial design can essentially be conceptualized as being comprised of two separate single factor experiments. As such, both **simple** and **complex comparisons** can be conducted contrasting different means or sets of means that represent the levels of each of the factors. Such comparisons involve contrasting within a specific factor the **marginal means** (i.e., the means of the p rows and the means of the q columns). In the case of Example 21.1, a simple comparison can be conducted in which two of the three levels of Factor B are compared with one another (i.e., $\bar{X}_{B_1}$ versus $\bar{X}_{B_2}$), or a complex comparison in which a composite mean involving two levels of Factor B is compared with the mean of the third level of Factor B (i.e., $(\bar{X}_{B_2} + \bar{X}_{B_3})/2$ versus $\bar{X}_{B_1}$). If Factor B has four levels, a complex comparison contrasting two sets of composite means (each set representing a composite mean of two of the four levels) can be conducted (i.e., $(\bar{X}_{B_1} + \bar{X}_{B_2})/2$ versus $(\bar{X}_{B_3} + \bar{X}_{B_4})/2$). Since there are only two levels of Factor A, no additional comparisons are possible involving the means of the levels of that factor (i.e., the omnibus F value for Factor A represents the comparison $\bar{X}_{A_1}$ versus $\bar{X}_{A_2}$). As is the case for the **single-factor between-subjects analysis of variance,** in designs in which one or both of the factors are comprised of more than three levels, it is possible to conduct an omnibus F test comparing the means of three or more of the levels of a specific factor. In addition to all of the aforementioned comparisons, within a given level of a specific factor simple and complex comparisons can be conducted that contrast the means of specific groups that are a combination of both factors (i.e., a simple comparison such as $\bar{X}_{AB_{11}}$ versus $\bar{X}_{AB_{12}}$, or a complex comparison such as $\bar{X}_{AB_{11}}$ versus $(\bar{X}_{AB_{12}} + \bar{X}_{AB_{13}})/2$).[8] It is worth reiterating that, whenever possible, comparisons should be planned prior to the data collection phase of a study, and that any comparisons which are conducted should address important theoretical and/or practical questions that underlie the hypotheses under study. In addition, the total number of comparisons that are conducted should be limited in number, and should not be redundant with respect to the information they provide.

c) **A significant interaction is present with or without one or more of the main effects being significant** As noted previously, when the interaction is significant the main effects are essentially rendered meaningless, since any main effects will have to be qualified in reference to the levels of a second factor. Thus, any comparison which involves the levels of a specific factor (e.g., $\bar{X}_{B_1}$ versus $\bar{X}_{B_2}$) will reflect both the contribution of that factor, as well as the interaction between that factor and the second factor. For this reason, the most logical strategy to employ if a significant interaction is obtained is to test for what are referred to as **simple effects.** A test of a simple effect is essentially an analysis of variance evaluating all of the levels of one factor across only one level of the other factor. In the case of Example 21.1, two simple effects can be evaluated for Factor B. Specifically, an F test can be conducted which evaluates the scores of subjects on Factor B, but only for those subjects who serve under Level 1 of Factor A (i.e., an F ratio is computed for Groups AB_{11}, AB_{12}, and AB_{13}). A second simple effect for Factor B can be evaluated by contrasting the scores of subjects on Factor B, but only for those subjects who serve under Level 2 of Factor A (i.e., Groups AB_{21}, AB_{22}, and AB_{23}). In the case of Factor A, there are three possible simple effects that can be evaluated. Specifically, separate F tests can be conducted which evaluate the scores of subjects on Factor A for only those subjects who serve under: a) Level 1 of Factor B (i.e., Groups AB_{11} and AB_{21}); b) Level 2 of Factor B (i.e., Groups AB_{12} and AB_{22}); and c) Level 3 of Factor B (i.e., Groups AB_{13} and AB_{23}). In the event that one or more of the simple effects is significant, additional simple and complex comparisons contrasting specific groups within a given level of a factor can be conducted (e.g., a simple comparison such as $\bar{X}_{AB_{11}}$ versus $\bar{X}_{AB_{12}}$ or a complex comparison such as $\bar{X}_{AB_{11}}$ versus $(\bar{X}_{AB_{12}} + \bar{X}_{AB_{13}})/2$).

Description of analytical procedures (Including the following comparison procedures that are described for the **single-factor between-subjects analysis of variance** which in this section are described in reference to the **between-subjects factorial analysis of variance: Test 21a: Multiple *t* tests/Fisher's LSD test** (which is equivalent to **linear contrasts**); **Test 21b: The Bonferroni–Dunn test; Test 21c: Tukey's HSD test; Test 21d: The Newman–Keuls test; Test 21e: The Scheffé test; Test 21f: The Dunnett test**)

Comparisons between the marginal means The equations that are employed in conducting simple and complex comparisons involving the marginal means are basically the same equations that are employed for conducting comparisons for the **single-factor between-subjects analysis of variance.** Thus, in comparing two marginal means or two sets of marginal means (in the case of complex comparisons), **linear contrasts** can be conducted when no attempt is made to control the value of α_{FW} (which will generally be the case for planned comparisons).[9] In the case of either planned or unplanned comparisons where the value of α_{FW} is controlled, any of the multiple comparison procedures discussed under the **single-factor between-subjects analysis of variance** can be employed (i.e., **The Bonferroni–Dunn test, Tukey's HSD test, The Newman–Keuls test, The Scheffé test,**

and **The Dunnett test**). The only difference in employing any of the latter comparison procedures with a factorial design is that the sample size employed in a comparison equation will reflect the number of subjects in each of the levels of the relevant factor. Thus, any comparison involving the marginal means of Factor A will involve nq subjects per group (in Example 21.1, $nq = (3)(3) = 9$), and any comparison involving the marginal means of Factor B will involve np subjects per group (in Example 21.1, $np = (3)(2) = 6$).

As an example, assume we want to compare the scores of subjects on two of the levels of Factor B — specifically Level 1 versus Level 3 (i.e., $\bar{X}_{B_1}$ versus $\bar{X}_{B_3}$). If no attempt is made to control the value of α_{FW}, Equations 21.28–21.30 (which are the analogs of Equations 16.17–16.19 employed for conducting **linear contrasts** for the **single-factor between-subjects analysis of variance**) are employed to conduct a **linear contrast** comparing the two levels of Factor B (which within the framework of the comparison are conceptualized as two groups). Note that in Equation 21.28 the value np represents the number of subjects who served under each level of Factor B, and $[\Sigma(c_{B_k})(\bar{X}_{B_k})]^2$ will equal the squared difference between the means of the two levels of Factor B that are being compared (i.e., in the case of the comparison under discussion, $[\Sigma(c_{B_k})(\bar{X}_{B_k})]^2 = (\bar{X}_{B_1} - \bar{X}_{B_3})^2$).

$$SS_{B\ \text{comp}} = \frac{np\left[\Sigma(c_{B_k})(\bar{X}_{B_k})\right]^2}{\Sigma c_{B_k}^2} \qquad \textbf{(Equation 21.28)}$$

$$MS_{B\ \text{comp}} = \frac{SS_{B\ \text{comp}}}{df_{B\ \text{comp}}} \qquad \textbf{(Equation 21.29)}$$

$$F_{B\ \text{comp}} = \frac{MS_{B\ \text{comp}}}{MS_{WG}} \qquad \textbf{(Equation 21.30)}$$

The data from Example 21.1 are now employed in Equations 21.28–21.30 to conduct the comparison $\bar{X}_{B_1}$ versus $\bar{X}_{B_3}$. Note that since Levels 1 and 3 of Factor B constitute the groups that are involved in the comparison, the coefficients for the comparison are $c_{B_1} = +1$, $c_{B_2} = 0$, $c_{B_3} = -1$. Thus, $\Sigma c_{B_k}^2 = 2$ and $[\Sigma(c_{B_k})(\bar{X}_{B_k})]^2 = (\bar{X}_{B_1} - \bar{X}_{B_3})^2 = (6.5 - 2.5)^2 = (4)^2 = 16$. Substituting the appropriate values in Equation 21.28, the value $SS_{B\ \text{comp}} = 48$ is computed.

$$SS_{B\ \text{comp}} = \frac{(3)(2)(4)^2}{2} = 48$$

Since all **linear contrasts** represent a **single degree of freedom comparison**, $df_{B\ \text{comp}} = 1$. Employing Equations 21.29 and 21.30 the values $MS_{B\ \text{comp}} = 48$ and $F_{B_{\text{comp}}} = 48$ are computed. Note that the value $MS_{WG} = 1$ computed for the omnibus F test is employed in the denominator of Equation 21.30.

$$MS_{B\ \text{comp}} = \frac{48}{1} = 48$$

$$F_{B\ \text{comp}} = \frac{48}{1} = 48$$

The value $F_{B\ \text{comp}} = 48$ is evaluated with **Table A10**. Employing the latter table, the appropriate degrees of freedom for the numerator and denominator are $df_{\text{num}} = df_{B\ \text{comp}} = 1$

and $df_{den} = df_{WG} = 12$. For $df_{num} = 1$ and $df_{den} = 12$, the tabled critical .05 and .01 values are $F_{.05} = 4.75$ and $F_{.01} = 9.33$. Since the obtained value $F_{comp} = 48$ is greater than the aforementioned critical values, the nondirectional alternative hypothesis H_1: $u_{B_1} \neq \mu_{B_3}$ is supported at both the .05 and .01 levels.

Equations 21.31–21.33 are employed to evaluate comparisons involving the levels of Factor A.

$$SS_{A\ comp} = \frac{nq[\Sigma(c_{A_j})(\bar{X}_{A_j})]^2}{\Sigma c_{A_j}^2} \quad \textbf{(Equation 21.31)}$$

$$MS_{A\ comp} = \frac{SS_{A\ comp}}{df_{A\ comp}} \quad \textbf{(Equation 21.32)}$$

$$F_{A\ comp} = \frac{MS_{A\ comp}}{MS_{WG}} \quad \textbf{(Equation 21.33)}$$

Note that in Equation 21.31 nq represents the sample size, which in this case is the number of subjects who serve in each level of Factor A. The value $[\Sigma(c_{A_j})(\bar{X}_{A_j})]^2$ is equal to $(\bar{X}_{A_x} - \bar{X}_{A_y})^2$, which is the squared difference between the two means involved in the comparison (where x and y represent the levels of Factor A that are employed in the comparison).

As is the case with comparisons conducted for a **single-factor between-subjects analysis of variance**, a CD value can be computed for any comparison. Recollect that a CD value represents the minimum required difference in order for two means to differ significantly from one another. To demonstrate this, two CD values will be computed for the comparison $\bar{X}_{B_1}$ versus $\bar{X}_{B_3}$. Specifically, CD_{LSD} and CD_S will be computed. CD_{LSD}, which is the CD value associated with the **linear contrast** that is conducted with Equations 21.28–21.30, is the lowest possible difference that can be computed with any of the available comparison procedures. CD_S (the value for the **Scheffé test**), on the other hand, computes the largest CD value from the methods that are available. If the obtained difference for a comparison is less than CD_{LSD}, the null hypothesis will be retained, whereas if it is larger than CD_S it will be rejected. For the purpose of this discussion, it will be assumed that an obtained difference that is larger than CD_{LSD} but less than CD_S will be relegated to the **suspend judgement** category.[10]

Equation 21.34 is employed to compute the value $CD_{LSD} = 1.25$ for the **simple comparison** $\bar{X}_{B_1}$ versus $\bar{X}_{B_3}$, for $\alpha = .05$. In point of fact, the CD_{LSD} value computed with Equation 21.34 applies to all three simple comparisons that can be conducted with respect to Factor B (i.e., $\bar{X}_{B_1} - \bar{X}_{B_2} = 6.5 - 5.5 = 1$; $\bar{X}_{B_1} - \bar{X}_{B_3} = 6.5 - 2.5 = 4$; $\bar{X}_{B_2} - \bar{X}_{B_3} = 5.5 - 2.5 = 3$).

$$CD_{LSD} = \sqrt{F_{(1,WG)}}\sqrt{\frac{2MS_{WG}}{np}} = \sqrt{4.75}\sqrt{\frac{(2)(1)}{(3)(2)}} = 1.25 \quad \textbf{(Equation 21.34)}$$

In order to differ significantly at the .05 level, the means of any two levels of Factor B must differ from one another by at least 1.25 units. Thus, the differences $\bar{X}_{B_1} - \bar{X}_{B_3} = 4$ and $\bar{X}_{B_2} - \bar{X}_{B_3} = 3$ are significant, while the difference $\bar{X}_{B_1} - \bar{X}_{B_2} = 1$ is not.

Note that Equation 21.34 is identical to Equation 16.24 employed for computing the CD_{LSD} value for the **single-factor between-subjects analysis of variance**, except for the fact that in Equation 21.34 np subjects are employed per group/level of Factor B. In Equation 21.34, the value $F_{(1,WG)} = 4.75$ is the tabled critical .05 F value for df_{num} and df_{den}, which represent the degrees of freedom associated with the $F_{B\ \text{comp}}$ value computed with Equation 21.30.

Equations 21.35 and 21.36, which are analogous to Equation 16.25 (which is the generic equation for both simple and complex comparisons for CD_{LSD} for the **single-factor between-subjects analysis of variance**), are respectively the generic equations for Factors A and B for computing CD_{LSD}.

$$CD_{LSD} = \sqrt{F_{(1,WG)}}\sqrt{\frac{(\Sigma c_{A_j}^2)(MS_{WG})}{nq}} \qquad \textbf{(Equation 21.35)}$$

$$CD_{LSD} = \sqrt{F_{(1,WG)}}\sqrt{\frac{(\Sigma c_{B_k}^2)(MS_{WG})}{np}} \qquad \textbf{(Equation 21.36)}$$

At this point the **Scheffé test** will be employed to conduct the simple comparison $\bar{X}_{B_1}$ versus $\bar{X}_{B_3}$. Equation 21.37, which is analogous to Equation 16.32 (which is the equation for simple comparisons for CD_S for the **single-factor between-subjects analysis of variance**), is employed to compute the value $CD_S = 1.61$, with $\alpha_{FW} = .05$. The value $F_{(B,WG)} = 3.89$ used in Equation 21.37 is the tabled critical .05 F value employed in evaluating the main effect for Factor B in the omnibus F test.

(Equation 21.37)

$$CD_S = \sqrt{(q-1)F_{df_{(B,WG)}}}\sqrt{\frac{2MS_{WG}}{np}} = \sqrt{(3-1)(3.89)}\sqrt{\frac{(2)(1)}{(3)(2)}} = 1.61$$

Thus, in order to differ significantly at the .05 level, the means of any two levels of Factor B must differ from one another by at least 1.61 units. As is the case when $CD_{LSD} = 1.25$ is computed, the differences $\bar{X}_{B_1} - \bar{X}_{B_3} = 4$ and $\bar{X}_{B_2} - \bar{X}_{B_3} = 3$ are significant, while the difference $\bar{X}_{B_1} - \bar{X}_{B_2} = 1$ is not.

Equations 21.38 and 21.39, which are analogous to Equation 16.33 (which is the generic equation for both simple and complex comparisons for CD_S for the **single-factor between-subjects analysis of variance**), are respectively the generic equations for Factors A and B for computing CD_S. Note that in conducting comparisons involving the levels of Factor A, the value $F_{(A,WG)}$ employed in Equation 21.38 is the tabled critical F value at the prespecified level of significance employed in evaluating the main effect for Factor A in the omnibus F test.

$$CD_S = \sqrt{(p-1)F_{df_{(A,WG)}}}\sqrt{\frac{(\Sigma c_{A_j}^2)(MS_{WG})}{nq}} \qquad \textbf{(Equation 21.38)}$$

$$CD_S = \sqrt{(q-1)F_{df_{(B,WG)}}}\sqrt{\frac{(\Sigma c_{B_k}^2)(MS_{WG})}{np}} \qquad \textbf{(Equation 21.39)}$$

In closing the discussion of the **Scheffé test**, it should be noted that since Equation 21.37 only takes into account those comparisons that are possible which involve the levels of Factor B, it may not be viewed as imposing adequate control over α_{FW} if one intends to conduct additional comparisons involving the levels of Factor A and/or specific groups that are a combination of both factors. Because of this, some sources make a distinction between the **familywise error rate** (α_{FW}) and the **experimentwise error rate**. Although in the case of a single factor experiment the two values will be identical, in a multifactor experiment a **familywise error rate** can be computed for comparisons within each factor, as well as for comparisons between groups that are based on combinations of the factors. The **experimentwise error rate** will be a composite error rate which will be the result of combining all of the **familywise error rates**. Thus, in the above example if one intends to conduct additional comparisons involving the levels of Factor A and/or groups that are combinations of both factors, one can argue that the **Scheffé test** as employed does not impose sufficient control over the value of the **experimentwise error rate**. Probably the simplest way to deal with such a situation is to conduct a more conservative test in evaluating any null hypotheses involving the levels of Factor A, Factor B, or groups that are combinations of both factors (i.e., evaluate a null hypothesis at the .01 level instead of at the .05 level).

Evaluation of an omnibus hypothesis involving more than two marginal means If the interaction is not significant, it is conceivable that a researcher may wish to conduct an F test on three or more marginal means in a design where the factor involved has four or more levels. In other words, if in Example 21.1 there were four levels on Factor B instead of three, one might want to evaluate the null hypothesis H_0: $\mu_{B_1} = \mu_{B_2} = \mu_{B_3}$. The logic that is employed in conducting such an analysis for the **single-factor between-subjects analysis of variance** can be extended to a factorial design. Specifically, in the case of a 2 × 4 design, a **between-subjects factorial analysis of variance** employing all of the data is conducted initially. Upon determining that the interaction is not significant, a **single-factor between-subjects analysis of variance** can then be conducted employing only the data for the three levels of Factor B in which the researcher is interested (i.e., B_1, B_2, and B_3). The following F ratio is computed: $F_{(B_1/B_2/B_3)} = MS_{BG_{(B_1/B_2/B_3)}}/MS_{WG}$. Note that the mean square value in the numerator is based on the between-groups variability in the **single-factor between-subjects analysis of variance** that involves only the data for levels B_1, B_2, and B_3 of Factor B. The degrees of freedom associated with the numerator of the F ratio is 2, since it is based on the number of levels of Factor B evaluated with the **single-factor between-subjects analysis of variance** (i.e., $df_{(B_1/B_2/B_3)} = 3 - 1 = 2$). The mean square and degrees of freedom for the denominator of the F ratio are the within-groups mean square and degrees of freedom computed for the **between-subjects factorial analysis of variance** when the full set of data is employed (i.e., the data for all four levels of Factor B). For further clarification of the aforementioned procedure the reader should review Section VI of the **single-factor between-subjects analysis of variance.**

Comparisons between specific groups that are a combination of both factors The procedures employed for comparing the marginal means can also be employed to evaluate differences between specific groups that are a combination of both factors (e.g., a comparison such as $\bar{X}_{AB_{11}}$ versus $\bar{X}_{AB_{12}}$). Such differences are most likely to be of interest when an interaction is present. It should be noted that these are not the only types of comparisons that can provide more specific information regarding the nature of an interaction. A more comprehensive discussion of further analysis of an interaction can be found in books that specialize in the analysis of variance. Keppel (1991), among others, provides an excellent discussion of this general subject.

In comparing specific groups with one another, the same equations are essentially employed that are used for the comparison of marginal means, except for the fact that the equations must be modified in order to accommodate the sample size of the groups. Both simple and complex comparisons can be conducted. As an example, let us assume we want to conduct a **linear contrast** for the simple comparison $\bar{X}_{AB_{11}}$ versus $\bar{X}_{AB_{12}}$. Equation 21.40 is employed for conducting such a comparison. Note that the latter equation has the same basic structure as Equations 21.28 and 21.31, but is based on the sample size of n, which is the sample size of each of the $p \times q$ groups.

$$SS_{\text{comp}} = \frac{n[\Sigma(c_{AB_{jk}})(\bar{X}_{AB_{jk}})]^2}{\Sigma c^2_{AB_{jk}}} \quad \textbf{(Equation 21.40)}$$

In Equation 21.40 the value $[\Sigma(c_{AB_{jk}})(\bar{X}_{AB_{jk}})]^2$ is equal to the squared difference between the means of the two groups that are being compared (i.e., for the comparison under discussion it yields the same value as $(\bar{X}_{AB_{11}} - \bar{X}_{AB_{12}})^2$). Note that since only two of the $p \times q = 6$ groups are involved in the comparison, the coefficients for the comparison are $c_{AB_{11}} = +1$, $c_{AB_{12}} = -1$, and $c_{AB_{jk}} = 0$ for the remaining four groups. Thus, $[\Sigma(c_{AB_{jk}})(\bar{X}_{AB_{jk}})]^2 = (\bar{X}_{AB_{11}} - \bar{X}_{AB_{12}})^2 = (10 - 7)^2 = (3)^3 = 9$ and $\Sigma c^2_{AB_{jk}} = 2$. Substituting the appropriate values in Equation 21.40, the value $SS_{\text{comp}} = 13.5$ is computed.

$$SS_{\text{comp}} = \frac{(3)(3)^2}{2} = 13.5$$

Employing Equations 16.18 and 16.19, the values $MS_{\text{comp}} = 13.5$ and $F_{\text{comp}} = 13.5$ are computed.

$$MS_{\text{comp}} = \frac{SS_{\text{comp}}}{df_{\text{comp}}} = \frac{13.5}{1} = 13.5$$

$$F_{\text{comp}} = \frac{MS_{\text{comp}}}{MS_{WG}} = \frac{13.5}{1} = 13.5$$

Employing **Table A10**, the appropriate degrees of freedom for the numerator and denominator are $df_{\text{num}} = df_{\text{comp}} = 1$ (since the comparison is a **single degree of freedom comparison**) and $df_{\text{den}} = df_{WG} = 12$. For $df_{\text{num}} = 1$ and $df_{\text{den}} = 12$, the tabled critical .05 and .01 values are $F_{.05} = 4.75$ and $F_{.01} = 9.33$. Since the obtained value $F_{\text{comp}} = 13.5$ is greater than the aforementioned critical values, the nondirectional alternative hypothesis H_1: $\mu_{AB_{11}} \neq \mu_{AB_{12}}$ is supported at both the .05 and .01 levels.

CD_{LSD} and CD_S values will now be computed for the above comparison, for $\alpha = .05$. CD_{LSD} is computed with Equation 21.41 (which is identical to Equation 16.24 employed to compute CD_{LSD} for the **single-factor between-subjects analysis of variance**). Note that the sample size employed in Equation 21.41 is $n = n_{AB_{jk}} = 3$. Substituting the appropriate values in Equation 21.41, the value $CD_{LSD} = 1.78$ is computed.

$$CD_{LSD} = \sqrt{F_{(1,WG)}}\sqrt{\frac{2MS_{WG}}{n}} = \sqrt{4.75}\sqrt{\frac{(2)(1)}{3}} = 1.78 \quad \textbf{(Equation 21.41)}$$

Since in order to differ significantly at the .05 level the means of any two groups must differ from one another by at least 1.78 units, the difference $\bar{X}_{AB_{11}} - \bar{X}_{AB_{12}} = 3$ is significant. If we conduct comparisons for all 15 possible differences between pairs of groups (i.e., all simple comparisons), any difference that is equal to or greater than 1.78 units is significant at the .05 level.[11] Recollect, though, that since the computation of a CD_{LSD} value does not control the value of α_{FW}, the **per comparison Type I error rate** will equal .05.

Equation 21.42 (which is analogous to Equation 16.25 employed for the **single-factor between-subjects analysis of variance**) is the generic form of Equation 21.41 that can be employed for both simple and complex comparisons.

$$CD_{LSD} = \sqrt{F_{(1,WG)}} \sqrt{\frac{(\Sigma c^2_{AB_{jk}})(MS_{WG})}{n}} \qquad \textbf{(Equation 21.42)}$$

CD_S is computed with Equation 21.43 (which is analogous to Equation 16.32 employed to compute CD_S for the **single-factor between-subjects analysis of variance**). Substituting the appropriate values in Equation 21.43, the value $CD_S = 3.22$ is computed. The value $F_{df_{(BG,WG)}} = 3.11$ used in Equation 21.43 is the tabled critical .05 F value for $df_{\text{num}} = df_{BG} = pq - 1 = (2)(3) - 1 = 5$ and $df_{\text{den}} = df_{WG} = 12$ employed in the omnibus F test.

$$CD_S = \sqrt{(pq - 1)F_{df_{(BG,WG)}}} \sqrt{\frac{2MS_{WG}}{n}} \qquad \textbf{(Equation 21.43)}$$

$$= \sqrt{[(2)(3) - 1](3.11)} \sqrt{\frac{(2)(1)}{3}} = 3.22$$

Thus, in order for any pair of means to differ significantly, the difference between the two means must be equal to or greater than 3.22 units. Since the difference $\bar{X}_{AB_{11}} - \bar{X}_{AB_{12}} = 3$ is less than $CD_S = 3.22$, the null hypothesis cannot be rejected if the **Scheffé test** is employed. Thus, the nondirectional alternative hypothesis H_1: $\mu_{AB_{11}} \neq \mu_{AB_{12}}$ is not supported.

Equation 21.44 (which is analogous to Equation 16.33 employed for the **single-factor between-subjects analysis of variance**) is the generic form of Equation 21.43 that can be employed for both simple and complex comparisons.

$$CD_S = \sqrt{(pq - 1)F_{df_{(BG,WG)}}} \sqrt{\frac{(\Sigma c^2_{AB_{jk}})(MS_{WG})}{n}} \qquad \textbf{(Equation 21.44)}$$

Since the **linear contrast** procedure yields a significant difference and the **Scheffé test** does not, one might want to **suspend judgement** with respect to the comparison $\bar{X}_{AB_{11}}$ versus $\bar{X}_{AB_{12}}$ until a replication study is conducted. However, it is certainly conceivable that many researchers might consider the **Scheffé test** to be too conservative a procedure. Thus, one might elect to use a less conservative procedure such as **Tukey's HSD test**. Equation 21.45 (which is analogous to Equation 16.31 employed to compute CD_{HSD} for the **single-factor between-subjects analysis of variance**) is employed to compute CD_{HSD}. The value $q_{(pq, df_{WG})} = 4.75$ in Equation 21.45 is the value of the **Studentized range statistic** in **Table A13 (Table of the Studentized Range Statistic)** in the **Appendix** for $k = pq = 6$ and $df_{WG} = 12$.

$$CD_{HSD} = q_{(pq, df_{WG})} \sqrt{\frac{MS_{WG}}{n}} = 4.75\sqrt{\frac{1}{3}} = 2.74 \qquad \textbf{(Equation 21.45)}$$

Since $CD_{HSD} = 2.74$ is less than $\bar{X}_{AB_{11}} - \bar{X}_{AB_{12}} = 3$, we can conclude that the difference between the groups is significant.[12]

The computation of a confidence interval for a comparison The same procedure described for computing a confidence interval for a comparison for the **single-factor between-**

subjects analysis of variance can also be employed for the **between-subjects factorial analysis of variance**. Specifically, the following procedure is employed for computing a confidence interval for any of the methods described in this section: The obtained CD value is added to and subtracted from the obtained difference between the two means (or sets of means in the case of a complex comparison). The resulting range of values defines the confidence interval. The 95% confidence interval will be associated with a computed $CD_{.05}$ value, and the 99% confidence interval will be associated with a computed $CD_{.01}$ value. To illustrate the computation of a confidence interval, the 95% confidence interval for the value $CD_{HSD} = 2.74$ computed for the comparison $\bar{X}_{AB_{11}}$ versus $\bar{X}_{AB_{12}}$ is demonstrated below.

$$CI_{.95} = (\bar{X}_{AB_{11}} - \bar{X}_{AB_{12}}) \pm CD_{HSD} = 3 \pm 2.74$$

Thus, the researcher can be 95% sure that the mean of the population represented by Group AB_{11} is between .26 and 5.74 units larger than the mean of the population represented by Group AB_{12}. This result can be stated symbolically as follows: $.26 \leq (\mu_{AB_{11}} - \mu_{AB_{12}}) \leq 5.74$.

Analysis of simple effects Earlier in this section it is noted that the most logical strategy to employ when a significant interaction is detected is to initially test for what is referred to as **simple effects.** A test of a **simple effect** is essentially an analysis of variance evaluating all of the levels of one factor across only one level of the other factor. The analysis of simple effects will be illustrated with Example 21.1 by evaluating the simple effects of Factor B. Specifically, an F test will be conducted to evaluate the scores of subjects on the three levels of Factor B, but only the nq subjects who served under Level 1 of Factor A (i.e., an F ratio will be computed evaluating the Groups AB_{11}, AB_{12}, and AB_{13}). This represents the analysis of the simple effect of Factor B at level A_1. An analysis of a second simple effect (which represents the analysis of the simple effect of Factor B at level A_2) will evaluate the scores of subjects on the three levels of Factor B, but only the nq subjects who served under Level 2 of Factor A (i.e., Groups AB_{21}, AB_{22}, and AB_{23}).

Although it will not be done in reference to Example 21.1, since an interaction is present a comprehensive analysis of the data would also involve evaluating the simple effects of Factor A. There are three simple effects of Factor A that can be evaluated, each one involving comparing the scores of subjects on Factor A, but employing only the np subjects who served under one of the three levels of Factor B. The three simple effects of Factor A involve the following contrasts: 1) The simple effect of Factor A at Level B_1: $\bar{X}_{AB_{11}}$ versus $\bar{X}_{AB_{21}}$; 2) The simple effect of Factor A at Level B_2: $\bar{X}_{AB_{12}}$ versus $\bar{X}_{AB_{22}}$; and 3) The simple effect of Factor A at Level B_3: $\bar{X}_{AB_{13}}$ versus $\bar{X}_{AB_{23}}$.[13]

In order to evaluate a simple effect, it is necessary to initially compute a sum of squares for the specific effect. Thus, in evaluating the simple effects of Factor B it is necessary to compute a sum of squares for Factor B at Level 1 of Factor A ($SS_{B \text{ at } A_1}$) and a sum of squares for Factor B at Level 2 of Factor A ($SS_{B \text{ at } A_2}$). Upon computing all of the sums of squares for the simple effects for a specific factor, F ratios are computed for each of the simple effects by dividing the mean square for a simple effect (which is obtained by dividing the simple effect sum of squares by its degrees of freedom) by the within-groups mean square derived for the factorial analysis of variance. This procedure will now be demonstrated for the simple effects of Factor B.

Equation 21.46 is employed to compute the sum of squares for each of the simple effects. If $\Sigma X_{AB_{j.}}$ represents the sum of the scores on Level j of Factor A of subjects who serve under a specific level of Factor B, the notation $\Sigma[(\Sigma X_{AB_{j.}})^2/n]$ in Equation 21.46

indicates that the sum of the scores for each level of Factor B at a given level of Factor A is squared, divided by n, and the q squared sums are summed. The notation $(\Sigma\Sigma X_{AB_{j.}})^2$ represents the square of the sum of scores of the nq subjects who serve under the specified level of Factor A.[14]

$$SS_{B \text{ at } A_j} = \Sigma \left[\frac{(\Sigma X_{AB_{j.}})^2}{n} \right] - \frac{(\Sigma\Sigma X_{AB_{j.}})^2}{nq} \qquad \textbf{(Equation 21.46)}$$

$$SS_{B \text{ at } A_1} = \Sigma \left[\frac{(\Sigma X_{AB_{1.}})^2}{n} \right] - \frac{(\Sigma\Sigma X_{AB_{1.}})^2}{nq} = \left[\frac{(30)^2 + (21)^2 + (12)^2}{3} \right] - \frac{(63)^2}{(3)(3)} = 54$$

$$SS_{B \text{ at } A_2} = \Sigma \left[\frac{(\Sigma X_{AB_{2.}})^2}{n} \right] - \frac{(\Sigma\Sigma X_{AB_{2.}})^2}{nq} = \left[\frac{(9)^2 + (12)^2 + (3)^2}{3} \right] - \frac{(24)^2}{(3)(3)} = 14$$

Table 21.6 summarizes the analysis of variance for the simple effects of Factor B. Note that for each of the simple effects, the degrees of freedom for the effect is $df_{B \text{ at } A_j} = q - 1 = 3 - 1 = 2$ (which equals df_B employed for the **between-subjects factorial analysis of variance**). The mean square for each simple effect is obtained by dividing the sum of squares for the simple effect by its degrees of freedom. The F value for each simple effect is obtained by dividing the mean square for the simple effect by $MS_{WG} = 1$ computed for the factorial analysis of variance. Thus, $F_{B \text{ at } A_1} = 27/1 = 27$ and $F_{B \text{ at } A_2} = 7/1 = 7$.

Table 21.6 Analysis of Simple Effects of Factor B

Source of variation	*SS*	*df*	*MS*	*F*
B at A_1	54	2	27	27
B at A_2	14	2	7	7
Within-groups	12	12	1	

Employing **Table A10**, the degrees of freedom used in evaluating each of the simple effects are $df_{\text{num}} = df_{B \text{ at } A_j} = 2$ and $df_{\text{den}} = df_{WG} = 12$. Since both of the obtained values $F_{B \text{ at } A_1} = 27$ and $F_{B \text{ at } A_2} = 7$ are greater than $F_{.05} = 3.89$ and $F_{.01} = 6.93$ (which are the tabled critical values for $df_{\text{num}} = 2$ and $df_{\text{den}} = 12$), each of the simple effects is significant at both the .05 and .01 levels. On the basis of the result of the analysis of the simple effects of Factor B, we can conclude that within each level of Factor A there is at least one simple or complex comparison involving the levels of Factor B that is significant.

As noted earlier, when one or more of the simple effects is significant, additional simple and complex comparisons contrasting specific groups can be conducted. Thus, for Level 1 of Factor A, simple comparisons between $\bar{X}_{AB_{11}}$, $\bar{X}_{AB_{12}}$, and $\bar{X}_{AB_{13}}$, as well as complex comparisons (such as $\bar{X}_{AB_{11}}$ versus $(\bar{X}_{AB_{12}} + \bar{X}_{AB_{13}})/2$) can clarify the locus of the significant simple effect.

If the homogeneity of variance assumption of the **between-subjects factorial analysis of variance** (which is discussed in the next section) is violated, in computing the F ratios for the simple effects a researcher can justify employing a MS_{WG} value that is just based on the groups involved in analyzing a specific simple effect, instead of the value of MS_{WG} computed for the factorial analysis of variance. If the latter is done, the within-groups degrees of freedom employed in the analysis of the simple effects of Factor B becomes $df_{WG} = q(n - 1)$ instead of $df_{WG} = pq(n - 1)$. Since the within-groups degrees of freedom is smaller if $df_{WG} = q(n - 1)$ is employed, the test will be less powerful than a test employing $df_{WG} = pq(n - 1)$. The loss of power can be offset, however, if the new value for MS_{WG} is lower than the value derived for the omnibus F test.[15]

The reader should take note of the fact that the variability within each of the simple effects is the result of contributions from both the main effect on the factor for which the simple effect is being evaluated (Factor B in our example), as well as any interaction between the two factors. For this reason, the total of the sum of squares for each of the simple effects for a given factor will be equal to the interaction sum of squares (SS_{AB}) plus the sum of squares for that factor (SS_B). This can be confirmed by the fact that in our example the following is true:

$$[(SS_{B \text{ at } A_1} = 54) + (SS_{B \text{ at } A_2} = 14)] = [(SS_{AB} = 16) + (SS_B = 52)] = 68$$

It should be noted that analysis of simple effects in and of itself cannot provide definitive evidence with regard to the presence or absence of an interaction. In point of fact, it is possible for only one of two simple effects to be significant, and yet the value of F_{AB} computed for the factorial analysis of variance may not be significant. For a full clarification of this issue the reader should consult Keppel (1991).

2. Evaluation of the homogeneity of variance assumption of the between-subjects factorial analysis of variance The homogeneity of variance assumption discussed in reference to the **single-factor between-subjects analysis of variance** is also an assumption of the **between-subjects factorial analysis of variance.** Since both tests employ the same protocol in evaluating this assumption, prior to reading this section the reader should review the relevant material for evaluating the homogeneity of variance assumption (through use of **Hartley's F_{max} test (Test 8a))** in Section VI of the **single-factor between-subjects analysis of variance** (as well as the material on **Hartley's F_{max} test** in Section VI of the ***t* test for two independent samples**).

In the case of the **between-subjects factorial analysis of variance**, evaluation of the homogeneity of variance assumption requires the researcher to compute the estimated population variances for each of the pq groups The latter values are computed with Equation I.5. As it turns out, the value of the estimated population variance for all six groups equals $\tilde{s}^2_{AB_{jk}} = 1$. This can be demonstrated below for Group AB_{11}.

$$\tilde{s}^2_{AB_{11}} = \frac{\Sigma X^2_{AB_{11}} - \dfrac{(\Sigma X_{AB_{11}})^2}{n_{AB_{11}}}}{n_{AB_{11}} - 1} = \frac{302 - \dfrac{(30)^2}{3}}{3 - 1} = 1$$

Upon determining that the value of both the largest and smallest of the estimated population variances equals 1, Equation 16.37 is employed to compute the value of the F_{max} statistic. Employing Equation 16.37, the value $F_{max} = 1$ is computed.

$$F_{max} = \frac{\tilde{s}^2_L}{\tilde{s}^2_S} = \frac{1}{1} = 1$$

In order to reject the null hypothesis (H_0: $\sigma^2_L = \sigma^2_S$) and thus conclude that the homogeneity of variance assumption is violated, the obtained F_{max} value must be equal to or greater than the tabled critical value at the prespecified level of significance. Employing **Table A9 (Table of the F_{max} Distribution)** in the **Appendix**, we determine that the tabled critical F_{max} values for $n = n_{AB_{jk}} = 3$ and $k = pq = 6$ groups are $F_{max_{.05}} = 266$ and $F_{max_{.01}} = 1362$. Since the obtained value $F_{max} = 1$ is less than $F_{max_{.05}} = 266$, the null hypothesis cannot be rejected. In other words, the alternative hypothesis indicating the presence of heterogeneity of variance is not supported.

In instances where the homogeneity of variance assumption is violated, the researcher should employ one of the strategies recommended for heterogeneity of variance that are discussed in Section VI of the **single-factor between-subjects analysis of variance.** The simplest strategy is to use a more conservative test (i.e., employ a lower α level) in evaluating the three sets of hypotheses for the factorial analysis of variance.[16]

3. Computation of the power of the between-subjects factorial analysis of variance Prior to reading this section the reader should review the procedure described for computing the power of the **single-factor between-subjects analysis of variance**, since the latter procedure can be generalized to the **between-subjects factorial analysis of variance.** In determining the appropriate sample size for a factorial design, a researcher must consider the predicted effect size for each of the factors, as well as the magnitude of any predicted interactions. Thus, in the case of Example 21.1, prior to the experiment a separate power analysis can be conducted with respect to the main effect for Factor A, the main effect for Factor B, and the interaction between the two factors. The sample size the researcher should employ will be the largest of the sample sizes derived from analyzing the predicted effects associated with the two factors and the interaction. As is the case for the **single-factor between-subjects analysis of variance,** such an analysis will require the researcher to estimate the means of all of the experimental groups, as well as the value of error/within-groups variability (i.e., σ^2_{WG}).

Equation 21.47, which contains the same basic elements that comprise Equation 16.38, is the general equation that is employed for determining the minimum sample size necessary in order to achieve a specified power with regard to either of the main effects or the interaction.

$$\phi = \sqrt{(\text{number of observations})\left[\frac{\Sigma d^2}{(df_{\text{effect}} + 1)(\sigma^2_{WG})}\right]} \qquad \textbf{(Equation 21.47)}$$

The following should be noted with respect to Equation 21.47:

a) The value employed for the number of observations will equal nq for Factor A, np for Factor B, and n for the interaction.

b) Σd^2 represents the sum of the squared deviation scores. This value is obtained as follows: 1) For **Factor A,** p deviation scores are computed by subtracting the estimated grand mean (μ_G) from each of the estimated means of the levels of Factor A (i.e., $d_{A_j} = \mu_{A_j} - \mu_G$). Σd^2, the sum of the squared deviation scores, is obtained by squaring the p deviation scores and summing the resulting values; 2) For **Factor B,** q deviation scores are computed by subtracting the estimated grand mean from each of the estimated means of the levels of Factor B (i.e., $d_{B_k} = \mu_{B_k} - \mu_G$). The sum of the squared deviation scores is obtained by squaring the q deviation scores and summing the resulting values; and 3) For the interaction, pq deviation scores are computed — one for each of the groups. A deviation score is computed for each group by employing the following equation: $d_{AB_{jk}} = \mu_{AB_{jk}} - \mu_{A_j} - \mu_{B_k} + \mu_G$. The latter equation indicates the following: The mean of the group is estimated ($\mu_{AB_{jk}}$), after which both the estimated mean of the level of Factor A the group serves under (μ_{A_j}) and the estimated mean of the level of Factor B the group serves under (μ_{B_k}) are subtracted from the estimated mean of the group. The estimated grand mean (μ_G) is then added to this result. The resulting value represents the deviation score for that group. Upon computing a deviation score for each of the pq groups, the pq deviation scores are squared, after which the resulting squared deviation scores are summed. The resulting value equals Σd^2.

c) $(df_{\text{effect}} + 1)$ for Factor A equals $df_A + 1 = p$. $(df_{\text{effect}} + 1)$ for Factor B equals $df_B + 1 = q$. $(df_{\text{effect}} + 1)$ for the interaction equals $df_{AB} + 1 = (p - 1)(q - 1) + 1$.

d) σ^2_{WG} is the estimate of the population variance for any one of the pq groups (which are assumed to have equal variances if the homogeneity of variance assumption is true). If a power analysis is conducted after the data collection phase of a study, it is logical to employ MS_{WG} as the estimate of σ^2_{WG}.[17]

To illustrate the use of Equation 21.47, the power of detecting the main effect on Factor B will be computed. Let us assume that based on previous research, prior to evaluating the data we estimate the following values: $\mu_{B_1} = 7$, $\mu_{B_2} = 5$, $\mu_{B_3} = 3$. Since we know that $\mu_G = (\mu_{B_1} + \mu_{B_2} + \mu_{B_3})/q$, $\mu_G = (7 + 5 + 3)/3 = 5$.[18] It will also be assumed that the estimated value for error variability is $\sigma^2_{WG} = 1.5$. The relevant values are now substituted in Equation 21.47.

$$\phi = \sqrt{np\left[\frac{\Sigma(\mu_{B_k} - \mu_G)^2}{(q)(\sigma^2_{WG})}\right]} = \sqrt{n(2)\left[\frac{(7-5)^2 + (5-5)^2 + (3-5)^2}{(3)(1.5)}\right]} = 1.89\sqrt{n}$$

If we employ $n = 3$ subjects per groups (as is the case in Example 21.1), the value $\phi = 3.27$ is computed: $\phi = 1.89\sqrt{3} = 3.27$. Employing **Table A15 (Graphs of the Power Function for the Analysis of Variance)** in the **Appendix**, we use the set of power curves for $df_{\text{num}} = df_{\text{effect}} = df_B = 2$, and within that set employ the curve for $df_{\text{den}} = df_{WG} = 12$, for $\alpha = .05$. Since a perpendicular line erected from the value $\phi = 3.27$ on the abscissa to the curve for $df_{WG} = 12$ is beyond the highest point on the curve, the power of the test for the estimated effect on Factor B will be 1 if $n = 3$ subjects are employed per group. Thus, there is a 100% likelihood that an effect equal to or larger than the one stipulated by the values employed in Equation 21.47 will be detected.

Although it will not be demonstrated here, to conduct a thorough power analysis it is necessary to also determine the minimum required sample sizes required to achieve what a researcher would consider to be the minimum acceptable power for identifying the estimated effects for Factor A and the interaction. The largest of the values computed for n for each of the three power analyses is the sample size that should be employed for each of the pq groups in the study. For a more comprehensive discussion on computing the power of the **between-subjects factorial analysis of variance** the reader should consult Cohen (1988).

4. Measure of magnitude of treatment effect for the between-subjects factorial analysis of variance: Omega squared (Test 21g) Prior to reading this section the reader should review the discussion of magnitude of treatment effect in Section VI of both the ***t* test for two independent samples** and the **single-factor between-subjects analysis of variance**. The latter discussions note that the computation of an omnibus F value only provides a researcher with information regarding whether the null hypothesis can be rejected — i.e., whether a significant treatment effect exists. An F value (as well as the level of significance with which it is associated), however, does not provide the researcher with any information regarding the size of any treatment effect that is present. As is noted in earlier discussions of treatment effect, the latter is defined as the proportion of the variability on the dependent variable that is associated with the experimental treatments. The **omega squared** statistic ($\tilde{\omega}^2$) is the most commonly used measure of association computed for the **between-subjects factorial analysis of variance**. Although there is some disagreement with respect to which variance components should be employed in computing $\tilde{\omega}^2$, Equation 21.48, is recommended in most sources.[19] Equation 21.48 is a generic equation for computing the **omega squared** statistic for both the main effects and the interaction of the **between-subjects factorial analysis of variance.**

$$\tilde{\omega}^2 = \frac{SS_{\text{effect}} - (df_{\text{effect}})(MS_{WG})}{SS_T + MS_{WG}} \qquad \textbf{(Equation 21.48)}$$

Based on Equation 21.48, Equations 21.49–21.51 are respectively employed for computing the **omega squared** statistic for Factors A and B and the interaction.

$$\tilde{\omega}^2_A = \frac{SS_A - (df_A)(MS_{WG})}{SS_T + MS_{WG}} = \frac{84.5 - (1)(1)}{164.5 + 1} = .50 \qquad \textbf{(Equation 21.49)}$$

$$\tilde{\omega}^2_B = \frac{SS_B - (df_B)(MS_{WG})}{SS_T + MS_{WG}} = \frac{52 - (2)(1)}{164.5 + 1} = .30 \qquad \textbf{(Equation 21.50)}$$

$$\tilde{\omega}^2_{AB} = \frac{SS_{AB} - (df_{AB})(MS_{WG})}{SS_T + MS_{WG}} = \frac{16 - (2)(1)}{164.5 + 1} = .08 \qquad \textbf{(Equation 21.51)}$$

The result of the above analysis indicates that 50% of the variability on the dependent variable is associated with Factor A (Humidity), 30% with Factor B (Temperature), and 8% with their interaction. Thus, 50% + 30% + 8% = 88% of the variability on the dependent variable (problem solving scores) is associated with variability on the two factors/independent variables and the interaction between them. It should be noted that although in some instances a small value for $\tilde{\omega}^2$ may indicate that the contribution of a factor or the interaction is trivial, this will not always be the case. Thus, in the example under discussion, although the value $\tilde{\omega}^2_{AB} = .08$ is small relative to the $\tilde{\omega}^2$ values computed for the main effects, inspection of the data clearly indicates that in order to understand the influence of temperature on problem solving scores, it is imperative that one take into account the level of humidity, and vice versa.

5. Computation of a confidence interval for the mean of a population represented by a group The same procedure employed to compute a confidence interval for a treatment population for the **single-factor between-subjects analysis of variance** can be employed with the **between-subjects factorial analysis of variance** to compute a confidence interval for the mean of any population represented by the pq groups. Although it will not be demonstrated here, the computational procedure requires that the appropriate values be substituted in Equation 16.44. In the event a researcher wants to compute a confidence interval for the mean of one of the levels of any of the factors, the number of subjects in the denominator of the radical of Equation 16.44 is based on the number of subjects who served within each level of the relevant factor (i.e., nq in the case of Factor A and np in the case of Factor B).

6. Additional analysis of variance procedures for factorial designs Section IX (which follows the full discussion of the **between-subjects factorial analysis of variance**) provides a description of the following additional factorial analysis of variance procedures: a) **The factorial analysis of variance for a mixed design (Test 21h)**; and b) **The within-subjects factorial analysis of variance (Test 21i)**. The discussion of each of the aforementioned analysis of variance procedures includes a description of the design for which it is employed, an example involving the same variables which are employed in Example 21.1, the computational equations for computing the appropriate F ratios, and computation of the F ratios for the relevant example.

VII. Additional Discussion of the Between-Subjects Factorial Analysis of Variance

1. Theoretical rationale underlying the between-subjects factorial analysis of variance As noted in Section IV, as is the case for the **single-factor between-subjects analysis of variance**, the total variability for the **between-subjects factorial analysis of variance** can be divided into **between-groups variability** and **within-groups variability**. Although it is not required in order to compute the F ratios, the value MS_{BG} (which represents between-groups variability) can be used to represent the variance of the means of the pq groups. MS_{BG} can be computed with the equation $MS_{BG} = SS_{BG}/df_{BG}$. As noted earlier, between-groups variability is comprised of the following elements: a) **Variability attributable to Factor A** (represented by the notation MS_A), which represents the variance of the means of the p levels of Factor A; b) **Variability attributable to Factor B** (represented by the notation MS_B), which represents the variance of the means of the q levels of Factor B; and c) **Variability attributable to any interaction that is present between Factors A and B** (represented by the notation MS_{AB}), which is a measure of variance that represents whatever remains of between-groups variability after the contributions of the main effects of Factors A and B have been subtracted from between-groups variability.

In computing the three F ratios for the **between-subjects factorial analysis of variance**, the values MS_A, MS_B, and MS_{AB} are contrasted with MS_{WG}, which serves as a baseline measure of error variability. In other words, MS_{WG} represents experimental error which results from factors that are beyond an experimenter's control. As is the case for the **single-factor between-subjects analysis of variance**, in the **between-subjects factorial analysis of variance**, MS_{WG} is the normal amount of variability that is expected between the scores of different subjects who serve under the same experimental condition. Thus, MS_{WG} represents the average of the variances computed for each of the pq groups. As long as any of the elements that comprise between-groups variability (MS_A, MS_B, or MS_{AB}) are approximately the same value as within-groups variability (MS_{WG}), the experimenter can attribute variability on a between-groups component to experimental error. When, however, any of the components that comprise between-groups variability are substantially greater than MS_{WG}, it indicates that something over and above error variability is contributing to that element of variability. In such a case it is assumed that the inflated level of variability for the between-groups component is the result of a treatment effect.

2. Definitional equations for the between-subjects factorial analysis of variance In the description of the computational protocol for the **between-subjects factorial analysis of variance** in Section IV, Equations 21.9–21.14 are employed to compute the values SS_T, SS_{BG}, SS_A, SS_B, SS_{AB}, and SS_{WG}. The latter set of computational equations were employed, since they allow for the most efficient computation of the sum of squares values. As noted in Section IV, computational equations are derived from definitional equations which reveal the underlying logic involved in the derivation of the sums of squares.

As noted previously, the total sum of squares (SS_T) can be broken down into two elements, the between-groups sum of squares (SS_{BG}) and the within-groups sum of squares (SS_{WG}). The contribution of any single subject's score to the total variability in the data can be expressed in terms of a between-groups component and a within-groups component. When the between-groups component and the within-groups component are added, the sum reflects that subject's total contribution to the overall variability in the data. Furthermore, the between-groups sum of squares can be broken down into three elements: a sum of squares for Factor A (SS_A), a sum of squares for Factor B (SS_B), and an interaction sum of squares (SS_{AB}). The contribution of any single subject's score to between-groups variability in the data can be expressed in terms of an A, a B, and an AB component. When the A, B, and AB components for a given subject are added, the sum reflects that subject's total contribution to between-groups variability in the data. The aforementioned information is reflected in the definitional equations which will now be described for computing the sums of squares.

Equation 21.52 is the definitional equation for the **total sum of squares.**[20] In Equation 21.52 the notation X_{ijk} is employed to represent the score of the ith subject in the group that serves under Level j of Factor A and Level k of Factor B.[21] When the notation $\sum_{k=1}^{q} \sum_{j=1}^{p} \sum_{i=1}^{n}$ precedes a term in parentheses, it indicates that the designated operation should be carried out for all $N = npq$ subjects.[22]

$$SS_T = \sum_{k=1}^{q} \sum_{j=1}^{p} \sum_{i=1}^{n} (X_{ijk} - \bar{X}_T)^2 \qquad \textbf{(Equation 21.52)}$$

In employing Equation 21.52 to compute SS_T, the grand mean ($\bar{X}_T$) is subtracted from each of the $N = npq$ scores and each of the N difference scores is squared. The total sum of squares (SS_T) is the sum of the N squared difference scores. Equation 21.52 is computationally equivalent to Equation 21.9.

Equation 21.53 is the definitional equation for the **between-groups sum of squares.**

$$SS_{BG} = n \sum_{k=1}^{q} \sum_{j=1}^{p} (\bar{X}_{AB_{jk}} - \bar{X}_T)^2 \qquad \textbf{(Equation 21.53)}$$

In employing Equation 21.53 to compute SS_{BG}, the following operations are carried out for each of the pq groups. The grand mean ($\bar{X}_T$) is subtracted from the group mean ($\bar{X}_{AB_{jk}}$). The difference score is squared, and the squared difference score is multiplied by the number of subjects in the group (n). After this is done for all pq groups, the values that have been obtained for each group as a result of multiplying the squared difference score by the number of subjects in a group are summed. The resulting value represents the between-groups sum of squares (SS_{BG}). Equation 21.53 is computationally equivalent to Equation 21.10.

Equation 21.54 is the definitional equation for the **sum of squares for Factor A.**

$$SS_A = nq \sum_{j=1}^{p} (\bar{X}_{A_j} - \bar{X}_T)^2 \qquad \textbf{(Equation 21.54)}$$

In employing Equation 21.54 to compute SS_A, the following operations are carried out for each of the p levels of Factor A. The grand mean ($\bar{X}_T$) is subtracted from the mean of that level of Factor A ($\bar{X}_{A_j}$). The difference score is squared, and the squared difference score is multiplied by the number of subjects in the level ($n_{A_j} = nq$). After this is done for all p levels of Factor A, the values that have been obtained for each level as a result of multiplying the squared difference score by the number of subjects in a level are summed. The resulting value represents the sum of squares for Factor A (SS_A). Equation 21.54 is computationally equivalent to Equation 21.11.

Equation 21.55 is the definitional equation for the **sum of squares for Factor B.**

$$SS_B = np \sum_{k=1}^{q} (\bar{X}_{B_k} - \bar{X}_T)^2 \qquad \textbf{(Equation 21.55)}$$

In employing Equation 21.55 to compute SS_B, the following operations are carried out for each of the q levels of Factor B. The grand mean ($\bar{X}_T$) is subtracted from the mean of that level of Factor B ($\bar{X}_{B_k}$). The difference score is squared, and the squared difference score is multiplied by the number of subjects in the level ($n_{B_k} = np$). After this is done for all q levels of Factor B, the values that have been obtained for each level as a result of multiplying the squared difference score by the number of subjects in a level are summed. The resulting value represents the sum of squares for Factor B (SS_B). Equation 21.55 is computationally equivalent to Equation 21.12.

Equation 21.56 is the definitional equation for the **interaction sum of squares.**

$$SS_{AB} = n\sum_{k=1}^{q}\sum_{j=1}^{p}(\bar{X}_{AB_{jk}} - \bar{X}_{A_j} - \bar{X}_{B_k} + \bar{X}_T)^2 \qquad \textbf{(Equation 21.56)}$$

In employing Equation 21.56 to compute SS_{AB}, the following operations are carried out for each of the pq groups. The mean of the level of Factor A the group represents ($\bar{X}_{A_j}$) and the mean of the level of Factor B the group represents ($\bar{X}_{B_k}$) are subtracted from the mean of the group ($\bar{X}_{AB_{jk}}$), and the grand mean ($\bar{X}_T$) is added to the resulting value. The result of the aforementioned operation is squared, and the squared difference score is multiplied by the number of subjects in that group ($n = n_{AB_{jk}}$). After this is done for all pq groups, the values that have been obtained for each group are summed, and the resulting value represents the sum of squares for the interaction (SS_{AB}).[23] Equation 21.56 is computationally equivalent to Equation 21.13.

Equation 21.57 is the definitional equation for the **within-groups sum of squares.**

$$SS_{WG} = \sum_{k=1}^{q}\sum_{j=1}^{p}\sum_{i=1}^{n}(X_{ijk} - \bar{X}_{AB_{jk}})^2 \qquad \textbf{(Equation 21.57)}$$

In employing Equation 21.57 to compute SS_{WG}, the following operations are carried out for each of the pq groups. The group mean ($\bar{X}_{AB_{jk}}$) is subtracted from each score in the group. The difference scores are squared, after which the sum of the squared difference scores is obtained. The sum of the sum of the squared difference scores for all pq groups represents the within-groups sum of squares. Equation 21.57 is computationally equivalent to Equation 21.14.

3. Unequal sample sizes The equations presented in this book for the **between-subjects factorial analysis of variance** assume there is an equal number of subjects in each of the pq groups (i.e., the value of $n_{AB_{jk}}$ is equal for each group). When the number of subjects per group is not equal, most sources recommend that adjusted sum of squares and sample size values be employed in conducting the analysis of variance. One approach to dealing with unequal sample sizes, which is generally referred to as the **unweighted means procedure,** employs the harmonic mean of the sample sizes of the pq groups to represent the value of $n = n_{AB_{jk}}$.[24] Based on the computed value of the harmonic mean (which will be designated $\bar{n}_h$), the sample size of each row and column, as well as the total sample size are adjusted as follows: $n_{A_j} = (\bar{n}_h)(q)$, $n_{B_k} = (\bar{n}_h)(p)$, $N = (\bar{n}_h)(p)(q)$. In addition, the $\Sigma X_{AB_{jk}}$ score of each group is adjusted by multiplying the mean of the group derived from the original data by the value computed for the harmonic mean (i.e., $(\bar{X}_{AB_{jk}})(\bar{n}_h)$ = Adjusted value of $\Sigma X_{AB_{jk}}$). Employing the adjusted $\Sigma X_{AB_{jk}}$ values, the value of ΣX_T and the values of the sums of the rows (ΣX_{A_j}) and columns (ΣX_{B_k}) are adjusted accordingly. The adjusted values of ΣX_T,

$\Sigma X_{AB_{jk}}$, ΣX_{A_j}, ΣX_{B_k}, $n_{AB_{jk}}$, n_{A_j}, n_{B_k}, and N, are substituted in Equations 21.9–21.13 to compute the values SS_T, SS_{BG}, SS_A, SS_B, and SS_{AB}. The value of SS_{WG}, on the other hand, is a pooled within-groups sum of squares that is based on the original unadjusted data. Thus, employing the original unadjusted values of $\Sigma X_{AB_{jk}}$ and $n_{AB_{jk}}$, the sum of squares is computed for each of the pq groups employing the following equation: $\Sigma X^2_{AB_{jk}} - [(\Sigma X_{AB_{jk}})^2/n_{AB_{jk}}]$. The sum of the pq sum of squares values represents the pooled within-groups sum of squares. This later value is, in fact, computed in Endnote 4. The values MS_A, MS_B, and MS_{AB} are computed with Equations 21.15–21.17, by dividing the relevant sum of squares value by the appropriate degrees of freedom. The degrees of freedom for the aforementioned mean square values are computed with Equations 21.19–21.21. Although the value MS_{WG} is computed with Equation 21.18, the value df_{WG} in the denominator of Equation 21.18 is a pooled within-groups degrees of freedom. The latter degrees of freedom value is determined by computing the value $(n_{AB_{jk}} - 1)$ for each group, and summing the $(n_{AB_{jk}} - 1)$ values for each of the pq groups. When the resulting degrees of freedom value is divided into the pooled within-groups sum of squares, it yields the value MS_{WG} that is employed in computing the F ratios. Equations 21.25–21.27 are employed to compute the values F_A, F_B, and F_{AB}. Keppel (1991) notes that since the F values derived by the method described in this section may underestimate the likelihood of committing a Type I error, it is probably prudent to employ a lower tabled probability to represent the prespecified level of significance — i.e., employ the tabled critical $F_{.01}$ value to represent the tabled critical $F_{.05}$ value. Alternative methods for dealing with unequal sample sizes are described in Keppel (1991), Kirk (1982), and Winer *et al.* (1991).

4. Final comments on the between-subjects factorial analysis of variance

a) Fixed-effects versus random-effects versus mixed-effects models In Section VII of the **single-factor between-subjects analysis of variance** it is noted that one assumption underlying the analysis of variance is whether or not the levels of an independent variable are **fixed** or **random**. Whereas a **fixed-effects model** assumes that the levels of an independent variable are the same levels that will be employed in any attempted replication of the experiment, a **random-effects model** assumes that the levels have been randomly selected from the overall population of all possible levels that can be employed for the independent variable. The computational procedures for all of the analysis of variance procedures described in this book assume a fixed-effects model for all factors.

In the case of factorial designs it is also possible to have a **mixed-effects model**, which is a combination of a fixed-effects and a random-effects model. Specifically, in the case of a two-factor design, a mixed-effects model assumes that one of the factors is based on a fixed-effects model while the second factor is based on a random-effects model. When there are three or more factors, a mixed-effects model assumes that one or more of the factors is based on a fixed-effects model and one or more of the factors is based on a random-effects model. Texts that specialize in the analysis of variance provide in-depth discussions of this general subject, as well as describing the modified equations that are appropriate for evaluating factorial designs that are based on random- and mixed-effects models.

b) Nested factors/hierarchical designs and designs involving more than two factors In designing experiments that involve two or more factors, it is possible to employ what are referred to as **nested factors**. Nesting is present in an experimental design when different levels of one factor do not occur at all levels of another factor. To illustrate nesting, let us assume that a researcher wants to evaluate two teaching methods (which will comprise Factor A) in 10 different classes, each of which is unique with respect to the ethnic makeup of its students. The 10 different classes will comprise Factor B. Five of the classes (B_1 ... B_5) are taught by teaching method A_1 and the other five classes (B_6 ... B_{10}) by teaching method A_2. In such a case Factor B is nested under Factor A, since each level of Factor B serves under only one level of Factor A. Figure 21.3 outlines the design.[25]

A_1					A_2				
B_1	B_2	B_3	B_4	B_5	B_6	B_7	B_8	B_9	B_{10}

Figure 21.3 Example of a Nested Design

Winer *et al.* (1991) note that in the above described design it is not possible to evaluate whether there is an interaction between the two factors. The reason for this is that in order to test for an interaction it is necessary that all levels of Factor A must occur under all of the levels of Factor B. When the latter is true (as is the case in Example 21.1), the two factors are said to be **crossed.** The term **hierarchical design** is often employed to describe designs in which there are two or more nested factors. It is also possible to have a **partially hierarchical design,** in which at least one factor is nested and at least one factor is crossed. Since the statistical model upon which nested designs are based differs from the model that has been employed for the **between-subjects factorial analysis of variance**, the analysis of such designs requires the use of different equations.

As noted earlier, a **between-subjects factorial analysis of variance** (as well as a **factorial analysis of variance for a mixed design** and a **within-subjects factorial analysis of variance** which are discussed in Section IX) can involve more than two factors. To further complicate matters, designs with three of more factors can involve some factors that are nested and others that are crossed, plus the fact that a fixed-effects model may be assumed for some factors and a random-effects model for others. Honeck *et al.* (1983) is an excellent source for deriving the appropriate equations for the use of the analysis of variance with experimental designs involving nesting and/or the use of more than two factors. Other sources on this subject are Keppel (1991), Kirk (1982), and Winer *et al.* (1991).

VIII. Additional Examples Illustrating the Use of the Between-Subjects Factorial Analysis of Variance

Examples 21.2 and 21.3 are two additional examples that can be evaluated with the **between-subjects factorial analysis of variance.** Since the data for both examples are identical to that employed in Example 21.1, they yield the same result. Note that whereas in Example 21.1 both Factor A and Factor B are manipulated independent variables, in Example 21.2 Factor B is manipulated while Factor A is nonmanipulated (i.e., is a subject/attribute independent variable). In Example 21.3 both factors are nonmanipulated independent variables.

Example 21.2 *A study is conducted in order to evaluate the impact of gender (to be designated as Factor* A) *and anxiety level (to be designated as Factor* B) *on affiliation. The experimenter employs a* 2 × 3 *between-subjects (completely-randomized) factorial design. The two levels that comprise Factor* A *are* A_1*: Male;* A_2*: Female. The three levels that comprise Factor* B *are* B_1*: Low Anxiety;* B_2*: Moderate Anxiety;* B_3*: High Anxiety. Each of nine males and nine females is randomly assigned to one of three experimental conditions. All of the subjects are told they are participants in a learning experiment which will require them to learn lists of words. Subjects in the low anxiety condition are told that there will be no consequences for poor performance in the experiment. Subjects in the moderate anxiety condition are told if they perform below a certain level they will have to drink a distasteful beverage. Subjects in the high anxiety condition are told if they perform below a certain level they will be given a painful electric shock. All subjects are told that while waiting to be* *tested they can either wait by themselves or with other people. Each subject is asked to designate the number of people he or she would like in the room with him or her while waiting to be tested. This latter measure is employed to represent the dependent variable of affiliation. The experimenter assumes that the higher a subject is in affiliation, the more people the subject will want to be with while waiting. The affiliation scores of the three subjects in each of the six experimental groups/conditions (which result from the combinations of the levels that comprise the two factors) follow:* **Group AB_{11}**: *Male/Low anxiety* (11, 9, 10); **Group AB_{12}**: *Male/Moderate anxiety* (7, 8, 6); **Group AB_{13}**: *Male/High anxiety* (5, 4, 3); **Group AB_{21}**: *Female/Low anxiety* (2, 4, 3); **Group AB_{22}**: *Female/Moderate anxiety* (4, 5, 3); **Group AB_{23}**: *Female/High anxiety* (0, 1, 2). *Do the data indicate that either gender or anxiety level influence affiliation?*

Example 21.3 *A study is conducted in order to evaluate if there is a relationship between ethnicity (to be designated as Factor* A*) and socioeconomic class (to be designated as Factor* B*), and the number of times a year a person visits a doctor. The experimenter employs a* 2×3 *between-subjects (completely-randomized) factorial design. The two levels that comprise Factor* A *are* A_1*: Caucasian;* A_2*: Afro-American. The three levels that comprise Factor* B *are* B_1*: Lower socioeconomic class;* B_2*: Middle socioeconomic class;* B_3*: Upper socioeconomic class. Based on their occupation and income, each of nine Caucasians and nine Afro-Americans is categorized with respect to whether he or she is a member of the lower, middle, or upper socioeconomic class. Upon doing this the experimenter determines the number of times during the past year each of the subjects has visited a doctor. This latter measure represents the dependent variable in the study. The number of visits for the three subjects in each of the six experimental groups/conditions (which result from the combinations of the levels that comprise the two factors) follow:* **Group AB_{11}**: *Caucasian/Lower socioeconomic class* (11, 9, 10); **Group AB_{12}**: *Caucasian/Middle socioeconomic class* (7, 8, 6); **Group AB_{13}**: *Caucasian/Upper socioeconomic class* (5, 4, 3); **Group AB_{21}**: *Afro-American/Lower socioeconomic class* (2, 4, 3); **Group AB_{22}**: *Afro-American/Middle socioeconomic class* (4, 5, 3); **Group AB_{23}**: *Afro-American/Upper socioeconomic class* (0, 1, 2). *Do the data indicate that either ethnicity or socioeconomic class is related to how often a person visits a doctor?*

IX. Addendum

Discussion of additional analysis of variance procedures for factorial designs

1. Test 21h: The factorial analysis of variance for a mixed design A **mixed factorial design** involves two or more independent variables/factors in which at least one of independent variables is measured between-subjects (different subjects serve under each of the levels of that independent variable) and at least one of the independent variables is measured within-subjects (the same subjects or matched sets of subjects serve under all of the levels of that independent variable). Although the **factorial analysis of variance for a mixed design** can be used with designs involving more than two factors, the computational protocol to be described in this section will be limited to the two-factor experiment. For purposes of illustration it will be assumed that Factor A is measured between-subjects (i.e., different subjects serve in each of the p levels of Factor A), and that Factor B is measured within-subjects (i.e., all subjects are measured on each of the q levels of Factor B). Since one of the factors is measured within-subjects, a **mixed factorial design** requires a fraction of the subjects that are needed to evaluate the same set of hypotheses with a **between-subjects factorial design** (assuming both designs employ the same number of scores in each of the pq experimental conditions). To be more specific, the fraction of subjects required is 1 divided by the number of levels of the within-subjects factor (i.e., $1/q$ if Factor B is the within-subjects factor). The advantages as well as the disadvantages of a within-subjects analysis (which are discussed under the **t test for two dependent samples** and the **single-factor within-subjects analysis of variance**) also apply to the within-subjects factor that is evaluated with the **factorial analysis of variance for a mixed design**. Probably the most notable advantage associated with the within-subjects factor is that it allows for a more powerful test of an alternative hypothesis when contrasted with the between-subjects factor. Example 21.4 is employed to illustrate the use of the **factorial analysis of variance for a mixed design.**

Example 21.4 *A study is conducted in order to evaluate the effect of humidity (to be designated as Factor* A) *and temperature (to be designated as Factor* B) *on mechanical problem solving ability. The experimenter employs a* 2 × 3 *mixed factorial design. The two levels that comprise Factor* A *are* A_1*: Low humidity;* A_2*: High humidity. The three levels that comprise Factor* B *are* B_1*: Low temperature;* B_2*: Moderate temperature;* B_3*: High temperature. The study employs six subjects, three of whom are randomly assigned to Level* 1 *of Factor* A *and three of whom are randomly assigned to Level* 2 *of Factor* A. *Each subject is exposed to all three levels of Factor* B. *The order of presentation of the levels of Factor* B *is completely counterbalanced within the six subjects. The number of mechanical problems solved by the subjects in the six experimental conditions (which result from combinations of the levels of the two factors) follow:* **Condition AB_{11}**: *Low humidity/Low temperature* (11, 9, 10); **Condition AB_{12}**: *Low humidity/Moderate temperature* (7, 8, 6); **Condition AB_{13}**: *Low humidity/High temperature* (5, 4, 3); **Condition AB_{21}**: *High humidity/Low temperature* (2, 4, 3); **Condition AB_{22}**: *High humidity/Moderate temperature* (4, 5, 3); **Condition AB_{23}**: *High humidity/High temperature* (0, 1, 2). *Do the data indicate that either humidity or temperature influences mechanical problem solving ability?*

The data for Example 21.4 are summarized in Table 21.7.

Table 21.7 Data for Example 21.4 for Evaluation with the Factorial Analysis of Variance for a Mixed Design

	A_1			
	B_1	B_2	B_3	Subject sums (ΣS_i)
Subject 1	11	7	5	$\Sigma S_1 = 23$
Subject 2	9	8	4	$\Sigma S_2 = 21$
Subject 3	10	6	3	$\Sigma S_3 = 19$
Condition sums	$\Sigma X_{AB_{11}} = 30$ $\Sigma X^2_{AB_{11}} = 302$	$\Sigma X_{AB_{12}} = 21$ $\Sigma X^2_{AB_{12}} = 149$	$\Sigma X_{AB_{13}} = 12$ $\Sigma X^2_{AB_{13}} = 50$	$\Sigma X_{A_1} = 63$

	A_2			
	B_1	B_2	B_3	Subject sums (ΣS_i)
Subject 4	2	4	0	$\Sigma S_4 = 6$
Subject 5	4	5	1	$\Sigma S_5 = 10$
Subject 6	3	3	2	$\Sigma S_6 = 8$
Condition sums	$\Sigma X_{AB_{21}} = 9$ $\Sigma X^2_{AB_{21}} = 29$	$\Sigma X_{AB_{22}} = 12$ $\Sigma X^2_{AB_{22}} = 50$	$\Sigma X_{AB_{23}} = 3$ $\Sigma X^2_{AB_{23}} = 5$	$\Sigma X_{A_2} = 24$
	$\Sigma X_{B_1} = 39$	$\Sigma X_{B_2} = 33$	$\Sigma X_{B_3} = 15$	$\Sigma X_T = 87$ $\Sigma X^2_T = 585$

Examination of Table 21.7 reveals that since the data employed for Example 21.4 are identical to that employed for Example 21.1, the summary values for the rows, columns, and pq experimental conditions are identical to those in Table 21.1. Thus, the following values in Table 21.7 are identical to those obtained in Table 21.1: $n_{A_j} = 9$ and the values computed for ΣX_{A_j} and $\Sigma X_{A_j}^2$ for each of the levels of Factor A; $n_{B_k} = 6$ and the values computed for ΣX_{B_k} and $\Sigma X_{B_k}^2$ for each of the levels of Factor B; $n_{AB_{jk}} = n = 3$ and the values computed for $\Sigma X_{AB_{jk}}$ and $\Sigma X_{AB_{jk}}^2$ for each of the pq experimental conditions that result from combinations of the levels of the two factors; $N = npq = 18$; $\Sigma X_T = 87$; $\Sigma X_T^2 = 585$.

Note that in both the **between-subjects factorial analysis of variance** and the **factorial analysis of variance for a mixed design**, the value $n_{AB_{jk}} = n = 3$ represents the number of scores in each of the pq experimental conditions. In the case of the **factorial analysis of variance for a mixed design**, the value $N = npq = 18$ represents the total number of scores in the set of data. Note, however, that the latter value does not represent the total number of subjects employed in the study, as it does in the case of the **between-subjects factorial analysis of variance**. The number of subjects employed for a **factorial analysis of variance for a mixed design** will always be the value of n multiplied by the number of levels of the between-subjects factor. Thus, in Example 21.4 the number of subjects is $np = (3)(2) = 6$.[26]

As is the case for the **between-subjects factorial analysis of variance**, the following three F ratios are computed for the **factorial analysis of variance for a mixed design**: F_A, F_B, F_{AB}. The equations required for computing the F ratios are summarized in Table 21.8. Table 21.9 summarizes the computations for the **factorial analysis of variance for a mixed design** when it is employed to evaluate Example 21.4. In order to compute the F ratios for the **factorial analysis of variance for a mixed design** it is required that the following summary values (which are also computed for the **between-subjects factorial analysis of variance**) be computed: $[XS]$, $[T]$, $[A]$, $[B]$, and $[AB]$. Since the summary values computed in Table 21.7 are identical to those computed in Table 21.1 (for Example 21.1), the same summary values are employed in Tables 21.8 and 21.9 to compute the values $[XS]$, $[T]$, $[A]$, $[B]$, and $[AB]$ (which are respectively computed with Equations 21.4–21.8). Thus: $[XS] = 585$, $[T] = 420.5$, $[A] = 505$, $[B] = 472.5$, $[AB] = 573$. Since the same set of data and the same equations are employed for the **factorial analysis of variance for a mixed design** and the **between-subjects factorial analysis of variance**, both analysis of variance procedures yield identical values for $[XS]$, $[T]$, $[A]$, $[B]$, and $[AB]$. Inspection of Table 21.8 also reveals that the **factorial analysis of variance for a mixed design** and the **between-subjects factorial analysis of variance** employ the same equations to compute the values SS_A, SS_B, SS_{AB}, SS_T, MS_A, MS_B, and MS_{AB}.

In order to compute a number of additional sum of squares values for the **factorial analysis of variance for a mixed design**, it is necessary to compute the element $[AS]$ (which is not computed for the **between-subjects factorial analysis of variance**). $[AS]$, which is computed with Equation 21.58, is employed in Tables 21.8 and 21.9 to compute the following values: $SS_{\text{Between-subjects}}$, $SS_{\text{Subjects } WG}$, $SS_{\text{Within-subjects}}$, $SS_{B \times \text{subjects } WG}$.

$$[AS] = \sum_{i=1}^{np} \left[\frac{(\Sigma S_i)^2}{q} \right] \qquad \textbf{(Equation 21.58)}$$

The notation $\Sigma_{i=1}^{np}[(\Sigma S_i)^2/q]$ in Equation 21.58 indicates that for each of the $np = 6$ subjects, the score of the subject is squared and divided by q. The resulting values obtained for the np subjects are summed, yielding the value $[AS]$. Employing Equation 21.58, the value $[AS] = 510.33$ is computed.

$$[AS] = \frac{(23)^2}{3} + \frac{(21)^2}{3} + \frac{(19)^2}{3} + \frac{(6)^2}{3} + \frac{(10)^2}{3} + \frac{(8)^2}{3} = 510.33$$

Table 21.8 Summary Table of Equations for the Factorial Analysis of Variance for a Mixed Design

Source of variation	SS	df	MS	F
Between-subjects	$[AS]-[T]$	$np-1$		
A	$[A]-[T]$	$p-1$	$\frac{SS_A}{df_A}$	$F_A=\frac{MS_A}{MS_{\text{Subjects } WG}}$
Subjects *WG*	$[AS]-[A]$	$p(n-1)$	$\frac{SS_{\text{Subjects } WG}}{df_{\text{Subjects } WG}}$	
Within-subjects	$[XS]-[AS]$	$np(q-1)$		
B	$[B]-[T]$	$q-1$	$\frac{SS_B}{df_B}$	$F_B=\frac{MS_B}{MS_{B\times\text{subjects } WG}}$
AB	$[AB]-[A]-[B]+[T]$	$(p-1)(q-1)$	$\frac{SS_{AB}}{df_{AB}}$	$F_{AB}=\frac{MS_{AB}}{MS_{B\times\text{subjects } WG}}$
B×subjects *WG*	$[XS]-[AB]-[AS]+[A]$	$p(q-1)(n-1)$	$\frac{SS_{B\times\text{subjects } WG}}{df_{B\times\text{subjects } WG}}$	
Total	$[XS]-[T]$	$N-1=npq-1$		

Table 21.9 Summary Table of Computations for Example 21.4

Source of variation	SS	df	MS	F
Between-subjects	510.33−420.5=89.83	(3)(2)−1=5		
A	505−420.5=84.5	2−1=1	$MS_A=\frac{84.5}{1}=84.5$	$F_A=\frac{84.5}{1.33}=63.53$
Subjects *WG*	510.33−505=5.33	2(3−1)=4	$MS_{\text{Subjects } WG}=\frac{5.33}{4}=1.33$	
Within-subjects	585−510.33=74.67	(3)(2)(3−1)=12		
B	472.5−420.5=52	3−1=2	$MS_B=\frac{52}{2}=26$	$F_B=\frac{26}{.83}=31.33$
AB	573−505−472.5+420.5=16	(2−1)(3−1)=2	$MS_{AB}=\frac{16}{2}=8$	$F_{AB}=\frac{8}{.83}=9.64$
B×subjects *WG*	585−573−510.33+505=6.67	2(3−1)(3−1)=8	$MS_{B\times\text{subjects } WG}=\frac{6.67}{8}=.83$	
Total	585−420.5=164.5	18−1=(3)(2)(3)−1=17		

The reader should take note of the following relationships in Tables 21.8 and 21.9:

$$SS_{\text{Between-subjects}} = SS_A + SS_{\text{Subjects } WG}$$

$$SS_{\text{Within-subjects}} = SS_B + SS_{AB} + SS_{B \times \text{ subjects } WG}$$

$$SS_T = SS_{\text{Between-subjects}} + SS_{\text{Within-subjects}}$$

$$df_{\text{Between-subjects}} = df_A + df_{\text{Subjects } WG}$$

$$df_{\text{Within-subjects}} = df_B + df_{AB} + df_{B \times \text{ subjects } WG}$$

$$df_T = df_{\text{Between-subjects}} + df_{\text{Within-subjects}}$$

Inspection of Table 21.2 and Tables 21.8/21.9 reveals that if a **between-subjects factorial analysis of variance** and a **factorial analysis of variance for a mixed design** are employed with the same set of data, identical values are computed for the following: SS_A, SS_B, SS_{AB}, SS_T, df_A, df_B, df_{AB}, df_T, MS_A, MS_B, MS_{AB}.

In Table 21.9 the error term $MS_{\text{Subjects } WG} = 1.33$ (employed in computing the value $F_A = 63.53$) is identical to the value MS_{WG} which would be obtained if Factor B was not taken into account in Example 21.4, and the data on Factor A were evaluated with a **single-factor between-subjects analysis of variance.** The error term $MS_{B \times \text{ subjects } WG} = .83$, employed in computing the values $F_B = 31.33$ and $F_{AB} = 9.64$ is analogous to the error term employed for the **single-factor within-subjects analysis of variance.** For a thorough discussion of the derivation of the error terms for the **factorial analysis of variance for a mixed design,** the reader should consult books which discuss analysis of variance procedures in greater detail (e.g., Keppel (1991) and Winer *et al.* (1991)).

The following tabled critical values derived from **Table A10** are employed in evaluating the three F ratios computed for Example 21.4: a) **Factor A:** For $df_{\text{num}} = df_A = 1$ and $df_{\text{den}} = df_{\text{Subjects } WG} = 4$, $F_{.05} = 7.71$ and $F_{.01} = 21.20$; b) **Factor B:** For $df_{\text{num}} = df_B = 2$ and $df_{\text{den}} = df_{B \times \text{ subjects } WG} = 8$, $F_{.05} = 4.46$ and $F_{.01} = 8.65$; and c) **AB interaction:** For $df_{\text{num}} = df_{AB} = 2$ and $df_{\text{den}} = df_{B \times \text{ subjects } WG} = 8$, $F_{.05} = 4.46$ and $F_{.01} = 8.65$.

The identical null and alternative hypotheses that are evaluated in Section III of the **between-subjects factorial analysis of variance** are evaluated in the **factorial analysis of variance for a mixed design.** In order to reject the null hypothesis in reference to a computed F ratio, the obtained F value must be equal to or greater than the tabled critical value at the prespecified level of significance. Since the computed value $F_A = 63.53$ is greater than $F_{.05} = 7.71$ and $F_{.01} = 21.20$, the alternative hypothesis for Factor A is supported at both the .05 and .01 levels. Since the computed value $F_B = 31.33$ is greater than $F_{.05} = 4.46$ and $F_{.01} = 8.65$, the alternative hypothesis for Factor B is supported at both the .05 and .01 levels. Since the computed value $F_{AB} = 9.64$ is greater than $F_{.05} = 4.46$ and $F_{.01} = 8.65$, the alternative hypothesis for an interaction between Factors A and B is supported at both the .05 and .01 levels.

The analysis of the data for Example 21.4 allows the researcher to conclude that both humidity (Factor A) and temperature (Factor B) have a significant impact on problem solving scores. However, as is the case when the same set of data is evaluated with a **between-subjects factorial analysis of variance,** the relationships depicted by the main effects must be qualified because of the presence of a significant interaction. Although the comparison procedures following the computation of the omnibus F ratios (as well as the other analytical procedures for determining power, effect size, etc.) described in Section VI of the **between-subjects factorial analysis of variance** can be extended to the **factorial analysis of variance for a mixed design,** they will not be described in this book. For a full description of such procedures, the reader should consult texts that discuss analysis of variance procedures in greater detail (e.g., Keppel (1991) and Winer *et al.* (1991)).

2. Test 21i: The within-subjects factorial analysis of variance A **within-subjects factorial design** involves two or more factors, and all subjects are measured on each of the levels of all of the factors. The **within-subjects factorial analysis of variance** (also known as a **repeated-measures factorial analysis of variance**) is an extension of the **single-factor within-subjects analysis of variance** to experiments involving two or more independent variables/factors. Although the **within-subjects factorial analysis of variance** can be used with designs involving more than two factors, the computational protocol to be described in this section will be limited to the two factor experiment. Within the framework of the **within-subjects factorial design**, each subject contributes pq scores (which result from the combinations of the levels that comprise the two factors). Since subjects serve under all pq experimental conditions, a **within-subjects factorial design** requires a fraction of the subjects that are needed to evaluate the same set of hypotheses with either the **between-subjects factorial design** or the **mixed factorial design** (assuming a given design employs the same number of scores in each of the pq experimental conditions). To be more specific, only $1/pq^{th}$ of the subjects are required for a **within-subjects factorial design** in contrast to a **between-subjects factorial design.** The requirement of fewer subjects and the fact that a within-subjects analysis provides for a more powerful test of an alternative hypothesis than a between-subjects analysis, must to be weighed against the fact that it is often impractical or impossible to have subjects serve in multiple experimental conditions. In addition, a within-subjects analysis of variance is more sensitive to violations of its assumptions than a between-subjects analysis of variance. Example 21.5 is employed to illustrate the **within-subjects factorial analysis of variance.**

Example 21.5 *A study is conducted in order to evaluate the effect of humidity (to be designated as Factor* A) *and temperature (to be designated as Factor* B) *on mechanical problem solving ability. The experimenter employs a* 2 × 3 *within-subjects factorial design. The two levels that comprise Factor* A *are* A_1*: Low humidity;* A_2*: High humidity. The three levels that comprise Factor* B *are* B_1*: Low temperature;* B_2*: Moderate temperature;* B_3*: High temperature. The study employs three subjects, all of whom serve under the two levels of Factor A and the three levels of Factor* B. *The order of presentation of the combinations of the two factors is incompletely counterbalanced.*[27] *The number of mechanical problems solved by the subjects in the six experimental conditions (which result from combinations of the levels of the two factors) follow:* **Condition AB_{11}**: *Low humidity/Low temperature* (11, 9, 10); **Condition AB_{12}**: *Low humidity/Moderate temperature* (7, 8, 6); **Condition AB_{13}**: *Low humidity/High temperature* (5, 4, 3); **Condition AB_{21}**: *High humidity/Low temperature* (2, 4, 3); **Condition AB_{22}**: *High humidity/Moderate temperature* (4, 5, 3); **Condition AB_{23}**: *High humidity/High temperature* (0, 1, 2). *Do the data indicate that either humidity or temperature influences mechanical problem solving ability?*

The data for Example 21.5 are summarized in Tables 21.10–21.12. In Table 21.11 $S_{i_{A_j}}$ represents the score of Subject i under Level j of Factor A. In Table 21.12 $S_{i_{B_k}}$ represents the score of Subject i under Level k of Factor B.

Examination of Tables 21.10–21.12 reveals that since the data employed for Example 21.5 are identical to that employed for Examples 21.1 and 21.4, the summary values for the rows, columns, and pq experimental conditions are identical to those in Tables 21.1 and 21.7. Thus, the following values in Tables 21.10–21.12 are identical to those obtained in the tables for Examples 21.1 and 21.4: $n_{A_j} = 9$ and the values computed for ΣX_{A_j} and $\Sigma X^2_{A_j}$ for each of the levels of Factor A; $n_{B_k} = 6$ and the values computed for ΣX_{B_k} and $\Sigma X_{B_k}^2$ for each of the levels of Factor B; $n_{AB_{jk}} = n = 3$ and the values computed for $\Sigma X_{AB_{jk}}$ and $\Sigma X_{AB_{jk}}^2$ for each of the pq experimental conditions that result from combinations of the levels of the two factors; $N = npq = 18$; $\Sigma X_T = 87$; $\Sigma X_T^2 = 585$.

Table 21.10 Data for Example 21.5 for Evaluation with the Within-Subjects Factorial Analysis of Variance

	A_1			A_2			Subject sums
	B_1	B_2	B_3	B_1	B_2	B_3	(ΣS_i)
Subject 1	11	7	5	2	4	0	$\Sigma S_1 = 29$
Subject 2	9	8	4	4	5	1	$\Sigma S_2 = 31$
Subject 3	10	6	3	3	3	2	$\Sigma S_3 = 27$
Condition Sums	$\Sigma X_{AB_{11}} = 30$	$\Sigma X_{AB_{12}} = 21$	$\Sigma X_{AB_{13}} = 12$	$\Sigma X_{AB_{21}} = 9$	$\Sigma X_{AB_{22}} = 12$	$\Sigma X_{AB_{23}} = 3$	$\Sigma X_T = 87$
	$\Sigma X_{AB_{11}}^2 = 302$	$\Sigma X_{AB_{12}}^2 = 149$	$\Sigma X_{AB_{13}}^2 = 50$	$\Sigma X_{AB_{21}}^2 = 29$	$\Sigma X_{AB_{22}}^2 = 50$	$\Sigma X_{AB_{23}}^2 = 5$	$\Sigma X_T^2 = 585$

Table 21.11 Scores of Subjects on Levels of Factor A for Example 21.5

	A_1	A_2	Subject sums (ΣS_i)
Subject 1	$S_{1_{A_1}} = 23$	$S_{1_{A_2}} = 6$	$\Sigma S_1 = 29$
Subject 2	$S_{2_{A_1}} = 21$	$S_{2_{A_2}} = 10$	$\Sigma S_2 = 31$
Subject 3	$S_{3_{A_1}} = 19$	$S_{3_{A_2}} = 8$	$\Sigma S_3 = 27$
Sums for levels of Factor A	$\Sigma X_{A_1} = 63$	$\Sigma X_{A_2} = 24$	$\Sigma X_T = 87$

Table 21.12 Scores of Subjects on Levels of Factor B for Example 21.5

	B_1	B_2	B_3	Subject sums (ΣS_i)
Subject 1	$S_{1_{B_1}} = 13$	$S_{1_{B_2}} = 11$	$S_{1_{B_3}} = 5$	$\Sigma S_1 = 29$
Subject 2	$S_{2_{B_1}} = 13$	$S_{2_{B_2}} = 13$	$S_{2_{B_3}} = 5$	$\Sigma S_2 = 31$
Subject 3	$S_{3_{B_1}} = 13$	$S_{3_{B_2}} = 9$	$S_{3_{B_3}} = 5$	$\Sigma S_3 = 27$
Sums for levels of Factor B	$\Sigma X_{B_1} = 39$	$\Sigma X_{B_2} = 33$	$\Sigma X_{B_3} = 15$	$\Sigma X_T = 87$

Note that in the **within-subjects factorial analysis of variance**, the **between-subjects factorial analysis of variance**, and the **factorial analysis of variance for a mixed design**, the value $n_{AB_{jk}} = n = 3$ represents the number of scores in each of the pq experimental conditions. In the case of the **within-subjects factorial analysis of variance**, the value $N = npq = 18$ represents the total number of scores in the set of data. Note, however, that the latter value does not represent the total number of subjects employed in the study as it does in the case of the **between-subjects factorial analysis of variance**. The number of subjects employed for a **within-subjects factorial analysis of variance** will always be the value of $n = n_{AB_{jk}}$. Thus, in Example 21.5 the number of subjects is $n = n_{AB_{jk}} = 3$.

As is the case for the **between-subjects factorial analysis of variance** and the **factorial analysis of variance for a mixed design**, the following three F ratios are computed for the **within-subjects factorial analysis of variance:** F_A, F_B, F_{AB}. The equations required for computing the F ratios are summarized in Table 21.13. Table 21.14 summarizes the computations for the **within-subjects factorial analysis of variance** when it is employed to evaluate Example 21.5. In order to compute the F ratios for the **within-subjects factorial analysis of variance**, it is required that the following summary values (which are also computed for the **between-subjects factorial analysis of variance** and the **factorial analysis of variance for a mixed design**) be computed: $[XS]$, $[T]$, $[A]$, $[B]$, and $[AB]$. Since the summary values computed in Tables 21.10–21.12 are identical to those computed in Tables 21.1 and 21.7 (for Example 21.1 and Example 21.4), the same summary values are employed in Tables 21.13 and 21.14 to compute the values $[XS]$, $[T]$, $[A]$, $[B]$, and $[AB]$ (which are respectively computed with Equations 21.4–21.8). Thus: $[XS] = 585, [T] = 420.5$, $[A] = 505$, $[B] = 472.5$, $[AB] = 573$. Since the same set of data and the same equations are employed for the **within-subjects factorial analysis of variance**, the **between-subjects factorial analysis of variance**, and the **factorial analysis of variance for a mixed design**, all three analysis of variance procedures yield identical values for $[XS]$, $[T]$, $[A]$, $[B]$, and $[AB]$. Inspection of Table 21.13 also reveals that the **within-subjects factorial analysis of variance**, the **between-subjects factorial analysis of variance**, and the **factorial analysis of variance for a mixed design** employ the same equations to compute the values SS_A, SS_B, SS_{AB}, SS_T, MS_A, MS_B, and MS_{AB}.

In order to compute a number of additional sum of squares values for the **within-subjects factorial analysis of variance**, it is necessary to compute the following three elements which are not computed for the **between-subjects factorial analysis of variance:** $[S]$, $[AS]$, and $[BS]$.

$[S]$, which is computed with Equation 21.59, is employed in Tables 21.13 and 21.14 to compute the following values: $SS_{\text{Between-subjects}}$, $SS_{\text{Within-subjects}}$, $SS_{A \times \text{subjects}}$, $SS_{B \times \text{subjects}}$, $SS_{AB \times \text{subjects}}$.

$$[S] = \sum_{i=1}^{n} \left[\frac{(\Sigma S_i)^2}{pq} \right] \qquad \textbf{(Equation 21.59)}$$

The notation $\Sigma_{i=1}^{n}[(\Sigma S_i)^2/pq]$ in Equation 21.59 indicates that for each of the $n = 3$ subjects, the sum of that subject's three scores (i.e., ΣS_i) is squared and divided by pq. The resulting values obtained for the $n = 3$ subjects are summed, yielding the value $[S]$. Employing Equation 21.59, the value $[S] = 421.83$ is computed.

$$[S] = \frac{(29)^2}{6} + \frac{(31)^2}{6} + \frac{(27)^2}{6} = 421.83$$

$[AS]$, which is computed with Equation 21.60, is employed in Tables 21.13 and 21.14 to compute the following values: $SS_{A \times \text{subjects}}$, $SS_{AB \times \text{subjects}}$.

$$[AS] = \sum_{i=1}^{n} \sum_{j=1}^{p} \left[\frac{(\Sigma S_{i_{A_j}})^2}{q} \right] \qquad \textbf{(Equation 21.60)}$$

The notation $\Sigma_{i=1}^{n}\Sigma_{j=1}^{p}[(\Sigma S_{i_{A_j}})^2/q]$ in Equation 21.60 indicates that each of the $p = 2$ $S_{i_{A_j}}$ scores of the $n = 3$ subjects is squared and divided by $q = 3$. The resulting $np = 6$ values are summed, yielding the value $[AS]$. Employing Equation 21.60, the value $[AS] = 510.33$ is computed (which is the same value computed for $[AS]$ when the same set of data is evaluated with the **factorial analysis of variance for a mixed design**).

$$[AS] = \frac{(23)^2}{3} + \frac{(6)^2}{3} + \frac{(21)^2}{3} + \frac{(10)^2}{3} + \frac{(19)^2}{3} + \frac{(8)^2}{3} = 510.33$$

$[BS]$, which is computed with Equation 21.61, is employed in Tables 21.13 and 21.14 to compute the following values: $SS_{B \times \text{subjects}}$, $SS_{AB \times \text{subjects}}$.

$$[BS] = \sum_{i=1}^{n} \sum_{k=1}^{q} \left[\frac{(\Sigma S_{i_{B_k}})^2}{p} \right] \qquad \textbf{(Equation 21.61)}$$

The notation $\Sigma_{i=1}^{n} \Sigma_{k=1}^{q} [(\Sigma S_{i_{B_k}})^2/p]$ in Equation 21.61 indicates that each of the $q = 3$ $S_{i_{B_k}}$ scores of the $n = 3$ subjects is squared and divided by $p = 2$. The resulting $nq = 9$ values are summed, yielding the value $[BS]$. Employing Equation 21.61, the value $[BS] = 476.5$ is computed.

$$[BS] = \frac{(13)^2}{2} + \frac{(11)^2}{2} + \frac{(5)^2}{2} + \frac{(13)^2}{2} + \frac{(13)^2}{2} + \frac{(5)^2}{2} + \frac{(13)^2}{2} + \frac{(9)^2}{2} + \frac{(5)^2}{2} = 476.5$$

The reader should take note of the following relationships in Tables 21.13 and 21.14:

$$SS_{\text{Within-subjects}} = SS_A + SS_B + SS_{AB} + SS_{A \times \text{subjects}} + SS_{B \times \text{subjects}} + SS_{AB \times \text{subjects}}$$

$$SS_T = SS_{\text{Between-subjects}} + SS_{\text{Within-subjects}}$$

$$df_{\text{Within-subjects}} = df_A + df_B + df_{AB} + df_{A \times \text{subjects}} + df_{B \times \text{subjects}} + df_{AB \times \text{subjects}}$$

$$df_T = df_{\text{Between-subjects}} + df_{\text{Within-subjects}}$$

Inspection of Table 21.2, Tables 21.13/21.14, and Tables 21.8/21.9 reveals that if a **between-subjects factorial analysis of variance**, a **within-subjects factorial analysis of variance**, and a **factorial analysis of variance for a mixed design** are employed with the same set of data, identical values are computed for the following: SS_A, SS_B, SS_{AB}, SS_T, df_A, df_B, df_{AB}, df_T, MS_A, MS_B, MS_{AB}.

In Table 21.14, the error term $MS_{A \times \text{subjects}} = 2$, employed in computing the value $F_A = 42.25$, is analogous to the error term that would be obtained if in evaluating the data for Example 21.5, Factor B was not taken into account, and the data on Factor A were evaluated with a **single-factor within-subjects analysis of variance**. The error term $MS_{B \times \text{subjects}} = .67$, employed in computing the value $F_B = 38.81$, is analogous to the error term that would be obtained if in evaluating the data for Example 21.5, Factor A was not taken into account, and the data on Factor B were evaluated with a **single-factor within-subjects analysis of variance**. The value $MS_{AB \times \text{subjects}} = 1$, employed in computing the value $F_{AB} = 8$, is a measure of error variability specific to the AB interaction for the **within-subjects factorial analysis of variance**. For a thorough discussion of the derivation of the error terms for the **within-subjects factorial analysis of variance**, the reader should consult books which discuss analysis of variance procedures in greater detail (e.g., Keppel (1991) and Winer *et al.* (1991)).

The following tabled critical values derived from **Table A10** are employed in evaluating the three F ratios computed for Example 21.5: a) **Factor A**: For $df_{\text{num}} = df_A = 1$ and $df_{\text{den}} = df_{A \times \text{subjects}} = 2$, $F_{.05} = 18.51$ and $F_{.01} = 98.50$; b) **Factor B**: For $df_{\text{num}} = df_B = 2$ and $df_{\text{den}} = df_{B \times \text{subjects}} = 4$, $F_{.05} = 6.94$ and $F_{.01} = 18.00$; and c) **AB interaction**: For $df_{\text{num}} = df_{AB} = 2$ and $df_{\text{den}} = df_{AB \times \text{subjects}} = 4$, $F_{.05} = 6.94$ and $F_{.01} = 18.00$.

Table 21.13 Summary Table of Equations for the Within-Subjects Factorial Design

Source of variation	SS	df	MS	F
Between-subjects	$[S]-[T]$	$n-1$		
Within-subjects	$[XS]-[S]$	$n(pq-1)$		
A	$[A]-[T]$	$p-1$	$\frac{SS_A}{df_A}$	$F_A = \frac{MS_A}{MS_{A \times \text{subjects}}}$
B	$[B]-[T]$	$q-1$	$\frac{SS_B}{df_B}$	$F_B = \frac{MS_B}{MS_{B \times \text{subjects}}}$
AB	$[AB]-[A]-[B]+[T]$	$(p-1)(q-1)$	$\frac{SS_{AB}}{df_{AB}}$	$F_{AB} = \frac{MS_{AB}}{MS_{AB \times \text{subjects}}}$
$A \times$subjects	$[AS]-[A]-[S]+[T]$	$(p-1)(n-1)$	$\frac{SS_{A \times \text{subjects}}}{df_{A \times \text{subjects}}}$	
$B \times$subjects	$[BS]-[B]-[S]+[T]$	$(q-1)(n-1)$	$\frac{SS_{B \times \text{subjects}}}{df_{B \times \text{subjects}}}$	
$AB \times$subjects	$[XS]-[AB]-[AS]-[BS]$ $+[A]+[B]+[S]-[T]$	$(p-1)(q-1)(n-1)$	$\frac{SS_{AB \times \text{subjects}}}{df_{AB \times \text{subjects}}}$	
Total	$[XS]-[T]$	$N-1 = npq-1$		

Table 21.14 Summary Table of Computations for Example 21.5

Source of variation	SS	df	MS	F
Between-subjects	421.83−420.5=1.33	3−1=2		
Within-subjects	585−421.83=163.17	3[(2)(3)−1]=15		
A	505−420.5=84.5	2−1=1	$MS_A = \frac{84.5}{1} = 84.5$	$F_A = \frac{84.5}{2} = 42.25$
B	472.5−420.5=52	3−1=2	$MS_B = \frac{52}{2} = 26$	$F_B = \frac{26}{.67} = 38.81$
AB	573−505−472.5+420.5=16	(2−1)(3−1)=2	$MS_{AB} = \frac{16}{2} = 8$	$F_{AB} = \frac{8}{1} = 8$
$A \times$subjects	510.33−505−421.83 +420.5=4	(2−1)(3−1)=2	$MS_{A \times \text{subjects}} = \frac{4}{2} = 2$	
$B \times$subjects	476.5−472.5−421.83 +420.5=2.67	(3−1)(3−1)=4	$MS_{B \times \text{subjects}} = \frac{2.67}{4} = .67$	
$AB \times$subjects	585−573−510.33−476.5+505 +472.5+421.83−420.5=4	(2−1)(3−1)(3−1)=4	$MS_{AB \times \text{subjects}} = \frac{4}{4} = 1$	
Total	585−420.5=164.5	18−1=(3)(2)(3)−1=17		

The identical null and alternative hypotheses that are evaluated in Section III of the **between-subjects factorial analysis of variance** are evaluated in the **within-subjects factorial analysis of variance**. In order to reject the null hypothesis in reference to a computed F ratio, the obtained F value must be equal to or greater than the tabled critical value at the prespecified level of significance. Since the computed value $F_A = 42.25$ is greater than $F_{.05} = 18.51$, the alternative hypothesis for Factor A is supported, but only at the .05 level. Since the computed value $F_B = 38.81$ is greater than $F_{.05} = 6.94$ and $F_{.01} = 18.00$, the alternative hypothesis for Factor B is supported at both the .05 and .01 levels. Since the computed value $F_{AB} = 8$ is greater than $F_{.05} = 6.94$, the alternative hypothesis for an interaction between Factors A and B is supported, but only at the .05 level.[28]

The analysis of the data for Example 21.5 allows the researcher to conclude that both humidity (Factor A) and temperature (Factor B) have a significant impact on problem solving scores. However, as is the case when the same set of data is evaluated with a **between-subjects factorial analysis of variance**, the relationships depicted by the main effects must be qualified because of the presence of a significant interaction. Although the comparison procedures following the computation of the omnibus F ratios (as well as the other analytical procedures for determining power, effect size, etc.) described in Section VI of the **between-subjects factorial analysis of variance** can be extended to the **within-subjects factorial analysis of variance**, they will not be described in this book. For a full description of such procedures, the reader should consult texts that discuss analysis of variance procedures in greater detail (e.g., Keppel (1991) and Winer *et al.* (1991)).

References

Cohen, J. (1988). **Statistical power analysis for the behavioral sciences** (2nd ed.). New York: Academic Press.

Honeck, R. P., Kibler, C. T., and Sugar, J. (1983). **Experimental design and analysis: A systematic approach.** Lanham, MD: University Press of America.

Howell, D. C. (1992). **Statistical methods for psychology** (3rd ed.). Boston: PWS–Kent Publishing Company.

Keppel, G. (1991). **Design and analysis: A researcher's handbook** (3rd ed.). Englewood Cliffs, NJ: Prentice Hall.

Keppel, G., Saufley, W. H., and Tokunaga, H. (1992). **Introduction to design and analysis: A student's handbook** (2nd ed.). New York: W. H. Freeman and Company.

Kirk, R. E. (1982). **Experimental design: Procedures for the behavioral sciences** (2nd ed.). Belmont, CA: Brooks/Cole Publishing Company.

Maxwell, S. E. and Delaney, H. D. (1990). **Designing experiments and analyzing data.** Belmont, CA: Wadsworth Publishing Company.

Myers, J. L. and Well, A. D. (1991). **Research design and statistical analysis.** New York: Harper Collins.

Winer, B. J., Brown, D. R., and Michels, K. M. (1991). **Statistical principles in experimental design** (3rd ed.). New York: McGraw–Hill Publishing Company.

Endnotes

1. A **main effect** refers to the effect of one independent variable on the dependent variable, while ignoring the effect any of the other independent variables have on the dependent variable.

2. Although it is possible to conduct a directional analysis, such an analysis will not be described with respect to a factorial analysis of variance. A discussion of a directional

analysis when an independent variable is comprised of two levels can be found under the **t test for two independent samples.** In addition, a discussion of one-tailed F values can be found in Section VI of the latter test under the discussion of the **Hartley's F_{max} test for homogeneity of variance/F test for two population variances.** A discussion of the evaluation of a directional alternative hypothesis when there are two or more groups can be found in Section VII of the **chi-square goodness-of-fit test (Test 5).** Although the latter discussion is in reference to analysis of a k independent samples design involving categorical data, the general principles regarding the analysis of a directional alternative hypothesis are applicable to the analysis of variance.

3. The notational system employed for the factorial analysis of variance procedures described in this chapter is based on Keppel (1991).

4. The value $SS_{WG} = 12$ can also be computed employing the following equation:

$$SS_{WG} = \sum_{k=1}^{q} \sum_{j=1}^{p} \left[\Sigma X_{AB_{jk}}^2 - \frac{(\Sigma X_{AB_{jk}})^2}{n_{AB_{jk}}} \right]$$

$$= \left[302 - \frac{(30)^2}{3}\right] + \left[29 - \frac{(9)^2}{3}\right] + \left[149 - \frac{(21)^2}{3}\right] + \left[50 - \frac{(12)^2}{3}\right] + \left[50 - \frac{(12)^2}{3}\right] + \left[5 - \frac{(3)^2}{3}\right]$$

$$= 2 + 2 + 2 + 2 + 2 + 2 = 12$$

Note that in the above equation a within-groups sum of squares is computed for each of the $pq = 6$ groups, and $SS_{WG} = 12$ represents the sum of the six sum of squares values.

5. This averaging protocol only applies when there is an equal number of subjects in the groups represented in the specific row or column for which an average is computed.

6. If the factor represented on the abscissa is comprised of two levels (as is the case in Figure 21.1a), when no interaction is present the lines representing the different levels of the second factor will be parallel to one another by virtue of being equidistant from one another. When the abscissa factor is comprised of more than two factors, the lines can be equidistant but not parallel when no interaction is present.

7. As noted earlier, the fact that the lines are parallel to one another is not a requirement if no interaction is present when the abscissa factor is comprised of three or more levels.

8. If no interaction is present, such comparisons should yield results that are consistent with those obtained when the means of the levels of that factor are contrasted.

9. As noted in Section VI of the **single-factor between-subjects analysis of variance**, a **linear contrast** is equivalent to **multiple t tests/Fisher's LSD test.**

10. Many researchers would elect to employ a comparison procedure that is less conservative than the **Scheffé test**, and thus would not require as large a value as CD_S in order to reject the null hypothesis.

11. The number of pairwise comparisons is $[k(k - 1)]/2 = [6(6 - 1)]/2 = 15$, where $k = pq = (2)(3) = 6$ represents the number of groups.

12. If **Tukey's HSD test** is employed to contrast pairs or sets of marginal means for Factors A and B, the values $q_{(A, df_{WG})}$ and $q_{(B, df_{WG})}$ are respectively employed from **Table A13**. The sample sizes used in Equation 21.45 for Factors A and B are respectively nq and np.

13. When there are only two levels involved in analyzing the simple effects of a factor (as is the case with Factor A), the procedure to be described in this section will yield an F value for a simple effect that is equivalent to the F_{comp} value that can be computed by comparing the two groups employing the **linear contrast** procedure described earlier (i.e., the procedure in which Equation 21.40 is employed to compute SS_{comp}).

14. The equation for computing the sum of squares for each of the simple effects of Factor A is noted below.

$$SS_{A \text{ at } B_k} = \sum \left[\frac{(\Sigma X_{AB_{.k}})^2}{n} \right] - \frac{(\Sigma\Sigma X_{AB_{.k}})^2}{np}$$

If $\Sigma X_{AB_{.k}}$ represents the sum of the scores on Level k of Factor B of subjects who serve under a specific level of Factor A, the notation $\Sigma[(\Sigma X_{AB_{.k}})^2/n]$ in the above equation indicates that the sum of the scores for each level of Factor A at a given level of Factor B is squared, divided by n, and the p squared sums are summed. The notation $(\Sigma\Sigma X_{AB_{.k}})^2$ represents the square of the sum of scores of the np subjects who serve under the specified level of Factor B.

15. In the case of the simple effects of Factor A, the modified degrees of freedom value is $df_{WG} = p(n - 1)$.

16. The fact that in the example under discussion the tabled critical values employed for evaluating $F_{\max}$ are extremely large is due to the small value of n. However, under the discussion of homogeneity of variance under the **single-factor between-subjects analysis of variance**, it is noted that Keppel (1991) suggests employing a more conservative test anytime the value of $F_{\max} \geq 3$.

17. The fact that MS_{WG} is an unbiased estimate of σ^2_{WG} can be confirmed by the fact that in the discussion of the homogeneity of variance assumption in the previous section, it is noted that the estimated population variance of each group is $\tilde{s}^2_{AB_{jk}} = 1$. The latter value is equivalent to the value $MS_{WG} = 1$ computed for the factorial analysis of variance.

18. The procedure described in this section assumes there is an equal number of subjects in each group. If the latter is true, it is also the case for Example 21.1 that $\mu_G = (\mu_{A_1} + \mu_{A_2})/2$ and $\mu_G = (\mu_{AB_{11}} + \mu_{AB_{12}} + \mu_{AB_{13}} + \mu_{AB_{21}} + \mu_{AB_{22}} + \mu_{AB_{23}})/6$.

19. Keppel (1991) provides a discussion of an alternative approach for computing the **omega squared** statistic for the **between-subjects factorial analysis of variance.**

20. For a clarification of the use of multiple summation signs, the reader should review Endnote 61 under the **single-factor between-subjects analysis of variance** and Endnote 19 under the **single-factor within-subjects analysis of variance.**

21. The notation X_{ijk} is a simpler form of the notation $X_{i_{AB_{jk}}}$, which is more consistent with the notational format used throughout the discussion of the **between-subjects factorial analysis of variance.**

22. The notation $\sum_{k=1}^{q}\sum_{j=1}^{p}\sum_{i=1}^{n}X_{ijk}$ is an alternative way of writing $\sum X_T$. $\sum_{k=1}^{q}\sum_{j=1}^{p}\sum_{i=1}^{n}X_{ijk}$ indicates that the scores of each of the $n = n_{AB_{jk}}$ subjects in each of the pq groups are summed.

23. Since the interaction sum of squares is comprised of whatever remains of between-groups variability after the contributions of the main effects for Factor A and Factor B have been removed, Equation 21.56 can be derived from the equation noted below which subtracts Equations 21.54 and 21.55 from Equation 21.53.

$$SS_{AB} = n\sum_{k=1}^{q}\sum_{j=1}^{p}(\bar{X}_{AB_{jk}} - \bar{X}_T)^2 - nq\sum_{j=1}^{p}(\bar{X}_{A_j} - \bar{X}_T)^2 - np\sum_{k=1}^{q}(\bar{X}_{B_k} - \bar{X}_T)^2$$

24. The computation of the harmonic mean is described in Section VI of the ***t* test for two independent samples.**

25. Some sources note that the subjects employed in such an experiment (or for that matter any experiment involving independent samples) are nested within the level of the factor to which they are assigned, since each subject serves under only one level of that factor.

26. The computational procedure for the **factorial analysis of variance for a mixed design** assumes that there is an equal number of subjects in each of the levels of the between-subjects factor. When the latter is not true, adjusted equations should be employed which can be found in books that describe the **factorial analysis of variance for a mixed design** in greater detail.

27. There are 12 possible presentation orders involving combinations of the two factors ($p!q! = 3!2! = 12$). The sequences for presentation of the levels of both factors are determined in the following manner: If A_1 is followed by A_2, presentation of the levels of Factor B can be in the six following sequences: 123, 132, 213, 231, 312, 321. If A_2 is followed by A_1, presentation of the levels of Factor B can be in the same six sequences noted previously. Thus, there are a total of 12 possible sequence combinations. Since there are only six subjects in Example 21.5, only six of the 12 possible sequence combinations can be employed.

28. If Factors A and B are both within-subjects factors and a significant effect is present for the main effects and the interaction, the **within-subjects factorial analysis of variance** would be the most likely of the three factorial analysis of variance procedures discussed to yield significant F ratios. The F_A, F_B, and F_{AB} values obtained in Examples 21.1 and 21.4 are significant at both the .05 and .01 levels when the data are respectively evaluated with a **between-subjects factorial analysis of variance** and a **factorial analysis of variance for a mixed design.** However, when Example 21.5 is evaluated with the **within-subjects factorial analysis of variance**, although F_B is significant at both the .05 and .01 levels, F_A and F_{AB} are only significant at the .05 level. This latter result can be attributed to the fact that the data set employed for the three examples is hypothetical, and is not based on the scores of actual subjects who were evaluated within the framework of a within-subjects factorial design. In point of fact, in the case of the **within-subjects factorial analysis of variance**, the lower value for df_{den} employed for

a specific effect (in contrast to the values of df_{den} employed for the **between-subjects factorial analysis of variance** and the **factorial analysis of variance for a mixed design**) will be associated with a tabled critical F value that is larger than the values employed for the latter two tests. Thus, unless there is an actual correlation between subjects' scores under different conditions (which should be the case if a variable is measured within-subjects), the loss of degrees of freedom will nullify the increase in power associated with the **within-subjects factorial analysis of variance** (assuming the data are derived from the appropriate design). The superior power of the **within-subjects factorial analysis of variance** derives from the smaller MS error terms employed in evaluating the main effects and interaction.

Measures of Association/Correlation

Test 22: The Pearson Product-Moment Correlation Coefficient

Test 23: Spearman's Rank-Order Correlation Coefficient

Test 24: Kendall's Tau

Test 25: Kendall's Coefficient of Concordance

Test 26: Goodman and Kruskal's Gamma

Test 22

The Pearson Product-Moment Correlation Coefficient
(Parametric Measure of Association/Correlation Employed with Interval/Ratio Data)

I. Hypothesis Evaluated with Test and Relevant Background Information

The **Pearson product-moment correlation coefficient** is one of a number of measures of correlation or association discussed in this book. Measures of correlation are not inferential statistical tests, but are, instead, descriptive statistical measures which represent the degree of relationship between two or more variables. Upon computing a measure of correlation, it is common practice to employ one or more inferential statistical tests in order to evaluate one or more hypotheses concerning the correlation coefficient. The hypothesis stated below is the most commonly evaluated hypothesis for the **Pearson product-moment correlation coefficient.**

Hypothesis evaluated with test In the underlying population represented by a sample, is the correlation between subjects' scores on two variables some value other than zero? The latter hypothesis can also be stated in the following form: In the underlying population represented by the sample, is there a significant linear relationship between the two variables?

Relevant background information on test Developed by Pearson (1896, 1900), the **Pearson product-moment correlation coefficient** is employed with interval/ratio data to determine the degree to which two variables covary (i.e., vary in relationship to one another). Any measure of correlation/association that assesses the degree of relationship between two variables is referred to as a **bivariate** measure of association. In evaluating the extent to which two variables covary, the **Pearson product-moment correlation coefficient** determines the degree to which a linear relationship exists between the variables. One variable (usually designated as the X variable) is referred to as the **predictor variable**, since if indeed a linear relationship does exist between the two variables, a subject's score on the predictor variable can be used to predict the subject's score on the second variable. The latter variable, which is referred to as the **criterion variable**, is usually designated as the Y variable.[1] The degree of accuracy with which a researcher will be able to predict a subject's score on the criterion variable from the subject's score on the predictor variable will depend upon the strength of the linear relationship between the two variables. The use of correlational data for predictive purposes is summarized under the general subject of **regression analysis** (or more formally, **linear regression analysis**, since when prediction is discussed in reference to the **Pearson product-moment correlation coefficient** it is based on the degree of linear relationship between the two variables). A full discussion of regression analysis can be found in Section VI.

The statistic computed for the **Pearson product-moment correlation coefficient** is represented by the letter r. r is an estimate of ρ (the Greek letter **rho**), which is the correlation between the two variables in the underlying population. r can assume any value within the range of -1 to $+1$ (i.e., $-1 \leq r \leq +1$). Thus, the value of r can never be less than -1 (i.e., r cannot equal -1.2, -50, etc.) or be greater than $+1$ (i.e., r cannot equal 1.2, 50, etc.). The **absolute value** of r (i.e., $|r|$) indicates the **strength** of the relationship between the two variables. As the absolute value of r approaches 1, the degree of linear relationship between the variables becomes stronger, achieving the maximum when $|r| = 1$ (i.e., when r equals either $+1$ or -1). The closer the absolute value of r is to 1, the more accurately a researcher will be able to predict a subject's score on one variable from the subject's score on the other variable. The closer the absolute value of r is to 0, the weaker the linear relationship between the two variables. As the absolute value of r approaches 0, the degree of accuracy with which a researcher can predict a subject's score on one variable from the other variable decreases, until finally, when $r = 0$ there is no predictive relationship between the two variables. To state it another way, when $r = 0$ the use of the correlation coefficient to predict a subject's X score from the subject's Y score (or vice versa) will not be any more accurate than a prediction that is based on some random process (i.e., a prediction that is based purely on chance).

The **sign** of r indicates the nature or **direction** of the linear relationship that exists between the two variables. A positive sign indicates a **direct** linear relationship, whereas a negative sign indicates an **indirect** (or **inverse**) linear relationship. A direct linear relationship is one in which a change on one variable is associated with a change on the other variable in the same direction (i.e., an increase on one variable is associated with an increase on the other variable, and a decrease on one variable is associated with a decrease on the other variable). When there is a direct relationship, subjects who have a high score on one variable will have a high score on the other variable, and subjects who have a low score on one variable will have a low score on the other variable. The closer a positive value of r is to $+1$, the stronger the direct relationship between the two variables; whereas the closer a positive value of r is to 0, the weaker the direct relationship between the variables. Thus, when r is close to $+1$, most subjects who have a high score on one variable will have a comparably high score on the second variable, and most subjects who have a low score on one variable will have a comparably low score on the second variable. As the value of r approaches 0, the consistency of the general pattern described by a positive correlation deteriorates, until finally, when $r = 0$ there will be no consistent pattern that allows one to predict at above chance a subject's score on one variable if one knows the subject's score on the other variable.

An indirect/inverse relationship is one in which a change on one variable is associated with a change on the other variable in the opposite direction (i.e., an increase on one variable is associated with a decrease on the other variable, and a decrease on one variable is associated with an increase on the other variable). When there is an indirect linear relationship, subjects who have a high score on one variable will have a low score on the other variable, and vice versa. The closer a negative value of r is to -1, the stronger the indirect relationship between the two variables, whereas the closer a negative value of r is to 0, the weaker the indirect relationship between the variables. Thus, when r is close to -1, most subjects who have a high score on one variable will have a comparably low score on the second variable (i.e., as extreme a score in the opposite direction), and most subjects who have a low score on one variable will have a comparably high score on the second variable. As the value of r approaches 0, the consistency of the general pattern described by a negative correlation deteriorates, until finally, when $r = 0$ there will be no consistent pattern that allows one to predict at above chance a subject's score on one variable if one knows the subject's score on the other variable.

The use of the **Pearson product-moment correlation coefficient** assumes that a linear function best describes the relationship between the two variables. If, however, the relationship between the variables is better described by a curvilinear function, the value of

r computed for a set of data may not indicate the actual extent of the relationship between the variables. In view of this, when a computed r value is equal to or close to 0, a researcher should always rule out the possibility that the two variables are related curvilinearly. One quick way of assessing the likelihood of the latter is to construct a **scatterplot** of the data. A scatterplot, which is described in Section VI, displays the data for a correlational analysis in a graphical format.

It is important to note that correlation does not imply causation. Consequently, if there is a strong correlation between two variables (i.e., the absolute value of r is close to 1), a researcher is not justified in concluding that one variable causes the other variable. Although it is possible that when a strong correlation exists one variable may, in fact, cause the other variable, the information employed in computing the **Pearson product-moment correlation coefficient** does not allow a researcher to draw such a conclusion. This is the case, since extraneous variables which have not been taken into account by the researcher can be responsible for the observed correlation between the two variables.

The **Pearson product-moment correlation coefficient** is based on the following assumptions: a) The sample of n subjects for which the value r is computed is randomly selected from the population it represents; b) The level of measurement upon which each of the variables is based is interval or ratio; c) The two variables have a **bivariate normal distribution**. The assumption of bivariate normality states that each of the variables and the linear combination of the two variables are normally distributed. With respect to the latter, if every possible pair of data points are plotted on a three-dimensional plane, the resulting surface (which will look like a mountain with a rounded peak) will be a three-dimensional normal distribution (i.e., a three-dimensional structure in which any cross-section is a standard normal distribution). Another characteristic of a bivariate normal distribution is that for any given value of the X variable, the scores on the Y variable will be normally distributed, and for any given value of the Y variable, the scores on the X variable will be normally distributed. In conjunction with the latter, the variances for the Y variable will be equal for each of the possible values of the X variable, and the variances for the X variable will be equal for each of the possible values of the Y variable; d) Related to the bivariate normality assumption is the assumption of **homoscedasticity**. Homoscedasticity exists in a set of data if the relationship between the X and Y variables is of equal strength across the whole range of both variables. Tabachnick and Fidell (1989) note that when the assumption of bivariate normality is met, the two variables will be homoscedastic. The concept of homoscedasticity is discussed in Section VII; and e) Another assumption of the **Pearson product-moment correlation coefficient**, referred to as **nonautoregression**, is discussed in many books on business and economics. This latter assumption, which is discussed within the framework of a special case of correlation referred to as **autocorrelation**, is only likely to be violated when pairs of numbers that are derived from a series of n numbers are correlated with one another. A discussion of autocorrelation can be found in Section VII.

II. Example

Example 22.1 *A psychologist conducts a study employing a sample of five children to determine whether there is a statistical relationship between the number of ounces of sugar a ten-year-old child eats per week (which will represent the X variable) and the number of cavities in a child's mouth (which will represent the Y variable). The two scores (ounces of sugar consumed per week and number of cavities) obtained for each of the five children follow:* **Child 1** (20, 7); **Child 2** (0, 0); **Child 3** (1, 2); **Child 4** (12, 5); **Child 5** (3, 3). *Is there a significant correlation between sugar consumption and the number of cavities?*

III. Null versus Alternative Hypotheses

Upon computing the **Pearson product-moment correlation coefficient**, it is common practice to determine whether the obtained absolute value of the correlation coefficient is large enough to allow a researcher to conclude that the underlying population correlation coefficient between the two variables is some value other than zero. Section V describes how the latter hypothesis, which is stated below, can be evaluated through use of tables of critical r values or through use of an inferential statistical test that is based on the t distribution.[2]

Null hypothesis H_0: $\rho = 0$

(In the underlying population the sample represents, the correlation between the scores of subjects on Variable X and Variable Y equals 0.)

Alternative hypothesis H_1: $\rho \neq 0$

(In the underlying population the sample represents, the correlation between the scores of subjects on Variable X and Variable Y equals some value other than 0. This is a **non-directional alternative hypothesis**, and it is evaluated with a **two-tailed test**. Either a significant positive r value or a significant negative r value will provide support for this alternative hypothesis. In order to be significant, the obtained absolute value of r must be equal to or greater than the tabled critical two-tailed r value at the prespecified level of significance.)

or

H_1: $\rho > 0$

(In the underlying population the sample represents, the correlation between the scores of subjects on Variable X and Variable Y equals some value greater than 0. This is a **directional alternative hypothesis**, and it is evaluated with a **one-tailed test**. Only a significant positive r value will provide support for this alternative hypothesis. In order to be significant (in addition to the requirement of a positive r value), the obtained absolute value of r must be equal to or greater than the tabled critical one-tailed r value at the prespecified level of significance.)

or

H_1: $\rho < 0$

(In the underlying population the sample represents, the correlation between the scores of subjects on Variable X and Variable Y equals some value less than 0. This is a **directional alternative hypothesis**, and it is evaluated with a **one-tailed test**. Only a significant negative r value will provide support for this alternative hypothesis. In order to be significant (in addition to the requirement of a negative r value), the obtained absolute value of r must be equal to or greater than the tabled critical one-tailed r value at the prespecified level of significance.)

Note: Only one of the above noted alternative hypotheses is employed. If the alternative hypothesis the researcher selects is supported, the null hypothesis is rejected.

IV. Test Computations

Table 22.1 summarizes the data for Example 22.1. The following should be noted with respect to Table 22.1: a) The number of subjects is $n = 5$. Each subject has an X score and a Y score, and thus there are five X scores and five Y scores; b) ΣX, ΣX^2, ΣY, and ΣY^2

respectively represent the sum of the five subjects' scores on the X variable, the sum of the five subjects' squared scores on the X variable, the sum of the five subjects' scores on the Y variable, and the sum of the five subjects' squared scores on the Y variable; and c) An XY score is obtained for each subject by multiplying a subject's X score by the subject's Y score. ΣXY represents the sum of the five subjects' XY scores.

Table 22.1 Summary of Data for Example 22.1

Subject	X	X^2	Y	Y^2	XY
1	20	400	7	49	140
2	0	0	0	0	0
3	1	1	2	4	2
4	12	144	5	25	60
5	3	9	3	9	9
	$\Sigma X = 36$	$\Sigma X^2 = 554$	$\Sigma Y = 17$	$\Sigma Y^2 = 87$	$\Sigma XY = 211$

Although they are not required for computing the value of r, the mean score ($\bar{X}$ and $\bar{Y}$) and the estimated population standard deviation ($\tilde{s}_X$ and $\tilde{s}_Y$) for each of the variables are computed (the latter values are computed with Equation I.8). These values are employed in Section VI to derive **regression equations**, which are used to predict a subject's score on one variable from the subject's score on the other variable.

$$\bar{X} = \frac{\Sigma X}{n} = \frac{36}{5} = 7.2 \qquad \bar{Y} = \frac{\Sigma Y}{n} = \frac{17}{5} = 3.4$$

$$\tilde{s}_X = \sqrt{\frac{\Sigma X^2 - \frac{(\Sigma X)^2}{n}}{n - 1}} = \sqrt{\frac{554 - \frac{(36)^2}{5}}{5 - 1}} = 8.58$$

$$\tilde{s}_Y = \sqrt{\frac{\Sigma Y^2 - \frac{(\Sigma Y)^2}{n}}{n - 1}} = \sqrt{\frac{87 - \frac{(17)^2}{5}}{5 - 1}} = 2.70$$

Equation 22.1 (which is identical to Equation 12.7, except for the fact that the notations X and Y are used in place of X_1 and X_2) is employed to compute the value of r.[3]

$$r = \frac{\Sigma XY - \frac{(\Sigma X)(\Sigma Y)}{n}}{\sqrt{\left[\Sigma X^2 - \frac{(\Sigma X)^2}{n}\right]\left[\Sigma Y^2 - \frac{(\Sigma Y)^2}{n}\right]}} \qquad \textbf{(Equation 22.1)}$$

Substituting the appropriate values in Equation 22.1, the value $r = .995$ is computed.

$$r = \frac{211 - \frac{(36)(17)}{5}}{\sqrt{\left[554 - \frac{(36)^2}{5}\right]\left[87 - \frac{(17)^2}{5}\right]}} = .955$$

The numerator of Equation 22.1, which is referred to as the **sum of products** (which is summarized with the notation SP_{XY}), will determine the sign of r. If the numerator is a negative value, r will be a negative number. If the numerator is a positive value, r will be a positive number. If the numerator equals zero, r will equal zero. In the case of Example 22.1, $SP_{XY} = \Sigma XY - [(\Sigma X)(\Sigma Y)/n] = 211 - [(36)(17)/5] = 88.6$. The denominator of Equation 22.1 is the square root of the product of the sum of squares of the X **scores** (which is summarized with the notation SS_X), and the **sum of squares of the** Y **scores** (which is summarized with the notation SS_Y). Thus, $SS_X = \Sigma X^2 - [(\Sigma X)^2/n] = 554 - [(36)^2/5] = 294.8$ and $SS_Y = \Sigma Y^2 - [(\Sigma Y)^2/n] = 87 - [(17)^2/5] = 29.2$. The aforementioned sum of squares values represent the numerator of the equation for computing the estimated population standard deviation of the X and Y scores (i.e., Equation I.8). Employing the notation for the sum of products and the sums of squares, the equation for the **Pearson product-moment correlation coefficient** can be expressed as follows: $r = SP_{XY}/\sqrt{SS_X SS_Y}$.

The reader should take note of the fact that each of the sum of squares values must be a positive number. If either of the sum of squares values is a negative number, it indicates that a computational error has been made. The only time a sum of squares value will equal zero, will be if all of the subjects have the identical score on the variable for which the sum of squares is computed. Anytime one or both of the sum of squares values equals zero, Equation 22.1 will be insoluble. It is noted in Section I that the computed value of r must fall within the range $-1 \leq r \leq +1$. Consequently, if the value of r is less than -1 or greater than $+1$, it indicates that a computational error has been made.

V. Interpretation of the Test Results

The obtained value $r = .995$ is evaluated with **Table A16 (Table of Critical Values for Pearson** r**)** in the **Appendix**. The degrees of freedom employed for evaluating the significance of r are computed with Equation 22.2.

$$df = n - 2 \qquad \textbf{(Equation 22.2)}$$

Employing Equation 22.2, the value $df = 5 - 3 = 2$ is computed. Using **Table A16**, it can be determined that the tabled critical two-tailed r values at the .05 and .01 levels of significance are $r_{.05} = .878$ and $r_{.01} = .959$, and the tabled critical one-tailed r values at the .05 and .01 levels of significance are $r_{.05} = .805$ and $r_{.01} = .934$.

The following guidelines are employed in evaluating the null hypothesis H_0: $\rho = 0$.

a) If the nondirectional alternative hypothesis H_1: $\rho \neq 0$ is employed, the null hypothesis can be rejected if the obtained absolute value of r is equal to or greater than the tabled critical two-tailed value at the prespecified level of significance.

b) If the directional alternative hypothesis H_1: $\rho > 0$ is employed, the null hypothesis can be rejected if the sign of r is positive, and the value of r is equal to or greater than the tabled critical one-tailed value at the prespecified level of significance.

c) If the directional alternative hypothesis H_1: $\rho < 0$ is employed, the null hypothesis can be rejected if the sign of r is negative, and the absolute value of r is equal to or greater than the tabled critical one-tailed value at the prespecified level of significance.

Employing the above guidelines, the nondirectional alternative hypothesis H_1: $\rho \neq 0$ is supported at the .05 level, since the computed value $r = .955$ is greater than the tabled critical two-tailed value $r_{.05} = .878$. It is not, however, supported at the .01 level, since $r = .955$ is less than the tabled critical two-tailed value $r_{.01} = .959$.

The directional alternative hypothesis H_1: $\rho > 0$ is supported at both the .05 and .01 levels, since the computed value $r = .955$ is a positive number that is greater than the tabled critical one-tailed values $r_{.05} = .805$ and $r_{.01} = .934$.

The directional alternative hypothesis H_1: $\rho < 0$ is not supported, since the computed value $r = .955$ is a positive number. In order for the alternative hypothesis H_1: $\rho < 0$ to be supported, the computed value of r must be a negative number (as well as the fact that the absolute value of r must be equal to or greater than the tabled critical one-tailed value at the prespecified level of significance).

It may seem surprising that such a large correlation (i.e., an r value that almost equals 1) is not significant at the .01 level. Inspection of **Table A16** reveals that when the sample size is small (as is the case in Example 22.1), the tabled critical r values are relatively large. The large critical values reflect the fact that the smaller the sample size, the higher likelihood of sampling error resulting in a spuriously inflated correlation. At this point it is worth noting that there are a number of factors which can dramatically influence the value of r, and such factors are much more likely to distort the computed value of a correlation coefficient when the sample size is small. The following are among those factors that can dramatically influence the value of r: a) If the range of scores on either the X or Y variable is restricted, the absolute value of r will be reduced; b) A correlation based on a sample which is characterized by the presence of extreme scores on one or both of the variables (even though the scores are not extreme enough to be considered **outliers,** which are atypically extreme scores) may be spuriously high (i.e., the absolute value of r will be higher than the absolute value of ρ in the underlying population); and c) The presence of one or more **outliers** can grossly distort the absolute value of r, or even affect the sign of r.

Further examination of **Table A16** reveals that as the value of n increases, the tabled critical values at a given level of significance decrease, until finally when n is quite large the tabled critical values are quite low. What this translates into is that when the sample size is extremely large, an absolute r value that is barely above zero will be statistically significant. Keep in mind, however, that the alternative hypothesis that is evaluated only stipulates that the underlying population correlation is some value other than zero. The distinction between statistical versus practical significance (which is discussed in Section VI of the ***t* test for two independent samples (Test 8)**) is germane to this discussion, in that a small correlation may be statistically significant, yet not be of any practical and/or theoretical value. It should be noted, however, that in many instances a significant correlation which is close to zero may be of practical and/or theoretical significance.

Test 22a: Test of significance for a Pearson product-moment correlation coefficient In the event a researcher does not have access to **Table A16**, Equation 22.3, which employs the t distribution, provides an alternative way of evaluating the null hypothesis H_0: $\rho = 0$.

$$t = \frac{r\sqrt{n - 2}}{\sqrt{1 - r^2}} \qquad \textbf{(Equation 22.3)}$$

Substituting the appropriate values in Equation 22.3, the value $t = 5.58$ is computed.

$$t = \frac{.955\sqrt{5 - 2}}{\sqrt{1 - (.955)^2}} = 5.58$$

The computed value $t = 5.58$ is evaluated with **Table A2 (Table of Student's *t* Distribution)** in the **Appendix.** The degrees of freedom employed in evaluating Equation 22.3 are $df = n - 2$. Thus, $df = 5 - 2 = 3$. For $df = 3$, the tabled critical two-tailed .05 and .01 values are $t_{.05} = 3.18$ and $t_{.01} = 5.84$, and the tabled critical one-tailed .05 and .01 values are $t_{.05} = 2.35$ and $t_{.01} = 4.54$. Since the sign of the t value computed with Equation 22.3 will always be the same as the sign of r, the guidelines described earlier in reference to **Table A16** for evaluating an r value can also be applied in evaluating the t value computed with Equation 22.3 (i.e., substitute t in place of r in the text of the guidelines for evaluating r).

Employing the guidelines, the nondirectional alternative hypothesis H_1: $\rho \neq 0$ is supported at the .05 level, since the computed value $t = 5.58$ is greater than the tabled critical two-tailed value $t_{.05} = 3.18$. It is not, however, supported at the .01 level, since $t = 5.58$ is less than the tabled critical two-tailed value $t_{.01} = 5.84$.

The directional alternative hypothesis H_1: $\rho > 0$ is supported at both the .05 and .01 levels, since the computed value $t = 5.58$ is a positive number that is greater than the tabled critical one-tailed values $t_{.05} = 2.35$ and $t_{.01} = 4.54$.

The directional alternative hypothesis H_1: $\rho < 0$ is not supported, since the computed value $t = 5.58$ is a positive number. In order for the alternative hypothesis H_1: $\rho < 0$ to be supported, the computed value of t must be a negative number (as well as the fact that the absolute value of t must be equal to or greater than the tabled critical one-tailed value at the prespecified level of significance).

Note that the results obtained through use of Equation 22.3 are consistent with those that are obtained when **Table A16** is employed.[4] A summary of the analysis of Example 22.1 follows: It can be concluded that there is a significant positive correlation between the number of ounces of sugar a ten-year-old child eats and the number of cavities in a child's mouth. This result can be summarized as follows (if it is assumed the nondirectional alternative hypothesis H_1: $\rho \neq 0$ is employed): $r = .955$, $p < .05$.

The coefficient of determination The square of a computed r value (i.e., r^2) is referred to as the **coefficient of determination.** r^2 represents the proportion of variance on one variable that can be accounted for by variance on the other variable.[5] The use of the term "accounted for" in the previous sentence should not be interpreted as indicating that a cause–effect relationship exists between the two variables. As noted in Section I, a substantial correlation between two variables does not allow one to conclude that one variable causes the other.

For Example 22.1 the coefficient of determination is computed to be $r^2 = (.955)^2 = .912$, which expressed as a percentage is 91.2%.[6] This indicates that 91.2% of the variation on the X variable can be accounted for on the basis of variability on the Y variable (or vice versa). Although it is possible that X causes Y (or that Y causes X), it is also possible that one or more extraneous variables which are related to X and/or Y, which have not been taken into account in the analysis, are the real reason for the strong relationship between the two variables. In order to demonstrate that the amount of sugar a child eats is the direct cause of the number of cavities he or she develops, a researcher would be required to conduct an experiment in which the amount of sugar consumed is a manipulated independent variable, and the number of cavities is the dependent variable. As noted in the **Introduction** of the book, an experiment in which the independent variable is directly manipulated by the researcher is often referred to as a **true experiment.** If a researcher conducts a true experiment to evaluate the relationship between the amount of sugar eaten and the number of cavities, such a study would require randomly assigning a representative sample of young children to two or more groups. By virtue of random assignment, it would be assumed that the resulting groups are comparable to one another. Each of the groups would be differentiated from one another on the basis of the amount of sugar the children within a group consume. Since the independent variable is manipulated, the amount of sugar consumed by each group is under the direct control of the experimenter. Any observed differences on the dependent variable between the groups at some later point in time could be attributed to the manipulated independent variable. Thus, if, in fact, significant group differences with respect to the number of cavities are observed, the researcher would have a reasonable basis for concluding that sugar consumption is responsible for such differences.

Whereas the correlational study represented by Example 22.1 is not able to control for potentially confounding variables, the true experiment described above is able control for such variables. Common sense suggests, however, that practical and ethical considerations would make it all but impossible to conduct the sort of experiment described above. Realistically, in a democratic society a researcher cannot force a parent to feed her child a specified amount of sugar if the parent is not naturally inclined to do so. Even if a researcher discovers that through the use of monetary incentives she can persuade some parents to feed their children different amounts of sugar than they deem prudent, the latter sort of inducement would most likely compromise a researcher's ability to randomly assign subjects to groups, not to mention the fact that it would be viewed as unethical by many people. Consequently, if a researcher is inclined to conduct a study evaluating the relationship between sugar consumption and the number of cavities, it is highly unlikely that sugar consumption would be employed as a manipulated independent variable. In order to assess what, if any, relationship there is between the two variables, it is much more likely that a researcher would solicit parents whose children ate large versus moderate versus small amounts of sugar, and use the latter as a basis for defining her groups. In such a study, the amount of sugar consumed would be a nonmanipulated independent variable (since it represents a preexisting subject characteristic). The information derived from this type of study (which is commonly referred to as an **ex post facto study** or a **natural experiment**) is correlational in nature. This is the case, since in any study in which the independent variable is not manipulated by the experimenter, one is not able to effectively control for the influence of potentially confounding variables. Thus, if, in fact, differences are observed between two or more groups in an ex post facto study, although such differences may be due to the independent variable, they can also be due to extraneous variables. Consequently, in the case of the example under discussion, any observed differences in the number of cavities between two or more groups can be due to extraneous factors such as maternal prenatal health care, different home environments, dietary elements other than sugar, socioeconomic and/or educational differences between the families that comprise the different groups, etc.

VI. Additional Analytical Procedures for the Pearson Product-Moment Correlation Coefficient and/or Related Tests

1. Derivation of a regression line The obtained value $r = .955$ suggests a strong direct relationship between sugar consumption (X) and the number of cavities (Y). The high positive value of the correlation coefficient suggests that as the number of ounces of sugar consumed increases, there is a corresponding increase in the number of cavities. This is confirmed in Figure 22.1 which is a **scatterplot** of the data for Example 22.1. A scatterplot depicts the data employed in a correlational analysis in a graphical format. Each subject's two scores are represented by a single point on the scatterplot. The point which depicts a subject's two scores is arrived at by moving horizontally on the abscissa (X-axis) the number of units that corresponds to the subject's X score, and moving vertically on the ordinate (Y-axis) the number of units that corresponds to the subject's Y score.

Employing the scatterplot, one can visually estimate the straight line that comes closest to passing through all of the data points. This line is referred to as the **regression line** (also known as the **line of best fit**). In actuality, there are two regression lines. The line which is most commonly determined is **the regression line of Y on X**. The latter line is employed to predict a subject's Y score (which represents the criterion variable) by employing the subject's X score (which represents the predictor variable). The second regression line, **the regression line of X on Y**, allows one to predict a subject's X score by employing the subject's Y score. As will be noted later in this discussion, the only time the two regression lines will be identical is when the absolute value of r equals 1. Because X is usually designated as the predictor variable and Y as the criterion variable, the **regression line of Y on X** is the more commonly determined of the two regression lines.

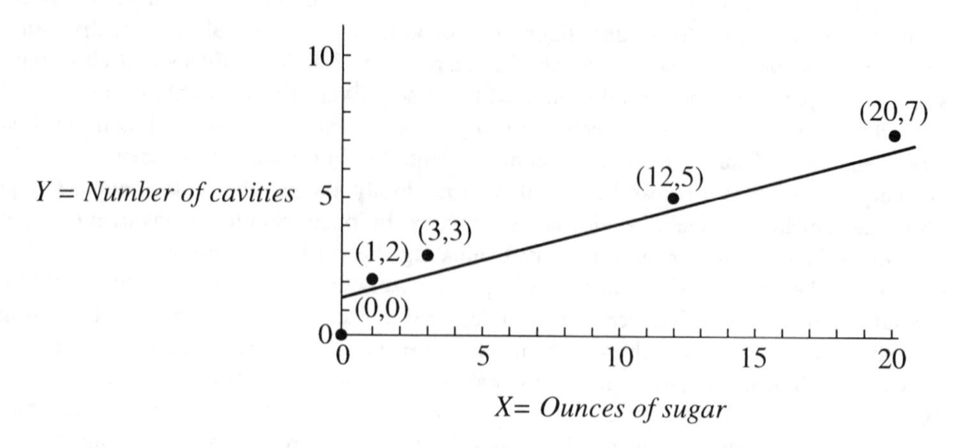

Figure 22.1 Scatterplot for Example 22.1

The **regression line of *Y* on *X*** (which, along with the **regression line of *X* on Y**, is determined mathematically later in this section) has been inserted in Figure 22.1. Note that the line is positively sloped — i.e., the lowest part of the line is on lower left of the graph with the line slanting upward to the right. A line that is positively sloped reflects the fact that a change on one variable in a specific direction is accompanied by a change in the other variable in the same direction. A positive correlation will always result in a positively sloped regression line. A negative correlation, on the other hand, will always result in a negatively sloped regression line. In a negatively sloped regression line, the upper part of the line is at the left of the graph and the line slants downward as one moves to the right. A line that is negatively sloped reflects the fact that a change on one variable in a specific direction is accompanied by a change in the other variable in the opposite direction.

Whereas the slope of the regression line indicates whether a computed r value is a positive or negative number, the magnitude of the absolute value of r reflects how close the n data points fall in relation to the regression line. When $r = +1$ or $r = -1$, all of the data points fall on the regression line. As the absolute value of r deviates from 1 and moves toward 0, the data points deviate further and further from the regression line. Figure 22.2 depicts a variety of hypothetical regression lines, which are presented to illustrate the relationship between the sign and absolute value of r and the regression line.

In Figure 22.2 the regression lines (a), (b), (c), and (d) are positively sloped, and are thus associated with a positive correlation. Lines (e), (f), (g) and (h), on the other hand, are negatively sloped, and are associated with a negative correlation. Note that in each graph, the closer the data points are to the regression line, the closer the absolute value of r is to one. Thus, in graphs (a)–(h), the strength of the correlation (i.e., maximum, strong, moderate, weak) is a function of how close the data points are to the regression line.

The use of the terms strong, moderate, and weak in relation to specific values of correlation coefficients is somewhat arbitrary. For the purpose of discussion the following rough guidelines will be employed for designating the strength of a correlation coefficient: a) If $|r| \geq .7$, a correlation is considered to be strong; b) If $.3 \leq |r| < .7$, a correlation is considered to be moderate; and c) If $|r| < .3$, a correlation is considered to be weak. In point of fact, most statistically significant correlations in the scientific literature are in the weak to moderate range. As noted earlier, although such correlations are not always of practical and/or theoretical importance, there are many instances where they are.

Graphs (i) and (j) in Figure 22.2 depict data which result in a correlation of zero, since in both instances the distribution of data points is random, and consequently a straight line

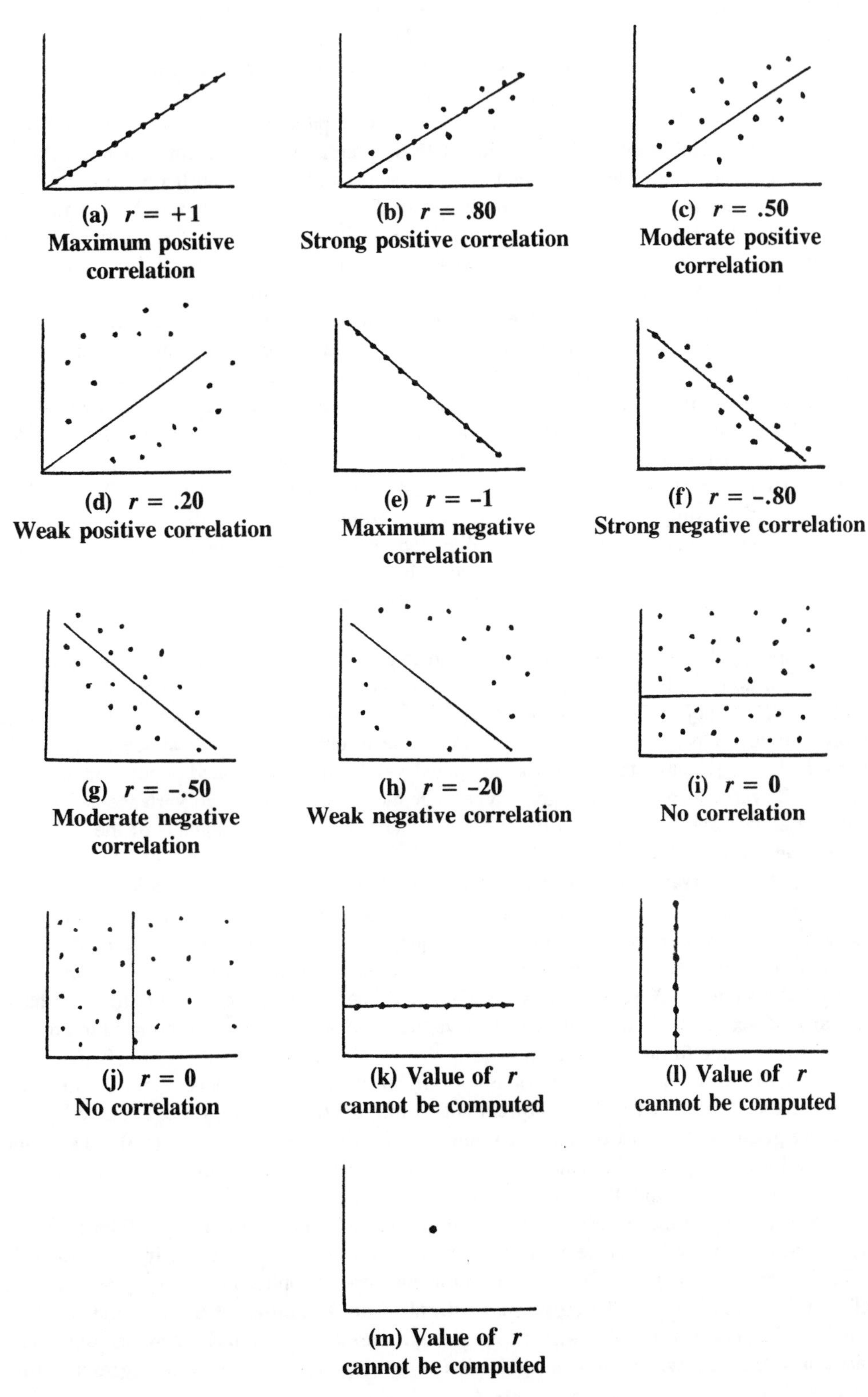

Figure 22.2 Hypothetical Regression Lines

cannot be used to describe the relationship between the two variables with any degree of accuracy. Whenever the **Pearson-moment correlation coefficient** equals zero, the regression line will be parallel to either the X-axis (as in Graph (i)) or the Y-axis (as in Graph (j)), depending upon which regression line is drawn.

Two other instances in which the regression line is parallel to the X-axis or the Y-axis are depicted in Graphs (k) and (l). Both of these graphs depict data for which a value of r cannot be computed. The data depicted in graphs (k) and (l) illustrate that in order to compute a coefficient of correlation, there must be variability on both the X and the Y variables. Specifically, in Graph (k) the regression line is parallel to the X-axis. The configuration of the data upon which this graph is based indicates that although there is variability with respect to subjects' scores on the X variable, there is no variability with respect to their scores on the Y variable — i.e., all of the subjects obtain the identical score on the Y variable. As a result of the latter, the computed value for the estimated population variance for the Y variable will equal zero. When, in fact, the value of the variance for the Y variable equals zero, it is directly related to the fact that the sum of squares of the Y scores equals zero (i.e., $SS_Y = \Sigma Y^2 - [(\Sigma Y)^2/n] = 0$). The sum of squares of the Y scores (which is the numerator of the equation for computing the estimated population variance of the Y scores) is, as is noted earlier, one of the elements that comprises the denominator of Equation 22.1. Consequently, if the sum of squares of the Y scores equals zero, the latter equation becomes insoluble. Note that if the regression line depicted in Graph (k) is employed to predict a subject's Y score from the subject's X score, all subjects are predicted to have the same score. If, in fact, all subjects have the same Y score, there is no need to employ the regression line to make a prediction.

Graph (l) illustrates a regression line that is parallel to the Y-axis. The configuration of the data upon which the latter graph is based indicates that although there is variability with respect to subjects' scores on the Y variable, there is no variability with respect to their scores on the X variable — i.e., all of the subjects obtain the identical score on the X variable. As a result of the latter, the computed value for the estimated population variance for the X variable will equal zero. When, in fact, the value of the variance for the X variable equals zero, it is directly related to the fact that the sum of squares of the X scores equals zero (i.e., $SS_X = \Sigma X^2 - [(\Sigma X)^2/n] = 0$). The sum of squares of the X scores (which is the numerator of the equation for computing the estimated population variance of the X scores) is, as is noted earlier, one of the elements that comprises the denominator of Equation 22.1. Consequently, if the sum of squares of the X scores equals zero, the latter equation becomes insoluble. Note that if the regression line depicted in Graph (l) is employed to predict a subject's X score from the subject's Y score, all subjects are predicted to have the same X score. If, in fact, all subjects have the same X score, there is no need to employ the regression line to make a prediction.

If both the X and Y variable have no variability (i.e., all subjects obtain the identical score on the X variable, and all subjects obtain the identical score on the Y variable), the resulting graph will consist of a single point (which is the case for Graph (m)). Thus, the single point in Graph (m) indicates that each of the n subjects in a sample obtains identical scores on both the X and Y variables.

At this point in the discussion, the role of the slope of a regression line will be clarified. The slope of a line indicates the number of units the Y variable will change if the X variable is incremented by one unit. This definition for the slope is applicable to the **regression line of *Y* on *X***. The slope of the **regression line of *X* on *Y***, on the other hand, indicates the number of units the X variable will change if the Y variable is incremented by one unit. The discussion to follow will employ the definition of the slope in reference to the **regression line of *Y* on *X***.

A line with a large positive slope or large negative slope is inclined in an upward direction away from the X-axis — i.e., like a hill with a high grade. The more the magnitude of the positive slope increases, the more the line approaches being parallel to the Y-axis. A line with a small positive slope or small negative slope has a minimal inclination in relation to the X-axis — i.e., like a hill with a low grade. The smaller the slope of a line, the more the line approaches being parallel to the X-axis. The graphs in Figure 22.3 reflect the following degrees of slope: Graphs (a) and (b) respectively depict lines with a large positive slope and a large negative slope; Graphs (c) and (d) respectively depict lines with a moderate positive slope and a moderate negative slope (i.e., the severity of the angle in relation to the X-axis is in between that of a line with a large slope and a small slope); Graphs (e) and (f) respectively depict lines with a small positive slope and a small negative slope.

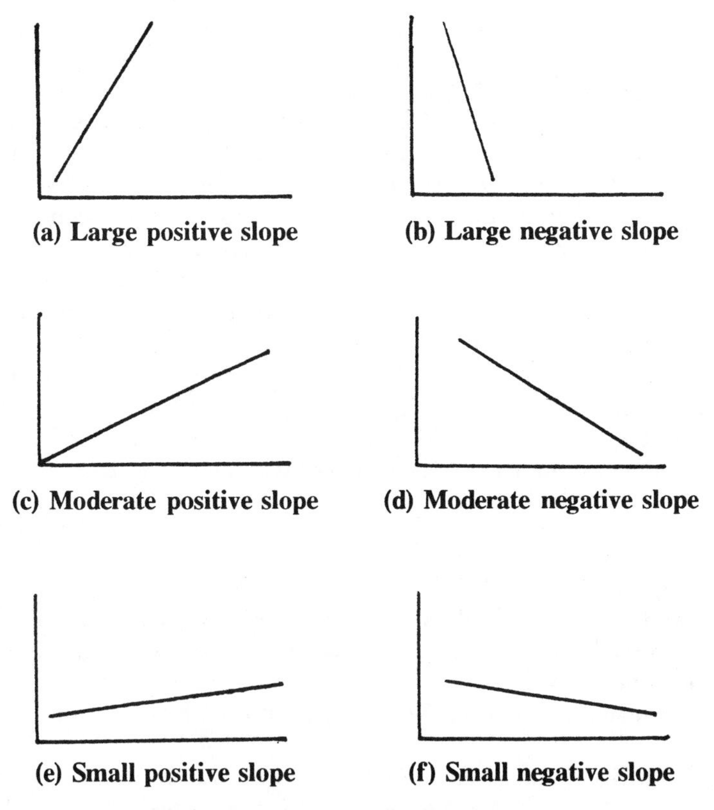

Figure 22.3 Hypothetical Regression Lines

It is important to keep in mind that although the slope of the **regression line of Y on X** plays a role in determining the specific value of Y that is predicted from the value of X, the magnitude of the slope is not related to the magnitude of the absolute value of the coefficient of correlation. A regression line with a large slope can be associated with a correlation coefficient that has a large, moderate, or small absolute value. In the same respect, a regression line with a small slope can be associated with a correlation coefficient that has a large, moderate, or small absolute value. Thus, the accuracy of a prediction is not a function of the slope of the regression line. Instead, it is a function of how far removed the data points are from the regression line. To illustrate this point, let us assume that a regression line which has a large positive slope (such as Graph (a) in Figure 22.3) is being used to predict Y scores for a set of X scores that are one unit apart from one another. As the magnitude of an X score increases by one unit, there is a sizeable increase in the Y score predicted for each subsequent value of X. In the opposite respect, if the regression line has a small positive slope (such as Graph (e) in Figure 22.3), as the magnitude of an X score increases by one unit, there is a minimal increase in the Y score predicted for each subsequent value of X. It is important to note, however, that in both of the aforementioned examples, regardless of whether the slope of the regression line is large or small, the accuracy of the predicted Y scores will not be affected by the magnitude of the slope of the line. Consequently, for any of the regression lines depicted in Figure 22.3, the n data points can fall on, close to, or be far removed from the regression line.

Mathematical derivation of the regression line[7] The most accurate way to determine the regression line is to compute, through use of a procedure referred to as the **method of least squares**, the equation of the straight line that comes closest to passing through all of the data points. As noted earlier, in actuality there are two regression lines — **the regression line of Y on X** (which is employed to predict a subject's Y score by employing the subject's X score), and **the regression line of X on Y** (which is employed to predict a subject's X score by employing the subject's Y score). The equations for the two regression lines will always be different, except when the absolute value of r equals 1. When $|r| = 1$, the two regression lines are identical (both visually and algebraically). The reason why the two regression lines are always different (except when $|r| = 1$) is because the **regression line of Y on X** is based on the equation that results in the **minimum squared distance** of all the data points from the line, when the distance of the points from the line is measured **vertically** (i.e., ↑ or ↓). On the other hand, the **regression line of X on Y** is based on the minimum squared distance of the data points from the line, when the distance of the points from the line is measured **horizontally** (i.e., → or ←).[8] Since when $|r| = 1$ all the points fall on the regression line, both the vertical and horizontal squared distance for each data point equals zero. Consequently, when $|r| = 1$ the two regression lines are identical.

The **regression line of Y on X** is determined with Equation 22.4.

$$Y' = a_Y + b_Y X \qquad \textbf{(Equation 22.4)}$$

Where: Y' represents the predicted Y score for a subject
X represents the subject's X score that is used to predict the value Y'
a_Y represents the Y intercept, which is the point at which the regression line crosses the Y-axis
b_Y represents the slope of the **regression line of Y on X**

In order to derive Equation 22.4, the values of b_Y and a_Y must be computed. Either Equation 22.5 or Equation 22.6 can be employed to compute the value b_Y. The latter equations are employed below to compute the value $b_Y = .30$.

$$b_Y = \frac{SP_{XY}}{SS_X} = \frac{\Sigma XY - \frac{(\Sigma X)(\Sigma Y)}{n}}{\Sigma X^2 - \frac{(\Sigma X)^2}{n}} = \frac{211 - \frac{(36)(17)}{5}}{554 - \frac{(36)^2}{5}} = .30 \qquad \textbf{(Equation 22.5)}$$

$$b_Y = r\left(\frac{\tilde{s}_Y}{\tilde{s}_X}\right) = (.955)\left(\frac{2.70}{8.58}\right) = .30 \qquad \textbf{(Equation 22.6)}$$

Equation 22.7 is employed to compute the value a_Y. The latter equation is employed below to compute the value $a_Y = 1.24$.

$$a_Y = \bar{Y} - b_Y\bar{X} = 3.4 - (.30)(7.2) = 1.24 \qquad \textbf{(Equation 22.7)}$$

Substituting the values $a_Y = 1.24$ and $b_Y = .30$ in Equation 22.4, we determine that the equation for **regression line of *Y* on *X*** is $Y' = 1.24 + .3X$. Since two points can be used to construct a straight line, we can select two values for X and substitute each value in Equation 22.4, and solve for the values that would be predicted for Y'. Each set of values that is comprised of an X score and the resulting Y' value will represent one point on the regression line. Thus, if we plot any two points derived in this manner and connect them, the resulting line is the **regression line of *Y* on *X***. To demonstrate this, if the value $X = 0$ is substituted in the regression equation, it yields the value $Y' = 1.24$ (which equals the value of a_Y): $Y' = 1.24 - (.30)(0) = 1.24$. Thus, the first point that will be employed in constructing the regression line is (0, 1.24). If we next substitute the value $X = 5$ in the regression equation, it yields the value $Y' = 2.74$: $Y' = 1.24 + (.30)(5) = 2.74$. Thus, the second point to be used in constructing the regression line is (5, 2.74). The regression line that results from connecting the points (0, 1.24) and (5, 2.74) is displayed in Figure 22.4.

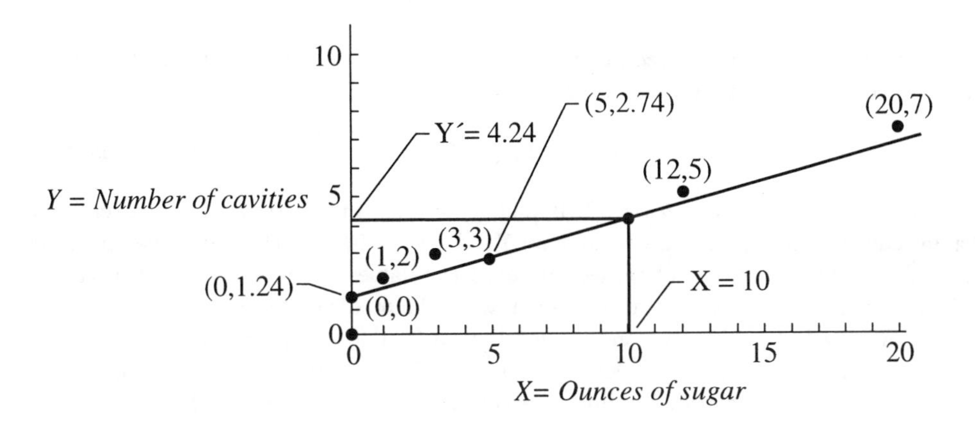

Figure 22.4 Regression Line of *Y* on *X* for Example 22.1

If the researcher wants to predict a subject's score on the Y variable by employing the subject's score on the X variable, the predicted value Y' can be derived either from the regression equation or from Figure 22.4. If the regression equation is employed, the value Y' is derived by substituting a subject's X score in the equation (which is the same procedure that is employed to determine the two points that are used to construct the regression line). Thus, if a child consumes ten ounces of sugar per week, employing $X = 10$ in the regression equation, the predicted number of cavities for the child is $Y' = 1.24 + (.30)(10) = 4.24$. In using Figure 22.4 to predict the value of Y', we identify the point on the X-axis which corresponds to the subject's score on the X variable. A perpendicular line is erected from that point until it intersects the regression line. At the point the perpendicular line intersects the regression line, a second perpendicular line is dropped to the Y-axis. The point at which the latter perpendicular line intersects the Y-axis corresponds to the predicted value Y'. This procedure, which is illustrated in Figure 22.4, yields the same value $Y' = 4.24$, which is obtained when the regression equation is employed.

The **regression line of *X* on *Y*** is determined with Equation 22.8.

$$X' = a_X + b_X Y \qquad \textbf{(Equation 22.8)}$$

Where: X' represents the predicted X score for a subject
Y represents the subject's Y score which is used to predict the value X'
a_X represents the X intercept, which is the point at which the regression line crosses the X-axis
b_X represents the slope of the **regression line of *X* on *Y***

In order to derive Equation 22.8, the values of b_X and a_X must be computed. Either Equation 22.9 or Equation 22.10 can be employed to compute the value b_X. The latter equations are employed below to compute the value $b_X = 3.03$.

$$b_X = \frac{SP_{XY}}{SS_Y} = \frac{\Sigma XY - \frac{(\Sigma X)(\Sigma Y)}{n}}{\Sigma Y^2 - \frac{(\Sigma Y)^2}{n}} = \frac{211 - \frac{(36)(17)}{5}}{87 - \frac{(17)^2}{5}} = 3.03 \qquad \textbf{(Equation 22.9)}$$

$$b_X = r\left(\frac{\tilde{s}_X}{\tilde{s}_Y}\right) = (.955)\left(\frac{8.58}{2.70}\right) = 3.03 \qquad \textbf{(Equation 22.10)}$$

Equation 22.11 is employed to compute the value a_X. The latter equation is employed below to compute the value $a_X = -3.10$.

$$a_X = \bar{X} - b_X \bar{Y} = 7.2 - (3.03)(3.4) = -3.10 \qquad \textbf{(Equation 22.11)}$$

Substituting the values $a_X = -3.10$ and $b_X = 3.03$ in Equation 22.8, we determine that the equation for **regression line of *X* on *Y*** is $X' = -3.10 + 3.03Y$. Since two points can be used to construct a straight line, we can select two values for Y and substitute each value in Equation 22.8, and solve for the values that would be predicted for X'. Each set of values that is comprised of an Y score and the resulting X' value will represent one point on the regression line. Thus, if we plot any two points derived in this manner and connect them, the resulting line is the **regression line of *X* on *Y*.** To demonstrate this, if the value $Y = 0$ is substituted in the regression equation, it yields the value $X' = -3.10$ (which equals the value of a_X): $X' = -3.10 + (3.03)(0) = -3.10$. Thus, the first point that will be employed in constructing the regression line is (−3.10, 0). If we next substitute the value $Y = 5$ in the regression equation, it yields the value $X' = 12.05$: $X' = -3.10 + (3.03)(5) = 12.05$. Thus, the second point to be used in constructing the regression line is (12.05, 5). The regression line that results from connecting the points (−3.10, 0) and (12.05, 5) is displayed in Figure 22.5. Note that since the value $Y = 0$ results in a negative X value, the X-axis in Figure 22.5 must be extended to the left of the origin in order to accommodate the value $X = -3.10$.

If the researcher wants to predict a subject's score on the X variable by employing the subject's score on the Y variable, the predicted value X' can be derived either from the regression equation or from Figure 22.5. If the regression equation is employed, the value X' is derived by substituting a subject's Y score in the equation (which is the same procedure that is employed to determine the two points that are used to construct the regression line). Thus, if a child has four cavities, employing $Y = 4$ in the regression equation, the predicted number of ounces of sugar the child eats per week is $X' = -3.10 + (3.03)(4) = 9.02$. In using Figure 22.5 to predict the value of X', we identify the point on the Y-axis which corresponds to the subject's score on the Y variable. A perpendicular line is erected

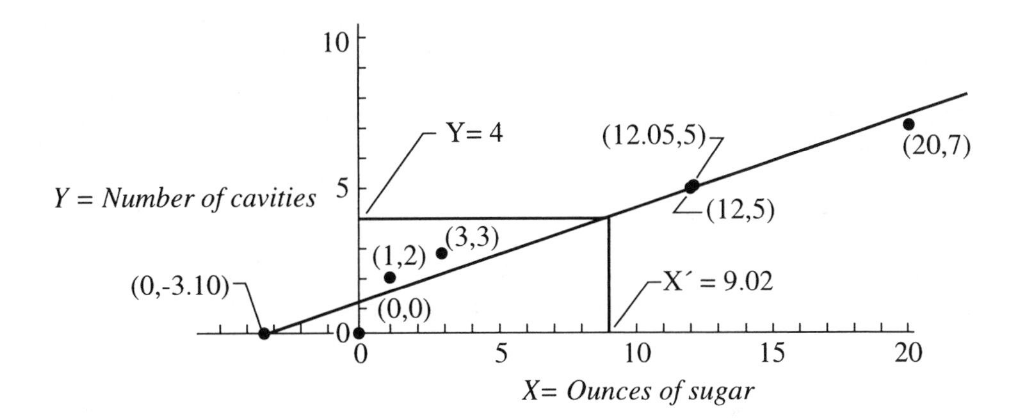

Figure 22.5 Regression line of *X* on *Y* for Example 22.1

from that point until it intersects the regression line. At the point the perpendicular line intersects the regression line, a second perpendicular line is dropped to the X-axis. The point at which the latter perpendicular line intersects the X-axis corresponds to the predicted value X'. This procedure, which is illustrated in Figure 22.5, yields the same value $X' = 9.02$, which is obtained when the regression equation is employed.

The protocol described in this section for deriving a regression equation does not provide any information regarding the accuracy of prediction that will result from such an equation. The **standard error of estimate**, which is discussed in the next section, is used as an index of accuracy in regression analysis. The standard error of estimate is a function of a set of n deviation scores that are referred to as **residuals.** A **residual** is the difference between the predicted value of the criterion variable for a subject (i.e., Y' or X'), and a subject's actual score on the criterion variable (i.e., Y or X). A discussion of the role of residuals in regression analysis can be found in Section VII.

In closing the discussion of the derivation of a regression line, it is important to emphasize that in some instances where the value of r is equal to or close to zero, there may actually be a curvilinear relationship between the two variables. When the absolute value of r is such that there is a weak to moderate relationship between the variables, if, in fact, a curvilinear function best describes the relationship between the variables, it will provide a more accurate basis for prediction than will the straight line derived through use of the method of least squares. One advantage of constructing a scatterplot is that it allows a researcher to visually assess whether or not a curvilinear function is more appropriate than a straight line in describing the relationship between the variables. If the latter is true, the researcher should derive the equation for the appropriate curve. Although the derivation of equations for curvilinear functions will not be described in this book, it is discussed in many books on correlation and regression.

2. The standard error of estimate The **standard error of estimate** is a standard deviation of the distribution of error scores employed in regression analysis. More specifically, it is an index of the difference between the predicted versus the actual value of the criterion variable. The standard error of estimate for the **regression line of *Y* on *X*** (which is represented by the notation $s_{Y.X}$) represents the standard deviation of the values of Y for a specific value of X. The standard error of estimate for the **regression line of *X* on *Y*** (which is represented by the notation $s_{X.Y}$) represents the standard deviation of the values of

X for a specific value of Y. Thus, in Example 22.1, $s_{Y.X}$ represents the standard deviation for the number of cavities of any subject whose weekly sugar consumption is equal to a specific number of ounces. $s_{X.Y}$, on the other hand, represents the standard deviation for the number of ounces of sugar consumed by any subject who has a specific number of cavities.

The standard error of estimate can be employed to compute a confidence interval for the predicted value of Y (or X). The larger the value of a standard error of estimate, the larger will be the range of values that define the confidence interval, and consequently, the less likely it is that the predicted value Y' (or X') will equal or be close to the actual score of a given subject on that variable.

Equations 22.12 and 22.13 are respectively employed to compute the values $s_{Y.X}$ and $s_{X.Y}$ (which are estimates of the underlying population parameters $\sigma_{Y.X}$ and $\sigma_{X.Y}$).

$$s_{Y.X} = \tilde{s}_Y \sqrt{\left[\frac{n-1}{n-2}\right][1 - r^2]} \qquad \textbf{(Equation 22.12)}$$

$$s_{X.Y} = \tilde{s}_X \sqrt{\left[\frac{n-1}{n-2}\right][1 - r^2]} \qquad \textbf{(Equation 22.13)}$$

As the size of the sample increases, the value $(n - 1)/(n - 2)$ in the radical of Equations 22.12 and 22.13 approaches 1, and thus for large sample sizes the equations simplify to $s_{Y.X} = \tilde{s}_Y\sqrt{1 - r^2}$ and $s_{X.Y} = \tilde{s}_X\sqrt{1 - r^2}$. Note, however, that for small sample sizes the latter equations underestimate the values of $s_{Y.X}$ and $s_{X.Y}$.

Equations 22.12 and 22.13 are employed to compute the values $s_{Y.X} = .92$ and $s_{X.Y} = 2.94$.[9]

$$s_{Y.X} = 2.70\sqrt{\left[\frac{5-1}{5-2}\right][1 - (.955)^2]} = .92$$

$$s_{X.Y} = 8.58\sqrt{\left[\frac{5-1}{5-2}\right][1 - (.955)^2]} = 2.94$$

3. Computation of a confidence interval for the value of the criterion variable[10] It turns out that Equations 22.12 and 22.13 are not unbiased estimates of error throughout the full range of values the criterion variable may assume. What this translates into is that if a researcher wants to compute a confidence interval with respect to a specific subject's score on the criterion variable, in the interest of complete accuracy an adjusted standard error of estimate value should be employed. The adjusted standard error of estimate values will be designated $\tilde{s}_{Y.X}$ and $\tilde{s}_{X.Y}$. The values computed for $\tilde{s}_{Y.X}$ and $\tilde{s}_{X.Y}$ will always be larger than the values computed for $s_{Y.X}$ and $s_{X.Y}$. The larger the deviation between a subject's score on the predictor variable and the mean score for the predictor variable, the greater the difference between the values $s_{Y.X}$ versus $\tilde{s}_{Y.X}$ and $s_{X.Y}$ versus $\tilde{s}_{X.Y}$. Equations 22.14 and 22.15 are employed to compute the values $\tilde{s}_{Y.X}$ and $\tilde{s}_{X.Y}$.[11] In the latter equations, the values X and Y respectively represent the X and Y scores of the specific subject for whom the standard error of estimate is computed.

$$\tilde{s}_{Y.X} = s_{Y.X}\sqrt{1 + \frac{1}{n} + \frac{(X - \bar{X})^2}{SS_X}} \qquad \textbf{(Equation 22.14)}$$

$$\tilde{s}_{X.Y} = s_{X.Y}\sqrt{1 + \frac{1}{n} + \frac{(Y - \bar{Y})^2}{SS_Y}}$$ **(Equation 22.15)**

At this point two confidence intervals will be computed employing the values $\tilde{s}_{Y.X}$ and $\tilde{s}_{X.Y}$. The two confidence intervals will be in reference to the two subjects for whom the values $Y' = 4.24$ and $X' = 9.02$ are predicted in the previous section (employing Equations 22.4 and 22.8). Initially, the use of Equation 22.14 will be demonstrated to compute a confidence interval for the subject who consumes 10 ounces of sugar (i.e., $X = 10$), and (through use of Equation 22.4) is predicted to have $Y' = 4.24$ cavities. Equation 22.16 is employed to compute a confidence interval for the predicted value of Y.

$$CI_{(1 - \alpha)} = Y' \pm (t_{\alpha/2})(\tilde{s}_{Y.X})$$ **(Equation 22.16)**

Where: $t_{\alpha/2}$ represents the tabled critical two-tailed value in the t distribution, for $df = n - 2$, below which a proportion equal to $[1 - (\alpha/2)]$ of the cases falls. If the percentage of the distribution that falls within the confidence interval is subtracted from 1, it will equal the value of α.

In the computation of a confidence interval, the predicted value Y' can be conceptualized as the mean value in a population of scores on the Y variable for a specific subject. When the sample size employed for the analysis is large (i.e., $n > 100$), one can assume that the shape of such a distribution for each subject will be normal, and in such a case the relevant tabled critical two-tailed z value (i.e., $z_{\alpha/2}$) can be employed in Equation 22.16 in place of the relevant tabled critical t value. For smaller sample sizes (as is the case for Example 22.1), however, the t distribution provides a more accurate approximation of the underlying population distribution. Use of the normal distribution with small sample sizes underestimates the range of values that define a confidence interval. Inspection of Equation 22.16 reveals that the range of values computed for a confidence interval is a function of the magnitude of the standard error of estimate and the tabled critical t value (the magnitude of the latter being inversely related to the sample size).

In order to use Equation 22.16, the value $\tilde{s}_{Y.X}$ must be computed with Equation 22.14. Employing Equation 22.14, the value $\tilde{s}_{Y.X} = 1.02$ is computed. Note that the latter value is slightly larger than the value $s_{Y.X} = .92$ computed with Equation 22.12.

$$\tilde{s}_{Y.X} = .92\sqrt{1 + \frac{1}{5} + \frac{(10 - 7.2)^2}{294.8}} = 1.02$$

To demonstrate the use of Equation 22.16, the 95% confidence interval will be computed. The value $t_{.05} = 3.18$ is employed to represent $t_{\alpha/2}$, since in **Table A2** it is the tabled critical two-tailed .05 t value for $df = 3$. The appropriate values are now substituted in Equation 22.16 to compute the 95% confidence interval.

$$CI_{.95} = 4.24 \pm (3.18)(1.02) = 4.24 \pm 3.24$$

This result indicates that the researcher can be 95% confident that the number of cavities the subject actually has falls within the range 1.00 and 7.48 (i.e., $1.00 \leq Y \leq 7.48$).

A confidence interval will now be computed for the subject who has 4 cavities (i.e., $Y = 4$), and (through use of Equation 22.8) is predicted to eat 9.02 ounces of sugar. Equation 22.17 is employed to compute a confidence interval for the value of X.

$$CI_{(1 - \alpha)} = X' \pm (t_{\alpha/2})(\tilde{s}_{X.Y})$$ **(Equation 22.17)**

In order to use Equation 22.17 the value $\tilde{s}_{X.Y}$ must be computed with Equation 22.15. Employing Equation 22.15, the value $\tilde{s}_{X.Y} = 3.24$ is computed. Note that the latter value is slightly larger than the value $s_{X.Y} = 2.94$ computed with Equation 22.13.

$$\tilde{s}_{X.Y} = 2.94\sqrt{1 + \frac{1}{5} + \frac{(4 - 3.4)^2}{29.2}} = 3.24$$

As is done in the previous example, the 95% confidence interval will be computed. Thus, the values $t_{.05} = 3.18$ and $\tilde{s}_{X.Y} = 3.24$ are substituted in Equation 22.17.

$$CI_{.95} = 9.02 \pm (3.18)(3.24) = 9.02 \pm 10.30$$

This result indicates that the researcher can be 95% confident that the number of ounces of sugar the subject actually eats falls within the range −1.28 and 19.32 (i.e., $-1.28 \le X \le 19.32$). Since it is impossible to have a negative number of ounces of sugar, the result translates into between 0 and 19.32 ounces of sugar.

4. Computation of a confidence interval for a Pearson product-moment correlation coefficient In order to compute a confidence interval for a computed value of the **Pearson product-moment correlation coefficient,** it is necessary to employ a procedure developed by Fisher (1921) referred to as **Fisher's z_r (or z) transformation.** The latter procedure transforms an r value to a scale that is based on the normal distribution. The rationale behind the use of **Fisher's z_r transformation** is that although the theoretical sampling distribution of the correlation coefficient can be approximated by the normal distribution when the value of a population correlation is equal to zero, as the value of the population correlation deviates from zero the sampling distribution becomes more and more skewed. Thus, in computing confidence intervals (as well as in testing hypotheses involving one or more populations in which a hypothesized population correlation is some value other than zero), **Fisher's z_r transformation** is required to transform a skewed sampling distribution into a normalized format.

Equation 22.18 is employed to convert an r value into a **Fisher transformed value,** which is represented by the notation z_r.

$$z_r = \frac{1}{2}\log_e\left[\frac{1 + r}{1 - r}\right] \qquad \textbf{(Equation 22.18)}$$

Where: $\log_e$ represents the natural logarithm of a number

Although logarithmic values can be computed with a function key on most scientific calculators, if one does not have access to a calculator, **Table A17 (Table of Fisher's z_r Transformation)** in the **Appendix** provides an alternative way of deriving the Fisher transformed values. The latter table contains the z_r values that correspond to specific values of r. The reader should take note of the fact that in employing Equation 22.18 or **Table A17,** the sign assigned to a z_r value is always the same as the sign of the r value upon which it is based. Thus, a positive r value will always be associated with a positive z_r value, and a negative r value will always be associated with a negative z_r value. When $r = 0$, z_r will also equal zero.

Equation 22.19 is employed to compute the confidence interval for a computed r value.

$$CI_{z_{r(1-\alpha)}} = z_r \pm (z_{\alpha/2})\sqrt{\frac{1}{n - 3}} \qquad \textbf{(Equation 22.19)}$$

Where: $z_{\alpha/2}$ represents the tabled critical two-tailed value in the normal distribution below which a proportion equal to $[1 - (\alpha/2)]$ of the cases fall. If the percentage of the distribution that falls within the confidence interval is subtracted from 1, it will equal the value of α.

The value $\sqrt{1/(n - 3)}$ in Equation 22.19 represents the standard error of z_r. In employing Equation 22.19 to compute the 95% confidence interval, the product of the tabled critical two-tailed .05 z value and the standard error of z_r are added to and subtracted from the Fisher transformed value for the computed r value. The two resulting values, which represent z_r values, are then reconverted into correlation coefficients through use of **Table A17** or by reconfiguring Equation 22.18 to solve for r. Use of **Table A17** for the latter is accomplished by identifying the r values which correspond to the computed z_r values. The resulting r values derived from the table identify the limits that define the 95% confidence interval.

Equation 22.19 will now be used to compute the 95% confidence interval for $r = .955$. From **Table A17** it is determined that the Fisher transformed value which corresponds to $r = .955$ is $z_r = 1.886$. The appropriate values are now substituted in Equation 22.19.

$$CI_{z_{r(.95)}} = 1.886 \pm (1.96)\sqrt{\frac{1}{5 - 3}} = 1.886 \pm 1.386$$

Subtracting from and adding 1.386 to 1.886, yields the values .5 and 3.272. The latter values are now converted into r values through use of **Table A17**. By interpolating, we can determine that a z_r value of .5 corresponds to the value $r = .462$, which will define the lower limit of the confidence interval. Since the value $z_r = 3.272$ is substantially above the z value that corresponds to the largest tabled r value, it will be associated with the value $r = 1$. Thus, we can be 95% confident that the true value of the population correlation falls between .462 and 1. Symbolically, this can be written as follows: $.462 \leq \rho \leq 1$. Note that because of the small sample size employed in the experiment, the range of values that define the confidence interval is quite large.

If the 99% confidence interval is computed, the tabled critical two-tailed .01 value $z_{.01} = 2.58$ is employed in Equation 22.19 in place of $z_{.05} = 1.96$. As is always the case in computing a confidence interval, the range of values that defines a 99% confidence interval will be larger than the range which defines a 95% confidence interval.

5. Test 22b: Test for evaluating the hypothesis that the true population correlation is a specific value other than zero In certain instances a researcher may want to evaluate whether an obtained correlation could have come from a population in which the true correlation between two variables is a specific value other than zero. The null and alternative hypotheses that are evaluated under such conditions are as follows.

$$H_0\text{: } \rho = \rho_0$$

(In the underlying population the sample represents, the correlation between the scores of subjects on Variable X and Variable Y equals ρ_0.)

$$H_1\text{: } \rho \neq \rho_0$$

(In the underlying population the sample represents, the correlation between the scores of subjects on Variable X and Variable Y equals some value other than ρ_0. The alternative hypothesis as stated is **nondirectional**, and is evaluated with a **two-tailed test**. It is also possible to state the alternative hypothesis directionally (H_1: $\rho > \rho_0$ or H_1: $\rho < \rho_0$), in which case it is evaluated with a one-tailed test.)

Equation 22.20 is employed to evaluate the null hypothesis H_0: $\rho = \rho_0$.

$$z = \frac{z_r - z_{\rho_0}}{\sqrt{\frac{1}{(n - 3)}}}$$ **(Equation 22.20)**

Where: z_r represents the Fisher transformed value of the computed value of r
z_{ρ_0} represents the Fisher transformed value of ρ_0, the hypothesized population correlation

Equation 22.20 will now be employed in reference to Example 22.1. Let us assume that we want to evaluate whether the true population correlation between the number of ounces of sugar consumed and the number of cavities is .80. Thus, the null hypothesis is H_0: $\rho = .80$, and the nondirectional alternative hypothesis is H_1: $\rho \neq .80$.

By employing **Table A17** (or Equation 22.18), we determine that the corresponding z_r values for the obtained correlation coefficient $r = .955$ and the hypothesized population correlation coefficient $\rho = .80$ are respectively $z_r = 1.886$ and $z_\rho = 1.099$ (the notation z_ρ is employed in place of z_r whenever the relevant element in an equation identifies a population correlation). Substituting the Fisher transformed values in Equation 22.20, the value $z = 1.11$ is computed.[12]

$$z = \frac{1.886 - 1.099}{\sqrt{\frac{1}{(5 - 3)}}} = 1.11$$

The computed value $z = 1.11$ is evaluated with **Table A1 (Table of the Normal Distribution)** in the **Appendix**. In order to reject the null hypothesis, the obtained absolute value of z must be equal to or greater than the tabled critical two-tailed value at the prespecified level of significance. Since $z = 1.11$ is less than the tabled critical two-tailed values $z_{.05} = 1.96$ and $z_{.01} = 2.58$, the null hypothesis cannot be rejected at either the .05 or .01 level. Thus, the null hypothesis that the true population correlation equals .80 is retained.

If the alternative hypothesis is stated directionally, in order to reject the null hypothesis the obtained absolute value of z must be equal to or greater than the tabled critical one-tailed value at the prespecified level of significance (i.e., $z_{.05} = 1.65$ or $z_{.01} = 2.33$). Since $z = 1.11$ is less than $z_{.05} = 1.65$, the directional alternative hypothesis H_0: $\rho > .80$ is not supported. Note that the sign of the value of z computed with Equation 22.20 will be positive when the computed value of r is greater than the hypothesized value ρ_0, and negative when the computed value of r is less than the hypothesized value ρ_0.

6. Computation of power for the Pearson product-moment correlation coefficient Prior to collecting correlational data a researcher can determine the likelihood of detecting a population correlation of a specific magnitude if a specific value of n is employed. As a result of such a power analysis, one can determine the minimum required sample size in order to detect a prespecified population correlation. To illustrate the computation of power, let us assume that prior to collecting the data for 25 subjects a researcher wants to determine the power associated with the analysis if the value of the population correlation he wants to detect is $\rho = .40$.[13] It will be assumed that a nondirectional analysis is conducted, with $\alpha = .05$.

Equation 22.21 (which is described in Guenther (1965)) is employed to compute the power of the analysis.

$$\delta = |z_{\rho_0} - z_{\rho_1}|\sqrt{n - 3} \qquad \textbf{(Equation 22.21)}$$

Where: z_{ρ_0} is the **Fisher transformed value** of the population correlation stipulated in the null hypothesis, and z_{ρ_1} is the **Fisher transformed value** of the population correlation the researcher wants to detect

Table A17 in the **Appendix** reveals that the Fisher transformed value associated with $\rho = 0$ is $z_\rho = z_{\rho_0} = 0$, and thus when the null hypothesis H_0: $\rho = 0$ is employed (which it will be assumed is the case), Equation 22.21 reduces to $\delta = z_{\rho_1}\sqrt{n - 3}$.

Employing **Table A17,** we determine that the Fisher transformed value for the population correlation of $\rho = .40$ is $z_\rho = .424$. Substituting the appropriate values in Equation 22.21, the value $\delta = 1.99$ is computed.

$$\delta = |0 - .424|\sqrt{25 - 3} = 1.99$$

The obtained value $\delta = 1.99$ is evaluated with **Table A3 (Power Curves for Student's *t* Distribution)** in the **Appendix.** A full discussion on the use of **Table A3** (which is employed to evaluate the power of a number of different types of ***t* tests**) can be found in Section VI of the **single-sample *t* test (Test 2).** Employing the power curve for $df = \infty$ in **Table A3-C** (the appropriate table for a nondirectional/two-tailedanalysis, with $\alpha = .05$), we determine the power of the correlational analysis to be approximately .52.[14] Thus, if the underlying population correlation is $\rho = .40$ and a sample size of $n = 25$ is employed, the likelihood of the researcher rejecting the null hypothesis is only .52. If this value is deemed too small, the researcher can substitute larger values of n in Equation 22.21 until a value is computed for δ that is associated with an acceptable level of power.

Equation 22.21 can also be employed if the value stated in the null hypothesis is some value other than $\rho = 0$. Assume that a number of studies suggest that the population correlation between two variables is $\rho = .60$. A researcher who has reason to believe that the latter value may overestimate the true population correlation wants to compute the power of a correlational analysis to determine if the true population correlation is, in fact, $\rho = .40$. In this example (for which the value $n = 25$ will be employed) H_0: $\rho = .60$. Since the researcher believes the true population correlation may be less than .60, the alternative hypothesis is stated directionally. Thus, H_1: $\rho < .60$.

Employing **Table A17**, we determine the Fisher transformed value for $\rho = .60$ is $z_\rho = .693$, and from the previous analysis we know that for $\rho = .40$, $z_\rho = .424$. Substituting the appropriate values in Equation 22.21, the value $\delta = 1.26$ is computed.

$$\delta = |.693 - .424|\sqrt{25 - 3} = 1.26$$

Employing the power curve for $df = \infty$ in **Table A3-D** (i.e., the curves for the one-tailed .05 value), we determine the power of the analysis to be approximately .37. Thus, if the underlying population correlation is $\rho = .40$ and a sample size of $n = 25$ is employed, the likelihood of the researcher rejecting the null hypothesis H_0: $\rho = .60$ is only .37.

7. Test 22c: Test for evaluating a hypothesis on whether there is a significant difference between two independent correlations There are occasions when a researcher will compute a correlation between the same two variables for two independent samples. In the event the correlation coefficients obtained for the two samples are not equal, the researcher may wish to determine whether the difference between the two correlations is statistically significant. The null and alternative hypotheses that are evaluated under such conditions are as follows.

$$H_0\colon \rho_1 = \rho_2$$

(In the underlying populations represented by the two samples, the correlation between the two variables is equal.)

$$H_1\colon \rho_1 \neq \rho_2$$

(In the underlying populations represented by the two samples, the correlation between the two variables is not equal. The alternative hypothesis as stated is **nondirectional**, and is evaluated with a **two-tailed test**. It is also possible to state the alternative hypothesis directionally ($H_1\colon \rho_1 > \rho_2$ or $H_1\colon \rho_1 < \rho_2$), in which case it is evaluated with a one-tailed test.)

To illustrate, let us assume that in Example 22.1 the correlation of $r = .955$ between the number of ounces of sugar eaten per week and the number of cavities is based on a sample of five-ten-year old boys (to be designated Sample 1). Let us also assume that the researcher evaluates a sample of five-ten-year old girls (to be designated Sample 2), and determines that in this second sample the correlation between the number of ounces of sugar eaten per week and the number of cavities is $r = .765$. Equation 22.22 can be employed to determine whether or not the difference between $r_1 = .955$ and $r_2 = .765$ is significant.

$$z = \frac{z_{r_1} - z_{r_2}}{\sqrt{\dfrac{1}{n_1 - 3} + \dfrac{1}{n_2 - 3}}} \qquad \textbf{(Equation 22.22)}$$

Where: z_{r_1} represents the Fisher transformed value of the computed value of r_1 for Sample 1

z_{r_2} represents the Fisher transformed value of the computed value of r_2 for Sample 2

n_1 and n_2 are respectively the number of subjects in Sample 1 and Sample 2

Since there are five subjects in both samples, $n_1 = n_2 = 5$. From the analysis in the previous section we already know that the Fisher transformed value of $r_1 = .955$ is $z_{r_1} = 1.886$. For the female sample, employing **Table A17** we determine that the Fisher transformed value of $r_2 = .765$ is $z_{r_2} = 1.008$. When the appropriate values are substituted in Equation 22.22, they yield the value $z = .878$.

$$z = \frac{1.886 - 1.008}{\sqrt{\dfrac{1}{5 - 3} + \dfrac{1}{5 - 3}}} = .878$$

The value $z = .878$ is evaluated with **Table A1**. In order to reject the null hypothesis, the obtained absolute value of z must be equal to or greater than the tabled critical two-tailed value at the prespecified level of significance. Since $z = .878$ is less than the tabled critical two-tailed values $z_{.05} = 1.96$ and $z_{.01} = 2.58$, the nondirectional alternative hypothesis $H_1\colon \rho_1 \neq \rho_2$ is not supported at either the .05 or .01 level. Thus, we retain the null hypothesis that there is an equal correlation between the two variables in each of the populations represented by the samples.

If the alternative hypothesis is stated directionally, in order to reject the null hypothesis the obtained absolute value of z must be equal to or greater than the tabled critical one-tailed value at the prespecified level of significance (i.e., $z_{.05} = 1.65$ or $z_{.01} = 2.33$). The sign of z will be positive when $r_1 > r_2$, and thus can only support the alternative hypothesis

H_1: $\rho_1 > \rho_2$. The sign of z will be negative when $r_1 < r_2$, and thus can only support the alternative hypothesis H_1: $\rho_1 < \rho_2$. Since $z = .878$ is less than $z_{.05} = 1.65$, the directional alternative hypothesis H_1: $\rho_1 > \rho_2$ is not supported.

Edwards (1984) notes that when the null hypothesis is retained, since the analysis suggests that the two samples represent a single population, Equation 22.23 can be employed to provide a weighted estimate of the common population correlation.

$$\bar{z}_r = \frac{(n_1 - 3)z_{r_1} + (n_2 - 3)z_{r_2}}{(n_1 - 3) + (n_2 - 3)} \qquad \textbf{(Equation 22.23)}$$

Substituting the data in Equation 22.23, the Fisher transformed value $\bar{z}_r = 1.447$ is computed.

$$\bar{z}_r = \frac{(5 - 3)(1.886) + (5 - 3)(1.008)}{(5 - 3) + (5 - 3)} = 1.447$$

Employing **Table A17**, we determine that the Fisher transformed value $z_r = 1.447$ corresponds to the value $r = .895$. Thus, $r = .895$ can be employed as the best estimate of the common population correlation. Note that the estimated common population correlation computed with Equation 22.23 is not the same value that is obtained if, instead, one calculates the weighted average of the two correlations (which, since the sample sizes are equal, is the average of the two correlations: $(.955 + .765)/2 = .86$). The fact that the weighted average of the two correlations yields a different value from the result obtained with Equation 22.23 can be attributed to fact that the theoretical sampling distribution of the correlation coefficient becomes more skewed as the absolute value of r approaches 1. The procedure and tables for computing the power of the test comparing two independent correlation coefficients can be found in Cohen (1988).

8. Test 22d: Test for evaluating a hypothesis on whether *k* independent correlations are homogeneous **Test 22c** can be extended to determine whether more than two independent correlation coefficients are homogeneous (in other words, can be viewed as representing the same population correlation, ρ). The null and alternative hypotheses that are evaluated under such conditions are as follows.

$$H_0\text{: } \rho_1 = \rho_2 = \cdots = \rho_k$$

(In the underlying populations represented by the k samples, the correlation between the two variables is equal.)

$$H_1\text{: Not } H_0$$

(In the underlying populations represented by the k samples, the correlation between the two variables is not equal in at least two of the populations. The alternative hypothesis as stated is **nondirectional**.)

To illustrate, let us assume that the correlation between the number of ounces of sugar eaten per week and the number of cavities is computed for three independent samples, each sample consisting of five children living in different parts of the country. The values of the correlations obtained for the three samples are as follows: $r_1 = .955$, $r_2 = .765$, $r_3 = .845$.

Equation 22.24 is employed to determine whether the $k = 3$ sample correlations are homogeneous. In Equation 22.24, wherever the summation sign $\sum_{i=1}^{k}$ appears it indicates that the operation following the summation sign is carried out for each of the $k = 3$ samples, and the resulting $k = 3$ values are summed.

$$\chi^2 = \sum_{i=1}^{k} \left[(n_i - 3) z_{r_i}^2 \right] - \frac{\left[\sum_{i=1}^{k} (n_i - 3) z_{r_i} \right]^2}{\sum_{i=1}^{k} (n_i - 3)} \qquad \textbf{(Equation 22.24)}$$

Since there are five subjects in each sample, $n_1 = n_2 = n_3 = 5$. From the analysis in the previous section, we already know that the Fisher transformed values of $r_1 = .955$ and $r_2 = .765$ are respectively $z_{r_1} = 1.886$ and $z_{r_2} = 1.008$. Employing **Table A17**, we determine for Sample 3 the Fisher transformed value of $r_3 = .845$ is $z_{r_3} = 1.238$. When the appropriate values are substituted in Equation 22.24, they yield the value $\chi^2 = .83$.

$$\chi^2 = [(5 - 3)(1.886)^2 + (5 - 3)(1.008)^2 + (5 - 3)(1.238)^2] - \frac{[(5-3)(1.886) + (5-3)(1.008) + (5-3)(1.238)]^2}{(5-3) + (5-3) + (5-3)} = .83$$

The value $\chi^2 = .83$ is evaluated with **Table A4 (Table of the Chi-Square Distribution)** in the **Appendix**. The degrees of freedom employed in evaluating the obtained chi-square value are $df = k - 1$. Thus, for the above example, $df = 3 - 1 = 2$. In order to reject the null hypothesis, the obtained value of χ^2 must be equal to or greater than the tabled critical value at the prespecified level of significance. Since $\chi^2 = .83$ is less than $\chi^2_{.05} = 5.99$ and $\chi^2_{.01} = 9.21$ (which are the tabled critical .05 and .01 values for $df = 2$ when a nondirectional alternative hypothesis is employed), the null hypothesis cannot be rejected at either the .05 or .01 level. Thus, we retain the null hypothesis that in the underlying populations represented by the $k = 3$ samples the correlations between the two variables are equal.[15]

Edwards (1984) notes that when the null hypothesis is retained, since the analysis suggests that the k samples represent a single population, Equation 22.25 can be employed to provide a weighted estimate of the common population correlation.[16]

$$\bar{z}_r = \frac{\sum_{i=1}^{k} (n_i - 3) z_{r_i}}{\sum_{i=1}^{k} (n_i - 3)} \qquad \textbf{(Equation 22.25)}$$

Substituting the data in Equation 22.25, the Fisher transformed value $\bar{z}_r = 1.377$ is computed.

$$\bar{z}_r = \frac{(5 - 3)(1.886) + (5 - 3)(1.008) + (5 - 3)(1.238)}{(5 - 3) + (5 - 3) + (5 - 3)} = 1.377$$

Employing **Table A17**, we determine that the Fisher transformed value $z_r = 1.377$ corresponds to the value $r = .88$. Thus, $r = .88$ can be employed as the best estimate of the common population correlation. Note that, as is the case when the same analysis is conducted for $k = 2$ samples, the value obtained for the common population correlation (using Equation 22.25) is not the same as the value that is obtained if the weighted average of the three correlation coefficients is computed (i.e., $(.955 + .765 + .845)/3 = .855$).

9. Test 22e: Test for evaluating the null hypothesis H_0: $\rho_{XZ} = \rho_{YZ}$ There are instances when a researcher may want to evaluate if, within a specific population, one variable (X) has the same correlation with some criterion variable (Z) as does another variable (Y). The null and alternative hypotheses which are evaluated in such a situation are as follows.

$$H_0\text{: } \rho_{XZ} = \rho_{YZ}$$

(In the underlying population represented by the sample, the correlation between variables X and Z is equal to the correlation between variables Y and Z.)

$$H_1\text{: } \rho_{XZ} \neq \rho_{YZ}$$

(In the underlying population represented by the sample, the correlation between variables X and Z is not equal to the correlation between variables Y and Z. The alternative hypothesis as stated is **nondirectional**, and is evaluated with a **two-tailed test**. It is also possible to state the alternative hypothesis directionally (H_1: $\rho_{XZ} > \rho_{YZ}$ or H_1: $\rho_{XZ} < \rho_{YZ}$), in which case it is evaluated with a one-tailed test.)

To illustrate how one can evaluate the null hypothesis H_0: $\rho_{XZ} = \rho_{YZ}$, let us assume that the correlation between the number of ounces of sugar eaten per week and the number of cavities is computed for five subjects, and $r = .955$. Let us also assume that for the same five subjects we determine that the correlation between the number of ounces of salt eaten per week and the number of cavities is $r = .52$. We want to determine whether there is a significant difference in the correlation between the number of ounces of sugar eaten per week and the number of cavities versus the number of ounces of salt eaten per week and the number of cavities. Let us also assume that for the sample of five subjects, the correlation between the number of ounces of sugar eaten per week and the number of ounces of salt eaten per week is $r = .37$.

In the above example, within the framework of the hypothesis being evaluated we have two predictor variables — the number of ounces of sugar eaten per week and the number of ounces of salt eaten per week. These two predictor variables will respectively represent the X and Y variables in the analysis to be described. The number of cavities, which is the criterion variable, will be designated as the Z variable. Thus, $r_{XZ} = .955$, $r_{YZ} = .52$, $r_{XY} = .37$.

The test statistic for evaluating the null hypothesis, which is based on the t distribution, is computed with Equation 22.26. A more detailed description of the test statistic can be found in Steiger (1980), who notes that Equation 22.26 provides a superior test of the hypothesis being evaluated when compared with an alternative procedure developed by Hotelling (1940) (which is described in Lindeman *et al.* (1980)).

(Equation 22.26)

$$t = (r_{YZ} - r_{XZ}) \sqrt{\frac{(n - 1)(1 + r_{XY})}{2\left[\dfrac{(n - 1)}{(n - 3)}\right]\left[1 - r_{YZ}^2 - r_{XZ}^2 - r_{XY}^2 + 2r_{YZ}r_{XZ}r_{XY}\right] + \left[\dfrac{r_{YZ} + r_{XZ}}{2}\right]^2 \left[1 - r_{XY}\right]^3}}$$

Substituting the values $n = 5$, $r_{XZ} = .955$, $r_{YZ} = .52$, and $r_{XY} = .37$ in Equation 22.26, the value $t = -1.78$ is computed.

$$t = (.52 - .955) \sqrt{\frac{(5-1)(1+.37)}{2\left[\frac{(5-1)}{(5-3)}\right]\left[1-(.52)^2-(.955)^2-(.37)^2+2(.52)(.955)(.37)\right]+\left[\frac{(.52+.955)}{2}\right]^2\left[1-(.37)\right]^3}} = -1.78$$

The value $t = -1.78$ is evaluated with **Table A2**. The degrees of freedom employed in evaluating the obtained t value are $df = n - 3$. Thus, for the above example, $df = 5 - 3 = 2$. In order to reject the null hypothesis, the obtained absolute value of t must be equal to or greater than the tabled critical value at the prespecified level of significance. Since the absolute value $t = 1.78$ is less than $t_{.05} = 4.30$ and $t_{.01} = 9.93$ (which are the tabled critical two-tailed values for $df = 2$), the null hypothesis cannot be rejected at either the .05 or .01 level. Thus, we retain the null hypothesis that the population correlation between variables X and Z is equal to the population correlation between variables Y and Z.

If the alternative hypothesis is stated directionally, in order to reject the null hypothesis, the obtained absolute value of t must be equal to or greater than the tabled critical one-tailed value at the prespecified level of significance (which for $df = 2$ are $t_{.05} = 2.92$ or $t_{.01} = 6.97$). The sign of t must be positive if H_1: $\rho_{XZ} < \rho_{YZ}$, and must be negative if H_1: $\rho_{XZ} > \rho_{YZ}$. Since the absolute value $t = 1.78$ is less than $t_{.05} = 2.92$, the directional alternative hypothesis H_1: $\rho_{XZ} > \rho_{YZ}$ is not supported.

In the event the t value obtained with Equation 22.26 is significant, it indicates that the predictor variable which correlates highest with the criterion variable (i.e., the one with the highest absolute value) is the best predictor of subjects' scores on the latter variable. It should be noted that because the analysis discussed in this section represents a dependent samples analysis (since all three correlations are based on the same sample), Equation 22.22 (the equation for contrasting two independent correlations) is not appropriate to use to evaluate the null hypothesis H_0: $\rho_{XZ} = \rho_{YZ}$.

10. Tests for evaluating a hypothesis regarding one or more regression coefficients A number of tests have been developed that evaluate hypotheses concerning the slope of a regression line (which is also referred to as a **regression coefficient**). This section will present a brief description of such tests. In the statement of the null and alternative hypotheses of tests concerning a regression coefficient, the notation β is employed to represent the slope of the line in the underlying population represented by a sample. Thus, β_Y is the population regression coefficient of the **regression line of *Y* on *X***, and β_X is the population regression coefficient of the **regression line of *X* on *Y***.

Test 22f: Test for evaluating the null hypothesis H_0: $\beta = 0$ A test of significance can be conducted to evaluate the hypothesis of whether in the underlying population, the value of the slope of a regression line is equal to zero. The null hypotheses that can be evaluated in reference to the two regression lines are H_0: $\beta_Y = 0$ and H_0: $\beta_X = 0$. In point of fact, the test of the generic null hypothesis H_0: $\beta = 0$ will always yield the same result as that obtained when the null hypothesis H_0: $\rho = 0$ is evaluated using **Test 22a** (which employs Equation 22.3). This is the case, since whenever $\rho = 0$ the slope of a regression line in the underlying population will also equal zero. Equations 22.27 and 22.28 are respectively employed to evaluate the null hypotheses H_0: $\beta_Y = 0$ and H_0: $\beta_X = 0$. The equations are employed below with the data for Example 22.1, and in both instances yield the same t value as that obtained when the null hypothesis H_0: $\rho = 0$ is evaluated with Equation 22.3. The slight discrepancies between the t values computed with Equations 22.3, 22.27, and 22.28 are the result of rounding off error.

$$t = \frac{(b_Y)(\tilde{s}_X)\sqrt{n-1}}{s_{Y.X}} = \frac{(.30)(8.58)\sqrt{5-1}}{.92} = 5.60 \qquad \textbf{(Equation 22.27)}$$

$$t = \frac{(b_X)(\tilde{s}_Y)\sqrt{n-1}}{s_{X.Y}} = \frac{(3.03)(2.70)\sqrt{5-1}}{2.94} = 5.57 \qquad \textbf{(Equation 22.28)}$$

The t values computed with Equations 22.27 and 22.28 are evaluated with **Table A2**. The degrees of freedom employed in evaluating the obtained t values are $df = n - 2$. Since both t values are identical to the t value computed with Equation 22.3 and the same degrees of freedom are employed, interpretation of the t values leads to the same conclusions —

except, in this case the conclusions are in reference to the regression coefficients β_Y and β_X. Thus, the nondirectional alternative hypotheses H_1: $\beta_Y \neq 0$ and H_1: $\beta_X \neq 0$ are supported at the .05 level, since (for $df = 3$) the computed t values are greater than the tabled critical two-tailed value $t_{.05} = 3.18$. The directional alternative hypotheses H_1: $\beta_Y > 0$ and H_1: $\beta_X > 0$ are supported at both the .05 and .01 levels, since the sign of t (as well as the sign of each of the regression coefficients) is positive, and the computed t values are greater than the tabled critical one-tailed values $t_{.05} = 2.35$ and $t_{.01} = 4.54$.

Equations 22.29 and 22.30 can respectively be employed to compute confidence intervals for the regression coefficients β_Y and β_X. The computation of the 95% confidence interval for the two regression coefficients is demonstrated below. The value $t_{.05} = 3.18$ (which is also employed in computing the confidence intervals derived with Equations 22.16 and 22.17) is employed in Equations 22.29 and 22.30 to represent $t_{\alpha/2}$, since in **Table A2** it is the tabled critical two-tailed .05 t value for $df = 3$ (which is computed with $df = n - 2$).

(Equation 22.29)

$$CI_{\beta_{Y(1-\alpha)}} = b_Y \pm (t_{\alpha/2})\left[\frac{s_{Y.X}}{(\tilde{s}_X)\sqrt{n-1}}\right] = .30 \pm (3.18)\left[\frac{.92}{8.58\sqrt{5-1}}\right] = .30 \pm .17$$

(Equation 22.30)

$$CI_{\beta_{X(1-\alpha)}} = b_X \pm (t_{\alpha/2})\left[\frac{s_{X.Y}}{(\tilde{s}_Y)\sqrt{n-1}}\right] = 3.03 \pm (3.18)\left[\frac{2.94}{2.70\sqrt{5-1}}\right] = 3.03 \pm 1.73$$

The above results indicate the following: a) There is a 95% likelihood that the population regression coefficient β_Y falls within the range .13 and .47 (i.e., $.13 \leq \beta_Y \leq .47$); and b) There is a 95% likelihood that the population regression coefficient β_X falls within the range 1.30 and 4.76 (i.e., $1.30 \leq \beta_X \leq 4.76$). Since the nondirectional alternative hypotheses H_1: $\beta_Y \neq 0$ and H_1: $\beta_X \neq 0$ are supported at the .05 level, it logically follows that the value zero will not fall within the range that defines either of the confidence intervals.

Test 22g: Test for evaluating the null hypothesis H_0: $\beta_1 = \beta_2$ A test of significance can be conducted to evaluate whether the slopes of two regression lines obtained from two independent samples are equal to one another. As is the case with **Test 22c**, it is assumed that the correlations for the independent samples are for the same two variables. The null hypotheses evaluated by the test for the **regression lines of *Y* on *X*** and the **regression lines *X* on *Y*** are respectively H_0: $\beta_{Y_1} = \beta_{Y_2}$ and H_0: $\beta_{X_1} = \beta_{X_2}$ (where β_{Y_i} represents the slope of the **regression line of *Y* on *X*** in the underlying population represented by Sample i, and β_{X_i} represents the slope of the **regression line of *X* on *Y*** in the underlying population represented by Sample i). As a result of evaluating the null hypothesis in reference to two independent **regression lines of *Y* on *X***, a researcher can determine if the degree of change on the Y variable when the X variable is incremented by one unit is equivalent in the two samples. In the case of two independent **regression lines of *X* on *Y***, a researcher can determine if the degree of change on the X variable when the Y variable is incremented by one unit is equivalent in the two samples. It should be noted that the test which is employed to evaluate the generic null hypothesis H_0: $\beta_1 = \beta_2$ is not equivalent to **Test 22c**, which evaluates the null hypothesis H_0: $\rho_1 = \rho_2$. This is the case since (as is illustrated in Figure 22.2) it is entirely possible for the regression lines associated with two independent correlations to have identical slopes, yet be associated with dramatically different correlations.

Equations 22.31 and 22.32 are employed to evaluate the null hypotheses H_0: $\beta_{Y_1} = \beta_{Y_2}$ and H_0: $\beta_{X_1} = \beta_{X_2}$. The t values computed with Equations 22.31 and 22.32 are evaluated with **Table A2**. The degrees of freedom employed in evaluating the obtained t values are $df = (n_1 - 2) + (n_2 - 2) = n_1 + n_2 - 4$.[17]

$$t = \frac{b_{Y_1} - b_{Y_2}}{\sqrt{\dfrac{s^2_{Y.X_1}}{(\tilde{s}^2_{X_1})(n_1 - 1)} + \dfrac{s^2_{Y.X_2}}{(\tilde{s}^2_{X_2})(n_2 - 1)}}}$$ **(Equation 22.31)**

$$t = \frac{b_{X_1} - b_{X_2}}{\sqrt{\dfrac{s^2_{X.Y_1}}{(\tilde{s}^2_{Y_1})(n_1 - 1)} + \dfrac{s^2_{X.Y_2}}{(\tilde{s}^2_{Y_2})(n_2 - 1)}}}$$ **(Equation 22.32)**

Equations 22.31 and 22.32 can be employed to evaluate the regression coefficients associated with the two independent correlations described within the framework of the example employed to demonstrate **Test 22c**. If, for instance, the regression coefficient b_{Y_1} computed for boys (who will represent Sample 1) is larger than the regression coefficient b_{Y_2} computed for girls (who will represent Sample 2), Equation 22.31 can be employed to evaluate the null hypothesis H_0: $\beta_{Y_1} = \beta_{Y_2}$. Although the full analysis will not be done here, the following values would be used in Equation 22.31 for Sample 1/boys (whose data are the same as that employed in Example 22.1): $n_1 = 5$, $b_{Y_1} = .30$, $\tilde{s}^2_{X_1} = (8.58)^2 = 73.62$, $s^2_{Y.X_1} = (.92)^2 = .85$. If upon substituting the analogous values for a sample of five girls in Equation 22.31 the resulting t value is significant, the null hypothesis H_0: $\beta_{Y_1} = \beta_{Y_2}$ is rejected. The number of degrees of freedom employed for the analysis are $df = 5 + 5 - 4 = 6$. The tabled critical .05 and .01 two-tailed and one-tailed t values for $df = 6$ are respectively $t_{.05} = 2.45$ and $t_{.01} = 3.71$, and $t_{.05} = 1.94$ and $t_{.01} = 3.14$. If the nondirectional alternative hypothesis H_1: $\beta_{Y_1} \neq \beta_{Y_2}$ is employed, in order to be significant the obtained absolute value of t must be equal to or greater than the tabled critical two-tailed value at the prespecified level of significance. If the directional alternative hypothesis H_1: $\beta_{Y_1} > \beta_{Y_2}$ is employed, in order to be significant the computed t value must be a positive number that is equal to or greater than the tabled critical one-tailed value at the prespecified level of significance. If the directional alternative hypothesis H_1: $\beta_{Y_1} < \beta_{Y_2}$ is employed, in order to be significant the computed t value must be a negative number that has an absolute value which is equal to or greater than the tabled critical value at the prespecified level of significance.[18]

11. Additional correlational procedures At the conclusion of the discussion of the **Pearson product-moment correlation coefficient**, an **Addendum** (Section IX) has been included which provides a description of the following additional correlational procedures that are directly or indirectly related to the **Pearson product-moment correlation coefficient:** a) **Test 22h: The point-biserial correlation coefficient;** b) **Test 22i: The biserial correlation coefficient;** c) **Test 22j: The tetrachoric correlation coefficient** ; d) **Test 22k: The multiple correlation coefficient;** e) **Test 22l: The partial correlation coefficient;** f) **Test 22m: The semi-partial correlation coefficient.**

VII. Additional Discussion of the Pearson Product-Moment Correlation Coefficient

1. The definitional equation for the Pearson product-moment correlation coefficient Although more computationally tedious than Equation 22.1, Equation 22.33 is a conceptually more meaningful equation for computing the **Pearson product-moment correlation coefficient.** Unlike Equation 22.1, which allows for the quick computation of r, Equation 22.33 reveals the fact that Pearson conceptualized the product-moment coefficient of correlation as the average of the products of the paired z scores of subjects on the X and Y variables.

$$r = \frac{\sum_{i=1}^{n} z_{X_i} z_{Y_i}}{n - 1} \qquad \textbf{(Equation 22.33)}$$

Where: $z_{X_i} = (X_i - \bar{X})/\tilde{s}_X$ and $z_{Y_i} = (Y_i - \bar{Y})/\tilde{s}_Y$, with X_i and Y_i representing the scores of the ith subject on the X and Y variables

As noted above, the correlation coefficient is, in actuality, the mean of the product of each subject's X and Y scores, when the latter are expressed as z scores. Since the computed r value represents an average score, many books employ n as the denominator of Equation 22.33 instead of $(n - 1)$. In point of fact, n can be employed as the denominator of Equation 22.33 if in computing the z_{X_i} and z_{Y_i} scores, the sample standard deviations s_X and s_Y (computed with Equation I.7) are employed in place of the estimated population standard deviations $\tilde{s}_X$ and $\tilde{s}_Y$ (computed with Equation I.8). When the estimated population standard deviations are employed, however, $(n - 1)$ is the appropriate value to employ in the denominator of Equation 22.33.

In employing Equation 22.33 to compute the value of r, initially the mean and estimated population standard deviation of the X and Y scores must be computed. Each X score is then converted into a z score by employing the equation for converting a raw score into a z score (i.e., $z_{X_i} = (X_i - \bar{X})/\tilde{s}_X$). Each Y score is also converted into a z score using the same equation with reference to the Y variable (i.e., $z_{Y_i} = (Y_i - \bar{Y})/\tilde{s}_Y$).[19] The product of each subject's z_{X_i} and z_{Y_i} score is obtained, and the sum of the products for the n subjects is computed. The latter sum is divided by $(n - 1)$, yielding the value of r which represents an average of the sum of the products.[20] The value of r computed with Equation 22.33 will be identical to that computed with Equation 22.1.

The computation of the value $r = .955$ with Equation 22.33 is demonstrated in Table 22.2. In deriving the z_{X_i} and z_{Y_i} scores, the following summary values are employed: $\bar{X} = 7.2$, $\tilde{s}_X = 8.58$, $\bar{Y} = 3.4$, $\tilde{s}_Y = 2.70$. Equation 22.33 yields the value $r = 3.818/4 = .955$ when the sum of the products in the last column of Table 22.2 is divided by $(n - 1)$.

Table 22.2 Computation of r with Equation 22.33

X_i	$z_{X_i} = \frac{X_i - \bar{X}}{\tilde{s}_X}$	Y	$z_{Y_i} = \frac{Y_i - \bar{Y}}{\tilde{s}_Y}$	$z_{X_i} z_{Y_i}$
20	1.49	7	1.33	1.982
0	−.84	0	−1.26	1.058
1	−.72	2	−.52	.374
12	.56	5	.59	.330
3	−.49	3	−.15	.074
				$\Sigma z_{X_i} z_{Y_i} = 3.818$

2. Residuals In Section VI it is noted that a **residual** is the difference between the predicted value of a subject's score on the criterion variable and the subject's actual score on the criterion variable. Thus, a residual indicates the amount of error between a subject's actual and predicted scores. If e_i represents the amount of error for the ith subject, the residual for a subject can be defined as follows (assuming the **regression line of Y on X** is employed): $e_i = (Y_i - Y_i')$. In the least squares regression model, the sum of the residuals will always equal zero. Thus, $\sum_{i=1}^{n} e_i = \sum_{i=1}^{n}(Y_i - Y_i') = 0$. Since the sum of the residuals equals zero, the average of the residuals will also equal zero (i.e., $\bar{e}_i = (\sum_{i=1}^{n} e_i)/n = 0$). The latter reflects the fact that for some of the subjects the predicted value of Y_i' will be larger than Y_i, while for other subjects the predicted value of Y_i' will be smaller than Y_i (of course, for some subjects Y_i' may equal Y_i). It should be noted that if the sum of the squared distances of the data points from the regression line is not the minimum possible value, the sum of the residuals will be some value other than zero.

$\sum_{i=1}^{n} e_i^2 = \sum_{i=1}^{n}(Y_i - Y_i')^2$ (which is the sum of the squared residuals) provides an index of the accuracy of prediction that results from use of the regression equation. When the sum of the squared residuals is small, prediction will be accurate but when it is large, prediction will be inaccurate. When the sum of the squared residuals equals zero (which will only be the case when $|r| = 1$), prediction will be perfect. The latter statement, however, only applies to the scores of subjects in the sample employed in the study. It does not ensure that prediction will be perfect for other members of the underlying population the sample represents. The accuracy of prediction for the population will depend upon the degree to which the derived regression equation is an accurate estimate of the actual regression equation in the underlying population.

The partitioning of the variation on the criterion variable in the least squares regression model can be summarized by Equation 22.34.

$$\sum_{i=1}^{n}(Y_i - \bar{Y})^2 = \sum_{i=1}^{n}(Y_i' - \bar{Y})^2 + \sum_{i=1}^{n}(Y_i - Y_i')^2 \qquad \textbf{(Equation 22.34)}$$

Total variation = Explained variation + Error variation

Note that in Equation 22.34, the error (unexplained) variation is the sum of the squared residuals. When $|r| = 1$, $\sum_{i=1}^{n}(Y_i - Y_i')^2 = 0$, which as noted earlier results in perfect prediction. When, on the other hand, $r = 0$, $\sum_{i=1}^{n}(Y_i' - \bar{Y})^2 = 0$, and thus the value $\bar{Y}$ will be the predicted value of Y' for each subject (since using Equations 22.6 and 22.7, if $r = 0$, $b_Y = 0$ and $a_Y = \bar{Y}$. If $a_Y = \bar{Y}$, the value of Y' computed with Equation 22.4 is $Y' = \bar{Y}$).

Through use of the residuals, the **coefficient of determination** can be expressed with Equation 22.35.

$$r^2 = \frac{\text{Explained variation}}{\text{Total variation}} = \frac{\sum_{i=1}^{n}(Y_i' - \bar{Y})^2}{\sum_{i=1}^{n}(Y_i - \bar{Y})^2} \qquad \textbf{(Equation 22.35)}$$

Equation 22.36 (which is the square root of Equation 22.35) represents an alternative (albeit more tedious) way of computing the correlation coefficient.

$$r = \pm\sqrt{\frac{\sum_{i=1}^{n}(Y_i' - \bar{Y})^2}{\sum_{i=1}^{n}(Y_i - \bar{Y})^2}} \qquad \textbf{(Equation 22.36)}$$

The value $s_{Y.X}^2$ is the **residual variance** (i.e., the variance of the residuals). The residual variance (which is the square of the value $s_{Y.X}$ computed with Equation 22.12) can be defined by Equation 22.37. The denominator of Equation 22.37 represents the degrees of freedom employed in the analysis.

$$s_{Y.X}^2 = \frac{\sum_{i=1}^{n} (Y_i - Y_i')^2}{n - 2} \qquad \text{(Equation 22.37)}$$

Equation 22.38, which is the square root of Equation 22.37, is an alternative (albeit more tedious) way of computing the **standard error of estimate**. Inspection of Equation 22.38 reveals that the greater the sum of the squared residuals, the greater the value of $s_{Y.X}$.

$$s_{Y.X} = \sqrt{\frac{\sum_{i=1}^{n} (Y_i - Y_i')^2}{n - 2}} \qquad \text{(Equation 22.38)}$$

Everything that has been said about the residuals with reference to the **regression line of *Y* on *X*** can be generalized to the **regression line of *X* on *Y*** (in which case the residual for each subject is represented by $e_i = (X_i - X_i')$). Thus, all of the equations described in this section can be generalized to the second regression line by employing the values X_i, X_i', and $\bar{X}$ place of Y_i, Y_i', and $\bar{Y}$.

3. Covariance In Section IV it is noted that the numerator of Equation 22.1 is referred to as the **sum of products**. When the sum of products is divided by $(n - 1)$, the resulting value represents a measure that is referred to as the **covariance**. Equation 22.39 is the computational equation for the covariance.

$$\text{cov}_{XY} = \frac{\Sigma XY - \frac{(\Sigma X)(\Sigma Y)}{n}}{n - 1} \qquad \text{(Equation 22.39)}$$

Equation 22.40 is the definitional equation of the covariance, which reveals the fact that covariance is an index of the degree to which two variables covary (i.e., vary in relation to one another).

$$\text{cov}_{XY} = \frac{\sum_{i=1}^{n} \left[(X_i - \bar{X})(Y_i - \bar{Y})\right]}{n - 1} \qquad \text{(Equation 22.40)}$$

Each subject's contribution to the covariance is computed as follows: The difference between a subject's score on the X variable and the mean of the X variable, and the difference between a subject's score on the Y variable and the mean of the Y variable are computed. The two resulting deviation scores are multiplied together. The resulting product represents that subject's contribution to the covariance. Upon obtaining a product for all n subjects, the sum of the n products (which is the numerator of Equation 22.1) is divided by $(n - 1)$. The resulting value represents the covariance, which it can be seen is essentially the average of the products of the deviation scores. The reason why the sum of the products is divided by $(n - 1)$ instead of n is because (as is also the case in computing the variance) division by the latter value provides a biased estimate of the population covariance. In the event one is computing a covariance for a sample and not using it as an estimate of the underlying population covariance, n is employed as the denominator of Equations 22.39 and 22.40.

Inspection of Equation 22.40 reveals that subjects who are above the mean on both variables or below the mean on both variables will contribute a positive product to the covariance. On the other hand, subjects who are above the mean on one of the variables but below the mean on the other variable will contribute a negative product to the covariance. If all or most of the subjects contribute positive products, the covariance will be a positive number. Since the value of r is a direct function of the sign of the covariance (which is a function of the sum of products), the resulting correlation coefficient will be a positive number. If all or most of the subjects contribute negative products, the covariance will be a negative number, and consequently the resulting correlation coefficient will also be negative. When among the n subjects the distribution of negative and positive products is such that they sum to zero, the sum of products will equal zero resulting in zero covariance, and r will equal zero. If for one of the two variables all subjects obtain the identical score, each subject will yield a product of zero, resulting in zero covariance (since the sum of products will equal zero). However, as noted in Section VI, since the value for the sum of squares will equal zero for a variable on which all subjects have the same score, Equation 22.1 becomes insoluble when all subjects have the same score on either of the variables. Based on what has been said with respect to the relationship between the sum of products and the covariance, the computation of r can be summarized by either of the following equations: $r = \text{cov}_{XY}/(\tilde{s}_X \tilde{s}_Y)$ and $r = SP_{XY}/\sqrt{SS_X SS_Y}$.

4. The homoscedasticity assumption of the Pearson product-moment correlation coefficient It is noted in Section I that one of the assumptions underlying the **Pearson product-moment correlation coefficient** is a condition referred to as **homoscedasticity (homo** means same and **scedastic** means scatter). Homoscedasticity exists in a set of data if the relationship between the X and Y variables is of equal strength across the whole range of both variables. Data that are not homoscedastic are **heteroscedastic**. When data are homoscedastic the accuracy of a prediction based on the regression line will be consistent across the full range of both variables. To illustrate, if data are homoscedastic and a strong positive correlation is computed between X and Y, the strong positive correlation will exist across all values of both variables. However, if for high values of X the correlation between X and Y is a strong positive one, but the strength of this relationship decreases as the value of X decreases, the data are heteroscedastic. As a general rule, if the distribution of one or both of the variables employed in a correlation is saliently skewed, the data are likely to be heteroscedastic. When, however, the data for both variables are distributed normally, the data will be homoscedastic.

Figure 22.6 presents two regression lines (which it will be assumed represent the **regression line of *Y* on *X***) and the accompanying data points. Note that in Figure 22.6a, which represents homoscedastic data, the distance of the data points from the regression line is about the same along the entire length of the line. Figure 22.6b, on the other hand, represents heteroscedastic data, since the data are not dispersed evenly along the regression line. Specifically, in Figure 22.6b the data points are close to the line for high values of X, yet as the value of X decreases, the data points become further removed from the line. Thus, the strength of the positive correlation is much greater for high values of X than it is for low values. This translates into the fact that a subject's Y score can be predicted with a greater degree of accuracy if the subject has a high score on the X variable as opposed to a low score. Directly related to this is the fact that the value of the standard error of estimate computed with Equation 22.12 ($s_{Y.X}$) will not be a representative measure of error variability for all values of X. Specifically, when $\tilde{s}_{Y.X}$ computed with Equation 22.14 (which is a function of the value $s_{Y.X}$) is employed to predict (within the framework of Equation 22.16) the Y score of a subject, the value of $\tilde{s}_{Y.X}$ will be larger for subjects who have a low score on the X variable, and thus the confidence interval (computed with Equation 22.16) associated with the predicted scores of such subjects will be larger than the confidence interval for subjects who have a high score on the X variable.

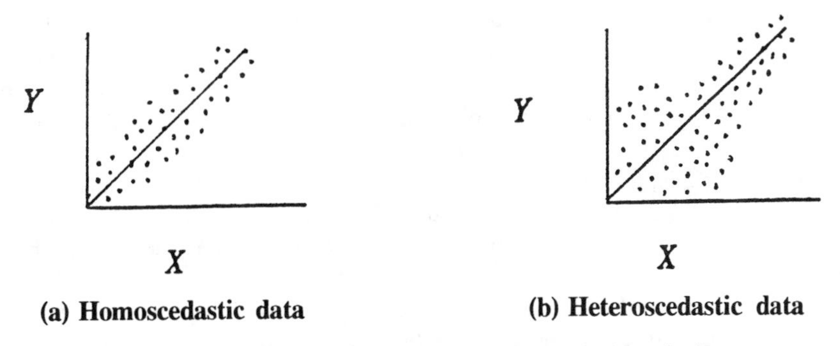

(a) Homoscedastic data **(b) Heteroscedastic data**

Figure 22.6 Homoscedastic Versus Heteroscedastic Data

5. The phi coefficient as a special case of the Pearson product-moment correlation coefficient A number of the correlational procedures discussed in this book represent special cases of the **Pearson product-moment correlation coefficient**. One of the procedures, the **phi coefficient (Test 11g)**, is described in Section VI of the **chi-square test for** $r \times c$ **tables (Test 11)**. Another of the procedures, the **point-biserial correlation coefficient (Test 22h)**, is described in Section IX; while a third procedure, the **Spearman rank-order correlation coefficient (Test 23)** is discussed later in the book.

In this section it will be demonstrated how the **phi coefficient** (ϕ) can be computed with Equation 22.1. In the discussion of the latter measure of association, it is noted that the value of **phi** is equivalent to the value of the **Pearson product-moment correlation coefficient** that will be obtained if the scores 0 and 1 are employed with reference to two dichotomous variables in a 2×2 contingency table. Using the data for Examples 11.1/11.2 (which employ a 2×2 contingency table), the scores 0 and 1 are employed for each of the categories on the two variables. Table 22.3 summarizes the data.

Table 22.3 Summary of Data for Examples 11.1/11.2

		X variable		Row sums
		0	1	
Y variable	0	$a = 30$	$b = 70$	100
	1	$c = 60$	$d = 40$	100
Column sums		90	110	Total = 200

Table 22.3 reveals the following: 30 subjects have both an X score and a Y score of 0; 70 subjects have an X score of 1 and a Y score of 0; 60 subjects have an X score of 0 and a Y score of 1; 40 subjects have both an X score and a Y score of 1. Employing this information we can determine $\Sigma X = 110$, $\Sigma X^2 = 110$, $\Sigma Y = 100$, $\Sigma Y^2 = 100$, $\Sigma XY = 40$. Substituting these values in Equation 22.1, the value $r = .30$ is computed. The latter value is identical to $\phi = .30$ computed for Examples 11.1/11.2 with Equations 11.17 and 11.18.

$$r = \frac{40 - \dfrac{(110)(100)}{200}}{\sqrt{\left[110 - \dfrac{(110)^2}{200}\right]\left[100 - \dfrac{(100)^2}{200}\right]}} = .30$$

6. Autocorrelation/serial correlation In Section VII of the **single-sample runs test (Test 7)** a number of procedures are discussed that are employed in determining whether the ordering of a series of numbers is random. Among the procedures that are briefly discussed is

autocorrelation, which is also known as **serial correlation**. Autocorrelation and the derivation of the corresponding regression equations (referred to as **autoregression**) are complex subjects that are primarily discussed in books that deal with statistical applications in business and economics. The most basic methodology that can be employed for autocorrelation is to pair each of the numbers in a series of n numbers with the number that follows it in the series. Upon doing this, the **Pearson product-moment correlation coefficient** between the resulting $(n - 1)$ pairs of numbers is computed. It is also possible to pair each number with the number whose ordinal position is some value other than one digit after it. In other words, each number can be paired with the number that is two digits after it, three digits after it, etc., or, with the number that is one, two, three, etc. digits before it in the series. In autocorrelation the number of digits that separate two values that are paired with one another is referred to as the **lag value**. In the example to be employed in this section the **lag +1** will be used, since each number will be paired with the number that is one ordinal position above it in the series. If, instead, each number is paired with the number that falls two ordinal positions above it in the series, the lag value is +2. If, on the other hand, each number is paired with the number that precedes it by one ordinal position in the series, the lag value is −1. The higher the absolute value of the lag value, the fewer the number of pairs that will be employed in computing the correlation. Thus, if in a series of ten digits each number is paired with the number that is above it by two ordinal positions, there will only be $n - 2 = 8$ pairs of X and Y scores. This is the case, since the first two numbers in the series can only be X scores, and the last two numbers in the series can only be Y scores. Regardless of the lag value employed in an autocorrelation, if the sequence of numbers in a series is random, the computed value of the correlation coefficient should equal zero.

One variant of the methodology described in this section (which is referred to as **noncircular serial correlation**) is a procedure referred to as **circular serial correlation**. In circular serial correlation every number in a series of n numbers is paired with another number, including any numbers in the series that do not have a number following it. Numbers that are not followed by any numbers are sequentially paired with the numbers at the beginning of the series. Thus, if the lag value is +1, the last number in the series is paired with the first number in the series. If the lag value is +2, the $(n - 1)^{th}$ number is paired with the first number in the series, and the nth number is paired with the second number in the series.

Table 22.4 Arrangement of Numbers for Autocorrelation

Column A	Column B
	4
4	3
3	5
5	2
2	1
1	3
3	2
2	1
1	1
1	2
2	

To illustrate autocorrelation, the following ten digit series of numbers will be evaluated: 4, 3, 5, 2, 1, 3, 2, 1, 1, 2. In Table 22.4 the ten digits are arranged sequentially from top to bottom in Column A. The same ten digits are arranged sequentially in Column B, except for the fact that they are arranged so that each digit in Column B is adjacent to the digit in

Column A that directly precedes it in the series. If each pair of adjacent values is treated as a set of scores, the value in Column A can be designated as an X score, and the value in Column B can be designated as a Y score. If the latter is done, each of the ten digits in the series will at some point be designated as both an X score and a Y score, except for the first digit which will only be an X score and the last digit which will only be a Y score.

Table 22.5, which contains the nine pairs of digits in Table 22.4, summarizes the required values for computing the **Pearson product-moment correlation coefficient.** Note that the value $n = 9$ is employed in computing the value of r, since that is the number of sets of paired scores.

Employing Equation 22.1, the value $r = .28$ is computed for the correlation coefficient.

$$r = \frac{53 - \dfrac{(22)(20)}{9}}{\sqrt{\left[70 - \dfrac{(22)^2}{9}\right]\left[58 - \dfrac{(20)^2}{9}\right]}} = .28$$

Table 22.5 Data for Autocorrelation

X	X^2	Y	Y^2	XY
4	16	3	9	12
3	9	5	25	15
5	25	2	4	10
2	4	1	1	2
1	1	3	9	3
3	9	2	4	6
2	4	1	1	2
1	1	1	1	1
1	1	2	4	2
$\Sigma X = 22$	$\Sigma X^2 = 70$	$\Sigma Y = 20$	$\Sigma Y^2 = 58$	$\Sigma XY = 53$

If the usual criteria for evaluating an r value are employed, the degrees of freedom for the analysis are $df = n - 2 = 9 - 2 = 7$. In **Table A16**, the tabled critical two-tailed .05 and .01 values for $df = 7$ that are employed to evaluate the nondirectional alternative hypothesis H_1: $\rho \neq 0$ are $r_{.05} = .666$ and $r_{.01} = .798$. Since the computed value $r = .28$ is less than $r_{.05} = .666$, the null hypothesis H_0: $\rho = 0$ cannot be rejected. Thus, the data do not indicate that the underlying population correlation is some value other than zero.

In point of fact, the tabled critical values in **Table A16** are not the most appropriate values for evaluating the value $r = .28$. This is the case, since the sampling distribution for a serial correlation coefficient is not identical to the sampling distribution upon which the values in **Table A16** are based. The sampling distribution upon which **Table A16** is based assumes that the n pairs of scores are independent of one another. Since in Table 22.5 all of the digits in the series (with the exception of the last digit) represent both an X and a Y variable, the latter assumption is violated. Because the pairs are not independent, the residuals derived from the data may also not be independent (independence of the residuals is an underlying assumption of the least squares regression model).[21] Although not necessarily the case with pseudorandom numbers (i.e., a series of random numbers generated with a computer algorithm), it is common in autocorrelated data in business and economics for residuals to be dependent on one another. Most commonly, in the latter disciplines there is a positive correlation between residuals. When the latter is true, residuals of the same sign occur in clusters — i.e., residuals for adjacent pairs have the identical sign. When there is a negative autocorrelation, adjacent residuals tend to alternate between a positive and negative sign.

Because of the fact that the residuals may not be independent, a sampling distribution other than the one upon which the critical values in **Table A16** are based should be employed to evaluate the value $r = .28$ computed with Equation 22.1. Anderson (1942) demonstrated that in the sampling distribution for a serial correlation, the absolute value of a critical value at a prespecified level of significance is smaller than the corresponding critical value in **Table A16**. Furthermore, the limits that define a critical value at a prespecified level of significance are asymmetrical (i.e., the absolute value of a critical value will not be identical for a positive versus a negative r value). Anderson (1942) computed the critical two-tailed .05 and .01 values of r for values of n between 5 and 75 for lag +1. For large sample sizes he determined that Equation 22.41 (which employs the normal distribution) can be used to provide a good approximation of the critical values of r when the lag value is +1.[22]

$$r = \pm z \sqrt{\frac{n-2}{(n-1)^2}} - \frac{1}{n-1} \qquad \textbf{(Equation 22.41)}$$

Where: z represents the tabled critical value in the normal distribution that corresponds to the prespecified level of significance employed in evaluating r

n represents the total number of numbers in the series. Note that n is not the number of pairs of numbers employed in computing the coefficient of correlation.

Employing the values derived by Anderson (1942) for the exact sampling distribution of the serial correlation coefficient (which is not reproduced in this book), it can be determined (for $\alpha = .05$ and $n = 10$) that in order to reject the null hypothesis H_0: $\rho = 0$, the computed value of r must be equal to or greater than $r_{.05} = .360$ or equal to or less than $r_{.05} = -.564$. (The value of n used in Anderson's (1942) table represents the total number of digits in the series and not the number of pairs of digits employed in computing the correlation.) Since $r = .28$ is less than $r_{.05} = .360$, the null hypothesis can be retained. Thus, regardless of whether one employs **Table A16** or Anderson's (1942) critical values, the null hypothesis is retained. Nevertheless, the difference between the critical values in the two tables is substantial.

Use of Anderson's (1942) tables and/or Equation 22.41 provide for a more powerful test of the alternative hypothesis H_0: $\rho \neq 0$ than do the critical values in **Table A16**. Although the degree of discrepancy between a critical value in **Table A16** and a critical value computed with Equation 22.41 decreases as the size of n increases, even for large sample sizes the absolute values in **Table A16** are noticeably higher. It should be noted that use of Equation 22.41 with small samples yields absolute critical values that are too high.

It is noted in the discussion of tests of randomness under the **single-sample runs test**, that it is not uncommon for two or more of the available tests for determining randomness to yield conflicting results. Although autocorrelation is not considered to be among the most rigorous tests for randomness, if one conducts multiple autocorrelations on a series (i.e., for the lag values +1, +2, +3, etc. and −1, −2, −3, etc.), and all or most lead to retention of the null hypothesis, such a protocol will provide a more authoritative analysis with respect to randomness than will the single analysis for lag +1 conducted in this section. It should be noted that if for a series of n numbers (where the value of n is large) an autocorrelation is conducted for every possible positive and negative lag value, just by chance it is expected that some of the computed serial correlations will be significant. Whatever prespecified alpha value the researcher employs will determine the proportion of significant correlations that can be obtained which will still allow one to retain the null hypothesis H_0: $\rho = 0$.

One limitation of autocorrelation as a test of randomness should be noted. Assume that a researcher is evaluating a series in which in any trial a number can assume any one of $k = 5$ possible values. For instance, in the example employed in this section it is assumed that the integer values 1, 2, 3, 4, 5 are the only possible values that can occur. In a truly random series of reasonable length, each of the five digits would be expected to occur approximately the same number of times. Yet, it is entirely possible to have a series of numbers in which one or more of the integer values does not even occur one time, yet the resulting autocorrelation is $r = 0$. For instance, a computer can be programmed to generate a series of 1000 pseudorandom numbers employing the integer values 1, 2, 3 ,4, 5. Yet it is theoretically possible for the computer algorithm to generate 1000 digits, all of which are either 1 or 2. If the autocorrelation between the values of 1 and 2 that are generated is zero, it will suggest the sequence of numbers is random. Although it may be a random sequence for a population in which the only values the numbers may assume are the integer values 1 or 2, it is not a random series for a population in which the numbers may assume an integer value between 1 and 5. Whereas most of the other tests that are employed in evaluating randomness will identify this problem, autocorrelation will not.

Research in such fields as economics, business, and political science often employs autocorrelation for **time series analysis**, which is a methodology for studying the sequential progression of events. The results of a time series analysis can be useful in predicting future values for such variables as stock prices, sales revenues, crop yields, crime rates, weather, etc. Such predictions are predicated on the fact that significant data based on autocorrelation indicate sequential dependence with respect to the variable of interest. As is the case with Example 21.1, a regression equation that is derived from data that are autocorrelated is employed in making predictions. Use of a regression equation in this context is referred to as **autoregression**. Sources that discuss autoregression note that derivation of a regression equation through use of the method of least squares as described in Section VI will underestimate error variability, and consequently will not provide the most accurate basis for prediction. For this reason when autocorrelation is employed with a set of data, alternative procedures are recommended for making predictions, as well as for evaluating the null hypothesis H_0: $\rho = 0$. A procedure recommended in most sources is one developed by Durbin and Watson (1950, 1951, 1971). Among the sources that describe the **Durbin–Watson test** (which is only appropriate for a lag value of +1) are Chou (1989), Montgomery and Peck (1992), and Netter *et al.* (1983). The latter sources also describe other alternative approaches for autoregression.

VIII. Additional Examples Illustrating the Use of the Pearson Product-Moment Correlation Coefficient

Two additional examples that can be evaluated with the **Pearson product-moment correlation coefficient** are presented in this section. Since the data for Examples 22.2 and 22.3 are identical to the data employed in Example 22.1, they yield the same result.

Example 22.2 *The editor of an automotive magazine conducts a survey to see whether it is possible to predict the number of traffic citations one receives for speeding based on how often a person changes his or her motor oil. The responses of five subjects on the two variables follow. (For each subject, the first score represents the number of oil changes (which represents the X variable), and the second score the number of traffic citations (which represents the Y variable).)* **Subject 1** (20, 7); **Subject 2** (0, 0); **Subject 3** (1, 2); **Subject 4** (12, 5); **Subject 5** (3, 3). *Do the data indicate there is a significant correlation between the two variables?*

Example 22.3 *A pediatrician speculates that the length of time an infant is breast fed may be related to how often a child becomes ill. In order to answer the question, the pediatrician obtains the following two scores for five three-year-old children: The number of months the child was breast fed (which represents the X variable) and the number of times the child was brought to the pediatrician's office during the current year (which represents the Y variable). The scores for the five children follow:* **Child 1** (20, 7); **Child 2** (0, 0); **Child 3** (1, 2); **Child 4** (12, 5); **Child 5** (3, 3). *Do the data indicate that the length of time a child is breast fed is related to the number of times a child is brought to the pediatrician?*

IX. Addendum

Three bivariate correlational measures that are related to the **Pearson product-moment correlation coefficient** are described in the **Addendum**. Each of the correlation coefficients to be described assumes that the scores on at least one of the variables can be expressed within the format of interval/ratio data, and that the underlying distribution of these scores is continuous and normal. Two of the correlational procedures assume that the underlying interval/ratio scores on one or both of the variables have been converted into a dichotomous (two category) format. A brief description of the three procedures follows:

The point-biserial correlation coefficient (Test 22h) The **point-biserial correlation coefficient** (r_{pb}) (which is a special case of the **Pearson product-moment correlation coefficient**) is employed if one variable is expressed as interval/ratio data, and the other variable is represented by a dichotomous nominal/categorical scale (i.e., two categories).

The biserial correlation coefficient (Test 22i) The **biserial correlation coefficient** (r_b) is employed if both variables are based on an interval/ratio scale, but the scores on one of the variables have been transformed into a dichotomous nominal/categorical scale. It provides an estimate of the value that would be obtained for the **Pearson product-moment correlation coefficient**, if instead of the dichotomized variable one employed the scores on the underlying interval/ratio scale which the latter variable represents.

The tetrachoric correlation coefficient (Test 22j) The **tetrachoric correlation coefficient** (r_{tet}) is employed if both variables are based on an interval/ratio scale, but the scores on both of the variables have been transformed into a dichotomous nominal/categorical scale. It provides an estimate of the value that would be obtained for the **Pearson product-moment correlation coefficient**, if instead of the dichotomized variables one employed the scores on the underlying interval/ratio scales that the latter variables represent.

In the latter part of the **Addendum**, three **multivariate** correlational procedures are described. The use of the term multivariate implies that data for three or more variables are employed in the analysis. The procedures that are described are extensions of the **Pearson product-moment correlation coefficient** to an analysis involving three or more variables. All of the procedures (each of which is discussed in reference to an analysis involving three variables) assume that all of the variables are measured on an interval/ratio scale. A brief description of the procedures to be described follows.

The multiple correlation coefficient (Test 22k) The **multiple correlation coefficient** (R) is a correlation between a criterion variable and a linear combination of two or more predictor variables.

The partial correlation coefficient (Test 22l) The **partial correlation coefficient** (e.g., $r_{YX_1 . X_2}$) measures the degree of association between two variables, after any linear association one or more additional variables has with the two variables has been removed.

The semi-partial correlation coefficient (Test 22m) The **semi-partial correlation coefficient** (or **part correlation coefficient**) (e.g., $r_{Y(X_1 . X_2)}$) measures the degree of association between two variables, with the linear association of one or more other variables removed from only one of the two variables that are being correlated with one another.

1. Bivariate measures of correlation that are related to the Pearson-product moment correlation coefficient

Test 22h: The point-biserial correlation coefficient (r_{pb}) As noted earlier, the **point-biserial correlation coefficient** represents a special case of the **Pearson product-moment correlation coefficient.** The **point-biserial correlation coefficient** is employed if one variable is expressed as interval/ratio data, and the other variable is represented by a dichotomous nominal/categorical scale. Examples of variables that constitute a dichotomous nominal/categorical scale are **male** versus **female** and **employed** versus **unemployed**. In using the **point-biserial correlation coefficient,** it is assumed that the dichotomous variable is not based on an underlying continuous interval/ratio distribution. If, in fact, the dichotomous variable is based on the latter type of distribution, the **biserial correlation coefficient (Test 22i)** is the appropriate measure to employ. Examples of variables that are expressed in a dichotomous format, but which are based on an underlying continuous interval/ratio distribution are **pass** versus **fail** and **above average intelligence** versus **below average intelligence**. Obviously, not everyone who passes (or fails) a test or a course performs at the same level. In the same respect, the distribution of intelligence of people who are **above average** or **below average** is not uniform. There are, of course, variables with respect to which it can be argued whether they are based on an underlying continuous distribution (such as perhaps handedness, which will be employed as a dichotomous variable in the example to be presented in this section). As is the case with the **Pearson product-moment correlation coefficient,** the range of values within which r_{pb} can fall are $-1 \leq r_{pb} \leq +1$. Example 22.4 will be employed to illustrate the use of the **point-biserial correlation coefficient.**

Example 22.4 *A study is conducted to determine whether there is a correlation between handedness and eye-hand coordination. Five right-handed and five left-handed subjects are administered a test of eye-hand coordination. The test scores of the subjects follow (the higher a subject's score, the better his or her eye-hand coordination):* **Right-handers:** 11, 1, 0, 2, 0; **Left-handers:** 11, 11, 5, 8, 4. *Is there a statistical relationship between handedness and eye-hand coordination?*

In the analysis handedness will represent the X variable, and the eye-hand coordination test scores will represent the Y variable. With respect to handedness (which is a dichotomous variable), all right-handed subjects will be assigned a score of 1 on the X variable, and all left-handed subjects will be assigned a score of 0. Table 22.6 summarizes the data for the ten subjects employed in the study.

Table 22.6 Data for Example 22.4

Subject	X	X^2	Y	Y^2	XY
1	1	1	11	121	11
2	1	1	1	1	1
3	1	1	0	0	0
4	1	1	2	4	2
5	1	1	0	0	0
6	0	0	11	121	0
7	0	0	11	121	0
8	0	0	5	25	0
9	0	0	8	64	0
10	0	0	4	16	0
	$\Sigma X = 5$	$\Sigma X^2 = 5$	$\Sigma Y = 53$	$\Sigma Y^2 = 473$	$\Sigma XY = 14$

Since the **point-biserial correlation coefficient** is a special case of the **Pearson product-moment correlation coefficient,** Equation 22.42 (which is identical to Equation 22.1) is employed to compute r_{pb}. Employing Equation 22.42, the value $r_{pb} = -.57$ is computed.

$$r_{pb} = \frac{\Sigma XY - \dfrac{(\Sigma X)(\Sigma Y)}{n}}{\sqrt{\left[\Sigma X^2 - \dfrac{(\Sigma X)^2}{n}\right]\left[\Sigma Y^2 - \dfrac{(\Sigma Y)^2}{n}\right]}}$$

(Equation 22.42)

$$= \frac{14 - \dfrac{(5)(53)}{10}}{\sqrt{\left[5 - \dfrac{(5)^2}{10}\right]\left[473 - \dfrac{(53)^2}{10}\right]}} = -.57$$

Equation 22.43 is an alternative equation for computing the **point-biserial correlation coefficient.**

$$r_{pb} = \left[\frac{\bar{Y}_1 - \bar{Y}_0}{\tilde{s}_Y}\right]\sqrt{p_0 p_1}\sqrt{\frac{n}{n-1}}$$

(Equation 22.43)

$$= \left[\frac{2.8 - 7.8}{4.62}\right]\sqrt{(.5)(.5)}\sqrt{\frac{10}{10-1}} = -.57$$

Where: $\bar{Y}_0$ and $\bar{Y}_1$ are respectively the average scores on the Y variable for subjects who are categorized 0 versus 1 on the X variable
p_0 equals the proportion of subjects with an X score of 0
p_1 equals the proportion of subjects with an X score of 1

In employing Equation 22.43, $\bar{Y}_1 = 2.8$ and $\bar{Y}_0 = 7.8$. The value $\tilde{s}_Y = 4.62$, which represents the unbiased estimate of the population standard deviation for the Y variable (which is computed with Equation I.8), is computed below.

$$\tilde{s}_Y = \sqrt{\frac{\Sigma Y^2 - \dfrac{(\Sigma Y)^2}{n}}{n-1}} = \sqrt{\frac{473 - \dfrac{(53)^2}{10}}{10-1}} = 4.62$$

It should be noted that some sources employ Equation 22.44 to compute the value of the **point-biserial correlation coefficient.**

$$r_{pb} = \left[\frac{\bar{Y}_1 - \bar{Y}_0}{s_Y}\right]\sqrt{p_0 p_1} = \left[\frac{2.8 - 7.8}{4.38}\right]\sqrt{(.5)(.5)} = -.57 \quad \textbf{(Equation 22.44)}$$

Note that Equation 22.44 employs the sample standard deviation (computed with Equation I.7 — i.e., $s_Y = \sqrt{[\Sigma Y^2 - ((\Sigma Y)^2/n)]/n}$), which is a biased estimate of the population standard deviation. For Example 22.2, $s_Y = 4.38$. When $s_Y = 4.38$ is substituted in the Equation 22.44, it yields the value $r_{pb} = -.57$.

The reader should take note of the fact that the sign of r_{pb} is irrelevant unless the categories on the dichotomized variable are ordered (which is not the case for Example 22.4). The reason for employing the absolute value of r_{pb} is that the use of the scores 0 and 1 for the two categories is arbitrary, and does not indicate that one category is superior to the other. (If all right-handed subjects are assigned a score of 0 on the X variable and all left-handed subjects are assigned a score of 1, the value computed for $r_{pb} = +.57$, which is the same absolute value computed for the data in Table 22.6.) Since the categories are not ordered, from this point on in the discussion the absolute value $r_{pb} = .57$ will be employed to represent the value of r_{pb}. In the event the categories are ordered, the score 1 should be employed for the category associated with higher performance/quality, and the score 0 should be employed for the category associated with lower performance/quality. In all likelihood, if the categories are ordered they are likely to be based on an underlying continuous distribution, and in such a case the appropriate correlational measure to employ is the **biserial correlation coefficient.**

The square of the **point-biserial correlation coefficient** represents the **coefficient of determination**, which as noted in Section VI indicates the amount of variability on the Y variable that can be accounted for by variability on the X variable. Since, $r_{pb}^2 = (.57)^2 = .325$, 32.5% of the variability on the test of eye-hand coordination can be accounted for on the basis of a person's handedness.

The data employed for Example 22.4 are identical to that employed for Example 8.1 (which is used to illustrate the ***t* test for two independent samples**). In point of fact, the **point-biserial correlation coefficient** can be employed to measure the magnitude of treatment effect in an experiment that has been evaluated with a ***t* test for two independent samples**, if the grouping of the subjects is conceptualized as the dichotomous variable. Thus, in Example 8.1, if each subject in Group 1 (Drug Group) is assigned an X score of 1, and each subject in Group 2 (Placebo Group) is assigned an X score of 0, the data for the experiment can be summarized with Table 22.6. Since analysis of the data in Table 22.6 yields $r_{pb} = .57$ and $r_{pb}^2 = .325$, the researcher can conclude that 32.5% of the variability on the dependent variable (the depression ratings for subjects) can be accounted for on the basis of which group a subject is a member.

In Section VI of the ***t* test for two independent samples**, the measure of association that is employed to measure the magnitude of treatment effect is the **omega squared** ($\tilde{\omega}^2$) statistic. The value of **omega squared** computed for Example 8.1 is $\tilde{\omega}^2 = .22$. Since $\tilde{\omega}^2$ is interpreted in the same manner as r_{pb}^2, the value $\tilde{\omega}^2 = .22$ indicates that 22% of the variability on the dependent variable can be accounted for on the basis of which group a subject is a member. Obviously, the latter value is lower than the value $r_{pb}^2 = .325$ computed in this section. The discrepancy between the two values will be discussed further later in this section. In point of fact, $r_{pb}^2 = .325$ is equivalent to the **eta squared** ($\tilde{\eta}^2$) statistic, which is an alternative measure of association that some sources employ in assessing the magnitude of treatment effect for the ***t* test for two independent samples**. Marascuilo and Serlin (1988) note that both r_{pb}^2 and $\tilde{\eta}^2$ represent a **correlation ratio**. The **correlation ratio**, which can be defined within the framework of an analysis of variance, is the ratio of the explained sum of squares over the total sum of squares. To clarify the meaning of a correlation ratio, let us assume that in lieu of the ***t* test for two independent samples**, the data for Example 8.1 are evaluated with the **single-factor between-subjects analysis of variance (Test 16)**. Table 22.7 is the summary table of the analysis of variance for Example 8.1.

Table 22.7 Summary Table of Analysis of Variance for Example 8.1

Source of variation	*SS*	*df*	*MS*	*F*
Between-groups	62.5	1	62.5	3.86
Within-groups	129.6	8	16.2	
Total	**192.1**	**9**		

Within the framework of the **single-factor between-subjects analysis of variance**, the **correlation ratio** (which is computed with Equation 16.42) is $\tilde{\eta}^2 = SS_{BG}/SS_T$. Since both $\tilde{\eta}^2$ and r_{pb}^2, represent the correlation ratio, $\tilde{\eta}^2 = r_{pb}^2 = SS_{BG}/SS_T$. Thus, for Example 8.1, $\tilde{\eta}^2 = r_{pb}^2 = 62.5/192.1 = .325$.

Equation 22.45 can also be employed to compute the value $\tilde{\eta}^2 = r_{pb}^2$ (where $t = \sqrt{F} = \sqrt{3.86} = 1.964$).

$$\tilde{\eta}^2 = r_{pb}^2 = \frac{t^2}{t^2 + df} = \frac{(1.964)^2}{(1.964)^2 + 8} = .325 \qquad \textbf{(Equation 22.45)}$$

Where: $df = n_1 + n_2 - 2$ (which is the degrees of freedom for the **t test for two independent samples**)

The fact that different measures of magnitude of treatment effect may not yield the same value is discussed in Section VI of the **single-factor between-subjects analysis of variance**. In the latter discussion it is noted that the computed value $\tilde{\eta}^2$ is a biased estimate of the underlying population parameter η^2, and an adjusted value (which is less biased) can be computed with Equation 16.43. The latter value is now computed for Example 8.1: Adjusted $\tilde{\eta}^2 = 1 - [MS_{WG}/MS_T] = 1 - [16.2/21.34] = .24$. (Where, $MS_T = SS_T/df_T = 192.1/9 = 21.34$.) Note that the value Adjusted $\tilde{\eta}^2 = .24$ is closer to the value $\tilde{\omega}^2 = .22$ than the previously computed value $\tilde{\eta}^2 = r_{pb}^2 = .325$.

Test 22h-a: Test of significance for a point biserial correlation coefficient The null hypothesis H_0: $\rho_{pb} = 0$ can be evaluated with Equation 22.46 (which is identical to Equation 22.3, which is employed to evaluate the same hypothesis with reference to the **Pearson product-moment correlation coefficient**). As is the case for Equation 22.3, the degrees of freedom for Equation 22.46 are $df = n - 2$. Employing Equation 22.46, the value $t = 1.96$ is computed.

$$t = \frac{r_{pb}\sqrt{n - 2}}{\sqrt{1 - r_{pb}^2}} = \frac{.57\sqrt{10 - 2}}{\sqrt{1 - (.57)^2}} = 1.96 \qquad \textbf{(Equation 22.46)}$$

It will be assumed that the nondirectional alternative hypothesis H_1: $\rho_{pb} \neq 0$ is evaluated. Employing **Table A2**, for $df = 10 - 2 = 8$, the tabled critical two-tailed .05 and .01 values are $t_{.05} = 2.31$ and $t_{.01} = 3.36$. Since the obtained value $t = 1.96$ is less than both of the aforementioned critical values, the null hypothesis H_0: $\rho_{pb} = 0$ cannot be rejected.

Since the value of the **point-biserial correlation coefficient** is a direct function of the difference between $\bar{Y}_0$ and $\bar{Y}_1$, a significant difference between the latter two mean values indicates that the absolute value of the correlation between the two variables is significantly above zero. Thus, an alternative way of evaluating the null hypothesis H_0: $\rho_{pb} = 0$ is to conduct a **t test for two independent samples** contrasting the two mean values $\bar{Y}_0$ and $\bar{Y}_1$. The fact that the latter analysis will yield a result that is equivalent to that obtained with Equation 22.46 can be confirmed by the fact that the value $t = 1.96$ computed above with Equation 22.46 is identical to the absolute t value computed for the same set of data with Equation 8.1 (for Example 8.1). Sources that provide additional discussion of the **point-biserial correlation coefficient** are Guilford (1965), Lindeman *et al.* (1980), and McNemar (1969).

Test 22i: The biserial correlation coefficient (r_b) The **biserial correlation coefficient** is employed if both variables are based on an interval/ratio scale, but the scores on one of the variables have been transformed into a dichotomous nominal/categorical scale. An example of a situation where an interval/ratio variable would be expressed as a dichotomous variable is a test based on a normally distributed interval/ratio scale for which the only information available is whether a subject has passed or failed the test. The value computed for the **biserial correlation coefficient** represents an estimate of the value that would be obtained for the **Pearson product-moment correlation coefficient**, if instead of employing a dichotomized variable one had employed the scores on the underlying interval/ratio scale.

The **biserial correlation coefficient** is based on the assumption that the underlying distribution for both of the variables is continuous and normal. Since the accuracy of r_b is highly dependent upon the assumption of normality, it should not be employed unless there is empirical evidence to indicate that the distribution underlying the dichotomous variable is normal. If the underlying distribution of the dichotomous variable deviates substantially from normality, the computed value of r_b will not be an accurate approximation of the underlying population correlation r_b estimates. One consequence of the normality assumption being violated is that under certain conditions the absolute value computed for r_b may exceed 1. In point of fact, Lindeman *et al.* (1980) note that the theoretical limits of r_b are $-\infty \leq r_b \leq +\infty$.

In contrast to r_{pb}, the sign of the **biserial correlation coefficient** should be taken into account, since it clarifies the nature of the relationship between the two variables. This is the case, since the dichotomous variable will involve two ordered categories. In assigning scores to subjects on the ordered dichotomized variable, the score 1 should be employed for the category associated with higher performance/quality, and the score 0 should be employed for the category associated with lower performance/quality.

In order to illustrate the computation of the **biserial correlation coefficient**, let us assume a researcher wants to determine whether there is a statistical relationship between intelligence and eye-hand coordination. Ten subjects are categorized with respect to both variables. Although the evaluation of each subject's intelligence is based on an interval/ratio intelligence test score, we will assume that the only information available to the researcher is whether an individual is above or below average in intelligence. In view of this, intelligence, which will be designated the X variable, will have to be represented as a dichotomous variable. Subjects who are above average in intelligence will be assigned a score of 1, and subjects who are below average in intelligence will be assigned a score of 0. The scores on the eye-hand coordination test will represent the Y variable. Example 22.5, which employs the same set of data as Example 22.4, summarizes the above described experiment.

Example 22.5 *A study is conducted to determine whether there is a correlation between intelligence and eye-hand coordination. Five subjects who are above average in intelligence and five subjects who are below average in intelligence are administered a test of eye-hand coordination. The test scores of the subjects follow (the higher a subject's score, the better his or her eye-hand coordination):* **Above average intelligence**: 11, 1, 0, 2, 0; **Below average intelligence**: 11, 11, 5, 8, 4. *Is there a statistical relationship between intelligence and eye-hand coordination?*

The **biserial correlation coefficient** can be computed with either Equation 22.47 or 22.48. It can also be computed with Equation 22.49 if r_{pb} has been computed for same set of data. Note that except for h, all of the terms in the aforementioned equations are also employed in computing the **point-biserial correlation coefficient**. The value h represents the height (known more formally as the **ordinate**) of the standard normal distribution at the point which divides the proportions p_0 and p_1. Specifically, employing **Table A1** the z value is identified that delineates the point on the normal curve that a proportion of the cases corresponding to the smaller of the two proportions p_0 versus p_1 falls above and the larger of the two proportions falls below. The tabled value of h associated with that z value is employed in whatever equation one employs for computing r_b. If, as is the case in our example, $p_0 = p_1 = .5$, the value of z will equal zero, and thus the corresponding value of $h = .3989$. When $h = .3989$ and the other appropriate values employed for Example 22.5 (which are summarized in Table 22.6) are substituted in Equations 22.47–22.49, the value $r_b = -.71$ is computed.

(Equation 22.47)

$$r_b = \left[\frac{\bar{Y}_1 - \bar{Y}_0}{\tilde{s}_Y}\right]\left[\frac{p_0 p_1}{h}\right]\sqrt{\frac{n}{n-1}} = \left[\frac{2.8 - 7.8}{4.62}\right]\left[\frac{(.5)(.5)}{.3989}\right]\sqrt{\frac{10}{10-1}} = -.71$$

$$r_b = \left[\frac{\bar{Y}_1 - \bar{Y}_0}{s_Y}\right]\left[\frac{p_0 p_1}{h}\right] = \left[\frac{2.8 - 7.8}{4.38}\right]\left[\frac{(.5)(.5)}{.3989}\right] = -.71 \quad \textbf{(Equation 22.48)}$$

$$r_b = \frac{r_{pb}\sqrt{p_0 p_1}}{h} = \frac{(-.57)\sqrt{(.5)(.5)}}{.3989} = -.71 \quad \textbf{(Equation 22.49)}$$

Note that for the same set of data, the absolute value $r_b = .71$ is larger than the absolute value $r_{pb} = .57$ computed for the **point-biserial correlation coefficient**. In point of fact, for the same set of data (except when $r_b = r_{pb} = 0$) the absolute value of r_b will always be larger than the absolute value of r_{pb}, since $\sqrt{p_0 p_1}/h$ will always be larger than 1. The closer together the values p_0 and p_1, the less the discrepancy between the values of r_b and r_{pb}. If there is reason to believe that the normality assumption for the dichotomous variable has been violated, most sources recommend computing r_{pb} instead of r_b, since r_b may be a spuriously inflated estimate of the underlying population correlation. When the latter is taken into consideration, along with the fact that by dichotomizing a continuous variable one sacrifices valuable information, it can be understood why the **biserial correlation coefficient** is infrequently employed within the framework of research.

Guilford (1965) notes that, given the normality assumption has not been violated, those conditions which optimize the likelihood of r_b providing a good estimate of the underlying population parameter ρ_b are as follows: a) The value of n is large; and b) The values of p_0 and p_1 are close together. It should be noted that (as is the case for Pearson r and r_{pb}) if the relationship between two variables is nonlinear, the computed value of r_b will only represent the degree of linear relationship between the variables.

Test 22i-a: Test of significance for a biserial correlation coefficient Lindeman *et al.* (1980) note that the null hypothesis H_0: $\rho_b = 0$ can be evaluated with Equation 22.50. Although the latter equation, which is based on the normal distribution, assumes a large sample size, it is employed to evaluate the value $r_b = -.71$ computed for Example 22.5. Note that the sign of z will always be the same as the sign of r_b.

$$z = \frac{h r_b}{\sqrt{\frac{p_0 p_1}{n}}} = \frac{(.3989)(-.71)}{\sqrt{\frac{(.5)(.5)}{10}}} = -1.79 \quad \textbf{(Equation 22.50)}$$

Employing **Table A1**, the tabled critical two-tailed values are $z_{.05} = 1.96$ and $z_{.01} = 2.58$, and the tabled critical one-tailed values are $z_{.05} = 1.65$ and $z_{.01} = 2.33$. The nondirectional alternative hypothesis H_1: $\rho_b \neq 0$ is not supported, since the absolute value $z = 1.79$ is less than the tabled critical two-tailed value $z_{.05} = 1.96$. However, the directional alternative hypothesis H_1: $\rho_b < 0$ is supported at the .05 level, since $z = -1.79$ is a negative number with an absolute value that is greater than the tabled critical one-tailed value $z_{.05} = 1.65$. The moderately strong negative correlation between the two variables indicates that subjects who score below average on the intelligence test perform better on the test of eye-hand coordination than do subjects who score above average on the intelligence test. The latter can be confirmed by the fact that $\bar{Y}_1 = 2.8$ is less than $\bar{Y}_0 = 7.8$.[23]

As is the case for the **point-biserial correlation coefficient**, since the value of the **biserial correlation coefficient** is a direct function of the difference between $\bar{Y}_0$ and $\bar{Y}_1$, a significant difference between the two mean values indicates that the absolute value of the correlation between the two variables is significantly above zero. Thus, an alternative way of evaluating the null hypothesis H_0: $\rho_b = 0$ is to contrast the means $\bar{Y}_0$ and $\bar{Y}_1$ with a ***t* test for two independent samples**. However, the result obtained with Equation 22.50 will not necessarily be consistent with the result obtained if the ***t* test for two independent samples** is employed to contrast $\bar{Y}_0$ versus $\bar{Y}_1$ (especially if the sample size is small). In point of fact, use of the ***t* test for two independent samples** to contrast $\bar{Y}_0$ versus $\bar{Y}_1$ assumes the use of r_{pb} as a measure of association. Within the context of employing the t test, the correlational example under discussion can be conceptualized as a study in which intelligence represents the independent variable and eye-hand coordination the dependent variable. The independent variable, which is nonmanipulated, is comprised of the two levels, **above average intelligence** versus **below average intelligence.** Sources that provide additional discussion of the **biserial correlation coefficient** are Guilford (1965), Lindeman *et al.* (1980), and McNemar (1969).

Test 22j: The tetrachoric correlation coefficient (r_{tet}) The **tetrachoric correlation coefficient** is employed if both variables are based on an interval/ratio scale, but the scores on both of the variables have been transformed into a dichotomous nominal/categorical scale. The value computed for the **tetrachoric correlation coefficient** represents an estimate of the value one would obtain for the **Pearson product-moment correlation coefficient**, if instead of employing dichotomized variables, one had used the scores on the underlying interval/ratio scales. The **tetrachoric correlation coefficient** (which was developed by Karl Pearson (1901)) is based on the assumption that the underlying distribution for both of the variables is continuous and normal. Among others, Cohen and Cohen (1983) note that caution should be employed in using both the **tetrachoric and biserial correlation coefficients**, since both measures are based on hypothetical underlying distributions that are not directly observed. Since the accuracy of r_{tet} is highly dependent upon the assumption of normality, it should not be employed unless there is empirical evidence to indicate that the distributions underlying the dichotomous variables are normal.

Since the magnitude of the standard error of estimate of r_{tet} is large relative to the standard error of estimate of Pearson r, in order to provide a reasonable estimate of r the sample size employed for computing r_{tet} should be quite large. Guilford (1965) and Lindeman *et al.* (1980) state that the value of n employed in computing r_{tet} should be at least two times that which would be employed to compute r. As is the case for the **Pearson product-moment correlation coefficient,** the following apply to the **tetrachoric correlation coefficient:** a) The range of values within which r_{tet} can fall is $-1 \leq r_{tet} \leq +1$; and b) If the relationship between two variables is nonlinear, the computed value of r_{tet} will only represent the degree of linear relationship between the variables.

Earlier in this section (as well as in the discussion of the **chi-square test for $r \times c$ tables**) it is noted that the **phi coefficient (ϕ)** is also employed as a measure of association for a 2×2 contingency table involving two dichotomous variables. The basic difference between r_{tet} and ϕ, is that the latter measure is employed with two genuinely dichotomous variables (i.e., variables that are not based on an underlying distribution involving an interval/ratio scale). Cohen and Cohen (1983) and McNemar (1969) note that the value of

r_{tet} computed for a 2×2 contingency table will always be larger than the value of ϕ computed for the same data.[24]

A number of reasons account for the fact that the **tetrachoric correlation coefficient** is infrequently employed within the framework of research. One reason is that, in most instances, data on variables that represent an interval/ratio scale are available in the latter format, and thus there is no need to convert it into a dichotomous format. Another reason is the reluctance of researchers to accept the normality assumption with respect to variables for which only dichotomous information is available.

Without the aid of a computer or special tables (which can be found in Guilford (1965) and Lindeman *et al.* (1980)), the computation of the exact value of r_{tet} is both time consuming and tedious. There are, however, two equations that have been developed which provide reasonably good approximations of r_{tet} under most conditions. These equations will be employed to evaluate to Example 22.6.

Example 22.6 *Two hundred subjects are asked whether they* **Agree** *(which will be assigned a score of* 1) *or* **Disagree** *(which will be assigned a score of* 0) *with the following two statements:* **Question 1**: *I believe that abortion should be legal.* **Question 2**: *I believe that murderers should be executed. The responses of the* 200 *subjects are summarized in* Table 22.8. *Is there a statistical relationship between subjects' responses to the two questions?*

Table 22.8 Summary of Data for Example 22.6

		X variable Question 1		
		0 Disagree	1 Agree	Row Sums
Y variable Question 2	0 Disagree	$a = 30$	$b = 70$	100
	1 Agree	$c = 60$	$d = 40$	100
Column Sums		90	110	Total = 200

Subjects' responses to Question 1 will represent the X variable, and their responses to Question 2 will represent the Y variable. The use of the **tetrachoric correlation coefficient** in evaluating the data is based on the assumption that the permissible responses **Agree** versus **Disagree** represent two points that lie on a continuous scale. It will be assumed that if subjects are allowed to present their opinions to the questions with more precision, their responses can be quantified on an interval/ratio scale, and that the overall distribution of these responses in the underlying population will be normal. Thus, the responses of 0 and 1 on each variable are the result of dichotomizing information that is based on an underlying interval/ratio scale.

Equations 22.51 and 22.52 can be employed to compute reasonably good approximations of the value of r_{tet}. Lindeman *et al.* (1980) note that Equation 22.51 provides a good approximation of r_{tet} when $p_0 = p_1 = .5$ for both of the dichotomous variables. In other words, for the X variable both $p_{0_X} = (a + c)/n$ and $p_{1_X} = (b + d)/n$ will equal .5, and for the Y variable both $p_{0_Y} = (a + b)/n$ and $p_{1_Y} = (c + d)/n$ will equal .5 (where $n = a + b + c + d$). Equation 22.52, on the other hand, is recommended when the values of p_0 and p_1 are not equal. As the discrepancy between p_0 and p_1 increases, the less accurate the approximation provided by Equation 22.52 becomes.

In both Equations 22.51 and 22.52, a and d will always represent the frequencies of cells in which subjects provide the same response for both variables/questions, and b and c will always represent the frequencies of cells in which subjects provide opposite responses for the two variables/questions. Inspection of Table 22.8 (which is identical to Table 22.3, which is employed to illustrate that ϕ is a special case of r) indicates the following: 1) $a = 30$ subjects respond **Disagree** to both questions; 2) $d = 40$ subjects respond **Agree** to both questions; 3) $b = 70$ subjects respond **Agree** to Question 1 and **Disagree** to Question 2; and 4) $c = 60$ subjects respond **Disagree** to Question 1 and **Agree** to Question 2.

The configuration of the data is such that $p_0 = p_1$ for the Y variable, and the relationship is closely approximated for the X variable. Specifically, for the Y variable, $p_{0_Y} = (a + b)/n = (30 + 70)/200 = .5$ and $p_{1_Y} = (c + d)/n = (60 + 40)/200 = .5$. In the case of the X variable, $p_{0_X} = (a + c)/n = (30 + 60)/200 = .45$ and $p_{1_X} = (b + d)/n = (70 + 40)/200 = .55$.

The appropriate values are substituted in Equations 22.51 and 22.52 below. The trigonometric functions in each of the equations can be easily calculated with one keystroke on most scientific calculators.

(Equation 22.51)

$$r_{\text{tet}} = \sin\left[90^\circ\left(\frac{a+d-b-c}{n}\right)\right] = \sin\left[90^\circ\left(\frac{30+40-70-60}{200}\right)\right] = \sin\ -27^\circ = -.45$$

(Equation 22.52)

$$r_{\text{tet}} = \cos\left(\frac{180^\circ}{1+\sqrt{\dfrac{ad}{bc}}}\right) = \cos\left(\frac{180^\circ}{1+\sqrt{\dfrac{(30)(40)}{(70)(60)}}}\right) = \cos\ 117.30^\circ = -.46$$

Since for both variables the condition $p_0 = p_1 = .5$ is present or approximated, the two equations result in almost identical values. The negative sign in front of the correlation coefficient reflects the fact that subjects who are in one response category on one variable are more likely to be in the other response category on the other variable. A positive correlation would indicate that subjects tend to be in the same response category on both variables. Note that the absolute value $r_{\text{tet}} = .45$ (or .46) is greater than the value $\phi = .30$ obtained for the same set of data. This is consistent with what was noted earlier — that the value of r_{tet} computed for a 2 × 2 table will always be larger than the value of ϕ computed for the same data. The reader should take note of the fact, however, that unlike r_{tet}, ϕ is always expressed as a positive number. Thus, in comparing the two values for the same set of data, the absolute value of r_{tet} should be employed.

Test 22j-a: Test of significance for a tetrachoric correlation coefficient In order to evaluate the null hypothesis H_0: $\rho_{\text{tet}} = 0$, the standard error of estimate of r_{tet} must first be computed employing Equation 22.53. In the latter equation, the values h_X and h_Y are the height (ordinate) of the standard normal distribution at the point for each of the variables which divides the proportions p_0 and p_1. The protocol for determining the ordinate for each of the variables is identical to that employed for determining the ordinate for the **biserial correlation coefficient**. Employing **Table A1** for both the X and the Y variables, the z value is identified which delineates the point on the normal curve that a proportion of cases corresponding to the smaller of the two proportions p_0 versus p_1 falls above and the larger of the two proportions falls below. The corresponding ordinate is then determined. Thus, in the case of the X variable, $h_X = .3958$ (since 45% of cases fall above the corresponding value $z = .128$). In the case of the Y variable, $h_Y = .3989$ (since 50% of cases fall above the corresponding value $z = 0$). When the appropriate values are substituted in Equation 22.53, the value $s_{r_{tet}}$ .111 is computed.

$$s_{r_{tet}} = \frac{\sqrt{p_{0_X} p_{1_X} p_{0_Y} p_{1_Y}}}{h_X h_Y \sqrt{n}} = \frac{\sqrt{(.45)(.55)(.5)(.5)}}{(.3958)(.3989)\sqrt{200}} = .111 \quad \textbf{(Equation 22.53)}$$

The value $s_{r_{tet}} = .111$ is substituted in Equation 22.54, which is employed to evaluate the null hypothesis H_0: $r_{tet} = 0$. Use of the normal distribution in evaluating the null hypothesis assumes that the computation of r_{tet} is based on a large sample size (since, as noted earlier, r_{tet} will be extremely unreliable if it is based on a small sample). Employing Equation 22.54, the value $z = -4.14$ is computed. Note that the sign of z will always be the same as the sign of r_{tet}.

$$z = \frac{r_{tet}}{s_{r_{tet}}} = \frac{-.46}{.111} = -4.14 \quad \textbf{(Equation 22.54)}$$

It will be assumed that the nondirectional alternative hypothesis H_1: $\rho_{tet} \neq 0$ is evaluated. Employing **Table A1**, it is determined that the tabled critical two-tailed .05 and .01 values are $z_{.05} = 1.96$ and $z_{.01} = 2.58$. Since the obtained absolute value $z = 4.14$ is greater than both of the aforementioned critical values, the nondirectional alternative hypothesis H_1: $\rho_{tet} \neq 0$ is supported at both the .05 and .01 levels. Additional discussion of the **tetrachoric correlation coefficient** can be found in Guilford (1965), Lindeman *et al.* (1980), and McNemar (1969).

2. Multiple regression analysis

General Introduction to multiple regression analysis **Multiple regression analysis** is the term employed to describe the use of correlation and regression with designs involving more than two variables. Such analysis, which is considerably more complex than bivariate analysis, involves laborious computational procedures that make it all but impractical to conduct without the aid of a computer. Although this section will provide the reader with an overview of **multiple regression analysis**, the discussion to follow is not intended to provide comprehensive coverage of the subject. For a more thorough discussion of multiple regression, the reader should consult sources on multivariate analysis (e.g., Marascuilo and Levin (1983), Stevens (1986), and Tabachnick and Fidell (1989)).

In contrast to simple linear regression, where scores on one predictor variable are employed to predict the scores on a criterion variable, in multiple regression analysis a researcher attempts to increase the accuracy of prediction through the use of multiple predictor variables. By employing multiple predictor variables, one can often account for a greater amount of the variability on the criterion variable than will be the case if only one of the predictor variables is employed. Thus, the major goal of multiple regression analysis is to identify a limited number of predictor variables which optimize one's ability to predict scores on a criterion variable.

Since researchers generally want the simplest possible predictive model (as well as the fact that from a time and cost perspective, a model that involves a limited number of variables is less costly and easier to implement), it is unusual to find a multiple regression model that involves more than five predictor variables. Two additional factors which limit the number of predictor variables derived in multiple regression analysis follow: a) Once a limited number of predictor variables has been identified that explains a relatively large proportion of the variability on the criterion variable, it becomes increasingly unlikely that any new predictor variables which are identified will result in a significant increase in predictive power; and b) Although the researcher wants to identify predictor variables that are highly correlated with the criterion variable, he also wants to make sure that the predictor variables employed account for different proportions of the variability on the criterion variable. In order to accomplish the latter, none of the predictor variables should be highly correlated with one another, since if the latter is true the variables will be redundant with respect to the variation on the criterion variable they explain. As a general rule, it is difficult to find a large number of predictor variables that are highly correlated with a criterion variable, yet not correlated with one another. The term **multicollinearity** is used to describe a situation where predictor variables have a high intercorrelation with one another. When multicollinearity exists, the reliability of multiple regression analysis may be severely compromised.

Within the framework of multiple regression analysis there are a variety of strategies that are employed in selecting predictor variables. Among the strategies that are available are the following: a) **Forward selection** — In **forward selection** predictor variables are determined one at a time with respect to the order of their contribution in explaining variability on the criterion variable; b) **Backward selection** — In **backward selection** the researcher starts with a large pool of predictor variables and, starting with the smallest contributor, eliminates them one at a time based on whether or not they make a significant contribution to the predictive model; c) **Stepwise regression** — **Stepwise regression** is a combination of the forward and backward selection methods. In stepwise regression one employs the forward selection method, but upon adding each new predictor variable, all of the remaining predictor variables from the original pool are reexamined to determine whether they should be retained in the predictive model; d) **Hierarchical regression** — Whereas in the three previous methods statistical considerations dictate which predictor variables are included in the model, in **hierarchical regression** the researcher determines which variables should be included in the model. The latter determination is based on such factors as logic, theory, results of prior research, and cost; and e) **Standard or direct regression** — In **standard/direct regression** all available predictor variables are included in the model, including those that only explain a minimal amount of variability on the criterion variable. This type of regression may be employed when a researcher wants to explore for theoretical or other reasons the relationship between a large set of predictor variables and a criterion variable. Standard/direct regression is atypical when compared with the other methods of regression analysis, in that it is more likely to result in a large number of predictor variables.

Upon conducting a multiple regression analysis it is recommended that the resulting regression model be **cross-validated**. Minimally, this means replicating the results of the analysis on two subsamples, each representing a different half of the original sample. An even more desirable strategy is replicating the results on one or more independent samples that are representative of the population to which one wishes to apply the model. By cross-validating a model one can demonstrate that it generates consistent results, and will thus be of practical value in making predictions among members of the reference population upon which the model is based.

Although multiple regression analysis may result in a mechanism for reasonably accurate predictions, it does not provide sufficient control over the variables under study to allow a researcher to draw conclusions with regard to cause and effect. As is the case with bivariate correlation, multivariate correlation is not immune to the potential impact of extraneous variables that may be critical in understanding the causal relationship between the variables under study. In order to demonstrate cause and effect one must employ the experimental method (specifically, through use of the **true experiment**, demonstrate a causal connection between scores on a dependent variable and a manipulated independent variable). It should be noted that there is a procedure called **path analysis** (which will not be described in this book) that employs correlational information to evaluate causal relationships between variables. Statisticians are not in agreement, however, with respect to what role path analysis (as well as a number of related procedures) should play in making judgements with regard to cause and effect in correlational research.

Multiple regression analysis has a number of assumptions which if violated can compromise the reliability of the results. These assumptions (which concern normality, linearity, homoscedasticity, and independence of the residuals), as well as the impact of **outliers** on a multiple regression analysis, are discussed in books that provide comprehensive coverage of the subject.

Computational procedures for multiple regression analysis involving three variables

Test 22k: The multiple correlation coefficient Within the framework of multiple regression analysis, a researcher is able to compute a correlation coefficient between the criterion variable and the k predictor variables (where $k \geq 2$). The computed **multiple correlation coefficient** is represented by the notation R. A computed value of R must fall within the range 0 to +1 (i.e., $0 \leq R \leq +1$). Unlike an r value computed for two variables, the **multiple correlation coefficient** cannot be a negative number. The closer the value of R is to 1, the stronger the linear relationship between the criterion variable and the k predictor variables, whereas the closer it is to 0, the weaker the linear relationship. Because of the complexity of the computations involved, the discussion of the computational procedures for multiple regression analysis will be restricted to designs involving two predictor variables (which in the examples to be discussed will be designated X_1 and X_2), and a criterion variable (which will be designated Y). When $k = 2$, Equation 22.55 is employed to compute the value of R. The notation $R_{Y.X_1X_2}$ represents the **multiple correlation coefficient** between the criterion variable Y and the linear combination of two predictor variables X_1 and X_2.

$$R_{Y.X_1X_2} = \sqrt{\frac{r_{YX_1}^2 + r_{YX_2}^2 - 2r_{YX_1}r_{YX_2}r_{X_1X_2}}{1 - r_{X_1X_2}^2}} \qquad \textbf{(Equation 22.55)}$$

The following example will be employed to illustrate the use of Equation 22.55. Assume that the following correlation coefficients have been computed: a) The correlation between the number of ounces of sugar a child eats (which will represent predictor variable X_1) and the number of cavities (which will represent the criterion variable Y) is $r_{YX_1} = .955$; b) The correlation between the number of ounces of salt a child eats (which will represent predictor variable X_2) and the number of cavities is $r_{YX_2} = .52$; and c) The correlation between the number of ounces of sugar a child eats and the number of ounces of salt a child eats is $r_{X_1X_2} = .37$. The three above noted correlations are based on the following data: a) Table 22.1 lists the number of ounces of sugar a child eats (X_1) and the number of cavities (Y); and b) The following values, which are used to compute the correlations $r_{YX_2} = .52$ and $r_{X_1X_2} = .37$, are employed to represent the number of ounces of salt eaten (X_2) by the five subjects in Table 22.1: 4, 1, 1, 3, 6. Employing the latter set of five scores, the mean and estimated population standard deviation for variable X_2 are $\bar{X}_2 = 15/5 = 3$ and $\tilde{s}_{X_2} = [63-((15)^2/5)]/4 = 2.12$.

Substituting the correlations $r_{YX_1} = .955$, $r_{YX_2} = .52$, and $r_{X_1X_2} = .37$ in Equation 22.55, the multiple correlation coefficient $R_{Y.X_1X_2} = .972$ is computed.

$$R_{Y.X_1X_2} = \sqrt{\frac{(.955)^2 + (.52)^2 - 2(.955)(.52)(.37)}{1 - (.37)^2}} = \sqrt{.944} = .972$$

Note that the value $R_{Y.X_1X_2} = .972$ is larger than either value that is computed when each of the predictor variables is correlated separately with the criterion variable. Of course, the value of $R_{Y.X_1X_2}$ can be only minimally above $r_{YX_1} = .955$, since the maximum value R can attain is 1.

The coefficient of multiple determination R^2, which is the square of the **multiple correlation coefficient,** is referred to as the **coefficient of multiple determination.** The **coefficient of multiple determination** indicates the proportion of variance on the criterion variable that can be accounted for on the basis of variability on the k predictor variables. In our example $R^2_{Y.X_1X_2} = (.972)^2 = .944$. In point of fact, R^2 is a biased measure of P^2, which is the population parameter it is employed to estimate (P is the upper case Greek letter **rho**). The degree to which the computed value of R^2 is a biased estimate of P^2 will be a function of the sample size and the number of predictor variables employed in the analysis. The value of R^2 will be spuriously inflated when the sample size is close in value to the number of predictor variables employed in the analysis.[25] For this reason, sources emphasize that the number of subjects employed in a multiple regression analysis should always be substantially larger than the number of predictor variables. For example, Marascuilo and Levin (1983) recommend that the value of n should be at least ten times the value of k.[26]

One way of correcting for bias resulting from a small sample size is to employ Equation 22.56 to compute $\tilde{R}^2$, which is a relatively unbiased estimate of P^2. The value $\tilde{R}^2$ is commonly referred to as a "shrunken" estimate of the **coefficient of multiple determination.**[27]

$$\tilde{R}^2 = 1 - \frac{(1 - R^2)(n - 1)}{n - k - 1} \qquad \textbf{(Equation 22.56)}$$

It can be seen below that substituting the values from our example in Equation 22.56 yields the value $\tilde{R}^2_{Y.X_1X_2} = .89$, which is lower than $R^2_{Y.X_1X_2} = .944$ computed with Equaion 22.55 ($R^2_{Y.X_1X_2} = .944$ is the value in the radical of Equation 22.55 prior to computing the square root).

$$\tilde{R}^2_{Y.X_1X_2} = 1 - \frac{[1 - (.944)][5 - 1]}{5 - 2 - 1} = .89$$

Test 22k-a: Test of significance for a multiple correlation coefficient Equation 22.57 is employed to evaluate the null hypothesis H_0: $\mathrm{P}^2 = 0$. If the latter null hypothesis is supported the value of P will also equal zero, and consequently the null hypothesis H_0: $\mathrm{P} = 0$ is also supported.

$$F = \frac{(n - k - 1)R^2}{k(1 - R^2)} \qquad \textbf{(Equation 22.57)}$$

The computed value $R^2_{Y.X_1X_2} = .944$, as well as the shrunken estimate $\tilde{R}^2_{Y.X_1X_2} = .89$, are substituted in Equation 22.57 below. If one has to choose which of the two values to employ, researchers would probably consider it more prudent to employ the shrunken estimate $\tilde{R}^2_{Y.X_1X_2} = .89$ (especially when the sample size is small).

$$F = \frac{(5 - 2 - 1)(.944)}{2(1 - .944)} = 16.86$$

$$F = \frac{(5 - 2 - 1)(.89)}{2(1 - .89)} = 8.09$$

The computed F value is evaluated with **Table A10 (Table of the F Distribution)** in the **Appendix.** In order to reject the null hypothesis, the F value must be equal to or greater than the tabled critical value at the prespecified level of significance. The degrees of freedom employed for the analysis are $df_{\text{num}} = k$ and $df_{\text{den}} = n - k - 1$. Thus, for our example $df_{\text{num}} = 2$ and $df_{\text{den}} = 5 - 2 - 1 = 2$. In **Table A10**, for $df_{\text{num}} = 2$ and $df_{\text{den}} = 2$, the tabled critical .05 and .01 values are $F_{.05} = 19.00$ and $F_{.01} = 99.00$. Since both of the obtained values $F = 16.86$ and $F = 8.09$ are less than $F_{.05} = 19.00$, regardless of whether R^2 or $\tilde{R}^2$ is computed, the null hypothesis cannot be rejected. Thus, in spite of the fact that the obtained value of R is close to 1, the data still do not allow one to conclude that the population multiple correlation coefficient is some value other than zero. The lack of significance for such a large R value can be attributed to the small sample size.

The multiple regression equation A major goal of multiple regression analysis is to derive a **multiple regression equation** that utilizes scores on the k predictor variables to predict the scores of subjects on the criterion variable. Equation 22.58 is the general form of the multiple regression equation.

$$Y' = a + b_1X_1 + b_2X_2 + \cdots + b_kX_k \qquad \textbf{(Equation 22.58)}$$

Note that the multiple regression equation contains a **regression coefficient** (b_i) for each of the predictor variables ($X_1, X_2, ..., X_k$) and a **regression constant** (a). In contrast to the regression line employed in simple linear regression, the multiple regression equation describes a **regression plane** that provides the best fit through a set of data points that exists in a **multidimensional space.** The values computed for the regression equation minimize the sum of the squared residuals, which in the case of multiple regression are the sum of the squared distances of all the data points from the regression plane.

Equations 22.59 and 22.60 can be employed to determine the values of the regression coefficients b_1 and b_2, which are the coefficients for predictor variables X_1 and X_2. Each of the regression coefficients indicates the amount of change on the criterion variable Y that will be associated with a one unit change on that predictor variable if the effect of the second predictor variable is held constant.

$$b_1 = \left[\frac{\tilde{s}_Y}{\tilde{s}_{X_1}}\right]\left[\frac{r_{YX_1} - r_{YX_2}r_{X_1X_2}}{1 - r^2_{X_1X_2}}\right] \qquad \textbf{(Equation 22.59)}$$

$$b_2 = \left[\frac{\tilde{s}_Y}{\tilde{s}_{X_2}}\right]\left[\frac{r_{YX_2} - r_{YX_1}r_{X_1X_2}}{1 - r^2_{X_1X_2}}\right] \qquad \textbf{(Equation 22.60)}$$

Equation 22.61 is employed to compute the regression constant a (which is analogous to the Y intercept computed in simple linear regression).

$$a = \bar{Y} - b_1\bar{X}_1 - b_2\bar{X}_2 \qquad \textbf{(Equation 22.61)}$$

The multiple regression equation will now be computed. From earlier discussion we know the following values: $\bar{X}_1 = 7.2$, $\tilde{s}_{X_1} = 8.58$, $\bar{X}_2 = 3$, $\tilde{s}_{X_2} = 2.12$, $\bar{Y} = 3.4$, $\tilde{s}_Y = 2.70$, $r_{YX_1} = .955$, $r_{YX_2} = .52$, $r_{X_1X_2} = .37$. Substituting the appropriate values in Equations 22.59–22.61, the multiple regression equation is determined below.

$$b_1 = \left[\frac{2.70}{8.58}\right]\left[\frac{.955 - (.52)(.37)}{1 - (.37)^2}\right] = .278$$

$$b_2 = \left[\frac{2.70}{2.12}\right]\left[\frac{.52 - (.955)(.37)}{1 - (.37)^2}\right] = .246$$

$$a = 3.4 - (.278)(7.2) - (.246)(3) = .660$$

$$Y' = .660 + .278X_1 + .246X_2$$

To illustrate the application of the multiple regression equation, when the appropriate values are substituted in Equation 22.58, a child who consumes 4 ounces of sugar (X_1) and 2 ounces of salt (X_2) per week is predicted to have 2.264 cavities.

$$Y' = .660 + (.278)(4) + (.246)(2) = 2.264$$

The standard error of multiple estimate As is the case with simple linear regression, a **standard error of estimate** can be computed which can be employed to determine how accurately the multiple regression equation will predict a subject's score on the criterion variable. Employing this error term, which in the case of multiple regression is referred to as the **standard error of multiple estimate,** one can compute a confidence interval for a predicted score. The standard error of multiple estimate will be represented by the notation $s_{Y.X_1X_2}$. Equation 22.62 is employed to compute $s_{Y.X_1X_2}$ if R^2 (the biased estimate of P^2) is used to represent the **coefficient of multiple determination.**[28] If, on the other hand, the unbiased estimate $\tilde{R}^2$ is employed to represent the **coefficient of multiple determination,** Equation 22.63 can be used to compute $s_{Y.X_1X_2}$. The two equations are employed below to compute the value $s_{Y.X_1X_2} = .90$.

(Equation 22.62)

$$s_{Y.X_1X_2} = \tilde{s}_Y\sqrt{\left[\frac{n - 1}{n - k - 1}\right](1 - R^2_{Y.X_1X_2})} = (2.70)\sqrt{\left[\frac{5 - 1}{5 - 2 - 1}\right](1 - .944)} = .90$$

$$s_{Y.X_1X_2} = \tilde{s}_Y\sqrt{1 - \tilde{R}^2_{Y.X_1X_2}} = (2.70)\sqrt{1 - .89} = .90 \qquad \textbf{(Equation 22.63)}$$

Computation of a confidence interval for Y' Equation 22.64 (which is analogous to Equation 22.16) can be employed to compute a confidence interval for the predicted value Y'. The value $s_{Y.X_1X_2} = .90$ is employed in Equation 22.64 to compute the 95% confidence interval for the subject who is predicted to have $Y' = 2.264$ cavities. Also employed in the latter equation is the tabled critical two-tailed .05 t value $t_{.05} = 4.30$ (for $df = n - k - 1 = 5 - 2 - 1 = 2$), which delineates to the 95% confidence interval.

(Equation 22.64)

$$CI_{(1-\alpha)} = Y' \pm (t_{\alpha/2})(s_{Y.X_1X_2}) = 2.264 \pm (4.30)(.90) = 2.264 \pm 3.87$$

This result indicates that the researcher can be 95% confident that the number of cavities the subject actually has falls within the range −1.606 and 6.134 (i.e., $-1.606 \le Y \le 6.134$). Since a person cannot have a negative number of cavities, the latter result indicates that the person will have between 0 and 6.134 cavities.

Evaluation of the relative importance of the predictor variables If the result of a multiple regression analysis is significant, a researcher will want to assess the relative importance of the predictor variables in explaining variability on the criterion variable. It should be noted that although the value computed for the **multiple correlation coefficient** is not significant for the example under discussion, within the framework of the discussion of the material in this section it will be assumed that it is. Intuitively, it might appear that one can evaluate the relative importance of the predictor variables based on the relative magnitude of the regression coefficients. However, since the different predictor variables represent different units of measurement, comparison of the regression coefficients will not allow a researcher to make such an estimate. One approach to solving this problem is to standardize each of the variables so that scores on all of the variables are based on standard normal distributions. As a result of standardizing all of the variables, Equation 22.65 (which is referred to as the **standardized multiple regression equation**) becomes the general form of the multiple regression equation.

$$z_{Y'} = \beta_1 z_1 + \beta_2 z_2 + \cdots + \beta_k z_k \qquad \textbf{(Equation 22.65)}$$

In Equation 22.65 the predicted value Y', as well as the scores on the predictor variables X_1 and X_2, are expressed as standard deviation scores (i.e., $z_{Y'}$, z_1, z_2). The standardized equivalent of a regression coefficient, referred to as a **beta weight**, is represented by the notation β_i. Since the regression constant a will always equal zero in the standardized multiple regression equation, it is not included. When there are two predictor variables, Equations 22.66 and 22.67 can be employed to compute the values of β_1 and β_2. Note that each equation is expressed in two equivalent forms, one form employing the regression coefficients and the relevant estimated population standard deviations, and the other form employing the correlations between the three variables.

$$\beta_1 = b_1\left[\frac{\tilde{s}_{X_1}}{\tilde{s}_Y}\right] = \frac{r_{YX_1} - r_{YX_2} r_{X_1X_2}}{1 - r^2_{X_1X_2}} \qquad \textbf{(Equation 22.66)}$$

$$\beta_2 = b_2\left[\frac{\tilde{s}_{X_2}}{\tilde{s}_Y}\right] = \frac{r_{YX_2} - r_{YX_1} r_{X_1X_2}}{1 - r^2_{X_1X_2}} \qquad \textbf{(Equation 22.67)}$$

Substituting the appropriate values in Equations 22.66 and 22.67, the values $\beta_1 = .883$ and $\beta_2 = .193$ are computed.

$$\beta_1 = .278\left[\frac{8.58}{2.70}\right] = \frac{.955 - (.52)(.37)}{[1 - (.37)^2]} = .883$$

$$\beta_2 = .246\left[\frac{2.12}{2.70}\right] = \frac{.52 - (.955)(.37)}{[1 - (.37)^2]} = .193$$

Thus, the standardized multiple regression equation is as follows: $z_{Y'} = .883z_1 + .193z_2$. The value $\beta_1 = .883$ indicates that an increase of one standard deviation unit on variable X_1 is associated with an increase of .883 standard deviation units on the criterion variable (if the predictor variable X_2 remains at a fixed value). In the same respect the value $\beta_2 = .193$ indicates that an increase of one standard deviation unit on variable X_2 is associated with an increase of .193 standard deviation units on the criterion variable (if the predictor variable X_1 remains at a fixed value).

When there are two predictor variables, Equation 22.68 employs the standardized beta weights to provide an alternative method for computing the value $R^2_{Y.X_1X_2}$. The value derived with Equation 22.68 will be equivalent to the square of the value computed with Equation 22.55.

$$R^2_{Y.X_1X_2} = \beta_1 r_{YX_1} + \beta_2 r_{YX_2}$$

(Equation 22.68)

$$R^2_{Y.X_1X_2} = (.883)(.955) + (.193)(.52) = .944$$

Some sources note that the absolute values of the beta weights reflect the rank-ordering of the predictor variables with respect to the role they play in accounting for variability on the criterion variable. Kachigan (1986) suggests that by dividing the square of a larger beta weight by the square of a smaller beta weight, a researcher can determine the relative influence of two predictor variables on the criterion variable. The problem with the latter approach (as Kachigan (1986) himself notes) is that beta weights do not allow a researcher to separate the joint contribution of two or more predictor variables. The fact that predictor variables are usually correlated with one another (as is the case in the example under discussion), makes it difficult to determine how much variability on the criterion variable can be accounted for by any single predictor variable in and of itself. In view of this, Howell (1992) and Marascuilo and Serlin (1988) note that statisticians are not in agreement with respect to what, if any, methodology is appropriate for determining the precise amount of variability attributable to each of the predictor variables.

Evaluating the significance of a regression coefficient In conducting a multiple regression analysis it is common practice to evaluate whether each of the regression coefficients is statistically significant. Since the unstandardized/raw score coefficients and standardized coefficients are linear transformations of one another, a statistical test on either set of coefficients will yield the same result. The null hypothesis that is evaluated is that the true value of the regression coefficient in the population equals zero. Thus, H_0: $\mathrm{B}_i = 0$ (where B_i, which is the upper case Greek letter **beta**, represents the value of the coefficient for the ith predictor variable in the underlying population). For the purpose of discussion it will be assumed that the aforementioned null hypothesis is stated in reference to the unstandardized coefficients. In order to evaluate the null hypothesis H_0: $\mathrm{B}_i = 0$, a standard error of estimate must be computed for a coefficient. When there are two predictor variables, the standard error of estimate for an unstandardized coefficient (represented by the notation s_{b_i}) is computed with Equation 22.69.[29]

$$s_{b_i} = \frac{\tilde{s}_Y}{\tilde{s}_{X_i}} \sqrt{\frac{1 - R^2_{Y.X_1X_2}}{(1 - r^2_{X_1X_2})(n - k - 1)}}$$

(Equation 22.69)

Employing Equation 22.69, the values $s_{b_1} = .057$ and $s_{b_2} = .229$ are computed.

$$s_{b_1} = \frac{2.70}{8.58} \sqrt{\frac{1 - .944}{[1 - (.37)^2][5 - 2 - 1]}} = .057$$

$$s_{b_2} = \frac{2.70}{2.12} \sqrt{\frac{1 - .944}{[1 - (.37)^2][5 - 2 - 1]}} = .229$$

Each of the values $s_{b_1} = .057$ and $s_{b_2} = .229$ can be substituted in Equation 22.70, which employs the t distribution to evaluate the null hypothesis H_0: $\mathrm{B}_i = 0$.[30] In the analysis to be described it will be assumed that the nondirectional alternative hypothesis H_1: $\mathrm{B}_i \neq 0$ is evaluated for each regression coefficient. The degrees of freedom employed in evaluating a t value computed with Equation 22.70 are $df = n - k - 1$. Thus, $df = 5 - 2 - 1 = 2$. If the null hypothesis cannot be rejected for a specific coefficient, the researcher can conclude that the predictor variable in question will not be of any use in predicting scores on the criterion variable.

$$t_{b_i} = \frac{b_i}{s_{b_i}} \qquad \textbf{(Equation 22.70)}$$

Employing Equation 22.70, the null hypotheses H_0: $B_1 = 0$ and H_0: $B_2 = 0$ are evaluated.[31]

$$t_{b_1} = \frac{.278}{.057} = 4.88 \qquad t_{b_2} = \frac{.246}{.229} = 1.07$$

Employing **Table A2**, for $df = 2$, the tabled critical .05 and .01 t values are $t_{.05} = 4.30$ and $t_{.01} = 9.93$.[32] Since the value $t_{b_1} = 4.88$ is greater than $t_{.05} = 4.30$, the nondirectional alternative hypothesis H_1: $B_1 \neq 0$ is supported at the .05 level. It is not supported at the .01 level, since $t_{b_1} = 4.88$ is less than $t_{.01} = 9.93$. Since $t_{b_2} = 1.07$ is less than $t_{.05} = 4.30$, the nondirectional alternative hypothesis H_1: $B_2 \neq 0$ is not supported. Thus, we can conclude that whereas predictor variable X_1 (sugar consumption) contributes significantly in predicting variability on the criterion variable (number of cavities), predictor variable X_2 (salt consumption) does not. Consequently, the latter predictor variable can be removed from the analysis. It should be noted that when a researcher elects to eliminate a predictor variable from the analysis, a new regression equation should be derived which just involves the data for the remaining predictor variable(s).

Computation of a confidence interval for a regression coefficient Equation 22.71 can be employed to compute a confidence interval for a regression coefficient.

$$CI_{b_{i(1-\alpha)}} = b_i \pm (t_{\alpha/2})(s_{b_i}) \qquad \textbf{(Equation 22.71)}$$

Using Equation 22.71, the 95% confidence intervals for the two regression coefficients are computed. The t value employed in Equation 22.71 is $t_{.05} = 4.30$, which is the tabled critical two-tailed .05 t value for $df = n - k - 1 = 5 - 2 - 1 = 2$.

$$CI_{.95_{b_1}} = .278 \pm (4.30)(.057) = .278 \pm .245$$

$$CI_{.95_{b_2}} = .246 \pm (4.30)(.229) = .246 \pm .985$$

Thus, the range of values in which the researcher can be 95% sure the true values of the coefficients lie are as follows: $.033 \leq B_1 \leq .523$ and $-.739 \leq B_2 \leq 1.231$.

Partial and semipartial correlation Within the framework of multiple regression analysis there are a number of other types of correlation coefficients which can be computed that can further clarify the nature of the relationship between predictor variables and a criterion variable. In evaluating the relationship between a criterion variable and a single predictor variable, it is not uncommon that the correlation is influenced by a third variable. As an example, the relationship between frequency of violent crimes (which will represent the criterion variable Y) and level of stress (which will represent the predictor variable X_1) is undoubtedly influenced by extraneous variables such as social class, which if included in a multiple regression analysis can represent a second potential predictor variable X_2. In some instances, by measuring the influence of a third variable (in this case social class), a researcher will be better able to understand the nature of the relationship between the other two variables (violent crime and stress). By allowing the third variable to serve in the role of a meditating variable, it often increases the researcher's ability to predict the scores of subjects on the criterion variable. In other instances, however, the researcher may view the contribution of a third variable as interfering with the study of the relationship between the other two variables. Thus, if a researcher wants to obtain a "purer" measure of the relationship between violent crime and stress, he might want to eliminate the influence of social class from the analysis. In instances where one wants to control for the influence of an extraneous variable, the latter variable is viewed as a **nuisance variable.** Fortunately, correlational procedures have been developed which allow researchers to statistically control for the influence of extraneous variables. Two of these procedures, **partial correlation** and **semipartial correlation** (also referred to as **part correlation**), are described in this section.

Test 22l: The partial correlation coefficient A **partial correlation coefficient** allows a researcher to measure the degree of association between two variables, after any linear association one or more additional variables has with the other two variables has been removed. Partial correlation reverses that which multiple correlation accomplishes. Whereas multiple correlation combines variables in order to assess their cumulative effect, partial correlation removes the effects of variables in order to determine what effect remains when one or more of the variables have been eliminated.

In the case of two predictor variables X_1 and X_2 and the criterion variable Y, the **partial correlation coefficient** $r_{YX_1.X_2}$ represents the correlation between Y and X_1 after any linear association that X_2 has with either Y or X_1 has been removed. It can also be stated that the **partial correlation coefficient** $r_{YX_1.X_2}$ represents the correlation between Y and X_1 if X_2 is held constant. By computing the **partial correlation coefficient** $r_{YX_1.X_2}$, one is able to have a "purer" measure of the relationship between a criterion variable Y and the predictor variable X_1.

When there are three variables, it is possible to compute the following **partial correlation coefficients:** $r_{YX_1.X_2}$, $r_{YX_2.X_1}$, $r_{X_1X_2.Y}$. Equation 22.72 is the general equation for computing a **partial correlation coefficient** involving three variables (where A, B, and C represent the three variables). The notation $r_{AB.C}$ represents the correlation between A and B, after any linear relationship C has with A and B has been removed.

$$r_{AB.C} = \frac{r_{AB} - r_{AC}r_{BC}}{\sqrt{(1 - r_{AC}^2)(1 - r_{BC}^2)}} \qquad \textbf{(Equation 22.72)}$$

Employing Equation 22.72, the **partial correlation coefficients** $r_{YX_1.X_2}$ (which is the partial correlation of Y and X_1, with the effect of X_2 removed) and $r_{YX_2.X_1}$ (which is the partial correlation of Y and X_2, with the effect of X_1 removed) will be computed. When the appropriate correlations from the example involving the relationship between sugar and salt consumption and the number of cavities are substituted in Equation 22.72, the **partial correlation coefficients** $r_{YX_1.X_2} = .96$ and $r_{YX_2.X_1} = .60$ are computed.

$$r_{YX_1.X_2} = \frac{r_{YX_1} - r_{YX_2}r_{X_1X_2}}{\sqrt{(1 - r_{YX_2}^2)(1 - r_{X_1X_2}^2)}} = \frac{.955 - (.52)(.37)}{\sqrt{[1 - (.52)^2][1 - (.37)^2]}} = .96$$

$$r_{YX_2.X_1} = \frac{r_{YX_2} - r_{YX_1}r_{X_1X_2}}{\sqrt{(1 - r_{YX_1}^2)(1 - r_{X_1X_2}^2)}} = \frac{.52 - (.955)(.37)}{\sqrt{[1 - (.955)^2][1 - (.37)^2]}} = .60$$

The value $r_{YX_1 \cdot X_2} = .96$ indicates that the correlation between sugar consumption and the number of cavities, with salt consumption removed, is .96. The value $r_{YX_2 \cdot X_1} = .60$ indicates that the correlation between salt consumption and the number of cavities, with sugar consumption removed, is .60. Although the partial correlation between two variables is generally (but not always) smaller than the **zero order correlation** (which is the term that refers to the correlation between the two variables before the effect of the third variable has been removed), this is not the case in the example under discussion (since the partial correlations $r_{YX_1 \cdot X_2} = .96$ and $r_{YX_2 \cdot X_1} = .60$ are larger than the **zero order correlations** $r_{YX_1} = .955$ and $r_{YX_2} = .52$). When a partial correlation is substantially different from the corresponding zero order correlation (especially when the absolute value of the partial correlation is substantially larger), it may indicate the presence of a **suppressor variable.** A **suppressor variable** is a predictor variable that can improve prediction on the criterion variable by suppressing variance that is irrelevant to predicting the criterion variable. In a set of three variables, a suppressor variable is a predictor variable (X_i) that has a low correlation with the criterion variable (Y), but a high correlation with the other predictor variable (X_j). By virtue of the latter, inclusion of the suppressor variable in the analysis may result in a multiple correlation coefficient ($R_{Y.X_iX_j}$) that has a larger absolute value (or even a different sign) than the zero order correlation coefficient between the criterion variable and the suppressor variable (r_{YX_i}). It should be noted that suppressor variables can create major problems in interpreting the results of a multiple regression analysis. For a more detailed discussion of suppressor variables the reader is referred to Cohen and Cohen (1983).

The square of a partial correlation coefficient represents the proportion of variability explained on the criterion variable by one of the predictor variables, after removing any linear effects of the other predictor variable from the other two variables. In the case of $r_{YX_1 \cdot X_2} = .96$, $r^2_{YX_1 \cdot X_2} = (.96)^2 = .92$. Thus, 92% of the variability on the criterion variable can be accounted for on the basis of the predictor variable X_1, when the linear effects of variable X_2 are removed from the other two variables. In the case of $r_{YX_2 \cdot X_1} = .60$, $r^2_{YX_2 \cdot X_1} = (.60)^2 = .36$. Thus, 36% of the variability on the criterion variable can be accounted for on the basis of the predictor variable X_2, when the linear effects of variable X_1 are removed from the other two variables.

Test 22l-a: Test of significance for a partial correlation coefficient The null hypothesis H_0: $\rho_p = 0$ can be evaluated with Equation 22.73 (where ρ_p represents the population **partial correlation coefficient**).[33]

$$t = \frac{r_p\sqrt{n - v}}{\sqrt{1 - r_p^2}}$$ **(Equation 22.73)**

Where: r_p is the **partial correlation coefficient**
v is the total number of variables employed in the analysis

When there are two predictor variables and one criterion variable, the total number of variables employed in the analysis is 3, and thus $\sqrt{n - v} = \sqrt{n - 3}$. The value $n - v = n - 3$ represents the number of degrees of freedom employed for the analysis.[34] Employing Equation 22.73, the null hypothesis is evaluated in reference to the partial correlation coefficients $r_{YX_1 \cdot X_2} = .96$ and $r_{YX_2 \cdot X_1} = .60$.

$$t = \frac{(.96)\sqrt{5 - 3}}{\sqrt{1 - (.96)^2}} = 4.85 \qquad t = \frac{(.60)\sqrt{5 - 3}}{\sqrt{1 - (.60)^2}} = 1.06$$

It will be assumed that the nondirectional alternative hypothesis H_1: $\rho_p \neq 0$ is evaluated. Employing **Table A2**, for $df = 2$ (since $df = 5 - 3 = 2$), the tabled critical two-tailed .05 and .01 values are $t_{.05} = 4.30$ and $t_{.01} = 9.93$. Since the obtained value $t = 4.85$ is greater than $t_{.05} = 4.30$, the nondirectional alternative hypothesis H_1: $\rho_{YX_1 . X_2} \neq 0$ is supported at the .05 level (but not at the .01 level). Since the obtained value $t = 1.06$ is less than $t_{.05} = 4.30$, the nondirectional alternative hypothesis H_1: $\rho_{YX_2 . X_1} \neq 0$ is not supported.

When there are two predictor variables, a partial correlation can obviously only eliminate the effect of one other predictor variable. This kind of partial correlation is often referred to as a **first-order partial correlation.** When there are more than two predictor variables, it is possible to compute **higher-order partial correlations** in which the effects of two or more predictor variables are eliminated. Thus, a **second-order partial correlation** is one in which the effects of two predictor variables are eliminated. A discussion of higher-order partial correlation can be found in Cohen and Cohen (1983), Hays (1994), and Marascuilo and Levin (1983).

Test 22m: The semipartial correlation coefficient A **semipartial (or part) correlation coefficient** measures the degree of association between two variables, with the linear association of one or more other variables removed from only one of the two variables that are being correlated with one another. In the case of two predictor variables and a criterion variable, a semipartial correlation coefficient measures the degree of association between two variables, with the influence of the third variable removed from only one of the two variables that are being correlated with one another. Thus, the **semipartial correlation coefficient** $r_{Y(X_1 . X_2)}$ represents the correlation between Y and X_1 after any linear association that X_2 has with X_1 has been removed.

When there are three variables, it is possible to compute the following six **semipartial correlation coefficients:** $r_{Y(X_1 . X_2)}$, $r_{Y(X_2 . X_1)}$, $r_{X_1(Y . X_2)}$, $r_{X_1(X_2 . Y)}$, $r_{X_2(Y . X_1)}$, $r_{X_2(X_1 . Y)}$. Equation 22.74 is the general equation for computing a **semipartial correlation coefficient** involving three variables (where A, B, and C represent the three variables). The notation $r_{A(B . C)}$ represents the correlation between A and B, after any linear relationship that C has with B has been removed.

$$r_{A(B . C)} = \frac{r_{AB} - r_{AC} r_{BC}}{\sqrt{1 - r_{BC}^2}} \qquad \textbf{(Equation 22.74)}$$

Employing Equation 22.74, the **semipartial correlation coefficients** $r_{Y(X_1 . X_2)} = .82$ and $r_{Y(X_2 . X_1)} = .18$ are computed.

$$r_{Y(X_1 . X_2)} = \frac{r_{YX_1} - r_{YX_2} r_{X_1 X_2}}{\sqrt{1 - r_{X_1 X_2}^2}} = \frac{.955 - (.52)(.37)}{\sqrt{1 - (.37)^2}} = .82$$

$$r_{Y(X_2 . X_1)} = \frac{r_{YX_2} - r_{YX_1} r_{X_1 X_2}}{\sqrt{1 - r_{X_1 X_2}^2}} = \frac{.52 - (.955)(.37)}{\sqrt{1 - (.37)^2}} = .18$$

The square of a semipartial correlation coefficient represents the proportion of variability explained on one of the variables by a second variable, after removing the linear effect of a third variable from the second variable. In the case of $r_{Y(X_1.X_2)} = .82$, $r^2_{Y(X_1.X_2)} = (.82)^2 = .67$. Thus, 67% of the variability on Y can be accounted for on the basis of X_1 when the linear effect of X_2 is removed from X_1. In the case of $r_{Y(X_2.X_1)} = .18$, $r^2_{Y(X_2.X_1)} = (.18)^2 = .03$. Thus, only 3% of the variability on Y can be accounted for on the basis of X_2 when the linear effect of X_1 is removed from X_2.

Marascuilo and Serlin (1983) note that although it is theoretically possible for the two values to be equal (in reference to the same variables), a **semipartial correlation coefficient** will have a smaller absolute value than a **partial correlation coefficient.** The latter can be confirmed by the fact that the **partial correlation coefficient** $r_{YX_1.X_2} = .96$ is larger than the **semipartial correlation coefficient** $r_{Y(X_1.X_2)} = .82$, and the **partial correlation coefficient** $r_{YX_2.X_1} = .60$ is larger than the **semipartial correlation coefficient** $r_{Y(X_2.X_1)} = .18$.

Test 22m-a: Test of significance for a semipartial correlation coefficient The null hypothesis H_0: $\rho_{sp} = 0$ can be evaluated with Equation 22.75 (where ρ_{sp} represents the population **semipartial correlation coefficient**).[35]

$$t = \frac{r_{sp}\sqrt{n - v}}{\sqrt{1 - r^2_{sp}}} \qquad \textbf{(Equation 22.75)}$$

Where: r_{sp} is the **semipartial correlation coefficient**
v is the total number of variables employed in the analysis

When there are two predictor variables and one criterion variable, the total number of variables employed in the analysis is 3, and thus $\sqrt{n - v} = \sqrt{n - 3}$.[36] The value $n - v = n - 3$ represents the number of degrees of freedom employed for the analysis. Employing Equation 22.73, the null hypothesis is evaluated in reference to the **semipartial correlation coefficients** $r_{Y(X_1.X_2)} = .82$ and $r_{Y(X_2.X_1)} = .18$.

$$t = \frac{(.82)\sqrt{5 - 3}}{\sqrt{1 - (.82)^2}} = 2.03$$

$$t = \frac{(.18)\sqrt{5 - 3}}{\sqrt{1 - (.18)^2}} = .26$$

It will be assumed that the nondirectional alternative hypothesis H_1: $\rho_{sp} \neq 0$ is evaluated. Employing **Table A2**, for $df = 2$ (since $df = 5 - 3 = 2$), the tabled critical two-tailed .05 and .01 values are $t_{.05} = 4.30$ and $t_{.01} = 9.93$. Since both of the obtained t values are less than $t_{.05} = 4.30$, the nondirectional alternative hypotheses H_1: $\rho_{Y(X_1.X_2)} \neq 0$ and H_1: $\rho_{Y(X_2.X_1)} \neq 0$ are not supported.

Final comments on multiple regression analysis Figure 22.7 (based on Cohen and Cohen (1983)), which is known as a **Venn diagram**, provides a visual summary of the proportion of variance represented by **zero order, multiple, partial,** and **semipartial correlation coefficients** (when there are two predictor variables and a criterion variable). Each of the three circles represents the variance of one of the three variables. Areas of overlap between circles represent shared variance between variables.

Zero order correlations: $r^2_{Y.X_1} = d + e \qquad r^2_{Y.X_2} = e + f \qquad r^2_{X_1.X_2} = e + b$

Multiple correlation: $R^2_{Y.X_1X_2} = d + e + f$

Partial correlations: $r^2_{YX_1.X_2} = \dfrac{d}{d + g} \qquad r^2_{YX_2.X_1} = \dfrac{f}{f + g}$

Semipartial correlations: $r^2_{Y(X_1.X_2)} = d \qquad r^2_{Y(X_2.X_1)} = f$

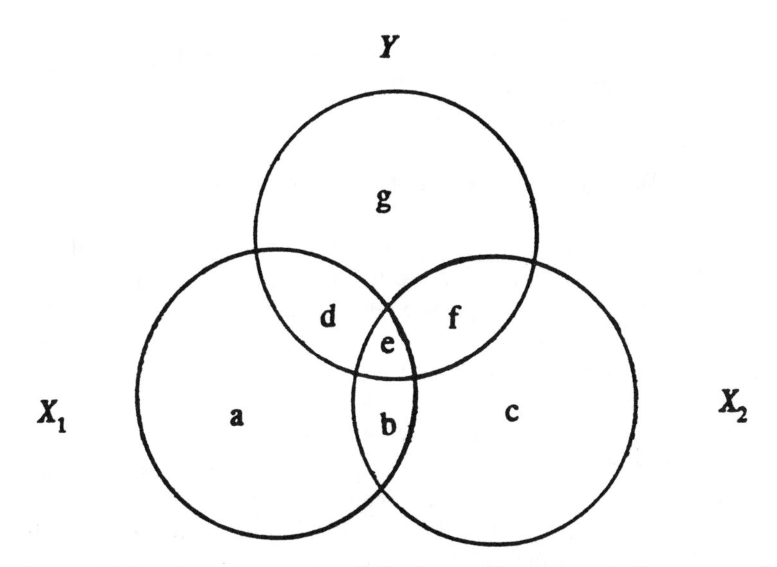

Figure 22.7 Venn Diagram of Variance Components Represented by Squared Correlation Coefficients

Sources that provide comprehensive coverage of the general subject of multiple regression analysis (e.g., Cohen and Cohen (1983) and Marascuilo and Levin (1983)) describe additional analytical procedures (many of which involve matrix algebra), as well as covering other issues that are relevant to the interpretation of a such an analysis. It should also be noted that it is possible to conduct curvilinear multiple regression analysis, in which case a multiple regression equation is derived that uses a curvilinear combination of the predictor variables to predict scores on the criterion variable. As noted earlier, because of the tedious computations involved, multiple regression analysis is generally not practical to employ unless one has access to the appropriate computer software.

References

Anderson, R. L. (1942). Distribution of the serial correlation coefficient. **Annals of** Mathematical Statistics, 13, 1–13.

Bennett, C. A. and Franklin, N. L. (1954). **Statistical analysis in chemistry and the chemical industry**. New York: John Wiley and Sons, Inc.

Chou, Y. (1989). **Statistical analysis for business and economics**. New York: Elsevier.

Cohen, J. (1988). **Statistical power analysis** (2nd ed.). Hillsdale, NJ: Erlbaum.

Cohen, J. and Cohen, P. (1983). **Applied multiple regression/correlation analysis for the behavioral sciences** (2nd ed.). Hillsdale, NJ: Lawrence Erlbaum Associates, Publishers.

David, F. N. (1938). **Tables of the ordinates and probability integral of the distribution of the correlation coefficient in small samples.** Cambridge: University Press.

Durbin, J. and Watson, G. S. (1950). Testing for serial correlation in least squares regression I. **Biometrika**, 37, 409–438.

Durbin, J. and Watson, G. S. (1951). Testing for serial correlation in least squares regression II. **Biometrika**, 38, 159–178.

Durbin, J. and Watson, G. S. (1971). Testing for serial correlation in least squares regression III. **Biometrika**, 58, 1–19.

Edwards, A. L. (1984). **An introduction to linear regression and correlation** (2nd ed.). New York: W. H. Freeman and Company.

Fisher, R. A. (1921). On the "probable error" of a coefficient of correlation deduced from a small sample. **Metron**, 1, Part 4, 3–32.

Guenther, W. C. (1965). **Concepts of statistical inference.** New York: McGraw–Hill Book Company.

Guilford, J. P. (1965). **Fundamental statistics in psychology and education** (4th ed.). New York: McGraw–Hill Book Company.

Hays, W. L. (1994). **Statistics** (5th ed.). Fort Worth: Harcourt Brace College Publishers.

Hotelling, H. (1940) The selection of variates for use in prediction with some comments on the general problem of nuisance parameters. **Annals of Mathematical Statistics**, 11, 271–283.

Howell, D. C. (1992). **Statistical methods for psychology** (3rd ed.). Boston: PWS–Kent Publishing Company.

Kachigan, S. K. (1986). **Statistical analysis.** New York: Radius Press.

Lindeman, R. H., Merenda, P. F., and Gold, R. Z. (1980). **Introduction to bivariate and multivariate analysis.** Glenview, IL: Scott, Foresman and Company.

Marascuilo, L. A. and Levin, J. R. (1983). **Multivariate statistics in the social sciences.** Monterey, CA: Brooks/Cole Publishing Company.

Marascuilo, L. A. and Serlin, R. C. (1988). **Statistical methods for the social and behavioral sciences.** New York: W. H. Freeman and Company.

McNemar, Q. (1969). **Psychological statistics** (4th ed.). New York: John Wiley and Sons, Inc.

Montgomery, D. C. and Peck, E. A. (1992). **Introduction to linear regression analysis** (2nd ed.). New York: John Wiley and Sons, Inc.

Netter, J., Wasserman, W., and Kutner, M. H. (1983). **Applied linear regression models** (3rd ed.). Homewood, IL: Richard D. Irwin, Inc.

Palumbo, D. J. (1977). **Statistics in political and social science** (Revised ed.). New York: Columbia University Press.

Pearson, K. (1896). Mathematical contributions to the theory of evolution — III. Regression, heredity and panmixia. **Philosophical Transactions of the Royal Society of London**, Series A 187, 253–318.

Pearson, K. (1900). On the criterion that a given system of deviations from the probable in the case of a correlated system of variables is such that it can reasonably be supposed to have arisen in a random sampling. **Philosophical Magazine**, 5, 157–175.

Pearson, K. (1901). On the correlation of characters not quantitatively measured. **Philosophical Transactions of the Royal Society** (Series A), 195, 1–47.

Steiger, J. H. (1980) Tests for comparing elements of a correlation matrix. **Psychological Bulletin**, 87, 245–251.

Stevens, J. (1986). **Applied multivariate statistics for the social sciences.** Hillsdale, NJ: Lawrence Erlbaum Associates, Publishers.

Tabachnick, B. G. and Fidell, L. S. (1989). **Using multivariate statistics.** New York: Harper Collins Publishers.

Endnotes

1. It is also possible to designate the Y variable as the predictor variable and the X variable as the criterion variable. The use of the Y variable as the predictor variable is discussed in Section VI.

2. It should be noted that when the joint distribution of two variables is bivariate normal, only a linear relationship can exist between the variables. As a result of the latter, whenever the population correlation between two bivariate normally distributed variables equals zero, one can conclude that the variables are statistically independent of one another. Under such conditions the null hypothesis H_0: $\rho = 0$ stated in this section is equivalent to the null hypothesis that the two variables are independent of one another. On the other hand, it is possible for each of two variables to be normally distributed, yet the joint distribution of the two variables not be bivariate normal. When the latter is true, it possible to compute the value $r = 0$, and at the same time have two variables that are statistically dependent upon one another. Statistical dependence in such a case will be the result of the fact that the variables are related to one another curvilinearly.

3. Howell (1992) notes that the value of r computed with Equation 22.1 is a biased estimate of the underlying population parameter ρ. The degree to which the computed value of r is biased is inversely related to the size of the sample employed in computing the correlation coefficient. For this reason, when one employs correlational data within the framework of research, it is always recommended that a reasonably large sample size be employed.

   One way of correcting for bias resulting from a small sample size is to employ the equation noted below to compute the value $\tilde{r}$, which represents a relatively unbiased estimate of the population parameter ρ. The value $\tilde{r}$ is referred to as a "shrunken" or "adjusted" estimate of the population correlation. The computation of $\tilde{r}$ (the absolute value of which will always be less than r) is demonstrated for Example 22.1.

$$\tilde{r} = \sqrt{1 - \frac{(1 - r^2)(n - 1)}{n - 2}} = \sqrt{1 - \frac{[1 - (.955)^2][5 - 1]}{5 - 2}} = .940$$

   Thus, in the case of Example 22.1, $\tilde{r} = .940$ provides a better estimate of the true population correlation than $r = .955$ (although even if $\tilde{r}$ is employed, $n = 5$ is an absurdly low sample size to employ within the framework of serious research). Since most sources use the computed value of r rather than $\tilde{r}$, the former value will be employed as the estimate of the population correlation throughout the discussion of the **Pearson product-moment correlation coefficient**.

4. An alternative form of Equation 22.3 that is based on the relationship $t^2 = F$ (described in Section VII of the **single-factor between-subjects analysis of variance**), which yields equivalent results, employs the F distribution. The equation employing the F distribution for evaluating the significance of r is noted below.

$$F = \frac{r^2(n - 2)}{1 - r^2}$$

   Employing the above equation with Example 22.1, the value $F = 31.10$ is computed.

$$F = \frac{(.955)^2(5 - 3)}{1 - (.955)^2} = 31.10$$

The computed value $F = 31.10$ (which is equivalent to $(t = 5.58)^2$ if rounding off error is ignored) is evaluated with **Table A10**. The degrees of freedom employed in evaluating the above equation are $df_{\text{num}} = 1$, $df_{\text{den}} = n - 2$. Thus, $df_{\text{num}} = 1$ and $df_{\text{den}} = 3$. It is determined that the tabled critical .05 and .01 two-tailed values are $F_{.05} = 10.13$ and $F_{.01} = 34.12$, and the tabled critical .05 and .01 one-tailed values are $F_{.05} = 5.54$ and $F_{.01} = 20.61$ (The latter values, which are not in **Table A10**, were obtained by squaring the tabled critical one-tailed values $t_{.05} = 2.35$ and $t_{.01} = 4.54$. A full discussion of one-tailed F values can be found in Section VI of the ***t* test for two independent samples** under the discussion of homogeneity of variance.). The same guidelines for interpreting a computed t value with Equation 22.3 are employed to interpret the computed F value, with one exception in reference to the directional alternative hypothesis H_1: $\rho < 0$. Since the value of F will always be a positive number, if the directional alternative H_1: $\rho < 0$ is employed, in order to reject the null hypothesis the value of F must be equal to or greater than the tabled critical one-tailed F value at the prespecified level of significance. However, the sign of r must be negative. When the F distribution is employed to evaluate the null hypothesis H_0: $\rho = 0$, it results in identical conclusions to those reached when Equation 22.3 is employed. Specifically, the nondirectional alternative hypothesis H_1: $\rho \neq 0$ is supported at the .05 level, since $F = 31.10$ is greater than the tabled critical two-tailed value $F_{.05} = 10.13$. The directional alternative hypothesis H_1: $\rho > 0$ is supported at both the .05 and .01 levels, since r is a positive number, and $F = 31.10$ is greater than the tabled critical one-tailed values $F_{.05} = 5.54$ and $F_{.01} = 20.61$.

Marascuilo and Serlin (1988) note that the following equation employing the normal distribution can also be used with large sample sizes to evaluate the null hypothesis H_0: $\rho = 0$: $z = r\sqrt{n - 1}$. If applied to Example 22.1, the value $z = (.955)\sqrt{5 - 1} = 1.91$ is computed. The value $z = 1.91$ only supports the directional alternative hypothesis H_1: $\rho > 0$ at the .05 level, since $z = 1.91$ is greater than the tabled critical one-tailed value $z_{.05} = 1.65$ in **Table A1** in the **Appendix**. The nondirectional alternative hypothesis H_1: $\rho \neq 0$ is not supported, since $z = 1.91$ is less than the tabled critical two-tailed value $z_{.05} = 1.96$. The latter result indicates that when employed with a small sample size, the normal approximation provides a more conservative test of an alternative hypothesis than does Equation 22.3.

5. The value $(1 - r^2)$ is often referred to as the **coefficient of nondetermination**, since it represents the proportion of variance that the two variables do not hold in common with one another.

6. The reader should keep in mind that for illustrative purposes the sample size employed for Example 22.1 is very small. Consequently, the values r and r^2 are, in all likelihood, not accurate estimates of the corresponding underlying population parameters ρ and ρ^2.

7. In the interest of complete accuracy, it should be noted that the following distinction is often made in discussing the underlying theories behind correlation and regression. Marascuilo and Levin (1983, p. 24) note that in correlation theory neither of the variables is stipulated as the predictor variable (referred to in some sources as the independent variable) or the criterion variable (referred to in some sources as the

dependent variable). Because of this, it is possible to derive two lines of best fit. In the case of one of the lines, Y is stipulated as the criterion variable, and in the case of the other line, X is stipulated as the criterion variable. In regression theory it is assumed that one of the variables (usually X) is stipulated as the predictor variable and the other (usually Y) as the criterion variable. The assumption of a joint bivariate normal distribution (noted in Section I) is directly related to the distinction between the two theories, in that the assumption of bivariate normality only applies to correlation theory. The aforementioned difference between correlation and regression theory is of no consequence to the computational procedures described in Sections V and VI.

8. An equation that is based on the minimum squared distance of all the points from the line reflects the fact that if the distance of each data point from the line is measured, and the resulting value is squared, the sum of the squared values for the n data points is the lowest possible value that can be obtained for that set of data.

9. The values $s_{Y.X}$ and $s_{X.Y}$ can also be computed with the equations noted below:

$$s_{Y.X} = \sqrt{\frac{SS_Y - \dfrac{(SP_{XY})^2}{SS_X}}{n - 2}} = \sqrt{\frac{29.2 - \dfrac{(88.6)^2}{294.8}}{5 - 2}} = .92$$

$$s_{X.Y} = \sqrt{\frac{SS_X - \dfrac{(SP_{XY})^2}{SS_Y}}{n - 2}} = \sqrt{\frac{294.8 - \dfrac{(88.6)^2}{29.2}}{5 - 2}} = 2.94$$

10. The reader may find it useful to review the discussion of confidence intervals in Section VI of the **single-sample *t* test** before reading this section.

11. The term SS_X in Equation 21.14 may also be written in the form $SS_X = (n - 1)\tilde{s}_X^2$, and the term SS_Y in Equation 22.15 may also be written in the form $SS_Y = (n - 1)\tilde{s}_Y^2$.

12. Equation 22.20 can also be written in the form: $z = (z_r - z_{\rho_0})\sqrt{n - 3}$. Thus, $z = (1.886 - 1.099)\sqrt{5 - 3} = 1.11$.

13. The value $n = 5$ employed in Example 22.1 is not used, since the method to be described is recommended when $n \geq 25$. For smaller sample sizes, tables in Cohen (1988) derived by David (1938) can be employed.

14. For the analysis described in this section the $df = \infty$ curve is employed for the relevant set of power curves, since **Fisher's z_r transformation** is based on the normal distribution.

15. Equation 22.24 can also be employed to evaluate the hypothesis of whether there is a significant difference between $k = 2$ independent correlations — i.e., the same hypothesis evaluated with Equation 22.22. When $k = 2$, the result obtained with Equation 22.24 will be equivalent to the result obtained with Equation 22.22. Specifically, the square of the obtained value of z obtained with Equation 22.22 will equal the value of χ^2 obtained with Equation 22.24. Thus, if the data employed in

Equation 22.22 are employed in Equation 22.24, the obtained value of chi-square equals $\chi^2 = z^2 = (.878)^2 = .771$.

$$\chi^2 = [(5-3)(1.886)^2 + (5-3)(1.008)^2] - \frac{[(5-3)(1.886) + (5-3)(1.008)]^2}{(5-3) + (5-3)} = .771$$

16. When $k = 2$, Equation 22.25 is equivalent to Equation 22.23.

17. If homogeneity of variance is assumed for the two samples, a pooled error variance can be computed as follows:

$$s^2_{Y.X} = \frac{(n_1 - 2)(s^2_{Y.X_1}) + (n_2 - 2)(s^2_{Y.X_2})}{n_1 + n_2 - 4}$$

$$s^2_{X.Y} = \frac{(n_1 - 2)(s^2_{X.Y_1}) + (n_2 - 2)(s^2_{X.Y_2})}{n_1 + n_2 - 4}$$

The computed value $s^2_{Y.X}$ is used in place of both $s^2_{Y.X_1}$ and $s^2_{Y.X_2}$ in Equation 22.31, and the computed value $s^2_{X.Y}$ is used in place of both $s^2_{X.Y_1}$ and $s^2_{X.Y_2}$ in Equation 22.32.

18. Marascuilo and Serlin (1988) describe how the procedure described in this section can be extended to the evaluation of a hypothesis contrasting three or more regression coefficients.

19. The equations for z_{X_i} and z_{Y_i} are analogous to Equation I.9 in the **Introduction.**

20. The sum of products within this context is not the same as the sum of products that represents the numerator of Equation 22.1.

21. One way of avoiding the problem of dependent pairs is to form pairs in which no digit is used for more than one pair. In other words, the first two digits in the series represent the X and Y variables for first pair, the third and fourth digits in the series represent the X and Y variables for second pair, and so on. Although use of the latter methodology really does not conform to the definition of autocorrelation, if it is employed one can justify employing the critical values in **Table A16**.

22. A discussion of the derivation of Equation 22.41 can be found in Bennett and Franklin (1954).

23. The reader should take note of the fact that the data for Example 22.5 are fictitious, and in reality the result of the analysis in this section is probably not consistent with actual studies that have been conducted which evaluate the relationship between intelligence and eye-hand coordination.

24. Although the **phi coefficient** is described in the book as a measure of association for the **chi-square test for $r \times c$ tables** (specifically, for 2×2 tables), it is also employed in psychological testing as a measure of association for 2×2 tables in order to evaluate the consistency of n subjects' responses to two questions. The latter type of analysis is essentially a dependent samples analysis for a 2×2 table, which, in fact, is the general model for which the **McNemar test (Test 15)** is employed.

25. It is also the case that the greater the number of predictor variables in a set of data involving a fixed number of subjects, the larger the value of R^2.

26. This principle has obviously not been adhered to in the example under discussion in order to minimize computations.

27. Tabachnick and Fidell (1989) note that for small sample sizes some sources recommend an even more severe adjustment than that which results from using Equation 22.56.

28. The equation noted below is equivalent to Equation 22.62.

$$s_{Y.X_1X_2} = \sqrt{\frac{SS_Y(1 - R^2_{Y.X_1X_2})}{n - k - 1}} = \sqrt{\frac{(29.2)(1 - .944)}{5 - 2 - 1}} = .90$$

29. The following should be noted with respect to Equation 22.69: a) When the sample size is small and/or the number of subjects is not substantially larger than the number of predictor variables, the "shrunken" estimate $\tilde{R}^2$ (computed with Equation 22.56) should be employed in Equation 22.69; b) When there are more than two predictor variables, the multiple correlation coefficient for the k variables is employed in the numerator of the radical of Equation 22.69 in place of $R^2_{Y.X_1X_2}$. The value $r^2_{X_1X_2}$ in the denominator of the radical of Equation 22.69 is replaced by the squared multiple correlation coefficient of variable i with all of the remaining predictor variables. Thus, if there are three predictor variables and s_{b_1} is computed, the values employed in the numerator and denominator of the radical are respectively $R^2_{Y.X_1X_2X_3}$ and $R^2_{X_1.X_2X_3}$.

30. Howell (1992) cites sources who argue that the t distribution does not provide a precise approximation of the underlying sampling distribution for the standard error of estimate of the coefficients. On the basis of this he states that caution should be employed in interpreting the results of the t test.

31. The same results are obtained if the analysis is done employing the standardized regression coefficients. This is demonstrated below employing the appropriate equations for the standardized coefficients. The minimal discrepancy between the values t_{β_1} and t_{b_1} is due to rounding off error.

$$s_{\beta_i} = \sqrt{\frac{1 - R^2_{Y.X_1X_2}}{(1 - r^2_{X_1X_2})(n - k - 1)}} = \sqrt{\frac{1 - .944}{[1 - (.37)^2][5 - 2 - 1]}} = .180$$

$$t_{\beta_i} = \frac{\beta_i}{s_{\beta_i}} \qquad t_{\beta_1} = \frac{.883}{.180} = 4.91 \qquad t_{\beta_2} = \frac{.193}{.180} = 1.07$$

32. Marascuilo and Levin (1983) and Marascuilo and Serlin (1988) recommend that in order to control the Type I error rate, a more conservative t value should be employed when the number of regression coefficients evaluated is greater than one. These sources describe the use of the **Bonferroni–Dunn** and **Scheffé procedures** (which are described in reference to multiple comparisons for analysis of variance procedures) in adjusting the t value.

33. The computed value r_p can also be evaluated through use of the critical values in **Table A16** (for $df = n - v$).

34. Note that when a simple bivariate/zero order correlation is computed, $n - v = n - 2$, and thus Equation 22.73 becomes identical to Equation 22.3 (which is used to evaluate the significance of the zero order correlation coefficient $r_{X_1X_2}$).

35. The computed value r_{sp} can also be evaluated through use of the critical values in **Table A16** (for $df = n - v$).

36. As is the case for Equation 22.73, Equation 22.75 becomes identical to Equation 22.3 when $n - v = n - 2$.

Test 23

Spearman's Rank-Order Correlation Coefficient

(Nonparametric Measure of Association/Correlation Employed with Ordinal Data)

I. Hypothesis Evaluated with Test and Relevant Background Information

Spearman's rank-order correlation coefficient is one of a number of measures of correlation or association discussed in this book. Measures of correlation are not inferential statistical tests, but are, instead, descriptive statistical measures which represent the degree of relationship between two or more variables. Upon computing a measure of correlation, it is common practice to employ one or more inferential statistical tests in order to evaluate one or more hypotheses concerning the correlation coefficient. The hypothesis stated below is the most commonly evaluated hypothesis for **Spearman's rank-order correlation coefficient.**

Hypothesis evaluated with test In the underlying population represented by a sample, is the correlation between subjects' scores on two variables some value other than zero? The latter hypothesis can also be stated in the following form: In the underlying population represented by the sample, is there a significant **monotonic** relationship between the two variables? It is important to note that the nature of the relationship described by **Spearman's rank-order correlation coefficient** is based on an analysis of two sets of ranks.

Relevant background information on test Prior to reading the material in this section the reader should review the general discussion of correlation in Section I of the **Pearson product-moment correlation coefficient (Test 22).** Developed by Spearman (1904), **Spearman's rank-order correlation coefficient** is a bivariate measure of correlation/association that is employed with rank-order data. The population parameter estimated by the correlation coefficient will be represented by the notation ρ_S (where ρ is the lower case Greek letter **rho**). The sample statistic computed to estimate the value of ρ_S will be represented by the notation r_S. In point of fact, **Spearman's rank-order correlation coefficient** is a special case of the **Pearson product-moment correlation coefficient,** when the latter measure is computed for two sets of ranks. The relationship between **Spearman's rank-order correlation coefficient** and the **Pearson product-moment correlation coefficient** is discussed in Section VI.

As is the case for the **Pearson product-moment correlation coefficient, Spearman's rank-order correlation coefficient** can be employed to evaluate data for n subjects, each of whom has contributed a score on two variables (designated as the X and Y variables). Within each of the variables, the n scores are rank-ordered. **Spearman's rank-order correlation coefficient** is also commonly employed to evaluate the degree of agreement between the rankings of $m = 2$ judges for n subjects/objects.

In computing **Spearman's rank-order correlation coefficient** one of the following is true with regard to the rank-order data that are evaluated: a) The data for both variables are in a rank-order format, since it is the only format for which data are available; b) The original data are in a rank-order format for one variable and in an interval/ratio format for the second variable. In such an instance, data on the second variable are converted to a rank-order format in order that both sets of data represent the same level of measurement; and c) The data for both variables have been transformed into a rank order-format from an interval/ratio format, since the researcher has reason to believe that one or more of the assumptions underlying the **Pearson product-moment correlation coefficient** (which is the analogous parametric correlational procedure employed for interval/ratio data) has been saliently violated. It should be noted that since information is sacrificed when interval/ratio data are transformed into a rank-order format, some researchers may elect to employ the **Pearson product-moment correlation coefficient** rather than **Spearman's rank-order correlation coefficient**, even when there is reason to believe that one or more of the assumptions of the former measure has been violated.

Spearman's rank-order correlation coefficient determines the degree to which a **monotonic** relationship exists between two variables. A monotonic relationship can be described as **monotonic increasing** (which is associated with a positive correlation) or **monotonic decreasing** (which is associated with a negative correlation). A relationship between two variables is monotonic increasing, if an increase in the value of one variable is always accompanied by an increase in the value of the other variable. A relationship between two variables is monotonic decreasing, if a decrease in the value of one variable is always accompanied by an decrease in the value of the other variable. Based on the above definitions, a positively sloped straight line represents an example of a monotonic increasing function, while a negatively sloped straight line represents an example of a monotonic decreasing function. In addition to the aforementioned linear functions, curvilinear functions can also be monotonic. For instance, the function $Y = X^2$ depicted in Figure 23.1 represents an example of a monotonic increasing function, since an increase in the X variable always results in an increase in Y variable. It should be noted that when the interval/ratio scores on two variables are monotonically related to one another, a linear function can be employed to describe the relationship between the rank-orderings of the two variables. This latter fact is demonstrated in Section VI.

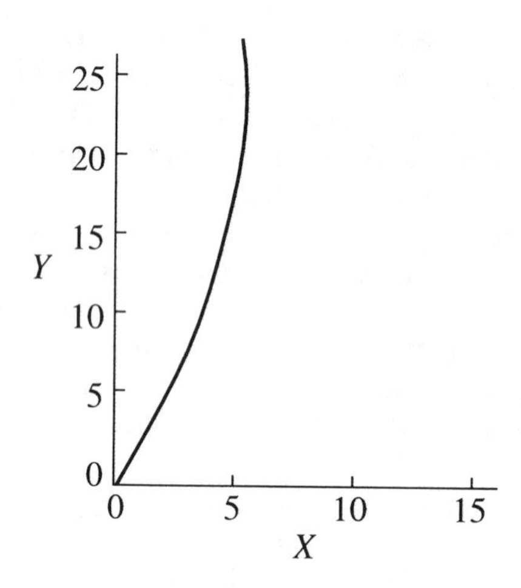

Figure 23.1 Monotonic Increasing Relationship ($Y = X^2$)

The same general guidelines that are described for interpreting the value of the **Pearson product-moment correlation coefficient** can be applied to **Spearman's rank-order correlation coefficient**. Thus, the range of values r_S can assume is defined by the limits -1 to $+1$ (i.e., $-1 \leq r_S \leq +1$). The absolute value of r_S (i.e., $|r_S|$) indicates the strength of the relationship between the two variables. As the absolute value of r_S approaches 1, the strength of the monotonic relationship increases, being the strongest when r_S equals either +1 or −1. The closer the absolute value of r_S is to 0, the weaker the monotonic relationship between the two variables, and when $r_S = 0$, no monotonic relationship is present. The sign of r_S indicates the direction of the monotonic relationship (i.e., positive/increasing monotonic versus negative/decreasing monotonic). As is the case for the **Pearson product-moment correlation coefficient**, a positive correlation indicates that an increase (decrease) on one variable is associated with an increase (decrease) on the other variable. A negative correlation indicates that an increase (decrease) on one variable is associated with a decrease (increase) on the other variable.

It is important to note that correlation does not imply causation. Consequently, if there is a strong correlation between two variables (i.e., the absolute value of r_S is close to 1), a researcher is not justified in concluding that one variable causes the other variable. Although it is possible that when a strong correlation exists one variable may, in fact, cause the other variable, the information employed in computing **Spearman's rank-order correlation coefficient** does not allow a researcher to draw such a conclusion. This is the case, since extraneous variables which have not been taken into account by the researcher can be responsible for the observed correlation between the two variables.

II. Example

Example 23.1 is identical to Example 22.1 (which is evaluated with the **Pearson product-moment correlation coefficient**). In evaluating Example 23.1 it will be assumed that the ratio data are rank-ordered, since one or more of the assumptions of the **Pearson product-moment correlation coefficient** has been saliently violated.[1]

Example 23.1 *A psychologist conducts a study employing a sample of five children to determine whether there is a statistical relationship between the number of ounces of sugar a ten-year-old child eats per week (which will represent the X variable) and the number of cavities in a child's mouth (which will represent the Y variable). The two scores (ounces of sugar consumed per week and number of cavities) obtained for each of the five children follow:* **Child 1** (20, 7); **Child 2** (0, 0); **Child 3** (1, 2); **Child 4** (12, 5); **Child 5** (3, 3). *Is there a significant correlation between sugar consumption and the number of cavities?*

III. Null versus Alternative Hypotheses

Upon computing **Spearman's rank-order correlation coefficient**, it is common practice to determine whether the obtained absolute value of the correlation coefficient is large enough to allow a researcher to conclude that the underlying population correlation coefficient between the two variables is some value other than zero. Section V describes how the latter hypothesis, which is stated below, can be evaluated through use of tables of critical r_S values or through use of an inferential statistical test that is based on either the t or z distributions.

Null hypothesis H_0: $\rho_S = 0$

(In the underlying population the sample represents, the correlation between the scores of subjects on Variable X and Variable Y equals 0.)

Alternative hypothesis H_1: $\rho_S \neq 0$

(In the underlying population the sample represents, the correlation between the scores of subjects on Variable X and Variable Y equals some value other than 0. This is a **non-directional alternative hypothesis**, and it is evaluated with a **two-tailed test**. Either a significant positive r_S value or a significant negative r_S value will provide support for this alternative hypothesis. In order to be significant, the obtained absolute value of r_S must be equal to or greater than the tabled critical two-tailed r_S value at the prespecified level of significance.)

or

$$H_1\text{: } \rho_S > 0$$

(In the underlying population the sample represents, the correlation between the scores of subjects on Variable X and Variable Y equals some value greater than 0. This is a **directional alternative hypothesis**, and it is evaluated with a **one-tailed test**. Only a significant positive r_S value will provide support for this alternative hypothesis. In order to be significant (in addition to the requirement of a positive r_S value), the obtained absolute value of r_S must be equal to or greater than the tabled critical one-tailed r_S value at the prespecified level of significance.)

or

$$H_1\text{: } \rho_S < 0$$

(In the underlying population the sample represents, the correlation between the scores of subjects on Variable X and Variable Y equals some value less than 0. This is a **directional alternative hypothesis**, and it is evaluated with a **one-tailed test**. Only a significant negative r_S value will provide support for this alternative hypothesis. In order to be significant (in addition to the requirement of a negative r_S value), the obtained absolute value of r_S must be equal to or greater than the tabled critical one-tailed r_S value at the prespecified level of significance.)

Note: Only one of the above noted alternative hypotheses is employed. If the alternative hypothesis the researcher selects is supported, the null hypothesis is rejected.[2]

IV. Test Computations

Table 23.1 summarizes the data for Example 23.1. The following should be noted with respect to Table 23.1: a) The number of subjects is $n = 5$. Each subject has an X score and a Y score, and thus there are five X scores and five Y scores; b) The rankings of the five subjects' scores on the X and Y variables are respectively recorded in the columns labelled R_X and R_Y; c) The column labelled $d = R_X - R_Y$ contains a difference score for each subject, which is obtained by subtracting a subject's rank on the Y variable from the subject's rank on the X variable; and d) The column labelled d^2 contains the square of each subject's difference score.

Table 23.1 Summary of Data for Example 23.1

Subject	X	R_X	Y	R_Y	$d = R_X - R_Y$	d^2
1	20	5	7	5	0	0
2	0	1	0	1	0	0
3	1	2	2	2	0	0
4	12	4	5	4	0	0
5	3	3	3	3	0	0
					$\Sigma d = 0$	$\Sigma d^2 = 0$

The ranking protocol employed in Table 23.1 is identical to that employed for the **Mann–Whitney *U* test (Test 9)**. Whereas in the case of the latter test the scores of subjects are ranked within each group, in the computation of **Spearman's rho** the scores of the $n = 5$ subjects are ranked within each of the variables. Thus, in Table 23.1 the five subjects' X scores are ranked such that a rank of 1 is assigned to the lowest score on the X variable, a rank of 2 is assigned to the next lowest score on the X variable, and so on until a rank of 5 is assigned to the highest score on the X variable. The identical ranking procedure is employed with respect to the Y scores (i.e., a rank of 1 is assigned to the lowest score on the Y variable, a rank of 2 is assigned to the next lowest score on the Y variable, and so on until a rank of 5 is assigned to the highest score on the Y variable). In the event of tied scores (which do not occur in Example 23.1), as is the case for other rank-order procedures, the average of the ranks involved is assigned to all scores tied for a given rank.

It should be noted that it is permissible to reverse the ranking protocol described above. Specifically, for each variable a rank of 1 can be assigned to the highest score on that variable and a rank of 5 to the lowest score on that variable. Employing this alternative ranking protocol will yield the identical value for r_S as the one yielded by the ranking protocol employed in Table 23.1. It should be emphasized that regardless of which ranking protocol is employed, the same protocol must be employed for both variables. The protocol of assigning the lowest rank to the lowest score and the highest rank to the highest score is employed in Example 23.1, since it allows for easiest interpretation of the results of the study.

In Column 6 of Table 23.1, the sum of the difference scores is computed to be $\Sigma d = 0$. In point fact, Σd will always equal zero and if Σd is some value other than zero, it indicates that an error has been made in the rankings and/or computations. In the last column of Table 23.1, the sum of the squared difference scores ($\Sigma d^2 = 0$) is computed. This latter value (which will only equal zero when $r_S = 1$) and the value of n are employed in Equation 23.1, which is the equation for computing **Spearman's rank-order correlation coefficient.**[3]

$$r_S = 1 - \frac{6\Sigma d^2}{n(n^2 - 1)} \qquad \textbf{(Equation 23.1)}$$

Substituting the appropriate values in Equation 23.1, the value $r_S = 1$ is computed.

$$r_S = 1 - \frac{(6)(0)}{5[(5)^2 - 1]} = 1$$

V. Interpretation of the Test Results

The obtained value $r_S = 1$ is evaluated with **Table A18 (Table of Critical Values for Spearman's Rho)** in the **Appendix**. The critical values in **Table A18** are listed in reference to n.[4] Employing **Table A18**, it can be determined that the tabled critical two-tailed r_S value at the .05 level of significance is $r_{S_{.05}} = 1$. Because of the small sample size, it is not possible to evaluate the nondirectional null hypothesis at the .01 level. The tabled critical one-tailed r_S values at the .05 and .01 levels of significance are $r_{S_{.05}} = .90$ and $r_{S_{.01}} = 1$.

The following guidelines are employed in evaluating the null hypothesis H_0: $\rho_S = 0$.

a) If the nondirectional alternative hypothesis H_1: $\rho_S \neq 0$ is employed, the null hypothesis can be rejected if the obtained absolute value of r_S is equal to or greater than the tabled critical two-tailed value at the prespecified level of significance.

b) If the directional alternative hypothesis H_1: $\rho_S > 0$ is employed, the null hypothesis can be rejected if the sign of r_S is positive, and the value of r_S is equal to or greater than the tabled critical one-tailed value at the prespecified level of significance.

c) If the directional alternative hypothesis H_1: $\rho_S < 0$ is employed, the null hypothesis can be rejected if the sign of r_S is negative, and the absolute value of r_S is equal to or greater than the tabled critical one-tailed value at the prespecified level of significance.

Employing the above guidelines, the nondirectional alternative hypothesis H_1: $\rho_s \neq 0$ is supported at the .05 level, since the computed value $r_S = 1$ is equal to the tabled critical two-tailed value $r_{S_{.05}} = 1$. The directional alternative hypothesis H_1: $\rho_s > 0$ is supported at both the .05 and .01 levels, since the computed value $r_S = 1$ is a positive number that is equal to or greater than the tabled critical one-tailed values $r_{S_{.05}} = .90$ and $r_{S_{.01}} = 1$. The directional alternative hypothesis H_1: $\rho_S < 0$ is not supported, since the computed value $r_S = 1$ is a positive number.

When the **Pearson product-moment correlation coefficient** is employed to evaluate the same set of data (i.e., the ratio scores of subjects are correlated with one another), the nondirectional alternative hypothesis (i.e., H_1: $\rho \neq 0$) is also supported at only the .05 level, and the directional alternative hypothesis (i.e., H_1: $\rho > 0$ is supported at both the .05 and .01 levels. Thus, in this instance the two correlation coefficients yield comparable results. (However, since **Pearson** r is the more powerful of the two correlational procedures, it is more likely to result in rejection of the null hypothesis at a given level of significance when applied to the same set of data.)

Test 23a: Test of significance for Spearman's rank-order correlation coefficient In the event a researcher does not have access to **Table A18**, Equation 23.2 which employs the t distribution, provides an alternative way of evaluating the null hypothesis H_0: $\rho_S = 0$. Most sources that recommend Equation 23.2 state that it provides a reasonably good approximation of the underlying sampling distribution when $n > 10$.

$$t = \frac{r_S\sqrt{n - 2}}{\sqrt{1 - r_S^2}} \qquad \textbf{(Equation 23.2)}$$

The t value computed with Equation 23.2 is evaluated with **Table A2 (Table of Student's t Distribution)** in the **Appendix**. The degrees of freedom employed are $df = n - 2$. Thus, in the case of Example 23.1, $df = 5 - 2 = 3$. For $df = 3$, the tabled critical two-tailed .05 and .01 values are $t_{.05} = 3.18$ and $t_{.01} = 5.84$, and the tabled critical one-tailed .05 and .01 values are $t_{.05} = 2.35$ and $t_{.01} = 4.54$. Since the sign of the t value computed with Equation 23.2 will always be the same as the sign of r_S, the guidelines described earlier in reference to **Table A18** for evaluating an r_S value can also be applied in evaluating the t value computed with Equation 23.2 (i.e., substitute t in place of r_S in the text of the guidelines for evaluating r_S).

Inspection of Equation 23.2 reveals that if the absolute value of r_S equals 1, the term $\sqrt{1 - r^2}$ will equal zero, thus rendering the equation insoluble (i.e., $t = [(1)\sqrt{5-2}]/\sqrt{1-(1)^2}$ = ?). Consequently, Equation 23.2 cannot be applied to Example 23.1.

Equation 23.3, which employs the normal distribution, is an alternative equation for evaluating the significance of r_S. When the sample size is large (approximately 200 or greater), Equation 23.3 will yield a result that is equivalent to that obtained with Equation 23.2.[5]

$$z = r_S\sqrt{n - 1} \qquad \textbf{(Equation 23.3)}$$

Although the sample size in Example 23.1 is well below the minimum size recommended for Equation 23.3, the appropriate values will be substituted in the latter equation in order to demonstrate its application. Substituting the values $r_S = 1$ and $n = 5$ in Equation 23.3, the value $z = 2.00$ is computed.

$$z = (1)\sqrt{5 - 1} = 2$$

The computed value $z = 2.00$ is evaluated with **Table A1 (Table of the Normal Distribution)** in the **Appendix**. In the latter table, the tabled critical two-tailed .05 and .01 values are $z_{.05} = 1.96$ and $z_{.01} = 2.58$, and the tabled critical one-tailed .05 and .01 values are $z_{.05} = 1.65$ and $z_{.01} = 2.33$. Since the sign of the z value computed with Equation 23.3 will always be the same as the sign of r_S, the guidelines described earlier in reference to **Table A18** for evaluating an r_S value can also be applied in evaluating the z value computed with Equation 23.3 (i.e., substitute z in place of r_S in the text of the guidelines for evaluating r_S).

Employing the guidelines, the nondirectional alternative hypothesis H_1: $\rho_S \neq 0$ is supported at the .05 level, since the computed value $z = 2.00$ is greater than the tabled critical two-tailed value $z_{.05} = 1.96$. It is not, however, supported at the .01 level, since $z = 2.00$ is less than the tabled critical two-tailed value $z_{.01} = 2.58$.

The directional alternative hypothesis H_1: $\rho_S > 0$ is supported at the .05 level, since the computed value $z = 2.00$ is a positive number that is greater than the tabled critical one-tailed value $z_{.05} = 1.65$. It is not, however, supported at the .01 level, since $z = 2.00$ is less than the tabled critical one-tailed value $z_{.01} = 2.33$.

The directional alternative hypothesis H_1: $\rho < 0$ is not supported, since the computed value $z = 2.00$ is a positive number. In order for the alternative hypothesis H_1: $\rho_S < 0$ to be supported, the computed value of z must be a negative number (as well as the fact that the absolute value of z must be equal to or greater than the tabled critical one-tailed value at the prespecified level of significance). Note that the results obtained through use of Equation 23.3 are reasonably consistent with those that are obtained when **Table A18** is employed.[6]

A summary of the analysis of Example 23.1 follows: It can be concluded that there is a significant monotonic increasing/positive relationship between the number of ounces of sugar a ten-year-old child eats and the number of cavities in a child's mouth. This result can be summarized as follows (if it is assumed the nondirectional alternative hypothesis H_1: $\rho_S \neq 0$ is employed): $r_S = 1, p < .05$.

VI. Additional Analytical Procedures for Spearman's Rank-Order Correlation Coefficient and/or Related Tests

1. Tie correction for Spearman's rank-order correlation coefficient When one or more ties are present in a set of data, many sources recommend that the r_S value computed with Equation 23.1 be adjusted. The reason for this is that when ties are present, Equation 23.1 spuriously inflates the absolute value of r_S. In practice, most of the time that ties are present the effect on the value of r_S will be minimal (unless the number of ties is excessive). The tie correction procedure to be demonstrated in this section will employ the data summarized in Table 23.2. Assume that the data are for the same variables evaluated in Example 23.1, except for the fact that a different set of subjects is employed with $n = 10$.

Employing Equation 23.1, it is determined that the value of **Spearman's rho** without employing a tie correction is $r_S = .764$.

$$r_S = 1 - \frac{6(39)}{10[(10)^2 - 1)]} = .764$$

Table 23.2 Data Employed with Tie Correction Procedure

Subject	X	R_X	Y	R_Y	$d = R_X - R_Y$	d^2
1	0	1.5	1	2	−.5	.25
2	0	1.5	0	1	.5	.25
3	2	3	2	3.5	−.5	.25
4	4	4	2	3.5	.5	.25
5	8	6	8	9.5	−3.5	12.25
6	8	6	8	9.5	−3.5	12.25
7	8	6	3	5	1	1
8	13	8	4	6	2	4
9	16	9.5	6	7	2.5	6.25
10	16	9.5	7	8	1.5	2.25
					$\Sigma d = 0$	$\Sigma d^2 = 39$

The tie correction will now be introduced. In the example under discussion there are $s = 3$ sets of ties involving the ranks of subjects' X scores (Subjects 1 and 2; Subjects 5, 6, and 7; Subjects 9 and 10), and $s = 2$ sets of ties involving the ranks of subjects' Y scores (Subjects 3 and 4; Subjects 5 and 6). Equation 23.8 is employed to compute the tie corrected **Spearman's rank-order correlation coefficient**, which will be represented by the notation r_{S_c}. Note that the values Σx^2 and Σy^2 in Equation 23.8 are computed with Equations 23.6 and 23.7, and that Equations 23.6 and 23.7 are respectively based on the values T_X and T_Y, which are computed with Equations 23.4 and 23.5. In Equation 23.4, $t_{i_{(X)}}$ represents the number of X scores that are tied for a given rank. In Equation 23.5, $t_{i_{(Y)}}$ represents the number of Y scores that are tied for a given rank. The notations $\Sigma_{i=1}^{s}\left(t_{i_{(X)}}^3 - t_{i_{(X)}}\right)$ and $\Sigma_{i=1}^{s}\left(t_{i_{(Y)}}^3 - t_{i_{(Y)}}\right)$ indicate that the following is done with respect to each of the variables: a) For each set of ties, the number of ties in the set is subtracted from the cube of the number of ties in that set; and b) The sum of all the values computed in part a) is obtained for that variable.

When the data from Table 23.2 are substituted in Equations 23.4–23.8, the tie corrected value $r_{S_c} = .758$ is computed.

$$T_X = \sum_{i=1}^{s}\left(t_{i_{(X)}}^3 - t_{i_{(X)}}\right) = [(2)^3 - 2] + [(3)^3 - 3] + [(2)^3 - 2] = 36 \quad \textbf{(Equation 23.4)}$$

$$T_Y = \sum_{i=1}^{s}\left(t_{i_{(Y)}}^3 - t_{i_{(Y)}}\right) = [(2)^3 - 2] + [(2)^3 - 2] = 12 \quad \textbf{(Equation 23.5)}$$

$$\Sigma x^2 = \frac{n^3 - n - T_X}{12} = \frac{(10)^3 - 10 - 36}{12} = 79.5 \quad \textbf{(Equation 23.6)}$$

$$\Sigma y^2 = \frac{n^3 - n - T_Y}{12} = \frac{(10)^3 - 10 - 12}{12} = 81.5 \quad \textbf{(Equation 23.7)}$$

$$r_{S_c} = \frac{\Sigma x^2 + \Sigma y^2 - \Sigma d^2}{2\sqrt{\Sigma x^2 \Sigma y^2}} = \frac{79.5 + 81.5 - 39}{2\sqrt{(79.5)(81.5)}} = .758 \quad \textbf{(Equation 23.8)}$$

Thus, by employing the tie correction, the value of **rho** is reduced from the uncorrected value of $r_S = .764$ to $r_{S_c} = .758$. As noted earlier, the correction is minimal.

2. Spearman's rank-order correlation coefficient as a special case of the Pearson product-moment correlation coefficient Although the procedure described in the previous section for dealing with ties is the one recommended in most sources, in actuality, an alternative and at times more computationally efficient procedure can be employed. In Section I it is noted that **Spearman's rank-order correlation coefficient** is a special case of the **Pearson product-moment correlation coefficient.** In point of fact, if the **Pearson product-moment correlation coefficient** is computed for the rank-orders in a set of interval/ratio data, the computed r value will be identical to the value computed for r_S with Equation 23.1. This is demonstrated below for Example 23.1, where Equation 22.1 (the equation for computing the **Pearson product-moment correlation coefficient**) is employed to compute the value $r = r_S = 1$. Table 23.3 summarizes the values that are substituted in Equation 22.1. Note that the ranks R_X and R_Y employed in Table 23.1 are used in Table 23.3 to represent the scores on the X and Y variables.[7]

$$r = r_S = \frac{55 - \dfrac{(15)(15)}{5}}{\sqrt{\left[55 - \dfrac{(15)^2}{10}\right]\left[55 - \dfrac{(15)^2}{10}\right]}} = 1$$

Table 23.3 Summary of Data for Example 23.1 for Evaluation with Equation 22.1

Subject	X	X^2	Y	Y^2	XY
1	5	25	5	25	25
2	1	1	1	1	1
3	2	4	2	4	4
4	4	16	4	16	16
5	3	9	3	9	9
	$\Sigma X = 15$	$\Sigma X^2 = 55$	$\Sigma Y = 15$	$\Sigma Y^2 = 55$	$\Sigma XY = 55$

Table 23.4 Summary of Data in Table 23.2 for Evaluation with Equation 22.1

Subject	X	X^2	Y	Y^2	XY
1	1.5	2.25	2	4	3
2	1.5	2.25	1	1	1.5
3	3	9	3.5	12.25	10.5
4	4	16	3.5	12.25	14
5	6	36	9.5	90.25	57
6	6	36	9.5	90.25	57
7	6	36	5	25	30
8	8	64	6	36	48
9	9.5	90.25	7	49	66.5
10	9.5	90.25	8	64	76
	$\Sigma X = 55$	$\Sigma X^2 = 382$	$\Sigma Y = 55$	$\Sigma Y^2 = 384$	$\Sigma XY = 363.5$

When there are no ties present in the data, Equations 23.1 and 22.1 will always yield the identical value for r_S. However, anytime there is at least one set of ties, the values yielded by the two equations will not be identical. In point of fact, Howell (1992) notes that when ties are present in the data, the r_S value computed with Equation 22.1 will be equivalent to the tie corrected value r_{S_c} computed with Equation 23.8. When there are no ties present in the data, it is clearly more efficient to employ Equation 23.1 than it is to employ

Equation 22.1. However, when ties are present, it can be argued that use of Equation 22.1 is more computationally efficient than Equation 23.8. To demonstrate the equivalency of Equation 22.1 and Equation 23.8, Equation 22.1 is employed below with the rank-orders in Table 23.2. Table 23.4 summarizes the values that are substituted in Equation 22.1. The value $r = .758$ obtained with Equation 22.1 is identical to the value $r_{S_c} = .758$ obtained with Equation 23.8.

$$r = \frac{363.5 - \frac{(55)(55)}{10}}{\sqrt{\left[382 - \frac{(55)^2}{10}\right]\left[384 - \frac{(55)^2}{10}\right]}} = .758$$

3. **Regression analysis and Spearman's rank-order correlation coefficient** When **Spearman's rank-order correlation coefficient** is computed for a set of data, a researcher may also want to derive the mathematical function that best allows one to predict a subject's score on one variable through use of the subject's score on the second variable. To do this requires the use of regression analysis which, as noted in Section VI of the **Pearson product-moment correlation coefficient**, is a general term that describes statistical procedures which determine the mathematical function that best describes the relationship between two or more variables. One type of regression analysis that falls within the general category of **nonparametric regression analysis** is referred to as **monotonic regression analysis**. The latter type of analysis is based on the fact that if two variables (which are represented by interval/ ratio data) are monotonically related to one another, the rankings on the variables will be linearly related to one another. This can be illustrated in reference to Example 23.1 through use of Figure 23.2. Whereas Figure 22.1 (in Section VI of the **Pearson product-moment correlation coefficient**) represents a scatterplot of the five pairs of ratio scores for Examples 22.1/23.1, Figure 23.2 is a scatterplot of the five pairs of ranks on the two variables. Note that the scatterplot is such that one can draw a positively sloped straight line which passes through all of the data points. The only time all of the data points will fall on the regression line is when the absolute value of the correlation between the two variables equals 1. Although some data points may fall on the line when an imperfect monotonic relationship is present, the others will not. The stronger the monotonic relationship, the closer the proximity of the data points to the line.

As is noted in Section VI of the **Pearson product-moment correlation coefficient**, the most commonly employed method of regression analysis is the method of least squares (which is a linear regression procedure that derives the straight line which provides the best fit for a set of data). Although visual inspection of Figures 22.1 and 23.2 suggests a strong monotonic increasing relationship between the two variables (i.e., an increase in the number of ounces of sugar consumed is associated with an increase in the number of cavities), it does not allow one to precisely determine whether the function that best describes the relationship is a straight line or a monotonic curve. In order to determine the latter, it is necessary to contrast the predictive accuracy of the method of least squares with some alternative form of regression analysis. Conover (1980), who provides a bibliography on the general subject of monotonic regression analysis, describes its application in deriving a curve for a set of rank-ordered data. Marascuilo and McSweeney (1977) and Sprent (1989) also discuss the subject of monotonic regression analysis. In addition to sources on nonparametric statistics that discuss monotonic regression, many books on correlation and regression describe procedures for deriving different types of curvilinear functions. Daniel (1990) discusses a number of different approaches to nonparametric regression analysis, which derive the straight line that best describes the relationship between the interval/ratio scores on the two variables. These

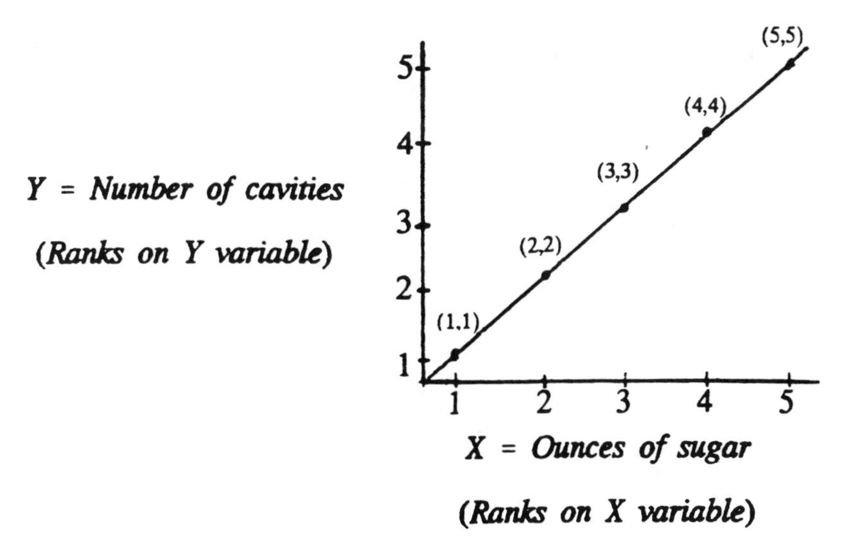

Figure 23.2 Scatterplot of Ranks for Example 23.1

latter types of regression analysis (which employ the median instead of the mean as a reference point) are recommended when there is reason to believe that one or more of the assumptions underlying the method of least squares is saliently violated. Among those procedures Daniel (1990) describes are the **Brown-Mood method** (Brown and Mood (1951), and Mood (1950)), and a methodology developed by Theil (1950). Daniel (1990) also provides a comprehensive bibliography on the subject of nonparametric regression analysis.

4. Partial rank correlation The computation of a **partial correlation coefficient**, described in Section IX of the **Pearson product-moment correlation coefficient**, can be extended to **Spearman's rank-order correlation coefficient**. Thus, when the rank-orders for three variables are evaluated, Equation 22.72 can be employed to compute a **partial correlation coefficient** for **Spearman's rho** (employing the relevant r_S values in the equation). Conover (1980) and Daniel (1990) discuss the computation of a **partial correlation coefficient** in reference to **Spearman's rho**.

VII. Additional Discussion of Spearman's Rank-Order Correlation Coefficient

1. The relationship between Kendall's coefficient of concordance (Test 25), Spearman's rank-order correlation coefficient, and the Friedman two-way analysis of variance by ranks (Test 19) **Kendall's coefficient of concordance**, which is discussed later in the book, is a measure of association that allows a researcher to evaluate the degree of agreement between m sets of ranks on n subjects/objects. In point of fact, **Kendall's coefficient of concordance** is linearly related to **Spearman's rank-order correlation coefficient.**[8] The underlying statistical model upon which **Kendall's coefficient of concordance** is based is identical to the model for the **Friedman two-way analysis of variance by ranks**. As a result of this, the **Friedman two-way analysis of variance by ranks** can be employed to determine whether the value of the **coefficient of concordance** is significant. In point of fact, the **Friedman two-way analysis of variance by ranks** can also be used to determine whether the value of **Spearman's rho** is significant. This will be illustrated with Example 23.2, which represents a type of problem that is commonly evaluated with **Spearman's rank-order correlation coefficient** (as well as **Kendall's coefficient of concordance** when there are more than two sets of ranks). In Example 23.2, $n = 10$ films (i.e., objects/subjects) are rank-ordered by $m = 2$ judges, and a determination is made with respect to the degree of agreement among the rankings of the judges.

Example 23.2 *In order to determine whether two critics agree with one another in their evaluation of movies, a newspaper editor asks the two critics to rank-order ten movies (assigning a rank of* 1 *to the best movie, a rank of* 2 *to the next best movie, etc.).* Table 23.5 *summarizes the data for the study. Is there a significant association between the two sets of ranks?*

Table 23.5 Summary of Data for Example 23.2

	Critic 1	Critic 2		
Movie	R_X	R_Y	$d = R_X - R_Y$	d^2
1	7	10	−3	9
2	1	2	−1	1
3	8	6	2	4
4	10	8	2	4
5	9	7	2	4
6	6	4	2	4
7	5	9	−4	16
8	2.5	3	−.5	.25
9	2.5	1	1.5	2.25
10	4	5	−1	1
			$\Sigma d = 0$	$\Sigma d^2 = 45.5$

Note that in Table 23.5 each of the $n = 10$ rows represents one of the ten movies, instead of representing $n = 10$ subjects (as is the case in Example 23.1). The ranks of Critic 1 are represented in the column labelled R_X, and the ranks of Critic 2 are represented in the column labelled R_Y. Note that Critic 1 places Movies 8 and 9 in a tie for the second best movie. Thus (employing the protocol for tied ranks described in Section IV of the **Mann–Whitney *U* test**), the two ranks involved (2 and 3) are averaged ($(2 + 3)/2 = 2.5$), and each of the movies is assigned the average rank of 2.5.

Employing Equation 23.1, the value $r_S = .724$ is computed. The tie corrected value $r_{S_c} = .723$ (for which the calculations are not shown) is almost identical.

$$r = 1 - \frac{(6)(45.5)}{10[(10)^2 - 1]} = .724$$

Employing **Table A18**, it is determined that for $n = 10$, the tabled critical two-tailed .05 and .01 values are $r_{S_{.05}} = .648$ and $r_{S_{.01}} = .794$, and the tabled critical one-tailed .05 and .01 values are $r_{S_{.05}} = .564$ and $r_{S_{.01}} = .745$. Employing the aforementioned critical values, the nondirectional alternative hypothesis H_1: $\rho_S \neq 0$ and the directional alternative hypothesis H_1: $\rho_S > 0$ are supported at the .05 level, since the computed value $r_S = .724$ is greater than the tabled critical two-tailed value $r_{S_{.05}} = .648$ and the tabled critical one-tailed value $r_{S_{.05}} = .564$. The alternative hypotheses are not supported at the .01 level, since $r_S = .724$ is less than the tabled critical two-tailed value $r_{S_{.01}} = .794$ and the tabled critical one-tailed value $r_{S_{.01}} = .745$.

If Equation 23.2 is employed to evaluate the null hypothesis H_0: $\rho_S = 0$, the value $t = 2.97$ is computed.

$$t = \frac{(.724)\sqrt{10 - 2}}{\sqrt{1 - (.724)^2}} = 2.97$$

Employing **Table A2**, it is determined that for $df = 10 - 2 = 8$, the tabled critical two-tailed .05 and .01 values are $t_{.05} = 2.31$ and $t_{.01} = 3.36$, and the tabled critical one-tailed .05 and .01 values are $t_{.05} = 1.86$ and $t_{.01} = 2.90$. Employing the aforementioned critical values, the nondirectional alternative hypothesis H_1: $\rho_S \neq 0$ is supported at the .05 level, since the computed value $t = 2.97$ is greater than the tabled critical two-tailed value $t_{.05} = 2.31$. It is not supported at the .01 level, since $t = 2.97$ is less than $t_{.01} = 3.36$. The directional alternative hypothesis H_1: $\rho_S > 0$ is supported at both the .05 and .01 levels, since the computed value $t = 2.97$ is a positive number (since $r_S = .724$ is a positive number) that is greater than the tabled critical one-tailed values $t_{.05} = 1.86$ and $t_{.01} = 2.90$.

If Equation 23.3 is employed to evaluate the null hypothesis H_0: $\rho_S = 0$, the value $z = 2.17$ is computed.

$$z = (.724)\sqrt{10 - 1} = 2.17$$

Employing **Table A1**, it is determined that the computed value $z = 2.17$ is greater than the tabled critical two-tailed value $z_{.05} = 1.96$ and the tabled critical one-tailed value $z_{.05} = 1.65$, but less than the tabled critical two-tailed value $z_{.01} = 2.58$ and the tabled critical one-tailed value $z_{.01} = 2.33$. Thus, both the nondirectional alternative hypothesis H_1: $\rho_S \neq 0$ and the directional alternative hypothesis H_1: $\rho_S > 0$ are supported at the .05 level, but not at the .01 level. Note that identical conclusions are reached with **Table A18** and Equation 23.3, but the latter conclusions are not identical to those obtained with Equation 23.2 (where the directional alternative hypothesis H_1: $\rho_S > 0$ is also supported at the .01 level). As noted in Section V, the conclusions based on use of **Table A18**, Equation 23.2, and Equation 23.3 will not always be in total agreement.

It is noted earlier in this section that the **Friedman two-way analysis of variance by ranks** can be employed to determine whether the value of **Spearman's rho** is significant. This will now be illustrated in reference to Example 23.2. The data for Example 23.2 are rearranged in Table 23.6 to conform to the test model for the **Friedman two-way analysis of variance by ranks.** Note that the rows and columns employed in Table 23.5 are reversed in Table 23.6. When Table 23.6 is employed within the framework of the **Friedman test** model, the two critics represent $n = 2$ subjects, and the 10 ranks represent $k = 10$ levels of a within-subjects/repeated-measures independent variable.

Table 23.6 Data for Example 23.2 Formatted for Analysis with the Friedman Two-Way Analysis of Variance by Ranks

Movie	1	2	3	4	5	6	7	8	9	10
Critic 1	7	1	8	10	9	6	5	2.5	2.5	4
Critic 2	10	2	6	8	7	4	9	3	1	5
ΣR_j	17	3	14	18	16	10	14	5.5	3.5	9
$(\Sigma R_j)^2$	289	9	196	324	256	100	196	30.25	12.25	81

From the summary information in Table 23.6, the value $\Sigma_{j=1}^{k}(\Sigma R_j)^2 = 1493.5$ is computed.

$$\sum_{j=1}^{k} (\Sigma R_j)^2 = 289 + 9 + 196 + 324 + 256 + 100 + 196 + 30.25 + 12.25 + 81 = 1493.5$$

Employing the above value, along with the other appropriate values in Equation 19.1 (the equation for the **Friedman two-way analysis of variance by ranks**), the value $\chi_r^2 = 15.46$ is computed.[9]

$$\chi_r^2 = \frac{12}{nk(k+1)} \sum_{j=1}^{k} (\Sigma R_j)^2 - 3n(k+1)$$

$$= \left[\frac{12}{(2)(10)(10+1)}\right][1493.5] - (3)(2)(10+1) = 15.46$$

The value $\chi_r^2 = 15.46$ is evaluated with **Table A4 (Table of the Chi-Square Distribution) in the Appendix**. For $df = k - 1 = 10 - 1 = 9$, the tabled critical two-tailed .05 and .01 values are $\chi^2_{.05} = 16.92$ and $\chi^2_{.01} = 21.67$, and the tabled critical one-tailed .05 and .01 values are $\chi^2_{.05} = 14.68$ and $\chi^2_{.01} = 19.50$ (the latter value is interpolated).[10] Employing the aforementioned critical values, the null hypothesis for the **Friedman two-way analysis of variance by ranks** (H_0: $\theta_1 = \theta_2 = \cdots = \theta_{10}$) can be rejected at the .05 level, but only if a one-tailed analysis is conducted (since $\chi_r^2 = 15.46$ is greater than the tabled critical one-tailed value $\chi^2_{.05} = 14.68$).[11] The result falls short of being significant at the .05 level for a two-tailed analysis, since $\chi_r^2 = 15.46$ is less than the tabled critical two-tailed value $\chi^2_{.05} = 16.92$. Rejection of the null hypothesis for the **Friedman two-way analysis of variance by ranks** is commensurate with rejection of the null hypothesis H_0: $\rho_S = 0$ for **Spearman's rank-order correlation coefficient**. In actuality, the result derived employing the **Friedman two-way analysis of variance by ranks** is similar, but not identical, to the analysis of **Spearman's rho** with **Table A18**, Equation 23.2, and Equation 23.3 (which, as noted earlier, are not in themselves in total agreement). The slight discrepancy between the results of the **Friedman test** and the more commonly employed methods for assessing the significance of **Spearman's rho** can be attributed to the fact that the test statistics based on the t, normal, and chi-square distributions are large sample approximations, which in the case of Example 23.2 are employed with a small sample size. It was also noted earlier, that the values in **Table A18** are approximations of the exact values in the underlying sampling distribution.

2. Power efficiency of Spearman's rank-order correlation coefficient Daniel (1990) and Siegel and Castellan (1988) note that (for large sample sizes) the power efficiency of **Spearman's rank-order correlation coefficient** relative to the **Pearson product-moment correlation coefficient** is approximately .91. The value .91 indicates that if a test of significance is employed to evaluate the null hypothesis of a zero population correlation, use of the **Pearson product-moment correlation coefficient** requires 91% of the subjects to detect a specific effect size (i.e., reject a false null hypothesis) when compared with **Spearman's rank-order correlation coefficient.**[12]

3. Brief discussion of Kendall's Tau (Test 24): An alternative measure of association for two sets of ranks **Kendall's tau** is an alternative measure of association that can be employed to evaluate two sets of ranks. Although **Spearman's rho** and **Kendall's tau** can be employed to measure the degree of association for the same set of data, **Spearman's rho** is the more commonly described of the two measures (primarily because it requires fewer computations). A comparative discussion of **Spearman's rho** and **Kendall's tau** can be found in Section I of the latter test.

VIII. Additional Examples Illustrating the Use of Spearman's Rank-Order Correlation Coefficient

If a researcher elects to rank-order the scores of subjects in any of the examples for which the **Pearson product-moment correlation coefficient** is employed, a value can be computed for **Spearman's rank-order correlation coefficient.** Thus, as is the case for Example 22.1, the data for Examples 22.2 and 22.3 can be rank-ordered and evaluated with **Spearman's rho.** Since the rankings for the latter two examples are identical to the rankings for Example 23.1, all three examples yield the identical result. Since **Kendall's tau** and **Spearman's rho** can be employed to evaluate the same data, Example 24.1, as well as the data set presented in Table 24.4, can also be evaluated with **Spearman's rho.**

References

Brown, G. M. and Mood, A. M. (1951). "On median tests for linear hypotheses," Jerzy Neyman (ed.), **Proceedings of the Second Berkeley Symposium on Mathematical Statistics and Probability.** Berkeley and Los Angeles: The University of California Press, 159–166.

Conover, W. J. (1980). **Practical nonparametric statistics** (2nd ed.). New York: John Wiley and Sons.

Daniel, W. W. (1990). **Applied nonparametric statistics** (2nd ed.). Boston: PWS–Kent Publishing Company.

Edwards, A. L. (1984). **An introduction to linear regression and correlation** (2nd ed.). New York: W. H. Freeman and Company.

Howell, D. C. (1992). **Statistical methods for psychology** (3rd ed.). Boston: PWS–Kent Publishing Company.

Lindeman, R. H., Merenda, P. F., and Gold, R. Z. (1980). **Introduction to bivariate and multivariate analysis.** Glenview, IL: Scott, Foresman and Company.

Marascuilo, L. A. and McSweeney, M. (1977). **Nonparametric and distribution-free methods for the social sciences.** Monterey, CA: Brooks/Cole Publishing Company.

Mood, A. M. (1950). **Introduction to the theory of statistics.** New York: McGraw–Hill Book Company.

Olds, E. G. (1938). Distribution of sum of squares of rank differences for small numbers of individuals. **Annals of Mathematical Statistics,** 9, 133–148.

Olds, E. G. (1949). The 5% significance levels of sums of squares of rank differences and a correlation. **Annals of Mathematical Statistics,** 20, 117–119.

Siegel, S. and Castellan, N. J., Jr. (1988). **Nonparametric statistics for the behavioral sciences** (2nd ed.). New York: McGraw–Hill Book Company.

Spearman, C. (1904). The proof and measurement of association between two things. **American Journal of Psychology,** 15, 72–101.

Sprent, P. (1989). **Applied nonparametric statistical methods.** London: Chapman and Hall.

Theil, H. (1950). A rank–invariant method of linear and polynomial regression analysis III. **Nederl. Akad. Wetensch. Proc., Series A,** 53, 1397–1412.

Zar, J. H. (1972). Significance testing of Spearman rank correlation coefficient. **Journal of the American Statistical Association,** 67, 578–580.

Endnotes

1. It should be noted that although the scores of subjects in Example 23.1 are ratio data, in most instances when **Spearman's rank-order correlation coefficient** is employed it is more likely that the original data for both variables are in a rank-order format. As is noted in Section I, conversion of ratio data to a rank-order format (which is done in Section IV with respect to Example 23.1) is most likely to occur when a researcher has reason to believe that one or more of the underlying assumptions of the **Pearson product-moment correlation coefficient** is saliently violated. Example 23.2 in Section VI represents a study involving two variables that are originally in a rank-order format for which **Spearman's rho** is computed.

2. Some sources employ the following statements as the null hypothesis and the nondirectional alternative hypothesis for **Spearman's rank-order correlation coefficient: Null hypothesis:** H_0: Variables X and Y are independent of one another; **Nondirectional alternative hypothesis:** H_1: Variables X and Y are not independent of one another.
   It is, in fact, true that if in the underlying population the two variables are independent, the value of ρ_s will equal zero. However, the fact that $\rho_s = 0$ in and of itself does not ensure that the variables are independent of one another. Thus, it is conceivable that in a population in which the correlation between X and Y is $\rho_s = 0$, a nonmonotonic curvilinear function can be employed to describe the relationship between the variables.

3. Daniel (1990) notes that the computed value of r_S is not an unbiased estimate of ρ_S.

4. The reader may find slight discrepancies in the critical values listed for **Spearman's rho** in the tables published in different books. The differences are due to the fact that separate tables derived by Olds (1938, 1949) and Zar (1972), which are not identical, are employed in different sources. Howell (1992) notes that the tabled critical values noted in various sources are approximations and not exact values.

5. The minimum sample size for which Equation 23.3 is recommended varies depending upon which source one consults. Some sources recommend the use of Equation 23.3 for values as low as $n = 25$, whereas others state that n should equal at least 100.

6. The results obtained through use of **Table A18**, Equation 23.2, and Equation 23.3 will not always be in total agreement with one another. In instances where the different methods for evaluating significance do not agree, there will usually not be a major discrepancy between them. In the final analysis, the larger the sample size the more likely it is that the methods will be consistent with one another.

7. The following will always be true when Equation 22.1 is employed in computing **Pearson** r (and r_S), and the rank-orders are employed to represent the scores on the X and Y variables: $\Sigma X = \Sigma Y$ and $\Sigma X^2 = \Sigma Y^2$ (however, the latter will only be true if there are no ties).

8. The relationship between **Spearman's rank-order correlation coefficient** and **Kendall's coefficient of concordance** is discussed in greater detail in Section VII of the latter test. In the latter discussion, it is noted that although when there are two sets of ranks the values computed for **Spearman's rho** and **Kendall's coefficient of concordance** will not be identical, one value can be converted into the other through use of Equation 25.7.

9. If the tie correction for the **Friedman two-way analysis of variance by ranks** is employed, the computed value of χ_r^2 will be slightly higher.

10. The tabled critical two-tailed .05 and .01 chi-square values represent the chi-square values at the 95th and 99th percentiles, and the tabled critical one-tailed .05 and .01 chi-square values represent the chi-square values at the 90th and 98th percentiles.

11. In the discussion of the **Friedman two-way analysis of variance by ranks,** it is assumed that a nondirectional analysis is always conducted for the latter test. A directional/one-tailed analysis is used here in order to employ probability values that are comparable to the one-tailed values employed in evaluating **Spearman's rho.** Within the **Friedman test model,** when $k = 10$ the usage of the term one-tailed analysis is really not meaningful. For a clarification of this issue (i.e., conducting a directional analysis when $k > 2$), the reader should read the discussion on the directionality of the **chi-square goodness-of-fit test (Test 5)** in Section VII of the latter test (which can be generalized to the **Friedman test).**

12. The concept of power efficiency is discussed in Section VII of the **Mann–Whitney *U* test.**

Test 24

Kendall's Tau

(Nonparametric Measure of Association/Correlation Employed with Ordinal Data)

I. Hypothesis Evaluated with Test and Relevant Background Information

Kendall's tau is one of a number of measures of correlation or association discussed in this book. Measures of correlation are not inferential statistical tests, but are, instead, descriptive statistical measures which represent the degree of relationship between two or more variables. Upon computing a measure of correlation, it is common practice to employ one or more inferential statistical tests in order to evaluate one or more hypotheses concerning the correlation coefficient. The hypothesis stated below is the most commonly evaluated hypothesis for **Kendall's tau.**

Hypothesis evaluated with test In the underlying population represented by a sample, is the correlation between subjects' scores on two variables some value other than zero? The latter hypothesis can also be stated in the following form: In the underlying population represented by the sample, is there a significant **monotonic** relationship between the two variables?[1] It is important to note that the nature of the relationship described by **Kendall's tau** is based on an analysis of two sets of ranks.

Relevant background information on test Prior to reading the material in this section the reader should review the general discussion of correlation in Section I of the **Pearson product-moment correlation coefficient (Test 22)**, and the material in Section I of **Spearman's rank-order correlation coefficient (Test 23)** (which also evaluates whether a monotonic relationship exists between two sets of ranks). Developed by Kendall (1938), **tau** is a bivariate measure of correlation/association that is employed with rank-order data. The population parameter estimated by the correlation coefficient will be represented by the notation τ (which is the lower case Greek letter **tau**). The sample statistic computed to estimate the value of τ will be represented by the notation $\tilde{\tau}$. As is the case with **Spearman's rank-order correlation coefficient, Kendall's tau** can be employed to evaluate data in which a researcher has scores for n subjects/objects on two variables (designated as the X and Y variables), both of which are rank-ordered. **Kendall's tau** is also commonly employed to evaluate the degree of agreement between the rankings of $m = 2$ judges for n subjects/objects.

As is the case with **Spearman's rho**, the range of possible values **Kendall's tau** can assume is defined by the limits -1 to $+1$ (i.e., $-1 \leq \tilde{\tau} \leq +1$). Although **Kendall's tau** and **Spearman's rho** share certain properties in common with one another, they employ a different logic with respect to how they evaluate the degree of association between two variables. **Kendall's tau** measures the degree of agreement between two sets of ranks with respect to the relative ordering of all possible pairs of subjects/objects. One set of ranks represents the ranks on the X variable, and the other set represents the ranks on the Y variable. Specifically, assume data are in the form of the following two pairs of observations expressed in a rank-order format: a) (R_{X_i}, R_{Y_i}) (which respectively represent the ranks on

Variables X and Y for the ith subject/object; and b) (R_{X_j}, R_{Y_j}) (which respectively represent the ranks on Variables X and Y for the jth subject/object. If the sign/direction of the difference $(R_{X_i} - R_{X_j})$ is the same as the sign/direction of the difference $(R_{Y_i} - R_{Y_j})$, a pair of ranks is said to be **concordant** (i.e., in agreement). If the sign/direction of the difference $(R_{X_i} - R_{X_j})$ is not the same as the sign/direction of the difference $(R_{Y_i} - R_{Y_j})$, a pair of ranks is said to be **discordant** (i.e., disagree). If $(R_{X_i} - R_{X_j})$ and/or $(R_{Y_i} - R_{Y_j})$ result in the value zero, a pair of ranks is neither concordant or discordant (and is conceptualized within the framework of a tie which is discussed in Section VI). **Kendall's tau** is a proportion that represents the difference between the proportion of concordant pairs of ranks less the proportion of discordant pairs of ranks. The computed value of **tau** will equal +1 when there is complete agreement among the rankings (i.e., all of the pairs of ranks are concordant), and will equal –1 when there is complete disagreement among the rankings (i.e., all of the pairs of ranks are discordant).

As a result of the different logic involved in computing **Kendall's tau** and **Spearman's rho**, the two measures have different underlying scales, and because of this it is not possible to determine the exact value of one measure if the value of the other measure is known. As a general rule, however, the computed absolute value of $\tilde{\tau}$ will always be less than the computed absolute value of r_S for a set of data, and as the sample size increases the ratio $\tilde{\tau}/r_S$ approaches the value .67.[2] Siegel and Castellan (1988) note the following inequality can be employed to describe the relationship between r_S and $\tilde{\tau}$: $-1 \leq (3\tilde{\tau} - 2r_S) \leq 1$.

In spite of the differences between **Kendall's tau** and **Spearman's rho**, the two statistics employ the same amount of information, and because of this are equally likely to detect a significant effect in a population. Thus, although for the same set of data different values will be computed for r_S and $\tilde{\tau}$ (unless, as noted in Endnote 2, the correlation between the two variables is +1 or –1), the two measures will essentially result in the same conclusions with respect to whether or not the underlying population correlation equals zero. The comparability of $\tilde{\tau}$ and r_S is discussed in more detail in Section V.

In contrast to **Kendall's tau**, **Spearman's rho** is more commonly discussed in statistics books as a bivariate measure of correlation for ranked data. Two reasons for this are as follows: a) The computations required for computing **tau** are more tedious than those required for computing **rho**; and b) When a sample is derived from a bivariate normal distribution (which is discussed in Section I of the **Pearson product-moment correlation coefficient**), the computed value r_S will generally provide a reasonably good approximation of **Pearson** r, whereas the value of $\tilde{\tau}$ will not. Since r_S provides a good estimate of r, r_S^2 can be employed to represent the **coefficient of determination** (i.e., a measure of the proportion of variability on one variable than can be accounted for by variability on the other variable).[3] One commonly cited advantage of **tau** over **rho** is that $\tilde{\tau}$ is an unbiased estimate of the population parameter τ, whereas the value computed for r_S is not an unbiased estimate of the population parameter ρ_S. Lindeman *et al.* (1980) note another advantage of **tau** is that unlike **rho**, the sampling distribution of **tau** approaches normality very quickly. Because of this, the normal distribution provides a good approximation of the exact sampling distribution of **tau** for small sample sizes. In contrast, a large sample size is required in order to employ the normal distribution to approximate the exact sampling distribution of **rho**.

II. Example

Example 24.1 *Two psychiatrists, Dr. X and Dr. Y, rank-order ten patients with respect to their level of psychological disturbance (assigning a rank of* 1 *to the least disturbed patient and a rank of 10 to the most disturbed patient). The rankings of the two psychiatrists (along* *with additional information that allows the value of* **Spearman's rho** *to be computed for the same set of data) are presented in* Table 24.1. *Is there a significant correlation between the rank-orders assigned to the patients by the two doctors?*

III. Null versus Alternative Hypotheses

Upon computing **Kendall's tau**, it is common practice to determine whether the obtained absolute value of the correlation coefficient is large enough to allow a researcher to conclude that the underlying population correlation coefficient between the two variables is some value other than zero. Section V describes how the latter hypothesis, which is stated below, can be evaluated through use of tables of critical $\tilde{\tau}$ values or through use of an inferential statistical test that is based on the normal distribution.

Null hypothesis H_0: $\tau = 0$

(In the underlying population the sample represents, the correlation between the scores of subjects on Variable X and Variable Y equals 0.)

Alternative hypothesis H_1: $\tau \neq 0$

(In the underlying population the sample represents, the correlation between the scores of subjects on Variable X and Variable Y equals some value other than 0. This is a **non-directional alternative hypothesis**, and it is evaluated with a **two-tailed test**. Either a significant positive $\tilde{\tau}$ value or a significant negative $\tilde{\tau}$ value will provide support for this alternative hypothesis. In order to be significant, the obtained absolute value of $\tilde{\tau}$ must be equal to or greater than the tabled critical two-tailed $\tilde{\tau}$ value at the prespecified level of significance.)

or

H_1: $\tau > 0$

(In the underlying population the sample represents, the correlation between the scores of subjects on Variable X and Variable Y equals some value greater than 0. This is a **directional alternative hypothesis**, and it is evaluated with a **one-tailed test**. Only a significant positive $\tilde{\tau}$ value will provide support for this alternative hypothesis. In order to be significant (in addition to the requirement of a positive $\tilde{\tau}$ value), the obtained absolute value of $\tilde{\tau}$ must be equal to or greater than the tabled critical one-tailed $\tilde{\tau}$ value at the prespecified level of significance.)

or

H_1: $\tau < 0$

(In the underlying population the sample represents, the correlation between the scores of subjects on Variable X and Variable Y equals some value less than 0. This is a **directional alternative hypothesis**, and it is evaluated with a **one-tailed test**. Only a significant negative $\tilde{\tau}$ value will provide support for this alternative hypothesis. In order to be significant (in addition to the requirement of a negative $\tilde{\tau}$ value), the obtained absolute value of $\tilde{\tau}$ must be equal to or greater than the tabled critical one-tailed $\tilde{\tau}$ value at the prespecified level of significance.)

Note: Only one of the above noted alternative hypotheses is employed. If the alternative hypothesis the researcher selects is supported, the null hypothesis is rejected.[4]

IV. Test Computations

The data for Example 24.1 are summarized in Table 24.1. Although the last two columns of Table 24.1 are not necessary to compute the value of **Kendall's tau**, they are included to allow for the computation of **Spearman's rank-order correlation coefficient** for the same set of data (which is done is Section V).

Table 24.1 Data for Example 24.1

Patient	Rankings of Dr. X R_{X_i}	Rankings of Dr. Y R_{Y_i}	$d_i = R_{X_i} - R_{Y_i}$	d_i^2
1	7	10	-3	9
2	1	2	-1	1
3	8	6	2	4
4	10	8	2	4
5	9	7	2	4
6	6	4	2	4
7	5	9	-4	16
8	3	3	0	0
9	2	1	1	1
10	4	5	-1	1
			$\Sigma d_i = 0$	$\Sigma d_i^2 = 44$

Equation 24.1 is employed to compute the value of **Kendall's tau**

$$\tilde{\tau} = \frac{n_C - n_D}{\left[\dfrac{n(n-1)}{2}\right]} \qquad \textbf{(Equation 24.1)}$$

Where: n_C is the number of concordant pairs of ranks
n_D is the number of discordant pairs of ranks
$[n(n-1)]/2$ is the total number of possible pairs of ranks

In order to employ Equation 24.1 to compute the value of **Kendall's tau**, it is necessary to determine the number of concordant versus discordant pairs of ranks. In order to do this, the data are recorded in the format employed in Table 24.2.

The first row of Table 24.2 consists of the identification number of each subject. The order in which subjects are listed is based on their rank-order on the X variable (i.e., R_{X_i}). The latter set of ranks are recorded in the second row of the table. The third row lists each subject's rank-order on the Y variable (i.e., R_{Y_i}). Inspection of Table 24.2 reveals that no ties are present in the data on either the X or the Y variable. The protocol to be described in this section assumes that there are no ties. The protocol for handling ties is described in Section VI. The portion of Table 24.2 that lies below the double line consists of cells in which there is either an entry of **C** or **D** (except for the number value to the left of each row). This part of the table provides information with regard to the concordant versus discordant pairs of observations for the two sets of ranks. Specifically, in each of the rows of the table that fall below the double line, the number to the left of a row is the R_{Y_i} value (i.e., rank on the Y variable) of the subject represented by the column in which that value appears. Within each row, the R_{Y_i} value is compared with those R_{Y_j} values that fall in the columns to its right. In any instance where an R_{Y_j} value in a column is larger than the R_{Y_i} value for the row, a **C** is recorded in the cell that is the intersection of that row and column. In any instance where an R_{Y_j} value in a column is smaller than the R_{Y_i} value for the row, a **D** is recorded in the cell that is the intersection of that row and column. The presence of a **C** in a cell indicates a **concordant** pair of observations, since the ordering of the ranks on both the X and Y variables for that pair of observations is in the same direction. The presence of a **D** in a cell indicates a **discordant** pair of observations, since the ordering of the ranks on both the X and Y variables for that pair of observations is in the opposite direction.

Table 24.2 Computational Table for Kendall's Tau

Subject	**2**	**9**	**8**	**10**	**7**	**6**	**1**	**3**	**5**	**4**	ΣC	ΣD
R_{X_i}	1	2	3	4	5	6	7	8	9	10		
R_{Y_i}	2	1	3	5	9	4	10	6	7	8		
	2	D	C	C	C	C	C	C	C	C	8	1
		1	C	C	C	C	C	C	C	C	8	0
			3	C	C	C	C	C	C	C	7	0
				5	C	D	C	C	C	C	5	1
					9	D	C	D	D	D	1	4
						4	C	C	C	C	4	0
							10	D	D	D	0	3
								6	C	C	2	0
									7	C	1	0
										8	0	0
											$\Sigma\Sigma C = n_C = 36$	$\Sigma\Sigma D = n_D = 9$

The last two columns of Table 24.2 contain the number of concordant (ΣC) versus discordant (ΣD) pairs of observations in each row. The value $\Sigma\Sigma C = n_C = 36$ in the last row of Table 24.2 is the sum of the column labelled ΣC. The value $\Sigma\Sigma C = n_C = 36$ represents the total number of **C** entries in the table, which is the total number of concordant pairs of observations in the data. The value $\Sigma\Sigma D = n_D = 9$ is the sum of the column labelled ΣD. The value $\Sigma\Sigma D = n_D = 9$ represents the total number of **D** entries in the table, which is the total number of discordant pairs of observations in the data (which are also referred to as **inversions**).

Substituting the values $n_C = 36$, $n_D = 9$, and $n = 10$ in Equation 24.1, the value $\tilde{\tau} = .60$ is computed.

$$\tilde{\tau} = \frac{36 - 9}{\left[\dfrac{10(10 - 1)}{2}\right]} = .60$$

The reader should note the following with respect to the sign of $\tilde{\tau}$: a) When $n_C > n_D$, the sign of $\tilde{\tau}$ will be positive; b) When $n_D > n_C$, the sign of $\tilde{\tau}$ will be negative; and c) When $n_C = n_D$, $\tilde{\tau}$ will equal zero.

Some sources employ the notation S to represent the value $(n_C - n_D)$ in the numerator of Equation 24.1. For Example 24.1, $S = 36 - 9 = 27$. If the value S is employed, the value of $\tilde{\tau}$ can be computed with Equation 24.2 .

$$\tilde{\tau} = \frac{2S}{n(n - 1)} = \frac{(2)(27)}{(10)(10 - 1)} = .60 \qquad \textbf{(Equation 24.2)}$$

Equation 24.3 can also be employed to compute $\tilde{\tau}$.

$$\tilde{\tau} = 1 - \frac{4n_D}{n(n-1)} = 1 - \frac{(4)(9)}{10(10-1)} = .60 \qquad \textbf{(Equation 24.3)}$$

If there are no tied ranks present in the data, Equation 24.3 can be employed to compute **tau** in conjunction with a less tedious method than the one based on use of Table 24.2. The alternative method (which becomes impractical when the sample size is large) involves the use of Figure 24.1.

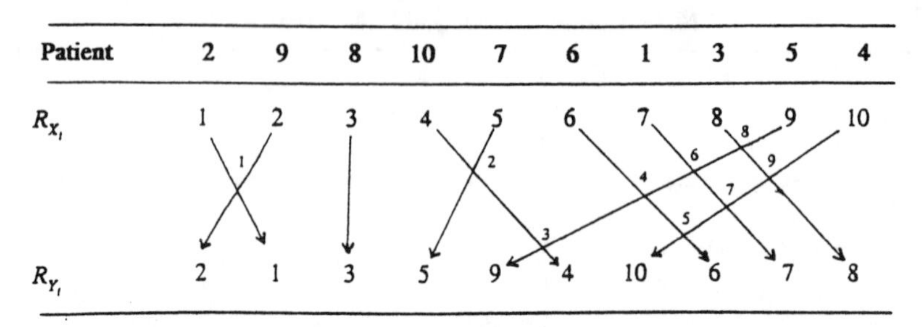

Figure 24.1 Visual Representation of Discordant Pairs of Ranks for Example 24.1

In Figure 24.1 the values of R_{X_i} and R_{Y_i} are recorded as they appear in the second and third rows of Table 24.2. Lines are drawn to connect each of the $n = 10$ corresponding values of R_{X_i} and R_{Y_i}. The total number of intersecting points in the diagram represents the number of discordant pairs of ranks or inversions in the data — i.e., the value of n_D. The value $n_D = 9$ along with the value $n = 10$ are substituted in Equation 24.3 to compute the value $\tilde{\tau} = .60$. Although not required for use in the latter equation, the number of concordant ranks (which along with n_D are employed in Equation 24.1) can be computed as follows: $n_C = [n(n-1)/2] - n_D$.

V. Interpretation of the Test Results

The obtained value $\tilde{\tau} = .60$ is evaluated with **Table A19 (Table of Critical Values for Kendall's Tau)** in the **Appendix**. Note that **Table A19** lists critical values for both **tau** and S.[5] Table 24.3 lists the tabled critical two-tailed and one-tailed .05 and.01 values for **tau** and S for $n = 10$.

Table 24.3 Exact Tabled Critical Values for $\tilde{\tau}$ and S for $n = 10$

	$\tilde{\tau}_{.05}/S_{.05}$	$\tilde{\tau}_{.01}/S_{.01}$
Two-tailed values	$\tilde{\tau} = .511$ $S = 23$	$\tilde{\tau} = .644$ $S = 29$
One-tailed values	$\tilde{\tau} = .467$ $S = 21$	$\tilde{\tau} = .600$ $S = 27$

The following guidelines are employed in evaluating the null hypothesis.

a) If the nondirectional alternative hypothesis H_1: $\tau \neq 0$ is employed, the null hypothesis can be rejected if the obtained absolute value of $\tilde{\tau}$ (or S) is equal to or greater than the tabled critical two-tailed value at the prespecified level of significance.

b) If the directional alternative hypothesis H_1: $\tau > 0$ is employed, the null hypothesis can be rejected if the sign of $\tilde{\tau}$ (or S) is positive, and the obtained value of $\tilde{\tau}$ (or S) is equal to or greater than the tabled critical one-tailed value at the prespecified level of significance.

c) If the directional alternative hypothesis H_1: $\tau < 0$ is employed, the null hypothesis can be rejected if the sign of $\tilde{\tau}$ (or S) is negative, and the obtained absolute value of $\tilde{\tau}$ (or S) is equal to or greater than the tabled critical one-tailed value at the prespecified level of significance.

Employing the above guidelines, the nondirectional alternative hypothesis H_1: $\tau \neq 0$ is supported at the .05 level, since the computed value $\tilde{\tau} = .60$ $(S = 27)$ is greater than the tabled critical two-tailed value $\tilde{\tau}_{.05} = .511$ $(S_{.05} = 23)$. It is not supported at the .01 level, since $\tilde{\tau} = .60$ $(S = 27)$ is less than the tabled critical two-tailed value $\tilde{\tau}_{.01} = .644$ $(S_{.01} = 29)$.

The directional alternative hypothesis H_1: $\tau > 0$ is supported at the .05 level, since the computed value $\tilde{\tau} = .60$ is a positive number that is greater than the tabled critical one-tailed value $\tilde{\tau}_{.05} = .467$ $(S_{.05} = 21)$. It is also supported at the .01 level, since $\tilde{\tau} = .60$ is equal to the tabled critical one-tailed value $\tilde{\tau}_{.01} = .600$ $(S_{.01} = 27)$.

The directional alternative hypothesis H_1: $\tau < 0$ is not supported, since the computed value $\tilde{\tau} = .60$ (S) is a positive number.

Test 24a: Test of significance for Kendall's tau When $n > 10$, the normal distribution provides an excellent approximation of the sampling distribution of **tau**. Equation 24.4 is the normal approximation for evaluating the null hypothesis H_0: $\tau = 0$.

$$z = \frac{3\tilde{\tau}\sqrt{n(n-1)}}{\sqrt{2(2n+5)}} \qquad \textbf{(Equation 24.4)}$$

In view of the fact that the sample size $n = 10$ employed in Example 24.1 is just one subject below the minimum value generally recommended for use with Equation 24.4, the normal approximation will still provide a reasonably good approximation of the exact sampling distribution. When the appropriate values from Example 24.1 are substituted in Equation 24.4, the value $z = 2.41$ is computed.

$$z = \frac{(3)(.60)\sqrt{10(10-1)}}{\sqrt{2\,[(2)(10)+5]}} = 2.41$$

Equations 24.5 and 24.6 are alternative equations for computing the value of z that yield the identical result.[6]

$$z = \frac{\tilde{\tau}}{\sqrt{\dfrac{2(2n+5)}{9n(n-1)}}} = \frac{.60}{\sqrt{\dfrac{(2)[(2)(10)+5]}{(9)(10)(10-1)}}} = 2.41 \qquad \textbf{(Equation 24.5)}$$

$$z = \frac{S}{\sqrt{\dfrac{n(n-1)(2n+5)}{18}}} = \frac{27}{\sqrt{\dfrac{(10)(10-1)[(2)(10)+5]}{18}}} = 2.41 \qquad \textbf{(Equation 24.6)}$$

The computed value $z = 2.41$ is evaluated with **Table A1 (Table of the Normal Distribution)** in the **Appendix**. In the latter table, the tabled critical two-tailed .05 and .01 values are $z_{.05} = 1.96$ and $z_{.01} = 2.58$, and the tabled critical one-tailed .05 and .01 values are $z_{.05} = 1.65$ and $z_{.01} = 2.33$. Since the sign of the z value computed with Equations 24.4–24.6 will always be the same as the sign of $\tilde{\tau}$ (and S), the guidelines that are described earlier in this section for evaluating a $\tilde{\tau}$ (or S) value can also be applied in evaluating the z value computed with Equations 24.4–24,6 (i.e., substitute z in place of $\tilde{\tau}$ (or S) in the text of the guidelines for evaluating $\tilde{\tau}$ (or S)).

Employing the guidelines, the nondirectional alternative hypothesis H_1: $\tau \neq 0$ is supported at the .05 level, since the computed value $z = 2.41$ is greater than the tabled critical two-tailed value $z_{.05} = 1.96$. It is not, however, supported at the .01 level, since $z = 2.41$ is less than the tabled critical two-tailed value $z_{.01} = 2.58$.

The directional alternative hypothesis H_1: $\tau > 0$ is supported at both the .05 and .01 levels, since the computed value $z = 2.41$ is a positive number that is greater than the tabled critical one-tailed values $z_{.05} = 1.65$ and $z_{.01} = 2.33$.

The directional alternative hypothesis H_1: $\tau < 0$ is not supported, since the computed value $z = 2.41$ is a positive number. In order for the alternative hypothesis H_1: $\tau < 0$ to be supported, the computed value of z must be a negative number (as well as the fact that the absolute value of z must be equal to or greater than the tabled critical one-tailed value at the prespecified level of significance).

Note that the results for the normal approximation are identical to those obtained when the exact values of the sampling distribution of **tau** are employed. A summary of the analysis of Example 24.1 follows: It can be concluded that there is a significant monotonic increasing/positive relationship between the rankings of the two judges.[7] The result of the analysis (based on the critical values in Table 24.3 and the normal approximation) can be summarized as follows (if it is assumed the nondirectional alternative hypothesis H_1: $\tau \neq 0$ is employed): $\tilde{\tau} = .60$, $p < .05$.

It is noted in Section I that if both **Kendall's tau** and **Spearman's rho** are computed for the same set of data, the two measures will result in essentially the same conclusions with respect to whether the value of the underlying population correlation equals zero. In order to demonstrate this, employing the relevant values from Table 24.1 in Equation 23.1, the value $r_S = .733$ is computed for Example 24.1.[8]

$$r_S = 1 - \frac{6\Sigma d^2}{n(n^2 - 1)} = 1 - \frac{6(44)}{10[(10)^2 - 1]} = .733$$

Note that the values $\tilde{\tau} = .60$ and $r_S = .733$ computed for Example 24.1 are not identical to one another, and that as noted in Section I, the absolute value of $\tilde{\tau}$ is less than the absolute value of r_S. It is also the case that the inequality $-1 \leq (3\tilde{\tau} - 2r_S) \leq 1$ (which is noted in Section I) is substantiated, since $(3)(.60) - (2)(.733) = .334$ (which falls within the range -1 to $+1$).

The computed value $r_S = .733$ is evaluated with **Table A18 (Table of Critical Values for Spearman's Rho)** in the **Appendix**. Employing the latter table, it is determined that for $n = 10$, the tabled critical two-tailed .05 and .01 values are $r_{S_{.05}} = .648$ and $r_{S_{.01}} = .794$, and the tabled critical one-tailed .05 and .01 values are $r_{S_{.05}} = .564$ and $r_{S_{.01}} = .745$. Employing the aforementioned critical values, the nondirectional alternative hypothesis H_1: $\rho_s \neq 0$ and the directional alternative hypothesis H_1: $\rho_s > 0$ are supported at the .05 level, since the computed value $r_S = .733$ is greater than the tabled critical two-tailed value $r_{S_{.05}} = .648$ and the tabled critical one-tailed value $r_{S_{.05}} = .564$. The alternative hypotheses are not supported at the .01 level, since $r_S = .733$ is less than the tabled critical two-tailed value $r_{S_{.01}} = .794$ and the tabled critical one-tailed value $r_{S_{.01}} = .745$. This result is almost identical to that obtained when **Kendall's tau** is employed (although in the analysis for **Kendall's tau**, the directional alternative hypothesis H_1: $\tau > 0$ is supported at the .01 level).

If the computed value $r_S = .733$ is evaluated with Equation 23.2, the value $t = 3.05$ is computed.

$$t = \frac{r_S\sqrt{n - 2}}{\sqrt{1 - r_S^2}} = \frac{.733\sqrt{10 - 2}}{\sqrt{[1 - (.733)^2]}} = 3.05$$

The t value computed with Equation 23.2 is evaluated with **Table A2 (Table of Student's t Distribution)** in the **Appendix**. The degrees of freedom employed are $df = n - 2$. Employing **Table A2**, it is determined that for $df = 10 - 2 = 8$, the tabled critical two-tailed .05 and .01 values are $t_{.05} = 2.31$ and $t_{.01} = 3.36$, and the tabled critical one-tailed .05 and .01 values are $t_{.05} = 1.86$ and $t_{.01} = 2.90$. Employing the aforementioned critical values, the nondirectional alternative hypothesis H_1: $\rho_s \neq 0$ is supported at the .05 level, since the computed value $t = 3.05$ is greater than the tabled critical two-tailed value $t_{.05} = 2.31$. It is not supported at the .01 level, since $t = 3.05$ is less than $t_{.01} = 3.36$. The directional alternative hypothesis H_1: $\rho_s > 0$ is supported at both the .05 and .01 levels, since the computed value $t = 3.05$ is a positive number (since $r_S = .733$ is a positive number) that is greater than the tabled critical one-tailed values $t_{.05} = 1.86$ and $t_{.01} = 2.90$. This result is identical to that obtained when **Kendall's tau** is employed.

The slight discrepancies between the various methods for assessing the significance of $\tilde{\tau}$ and r_S can be attributed to the fact that the values in **Table A18** and the result of Equation 23.2 are approximations of the exact sampling distribution of **Spearman's rho**, as well as the fact that the use of the normal distribution for assessing the significance of **tau** also represents an approximation of an exact sampling distribution. However, for the most part, regardless of whether one elects to compute $\tilde{\tau}$ or r_S as the measure of association for Example 24.1, it will be concluded that the population correlation is some value other than zero, and the latter conclusion will be reached irrespective of whether a nondirectional or directional alternative hypothesis is employed.

VI. Additional Analytical Procedures for Kendall's Tau and/or Related Tests

1. Tie correction for Kendall's tau When one or more ties are present in a set of data it is necessary to employ a tie correction in order to compute the value of $\tilde{\tau}$. To illustrate how ties are handled, let us assume that Table 24.4 summarizes the data for Example 24.1. Note that in contrast to Table 24.2, the data in Table 24.4 are characterized by the presence of ties on both the X and Y variables. Specifically, Subjects 2 and 9 are tied for the first ordinal position on the X variable, Subjects 8 and 10 are tied for the third ordinal position on the Y variable, and Subjects 6 and 7 are tied for the ninth ordinal position on the Y variable.

As is the case for Table 24.2, the entries C and D are employed in the cells of Table 24.4 to indicate concordant versus discordant pairs of ranks. There are, however, three cells in Table 24.4 that involve tied ranks in which the cell entry is **0**. Note that if the R_{Y_i} value for a row is equal to a R_{Y_j} that falls in a column to its right, a **0** is written in the cell that is the intersection of that row and column. Since, however, this protocol only takes into account ties on the Y variable, it will not allow one to identify all of the cells in the table for which **0** is the appropriate entry. In point of fact, a **0** entry should also appear in any cell that involves a pair of tied observations on the X variable, even though the two ranks on the Y variable with which the X variable pair is being contrasted are not tied. In the above example there is just one set of ties on the X variable (the rank of 1.5 for Subjects 2 and 9). Note that the cell identified with an asterisk in the upper left of the table has a **0** entry, even though the rank-order directly to the right of the value $R_{Y_2} = 2$ (which is the rank of Subject 2 on the Y variable) is $R_{Y_9} = 1$ (which is the rank of Subject 9 on the Y variable). If the protocol described for Table 24.2 is employed, since the rank-order $R_{Y_9} = 1$ to the right of $R_{Y_2} = 2$ is less than the latter value, a **D** should be placed in that cell. The reason for employing a **0** in the cell is that if the arrangement of the ranks on the X variable is reversed, with Subject 9 listed first and Subject 2 listed second, the value of R_{Y_i} for that row will be $R_{Y_9} = 1$, and the first rank/R_{Y_j} value that it will be compared with will be $R_{Y_2} = 2$. If the

Table 24.4 Computational Table for Kendall's Tau Involving Ties

Subject	2	9	8	10	7	6	1	3	5	4	ΣC	ΣD
R_{X_i}	1.5	1.5	3	4	5	6	7	8	9	10		
R_{Y_i}	2	1	3.5	3.5	9.5	9.5	5	6	7	8		
	2	0*	C	C	C	C	C	C	C	C	8	0
		1	C	C	C	C	C	C	C	C	8	0
			3.5	0	C	C	C	C	C	C	6	0
				3.5	C	C	C	C	C	C	6	0
					9.5	0	D	D	D	D	0	4
						9.5	D	D	D	D	0	4
							5	C	C	C	3	0
								6	C	C	2	0
									7	C	1	0
										8	0	0
											$\Sigma\Sigma C = n_C = 34$	$\Sigma\Sigma D = n_D = 8$

latter arrangement is employed in conjunction with the protocol described for Table 24.2, the appropriate entry for the cell under discussion is a **C**. Thus, whenever a different arrangement of the tied ranks on the X variable will result in a different letter entry for a cell (i.e., **C** versus **D**), that cell is assigned a **0**.

A general protocol for determining whether a cell should be assigned a **0** to represent a tie can be summarized as follows: a) If the R_{Y_i} value at the left of a row is tied with an R_{Y_j} value that falls in a column to its right, a **0** should be placed in the cell that is the intersection of that row and column; and b) If there is a tie between the values of R_{X_i} and R_{X_j} that fall directly above the values of R_{Y_i} and R_{Y_j} being compared, a **0** should be placed in the cell that is the intersection of the row and column the values R_{Y_i} and R_{Y_j} (as well as R_{X_j}) appear.

It should be noted that when there are no ties present in the data, $n_C + n_D = [n(n-1)]/2$. Thus, in the case of Table 24.2, $(n_C = 36) + (n_D = 9) = [(10)(10-1)]/2 = 45$. When, on the other hand, ties are present in the data, since entries of **0** are not counted as either concordant or discordant pairs, $n_C + n_D \neq [n(n-1)]/2$. The latter can be confirmed by the fact that in Table 24.4, $(n_C = 34) + (n_D = 8) \neq [(10)(10-1)]/2$.

The computation of the value of **Kendall's tau** using the tie correction will now be described. In the example under discussion there is $s = 1$ set of ties involving the ranks of subjects' X scores (Subjects 2 and 9), and $s = 2$ sets of ties involving the ranks of subjects' Y scores (Subjects 8 and 10; Subjects 6 and 7). Equation 24.9 is employed to compute the tie corrected value of **Kendall's tau,** which will be represented by the notation $\tilde{\tau}_c$. Note that the values ΣT_X and ΣT_Y in Equation 24.9 are computed with Equations 24.7 and 24.8. In Equation 24.7, $t_{i_{(X)}}$ represents the number of X scores that are tied for a given rank. In Equation 24.8, $t_{i_{(Y)}}$ represents the number of Y scores that are tied for a given rank. The notations $\Sigma_{i=1}^{s}(t_{i_{(X)}}^2 - t_{i_{(X)}})$ and $\Sigma_{i=1}^{s}(t_{i_{(Y)}}^2 - t_{i_{(Y)}})$ indicate that the following is done with respect to each of the variables: a) For each set of ties, the number of ties in the set is subtracted from the square of the number of ties in that set; and b) The sum of all the values computed in part a) is obtained for that variable.

When the data from Table 24.4 are substituted in Equations 24.7–24.9, the tie corrected value $\tilde{\tau}_c = .598$ is computed.[9]

$$T_X = \sum_{i=1}^{s} \left(t_{i_{(X)}}^2 - t_{i_{(X)}} \right) = [(2)^2 - 2] = 2 \qquad \textbf{(Equation 24.7)}$$

$$T_Y = \sum_{i=1}^{s} \left(t_{i_{(Y)}}^2 - t_{i_{(Y)}} \right) = [(2)^2 - 2] + [(2)^2 - 2] = 4 \qquad \textbf{(Equation 24.8)}$$

$$\tilde{\tau}_c = \frac{2(n_C - n_D)}{\sqrt{n(n-1) - T_X}\,\sqrt{n(n-1) - T_Y}} \qquad \textbf{(Equation 24.9)}$$

$$= \frac{(2)(34 - 8)}{\sqrt{10(10-1) - 2}\,\sqrt{10(10-1) - 4}} = .598$$

2. Regression analysis and Kendall's tau As noted in the discussion of **Spearman's rank-order correlation coefficient**, regression analysis procedures have been developed for rank-order data. Sources for nonparametric regression analysis (including monotonic regression analysis) are cited in Section VI of the latter test.

3. Partial rank correlation The computation of a **partial correlation coefficient** (described in Section IX of the **Pearson product-moment correlation coefficient**, and discussed briefly in Section VI of **Spearman's rank-order correlation coefficient**), can be extended to **Kendall's tau**. Thus, when the rank-orders for three variables are evaluated, Equation 22.72 can be employed to compute one or more **partial correlation coefficients** for **Kendall's tau** (employing the relevant values of $\tilde{\tau}$ in the equation). Conover (1980), Daniel (1990), Marascuilo and McSweeney (1977), and Siegel and Castellan (1988) discuss the computation of a **partial correlation coefficient** in reference to **Kendall's tau**. It should be noted that the partial rank-order correlation coefficient for **Kendall's tau** employs a different sampling distribution than the one that is employed for evaluating $\tilde{\tau}$. Tables for the appropriate sampling distribution can be found in Daniel (1990) and Siegel and Castellan (1988).

4. Sources for computing a confidence interval for Kendall's tau A procedure (attributed to Noether (1967)) for deriving a confidence interval for **Kendall's tau** is described in Daniel (1990).

VII. Additional Discussion of Kendall's Tau

1. Power efficiency of Kendall's tau Daniel (1990) and Siegel and Castellan (1988) note that (for large sample sizes) the power efficiency of **Kendall's tau** relative to the **Pearson product-moment correlation coefficient** is approximately .91 (which is the identical value of the power efficiency of **Spearman's rho** relative to **Pearson** *r*). The value .91 indicates that if a test of significance is employed to evaluate the null hypothesis of a zero population correlation, use of the **Pearson product-moment correlation coefficient** requires 91% of the subjects to detect a specific effect size (i.e., reject a false null hypothesis) when compared with **Kendall's tau.**[10]

2. Kendall's coefficient of agreement **Kendall's coefficient of agreement** is another measure of association that allows a researcher to evaluate the degree of agreement between m sets of ranks on n subjects/objects. The latter measure, which is described in Siegel and Castellan (1988), is essentially an extension of **Kendall's tau** to more than two sets of ranks. The relationship between **Kendall's tau** and **Kendall's coefficient of agreement** is analogous to the relationship between **Spearman's rho** and **Kendall's coefficient of concordance (Test 25).**

VIII. Additional Examples Illustrating the Use of Kendall's Tau

Since **Spearman's rho** and **Kendall's tau** can be employed to evaluate the same data, Examples 23.1 and 23.2, as well as the data set presented in Tables 23.2/23.4, can be evaluated with **Kendall's tau**. It is also the case, that if a researcher elects to rank-order the scores of subjects in any of the examples for which the **Pearson product-moment correlation coefficient** is employed, a value can be computed for **Kendall's tau**. To illustrate this, Example 22.1 (which is identical to Example 23.1) will be evaluated with **Kendall's tau**. The rank-orders of the scores of subjects on the X and Y variables in Examples 22.1/23.1 are arranged in Figure 24.2. The arrangement of the ranks in Figure 24.2 allows for use of the protocol for determining the number of discordant pairs of ranks that is described in reference to Figure 24.1. Since none of the vertical lines intersect, the number of pairs of discordant ranks is $n_D = 0$. Since each subject has the identical rank on both the X and the Y variables, all of the pairs of ranks are concordant. The total number of pairs of ranks is $[(5)(5-1)]/2 = 10$, which is also the value of n_C.

Subject	2	3	5	4	1
R_{X_i}	1	2	3	4	5
	↓	↓	↓	↓	↓
R_{Y_i}	1	2	3	4	5

Figure 24.2 Visual Representation of Discordant Pairs of Ranks for Examples 22.1/23.1

Employing the values $n = 5$ and $n_D = 0$ in Equation 24.3, the value $\tilde{\tau} = 1$ is computed. The same value can also be computed with either Equation 24.1 or Equation 24.2, if the values $n_C = 10$ and/or $S = 10$ are employed in the aforementioned equations

$$\tilde{\tau} = 1 - \frac{(4)(0)}{5(5-1)} = 1$$

$\tilde{\tau} = 1$ is identical to the value $r_S = 1$ computed for the same set of data. As noted in Section I, when there is a perfect positive or negative correlation between the variables, identical values are computed for $\tilde{\tau}$ and r_S.

References

Conover, W. J. (1980). **Practical nonparametric statistics** (2nd ed.). New York: John Wiley and Sons.

Daniel, W. W. (1990). **Applied nonparametric statistics** (2nd ed.). Boston: PWS–Kent Publishing Company.

Howell, D. C. (1992). **Statistical methods for psychology** (3rd ed.). Boston: PWS–Kent Publishing Company.

Kendall, M. G. (1938). A new measure of rank correlation. **Biometrika**, 30, 81–93.

Kendall, M. G. (1952). **The advanced theory of statistics** (Vol. 1). London: Charles Griffin and Co. Ltd.

Kendall, M. G. (1970). **Rank correlation methods** (4th ed.). London: Charles Griffin and Co. Ltd.

Lindeman, R. H., Meranda, P. F., and Gold, R. Z. (1980). **Introduction to bivariate and multivariate analysis**. Glenview, IL: Scott, Foresman and Company.

Marascuilo, L. A. and McSweeney, M. (1977). **Nonparametric and distribution-free methods for the social sciences.** Monterey, CA: Brooks/Cole Publishing Company.

Noether, G. E. (1967). **Elements of nonparametric statistics.** New York: John Wiley and Sons.

Siegel, S. and Castellan, N. J., Jr. (1988). **Nonparametric statistics for the behavioral sciences** (2nd ed.). New York: McGraw–Hill Book Company.

Sprent, P. (1989). **Applied nonparametric statistics.** London: Chapman and Hall.

Endnotes

1. A discussion of monotonic relationships can be found in Section I of **Spearman's rank-order correlation coefficient.**

2. The exception to this is that when the computed value of $\tilde{\tau}$ is either +1 or −1, the identical value will be computed for r_S.

3. The **coefficient of determination** is discussed in Section V of the **Pearson product-moment correlation coefficient.**

4. a) Some sources employ the following statements as the null hypothesis and the nondirectional alternative hypothesis for **Kendall's tau: Null hypothesis:** H_0: Variables X and Y are independent of one another; **Nondirectional alternative hypothesis:** H_1: Variables X and Y are not independent of one another.

   It is, in fact, true that if in the underlying population the two variables are independent, the value of τ will equal zero. However, the fact that $\tau = 0$ in and of itself does not ensure that the variables are independent of one another. Thus, it is conceivable that in a population in which the correlation between X and Y is $\tau = 0$, a nonmonotonic curvilinear function can be employed to describe the relationship between the variables.

   b) Note that in Example 24.1 the scores of subjects (who are the patients) on the X and Y variables are the respective ranks assigned to the subjects/patients by Dr. X and Dr. Y. Thus, the null hypothesis can also be stated as follows: In the underlying population the sample of subjects/patients represents, the correlation between the rankings of Dr. X and Dr. Y equals 0.

5. If either of the two values $\tilde{\tau}$ or S is known, Equation 24.2 can be employed to compute the other value. Some sources only list critical values for one of the two values $\tilde{\tau}$ or S.

6. The following should be noted with respect to Equations 24.5 and 24.6: a) The denominator of Equation 24.5 is the standard deviation of the sampling distribution of the normal approximation of **tau**; and b) Based on a recommendation by Kendall (1970), Marascuilo and McSweeney (1977) (who employ Equation 24.6) describe the use of a correction for continuity for the normal approximation. In employing the correction for continuity with Equation 24.6, when S is a positive number, the value 1 is subtracted from S, and when S is a negative number, the value 1 is added to S. The correction for continuity (which is not employed by most sources) reduces the absolute value of z, thus resulting in a more conservative test. The rationale for employing a correction for continuity for a normal approximation of a sampling distribution is discussed in Section VI of the **Wilcoxon signed-ranks test (Test 4).**

7. Howell (1992) notes that the value $\tilde{\tau} = .60$ indicates that if a pair of subjects are randomly selected, the likelihood that the pair will be ranked in the same order is .60 higher than the likelihood that they will be ranked in the reverse order.

8. The data for Examples 24.1 and 23.2 are identical, except for the fact that in the latter example there is a tie for the X score in the second ordinal position which involves the X scores in the eighth and ninth rows.

9. It should be noted that if Equation 24.1 is employed to compute the value of $\tilde{\tau}$ for the data in Table 24.4, the value $\tilde{\tau} = .578$ is computed. As noted in the text, because of the presence of ties, $n_C + n_D \neq [n(n - 1)]/2$.

$$\tilde{\tau} = \frac{34 - 8}{\left[\frac{10(10 - 1)}{2}\right]} = .578$$

10. The concept of power efficiency is discussed in Section VII of the **Mann-Whitney *U* test (Test 9)**.

Test 25

Kendall's Coefficient of Concordance
(Nonparametric Measure of Association/Correlation Employed with Ordinal Data)

I. Hypothesis Evaluated with Test and Relevant Background Information

Kendall's coefficient of concordance is one of a number of measures of correlation or association discussed in this book. Measures of correlation are not inferential statistical tests, but are, instead, descriptive statistical measures which represent the degree of relationship between two or more variables. Upon computing a measure of correlation, it is common practice to employ one or more inferential statistical tests in order to evaluate one or more hypotheses concerning the correlation coefficient. The hypothesis stated below is the most commonly evaluated hypothesis for **Kendall's coefficient of concordance.**

Hypothesis evaluated with test In the underlying population represented by a sample, is the correlation between m sets of ranks some value other than zero? The latter hypothesis can also be stated in the following form: In the underlying population represented by the sample, are m sets of ranks independent of one another?

Relevant background information on test Developed independently by Kendall and Babington–Smith (1939) and Wallis (1939), **Kendall's coefficient of concordance** is a measure of correlation/association that is employed for three or more sets of ranks. Specifically, **Kendall's coefficient of concordance** is a measure that allows a researcher to evaluate the degree of agreement between m sets of ranks for n subjects/objects. The population parameter estimated by the correlation coefficient will be represented by the notation W. The sample statistic computed to estimate the value of W will be represented by the notation $\tilde{W}$. The range of possible values within which **Kendall's coefficient of concordance** may fall is $0 \leq \tilde{W} \leq +1$. When there is complete agreement among all m sets of ranks, the value of $\tilde{W}$ will equal 1.[1] When, on the other hand, there is no pattern of agreement among the m sets of ranks, $\tilde{W}$ will equal 0. The value of $\tilde{W}$ cannot be a negative number, since when there are more than two sets of ranks it is not possible to have complete disagreement among all the sets. Because of this, it becomes meaningless to use a negative correlation to describe the degree of association in the data when $m \geq 3$.

It is important to note that **Kendall's coefficient of concordance** is related to both **Spearman's rank-order correlation coefficient (Test 23)** and **Friedman's two-way analysis of variance by ranks (Test 19)**. Specifically: a) The computed value of $\tilde{W}$ for m sets of ranks is linearly related to the average value of **Spearman's rho** that can be computed for all possible pairs of ranks. The relationship between **Kendall's coefficient of concordance** and **Spearman's rank-order correlation coefficient** is discussed in greater detail in Section VII. It should be noted that although **Kendall's coefficient of concordance** can be computed for two sets of ranks, in practice it is not. The latter can be attributed to the fact that in contrast to **Spearman's rho** and **Kendall's tau** (which are the measures of association that are employed with two sets of ranks), the value of $\tilde{W}$ cannot be a negative number (which in the case of **Spearman's rho** and **Kendall's tau** indicates the presence of an inverse relationship). Because the measures of association that are employed with two sets of ranks can assume a negative value, $\tilde{W}$ is not directly comparable to them;[2] and b) Although they were developed independently, **Kendall's coefficient of concordance** and **Friedman's two-way analysis of variance by ranks** are based on the same mathematical model. Because of this, for a given set of data the values computed for χ_r^2 (which is the **Friedman** test statistic) and $\tilde{W}$ can be algebraically derived from one another. The relationship between **Kendall's coefficient of concordance** and **Friedman's two-way analysis of variance by ranks** is discussed in Section VII.

II. Example

Example 25.1 *Six instructors at an art institute rank four students with respect to artistic ability. A rank of* 1 *is assigned to the student with the highest level of ability and a rank of* 4 *to the student with the lowest level of ability. The rankings of the six instructors for the four students are summarized in* Table 25.1. *Is there a significant association between the rank-orders assigned to the four students by the six instructors?*

Table 25.1 Data for Example 25.1

	Student				
Instructor	**1**	**2**	**3**	**4**	**Totals**
1	3	2	1	4	
2	3	2	1	4	
3	3	2	1	4	
4	4	2	1	3	
5	3	2	1	4	
6	4	1	2	3	
ΣR_j	20	11	7	22	$T = 60$
$(\Sigma R_j)^2$	400	121	49	484	$U = 1054$

III. Null versus Alternative Hypotheses

Upon computing **Kendall's coefficient of concordance**, it is common practice to determine whether the obtained value of the correlation coefficient is large enough to allow a researcher to conclude that the underlying population correlation coefficient between the m sets of ranks is some value other than zero. Section V describes how the latter hypothesis, which is stated below, can be evaluated through use of tables of critical $\tilde{W}$ values or through use of an inferential statistical test that is based on the chi-square distribution.

Null hypothesis H_0: $W = 0$

(In the underlying population the sample represents, the correlation between the $m = 6$ sets of ranks equals 0.)

Alternative hypothesis H_1: $W \neq 0$

(In the underlying population the sample represents, the correlation between the $m = 6$ sets of ranks equals some value other than 0. This is equivalent to stating that the $m = 6$ sets of ranks are not independent of one another. When there are more than two sets of ranks, the alternative hypothesis will always be stated **nondirectionally**.[3] In order to be significant, the obtained value of $\tilde{W}$ must be equal to or greater than the tabled critical value of $\tilde{W}$ at the prespecified level of significance.)

IV. Test Computations

The data for Example 25.1 are summarized in Table 25.1. Note that in Table 25.1 there are $m = 6$ instructors, who are represented by the six rows, and $n = 4$ students who are represented by the four columns.

The summary values $T = 60$ and $U = 1054$ in Table 25.1 are computed as follows.

$$T = \sum_{j=1}^{n} (\Sigma R_j) = \Sigma R_1 + \Sigma R_2 + \Sigma R_3 + \Sigma R_4 = 20 + 11 + 7 + 22 = 60$$

$$U = \sum_{j=1}^{n} (\Sigma R_j)^2 = (\Sigma R_1)^2 + (\Sigma R_2)^2 + (\Sigma R_3)^2 + (\Sigma R_4)^2$$

$$= (20)^2 + (11)^2 + (7)^2 + (22)^2 = 400 + 121 + 49 + 484 = 1054$$

The **coefficient of concordance** is a ratio of the variance of the sums of ranks for the subjects (i.e., the variance of the ΣR_j values) divided by the maximum possible value that can be computed for the variance of the sums of the ranks (for the relevant values of m and n). Equation 25.1 summarizes the definition of $\tilde{W}$.

$$\tilde{W} = \frac{\text{Variance of } \Sigma R_j \text{ values}}{\text{Maximum possible variance for } \Sigma R_j \text{ values for relevant values of } m \text{ and } n} \qquad \textbf{(Equation 25.1)}$$

The variance of the ΣR_j values (which is represented by the notation S) is computed with Equation 25.2.

$$S = \frac{nU - (T)^2}{n} \qquad \textbf{(Equation 25.2)}$$

Substituting the appropriate values from Example 25.1 in Equation 25.2, the value $S = 154$ is computed.

$$S = \frac{(4)(1054) - (60)^2}{4} = 154$$

$\tilde{W}$ is computed with Equation 25.3. The denominator of Equation 25.3 (which for Example 25.1 equals 180) represents the maximum possible value that can be computed for the variance of the sums of the ranks. The only time the value of S will equal the value of the denominator of Equation 25.3 (thus resulting in the value $\tilde{W} = 1$) will be when there is perfect agreement among the m judges with respect to their rankings of the n subjects.

$$\tilde{W} = \frac{S}{\left(\frac{m^2 n (n^2 - 1)}{12}\right)} \qquad \textbf{(Equation 25.3)}$$

Substituting the appropriate values in Equation 25.3, the value $\tilde{W} = .856$ is computed.

$$\tilde{W} = \frac{154}{\left(\frac{(6)^2(4)[(4)^2 - 1]}{12}\right)} = .856$$

Equation 25.4 is an alternative computationally quicker equation for computing the value of $\tilde{W}$. Equation 25.4, however, does not allow for the direct computation of S. The latter fact is noted, since some of the tables employed to evaluate whether $\tilde{W}$ is significant list critical values for S rather than critical values for $\tilde{W}$.

(Equation 25.4)

$$\tilde{W} = \frac{12U - 3m^2n(n+1)^2}{m^2n(n^2-1)} = \frac{(12)(1054) - (3)(6)^2(4)(4+1)^2}{(6)^2(4)[(4)^2-1]} = .856$$

The fact the value of $\tilde{W}$ is close to 1 indicates that there is a high degree of agreement among the six instructors with respect to how they rank the four students.

V. Interpretation of the Test Results

The obtained value $\tilde{W} = .856$ is evaluated with **Table A20 (Table of Critical Values for Kendall's Coefficient of Concordance)** in the **Appendix**. Note that Table **A20** lists critical values for both $\tilde{W}$ and S. The S values in **Table A20** are extracted from Friedman (1940), and the values of $\tilde{W}$ were computed by substituting the appropriate value of S in Equation 25.3. In order to reject the null hypothesis, the computed value of $\tilde{W}$ (or S) must be equal to or greater than the tabled critical value at the prespecified level of significance. For $m = 6$ and $n = 4$, the tabled critical .05 and .01 values for $\tilde{W}$ (S) in **Table A20** are $\tilde{W}_{.05} = .421$ $(S_{.05} = 75.7)$ and $\tilde{W}_{.01} = .553$ $(S_{.01} = 99.5)$. Since the computed value $\tilde{W} = .856$ $(S = 154)$ is greater than all of the aforementioned critical values, the alternative hypothesis H_1: $W \neq 0$ is supported at both the .05 and .01 levels.

Test 25a: Test of significance for Kendall's coefficient of concordance When exact tables for $\tilde{W}$ (or S) are not available, the chi-square distribution provides a reasonably good approximation of the sampling distribution of $\tilde{W}$. The chi-square approximation of the sampling distribution of $\tilde{W}$ is computed with Equation 25.5. The degrees of freedom employed for Equation 25.5 are $df = n - 1$.

$$\chi^2 = m(n-1)\tilde{W} \qquad \textbf{(Equation 25.5)}$$

When the appropriate values from Example 25.1 are substituted in Equation 25.5, the value $\chi^2 = 15.41$ is computed.

$$\chi^2 = (6)(4-1)(.856) = 15.41$$

The value $\chi^2 = 15.41$ is evaluated with **Table A4 (Table of the Chi-Square Distribution)** in the **Appendix**. In order to reject the null hypothesis, the obtained value of χ^2 must be equal to or greater than the tabled critical value at the prespecified level of significance. For $df = 4 - 1 = 3$, the tabled critical values are $\chi^2_{.05} = 7.81$ and $\chi^2_{.01} = 11.34$ (which are the chi-square values at the 95th and 99th percentiles). Since $\chi^2 = 15.41$ is greater than both of the aforementioned critical values, the alternative hypothesis H_1: $W \neq 0$ is supported at both the .05 and .01 levels.

For small sample sizes, the exact sampling distribution of the **Friedman two-way analysis of variance by ranks** (which, as noted in Section I, is mathematically equivalent to **Kendall's coefficient of concordance**) can be employed to evaluate the significance of $\tilde{W}$. In addition, when the values of m and n are reasonably small, some sources (e.g., Marascuilo and McSweeney (1977) and Siegel and Castellan (1988)) evaluate the significance of $\tilde{W}$ by employing an adjusted chi-square value (discussed in Section VII of the **Friedman two-way analysis of variance by ranks**) which represents an exact value for the underlying sampling distribution. For $m = 6$ and $n = 4$, the adjusted chi-square .05 and .01 critical values are $\chi^2_{r_{.05}} = 7.60$ and $\chi^2_{r_{.01}} = 10.00$ (which are reasonably close to the values

$\chi^2_{.05} = 7.81$ and $\chi^2_{.01} = 11.34$). Since the computed value $\chi^2 = 15.41$ is greater than both of the aforementioned critical values, the alternative hypothesis H_1: $W \neq 0$ is supported at both the .05 and .01 levels. Thus, regardless of which tables are employed to evaluate the results of Example 25.1, the alternative hypothesis H_1: $W \neq 0$ is supported at both the .05 and .01 levels. Consequently, one can conclude there is a significant association among the six instructors with respect to how they rank the four students.

VI. Additional Analytical Procedures for Kendall's Coefficient of Concordance and/or Related Tests

Tie correction for Kendall's coefficient of concordance When ties are present in a set of data, some sources recommend that the value of $\tilde{W}$ computed with Equations 25.3/25.4 be adjusted. Unless there is an excessive number of ties, the difference between the value of $\tilde{W}$ computed with Equations 25.3/25.4 and the value computed with the tie correction will be minimal. The tie correction, which results in a slight increase in the value of $\tilde{W}$, will be illustrated with Example 25.2.

Example 25.2 *Four judges rank four contestants in a beauty contest. The judges are told to assign the most beautiful contestant a rank of* 1 *and the least beautiful contestant a rank of* 4. *The rank-orders of the four judges are summarized in* Table 25.2. *Is there a significant association between the rank-orders assigned to the four contestants by the four judges?*

Table 25.2 Data for Example 25.2

	Contestant				
Judge	**1**	**2**	**3**	**4**	**Totals**
1	1	3	3	3	
2	1	4	2	3	
3	2	3	1	4	
4	1.5	1.5	3.5	3.5	
ΣR_j	5.5	11.5	9.5	13.5	$T = 40$
$(\Sigma R_j)^2$	30.25	132.25	90.25	182.25	$U = 435$

In Example 25.2, there are $m = 4$ sets of ranks/judges and $n = 4$ subjects/contestants who are ranked. Inspection of Table 25.2 reveals that Judges 1 and 4 employ tied ranks. As is the case with other rank-order tests described in the book, subjects who are tied for a specific rank are assigned the average of the ranks that are involved. Judge 1 assigns a rank of 1 to Contestant 1, and places the other three contestants in a tie for the next ordinal position. Thus, Contestants 2, 3, and 4 are all assigned a rank of 3, which is the average of the three ranks involved (i.e., $(2 + 3 + 4)/3 = 3$). Judge 4 places Contestants 1 and 2 in a tie for the first and second ordinal positions, and Contestants 3 and 4 in a tie for the third and fourth ordinal positions. Thus, the contestants evaluated by Judge 4 are assigned ranks that are the average of those ranks for which they are tied (i.e., $(1 + 2)/2 = 1.5$ and $(3 + 4)/2 = 3.5$).

Equation 25.6 (which is the tie corrected version of Equation 25.4) is employed to compute the tie corrected value of **Kendall's coefficient of concordance**, which will be represented by the notation $\tilde{W}_c$.

$$\tilde{W}_c = \frac{12U - 3m^2n(n+1)^2}{m^2n(n^2-1) - m\sum_{i=1}^{m}\left[\sum_{a=1}^{s}(t_a^3 - t_a)\right]} \qquad \textbf{(Equation 25.6)}$$

The notation $\sum_{i=1}^{m}\left[\sum_{a=1}^{s}(t_a^3 - t_a)\right]$ in Equation 25.6 indicates the following: a) Within each set of ranks, for each set of ties that is present the number of ties in the set is subtracted from the cube of the number of ties in that set; b) The sum of all the values computed in part a) is obtained for that set of ranks; and c) The sum of the values computed in part b) is computed for the m sets of ranks.

In the case of Example 25.2, Judge 1 has $s = 1$ set of ties involving three contestants. Thus, for Judge 1, $\sum_{a=1}^{s}(t_a^3 - t_a) = [(3)^3 - 3] = 24$. Since Judges 2 and 3 do not employ any ties, the latter two judges will not contribute to the tie correction, and thus the value of $\sum_{a=1}^{s}(t_a^3 - t_a)$ will equal 0 for both of the aforementioned judges. Judge 4 has $s = 2$ sets of ties, each set involving two contestants. Thus, for Judge 4, $\sum_{a=1}^{s}(t_a^3 - t_a) = [(2)^3 - 2] + [(2)^3 - 2] = 12$. We can now determine the value $\sum_{i=1}^{m}\left[\sum_{a=1}^{s}(t_a^3 - t_a)\right] = 36$, which is employed in Equation 25.6.

$$\sum_{i=1}^{m}\left[\sum_{a=1}^{s}(t_a^3 - t_a)\right] = 24 + 0 + 0 + 12 = 36$$

When the appropriate values are substituted in Equation 25.6, the tie corrected value $\tilde{W}_c = .51$ is computed for Example 25.2.[4]

$$\tilde{W}_c = \frac{(12)(435) - (3)(4)^2(4)(4+1)^2}{(4)^2(4)[(4)^2 - 1] - (4)(36)} = .51$$

It can be seen below that when Equation 25.4 (which does not employ the tie correction) is employed to compute $\tilde{W}$, the value $\tilde{W} = .44$ is obtained. Note that the latter value is less than $\tilde{W}_c = .51$. The computed correlation $\tilde{W}_c = .51$ (as well as $\tilde{W} = .44$) indicates a moderate degree of association between the four sets of ranks.

$$\tilde{W} = \frac{(12)(435) - (3)(4)^2(4)(4+1)^2}{(4)^2(4)[(4)^2 - 1]} = .44$$

Note that in **Table A20** the tabled critical .05 and.01 values for $m = 4$ and $n = 4$ are $\tilde{W}_{.05} = .619$ and $\tilde{W}_{.01} = .768$. Since both $\tilde{W}_c = .51$ and $\tilde{W} = .44$ are less than $\tilde{W}_{.05} = .619$, the null hypothesis H_0: $W = 0$ cannot be rejected.

VII. Additional Discussion of Kendall's Coefficient of Concordance

1. Relationship between Kendall's coefficient of concordance and Spearman's rank-order correlation coefficient The relationship between **Kendall's coefficient of concordance** and **Spearman's rank-order correlation coefficient** is as follows: If for data consisting of m sets of ranks a value for **Spearman's rho** is computed for every possible pair consisting of two sets of ranks (i.e., if $m = 3$, $r_{S_{12}}, r_{S_{13}}, r_{S_{23}}$), the average of all the r_S values (to be designated $\bar{r}_S$) is a linear function of the value of $\tilde{W}$ computed for the data. Equation 25.7 defines the exact relationship between **Spearman's rho** and $\tilde{W}$ for the same set of data.

$$\bar{r}_S = \frac{m\tilde{W} - 1}{m - 1} \qquad \textbf{(Equation 25.7)}$$

The above relationship will be demonstrated employing the data in Table 25.3 (which we will assume is a revised set of data for Example 25.1, in which $m = 3$ and $n = 3$).

Table 25.3 Data for Use in Equation 25.7

	Student			
Instructor	**1**	**2**	**3**	**Totals**
1	3	1	2	
2	1	2	3	
3	3	2	1	
ΣR_j	7	5	6	$T = 18$
$(\Sigma R_j)^2$	49	25	36	$U = 110$

Substituting the appropriate values in Equation 25.4, the value $\tilde{W} = .111$ is computed. The latter value indicates a weak degree of association between the three sets of ranks.[5]

$$\tilde{W} = \frac{(12)(110) - (3)(3)^2(3)(3 + 1)^2}{(3)^2(3)[(3)^2 - 1]} = .111$$

Substituting $\tilde{W} = .111$ in Equation 25.7, the value $\bar{r}_S = -.333$ is computed.

$$\bar{r}_S = \frac{(3)(.111) - 1}{3 - 1} = -.333$$

We will now confirm that $\bar{r}_S = -.333$. Equation 23.1 is employed to compute the r_S values for the 3 pairs of ranks (i.e., $r_{S_{12}}$ for the ranks of Instructor 1 versus Instructor 2; $r_{S_{13}}$ for the ranks of Instructor 1 versus Instructor 3; $r_{S_{23}}$ for the ranks of Instructor 2 versus Instructor 3). The resulting values are $r_{S_{12}} = -.5$, $r_{S_{13}} = .5$, and $r_{S_{23}} = -1$. The average of the values of the three pairs of ranks is $\bar{r}_S = [(-.5) + .5 + (-1)]/3 = -.333$, thus confirming the result obtained with Equation 25.7. It should be noted that when Equation 25.7 is employed to compute the value of $\bar{r}_S$, the range of values within which $\bar{r}_S$ can fall is defined by the following limits: $[-1/(m - 1)] \leq \bar{r}_S \leq +1$. When $m = 3$, as is the case in the example under discussion, the minimum possible value $\bar{r}_S$ can assume is $-1/(3 - 1) = -.5$. Note that even though the sign of $\tilde{W}$ cannot be negative, Equation 25.7 can convert a positive $\tilde{W}$ value into either a positive or negative $\bar{r}_S$ value.

The relationship described by Equation 25.7 can also be demonstrated for any of the examples employed in illustrating **Spearman's rank-order correlation coefficient**, where $m = 2$. To illustrate, in the case of Example 23.2 the value $r_S = .72$ is computed for two sets of ranks. When the relevant values from Example 23.2 (which are summarized in Table 23.6)[6] are substituted in Equation 25.4, the value $\tilde{W} = .86$ is computed. Note that in Example 23.2, $m = 2$ and $n = 10$.

$$\tilde{W} = \frac{(12)(1493.5) - (3)(2)^2(10)(10 + 1)^2}{(2)^2(10)[(10)^2 - 1]} = .86$$

Substituting $\tilde{W} = .86$ in Equation 25.7 yields the value $\bar{r}_S = .72$, which equals $r_S = .72$ computed with Equation 23.1.

$$\bar{r}_S = \frac{(2)(.86) - 1}{2 - 1} = .72$$

Thus, when $m = 2$, the value of $\bar{r}_S$ will equal r_S, since the average of a single value (based on one pair of ranks) is that value.

2. Relationship between Kendall's coefficient of concordance and the Friedman two-way analysis of variance by ranks In Section I it is noted that **Kendall's coefficient of concordance** and the **Friedman two-way analysis of variance by ranks** are based on the same mathematical model. Equation 25.8 defines the relationship between the computed value of $\tilde{W}$ and χ_r^2. The chi-square value (χ_r^2) in Equation 25.8 can be employed to represent the test statistic for the **Friedman two-way analysis of variance by ranks** (which is more commonly computed with Equation 19.1). Note that Equation 25.8 is identical to Equation 25.5.

$$\chi_r^2 = m(n - 1)\tilde{W} \qquad \textbf{(Equation 25.8)}$$

Equation 25.9, which is the algebraic transposition of Equation 25.8, provides an alternative way of computing the value $\tilde{W}$.

$$\tilde{W} = \frac{\chi_r^2}{m(n - 1)} \qquad \textbf{(Equation 25.9)}$$

In order to employ Equation 25.9 to compute the value of $\tilde{W}$, it is necessary to evaluate the data for m sets of ranks on n subjects/objects with the **Friedman two-way analysis of variance by ranks.** To illustrate the equivalence of **Kendall's coefficient of concordance** and the **Friedman two-way analysis of variance by ranks**, consider Example 25.3 which employs the same variables employed in Example 19.1 (which is used to illustrate the **Friedman two-way analysis of variance by ranks**). Note that in Example 25.3 there are $m = 6$ judges (who are represented by the six subjects) and $n = 3$ objects (which are represented by the three levels of noise).

Example 25.3 *Six subjects rank three levels of noise (based on the presence or absence of different types of music) with respect to the degree they believe each level of noise will disrupt one's ability to learn a list of nonsense syllables. The subjects are instructed to assign a rank of* 1 *to the most disruptive level of noise and a rank of* 3 *to the least disruptive level of noise.* Table 25.4 *summarizes the rankings of the subjects. Is there a significant association between the rank-orders assigned to the three levels of noise by the six subjects?*

Employing Equation 25.4, the value $\tilde{W} = .92$ is computed.[7] The value $\tilde{W} = .92$ indicates a strong degree of association between the six sets of ranks.

$$\tilde{W} = \frac{(12)(498.5) - (3)(6)^2(3)(3 + 1)^2}{(6)^2(3)[(3)^2 - 1]} = .92$$

Table 25.4 Data for Example 25.3

	Type of noise			
Subject	**No noise**	**Classical music**	**Rock music**	**Totals**
1	3	2	1	
2	3	2	1	
3	3	2	1	
4	3	2	1	
5	3	2	1	
6	3	1.5	1.5	
ΣR_j	18	11.5	6.5	$T = 36$
$(\Sigma R_j)^2$	324	132.25	42.25	$U = 498.5$

It happens to be the case that the configuration of ranks in Example 25.3 is identical to the configuration of ranks employed in Example 19.1. When the **Friedman two-way analysis of variance by ranks** is employed to evaluate the same six sets of ranks, the value $\chi_r^2 = 11.08$ is computed. The reader should take note of the fact that when the data are evaluated with Equation 19.1 in Section IV of the **Friedman test,** k is employed to represent the number of levels of the independent variable and n is employed to represent the number of subjects. In Table 19.1, the three columns of R_j values represent the $k = 3$ levels of the independent variable, and the six rows represent the $n = 6$ subjects. In the model employed for **Kendall's coefficient of concordance**, the value of n corresponds to the value employed for k in the **Friedman model**, and thus, $n = k = 3$. The value of m in the **Kendall model** corresponds to the value employed for n in the **Friedman model**, and thus, $m = n = 6$. The equations used in this section employ notations that are consistent with the **Kendall model.**

When the value $\chi_r^2 = 11.08$ is substituted in Equation 25.9, the value $\tilde{W} = .92$ is computed.

$$\tilde{W} = \frac{11.08}{(6)(3 - 1)} = .92$$

In the same respect, if $\tilde{W} = .92$ is substituted in Equations 25.8/25.5, it yields the value $\chi_r^2 = 11.08$.[8]

$$\chi_r^2 = (6)(3 - 1)(.92) = 11.08$$

Since the value of $\tilde{W}$ can be computed for the **Friedman test** model, **Kendall's coefficient of concordance** can be employed as a measure of effect size for a within-subjects design (involving data that are rank-ordered) with an independent variable that has three or more levels. The closer the value of $\tilde{W}$ is to 1, the stronger the relationship between the independent and dependent variables. Consequently, the value $\tilde{W} = .92$ computed for Example 19.1 (as well as Example 25.3), indicates there is a strong degree of association between the independent variable (noise) and the dependent variable (the rank-ordering on number of nonsense syllables recalled).

VIII. Additional Examples Illustrating the Use of Kendall's Coefficient of Concordance

Examples 25.4 and 25.5 are two additional examples that can be evaluated with **Kendall's coefficient of concordance.** Example 25.4 addresses the same question evaluated by Example 23.2, but in Example 25.4 the values $m = 6$ and $n = 4$ are employed in place of the values $m = 2$ and $n = 10$ employed in Example 23.2. Since Examples 25.4 and 25.5 employ the same data as Example 25.1, they yield the same result.

Example 25.4 *In order to determine whether critics agree with one another in their evaluation of movies, a newspaper editor asks six critics to rank four movies (assigning a rank of* 1 *to the best movie, a rank of* 2 *to the next best movie, etc.).* Table 25.5 *summarizes the data for the study. Is there a significant association between the six sets of ranks?*

Example 25.5 *Four members of a track team are ranked by the head coach with respect to their ability on six track and field events. For each event, the coach assigns a rank of* 1 *to the athlete who is best at the event and a rank of* 4 *to the athlete who is worst at the event.* Table 25.6 *summarizes the data for the study. Is there a significant association between the rank-orders assigned to the athletes on the six events?*

Table 25.5 Data for Example 25.4

	Movie				
Critic	**1**	**2**	**3**	**4**	**Totals**
1	3	2	1	4	
2	3	2	1	4	
3	3	2	1	4	
4	4	2	1	3	
5	3	2	1	4	
6	4	1	2	3	
ΣR_j	20	11	7	22	$T = 60$
$(\Sigma R_j)^2$	400	121	49	484	$U = 1054$

Table 25.6 Data for Example 25.5

	Athlete				
Event	**1**	**2**	**3**	**4**	**Totals**
Sprint	3	2	1	4	
1500 meters	3	2	1	4	
Pole vault	3	2	1	4	
Long jump	4	2	1	3	
Shot put	3	2	1	4	
400 meters	4	1	2	3	
ΣR_j	20	11	7	22	$T = 60$
$(\Sigma R_j)^2$	400	121	49	484	$U = 1054$

Note that in Example 25.5, even though one judge (the coach) is employed, the judge generates six sets of ranks. If there is a significant association between the six sets of ranks, it indicates that the athletes are perceived to be consistent with respect to performance on the six events.

References

Conover, W. J. (1980). **Practical nonparametric statistics** (2nd ed.). New York: John Wiley and Sons.

Daniel, W. W. (1990). **Applied nonparametric statistics** (2nd ed.). Boston: PWS–Kent Publishing Company.

Friedman, M. (1940). A comparison of alternative tests of significance for the problem of *m* rankings, **Annals of Mathematical Statistics**, 11, 86–92.

Kendall, M. G. and Babington–Smith, B. (1939). The problem of *m* rankings. **Annals of Mathematical Statistics**, 10, 275–287.

Kendall, M. G. (1970). **Rank correlation methods** (4th ed.). London: Charles Griffin and Co. Ltd.

Lindeman, R. H., Meranda, P. F., and Gold, R. Z. (1980). **Introduction to bivariate and multivariate analysis**. Glenview, IL: Scott, Foresman and Company.

Marascuilo, L. A. and McSweeney, M. (1977). **Nonparametric and distribution-free methods for the social sciences**. Monterey, CA: Brooks/Cole Publishing Company.

Siegel, S. and Castellan, N. J., Jr. (1988). **Nonparametric statistics for the behavioral sciences** (2nd ed.). New York: McGraw–Hill Book Company.

Sprent, P. (1989). **Applied nonparametric statistics**. London: Chapman and Hall.

Wallis, W. A. (1939). The correlation ratio for ranked data. **Journal of the American Statistical Association,** 34, 533–538.

Endnotes

1. Siegel and Castellan (1988) emphasize the fact that a correlation equal to or close to 1 does not in itself indicate that the rankings are correct. A high correlation only indicates that there is agreement among the m sets of ranks. It is entirely possible that there can be complete agreement among two or more sets of ranks, but that all of the rankings are, in fact, incorrect. In other words, the ranks may not reflect what is actually true with regard to the subjects/objects that are evaluated.

2. In point of fact, if the values of r_S and $\tilde{W}$ are computed for $m = 2$ sets of ranks, when the computed values for r_S are respectively 1, −1, and 0 the computed values of $\tilde{W}$ will respectively be 1, 0, and .5. The latter sets of values can be obtained through use of Equation 25.7, which is presented in Section VII.

3. Some sources state that the alternative hypothesis is **directional**, since $\tilde{W}$ can only be a positive value. Related to this is the fact that only the upper tail of the chi-square distribution (which is discussed in Section V) is employed in approximating the exact sampling distribution of $\tilde{W}$. In the final analysis, it becomes academic whether one elects to identify the alternative hypothesis as directional or nondirectional.

4. The tie corrected version of Equation 25.3 is noted below:

$$\tilde{W} = \frac{S}{\left(\dfrac{m^2 n(n^2 - 1) - m\sum_{i=1}^{m}\left[\sum_{a=1}^{s}(t_a^3 - t_a)\right]}{12}\right)} = \frac{35}{\left(\dfrac{(4)^2(4)[(4)^2 - 1] - (4)(36)}{12}\right)} = .51$$

5. Note that for $m = 3$ and $n = 3$, no tabled critical values are listed in **Table A20**. This is the case, since critical values cannot be computed for values of m and n that fall below specific minimum values. If Equation 25.5 is employed to evaluate $\tilde{W} = .111$, it yields the following result: $\chi^2 = (3)(3 - 1)(.111) = .666$. Since $\chi^2 = .666$ is less than the tabled critical two-tailed value (for $df = 2$) $\chi^2_{.05} = 5.99$, the obtained value $\tilde{W} = .111$ is not significant. In point of fact, even if the maximum possible value $\tilde{W} = 1$ is substituted in Equation 25.5, it yields the value $\chi^2 = 6$, which is barely above $\chi^2_{.05} = 5.99$. Since the chi-square distribution provides an approximation of the exact sampling distribution, in this instance it would appear that the tabled value $\chi^2_{.05} = 5.99$ is a little too high, and in actuality is associated with a Type I error rate that is slightly above .05.

6. The summary of the data for Example 23.2 in Table 23.6 provides the necessary values required to compute the value of $\tilde{W}$. The latter values are not computed in Table 23.5, which (employing a different format) also summarizes the data for Example 23.2.

7. Although there is one set of ties in the data, the tie correction described in Section VI is not employed for Example 25.3.

8. The exact value $\chi_r^2 = 11.08$ is computed if the value $\tilde{W} = .9236$ (which carries the computation of $\tilde{W}$ to four decimal places) is employed in Equations 25.8/25.5.

Test 26

Goodman and Kruskal's Gamma
(Nonparametric Measure of Association/Correlation Employed with Ordinal Data)

I. Hypothesis Evaluated with Test and Relevant Background Information

Goodman and Kruskal's gamma is one of a number of measures of correlation or association discussed in this book. Measures of correlation are not inferential statistical tests, but are, instead, descriptive statistical measures which represent the degree of relationship between two or more variables. Upon computing a measure of correlation, it is common practice to employ one or more inferential statistical tests in order to evaluate one or more hypotheses concerning the correlation coefficient. The hypothesis stated below is the most commonly evaluated hypothesis for **Goodman and Kruskal's gamma**.

Hypothesis evaluated with test In the underlying population represented by a sample, is the correlation between subjects' scores on two variables some value other than zero?

Relevant background information on test Prior to reading the material in this section the reader should review the general discussion of correlation in Section I of the **Pearson product-moment correlation coefficient (Test 22)**, as well as the material in Section I of **Kendall's tau (Test 24)**. Developed by Goodman and Kruskal (1954, 1959, 1963, 1972), **gamma** is a bivariate measure of correlation/association that is employed with rank-order data which is summarized within the format of an **ordered contingency table**. The population parameter estimated by the correlation coefficient will be represented by the notation γ (which is the lower case Greek letter **gamma**). The sample statistic computed to estimate the value of γ will be represented by the notation G. As is the case with **Spearman's rank-order correlation coefficient (Test 23)** and **Kendall's tau**, **Goodman and Kruskal's gamma** can be employed to evaluate data in which a researcher has scores for n subjects/objects on two variables (designated as the X and Y variables), both of which have been rank-ordered. However, in contrast to **Spearman's rho** and **Kendall's tau**, computation of **gamma** is recommended when there are many ties in a set of data, and thus it becomes more efficient to summarize the data within the format of an ordered $r \times c$ contingency table.

An ordered $r \times c$ contingency table consists of $r \times c$ cells, and is comprised of r rows and c columns.[1] In the model employed for **Goodman and Kruskal's gamma**, each of the rows in the contingency table represents one of the r levels of the X variable, and each of the columns represents one of the c levels of the Y variable (or vice versa). Since the contingency table that is employed to summarize the data is ordered, the categories for both the row and the column variables are arranged sequentially with respect to magnitude/ordinal position. To be more specific, the first row in the table represents the category that is lowest in magnitude on the X variable and the rth row represents the category that is highest in magnitude on the X variable. In the same respect, the first column represents the category that is lowest in magnitude on the Y variable and the cth column represents the category that is highest in magnitude on the Y variable.[2] Recorded within each of the

$r \times c$ cells of the contingency table are the number of subjects whose categorization on the X and Y variables corresponds to the row and column of a specific cell.

The value of **gamma** computed for a set of data represents the difference $p(C) - p(D)$, where: a) $p(C)$ is the probability that the ordering of the scores on the row and column variables for a pair of subjects is concordant (i.e., in agreement); and b) $p(D)$ is the probability that the ordering of the scores on the row and column variables for a pair of subjects is discordant (i.e., disagree).

To illustrate, if a subject is categorized on the lowest level of the row variable and the highest level of the column variable, that subject is concordant with respect to ordering when compared with any other subject who is assigned to a lower category on the row variable than he is on the column variable. On the other hand, that subject is discordant with respect to ordering when compared with another subject who is assigned to a higher category on the row variable than he is on the column variable. For a more thorough discussion of the concepts of concordance and discordance the reader should review Section I of **Kendall's tau.**

The range of possible values within which a computed value of **gamma** may fall is $-1 \leq G \leq +1$. As is the case for **Kendall's tau,** a positive value of G indicates that the number of concordant pairs in a set of data is greater than the number of discordant pairs, while a negative value indicates that the number of discordant pairs is greater than the number of concordant pairs. The computed value of G will equal 1 when the ordering of scores for all of the pairs of subjects in a set of data is concordant, and will equal -1 when the ordering of scores for all of the pairs of subjects is discordant. When $G = 0$ the number of concordant and discordant pairs of subjects in a set of data is equal.

Since **Goodman and Kruskal's gamma** and **Kendall's tau** both involve evaluating pairs of scores with respect to concordance versus discordance, the two measures of association are related to one another. Marascuilo and McSweeney (1977), who provide a detailed discussion on the nature of the relationship between **gamma** and **tau**, note that if $\tilde{\tau}$ and G are computed for the same set of data, as the number of pairs of ties increase, the absolute value computed for G will become increasingly larger relative to the absolute value of $\tilde{\tau}$. As a result of the latter, researchers who want to safeguard against obtaining an inflated value for the degree of association between the two variables may prefer to compute $\tilde{\tau}$ for a set of rank-order data, as opposed to computing the value of G.

It should be noted that **Yule's *Q* (Test 11i)** (which is one of a number of measures of association that can only be employed to evaluate a 2×2 contingency table) represents a special case of **Goodman and Kruskal's gamma.** Although **gamma** can be employed with a 2×2 contingency table, it is typically employed with ordered contingency tables in which there are at least three levels on either the row or column variable. A more detailed discussion of the relationship between **Yule's *Q*** and **Goodman and Kruskal's gamma** can be found in Section VII.

II. Example

Example 26.1 *A researcher wants to determine whether a relationship exists between a person's weight (which will be designated as the X variable) and birth order (which will be designated as the Y variable). Upon determining the weight and birth order of* 300 *subjects, each subject is categorized with respect to one of three weight categories and one of four birth order categories. Specifically, the following three categories are employed with respect to weight:* **below average, average, above average.** *The following four categories are employed with respect to birth order:* **first born, second born, third born, fourth born and all subsequent birth orders.** Table 26.1 *(which is a* 3×4 *ordered contingency table, with* $r = 3$ *and* $c = 4$*) summarizes the data. Do the data indicate there is a significant association between a person's weight and birth order?*

Table 26.1 Summary of Data for Example 26.1

		Birth order				Row sums
		1st born	2nd born	3rd born	4th born+	
	Below average	70	15	10	5	100
Weight	Average	10	60	20	10	100
	Above average	10	15	35	40	100
Column sums		90	90	65	55	300

III. Null versus Alternative Hypotheses

Upon computing **Goodman and Kruskal's gamma**, it is common practice to determine whether the obtained absolute value of the correlation coefficient is large enough to allow a researcher to conclude that the underlying population correlation coefficient between the two variables is some value other than zero. Section V describes how the latter hypothesis, which is stated below, can be evaluated through use of an inferential statistical test that is based on the normal distribution.

Null hypothesis H_0: $\gamma = 0$

(In the underlying population the sample represents, the correlation between the scores/categorization of subjects on Variable X and Variable Y equals 0.)

Alternative hypothesis H_1: $\gamma \neq 0$

(In the underlying population the sample represents, the correlation between the scores/categorization of subjects on Variable X and Variable Y equals some value other than 0. This is a **nondirectional alternative hypothesis**, and it is evaluated with a **two-tailed test**. Either a significant positive G value or a significant negative G value will provide support for this alternative hypothesis. In order to be significant, the obtained absolute value of G must be equal to or greater than the tabled critical two-tailed G value at the prespecified level of significance.)

or

H_1: $\gamma > 0$

(In the underlying population the sample represents, the correlation between the scores/categorization of subjects on Variable X and Variable Y equals some value greater than 0. This is a **directional alternative hypothesis**, and it is evaluated with a **one-tailed test**. Only a significant positive G value will provide support for this alternative hypothesis. In order to be significant (in addition to the requirement of a positive G value), the obtained absolute value of G must be equal to or greater than the tabled critical one-tailed G value at the prespecified level of significance.)

or

H_1: $\gamma < 0$

(In the underlying population the sample represents, the correlation between the scores/categorization of subjects on Variable X and Variable Y equals some value less than 0. This is a **directional alternative hypothesis**, and it is evaluated with a **one-tailed test**. Only a significant negative G value will provide support for this alternative hypothesis. In order to be significant (in addition to the requirement of a negative G value), the obtained absolute value of G must be equal to or greater than the tabled critical one-tailed G value at the prespecified level of significance.)

Note: Only one of the above noted alternative hypotheses is employed. If the alternative hypothesis the researcher selects is supported, the null hypothesis is rejected.[3]

IV. Test Computations

In order to compute the value of G it is necessary to determine the number of pairs of subjects who are concordant with respect to the ordering of their scores on the X and Y variables (which will be represented by the notation n_C), and the number of pairs of subjects who are discordant with respect to the ordering of their scores on the X and Y variables (which will be represented by the notation n_D). Upon computing the values of n_C and n_D, Equation 26.1 is employed to compute the value of **Goodman and Kruskal's gamma.**

$$G = \frac{n_C - n_D}{n_C + n_D} \qquad \textbf{(Equation 26.1)}$$

The determination of the values of n_C and n_D is based on an analysis of the frequencies in the ordered contingency table (i.e., Table 26.1). Each of the cells in the table will be identified by two digits. The first digit will represent the row within which the cell falls, and the second digit will represent the column within which the cell falls. Thus, Cell_{ij} is the cell in the ith row and j^{th} column. As an example, since it is in both the first row and first column, the cell in the upper left hand corner of Table 26.1 is Cell_{11}. The number of subjects within each cell is identified by the notation n_{ij}. Thus, in the case of Cell_{11}, $n_{11} = 70$.

The protocol for determining the values of n_C and n_D will now be described. The following procedure is employed to determine the value of n_C.

a) Begin with the cell in the **upper left hand corner of the table** (i.e., Cell_{11}). Determine the frequency of that cell (which will be referred to as the target cell), and multiply the frequency by the sum of the frequencies of **all other cells in the table that fall both below it and to the right of it.** In Table 26.1, the following six cells meet the criteria of being both below and to the right of Cell_{11}: Cell_{22}, Cell_{23}, Cell_{23}, Cell_{32}, Cell_{33}, Cell_{34}. Note that although Cell_{21} and Cell_{31} are below Cell_{11}, they are not to the right of it, and although Cell_{12}, Cell_{13}, and Cell_{14} fall to the right of Cell_{11}, they do not fall below it. Any subject who falls within a cell that is both below and to the right of Cell_{11} will form a concordant pair with any subject in Cell_{11}. The rationale for this is as follows: Assume that the values R_{X_i} and R_{Y_i} represent the score/ranking/category of Subject i on the X and Y variables, and that R_{X_j} and R_{Y_j} represent the score/ranking/category of Subject j on the X and Y variables. Assume that Subject i is a subject in the target cell, and that Subject j is a subject in a cell that falls below and to the right of the target cell. We can state that the sign of the difference $(R_{X_i} - R_{X_j})$ will be the same as the sign of the difference $(R_{Y_i} - R_{Y_j})$ when the scores of any subject in the target cell are compared with any subject in a cell that falls below and to the right of the target cell. When for any pair of subjects the signs of the differences $(R_{X_i} - R_{X_j})$ and $(R_{Y_i} - R_{Y_j})$ are identical, that pair of subjects is concordant with respect to their ordering on the two variables.

To illustrate, each of the 70 subjects in Cell_{11} has a rank of 1 on both of the variables, and each of the 60 subjects in Cell_{22} (which is one of the cells below and to the right of Cell_{11}) has a rank of 2 on both of the variables. Any pair of subjects that is formed by employing one subject from Cell_{11} and one subject from Cell_{22} will be concordant with respect to their ordering on the two variables, since for each pair the sign of the difference between the ranks on both variables will be negative (i.e., $(R_{X_i} - R_{X_j}) = (1 - 2) = -1$ and $(R_{Y_i} - R_{Y_j}) = (1 - 2) = -1$). If, on the other hand, we compare the ranks on both variables for any subject who is in the target cell with the ranks on both variables for any subject who is in a cell that is not below and to the right of the target cell, $(R_{X_i} - R_{X_j})$ and $(R_{Y_i} - R_{Y_j})$ will have different signs or will equal zero.

The expression which summarizes the product of the frequency of Cell_{11} and the sum of the frequencies of all the cells that fall both below and to the right of it is as follows: $n_{11}(n_{22} + n_{23} + n_{24} + n_{32} + n_{33} + n_{34})$. Substituting the appropriate frequencies from Table 26.1, we obtain $70(60 + 20 + 10 + 15 + 35 + 40) = (70)(180) = 12600$. This latter value will be designated as **Product 1**.

b) The same procedure employed with Cell_{11} is applied to all remaining cells. Moving to the right in Row 1, the procedure is next employed with Cell_{12}. **Product 2**, which represents the product for the second target cell, can be summarized by the expression $n_{12}(n_{23} + n_{24} + n_{33} + n_{34})$, since Cell_{23}, Cell_{24}, Cell_{33}, and Cell_{34} are the only cells that fall both below and to the right of Cell_{12}. Thus, **Product 2** will equal $15(20 + 10 + 35 + 40) = (15)(105) = 1575$.

c) Upon computing **Product 2**, products for the two remaining cells in Row 1 are computed, after which products are computed for each of the cells in Rows 2 and 3. The computation of the products for all 12 cells in the ordered contingency table is summarized in Table 26.2. Note that since many of the cells have no cell that falls both below and to the right of them, the value that the frequency of these cells will be multiplied by will equal zero, and thus the resulting product will equal zero. The value of n_C is the sum of all the products in Table 26.2. For Example 26.1, $n_C = 20875$.

Table 26.2 Computation of n_C for Example 26.1

Cell	Computation		Value	Product
Cell_{11} :	70 (60 + 20 + 10 + 15 + 35 + 40)	=	12600	**Product 1**
Cell_{12} :	15 (20 + 10 + 35 + 40)	=	1575	**Product 2**
Cell_{13} :	10 (10 + 40)	=	500	**Product 3**
Cell_{14} :	5 (0)	=	0	**Product 4**
Cell_{21} :	10 (15 + 35 + 40)	=	900	**Product 5**
Cell_{22} :	60 (35 + 40)	=	4500	**Product 6**
Cell_{23} :	20 (40)	=	800	**Product 7**
Cell_{24} :	10 (0)	=	0	**Product 8**
Cell_{31} :	10 (0)	=	0	**Product 9**
Cell_{32} :	15 (0)	=	0	**Product 10**
Cell_{33} :	35 (0)	=	0	**Product 11**
Cell_{34} :	40 (0)	=	0	**Product 12**
	n_C = Sum of products	=	**20875**	

Upon computing the value of n_C, the following protocol is employed to compute the value of n_D.

a) Begin with the cell in the **upper right hand corner of the table** (i.e., Cell_{14}). Determine the frequency of that cell, and multiply the frequency by the sum of the frequencies of **all other cells in the table that fall both below it and to the left of it**. In Table 26.1, the following six cells meet the criteria of being both below and to the left of Cell_{14}: Cell_{21}, Cell_{22}, Cell_{23}, Cell_{31}, Cell_{32}, Cell_{33}. Note that although Cell_{24} and Cell_{34} are below Cell_{14}, they are not to the left of it, and although Cell_{11}, Cell_{12}, and Cell_{13} fall to the left of Cell_{14}, they do not fall below it. Any subject who falls within a cell that is both below and to the left of Cell_{14} will form a discordant pair with any subject in Cell_{14}. The general rule that can be stated with respect to discordant pairs is as follows (if we assume that Subject i is a subject in the target cell, and Subject j is a subject in some other cell): The sign of the difference $(R_{X_i} - R_{X_j})$ will be different than the sign of the difference $(R_{Y_i} - R_{Y_j})$ when the scores of any subject in the target cell are compared with any subject in a cell that falls below and to the left of the target cell. When for any pair of subjects the signs of the differences $(R_{X_i} - R_{X_j})$ and $(R_{Y_i} - R_{Y_j})$ are different, that pair of subjects is discordant with respect to their ordering on the two variables.

To illustrate, each of the 5 subjects in Cell_{14} has a rank of 1 on weight and a rank of 4 on birth order. Each of the 20 subjects in Cell_{23} (which is one of the cells below and to the left of Cell_{14}) has a rank of 2 on weight and a rank of 3 on birth order. Any pair of subjects that is formed by employing one subject from Cell_{14} and one subject from Cell_{23}, will be discordant with respect to the ordering of the ranks of the subjects on the two variables, since for each pair the signs of the difference between the ranks on both variables will be different (i.e., $(R_{X_i} - R_{X_j}) = (1 - 2) = -1$ and $(R_{Y_i} - R_{Y_j}) = (4 - 3) = +1$). If, on the other hand, we compare the ranks on both variables for any subject who is in the target cell with the ranks on both variables for any subject who is in a cell that is not below and to the left of the target cell, $(R_{X_i} - R_{X_j})$ and $(R_{Y_i} - R_{Y_j})$ will have the same sign or will equal zero.

The expression which summarizes the product of the frequency of Cell_{14} and the sum of the frequencies of all cells that fall both below it and to the left of it is as follows: $n_{14}(n_{21} + n_{22} + n_{23} + n_{31} + n_{32} + n_{33})$. Substituting the appropriate frequencies, we obtain $5(10 + 60 + 20 + 10 + 15 + 35) = (5)(150) = 750$. As is the case in determining the number of concordant pairs, we will designate the product for the first cell that is analyzed as **Product 1**.

b) The same procedure employed with Cell_{14} is applied to all remaining cells. Moving to the left in Row 1, the procedure is next employed with Cell_{13}. **Product 2**, which represents the product for the second target cell, can be summarized by the expression $n_{13}(n_{21} + n_{22} + n_{31} + n_{32})$, since Cell_{21}, Cell_{22}, Cell_{31}, and Cell_{32} are the only cells that fall both below and to the left of Cell_{13}. Thus, **Product 2** will equal $10(10 + 60 + 10 + 15) = (10)(95) = 950$.

c) Upon computing **Product 2**, products for the two remaining cells in Row 1 are computed, after which products are computed for each of the cells in Rows 2 and 3. The computation of the products for all 12 cells in the ordered contingency table is summarized in Table 26.3. Note that since many of the cells have no cell that falls both below and to the left of them, the value that the frequency of such cells will be multiplied by will equal zero, and thus the resulting product will equal zero. The value of n_D will be the sum of all the products in Table 26.3. For Example 26.1, $n_D = 3700$.

Table 26.3 Computation of n_D for Example 26.1

Cell$_{14}$:	5 (10 + 60 + 20 + 10 + 15 + 35)	=	750	**Product 1**
Cell$_{13}$:	10 (10 + 60 + 10 + 5)	=	950	**Product 2**
Cell$_{12}$:	15 (10 + 10)	=	300	**Product 3**
Cell$_{11}$:	70 (0)	=	0	**Product 4**
Cell$_{24}$:	10 (10 + 15 + 35)	=	600	**Product 5**
Cell$_{23}$:	20 (10 + 15)	=	500	**Product 6**
Cell$_{22}$:	60 (10)	=	600	**Product 7**
Cell$_{21}$:	10 (0)	=	0	**Product 8**
Cell$_{34}$:	40 (0)	=	0	**Product 9**
Cell$_{33}$:	35 (0)	=	0	**Product 10**
Cell$_{32}$:	15 (0)	=	0	**Product 11**
Cell$_{31}$:	10 (0)	=	0	**Product 12**

n_D = **Sum of products = 3700**

Substituting the values $n_C = 20875$ and $n_D = 3700$ in Equation 26.1, the value $G = .70$ is computed. Note that the value of G is positive, since the number of concordant pairs is greater than the number of discordant pairs.

$$G = \frac{20875 - 3700}{20875 + 3700} = .70$$

The value $G = .70$ can also be computed employing the definition of **gamma** presented in Section I. Specifically:

$$G = p(C) - p(D) = \frac{20875}{24575} - \frac{3700}{24575} = .70$$

In the above equation the value 24575 is the total number of pairs (to be designated n_T), which is the denominator of Equation 26.1. Thus, $p(C) = n_C/n_T$ and $p(D) = n_D/n_T$.

Since the computed value $G = .70$ is close to 1, it indicates the presence of a strong positive/direct relationship between the two variables. Specifically, it suggests that the higher the rank of a subject's weight category, the higher the rank of the subject's birth order category.

V. Interpretation of the Test Results

Test 26a: Test of significance for Goodman and Kruskal's gamma When the sample size is relatively large (which will generally be the case when **gamma** is computed), the computed value of G can be evaluated with Equation 26.2. To be more specific, Equation 26.2 (which employs the normal distribution) is employed to evaluate the null hypothesis H_0: $\gamma = 0$.[4] The sign of the z value computed with Equation 26.2 will be the same as the sign of the value computed for G.

$$z = G\sqrt{\frac{n_C + n_D}{N(1 - G^2)}} \qquad \textbf{(Equation 26.2)}$$

Where: N is the total number of subjects for whom scores are recorded in the ordered contingency table

When the appropriate values from Example 26.1 are substituted in Equation 26.2, the value $z = 8.87$ is computed.

$$z = .70\sqrt{\frac{20875 + 3700}{300[1 - (.70)^2]}} = 8.87$$

Equation 26.3 is an alternative equation for computing the value of z. The denominator of Equation 26.3 represents the **standard error** of the G statistic (which will be represented by the notation SE_G). In Section VI, SE_G is employed to compute a confidence interval for **gamma**.

(Equation 26.3)

$$z = \frac{G}{SE_G} = \frac{G}{\left[\dfrac{1}{\sqrt{\dfrac{n_C + n_D}{N(1 - G^2)}}}\right]} = \frac{.70}{\left[\dfrac{1}{\sqrt{\dfrac{20875 + 3700}{(300)[1 - (.70)^2]}}}\right]} = \frac{.70}{.0789} = 8.87$$

The computed value $z = 8.87$ is evaluated with **Table A1 (Table of the Normal Distribution)** in the **Appendix.**[5] In the latter table, the tabled critical two-tailed .05 and .01 values are $z_{.05} = 1.96$ and $z_{.01} = 2.58$, and the tabled critical one-tailed .05 and .01 values are $z_{.05} = 1.65$ and $z_{.01} = 2.33$.

The following guidelines are employed in evaluating the null hypothesis.

a) If the nondirectional alternative hypothesis H_1: $\gamma \neq 0$ is employed, the null hypothesis can be rejected if the obtained absolute value of z is equal to or greater than the tabled critical two-tailed value at the prespecified level of significance.

b) If the directional alternative hypothesis H_1: $\gamma > 0$ is employed, the null hypothesis can be rejected if the sign of z is positive, and the value of z is equal to or greater than the tabled critical one-tailed value at the prespecified level of significance.

c) If the directional alternative hypothesis H_1: $\gamma < 0$ is employed, the null hypothesis can be rejected if the sign of z is negative, and the absolute value of z is equal to or greater than the tabled critical one-tailed value at the prespecified level of significance.

Employing the above guidelines, the nondirectional alternative hypothesis H_1: $\gamma \neq 0$ is supported at both the .05 and .01 levels, since the computed value $z = 8.87$ is greater than the tabled critical two-tailed values $z_{.05} = 1.96$ and $z_{.01} = 2.58$. The directional alternative hypothesis H_1: $\gamma > 0$ is supported at both the .05 and .01 levels, since the computed value $z = 8.87$ is a positive number that is greater than the tabled critical one-tailed values $z_{.05} = 1.65$ and $z_{.01} = 2.33$. The directional alternative hypothesis H_1: $\gamma < 0$ is not supported, since the computed value $z = 8.87$ is a positive number.

A summary of the analysis of Example 26.1 follows: It can be concluded that there is a significant positive relationship between weight and birth order.

VI. Additional Analytical Procedures for Goodman and Kruskal's Gamma and/or Related Tests

1. The computation of a confidence interval for the value of Goodman and Kruskal's gamma Equation 26.4 is employed to compute a confidence interval for a computed value of **gamma**.

$$CI_{(1-\alpha)} = G \pm (z_{\alpha/2})(SE_G) \qquad \textbf{(Equation 26.4)}$$

Where: $z_{\alpha/2}$ represents the tabled critical two-tailed value in the normal distribution below which a proportion equal to $[1 - (\alpha/2)]$ of the cases falls. If the percentage of the distribution that falls within the confidence interval is subtracted from 1, it will equal the value of α.

Equation 26.4 will be employed to compute the 95% confidence interval for **gamma**. Along with the tabled critical two-tailed .05 value $z_{.05} = 1.96$, the following values computed for Example 26.1 are substituted in Equation 26.4: $G = .70$ and $SE_G = .0789$ (which is the computed value of the denominator of Equation 26.3, which as noted in Section V represents the **standard error** of G).

$$CI_{.95} = .70 \pm (1.96)(.0789) = .70 \pm .15$$

Subtracting from and adding .15 to .70, yields the values .55 and .85. Thus, the researcher can be 95% confident that the true value of **gamma** in the underlying population falls between .55 and 85. Symbolically, this can be written as follows: $.55 \leq \gamma \leq .85$.

2. Test 26b: Test for evaluating the null hypothesis H_0: $\gamma_1 = \gamma_2$ Marascuilo and McSweeney (1977) note that Equation 26.5 can be employed to determine whether or not there is a significant difference between two independent values of **gamma**. Use of Equation 26.5 assumes that the following conditions have been met: a) The sample size in each of two ordered contingency tables is large enough for evaluation with the normal approximation; b) The values of r and c are identical in the two ordered contingency tables; and c) The same row and column categories are employed in the two ordered contingency tables.

$$z = \frac{G_1 - G_2}{\sqrt{SE_{G_1} + SE_{G_2}}} \qquad \textbf{(Equation 26.5)}$$

Where: G_1 and G_2 are the computed values of **gamma** for the two ordered contingency tables, and SE_{G_1} and SE_{G_2} are the computed values of the **standard error** for the two values of **gamma**

To illustrate the use of Equation 26.5, assume that the study described in Example 26.1 is replicated with a different sample comprised of $N = 600$ subjects. The obtained value of **gamma** for the sample is $G = .50$, with $SE_G = .0438$. By employing the values $G_1 = .70$, $SE_{G_1} = .0789$, $G_2 = .50$, and $SE_{G_2} = .0438$ in Equation 26.5, the researcher can evaluate the null hypothesis H_0: $\gamma_1 = \gamma_2$. Substituting the appropriate values in Equation 26.5 yields the value $z = .57$.

$$z = \frac{.7 - .5}{\sqrt{.0789 + .0438}} = .57$$

The same guidelines described for evaluating the alternative hypotheses H_1: $\gamma \neq 0$, H_1: $\gamma > 0$, and H_1: $\gamma < 0$ are respectively employed for evaluating the alternative hypotheses H_1: $\gamma_1 \neq \gamma_2$, H_1: $\gamma_1 > \gamma_2$, and H_1: $\gamma_1 < \gamma_2$. The nondirectional alternative hypothesis H_1: $\gamma_1 \neq \gamma_2$ is not supported, since the computed value $z = .57$ is less than the tabled critical two-tailed value $z_{.05} = 1.96$. The directional alternative hypothesis H_1: $\gamma_1 > \gamma_2$ is not supported, since the computed value $z = .57$ is less than the tabled critical one-tailed value $z_{.05} = 1.65$. The directional alternative hypothesis H_1: $\gamma_1 < \gamma_2$ is not supported, since the computed value $z = .57$ is a positive number. The fact that the difference $\gamma_1 - \gamma_2 = .7 - .5 = .2$ (which is reasonably large) is not significant, can be attributed to the fact that both samples have relatively large standard errors.

3. Sources for computing a partial correlation coefficient for Goodman and Kruskal's gamma A procedure developed by Davis (1967) for computing a partial correlation for **gamma** is described in Marascuilo and McSweeney (1977).

VII. Additional Discussion of Goodman and Kruskal's Gamma

1. Relationship between Goodman and Kruskal's gamma and Yule's Q In Section I it is noted that **Yule's Q** is a special case of **Goodman and Kruskal's gamma**. To illustrate this, assume that the four cells in Tables 11.2/11.3 (for which **Yule's Q** is computed) represent a 2 × 2 contingency table in which the cells on both the row and the column variables are ordered. If the procedure described for determining concordant pairs is employed with the data in Tables 11.2/11.3, the only cell that will generate a product other than zero is Cell_{11} (which corresponds to **Cell a** within the framework of the notation used for a 2 × 2 contingency table). Specifically, the product for Cell_{11}, which will correspond to the value of n_C, is $(n_{11})(n_{12}) = (30)(40) = 1200$. In the same respect if the procedure described for determining discordant pairs is employed, the only cell that will generate a product other than zero is Cell_{12} (which corresponds to **Cell b** within the framework of the notation used for a 2 × 2 contingency table). Specifically, the product for Cell_{12}, which will correspond to the value of n_D, is $(n_{12})(n_{21}) = (70)(60) = 4200$. When the values $n_C = 1200$ and $n_D = 4200$ are substituted in Equation 26.1, $G = (1200 - 4200)/(1200 + 4200) = -.56$. Note that this result is identical to that obtained when Equation 11.20 is employed to compute **Yule's Q** for the same set of data: $Q = (ad - bc)/(ad + bc) = [(30)(40) - (70)(60)]/[(30)(40) + (70)(60)] = -.56$. It should be noted that unlike **gamma**, which is only employed with ordered contingency tables, **Yule's Q** can be employed with both ordered and unordered 2 × 2 contingency tables.

2. Somers' delta as an alternative measure of association for an ordered contingency table Somers (1962) has developed an alternative measure of association for ordered contingency tables referred to as **delta** (which is represented by the upper case Greek letter Δ). Siegel and Castellan (1988) identify **delta** as an **asymmetrical measure of association** (as opposed to a **symmetrical measure of association**). An asymmetrical measure of association is employed when one variable is distinguished in a meaningful way from the other variable (e.g., within the context of the study, one variable is more important than the other or one variable represents an independent variable and the other a dependent variable). Within this framework, **gamma** is viewed as a symmetrical measure of association, since it does not assume a meaningful distinction between the variables within the context noted above. A full discussion of **Somers' delta** can be found in Siegel and Castellan (1988).

VIII. Additional Examples Illustrating the Use of Goodman and Kruskal's Gamma

Examples 26.2 and 26.3 are two additional examples that can be evaluated with **Goodman and Kruskal's gamma**. Since Examples 26.2 and 26.3 employ the same data as Example 26.1, they yield the same result. Example 26.4 describes the identical study described by Example 26.1, but uses a different configuration of data in order to illustrate the computation of a negative value for **gamma**.

Example 26.2 *A consumer group conducts a survey in order to determine whether a relationship exists between customer satisfaction and the price a person pays for an automobile. Each of 300 individuals who has purchased a new vehicle within the past year is classified in one of four categories based on the purchase price of one's automobile. Each subject is also classified in one of three categories with respect to how satisfied he or she is with one's automobile. The results are summarized in* Table 26.4. *Do the data indicate there is a relationship between the price of an automobile and degree of satisfaction?*

Example 26.3 *A panel of psychiatrists wants to determine whether a relationship exists between the number of years a patient is in psychotherapy and the degree of change in a patient's behavior. Each of 300 patients is categorized with respect to one of four time periods during which he or she is in psychotherapy, and one of three categories with respect to the change in behavior he or she has exhibited since initiating therapy. Specifically, the following four categories are employed with respect to psychotherapy duration:* **less than one year; one to two years; two to three years; more than three years.** *The following three* *categories are employed with respect to changes in behavior:* **deteriorated (–), no change, improved (+)**. Table 26.5 *summarizes the data. Do the data indicate there is an association between the amount of time a patient is in psychotherapy and the degree to which he or she changes?*[6]

Table 26.4 Summary of Data for Example 26.2

		Purchase price				**Row sums**
		Under $10,000	**$10,000 to $18,000**	**$18,001 to $30,000**	**More than $30,000**	
Level of satisfaction	**Below average**	70	15	10	5	**100**
	Average	10	60	20	10	**100**
	Above average	10	15	35	40	**100**
Column sums		**90**	**90**	**65**	**55**	**300**

Table 26.5 Summary of Data for Example 26.3

		Number of years in psychotherapy				**Row sums**
		Less than one year	**One to two years**	**Two to three years**	**More than three years**	
Amount of change	**–**	70	15	10	5	**100**
	No change	10	60	20	10	**100**
	+	10	15	35	40	**100**
Column sums		**90**	**90**	**65**	**55**	**300**

Example 26.4 *A researcher wants to determine whether a relationship exists between a person's weight and birth order. Upon determining the weight and birth order of* 300 *subjects, each subject is categorized with respect to one of three weight categories and one of four birth order categories. Specifically, the following three categories are employed with respect to weight:* **below average, average, above average**. *The following four categories are employed with respect to birth order:* **first born, second born, third born, fourth born and all subsequent birth orders**. Table 26.6 *summarizes the data. Do the data indicate there is a significant association between a person's weight and birth order?*

Table 26.6 Summary of Data for Example 26.4

		Birth order				**Row sums**
		1st born	**2nd born**	**3rd born**	**4th born+**	
Weight	**Below average**	5	10	15	70	**100**
	Average	10	20	60	10	**100**
	Above average	40	35	15	10	**100**
Column sums		**55**	**65**	**90**	**90**	**300**

Inspection of the data reveals that the cell frequencies in Table 26.6 are the mirror image of those employed in Table 26.1. By virtue of employing the same frequencies in an inverted format, the values of n_C and n_D for Table 26.6 are the reverse of those obtained for Table 26.1. Thus, for Table 26.6, $n_C = 3700$ and $n_D = 20875$. Consequently, employing

Equation 26.1, $G = (3700 - 20875)/(3700 + 20875) = -.70$. Because the same configuration of data is employed in an inverted format, the value $G = -.70$ computed for Table 26.6 is the same absolute value computed for Table 26.1. Note that the negative correlation $G = -.70$ indicates that a subject's birth order is inversely related to his weight. Specifically, subjects in a low birth order category are more likely to be above average in weight, while subjects in a high birth order category are more likely to be below average in weight.

References

Daniel, W. W. (1990). **Applied nonparametric statistics** (2nd ed.). Boston: PWS–Kent Publishing Company.

Davis, J. A. (1967). A partial coefficient for Goodman and Kruskal's gamma. **Journal of the American Statistical Association**, 62, 189–193.

Goodman, L. A. and Kruskal, W. H. (1954). Measures of association for cross-classification. **Journal of the American Statistical Association**, 49, 732–764.

Goodman, L. A. and Kruskal, W. H. (1959). Measures of association for cross-classification II: Further discussion and references. **Journal of the American Statistical Association**, 54, 123–163.

Goodman, L. A. and Kruskal, W. H. (1963). Measures of association for cross-classification III: Approximate sample theory. **Journal of the American Statistical Association**, 58, 310–364.

Goodman, L. A. and Kruskal, W. H. (1972). Measures of association for cross-classification IV: Simplification for asymptotic variances. **Journal of the American Statistical Association**, 67, 415–421.

Marascuilo, L. A. and McSweeney, M. (1977). **Nonparametric and distribution-free methods for the social sciences.** Monterey, CA: Brooks/Cole Publishing Company.

Ott, R. L, Larson, R., Rexroat, C., and Mendenhall, W. (1992). **Statistics: A tool for the social sciences** (5th ed.). Boston: PWS–Kent Publishing Company.

Siegel, S. and Castellan, N. J., Jr. (1988). **Nonparametric statistics for the behavioral sciences** (2nd ed.). New York: McGraw–Hill Book Company.

Somers, R. H. (1962). A new asymmetric measure of association for ordinal variables. **American Sociological Review**, 27, 799–811.

Endnotes

1. The general model for an $r \times c$ contingency table (which is summarized in Table 11.1) is discussed in Section I of the **chi-square test for $r \times c$ tables (Test 11).**

2. **Gamma** can also be computed if the ordering is reversed — i.e. within both variables, the first row/column represents the category with the highest magnitude, and the last row/column represents the category with the lowest magnitude.

3. Some sources employ the following statements as the null hypothesis and the nondirectional alternative hypothesis for **Goodman and Kruskal's gamma: Null hypothesis:** H_0: Variables X and Y are independent of one another; **Nondirectional alternative hypothesis:** H_1: Variables X and Y are not independent of one another.

   It is, in fact, true that if in the underlying population the two variables are independent, the value of γ will equal zero. However, Siegel and Castellan (1988) note that if $\gamma = 0$, the latter does not in and of itself ensure that the two variables are independent of one another (unless the contingency table is a 2×2 table).

4. Equation 26.2 can also be written in the following form:

$$z = (G - \gamma)\sqrt{\frac{n_C + n_D}{N(1 - G^2)}}$$

In the above equation, γ represents the value of **gamma** stated in the null hypothesis. When the latter value equals zero, the above equation reduces to Equation 26.2. When some value other than zero is stipulated for **gamma** in the null hypothesis, the equation noted above can be employed to evaluate the null hypothesis H_0: $\gamma = \gamma_0$ (where γ_0 represents the value stipulated for the population correlation).

5. Sources that discuss the evaluation of the null hypothesis H_0: $\gamma = 0$ note that the normal approximation computed with Equations 26.2/26.3 tends to be overly conservative. Consequently, the likelihood of committing a Type I error (i.e., rejecting H_0 when it is true) is actually less than the value of alpha employed in the analysis.

6. It could be argued that it might be more appropriate to employ **Somers' delta** (which is briefly discussed in Section VII) rather than **gamma** as a measure of association for Example 26.3. The use of **delta** could be justified, if within the framework of a study the number of years of therapy represents an independent variable and the amount of change represents the dependent variable. In point of fact, depending upon how one conceptualizes the relationship between the two variables, one could also argue for the use of **delta** as a measure of association for Example 26.1. In the final analysis, it will not always be clear whether it is more appropriate to employ **gamma** or **delta** as a measure of association for an ordered contingency table.

Appendix: Tables

Acknowledgments and Sources for Tables in Appendix

Table A1 Table of the Normal Distribution

Reprinted with permission of CRC Press, Boca Raton, Florida from W. H. Beyer (1968), **CRC Handbook of Tables for Probability and Statistics** (2nd ed.), Table II.1 (The normal probability function and related functions), pp. 127–134.

Table A2 Table of Student's *t* Distribution

Reprinted with permission from Table 12 (Percentage points for the t distribution) in E. S. Pearson and H. O. Hartley, eds. (1970), **Biometrika Tables for Statisticians** (3rd ed., Volume I). New York: Cambridge University Press. Reproduced with kind permission of **Biometrika** trustees.

Table A3 Power Curves for Student's *t* Distribution

Reprinted with permission of Addison–Wesley Longman Publishing Company, Inc. from Table 2.2 (Graphs of the operating characteristics of Student's t test) in D. B. Owen (©1962), **Handbook of Statistical Tables**. Reading, MA: Addison–Wesley, pp. 32–35.

Table A4 Table of the Chi-Square Distribution

Reprinted with permission from Table 8 (Percentage points of the χ^2 distribution) in E. S. Pearson and H. O. Hartley, eds. (1970), **Biometrika Tables for Statisticians** (3rd ed., Volume I). New York: Cambridge University Press. Reproduced with kind permission of **Biometrika** trustees.

Table A5 Table of Critical *T* Values for Wilcoxon's Signed Ranks and Matched-Pairs Signed-Ranks Tests

Material from Table II in F. Wilcoxon, S. K. Katti and R. A. Wilcox (1963), **Critical Values and Probability Levels for the Wilcoxon Rank Sum Test and the Wilcoxon Signed Rank Test**. Copyright © 1963, American Cyanamid Company, Lederle Laboratories Division. All rights reserved and reprinted with permission.

Table A6 Table of the Binomial Distribution, Individual Probabilities

Reprinted with permission of CRC Press, Boca Raton, Florida from W. H. Beyer (1968), **CRC Handbook of Tables for Probability and Statistics** (2nd ed.), Table III.1 (Individual terms, binomial distribution), pp. 182–193.

Table A7 Table of the Binomial Distribution, Cumulative Probabilities

Reprinted with permission of CRC Press, Boca Raton, Florida from W. H. Beyer (1968), **CRC Handbook of Tables for Probability and Statistics** (2nd ed.), Table III.2 (Cumulative terms, binomial distribution), pp. 194–205.

Table A8 Table of Critical Values for the Single-Sample Runs Test

Reprinted with permission of Institute of Mathematical Statistics, Hayward, CA from the following: Portions of **Table II** on pp. 83–87 from: F. S. Swed and C. Eisenhart (1943). Tables for testing randomness of grouping in a sequence of alternatives, **Annals**

of Mathematical Statistics, 14, 66–87.

Table A9 Table of the F_{max} Distribution
Reprinted with permission from Table 31 (Percentage points of the ratio s_{max}^2/s_{min}^2) in E. S. Pearson and H. O. Hartley, eds. (1970). **Biometrika Tables for Statisticians** (3rd ed., Volume 1). New York: Cambridge University Press. Reproduced with the kind permission of the **Biometrika** trustees.

Table A10 Table of the *F* Distribution
Reprinted with permission from Table 18 (Percentage points of the F-distribution (variance ratio)) in E. S. Pearson and H. O. Hartley (eds.) (1970), **Biometrika Tables for Statisticians** (3rd ed., Volume 1). New York: Cambridge University Press. Reproduced with the kind permission of the **Biometrika** trustees. Table reproduced with permission of CRC Press, Boca Raton, Florida from W. H. Beyer (1968), **CRC Handbook of Tables for Probability and Statistics** (2nd ed.), Table VI.1 (Percentage Points, F Distribution), pp. 304–310.

Table A11 Table of Critical Values for Mann-Whitney *U* Statistic
Reprinted with permission of Indiana University from D. Auble (1953), Extended Tables for the Mann–Whitney Statistic. **Bulletin of the Institute of Educational Research at Indiana University** Vol. 1, No. 2. Table reproduced with permission of CRC Press, Boca Raton, Florida from W. H. Beyer (1968), **CRC Handbook of Tables for Probability and Statistics** (2nd ed.), Table X.4 (Critical Values of U in the Wilcoxon (Mann–Whitney) Two-Sample Statistic), pp. 405–408.

Table A12 Table of Sandler's *A* Statistic
Reprinted with permission of British Psychological Society and Joseph Sandler from J. Sandler (1955), A test of the significance of difference between the means of correlated measures based on a simplification of Student's t. **British Journal of Psychology**, 46, pp. 225–226.

Table A13 Table of the Studentized Range Statistic
Reprinted with permission from Table 29 (Percentage points of the studentized range) in E. S. Pearson and H. O. Hartley, eds. (1970), **Biometrika Tables for Statisticians** (3rd ed., Volume 1). New York: Cambridge University Press. Reproduced with the kind permission of the **Biometrika** trustees.

Table A14 Table of Dunnett's Modified *t* Statistic for a Control Group Comparison
Two-tailed values: Reprinted with permission of the Biometric Society, Alexandria, VA from: C. W. Dunnett (1964), New tables for multiple comparisons with a control. **Biometrics**, 20, pp. 482–491.
One-tailed values: Reprinted with permission of the American Statistical Association, Alexandria, VA from: C. W. Dunnett (1955). A multiple comparison procedure for comparing several treatments with a control. **Journal of the American Statistical Association**, 50, pp. 1096–1121.

Table A15 Graphs of the Power Function for the Analysis of Variance
Reprinted with permission of **Biometrika** from E. S. Pearson and H. O. Hartley (1951), Charts of the power function for analysis of variance tests, derived from the non-central F distribution, **Biometrika**, 38, pp. 112–130.

Table A16 Table of Critical Values for Pearson r

Reprinted with permission from Table 13 (Percentage points for the distribution of the correlation coefficient, r, when $\rho = 0$) in E. S. Pearson and H. O. Hartley, eds. (1970), **Biometrika Tables for Statisticians** (3rd ed., Volume 1). New York: Cambridge University Press. Reproduced with the kind permission of the **Biometrika** trustees.

Table A17 Table of Fisher's z_r Transformation

Reprinted with permission from Table 14 (The z-transformation of the correlation coefficient, $z = \tanh^{-1} r$) in E. S. Pearson and H. O. Hartley, eds. (1970), **Biometrika Tables for Statisticians** (3rd ed., Volume 1). New York: Cambridge University Press. Reproduced with the kind permission of the **Biometrika** trustees.

Table A18 Table of Critical Values for Spearman's Rho

Reprinted with permission of the American Statistical Association, Alexandria, VA from: J. H. Zar (1972), Significance testing of the Spearman rank correlation coefficient. **Journal of the American Statistical Association**, 67, pp. 578–580 (Table 1, p. 579).

Table A19 Table of Critical Values for Kendall's Tau

Reprinted with permission of Blackwell Publishers and Statistica Neerlandica, from Table III in L. Kaarsemaker and A. van Wijngaarden (1953), Tables for use in rank correlation. **Statistica Neerlandica,** 7, pp. 41–54 (Copyright: The Netherlands Statistical Society (VVS)).

Table A20 Table of Critical Values for Kendall's Coefficient of Concordance

Reprinted with permission of Institute of Mathematical Statistics, Hayward, CA from the following: M. Friedman (1940), A comparison of alternative tests of significance for the problem of m rankings. **Annals of Mathematical Statistics**, 11, 86–92 (Table III, p. 91).

Table A1 Table of the Normal Distribution

x	$p(\mu$ to $z)$	$p(z$ to tail)	ordinate	x	$p(\mu$ to $z)$	$p(z$ to tail)	ordinate
.00	.0000	.5000	.3989	.45	.1736	.3264	.3605
.01	.0040	.4960	.3989	.46	.1772	.3228	.3589
.02	.0080	.4920	.3989	.47	.1808	.3192	.3572
.03	.0120	.4880	.3988	.48	.1844	.3156	.3555
.04	.0160	.4840	.3986	.49	.1879	.3121	.3538
.05	.0199	.4801	.3984	.50	.1915	.3085	.3521
.06	.0239	.4761	.3982	.51	.1950	.3050	.3503
.07	.0279	.4721	.3980	.52	.1985	.3015	.3485
.08	.0319	.4681	.3977	.53	.2019	.2981	.3467
.09	.0359	.4641	.3973	.54	.2054	.2946	.3448
.10	.0398	.4602	.3970	.55	.2088	.2912	.3429
.11	.0438	.4562	.3965	.56	.2123	.2877	.3410
.12	.0478	.4522	.3961	.57	.2157	.2843	.3391
.13	.0517	.4483	.3956	.58	.2190	.2810	.3372
.14	.0557	.4443	.3951	.59	.2224	.2776	.3352
.15	.0596	.4404	.3945	.60	.2257	.2743	.3332
.16	.0636	.4364	.3939	.61	.2291	.2709	.3312
.17	.0675	.4325	.3932	.62	.2324	.2676	.3292
.18	.0714	.4286	.3925	.63	.2357	.2643	.3271
.19	.0753	.4247	.3918	.64	.2389	.2611	.3251
.20	.0793	.4207	.3910	.65	.2422	.2578	.3230
.21	.0832	.4168	.3902	.66	.2454	.2546	.3209
.22	.0871	.4129	.3894	.67	.2486	.2514	.3187
.23	.0901	.4090	.3885	.68	.2517	.2483	.3166
.24	.0948	.4052	.3876	.69	.2549	.2451	.3144
.25	.0987	.4013	.3867	.70	.2580	.2420	.3123
.26	.1026	.397	.3857	.71	.2611	.2389	.3101
.27	.1064	.3936	.3847	.72	.2642	.2358	.3079
.28	.1103	.3897	.3836	.73	.2673	.2327	.3056
.29	.1141	.3859	.3825	.74	.2704	.2296	.3034
.30	.1179	.3821	.3814	.75	.2734	.2266	.3011
.31	.1217	.3783	.3802	.76	.2764	.2236	.2989
.32	.1255	.3745	.3790	.77	.2794	.2206	.2966
.33	.1293	.3707	.3778	.78	.2823	.2177	.2943
.34	.1331	.3669	.3765	.79	.2852	.2148	.2920
.35	.1368	.3632	.3752	.80	.2881	.2119	.2897
.36	.1406	.3594	.3739	.81	.2910	.2090	.2874
.37	.1443	.3557	.3725	.82	.2939	.2061	.2850
.38	.1480	.3520	.3712	.83	.2967	.2033	.2827
.39	.1517	.3483	.3697	.84	.2995	.2005	.2803
.40	.1554	.3446	.3683	.85	.3023	.1977	.2780
.41	.1591	.3409	.3668	.86	.3051	.1949	.2756
.42	.1628	.3372	.3653	.87	.3078	.1922	.2732
.43	.1664	.3336	.3637	.88	.3106	.1894	.2709
.44	.1700	.3300	.3621	.89	.3133	.1867	.2685

Table A1 Table of the Normal Distribution (continued)

x	$p(\mu$ to $z)$	$p(z$ to tail)	ordinate
.90	.3159	.1841	.2661
.91	.3186	.1814	.2637
.92	.3212	.1788	.2613
.93	.3238	.1762	.2589
.94	.3264	.1736	.2565
.95	.3289	.1711	.2541
.96	.3315	.1685	.2516
.97	.3340	.1660	.2492
.98	.3365	.1635	.2468
.99	.3389	.1611	.2444
1.00	.3413	.1587	.2420
1.01	.3438	.1562	.2396
1.02	.3461	.1539	.2371
1.03	.3485	.1515	.2347
1.04	.3508	.1492	.2323
1.05	.3531	.1469	.2299
1.06	.3554	.1446	.2275
1.07	.3577	.1423	.2251
1.08	.3599	.1401	.2227
1.09	.3621	.1379	.2203
1.10	.3643	.1357	.2179
1.11	.3665	.1335	.2155
1.12	.3686	.1314	.2131
1.13	.3708	.1292	.2107
1.14	.3729	.1271	.2083
1.15	.3749	.1251	.2059
1.16	.3770	.1230	.2036
1.17	.3790	.1210	.2012
1.18	.3810	.1190	.1989
1.19	.3830	.1170	.1965
1.20	.3849	.1151	.1942
1.21	.3869	.1131	.1919
1.22	.3888	.1112	.1895
1.23	.3907	.1093	.1872
1.24	.3925	.1075	.1849
1.25	.3944	.1056	.1826
1.26	.3962	.1038	.1804
1.27	.3980	.1020	.1781
1.28	.3997	.1003	.1758
1.29	.3015	.0985	.1736
1.30	.3032	.0968	.1714
1.31	.3049	.0951	.1691
1.32	.3066	.0934	.1669
1.33	.3082	.0918	.1447
1.34	.3099	.0901	.1626

x	$p(\mu$ to $z)$	$p(z$ to tail)	ordinate
1.35	.4115	.0885	.1604
1.36	.4131	.0869	.1582
1.37	.4147	.0853	.1561
1.38	.4162	.0838	.1539
1.39	.4177	.0823	.1518
1.40	.4192	.0808	.1497
1.41	.4207	.0793	.1476
1.42	.4222	.0778	.1456
1.43	.4236	.0764	.1435
1.44	.4251	.0749	.1415
1.45	.4265	.0735	.1394
1.46	.4279	.0721	.1374
1.47	.4292	.0708	.1354
1.48	.4306	.0694	.1334
1.49	.4319	.0681	.1315
1.50	.4332	.0668	.1295
1.51	.4345	.0655	.1276
1.52	.4357	.0643	.1257
1.53	.4370	.0630	.1238
1.54	.4382	.0618	.1219
1.55	.4394	.0606	.1200
1.56	.4406	.0594	.1182
1.57	.4418	.0582	.1163
1.58	.4429	.0571	.1145
1.59	.4441	.0559	.1127
1.60	.4452	.0548	.1109
1.61	.4463	.0537	.1092
1.62	.4474	.0526	.1074
1.63	.4484	.0516	.1057
1.64	.4495	.0505	.1040
1.65	.4505	.0495	.1023
1.66	.4515	.0485	.1006
1.67	.4525	.0475	.0989
1.68	.4535	.0465	.0973
1.69	.4545	.0455	.0957
1.70	.4554	.0446	.0940
1.71	.4564	.0436	.0925
1.72	.4573	.0427	.0909
1.73	.4582	.0418	.0893
1.74	.4591	.0409	.0878
1.75	.4599	.0401	.0863
1.76	.4608	.0392	.0848
1.77	.4616	.0384	.0833
1.78	.4625	.0375	.0818
1.79	.4633	.0367	.0804

Table A1 Table of the Normal Distribution (continued)

x	$p(\mu$ to $z)$	$p(z$ to tail)	ordinate
1.80	.4641	.0359	.0790
1.81	.4649	.0351	.0775
1.82	.4656	.0344	.0761
1.83	.4664	.0336	.0748
1.84	.4671	.0329	.0734
1.85	.4678	.0322	.0721
1.86	.4686	.0314	.0707
1.87	.4693	.0307	.0694
1.88	.4699	.0301	.0681
1.89	.4706	.0294	.0669
1.90	.4713	.0287	.0656
1.91	.4719	.0281	.0644
1.92	.4726	.0274	.0632
1.93	.4732	.0268	.0620
1.94	.4738	.0262	.0608
1.95	.4744	.0256	.0596
1.96	.4750	.0250	.0584
1.97	.4756	.0244	.0573
1.98	.4761	.0239	.0562
1.99	.4767	.0233	.0551
2.00	.4772	.0228	.0540
2.01	.4778	.0222	.0529
2.02	.4783	.0217	.0519
2.03	.4788	.0212	.0508
2.04	.4793	.0207	.0498
2.05	.4798	.0202	.0488
2.06	.4803	.0197	.0478
2.07	.4808	.0192	.0468
2.08	.4812	.0188	.0459
2.09	.4817	.0183	.0449
2.10	.4821	.0179	.0440
2.11	.4826	.0174	.0431
2.12	.4830	.0170	.0422
2.13	.4834	.0166	.0413
2.14	.4838	.0162	.0404
2.15	.4842	.0158	.0396
2.16	.4846	.0154	.0387
2.17	.4850	.0150	.0379
2.18	.4854	.0146	.0371
2.19	.4857	.0143	.0363
2.20	.4861	.0139	.0355
2.21	.4864	.0136	.0347
2.22	.4868	.0132	.0339
2.23	.4871	.0129	.0332
2.24	.4875	.0125	.0325

x	$p(\mu$ to $z)$	$p(z$ to tail)	ordinate
2.25	.4878	.0122	.0317
2.26	.4881	.0119	.0310
2.27	.4884	.0116	.0303
2.28	.4887	.0113	.0297
2.29	.4890	.0110	.0290
2.30	.4893	.0107	.0283
2.31	.4896	.0104	.0277
2.32	.4898	.0102	.0270
2.33	.4901	.0099	.0264
2.34	.4904	.0096	.0258
2.35	.4906	.0094	.0252
2.36	.4909	.0091	.0246
2.37	.4911	.0089	.0241
2.38	.4913	.0087	.0235
2.39	.4916	.0084	.0229
2.40	.4918	.0082	.0224
2.41	.4920	.0080	.0219
2.42	.4922	.0078	.0213
2.43	.4925	.0075	.0208
2.44	.4927	.0073	.0203
2.45	.4929	.0071	.0198
2.46	.4931	.0069	.0194
2.47	.4932	.0068	.0189
2.48	.4934	.0066	.0184
2.49	.4936	.0064	.0180
2.50	.4938	.0062	.0175
2.51	.4940	.0060	.0171
2.52	.4941	.0059	.0167
2.53	.4943	.0057	.0163
2.54	.4945	.0055	.0158
2.55	.4946	.0054	.0155
2.56	.4948	.0052	.0151
2.57	.4949	.0051	.0147
2.58	.4951	.0049	.0143
2.59	.4952	.0048	.0139
2.60	.4953	.0047	.0136
2.61	.4955	.0045	.0132
2.62	.4956	.0044	.0129
2.63	.4957	.0043	.0126
2.64	.4959	.0041	.0122
2.65	.4960	.0040	.0119
2.66	.4961	.0039	.0116
2.67	.4962	.0038	.0113
2.68	.4963	.0037	.0110
2.69	.4964	.0036	.0107

Table A1 Table of the Normal Distribution (continued)

x	$p(\mu$ to $z)$	$p(z$ to tail$)$	ordinate
2.70	.4965	.0035	.0104
2.71	.4966	.0034	.0101
2.72	.4967	.0033	.0099
2.73	.4968	.0032	.0096
2.74	.4969	.0031	.0093
2.75	.4970	.0030	.0091
2.76	.4971	.0029	.0088
2.77	.4972	.0028	.0086
2.78	.4973	.0027	.0084
2.79	.4974	.0026	.0081
2.80	.4974	.0026	.0079
2.81	.4975	.0025	.0077
2.82	.4976	.0024	.0075
2.83	.4977	.0023	.0073
2.84	.4977	.0023	.0071
2.85	.4978	.0022	.0069
2.86	.4979	.0021	.0067
2.87	.4979	.0021	.0065
2.88	.4980	.0020	.0063
2.89	.4981	.0019	.0061
2.90	.4981	.0019	.0060
2.91	.4982	.0018	.0058
2.92	.4982	.0018	.0056
2.93	.4983	.0017	.0055
2.94	.4984	.0016	.0053
2.95	.4984	.0016	.0051
2.96	.4985	.0015	.0050
2.97	.4985	.0015	.0048
2.98	.4986	.0014	.0047
2.99	.4986	.0014	.0046
3.00	.4987	.0013	.0044
3.01	.4987	.0013	.0043
3.02	.4987	.0013	.0042
3.03	.4988	.0012	.0040
3.04	.4988	.0012	.0039
3.05	.4989	.0011	.0038
3.06	.4989	.0011	.0037
3.07	.4989	.0011	.0036
3.08	.4990	.0010	.0035
3.09	.4990	.0010	.0034
3.10	.4990	.0010	.0033
3.11	.4991	.0009	.0032
3.12	.4991	.0009	.0031
3.13	.4991	.0009	.0030
3.14	.4992	.0008	.0029
3.15	.4992	.0008	.0028
3.16	.4992	.0008	.0027
3.17	.4992	.0008	.0026
3.18	.4993	.0007	.0025
3.19	.4993	.0007	.0025
3.20	.4993	.0007	.0024
3.21	.4993	.0007	.0023
3.22	.4994	.0006	.0022
3.23	.4994	.0006	.0022
3.24	.4994	.0006	.0021
3.25	.4994	.0006	.0020
3.26	.4994	.0006	.0020
3.27	.4995	.0005	.0019
3.28	.4995	.0005	.0018
3.29	.4995	.0005	.0018
3.30	.4995	.0005	.0017
3.31	.4995	.0005	.0017
3.32	.4995	.0005	.0016
3.33	.4996	.0004	.0016
3.34	.4996	.0004	.0015
3.35	.4996	.0004	.0015
3.36	.4996	.0004	.0014
3.37	.4996	.0004	.0014
3.38	.4996	.0004	.0013
3.39	.4997	.0003	.0013
3.40	.4997	.0003	.0012
3.41	.4997	.0003	.0012
3.42	.4997	.0003	.0012
3.43	.4997	.0003	.0011
3.44	.4997	.0003	.0011
3.45	.4997	.0003	.0010
3.46	.4997	.0003	.0010
3.47	.4997	.0003	.0010
3.48	.4997	.0003	.0009
3.49	.4998	.0002	.0009
3.50	.4998	.0002	.0009
3.51	.4998	.0002	.0008
3.52	.4998	.0002	.0008
3.53	.4998	.0002	.0008
3.54	.4998	.0002	.0008
3.55	.4998	.0002	.0007
3.56	.4998	.0002	.0007
3.57	.4998	.0002	.0007
3.58	.4998	.0002	.0007
3.59	.4998	.0002	.0006

Table A1 Table of the Normal Distribution (continued)

x	$p(\mu$ to $z)$	$p(z$ to tail)	ordinate
3.60	.4998	.0002	.0006
3.61	.4998	.0002	.0006
3.62	.4999	.0001	.0006
3.63	.4999	.0001	.0005
3.64	.4999	.0001	.0005
3.65	.4999	.0001	.0005
3.66	.4999	.0001	.0005
3.67	.4999	.0001	.0005
3.68	.4999	.0001	.0005
3.69	.4999	.0001	.0004
3.70	.4999	.0001	.0004
3.71	.4999	.0001	.0004
3.72	.4999	.0001	.0004
3.73	.4999	.0001	.0004
3.74	.4999	.0001	.0004
3.75	.4999	.0001	.0004
3.76	.4999	.0001	.0003
3.77	.4999	.0001	.0003
3.78	.4999	.0001	.0003
3.79	.4999	.0001	.0003

x	$p(\mu$ to $z)$	$p(z$ to tail)	ordinate
3.80	.4999	.0001	.0003
3.81	.4999	.0001	.0003
3.82	.4999	.0001	.0003
3.83	.4999	.0001	.0003
3.84	.4999	.0001	.0003
3.85	.4999	.0001	.0002
3.86	.4999	.0001	.0002
3.87	.4999	.0001	.0002
3.88	.4999	.0001	.0002
3.89	1.0000	.0000	.0002
3.90	1.0000	.0000	.0002
3.91	1.0000	.0000	.0002
3.92	1.0000	.0000	.0002
3.93	1.0000	.0000	.0002
3.94	1.0000	.0000	.0002
3.95	1.0000	.0000	.0002
3.96	1.0000	.0000	.0002
3.97	1.0000	.0000	.0002
3.98	1.0000	.0000	.0001
3.99	1.0000	.0000	.0001
4.00	1.0000	.0000	.0001

Table A2 Table of Student's *t* Distribution

Two-tailed	.80	.50	.20	.10	.05	.02	.01	.001
One-tailed	.40	.25	.10	.05	.025	.01	.005	.0005
df / *p*	.60	.75	.90	.95	.975	.99	.995	.9995
1	.325	1.000	3.078	6.314	12.706	31.821	63.657	636.619
2	.289	.816	1.886	2.920	4.303	6.965	9.925	31.598
3	.277	.765	1.638	2.353	3.182	4.541	5.841	12.924
4	.271	.741	1.533	2.132	2.776	3.747	4.604	8.610
5	.267	.727	1.476	2.015	2.571	3.365	4.032	6.869
6	.265	.718	1.440	1.943	2.447	3.143	3.707	5.959
7	.263	.711	1.415	1.895	2.365	2.998	3.499	5.408
8	.262	.706	1.397	1.860	2.306	2.896	3.355	5.041
9	.261	.703	1.383	1.833	2.262	2.821	3.250	4.781
10	.260	.700	1.372	1.812	2.228	2.764	3.169	4.587
11	.260	.697	1.363	1.796	2.201	2.718	3.106	4.437
12	.259	.695	1.356	1.782	2.179	2.681	3.055	4.318
13	.259	.694	1.350	1.771	2.160	2.650	3.012	4.221
14	.258	.692	1.345	1.761	2.145	2.624	2.977	4.140
15	.258	.691	1.341	1.753	2.131	2.602	2.947	4.073
16	.258	.690	1.337	1.746	2.120	2.583	2.921	4.015
17	.257	.689	1.333	1.740	2.110	2.567	2.898	3.965
18	.257	.688	1.330	1.734	2.101	2.552	2.878	3.922
19	.257	.688	1.328	1.729	2.093	2.539	2.861	3.883
20	.257	.687	1.325	1.725	2.086	2.528	2.845	3.850
21	.257	.686	1.323	1.721	2.080	2.518	2.831	3.819
22	.256	.686	1.321	1.717	2.074	2.508	2.819	3.792
23	.256	.685	1.319	1.714	2.069	2.500	2.807	3.767
24	.256	.685	1.318	1.711	2.064	2.492	2.797	3.745
25	.256	.684	1.316	1.708	2.060	2.485	2.787	3.725
26	.256	.684	1.315	1.706	2.056	2.479	2.779	3.707
27	.256	.684	1.314	1.703	2.052	2.473	2.771	3.690
28	.256	.683	1.313	1.701	2.048	2.467	2.763	3.674
29	.256	.683	1.311	1.699	2.045	2.462	2.756	3.659
30	.256	.683	1.310	1.697	2.042	2.457	2.750	3.646
40	.255	.681	1.303	1.684	2.021	2.423	2.704	3.551
60	.254	.679	1.296	1.671	2.000	2.390	2.660	3.460
120	.254	.677	1.289	1.658	1.980	2.358	2.617	3.373
∞	.253	.674	1.282	1.645	1.960	2.326	2.576	3.291

Table A3 Power Curves for Student's *t* Distribution

Table A3-A (Two-Tailed .01 and One-Tailed .005 Values)

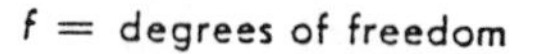

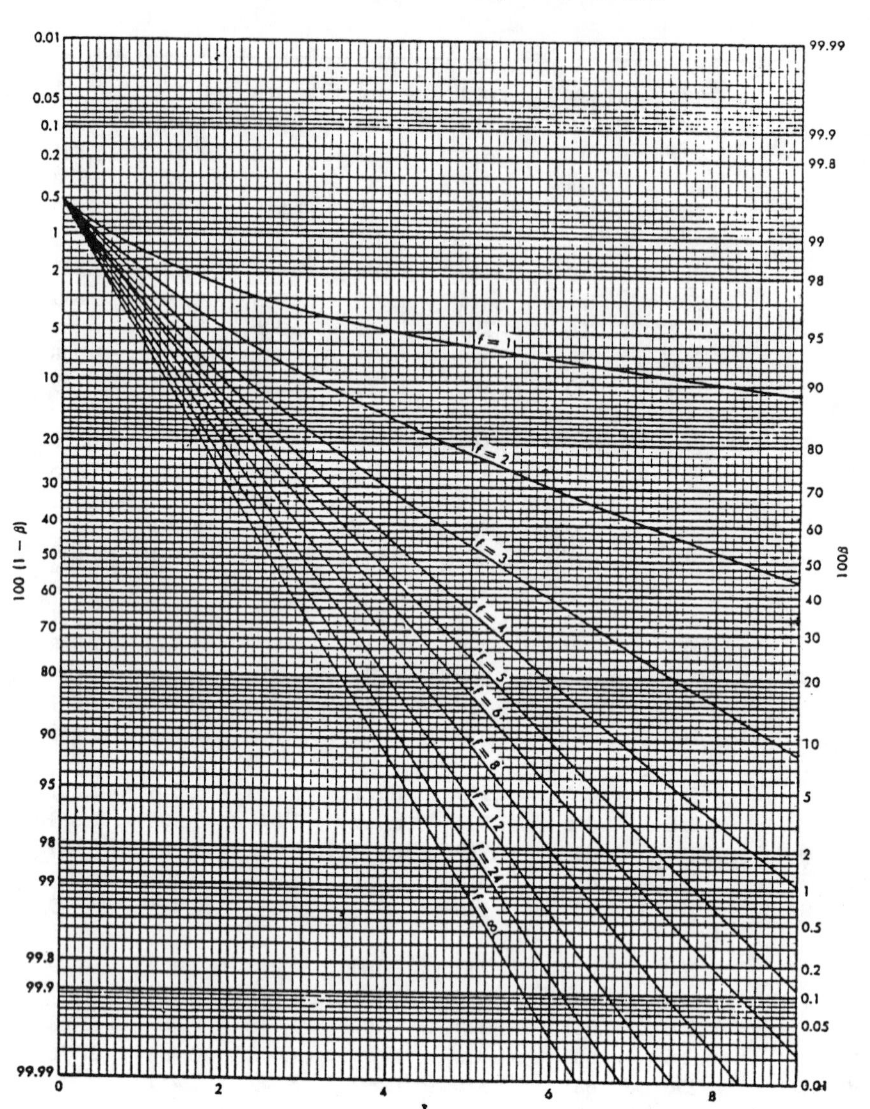

Table A3 Power Curves for Student's *t* Distribution (continued)

Table A3-B (Two-Tailed .02 and One-Tailed .01 Values)

Power

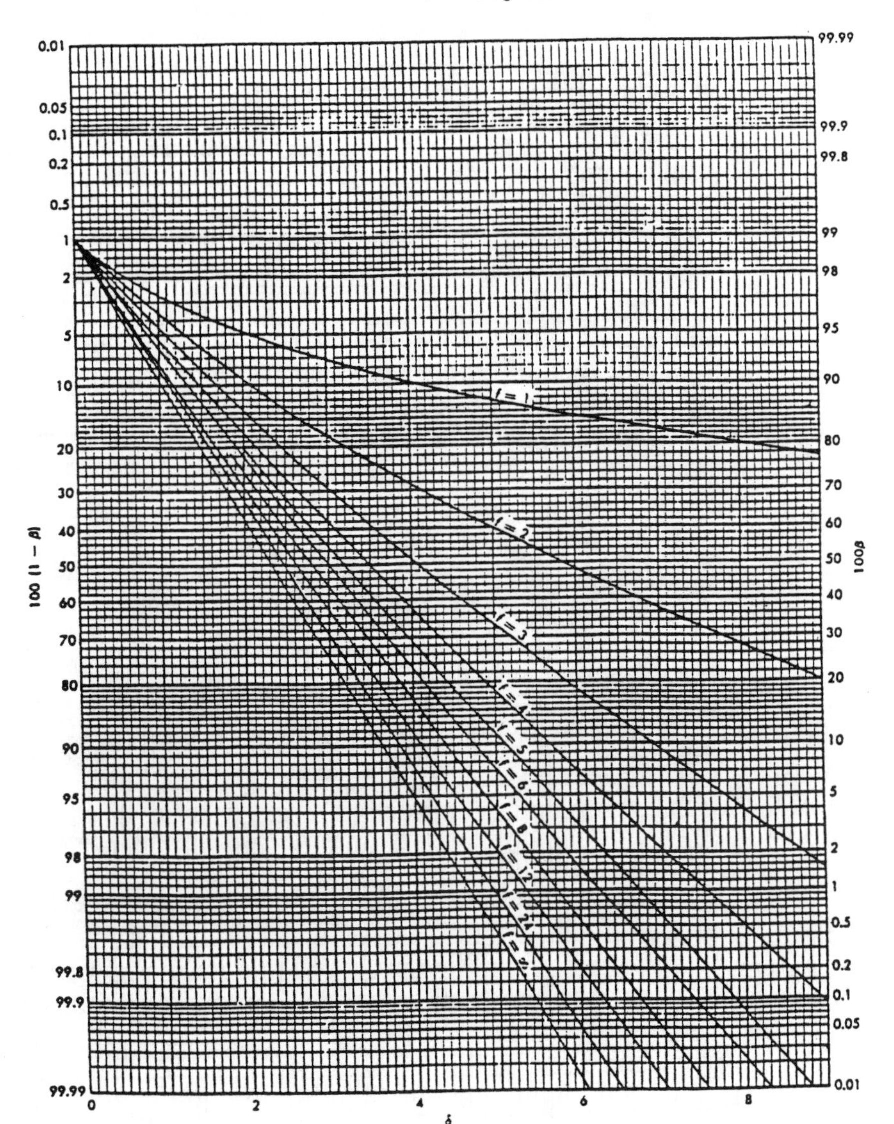

β

Table A3 Power Curves for Student's *t* Distribution (continued)

Table A3-C (Two-Tailed .05 and One-Tailed .025 Values)

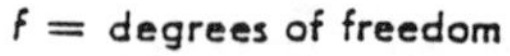

Power

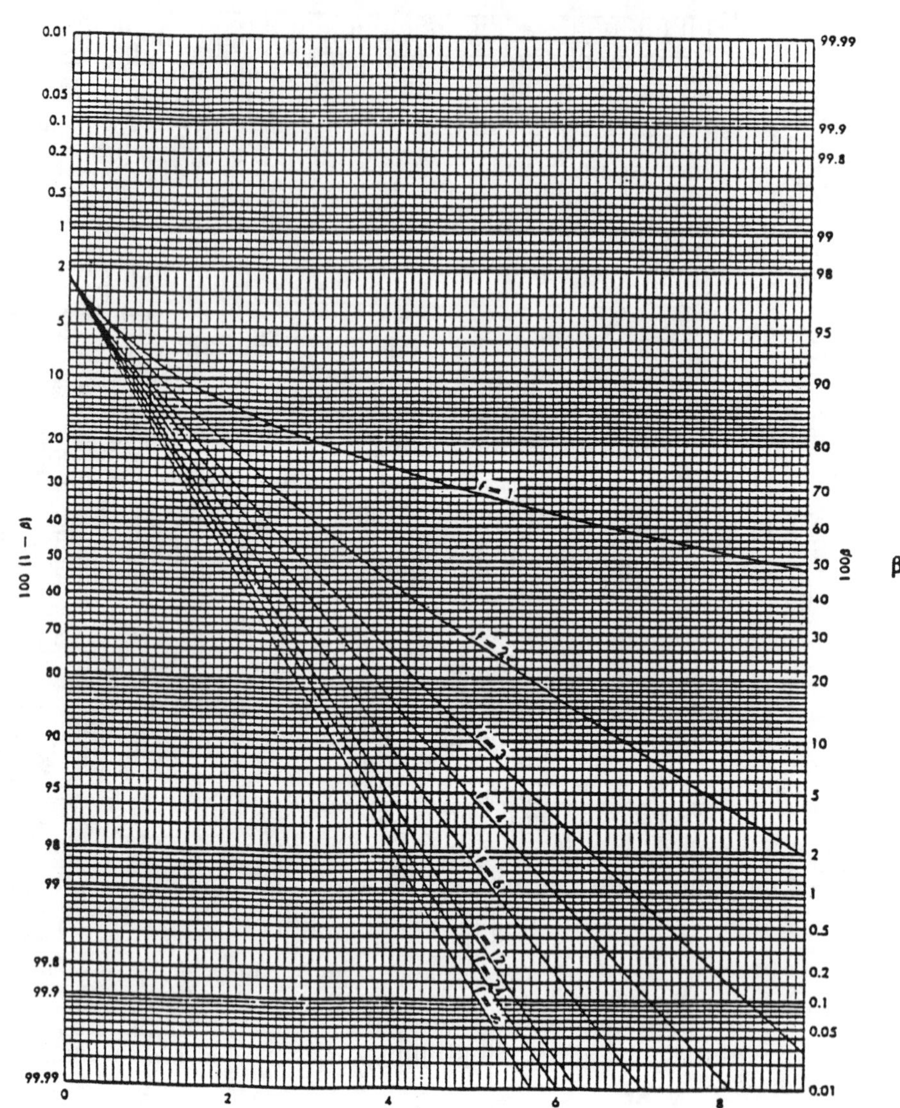

Table A3 Power Curves for Student's t Distribution (continued)

Table A3-D (Two-Tailed .10 and One-Tailed .05 Values)

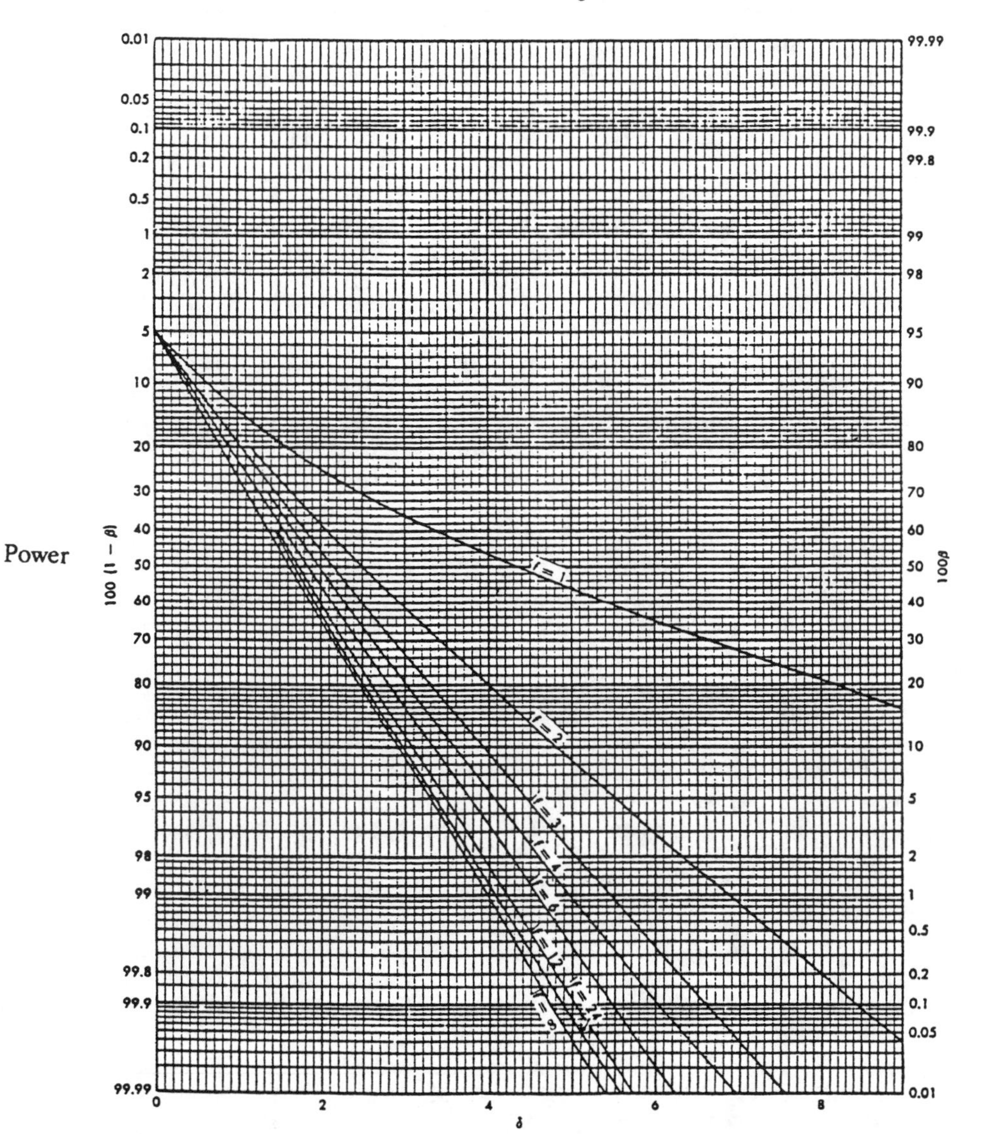

Table A4 Table of the Chi-Square Distribution

p / *df*	.005	.010	.025	.050	.100	.900	.950	.975	.990	.995	.999
1	.0393	.0157	.0982	.0393	.0158	2.71	3.84	5.02	6.63	7.88	10.83
2	.0100	.0201	.0506	.103	.211	4.61	5.99	7.38	9.21	10.60	13.82
3	.072	.115	.216	.352	.584	6.25	7.81	9.35	11.34	12.84	16.27
4	.0207	.297	.484	.711	1.064	7.78	9.49	11.14	13.28	14.86	18.47
5	.412	.554	.831	1.145	1.61	9.24	11.07	12.83	15.09	16.75	20.52
6	.676	.872	1.24	1.64	2.20	10.64	12.59	14.45	16.81	18.55	22.46
7	.989	1.24	1.69	2.17	2.83	12.02	14.07	16.01	18.48	20.28	24.32
8	1.34	1.65	2.18	2.73	3.49	13.36	15.51	17.53	20.09	21.96	26.13
9	1.73	2.09	2.70	3.33	4.17	14.68	16.92	19.02	21.67	23.59	27.88
10	2.16	2.56	3.25	3.94	4.87	15.99	18.31	20.48	23.21	25.19	29.59
11	2.60	3.05	3.82	4.57	5.58	17.28	19.68	21.92	24.72	26.76	31.26
12	3.07	3.57	4.40	5.23	6.30	18.55	21.03	23.34	26.22	28.30	32.91
13	3.57	4.11	5.01	5.89	7.04	19.81	22.36	24.74	27.69	29.82	34.53
14	4.07	4.66	5.63	6.57	7.79	21.06	23.68	26.12	29.14	31.32	36.12
15	4.60	5.23	6.26	7.26	8.55	22.31	25.00	27.49	30.58	32.80	37.70
16	5.14	5.81	6.91	7.96	9.31	23.54	26.30	28.85	32.00	34.27	39.25
17	5.70	6.41	7.56	8.67	10.09	24.77	27.59	30.19	33.41	35.72	40.79
18	6.26	7.01	8.23	9.39	10.86	25.99	28.87	31.53	34.81	37.16	42.31
19	6.84	7.63	8.91	10.12	11.65	27.20	30.14	32.85	36.19	38.58	43.82
20	7.43	8.26	8.59	10.85	12.44	28.41	31.41	34.17	37.57	40.00	43.32
21	8.03	8.90	10.28	11.59	13.24	29.62	32.67	35.48	38.93	41.40	46.80
22	8.64	9.54	10.98	12.34	14.04	30.81	33.92	36.78	40.29	42.80	48.27
23	9.26	10.20	11.69	13.09	14.85	32.01	35.17	38.08	41.64	44.18	49.73
24	9.89	10.86	12.40	13.85	15.66	33.20	36.42	39.36	42.98	45.56	51.18
25	10.52	11.52	13.12	14.61	16.47	34.38	37.65	40.65	44.31	46.93	52.62
26	11.16	12.20	13.84	15.38	17.29	35.56	38.89	41.92	45.64	48.29	54.05
27	11.81	12.88	14.57	16.15	18.11	36.74	40.11	43.19	46.96	49.64	55.48
28	12.46	13.56	15.31	16.93	18.94	37.92	41.34	44.46	48.28	50.99	56.89
29	13.21	14.26	16.05	17.71	19.77	39.09	42.56	45.72	49.59	52.34	58.30
30	13.79	14.95	16.79	18.49	20.60	40.26	43.77	46.98	50.89	53.67	59.70
40	20.71	22.16	24.43	26.51	29.05	51.80	55.76	59.34	63.69	66.77	73.40
50	27.99	29.71	32.36	34.76	37.69	63.17	67.50	71.42	76.15	79.49	86.66
60	35.53	37.48	40.48	43.19	46.46	74.40	79.08	83.30	88.38	91.95	99.61
70	43.28	45.44	48.76	51.74	55.33	85.53	90.53	95.02	100.43	104.22	112.32
80	51.17	53.54	57.15	60.39	64.28	96.58	101.88	106.63	112.33	116.32	124.84
90	59.20	61.75	65.65	69.13	73.29	107.56	113.15	118.14	124.12	128.30	137.21
100	67.33	70.06	74.22	77.93	82.36	118.50	124.34	129.56	135.81	140.17	149.45

Table A5 Table of Critical *T* Values for Wilcoxon's Signed-Ranks and Matched-Pairs Signed-Ranks Test

	One-tailed level of significance					One-tailed level of significance			
	.05	.025	.01	.005		.05	.025	.01	.005
	Two-tailed level of significance					Two-tailed level of significance			
	.10	.05	.02	.01		.10	.05	.02	.01
n					*n*				
5	0	–	–	–	28	130	116	101	91
6	2	0	–	–	29	140	126	110	100
7	3	2	0	–	30	151	137	120	109
8	5	3	1	0	31	163	147	130	118
9	8	5	3	1	32	175	159	140	128
10	10	8	5	3	33	187	170	151	138
11	13	10	7	5	34	200	182	162	148
12	17	13	9	7	35	213	195	173	159
13	21	17	12	9	36	227	208	185	171
14	25	21	15	12	37	241	221	198	182
15	30	25	19	15	38	256	235	211	194
16	35	29	23	19	39	271	249	224	207
17	41	34	27	23	40	286	264	238	220
18	47	40	32	27	41	302	279	252	233
19	53	46	37	32	42	319	294	266	247
20	60	52	43	37	43	336	310	281	261
21	67	58	49	42	44	353	327	296	276
22	75	65	55	48	45	371	343	312	291
23	83	73	62	54	46	389	361	328	307
24	91	81	69	61	47	407	378	345	322
25	100	89	76	68	48	426	396	362	339
26	110	98	84	75	49	446	415	379	355
27	119	107	92	83	50	466	434	397	373

Table A6 Table of the Binomial Distribution, Individual Probabilities

						π					
n	x	.05	.10	.15	.20	.25	.30	.35	.40	.45	.50
1	0	.9500	.9000	.8500	.8000	.7500	.7000	.6500	.6000	.5500	.5000
	1	.0500	.1000	.1500	.2000	.2500	.3000	.3500	.4000	.4500	.5000
2	0	.9025	.8100	.7225	.6400	.5625	.4900	.4225	.3600	.3025	.2500
	1	.0950	.1800	.2550	.3200	.3750	.4200	.4550	.4800	.4950	.5000
	2	.0025	.0100	.0225	.0400	.0625	.0900	.1225	.1600	.2025	.2500
3	0	.8574	.7290	.6141	.5120	.4219	.3430	.2746	.2160	.1664	.1250
	1	.1354	.2430	.3251	.3840	.4219	.4410	.4436	.4320	.4084	.3750
	2	.0071	.0270	.0574	.0960	.1406	.1890	.2389	.2880	.3341	.3750
	3	.0001	.0010	.0034	.0080	.0156	.0270	.0429	.0640	.0911	.1250
4	0	.8145	.6561	.5220	.4096	.3164	.2401	.1785	.1296	.0915	.0625
	1	.1715	.2916	.3685	.4096	.4219	.4116	.3845	.3456	.2995	.2500
	2	.0135	.0486	.0975	.1536	.2109	.2646	.3105	.3456	.3675	.3750
	3	.0005	.0036	.0115	.0256	.0469	.0756	.1115	.1536	.2005	.2500
	4	.0000	.0001	.0005	.0016	.0039	.0081	.0150	.0256	.0410	.0625
5	0	.7738	.5905	.4437	.3277	.2373	.1681	.1160	.0778	.0503	.0312
	1	.2036	.3280	.3915	.4096	.3955	.3602	.3124	.2592	.2059	.1562
	2	.0214	.0729	.1382	.2048	.2637	.3087	.3364	.3456	.3369	.3125
	3	.0011	.0081	.0244	.0512	.0879	.1323	.1811	.2304	.2757	.3125
	4	.0000	.0004	.0022	.0064	.0146	.0284	.0488	.0768	.1128	.1562
	5	.0000	.0000	.0001	.0003	.0010	.0024	.0053	.0102	.0185	.0312
6	0	.7351	.5314	.3771	.2621	.1780	.1176	.0754	.0467	.0277	.0156
	1	.2321	.3543	.3993	.3932	.3560	.3025	.2437	.1866	.1359	.0938
	2	.0305	.0984	.1762	.2458	.2966	.3241	.3280	.3110	.2780	.2344
	3	.0021	.0146	.0415	.0819	.1318	.1852	.2355	.2765	.3032	.3125
	4	.0001	.0012	.0055	.0154	.0330	.0595	.0951	.1382	.1861	.2344
	5	.0000	.0001	.0004	.0015	.0044	.0102	.0205	.0369	.0609	.0938
	6	.0000	.0000	.0000	.0001	.0002	.0007	.0018	.0041	.0083	.0156
7	0	.6983	.4783	.3206	.2097	.1335	.0824	.0490	.0280	.0152	.0078
	1	.2573	.3720	.3960	.3670	.3115	.2471	.1848	.1306	.0872	.0547
	2	.0406	.1240	.2097	.2753	.3115	.3177	.2985	.2613	.2140	.1641
	3	.0036	.0230	.0617	.1147	.1730	.2269	.2679	.2903	.2918	.2734
	4	.0002	.0026	.0109	.0287	.0577	.0972	.1442	.1935	.2388	.2734
	5	.0000	.0002	.0012	.0043	.0115	.0250	.0466	.0774	.1172	.1641
	6	.0000	.0000	.0001	.0004	.0013	.0036	.0084	.0172	.0320	.0547
	7	.0000	.0000	.0000	.0000	.0001	.0002	.0006	.0016	.0037	.0078
8	0	.6634	.4305	.2725	.1678	.1001	.0576	.0319	.0168	.0084	.0039
	1	.2793	.3826	.3847	.3355	.2670	.1977	.1373	.0896	.0548	.0312
	2	.0515	.1488	.2376	.2936	.3115	.2965	.2587	.2090	.1569	.1094
	3	.0054	.0331	.0839	.1468	.2076	.2541	.2786	.2787	.2568	.2188
	4	.0004	.0046	.0185	.0459	.0865	.1361	.1875	.2322	.2627	.2734
	5	.0000	.0004	.0026	.0092	.0231	.0467	.0808	.1239	.1719	.2188
	6	.0000	.0000	.0002	.0011	.0038	.0100	.0217	.0413	.0703	.1094
	7	.0000	.0000	.0000	.0001	.0004	.0012	.0033	.0079	.0164	.0312
	8	.0000	.0000	.0000	.0000	.0000	.0001	.0002	.0007	.0017	.0039

Table A6 Table of the Binomial Distribution, Individual Probabilities (continued)

n	x	π .05	.10	.15	.20	.25	.30	.35	.40	.45	.50
9	0	.6302	.3874	.2316	.1342	.0751	.0404	.0207	.0101	.0046	.0020
	1	.2985	.3874	.3679	.3020	.2253	.1556	.1004	.0605	.0339	.0176
	2	.0629	.1722	.2597	.3020	.3003	.2668	.2162	.1612	.1110	.0703
	3	.0077	.0446	.1069	.1762	.2336	.2668	.2716	.2508	.2119	.1641
	4	.0006	.0074	.0283	.0661	.1168	.1715	.2194	.2508	.2600	.2461
	5	.0000	.0008	.0050	.0165	.0389	.0735	.1181	.1672	.2128	.2461
	6	.0000	.0001	.0006	.0028	.0087	.0210	.0424	.0743	.1160	.1641
	7	.0000	.0000	.0000	.0003	.0012	.0039	.0098	.0212	.0407	.0703
	8	.0000	.0000	.0000	.0000	.0001	.0004	.0013	.0035	.0083	.0176
	9	.0000	.0000	.0000	.0000	.0000	.0000	.0001	.0003	.0008	.0020
10	0	.5987	.3487	.1969	.1074	.0563	.0282	.0135	.0060	.0025	.0010
	1	.3151	.3874	.3474	.2684	.1877	.1211	.0725	.0403	.0207	.0098
	2	.0746	.1937	.2759	.3020	.2816	.2335	.1757	.1209	.0763	.0439
	3	.0105	.0574	.1298	.2013	.2503	.2668	.2522	.2150	.1665	.1172
	4	.0010	.0112	.0401	.0881	.1460	.2001	.2377	.2508	.2384	.2051
	5	.0001	.0015	.0085	.0264	.0584	.1029	.1536	.2007	.2340	.2461
	6	.0000	.0001	.0012	.0055	.0162	.0368	.0689	.1115	.1596	.2051
	7	.0000	.0000	.0001	.0008	.0031	.0090	.0212	.0425	.0746	.1172
	8	.0000	.0000	.0000	.0001	.0004	.0014	.0043	.0106	.0229	.0439
	9	.0000	.0000	.0000	.0000	.0000	.0001	.0005	.0016	.0042	.0098
	10	.0000	.0000	.0000	.0000	.0000	.0000	.0000	.0001	.0003	.0010
11	0	.5688	.3138	.1673	.0859	.0422	.0198	.0088	.0036	.0014	.0004
	1	.3293	.3835	.3248	.2362	.1549	.0932	.0518	.0266	.0125	.0055
	2	.0867	.2131	.2866	.2953	.2581	.1998	.1395	.0887	.0513	.0269
	3	.0137	.0710	.1517	.2215	.2581	.2568	.2254	.1774	.1259	.0806
	4	.0014	.0158	.0536	.1107	.1721	.2201	.2428	.2365	.2060	.1611
	5	.0001	.0025	.0132	.0388	.0803	.1321	.1830	.2207	.2360	.2256
	6	.0000	.0003	.0023	.0097	.0268	.0566	.0985	.1471	.1931	.2256
	7	.0000	.0000	.0003	.0017	.0064	.0173	.0379	.0701	.1128	.1611
	8	.0000	.0000	.0000	.0002	.0011	.0037	.0102	.0234	.0462	.0806
	9	.0000	.0000	.0000	.0000	.0001	.0005	.0018	.0052	.0126	.0269
	10	.0000	.0000	.0000	.0000	.0000	.0000	.0002	.0007	.0021	.0054
	11	.0000	.0000	.0000	.0000	.0000	.0000	.0000	.0000	.0002	.0005
12	0	.5404	.2824	.1422	.0687	.0317	.0138	.0057	.0022	.0008	.0002
	1	.3413	.3766	.3012	.2062	.1267	.0712	.0368	.0174	.0075	.0029
	2	.0988	.2301	.2924	.2835	.2323	.1678	.1088	.0639	.0339	.0161
	3	.0173	.0852	.1720	.2362	.2581	.2397	.1954	.1419	.0923	.0537
	4	.0021	.0213	.0683	.1329	.1936	.2311	.2367	.2128	.1700	.1208
	5	.0002	.0038	.0193	.0532	.1032	.1585	.2039	.2270	.2225	.1934
	6	.0000	.0005	.0040	.0155	.0401	.0792	.1281	.1766	.2124	.2256
	7	.0000	.0000	.0006	.0033	.0115	.0291	.0591	.1009	.1489	.1934
	8	.0000	.0000	.0001	.0005	.0024	.0078	.0199	.0420	.0762	.1208
	9	.0000	.0000	.0000	.0001	.0004	.0015	.0048	.0125	.0277	.0537
	10	.0000	.0000	.0000	.0000	.0000	.0002	.0008	.0025	.0068	.0161
	11	.0000	.0000	.0000	.0000	.0000	.0000	.0001	.0003	.0010	.0029
	12	.0000	.0000	.0000	.0000	.0000	.0000	.0000	.0000	.0001	.0002

Table A6 Table of the Binomial Distribution, Individual Probabilities (continued)

n	x	.05	.10	.15	.20	π .25	.30	.35	.40	.45	.50
13	0	.5133	.2542	.1209	.0550	.0238	.0097	.0037	.0013	.0004	.0001
	1	.3512	.3672	.2774	.1787	.1029	.0540	.0259	.0113	.0045	.0016
	2	.1109	.2448	.2937	.2680	.2059	.1388	.0836	.0453	.0220	.0095
	3	.0214	.0997	.1900	.2457	.2517	.2181	.1651	.1107	.0660	.0349
	4	.0028	.0277	.0838	.1535	.2097	.2337	.2222	.1845	.1350	.0873
	5	.0003	.0055	.0266	.0691	.1258	.1803	.2154	.2214	.1989	.1571
	6	.0000	.0008	.0063	.0230	.0559	.1030	.1546	.1968	.2169	.2095
	7	.0000	.0001	.0011	.0058	.0186	.0442	.0833	.1312	.1775	.2095
	8	.0000	.0000	.0001	.0011	.0047	.0142	.0336	.0656	.1089	.1571
	9	.0000	.0000	.0000	.0001	.0009	.0034	.0101	.0243	.0495	.0873
	10	.0000	.0000	.0000	.0000	.0001	.0006	.0022	.0065	.0162	.0349
	11	.0000	.0000	.0000	.0000	.0000	.0001	.0003	.0012	.0036	.0095
	12	.0000	.0000	.0000	.0000	.0000	.0000	.0000	.0001	.0005	.0016
	13	.0000	.0000	.0000	.0000	.0000	.0000	.0000	.0000	.0000	.0001
14	0	.4877	.2288	.1028	.0440	.0178	.0068	.0024	.0008	.0002	.0001
	1	.3593	.3559	.2539	.1539	.0832	.0407	.0181	.0073	.0027	.0009
	2	.1229	.2570	.2912	.2501	.1802	.1134	.0634	.0317	.0141	.0056
	3	.0259	.1142	.2056	.2501	.2402	.1943	.1366	.0845	.0462	.0222
	4	.0037	.0349	.0998	.1720	.2202	.2290	.2022	.1549	.1040	.0611
	5	.0004	.0078	.0352	.0860	.1468	.1963	.2178	.2066	.1701	.1222
	6	.0000	.0013	.0093	.0322	.0734	.1262	.1759	.2066	.2088	.1833
	7	.0000	.0002	.0019	.0092	.0280	.0618	.1082	.1574	.1952	.2095
	8	.0000	.0000	.0003	.0020	.0082	.0232	.0510	.0918	.1398	.1833
	9	.0000	.0000	.0000	.0003	.0018	.0066	.0183	.0408	.0762	.1222
	10	.0000	.0000	.0000	.0000	.0003	.0014	.0049	.0136	.0312	.0611
	11	.0000	.0000	.0000	.0000	.0000	.0002	.0010	.0033	.0093	.0222
	12	.0000	.0000	.0000	.0000	.0000	.0000	.0001	.0005	.0019	.0056
	13	.0000	.0000	.0000	.0000	.0000	.0000	.0000	.0001	.0002	.0009
	14	.0000	.0000	.0000	.0000	.0000	.0000	.0000	.0000	.0000	.0001
15	0	.4633	.2059	.0874	.0352	.0134	.0047	.0016	.0005	.0001	.0000
	1	.3658	.3432	.2312	.1319	.0668	.0305	.0126	.0047	.0016	.0005
	2	.1348	.2669	.2856	.2309	.1559	.0916	.0476	.0219	.0090	.0032
	3	.0307	.1285	.2184	.2501	.2252	.1700	.1110	.0634	.0318	.0139
	4	.0049	.0428	.1156	.1876	.2252	.2186	.1792	.1268	.0780	.0417
	5	.0006	.0105	.0449	.1032	.1651	.2061	.2123	.1859	.1404	.0916
	6	.0000	.0019	.0132	.0430	.0917	.1472	.1906	.2066	.1914	.1527
	7	.0000	.0003	.0030	.0138	.0393	.0811	.1319	.1771	.2013	.1964
	8	.0000	.0000	.0005	.0035	.0131	.0348	.0710	.1181	.1647	.1964
	9	.0000	.0000	.0001	.0007	.0034	.0116	.0298	.0612	.1048	.1527
	10	.0000	.0000	.0000	.0001	.0007	.0030	.0096	.0245	.0515	.0916
	11	.0000	.0000	.0000	.0000	.0001	.0006	.0024	.0074	.0191	.0417
	12	.0000	.0000	.0000	.0000	.0000	.0001	.0004	.0016	.0052	.0139
	13	.0000	.0000	.0000	.0000	.0000	.0000	.0001	.0003	.0010	.0032
	14	.0000	.0000	.0000	.0000	.0000	.0000	.0000	.0000	.0001	.0005
	15	.0000	.0000	.0000	.0000	.0000	.0000	.0000	.0000	.0000	.0000

Table A7 Table of the Binomial Distribution, Cumulative Probabilities

n	x'	π .05	.10	.15	.20	.25	.30	.35	.40	.45	.50
2	1	.0975	.1900	.2775	.3600	.4375	.5100	.5775	.6400	.6975	.7500
	2	.0025	.0100	.0225	.0400	.0625	.0900	.1225	.1600	.2025	.2500
3	1	.1426	.2710	.3859	.4880	.5781	.6570	.7254	.7840	.8336	.8750
	2	.0072	.0280	.0608	.1040	.1562	.2160	.2818	.3520	.4252	.5000
	3	.0001	.0010	.0034	.0080	.0156	.0270	.0429	.0640	.0911	.1250
4	1	.1855	.3439	.4780	.5904	.6836	.7599	.8215	.8704	.9085	.9375
	2	.0140	.0523	.1095	.1808	.2617	.3483	.4370	.5248	.6090	.6875
	3	.0005	.0037	.0120	.0272	.0508	.0837	.1265	.1792	.2415	.3125
	4	.0000	.0001	.0005	.0016	.0039	.0081	.0150	.0256	.0410	.0625
5	1	.2262	.4095	.5563	.6723	.7627	.8319	.8840	.9222	.9497	.9688
	2	.0226	.0815	.1648	.2627	.3672	.4718	.5716	.6630	.7438	.8125
	3	.0012	.0086	.0266	.0579	.1035	.1631	.2352	.3174	.4069	.5000
	4	.0000	.0005	.0022	.0067	.0156	.0308	.0540	.0870	.1312	.1875
	5	.0000	.0000	.0001	.0003	.0010	.0024	.0053	.0102	.0185	.0312
6	1	.2649	.4686	.6229	.7379	.8220	.8824	.9246	.9533	.9723	.9844
	2	.0328	.1143	.2235	.3447	.4661	.5798	.6809	.7667	.8364	.8906
	3	.0022	.0158	.0473	.0989	.1694	.2557	.3529	.4557	.5585	.6562
	4	.0001	.0013	.0059	.0170	.0376	.0705	.1174	.1792	.2553	.3438
	5	.0000	.0001	.0004	.0016	.0046	.0109	.0223	.0410	.0692	.1094
	6	.0000	.0000	.0000	.0001	.0002	.0007	.0018	.0041	.0083	.0156
7	1	.3017	.5217	.6794	.7903	.8665	.9176	.9510	.9720	.9848	.9922
	2	.0444	.1497	.2834	.4233	.5551	.6706	.7662	.8414	.8976	.9375
	3	.0038	.0257	.0738	.1480	.2436	.3529	.4677	.5801	.6836	.7734
	4	.0002	.0027	.0121	.0333	.0706	.1260	.1998	.2898	.3917	.5000
	5	.0000	.0002	.0012	.0047	.0129	.0288	.0556	.0963	.1529	.2266
	6	.0000	.0000	.0001	.0004	.0013	.0038	.0090	.0188	.0357	.0625
	7	.0000	.0000	.0000	.0000	.0001	.0002	.0006	.0016	.0037	.0078
8	1	.3366	.5695	.7275	.8322	.8999	.9424	.9681	.9832	.9916	.9961
	2	.0572	.1869	.3428	.4967	.6329	.7447	.8309	.8936	.9368	.9648
	3	.0058	.0381	.1052	.2031	.3215	.4482	.5722	.6846	.7799	.8555
	4	.0004	.0050	.0214	.0563	.1138	.1941	.2936	.4059	.5230	.6367
	5	.0000	.0004	.0029	.0104	.0273	.0580	.1061	.1737	.2604	.3633
	6	.0000	.0000	.0002	.0012	.0042	.0113	.0253	.0498	.0885	.1445
	7	.0000	.0000	.0000	.0001	.0004	.0013	.0036	.0085	.0181	.0352
	8	.0000	.0000	.0000	.0000	.0000	.0001	.0002	.0007	.0017	.0039
9	1	.3698	.6126	.7684	.8658	.9249	.9596	.9793	.9899	.9954	.9980
	2	.0712	.2252	.4005	.5638	.6997	.8040	.8789	.9295	.9615	.9805
	3	.0084	.0530	.1409	.2618	.3993	.5372	.6627	.7682	.8505	.9102
	4	.0006	.0083	.0339	.0856	.1657	.2703	.3911	.5174	.6386	.7461
	5	.0000	.0009	.0056	.0196	.0489	.0988	.1717	.2666	.3786	.5000
	6	.0000	.0001	.0006	.0031	.0100	.0253	.0536	.0994	.1658	.2539
	7	.0000	.0000	.0000	.0003	.0013	.0043	.0112	.0250	.0498	.0898
	8	.0000	.0000	.0000	.0000	.0001	.0004	.0014	.0038	.0091	.0195
	9	.0000	.0000	.0000	.0000	.0000	.0000	.0001	.0003	.0008	.0020

Table A7 Table of the Binomial Distribution, Cumulative Probabilities (continued)

						π					
n	x'	.05	.10	.15	.20	.25	.30	.35	.40	.45	.50
10	1	.4013	.6513	.8031	.8926	.9437	.9718	.9865	.9940	.9975	.9990
	2	.0861	.2639	.4557	.6242	.7560	.8507	.9140	.9536	.9767	.9893
	3	.0115	.0702	.1798	.3222	.4744	.6172	.7384	.8327	.9004	.9453
	4	.0010	.0128	.0500	.1209	.2241	.3504	.4862	.6177	.7340	.8281
	5	.0001	.0016	.0099	.0328	.0781	.1503	.2485	.3669	.4956	.6230
	6	.0000	.0001	.0014	.0064	.0197	.0473	.0949	.1662	.2616	.3770
	7	.0000	.0000	.0001	.0009	.0035	.0106	.0260	.0548	.1020	.1719
	8	.0000	.0000	.0000	.0001	.0004	.0016	.0048	.0123	.0274	.0547
	9	.0000	.0000	.0000	.0000	.0000	.0001	.0005	.0017	.0045	.0107
	10	.0000	.0000	.0000	.0000	.0000	.0000	.0000	.0001	.0003	.0010
11	1	.4312	.6862	.8327	.9141	.9578	.9802	.9912	.9964	.9986	.9995
	2	.1019	.3026	.5078	.6779	.8029	.8870	.9394	.9698	.9861	.9941
	3	.0152	.0896	.2212	.3826	.5448	.6873	.7999	.8811	.9348	.9673
	4	.0016	.0185	.0694	.1611	.2867	.4304	.5744	.7037	.8089	.8867
	5	.0001	.0028	.0159	.0504	.1146	.2103	.3317	.4672	.6029	.7256
	6	.0000	.0003	.0027	.0117	.0343	.0782	.1487	.2465	.3669	.5000
	7	.0000	.0000	.0003	.0020	.0076	.0216	.0501	.0994	.1738	.2744
	8	.0000	.0000	.0000	.0002	.0012	.0043	.0122	.0293	.0610	.1133
	9	.0000	.0000	.0000	.0000	.0001	.0006	.0020	.0059	.0148	.0327
	10	.0000	.0000	.0000	.0000	.0000	.0000	.0002	.0007	.0022	.0059
	11	.0000	.0000	.0000	.0000	.0000	.0000	.0000	.0000	.0002	.0005
12	1	.4596	.7176	.8578	.9313	.9683	.9862	.9943	.9978	.9992	.9998
	2	.1184	.3410	.5565	.7251	.8416	.9150	.9576	.9804	.9917	.9968
	3	.0196	.1109	.2642	.4417	.6093	.7472	.8487	.9166	.9579	.9807
	4	.0022	.0256	.0922	.2054	.3512	.5075	.6533	.7747	.8655	.9270
	5	.0002	.0043	.0239	.0726	.1576	.2763	.4167	.5618	.6956	.8062
	6	.0000	.0005	.0046	.0194	.0544	.1178	.2127	.3348	.4731	.6128
	7	.0000	.0001	.0007	.0039	.0143	.0386	.0846	.1582	.2607	.3872
	8	.0000	.0000	.0001	.0006	.0028	.0095	.0255	.0573	.1117	.1938
	9	.0000	.0000	.0000	.0001	.0004	.0017	.0056	.0153	.0356	.0730
	10	.0000	.0000	.0000	.0000	.0000	.0002	.0008	.0028	.0079	.0193
	11	.0000	.0000	.0000	.0000	.0000	.0000	.0001	.0003	.0011	.0032
	12	.0000	.0000	.0000	.0000	.0000	.0000	.0000	.0000	.0001	.0002
13	1	.4867	.7458	.8791	.9450	.9762	.9903	.9963	.9987	.9996	.9999
	2	.1354	.3787	.6017	.7664	.8733	.9363	.9704	.9874	.9951	.9983
	3	.0245	.1339	.2704	.4983	.6674	.7975	.8868	.9421	.9731	.9888
	4	.0031	.0342	.0967	.2527	.4157	.5794	.7217	.8314	.9071	.9539
	5	.0003	.0065	.0260	.0991	.2060	.3457	.4995	.6470	.7721	.8666
	6	.0000	.0009	.0053	.0300	.0802	.1654	.2841	.4256	.5732	.7095
	7	.0000	.0001	.0013	.0070	.0243	.0624	.1295	.2288	.3563	.5000
	8	.0000	.0000	.0002	.0012	.0056	.0182	.0462	.0977	.1788	.2905
	9	.0000	.0000	.0000	.0002	.0010	.0040	.0126	.0321	.0698	.1334
	10	.0000	.0000	.0000	.0000	.0001	.0007	.0025	.0078	.0203	.0461
	11	.0000	.0000	.0000	.0000	.0000	.0001	.0003	.0013	.0041	.0112
	12	.0000	.0000	.0000	.0000	.0000	.0000	.0000	.0001	.0005	.0017
	13	.0000	.0000	.0000	.0000	.0000	.0000	.0000	.0000	.0000	.0001

Table A7 Table of the Binomial Distribution, Cumulative Probabilities (continued)

n	x'	.05	.10	.15	.20	π .25	.30	.35	.40	.45	.50
14	1	.5123	.7712	.8972	.9560	.9822	.9932	.9976	.9992	.9998	.9999
	2	.1530	.4154	.6433	.8021	.8990	.9525	.9795	.9919	.9971	.9991
	3	.0301	.1584	.3521	.5519	.7189	.8392	.9161	.9602	.9830	.9935
	4	.0042	.0441	.1465	.3018	.4787	.6448	.7795	.8757	.9368	.9713
	5	.0004	.0092	.0467	.1298	.2585	.4158	.5773	.7207	.8328	.9102
	6	.0000	.0015	.0115	.0439	.1117	.2195	.3595	.5141	.6627	.7880
	7	.0000	.0002	.0022	.0116	.0383	.0933	.1836	.3075	.4539	.6047
	8	.0000	.0000	.0003	.0024	.0103	.0315	.0753	.1501	.2586	.3953
	9	.0000	.0000	.0000	.0004	.0022	.0083	.0243	.0583	.1189	.2120
	10	.0000	.0000	.0000	.0000	.0003	.0017	.0060	.0175	.0426	.0898
	11	.0000	.0000	.0000	.0000	.0000	.0002	.0011	.0039	.0114	.0287
	12	.0000	.0000	.0000	.0000	.0000	.0000	.0001	.0006	.0022	.0065
	13	.0000	.0000	.0000	.0000	.0000	.0000	.0000	.0001	.0003	.0009
	14	.0000	.0000	.0000	.0000	.0000	.0000	.0000	.0000	.0000	.0001
15	1	.5367	.7941	.9126	.9648	.9866	.9953	.9984	.9995	.9999	1.0000
	2	.1710	.4510	.6814	.8329	.9198	.9647	.9858	.9948	.9983	.9995
	3	.0362	.1841	.3958	.6020	.7639	.8732	.9383	.9729	.9893	.9963
	4	.0055	.0556	.1773	.3518	.5387	.7031	.8273	.9095	.9576	.9824
	5	.0006	.0127	.0617	.1642	.3135	.4845	.6481	.7827	.8796	.9408
	6	.0001	.0022	.0168	.0611	.1484	.2784	.4357	.5968	.7392	.8491
	7	.0000	.0003	.0036	.0181	.0566	.1311	.2452	.3902	.5478	.6964
	8	.0000	.0000	.0006	.0042	.0173	.0500	.1132	.2131	.3465	.5000
	9	.0000	.0000	.0001	.0008	.0042	.0152	.0422	.0950	.1818	.3036
	10	.0000	.0000	.0000	.0001	.0008	.0037	.0124	.0338	.0769	.1509
	11	.0000	.0000	.0000	.0000	.0001	.0007	.0028	.0093	.0255	.0592
	12	.0000	.0000	.0000	.0000	.0000	.0001	.0005	.0019	.0063	.0176
	13	.0000	.0000	.0000	.0000	.0000	.0000	.0001	.0003	.0011	.0037
	14	.0000	.0000	.0000	.0000	.0000	.0000	.0000	.0000	.0001	.0005
	15	.0000	.0000	.0000	.0000	.0000	.0000	.0000	.0000	.0000	.0000

Table A8 Table of Critical Values for the Single-Sample Runs Test

Numbers listed are tabled critical two-tailed .05 and one-tailed .025 values.

n_2 \ n_1	2	3	4	5	6	7	8	9	10	11	12	13	14	15	16	17	18	19	20
2											2	2	2	2	2	2	2	2	2
											–	–	–	–	–	–	–	–	–
3					2	2	2	2	2	2	2	2	2	3	3	3	3	3	3
					–	–	–	–	–	–	–	–	–	–	–	–	–	–	–
4				2	2	2	3	3	3	3	3	3	3	3	4	4	4	4	4
				9	9	–	–	–	–	–	–	–	–	–	–	–	–	–	–
5			2	2	3	3	3	3	3	4	4	4	4	4	4	4	5	5	5
			9	10	10	11	11	–	–	–	–	–	–	–	–	–	–	–	–
6		2	2	3	3	3	3	4	4	4	4	5	5	5	5	5	5	6	6
		–	9	10	11	12	12	13	13	13	13	–	–	–	–	–	–	–	–
7		2	2	3	3	3	4	4	5	5	5	5	5	6	6	6	6	6	6
		–	–	11	12	13	13	14	14	14	14	15	15	15	–	–	–	–	–
8		2	3	3	3	4	4	5	5	5	6	6	6	6	6	7	7	7	7
		–	–	11	12	13	14	14	15	15	16	16	16	16	17	17	17	17	17
9		2	3	3	4	4	5	5	5	6	6	6	7	7	7	7	8	8	8
		–	–	–	13	14	14	15	16	16	16	17	17	18	18	18	18	18	18
10		2	3	3	4	5	5	5	6	6	7	7	7	7	8	8	8	8	9
		–	–	–	13	14	15	16	16	17	17	18	18	18	19	19	19	20	20
11		2	3	4	4	5	5	6	6	7	7	7	8	8	8	9	9	9	9
		–	–	–	13	14	15	16	17	17	18	19	19	19	20	20	20	21	21
12	2	2	3	4	4	5	6	6	7	7	7	8	8	8	9	9	9	10	10
	–	–	–	–	13	14	16	16	17	18	19	19	20	20	21	21	21	22	22
13	2	2	3	4	5	5	6	6	7	7	8	8	9	9	9	10	10	10	10
	–	–	–	–	–	15	16	17	18	19	19	20	20	21	21	22	22	23	23
14	2	2	3	4	5	5	6	7	7	8	8	9	9	9	10	10	10	11	11
	–	–	–	–	–	15	16	17	18	19	20	20	21	22	22	23	23	23	24
15	2	3	3	4	5	6	6	7	7	8	8	9	9	10	10	11	11	11	12
	–	–	–	–	–	15	16	18	18	19	20	21	22	22	23	23	24	24	25
16	2	3	4	4	5	6	6	7	8	8	9	9	10	10	11	11	11	12	12
	–	–	–	–	–	–	17	18	19	20	21	21	22	23	23	24	25	25	25
17	2	3	4	4	5	6	7	7	8	9	9	10	10	11	11	11	12	12	13
	–	–	–	–	–	–	17	18	19	20	21	22	23	23	24	25	25	26	26
18	2	3	4	5	5	6	7	8	8	9	9	10	10	11	11	12	12	13	13
	–	–	–	–	–	–	17	18	19	20	21	22	23	24	25	25	26	26	27
19	2	3	4	5	6	6	7	8	8	9	10	10	11	11	12	12	13	13	13
	–	–	–	–	–	–	17	18	20	21	22	23	23	24	25	26	26	27	27
20	2	3	4	5	6	6	7	8	9	9	10	10	11	12	12	13	13	13	14
	–	–	–	–	–	–	17	18	20	21	22	23	24	25	25	26	27	27	28

Table A9 Table of the F_{max} Distribution

The .05 critical values are in lightface type, and the .01 critical values are in **bold** type.

$n-1$ \ k	2	3	4	5	6	7	8	9	10	11	12
2	39	87.5	142	202	266	333	403	475	550	626	704
	199	**448**	**729**	**1036**	**1362**	**1705**	**2063**	**2432**	**2813**	**3204**	**3605**
3	15.4	27.8	39.2	50.7	62	72.9	83.5	93.9	104	114	124
	47.5	85	120	151	184	216*	249*	281*	310*	337*	361*
4	9.60	15.5	20.6	25.2	29.5	33.6	37.5	41.4	44.6	48.0	51.4
	23.2	**37**	**49**	**59**	**69**	**79**	**89**	**97**	**106**	**113**	**120**
5	7.15	10.8	13.7	16.3	18.7	20.8	22.9	24.7	26.5	28.2	29.9
	14.9	**22**	**28**	**33**	**38**	**42**	**46**	**50**	**54**	**57**	**60**
6	5.82	8.38	10.4	12.1	13.7	15.0	16.3	17.5	18.6	19.7	20.7
	11.1	**15.5**	**19.1**	**22**	**25**	**27**	**30**	**32**	**34**	**36**	**37**
7	4.99	6.94	8.44	9.70	10.8	11.8	12.7	13.5	14.3	15.1	15.8
	8.89	**12.1**	**14.5**	**16.5**	**18.4**	**20.**	**22**	**23**	**24**	**26**	**27**
8	4.43	6.00	7.18	8.12	9.03	9.78	10.5	11.1	11.7	12.2	12.7
	7.50	9.9	11.7	13.2	14.5	15.8	16.9	17.9	18.9	19.8	21
9	4.03	5.34	6.31	7.11	7.80	8.41	8.95	9.45	9.91	10.3	10.7
	6.54	**8.5**	**9.9**	**11.1**	**12.1**	**13.1**	**13.9**	**14.7**	**15.3**	**16.0**	**16.6**
10	3.72	4.85	5.67	6.34	6.92	7.42	7.87	8.28	8.66	9.01	9.34
	5.85	**7.4**	**8.6**	**9.6**	**10.4**	**11.1**	**11.8**	**12.4**	**12.9**	**13.4**	**13.9**
12	3.28	4.16	4.79	5.30	5.72	6.09	6.42	6.72	7.00	7.25	7.48
	4.91	**6.1**	**6.9**	**7.6**	**8.2**	**8.7**	**9.1**	**9.5**	**9.9**	**10.2**	**10.6**
15	2.86	3.54	4.01	4.37	4.68	4.95	5.19	5.40	5.59	5.77	5.93
	4.07	**4.9**	**5.5**	**6.0**	**6.4**	**6.7**	**7.1**	**7.3**	**7.5**	**7.8**	**8.0**
20	2.46	2.95	3.29	3.54	3.76	3.94	4.10	4.24	4.37	4.49	4.59
	3.32	**3.8**	**4.3**	**4.6**	**4.9**	**5.1**	**5.3**	**5.5**	**5.6**	**5.8**	**5.9**
30	2.07	2.40	2.61	2.78	2.91	3.02	3.12	3.21	3.29	3.36	3.39
	2.63	3.0	3.3	3.5	3.6	3.7	3.8	3.9	4.0	4.1	4.2
60	1.67	1.85	1.96	2.04	2.11	2.17	2.22	2.26	2.30	2.33	2.36
	1.96	**2.2**	**2.3**	**2.4**	**2.4**	**2.5**	**2.5**	**2.6**	**2.6**	**2.7**	**2.7**

* The third digit in these values is an approximation

Table A10 Table of the F Distribution

$F_{.95}$

$F_{.95}$

df_{den} \ df_{num}	1	2	3	4	5	6	7	8	9	10	12	15	20	24	30	40	60	120	∞
1	161.4	199.5	215.7	224.6	230.2	234.0	236.8	238.9	240.5	241.9	243.9	245.9	248.0	249.1	250.1	251.1	252.2	253.3	254.3
2	18.51	19.00	19.16	19.25	19.30	19.33	19.35	19.37	19.38	19.40	19.41	19.43	19.45	19.45	19.46	19.47	19.48	19.49	19.50
3	10.13	9.55	9.28	9.12	9.01	8.94	8.89	8.85	8.81	8.79	8.74	8.70	8.66	8.64	8.62	8.59	8.57	8.55	8.53
4	7.71	6.94	6.59	6.39	6.26	6.16	6.09	6.04	6.00	5.96	5.91	5.86	5.80	5.77	5.75	5.72	5.69	5.66	5.63
5	6.61	5.79	5.41	5.19	5.05	4.95	4.88	4.82	4.77	4.74	4.68	4.62	4.56	4.53	4.50	4.46	4.43	4.40	4.36
6	5.99	5.14	4.76	4.53	4.39	4.28	4.21	4.15	4.10	4.06	4.00	3.94	3.87	3.84	3.81	3.77	3.74	3.70	3.67
7	5.59	4.74	4.35	4.12	3.97	3.87	3.79	3.73	3.68	3.64	3.57	3.51	3.44	3.41	3.38	3.34	3.30	3.27	3.23
8	5.32	4.46	4.07	3.84	3.69	3.58	3.50	3.44	3.39	3.35	3.28	3.22	3.15	3.12	3.08	3.04	3.01	2.97	2.93
9	5.12	4.26	3.86	3.63	3.48	3.37	3.29	3.23	3.18	3.14	3.07	3.01	2.94	2.90	2.86	2.83	2.79	2.75	2.71
10	4.96	4.10	3.71	3.48	3.33	3.22	3.14	3.07	3.02	2.98	2.91	2.85	2.77	2.74	2.70	2.66	2.62	2.58	2.54
11	4.84	3.98	3.59	3.36	3.20	3.09	3.01	2.95	2.90	2.85	2.79	2.72	2.65	2.61	2.57	2.53	2.49	2.45	2.40
12	4.75	3.89	3.49	3.26	3.11	3.00	2.91	2.85	2.80	2.75	2.69	2.62	2.54	2.51	2.47	2.43	2.38	2.34	2.30
13	4.67	3.81	3.41	3.18	3.03	2.92	2.83	2.77	2.71	2.67	2.60	2.53	2.46	2.42	2.38	2.34	2.30	2.25	2.21
14	4.60	3.74	3.34	3.11	2.96	2.85	2.76	2.70	2.65	2.60	2.53	2.46	2.39	2.35	3.31	2.27	2.22	2.18	2.13
15	4.54	3.68	3.29	3.06	2.90	2.79	2.71	2.64	2.59	2.54	2.48	2.40	2.33	2.29	2.25	2.20	2.16	2.11	2.07
16	4.49	3.63	3.24	3.01	2.85	2.74	2.66	2.59	2.54	2.49	2.42	2.35	2.28	2.24	2.19	2.15	2.11	2.06	2.01
17	4.45	3.59	3.20	2.96	2.81	2.70	2.61	2.55	2.49	2.45	2.38	2.31	2.23	2.19	2.15	2.10	2.06	2.01	1.96
18	4.41	3.55	3.16	2.93	2.77	2.66	2.58	2.51	2.46	2.41	2.34	2.27	2.19	2.15	2.11	2.06	2.02	1.97	1.92
19	4.38	3.52	3.13	2.90	2.74	2.63	2.54	2.48	2.42	2.38	2.31	2.23	2.16	2.11	2.07	2.03	1.98	1.93	1.88
20	4.35	3.49	3.10	2.87	2.71	2.60	2.51	2.45	2.39	2.35	2.28	2.20	2.12	2.08	2.04	1.99	1.95	1.90	1.84
21	4.32	3.47	3.07	2.84	2.68	2.57	2.49	2.42	2.37	2.32	2.25	2.18	2.10	2.05	2.01	1.96	1.92	1.87	1.81
22	4.30	3.44	3.05	2.82	2.66	2.55	2.46	2.40	2.34	2.30	2.23	2.15	2.07	2.03	1.98	1.94	1.89	1.84	1.78
23	4.28	3.42	3.03	2.80	2.64	2.53	2.44	2.37	2.32	2.27	2.20	2.13	2.05	2.01	1.96	1.91	1.86	1.81	1.76
24	4.26	3.40	3.01	2.78	2.62	2.51	2.42	2.36	2.30	2.25	2.18	2.11	2.03	1.98	1.94	1.89	1.84	1.79	1.73
25	4.24	3.39	2.99	2.76	2.60	2.49	2.40	2.34	2.28	2.24	2.16	2.09	2.01	1.96	1.92	1.87	1.82	1.77	1.71
26	4.23	3.37	2.98	2.74	2.59	2.47	2.39	2.32	2.27	2.22	2.15	2.07	1.99	1.95	1.90	1.85	1.80	1.75	1.69
27	4.21	3.35	2.96	2.73	2.57	2.46	2.37	2.31	2.25	2.20	2.13	2.06	1.97	1.93	1.88	1.84	1.79	1.73	1.67
28	4.20	3.34	2.95	2.71	2.56	2.45	2.36	2.29	2.24	2.19	2.12	2.04	1.96	1.91	1.87	1.82	1.77	1.71	1.65
29	4.18	3.33	2.93	2.70	2.55	2.43	2.35	2.28	2.22	2.18	2.10	2.03	1.94	1.90	1.85	1.81	1.75	1.70	1.64
30	4.17	3.32	2.92	2.69	2.53	2.42	2.33	2.27	2.21	2.16	2.09	2.01	1.93	1.89	1.84	1.79	1.74	1.68	1.62
40	4.08	3.23	2.84	2.61	2.45	2.34	2.25	2.18	2.12	2.08	2.00	1.92	1.84	1.79	1.74	1.69	1.64	1.58	1.51
60	4.00	3.15	2.76	2.53	2.37	2.25	2.17	2.10	2.04	1.99	1.92	1.84	1.75	1.70	1.65	1.59	1.53	1.47	1.39
120	3.92	3.07	2.68	2.45	2.29	2.17	2.09	2.02	1.96	1.91	1.83	1.75	1.66	1.61	1.55	1.50	1.43	1.35	1.25
∞	3.84	3.00	2.60	2.37	2.21	2.10	2.01	1.94	1.88	1.83	1.75	1.67	1.57	1.52	1.46	1.39	1.32	1.22	1.00

Table A10 Table of the F Distribution (continued)

$F_{.975}$

df_{den} \ df_{num}	1	2	3	4	5	6	7	8	9	10	12	15	20	24	30	40	60	120	∞
1	647.8	799.5	864.2	899.6	921.8	937.1	948.2	956.7	963.3	968.6	976.7	984.9	993.1	997.2	1001	1006	1010	1014	1018
2	38.51	39.00	39.17	39.25	39.30	39.33	39.36	39.37	39.39	39.40	39.41	39.43	39.45	39.46	39.46	39.47	39.48	39.49	39.50
3	17.44	16.04	15.44	15.10	14.88	14.73	14.62	14.54	14.47	14.42	14.34	14.25	14.17	14.12	14.08	14.04	13.99	13.95	13.90
4	12.22	10.65	9.98	9.60	9.36	9.20	9.07	8.98	8.90	8.84	8.75	8.66	8.56	8.51	8.46	8.41	8.36	8.31	8.26
5	10.01	8.43	7.76	7.39	7.15	6.98	6.85	6.76	6.68	6.62	6.52	6.43	6.33	6.28	6.23	6.18	6.12	6.07	6.02
6	8.81	7.26	6.60	6.23	5.99	5.82	5.70	5.60	5.52	5.46	5.37	5.27	5.17	5.12	5.07	5.01	4.96	4.90	4.85
7	8.07	6.54	5.89	5.52	5.29	5.12	4.99	4.90	4.82	4.76	4.67	4.57	4.47	4.42	4.36	4.31	4.25	4.20	4.14
8	7.57	6.06	5.42	5.05	4.82	4.65	4.53	4.43	4.36	4.30	4.20	4.10	4.00	3.95	3.89	3.84	3.78	3.73	3.67
9	7.21	5.71	5.08	4.72	4.48	4.32	4.20	4.10	4.03	3.96	3.87	3.77	3.67	3.61	3.56	3.51	3.45	3.39	3.33
10	6.94	5.46	4.83	4.47	4.24	4.07	3.95	3.85	3.78	3.72	3.62	3.52	3.42	3.37	3.31	3.26	3.20	3.14	3.08
11	6.72	5.26	4.63	4.28	4.04	3.88	3.76	3.66	3.59	3.53	3.43	3.33	3.23	3.17	3.12	3.06	3.00	2.94	2.88
12	6.55	5.10	4.47	4.12	3.89	3.73	3.61	3.51	3.44	3.37	3.28	3.18	3.07	3.02	2.96	2.91	2.85	2.79	2.72
13	6.41	4.97	4.35	4.00	3.77	3.60	3.48	3.39	3.31	3.25	3.15	3.05	2.95	2.89	2.84	2.78	2.72	2.66	2.60
14	6.30	4.86	4.24	3.89	3.66	3.50	3.38	3.29	3.21	3.15	3.05	2.95	2.84	2.79	2.73	2.67	2.61	2.55	2.49
15	6.20	4.77	4.15	3.80	3.58	3.41	3.29	3.20	3.12	3.06	2.96	2.86	2.76	2.70	2.64	2.59	2.52	2.46	2.40
16	6.12	4.69	4.08	3.73	3.50	3.34	3.22	3.12	3.05	2.99	2.89	2.79	2.68	2.63	2.57	2.51	2.45	2.38	2.32
17	6.04	4.62	4.01	3.66	3.44	3.28	3.16	3.06	2.98	2.92	2.82	2.72	2.62	2.56	2.50	2.44	2.38	2.32	2.25
18	5.98	4.56	3.95	3.61	3.38	3.22	3.10	3.01	2.93	2.87	2.77	2.67	2.56	2.50	2.44	2.38	2.32	2.26	2.19
19	5.92	4.51	3.90	3.56	3.33	3.17	3.05	2.96	2.88	2.82	2.72	2.62	2.51	2.45	2.39	2.33	2.27	2.20	2.13
20	5.87	4.46	3.86	3.51	3.29	3.13	3.01	2.91	2.84	2.77	2.68	2.57	2.46	2.41	2.35	2.29	2.22	2.16	2.09
21	5.83	4.42	3.82	3.48	3.25	3.09	2.97	2.87	2.80	2.73	2.64	2.53	2.42	2.37	2.31	2.25	2.18	2.11	2.04
22	5.79	4.38	3.78	3.44	3.22	3.05	2.93	2.84	2.76	2.70	2.60	2.50	2.39	2.33	2.27	2.21	2.14	2.08	2.00
23	5.75	4.35	3.75	3.41	3.18	3.02	2.90	2.81	2.73	2.67	2.57	2.47	2.36	2.30	2.24	2.18	2.11	2.04	1.97
24	5.72	4.32	3.72	3.38	3.15	2.99	2.87	2.78	2.70	2.64	2.54	2.44	2.33	2.27	2.21	2.15	2.08	2.01	1.94
25	5.69	4.29	3.69	3.35	3.13	2.97	2.85	2.75	2.68	2.61	2.51	2.41	2.30	2.24	2.18	2.12	2.05	1.98	1.91
26	5.66	4.27	3.67	3.33	3.10	2.94	2.82	2.73	2.65	2.59	2.49	2.39	2.28	2.22	2.16	2.09	2.03	1.95	1.88
27	5.63	4.24	3.65	3.31	3.08	2.92	2.80	2.71	2.63	2.57	2.47	2.36	2.25	2.19	2.13	2.07	2.00	1.93	1.85
28	5.61	4.22	3.63	3.29	3.06	2.90	2.78	2.69	2.61	2.55	2.45	2.34	2.23	2.17	2.11	2.05	1.98	1.91	1.83
29	5.59	4.20	3.61	3.27	3.04	2.88	2.76	2.67	2.59	2.53	2.43	2.32	2.21	2.15	2.09	2.03	1.96	1.89	1.81
30	5.57	4.18	3.59	3.25	3.03	2.87	2.75	2.65	2.57	2.51	2.41	2.31	2.20	2.14	2.07	2.01	1.94	1.87	1.79
40	5.42	4.05	3.46	3.13	2.90	2.74	2.62	2.53	2.45	2.39	2.29	2.18	2.07	2.01	1.94	1.88	1.80	1.72	1.64
60	5.29	3.93	3.34	3.01	2.79	2.63	2.51	2.41	2.33	2.27	2.17	2.06	1.94	1.88	1.82	1.74	1.67	1.58	1.48
120	5.15	3.80	3.23	2.89	2.67	2.52	2.39	2.30	2.22	2.16	2.05	1.94	1.82	1.76	1.69	1.61	1.53	1.43	1.31
∞	5.02	3.69	3.12	2.79	2.57	2.41	2.29	2.19	2.11	2.05	1.94	1.83	1.71	1.64	1.57	1.48	1.39	1.27	1.00

Table A10 Table of the F Distribution (continued)

$F_{.99}$

df_{den} \ df_{num}	1	2	3	4	5	6	7	8	9	10	12	15	20	24	30	40	60	120	∞
1	4052	4999.5	5403	5625	5764	5859	5928	5982	6022	6056	6106	6157	6209	6235	6261	6287	6313	6339	6366
2	98.50	99.00	99.17	99.25	99.30	99.33	99.36	99.37	99.39	99.40	99.42	99.43	99.45	99.46	99.47	99.47	99.48	99.49	99.50
3	34.12	30.82	29.46	28.71	28.24	27.91	27.67	27.49	27.35	27.23	27.05	26.87	26.69	26.60	26.50	26.41	26.32	26.22	26.13
4	21.20	18.00	16.69	15.98	15.52	15.21	14.98	14.80	14.66	14.55	14.37	14.20	14.02	13.93	13.84	13.75	13.65	13.56	13.46
5	16.26	13.27	12.06	11.39	10.97	10.67	10.46	10.29	10.16	10.05	9.89	9.72	9.55	9.47	9.38	9.29	9.20	9.11	9.02
6	13.75	10.92	9.78	9.15	8.75	8.47	8.26	8.10	7.98	7.87	7.72	7.56	7.40	7.31	7.23	7.14	7.06	6.97	6.88
7	12.25	9.55	8.45	7.85	7.46	7.19	6.99	6.84	6.72	6.62	6.47	6.31	6.16	6.07	5.99	5.91	5.82	5.74	5.65
8	11.26	8.65	7.59	7.01	6.63	6.37	6.18	6.03	5.91	5.81	5.67	5.52	5.36	5.28	5.20	5.12	5.03	4.95	4.86
9	10.56	8.02	6.99	6.42	6.06	5.80	5.61	5.47	5.35	5.26	5.11	4.96	4.81	4.73	4.65	4.57	4.48	4.40	4.31
10	10.04	7.56	6.55	5.99	5.64	5.39	5.20	5.06	4.94	4.85	4.71	4.56	4.41	4.33	4.25	4.17	4.08	4.00	3.91
11	9.65	7.21	6.22	5.67	5.32	5.07	4.89	4.74	4.63	4.54	4.40	4.25	4.10	4.02	3.94	3.86	3.78	3.69	3.60
12	9.33	6.93	5.95	5.41	5.06	4.82	4.64	4.50	4.39	4.30	4.16	4.01	3.86	3.78	3.70	3.62	3.54	3.45	3.36
13	9.07	6.70	5.74	5.21	4.86	4.62	4.44	4.30	4.19	4.10	3.96	3.82	3.66	3.59	3.51	3.43	3.34	3.25	3.17
14	8.86	6.51	5.56	5.04	4.69	4.46	4.28	4.14	4.03	3.94	3.80	3.66	3.51	3.43	3.35	3.27	3.18	3.09	3.00
15	8.68	6.36	5.42	4.89	4.56	4.32	4.14	4.00	3.89	3.80	3.67	3.52	3.37	3.29	3.21	3.13	3.05	2.96	2.87
16	8.53	6.23	5.29	4.77	4.44	4.20	4.03	3.89	3.78	3.69	3.55	3.41	3.26	3.18	3.10	3.02	2.93	2.84	2.75
17	8.40	6.11	5.18	4.67	4.34	4.10	3.93	3.79	3.68	3.59	3.46	3.31	3.16	3.08	3.00	2.92	2.83	2.75	2.65
18	8.29	6.01	5.09	4.58	4.25	4.01	3.84	3.71	3.60	3.51	3.37	3.23	3.08	3.00	2.92	2.84	2.75	2.66	2.57
19	8.18	5.93	5.01	4.50	4.17	3.94	3.77	3.63	3.52	3.43	3.30	3.15	3.00	2.92	2.84	2.76	2.67	2.58	2.49
20	8.10	5.85	4.94	4.43	4.10	3.87	3.70	3.56	3.46	3.37	3.23	3.09	2.94	2.86	2.78	2.69	2.61	2.52	2.42
21	8.02	5.78	4.87	4.37	4.04	3.81	3.64	3.51	3.40	3.31	3.17	3.03	2.88	2.80	2.72	2.64	2.55	2.46	2.36
22	7.95	5.72	4.82	4.31	3.99	3.76	3.59	3.45	3.35	3.26	3.12	2.98	2.83	2.75	2.67	2.58	2.50	2.40	2.31
23	7.88	5.66	4.76	4.26	3.94	3.71	3.54	3.41	3.30	3.21	3.07	2.93	2.78	2.70	2.62	2.54	2.45	2.35	2.26
24	7.82	5.61	4.72	4.22	3.90	3.67	3.50	3.36	3.26	3.17	3.03	2.89	2.74	2.66	2.58	2.49	2.40	2.31	2.21
25	7.77	5.57	4.68	4.18	3.85	3.63	3.46	3.32	3.22	3.13	2.99	2.85	2.70	2.62	2.54	2.45	2.36	2.27	2.17
26	7.72	5.53	4.64	4.14	3.82	3.59	3.42	3.29	3.18	3.09	2.96	2.81	2.66	2.58	2.50	2.42	2.33	2.23	2.13
27	7.68	5.49	4.60	4.11	3.78	3.56	3.39	3.26	3.15	3.06	2.93	2.78	2.63	2.55	2.47	2.38	2.29	2.20	2.10
28	7.64	5.45	4.57	4.07	3.75	3.53	3.36	3.23	3.12	3.03	2.90	2.75	2.60	2.52	2.44	2.35	2.26	2.17	2.06
29	7.60	5.42	4.54	4.04	3.73	3.50	3.33	3.20	3.09	3.00	2.87	2.73	2.57	2.49	2.41	2.33	2.23	2.14	2.03
30	7.56	5.39	4.51	4.02	3.70	3.47	3.30	3.17	3.07	2.98	2.84	2.70	2.55	2.47	2.39	2.3)	2.21	2.11	2.01
40	7.31	5.18	4.31	3.83	3.51	3.29	3.12	2.99	2.89	2.80	2.66	2.52	2.37	2.29	2.20	2.11	2.02	1.92	1.80
60	7.08	4.98	4.13	3.65	3.34	3.12	2.95	2.82	2.72	2.63	2.50	2.35	2.20	2.12	2.03	1.94	1.84	1.73	1.60
120	6.85	4.79	3.95	3.48	3.17	2.96	2.79	2.66	2.56	2.47	2.34	2.19	2.03	1.95	1.86	1.76	1.66	1.53	1.38
∞	6.63	4.61	3.78	3.32	3.02	2.80	2.64	2.51	2.41	2.32	2.18	2.04	1.88	1.79	1.70	1.59	1.47	1.32	1.00

Table A10 Table of the *F* Distribution (continued)

$F_{.995}$

df_{den} \ df_{num}	1	2	3	4	5	6	7	8	9	10	12	15	20	24	30	40	60	120	∞
1	16211	20000	21615	22500	23056	23437	23715	23925	24091	24224	24426	24630	24836	24940	25044	25148	25253	25359	25465
2	198.5	199.0	199.2	199.2	199.3	199.3	199.4	199.4	199.4	199.4	199.4	199.4	199.4	199.5	199.5	199.5	199.5	199.5	199.5
3	55.55	49.80	47.47	46.19	45.39	44.84	44.43	44.13	43.88	43.69	43.39	43.08	42.78	42.62	42.47	42.31	42.15	41.99	41.83
4	31.33	26.28	24.26	23.15	22.46	21.97	21.62	21.35	21.14	20.97	20.70	20.44	20.17	20.03	19.89	19.75	19.61	19.47	19.32
5	22.78	18.31	16.53	15.56	14.94	14.51	14.20	13.96	13.77	13.62	13.38	13.15	12.90	12.78	12.66	12.53	12.40	12.27	12.14
6	18.63	14.54	12.92	12.03	11.46	11.07	10.79	10.57	10.39	10.25	10.03	9.81	9.59	9.47	9.36	9.24	9.12	9.00	8.88
7	16.24	12.40	10.88	10.05	9.52	9.16	8.89	8.68	8.51	8.38	8.18	7.97	7.75	7.65	7.53	7.42	7.31	7.19	7.08
8	14.69	11.04	9.60	8.81	8.30	7.95	7.69	7.50	7.34	7.21	7.01	6.81	6.61	6.50	6.40	6.29	6.18	6.06	5.95
9	13.61	10.11	8.72	7.96	7.47	7.13	6.88	6.69	6.54	6.42	6.23	6.03	5.83	5.73	5.62	5.52	5.41	5.30	5.19
10	12.83	9.43	8.08	7.34	6.87	6.54	6.30	6.12	5.97	5.85	5.66	5.47	5.27	5.17	5.07	4.97	4.86	4.75	4.64
11	12.23	8.91	7.60	6.88	6.42	6.10	5.86	5.68	5.54	5.42	5.24	5.05	4.86	4.76	4.65	4.55	4.44	4.34	4.23
12	11.75	8.51	7.23	6.52	6.07	5.76	5.52	5.35	5.20	5.09	4.91	4.72	4.53	4.43	4.33	4.23	4.12	4.01	3.90
13	11.37	8.19	6.93	6.23	5.79	5.48	5.25	5.08	4.94	4.82	4.64	4.46	4.27	4.17	4.07	3.97	3.87	3.76	3.65
14	11.06	7.92	6.68	6.00	5.56	5.26	5.03	4.86	4.72	4.60	4.43	4.25	4.06	3.96	3.86	3.76	3.66	3.55	3.44
15	10.80	7.70	6.48	5.80	5.37	5.07	4.85	4.67	4.54	4.42	4.25	4.07	3.88	3.79	3.69	3.58	3.48	3.37	3.26
16	10.58	7.51	6.30	5.64	5.21	4.91	4.69	4.52	4.38	4.27	4.10	3.92	3.73	3.64	3.54	3.44	3.33	3.22	3.11
17	10.38	7.35	6.16	5.50	5.07	4.78	4.56	4.39	4.25	4.14	3.97	3.79	3.61	3.51	3.41	3.31	3.21	3.10	2.98
18	10.22	7.21	6.03	5.37	4.96	4.66	4.44	4.28	4.14	4.03	3.86	3.68	3.50	3.40	3.30	3.20	3.10	2.99	2.87
19	10.07	7.09	5.92	5.27	4.85	4.56	4.34	4.18	4.04	3.93	3.76	3.59	3.40	3.31	3.21	3.11	3.00	2.89	2.78
20	9.94	6.99	5.82	5.17	4.76	4.47	4.26	4.09	3.96	3.85	3.68	3.50	3.32	3.22	3.12	3.02	2.92	2.81	2.69
21	9.83	6.89	5.73	5.09	4.68	4.39	4.18	4.01	3.88	3.77	3.60	3.43	3.24	3.15	3.05	2.95	2.84	2.73	2.61
22	9.73	6.81	5.65	5.02	4.61	4.32	4.11	3.94	3.81	3.70	3.54	3.36	3.18	3.08	2.98	2.88	2.77	2.66	2.55
23	9.63	6.73	5.58	4.95	4.54	4.26	4.05	3.88	3.75	3.64	3.47	3.30	3.12	3.02	2.92	2.82	2.71	2.60	2.48
24	9.55	6.66	5.52	4.89	4.49	4.20	3.99	3.83	3.69	3.59	3.42	3.25	3.06	2.97	2.87	2.77	2.66	2.55	2.43
25	9.48	6.60	5.46	4.84	4.43	4.15	3.94	3.78	3.64	3.54	3.37	3.20	3.01	2.92	2.82	2.72	2.61	2.50	2.38
26	9.41	6.54	5.41	4.79	4.38	4.10	3.89	3.73	3.60	3.49	3.33	3.15	2.97	2.87	2.77	2.67	2.56	2.45	2.33
27	9.34	6.49	5.36	4.74	4.34	4.06	3.85	3.69	3.56	3.45	3.28	3.11	2.93	2.83	2.73	2.63	2.52	2.41	2.25
28	9.28	6.44	5.32	4.70	4.30	4.02	3.81	3.65	3.52	3.41	3.25	3.07	2.89	2.79	2.69	2.59	2.48	2.37	2.29
29	9.23	6.40	5.28	4.66	4.26	3.98	3.77	3.61	3.48	3.38	3.21	3.04	2.86	2.76	2.66	2.56	2.45	2.33	2.24
30	9.18	6.35	5.24	4.62	4.23	3.95	3.74	3.58	3.45	3.34	3.18	3.01	2.82	2.73	2.63	2.52	2.42	2.30	2.18
40	8.83	6.07	4.98	4.37	3.99	3.71	3.51	3.35	3.22	3.12	2.95	2.78	2.60	2.50	2.40	2.30	2.18	2.06	1.93
60	8.49	5.79	4.73	4.14	3.76	3.49	3.29	3.13	3.01	2.90	2.74	2.57	2.39	2.29	2.19	2.08	1.96	1.83	1.69
120	8.18	5.54	4.50	3.92	3.55	3.28	3.09	2.93	2.81	2.71	2.54	2.37	2.19	2.09	1.98	1.87	1.75	1.61	1.43
∞	7.88	5.30	4.28	3.72	3.35	3.09	2.90	2.74	2.62	2.52	2.36	2.19	2.00	1.90	1.79	1.67	1.53	1.36	1.00

Table A11 Table of Critical Values for Mann–Whitney *U* Statistic

(Two–Tailed .05 Values)

n_2 \ n_1	1	2	3	4	5	6	7	8	9	10	11	12	13	14	15	16	17	18	19	20
1																				
2								0	0	0	0	1	1	1	1	1	2	2	2	2
3					0	1	1	2	2	3	3	4	4	5	5	6	6	7	7	8
4				0	1	2	3	4	4	5	6	7	8	9	10	11	11	12	13	13
5			0	1	2	3	5	6	7	8	9	11	12	13	14	15	17	18	19	20
6			1	2	3	5	6	8	10	11	13	14	16	17	19	21	22	24	25	27
7			1	3	5	6	8	10	12	14	16	18	20	22	24	26	28	30	32	34
8		0	2	4	6	8	10	13	15	17	19	22	24	26	29	31	34	36	38	41
9		0	2	4	7	10	12	15	17	20	23	26	28	31	34	37	39	42	45	48
10		0	3	5	8	11	14	17	20	23	26	29	33	36	39	42	45	48	52	55
11		0	3	6	9	13	16	19	23	26	30	33	37	40	44	47	51	55	58	62
12		1	4	7	11	14	18	22	26	29	33	37	41	45	49	53	57	61	65	69
13		1	4	8	12	16	20	24	28	33	37	41	45	50	54	59	63	67	72	76
14		1	5	9	13	17	22	26	31	36	40	45	50	55	59	64	67	74	78	83
15		1	5	10	14	19	24	29	34	39	44	49	54	59	64	70	75	80	85	90
16		1	6	11	15	21	26	31	37	42	47	53	59	64	70	75	81	86	92	98
17		2	6	11	17	22	28	34	39	45	51	57	63	67	75	81	87	93	99	105
18		2	7	12	18	24	30	36	42	48	55	61	67	74	80	86	93	99	106	112
19		2	7	13	19	25	32	38	45	52	58	65	72	78	85	92	99	106	113	119
20		2	8	13	20	27	34	41	48	55	62	69	76	83	90	98	105	112	119	127

(One–Tailed .05 Values)

n_2 \ n_1	1	2	3	4	5	6	7	8	9	10	11	12	13	14	15	16	17	18	19	20
1																			0	0
2					0	0	0	1	1	1	1	2	2	2	3	3	3	4	4	4
3			0	0	1	2	2	3	3	4	5	5	6	7	7	8	9	9	10	11
4			0	1	2	3	4	5	6	7	8	9	10	11	12	14	15	16	17	18
5		0	1	2	4	5	6	8	9	11	12	13	15	16	18	19	20	22	23	25
6		0	2	3	5	7	8	10	12	14	16	17	19	21	23	25	26	28	30	32
7		0	2	4	6	8	11	13	15	17	19	21	24	26	28	30	33	35	37	39
8		1	3	5	8	10	13	15	18	20	23	26	28	31	33	36	39	41	44	47
9		1	3	6	9	12	15	18	21	24	27	30	33	36	39	42	45	48	51	54
10		1	4	7	11	14	17	20	24	27	31	34	37	41	44	48	51	55	58	62
11		1	5	8	12	16	19	23	27	31	34	38	42	46	50	54	57	61	65	69
12		2	5	9	13	17	21	26	30	34	38	42	47	51	55	60	64	68	72	77
13		2	6	10	15	19	24	28	33	37	42	47	51	56	61	65	70	75	80	84
14		2	7	11	16	21	26	31	36	41	46	51	56	61	66	71	77	82	87	92
15		3	7	12	18	23	28	33	39	44	50	55	61	66	72	77	83	88	94	100
16		3	8	14	19	25	30	36	42	48	54	60	65	71	77	83	89	95	101	107
17		3	9	15	20	26	33	39	45	51	57	64	70	77	83	89	96	102	109	115
18		4	9	16	22	28	35	41	48	55	61	68	75	82	88	95	102	109	116	123
19	0	4	10	17	23	30	37	44	51	58	65	72	80	87	94	101	109	116	123	130
20	0	4	11	18	25	32	39	47	54	62	69	77	84	92	100	107	115	123	130	138

Table A11 Table of critical Values for Mann–Whitney U Statistic (continued)

(Two–Tailed .01 Values)

n_2 \ n_1	1	2	3	4	5	6	7	8	9	10	11	12	13	14	15	16	17	18	19	20
1																				
2																			0	0
3									0	0	0	1	1	1	2	2	2	2	3	3
4						0	0	1	1	2	2	3	3	4	5	5	6	6	7	8
5					0	1	1	2	3	4	5	6	7	7	8	9	10	11	12	13
6				0	1	2	3	4	5	6	7	9	10	11	12	13	15	16	17	18
7				0	1	3	4	6	7	9	10	12	13	15	16	18	19	21	22	24
8				1	2	4	6	7	9	11	13	15	17	18	20	22	24	26	28	30
9			0	1	3	5	7	9	11	13	16	18	20	22	24	27	29	31	33	36
10			0	2	4	6	9	11	13	16	18	21	24	26	29	31	34	37	39	42
11			0	2	5	7	10	13	16	18	21	24	27	30	33	36	39	42	45	48
12			1	3	6	9	12	15	18	21	24	27	31	34	37	41	44	47	51	54
13			1	3	7	10	13	17	20	24	27	31	34	38	42	45	49	53	56	60
14			1	4	7	11	15	18	22	26	30	34	38	42	46	50	54	58	63	67
15			2	5	8	12	16	20	24	29	33	37	42	46	51	55	60	64	69	73
16			2	5	9	13	18	22	27	31	36	41	45	50	55	60	65	70	74	79
17			2	6	10	15	19	24	29	34	39	44	49	54	60	65	70	75	81	86
18			2	6	11	16	21	26	31	37	42	47	53	58	64	70	75	81	87	92
19		0	3	7	12	17	22	28	33	39	45	51	56	63	69	74	81	87	93	99
20		0	3	8	13	18	24	30	36	42	48	54	60	67	73	79	86	92	99	105

(One–Tailed .01 Values)

n_2 \ n_1	1	2	3	4	5	6	7	8	9	10	11	12	13	14	15	16	17	18	19	20
1																				
2													0	0	0	0	0	0	1	1
3							0	0	1	1	1	2	2	2	3	3	4	4	4	5
4					0	1	1	2	3	3	4	5	5	6	7	7	8	9	9	10
5				0	1	2	3	4	5	6	7	8	9	10	11	12	13	14	15	16
6				1	2	3	4	6	7	8	9	11	12	13	15	16	18	19	20	22
7			0	1	3	4	6	7	9	11	12	14	16	17	19	21	23	24	26	28
8			0	2	4	6	7	9	11	13	15	17	20	22	24	26	28	30	32	34
9			1	3	5	7	9	11	14	16	18	21	23	26	28	31	33	36	38	40
10			1	3	6	8	11	13	16	19	22	24	27	30	33	36	38	41	44	47
11			1	4	7	9	12	15	18	22	25	28	31	34	37	41	44	47	50	53
12			2	5	8	11	14	17	21	24	28	31	35	38	42	46	49	53	56	60
13		0	2	5	9	12	16	20	23	27	31	35	39	43	47	51	55	59	63	67
14		0	2	6	10	13	17	22	26	30	34	38	43	47	51	56	60	65	69	73
15		0	3	7	11	15	19	24	28	33	37	42	47	51	56	61	66	70	75	80
16		0	3	7	12	16	21	26	31	36	41	46	51	56	61	66	71	76	82	87
17		0	4	8	13	18	23	28	33	38	44	49	55	60	66	71	77	82	88	93
18		0	4	9	14	19	24	30	36	41	47	53	59	65	70	76	82	88	94	100
19		1	4	9	15	20	26	32	38	44	50	56	63	69	75	82	88	94	101	107
20		1	5	10	16	22	28	34	40	47	53	60	67	73	80	87	93	100	107	114

Table A12 Table of Sandler's *A* Statistic

	One–tailed level of significance				
	.05	**.025**	**.01**	**.005**	**.0005**
	Two–tailed level of significance				
	.10	**.05**	**.02**	**.01**	**.001**
df = *n*–1					
1	.5125	.5031	.50049	.50012	.5000012
2	.412	.369	.347	.340	.334
3	.385	.324	.286	.272	.254
4	.376	.304	.257	.238	.211
5	.372	.293	.240	.218	.184
6	.370	.286	.230	.205	.167
7	.369	.281	.222	.196	.155
8	.368	.278	.217	.190	.146
9	.368	.276	.213	.185	.139
10	.368	.274	.210	.181	.134
11	.368	.273	.207	.178	.130
12	.368	.271	.205	.176	.126
13	.368	.270	.204	.174	.124
14	.368	.270	.202	.172	.121
15	.368	.269	.201	.170	.119
16	.368	.268	.200	.169	.117
17	.368	.268	.199	.168	.116
18	.368	.267	.198	.167	.114
19	.368	.267	.197	.166	.113
20	.368	.266	.197	.165	.112
21	.368	.266	.196	.165	.111
22	.368	.266	.196	.164	.110
23	.368	.266	.195	.163	.109
24	.368	.265	.195	.163	.108
25	.368	.265	.194	.162	.108
26	.368	.265	.194	.162	.107
27	.368	.265	.193	.161	.107
28	.368	.265	.193	.161	.106
29	.368	.264	.193	.161	.106
30	.368	.264	.193	.160	.105
40	.368	.263	.191	.158	.102
60	.369	.262	.189	.155	.099
120	.369	.261	.187	.153	.095
∞	.370	.260	.185	.151	.092

Table A13 Table of the Studentized Range Statistic

$q_{.95}$ (α = .05)

df_{error} \ k	2	3	4	5	6	7	8	9	10
1	17·97	26·98	32·82	37·08	40·41	43·12	45·40	47·36	49·07
2	6·08	8·33	9·80	10·88	11·74	12·44	13·03	13·54	13·99
3	4·50	5·91	6·82	7·50	8·04	8·48	8·85	9·18	9·46
4	3·93	5·04	5·76	6·29	6·71	7·05	7·35	7·60	7·83
5	3·64	4·60	5·22	5·67	6·03	6·33	6·58	6·80	6·99
6	3·46	4·34	4·90	5·30	5·63	5·90	6·12	6·32	6·49
7	3·34	4·16	4·68	5·06	5·36	5·61	5·82	6·00	6·16
8	3·26	4·04	4·53	4·89	5·17	5·40	5·60	5·77	5·92
9	3·20	3·95	4·41	4·76	5·02	5·24	5·43	5·59	5·74
10	3·15	3·88	4·33	4·65	4·91	5·12	5·30	5·46	5·60
11	3·11	3·82	4·26	4·57	4·82	5·03	5·20	5·35	5·49
12	3·08	3·77	4·20	4·51	4·75	4·95	5·12	5·27	5·39
13	3·06	3·73	4·15	4·45	4·69	4·88	5·05	5·19	5·32
14	3·03	3·70	4·11	4·41	4·64	4·83	4·99	5·13	5·25
15	3·01	3·67	4·08	4·37	4·59	4·78	4·94	5·08	5·20
16	3·00	3·65	4·05	4·33	4·56	4·74	4·90	5·03	5·15
17	2·98	3·63	4·02	4·30	4·52	4·70	4·86	4·99	5·11
18	2·97	3·61	4·00	4·28	4·49	4·67	4·82	4·96	5·07
19	2·96	3·59	3·98	4·25	4·47	4·65	4·79	4·92	5·04
20	2·95	3·58	3·96	4·23	4·45	4·62	4·77	4·90	5·01
24	2·92	3·53	3·90	4·17	4·37	4·54	4·68	4·81	4·92
30	2·89	3·49	3·85	4·10	4·30	4·46	4·60	4·72	4·82
40	2·86	3·44	3·79	4·04	4·23	4·39	4·52	4·63	4·73
60	2·83	3·40	3·74	3·98	4·16	4·31	4·44	4·55	4·65
120	2·80	3·36	3·68	3·92	4·10	4·24	4·36	4·47	4·56
∞	2·77	3·31	3·63	3·86	4·03	4·17	4·29	4·39	4·47

df_{error} \ k	11	12	13	14	15	16	17	18	19	20
1	50·59	51·96	53·20	54·33	55·36	56·32	57·22	58·04	58·83	59·56
2	14·39	14·75	15·08	15·38	15·65	15·91	16·14	16·37	16·57	16·77
3	9·72	9·95	10·15	10·35	10·52	10·69	10·84	10·98	11·11	11·24
4	8·03	8·21	8·37	8·52	8·66	8·79	8·91	9·03	9·13	9·23
5	7·17	7·32	7·47	7·60	7·72	7·83	7·93	8·03	8·12	8·21
6	6·65	6·79	6·92	7·03	7·14	7·24	7·34	7·43	7·51	7·59
7	6·30	6·43	6·55	6·66	6·76	6·85	6·94	7·02	7·10	7·17
8	6·05	6·18	6·29	6·39	6·48	6·57	6·65	6·73	6·80	6·87
9	5·87	5·98	6·09	6·19	6·28	6·36	6·44	6·51	6·58	6·64
10	5·72	5·83	5·93	6·03	6·11	6·19	6·27	6·34	6·40	6·47
11	5·61	5·71	5·81	5·90	5·98	6·06	6·13	6·20	6·27	6·33
12	5·51	5·61	5·71	5·80	5·88	5·95	6·02	6·09	6·15	6·21
13	5·43	5·53	5·63	5·71	5·79	5·86	5·93	5·99	6·05	6·11
14	5·36	5·46	5·55	5·64	5·71	5·79	5·85	5·91	5·97	6·03
15	5·31	5·40	5·49	5·57	5·65	5·72	5·78	5·85	5·90	5·96
16	5·26	5·35	5·44	5·52	5·59	5·66	5·73	5·79	5·84	5·90
17	5·21	5·31	5·39	5·47	5·54	5·61	5·67	5·73	5·79	5·84
18	5·17	5·27	5·35	5·43	5·50	5·57	5·63	5·69	5·74	5·79
19	5·14	5·23	5·31	5·39	5·46	5·53	5·59	5·65	5·70	5·75
20	5·11	5·20	5·28	5·36	5·43	5·49	5·55	5·61	5·66	5·71
24	5·01	5·10	5·18	5·25	5·32	5·38	5·44	5·49	5·55	5·59
30	4·92	5·00	5·08	5·15	5·21	5·27	5·33	5·38	5·43	5·47
40	4·82	4·90	4·98	5·04	5·11	5·16	5·22	5·27	5·31	5·36
60	4·73	4·81	4·88	4·94	5·00	5·06	5·11	5·15	5·20	5·24
120	4·64	4·71	4·78	4·84	4·90	4·95	5·00	5·04	5·09	5·13
∞	4·55	4·62	4·68	4·74	4·80	4·85	4·89	4·93	4·97	5·01

Table A13 Table of the Studentized Range Statistic (continued)

$q_{.99}$ **(α = .01)**

df_{error} \ k	2	3	4	5	6	7	8	9	10
1	90·03	135·0	164·3	185·6	202·2	215·8	227·2	237·0	245·6
2	14·04	19·02	22·29	24·72	26·63	28·20	29·53	30·68	31·69
3	8·26	10·62	12·17	13·33	14·24	15·00	15·64	16·20	16·69
4	6·51	8·12	9·17	9·96	10·58	11·10	11·55	11·93	12·27
5	5·70	6·98	7·80	8·42	8·91	9·32	9·67	9·97	10·24
6	5·24	6·33	7·03	7·56	7·97	8·32	8·61	8·87	9·10
7	4·95	5·92	6·54	7·01	7·37	7·68	7·94	8·17	8·37
8	4·75	5·64	6·20	6·62	6·96	7·24	7·47	7·68	7·86
9	4·60	5·43	5·96	6·35	6·66	6·91	7·13	7·33	7·49
10	4·48	5·27	5·77	6·14	6·43	6·67	6·87	7·05	7·21
11	4·39	5·15	5·62	5·97	6·25	6·48	6·67	6·84	6·99
12	4·32	5·05	5·50	5·84	6·10	6·32	6·51	6·67	6·81
13	4·26	4·96	5·40	5·73	5·98	6·19	6·37	6·53	6·67
14	4·21	4·89	5·32	5·63	5·88	6·08	6·26	6·41	6·54
15	4·17	4·84	5·25	5·56	5·80	5·99	6·16	6·31	6·44
16	4·13	4·79	5·19	5·49	5·72	5·92	6·08	6·22	6·35
17	4·10	4·74	5·14	5·43	5·66	5·85	6·01	6·15	6·27
18	4·07	4·70	5·09	5·38	5·60	5·79	5·94	6·08	6·20
19	4·05	4·67	5·05	5·33	5·55	5·73	5·89	6·02	6·14
20	4·02	4·64	5·02	5·29	5·51	5·69	5·84	5·97	6·09
24	3·96	4·55	4·91	5·17	5·37	5·54	5·69	5·81	5·92
30	3·89	4·45	4·80	5·05	5·24	5·40	5·54	5·65	5·76
40	3·82	4·37	4·70	4·93	5·11	5·26	5·39	5·50	5·60
60	3·76	4·28	4·59	4·82	4·99	5·13	5·25	5·36	5·45
120	3·70	4·20	4·50	4·71	4·87	5·01	5·12	5·21	5·30
∞	3·64	4·12	4·40	4·60	4·76	4·88	4·99	5·08	5·16

df_{error} \ k	11	12	13	14	15	16	17	18	19	20
1	253·2	260·0	266·2	271·8	277·0	281·8	286·3	290·4	294·3	298·0
2	32·59	33·40	34·13	34·81	35·43	36·00	36·53	37·03	37·50	37·95
3	17·13	17·53	17·89	18·22	18·52	18·81	19·07	19·32	19·55	19·77
4	12·57	12·84	13·09	13·32	13·53	13·73	13·91	14·08	14·24	14·40
5	10·48	10·70	10·89	11·08	11·24	11·40	11·55	11·68	11·81	11·93
6	9·30	9·48	9·65	9·81	9·95	10·08	10·21	10·32	10·43	10·54
7	8·55	8·71	8·86	9·00	9·12	9·24	9·35	9·46	9·55	9·65
8	8·03	8·18	8·31	8·44	8·55	8·66	8·76	8·85	8·94	9·03
9	7·65	7·78	7·91	8·03	8·13	8·23	8·33	8·41	8·49	8·57
10	7·36	7·49	7·60	7·71	7·81	7·91	7·99	8·08	8·15	8·23
11	7·13	7·25	7·36	7·46	7·56	7·65	7·73	7·81	7·88	7·95
12	6·94	7·06	7·17	7·26	7·36	7·44	7·52	7·59	7·66	7·73
13	6·79	6·90	7·01	7·10	7·19	7·27	7·35	7·42	7·48	7·55
14	6·66	6·77	6·87	6·96	7·05	7·13	7·20	7·27	7·33	7·39
15	6·55	6·66	6·76	6·84	6·93	7·00	7·07	7·14	7·20	7·26
16	6·46	6·56	6·66	6·74	6·82	6·90	6·97	7·03	7·09	7·15
17	6·38	6·48	6·57	6·66	6·73	6·81	6·87	6·94	7·00	7·05
18	6·31	6·41	6·50	6·58	6·65	6·73	6·79	6·85	6·91	6·97
19	6·25	6·34	6·43	6·51	6·58	6·65	6·72	6·78	6·84	6·89
20	6·19	6·28	6·37	6·45	6·52	6·59	6·65	6·71	6·77	6·82
24	6·02	6·11	6·19	6·26	6·33	6·39	6·45	6·51	6·56	6·61
30	5·85	5·93	6·01	6·08	6·14	6·20	6·26	6·31	6·36	6·41
40	5·69	5·76	5·83	5·90	5·96	6·02	6·07	6·12	6·16	6·21
60	5·53	5·60	5·67	5·73	5·78	5·84	5·89	5·93	5·97	6·01
120	5·37	5·44	5·50	5·56	5·61	5·66	5·71	5·75	5·79	5·83
∞	5·23	5·29	5·35	5·40	5·45	5·49	5·54	5·57	5·61	5·65

Table A14 Table of Dunnett's Modified *t* Statistic for a Control Group Comparison
Two-Tailed Values

The .05 critical values are in lightface type, and the .01 critical values are in **bold** type.

	k = number of treatment means, including control								
df_{error}	2	3	4	5	6	7	8	9	10
5	2.57	3.03	3.29	3.48	3.62	3.73	3.82	3.90	3.97
	4.03	**4.63**	**4.98**	**5.22**	**5.41**	**5.56**	**5.69**	**5.80**	**5.89**
6	2.45	2.86	3.10	3.26	3.39	3.49	3.57	3.64	3.71
	3.71	**4.21**	**4.51**	**4.71**	**4.87**	**5.00**	**5.10**	**5.20**	**5.28**
7	2.36	2.75	2.97	3.12	3.24	3.33	3.41	3.47	3.53
	3.50	**3.95**	**4.21**	**4.39**	**4.53**	**4.64**	**4.74**	**4.82**	**4.89**
8	2.31	2.67	2.88	3.02	3.13	3.22	3.29	3.35	3.41
	3.36	**3.77**	**4.00**	**4.17**	**4.29**	**4.40**	**4.48**	**4.56**	**4.62**
9	2.26	2.61	2.81	2.95	3.05	3.14	3.20	3.26	3.32
	3.25	**3.63**	**3.85**	**4.01**	**4.12**	**4.22**	**4.30**	**4.37**	**4.43**
10	2.23	2.57	2.76	2.89	2.99	3.07	3.14	3.19	3.24
	3.17	**3.53**	**3.74**	**3.88**	**3.99**	**4.08**	**4.16**	**4.22**	**4.28**
11	2.20	2.53	2.72	2.84	2.94	3.02	3.08	3.14	3.19
	3.11	**3.45**	**3.65**	**3.79**	**3.89**	**3.98**	**4.05**	**4.11**	**4.16**
12	2.18	2.50	2.68	2.81	2.90	2.98	3.04	3.09	3.14
	3.05	**3.39**	**3.58**	**3.71**	**3.81**	**3.89**	**3.96**	**4.02**	**4.07**
13	2.16	2.48	2.65	2.78	2.87	2.94	3.00	3.06	3.10
	3.01	**3.33**	**3.52**	**3.65**	**3.74**	**3.82**	**3.89**	**3.94**	**3.99**
14	2.14	2.46	2.63	2.75	2.84	2.91	2.97	3.02	3.07
	2.98	**3.29**	**3.47**	**3.59**	**3.69**	**3.76**	**3.83**	**3.88**	**3.93**
15	2.13	2.44	2.61	2.73	2.82	2.89	2.95	3.00	3.04
	2.95	**3.25**	**3.43**	**3.55**	**3.64**	**3.71**	**3.78**	**3.83**	**3.88**
16	2.12	2.42	2.59	2.71	2.80	2.87	2.92	2.97	3.02
	2.92	**3.22**	**3.39**	**3.51**	**3.60**	**3.67**	**3.73**	**3.78**	**3.83**
17	2.11	2.41	2.58	2.69	2.78	2.85	2.90	2.95	3.00
	2.90	**3.19**	**3.36**	**3.47**	**3.56**	**3.63**	**3.69**	**3.74**	**3.79**
18	2.10	2.40	2.56	2.68	2.76	2.83	2.89	2.94	2.98
	2.88	**3.17**	**3.33**	**3.44**	**3.53**	**3.60**	**3.66**	**3.71**	**3.75**
19	2.09	2.39	2.55	2.66	2.75	2.81	2.87	2.92	2.96
	2.86	**3.15**	**3.31**	**3.42**	**3.50**	**3.57**	**3.63**	**3.68**	**3.72**
20	2.09	2.38	2.54	2.65	2.73	2.80	2.86	2.90	2.95
	2.85	**3.13**	**3.29**	**3.40**	**3.48**	**3.55**	**3.60**	**3.65**	**3.69**
24	2.06	2.35	2.51	2.61	2.70	2.76	2.81	2.86	2.90
	2.80	**3.07**	**3.22**	**3.32**	**3.40**	**3.47**	**3.52**	**3.57**	**3.61**
30	2.04	2.32	2.47	2.58	2.66	2.72	2.77	2.82	2.86
	2.75	**3.01**	**3.15**	**3.25**	**3.33**	**3.39**	**3.44**	**3.49**	**3.52**
40	2.02	2.29	2.44	2.54	2.62	2.68	2.73	2.77	2.81
	2.70	**2.95**	**3.09**	**3.19**	**3.26**	**3.32**	**3.37**	**3.41**	**3.44**
60	2.00	2.27	2.41	2.51	2.58	2.64	2.69	2.73	2.77
	2.66	**2.90**	**3.03**	**3.12**	**3.19**	**3.25**	**3.29**	**3.33**	**3.37**
120	1.98	2.24	2.38	2.47	2.55	2.60	2.65	2.69	2.73
	2.62	**2.85**	**2.97**	**3.06**	**3.12**	**3.18**	**3.22**	**3.26**	**3.29**
∞	1.96	2.21	2.35	2.44	2.51	2.57	2.61	2.65	2.69
	2.58	**2.79**	**2.92**	**3.00**	**3.06**	**3.11**	**3.15**	**3.19**	**3.22**

Table A14 Table of Dunnett's Modified *t* Statistic for a Control Group Comparison (continued)

One–Tailed Values

	***k* = number of treatment means, including control**								
***df*$_{error}$**	**2**	**3**	**4**	**5**	**6**	**7**	**8**	**9**	**10**
5	2.02	2.44	2.68	2.85	2.98	3.08	3.16	3.24	3.30
	3.37	**3.90**	**4.21**	**4.43**	**4.60**	**4.73**	**4.85**	**4.94**	**5.03**
6	1.94	2.34	2.56	2.71	2.83	2.92	3.00	3.07	3.12
	3.14	**3.61**	**3.88**	**4.07**	**4.21**	**4.33**	**4.43**	**4.51**	**4.59**
7	1.89	2.27	2.48	2.62	2.73	2.82	2.89	2.95	3.01
	3.00	**3.42**	**3.66**	**3.83**	**3.96**	**4.07**	**4.15**	**4.23**	**4.30**
8	1.86	2.22	2.42	2.55	2.66	2.74	2.81	2.87	2.92
	2.90	**3.29**	**3.51**	**3.67**	**3.79**	**3.88**	**3.96**	**4.03**	**4.09**
9	1.83	2.18	2.37	2.50	2.20	2.68	2.75	2.81	2.86
	2.82	**3.19**	**3.40**	**3.55**	**3.66**	**3.75**	**3.82**	**3.89**	**3.94**
10	1.81	2.15	2.34	2.47	2.56	2.64	2.70	2.76	2.81
	2.76	**3.11**	**3.31**	**3.45**	**3.56**	**3.64**	**3.71**	**3.78**	**3.83**
11	1.80	2.13	2.31	2.44	2.53	2.60	2.67	2.72	2.77
	2.72	**3.06**	**3.25**	**3.38**	**3.48**	**3.56**	**3.63**	**3.69**	**3.74**
12	1.78	2.11	2.29	2.41	2.50	2.58	2.64	2.69	2.74
	2.68	**3.01**	**3.19**	**3.32**	**3.42**	**3.50**	**3.56**	**3.62**	**3.67**
13	1.77	2.09	2.27	2.39	2.48	2.55	2.61	2.66	2.71
	2.65	**2.97**	**3.15**	**3.27**	**3.37**	**3.44**	**3.51**	**3.56**	**3.61**
14	1.76	2.08	2.25	2.37	2.46	2.53	2.59	2.64	2.69
	2.62	**2.94**	**3.11**	**3.23**	**3.32**	**3.40**	**3.46**	**3.51**	**3.56**
15	1.75	2.07	2.24	2.36	2.44	2.51	2.57	2.62	2.67
	2.60	**2.91**	**3.08**	**3.20**	**3.29**	**3.36**	**3.42**	**3.47**	**3.52**
16	1.75	2.06	2.23	2.34	2.43	2.50	2.56	2.61	2.65
	2.58	**2.88**	**3.05**	**3.17**	**3.26**	**3.33**	**3.39**	**3.44**	**3.48**
17	1.74	2.05	2.22	2.33	2.42	2.49	2.54	2.59	2.64
	2.57	**2.86**	**3.03**	**3.14**	**3.23**	**3.30**	**3.36**	**3.41**	**3.45**
18	1.73	2.04	2.21	2.32	2.41	2.48	2.53	2.58	2.62
	2.55	**2.84**	**3.01**	**3.12**	**3.21**	**3.27**	**3.33**	**3.38**	**3.42**
19	1.73	2.03	2.20	2.31	2.40	2.47	2.52	2.57	2.61
	2.54	**2.83**	**2.99**	**3.10**	**3.18**	**3.25**	**3.31**	**3.36**	**3.40**
20	1.72	2.03	2.19	2.30	2.39	2.46	2.51	2.56	2.60
	2.53	**2.81**	**2.97**	**3.08**	**3.17**	**3.23**	**3.29**	**3.34**	**3.38**
24	1.71	2.01	2.17	2.28	2.36	2.43	2.48	2.53	2.57
	2.49	**2.77**	**2.92**	**3.03**	**3.11**	**3.17**	**3.22**	**3.27**	**3.31**
30	1.70	1.99	2.15	2.25	2.33	2.40	2.45	2.50	2.54
	2.46	**2.72**	**2.87**	**2.97**	**3.05**	**3.11**	**3.16**	**3.21**	**3.24**
40	1.68	1.97	2.13	2.23	2.31	2.37	2.42	2.47	2.51
	2.42	**2.68**	**2.82**	**2.92**	**2.99**	**3.05**	**3.10**	**3.14**	**3.18**
60	1.67	1.95	2.10	2.21	2.28	2.35	2.39	2.44	2.48
	2.39	**2.64**	**2.78**	**2.87**	**2.94**	**3.00**	**3.04**	**3.08**	**3.12**
120	1.66	1.93	2.08	2.18	2.26	2.32	2.37	2.41	2.45
	2.36	**2.60**	**2.73**	**2.82**	**2.89**	**2.94**	**2.99**	**3.03**	**3.06**
∞	1.64	1.92	2.06	2.16	2.23	2.29	2.34	2.38	2.42
	2.33	**2.56**	**2.68**	**2.77**	**2.84**	**2.89**	**2.93**	**2.97**	**3.00**

Table A15 Graphs of the Power Function for the Analysis of Variance (Fixed-Effects Model)

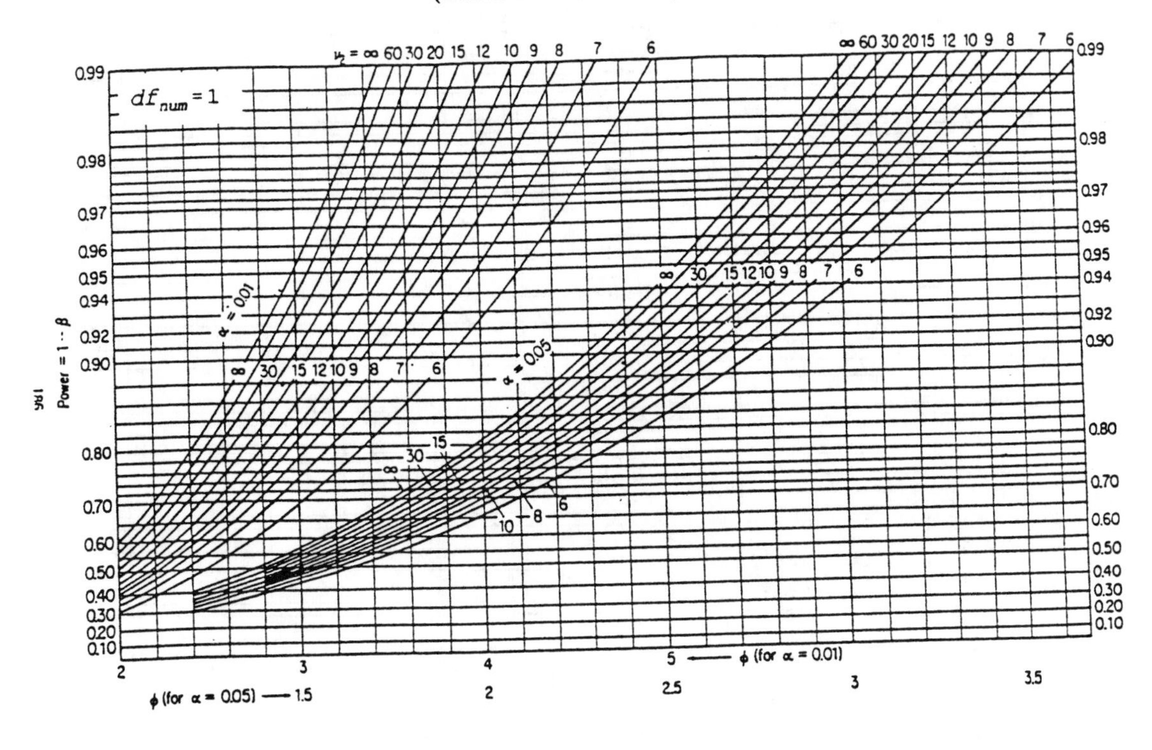

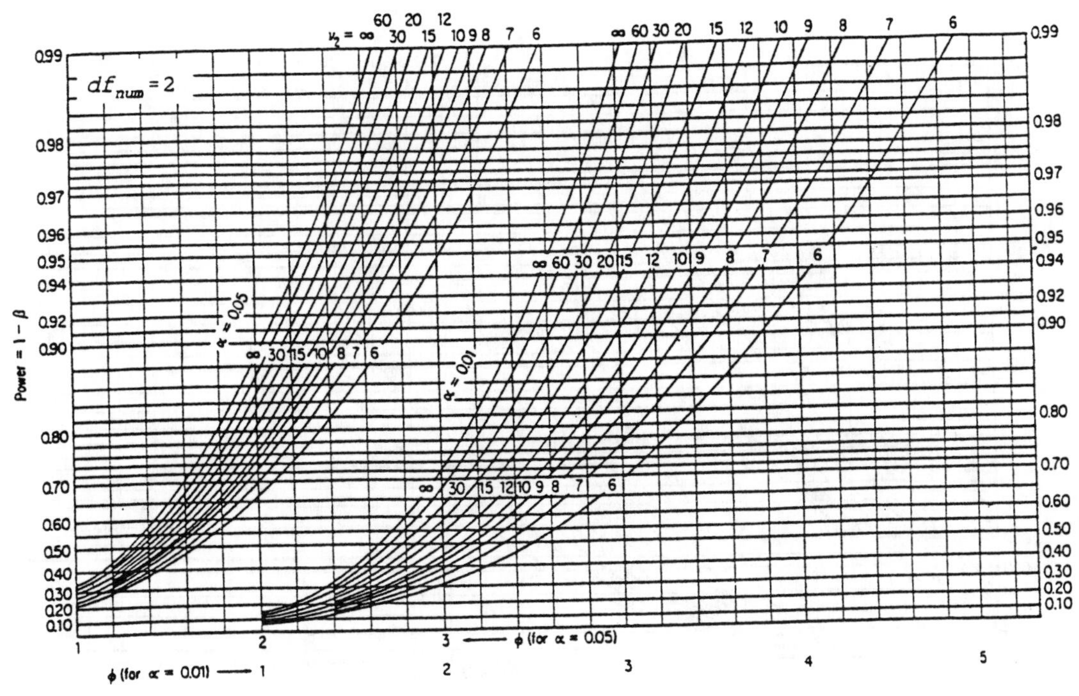

Table A15 Graphs of the Power Function for the Analysis of Variance (continued)

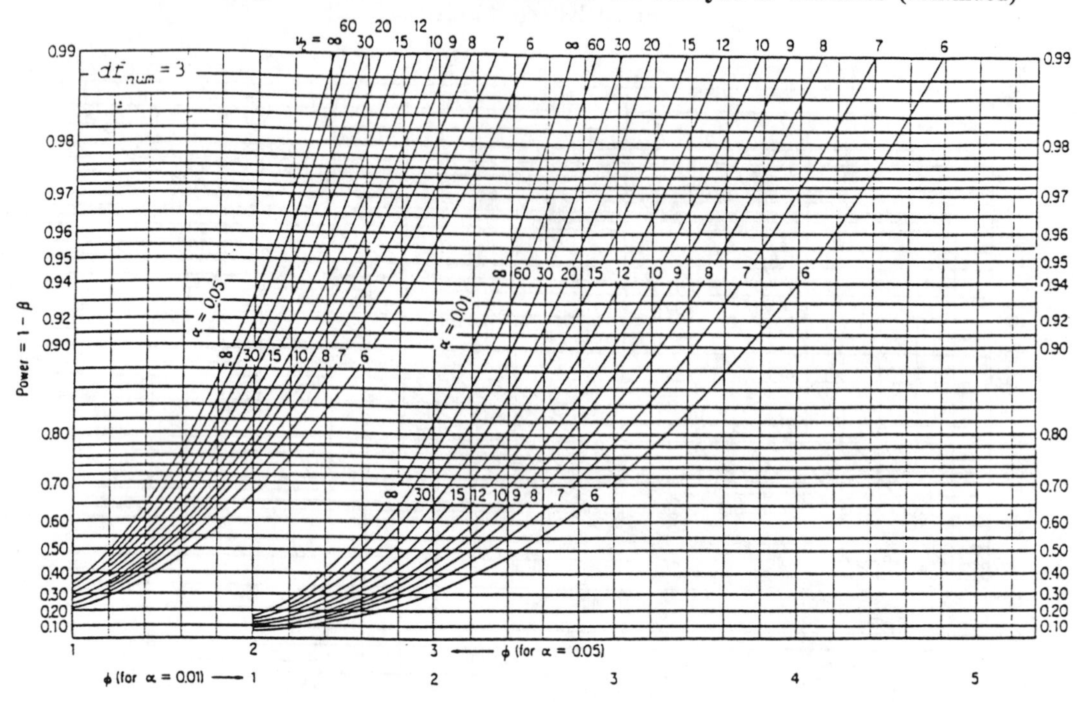

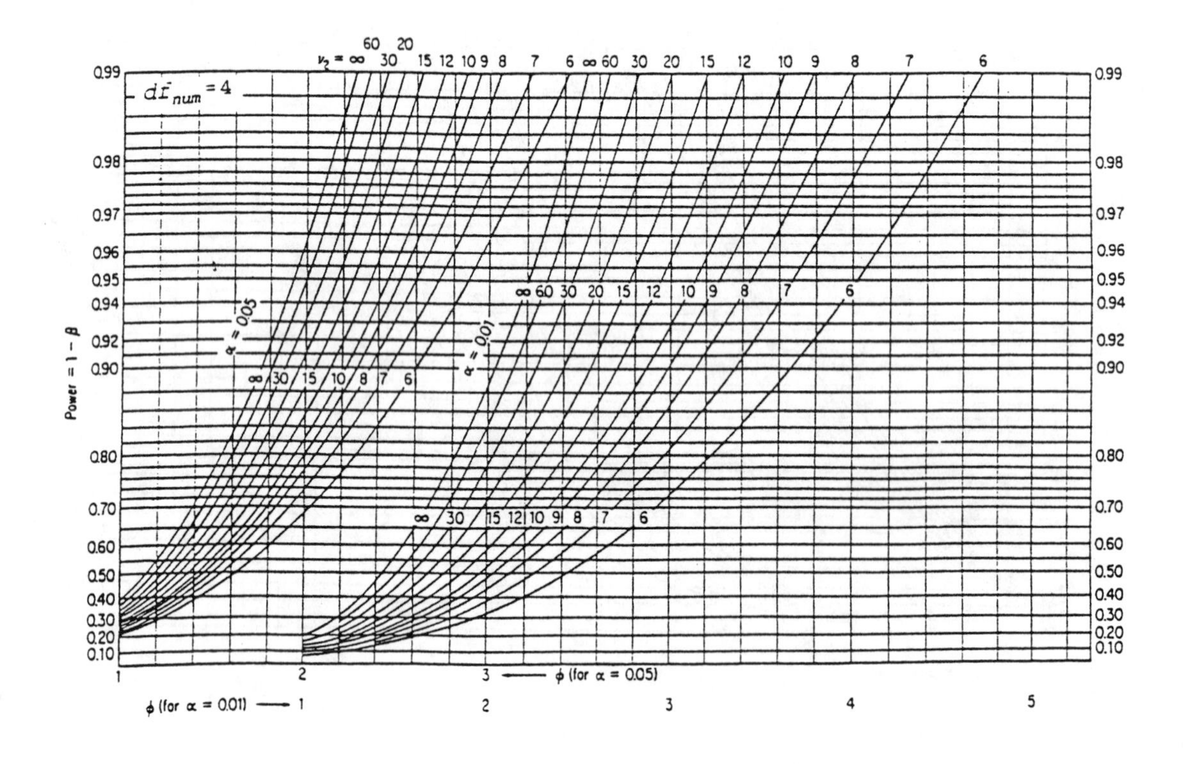

Table A15 Graphs of the Power Function for the Analysis of Variance (continued)

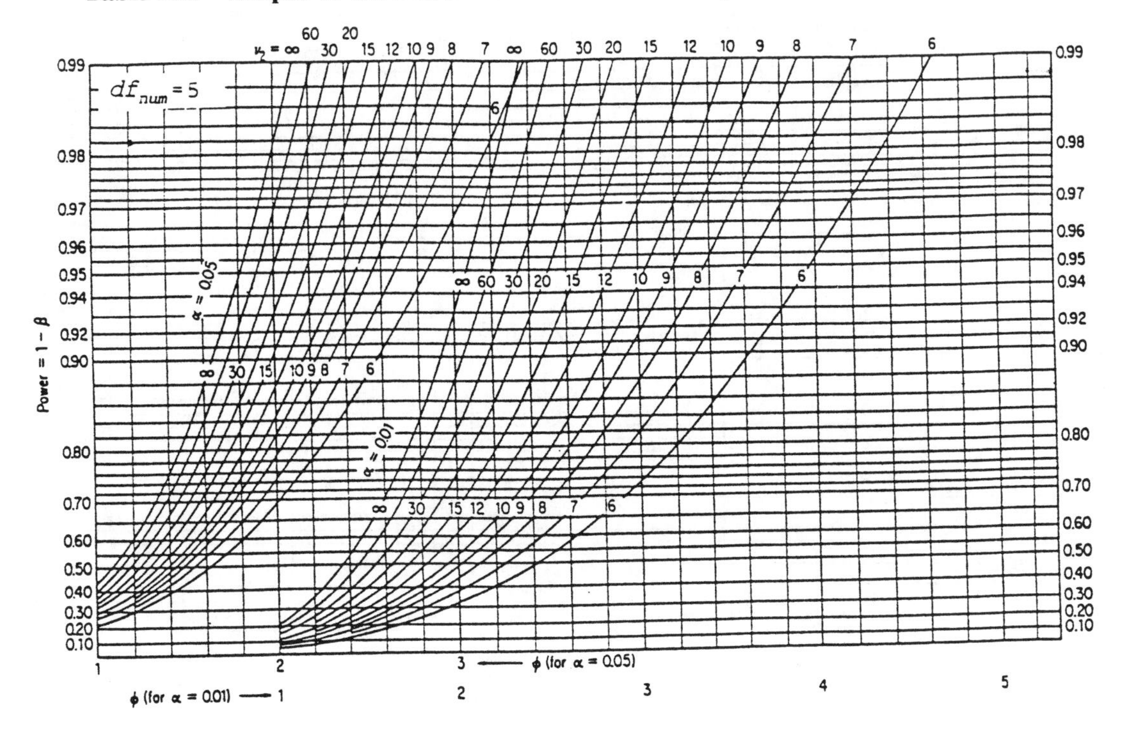

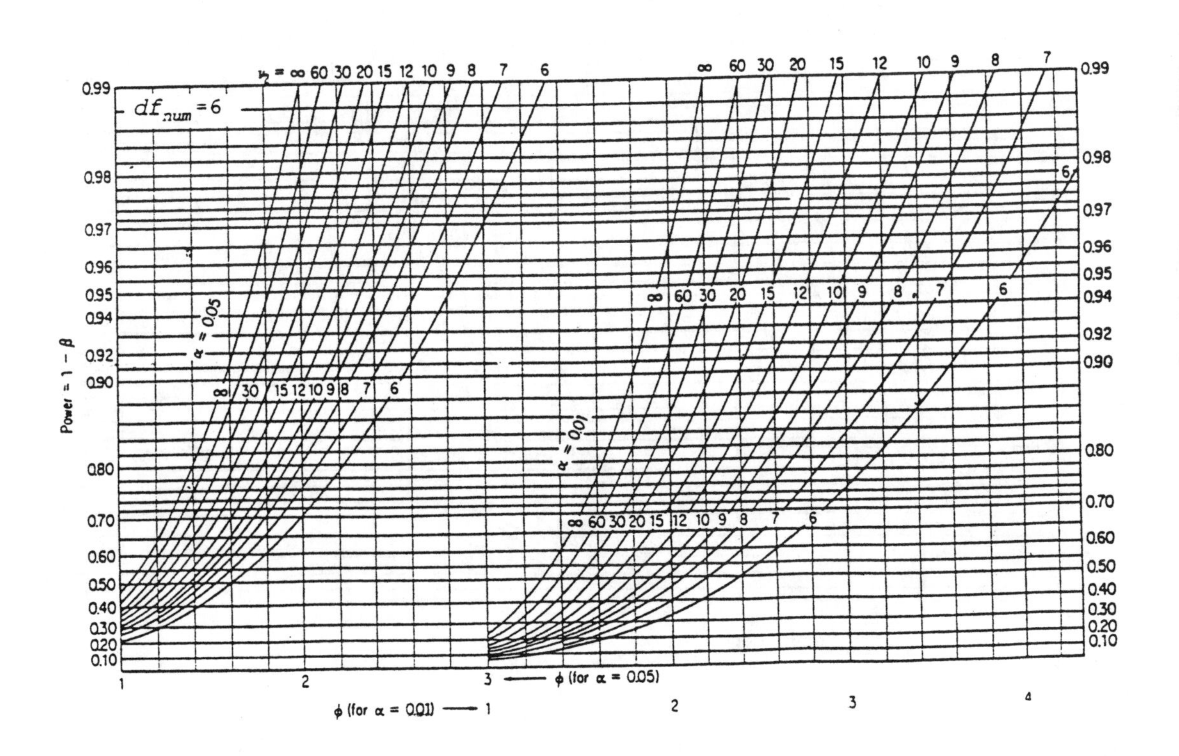

Table A15 Graphs of the Power Function for the Analysis of Variance (continued)

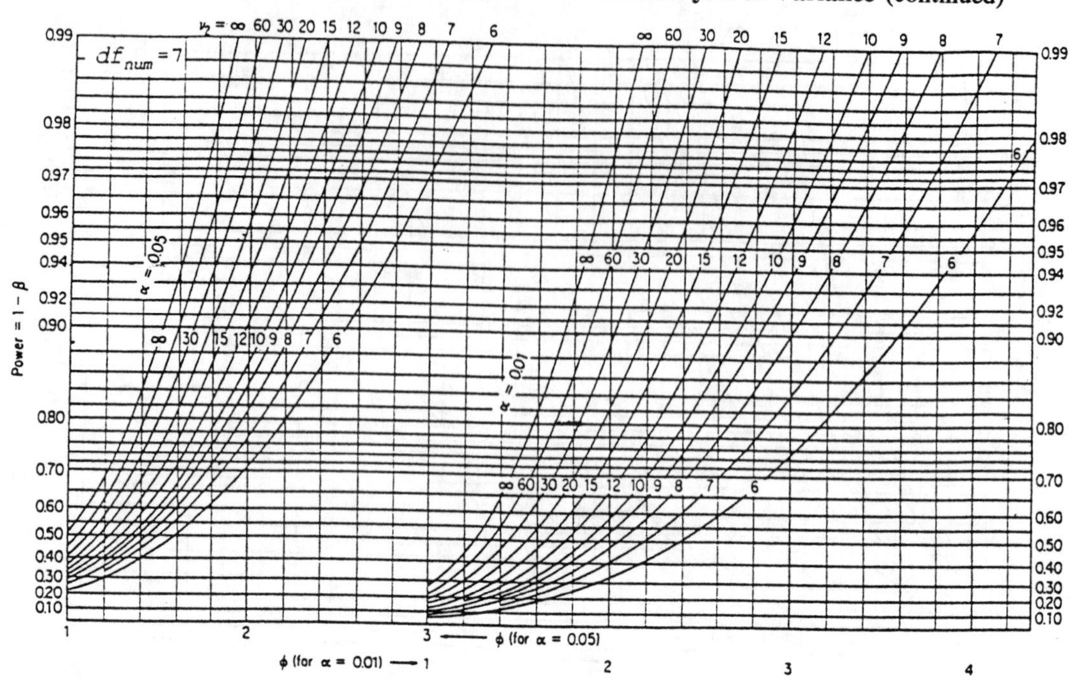

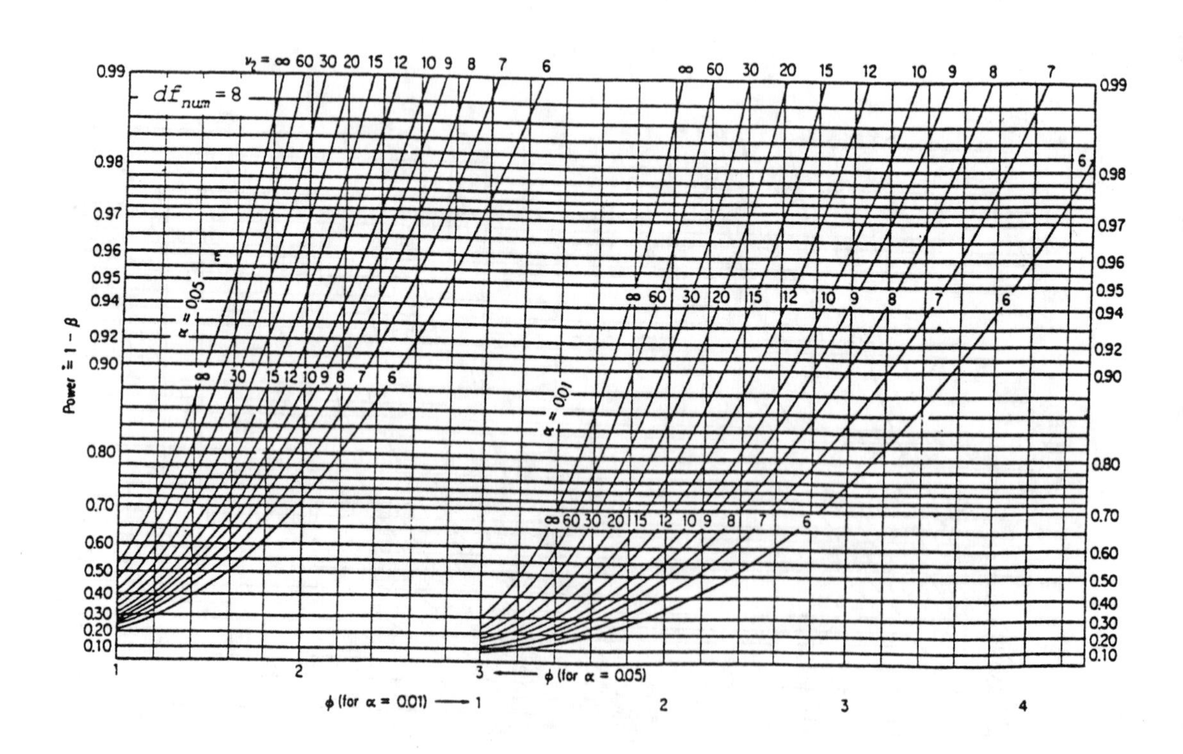

Table A16 Table of Critical Values for Pearson r

	One-tailed level of significance			
	.05	.025	.01	.005
	Two-tailed level of significance			
	.10	.05	.02	.01
$df = n-2$				
1	.988	.997	.9995	.9999
2	.900	.950	.980	.990
3	.805	.878	.934	.959
4	.729	.811	.882	.917
5	.669	.754	.833	.874
6	.622	.707	.789	.834
7	.582	.666	.750	.798
8	.549	.632	.716	.765
9	.521	.602	.685	.735
10	.497	.576	.658	.708
11	.476	.553	.634	.684
12	.458	.532	.612	.661
13	.441	.514	.592	.641
14	.426	.497	.574	.623
15	.412	.482	.558	.606
16	.400	.468	.542	.590
17	.389	.456	.528	.575
18	.378	.444	.516	.561
19	.369	.433	.503	.549
20	.360	.423	.492	.537
21	.352	.413	.482	.526
22	.344	.404	.472	.515
23	.337	.396	.462	.505
24	.330	.388	.453	.496
25	.323	.381	.445	.487
26	.317	.374	.437	.479
27	.311	.367	.430	.471
28	.306	.361	.423	.463
29	.301	.355	.416	.456
30	.296	.349	.409	.449
35	.275	.325	.381	.418
40	.257	.304	.358	.393
45	.243	.288	.338	.372
50	.231	.273	.322	.354
60	.211	.250	.295	.325
70	.195	.232	.274	.302
80	.183	.217	.256	.283
90	.173	.205	.242	.267
100	.164	.195	.230	.254

Table A17 Table of Fisher's z_r Transformation

r	z_r	r	z_r	r	z_r	r	z_r	r	z_r
.000	.000	.200	.203	.400	.424	.600	.693	.800	1.099
.005	.005	.205	.208	.405	.430	.605	.701	.805	1.113
.010	.010	.210	.213	.410	.436	.610	.709	.810	1.127
.015	.015	.215	.218	.415	.442	.615	.717	.815	1.142
.020	.020	.220	.224	.420	.448	.620	.725	.820	1.157
.025	.025	.225	.229	.425	.454	.625	.733	.825	1.172
.030	.030	.230	.234	.430	.460	.630	.741	.830	1.188
.035	.035	.235	.239	.435	.466	.635	.750	.835	1.204
.040	.040	.240	.245	.440	.472	.640	.758	.840	1.221
.045	.045	.245	.250	.445	.478	.645	.767	.845	1.238
.050	.050	.250	.255	.450	.485	.650	.775	.850	1.256
.055	.055	.255	.261	.455	.491	.655	.784	.855	1.274
.060	.060	.260	.266	.460	.497	.660	.793	.860	1.293
.065	.065	.265	.271	.465	.504	.665	.802	.865	1.313
.070	.070	.270	.277	.470	.510	.670	.811	.870	1.333
.075	.075	.275	.282	.475	.517	.675	.820	.875	1.354
.080	.080	.280	.288	.480	.523	.680	.829	.880	1.376
.085	.085	.285	.293	.485	.530	.685	.838	.885	1.398
.090	.090	.290	.299	.490	.536	.690	.848	.890	1.422
.095	.095	.295	.304	.495	.543	.695	.858	.895	1.447
.100	.100	.300	.310	.500	.549	.700	.867	.900	1.472
.105	.105	.305	.315	.505	.556	.705	.877	.905	1.499
.110	.110	.310	.321	.510	.563	.710	.887	.910	1.528
.115	.116	.315	.326	.515	.570	.715	.897	.915	1.557
.120	.121	.320	.332	.520	.576	.720	.908	.920	1.589
.125	.126	.325	.337	.525	.583	.725	.918	.925	1.623
.130	.131	.330	.343	.530	.590	.730	.929	.930	1.658
.135	.136	.335	.348	.535	.597	.735	.940	.935	1.697
.140	.141	.340	.354	.540	.604	.740	.950	.940	1.738
.145	.146	.345	.360	.545	.611	.745	.962	.945	1.783
.150	.151	.350	.365	.550	.618	.750	.973	.950	1.832
.155	.156	.355	.371	.555	.626	.755	.984	.955	1.886
.160	.161	.360	.377	.560	.633	.760	.996	.960	1.946
.165	.167	.365	.383	.565	.640	.765	1.008	.965	2.014
.170	.172	.370	.388	.570	.648	.770	1.020	.970	2.092
.175	.177	.375	.394	.575	.655	.775	1.033	.975	2.185
.180	.182	.380	.400	.580	.662	.780	1.045	.980	2.298
.185	.187	.385	.406	.585	.670	.785	1.058	.985	2.443
.190	.192	.390	.412	.590	.678	.790	1.071	.990	2.647
.195	.198	.395	.418	.595	.685	.795	1.085	.995	2.994

Table A18 Table of Critical Values for Spearman's Rho

	One-tailed level of significance			
	.05	.025	.01	.005
	Two-tailed level of significance			
	.10	.05	.02	.01
n				
4	1.000	–	–	–
5	.900	1.000	1.000	–
6	.829	.886	.943	1.000
7	.714	.786	.893	.929
8	.643	.738	.833	.881
9	.600	.700	.783	.833
10	.564	.648	.745	.794
11	.536	.618	.709	.755
12	.503	.587	.671	.727
13	.484	.560	.648	.703
14	.464	.538	.622	.675
15	.443	.521	.604	.654
16	.429	.503	.582	.635
17	.414	.485	.566	.615
18	.401	.472	.550	.600
19	.391	.460	.535	.584
20	.380	.447	.520	.570
21	.370	.435	.508	.556
22	.361	.425	.496	.544
23	.353	.415	.486	.532
24	.344	.406	.476	.521
25	.337	.398	.466	.511
26	.331	.390	.457	.501
27	.324	.382	.448	.491
28	.317	.375	.440	.483
29	.312	.368	.433	.475
30	.306	.362	.425	.467
35	.283	.335	.394	.433
40	.264	.313	.368	.405
45	.248	.294	.347	.382
50	.235	.279	.329	.363
60	.214	.255	.300	.331
70	.190	.235	.278	.307
80	.185	.220	.260	.287
90	.174	.207	.245	.271
100	.165	.197	.233	.257

Table A19 Table of Critical Values for Kendall's Tau

Critical values for both $\tilde{\tau}$ and S are listed in the table.

Two-tailed	.01		.02		.05		.10		.20	
One tailed	.005		.01		.025		.05		.10	
n	S	$\tilde{\tau}$	S	$\tilde{\tau}$	S	$\tilde{\tau}$	S	$\tilde{\tau}$	S	$\tilde{\tau}$
4	8	1.000	8	1.000	8	1.000	6	1.000	6	1.000
5	12	1.000	10	1.000	10	1.000	8	.800	8	.800
6	15	1.000	13	.867	13	.867	11	.733	9	.600
7	19	.905	17	.810	15	.714	13	.619	11	.524
8	22	.786	20	.714	18	.643	16	.571	12	.429
9	26	.722	24	.667	20	.556	18	.500	14	.389
10	29	.644	27	.600	23	.511	21	.467	17	.378
11	33	.600	31	.564	27	.491	23	.418	19	.345
12	38	.576	36	.545	30	.455	26	.394	20	.303
13	44	.564	40	.513	34	.436	28	.359	24	.308
14	47	.516	43	.473	37	.407	33	.363	25	.275
15	53	.505	49	.467	41	.390	35	.333	29	.276
16	58	.483	52	.433	46	.383	38	.317	30	.250
17	64	.471	58	.426	50	.368	42	.309	34	.250
18	69	.451	63	.412	53	.346	45	.294	37	.242
19	75	.439	67	.392	57	.333	49	.287	39	.228
20	80	.421	72	.379	62	.326	52	.274	42	.221
21	86	.410	78	.371	66	.314	56	.267	44	.210
22	91	.394	83	.359	71	.307	61	.264	47	.203
23	99	.391	89	.352	75	.296	65	.257	51	.202
24	104	.377	94	.341	80	.290	68	.246	54	.196
25	110	.367	100	.333	86	.287	72	.240	58	.193
26	117	.360	107	.329	91	.280	77	.237	61	.188
27	125	.356	113	.322	95	.271	81	.231	63	.179
28	130	.344	118	.312	100	.265	86	.228	68	.180
29	138	.340	126	.310	106	.261	90	.222	70	.172
30	145	.333	131	.301	111	.255	95	.218	75	.172
31	151	.325	137	.295	117	.252	99	.213	77	.166
32	160	.323	144	.290	122	.246	104	.210	82	.165
33	166	.314	152	.288	128	.242	108	.205	86	.163
34	175	.312	157	.280	133	.237	113	.201	89	.159
35	181	.304	165	.277	139	.234	117	.197	93	.156
36	190	.302	172	.273	146	.232	122	.194	96	.152
37	198	.297	178	.267	152	.228	128	.192	100	.150
38	205	.292	185	.263	157	.223	133	.189	105	.149
39	213	.287	193	.260	163	.220	139	.188	109	.147
40	222	.285	200	.256	170	.218	144	.185	112	.144

Table A20 Table of Critical Values for Kendall's Coefficient of Concordance

Critical values for both S and $\tilde{W}$ are listed. The values of $\tilde{W}$ in the table were computed by substituting the tabled S values in Equation 25.3.

	n									
m	3		4		5		6		7	
	Values at .05 level of significance									
	S	$\tilde{W}$	S	$\tilde{W}$	S	$\tilde{W}$	S	$\tilde{W}$	S	$\tilde{W}$
3					64.4	.716	103.9	.660	157.3	.624
4			49.5	.619	88.4	.552	143.3	.512	217.0	.484
5			62.6	.501	112.3	.449	182.4	.417	276.2	.395
6			75.7	.421	136.1	.378	221.4	.351	335.2	.333
8	48.1	.376	101.7	.318	183.7	.287	299.0	.267	453.1	.253
10	60.0	.300	127.8	.256	231.2	.231	376.7	.215	571.0	.204
15	89.8	.200	192.9	.171	349.8	.155	570.5	.145	864.9	.137
20	119.7	.150	258.0	.129	468.5	.117	764.4	.109	1158.7	.103
	Values at .01 level of significance									
3					75.6	.840	122.8	.780	185.6	.737
4			61.4	.768	109.3	.683	176.2	.629	265.0	.592
5			80.5	.644	142.8	.571	229.4	.524	343.8	.491
6			99.5	.553	176.1	.489	282.4	.448	422.6	.419
8	66.8	.522	137.4	.429	242.7	.379	388.3	.347	579.9	.324
10	85.1	.425	175.3	.351	309.1	.309	494.0	.282	737.0	.263
15	131.0	.291	269.8	.240	475.2	.211	758.2	.193	1129.5	.179
20	177.0	.221	364.2	.182	641.2	.160	1022.2	.146	1521.9	.136

Additional values for $n = 3$

	At .05 level		At .01 level	
m	S	$\tilde{W}$	S	$\tilde{W}$
9	54.0	.333	75.9	.469
12	71.9	.250	103.5	.359
14	83.8	.214	121.9	.311
16	95.8	.187	140.2	.274
18	107.7	.166	158.6	.245

Index

Q

R

S

T